Applications Index

SIXTH EDITION

Linear Algebra
and Its Applications

David C. Lay

University of Maryland–College Park

Steven R. Lay

Lee University

Judi J. McDonald

Washington State University

 Pearson

Library of Congress Cataloging in Publication Control Number: 2019044303

1 2019

SE: 978-0-13-585125-8
IE: 978-0-13-588280-1

To my wife, Lillian, and our children, Christina, Deborah, and Melissa, whose support, encouragement, and faithful prayers made this book possible.

David C. Lay

About the Authors

David C. Lay

As a founding member of the NSF-sponsored Linear Algebra Curriculum Study Group (LACSG), David Lay was a leader in the movement to modernize the linear algebra curriculum and shared those ideas with students and faculty through his authorship of the first four editions of this textbook. David C. Lay earned a B.A. from Aurora University (Illinois), and an M.A. and Ph.D. from the University of California at Los Angeles. David Lay was an educator and research mathematician for more than 40 years, mostly at the University of Maryland, College Park. He also served as a visiting professor at the University of Amsterdam, the Free University in Amsterdam, and the University of Kaiserslautern, Germany. He published more than 30 research articles on functional analysis and linear algebra. Lay was also a coauthor of several mathematics texts, including *Introduction to Functional Analysis* with Angus E. Taylor, *Calculus and Its Applications*, with L. J. Goldstein and D. I. Schneider, and *Linear Algebra Gems—Assets for Undergraduate Mathematics*, with D. Carlson, C. R. Johnson, and A. D. Porter.

David Lay received four university awards for teaching excellence, including, in 1996, the title of Distinguished Scholar-Teacher of the University of Maryland. In 1994, he was given one of the Mathematical Association of America's Awards for Distinguished College or University Teaching of Mathematics. He was elected by the university students to membership in Alpha Lambda Delta National Scholastic Honor Society and Golden Key National Honor Society. In 1989, Aurora University conferred on him the Outstanding Alumnus award. David Lay was a member of the American Mathematical Society, the Canadian Mathematical Society, the International Linear Algebra Society, the Mathematical Association of America, Sigma Xi, and the Society for Industrial and Applied Mathematics. He also served several terms on the national board of the Association of Christians in the Mathematical Sciences.

In October 2018, David Lay passed away, but his legacy continues to benefit students of linear algebra as they study the subject in this widely acclaimed text.

Steven R. Lay

Steven R. Lay began his teaching career at Aurora University (Illinois) in 1971, after earning an M.A. and a Ph.D. in mathematics from the University of California at Los Angeles. His career in mathematics was interrupted for eight years while serving as a missionary in Japan. Upon his return to the States in 1998, he joined the mathematics faculty at Lee University (Tennessee) and has been there ever since. Since then he has supported his brother David in refining and expanding the scope of this popular linear algebra text, including writing most of Chapters 8 and 9. Steven is also the author of three college-level mathematics texts: *Convex Sets and Their Applications, Analysis with an Introduction to Proof, and Principles of Algebra*.

In 1985, Steven received the Excellence in Teaching Award at Aurora University. He and David, and their father, Dr. L. Clark Lay, are all distinguished mathematicians, and in 1989, they jointly received the Outstanding Alumnus award from their alma mater, Aurora University. In 2006, Steven was honored to receive the Excellence in Scholarship Award at Lee University. He is a member of the American Mathematical Society, the Mathematics Association of America, and the Association of Christians in the Mathematical Sciences.

Judi J. McDonald

Judi J. McDonald became a co-author on the fifth edition, having worked closely with David on the fourth edition. She holds a B.Sc. in Mathematics from the University of Alberta, and an M.A. and Ph.D. from the University of Wisconsin. As a professor of Mathematics, she has more than 40 publications in linear algebra research journals and more than 20 students have completed graduate degrees in linear algebra under her supervision. She is an associate dean of the Graduate School at Washington State University and a former chair of the Faculty Senate. She has worked with the mathematics outreach project Math Central (http://mathcentral.uregina.ca/) and is a member of the second Linear Algebra Curriculum Study Group (LACSG 2.0).

Judi has received three teaching awards: two Inspiring Teaching awards at the University of Regina, and the Thomas Lutz College of Arts and Sciences Teaching Award at Washington State University. She also received the College of Arts and Sciences Institutional Service Award at Washington State University. Throughout her career, she has been an active member of the International Linear Algebra Society and the Association for Women in Mathematics. She has also been a member of the Canadian Mathematical Society, the American Mathematical Society, the Mathematical Association of America, and the Society for Industrial and Applied Mathematics.

Contents

Preface

The response of students and teachers to the first five editions of *Linear Algebra and Its Applications* has been most gratifying. This *Sixth Edition* provides substantial support both for teaching and for using technology in the course. As before, the text provides a modern elementary introduction to linear algebra and a broad selection of interesting classical and leading-edge applications. The material is accessible to students with the maturity that should come from successful completion of two semesters of college-level mathematics, usually calculus.

The main goal of the text is to help students master the basic concepts and skills they will use later in their careers. The topics here follow the recommendations of the original Linear Algebra Curriculum Study Group (LACSG), which were based on a careful investigation of the real needs of the students and a consensus among professionals in many disciplines that use linear algebra. Ideas being discussed by the second Linear Algebra Curriculum Study Group (LACSG 2.0) have also been included. We hope this course will be one of the most useful and interesting mathematics classes taken by undergraduates.

What's New in This Edition

The *Sixth Edition* has exiting new material, examples, and online resources. After talking with high-tech industry researchers and colleagues in applied areas, we added new topics, vignettes, and applications with the intention of highlighting for students and faculty the linear algebraic foundational material for machine learning, artificial intelligence, data science, and digital signal processing.

Content Changes

- Since matrix multiplication is a highly useful skill, we added new examples in Chapter 2 to show how matrix multiplication is used to identify patterns and scrub data. Corresponding exercises have been created to allow students to explore using matrix multiplication in various ways.

- In our conversations with colleagues in industry and electrical engineering, we heard repeatedly how important understanding abstract vector spaces is to their work. After reading the reviewers' comments for Chapter 4, we reorganized the chapter, condensing some of the material on column, row, and null spaces; moving Markov chains to the end of Chapter 5; and creating a new section on signal processing. We view signals as an infinite dimensional vector space and illustrate the usefulness of linear transformations to filter out unwanted "vectors" (a.k.a. noise), analyze data, and enhance signals.

- By moving Markov chains to the end of Chapter 5, we can now discuss the steady state vector as an eigenvector. We also reorganized some of the summary material on determinants and change of basis to be more specific to the way they are used in this chapter.

- In Chapter 6, we present pattern recognition as an application of orthogonality, and the section on linear models now illustrates how machine learning relates to curve fitting.

- Chapter 9 on optimization was previously available only as an online file. It has now been moved into the regular textbook where it is more readily available to faculty and students. After an opening section on finding optimal strategies to two-person zero-sum games, the rest of the chapter presents an introduction to linear programming—from two-dimensional problems that can be solved geometrically to higher dimensional problems that are solved using the Simplex Method.

Other Changes

- In the high-tech industry, where most computations are done on computers, judging the validity of information and computations is an important step in preparing and analyzing data. In this edition, students are encouraged to learn to analyze their own computations to see if they are consistent with the data at hand and the questions being asked. For this reason, we have added "Reasonable Answers" advice and exercises to guide students.

- We have added a list of projects to the end of each chapter (available online at `bit.ly/30IM8gT` and in MyLab Math). Some of these projects were previously available online and have a wide range of themes from using linear transformations to create art to exploring additional ideas in mathematics. They can be used for group work or to enhance the learning of individual students.

- Free-response writing exercises have been added to MyLab Math, allowing faculty to ask more sophisticated questions online and create a paperless class without losing the richness of discussing how concepts relate to each other and introductory proof writing.

- The electronic interactive textbook has been changed from Wolfram CDF to Wolfram Cloud format. This allows faculty and students to interact with figures and examples on a wider variety of electronic, devices, without the need to install the CDF Player.

- PowerPoint lecture slides have been updated to cover all sections of the text and cover them more thoroughly.

Distinctive Features

Early Introduction of Key Concepts

Many fundamental ideas of linear algebra are introduced within the first seven lectures, in the concrete setting of $\mathbb{R}^n$, and then gradually examined from different points of view. Later generalizations of these concepts appear as natural extensions of familiar ideas, visualized through the geometric intuition developed in Chapter 1. A major achievement of this text is that the level of difficulty is fairly even throughout the course.

A Modern View of Matrix Multiplication

Good notation is crucial, and the text reflects the way scientists and engineers actually use linear algebra in practice. The definitions and proofs focus on the columns of a matrix rather than on the matrix entries. A central theme is to view a matrix–vector product $A\mathbf{x}$ as a linear combination of the columns of A. This modern approach simplifies many arguments, and it ties vector space ideas into the study of linear systems.

Linear Transformations

Linear transformations form a "thread" that is woven into the fabric of the text. Their use enhances the geometric flavor of the text. In Chapter 1, for instance, linear transformations provide a dynamic and graphical view of matrix–vector multiplication.

Eigenvalues and Dynamical Systems

Eigenvalues appear fairly early in the text, in Chapters 5 and 7. Because this material is spread over several weeks, students have more time than usual to absorb and review these critical concepts. Eigenvalues are motivated by and applied to discrete and continuous dynamical systems, which appear in Sections 1.10, 4.8, and 5.9, and in five sections of Chapter 5. Some courses reach Chapter 5 after about five weeks by covering Sections 2.8 and 2.9 instead of Chapter 4. These two optional sections present all the vector space concepts from Chapter 4 needed for Chapter 5.

Orthogonality and Least-Squares Problems

These topics receive a more comprehensive treatment than is commonly found in beginning texts. The original Linear Algebra Curriculum Study Group has emphasized the need for a substantial unit on orthogonality and least-squares problems, because orthogonality plays such an important role in computer calculations and numerical linear algebra and because inconsistent linear systems arise so often in practical work.

Pedagogical Features

Applications

A broad selection of applications illustrates the power of linear algebra to explain fundamental principles and simplify calculations in engineering, computer science, mathematics, physics, biology, economics, and statistics. Some applications appear in separate sections; others are treated in examples and exercises. In addition, each chapter opens with an introductory vignette that sets the stage for some application of linear algebra and provides a motivation for developing the mathematics that follows.

A Strong Geometric Emphasis

Every major concept in the course is given a geometric interpretation, because many students learn better when they can visualize an idea. There are substantially more drawings here than usual, and some of the figures have never before appeared in a linear algebra text. Interactive versions of these figures, and more, appear in the electronic version of the textbook.

Examples

This text devotes a larger proportion of its expository material to examples than do most linear algebra texts. There are more examples than an instructor would ordinarily present in class. But because the examples are written carefully, with lots of detail, students can read them on their own.

Theorems and Proofs

Important results are stated as theorems. Other useful facts are displayed in tinted boxes, for easy reference. Most of the theorems have formal proofs, written with the beginner student in mind. In a few cases, the essential calculations of a proof are exhibited in a carefully chosen example. Some routine verifications are saved for exercises, when they will benefit students.

Practice Problems

A few carefully selected Practice Problems appear just before each exercise set. Complete solutions follow the exercise set. These problems either focus on potential trouble spots in the exercise set or provide a "warm-up" for the exercises, and the solutions often contain helpful hints or warnings about the homework.

Exercises

The abundant supply of exercises ranges from routine computations to conceptual questions that require more thought. A good number of innovative questions pinpoint conceptual difficulties that we have found on student papers over the years. Each exercise set is carefully arranged in the same general order as the text; homework assignments are readily available when only part of a section is discussed. A notable feature of the exercises is their numerical simplicity. Problems "unfold" quickly, so students spend little time on numerical calculations. The exercises concentrate on teaching understanding rather than mechanical calculations. The exercises in the *Sixth Edition* maintain the integrity of the exercises from previous editions, while providing fresh problems for students and instructors.

Exercises marked with the symbol ▇ are designed to be worked with the aid of a "matrix program" (a computer program, such as MATLAB, Maple, Mathematica, MathCad, or Derive, or a programmable calculator with matrix capabilities, such as those manufactured by Texas Instruments).

True/False Questions

To encourage students to read all of the text and to think critically, we have developed over 300 simple true/false questions that appear throughout the text, just after the computational problems. They can be answered directly from the text, and they prepare students for the conceptual problems that follow. Students appreciate these questions-after they get used to the importance of reading the text carefully. Based on class testing and discussions with students, we decided not to put the answers in the text. (The *Study Guide* tells the students where to find the answers to the odd-numbered questions.) An additional 150 true/false questions (mostly at the ends of chapters) test understanding of the material. The text does provide simple T/F answers to most of these supplementary exercises, but it omits the justifications for the answers (which usually require some thought).

Writing Exercises

An ability to write coherent mathematical statements in English is essential for all students of linear algebra, not just those who may go to graduate school in mathematics. The text includes many exercises for which a written justification is part of the answer. Conceptual exercises that require a short proof usually contain hints that help a student get started. For all odd-numbered writing exercises, either a solution is included at the back of the text or a hint is provided and the solution is given in the *Study Guide,* described below.

Projects

A list of projects (available online at `bit.ly/30IM8gT`) have been identified at the end of each chapter. They can be used by individual students or in groups. These projects provide the opportunity for students to explore fundamental concepts and applications in more detail. Two of the projects even encourage students to engage their creative side and use linear transformations to build artwork.

Reasonable Answers

Many of our students will enter a workforce where important decisions are being made based on answers provided by computers and other machines. The Reasonable Answers boxes and exercises help students develop an awareness of the need to analyze their answers for correctness and accuracy.

Computational Topics

The text stresses the impact of the computer on both the development and practice of linear algebra in science and engineering. Frequent Numerical Notes draw attention to issues in computing and distinguish between theoretical concepts, such as matrix inversion, and computer implementations, such as LU factorizations.

Acknowledgments

David Lay was grateful to many people who helped him over the years with various aspects of this book. He was particularly grateful to Israel Gohberg and Robert Ellis for more than fifteen years of research collaboration, which greatly shaped his view of linear algebra. And he was privileged to be a member of the Linear Algebra Curriculum Study Group along with David Carlson, Charles Johnson, and Duane Porter. Their creative ideas about teaching linear algebra have influenced this text in significant ways. He often spoke fondly of three good friends who guided the development of the book nearly from the beginning—giving wise counsel and encouragement—Greg Tobin, publisher; Laurie Rosatone, former editor; and William Hoffman, former editor.

Judi and Steven have been privileged to work on recent editions of Professor David Lay's linear algebra book. In making this revision, we have attempted to maintain the basic approach and the clarity of style that has made earlier editions popular with students and faculty. We thank Eric Schulz for sharing his considerable technological and pedagogical expertise in the creation of the electronic textbook. His help and encouragement were essential in bringing the figures and examples to life in the Wolfram Cloud version of this textbook.

Mathew Hudelson has been a valuable colleague in preparing the *Sixth Edition;* he is always willing to brainstorm about concepts or ideas and test out new writing and exercises. He contributed the idea for new vignette for Chapter 3 and the accompanying project. He has helped with new exercises throughout the text. Harley Weston has provided Judi with many years of good conversations about how, why, and who we appeal to when we present mathematical material in different ways. Katerina Tsatsomeros' artistic side has been a definite asset when we needed artwork to transform (the fish and the sheep), improved writing in the new introductory vignettes, or information from the perspective of college-age students.

We appreciate the encouragement and shared expertise from Nella Ludlow, Thomas Fischer, Amy Johnston, Cassandra Seubert, and Mike Manzano. They provided information about important applications of linear algebra and ideas for new examples and exercises. In particular, the new vignettes and material in Chapters 4 and 6 were inspired by conversations with these individuals.

We are energized by Sepideh Stewart and the other new Linear Algebra Curriculum Study Group (LACSG 2.0) members: Sheldon Axler, Rob Beezer, Eugene Boman, Minerva Catral, Guershon Harel, David Strong, and Megan Wawro. Initial meetings of this group have provided valuable guidance in revising the *Sixth Edition.*

We sincerely thank the following reviewers for their careful analyses and constructive suggestions:

Maila C. Brucal-Hallare, *Norfolk State University*
Kristen Campbell, *Elgin Community College*
Charles Conrad, *Volunteer State Community College*
R. Darrell Finney, *Wilkes Community College*
Xiaofeng Gu, *University of West Georgia*
Jeong Mi-Yoon, *University of Houston–Downtown*
Michael T. Muzheve, *Texas A&M U.–Kingsville*
Iason Rusodimos, *Perimeter C. at Georgia State U.*
Rebecca Swanson, *Colorado School of Mines*
Casey Wynn, *Kenyon College*
Taoye Zhang, *Penn State U.–Worthington Scranton*

Steven Burrow, *Central Texas College*
J. S. Chahal, *Brigham Young University*
Kevin Farrell, *Lyndon State College*
Chris Fuller, *Cumberland University*
Jeffrey Jauregui, *Union College*
Christopher Murphy, *Guilford Tech. C.C.*
Charles I. Odion, *Houston Community College*
Desmond Stephens, *Florida Ag. and Mech. U.*
Jiyuan Tao, *Loyola University–Maryland*
Amy Yielding, *Eastern Oregon University*
Houlong Zhuang, *Arizona State University*

We appreciate the proofreading and suggestions provided by John Samons and Jennifer Blue. Their careful eye has helped to minimize errors in this edition.

We thank Kristina Evans, Phil Oslin, and Jean Choe for their work in setting up and maintaining the online homework to accompany the text in MyLab Math, and for continuing to work with us to improve it. The reviews of the online homework done by Joan Saniuk, Robert Pierce, Doron Lubinsky and Adriana Corinaldesi were greatly appreciated. We also thank the faculty at University of California Santa Barbara, University of Alberta, Washington State University and the Georgia Institute of Technology for their feedback on the MyLab Math course. Joe Vetere has provided much appreciated technical help with the *Study Guide* and *Instructor's Solutions Manual.*

We thank Jeff Weidenaar, our content manager, for his continued careful, well-thought-out advice. Project Manager Ron Hampton has been a tremendous help guiding us through the production process. We are also grateful to Stacey Sveum and Rosemary Morton, our marketers, and Jon Krebs, our editorial associate, who have also contributed to the success of this edition.

Steven R. Lay and Judi J. McDonald

Get the *most* out of
MyLab Math

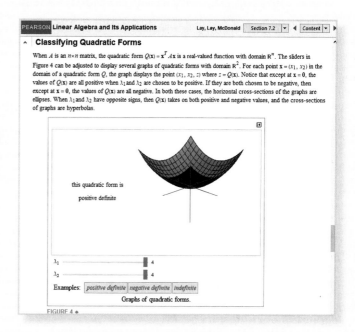

Supporting Instruction

MyLab Math provides resources to help you assess and improve student results and unparalleled flexibility to create a course tailored to you and your students.

PowerPoint® Lecture Slides

Fully editable PowerPoint slides are available for all sections of the text. The slides include definitions, theorems, examples and solutions. When used in the classroom, these slides allow the instructor to focus on teaching, rather than writing on the board. PowerPoint slides are available to students (within the Video and Resource Library in MyLab Math) so that they can follow along.

Sample Assignments

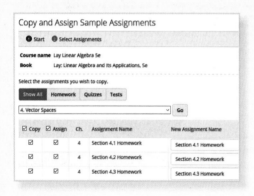

Sample Assignments are crafted to maximize student performance in the course. They make course set-up easier by giving instructors a starting point for each section.

Comprehensive Gradebook

The gradebook includes enhanced reporting functionality, such as item analysis and a reporting dashboard to enable you to efficiently manage your course. Student performance data are presented at the class, section, and program levels in an accessible, visual manner so you'll have the information you need to keep your students on track.

pearson.com/mylab/math

Resources for
Success

Instructor Resources

Online resources can be downloaded from MyLab Math or from www.pearson.com.

Instructor's Edition

ISBN 013588280X / 9780135882801
The instructor's edition includes brief answers to all exercises and provides instructor teaching and course structure suggestions, including sample syllabi.

Instructor's Solution Manual

Includes fully worked solutions to all exercises in the text and teaching notes for many sections.

PowerPoint® Lecture Slides

These fully editable lecture slides are available for all sections of the text.

Instructor's Technology Manuals

Each manual provides detailed guidance for integrating technology throughout the course, written by faculty who teach with the software and this text. Available For MATLAB, Maple, Mathematica, and Texas Instruments graphing calculators.

TestGen®

TestGen (www.pearsoned.com/testgen) enables instructors to build, edit, print, and administer tests using a computerized bank of questions developed to cover all the objectives of the text.

Student Resources

Additional resources to enhance student success. All resources can be downloaded from MyLab Math.

Study Guide

Provides detailed worked-out solutions to every third odd-numbered exercise. Also, a complete explanation is provided whenever an odd-numbered writing exercise has a Hint in the answers. Special subsections of the *Study Guide* suggest how to master key concepts of the course. Frequent "Warnings" identify common errors and show how to prevent them. MATLAB boxes introduce commands as they are needed. Appendixes in the *Study Guide* provide comparable information about Maple, Mathematica, and TI graphing calculators. Available within MyLab math and also available for purchase separately using ISBN 9780135851234.

Getting Started with Technology

A quick-start guide for students to the technology they may use in this course. Available for MATLAB, Maple, Mathematica, or Texas Instrument graphing calculators. Downloadable from MyLab Math.

Projects

Exploratory projects, written by experienced faculty members, invite students to discover applications of linear algebra.

pearson.com/mylab/math

A Note to Students

This course is potentially the most interesting and worthwhile undergraduate mathematics course you will complete. In fact, some students have written or spoken to us after graduation and said that they still use this text occasionally as a reference in their careers at major corporations and engineering graduate schools. The following remarks offer some practical advice and information to help you master the material and enjoy the course.

In linear algebra, the *concepts* are as important as the *computations*. The simple numerical exercises that begin each exercise set only help you check your understanding of basic procedures. Later in your career, computers will do the calculations, but you will have to choose the calculations, know how to interpret the results, analyze whether the results are reasonable, then explain the results to other people. For this reason, many exercises in the text ask you to explain or justify your calculations. A written explanation is often required as part of the answer. If you are working on questions in MyLab Math, keep a notebook for calculations and notes on what you are learning. For odd-numbered exercises in the textbook, you will find either the desired explanation or at least a good hint. You must avoid the temptation to look at such answers before you have tried to write out the solution yourself. Otherwise, you are likely to think you understand something when in fact you do not.

To master the concepts of linear algebra, you will have to read and reread the text carefully. New terms are in boldface type, sometimes enclosed in a definition box. A glossary of terms is included at the end of the text. Important facts are stated as theorems or are enclosed in tinted boxes, for easy reference. We encourage you to read the Preface to learn more about the structure of this text. This will give you a framework for understanding how the course may proceed.

In a practical sense, linear algebra is a language. You must learn this language the same way you would a foreign language—with daily work. Material presented in one section is not easily understood unless you have thoroughly studied the text and worked the exercises for the preceding sections. Keeping up with the course will save you lots of time and distress!

Numerical Notes

We hope you read the Numerical Notes in the text, even if you are not using a computer or graphing calculator with the text. In real life, most applications of linear algebra involve numerical computations that are subject to some numerical error, even though that error may be extremely small. The Numerical Notes will warn you of potential difficulties in using linear algebra later in your career, and if you study the notes now, you are more likely to remember them later.

If you enjoy reading the Numerical Notes, you may want to take a course later in numerical linear algebra. Because of the high demand for increased computing power, computer scientists and mathematicians work in numerical linear algebra to develop faster and more reliable algorithms for computations, and electrical engineers design faster and smaller computers to run the algorithms. This is an exciting field, and your first course in linear algebra will help you prepare for it.

Study Guide

To help you succeed in this course, we suggest that you use the *Study Guide* available in MyLab Math and for purchase in print (ISBN 9780135851234). Not only will it help you learn linear algebra, it also will show you how to study mathematics. At strategic points in your textbook, marginal notes will remind you to check that section of the *Study Guide* for special subsections entitled "Mastering Linear Algebra Concepts." There you will find suggestions for constructing effective review sheets of key concepts. The act of preparing the sheets is one of the secrets to success in the course, because you will construct *links between ideas*. These links are the "glue" that enables you to build a solid foundation for learning and *remembering* the main concepts in the course.

The *Study Guide* contains a detailed solution to more than a third of the odd-numbered exercises, plus solutions to all odd-numbered writing exercises for which only a hint is given in the Answers section of this book. The *Guide* is separate from the text because you must learn to write solutions by yourself, without much help. (We know from years of experience that easy access to solutions in the back of the text slows the mathematical development of most students.) The *Guide* also provides warnings of common errors and helpful hints that call attention to key exercises and potential exam questions.

If you have access to technology—MATLAB, Octave, Maple, Mathematica, or a TI graphing calculator—you can save many hours of homework time. The *Study Guide* is your "lab manual" that explains how to use each of these matrix utilities. It introduces new commands when they are needed. You will also find that most software commands you might use are easily found using an online search engine. Special matrix commands will perform the computations for you!

What you do in your first few weeks of studying this course will set your pattern for the term and determine how well you finish the course. Please read "How to Study Linear Algebra" in the *Study Guide* as soon as possible. Many students have found the strategies there very helpful, and we hope you will, too.

1 Linear Equations in Linear Algebra

Introductory Example

LINEAR MODELS IN ECONOMICS AND ENGINEERING

It was late summer in 1949. Harvard Professor Wassily Leontief was carefully feeding the last of his punched cards into the university's Mark II computer. The cards contained information about the U.S. economy and represented a summary of more than 250,000 pieces of information produced by the U.S. Bureau of Labor Statistics after two years of intensive work. Leontief had divided the U.S. economy into 500 "sectors," such as the coal industry, the automotive industry, communications, and so on. For each sector, he had written a linear equation that described how the sector distributed its output to the other sectors of the economy. Because the Mark II, one of the largest computers of its day, could not handle the resulting system of 500 equations in 500 unknowns, Leontief had distilled the problem into a system of 42 equations in 42 unknowns.

Programming the Mark II computer for Leontief's 42 equations had required several months of effort, and he was anxious to see how long the computer would take to solve the problem. The Mark II hummed and blinked for 56 hours before finally producing a solution. We will discuss the nature of this solution in Sections 1.6 and 2.6.

Leontief, who was awarded the 1973 Nobel Prize in Economic Science, opened the door to a new era in mathematical modeling in economics. His efforts at Harvard in 1949 marked one of the first significant uses of computers to analyze what was then a large-scale mathematical model. Since that time, researchers in many other fields have employed computers to analyze mathematical models. Because of the massive amounts of data involved, the models are usually *linear*; that is, they are described by *systems of linear equations*.

The importance of linear algebra for applications has risen in direct proportion to the increase in computing power, with each new generation of hardware and software triggering a demand for even greater capabilities. Computer science is thus intricately linked with linear algebra through the explosive growth of parallel processing and large-scale computations.

Scientists and engineers now work on problems far more complex than even dreamed possible a few decades ago. Today, linear algebra has more potential value for students in many scientific and business fields than any other undergraduate mathematics subject! The material in this text provides the foundation for further work in many interesting areas. Here are a few possibilities; others will be described later.

- *Oil exploration.* When a ship searches for offshore oil deposits, its computers solve thousands of separate systems of linear equations *every day*. The seismic data for the equations are obtained from underwater shock waves created by explosions from air guns. The waves bounce off subsurface rocks and are measured by geophones attached to mile-long cables behind the ship.

- *Linear programming.* Many important management decisions today are made on the basis of linear programming models that use hundreds of variables. The airline industry, for instance, employs linear programs that schedule flight crews, monitor the locations of aircraft, or plan the varied schedules of support services such as maintenance and terminal operations.

- *Electrical networks.* Engineers use simulation software to design electrical circuits and microchips involving millions of transistors. Such software

relies on linear algebra techniques and systems of linear equations.

- *Artificial intelligence.* Linear algebra plays a key role in everything from scrubbing data to facial recognition.

- *Signals and signal processing.* From a digital photograph to the daily price of a stock, important information is recorded as a signal and processed using linear transformations.

- *Machine learning.* Machines (specifically computers) use linear algebra to learn about anything from online shopping preferences to speech recognition.

Systems of linear equations lie at the heart of linear algebra, and this chapter uses them to introduce some of the central concepts of linear algebra in a simple and concrete setting. Sections 1.1 and 1.2 present a systematic method for solving systems of linear equations. This algorithm will be used for computations throughout the text. Sections 1.3 and 1.4 show how a system of linear equations is equivalent to a *vector equation* and to a *matrix equation*. This equivalence will reduce problems involving linear combinations of vectors to questions about systems of linear equations. The fundamental concepts of spanning, linear independence, and linear transformations, studied in the second half of the chapter, will play an essential role throughout the text as we explore the beauty and power of linear algebra.

1.1 Systems of Linear Equations

A **linear equation** in the variables $x_1, \ldots, x_n$ is an equation that can be written in the form

$$a_1 x_1 + a_2 x_2 + \cdots + a_n x_n = b \tag{1}$$

where b and the **coefficients** $a_1, \ldots, a_n$ are real or complex numbers, usually known in advance. The subscript n may be any positive integer. In textbook examples and exercises, n is normally between 2 and 5. In real-life problems, n might be 50 or 5000, or even larger.

The equations

$$4x_1 - 5x_2 + 2 = x_1 \quad \text{and} \quad x_2 = 2\left(\sqrt{6} - x_1\right) + x_3$$

are both linear because they can be rearranged algebraically as in equation (1):

$$3x_1 - 5x_2 = -2 \quad \text{and} \quad 2x_1 + x_2 - x_3 = 2\sqrt{6}$$

The equations

$$4x_1 - 5x_2 = x_1 x_2 \quad \text{and} \quad x_2 = 2\sqrt{x_1} - 6$$

are not linear because of the presence of $x_1 x_2$ in the first equation and $\sqrt{x_1}$ in the second.

A **system of linear equations** (or a **linear system**) is a collection of one or more linear equations involving the same variables—say, $x_1, \ldots, x_n$. An example is

$$
\begin{aligned}
2x_1 - x_2 + 1.5x_3 &= 8 \\
x_1 - 4x_3 &= -7
\end{aligned}
\tag{2}
$$

A **solution** of the system is a list $(s_1, s_2, \ldots, s_n)$ of numbers that makes each equation a true statement when the values $s_1, \ldots, s_n$ are substituted for $x_1, \ldots, x_n$, respectively. For instance, $(5, 6.5, 3)$ is a solution of system (2) because, when these values are substituted in (2) for x_1, x_2, x_3, respectively, the equations simplify to $8 = 8$ and $-7 = -7$.

The set of all possible solutions is called the **solution set** of the linear system. Two linear systems are called **equivalent** if they have the same solution set. That is, each solution of the first system is a solution of the second system, and each solution of the second system is a solution of the first.

Finding the solution set of a system of two linear equations in two variables is easy because it amounts to finding the intersection of two lines. A typical problem is

$$
\begin{aligned}
x_1 - 2x_2 &= -1 \\
-x_1 + 3x_2 &= 3
\end{aligned}
$$

The graphs of these equations are lines, which we denote by ℓ_1 and ℓ_2. A pair of numbers (x_1, x_2) satisfies *both* equations in the system if and only if the point (x_1, x_2) lies on both ℓ_1 and ℓ_2. In the system above, the solution is the single point $(3, 2)$, as you can easily verify. See Figure 1.

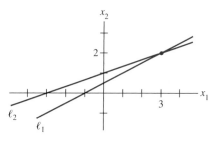

FIGURE 1 Exactly one solution.

Of course, two lines need not intersect in a single point—they could be parallel, or they could coincide and hence "intersect" at every point on the line. Figure 2 shows the graphs that correspond to the following systems:

$$
\begin{array}{ll}
\text{(a)} \quad
\begin{aligned}
x_1 - 2x_2 &= -1 \\
-x_1 + 2x_2 &= 3
\end{aligned}
&
\text{(b)} \quad
\begin{aligned}
x_1 - 2x_2 &= -1 \\
-x_1 + 2x_2 &= 1
\end{aligned}
\end{array}
$$

Figures 1 and 2 illustrate the following general fact about linear systems, to be verified in Section 1.2.

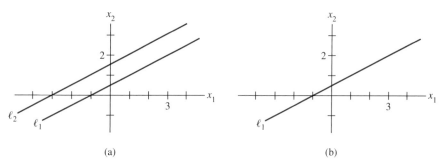

FIGURE 2 (a) No solution. (b) Infinitely many solutions.

A system of linear equations has

1. no solution, or
2. exactly one solution, or
3. infinitely many solutions.

A system of linear equations is said to be **consistent** if it has either one solution or infinitely many solutions; a system is **inconsistent** if it has no solution.

Matrix Notation

The essential information of a linear system can be recorded compactly in a rectangular array called a **matrix**. Given the system

$$
\begin{aligned}
x_1 - 2x_2 + x_3 &= 0 \\
2x_2 - 8x_3 &= 8 \\
5x_1 \qquad - 5x_3 &= 10
\end{aligned}
\tag{3}
$$

with the coefficients of each variable aligned in columns, the matrix

$$
\begin{bmatrix}
1 & -2 & 1 \\
0 & 2 & -8 \\
5 & 0 & -5
\end{bmatrix}
$$

is called the **coefficient matrix** (or **matrix of coefficients**) of the system (3), and the matrix

$$
\begin{bmatrix}
1 & -2 & 1 & 0 \\
0 & 2 & -8 & 8 \\
5 & 0 & -5 & 10
\end{bmatrix}
\tag{4}
$$

is called the **augmented matrix** of the system. (The second row here contains a zero because the second equation could be written as $0 \cdot x_1 + 2x_2 - 8x_3 = 8$.) An augmented matrix of a system consists of the coefficient matrix with an added column containing the constants from the respective right sides of the equations.

The **size** of a matrix tells how many rows and columns it has. The augmented matrix (4) above has 3 rows and 4 columns and is called a 3×4 (read "3 by 4") matrix. If m and n are positive integers, an $m \times n$ **matrix** is a rectangular array of numbers with m rows and n columns. (The number of rows always comes first.) Matrix notation will simplify the calculations in the examples that follow.

Solving a Linear System

This section and the next describe an algorithm, or a systematic procedure, for solving linear systems. The basic strategy is *to replace one system with an equivalent system (that is one with the same solution set) that is easier to solve.*

Roughly speaking, use the x_1 term in the first equation of a system to eliminate the x_1 terms in the other equations. Then use the x_2 term in the second equation to eliminate the x_2 terms in the other equations, and so on, until you finally obtain a very simple equivalent system of equations.

Three basic operations are used to simplify a linear system: Replace one equation by the sum of itself and a multiple of another equation, interchange two equations, and multiply all the terms in an equation by a nonzero constant. After the first example, you will see why these three operations do not change the solution set of the system.

EXAMPLE 1 Solve system (3).

SOLUTION The elimination procedure is shown here with and without matrix notation, and the results are placed side by side for comparison:

$$\begin{array}{rrrr} x_1 - 2x_2 + & x_3 = & 0 \\ 2x_2 - & 8x_3 = & 8 \\ 5x_1 \quad\quad - & 5x_3 = & 10 \end{array} \qquad \begin{bmatrix} 1 & -2 & 1 & 0 \\ 0 & 2 & -8 & 8 \\ 5 & 0 & -5 & 10 \end{bmatrix}$$

Keep x_1 in the first equation and eliminate it from the other equations. To do so, add -5 times equation 1 to equation 3. After some practice, this type of calculation is usually performed mentally:

$$\begin{array}{ll} -5 \cdot [\text{equation 1}] & \quad -5x_1 + 10x_2 - 5x_3 = 0 \\ +\, [\text{equation 3}] & \quad \underline{5x_1 \quad\quad - 5x_3 = 10} \\ \overline{[\text{new equation 3}]} & \quad 10x_2 - 10x_3 = 10 \end{array}$$

The result of this calculation is written in place of the original third equation:

$$\begin{array}{rrrr} x_1 - & 2x_2 + & x_3 = & 0 \\ & 2x_2 - & 8x_3 = & 8 \\ & 10x_2 - & 10x_3 = & 10 \end{array} \qquad \begin{bmatrix} 1 & -2 & 1 & 0 \\ 0 & 2 & -8 & 8 \\ 0 & 10 & -10 & 10 \end{bmatrix}$$

Now, multiply equation 2 by $\frac{1}{2}$ in order to obtain 1 as the coefficient for x_2. (This calculation will simplify the arithmetic in the next step.)

$$\begin{array}{rrrr} x_1 - & 2x_2 + & x_3 = & 0 \\ & x_2 - & 4x_3 = & 4 \\ & 10x_2 - & 10x_3 = & 10 \end{array} \qquad \begin{bmatrix} 1 & -2 & 1 & 0 \\ 0 & 1 & -4 & 4 \\ 0 & 10 & -10 & 10 \end{bmatrix}$$

Use the x_2 in equation 2 to eliminate the $10x_2$ in equation 3. The "mental" computation is

$$\begin{array}{ll} -10 \cdot [\text{equation 2}] & \quad -10x_2 + 40x_3 = -40 \\ +\, [\text{equation 3}] & \quad \underline{10x_2 - 10x_3 = \quad 10} \\ \overline{[\text{new equation 3}]} & \quad 30x_3 = -30 \end{array}$$

The result of this calculation is written in place of the previous third equation (row):

$$\begin{array}{rrrr} x_1 - 2x_2 + & x_3 = & 0 \\ x_2 - & 4x_3 = & 4 \\ & 30x_3 = & -30 \end{array} \qquad \begin{bmatrix} 1 & -2 & 1 & 0 \\ 0 & 1 & -4 & 4 \\ 0 & 0 & 30 & -30 \end{bmatrix}$$

Now, multiply equation 3 by $\frac{1}{30}$ in order to obtain 1 as the coefficient for x_3. (This calculation will simplify the arithmetic in the next step.)

$$\begin{array}{rrrr} x_1 - 2x_2 + & x_3 = & 0 \\ x_2 - & 4x_3 = & 4 \\ & x_3 = & -1 \end{array} \qquad \begin{bmatrix} 1 & -2 & 1 & 0 \\ 0 & 1 & -4 & 4 \\ 0 & 0 & 1 & -1 \end{bmatrix}$$

The new system has a *triangular* form (the intuitive term *triangular* will be replaced by a precise term in the next section):

$$\begin{aligned} x_1 - 2x_2 + x_3 &= 0 \\ x_2 - 4x_3 &= 4 \\ x_3 &= -1 \end{aligned} \qquad \begin{bmatrix} 1 & -2 & 1 & 0 \\ 0 & 1 & -4 & 4 \\ 0 & 0 & 1 & -1 \end{bmatrix}$$

Eventually, you want to eliminate the $-2x_2$ term from equation 1, but it is more efficient to use the x_3 in equation 3 first, to eliminate the $-4x_3$ and $+x_3$ terms in equations 2 and 1. The two "mental" calculations are

$$\begin{array}{ll} 4 \cdot [\text{equation 3}] & 4x_3 = -4 \\ + \,[\text{equation 2}] & x_2 - 4x_3 = 4 \\ \hline [\text{new equation 2}] & x_2 \qquad = 0 \end{array} \qquad \begin{array}{ll} -1 \cdot [\text{equation 3}] & -x_3 = 1 \\ + \,[\text{equation 1}] & x_1 - 2x_2 + x_3 = 0 \\ \hline [\text{new equation 1}] & x_1 - 2x_2 \qquad = 1 \end{array}$$

It is convenient to combine the results of these two operations:

$$\begin{aligned} x_1 - 2x_2 \quad &= 1 \\ x_2 \quad &= 0 \\ x_3 &= -1 \end{aligned} \qquad \begin{bmatrix} 1 & -2 & 0 & 1 \\ 0 & 1 & 0 & 0 \\ 0 & 0 & 1 & -1 \end{bmatrix}$$

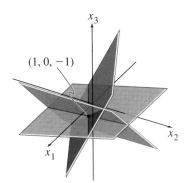

$(1, 0, -1)$

Each of the original equations determines a plane in three-dimensional space. The point $(1, 0, -1)$ lies in all three planes.

Now, having cleaned out the column above the x_3 in equation 3, move back to the x_2 in equation 2 and use it to eliminate the $-2x_2$ above it. Because of the previous work with x_3, there is now no arithmetic involving x_3 terms. Add 2 times equation 2 to equation 1 and obtain the system:

$$\begin{aligned} x_1 \qquad &= 1 \\ x_2 \qquad &= 0 \\ x_3 &= -1 \end{aligned} \qquad \begin{bmatrix} 1 & 0 & 0 & 1 \\ 0 & 1 & 0 & 0 \\ 0 & 0 & 1 & -1 \end{bmatrix}$$

The work is essentially done. It shows that the only solution of the original system is $(1, 0, -1)$. However, since there are so many calculations involved, it is a good practice to check the work. To verify that $(1, 0, -1)$ *is* a solution, substitute these values into the left side of the original system, and compute:

$$\begin{aligned} 1(1) - 2(0) + 1(-1) &= 1 - 0 - 1 = 0 \\ 2(0) - 8(-1) &= 0 + 8 = 8 \\ 5(1) - 5(-1) = 5 \quad &+ 5 = 10 \end{aligned}$$

The results agree with the right side of the original system, so $(1, 0, -1)$ is a solution of the system.   ∎

Example 1 illustrates how operations on equations in a linear system correspond to operations on the appropriate rows of the augmented matrix. The three basic operations listed earlier correspond to the following operations on the augmented matrix.

> **ELEMENTARY ROW OPERATIONS**
> **1.** (Replacement) Replace one row by the sum of itself and a multiple of another row.[1]
> **2.** (Interchange) Interchange two rows.
> **3.** (Scaling) Multiply all entries in a row by a nonzero constant.

[1] A common paraphrase of row replacement is "Add to one row a multiple of another row."

Row operations can be applied to any matrix, not merely to one that arises as the augmented matrix of a linear system. Two matrices are called **row equivalent** if there is a sequence of elementary row operations that transforms one matrix into the other.

It is important to note that row operations are *reversible*. If two rows are interchanged, they can be returned to their original positions by another interchange. If a row is scaled by a nonzero constant c, then multiplying the new row by $1/c$ produces the original row. Finally, consider a replacement operation involving two rows—say, rows 1 and 2—and suppose that c times row 1 is added to row 2 to produce a new row 2. To "reverse" this operation, add $-c$ times row 1 to (new) row 2 and obtain the original row 2. See Exercises 39–42 at the end of this section.

At the moment, we are interested in row operations on the augmented matrix of a system of linear equations. Suppose a system is changed to a new one via row operations. By considering each type of row operation, you can see that any solution of the original system remains a solution of the new system. Conversely, since the original system can be produced via row operations on the new system, each solution of the new system is also a solution of the original system. This discussion justifies the following statement.

> If the augmented matrices of two linear systems are row equivalent, then the two systems have the same solution set.

Though Example 1 is lengthy, you will find that after some practice, the calculations go quickly. Row operations in the text and exercises will usually be extremely easy to perform, allowing you to focus on the underlying concepts. Still, you must learn to perform row operations accurately because they will be used throughout the text.

The rest of this section shows how to use row operations to determine the size of a solution set, without completely solving the linear system.

Existence and Uniqueness Questions

Section 1.2 will show why a solution set for a linear system contains either no solutions, one solution, or infinitely many solutions. Answers to the following two questions will determine the nature of the solution set for a linear system.

To determine which possibility is true for a particular system, we ask two questions.

TWO FUNDAMENTAL QUESTIONS ABOUT A LINEAR SYSTEM

1. Is the system consistent; that is, does at least one solution *exist*?

2. If a solution exists, is it the *only* one; that is, is the solution *unique*?

These two questions will appear throughout the text, in many different guises. This section and the next will show how to answer these questions via row operations on the augmented matrix.

EXAMPLE 2   Determine if the following system is consistent:

$$
\begin{aligned}
x_1 - 2x_2 + x_3 &= 0 \\
2x_2 - 8x_3 &= 8 \\
5x_1 \qquad\quad - 5x_3 &= 10
\end{aligned}
$$

SOLUTION This is the system from Example 1. Suppose that we have performed the row operations necessary to obtain the triangular form

$$
\begin{aligned}
x_1 - 2x_2 + x_3 &= 0 \\
x_2 - 4x_3 &= 4 \\
x_3 &= -1
\end{aligned}
\qquad
\begin{bmatrix}
1 & -2 & 1 & 0 \\
0 & 1 & -4 & 4 \\
0 & 0 & 1 & -1
\end{bmatrix}
$$

At this point, we know x_3. Were we to substitute the value of x_3 into equation 2, we could compute x_2 and hence could determine x_1 from equation 1. So a solution exists; the system is consistent. (In fact, x_2 is uniquely determined by equation 2 since x_3 has only one possible value, and x_1 is therefore uniquely determined by equation 1. So the solution is unique.) ∎

EXAMPLE 3 Determine if the following system is consistent:

$$
\begin{aligned}
x_2 - 4x_3 &= 8 \\
2x_1 - 3x_2 + 2x_3 &= 1 \\
4x_1 - 8x_2 + 12x_3 &= 1
\end{aligned}
\tag{5}
$$

SOLUTION The augmented matrix is

$$
\begin{bmatrix}
0 & 1 & -4 & 8 \\
2 & -3 & 2 & 1 \\
4 & -8 & 12 & 1
\end{bmatrix}
$$

To obtain an x_1 in the first equation, interchange rows 1 and 2:

$$
\begin{bmatrix}
2 & -3 & 2 & 1 \\
0 & 1 & -4 & 8 \\
4 & -8 & 12 & 1
\end{bmatrix}
$$

To eliminate the $4x_1$ term in the third equation, add -2 times row 1 to row 3:

$$
\begin{bmatrix}
2 & -3 & 2 & 1 \\
0 & 1 & -4 & 8 \\
0 & -2 & 8 & -1
\end{bmatrix}
\tag{6}
$$

Next, use the x_2 term in the second equation to eliminate the $-2x_2$ term from the third equation. Add 2 times row 2 to row 3:

$$
\begin{bmatrix}
2 & -3 & 2 & 1 \\
0 & 1 & -4 & 8 \\
0 & 0 & 0 & 15
\end{bmatrix}
\tag{7}
$$

The augmented matrix is now in triangular form. To interpret it correctly, go back to equation notation:

$$
\begin{aligned}
2x_1 - 3x_2 + 2x_3 &= 1 \\
x_2 - 4x_3 &= 8 \\
0 &= 15
\end{aligned}
\tag{8}
$$

The equation $0 = 15$ is a short form of $0x_1 + 0x_2 + 0x_3 = 15$. This system in triangular form obviously has a built-in contradiction. There are no values of x_1, x_2, x_3 that satisfy (8) because the equation $0 = 15$ is never true. Since (8) and (5) have the same solution set, the original system is inconsistent (it has no solution). ∎

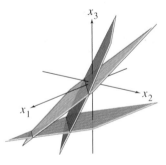

The system is inconsistent because there is no point that lies on all three planes.

Pay close attention to the augmented matrix in (7). Its last row is typical of an inconsistent system in triangular form.

Reasonable Answers

Once you have one or more solutions to a system of equations, remember to check your answer by substituting the solution you found back into the original equation. For example, if you found $(2, 1, -1)$ was a solution to the system of equations

$$
\begin{aligned}
x_1 - 2x_2 + x_3 &= 2 \\
x_1 \qquad\quad - 2x_3 &= -2 \\
x_2 + x_3 &= 3
\end{aligned}
$$

you could substitute your solution into the original equations to get

$$
\begin{aligned}
2 - 2(1) + (-1) &= -1 \neq 2 \\
2 \qquad\quad - 2(-1) &= 4 \neq -2 \\
1 + (-1) &= 0 \neq 3
\end{aligned}
$$

It is now clear that there must have been an error in your original calculations. If upon rechecking your arithmetic, you find the answer $(2, 1, 2)$, you can see that

$$
\begin{aligned}
2 - 2(1) + (2) &= 2 = 2 \\
2 \qquad\quad - 2(2) &= -2 = -2 \\
1 + 2 &= 3 = 3
\end{aligned}
$$

and you can now be confident you have a correct solution to the given system of equations.

Numerical Note

In real-world problems, systems of linear equations are solved by a computer. For a square coefficient matrix, computer programs nearly always use the elimination algorithm given here and in Section 1.2, modified slightly for improved accuracy.

The vast majority of linear algebra problems in business and industry are solved with programs that use *floating point arithmetic*. Numbers are represented as decimals $\pm .d_1 \cdots d_p \times 10^r$, where r is an integer and the number p of digits to the right of the decimal point is usually between 8 and 16. Arithmetic with such numbers typically is inexact, because the result must be rounded (or truncated) to the number of digits stored. "Roundoff error" is also introduced when a number such as $1/3$ is entered into the computer, since its decimal representation must be approximated by a finite number of digits. Fortunately, inaccuracies in floating point arithmetic seldom cause problems. The numerical notes in this book will occasionally warn of issues that you may need to consider later in your career.

Practice Problems

Throughout the text, practice problems should be attempted before working the exercises. Solutions appear after each exercise set.

1. State in words the next elementary row operation that should be performed on the system in order to solve it. [More than one answer is possible in (a).]

Practice Problems (Continued)

a.
$$x_1 + 4x_2 - 2x_3 + 8x_4 = 12$$
$$x_2 - 7x_3 + 2x_4 = -4$$
$$5x_3 - x_4 = 7$$
$$x_3 + 3x_4 = -5$$

b.
$$x_1 - 3x_2 + 5x_3 - 2x_4 = 0$$
$$x_2 + 8x_3 = -4$$
$$2x_3 = 3$$
$$x_4 = 1$$

2. The augmented matrix of a linear system has been transformed by row operations into the form below. Determine if the system is consistent.

$$\begin{bmatrix} 1 & 5 & 2 & -6 \\ 0 & 4 & -7 & 2 \\ 0 & 0 & 5 & 0 \end{bmatrix}$$

3. Is $(3, 4, -2)$ a solution of the following system?

$$5x_1 - x_2 + 2x_3 = 7$$
$$-2x_1 + 6x_2 + 9x_3 = 0$$
$$-7x_1 + 5x_2 - 3x_3 = -7$$

4. For what values of h and k is the following system consistent?

$$2x_1 - x_2 = h$$
$$-6x_1 + 3x_2 = k$$

1.1 Exercises

Solve each system in Exercises 1–4 by using elementary row operations on the equations or on the augmented matrix. Follow the systematic elimination procedure described in this section.

1.
$$x_1 + 5x_2 = 7$$
$$-2x_1 - 7x_2 = -5$$

2.
$$2x_1 + 4x_2 = -4$$
$$5x_1 + 7x_2 = 11$$

3. Find the point (x_1, x_2) that lies on the line $x_1 + 5x_2 = 7$ and on the line $x_1 - 2x_2 = -2$. See the figure.

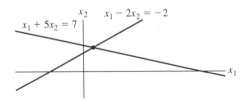

4. Find the point of intersection of the lines $x_1 - 5x_2 = 1$ and $3x_1 - 7x_2 = 5$.

Consider each matrix in Exercises 5 and 6 as the augmented matrix of a linear system. State in words the next two elementary row operations that should be performed in the process of solving the system.

5.
$$\begin{bmatrix} 1 & -4 & 5 & 0 & 7 \\ 0 & 1 & -3 & 0 & 6 \\ 0 & 0 & 1 & 0 & 2 \\ 0 & 0 & 0 & 1 & -5 \end{bmatrix}$$

6.
$$\begin{bmatrix} 1 & -6 & 4 & 0 & -1 \\ 0 & 2 & -7 & 0 & 4 \\ 0 & 0 & 1 & 2 & -3 \\ 0 & 0 & 3 & 1 & 6 \end{bmatrix}$$

In Exercises 7–10, the augmented matrix of a linear system has been reduced by row operations to the form shown. In each case, continue the appropriate row operations and describe the solution set of the original system.

7.
$$\begin{bmatrix} 1 & 7 & 3 & -4 \\ 0 & 1 & -1 & 3 \\ 0 & 0 & 0 & 1 \\ 0 & 0 & 1 & -2 \end{bmatrix}$$

8.
$$\begin{bmatrix} 1 & 1 & 2 & 0 \\ 0 & 1 & 7 & 0 \\ 0 & 0 & 2 & -2 \end{bmatrix}$$

9.
$$\begin{bmatrix} 1 & -1 & 0 & 0 & -4 \\ 0 & 1 & -3 & 0 & -7 \\ 0 & 0 & 1 & -3 & -1 \\ 0 & 0 & 0 & 0 & 4 \end{bmatrix}$$

10. $\begin{bmatrix} 1 & -2 & 0 & 3 & 0 \\ 0 & 1 & 0 & -4 & 0 \\ 0 & 0 & 1 & 0 & 0 \\ 0 & 0 & 0 & 1 & 0 \end{bmatrix}$

Solve the systems in Exercises 11–14.

11.
$$x_2 + 4x_3 = -4$$
$$x_1 + 3x_2 + 3x_3 = -2$$
$$3x_1 + 7x_2 + 5x_3 = 6$$

12.
$$x_1 - 3x_2 + 4x_3 = -4$$
$$3x_1 - 7x_2 + 7x_3 = -8$$
$$-4x_1 + 6x_2 + 2x_3 = 4$$

13.
$$x_1 \quad\quad - 3x_3 = 8$$
$$2x_1 + 2x_2 + 9x_3 = 7$$
$$x_2 + 5x_3 = -2$$

14.
$$x_1 - 3x_2 \quad\quad = 5$$
$$-x_1 + x_2 + 5x_3 = 2$$
$$x_2 + x_3 = 0$$

15. Verify that the solution you found to Exercise 11 is correct by substituting the values you obtained back into the original equations.

16. Verify that the solution you found to Exercise 12 is correct by substituting the values you obtained back into the original equations.

17. Verify that the solution you found to Exercise 13 is correct by substituting the values you obtained back into the original equations.

18. Verify that the solution you found to Exercise 14 is correct by substituting the values you obtained back into the original equations.

Determine if the systems in Exercises 19 and 20 are consistent. Do not completely solve the systems.

19.
$$x_1 \quad\quad + 3x_3 \quad\quad = 2$$
$$x_2 \quad\quad - 3x_4 = 3$$
$$-2x_2 + 3x_3 + 2x_4 = 1$$
$$3x_1 \quad\quad + 7x_4 = -5$$

20.
$$x_1 \quad\quad - 2x_4 = -3$$
$$2x_2 + 2x_3 \quad\quad = 0$$
$$x_3 + 3x_4 = 1$$
$$-2x_1 + 3x_2 + 2x_3 + x_4 = 5$$

21. Do the three lines $x_1 - 4x_2 = 1$, $2x_1 - x_2 = -3$, and $-x_1 - 3x_2 = 4$ have a common point of intersection? Explain.

22. Do the three planes $x_1 + 2x_2 + x_3 = 4$, $x_2 - x_3 = 1$, and $x_1 + 3x_2 = 0$ have at least one common point of intersection? Explain.

In Exercises 23–26, determine the value(s) of h such that the matrix is the augmented matrix of a consistent linear system.

23. $\begin{bmatrix} 1 & h & 4 \\ 3 & 6 & 8 \end{bmatrix}$ **24.** $\begin{bmatrix} 1 & h & -3 \\ -2 & 4 & 6 \end{bmatrix}$

25. $\begin{bmatrix} 1 & 3 & -2 \\ -4 & h & 8 \end{bmatrix}$ **26.** $\begin{bmatrix} 2 & -3 & h \\ -6 & 9 & 5 \end{bmatrix}$

In Exercises 27–34, key statements from this section are either quoted directly, restated slightly (but still true), or altered in some way that makes them false in some cases. Mark each statement True or False, and *justify* your answer. (If true, give the approximate location where a similar statement appears, or refer to a definition or theorem. If false, give the location of a statement that has been quoted or used incorrectly, or cite an example that shows the statement is not true in all cases.) Similar true/false questions will appear in many sections of the text and will be flagged with a **(T/F)** at the beginning of the question.

27. **(T/F)** Every elementary row operation is reversible.

28. **(T/F)** Elementary row operations on an augmented matrix never change the solution set of the associated linear system.

29. **(T/F)** A 5×6 matrix has six rows.

30. **(T/F)** Two matrices are row equivalent if they have the same number of rows.

31. **(T/F)** The solution set of a linear system involving variables $x_1, \ldots, x_n$ is a list of numbers $(s_1, \ldots, s_n)$ that makes each equation in the system a true statement when the values $s_1, \ldots, s_n$ are substituted for $x_1, \ldots, x_n$, respectively.

32. **(T/F)** An inconsistent system has more than one solution.

33. **(T/F)** Two fundamental questions about a linear system involve existence and uniqueness.

34. **(T/F)** Two linear systems are equivalent if they have the same solution set.

35. Find an equation involving g, h, and k that makes this augmented matrix correspond to a consistent system:
$$\begin{bmatrix} 1 & -4 & 7 & g \\ 0 & 3 & -5 & h \\ -2 & 5 & -9 & k \end{bmatrix}$$

36. Construct three different augmented matrices for linear systems whose solution set is $x_1 = -2$, $x_2 = 1$, $x_3 = 0$.

37. Suppose the system below is consistent for all possible values of f and g. What can you say about the coefficients c and d? Justify your answer.
$$x_1 + 3x_2 = f$$
$$cx_1 + dx_2 = g$$

38. Suppose a, b, c, and d are constants such that a is not zero and the system below is consistent for all possible values of f and g. What can you say about the numbers a, b, c, and d? Justify your answer.

$$ax_1 + bx_2 = f$$
$$cx_1 + dx_2 = g$$

In Exercises 39–42, find the elementary row operation that transforms the first matrix into the second, and then find the reverse row operation that transforms the second matrix into the first.

39.
$$\begin{bmatrix} 0 & -2 & 5 \\ 1 & 4 & -7 \\ 3 & -1 & 6 \end{bmatrix}, \begin{bmatrix} 1 & 4 & -7 \\ 0 & -2 & 5 \\ 3 & -1 & 6 \end{bmatrix}$$

40.
$$\begin{bmatrix} 1 & 3 & -4 \\ 0 & -2 & 6 \\ 0 & -5 & 9 \end{bmatrix}, \begin{bmatrix} 1 & 3 & -4 \\ 0 & 1 & -3 \\ 0 & -5 & 9 \end{bmatrix}$$

41.
$$\begin{bmatrix} 1 & -2 & 1 & 0 \\ 0 & 5 & -2 & 8 \\ 4 & -1 & 3 & -6 \end{bmatrix}, \begin{bmatrix} 1 & -2 & 1 & 0 \\ 0 & 5 & -2 & 8 \\ 0 & 7 & -1 & -6 \end{bmatrix}$$

42.
$$\begin{bmatrix} 1 & 2 & -5 & 0 \\ 0 & 1 & -3 & -2 \\ 0 & -3 & 9 & 5 \end{bmatrix}, \begin{bmatrix} 1 & 2 & -5 & 0 \\ 0 & 1 & -3 & -2 \\ 0 & 0 & 0 & -1 \end{bmatrix}$$

An important concern in the study of heat transfer is to determine the steady-state temperature distribution of a thin plate when the temperature around the boundary is known. Assume the plate shown in the figure represents a cross section of a metal beam, with negligible heat flow in the direction perpendicular to the plate. Let $T_1, \ldots, T_4$ denote the temperatures at the four interior nodes of the mesh in the figure. The temperature at a node is approximately equal to the average of the four nearest nodes—to the left, above, to the right, and below.[2] For instance,

$$T_1 = (10 + 20 + T_2 + T_4)/4, \quad \text{or} \quad 4T_1 - T_2 - T_4 = 30$$

43. Write a system of four equations whose solution gives estimates for the temperatures $T_1, \ldots, T_4$.

44. Solve the system of equations from Exercise 43. [*Hint:* To speed up the calculations, interchange rows 1 and 4 before starting "replace" operations.]

---

[2] See Frank M. White, *Heat and Mass Transfer* (Reading, MA: Addison-Wesley Publishing, 1991), pp. 145–149.

Solutions to Practice Problems

1. a. For "hand computation," the best choice is to interchange equations 3 and 4. Another possibility is to multiply equation 3 by $1/5$. Or, replace equation 4 by its sum with $-1/5$ times row 3. (In any case, do not use the x_2 in equation 2 to eliminate the $4x_2$ in equation 1. Wait until a triangular form has been reached and the x_3 terms and x_4 terms have been eliminated from the first two equations.)

   b. The system is in triangular form. Further simplification begins with the x_4 in the fourth equation. Use the x_4 to eliminate all x_4 terms above it. The appropriate step now is to add 2 times equation 4 to equation 1. (After that, move to equation 3, multiply it by $1/2$, and then use the equation to eliminate the x_3 terms above it.)

2. The system corresponding to the augmented matrix is

$$
\begin{aligned}
x_1 + 5x_2 + 2x_3 &= -6 \\
4x_2 - 7x_3 &= 2 \\
5x_3 &= 0
\end{aligned}
$$

The third equation makes $x_3 = 0$, which is certainly an allowable value for x_3. After eliminating the x_3 terms in equations 1 and 2, you could go on to solve for unique values for x_2 and x_1. Hence a solution exists, and it is unique. Contrast this situation with that in Example 3.

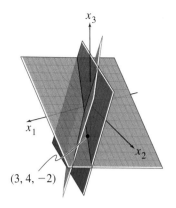

Since $(3, 4, -2)$ satisfies the first two equations, it is on the line of the intersection of the first two planes. Since $(3, 4, -2)$ does not satisfy all three equations, it does not lie on all three planes.

3. It is easy to check if a specific list of numbers is a solution. Set $x_1 = 3$, $x_2 = 4$, and $x_3 = -2$, and find that

$$5(3) - (4) + 2(-2) = 15 - 4 - 4 = 7$$
$$-2(3) + 6(4) + 9(-2) = -6 + 24 - 18 = 0$$
$$-7(3) + 5(4) - 3(-2) = -21 + 20 + 6 = 5$$

Although the first two equations are satisfied, the third is not, so $(3, 4, -2)$ is not a solution of the system. Notice the use of parentheses when making the substitutions. They are strongly recommended as a guard against arithmetic errors.

4. When the second equation is replaced by its sum with 3 times the first equation, the system becomes

$$2x_1 - x_2 = h$$
$$0 = k + 3h$$

If $k + 3h$ is nonzero, the system has no solution. The system is consistent for any values of h and k that make $k + 3h = 0$.

1.2 Row Reduction and Echelon Forms

This section refines the method of Section 1.1 into a row reduction algorithm that will enable us to analyze any system of linear equations.[1] By using only the first part of the algorithm, we will be able to answer the fundamental existence and uniqueness questions posed in Section 1.1.

The algorithm applies to any matrix, whether or not the matrix is viewed as an augmented matrix for a linear system. So the first part of this section concerns an arbitrary rectangular matrix and begins by introducing two important classes of matrices that include the "triangular" matrices of Section 1.1. In the definitions that follow, a *nonzero* row or column in a matrix means a row or column that contains at least one nonzero entry; a **leading entry** of a row refers to the leftmost nonzero entry (in a nonzero row).

DEFINITION

A rectangular matrix is in **echelon form** (or **row echelon form**) if it has the following three properties:

1. All nonzero rows are above any rows of all zeros.

2. Each leading entry of a row is in a column to the right of the leading entry of the row above it.

3. All entries in a column below a leading entry are zeros.

If a matrix in echelon form satisfies the following additional conditions, then it is in **reduced echelon form** (or **reduced row echelon form**):

4. The leading entry in each nonzero row is 1.

5. Each leading 1 is the only nonzero entry in its column.

---

[1] The algorithm here is a variant of what is commonly called *Gaussian elimination*. A similar elimination method for linear systems was used by Chinese mathematicians in about 250 B.C. The process was unknown in Western culture until the nineteenth century, when a famous German mathematician, Carl Friedrich Gauss, discovered it. A German engineer, Wilhelm Jordan, popularized the algorithm in an 1888 text on geodesy.

An **echelon matrix** (respectively, **reduced echelon matrix**) is one that is in echelon form (respectively, reduced echelon form). Property 2 says that the leading entries form an *echelon* ("steplike") pattern that moves down and to the right through the matrix. Property 3 is a simple consequence of property 2, but we include it for emphasis.

The "triangular" matrices of Section 1.1, such as

$$\begin{bmatrix} 2 & -3 & 2 & 1 \\ 0 & 1 & -4 & 8 \\ 0 & 0 & 0 & 5/2 \end{bmatrix} \text{ and } \begin{bmatrix} 1 & 0 & 0 & 29 \\ 0 & 1 & 0 & 16 \\ 0 & 0 & 1 & 3 \end{bmatrix}$$

are in echelon form. In fact, the second matrix is in reduced echelon form. Here are additional examples.

EXAMPLE 1   The following matrices are in echelon form. The leading entries (■) may have any nonzero value; the starred entries (∗) may have any value (including zero).

$$\begin{bmatrix} ■ & ∗ & ∗ & ∗ \\ 0 & ■ & ∗ & ∗ \\ 0 & 0 & 0 & 0 \\ 0 & 0 & 0 & 0 \end{bmatrix}, \begin{bmatrix} 0 & ■ & ∗ & ∗ & ∗ & ∗ & ∗ & ∗ & ∗ & ∗ \\ 0 & 0 & 0 & ■ & ∗ & ∗ & ∗ & ∗ & ∗ & ∗ \\ 0 & 0 & 0 & 0 & ■ & ∗ & ∗ & ∗ & ∗ & ∗ \\ 0 & 0 & 0 & 0 & 0 & ■ & ∗ & ∗ & ∗ & ∗ \\ 0 & 0 & 0 & 0 & 0 & 0 & 0 & 0 & ■ & ∗ \end{bmatrix}$$

The following matrices are in reduced echelon form because the leading entries are 1's, and there are 0's below *and above* each leading 1.

$$\begin{bmatrix} 1 & 0 & ∗ & ∗ \\ 0 & 1 & ∗ & ∗ \\ 0 & 0 & 0 & 0 \\ 0 & 0 & 0 & 0 \end{bmatrix}, \begin{bmatrix} 0 & 1 & ∗ & 0 & 0 & 0 & ∗ & ∗ & 0 & ∗ \\ 0 & 0 & 0 & 1 & 0 & 0 & ∗ & ∗ & 0 & ∗ \\ 0 & 0 & 0 & 0 & 1 & 0 & ∗ & ∗ & 0 & ∗ \\ 0 & 0 & 0 & 0 & 0 & 1 & ∗ & ∗ & 0 & ∗ \\ 0 & 0 & 0 & 0 & 0 & 0 & 0 & 0 & 1 & ∗ \end{bmatrix}$$

■

Any nonzero matrix may be **row reduced** (that is, transformed by elementary row operations) into more than one matrix in echelon form, using different sequences of row operations. However, the reduced echelon form one obtains from a matrix is unique. The following theorem is proved in Appendix A at the end of the text.

THEOREM 1

Uniqueness of the Reduced Echelon Form

Each matrix is row equivalent to one and only one reduced echelon matrix.

If a matrix A is row equivalent to an echelon matrix U, we call U **an echelon form** (or row echelon form) **of** A; if U is in reduced echelon form, we call U **the reduced echelon form of** A. [Most matrix programs and calculators with matrix capabilities use the abbreviation RREF for reduced (row) echelon form. Some use REF for (row) echelon form.]

Pivot Positions

When row operations on a matrix produce an echelon form, further row operations to obtain the reduced echelon form do not change the positions of the leading entries. Since the reduced echelon form is unique, *the leading entries are always in the same positions in any echelon form obtained from a given matrix.* These leading entries correspond to leading 1's in the reduced echelon form.

DEFINITION | A **pivot position** in a matrix A is a location in A that corresponds to a leading 1 in the reduced echelon form of A. A **pivot column** is a column of A that contains a pivot position.

In Example 1, the squares (■) identify the pivot positions. Many fundamental concepts in the first four chapters will be connected in one way or another with pivot positions in a matrix.

EXAMPLE 2   Row reduce the matrix A below to echelon form, and locate the pivot columns of A.

$$A = \begin{bmatrix} 0 & -3 & -6 & 4 & 9 \\ -1 & -2 & -1 & 3 & 1 \\ -2 & -3 & 0 & 3 & -1 \\ 1 & 4 & 5 & -9 & -7 \end{bmatrix}$$

SOLUTION   Use the same basic strategy as in Section 1.1. The top of the leftmost nonzero column is the first pivot position. A nonzero entry, or *pivot*, must be placed in this position. A good choice is to interchange rows 1 and 4 (because the mental computations in the next step will not involve fractions).

$$\begin{bmatrix} 1 & 4 & 5 & -9 & -7 \\ -1 & -2 & -1 & 3 & 1 \\ -2 & -3 & 0 & 3 & -1 \\ 0 & -3 & -6 & 4 & 9 \end{bmatrix}$$

Create zeros below the pivot, 1, by adding multiples of the first row to the rows below, and obtain matrix (1) below. The pivot position in the second row must be as far left as possible—namely in the second column. Choose the 2 in this position as the next pivot.

$$\begin{bmatrix} 1 & 4 & 5 & -9 & -7 \\ 0 & 2 & 4 & -6 & -6 \\ 0 & 5 & 10 & -15 & -15 \\ 0 & -3 & -6 & 4 & 9 \end{bmatrix} \tag{1}$$

Add $-5/2$ times row 2 to row 3, and add $3/2$ times row 2 to row 4.

$$\begin{bmatrix} 1 & 4 & 5 & -9 & -7 \\ 0 & 2 & 4 & -6 & -6 \\ 0 & 0 & 0 & 0 & 0 \\ 0 & 0 & 0 & -5 & 0 \end{bmatrix} \tag{2}$$

The matrix in (2) is different from any encountered in Section 1.1. There is no way to create a leading entry in column 3! (We can't use row 1 or 2 because doing so would destroy the echelon arrangement of the leading entries already produced.) However, if we interchange rows 3 and 4, we can produce a leading entry in column 4.

$$
\begin{bmatrix}
1 & 4 & 5 & -9 & -7 \\
0 & 2 & 4 & -6 & -6 \\
0 & 0 & 0 & -5 & 0 \\
0 & 0 & 0 & 0 & 0
\end{bmatrix}
\qquad \text{General form:} \qquad
\begin{bmatrix}
\blacksquare & * & * & * & * \\
0 & \blacksquare & * & * & * \\
0 & 0 & 0 & \blacksquare & * \\
0 & 0 & 0 & 0 & 0
\end{bmatrix}
$$

The matrix is in echelon form and thus reveals that columns 1, 2, and 4 of A are pivot columns.

$$
A =
\begin{bmatrix}
0 & -3 & -6 & 4 & 9 \\
-1 & -2 & -1 & 3 & 1 \\
-2 & -3 & 0 & 3 & -1 \\
1 & 4 & 5 & -9 & -7
\end{bmatrix}
\tag{3}
$$

A **pivot**, as illustrated in Example 2, is a nonzero number in a pivot position that is used as needed to create zeros via row operations. The pivots in Example 2 were 1, 2, and -5. Notice that these numbers are not the same as the actual elements of A in the highlighted pivot positions shown in (3).

With Example 2 as a guide, we are ready to describe an efficient procedure for transforming a matrix into an echelon or reduced echelon matrix. Careful study and mastery of this procedure now will pay rich dividends later in the course.

The Row Reduction Algorithm

The algorithm that follows consists of four steps, and it produces a matrix in echelon form. A fifth step produces a matrix in reduced echelon form. We illustrate the algorithm by an example.

EXAMPLE 3 Apply elementary row operations to transform the following matrix first into echelon form and then into reduced echelon form:

$$
\begin{bmatrix}
0 & 3 & -6 & 6 & 4 & -5 \\
3 & -7 & 8 & -5 & 8 & 9 \\
3 & -9 & 12 & -9 & 6 & 15
\end{bmatrix}
$$

SOLUTION

Step 1

Begin with the leftmost nonzero column. This is a pivot column. The pivot position is at the top.

$$
\begin{bmatrix}
0 & 3 & -6 & 6 & 4 & -5 \\
3 & -7 & 8 & -5 & 8 & 9 \\
3 & -9 & 12 & -9 & 6 & 15
\end{bmatrix}
$$

Pivot column

> ### Step 2
> Select a nonzero entry in the pivot column as a pivot. If necessary, interchange rows to move this entry into the pivot position.

Interchange rows 1 and 3. (We could have interchanged rows 1 and 2 instead.)

$$\begin{array}{c} \overset{\text{Pivot}}{} \\ \begin{bmatrix} 3 & -9 & 12 & -9 & 6 & 15 \\ 3 & -7 & 8 & -5 & 8 & 9 \\ 0 & 3 & -6 & 6 & 4 & -5 \end{bmatrix} \end{array}$$

> ### Step 3
> Use row replacement operations to create zeros in all positions below the pivot.

As a preliminary step, we could divide the top row by the pivot, 3. But with two 3's in column 1, it is just as easy to add -1 times row 1 to row 2.

$$\begin{array}{c} \overset{\text{Pivot}}{} \\ \begin{bmatrix} 3 & -9 & 12 & -9 & 6 & 15 \\ 0 & 2 & -4 & 4 & 2 & -6 \\ 0 & 3 & -6 & 6 & 4 & -5 \end{bmatrix} \end{array}$$

> ### Step 4
> Cover (or ignore) the row containing the pivot position and cover all rows, if any, above it. Apply steps 1–3 to the submatrix that remains. Repeat the process until there are no more nonzero rows to modify.

With row 1 covered, step 1 shows that column 2 is the next pivot column; for step 2, select as a pivot the "top" entry in that column.

$$\begin{array}{c} \overset{\text{Pivot}}{} \\ \begin{bmatrix} 3 & -9 & 12 & -9 & 6 & 15 \\ 0 & 2 & -4 & 4 & 2 & -6 \\ 0 & 3 & -6 & 6 & 4 & -5 \end{bmatrix} \\ \underset{\text{New pivot column}}{} \end{array}$$

For step 3, we could insert an optional step of dividing the "top" row of the submatrix by the pivot, 2. Instead, we add $-3/2$ times the "top" row to the row below. This produces

$$\begin{bmatrix} 3 & -9 & 12 & -9 & 6 & 15 \\ 0 & 2 & -4 & 4 & 2 & -6 \\ 0 & 0 & 0 & 0 & 1 & 4 \end{bmatrix}$$

When we cover the row containing the second pivot position for step 4, we are left with a new submatrix having only one row:

$$\begin{bmatrix} 3 & -9 & 12 & -9 & 6 & 15 \\ 0 & 2 & -4 & 4 & 2 & -6 \\ 0 & 0 & 0 & 0 & 1 & 4 \end{bmatrix}$$

Pivot

Steps 1–3 require no work for this submatrix, and we have reached an echelon form of the full matrix. If we want the reduced echelon form, we perform one more step.

> **Step 5**
>
> Beginning with the rightmost pivot and working upward and to the left, create zeros above each pivot. If a pivot is not 1, make it 1 by a scaling operation.

The rightmost pivot is in row 3. Create zeros above it, adding suitable multiples of row 3 to rows 2 and 1.

$$\begin{bmatrix} 3 & -9 & 12 & -9 & 0 & -9 \\ 0 & 2 & -4 & 4 & 0 & -14 \\ 0 & 0 & 0 & 0 & 1 & 4 \end{bmatrix} \begin{array}{l} \leftarrow \text{Row } 1 + (-6) \cdot \text{row } 3 \\ \leftarrow \text{Row } 2 + (-2) \cdot \text{row } 3 \end{array}$$

The next pivot is in row 2. Scale this row, dividing by the pivot.

$$\begin{bmatrix} 3 & -9 & 12 & -9 & 0 & -9 \\ 0 & 1 & -2 & 2 & 0 & -7 \\ 0 & 0 & 0 & 0 & 1 & 4 \end{bmatrix} \leftarrow \text{Row scaled by } \tfrac{1}{2}$$

Create a zero in column 2 by adding 9 times row 2 to row 1.

$$\begin{bmatrix} 3 & 0 & -6 & 9 & 0 & -72 \\ 0 & 1 & -2 & 2 & 0 & -7 \\ 0 & 0 & 0 & 0 & 1 & 4 \end{bmatrix} \leftarrow \text{Row } 1 + (9) \cdot \text{row } 2$$

Finally, scale row 1, dividing by the pivot, 3.

$$\begin{bmatrix} 1 & 0 & -2 & 3 & 0 & -24 \\ 0 & 1 & -2 & 2 & 0 & -7 \\ 0 & 0 & 0 & 0 & 1 & 4 \end{bmatrix} \leftarrow \text{Row scaled by } \tfrac{1}{3}$$

This is the reduced echelon form of the original matrix. ∎

The combination of steps 1–4 is called the **forward phase** of the row reduction algorithm. Step 5, which produces the unique reduced echelon form, is called the **backward phase**.

> **Numerical Note**
>
> In step 2 on page 17, a computer program usually selects as a pivot the entry in a column having the largest absolute value. This strategy, called **partial pivoting**, is used because it reduces roundoff errors in the calculations.

Solutions of Linear Systems

The row reduction algorithm leads directly to an explicit description of the solution set of a linear system when the algorithm is applied to the augmented matrix of the system.

Suppose, for example, that the augmented matrix of a linear system has been changed into the equivalent *reduced* echelon form

$$\begin{bmatrix} 1 & 0 & -5 & 1 \\ 0 & 1 & 1 & 4 \\ 0 & 0 & 0 & 0 \end{bmatrix}$$

There are three variables because the augmented matrix has four columns. The associated system of equations is

$$\begin{aligned} x_1 \quad\quad - 5x_3 &= 1 \\ x_2 + \; x_3 &= 4 \\ 0 &= 0 \end{aligned} \tag{4}$$

The variables x_1 and x_2 corresponding to pivot columns in the matrix are called **basic variables**.[2] The other variable, x_3, is called a **free variable**.

Whenever a system is consistent, as in (4), the solution set can be described explicitly by solving the *reduced* system of equations for the basic variables in terms of the free variables. This operation is possible because the reduced echelon form places each basic variable in one and only one equation. In (4), solve the first equation for x_1 and the second for x_2. (Ignore the third equation; it offers no restriction on the variables.)

$$\begin{cases} x_1 = 1 + 5x_3 \\ x_2 = 4 - x_3 \\ x_3 \text{ is free} \end{cases} \tag{5}$$

The statement "x_3 is free" means that you are free to choose any value for x_3. Once that is done, the formulas in (5) determine the values for x_1 and x_2. For instance, when $x_3 = 0$, the solution is $(1, 4, 0)$; when $x_3 = 1$, the solution is $(6, 3, 1)$. *Each different choice of x_3 determines a (different) solution of the system, and every solution of the system is determined by a choice of x_3.*

EXAMPLE 4   Find the general solution of the linear system whose augmented matrix has been reduced to

$$\begin{bmatrix} 1 & 6 & 2 & -5 & -2 & -4 \\ 0 & 0 & 2 & -8 & -1 & 3 \\ 0 & 0 & 0 & 0 & 1 & 7 \end{bmatrix}$$

SOLUTION   The matrix is in echelon form, but we want the reduced echelon form before solving for the basic variables. The row reduction is completed next. The symbol $\sim$ before a matrix indicates that the matrix is row equivalent to the preceding matrix.

$$\begin{bmatrix} 1 & 6 & 2 & -5 & -2 & -4 \\ 0 & 0 & 2 & -8 & -1 & 3 \\ 0 & 0 & 0 & 0 & 1 & 7 \end{bmatrix} \sim \begin{bmatrix} 1 & 6 & 2 & -5 & 0 & 10 \\ 0 & 0 & 2 & -8 & 0 & 10 \\ 0 & 0 & 0 & 0 & 1 & 7 \end{bmatrix}$$

$$\sim \begin{bmatrix} 1 & 6 & 2 & -5 & 0 & 10 \\ 0 & 0 & 1 & -4 & 0 & 5 \\ 0 & 0 & 0 & 0 & 1 & 7 \end{bmatrix} \sim \begin{bmatrix} 1 & 6 & 0 & 3 & 0 & 0 \\ 0 & 0 & 1 & -4 & 0 & 5 \\ 0 & 0 & 0 & 0 & 1 & 7 \end{bmatrix}$$

---

[2] Some texts use the term *leading variables* because they correspond to the columns containing leading entries.

There are five variables because the augmented matrix has six columns. The associated system now is

$$\begin{aligned}
x_1 + 6x_2 \quad\; + 3x_4 \quad\;\; &= 0 \\
x_3 - 4x_4 \quad\;\; &= 5 \\
x_5 &= 7
\end{aligned} \tag{6}$$

The pivot columns of the matrix are 1, 3, and 5, so the basic variables are x_1, x_3, and x_5. The remaining variables, x_2 and x_4, must be free. Solve for the basic variables to obtain the general solution:

$$\begin{cases}
x_1 = -6x_2 - 3x_4 \\
x_2 \text{ is free} \\
x_3 = 5 + 4x_4 \\
x_4 \text{ is free} \\
x_5 = 7
\end{cases} \tag{7}$$

Note that the value of x_5 is already fixed by the third equation in system (6). ∎

Parametric Descriptions of Solution Sets

The descriptions in (5) and (7) are *parametric descriptions* of solution sets in which the free variables act as parameters. *Solving a system* amounts to finding a parametric description of the solution set or determining that the solution set is empty.

Whenever a system is consistent and has free variables, the solution set has many parametric descriptions. For instance, in system (4), we may add 5 times equation 2 to equation 1 and obtain the equivalent system

$$\begin{aligned}
x_1 + 5x_2 \quad\;\; &= 21 \\
x_2 + x_3 &= \;\; 4
\end{aligned}$$

We could treat x_2 as a parameter and solve for x_1 and x_3 in terms of x_2, and we would have an accurate description of the solution set. However, to be consistent, we make the (arbitrary) convention of always using the free variables as the parameters for describing a solution set. (The answer section at the end of the text also reflects this convention.)

Whenever a system is inconsistent, the solution set is empty, even when the system has free variables. In this case, the solution set has *no* parametric representation.

Back-Substitution

Consider the following system, whose augmented matrix is in echelon form but is *not* in reduced echelon form:

$$\begin{aligned}
x_1 - 7x_2 + 2x_3 - 5x_4 + 8x_5 &= \;\; 10 \\
x_2 - 3x_3 + 3x_4 + \;\; x_5 &= -5 \\
x_4 - \;\; x_5 &= \;\;\; 4
\end{aligned}$$

A computer program would solve this system by back-substitution, rather than by computing the reduced echelon form. That is, the program would solve equation 3 for x_4 in terms of x_5 and substitute the expression for x_4 into equation 2, solve equation 2 for x_2, and then substitute the expressions for x_2 and x_4 into equation 1 and solve for x_1.

Our matrix format for the backward phase of row reduction, which produces the reduced echelon form, has the same number of arithmetic operations as back-substitution. But the discipline of the matrix format substantially reduces the likelihood of errors during hand computations. The best strategy is to use only the *reduced* echelon form to solve a system! The *Study Guide* that accompanies this text offers several helpful suggestions for performing row operations accurately and rapidly.

Numerical Note

In general, the forward phase of row reduction takes much longer than the backward phase. An algorithm for solving a system is usually measured in flops (or floating point operations). A **flop** is one arithmetic operation $(+, -, *, /)$ on two real floating point numbers.[3] For an $n \times (n + 1)$ matrix, the reduction to echelon form can take $2n^3/3 + n^2/2 - 7n/6$ flops (which is approximately $2n^3/3$ flops when n is moderately large—say, $n \geq 30$). In contrast, further reduction to reduced echelon form needs at most n^2 flops.

Existence and Uniqueness Questions

Although a nonreduced echelon form is a poor tool for solving a system, this form is just the right device for answering two fundamental questions posed in Section 1.1.

EXAMPLE 5 Determine the existence and uniqueness of the solutions to the system

$$
\begin{aligned}
3x_2 - 6x_3 + 6x_4 + 4x_5 &= -5 \\
3x_1 - 7x_2 + 8x_3 - 5x_4 + 8x_5 &= 9 \\
3x_1 - 9x_2 + 12x_3 - 9x_4 + 6x_5 &= 15
\end{aligned}
$$

SOLUTION The augmented matrix of this system was row reduced in Example 3 to

$$
\begin{bmatrix}
3 & -9 & 12 & -9 & 6 & 15 \\
0 & 2 & -4 & 4 & 2 & -6 \\
0 & 0 & 0 & 0 & 1 & 4
\end{bmatrix}
\tag{8}
$$

The basic variables are x_1, x_2, and x_5; the free variables are x_3 and x_4. There is no equation such as $0 = 1$ that would indicate an inconsistent system, so we could use back-substitution to find a solution. But the *existence* of a solution is already clear in (8). Also, the solution is *not unique* because there are free variables. Each different choice of x_3 and x_4 determines a different solution. Thus the system has infinitely many solutions. ■

When a system is in echelon form and contains no equation of the form $0 = b$, with b nonzero, every nonzero equation contains a basic variable with a nonzero coefficient. Either the basic variables are completely determined (with no free variables) or at least one of the basic variables may be expressed in terms of one or more free variables. In the former case, there is a unique solution; in the latter case, there are infinitely many solutions (one for each choice of values for the free variables).

These remarks justify the following theorem.

---

[3] Traditionally, a *flop* was only a multiplication or division because addition and subtraction took much less time and could be ignored. The definition of *flop* given here is preferred now, as a result of advances in computer architecture. See Golub and Van Loan, *Matrix Computations*, 2nd ed. (Baltimore: The Johns Hopkins Press, 1989), pp. 19–20.

THEOREM 2

Existence and Uniqueness Theorem

A linear system is consistent if and only if the rightmost column of the augmented matrix is *not* a pivot column—that is, if and only if an echelon form of the augmented matrix has *no* row of the form

$$[\,0 \;\; \cdots \;\; 0 \;\; b\,] \qquad \text{with } b \text{ nonzero}$$

If a linear system is consistent, then the solution set contains either (i) a unique solution, when there are no free variables, or (ii) infinitely many solutions, when there is at least one free variable.

The following procedure outlines how to find and describe all solutions of a linear system.

USING ROW REDUCTION TO SOLVE A LINEAR SYSTEM

1. Write the augmented matrix of the system.

2. Use the row reduction algorithm to obtain an equivalent augmented matrix in echelon form. Decide whether the system is consistent. If there is no solution, stop; otherwise, go to the next step.

3. Continue row reduction to obtain the reduced echelon form.

4. Write the system of equations corresponding to the matrix obtained in step 3.

5. Rewrite each nonzero equation from step 4 so that its one basic variable is expressed in terms of any free variables appearing in the equation.

Reasonable Answers

Remember that each augmented matrix corresponds to a system of equations. If you row reduce the augmented matrix $\begin{bmatrix} 1 & -2 & 1 & 2 \\ 1 & -1 & 2 & 5 \\ 0 & 1 & 1 & 3 \end{bmatrix}$ to get the matrix $\begin{bmatrix} 1 & 0 & 3 & 8 \\ 0 & 1 & 1 & 3 \\ 0 & 0 & 0 & 0 \end{bmatrix}$, the solution set is

$$\begin{cases} x_1 = 8 - 3x_3 \\ x_2 = 3 - x_3 \\ x_3 \text{ is free} \end{cases}$$

The system of equations corresponding to the original augmented matrix is

$$\begin{aligned} x_1 \;-\; 2x_2 \;+\;\;\; x_3 &= 2 \\ x_1 \;-\;\;\; x_2 \;+\; 2x_3 &= 5 \\ x_2 \;+\;\;\; x_3 &= 3 \end{aligned}$$

You can now check whether your solution is correct by substituting it into the original equations. Notice that you can just leave the free variables in the solution.

$$
\begin{array}{rcrcrcl}
(8-3x_3) & - & 2(3-x_3) & + & (x_3) & = & 8-3x_3-6+2x_3+x_3 = 2 \\
(8-3x_3) & - & (3-x_3) & + & 2(x_3) & = & 8-3x_3-3+x_3+2x_3 = 5 \\
& & (3-x_3) & + & (x_3) & = & 3-x_3+x_3 = 3
\end{array}
$$

You can now be confident you have a correct solution to the system of equations represented by the augmented matrix.

Practice Problems

1. Find the general solution of the linear system whose augmented matrix is

$$
\begin{bmatrix}
1 & -3 & -5 & 0 \\
0 & 1 & -1 & -1
\end{bmatrix}
$$

2. Find the general solution of the system

$$
\begin{array}{rcrcrcrcr}
x_1 & - & 2x_2 & - & x_3 & + & 3x_4 & = & 0 \\
-2x_1 & + & 4x_2 & + & 5x_3 & - & 5x_4 & = & 3 \\
3x_1 & - & 6x_2 & - & 6x_3 & + & 8x_4 & = & 2
\end{array}
$$

3. Suppose a 4×7 coefficient matrix for a system of equations has 4 pivots. Is the system consistent? If the system is consistent, how many solutions are there?

1.2 Exercises

In Exercises 1 and 2, determine which matrices are in reduced echelon form and which others are only in echelon form.

1. a. $\begin{bmatrix} 1 & 0 & 0 & 0 \\ 0 & 1 & 0 & 0 \\ 0 & 0 & 1 & 1 \end{bmatrix}$ **b.** $\begin{bmatrix} 1 & 0 & 1 & 0 \\ 0 & 0 & 1 & 0 \\ 0 & 0 & 0 & 1 \end{bmatrix}$

c. $\begin{bmatrix} 1 & 0 & 0 & 0 \\ 0 & 1 & 1 & 0 \\ 0 & 0 & 0 & 0 \\ 0 & 0 & 0 & 0 \end{bmatrix}$ **d.** $\begin{bmatrix} 1 & 1 & 0 & 1 & 1 \\ 0 & 2 & 0 & 2 & 2 \\ 0 & 0 & 0 & 3 & 3 \\ 0 & 0 & 0 & 0 & 4 \end{bmatrix}$

2. a. $\begin{bmatrix} 1 & 1 & 0 & 1 \\ 0 & 0 & 1 & 1 \\ 0 & 0 & 0 & 0 \end{bmatrix}$ **b.** $\begin{bmatrix} 1 & 0 & 0 & 0 \\ 0 & 1 & 0 & 0 \\ 0 & 0 & 1 & 1 \end{bmatrix}$

c. $\begin{bmatrix} 1 & 0 & 0 & 0 \\ 1 & 0 & 0 & 0 \\ 0 & 1 & 0 & 0 \\ 0 & 0 & 1 & 1 \end{bmatrix}$ **d.** $\begin{bmatrix} 0 & 1 & 1 & 1 & 1 \\ 0 & 0 & 2 & 2 & 2 \\ 0 & 0 & 0 & 0 & 3 \\ 0 & 0 & 0 & 0 & 0 \end{bmatrix}$

Row reduce the matrices in Exercises 3 and 4 to reduced echelon form. Circle the pivot positions in the final matrix and in the original matrix, and list the pivot columns.

3. $\begin{bmatrix} 1 & 2 & 3 & 4 \\ 4 & 5 & 6 & 7 \\ 6 & 7 & 8 & 9 \end{bmatrix}$ **4.** $\begin{bmatrix} 1 & 3 & 5 & 7 \\ 3 & 5 & 7 & 9 \\ 5 & 7 & 9 & 1 \end{bmatrix}$

5. Describe the possible echelon forms of a nonzero 2×2 matrix. Use the symbols ■, *, and 0, as in the first part of Example 1.

6. Repeat Exercise 5 for a nonzero 3×2 matrix.

Find the general solutions of the systems whose augmented matrices are given in Exercises 7–14.

7. $\begin{bmatrix} 1 & 3 & 4 & 7 \\ 3 & 9 & 7 & 6 \end{bmatrix}$ **8.** $\begin{bmatrix} 1 & 4 & 0 & 7 \\ 2 & 7 & 0 & 11 \end{bmatrix}$

9. $\begin{bmatrix} 0 & 1 & -6 & 5 \\ 1 & -2 & 7 & -4 \end{bmatrix}$ **10.** $\begin{bmatrix} 1 & -2 & -1 & 3 \\ 3 & -6 & -2 & 2 \end{bmatrix}$

11. $\begin{bmatrix} 3 & -4 & 2 & 0 \\ -9 & 12 & -6 & 0 \\ -6 & 8 & -4 & 0 \end{bmatrix}$ **12.** $\begin{bmatrix} 1 & -7 & 0 & 6 & 5 \\ 0 & 0 & 1 & -2 & -3 \\ -1 & 7 & -4 & 2 & 7 \end{bmatrix}$

13. $\begin{bmatrix} 1 & -3 & 0 & -1 & 0 & -2 \\ 0 & 1 & 0 & 0 & -4 & 1 \\ 0 & 0 & 0 & 1 & 9 & -4 \\ 0 & 0 & 0 & 0 & 0 & 0 \end{bmatrix}$

14. $\begin{bmatrix} 1 & 2 & -5 & -4 & 0 & -5 \\ 0 & 1 & -6 & -4 & 0 & 2 \\ 0 & 0 & 0 & 0 & 1 & 0 \\ 0 & 0 & 0 & 0 & 0 & 0 \end{bmatrix}$

You may find it helpful to review the information in the Reasonable Answers box from this section before answering Exercises 15–18.

15. Write down the equations corresponding to the augmented matrix in Exercise 9 and verify your answer to Exercise 9 is correct by substituting the solutions you obtained back into the original equations.

16. Write down the equations corresponding to the augmented matrix in Exercise 10 and verify your answer to Exercise 10 is correct by substituting the solutions you obtained back into the original equations.

17. Write down the equations corresponding to the augmented matrix in Exercise 11 and verify your answer to Exercise 11 is correct by substituting the solutions you obtained back into the original equations.

18. Write down the equations corresponding to the augmented matrix in Exercise 12 and verify your answer to Exercise 12 is correct by substituting the solutions you obtained back into the original equations.

Exercises 19 and 20 use the notation of Example 1 for matrices in echelon form. Suppose each matrix represents the augmented matrix for a system of linear equations. In each case, determine if the system is consistent. If the system is consistent, determine if the solution is unique.

19. a. $\begin{bmatrix} \blacksquare & * & * & * \\ 0 & \blacksquare & * & * \\ 0 & 0 & \blacksquare & 0 \end{bmatrix}$

b. $\begin{bmatrix} 0 & \blacksquare & * & * & * \\ 0 & 0 & \blacksquare & * & * \\ 0 & 0 & 0 & 0 & \blacksquare \end{bmatrix}$

20. a. $\begin{bmatrix} \blacksquare & * & * \\ 0 & \blacksquare & * \\ 0 & 0 & 0 \end{bmatrix}$

b. $\begin{bmatrix} \blacksquare & * & * & * & * \\ 0 & 0 & \blacksquare & * & * \\ 0 & 0 & 0 & \blacksquare & * \end{bmatrix}$

In Exercises 21 and 22, determine the value(s) of h such that the matrix is the augmented matrix of a consistent linear system.

21. $\begin{bmatrix} 2 & 3 & h \\ 4 & 6 & 7 \end{bmatrix}$ **22.** $\begin{bmatrix} 1 & -3 & -2 \\ 5 & h & -7 \end{bmatrix}$

In Exercises 23 and 24, choose h and k such that the system has (a) no solution, (b) a unique solution, and (c) many solutions. Give separate answers for each part.

23. $x_1 + hx_2 = 2$
$4x_1 + 8x_2 = k$

24. $x_1 + 3x_2 = 2$
$3x_1 + hx_2 = k$

In Exercises 25–34, mark each statement True or False **(T/F)**. Justify each answer.[4]

25. **(T/F)** In some cases, a matrix may be row reduced to more than one matrix in reduced echelon form, using different sequences of row operations.

26. **(T/F)** The echelon form of a matrix is unique.

27. **(T/F)** The row reduction algorithm applies only to augmented matrices for a linear system.

28. **(T/F)** The pivot positions in a matrix depend on whether row interchanges are used in the row reduction process.

29. **(T/F)** A basic variable in a linear system is a variable that corresponds to a pivot column in the coefficient matrix.

30. **(T/F)** Reducing a matrix to echelon form is called the *forward phase* of the row reduction process.

31. **(T/F)** Finding a parametric description of the solution set of a linear system is the same as *solving* the system.

32. **(T/F)** Whenever a system has free variables, the solution set contains a unique solution.

33. **(T/F)** If one row in an echelon form of an augmented matrix is $[0 \ 0 \ 0 \ 0 \ 5]$, then the associated linear system is inconsistent.

34. **(T/F)** A general solution of a system is an explicit description of all solutions of the system.

35. Suppose a 3×5 *coefficient* matrix for a system has three pivot columns. Is the system consistent? Why or why not?

36. Suppose a system of linear equations has a 3×5 *augmented* matrix whose fifth column is a pivot column. Is the system consistent? Why (or why not)?

37. Suppose the coefficient matrix of a system of linear equations has a pivot position in every row. Explain why the system is consistent.

38. Suppose the coefficient matrix of a linear system of three equations in three variables has a pivot in each column. Explain why the system has a unique solution.

39. Restate the last sentence in Theorem 2 using the concept of pivot columns: "If a linear system is consistent, then the solution is unique if and only if _____."

40. What would you have to know about the pivot columns in an augmented matrix in order to know that the linear system is consistent and has a unique solution?

41. A system of linear equations with fewer equations than unknowns is sometimes called an *underdetermined system.*

---

[4] True/false questions of this type will appear in many sections. Methods for justifying your answers were described before the True or False exercises in Section 1.1.

Suppose that such a system happens to be consistent. Explain why there must be an infinite number of solutions.

42. Give an example of an inconsistent underdetermined system of two equations in three unknowns.

43. A system of linear equations with more equations than unknowns is sometimes called an *overdetermined system*. Can such a system be consistent? Illustrate your answer with a specific system of three equations in two unknowns.

44. Suppose an $n \times (n + 1)$ matrix is row reduced to reduced echelon form. Approximately what fraction of the total number of operations (flops) is involved in the backward phase of the reduction when $n = 30$? when $n = 300$?

Suppose experimental data are represented by a set of points in the plane. An **interpolating polynomial** for the data is a polynomial whose graph passes through every point. In scientific work, such a polynomial can be used, for example, to estimate values between the known data points. Another use is to create curves for graphical images on a computer screen. One method for finding an interpolating polynomial is to solve a system of linear equations.

45. Find the interpolating polynomial $p(t) = a_0 + a_1 t + a_2 t^2$ for the data $(1, 12)$, $(2, 15)$, $(3, 16)$. That is, find a_0, a_1, and a_2 such that

$$a_0 + a_1(1) + a_2(1)^2 = 12$$
$$a_0 + a_1(2) + a_2(2)^2 = 15$$
$$a_0 + a_1(3) + a_2(3)^2 = 16$$

T 46. In a wind tunnel experiment, the force on a projectile due to air resistance was measured at different velocities:

Velocity (100 ft/sec)	0	2	4	6	8	10
Force (100 lb)	0	2.90	14.8	39.6	74.3	119

Find an interpolating polynomial for these data and estimate the force on the projectile when the projectile is traveling at 750 ft/sec. Use $p(t) = a_0 + a_1 t + a_2 t^2 + a_3 t^3 + a_4 t^4 + a_5 t^5$. What happens if you try to use a polynomial of degree less than 5? (Try a cubic polynomial, for instance.)[5]

---

[5] Exercises marked with the symbol **T** are designed to be worked with the aid of a "Matrix program" (a computer program, such as MATLAB, Maple, Mathematica, MathCad, Octave, or Derive, or a programmable calculator with matrix capabilities, such as those manufactured by Texas Instruments or Hewlett-Packard).

Solutions to Practice Problems

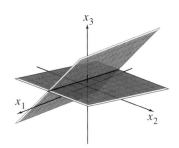

The general solution of the system of equations is the line of intersection of the two planes.

1. The reduced echelon form of the augmented matrix and the corresponding system are

$$\begin{bmatrix} 1 & 0 & -8 & -3 \\ 0 & 1 & -1 & -1 \end{bmatrix} \quad \text{and} \quad \begin{matrix} x_1 & - 8x_3 = -3 \\ x_2 - x_3 = -1 \end{matrix}$$

The basic variables are x_1 and x_2, and the general solution is

$$\begin{cases} x_1 = -3 + 8x_3 \\ x_2 = -1 + x_3 \\ x_3 \text{ is free} \end{cases}$$

Note: It is essential that the general solution describe each variable, with any parameters clearly identified. The following statement does *not* describe the solution:

$$\begin{cases} x_1 = -3 + 8x_3 \\ x_2 = -1 + x_3 \\ x_3 = 1 + x_2 \quad \text{Incorrect solution} \end{cases}$$

This description implies that x_2 and x_3 are *both* free, which certainly is not the case.

2. Row reduce the system's augmented matrix:

$$\begin{bmatrix} 1 & -2 & -1 & 3 & 0 \\ -2 & 4 & 5 & -5 & 3 \\ 3 & -6 & -6 & 8 & 2 \end{bmatrix} \sim \begin{bmatrix} 1 & -2 & -1 & 3 & 0 \\ 0 & 0 & 3 & 1 & 3 \\ 0 & 0 & -3 & -1 & 2 \end{bmatrix}$$

$$\sim \begin{bmatrix} 1 & -2 & -1 & 3 & 0 \\ 0 & 0 & 3 & 1 & 3 \\ 0 & 0 & 0 & 0 & 5 \end{bmatrix}$$

Solutions to Practice Problems (Continued)

This echelon matrix shows that the system is *inconsistent*, because its rightmost column is a pivot column; the third row corresponds to the equation $0 = 5$. There is no need to perform any more row operations. Note that the presence of the free variables in this problem is irrelevant because the system is inconsistent.

3. Since the coefficient matrix has four pivots, there is a pivot in every row of the coefficient matrix. This means that when the coefficient matrix is row reduced, it will *not* have a row of zeros, thus the corresponding row reduced augmented matrix can never have a row of the form $[0 \ 0 \ \cdots \ 0 \ b]$, where b is a nonzero number. By Theorem 2, the system is consistent. Moreover, since there are seven columns in the coefficient matrix and only four pivot columns, there will be three free variables resulting in infinitely many solutions.

1.3 Vector Equations

Important properties of linear systems can be described with the concept and notation of vectors. This section connects equations involving vectors to ordinary systems of equations. The term *vector* appears in a variety of mathematical and physical contexts, which we will discuss in Chapter 4, "Vector Spaces." Until then, *vector* will mean an *ordered list of numbers*. This simple idea enables us to get to interesting and important applications as quickly as possible.

Vectors in $\mathbb{R}^2$

A matrix with only one column is called a **column vector** or simply a **vector**. Examples of vectors with two entries are

$$\mathbf{u} = \begin{bmatrix} 3 \\ -1 \end{bmatrix}, \qquad \mathbf{v} = \begin{bmatrix} .2 \\ .3 \end{bmatrix}, \qquad \mathbf{w} = \begin{bmatrix} w_1 \\ w_2 \end{bmatrix}$$

where w_1 and w_2 are any real numbers. The set of all vectors with two entries is denoted by $\mathbb{R}^2$ (read "r-two"). The $\mathbb{R}$ stands for the real numbers that appear as entries in the vectors, and the exponent 2 indicates that each vector contains two entries.[1]

Two vectors in $\mathbb{R}^2$ are **equal** if and only if their corresponding entries are equal. Thus $\begin{bmatrix} 4 \\ 7 \end{bmatrix}$ and $\begin{bmatrix} 7 \\ 4 \end{bmatrix}$ are *not* equal, because vectors in $\mathbb{R}^2$ are *ordered pairs* of real numbers.

Given two vectors $\mathbf{u}$ and $\mathbf{v}$ in $\mathbb{R}^2$, their **sum** is the vector $\mathbf{u} + \mathbf{v}$ obtained by adding corresponding entries of $\mathbf{u}$ and $\mathbf{v}$. For example,

$$\begin{bmatrix} 1 \\ -2 \end{bmatrix} + \begin{bmatrix} 2 \\ 5 \end{bmatrix} = \begin{bmatrix} 1+2 \\ -2+5 \end{bmatrix} = \begin{bmatrix} 3 \\ 3 \end{bmatrix}$$

Given a vector $\mathbf{u}$ and a real number c, the **scalar multiple** of $\mathbf{u}$ by c is the vector $c\mathbf{u}$ obtained by multiplying each entry in $\mathbf{u}$ by c. For instance,

$$\text{if} \quad \mathbf{u} = \begin{bmatrix} 3 \\ -1 \end{bmatrix} \quad \text{and} \quad c = 5, \qquad \text{then} \quad c\mathbf{u} = 5\begin{bmatrix} 3 \\ -1 \end{bmatrix} = \begin{bmatrix} 15 \\ -5 \end{bmatrix}$$

---

[1] Most of the text concerns vectors and matrices that have only real entries. However, all definitions and theorems in Chapters 1–5, and in most of the rest of the text, remain valid if the entries are complex numbers. Complex vectors and matrices arise naturally, for example, in electrical engineering and physics.

The number c in $c\mathbf{u}$ is called a **scalar**; it is written in lightface type to distinguish it from the boldface vector $\mathbf{u}$.

The operations of scalar multiplication and vector addition can be combined, as in the following example.

EXAMPLE 1   Given $\mathbf{u} = \begin{bmatrix} 1 \\ -2 \end{bmatrix}$ and $\mathbf{v} = \begin{bmatrix} 2 \\ -5 \end{bmatrix}$, find $4\mathbf{u}$, $(-3)\mathbf{v}$, and $4\mathbf{u} + (-3)\mathbf{v}$.

SOLUTION

$$4\mathbf{u} = \begin{bmatrix} 4 \\ -8 \end{bmatrix}, \qquad (-3)\mathbf{v} = \begin{bmatrix} -6 \\ 15 \end{bmatrix}$$

and

$$4\mathbf{u} + (-3)\mathbf{v} = \begin{bmatrix} 4 \\ -8 \end{bmatrix} + \begin{bmatrix} -6 \\ 15 \end{bmatrix} = \begin{bmatrix} -2 \\ 7 \end{bmatrix}$$ ∎

Sometimes, for convenience (and also to save space), this text may write a column vector such as $\begin{bmatrix} 3 \\ -1 \end{bmatrix}$ in the form $(3, -1)$. In this case, the parentheses and the comma distinguish the vector $(3, -1)$ from the 1×2 row matrix $\begin{bmatrix} 3 & -1 \end{bmatrix}$, written with brackets and no comma. Thus

$$\begin{bmatrix} 3 \\ -1 \end{bmatrix} \neq \begin{bmatrix} 3 & -1 \end{bmatrix}$$

because the matrices have different shapes, even though they have the same entries.

Geometric Descriptions of $\mathbb{R}^2$

Consider a rectangular coordinate system in the plane. Because each point in the plane is determined by an ordered pair of numbers, *we can identify a geometric point* (a, b) *with the column vector* $\begin{bmatrix} a \\ b \end{bmatrix}$. So we may regard $\mathbb{R}^2$ as the set of all points in the plane. See Figure 1.

FIGURE 1 Vectors as points.   **FIGURE 2** Vectors with arrows.

The geometric visualization of a vector such as $\begin{bmatrix} 3 \\ -1 \end{bmatrix}$ is often aided by including an arrow (directed line segment) from the origin $(0, 0)$ to the point $(3, -1)$, as in Figure 2. In this case, the individual points along the arrow itself have no special significance.[2]

The sum of two vectors has a useful geometric representation. The following rule can be verified by analytic geometry.

---

[2] In physics, arrows can represent forces and usually are free to move about in space. This interpretation of vectors will be discussed in Section 4.1.

Parallelogram Rule for Addition

If **u** and **v** in $\mathbb{R}^2$ are represented as points in the plane, then **u** + **v** corresponds to the fourth vertex of the parallelogram whose other vertices are **u**, **0**, and **v**. See Figure 3.

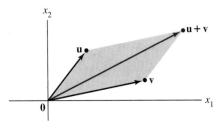

FIGURE 3 The parallelogram rule.

EXAMPLE 2 The vectors $\mathbf{u} = \begin{bmatrix} 2 \\ 2 \end{bmatrix}$, $\mathbf{v} = \begin{bmatrix} -6 \\ 1 \end{bmatrix}$, and $\mathbf{u} + \mathbf{v} = \begin{bmatrix} -4 \\ 3 \end{bmatrix}$ are displayed in Figure 4. ∎

FIGURE 4

The next example illustrates the fact that the set of all scalar multiples of one fixed nonzero vector is a line through the origin, $(0, 0)$.

EXAMPLE 3 Let $\mathbf{u} = \begin{bmatrix} 3 \\ -1 \end{bmatrix}$. Display the vectors **u**, 2**u**, and $-\frac{2}{3}\mathbf{u}$ on a graph.

SOLUTION See Figure 5, where **u**, $2\mathbf{u} = \begin{bmatrix} 6 \\ -2 \end{bmatrix}$, and $-\frac{2}{3}\mathbf{u} = \begin{bmatrix} -2 \\ 2/3 \end{bmatrix}$ are displayed. The arrow for 2**u** is twice as long as the arrow for **u**, and the arrows point in the same direction. The arrow for $-\frac{2}{3}\mathbf{u}$ is two-thirds the length of the arrow for **u**, and the arrows point in opposite directions. In general, the length of the arrow for $c\mathbf{u}$ is $|c|$ times the length of the arrow for **u**. [Recall that the length of the line segment from $(0, 0)$ to (a, b) is $\sqrt{a^2 + b^2}$.

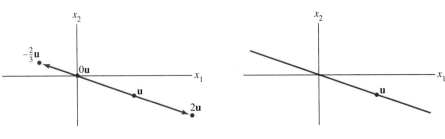

Typical multiples of **u**   The set of all multiples of **u**

FIGURE 5 ∎

Vectors in $\mathbb{R}^3$

Vectors in $\mathbb{R}^3$ are 3×1 column matrices with three entries. They are represented geometrically by points in a three-dimensional coordinate space, with arrows from the origin sometimes included for visual clarity. The vectors $\mathbf{a} = \begin{bmatrix} 2 \\ 3 \\ 4 \end{bmatrix}$ and $2\mathbf{a}$ are displayed in Figure 6.

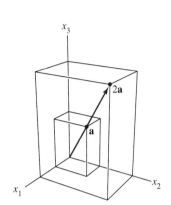

FIGURE 6
Scalar multiples.

Vectors in $\mathbb{R}^n$

If n is a positive integer, $\mathbb{R}^n$ (read "r-n") denotes the collection of all lists (or *ordered n-tuples*) of n real numbers, usually written as $n \times 1$ column matrices, such as

$$\mathbf{u} = \begin{bmatrix} u_1 \\ u_2 \\ \vdots \\ u_n \end{bmatrix}$$

The vector whose entries are all zero is called the **zero vector** and is denoted by $\mathbf{0}$. (The number of entries in $\mathbf{0}$ will be clear from the context.)

Equality of vectors in $\mathbb{R}^n$ and the operations of scalar multiplication and vector addition in $\mathbb{R}^n$ are defined entry by entry just as in $\mathbb{R}^2$. These operations on vectors have the following properties, which can be verified directly from the corresponding properties for real numbers. See Practice Problem 1 and Exercises 41 and 42 at the end of this section.

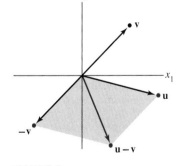

FIGURE 7
Vector subtraction.

> **Algebraic Properties of $\mathbb{R}^n$**
>
> For all $\mathbf{u}, \mathbf{v}, \mathbf{w}$ in $\mathbb{R}^n$ and all scalars c and d:
>
> (i) $\mathbf{u} + \mathbf{v} = \mathbf{v} + \mathbf{u}$ (v) $c(\mathbf{u} + \mathbf{v}) = c\mathbf{u} + c\mathbf{v}$
>
> (ii) $(\mathbf{u} + \mathbf{v}) + \mathbf{w} = \mathbf{u} + (\mathbf{v} + \mathbf{w})$ (vi) $(c + d)\mathbf{u} = c\mathbf{u} + d\mathbf{u}$
>
> (iii) $\mathbf{u} + \mathbf{0} = \mathbf{0} + \mathbf{u} = \mathbf{u}$ (vii) $c(d\mathbf{u}) = (cd)\mathbf{u}$
>
> (iv) $\mathbf{u} + (-\mathbf{u}) = -\mathbf{u} + \mathbf{u} = \mathbf{0}$,   (viii) $1\mathbf{u} = \mathbf{u}$
>   where $-\mathbf{u}$ denotes $(-1)\mathbf{u}$

For simplicity of notation, a vector such as $\mathbf{u} + (-1)\mathbf{v}$ is often written as $\mathbf{u} - \mathbf{v}$. Figure 7 shows $\mathbf{u} - \mathbf{v}$ as the sum of $\mathbf{u}$ and $-\mathbf{v}$.

Linear Combinations

Given vectors $\mathbf{v}_1, \mathbf{v}_2, \ldots, \mathbf{v}_p$ in $\mathbb{R}^n$ and given scalars $c_1, c_2, \ldots, c_p$, the vector $\mathbf{y}$ defined by

$$\mathbf{y} = c_1\mathbf{v}_1 + \cdots + c_p\mathbf{v}_p$$

is called a **linear combination** of $\mathbf{v}_1, \ldots, \mathbf{v}_p$ with **weights** $c_1, \ldots, c_p$. Algebraic Property (ii) above permits us to omit parentheses when forming such a linear combination. The weights in a linear combination can be any real numbers, including zero. For example, some linear combinations of vectors $\mathbf{v}_1$ and $\mathbf{v}_2$ are

$$\sqrt{3}\,\mathbf{v}_1 + \mathbf{v}_2, \quad \tfrac{1}{2}\mathbf{v}_1 \ (= \tfrac{1}{2}\mathbf{v}_1 + 0\mathbf{v}_2), \quad \text{and} \quad \mathbf{0} \ (= 0\mathbf{v}_1 + 0\mathbf{v}_2)$$

EXAMPLE 4 Figure 8 identifies selected linear combinations of $\mathbf{v}_1 = \begin{bmatrix} -1 \\ 1 \end{bmatrix}$ and $\mathbf{v}_2 = \begin{bmatrix} 2 \\ 1 \end{bmatrix}$. (Note that sets of parallel grid lines are drawn through integer multiples of $\mathbf{v}_1$ and $\mathbf{v}_2$.) Estimate the linear combinations of $\mathbf{v}_1$ and $\mathbf{v}_2$ that generate the vectors $\mathbf{u}$ and $\mathbf{w}$.

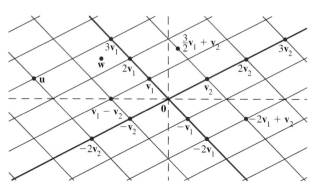

FIGURE 8 Linear combinations of $\mathbf{v}_1$ and $\mathbf{v}_2$.

SOLUTION The parallelogram rule shows that $\mathbf{u}$ is the sum of $3\mathbf{v}_1$ and $-2\mathbf{v}_2$; that is,

$$\mathbf{u} = 3\mathbf{v}_1 - 2\mathbf{v}_2$$

This expression for $\mathbf{u}$ can be interpreted as instructions for traveling from the origin to $\mathbf{u}$ along two straight paths. First, travel 3 units in the $\mathbf{v}_1$ direction to $3\mathbf{v}_1$, and then travel -2 units in the $\mathbf{v}_2$ direction (parallel to the line through $\mathbf{v}_2$ and $\mathbf{0}$). Next, although the vector $\mathbf{w}$ is not on a grid line, $\mathbf{w}$ appears to be about halfway between two pairs of grid lines, at the vertex of a parallelogram determined by $(5/2)\mathbf{v}_1$ and $(-1/2)\mathbf{v}_2$. (See Figure 9.) Thus a reasonable estimate for $\mathbf{w}$ is

$$\mathbf{w} = \tfrac{5}{2}\mathbf{v}_1 - \tfrac{1}{2}\mathbf{v}_2 \qquad \blacksquare$$

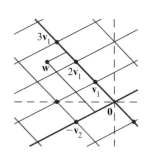

FIGURE 9

The next example connects a problem about linear combinations to the fundamental existence question studied in Sections 1.1 and 1.2.

EXAMPLE 5 Let $\mathbf{a}_1 = \begin{bmatrix} 1 \\ -2 \\ -5 \end{bmatrix}$, $\mathbf{a}_2 = \begin{bmatrix} 2 \\ 5 \\ 6 \end{bmatrix}$, and $\mathbf{b} = \begin{bmatrix} 7 \\ 4 \\ -3 \end{bmatrix}$. Determine whether $\mathbf{b}$ can be generated (or written) as a linear combination of $\mathbf{a}_1$ and $\mathbf{a}_2$. That is, determine whether weights x_1 and x_2 exist such that

$$x_1\mathbf{a}_1 + x_2\mathbf{a}_2 = \mathbf{b} \qquad (1)$$

If vector equation (1) has a solution, find it.

SOLUTION Use the definitions of scalar multiplication and vector addition to rewrite the vector equation which is the same as

$$\begin{bmatrix} x_1 \\ -2x_1 \\ -5x_1 \end{bmatrix} + \begin{bmatrix} 2x_2 \\ 5x_2 \\ 6x_2 \end{bmatrix} = \begin{bmatrix} 7 \\ 4 \\ -3 \end{bmatrix}$$

and

$$\begin{bmatrix} x_1 + 2x_2 \\ -2x_1 + 5x_2 \\ -5x_1 + 6x_2 \end{bmatrix} = \begin{bmatrix} 7 \\ 4 \\ -3 \end{bmatrix} \qquad (2)$$

The vectors on the left and right sides of (2) are equal if and only if their corresponding entries are both equal. That is, x_1 and x_2 make the vector equation (1) true if and only if x_1 and x_2 satisfy the system

$$\begin{aligned} x_1 + 2x_2 &= 7 \\ -2x_1 + 5x_2 &= 4 \\ -5x_1 + 6x_2 &= -3 \end{aligned} \qquad (3)$$

To solve this system, row reduce the augmented matrix of the system as follows:[3]

$$\begin{bmatrix} 1 & 2 & 7 \\ -2 & 5 & 4 \\ -5 & 6 & -3 \end{bmatrix} \sim \begin{bmatrix} 1 & 2 & 7 \\ 0 & 9 & 18 \\ 0 & 16 & 32 \end{bmatrix} \sim \begin{bmatrix} 1 & 2 & 7 \\ 0 & 1 & 2 \\ 0 & 16 & 32 \end{bmatrix} \sim \begin{bmatrix} 1 & 0 & 3 \\ 0 & 1 & 2 \\ 0 & 0 & 0 \end{bmatrix}$$

The solution of (3) is $x_1 = 3$ and $x_2 = 2$. Hence **b** is a linear combination of $\mathbf{a}_1$ and $\mathbf{a}_2$, with weights $x_1 = 3$ and $x_2 = 2$. That is,

$$3 \begin{bmatrix} 1 \\ -2 \\ -5 \end{bmatrix} + 2 \begin{bmatrix} 2 \\ 5 \\ 6 \end{bmatrix} = \begin{bmatrix} 7 \\ 4 \\ -3 \end{bmatrix} \qquad \blacksquare$$

Observe in Example 5 that the original vectors $\mathbf{a}_1$, $\mathbf{a}_2$, and **b** are the columns of the augmented matrix that we row reduced:

$$\begin{bmatrix} 1 & 2 & 7 \\ -2 & 5 & 4 \\ -5 & 6 & -3 \end{bmatrix}$$
$$\begin{array}{ccc} \uparrow & \uparrow & \uparrow \\ \mathbf{a}_1 & \mathbf{a}_2 & \mathbf{b} \end{array}$$

For brevity, write this matrix in a way that identifies its columns—namely

$$\begin{bmatrix} \mathbf{a}_1 & \mathbf{a}_2 & \mathbf{b} \end{bmatrix} \qquad (4)$$

It is clear how to write this augmented matrix immediately from vector equation (1), without going through the intermediate steps of Example 5. Take the vectors in the order in which they appear in (1) and put them into the columns of a matrix as in (4).

The discussion above is easily modified to establish the following fundamental fact.

---

[3] The symbol $\sim$ between matrices denotes row equivalence (Section 1.2).

A vector equation

$$x_1\mathbf{a}_1 + x_2\mathbf{a}_2 + \cdots + x_n\mathbf{a}_n = \mathbf{b}$$

has the same solution set as the linear system whose augmented matrix is

$$\begin{bmatrix} \mathbf{a}_1 & \mathbf{a}_2 & \cdots & \mathbf{a}_n & \mathbf{b} \end{bmatrix} \tag{5}$$

In particular, $\mathbf{b}$ can be generated by a linear combination of $\mathbf{a}_1, \ldots, \mathbf{a}_n$ if and only if there exists a solution to the linear system corresponding to the matrix (5).

One of the key ideas in linear algebra is to study the set of all vectors that can be generated or written as a linear combination of a fixed set $\{\mathbf{v}_1, \ldots, \mathbf{v}_p\}$ of vectors.

DEFINITION

If $\mathbf{v}_1, \ldots, \mathbf{v}_p$ are in $\mathbb{R}^n$, then the set of all linear combinations of $\mathbf{v}_1, \ldots, \mathbf{v}_p$ is denoted by Span $\{\mathbf{v}_1, \ldots, \mathbf{v}_p\}$ and is called the **subset of $\mathbb{R}^n$ spanned** (or **generated**) **by** $\mathbf{v}_1, \ldots, \mathbf{v}_p$. That is, Span $\{\mathbf{v}_1, \ldots, \mathbf{v}_p\}$ is the collection of all vectors that can be written in the form

$$c_1\mathbf{v}_1 + c_2\mathbf{v}_2 + \cdots + c_p\mathbf{v}_p$$

with $c_1, \ldots, c_p$ scalars.

Asking whether a vector $\mathbf{b}$ is in Span $\{\mathbf{v}_1, \ldots, \mathbf{v}_p\}$ amounts to asking whether the vector equation

$$x_1\mathbf{v}_1 + x_2\mathbf{v}_2 + \cdots + x_p\mathbf{v}_p = \mathbf{b}$$

has a solution, or, equivalently, asking whether the linear system with augmented matrix $\begin{bmatrix} \mathbf{v}_1 & \cdots & \mathbf{v}_p & \mathbf{b} \end{bmatrix}$ has a solution.

Note that Span $\{\mathbf{v}_1, \ldots, \mathbf{v}_p\}$ contains every scalar multiple of $\mathbf{v}_1$ (for example), since $c\mathbf{v}_1 = c\mathbf{v}_1 + 0\mathbf{v}_2 + \cdots + 0\mathbf{v}_p$. In particular, the zero vector must be in Span $\{\mathbf{v}_1, \ldots, \mathbf{v}_p\}$.

A Geometric Description of Span $\{\mathbf{v}\}$ and Span $\{\mathbf{u}, \mathbf{v}\}$

Let $\mathbf{v}$ be a nonzero vector in $\mathbb{R}^3$. Then Span $\{\mathbf{v}\}$ is the set of all scalar multiples of $\mathbf{v}$, which is the set of points on the line in $\mathbb{R}^3$ through $\mathbf{v}$ and $\mathbf{0}$. See Figure 10.

If $\mathbf{u}$ and $\mathbf{v}$ are nonzero vectors in $\mathbb{R}^3$, with $\mathbf{v}$ not a multiple of $\mathbf{u}$, then Span $\{\mathbf{u}, \mathbf{v}\}$ is the plane in $\mathbb{R}^3$ that contains $\mathbf{u}$, $\mathbf{v}$, and $\mathbf{0}$. In particular, Span $\{\mathbf{u}, \mathbf{v}\}$ contains the line in $\mathbb{R}^3$ through $\mathbf{u}$ and $\mathbf{0}$ and the line through $\mathbf{v}$ and $\mathbf{0}$. See Figure 11.

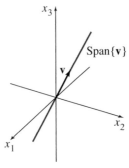

FIGURE 10 Span $\{\mathbf{v}\}$ as a line through the origin.

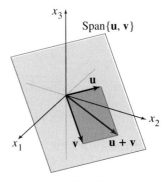

FIGURE 11 Span $\{\mathbf{u}, \mathbf{v}\}$ as a plane through the origin.

EXAMPLE 6   Let $\mathbf{a}_1 = \begin{bmatrix} 1 \\ -2 \\ 3 \end{bmatrix}$, $\mathbf{a}_2 = \begin{bmatrix} 5 \\ -13 \\ -3 \end{bmatrix}$,   and $\mathbf{b} = \begin{bmatrix} -3 \\ 8 \\ 1 \end{bmatrix}$.   Then Span $\{\mathbf{a}_1, \mathbf{a}_2\}$ is a plane through the origin in $\mathbb{R}^3$. Is $\mathbf{b}$ in that plane?

SOLUTION   Does the equation $x_1 \mathbf{a}_1 + x_2 \mathbf{a}_2 = \mathbf{b}$ have a solution? To answer this, row reduce the augmented matrix $[\, \mathbf{a}_1 \quad \mathbf{a}_2 \quad \mathbf{b} \,]$:

$$
\begin{bmatrix} 1 & 5 & -3 \\ -2 & -13 & 8 \\ 3 & -3 & 1 \end{bmatrix} \sim \begin{bmatrix} 1 & 5 & -3 \\ 0 & -3 & 2 \\ 0 & -18 & 10 \end{bmatrix} \sim \begin{bmatrix} 1 & 5 & -3 \\ 0 & -3 & 2 \\ 0 & 0 & -2 \end{bmatrix}
$$

The third equation is $0 = -2$, which shows that the system has no solution. The vector equation $x_1 \mathbf{a}_1 + x_2 \mathbf{a}_2 = \mathbf{b}$ has no solution, and so $\mathbf{b}$ is *not* in Span $\{\mathbf{a}_1, \mathbf{a}_2\}$.   ∎

Linear Combinations in Applications

The final example shows how scalar multiples and linear combinations can arise when a quantity such as "cost" is broken down into several categories. The basic principle for the example concerns the cost of producing several units of an item when the cost per unit is known:

$$
\left\{ \begin{matrix} \text{number} \\ \text{of units} \end{matrix} \right\} \cdot \left\{ \begin{matrix} \text{cost} \\ \text{per unit} \end{matrix} \right\} = \left\{ \begin{matrix} \text{total} \\ \text{cost} \end{matrix} \right\}
$$

EXAMPLE 7   A company manufactures two products. For $1.00 worth of product B, the company spends $.45 on materials, $.25 on labor, and $.15 on overhead. For $1.00 worth of product C, the company spends $.40 on materials, $.30 on labor, and $.15 on overhead. Let

$$
\mathbf{b} = \begin{bmatrix} .45 \\ .25 \\ .15 \end{bmatrix} \quad \text{and} \quad \mathbf{c} = \begin{bmatrix} .40 \\ .30 \\ .15 \end{bmatrix}
$$

Then $\mathbf{b}$ and $\mathbf{c}$ represent the "costs per dollar of income" for the two products.

a. What economic interpretation can be given to the vector $100\mathbf{b}$?

b. Suppose the company wishes to manufacture x_1 dollars worth of product B and x_2 dollars worth of product C. Give a vector that describes the various costs the company will have (for materials, labor, and overhead).

SOLUTION

a. Compute

$$
100\mathbf{b} = 100 \begin{bmatrix} .45 \\ .25 \\ .15 \end{bmatrix} = \begin{bmatrix} 45 \\ 25 \\ 15 \end{bmatrix}
$$

The vector $100\mathbf{b}$ lists the various costs for producing $100 worth of product B—namely $45 for materials, $25 for labor, and $15 for overhead.

b. The costs of manufacturing x_1 dollars worth of B are given by the vector $x_1\mathbf{b}$, and the costs of manufacturing x_2 dollars worth of C are given by $x_2\mathbf{c}$. Hence the total costs for both products are given by the vector $x_1\mathbf{b} + x_2\mathbf{c}$.   ∎

Practice Problems

1. Prove that $\mathbf{u} + \mathbf{v} = \mathbf{v} + \mathbf{u}$ for any $\mathbf{u}$ and $\mathbf{v}$ in $\mathbb{R}^n$.

2. For what value(s) of h will $\mathbf{y}$ be in Span$\{\mathbf{v}_1, \mathbf{v}_2, \mathbf{v}_3\}$ if

$$\mathbf{v}_1 = \begin{bmatrix} 1 \\ -1 \\ -2 \end{bmatrix}, \quad \mathbf{v}_2 = \begin{bmatrix} 5 \\ -4 \\ -7 \end{bmatrix}, \quad \mathbf{v}_3 = \begin{bmatrix} -3 \\ 1 \\ 0 \end{bmatrix}, \quad \text{and} \quad \mathbf{y} = \begin{bmatrix} -4 \\ 3 \\ h \end{bmatrix}$$

3. Let $\mathbf{w}_1, \mathbf{w}_2, \mathbf{w}_3, \mathbf{u}$, and $\mathbf{v}$ be vectors in $\mathbb{R}^n$. Suppose the vectors $\mathbf{u}$ and $\mathbf{v}$ are in Span $\{\mathbf{w}_1, \mathbf{w}_2, \mathbf{w}_3\}$. Show that $\mathbf{u} + \mathbf{v}$ is also in Span $\{\mathbf{w}_1, \mathbf{w}_2, \mathbf{w}_3\}$. [*Hint:* The solution requires the use of the definition of the span of a set of vectors. It is useful to review this definition before starting this exercise.]

1.3 Exercises

In Exercises 1 and 2, compute $\mathbf{u} + \mathbf{v}$ and $\mathbf{u} - 2\mathbf{v}$.

1. $\mathbf{u} = \begin{bmatrix} -1 \\ 2 \end{bmatrix}, \mathbf{v} = \begin{bmatrix} -3 \\ 3 \end{bmatrix}$

2. $\mathbf{u} = \begin{bmatrix} 3 \\ 2 \end{bmatrix}, \mathbf{v} = \begin{bmatrix} 2 \\ 3 \end{bmatrix}$

In Exercises 3 and 4, display the following vectors using arrows on an xy-graph: $\mathbf{u}, \mathbf{v}, -\mathbf{v}, -2\mathbf{v}, \mathbf{u} + \mathbf{v}, \mathbf{u} - \mathbf{v}$, and $\mathbf{u} - 2\mathbf{v}$. Notice that $\mathbf{u} - \mathbf{v}$ is the vertex of a parallelogram whose other vertices are $\mathbf{u}, \mathbf{0}$, and $-\mathbf{v}$.

3. $\mathbf{u}$ and $\mathbf{v}$ as in Exercise 1

4. $\mathbf{u}$ and $\mathbf{v}$ as in Exercise 2

In Exercises 5 and 6, write a system of equations that is equivalent to the given vector equation.

5. $x_1 \begin{bmatrix} 6 \\ -1 \\ 5 \end{bmatrix} + x_2 \begin{bmatrix} -3 \\ 4 \\ 0 \end{bmatrix} = \begin{bmatrix} 1 \\ -7 \\ -5 \end{bmatrix}$

6. $x_1 \begin{bmatrix} -2 \\ 3 \end{bmatrix} + x_2 \begin{bmatrix} 8 \\ 5 \end{bmatrix} + x_3 \begin{bmatrix} 1 \\ -6 \end{bmatrix} = \begin{bmatrix} 0 \\ 0 \end{bmatrix}$

Use the accompanying figure to write each vector listed in Exercises 7 and 8 as a linear combination of $\mathbf{u}$ and $\mathbf{v}$. Is every vector in $\mathbb{R}^2$ a linear combination of $\mathbf{u}$ and $\mathbf{v}$?

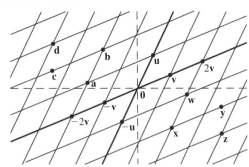

7. Vectors **a**, **b**, **c**, and **d**

8. Vectors **w**, **x**, **y**, and **z**

In Exercises 9 and 10, write a vector equation that is equivalent to the given system of equations.

9.
$$\begin{aligned} x_2 + 5x_3 &= 0 \\ 4x_1 + 6x_2 - x_3 &= 0 \\ -x_1 + 3x_2 - 8x_3 &= 0 \end{aligned}$$

10.
$$\begin{aligned} 4x_1 + x_2 + 3x_3 &= 9 \\ x_1 - 7x_2 - 2x_3 &= 2 \\ 8x_1 + 6x_2 - 5x_3 &= 15 \end{aligned}$$

In Exercises 11 and 12, determine if $\mathbf{b}$ is a linear combination of $\mathbf{a}_1, \mathbf{a}_2$, and $\mathbf{a}_3$.

11. $\mathbf{a}_1 = \begin{bmatrix} 1 \\ -2 \\ 0 \end{bmatrix}, \mathbf{a}_2 = \begin{bmatrix} 0 \\ 1 \\ 2 \end{bmatrix}, \mathbf{a}_3 = \begin{bmatrix} 5 \\ -6 \\ 8 \end{bmatrix}, \mathbf{b} = \begin{bmatrix} 2 \\ -1 \\ 6 \end{bmatrix}$

12. $\mathbf{a}_1 = \begin{bmatrix} 1 \\ -2 \\ 2 \end{bmatrix}, \mathbf{a}_2 = \begin{bmatrix} 0 \\ 5 \\ 5 \end{bmatrix}, \mathbf{a}_3 = \begin{bmatrix} 2 \\ 0 \\ 8 \end{bmatrix}, \mathbf{b} = \begin{bmatrix} -5 \\ 11 \\ -7 \end{bmatrix}$

In Exercises 13 and 14, determine if $\mathbf{b}$ is a linear combination of the vectors formed from the columns of the matrix A.

13. $A = \begin{bmatrix} 1 & -4 & 2 \\ 0 & 3 & 5 \\ -2 & 8 & -4 \end{bmatrix}, \mathbf{b} = \begin{bmatrix} 3 \\ -7 \\ -3 \end{bmatrix}$

14. $A = \begin{bmatrix} 1 & -2 & -6 \\ 0 & 3 & 7 \\ 1 & -2 & 5 \end{bmatrix}, \mathbf{b} = \begin{bmatrix} 11 \\ -5 \\ 9 \end{bmatrix}$

In Exercises 15 and 16, list five vectors in Span $\{\mathbf{v}_1, \mathbf{v}_2\}$. For each vector, show the weights on $\mathbf{v}_1$ and $\mathbf{v}_2$ used to generate the vector and list the three entries of the vector. Do not make a sketch.

15. $\mathbf{v}_1 = \begin{bmatrix} 7 \\ 1 \\ -6 \end{bmatrix}, \mathbf{v}_2 = \begin{bmatrix} -5 \\ 3 \\ 0 \end{bmatrix}$

16. $\mathbf{v}_1 = \begin{bmatrix} 3 \\ 0 \\ 2 \end{bmatrix}, \mathbf{v}_2 = \begin{bmatrix} -2 \\ 0 \\ 3 \end{bmatrix}$

17. Let $\mathbf{a}_1 = \begin{bmatrix} 1 \\ 4 \\ -2 \end{bmatrix}$, $\mathbf{a}_2 = \begin{bmatrix} -2 \\ -3 \\ 7 \end{bmatrix}$, and $\mathbf{b} = \begin{bmatrix} 4 \\ 1 \\ h \end{bmatrix}$. For what value(s) of h is $\mathbf{b}$ in the plane spanned by $\mathbf{a}_1$ and $\mathbf{a}_2$?

18. Let $\mathbf{v}_1 = \begin{bmatrix} 1 \\ 0 \\ -2 \end{bmatrix}$, $\mathbf{v}_2 = \begin{bmatrix} -3 \\ 1 \\ 8 \end{bmatrix}$, and $\mathbf{y} = \begin{bmatrix} h \\ -5 \\ -3 \end{bmatrix}$. For what value(s) of h is $\mathbf{y}$ in the plane generated by $\mathbf{v}_1$ and $\mathbf{v}_2$?

19. Give a geometric description of Span $\{\mathbf{v}_1, \mathbf{v}_2\}$ for the vectors
$$\mathbf{v}_1 = \begin{bmatrix} 8 \\ 2 \\ -6 \end{bmatrix} \text{ and } \mathbf{v}_2 = \begin{bmatrix} 12 \\ 3 \\ -9 \end{bmatrix}.$$

20. Give a geometric description of Span $\{\mathbf{v}_1, \mathbf{v}_2\}$ for the vectors in Exercise 16.

21. Let $\mathbf{u} = \begin{bmatrix} 2 \\ -1 \end{bmatrix}$ and $\mathbf{v} = \begin{bmatrix} 2 \\ 1 \end{bmatrix}$. Show that $\begin{bmatrix} h \\ k \end{bmatrix}$ is in Span $\{\mathbf{u}, \mathbf{v}\}$ for all h and k.

22. Construct a 3×3 matrix A, with nonzero entries, and a vector $\mathbf{b}$ in $\mathbb{R}^3$ such that $\mathbf{b}$ is *not* in the set spanned by the columns of A.

In Exercises 23–32, mark each statement True or False **(T/F)**. Justify each answer.

23. (T/F) Another notation for the vector $\begin{bmatrix} -4 \\ 3 \end{bmatrix}$ is $[-4 \ 3]$.

24. (T/F) Any list of five real numbers is a vector in $\mathbb{R}^5$.

25. (T/F) The points in the plane corresponding to $\begin{bmatrix} -2 \\ 5 \end{bmatrix}$ and $\begin{bmatrix} -5 \\ 2 \end{bmatrix}$ lie on a line through the origin.

26. (T/F) The vector $\mathbf{u}$ results when a vector $\mathbf{u} - \mathbf{v}$ is added to the vector $\mathbf{v}$.

27. (T/F) An example of a linear combination of vectors $\mathbf{v}_1$ and $\mathbf{v}_2$ is the vector $\frac{1}{2}\mathbf{v}_1$.

28. (T/F) The weights $c_1, \ldots, c_p$ in a linear combination $c_1\mathbf{v}_1 + \cdots + c_p\mathbf{v}_p$ cannot all be zero.

29. (T/F) The solution set of the linear system whose augmented matrix is $[\ \mathbf{a}_1 \ \ \mathbf{a}_2 \ \ \mathbf{a}_3 \ \ \mathbf{b} \ ]$ is the same as the solution set of the equation $x_1\mathbf{a}_1 + x_2\mathbf{a}_2 + x_3\mathbf{a}_3 = \mathbf{b}$.

30. (T/F) When $\mathbf{u}$ and $\mathbf{v}$ are nonzero vectors, Span $\{\mathbf{u}, \mathbf{v}\}$ contains the line through $\mathbf{u}$ and the origin.

31. (T/F) The set Span $\{\mathbf{u}, \mathbf{v}\}$ is always visualized as a plane through the origin.

32. (T/F) Asking whether the linear system corresponding to an augmented matrix $[\ \mathbf{a}_1 \ \ \mathbf{a}_2 \ \ \mathbf{a}_3 \ \ \mathbf{b} \ ]$ has a solution amounts to asking whether $\mathbf{b}$ is in Span $\{\mathbf{a}_1, \mathbf{a}_2, \mathbf{a}_3\}$.

33. Let $A = \begin{bmatrix} 1 & 0 & -4 \\ 0 & 3 & -2 \\ -2 & 6 & 3 \end{bmatrix}$ and $\mathbf{b} = \begin{bmatrix} 4 \\ 1 \\ -4 \end{bmatrix}$. Denote the columns of A by $\mathbf{a}_1, \mathbf{a}_2, \mathbf{a}_3$, and let $W = \text{Span} \{\mathbf{a}_1, \mathbf{a}_2, \mathbf{a}_3\}$.

a. Is $\mathbf{b}$ in $\{\mathbf{a}_1, \mathbf{a}_2, \mathbf{a}_3\}$? How many vectors are in $\{\mathbf{a}_1, \mathbf{a}_2, \mathbf{a}_3\}$?

b. Is $\mathbf{b}$ in W? How many vectors are in W?

c. Show that $\mathbf{a}_1$ is in W. [*Hint:* Row operations are unnecessary.]

34. Let $A = \begin{bmatrix} 2 & 0 & 6 \\ -1 & 8 & 5 \\ 1 & -2 & 1 \end{bmatrix}$, let $\mathbf{b} = \begin{bmatrix} 10 \\ 3 \\ 3 \end{bmatrix}$, and let W be the set of all linear combinations of the columns of A.

a. Is $\mathbf{b}$ in W?

b. Show that the third column of A is in W.

35. A mining company has two mines. One day's operation at mine 1 produces ore that contains 20 metric tons of copper and 550 kilograms of silver, while one day's operation at mine 2 produces ore that contains 30 metric tons of copper and 500 kilograms of silver. Let $\mathbf{v}_1 = \begin{bmatrix} 20 \\ 550 \end{bmatrix}$ and $\mathbf{v}_2 = \begin{bmatrix} 30 \\ 500 \end{bmatrix}$. Then $\mathbf{v}_1$ and $\mathbf{v}_2$ represent the "output per day" of mine 1 and mine 2, respectively.

a. What physical interpretation can be given to the vector $5\mathbf{v}_1$?

b. Suppose the company operates mine 1 for x_1 days and mine 2 for x_2 days. Write a vector equation whose solution gives the number of days each mine should operate in order to produce 150 tons of copper and 2825 kilograms of silver. Do not solve the equation.

T c. Solve the equation in (b).

36. A steam plant burns two types of coal: anthracite (A) and bituminous (B). For each ton of A burned, the plant produces 27.6 million Btu of heat, 3100 grams (g) of sulfur dioxide, and 250 g of particulate matter (solid-particle pollutants). For each ton of B burned, the plant produces 30.2 million Btu, 6400 g of sulfur dioxide, and 360 g of particulate matter.

a. How much heat does the steam plant produce when it burns x_1 tons of A and x_2 tons of B?

b. Suppose the output of the steam plant is described by a vector that lists the amounts of heat, sulfur dioxide, and particulate matter. Express this output as a linear combination of two vectors, assuming that the plant burns x_1 tons of A and x_2 tons of B.

T c. Over a certain time period, the steam plant produced 162 million Btu of heat, 23,610 g of sulfur dioxide, and 1623 g of particulate matter. Determine how many tons of each type of coal the steam plant must have burned. Include a vector equation as part of your solution.

37. Let $\mathbf{v}_1, \ldots, \mathbf{v}_k$ be points in $\mathbb{R}^3$ and suppose that for $j = 1, \ldots, k$ an object with mass m_j is located at point $\mathbf{v}_j$. Physicists call such objects *point masses*. The total mass of the system of point masses is

$$m = m_1 + \cdots + m_k$$

The *center of mass* (or *center of gravity*) of the system is

$$\overline{\mathbf{v}} = \frac{1}{m}[m_1\mathbf{v}_1 + \cdots + m_k\mathbf{v}_k]$$

Compute the center of gravity of the system consisting of the following point masses (see the figure):

Point	Mass
$\mathbf{v}_1 = (5, -4, 3)$	2 g
$\mathbf{v}_2 = (4, 3, -2)$	5 g
$\mathbf{v}_3 = (-4, -3, -1)$	2 g
$\mathbf{v}_4 = (-9, 8, 6)$	1 g

38. Let $\mathbf{v}$ be the center of mass of a system of point masses located at $\mathbf{v}_1, \ldots, \mathbf{v}_k$ as in Exercise 37. Is $\mathbf{v}$ in Span $\{\mathbf{v}_1, \ldots, \mathbf{v}_k\}$? Explain.

39. A thin triangular plate of uniform density and thickness has vertices at $\mathbf{v}_1 = (0, 1)$, $\mathbf{v}_2 = (8, 1)$, and $\mathbf{v}_3 = (2, 4)$, as in the figure below, and the mass of the plate is 3 g.

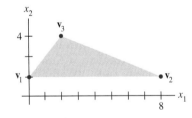

a. Find the (x, y)-coordinates of the center of mass of the plate. This "balance point" of the plate coincides with the center of mass of a system consisting of three 1-gram point masses located at the vertices of the plate.

b. Determine how to distribute an additional mass of 6 g at the three vertices of the plate to move the balance point of the plate to $(2, 2)$. [*Hint:* Let w_1, w_2, and w_3 denote the masses added at the three vertices, so that $w_1 + w_2 + w_3 = 6$.]

40. Consider the vectors $\mathbf{v}_1$, $\mathbf{v}_2$, $\mathbf{v}_3$, and $\mathbf{b}$ in $\mathbb{R}^2$, shown in the figure. Does the equation $x_1\mathbf{v}_1 + x_2\mathbf{v}_2 + x_3\mathbf{v}_3 = \mathbf{b}$ have a solution? Is the solution unique? Use the figure to explain your answers.

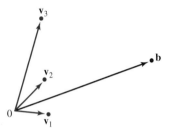

41. Use the vectors $\mathbf{u} = (u_1, \ldots, u_n)$, $\mathbf{v} = (v_1, \ldots, v_n)$, and $\mathbf{w} = (w_1, \ldots, w_n)$ to verify the following algebraic properties of $\mathbb{R}^n$.

a. $(\mathbf{u} + \mathbf{v}) + \mathbf{w} = \mathbf{u} + (\mathbf{v} + \mathbf{w})$

b. $c(\mathbf{u} + \mathbf{v}) = c\mathbf{u} + c\mathbf{v}$ for each scalar c

42. Use the vector $\mathbf{u} = (u_1, \ldots, u_n)$ to verify the following algebraic properties of $\mathbb{R}^n$.

a. $\mathbf{u} + (-\mathbf{u}) = (-\mathbf{u}) + \mathbf{u} = 0$

b. $c(d\mathbf{u}) = (cd)\mathbf{u}$ for all scalars c and d

The points $\begin{bmatrix} -4 \\ 3 \\ h \end{bmatrix}$ lie on a line that intersects the plane when $h = 5$.

Solutions to Practice Problems

1. Take arbitrary vectors $\mathbf{u} = (u_1, \ldots, u_n)$ and $\mathbf{v} = (v_1, \ldots, v_n)$ in $\mathbb{R}^n$, and compute

$$\mathbf{u} + \mathbf{v} = (u_1 + v_1, \ldots, u_n + v_n) \qquad \text{Definition of vector addition}$$
$$= (v_1 + u_1, \ldots, v_n + u_n) \qquad \text{Commutativity of addition in } \mathbb{R}$$
$$= \mathbf{v} + \mathbf{u} \qquad \text{Definition of vector addition}$$

2. The vector $\mathbf{y}$ belongs to Span $\{\mathbf{v}_1, \mathbf{v}_2, \mathbf{v}_3\}$ if and only if there exist scalars x_1, x_2, x_3 such that

$$x_1\begin{bmatrix} 1 \\ -1 \\ -2 \end{bmatrix} + x_2\begin{bmatrix} 5 \\ -4 \\ -7 \end{bmatrix} + x_3\begin{bmatrix} -3 \\ 1 \\ 0 \end{bmatrix} = \begin{bmatrix} -4 \\ 3 \\ h \end{bmatrix}$$

This vector equation is equivalent to a system of three linear equations in three unknowns. If you row reduce the augmented matrix for this system, you find that

$$\begin{bmatrix} 1 & 5 & -3 & -4 \\ -1 & -4 & 1 & 3 \\ -2 & -7 & 0 & h \end{bmatrix} \sim \begin{bmatrix} 1 & 5 & -3 & -4 \\ 0 & 1 & -2 & -1 \\ 0 & 3 & -6 & h-8 \end{bmatrix} \sim \begin{bmatrix} 1 & 5 & -3 & -4 \\ 0 & 1 & -2 & -1 \\ 0 & 0 & 0 & h-5 \end{bmatrix}$$

The system is consistent if and only if there is no pivot in the fourth column. That is, $h - 5$ must be 0. So $\mathbf{y}$ is in Span $\{\mathbf{v}_1, \mathbf{v}_2, \mathbf{v}_3\}$ if and only if $h = 5$.

Remember: The presence of a free variable in a system does not guarantee that the system is consistent.

3. Since the vectors $\mathbf{u}$ and $\mathbf{v}$ are in Span $\{\mathbf{w}_1, \mathbf{w}_2, \mathbf{w}_3\}$, there exist scalars c_1, c_2, c_3 and d_1, d_2, d_3 such that

$$\mathbf{u} = c_1\,\mathbf{w}_1 + c_2\,\mathbf{w}_2 + c_3\,\mathbf{w}_3 \quad \text{and} \quad \mathbf{v} = d_1\,\mathbf{w}_1 + d_2\,\mathbf{w}_2 + d_3\,\mathbf{w}_3.$$

Notice
$$\begin{aligned} \mathbf{u} + \mathbf{v} &= c_1\mathbf{w}_1 + c_2\mathbf{w}_2 + c_3\mathbf{w}_3 + d_1\mathbf{w}_1 + d_2\mathbf{w}_2 + d_3\mathbf{w}_3 \\ &= (c_1 + d_1)\,\mathbf{w}_1 + (c_2 + d_2)\,\mathbf{w}_2 + (c_3 + d_3)\,\mathbf{w}_3 \end{aligned}$$

Since $c_1 + d_1, c_2 + d_2$, and $c_3 + d_3$ are also scalars, the vector $\mathbf{u} + \mathbf{v}$ is in Span $\{\mathbf{w}_1, \mathbf{w}_2, \mathbf{w}_3\}$.

1.4 The Matrix Equation $A\mathbf{x} = \mathbf{b}$

A fundamental idea in linear algebra is to view a linear combination of vectors as the product of a matrix and a vector. The following definition permits us to rephrase some of the concepts of Section 1.3 in new ways.

DEFINITION

If A is an $m \times n$ matrix, with columns $\mathbf{a}_1, \ldots, \mathbf{a}_n$, and if $\mathbf{x}$ is in $\mathbb{R}^n$, then the **product of A and $\mathbf{x}$**, denoted by $A\mathbf{x}$, is **the linear combination of the columns of A using the corresponding entries in $\mathbf{x}$ as weights**; that is,

$$A\mathbf{x} = \begin{bmatrix} \mathbf{a}_1 & \mathbf{a}_2 & \cdots & \mathbf{a}_n \end{bmatrix} \begin{bmatrix} x_1 \\ \vdots \\ x_n \end{bmatrix} = x_1\mathbf{a}_1 + x_2\mathbf{a}_2 + \cdots + x_n\mathbf{a}_n$$

Note that $A\mathbf{x}$ is defined only if the number of columns of A equals the number of entries in $\mathbf{x}$.

EXAMPLE 1

a. $\begin{bmatrix} 1 & 2 & -1 \\ 0 & -5 & 3 \end{bmatrix} \begin{bmatrix} 4 \\ 3 \\ 7 \end{bmatrix} = 4\begin{bmatrix} 1 \\ 0 \end{bmatrix} + 3\begin{bmatrix} 2 \\ -5 \end{bmatrix} + 7\begin{bmatrix} -1 \\ 3 \end{bmatrix}$

$\qquad\qquad\qquad\qquad = \begin{bmatrix} 4 \\ 0 \end{bmatrix} + \begin{bmatrix} 6 \\ -15 \end{bmatrix} + \begin{bmatrix} -7 \\ 21 \end{bmatrix} = \begin{bmatrix} 3 \\ 6 \end{bmatrix}$

b. $\begin{bmatrix} 2 & -3 \\ 8 & 0 \\ -5 & 2 \end{bmatrix} \begin{bmatrix} 4 \\ 7 \end{bmatrix} = 4\begin{bmatrix} 2 \\ 8 \\ -5 \end{bmatrix} + 7\begin{bmatrix} -3 \\ 0 \\ 2 \end{bmatrix} = \begin{bmatrix} 8 \\ 32 \\ -20 \end{bmatrix} + \begin{bmatrix} -21 \\ 0 \\ 14 \end{bmatrix} = \begin{bmatrix} -13 \\ 32 \\ -6 \end{bmatrix}$ ∎

EXAMPLE 2 For $\mathbf{v}_1, \mathbf{v}_2, \mathbf{v}_3$ in $\mathbb{R}^m$, write the linear combination $3\mathbf{v}_1 - 5\mathbf{v}_2 + 7\mathbf{v}_3$ as a matrix times a vector.

SOLUTION Place $\mathbf{v}_1, \mathbf{v}_2, \mathbf{v}_3$ into the columns of a matrix A and place the weights 3, -5, and 7 into a vector $\mathbf{x}$. That is,

$$3\mathbf{v}_1 - 5\mathbf{v}_2 + 7\mathbf{v}_3 = \begin{bmatrix} \mathbf{v}_1 & \mathbf{v}_2 & \mathbf{v}_3 \end{bmatrix} \begin{bmatrix} 3 \\ -5 \\ 7 \end{bmatrix} = A\mathbf{x} \qquad \blacksquare$$

Section 1.3 showed how to write a system of linear equations as a vector equation involving a linear combination of vectors. For example, the system

$$\begin{aligned} x_1 + 2x_2 - x_3 &= 4 \\ -5x_2 + 3x_3 &= 1 \end{aligned} \tag{1}$$

is equivalent to

$$x_1 \begin{bmatrix} 1 \\ 0 \end{bmatrix} + x_2 \begin{bmatrix} 2 \\ -5 \end{bmatrix} + x_3 \begin{bmatrix} -1 \\ 3 \end{bmatrix} = \begin{bmatrix} 4 \\ 1 \end{bmatrix} \tag{2}$$

As in Example 2, the linear combination on the left side is a matrix times a vector, so that (2) becomes

$$\begin{bmatrix} 1 & 2 & -1 \\ 0 & -5 & 3 \end{bmatrix} \begin{bmatrix} x_1 \\ x_2 \\ x_3 \end{bmatrix} = \begin{bmatrix} 4 \\ 1 \end{bmatrix} \tag{3}$$

Equation (3) has the form $A\mathbf{x} = \mathbf{b}$. Such an equation is called a **matrix equation**, to distinguish it from a vector equation such as is shown in (2).

Notice how the matrix in (3) is just the matrix of coefficients of the system (1). Similar calculations show that any system of linear equations, or any vector equation such as (2), can be written as an equivalent matrix equation in the form $A\mathbf{x} = \mathbf{b}$. This simple observation will be used repeatedly throughout the text.

Here is the formal result.

THEOREM 3

If A is an $m \times n$ matrix, with columns $\mathbf{a}_1, \ldots, \mathbf{a}_n$, and if $\mathbf{b}$ is in $\mathbb{R}^m$, the matrix equation

$$A\mathbf{x} = \mathbf{b} \tag{4}$$

has the same solution set as the vector equation

$$x_1\mathbf{a}_1 + x_2\mathbf{a}_2 + \cdots + x_n\mathbf{a}_n = \mathbf{b} \tag{5}$$

which, in turn, has the same solution set as the system of linear equations whose augmented matrix is

$$\begin{bmatrix} \mathbf{a}_1 & \mathbf{a}_2 & \cdots & \mathbf{a}_n & \mathbf{b} \end{bmatrix} \tag{6}$$

Theorem 3 provides a powerful tool for gaining insight into problems in linear algebra, because a system of linear equations may now be viewed in three different but equivalent ways: as a matrix equation, as a vector equation, or as a system of linear equations. Whenever you construct a mathematical model of a problem in real life, you are free to choose whichever viewpoint is most natural. Then you may switch from one formulation of a problem to another whenever it is convenient. In any case, the matrix equation (4), the vector equation (5), and the system of equations are all solved in the same way—by row reducing the augmented matrix (6). Other methods of solution will be discussed later.

Existence of Solutions

The definition of $A\mathbf{x}$ leads directly to the following useful fact.

> The equation $A\mathbf{x} = \mathbf{b}$ has a solution if and only if $\mathbf{b}$ is a linear combination of the columns of A.

Section 1.3 considered the existence question, "Is $\mathbf{b}$ in Span $\{\mathbf{a}_1, \ldots, \mathbf{a}_n\}$?" Equivalently, "Is $A\mathbf{x} = \mathbf{b}$ consistent?" A harder existence problem is to determine whether the equation $A\mathbf{x} = \mathbf{b}$ is consistent *for all* possible $\mathbf{b}$.

EXAMPLE 3 Let $A = \begin{bmatrix} 1 & 3 & 4 \\ -4 & 2 & -6 \\ -3 & -2 & -7 \end{bmatrix}$ and $\mathbf{b} = \begin{bmatrix} b_1 \\ b_2 \\ b_3 \end{bmatrix}$. Is the equation $A\mathbf{x} = \mathbf{b}$ consistent for all possible b_1, b_2, b_3?

SOLUTION Row reduce the augmented matrix for $A\mathbf{x} = \mathbf{b}$:

$$\begin{bmatrix} 1 & 3 & 4 & b_1 \\ -4 & 2 & -6 & b_2 \\ -3 & -2 & -7 & b_3 \end{bmatrix} \sim \begin{bmatrix} 1 & 3 & 4 & b_1 \\ 0 & 14 & 10 & b_2 + 4b_1 \\ 0 & 7 & 5 & b_3 + 3b_1 \end{bmatrix}$$

$$\sim \begin{bmatrix} 1 & 3 & 4 & b_1 \\ 0 & 14 & 10 & b_2 + 4b_1 \\ 0 & 0 & 0 & b_3 + 3b_1 - \frac{1}{2}(b_2 + 4b_1) \end{bmatrix}$$

The third entry in column 4 equals $b_1 - \frac{1}{2}b_2 + b_3$. The equation $A\mathbf{x} = \mathbf{b}$ is *not* consistent for every $\mathbf{b}$ because some choices of $\mathbf{b}$ can make $b_1 - \frac{1}{2}b_2 + b_3$ nonzero. ∎

The reduced matrix in Example 3 provides a description of all $\mathbf{b}$ for which the equation $A\mathbf{x} = \mathbf{b}$ *is* consistent: The entries in $\mathbf{b}$ must satisfy

$$b_1 - \tfrac{1}{2}b_2 + b_3 = 0$$

This is the equation of a plane through the origin in $\mathbb{R}^3$. The plane is the set of all linear combinations of the three columns of A. See Figure 1.

The equation $A\mathbf{x} = \mathbf{b}$ in Example 3 fails to be consistent for all $\mathbf{b}$ because the echelon form of A has a row of zeros. If A had a pivot in all three rows, we would not care about the calculations in the augmented column because in this case an echelon form of the augmented matrix could not have a row such as $[\, 0 \quad 0 \quad 0 \quad 1 \,]$.

In the next theorem, the sentence "The columns of A span $\mathbb{R}^m$" means that *every* $\mathbf{b}$ in $\mathbb{R}^m$ is a linear combination of the columns of A. In general, a set of vectors $\{\mathbf{v}_1, \ldots, \mathbf{v}_p\}$ in $\mathbb{R}^m$ **spans** (or **generates**) $\mathbb{R}^m$ if every vector in $\mathbb{R}^m$ is a linear combination of $\mathbf{v}_1, \ldots, \mathbf{v}_p$—that is, if Span $\{\mathbf{v}_1, \ldots, \mathbf{v}_p\} = \mathbb{R}^m$.

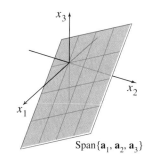

Span$\{\mathbf{a}_1, \mathbf{a}_2, \mathbf{a}_3\}$

FIGURE 1

The columns of
$A = [\, \mathbf{a}_1 \quad \mathbf{a}_2 \quad \mathbf{a}_3 \,]$ span a plane
through $\mathbf{0}$.

THEOREM 4

Let A be an $m \times n$ matrix. Then the following statements are logically equivalent. That is, for a particular A, either they are all true statements or they are all false.

a. For each $\mathbf{b}$ in $\mathbb{R}^m$, the equation $A\mathbf{x} = \mathbf{b}$ has a solution.

b. Each $\mathbf{b}$ in $\mathbb{R}^m$ is a linear combination of the columns of A.

c. The columns of A span $\mathbb{R}^m$.

d. A has a pivot position in every row.

Theorem 4 is one of the most useful theorems in this chapter. Statements (a), (b), and (c) are equivalent because of the definition of $A\mathbf{x}$ and what it means for a set of vectors to span $\mathbb{R}^m$. The discussion after Example 3 suggests why (a) and (d) are equivalent; a proof is given at the end of the section. The exercises will provide examples of how Theorem 4 is used.

Warning: Theorem 4 is about a *coefficient matrix*, not an augmented matrix. If an augmented matrix $[\,A \quad \mathbf{b}\,]$ has a pivot position in every row, then the equation $A\mathbf{x} = \mathbf{b}$ may or may not be consistent.

Computation of Ax

The calculations in Example 1 were based on the definition of the product of a matrix A and a vector $\mathbf{x}$. The following simple example will lead to a more efficient method for calculating the entries in $A\mathbf{x}$ when working problems by hand.

EXAMPLE 4 Compute $A\mathbf{x}$, where $A = \begin{bmatrix} 2 & 3 & 4 \\ -1 & 5 & -3 \\ 6 & -2 & 8 \end{bmatrix}$ and $\mathbf{x} = \begin{bmatrix} x_1 \\ x_2 \\ x_3 \end{bmatrix}$.

SOLUTION From the definition,

$$\begin{bmatrix} 2 & 3 & 4 \\ -1 & 5 & -3 \\ 6 & -2 & 8 \end{bmatrix}\begin{bmatrix} x_1 \\ x_2 \\ x_3 \end{bmatrix} = x_1\begin{bmatrix} 2 \\ -1 \\ 6 \end{bmatrix} + x_2\begin{bmatrix} 3 \\ 5 \\ -2 \end{bmatrix} + x_3\begin{bmatrix} 4 \\ -3 \\ 8 \end{bmatrix}$$

$$= \begin{bmatrix} 2x_1 \\ -x_1 \\ 6x_1 \end{bmatrix} + \begin{bmatrix} 3x_2 \\ 5x_2 \\ -2x_2 \end{bmatrix} + \begin{bmatrix} 4x_3 \\ -3x_3 \\ 8x_3 \end{bmatrix} \tag{7}$$

$$= \begin{bmatrix} 2x_1 + 3x_2 + 4x_3 \\ -x_1 + 5x_2 - 3x_3 \\ 6x_1 - 2x_2 + 8x_3 \end{bmatrix}$$

The first entry in the product $A\mathbf{x}$ is a sum of products (sometimes called a *dot product*), using the first row of A and the entries in $\mathbf{x}$. That is,

$$\begin{bmatrix} 2 & 3 & 4 \end{bmatrix}\begin{bmatrix} x_1 \\ x_2 \\ x_3 \end{bmatrix} = \begin{bmatrix} 2x_1 + 3x_2 + 4x_3 \end{bmatrix}$$

This matrix shows how to compute the first entry in $A\mathbf{x}$ directly, without writing down all the calculations shown in (7). Similarly, the second entry in $A\mathbf{x}$ can be calculated at once by multiplying the entries in the second row of A by the corresponding entries in $\mathbf{x}$ and then summing the resulting products:

$$\begin{bmatrix} -1 & 5 & -3 \end{bmatrix}\begin{bmatrix} x_1 \\ x_2 \\ x_3 \end{bmatrix} = \begin{bmatrix} -x_1 + 5x_2 - 3x_3 \end{bmatrix}$$

Likewise, the third entry in $A\mathbf{x}$ can be calculated from the third row of A and the entries in $\mathbf{x}$. ∎

Row–Vector Rule for Computing Ax

If the product $A\mathbf{x}$ is defined, then the ith entry in $A\mathbf{x}$ is the sum of the products of corresponding entries from row i of A and from the vector $\mathbf{x}$.

EXAMPLE 5

a. $\begin{bmatrix} 1 & 2 & -1 \\ 0 & -5 & 3 \end{bmatrix} \begin{bmatrix} 4 \\ 3 \\ 7 \end{bmatrix} = \begin{bmatrix} 1 \cdot 4 + 2 \cdot 3 + (-1) \cdot 7 \\ 0 \cdot 4 + (-5) \cdot 3 + 3 \cdot 7 \end{bmatrix} = \begin{bmatrix} 3 \\ 6 \end{bmatrix}$

b. $\begin{bmatrix} 2 & -3 \\ 8 & 0 \\ -5 & 2 \end{bmatrix} \begin{bmatrix} 4 \\ 7 \end{bmatrix} = \begin{bmatrix} 2 \cdot 4 + (-3) \cdot 7 \\ 8 \cdot 4 + 0 \cdot 7 \\ (-5) \cdot 4 + 2 \cdot 7 \end{bmatrix} = \begin{bmatrix} -13 \\ 32 \\ -6 \end{bmatrix}$

c. $\begin{bmatrix} 1 & 0 & 0 \\ 0 & 1 & 0 \\ 0 & 0 & 1 \end{bmatrix} \begin{bmatrix} r \\ s \\ t \end{bmatrix} = \begin{bmatrix} 1 \cdot r + 0 \cdot s + 0 \cdot t \\ 0 \cdot r + 1 \cdot s + 0 \cdot t \\ 0 \cdot r + 0 \cdot s + 1 \cdot t \end{bmatrix} = \begin{bmatrix} r \\ s \\ t \end{bmatrix}$ ∎

By definition, the matrix in Example 5(c) with 1's on the diagonal and 0's elsewhere is called an **identity matrix** and is denoted by I. The calculation in part (c) shows that $I\mathbf{x} = \mathbf{x}$ for every $\mathbf{x}$ in $\mathbb{R}^3$. There is an analogous $n \times n$ identity matrix, sometimes written as I_n. As in part (c), $I_n\mathbf{x} = \mathbf{x}$ for every $\mathbf{x}$ in $\mathbb{R}^n$.

Properties of the Matrix–Vector Product $A\mathbf{x}$

The facts in the next theorem are important and will be used throughout the text. The proof relies on the definition of $A\mathbf{x}$ and the algebraic properties of $\mathbb{R}^n$.

THEOREM 5

If A is an $m \times n$ matrix, $\mathbf{u}$ and $\mathbf{v}$ are vectors in $\mathbb{R}^n$, and c is a scalar, then:

a. $A(\mathbf{u} + \mathbf{v}) = A\mathbf{u} + A\mathbf{v}$;

b. $A(c\mathbf{u}) = c(A\mathbf{u})$.

PROOF For simplicity, take $n = 3$, $A = [\,\mathbf{a}_1 \quad \mathbf{a}_2 \quad \mathbf{a}_3\,]$, and $\mathbf{u}$, $\mathbf{v}$ in $\mathbb{R}^3$. (The proof of the general case is similar.) For $i = 1, 2, 3$, let u_i and v_i be the ith entries in $\mathbf{u}$ and $\mathbf{v}$, respectively. To prove statement (a), compute $A(\mathbf{u} + \mathbf{v})$ as a linear combination of the columns of A using the entries in $\mathbf{u} + \mathbf{v}$ as weights.

$$A(\mathbf{u} + \mathbf{v}) = [\,\mathbf{a}_1 \quad \mathbf{a}_2 \quad \mathbf{a}_3\,] \begin{bmatrix} u_1 + v_1 \\ u_2 + v_2 \\ u_3 + v_3 \end{bmatrix}$$

$$= (u_1 + v_1)\mathbf{a}_1 + (u_2 + v_2)\mathbf{a}_2 + (u_3 + v_3)\mathbf{a}_3 \qquad \text{Entries in } \mathbf{u} + \mathbf{v}$$

(Columns of A)

$$= (u_1\mathbf{a}_1 + u_2\mathbf{a}_2 + u_3\mathbf{a}_3) + (v_1\mathbf{a}_1 + v_2\mathbf{a}_2 + v_3\mathbf{a}_3)$$

$$= A\mathbf{u} + A\mathbf{v}$$

To prove statement (b), compute $A(c\mathbf{u})$ as a linear combination of the columns of A using the entries in $c\mathbf{u}$ as weights.

$$A(c\mathbf{u}) = [\,\mathbf{a}_1 \quad \mathbf{a}_2 \quad \mathbf{a}_3\,] \begin{bmatrix} cu_1 \\ cu_2 \\ cu_3 \end{bmatrix} = (cu_1)\mathbf{a}_1 + (cu_2)\mathbf{a}_2 + (cu_3)\mathbf{a}_3$$

$$= c(u_1\mathbf{a}_1) + c(u_2\mathbf{a}_2) + c(u_3\mathbf{a}_3)$$

$$= c(u_1\mathbf{a}_1 + u_2\mathbf{a}_2 + u_3\mathbf{a}_3)$$

$$= c(A\mathbf{u}) \qquad\qquad ∎$$

Numerical Note

To optimize a computer algorithm to compute $A\mathbf{x}$, the sequence of calculations should involve data stored in contiguous memory locations. The most widely used professional algorithms for matrix computations are written in Fortran, a language that stores a matrix as a set of columns. Such algorithms compute $A\mathbf{x}$ as a linear combination of the columns of A. In contrast, if a program is written in the popular language C, which stores matrices by rows, $A\mathbf{x}$ should be computed via the alternative rule that uses the rows of A.

PROOF OF THEOREM 4 As was pointed out after Theorem 4, statements (a), (b), and (c) are logically equivalent. So, it suffices to show (for an arbitrary matrix A) that (a) and (d) are either both true or both false. This will tie all four statements together.

Let U be an echelon form of A. Given $\mathbf{b}$ in $\mathbb{R}^m$, we can row reduce the augmented matrix $[\,A \quad \mathbf{b}\,]$ to an augmented matrix $[\,U \quad \mathbf{d}\,]$ for some $\mathbf{d}$ in $\mathbb{R}^m$:

$$[\,A \quad \mathbf{b}\,] \sim \cdots \sim [\,U \quad \mathbf{d}\,]$$

If statement (d) is true, then each row of U contains a pivot position and there can be no pivot in the augmented column. So $A\mathbf{x} = \mathbf{b}$ has a solution for any $\mathbf{b}$, and (a) is true. If (d) is false, the last row of U is all zeros. Let $\mathbf{d}$ be any vector with a 1 in its last entry. Then $[\,U \quad \mathbf{d}\,]$ represents an *inconsistent* system. Since row operations are reversible, $[\,U \quad \mathbf{d}\,]$ can be transformed into the form $[\,A \quad \mathbf{b}\,]$. The new system $A\mathbf{x} = \mathbf{b}$ is also inconsistent, and (a) is false. ∎

Practice Problems

1. Let $A = \begin{bmatrix} 1 & 5 & -2 & 0 \\ -3 & 1 & 9 & -5 \\ 4 & -8 & -1 & 7 \end{bmatrix}$, $\mathbf{p} = \begin{bmatrix} 3 \\ -2 \\ 0 \\ -4 \end{bmatrix}$, and $\mathbf{b} = \begin{bmatrix} -7 \\ 9 \\ 0 \end{bmatrix}$. It can be shown that $\mathbf{p}$ is a solution of $A\mathbf{x} = \mathbf{b}$. Use this fact to exhibit $\mathbf{b}$ as a specific linear combination of the columns of A.

2. Let $A = \begin{bmatrix} 2 & 5 \\ 3 & 1 \end{bmatrix}$, $\mathbf{u} = \begin{bmatrix} 4 \\ -1 \end{bmatrix}$, and $\mathbf{v} = \begin{bmatrix} -3 \\ 5 \end{bmatrix}$. Verify Theorem 5(a) in this case by computing $A(\mathbf{u} + \mathbf{v})$ and $A\mathbf{u} + A\mathbf{v}$.

3. Construct a 3×3 matrix A and vectors $\mathbf{b}$ and $\mathbf{c}$ in $\mathbb{R}^3$ so that $A\mathbf{x} = \mathbf{b}$ has a solution, but $A\mathbf{x} = \mathbf{c}$ does not.

1.4 Exercises

Compute the products in Exercises 1–4 using (a) the definition, as in Example 1, and (b) the row–vector rule for computing $A\mathbf{x}$. If a product is undefined, explain why.

1. $\begin{bmatrix} -4 & 2 \\ 1 & 6 \\ 0 & 1 \end{bmatrix} \begin{bmatrix} 3 \\ 1 \\ 7 \end{bmatrix}$

2. $\begin{bmatrix} 2 \\ 6 \\ -1 \end{bmatrix} \begin{bmatrix} 1 \\ -1 \end{bmatrix}$

3. $\begin{bmatrix} 6 & 5 \\ -4 & -3 \\ 7 & 6 \end{bmatrix} \begin{bmatrix} 1 \\ -3 \end{bmatrix}$

4. $\begin{bmatrix} 8 & 3 & 1 \\ 5 & 1 & 2 \end{bmatrix} \begin{bmatrix} 1 \\ 1 \\ 1 \end{bmatrix}$

In Exercises 5–8, use the definition of $A\mathbf{x}$ to write the matrix equation as a vector equation, or vice versa.

5. $\begin{bmatrix} 5 & 1 & -8 & 4 \\ -2 & -7 & 3 & -5 \end{bmatrix} \begin{bmatrix} 5 \\ -1 \\ 3 \\ -2 \end{bmatrix} = \begin{bmatrix} -8 \\ 16 \end{bmatrix}$

6. $\begin{bmatrix} 7 & -3 \\ 2 & 1 \\ 9 & -6 \\ -3 & 2 \end{bmatrix} \begin{bmatrix} -2 \\ -5 \end{bmatrix} = \begin{bmatrix} 1 \\ -9 \\ 12 \\ -4 \end{bmatrix}$

7. $x_1\begin{bmatrix} 4 \\ -1 \\ 7 \\ -4 \end{bmatrix} + x_2\begin{bmatrix} -5 \\ 3 \\ -5 \\ 1 \end{bmatrix} + x_3\begin{bmatrix} 7 \\ -8 \\ 0 \\ 2 \end{bmatrix} = \begin{bmatrix} 6 \\ -8 \\ 0 \\ -7 \end{bmatrix}$

8. $z_1\begin{bmatrix} 4 \\ -2 \end{bmatrix} + z_2\begin{bmatrix} -4 \\ 5 \end{bmatrix} + z_3\begin{bmatrix} -5 \\ 4 \end{bmatrix} + z_4\begin{bmatrix} 3 \\ 0 \end{bmatrix} = \begin{bmatrix} 4 \\ 13 \end{bmatrix}$

In Exercises 9 and 10, write the system first as a vector equation and then as a matrix equation.

9. $\begin{aligned} 3x_1 + x_2 - 5x_3 &= 9 \\ x_2 + 4x_3 &= 0 \end{aligned}$

10. $\begin{aligned} 8x_1 - x_2 &= 4 \\ 5x_1 + 4x_2 &= 1 \\ x_1 - 3x_2 &= 2 \end{aligned}$

Given A and $\mathbf{b}$ in Exercises 11 and 12, write the augmented matrix for the linear system that corresponds to the matrix equation $A\mathbf{x} = \mathbf{b}$. Then solve the system and write the solution as a vector.

11. $A = \begin{bmatrix} 1 & 2 & 4 \\ 0 & 1 & 5 \\ -2 & -4 & -3 \end{bmatrix}, \mathbf{b} = \begin{bmatrix} -2 \\ 2 \\ 9 \end{bmatrix}$

12. $A = \begin{bmatrix} 1 & 2 & 1 \\ -3 & -1 & 2 \\ 0 & 5 & 3 \end{bmatrix}, \mathbf{b} = \begin{bmatrix} 0 \\ 1 \\ -1 \end{bmatrix}$

13. Let $\mathbf{u} = \begin{bmatrix} 0 \\ 4 \\ 4 \end{bmatrix}$ and $A = \begin{bmatrix} 3 & -5 \\ -2 & 6 \\ 1 & 1 \end{bmatrix}$. Is $\mathbf{u}$ in the plane in $\mathbb{R}^3$

spanned by the columns of A? (See the figure.) Why or why not?

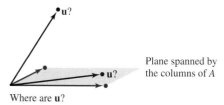

Where are **u**?

14. Let $\mathbf{u} = \begin{bmatrix} 2 \\ -3 \\ 2 \end{bmatrix}$ and $A = \begin{bmatrix} 5 & 8 & 7 \\ 0 & 1 & -1 \\ 1 & 3 & 0 \end{bmatrix}$. Is $\mathbf{u}$ in the subset

of $\mathbb{R}^3$ spanned by the columns of A? Why or why not?

15. Let $A = \begin{bmatrix} 2 & -1 \\ -6 & 3 \end{bmatrix}$ and $\mathbf{b} = \begin{bmatrix} b_1 \\ b_2 \end{bmatrix}$. Show that the equation

$A\mathbf{x} = \mathbf{b}$ does not have a solution for all possible $\mathbf{b}$, and describe the set of all $\mathbf{b}$ for which $A\mathbf{x} = \mathbf{b}$ *does* have a solution.

16. Repeat Exercise 15: $A = \begin{bmatrix} 1 & -3 & -4 \\ -3 & 2 & 6 \\ 5 & -1 & -8 \end{bmatrix}, \mathbf{b} = \begin{bmatrix} b_1 \\ b_2 \\ b_3 \end{bmatrix}$.

Exercises 17–20 refer to the matrices A and B below. Make appropriate calculations that justify your answers and mention an appropriate theorem.

$$A = \begin{bmatrix} 1 & 3 & 0 & 3 \\ -1 & -1 & -1 & 1 \\ 0 & -4 & 2 & -8 \\ 2 & 0 & 3 & -1 \end{bmatrix} \quad B = \begin{bmatrix} 1 & 3 & -2 & 2 \\ 0 & 1 & 1 & -5 \\ 1 & 2 & -3 & 7 \\ -2 & -8 & 2 & -1 \end{bmatrix}$$

17. How many rows of A contain a pivot position? Does the equation $A\mathbf{x} = \mathbf{b}$ have a solution for each $\mathbf{b}$ in $\mathbb{R}^4$?

18. Do the columns of B span $\mathbb{R}^4$? Does the equation $B\mathbf{x} = \mathbf{y}$ have a solution for each $\mathbf{y}$ in $\mathbb{R}^4$?

19. Can each vector in $\mathbb{R}^4$ be written as a linear combination of the columns of the matrix A above? Do the columns of A span $\mathbb{R}^4$?

20. Can every vector in $\mathbb{R}^4$ be written as a linear combination of the columns of the matrix B above? Do the columns of B span $\mathbb{R}^3$?

21. Let $\mathbf{v}_1 = \begin{bmatrix} 1 \\ 0 \\ -1 \\ 0 \end{bmatrix}, \mathbf{v}_2 = \begin{bmatrix} 0 \\ -1 \\ 0 \\ 1 \end{bmatrix}, \mathbf{v}_3 = \begin{bmatrix} 1 \\ 0 \\ 0 \\ -1 \end{bmatrix}$.

Does $\{\mathbf{v}_1, \mathbf{v}_2, \mathbf{v}_3\}$ span $\mathbb{R}^4$? Why or why not?

22. Let $\mathbf{v}_1 = \begin{bmatrix} 0 \\ 0 \\ -2 \end{bmatrix}, \mathbf{v}_2 = \begin{bmatrix} 0 \\ -3 \\ 8 \end{bmatrix}, \mathbf{v}_3 = \begin{bmatrix} 4 \\ -1 \\ -5 \end{bmatrix}$.

Does $\{\mathbf{v}_1, \mathbf{v}_2, \mathbf{v}_3\}$ span $\mathbb{R}^3$? Why or why not?

In Exercises 23–34, mark each statement True or False (**T/F**). Justify each answer.

23. (**T/F**) The equation $A\mathbf{x} = \mathbf{b}$ is referred to as a *vector equation.*

24. (**T/F**) Every matrix equation $A\mathbf{x} = \mathbf{b}$ corresponds to a vector equation with the same solution set.

25. (**T/F**) If the equation $A\mathbf{x} = \mathbf{b}$ is inconsistent, then $\mathbf{b}$ is not in the set spanned by the columns of A.

26. (**T/F**) A vector $\mathbf{b}$ is a linear combination of the columns of a matrix A if and only if the equation $A\mathbf{x} = \mathbf{b}$ has at least one solution.

27. (**T/F**) The equation $A\mathbf{x} = \mathbf{b}$ is consistent if the augmented matrix $[\ A\ \ \mathbf{b}\ ]$ has a pivot position in every row.

28. (**T/F**) If A is an $m \times n$ matrix whose columns do not span $\mathbb{R}^m$, then the equation $A\mathbf{x} = \mathbf{b}$ is inconsistent for some $\mathbf{b}$ in $\mathbb{R}^m$.

29. (**T/F**) The first entry in the product $A\mathbf{x}$ is a sum of products.

30. (**T/F**) Any linear combination vectors can always be written in the form $A\mathbf{x}$ for a suitable matrix A and vector $\mathbf{x}$.

31. (**T/F**) If the columns of an $m \times n$ matrix A span $\mathbb{R}^m$, then the equation $A\mathbf{x} = \mathbf{b}$ is consistent for each $\mathbf{b}$ in $\mathbb{R}^m$.

32. (**T/F**) The solution set of a linear system whose augmented matrix is $[\ \mathbf{a}_1\ \ \mathbf{a}_2\ \ \mathbf{a}_3\ \ \mathbf{b}\ ]$ is the same as the solution set of $A\mathbf{x} = \mathbf{b}$, if $A = [\ \mathbf{a}_1\ \ \mathbf{a}_2\ \ \mathbf{a}_3\ ]$.

33. (T/F) If A is an $m \times n$ matrix and if the equation $A\mathbf{x} = \mathbf{b}$ is inconsistent for some $\mathbf{b}$ in $\mathbb{R}^m$, then A cannot have a pivot position in every row.

34. (T/F) If the augmented matrix $[\, A \ \ \mathbf{b} \,]$ has a pivot position in every row, then the equation $A\mathbf{x} = \mathbf{b}$ is inconsistent.

35. Note that $\begin{bmatrix} 4 & -3 & 1 \\ 5 & -2 & 5 \\ -6 & 2 & -3 \end{bmatrix} \begin{bmatrix} -3 \\ -1 \\ 2 \end{bmatrix} = \begin{bmatrix} -7 \\ -3 \\ 10 \end{bmatrix}$. Use this fact

(and no row operations) to find scalars c_1, c_2, c_3 such that

$$\begin{bmatrix} -7 \\ -3 \\ 10 \end{bmatrix} = c_1 \begin{bmatrix} 4 \\ 5 \\ -6 \end{bmatrix} + c_2 \begin{bmatrix} -3 \\ -2 \\ 2 \end{bmatrix} + c_3 \begin{bmatrix} 1 \\ 5 \\ -3 \end{bmatrix}.$$

36. Let $\mathbf{u} = \begin{bmatrix} 7 \\ 2 \\ 5 \end{bmatrix}$, $\mathbf{v} = \begin{bmatrix} 3 \\ 1 \\ 3 \end{bmatrix}$, and $\mathbf{w} = \begin{bmatrix} 6 \\ 1 \\ 0 \end{bmatrix}$.

It can be shown that $3\mathbf{u} - 5\mathbf{v} - \mathbf{w} = \mathbf{0}$. Use this fact (and no row operations) to find x_1 and x_2 that satisfy the equation

$$\begin{bmatrix} 7 & 3 \\ 2 & 1 \\ 5 & 3 \end{bmatrix} \begin{bmatrix} x_1 \\ x_2 \end{bmatrix} = \begin{bmatrix} 6 \\ 1 \\ 0 \end{bmatrix}.$$

37. Let $\mathbf{q}_1, \mathbf{q}_2, \mathbf{q}_3$, and $\mathbf{v}$ represent vectors in $\mathbb{R}^5$, and let x_1, x_2, and x_3 denote scalars. Write the following vector equation as a matrix equation. Identify any symbols you choose to use.

$$x_1\mathbf{q}_1 + x_2\mathbf{q}_2 + x_3\mathbf{q}_3 = \mathbf{v}$$

38. Rewrite the (numerical) matrix equation below in symbolic form as a vector equation, using symbols $\mathbf{v}_1, \mathbf{v}_2, \dots$ for the vectors and $c_1, c_2, \dots$ for scalars. Define what each symbol represents, using the data given in the matrix equation.

$$\begin{bmatrix} -3 & 5 & -4 & 9 & 7 \\ 5 & 8 & 1 & -2 & -4 \end{bmatrix} \begin{bmatrix} -3 \\ 2 \\ 4 \\ -1 \\ 2 \end{bmatrix} = \begin{bmatrix} 8 \\ -1 \end{bmatrix}$$

39. Construct a 3×3 matrix, not in echelon form, whose columns span $\mathbb{R}^3$. Show that the matrix you construct has the desired property.

40. Construct a 3×3 matrix, not in echelon form, whose columns do *not* span $\mathbb{R}^3$. Show that the matrix you construct has the desired property.

STUDY GUIDE offers additional resources for mastering the concept of span.

41. Let A be a 3×2 matrix. Explain why the equation $A\mathbf{x} = \mathbf{b}$ cannot be consistent for all $\mathbf{b}$ in $\mathbb{R}^3$. Generalize your argument to the case of an arbitrary A with more rows than columns.

42. Could a set of three vectors in $\mathbb{R}^4$ span all of $\mathbb{R}^4$? Explain. What about n vectors in $\mathbb{R}^m$ when n is less than m?

43. Suppose A is a 4×3 matrix and $\mathbf{b}$ is a vector in $\mathbb{R}^4$ with the property that $A\mathbf{x} = \mathbf{b}$ has a unique solution. What can you say about the reduced echelon form of A? Justify your answer.

44. Suppose A is a 3×3 matrix and $\mathbf{b}$ is a vector in $\mathbb{R}^3$ with the property that $A\mathbf{x} = \mathbf{b}$ has a unique solution. Explain why the columns of A must span $\mathbb{R}^3$.

45. Let A be a 3×4 matrix, let $\mathbf{y}_1$ and $\mathbf{y}_2$ be vectors in $\mathbb{R}^3$, and let $\mathbf{w} = \mathbf{y}_1 + \mathbf{y}_2$. Suppose $\mathbf{y}_1 = A\mathbf{x}_1$ and $\mathbf{y}_2 = A\mathbf{x}_2$ for some vectors $\mathbf{x}_1$ and $\mathbf{x}_2$ in $\mathbb{R}^4$. What fact allows you to conclude that the system $A\mathbf{x} = \mathbf{w}$ is consistent? (*Note:* $\mathbf{x}_1$ and $\mathbf{x}_2$ denote vectors, not scalar entries in vectors.)

46. Let A be a 5×3 matrix, let $\mathbf{y}$ be a vector in $\mathbb{R}^3$, and let $\mathbf{z}$ be a vector in $\mathbb{R}^5$. Suppose $A\mathbf{y} = \mathbf{z}$. What fact allows you to conclude that the system $A\mathbf{x} = 4\mathbf{z}$ is consistent?

▣ In Exercises 47–50, determine if the columns of the matrix span $\mathbb{R}^4$.

47. $\begin{bmatrix} 7 & 2 & -5 & 8 \\ -5 & -3 & 4 & -9 \\ 6 & 10 & -2 & 7 \\ -7 & 9 & 2 & 15 \end{bmatrix}$ **48.** $\begin{bmatrix} 5 & -7 & -4 & 9 \\ 6 & -8 & -7 & 5 \\ 4 & -4 & -9 & -9 \\ -9 & 11 & 16 & 7 \end{bmatrix}$

49. $\begin{bmatrix} 12 & -7 & 11 & -9 & 5 \\ -9 & 4 & -8 & 7 & -3 \\ -6 & 11 & -7 & 3 & -9 \\ 4 & -6 & 10 & -5 & 12 \end{bmatrix}$

50. $\begin{bmatrix} 8 & 11 & -6 & -7 & 13 \\ -7 & -8 & 5 & 6 & -9 \\ 11 & 7 & -7 & -9 & -6 \\ -3 & 4 & 1 & 8 & 7 \end{bmatrix}$

▣ 51. Find a column of the matrix in Exercise 49 that can be deleted and yet have the remaining matrix columns still span $\mathbb{R}^4$.

▣ 52. Find a column of the matrix in Exercise 50 that can be deleted and yet have the remaining matrix columns still span $\mathbb{R}^4$. Can you delete more than one column?

Solutions to Practice Problems

1. The matrix equation

$$\begin{bmatrix} 1 & 5 & -2 & 0 \\ -3 & 1 & 9 & -5 \\ 4 & -8 & -1 & 7 \end{bmatrix} \begin{bmatrix} 3 \\ -2 \\ 0 \\ -4 \end{bmatrix} = \begin{bmatrix} -7 \\ 9 \\ 0 \end{bmatrix}$$

is equivalent to the vector equation

$$3\begin{bmatrix} 1 \\ -3 \\ 4 \end{bmatrix} - 2\begin{bmatrix} 5 \\ 1 \\ -8 \end{bmatrix} + 0\begin{bmatrix} -2 \\ 9 \\ -1 \end{bmatrix} - 4\begin{bmatrix} 0 \\ -5 \\ 7 \end{bmatrix} = \begin{bmatrix} -7 \\ 9 \\ 0 \end{bmatrix},$$

which expresses **b** as a linear combination of the columns of A.

2. $\mathbf{u} + \mathbf{v} = \begin{bmatrix} 4 \\ -1 \end{bmatrix} + \begin{bmatrix} -3 \\ 5 \end{bmatrix} = \begin{bmatrix} 1 \\ 4 \end{bmatrix}$

$A(\mathbf{u} + \mathbf{v}) = \begin{bmatrix} 2 & 5 \\ 3 & 1 \end{bmatrix}\begin{bmatrix} 1 \\ 4 \end{bmatrix} = \begin{bmatrix} 2 + 20 \\ 3 + 4 \end{bmatrix} = \begin{bmatrix} 22 \\ 7 \end{bmatrix}$

$A\mathbf{u} + A\mathbf{v} = \begin{bmatrix} 2 & 5 \\ 3 & 1 \end{bmatrix}\begin{bmatrix} 4 \\ -1 \end{bmatrix} + \begin{bmatrix} 2 & 5 \\ 3 & 1 \end{bmatrix}\begin{bmatrix} -3 \\ 5 \end{bmatrix}$

$= \begin{bmatrix} 3 \\ 11 \end{bmatrix} + \begin{bmatrix} 19 \\ -4 \end{bmatrix} = \begin{bmatrix} 22 \\ 7 \end{bmatrix}$

Remark: There are, in fact, infinitely many correct solutions to Practice Problem 3. When creating matrices to satisfy specified criteria, it is often useful to create matrices that are straightforward, such as those already in reduced echelon form. Here is one possible solution:

3. Let

$$A = \begin{bmatrix} 1 & 0 & 1 \\ 0 & 1 & 1 \\ 0 & 0 & 0 \end{bmatrix}, \mathbf{b} = \begin{bmatrix} 3 \\ 2 \\ 0 \end{bmatrix}, \text{ and } \mathbf{c} = \begin{bmatrix} 3 \\ 2 \\ 1 \end{bmatrix}.$$

Notice the reduced echelon form of the augmented matrix corresponding to $A\mathbf{x} = \mathbf{b}$ is

$$\begin{bmatrix} 1 & 0 & 1 & 3 \\ 0 & 1 & 1 & 2 \\ 0 & 0 & 0 & 0 \end{bmatrix},$$

which corresponds to a consistent system, and hence $A\mathbf{x} = \mathbf{b}$ has solutions. The reduced echelon form of the augmented matrix corresponding to $A\mathbf{x} = \mathbf{c}$ is

$$\begin{bmatrix} 1 & 0 & 1 & 3 \\ 0 & 1 & 1 & 2 \\ 0 & 0 & 0 & 1 \end{bmatrix},$$

which corresponds to an inconsistent system, and hence $A\mathbf{x} = \mathbf{c}$ does not have any solutions.

1.5 Solution Sets of Linear Systems

Solution sets of linear systems are important objects of study in linear algebra. They will appear later in several different contexts. This section uses vector notation to give explicit and geometric descriptions of such solution sets.

Homogeneous Linear Systems

A system of linear equations is said to be **homogeneous** if it can be written in the form $A\mathbf{x} = \mathbf{0}$, where A is an $m \times n$ matrix and $\mathbf{0}$ is the zero vector in $\mathbb{R}^m$. Such a system $A\mathbf{x} = \mathbf{0}$ *always* has at least one solution, namely $\mathbf{x} = \mathbf{0}$ (the zero vector in $\mathbb{R}^n$). This zero solution is usually called the **trivial solution**. For a given equation $A\mathbf{x} = \mathbf{0}$, the important question is whether there exists a **nontrivial solution**, that is, a nonzero vector $\mathbf{x}$ that satisfies $A\mathbf{x} = \mathbf{0}$. The Existence and Uniqueness Theorem in Section 1.2 (Theorem 2) leads immediately to the following fact.

> The homogeneous equation $A\mathbf{x} = \mathbf{0}$ has a nontrivial solution if and only if the equation has at least one free variable.

EXAMPLE 1 Determine if the following homogeneous system has a nontrivial solution. Then describe the solution set.

$$3x_1 + 5x_2 - 4x_3 = 0$$
$$-3x_1 - 2x_2 + 4x_3 = 0$$
$$6x_1 + x_2 - 8x_3 = 0$$

SOLUTION Let A be the matrix of coefficients of the system and row reduce the augmented matrix $[\,A \quad \mathbf{0}\,]$ to echelon form:

$$\begin{bmatrix} 3 & 5 & -4 & 0 \\ -3 & -2 & 4 & 0 \\ 6 & 1 & -8 & 0 \end{bmatrix} \sim \begin{bmatrix} 3 & 5 & -4 & 0 \\ 0 & 3 & 0 & 0 \\ 0 & -9 & 0 & 0 \end{bmatrix} \sim \begin{bmatrix} 3 & 5 & -4 & 0 \\ 0 & 3 & 0 & 0 \\ 0 & 0 & 0 & 0 \end{bmatrix}$$

Since x_3 is a free variable, $A\mathbf{x} = \mathbf{0}$ has nontrivial solutions (one for each nonzero choice of x_3). To describe the solution set, continue the row reduction of $[\,A \quad \mathbf{0}\,]$ to *reduced* echelon form:

$$\begin{bmatrix} 1 & 0 & -\frac{4}{3} & 0 \\ 0 & 1 & 0 & 0 \\ 0 & 0 & 0 & 0 \end{bmatrix} \qquad \begin{aligned} x_1 \quad\;\; - \tfrac{4}{3}x_3 &= 0 \\ x_2 \qquad\quad &= 0 \\ 0 &= 0 \end{aligned}$$

Solve for the basic variables x_1 and x_2 and obtain $x_1 = \frac{4}{3}x_3$, $x_2 = 0$, with x_3 free. As a vector, the general solution of $A\mathbf{x} = \mathbf{0}$ has the form

$$\mathbf{x} = \begin{bmatrix} x_1 \\ x_2 \\ x_3 \end{bmatrix} = \begin{bmatrix} \frac{4}{3}x_3 \\ 0 \\ x_3 \end{bmatrix} = x_3 \begin{bmatrix} \frac{4}{3} \\ 0 \\ 1 \end{bmatrix} = x_3 \mathbf{v}, \quad \text{where } \mathbf{v} = \begin{bmatrix} \frac{4}{3} \\ 0 \\ 1 \end{bmatrix}$$

Here x_3 is factored out of the expression for the general solution vector. This shows that every solution of $A\mathbf{x} = \mathbf{0}$ in this case is a scalar multiple of $\mathbf{v}$. The trivial solution is obtained by choosing $x_3 = 0$. Geometrically, the solution set is a line through $\mathbf{0}$ in $\mathbb{R}^3$. See Figure 1. ∎

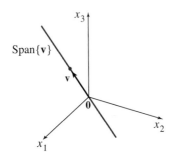

FIGURE 1

Notice that a nontrivial solution $\mathbf{x}$ can have some zero entries so long as not all of its entries are zero.

EXAMPLE 2 A single linear equation can be treated as a very simple system of equations. Describe all solutions of the homogeneous "system"

$$10x_1 - 3x_2 - 2x_3 = 0 \tag{1}$$

SOLUTION   There is no need for matrix notation. Solve for the basic variable x_1 in terms of the free variables. The general solution is $x_1 = .3x_2 + .2x_3$, with x_2 and x_3 free. As a vector, the general solution is

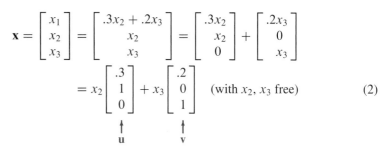

$$\mathbf{x} = \begin{bmatrix} x_1 \\ x_2 \\ x_3 \end{bmatrix} = \begin{bmatrix} .3x_2 + .2x_3 \\ x_2 \\ x_3 \end{bmatrix} = \begin{bmatrix} .3x_2 \\ x_2 \\ 0 \end{bmatrix} + \begin{bmatrix} .2x_3 \\ 0 \\ x_3 \end{bmatrix}$$

$$= x_2 \underset{\underset{\mathbf{u}}{\uparrow}}{\begin{bmatrix} .3 \\ 1 \\ 0 \end{bmatrix}} + x_3 \underset{\underset{\mathbf{v}}{\uparrow}}{\begin{bmatrix} .2 \\ 0 \\ 1 \end{bmatrix}} \quad \text{(with } x_2, x_3 \text{ free)} \tag{2}$$

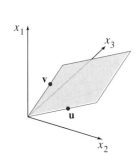

FIGURE 2

This calculation shows that every solution of (1) is a linear combination of the vectors $\mathbf{u}$ and $\mathbf{v}$, shown in (2). That is, the solution set is Span $\{\mathbf{u}, \mathbf{v}\}$. Since neither $\mathbf{u}$ nor $\mathbf{v}$ is a scalar multiple of the other, the solution set is a plane through the origin. See Figure 2.   ∎

Examples 1 and 2, along with the exercises, illustrate the fact that the solution set of a homogeneous equation $A\mathbf{x} = \mathbf{0}$ can always be expressed explicitly as Span $\{\mathbf{v}_1, \ldots, \mathbf{v}_p\}$ for suitable vectors $\mathbf{v}_1, \ldots, \mathbf{v}_p$. If the only solution is the zero vector, then the solution set is Span $\{\mathbf{0}\}$. If the equation $A\mathbf{x} = \mathbf{0}$ has only one free variable, the solution set is a line through the origin, as in Figure 1. A plane through the origin, as in Figure 2, provides a good mental image for the solution set of $A\mathbf{x} = \mathbf{0}$ when there are two or more free variables. Note, however, that a similar figure can be used to visualize Span $\{\mathbf{u}, \mathbf{v}\}$ even when $\mathbf{u}$ and $\mathbf{v}$ do not arise as solutions of $A\mathbf{x} = \mathbf{0}$. See Figure 11 in Section 1.3.

Parametric Vector Form

The original equation (1) for the plane in Example 2 is an *implicit* description of the plane. Solving this equation amounts to finding an *explicit* description of the plane as the set spanned by $\mathbf{u}$ and $\mathbf{v}$. Equation (2) is called a **parametric vector equation** of the plane. Sometimes such an equation is written as

$$\mathbf{x} = s\mathbf{u} + t\mathbf{v} \quad (s, t \text{ in } \mathbb{R})$$

to emphasize that the parameters vary over all real numbers. In Example 1, the equation $\mathbf{x} = x_3\mathbf{v}$ (with x_3 free), or $\mathbf{x} = t\mathbf{v}$ (with t in $\mathbb{R}$), is a parametric vector equation of a line. Whenever a solution set is described explicitly with vectors as in Examples 1 and 2, we say that the solution is in **parametric vector form**.

Solutions of Nonhomogeneous Systems

When a nonhomogeneous linear system has many solutions, the general solution can be written in parametric vector form as one vector plus an arbitrary linear combination of vectors that satisfy the corresponding homogeneous system.

EXAMPLE 3   Describe all solutions of $A\mathbf{x} = \mathbf{b}$, where

$$A = \begin{bmatrix} 3 & 5 & -4 \\ -3 & -2 & 4 \\ 6 & 1 & -8 \end{bmatrix} \quad \text{and} \quad \mathbf{b} = \begin{bmatrix} 7 \\ -1 \\ -4 \end{bmatrix}$$

SOLUTION Here A is the matrix of coefficients from Example 1. Row operations on $[A \quad \mathbf{b}]$ produce

$$\begin{bmatrix} 3 & 5 & -4 & 7 \\ -3 & -2 & 4 & -1 \\ 6 & 1 & -8 & -4 \end{bmatrix} \sim \begin{bmatrix} 1 & 0 & -\frac{4}{3} & -1 \\ 0 & 1 & 0 & 2 \\ 0 & 0 & 0 & 0 \end{bmatrix}, \qquad \begin{aligned} x_1 \quad - \tfrac{4}{3}x_3 &= -1 \\ x_2 \quad &= 2 \\ 0 &= 0 \end{aligned}$$

Thus $x_1 = -1 + \frac{4}{3}x_3$, $x_2 = 2$, and x_3 is free. As a vector, the general solution of $A\mathbf{x} = \mathbf{b}$ has the form

$$\mathbf{x} = \begin{bmatrix} x_1 \\ x_2 \\ x_3 \end{bmatrix} = \begin{bmatrix} -1 + \frac{4}{3}x_3 \\ 2 \\ x_3 \end{bmatrix} = \begin{bmatrix} -1 \\ 2 \\ 0 \end{bmatrix} + \begin{bmatrix} \frac{4}{3}x_3 \\ 0 \\ x_3 \end{bmatrix} = \underset{\underset{\mathbf{p}}{\uparrow}}{\begin{bmatrix} -1 \\ 2 \\ 0 \end{bmatrix}} + x_3 \underset{\underset{\mathbf{v}}{\uparrow}}{\begin{bmatrix} \frac{4}{3} \\ 0 \\ 1 \end{bmatrix}}$$

The equation $\mathbf{x} = \mathbf{p} + x_3\mathbf{v}$, or, writing t as a general parameter,

$$\mathbf{x} = \mathbf{p} + t\mathbf{v} \quad (t \text{ in } \mathbb{R}) \tag{3}$$

describes the solution set of $A\mathbf{x} = \mathbf{b}$ in parametric vector form. Recall from Example 1 that the solution set of $A\mathbf{x} = \mathbf{0}$ has the parametric vector equation

$$\mathbf{x} = t\mathbf{v} \quad (t \text{ in } \mathbb{R}) \tag{4}$$

[with the same $\mathbf{v}$ that appears in (3)]. Thus the solutions of $A\mathbf{x} = \mathbf{b}$ are obtained by adding the vector $\mathbf{p}$ to the solutions of $A\mathbf{x} = \mathbf{0}$. The vector $\mathbf{p}$ itself is just one particular solution of $A\mathbf{x} = \mathbf{b}$ [corresponding to $t = 0$ in (3)]. ∎

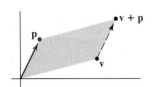

FIGURE 3

Adding $\mathbf{p}$ to $\mathbf{v}$ translates $\mathbf{v}$ to $\mathbf{v} + \mathbf{p}$.

To describe the solution set of $A\mathbf{x} = \mathbf{b}$ geometrically, we can think of vector addition as a *translation*. Given $\mathbf{v}$ and $\mathbf{p}$ in $\mathbb{R}^2$ or $\mathbb{R}^3$, the effect of adding $\mathbf{p}$ to $\mathbf{v}$ is to *move* $\mathbf{v}$ in a direction parallel to the line through $\mathbf{p}$ and $\mathbf{0}$. We say that $\mathbf{v}$ is **translated by $\mathbf{p}$** to $\mathbf{v} + \mathbf{p}$. See Figure 3. If each point on a line L in $\mathbb{R}^2$ or $\mathbb{R}^3$ is translated by a vector $\mathbf{p}$, the result is a line parallel to L. See Figure 4.

Suppose L is the line through $\mathbf{0}$ and $\mathbf{v}$, described by equation (4). Adding $\mathbf{p}$ to each point on L produces the translated line described by equation (3). Note that $\mathbf{p}$ is on the line in equation (3). We call (3) **the equation of the line through p parallel to v**. Thus *the solution set of $A\mathbf{x} = \mathbf{b}$ is a line through $\mathbf{p}$ parallel to the solution set of $A\mathbf{x} = \mathbf{0}$.* Figure 5 illustrates this case.

FIGURE 4

Translated line.

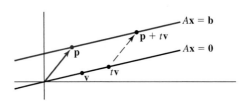

FIGURE 5 Parallel solution sets of $A\mathbf{x} = \mathbf{b}$ and $A\mathbf{x} = \mathbf{0}$.

The relation between the solution sets of $A\mathbf{x} = \mathbf{b}$ and $A\mathbf{x} = \mathbf{0}$ shown in Figure 5 generalizes to any *consistent* equation $A\mathbf{x} = \mathbf{b}$, although the solution set will be larger than a line when there are several free variables. The following theorem gives the precise statement. See Exercise 37 at the end of this section for a proof.

THEOREM 6

Suppose the equation $A\mathbf{x} = \mathbf{b}$ is consistent for some given $\mathbf{b}$, and let $\mathbf{p}$ be a solution. Then the solution set of $A\mathbf{x} = \mathbf{b}$ is the set of all vectors of the form $\mathbf{w} = \mathbf{p} + \mathbf{v}_h$, where $\mathbf{v}_h$ is any solution of the homogeneous equation $A\mathbf{x} = \mathbf{0}$.

Theorem 6 says that if $A\mathbf{x} = \mathbf{b}$ has a solution, then the solution set is obtained by translating the solution set of $A\mathbf{x} = \mathbf{0}$, using any particular solution $\mathbf{p}$ of $A\mathbf{x} = \mathbf{b}$ for the translation. Figure 6 illustrates the case in which there are two free variables. Even when $n > 3$, our mental image of the solution set of a consistent system $A\mathbf{x} = \mathbf{b}$ (with $\mathbf{b} \neq \mathbf{0}$) is either a single nonzero point or a line or plane not passing through the origin.

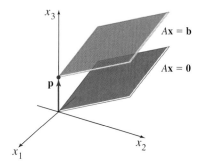

FIGURE 6 Parallel solution sets of $A\mathbf{x} = \mathbf{b}$ and $A\mathbf{x} = \mathbf{0}$.

Warning: Theorem 6 and Figure 6 apply only to an equation $A\mathbf{x} = \mathbf{b}$ that has at least one nonzero solution $\mathbf{p}$. When $A\mathbf{x} = \mathbf{b}$ has no solution, the solution set is empty.

The following algorithm outlines the calculations shown in Examples 1, 2, and 3.

WRITING A SOLUTION SET (OF A CONSISTENT SYSTEM) IN PARAMETRIC VECTOR FORM

1. Row reduce the augmented matrix to reduced echelon form.

2. Express each basic variable in terms of any free variables appearing in an equation.

3. Write a typical solution $\mathbf{x}$ as a vector whose entries depend on the free variables, if any.

4. Decompose $\mathbf{x}$ into a linear combination of vectors (with numeric entries) using the free variables as parameters.

Reasonable Answers

To verify that the solutions you found are indeed solutions to the homogeneous equation $A\mathbf{x} = \mathbf{0}$, simply multiply the matrix by each vector in your solution and check that the result is the zero vector. For example, if

$$A = \begin{bmatrix} 1 & -2 & 1 & 2 \\ 1 & -1 & 2 & 5 \\ 0 & 1 & 1 & 3 \end{bmatrix}, \text{ and you found the homogeneous solutions to}$$

Reasonable Answers (Continued)

be $x_3 \begin{bmatrix} -3 \\ -1 \\ 1 \\ 0 \end{bmatrix} + x_4 \begin{bmatrix} -8 \\ -3 \\ 0 \\ 1 \end{bmatrix}$, check $\begin{bmatrix} 1 & -2 & 1 & 2 \\ 1 & -1 & 2 & 5 \\ 0 & 1 & 1 & 3 \end{bmatrix} \begin{bmatrix} -3 \\ -1 \\ 1 \\ 0 \end{bmatrix} = \begin{bmatrix} 0 \\ 0 \\ 0 \end{bmatrix}$ and

$\begin{bmatrix} 1 & -2 & 1 & 2 \\ 1 & -1 & 2 & 5 \\ 0 & 1 & 1 & 3 \end{bmatrix} \begin{bmatrix} -8 \\ -3 \\ 0 \\ 1 \end{bmatrix} = \begin{bmatrix} 0 \\ 0 \\ 0 \end{bmatrix}$. Then $A \left(x_3 \begin{bmatrix} -3 \\ -1 \\ 1 \\ 0 \end{bmatrix} + x_4 \begin{bmatrix} -8 \\ -3 \\ 0 \\ 1 \end{bmatrix} \right)$

$= x_3 A \begin{bmatrix} -3 \\ -1 \\ 1 \\ 0 \end{bmatrix} + x_4 A \begin{bmatrix} -8 \\ -3 \\ 0 \\ 1 \end{bmatrix}$, which is equal to $x_3 \begin{bmatrix} 0 \\ 0 \\ 0 \end{bmatrix} + x_4 \begin{bmatrix} 0 \\ 0 \\ 0 \end{bmatrix} = \begin{bmatrix} 0 \\ 0 \\ 0 \end{bmatrix}$,

as desired.

If you are solving $A\mathbf{x} = \mathbf{b}$, then you can again verify that you have correct solutions by multiplying the matrix by each vector in your solutions. The product of A with the first vector (the one that is *not* part of the solution to the homogeneous equation) should be $\mathbf{b}$. The product of A with the remaining vectors (the ones that are part of the solution to the homogeneous equation) should of course be $\mathbf{0}$.

For example, to verify that $\begin{bmatrix} 2 \\ 1 \\ 1 \\ 2 \end{bmatrix} + x_3 \begin{bmatrix} -3 \\ -1 \\ 1 \\ 0 \end{bmatrix} + x_4 \begin{bmatrix} -8 \\ -3 \\ 0 \\ 1 \end{bmatrix}$ are solutions to

$A\mathbf{x} = \begin{bmatrix} 5 \\ 13 \\ 8 \end{bmatrix}$, check $\begin{bmatrix} 1 & -2 & 1 & 2 \\ 1 & -1 & 2 & 5 \\ 0 & 1 & 1 & 3 \end{bmatrix} \begin{bmatrix} 2 \\ 1 \\ 1 \\ 2 \end{bmatrix} = \begin{bmatrix} 5 \\ 13 \\ 8 \end{bmatrix}$, and use the

calculations from above. Notice $A \left(\begin{bmatrix} 2 \\ 1 \\ 1 \\ 2 \end{bmatrix} + x_3 \begin{bmatrix} -3 \\ -1 \\ 1 \\ 0 \end{bmatrix} + x_4 \begin{bmatrix} -8 \\ -3 \\ 0 \\ 1 \end{bmatrix} \right)$

$= A \begin{bmatrix} 2 \\ 1 \\ 1 \\ 2 \end{bmatrix} + x_3 A \begin{bmatrix} -3 \\ -1 \\ 1 \\ 0 \end{bmatrix} + x_4 A \begin{bmatrix} -8 \\ -3 \\ 0 \\ 1 \end{bmatrix}$, which is equal to $\begin{bmatrix} 5 \\ 13 \\ 8 \end{bmatrix} + x_3 \begin{bmatrix} 0 \\ 0 \\ 0 \end{bmatrix}$

$+ x_4 \begin{bmatrix} 0 \\ 0 \\ 0 \end{bmatrix} = \begin{bmatrix} 5 \\ 13 \\ 8 \end{bmatrix}$, as desired.

Practice Problems

1. Each of the following equations determines a plane in $\mathbb{R}^3$. Do the two planes intersect? If so, describe their intersection.

$$x_1 + 4x_2 - 5x_3 = 0$$
$$2x_1 - x_2 + 8x_3 = 9$$

2. Write the general solution of $10x_1 - 3x_2 - 2x_3 = 7$ in parametric vector form, and relate the solution set to the one found in Example 2.

3. Prove the first part of Theorem 6: Suppose that $\mathbf{p}$ is a solution of $A\mathbf{x} = \mathbf{b}$, so that $A\mathbf{p} = \mathbf{b}$. Let $\mathbf{v}_h$ be any solution to the homogeneous equation $A\mathbf{x} = \mathbf{0}$, and let $\mathbf{w} = \mathbf{p} + \mathbf{v}_h$. Show that $\mathbf{w}$ is a solution to $A\mathbf{x} = \mathbf{b}$.

1.5 Exercises

In Exercises 1–4, determine if the system has a nontrivial solution. Try to use as few row operations as possible.

1.
$$2x_1 - 5x_2 + 8x_3 = 0$$
$$-2x_1 - 7x_2 + x_3 = 0$$
$$4x_1 + 2x_2 + 7x_3 = 0$$

2.
$$x_1 - 3x_2 + 7x_3 = 0$$
$$-2x_1 + x_2 - 4x_3 = 0$$
$$x_1 + 2x_2 + 9x_3 = 0$$

3.
$$-3x_1 + 5x_2 - 7x_3 = 0$$
$$-6x_1 + 7x_2 + x_3 = 0$$

4.
$$-5x_1 + 7x_2 + 9x_3 = 0$$
$$x_1 - 2x_2 + 6x_3 = 0$$

In Exercises 5 and 6, follow the method of Examples 1 and 2 to write the solution set of the given homogeneous system in parametric vector form.

5.
$$x_1 + 3x_2 + x_3 = 0$$
$$-4x_1 - 9x_2 + 2x_3 = 0$$
$$- 3x_2 - 6x_3 = 0$$

6.
$$x_1 + 3x_2 - 5x_3 = 0$$
$$x_1 + 4x_2 - 8x_3 = 0$$
$$-3x_1 - 7x_2 + 9x_3 = 0$$

In Exercises 7–12, describe all solutions of $A\mathbf{x} = \mathbf{0}$ in parametric vector form, where A is row equivalent to the given matrix.

7. $\begin{bmatrix} 1 & 3 & -3 & 7 \\ 0 & 1 & -4 & 5 \end{bmatrix}$

8. $\begin{bmatrix} 1 & -2 & -9 & 5 \\ 0 & 1 & 2 & -6 \end{bmatrix}$

9. $\begin{bmatrix} 3 & -9 & 6 \\ -1 & 3 & -2 \end{bmatrix}$

10. $\begin{bmatrix} 1 & 3 & 0 & -4 \\ 2 & 6 & 0 & -8 \end{bmatrix}$

11. $\begin{bmatrix} 1 & -4 & -2 & 0 & 3 & -5 \\ 0 & 0 & 1 & 0 & 0 & -1 \\ 0 & 0 & 0 & 0 & 1 & -4 \\ 0 & 0 & 0 & 0 & 0 & 0 \end{bmatrix}$

12. $\begin{bmatrix} 1 & 5 & 2 & -6 & 9 & 0 \\ 0 & 0 & 1 & -7 & 4 & -8 \\ 0 & 0 & 0 & 0 & 0 & 1 \\ 0 & 0 & 0 & 0 & 0 & 0 \end{bmatrix}$

You may find it helpful to review the information in the Reasonable Answers box from this section before answering Exercises 13–16.

13. Verify that the solutions you found to Exercise 9 are indeed homogeneous solutions.

14. Verify that the solutions you found to Exercise 10 are indeed homogeneous solutions.

15. Verify that the solutions you found to Exercise 11 are indeed homogeneous solutions.

16. Verify that the solutions you found to Exercise 12 are indeed homogeneous solutions.

17. Suppose the solution set of a certain system of linear equations can be described as $x_1 = 5 + 4x_3$, $x_2 = -2 - 7x_3$, with x_3 free. Use vectors to describe this set as a line in $\mathbb{R}^3$.

18. Suppose the solution set of a certain system of linear equations can be described as $x_1 = 3x_4$, $x_2 = 8 + x_4$, $x_3 = 2 - 5x_4$, with x_4 free. Use vectors to describe this set as a line in $\mathbb{R}^4$.

19. Follow the method of Example 3 to describe the solutions of the following system in parametric vector form. Also, give a geometric description of the solution set and compare it to that in Exercise 5.

$$x_1 + 3x_2 + x_3 = 1$$
$$-4x_1 - 9x_2 + 2x_3 = -1$$
$$- 3x_2 - 6x_3 = -3$$

20. As in Exercise 19, describe the solutions of the following system in parametric vector form, and provide a geometric comparison with the solution set in Exercise 6.

$$x_1 + 3x_2 - 5x_3 = 4$$
$$x_1 + 4x_2 - 8x_3 = 7$$
$$-3x_1 - 7x_2 + 9x_3 = -6$$

21. Describe and compare the solution sets of $x_1 + 9x_2 - 4x_3 = 0$ and $x_1 + 9x_2 - 4x_3 = -2$.

22. Describe and compare the solution sets of $x_1 - 3x_2 + 5x_3 = 0$ and $x_1 - 3x_2 + 5x_3 = 4$.

In Exercises 23 and 24, find the parametric equation of the line through **a** parallel to **b**.

23. $\mathbf{a} = \begin{bmatrix} -2 \\ 0 \end{bmatrix}, \mathbf{b} = \begin{bmatrix} -5 \\ 3 \end{bmatrix}$ **24.** $\mathbf{a} = \begin{bmatrix} 3 \\ -4 \end{bmatrix}, \mathbf{b} = \begin{bmatrix} -7 \\ 8 \end{bmatrix}$

In Exercises 25 and 26, find a parametric equation of the line M through **p** and **q**. [*Hint: M* is parallel to the vector $\mathbf{q} - \mathbf{p}$. See the figure below.]

25. $\mathbf{p} = \begin{bmatrix} 2 \\ -5 \end{bmatrix}, \mathbf{q} = \begin{bmatrix} -3 \\ 1 \end{bmatrix}$ **26.** $\mathbf{p} = \begin{bmatrix} -6 \\ 3 \end{bmatrix}, \mathbf{q} = \begin{bmatrix} 0 \\ -4 \end{bmatrix}$

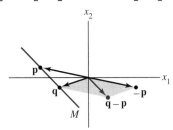

The line through **p** and **q**.

In Exercises 27–36, mark each statement True or False **(T/F)**. Justify each answer.

27. (**T/F**) A homogeneous equation is always consistent.

28. (**T/F**) If **x** is a nontrivial solution of $A\mathbf{x} = \mathbf{0}$, then every entry in **x** is nonzero.

29. (**T/F**) The equation $A\mathbf{x} = \mathbf{0}$ gives an explicit description of its solution set.

30. (**T/F**) The equation $\mathbf{x} = x_2\mathbf{u} + x_3\mathbf{v}$, with x_2 and x_3 free (and neither **u** nor **v** a multiple of the other), describes a plane through the origin.

31. (**T/F**) The homogeneous equation $A\mathbf{x} = \mathbf{0}$ has the trivial solution if and only if the equation has at least one free variable.

32. (**T/F**) The equation $A\mathbf{x} = \mathbf{b}$ is homogeneous if the zero vector is a solution.

33. (**T/F**) The equation $\mathbf{x} = \mathbf{p} + t\mathbf{v}$ describes a line through **v** parallel to **p**.

34. (**T/F**) The effect of adding **p** to a vector is to move the vector in a direction parallel to **p**.

35. (**T/F**) The solution set of $A\mathbf{x} = \mathbf{b}$ is the set of all vectors of the form $\mathbf{w} = \mathbf{p} + \mathbf{v}_h$, where $\mathbf{v}_h$ is any solution of the equation $A\mathbf{x} = \mathbf{0}$.

36. (**T/F**) The solution set of $A\mathbf{x} = \mathbf{b}$ is obtained by translating the solution set of $A\mathbf{x} = \mathbf{0}$.

37. Prove the second part of Theorem 6: Let **w** be any solution of $A\mathbf{x} = \mathbf{b}$, and define $\mathbf{v}_h = \mathbf{w} - \mathbf{p}$. Show that $\mathbf{v}_h$ is a solution of $A\mathbf{x} = \mathbf{0}$. This shows that every solution of $A\mathbf{x} = \mathbf{b}$ has the form $\mathbf{w} = \mathbf{p} + \mathbf{v}_h$, with **p** a particular solution of $A\mathbf{x} = \mathbf{b}$ and $\mathbf{v}_h$ a solution of $A\mathbf{x} = \mathbf{0}$.

38. Suppose $A\mathbf{x} = \mathbf{b}$ has a solution. Explain why the solution is unique precisely when $A\mathbf{x} = \mathbf{0}$ has only the trivial solution.

39. Suppose A is the 3×3 *zero* matrix (with all zero entries). Describe the solution set of the equation $A\mathbf{x} = \mathbf{0}$.

40. If $\mathbf{b} \neq \mathbf{0}$, can the solution set of $A\mathbf{x} = \mathbf{b}$ be a plane through the origin? Explain.

In Exercises 41–44, (a) does the equation $A\mathbf{x} = \mathbf{0}$ have a nontrivial solution and (b) does the equation $A\mathbf{x} = \mathbf{b}$ have at least one solution for every possible **b**?

41. A is a 3×3 matrix with three pivot positions.

42. A is a 3×3 matrix with two pivot positions.

43. A is a 3×2 matrix with two pivot positions.

44. A is a 2×4 matrix with two pivot positions.

45. Given $A = \begin{bmatrix} -2 & -6 \\ 7 & 21 \\ -3 & -9 \end{bmatrix}$, find one nontrivial solution of $A\mathbf{x} = \mathbf{0}$ by inspection. [*Hint:* Think of the equation $A\mathbf{x} = \mathbf{0}$ written as a vector equation.]

46. Given $A = \begin{bmatrix} 4 & -6 \\ -8 & 12 \\ 6 & -9 \end{bmatrix}$, find one nontrivial solution of $A\mathbf{x} = \mathbf{0}$ by inspection.

47. Construct a 3×3 nonzero matrix A such that the vector $\begin{bmatrix} 1 \\ 1 \\ 1 \end{bmatrix}$ is a solution of $A\mathbf{x} = \mathbf{0}$.

48. Construct a 3×3 nonzero matrix A such that the vector $\begin{bmatrix} 1 \\ -2 \\ 1 \end{bmatrix}$ is a solution of $A\mathbf{x} = \mathbf{0}$.

49. Construct a 2×2 matrix A such that the solution set of the equation $A\mathbf{x} = \mathbf{0}$ is the line in $\mathbb{R}^2$ through $(4, 1)$ and the origin. Then, find a vector **b** in $\mathbb{R}^2$ such that the solution set of $A\mathbf{x} = \mathbf{b}$ is *not* a line in $\mathbb{R}^2$ parallel to the solution set of $A\mathbf{x} = \mathbf{0}$. Why does this *not* contradict Theorem 6?

50. Suppose A is a 3×3 matrix and **y** is a vector in $\mathbb{R}^3$ such that the equation $A\mathbf{x} = \mathbf{y}$ does *not* have a solution. Does there exist a vector **z** in $\mathbb{R}^3$ such that the equation $A\mathbf{x} = \mathbf{z}$ has a unique solution? Discuss.

51. Let A be an $m \times n$ matrix and let **u** be a vector in $\mathbb{R}^n$ that satisfies the equation $A\mathbf{x} = \mathbf{0}$. Show that for any scalar c, the vector $c\mathbf{u}$ also satisfies $A\mathbf{x} = \mathbf{0}$. [That is, show that $A(c\mathbf{u}) = \mathbf{0}$.]

52. Let A be an $m \times n$ matrix, and let **u** and **v** be vectors in $\mathbb{R}^n$ with the property that $A\mathbf{u} = \mathbf{0}$ and $A\mathbf{v} = \mathbf{0}$. Explain why $A(\mathbf{u} + \mathbf{v})$ must be the zero vector. Then explain why $A(c\mathbf{u} + d\mathbf{v}) = \mathbf{0}$ for each pair of scalars c and d.

Solutions to Practice Problems

1. Row reduce the augmented matrix:

$$\begin{bmatrix} 1 & 4 & -5 & 0 \\ 2 & -1 & 8 & 9 \end{bmatrix} \sim \begin{bmatrix} 1 & 4 & -5 & 0 \\ 0 & -9 & 18 & 9 \end{bmatrix} \sim \begin{bmatrix} 1 & 0 & 3 & 4 \\ 0 & 1 & -2 & -1 \end{bmatrix}$$

$$\begin{aligned} x_1 \quad + 3x_3 &= 4 \\ x_2 - 2x_3 &= -1 \end{aligned}$$

Thus $x_1 = 4 - 3x_3$, $x_2 = -1 + 2x_3$, with x_3 free. The general solution in parametric vector form is

$$\begin{bmatrix} x_1 \\ x_2 \\ x_3 \end{bmatrix} = \begin{bmatrix} 4 - 3x_3 \\ -1 + 2x_3 \\ x_3 \end{bmatrix} = \underset{\mathbf{p}}{\begin{bmatrix} 4 \\ -1 \\ 0 \end{bmatrix}} + x_3 \underset{\mathbf{v}}{\begin{bmatrix} -3 \\ 2 \\ 1 \end{bmatrix}}$$

The intersection of the two planes is the line through $\mathbf{p}$ in the direction of $\mathbf{v}$.

2. The augmented matrix $\begin{bmatrix} 10 & -3 & -2 & 7 \end{bmatrix}$ is row equivalent to $\begin{bmatrix} 1 & -.3 & -.2 & .7 \end{bmatrix}$, and the general solution is $x_1 = .7 + .3x_2 + .2x_3$, with x_2 and x_3 free. That is,

$$\mathbf{x} = \begin{bmatrix} x_1 \\ x_2 \\ x_3 \end{bmatrix} = \begin{bmatrix} .7 + .3x_2 + .2x_3 \\ x_2 \\ x_3 \end{bmatrix} = \begin{bmatrix} .7 \\ 0 \\ 0 \end{bmatrix} + x_2 \begin{bmatrix} .3 \\ 1 \\ 0 \end{bmatrix} + x_3 \begin{bmatrix} .2 \\ 0 \\ 1 \end{bmatrix}$$

$$= \quad \mathbf{p} \quad + \quad x_2\mathbf{u} \quad + \quad x_3\mathbf{v}$$

The solution set of the nonhomogeneous equation $A\mathbf{x} = \mathbf{b}$ is the translated plane $\mathbf{p} + \text{Span}\,\{\mathbf{u}, \mathbf{v}\}$, which passes through $\mathbf{p}$ and is parallel to the solution set of the homogeneous equation in Example 2.

3. Using Theorem 5 from Section 1.4, notice

$$A(\mathbf{p} + \mathbf{v}_h) = A\mathbf{p} + A\mathbf{v}_h = \mathbf{b} + \mathbf{0} = \mathbf{b},$$

hence $\mathbf{p} + \mathbf{v}_h$ is a solution to $A\mathbf{x} = \mathbf{b}$.

1.6 Applications of Linear Systems

You might expect that a real-life problem involving linear algebra would have only one solution, or perhaps no solution. The purpose of this section is to show how linear systems with many solutions can arise naturally. The applications here come from economics, chemistry, and network flow.

A Homogeneous System in Economics

The system of 500 equations in 500 variables, mentioned in this chapter's introduction, is now known as a Leontief "input–output" (or "production") model.[1] Section 2.6 will examine this model in more detail, when more theory and better notation are available. For now, we look at a simpler "exchange model," also due to Leontief.

---

[1] See Wassily W. Leontief, "Input–Output Economics," *Scientific American*, October 1951, pp. 15–21.

Suppose a nation's economy is divided into many sectors, such as various manufacturing, communication, entertainment, and service industries. Suppose that for each sector we know its total output for one year and we know exactly how this output is divided or "exchanged" among the other sectors of the economy. Let the total dollar value of a sector's output be called the **price** of that output. Leontief proved the following result.

> There exist *equilibrium prices* that can be assigned to the total outputs of the various sectors in such a way that the income of each sector exactly balances its expenses.

The following example shows how to find the equilibrium prices.

EXAMPLE 1 Suppose an economy consists of the Coal, Electric (power), and Steel sectors, and the output of each sector is distributed among the various sectors as shown in Table 1, where the entries in a column represent the fractional parts of a sector's total output.

The second column of Table 1, for instance, says that the total output of the Electric sector is divided as follows: 40% to Coal, 50% to Steel, and the remaining 10% to Electric. (Electric treats this 10% as an expense it incurs in order to operate its business.) Since all output must be taken into account, the decimal fractions in each column must sum to 1.

Denote the prices (in dollar values) of the total annual outputs of the Coal, Electric, and Steel sectors by p_C, p_E, and p_S, respectively. If possible, find equilibrium prices that make each sector's income match its expenditures.

TABLE I   A Simple Economy

Coal	Electric	Steel	Purchased by
.0	.4	.6	Coal
.6	.1	.2	Electric
.4	.5	.2	Steel

(Distribution of Output from — Coal, Electric, Steel; Purchased by column on right)

SOLUTION A sector looks down a column to see where its output goes, and it looks across a row to see what it needs as inputs. For instance, the first row of Table 1 says that Coal receives (and pays for) 40% of the Electric output and 60% of the Steel output. Since the respective values of the total outputs are p_E and p_S, Coal must spend $.4p_E$ dollars for its share of Electric's output and $.6p_S$ for its share of Steel's output. Thus Coal's total expenses are $.4p_E + .6p_S$. To make Coal's income, p_C, equal to its expenses, we want

$$p_C = .4p_E + .6p_S \tag{1}$$

The second row of the exchange table shows that the Electric sector spends $.6p_C$ for coal, $.1p_E$ for electricity, and $.2p_S$ for steel. Hence the income/expense requirement for Electric is

$$p_E = .6p_C + .1p_E + .2p_S \tag{2}$$

Finally, the third row of the exchange table leads to the final requirement:

$$p_S = .4p_C + .5p_E + .2p_S \tag{3}$$

To solve the system of equations (1), (2), and (3), move all the unknowns to the left sides of the equations and combine like terms. [For instance, on the left side of (2), write $p_E - .1p_E$ as $.9p_E$.]

$$p_C - .4p_E - .6p_S = 0$$
$$-.6p_C + .9p_E - .2p_S = 0$$
$$-.4p_C - .5p_E + .8p_S = 0$$

Row reduction is next. For simplicity here, decimals are rounded to two places.

$$\begin{bmatrix} 1 & -.4 & -.6 & 0 \\ -.6 & .9 & -.2 & 0 \\ -.4 & -.5 & .8 & 0 \end{bmatrix} \sim \begin{bmatrix} 1 & -.4 & -.6 & 0 \\ 0 & .66 & -.56 & 0 \\ 0 & -.66 & .56 & 0 \end{bmatrix} \sim \begin{bmatrix} 1 & -.4 & -.6 & 0 \\ 0 & .66 & -.56 & 0 \\ 0 & 0 & 0 & 0 \end{bmatrix}$$

$$\sim \begin{bmatrix} 1 & -.4 & -.6 & 0 \\ 0 & 1 & -.85 & 0 \\ 0 & 0 & 0 & 0 \end{bmatrix} \sim \begin{bmatrix} 1 & 0 & -.94 & 0 \\ 0 & 1 & -.85 & 0 \\ 0 & 0 & 0 & 0 \end{bmatrix}$$

The general solution is $p_C = .94p_S$, $p_E = .85p_S$, and p_S is free. The equilibrium price vector for the economy has the form

$$\mathbf{p} = \begin{bmatrix} p_C \\ p_E \\ p_S \end{bmatrix} = \begin{bmatrix} .94p_S \\ .85p_S \\ p_S \end{bmatrix} = p_S \begin{bmatrix} .94 \\ .85 \\ 1 \end{bmatrix}$$

Any (nonnegative) choice for p_S results in a choice of equilibrium prices. For instance, if we take p_S to be 100 (or $100 million), then $p_C = 94$ and $p_E = 85$. The incomes and expenditures of each sector will be equal if the output of Coal is priced at $94 million, that of Electric at $85 million, and that of Steel at $100 million. ∎

Balancing Chemical Equations

Chemical equations describe the quantities of substances consumed and produced by chemical reactions. For instance, when propane gas burns, the propane (C_3H_8) combines with oxygen (O_2) to form carbon dioxide (CO_2) and water (H_2O), according to an equation of the form

$$(x_1)C_3H_8 + (x_2)O_2 \rightarrow (x_3)CO_2 + (x_4)H_2O \tag{4}$$

To "balance" this equation, a chemist must find whole numbers $x_1, \ldots, x_4$ such that the total numbers of carbon (C), hydrogen (H), and oxygen (O) atoms on the left match the corresponding numbers of atoms on the right (because atoms are neither destroyed nor created in the reaction).

A systematic method for balancing chemical equations is to set up a vector equation that describes the numbers of atoms of each type present in a reaction. Since equation (4) involves three types of atoms (carbon, hydrogen, and oxygen), construct a vector in $\mathbb{R}^3$ for each reactant and product in (4) that lists the numbers of "atoms per molecule," as follows:

$$C_3H_8: \begin{bmatrix} 3 \\ 8 \\ 0 \end{bmatrix}, \quad O_2: \begin{bmatrix} 0 \\ 0 \\ 2 \end{bmatrix}, \quad CO_2: \begin{bmatrix} 1 \\ 0 \\ 2 \end{bmatrix}, \quad H_2O: \begin{bmatrix} 0 \\ 2 \\ 1 \end{bmatrix} \begin{matrix} \leftarrow \text{Carbon} \\ \leftarrow \text{Hydrogen} \\ \leftarrow \text{Oxygen} \end{matrix}$$

To balance equation (4), the coefficients $x_1, \ldots, x_4$ must satisfy

$$x_1 \begin{bmatrix} 3 \\ 8 \\ 0 \end{bmatrix} + x_2 \begin{bmatrix} 0 \\ 0 \\ 2 \end{bmatrix} = x_3 \begin{bmatrix} 1 \\ 0 \\ 2 \end{bmatrix} + x_4 \begin{bmatrix} 0 \\ 2 \\ 1 \end{bmatrix}$$

To solve, move all the terms to the left (changing the signs in the third and fourth vectors):

$$x_1 \begin{bmatrix} 3 \\ 8 \\ 0 \end{bmatrix} + x_2 \begin{bmatrix} 0 \\ 0 \\ 2 \end{bmatrix} + x_3 \begin{bmatrix} -1 \\ 0 \\ -2 \end{bmatrix} + x_4 \begin{bmatrix} 0 \\ -2 \\ -1 \end{bmatrix} = \begin{bmatrix} 0 \\ 0 \\ 0 \end{bmatrix}$$

Row reduction of the augmented matrix for this equation leads to the general solution

$$x_1 = \tfrac{1}{4}x_4, \ x_2 = \tfrac{5}{4}x_4, \ x_3 = \tfrac{3}{4}x_4, \ \text{with } x_4 \text{ free}$$

Since the coefficients in a chemical equation must be integers, take $x_4 = 4$, in which case $x_1 = 1$, $x_2 = 5$, and $x_3 = 3$. The balanced equation is

$$C_3H_8 + 5O_2 \rightarrow 3CO_2 + 4H_2O$$

The equation would also be balanced if, for example, each coefficient were doubled. For most purposes, however, chemists prefer to use a balanced equation whose coefficients are the smallest possible whole numbers.

Network Flow

Systems of linear equations arise naturally when scientists, engineers, or economists study the flow of some quantity through a network. For instance, urban planners and traffic engineers monitor the pattern of traffic flow in a grid of city streets. Electrical engineers calculate current flow through electrical circuits. Economists analyze the distribution of products from manufacturers to consumers through a network of wholesalers and retailers. For many networks, the systems of equations involve hundreds or even thousands of variables and equations.

A *network* consists of a set of points called *junctions*, or *nodes*, with lines or arcs called *branches* connecting some or all of the junctions. The direction of flow in each branch is indicated, and the flow amount (or rate) is either shown or is denoted by a variable.

The basic assumption of network flow is that the total flow into the network equals the total flow out of the network and that the total flow into a junction equals the total flow out of the junction. For example, Figure 1 shows 30 units flowing into a junction through one branch, with x_1 and x_2 denoting the flows out of the junction through other branches. Since the flow is "conserved" at each junction, we must have $x_1 + x_2 = 30$. In a similar fashion, the flow at each junction is described by a linear equation. The problem of network analysis is to determine the flow in each branch when partial information (such as the flow into and out of the network) is known.

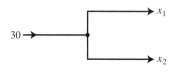

FIGURE 1

A junction or node.

EXAMPLE 2 The network in Figure 2 shows the traffic flow (in vehicles per hour) over several one-way streets in downtown Baltimore during a typical early afternoon. Determine the general flow pattern for the network.

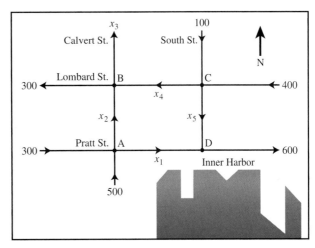

FIGURE 2 Baltimore streets.

SOLUTION   Write equations that describe the flow, and then find the general solution of the system. Label the street intersections (junctions) and the unknown flows in the branches, as shown in Figure 2. At each intersection, set the flow in equal to the flow out.

Intersection	Flow in		Flow out
A	$300 + 500$	$=$	$x_1 + x_2$
B	$x_2 + x_4$	$=$	$300 + x_3$
C	$100 + 400$	$=$	$x_4 + x_5$
D	$x_1 + x_5$	$=$	600

Also, the total flow into the network $(500 + 300 + 100 + 400)$ equals the total flow out of the network $(300 + x_3 + 600)$, which simplifies to $x_3 = 400$. Combine this equation with a rearrangement of the first four equations to obtain the following system of equations:

$$\begin{aligned} x_1 + x_2 \quad\quad\quad\quad\quad &= 800 \\ x_2 - x_3 + x_4 \quad\quad &= 300 \\ x_4 + x_5 &= 500 \\ x_1 \quad\quad\quad\quad\quad + x_5 &= 600 \\ x_3 \quad\quad\quad &= 400 \end{aligned}$$

Row reduction of the associated augmented matrix leads to

$$\begin{aligned} x_1 \quad\quad\quad\quad + x_5 &= 600 \\ x_2 \quad\quad - x_5 &= 200 \\ x_3 \quad\quad &= 400 \\ x_4 + x_5 &= 500 \end{aligned}$$

The general flow pattern for the network is described by

$$\begin{cases} x_1 = 600 - x_5 \\ x_2 = 200 + x_5 \\ x_3 = 400 \\ x_4 = 500 - x_5 \\ x_5 \text{ is free} \end{cases}$$

A negative flow in a network branch corresponds to flow in the direction opposite to that shown on the model. Since the streets in this problem are one way, none of the variables here can be negative. This fact leads to certain limitations on the possible values of the variables. For instance, $x_5 \leq 500$ because x_4 cannot be negative. Other constraints on the variables are considered in Practice Problem 2. ∎

Practice Problems

1. Suppose an economy has three sectors: Agriculture, Mining, and Manufacturing. Agriculture sells 5% of its output to Mining and 30% to Manufacturing, and retains the rest. Mining sells 20% of its output to Agriculture and 70% to Manufacturing, and retains the rest. Manufacturing sells 20% of its output to Agriculture and 30% to Mining, and retains the rest. Determine the exchange table for this economy, where the columns describe how the output of each sector is exchanged among the three sectors.

2. Consider the network flow studied in Example 2. Determine the possible range of values of x_1 and x_2. [*Hint:* The example showed that $x_5 \leq 500$. What does this imply about x_1 and x_2? Also, use the fact that $x_5 \geq 0$.]

1.6 Exercises

1. Suppose an economy has only two sectors, Goods and Services. Each year, Goods sells 80% of its output to Services and keeps the rest, while Services sells 70% of its output to Goods and retains the rest. Find equilibrium prices for the annual outputs of the Goods and Services sectors that make each sector's income match its expenditures.

2. Find another set of equilibrium prices for the economy in Example 1. Suppose the same economy used Japanese yen instead of dollars to measure the value of the various sectors' outputs. Would this change the problem in any way? Discuss.

3. Consider an economy with three sectors, Chemicals & Metals, Fuels & Power, and Machinery. Chemicals sells 30% of its output to Fuels and 50% to Machinery and retains the rest. Fuels sells 80% of its output to Chemicals and 10% to Machinery and retains the rest. Machinery sells 40% to Chemicals and 40% to Fuels and retains the rest.

   a. Construct the exchange table for this economy.

   b. Develop a system of equations that leads to prices at which each sector's income matches its expenses. Then write the augmented matrix that can be row reduced to find these prices.

T c. Find a set of equilibrium prices when the price for the Machinery output is 100 units.

4. Suppose an economy has four sectors, Agriculture (A), Energy (E), Manufacturing (M), and Transportation (T). Sector A sells 10% of its output to E and 25% to M and retains the rest. Sector E sells 30% of its output to A, 35% to M, and 25% to T and retains the rest. Sector M sells 30% of its output to A, 15% to E, and 40% to T and retains the rest. Sector T sells 20% of its output to A, 10% to E, and 30% to M and retains the rest.

   a. Construct the exchange table for this economy.

   T b. Find a set of equilibrium prices for the economy.

Balance the chemical equations in Exercises 5–10 using the vector equation approach discussed in this section.

5. Boron sulfide reacts violently with water to form boric acid and hydrogen sulfide gas (the smell of rotten eggs). The unbalanced equation is

   $$B_2S_3 + H_2O \rightarrow H_3BO_3 + H_2S$$

   [For each compound, construct a vector that lists the numbers of atoms of boron, sulfur, hydrogen, and oxygen.]

6. When solutions of sodium phosphate and barium nitrate are mixed, the result is barium phosphate (as a precipitate) and sodium nitrate. The unbalanced equation is

   $$Na_3PO_4 + Ba(NO_3)_2 \rightarrow Ba_3(PO_4)_2 + NaNO_3$$

   [For each compound, construct a vector that lists the numbers of atoms of sodium (Na), phosphorus, oxygen, barium, and nitrogen. For instance, barium nitrate corresponds to $(0, 0, 6, 1, 2)$.]

7. Alka-Seltzer contains sodium bicarbonate ($NaHCO_3$) and citric acid ($H_3C_6H_5O_7$). When a tablet is dissolved in water, the following reaction produces sodium citrate, water, and carbon dioxide (gas):

$$NaHCO_3 + H_3C_6H_5O_7 \rightarrow Na_3C_6H_5O_7 + H_2O + CO_2$$

8. The following reaction between potassium permanganate ($KMnO_4$) and manganese sulfate in water produces manganese dioxide, potassium sulfate, and sulfuric acid:

$$KMnO_4 + MnSO_4 + H_2O \rightarrow MnO_2 + K_2SO_4 + H_2SO_4$$

[For each compound, construct a vector that lists the numbers of atoms of potassium (K), manganese, oxygen, sulfur, and hydrogen.]

T 9. If possible, use exact arithmetic or rational format for calculations in balancing the following chemical reaction:

$$PbN_6 + CrMn_2O_8 \rightarrow Pb_3O_4 + Cr_2O_3 + MnO_2 + NO$$

T 10. The chemical reaction below can be used in some industrial processes, such as the production of arsene (AsH_3). Use exact arithmetic or rational format for calculations to balance this equation.

$$MnS + As_2Cr_{10}O_{35} + H_2SO_4$$
$$\rightarrow HMnO_4 + AsH_3 + CrS_3O_{12} + H_2O$$

11. Find the general flow pattern of the network shown in the figure. Assuming that the flows are all nonnegative, what is the largest possible value for x_3?

12. a. Find the general traffic pattern in the freeway network shown in the figure. (Flow rates are in cars/minute.)

b. Describe the general traffic pattern when the road whose flow is x_4 is closed.

c. When $x_4 = 0$, what is the minimum value of x_1?

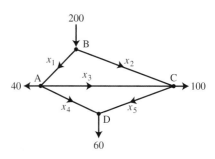

13. a. Find the general flow pattern in the network shown in the figure.

b. Assuming that the flow must be in the directions indicated, find the minimum flows in the branches denoted by x_2, x_3, x_4, and x_5.

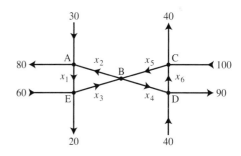

14. Intersections in England are often constructed as one-way "roundabouts," such as the one shown in the figure. Assume that traffic must travel in the directions shown. Find the general solution of the network flow. Find the smallest possible value for x_6.

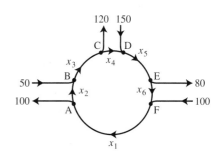

Solutions to Practice Problems

1. Write the percentages as decimals. Since all output must be taken into account, each column must sum to 1. This fact helps to fill in any missing entries.

Distribution of Output from			
Agriculture	**Mining**	**Manufacturing**	**Purchased by**
.65	.20	.20	Agriculture
.05	.10	.30	Mining
.30	.70	.50	Manufacturing

Solutions to Practice Problems (Continued)

2. Since $x_5 \leq 500$, the equations D and A for x_1 and x_2 imply that $x_1 \geq 100$ and $x_2 \leq 700$. The fact that $x_5 \geq 0$ implies that $x_1 \leq 600$ and $x_2 \geq 200$. So, $100 \leq x_1 \leq 600$, and $200 \leq x_2 \leq 700$.

1.7 Linear Independence

The homogeneous equations in Section 1.5 can be studied from a different perspective by writing them as vector equations. In this way, the focus shifts from the unknown solutions of $A\mathbf{x} = \mathbf{0}$ to the vectors that appear in the vector equations.

For instance, consider the equation

$$x_1 \begin{bmatrix} 1 \\ 2 \\ 3 \end{bmatrix} + x_2 \begin{bmatrix} 4 \\ 5 \\ 6 \end{bmatrix} + x_3 \begin{bmatrix} 2 \\ 1 \\ 0 \end{bmatrix} = \begin{bmatrix} 0 \\ 0 \\ 0 \end{bmatrix} \tag{1}$$

This equation has a trivial solution, of course, where $x_1 = x_2 = x_3 = 0$. As in Section 1.5, the main issue is whether the trivial solution is the *only one*.

DEFINITION

An indexed set of vectors $\{\mathbf{v}_1, \ldots, \mathbf{v}_p\}$ in $\mathbb{R}^n$ is said to be **linearly independent** if the vector equation

$$x_1 \mathbf{v}_1 + x_2 \mathbf{v}_2 + \cdots + x_p \mathbf{v}_p = \mathbf{0}$$

has only the trivial solution. The set $\{\mathbf{v}_1, \ldots, \mathbf{v}_p\}$ is said to be **linearly dependent** if there exist weights $c_1, \ldots, c_p$, not all zero, such that

$$c_1 \mathbf{v}_1 + c_2 \mathbf{v}_2 + \cdots + c_p \mathbf{v}_p = \mathbf{0} \tag{2}$$

Equation (2) is called a **linear dependence relation** among $\mathbf{v}_1, \ldots, \mathbf{v}_p$ when the weights are not all zero. An indexed set is linearly dependent if and only if it is not linearly independent. For brevity, we may say that $\mathbf{v}_1, \ldots, \mathbf{v}_p$ are linearly dependent when we mean that $\{\mathbf{v}_1, \ldots, \mathbf{v}_p\}$ is a linearly dependent set. We use analogous terminology for linearly independent sets.

EXAMPLE 1 Let $\mathbf{v}_1 = \begin{bmatrix} 1 \\ 2 \\ 3 \end{bmatrix}$, $\mathbf{v}_2 = \begin{bmatrix} 4 \\ 5 \\ 6 \end{bmatrix}$, and $\mathbf{v}_3 = \begin{bmatrix} 2 \\ 1 \\ 0 \end{bmatrix}$.

a. Determine if the set $\{\mathbf{v}_1, \mathbf{v}_2, \mathbf{v}_3\}$ is linearly independent.

b. If possible, find a linear dependence relation among $\mathbf{v}_1$, $\mathbf{v}_2$, and $\mathbf{v}_3$.

SOLUTION

a. We must determine if there is a nontrivial solution of equation (1) above. Row operations on the associated augmented matrix show that

$$\begin{bmatrix} 1 & 4 & 2 & 0 \\ 2 & 5 & 1 & 0 \\ 3 & 6 & 0 & 0 \end{bmatrix} \sim \begin{bmatrix} 1 & 4 & 2 & 0 \\ 0 & -3 & -3 & 0 \\ 0 & 0 & 0 & 0 \end{bmatrix}$$

Clearly, x_1 and x_2 are basic variables, and x_3 is free. Each nonzero value of x_3 determines a nontrivial solution of (1). Hence $\mathbf{v}_1, \mathbf{v}_2, \mathbf{v}_3$ are linearly dependent (and not linearly independent).

b. To find a linear dependence relation among $\mathbf{v}_1, \mathbf{v}_2$, and $\mathbf{v}_3$, completely row reduce the augmented matrix and write the new system:

$$\begin{bmatrix} 1 & 0 & -2 & 0 \\ 0 & 1 & 1 & 0 \\ 0 & 0 & 0 & 0 \end{bmatrix} \qquad \begin{aligned} x_1 \quad - 2x_3 &= 0 \\ x_2 + \ x_3 &= 0 \\ 0 &= 0 \end{aligned}$$

Thus $x_1 = 2x_3$, $x_2 = -x_3$, and x_3 is free. Choose any nonzero value for x_3—say, $x_3 = 5$. Then $x_1 = 10$ and $x_2 = -5$. Substitute these values into equation (1) and obtain

$$10\mathbf{v}_1 - 5\mathbf{v}_2 + 5\mathbf{v}_3 = \mathbf{0}$$

This is one (out of infinitely many) possible linear dependence relations among $\mathbf{v}_1$, $\mathbf{v}_2$, and $\mathbf{v}_3$. ∎

Linear Independence of Matrix Columns

Suppose that we begin with a matrix $A = [\, \mathbf{a}_1 \ \cdots \ \mathbf{a}_n \,]$ instead of a set of vectors. The matrix equation $A\mathbf{x} = \mathbf{0}$ can be written as

$$x_1\mathbf{a}_1 + x_2\mathbf{a}_2 + \cdots + x_n\mathbf{a}_n = \mathbf{0}$$

Each linear dependence relation among the columns of A corresponds to a nontrivial solution of $A\mathbf{x} = \mathbf{0}$. Thus we have the following important fact.

> The columns of a matrix A are linearly independent if and only if the equation $A\mathbf{x} = \mathbf{0}$ has *only* the trivial solution. (3)

EXAMPLE 2 Determine if the columns of the matrix $A = \begin{bmatrix} 0 & 1 & 4 \\ 1 & 2 & -1 \\ 5 & 8 & 0 \end{bmatrix}$ are linearly independent.

SOLUTION To study $A\mathbf{x} = \mathbf{0}$, row reduce the augmented matrix:

$$\begin{bmatrix} 0 & 1 & 4 & 0 \\ 1 & 2 & -1 & 0 \\ 5 & 8 & 0 & 0 \end{bmatrix} \sim \begin{bmatrix} 1 & 2 & -1 & 0 \\ 0 & 1 & 4 & 0 \\ 0 & -2 & 5 & 0 \end{bmatrix} \sim \begin{bmatrix} 1 & 2 & -1 & 0 \\ 0 & 1 & 4 & 0 \\ 0 & 0 & 13 & 0 \end{bmatrix}$$

At this point, it is clear that there are three basic variables and no free variables. So the equation $A\mathbf{x} = \mathbf{0}$ has only the trivial solution, and the columns of A are linearly independent. ∎

Sets of One or Two Vectors

A set containing only one vector—say, $\mathbf{v}$—is linearly independent if and only if $\mathbf{v}$ is not the zero vector. This is because the vector equation $x_1\mathbf{v} = \mathbf{0}$ has only the trivial solution when $\mathbf{v} \neq \mathbf{0}$. The zero vector is linearly dependent because $x_1\mathbf{0} = \mathbf{0}$ has many nontrivial solutions.

The next example will explain the nature of a linearly dependent set of two vectors.

EXAMPLE 3   Determine if the following sets of vectors are linearly independent.

a. $\mathbf{v}_1 = \begin{bmatrix} 3 \\ 1 \end{bmatrix}$, $\mathbf{v}_2 = \begin{bmatrix} 6 \\ 2 \end{bmatrix}$ b. $\mathbf{v}_1 = \begin{bmatrix} 3 \\ 2 \end{bmatrix}$, $\mathbf{v}_2 = \begin{bmatrix} 6 \\ 2 \end{bmatrix}$

SOLUTION

a. Notice that $\mathbf{v}_2$ is a multiple of $\mathbf{v}_1$, namely $\mathbf{v}_2 = 2\mathbf{v}_1$. Hence $-2\mathbf{v}_1 + \mathbf{v}_2 = \mathbf{0}$, which shows that $\{\mathbf{v}_1, \mathbf{v}_2\}$ is linearly dependent.

b. The vectors $\mathbf{v}_1$ and $\mathbf{v}_2$ are certainly *not* multiples of one another. Could they be linearly dependent? Suppose c and d satisfy

$$c\mathbf{v}_1 + d\mathbf{v}_2 = \mathbf{0}$$

If $c \neq 0$, then we can solve for $\mathbf{v}_1$ in terms of $\mathbf{v}_2$, namely $\mathbf{v}_1 = (-d/c)\mathbf{v}_2$. This result is impossible because $\mathbf{v}_1$ is *not* a multiple of $\mathbf{v}_2$. So c must be zero. Similarly, d must also be zero. Thus $\{\mathbf{v}_1, \mathbf{v}_2\}$ is a linearly independent set. ■

The arguments in Example 3 show that you can always decide *by inspection* when a set of two vectors is linearly dependent. Row operations are unnecessary. Simply check whether at least one of the vectors is a scalar times the other. (The test applies only to sets of *two* vectors.)

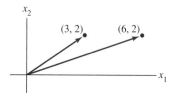

Linearly dependent

Linearly independent

FIGURE 1

A set of two vectors $\{\mathbf{v}_1, \mathbf{v}_2\}$ is linearly dependent if at least one of the vectors is a multiple of the other. The set is linearly independent if and only if neither of the vectors is a multiple of the other.

In geometric terms, two vectors are linearly dependent if and only if they lie on the same line through the origin. Figure 1 shows the vectors from Example 3.

Sets of Two or More Vectors

The proof of the next theorem is similar to the solution of Example 3. Details are given at the end of this section.

THEOREM 7

Characterization of Linearly Dependent Sets

An indexed set $S = \{\mathbf{v}_1, \ldots, \mathbf{v}_p\}$ of two or more vectors is linearly dependent if and only if at least one of the vectors in S is a linear combination of the others. In fact, if S is linearly dependent and $\mathbf{v}_1 \neq \mathbf{0}$, then some $\mathbf{v}_j$ (with $j > 1$) is a linear combination of the preceding vectors, $\mathbf{v}_1, \ldots, \mathbf{v}_{j-1}$.

Warning: Theorem 7 does *not* say that *every* vector in a linearly dependent set is a linear combination of the preceding vectors. A vector in a linearly dependent set may fail to be a linear combination of the other vectors. See Practice Problem 1(c).

EXAMPLE 4   Let $\mathbf{u} = \begin{bmatrix} 3 \\ 1 \\ 0 \end{bmatrix}$ and $\mathbf{v} = \begin{bmatrix} 1 \\ 6 \\ 0 \end{bmatrix}$. Describe the set spanned by $\mathbf{u}$ and $\mathbf{v}$,

and explain why a vector $\mathbf{w}$ is in Span $\{\mathbf{u}, \mathbf{v}\}$ if and only if $\{\mathbf{u}, \mathbf{v}, \mathbf{w}\}$ is linearly dependent.

SOLUTION   The vectors **u** and **v** are linearly independent because neither vector is a multiple of the other, and so they span a plane in $\mathbb{R}^3$. (See Section 1.3.) In fact, Span $\{\mathbf{u}, \mathbf{v}\}$ is the $x_1 x_2$-plane (with $x_3 = 0$). If **w** is a linear combination of **u** and **v**, then $\{\mathbf{u}, \mathbf{v}, \mathbf{w}\}$ is linearly dependent, by Theorem 7. Conversely, suppose that $\{\mathbf{u}, \mathbf{v}, \mathbf{w}\}$ is linearly dependent. By Theorem 7, some vector in $\{\mathbf{u}, \mathbf{v}, \mathbf{w}\}$ is a linear combination of the preceding vectors (since $\mathbf{u} \neq \mathbf{0}$). That vector must be **w**, since **v** is not a multiple of **u**. So **w** is in Span $\{\mathbf{u}, \mathbf{v}\}$. See Figure 2.   ■

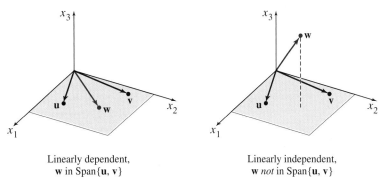

Linearly dependent,
w in Span$\{\mathbf{u}, \mathbf{v}\}$

Linearly independent,
w *not* in Span$\{\mathbf{u}, \mathbf{v}\}$

FIGURE 2   Linear dependence in $\mathbb{R}^3$.

Example 4 generalizes to any set $\{\mathbf{u}, \mathbf{v}, \mathbf{w}\}$ in $\mathbb{R}^3$ with **u** and **v** linearly independent. The set $\{\mathbf{u}, \mathbf{v}, \mathbf{w}\}$ will be linearly dependent if and only if **w** is in the plane spanned by **u** and **v**.

The next two theorems describe special cases in which the linear dependence of a set is automatic. Moreover, Theorem 8 will be a key result for work in later chapters.

THEOREM 8

If a set contains more vectors than there are entries in each vector, then the set is linearly dependent. That is, any set $\{\mathbf{v}_1, \ldots, \mathbf{v}_p\}$ in $\mathbb{R}^n$ is linearly dependent if $p > n$.

$$n \begin{bmatrix} \overset{\displaystyle p}{*} & * & * & * & * \\ * & * & * & * & * \\ * & * & * & * & * \end{bmatrix}$$

FIGURE 3

If $p > n$, the columns are linearly dependent.

PROOF   Let $A = [\, \mathbf{v}_1 \;\cdots\; \mathbf{v}_p \,]$. Then A is $n \times p$, and the equation $A\mathbf{x} = \mathbf{0}$ corresponds to a system of n equations in p unknowns. If $p > n$, there are more variables than equations, so there must be a free variable. Hence $A\mathbf{x} = \mathbf{0}$ has a nontrivial solution, and the columns of A are linearly dependent. See Figure 3 for a matrix version of this theorem.   ■

Warning: Theorem 8 says nothing about the case in which the number of vectors in the set does *not* exceed the number of entries in each vector.

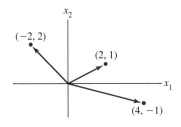

FIGURE 4

A linearly dependent set in $\mathbb{R}^2$.

EXAMPLE 5   The vectors $\begin{bmatrix} 2 \\ 1 \end{bmatrix}$, $\begin{bmatrix} 4 \\ -1 \end{bmatrix}$, $\begin{bmatrix} -2 \\ 2 \end{bmatrix}$ are linearly dependent by Theorem 8, because there are three vectors in the set and there are only two entries in each vector. Notice, however, that none of the vectors is a multiple of one of the other vectors. See Figure 4.   ■

THEOREM 9

If a set $S = \{\mathbf{v}_1, \ldots, \mathbf{v}_p\}$ in $\mathbb{R}^n$ contains the zero vector, then the set is linearly dependent.

PROOF By renumbering the vectors, we may suppose $\mathbf{v}_1 = \mathbf{0}$. Then the equation $1\mathbf{v}_1 + 0\mathbf{v}_2 + \cdots + 0\mathbf{v}_p = \mathbf{0}$ shows that S is linearly dependent. ∎

EXAMPLE 6 Determine by inspection if the given set is linearly dependent.

a. $\begin{bmatrix} 1 \\ 7 \\ 6 \end{bmatrix}, \begin{bmatrix} 2 \\ 0 \\ 9 \end{bmatrix}, \begin{bmatrix} 3 \\ 1 \\ 5 \end{bmatrix}, \begin{bmatrix} 4 \\ 1 \\ 8 \end{bmatrix}$ b. $\begin{bmatrix} 2 \\ 3 \\ 5 \end{bmatrix}, \begin{bmatrix} 0 \\ 0 \\ 0 \end{bmatrix}, \begin{bmatrix} 1 \\ 1 \\ 8 \end{bmatrix}$ c. $\begin{bmatrix} -2 \\ 4 \\ 6 \\ 10 \end{bmatrix}, \begin{bmatrix} 3 \\ -6 \\ -9 \\ 15 \end{bmatrix}$

SOLUTION

a. The set contains four vectors, each of which has only three entries. So the set is linearly dependent by Theorem 8.

b. Theorem 8 does not apply here because the number of vectors does not exceed the number of entries in each vector. Since the zero vector is in the set, the set is linearly dependent by Theorem 9.

c. Compare the corresponding entries of the two vectors. The second vector seems to be $-3/2$ times the first vector. This relation holds for the first three pairs of entries, but fails for the fourth pair. Thus neither of the vectors is a multiple of the other, and hence they are linearly independent. ∎

In general, you should read a section thoroughly *several* times to absorb an important concept such as linear independence. The notes in the *Study Guide* for this section will help you learn to form mental images of key ideas in linear algebra. For instance, the following proof is worth reading carefully because it shows how the definition of linear independence can be *used*.

PROOF OF THEOREM 7 (Characterization of Linearly Dependent Sets)
If some $\mathbf{v}_j$ in S equals a linear combination of the other vectors, then $\mathbf{v}_j$ can be subtracted from both sides of the equation, producing a linear dependence relation with a nonzero weight (-1) on $\mathbf{v}_j$. [For instance, if $\mathbf{v}_1 = c_2\mathbf{v}_2 + c_3\mathbf{v}_3$, then $\mathbf{0} = (-1)\mathbf{v}_1 + c_2\mathbf{v}_2 + c_3\mathbf{v}_3 + 0\mathbf{v}_4 + \cdots + 0\mathbf{v}_p$.] Thus S is linearly dependent.

Conversely, suppose S is linearly dependent. If $\mathbf{v}_1$ is zero, then it is a (trivial) linear combination of the other vectors in S. Otherwise, $\mathbf{v}_1 \neq \mathbf{0}$, and there exist weights $c_1, \ldots, c_p$, not all zero, such that

$$c_1\mathbf{v}_1 + c_2\mathbf{v}_2 + \cdots + c_p\mathbf{v}_p = \mathbf{0}$$

Let j be the largest subscript for which $c_j \neq 0$. If $j = 1$, then $c_1\mathbf{v}_1 = \mathbf{0}$, which is impossible because $\mathbf{v}_1 \neq \mathbf{0}$. So $j > 1$, and

$$c_1\mathbf{v}_1 + \cdots + c_j\mathbf{v}_j + 0\mathbf{v}_{j+1} + \cdots + 0\mathbf{v}_p = \mathbf{0}$$
$$c_j\mathbf{v}_j = -c_1\mathbf{v}_1 - \cdots - c_{j-1}\mathbf{v}_{j-1}$$
$$\mathbf{v}_j = \left(-\frac{c_1}{c_j}\right)\mathbf{v}_1 + \cdots + \left(-\frac{c_{j-1}}{c_j}\right)\mathbf{v}_{j-1} \quad ∎$$

Practice Problems

1. Let $\mathbf{u} = \begin{bmatrix} 3 \\ 2 \\ -4 \end{bmatrix}, \mathbf{v} = \begin{bmatrix} -6 \\ 1 \\ 7 \end{bmatrix}, \mathbf{w} = \begin{bmatrix} 0 \\ -5 \\ 2 \end{bmatrix}$, and $\mathbf{z} = \begin{bmatrix} 3 \\ 7 \\ -5 \end{bmatrix}$.

   a. Are the sets $\{\mathbf{u}, \mathbf{v}\}, \{\mathbf{u}, \mathbf{w}\}, \{\mathbf{u}, \mathbf{z}\}, \{\mathbf{v}, \mathbf{w}\}, \{\mathbf{v}, \mathbf{z}\}$, and $\{\mathbf{w}, \mathbf{z}\}$ each linearly independent? Why or why not?

  b. Does the answer to Part (a) imply that $\{\mathbf{u}, \mathbf{v}, \mathbf{w}, \mathbf{z}\}$ is linearly independent?

  c. To determine if $\{\mathbf{u}, \mathbf{v}, \mathbf{w}, \mathbf{z}\}$ is linearly dependent, is it wise to check if, say, $\mathbf{w}$ is a linear combination of $\mathbf{u}$, $\mathbf{v}$, and $\mathbf{z}$?

  d. Is $\{\mathbf{u}, \mathbf{v}, \mathbf{w}, \mathbf{z}\}$ linearly dependent?

2. Suppose that $\{\mathbf{v}_1, \mathbf{v}_2, \mathbf{v}_3\}$ is a linearly dependent set of vectors in $\mathbb{R}^n$ and $\mathbf{v}_4$ is a vector in $\mathbb{R}^n$. Show that $\{\mathbf{v}_1, \mathbf{v}_2, \mathbf{v}_3, \mathbf{v}_4\}$ is also a linearly dependent set.

1.7 Exercises

In Exercises 1–4, determine if the vectors are linearly independent. Justify each answer.

1. $\begin{bmatrix} 5 \\ 1 \\ 0 \end{bmatrix}, \begin{bmatrix} 7 \\ 2 \\ -6 \end{bmatrix}, \begin{bmatrix} -2 \\ -1 \\ 6 \end{bmatrix}$ **2.** $\begin{bmatrix} 0 \\ 0 \\ 2 \end{bmatrix}, \begin{bmatrix} 0 \\ 5 \\ -8 \end{bmatrix}, \begin{bmatrix} -3 \\ 4 \\ 1 \end{bmatrix}$

3. $\begin{bmatrix} 1 \\ -3 \end{bmatrix}, \begin{bmatrix} -3 \\ 6 \end{bmatrix}$ **4.** $\begin{bmatrix} -1 \\ 4 \end{bmatrix}, \begin{bmatrix} -2 \\ 8 \end{bmatrix}$

In Exercises 5–8, determine if the columns of the matrix form a linearly independent set. Justify each answer.

5. $\begin{bmatrix} 0 & -8 & 5 \\ 3 & -7 & 4 \\ -1 & 5 & -4 \\ 1 & -3 & 2 \end{bmatrix}$ **6.** $\begin{bmatrix} -4 & -3 & 0 \\ 0 & -1 & 4 \\ 1 & 0 & 3 \\ 5 & 4 & 6 \end{bmatrix}$

7. $\begin{bmatrix} 1 & 4 & -3 & 0 \\ -2 & -7 & 5 & 1 \\ -4 & -5 & 7 & 5 \end{bmatrix}$ **8.** $\begin{bmatrix} 1 & -3 & 3 & -2 \\ -3 & 7 & -1 & 2 \\ 0 & 1 & -4 & 3 \end{bmatrix}$

In Exercises 9 and 10, (a) for what values of h is $\mathbf{v}_3$ in Span $\{\mathbf{v}_1, \mathbf{v}_2\}$, and (b) for what values of h is $\{\mathbf{v}_1, \mathbf{v}_2, \mathbf{v}_3\}$ linearly *dependent*? Justify each answer.

9. $\mathbf{v}_1 = \begin{bmatrix} 1 \\ -3 \\ 2 \end{bmatrix}, \mathbf{v}_2 = \begin{bmatrix} -3 \\ 10 \\ -6 \end{bmatrix}, \mathbf{v}_3 = \begin{bmatrix} 2 \\ -7 \\ h \end{bmatrix}$

10. $\mathbf{v}_1 = \begin{bmatrix} 1 \\ -5 \\ -3 \end{bmatrix}, \mathbf{v}_2 = \begin{bmatrix} -2 \\ 10 \\ 6 \end{bmatrix}, \mathbf{v}_3 = \begin{bmatrix} 2 \\ -10 \\ h \end{bmatrix}$

In Exercises 11–14, find the value(s) of h for which the vectors are linearly *dependent*. Justify each answer.

11. $\begin{bmatrix} 1 \\ -1 \\ 4 \end{bmatrix}, \begin{bmatrix} 3 \\ -5 \\ 7 \end{bmatrix}, \begin{bmatrix} -1 \\ 5 \\ h \end{bmatrix}$ **12.** $\begin{bmatrix} 2 \\ -4 \\ 1 \end{bmatrix}, \begin{bmatrix} -6 \\ 7 \\ -3 \end{bmatrix}, \begin{bmatrix} 8 \\ h \\ 4 \end{bmatrix}$

13. $\begin{bmatrix} 1 \\ 5 \\ -3 \end{bmatrix}, \begin{bmatrix} -2 \\ -9 \\ 6 \end{bmatrix}, \begin{bmatrix} 3 \\ h \\ -9 \end{bmatrix}$ **14.** $\begin{bmatrix} 1 \\ -1 \\ 3 \end{bmatrix}, \begin{bmatrix} -5 \\ 7 \\ 8 \end{bmatrix}, \begin{bmatrix} 1 \\ 1 \\ h \end{bmatrix}$

Determine by inspection whether the vectors in Exercises 15–20 are linearly *independent*. Justify each answer.

15. $\begin{bmatrix} 5 \\ 1 \end{bmatrix}, \begin{bmatrix} 2 \\ 8 \end{bmatrix}, \begin{bmatrix} 1 \\ 3 \end{bmatrix}, \begin{bmatrix} -1 \\ 7 \end{bmatrix}$ **16.** $\begin{bmatrix} 4 \\ -2 \\ 6 \end{bmatrix}, \begin{bmatrix} 6 \\ -3 \\ 9 \end{bmatrix}$

17. $\begin{bmatrix} 3 \\ 5 \\ -1 \end{bmatrix}, \begin{bmatrix} 0 \\ 0 \\ 0 \end{bmatrix}, \begin{bmatrix} -6 \\ 5 \\ 4 \end{bmatrix}$ **18.** $\begin{bmatrix} 4 \\ 4 \end{bmatrix}, \begin{bmatrix} -1 \\ 3 \end{bmatrix}, \begin{bmatrix} 2 \\ 5 \end{bmatrix}, \begin{bmatrix} 8 \\ 1 \end{bmatrix}$

19. $\begin{bmatrix} -8 \\ 12 \\ -4 \end{bmatrix}, \begin{bmatrix} 2 \\ -3 \\ -1 \end{bmatrix}$ **20.** $\begin{bmatrix} 1 \\ 4 \\ -7 \end{bmatrix}, \begin{bmatrix} -2 \\ 5 \\ 3 \end{bmatrix}, \begin{bmatrix} 0 \\ 0 \\ 0 \end{bmatrix}$

In Exercises 21–28, mark each statement True or False (**T/F**). Justify each answer on the basis of a careful reading of the text.

21. (**T/F**) The columns of a matrix A are linearly independent if the equation $A\mathbf{x} = \mathbf{0}$ has the trivial solution.

22. (**T/F**) Two vectors are linearly dependent if and only if they lie on a line through the origin.

23. (**T/F**) If S is a linearly dependent set, then each vector is a linear combination of the other vectors in S.

24. (**T/F**) If a set contains fewer vectors than there are entries in the vectors, then the set is linearly independent.

25. (**T/F**) The columns of any 4×5 matrix are linearly dependent.

26. (**T/F**) If $\mathbf{x}$ and $\mathbf{y}$ are linearly independent, and if $\mathbf{z}$ is in Span $\{\mathbf{x}, \mathbf{y}\}$, then $\{\mathbf{x}, \mathbf{y}, \mathbf{z}\}$ is linearly dependent.

27. (**T/F**) If $\mathbf{x}$ and $\mathbf{y}$ are linearly independent, and if $\{\mathbf{x}, \mathbf{y}, \mathbf{z}\}$ is linearly dependent, then $\mathbf{z}$ is in Span $\{\mathbf{x}, \mathbf{y}\}$.

28. (**T/F**) If a set in $\mathbb{R}^n$ is linearly dependent, then the set contains more vectors than there are entries in each vector.

In Exercises 29–32, describe the possible echelon forms of the matrix. Use the notation of Example 1 in Section 1.2.

29. A is a 3×3 matrix with linearly independent columns.

30. A is a 2×2 matrix with linearly dependent columns.

31. A is a 4×2 matrix, $A = [\mathbf{a}_1 \quad \mathbf{a}_2]$, and $\mathbf{a}_2$ is not a multiple of $\mathbf{a}_1$.

32. A is a 4×3 matrix, $A = [\mathbf{a}_1 \quad \mathbf{a}_2 \quad \mathbf{a}_3]$, such that $\{\mathbf{a}_1, \mathbf{a}_2\}$ is linearly independent and $\mathbf{a}_3$ is not in Span $\{\mathbf{a}_1, \mathbf{a}_2\}$.

33. How many pivot columns must a 7×5 matrix have if its columns are linearly independent? Why?

34. How many pivot columns must a 5×7 matrix have if its columns span $\mathbb{R}^5$? Why?

35. Construct 3×2 matrices A and B such that $A\mathbf{x} = \mathbf{0}$ has only the trivial solution and $B\mathbf{x} = \mathbf{0}$ has a nontrivial solution.

36. a. Fill in the blank in the following statement: "If A is an $m \times n$ matrix, then the columns of A are linearly independent if and only if A has _____ pivot columns."

   b. Explain why the statement in (a) is true.

Exercises 37 and 38 should be solved *without performing row operations.* [*Hint:* Write $A\mathbf{x} = \mathbf{0}$ as a vector equation.]

37. Given $A = \begin{bmatrix} 2 & 3 & 5 \\ -5 & 1 & -4 \\ -3 & -1 & -4 \\ 1 & 0 & 1 \end{bmatrix}$, observe that the third column

is the sum of the first two columns. Find a nontrivial solution of $A\mathbf{x} = \mathbf{0}$.

38. Given $A = \begin{bmatrix} 4 & 1 & 6 \\ -7 & 5 & 3 \\ 9 & -3 & 3 \end{bmatrix}$, observe that the first column

plus twice the second column equals the third column. Find a nontrivial solution of $A\mathbf{x} = \mathbf{0}$.

Each statement in Exercises 39–44 is either true (in all cases) or false (for at least one example). If false, construct a specific example to show that the statement is not always true. Such an example is called a *counterexample* to the statement. If a statement is true, give a justification. (One specific example cannot explain why a statement is always true. You will have to do more work here than in Exercises 21–28.)

39. (T/F-C) If $\mathbf{v}_1, \ldots, \mathbf{v}_4$ are in $\mathbb{R}^4$ and $\mathbf{v}_3 = 2\mathbf{v}_1 + \mathbf{v}_2$, then $\{\mathbf{v}_1, \mathbf{v}_2, \mathbf{v}_3, \mathbf{v}_4\}$ is linearly dependent.

40. (T/F-C) If $\mathbf{v}_1, \ldots, \mathbf{v}_4$ are in $\mathbb{R}^4$ and $\mathbf{v}_3 = \mathbf{0}$, then $\{\mathbf{v}_1, \mathbf{v}_2, \mathbf{v}_3, \mathbf{v}_4\}$ is linearly dependent.

41. (T/F-C) If $\mathbf{v}_1$ and $\mathbf{v}_2$ are in $\mathbb{R}^4$ and $\mathbf{v}_2$ is not a scalar multiple of $\mathbf{v}_1$, then $\{\mathbf{v}_1, \mathbf{v}_2\}$ is linearly independent.

42. (T/F-C) If $\mathbf{v}_1, \ldots, \mathbf{v}_4$ are in $\mathbb{R}^4$ and $\mathbf{v}_3$ is *not* a linear combination of $\mathbf{v}_1, \mathbf{v}_2, \mathbf{v}_4$, then $\{\mathbf{v}_1, \mathbf{v}_2, \mathbf{v}_3, \mathbf{v}_4\}$ is linearly independent.

43. (T/F-C) If $\mathbf{v}_1, \ldots, \mathbf{v}_4$ are in $\mathbb{R}^4$ and $\{\mathbf{v}_1, \mathbf{v}_2, \mathbf{v}_3\}$ is linearly dependent, then $\{\mathbf{v}_1, \mathbf{v}_2, \mathbf{v}_3, \mathbf{v}_4\}$ is also linearly dependent.

44. (T/F-C) If $\mathbf{v}_1, \ldots, \mathbf{v}_4$ are linearly independent vectors in $\mathbb{R}^4$, then $\{\mathbf{v}_1, \mathbf{v}_2, \mathbf{v}_3\}$ is also linearly independent. [*Hint:* Think about $x_1\mathbf{v}_1 + x_2\mathbf{v}_2 + x_3\mathbf{v}_3 + 0 \cdot \mathbf{v}_4 = \mathbf{0}.$]

45. Suppose A is an $m \times n$ matrix with the property that for all $\mathbf{b}$ in $\mathbb{R}^m$ the equation $A\mathbf{x} = \mathbf{b}$ has at most one solution. Use the definition of linear independence to explain why the columns of A must be linearly independent.

46. Suppose an $m \times n$ matrix A has n pivot columns. Explain why for each $\mathbf{b}$ in $\mathbb{R}^m$ the equation $A\mathbf{x} = \mathbf{b}$ has at most one solution. [*Hint:* Explain why $A\mathbf{x} = \mathbf{b}$ cannot have infinitely many solutions.]

T In Exercises 47 and 48, use as many columns of A as possible to construct a matrix B with the property that the equation $B\mathbf{x} = \mathbf{0}$ has only the trivial solution. Solve $B\mathbf{x} = \mathbf{0}$ to verify your work.

47. $A = \begin{bmatrix} 8 & -3 & 0 & -7 & 2 \\ -9 & 4 & 5 & 11 & -7 \\ 6 & -2 & 2 & -4 & 4 \\ 5 & -1 & 7 & 0 & 10 \end{bmatrix}$

48. $A = \begin{bmatrix} 12 & 10 & -6 & -3 & 7 & 10 \\ -7 & -6 & 4 & 7 & -9 & 5 \\ 9 & 9 & -9 & -5 & 5 & -1 \\ -4 & -3 & 1 & 6 & -8 & 9 \\ 8 & 7 & -5 & -9 & 11 & -8 \end{bmatrix}$

T 49. With A and B as in Exercise 47 select a column $\mathbf{v}$ of A that was not used in the construction of B and determine if $\mathbf{v}$ is in the set spanned by the columns of B. (Describe your calculations.)

T 50. Repeat Exercise 49 with the matrices A and B from Exercise 48. Then give an explanation for what you discover, assuming that B was constructed as specified.

STUDY GUIDE offers additional resources for mastering the concept of linear independence.

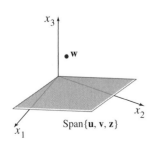

Span$\{\mathbf{u}, \mathbf{v}, \mathbf{z}\}$

Solutions to Practice Problems

1. a. Yes. In each case, neither vector is a multiple of the other. Thus each set is linearly independent.

   b. No. The observation in Part (a), by itself, says nothing about the linear independence of $\{\mathbf{u}, \mathbf{v}, \mathbf{w}, \mathbf{z}\}$.

   c. No. When testing for linear independence, it is usually a poor idea to check if one selected vector is a linear combination of the others. It may happen that the selected vector is not a linear combination of the others and yet the whole set of vectors is linearly dependent. In this practice problem, $\mathbf{w}$ is not a linear combination of $\mathbf{u}, \mathbf{v}$, and $\mathbf{z}$.

   d. Yes, by Theorem 8. There are more vectors (four) than entries (three) in them.

2. Applying the definition of linearly dependent to $\{\mathbf{v}_1, \mathbf{v}_2, \mathbf{v}_3\}$ implies that there exist scalars c_1, c_2, and c_3, not all zero, such that

$$c_1\mathbf{v}_1 + c_2\mathbf{v}_2 + c_3\mathbf{v}_3 = \mathbf{0}.$$

Adding $0\,\mathbf{v}_4 = \mathbf{0}$ to both sides of this equation results in

$$c_1\mathbf{v}_1 + c_2\mathbf{v}_2 + c_3\mathbf{v}_3 + 0\,\mathbf{v}_4 = \mathbf{0}.$$

Since c_1, c_2, c_3 and 0 are not *all* zero, the set $\{\mathbf{v}_1, \mathbf{v}_2, \mathbf{v}_3, \mathbf{v}_4\}$ satisfies the definition of a linearly dependent set.

1.8 Introduction to Linear Transformations

The difference between a matrix equation $A\mathbf{x} = \mathbf{b}$ and the associated vector equation $x_1\mathbf{a}_1 + \cdots + x_n\mathbf{a}_n = \mathbf{b}$ is merely a matter of notation. However, a matrix equation $A\mathbf{x} = \mathbf{b}$ can arise in linear algebra (and in applications such as computer graphics and signal processing) in a way that is not directly connected with linear combinations of vectors. This happens when we think of the matrix A as an object that "acts" on a vector $\mathbf{x}$ by multiplication to produce a new vector called $A\mathbf{x}$.

For instance, the equations

$$
\underset{A}{\begin{bmatrix} 4 & -3 & 1 & 3 \\ 2 & 0 & 5 & 1 \end{bmatrix}}
\underset{\mathbf{x}}{\begin{bmatrix} 1 \\ 1 \\ 1 \\ 1 \end{bmatrix}}
= \underset{\mathbf{b}}{\begin{bmatrix} 5 \\ 8 \end{bmatrix}}
\quad \text{and} \quad
\underset{A}{\begin{bmatrix} 4 & -3 & 1 & 3 \\ 2 & 0 & 5 & 1 \end{bmatrix}}
\underset{\mathbf{u}}{\begin{bmatrix} 1 \\ 4 \\ -1 \\ 3 \end{bmatrix}}
= \underset{\mathbf{0}}{\begin{bmatrix} 0 \\ 0 \end{bmatrix}}
$$

say that multiplication by A transforms $\mathbf{x}$ into $\mathbf{b}$ and transforms $\mathbf{u}$ into the zero vector. See Figure 1.

FIGURE 1 Transforming vectors via matrix multiplication.

From this new point of view, solving the equation $A\mathbf{x} = \mathbf{b}$ amounts to finding all vectors $\mathbf{x}$ in $\mathbb{R}^4$ that are transformed into the vector $\mathbf{b}$ in $\mathbb{R}^2$ under the "action" of multiplication by A.

The correspondence from $\mathbf{x}$ to $A\mathbf{x}$ is a *function* from one set of vectors to another. This concept generalizes the common notion of a function as a rule that transforms one real number into another.

A **transformation** (or **function** or **mapping**) T from $\mathbb{R}^n$ to $\mathbb{R}^m$ is a rule that assigns to each vector $\mathbf{x}$ in $\mathbb{R}^n$ a vector $T(\mathbf{x})$ in $\mathbb{R}^m$. The set $\mathbb{R}^n$ is called the **domain** of T, and $\mathbb{R}^m$

is called the **codomain** of T. The notation $T : \mathbb{R}^n \to \mathbb{R}^m$ indicates that the domain of T is $\mathbb{R}^n$ and the codomain is $\mathbb{R}^m$. For **x** in $\mathbb{R}^n$, the vector $T(\mathbf{x})$ in $\mathbb{R}^m$ is called the **image** of **x** (under the action of T). The set of all images $T(\mathbf{x})$ is called the **range** of T. See Figure 2.

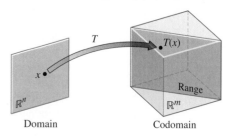

FIGURE 2 Domain, codomain, and range of $T : \mathbb{R}^n \to \mathbb{R}^m$.

The new terminology in this section is important because a dynamic view of matrix–vector multiplication is the key to understanding several ideas in linear algebra and to building mathematical models of physical systems that evolve over time. Such *dynamical systems* will be discussed in Sections 1.10, 4.8, and throughout Chapter 5.

Matrix Transformations

The rest of this section focuses on mappings associated with matrix multiplication. For each **x** in $\mathbb{R}^n$, $T(\mathbf{x})$ is computed as $A\mathbf{x}$, where A is an $m \times n$ matrix. For simplicity, we sometimes denote such a *matrix transformation* by $\mathbf{x} \mapsto A\mathbf{x}$. Observe that the domain of T is $\mathbb{R}^n$ when A has n columns and the codomain of T is $\mathbb{R}^m$ when each column of A has m entries. The range of T is the set of all linear combinations of the columns of A, because each image $T(\mathbf{x})$ is of the form $A\mathbf{x}$.

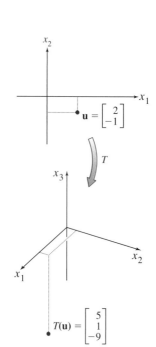

EXAMPLE 1 Let $A = \begin{bmatrix} 1 & -3 \\ 3 & 5 \\ -1 & 7 \end{bmatrix}$, $\mathbf{u} = \begin{bmatrix} 2 \\ -1 \end{bmatrix}$, $\mathbf{b} = \begin{bmatrix} 3 \\ 2 \\ -5 \end{bmatrix}$, $\mathbf{c} = \begin{bmatrix} 3 \\ 2 \\ 5 \end{bmatrix}$, and define a transformation $T : \mathbb{R}^2 \to \mathbb{R}^3$ by $T(\mathbf{x}) = A\mathbf{x}$, so that

$$T(\mathbf{x}) = A\mathbf{x} = \begin{bmatrix} 1 & -3 \\ 3 & 5 \\ -1 & 7 \end{bmatrix} \begin{bmatrix} x_1 \\ x_2 \end{bmatrix} = \begin{bmatrix} x_1 - 3x_2 \\ 3x_1 + 5x_2 \\ -x_1 + 7x_2 \end{bmatrix}$$

a. Find $T(\mathbf{u})$, the image of **u** under the transformation T.

b. Find an **x** in $\mathbb{R}^2$ whose image under T is **b**.

c. Is there more than one **x** whose image under T is **b**?

d. Determine if **c** is in the range of the transformation T.

SOLUTION

a. Compute

$$T(\mathbf{u}) = A\mathbf{u} = \begin{bmatrix} 1 & -3 \\ 3 & 5 \\ -1 & 7 \end{bmatrix} \begin{bmatrix} 2 \\ -1 \end{bmatrix} = \begin{bmatrix} 5 \\ 1 \\ -9 \end{bmatrix}$$

b. Solve $T(\mathbf{x}) = \mathbf{b}$ for **x**. That is, solve $A\mathbf{x} = \mathbf{b}$, or

$$\begin{bmatrix} 1 & -3 \\ 3 & 5 \\ -1 & 7 \end{bmatrix} \begin{bmatrix} x_1 \\ x_2 \end{bmatrix} = \begin{bmatrix} 3 \\ 2 \\ -5 \end{bmatrix} \tag{1}$$

Using the method discussed in Section 1.4, row reduce the augmented matrix:

$$
\begin{bmatrix} 1 & -3 & 3 \\ 3 & 5 & 2 \\ -1 & 7 & -5 \end{bmatrix} \sim \begin{bmatrix} 1 & -3 & 3 \\ 0 & 14 & -7 \\ 0 & 4 & -2 \end{bmatrix} \sim \begin{bmatrix} 1 & -3 & 3 \\ 0 & 1 & -.5 \\ 0 & 0 & 0 \end{bmatrix} \sim \begin{bmatrix} 1 & 0 & 1.5 \\ 0 & 1 & -.5 \\ 0 & 0 & 0 \end{bmatrix} \quad (2)
$$

Hence $x_1 = 1.5$, $x_2 = -.5$, and $\mathbf{x} = \begin{bmatrix} 1.5 \\ -.5 \end{bmatrix}$. The image of this $\mathbf{x}$ under T is the given vector $\mathbf{b}$.

c. Any $\mathbf{x}$ whose image under T is $\mathbf{b}$ must satisfy equation (1). From (2), it is clear that equation (1) has a unique solution. So there is exactly one $\mathbf{x}$ whose image is $\mathbf{b}$.

d. The vector $\mathbf{c}$ is in the range of T if $\mathbf{c}$ is the image of some $\mathbf{x}$ in $\mathbb{R}^2$, that is, if $\mathbf{c} = T(\mathbf{x})$ for some $\mathbf{x}$. This is just another way of asking if the system $A\mathbf{x} = \mathbf{c}$ is consistent. To find the answer, row reduce the augmented matrix:

$$
\begin{bmatrix} 1 & -3 & 3 \\ 3 & 5 & 2 \\ -1 & 7 & 5 \end{bmatrix} \sim \begin{bmatrix} 1 & -3 & 3 \\ 0 & 14 & -7 \\ 0 & 4 & 8 \end{bmatrix} \sim \begin{bmatrix} 1 & -3 & 3 \\ 0 & 1 & 2 \\ 0 & 14 & -7 \end{bmatrix} \sim \begin{bmatrix} 1 & -3 & 3 \\ 0 & 1 & 2 \\ 0 & 0 & -35 \end{bmatrix}
$$

The third equation, $0 = -35$, shows that the system is inconsistent. So $\mathbf{c}$ is *not* in the range of T. ∎

The question in Example 1(c) is a *uniqueness* problem for a system of linear equations, translated here into the language of matrix transformations: Is $\mathbf{b}$ the image of a *unique* $\mathbf{x}$ in $\mathbb{R}^n$? Similarly, Example 1(d) is an *existence* problem: Does there *exist* an $\mathbf{x}$ whose image is $\mathbf{c}$?

The next two matrix transformations can be viewed geometrically. They reinforce the dynamic view of a matrix as something that transforms vectors into other vectors. Section 2.7 contains other interesting examples connected with computer graphics.

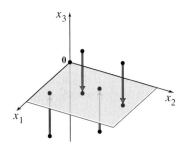

FIGURE 3

A projection transformation.

EXAMPLE 2 If $A = \begin{bmatrix} 1 & 0 & 0 \\ 0 & 1 & 0 \\ 0 & 0 & 0 \end{bmatrix}$, then the transformation $\mathbf{x} \mapsto A\mathbf{x}$ *projects* points in $\mathbb{R}^3$ onto the x_1x_2-plane because

$$
\begin{bmatrix} x_1 \\ x_2 \\ x_3 \end{bmatrix} \mapsto \begin{bmatrix} 1 & 0 & 0 \\ 0 & 1 & 0 \\ 0 & 0 & 0 \end{bmatrix} \begin{bmatrix} x_1 \\ x_2 \\ x_3 \end{bmatrix} = \begin{bmatrix} x_1 \\ x_2 \\ 0 \end{bmatrix}
$$

See Figure 3. ∎

sheep

sheared sheep

EXAMPLE 3 Let $A = \begin{bmatrix} 1 & 2 \\ 0 & 1 \end{bmatrix}$. The transformation $T : \mathbb{R}^2 \rightarrow \mathbb{R}^2$ defined by $T(\mathbf{x}) = A\mathbf{x}$ is called a **shear transformation**. It can be shown that if T acts on each point in the 2×2 square shown in Figure 4, then the set of images forms the sheared parallelogram. The key idea is to show that T maps line segments onto line segments (as shown in Exercise 35) and then to check that the corners of the square map onto the vertices of the parallelogram. For instance, the image of the point $\mathbf{u} = \begin{bmatrix} 0 \\ 2 \end{bmatrix}$ is

$T(\mathbf{u}) = \begin{bmatrix} 1 & 2 \\ 0 & 1 \end{bmatrix} \begin{bmatrix} 0 \\ 2 \end{bmatrix} = \begin{bmatrix} 4 \\ 2 \end{bmatrix}$, and the image of $\begin{bmatrix} 2 \\ 2 \end{bmatrix}$ is $\begin{bmatrix} 1 & 2 \\ 0 & 1 \end{bmatrix} \begin{bmatrix} 2 \\ 2 \end{bmatrix} = \begin{bmatrix} 6 \\ 2 \end{bmatrix}$. T deforms the square as if the top of the square were pushed to the right while the base is held fixed. Shear transformations appear in physics, geology, and crystallography. ∎

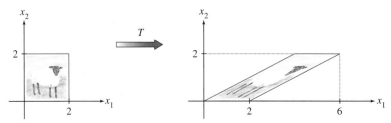

FIGURE 4 A shear transformation.

Linear Transformations

Theorem 5 in Section 1.4 shows that if A is $m \times n$, then the transformation $\mathbf{x} \mapsto A\mathbf{x}$ has the properties

$$A(\mathbf{u} + \mathbf{v}) = A\mathbf{u} + A\mathbf{v} \quad \text{and} \quad A(c\mathbf{u}) = cA\mathbf{u}$$

for all $\mathbf{u}, \mathbf{v}$ in $\mathbb{R}^n$ and all scalars c. These properties, written in function notation, identify the most important class of transformations in linear algebra.

DEFINITION

A transformation (or mapping) T is **linear** if

(i) $T(\mathbf{u} + \mathbf{v}) = T(\mathbf{u}) + T(\mathbf{v})$ for all $\mathbf{u}, \mathbf{v}$ in the domain of T;

(ii) $T(c\mathbf{u}) = cT(\mathbf{u})$ for all scalars c and all $\mathbf{u}$ in the domain of T.

Every matrix transformation is a linear transformation. Important examples of linear transformations that are not matrix transformations will be discussed in Chapters 4 and 5.

Linear transformations *preserve the operations of vector addition and scalar multiplication*. Property (i) says that the result $T(\mathbf{u} + \mathbf{v})$ of first adding $\mathbf{u}$ and $\mathbf{v}$ in $\mathbb{R}^n$ and then applying T is the same as first applying T to $\mathbf{u}$ and to $\mathbf{v}$ and then adding $T(\mathbf{u})$ and $T(\mathbf{v})$ in $\mathbb{R}^m$. These two properties lead easily to the following useful facts.

If T is a linear transformation, then

$$T(\mathbf{0}) = \mathbf{0} \tag{3}$$

and

$$T(c\mathbf{u} + d\mathbf{v}) = cT(\mathbf{u}) + dT(\mathbf{v}) \tag{4}$$

for all vectors $\mathbf{u}, \mathbf{v}$ in the domain of T and all scalars c, d.

Property (3) follows from condition (ii) in the definition, because $T(\mathbf{0}) = T(0\mathbf{u}) = 0T(\mathbf{u}) = \mathbf{0}$. Property (4) requires both (i) and (ii):

$$T(c\mathbf{u} + d\mathbf{v}) = T(c\mathbf{u}) + T(d\mathbf{v}) = cT(\mathbf{u}) + dT(\mathbf{v})$$

Observe that *if a transformation satisfies* (4) *for all* $\mathbf{u}, \mathbf{v}$ *and* c, d, *it must be linear.* (Set $c = d = 1$ for preservation of addition, and set $d = 0$ for preservation of scalar multiplication.) Repeated application of (4) produces a useful generalization:

$$T(c_1\mathbf{v}_1 + \cdots + c_p\mathbf{v}_p) = c_1T(\mathbf{v}_1) + \cdots + c_pT(\mathbf{v}_p) \tag{5}$$

In engineering and physics, (5) is referred to as a *superposition principle*. Think of $\mathbf{v}_1, \ldots, \mathbf{v}_p$ as signals that go into a system and $T(\mathbf{v}_1), \ldots, T(\mathbf{v}_p)$ as the responses of that system to the signals. The system satisfies the superposition principle if whenever an input is expressed as a linear combination of such signals, the system's response is *the same* linear combination of the responses to the individual signals. We will return to this idea in Chapter 4.

EXAMPLE 4   Given a scalar r, define $T : \mathbb{R}^2 \to \mathbb{R}^2$ by $T(\mathbf{x}) = r\mathbf{x}$. T is called a **contraction** when $0 \le r \le 1$ and a **dilation** when $r > 1$. Let $r = 3$, and show that T is a linear transformation.

SOLUTION   Let $\mathbf{u}, \mathbf{v}$ be in $\mathbb{R}^2$ and let c, d be scalars. Then

$$\begin{aligned}
T(c\mathbf{u} + d\mathbf{v}) &= 3(c\mathbf{u} + d\mathbf{v}) && \text{Definition of } T \\
&= 3c\mathbf{u} + 3d\mathbf{v} \\
&= c(3\mathbf{u}) + d(3\mathbf{v}) \\
&= cT(\mathbf{u}) + dT(\mathbf{v})
\end{aligned}$$

$\left.\vphantom{\begin{aligned}&\\&\\&\end{aligned}}\right\}$ Vector arithmetic

Thus T is a linear transformation because it satisfies (4). See Figure 5. ■

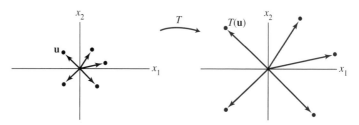

FIGURE 5 A dilation transformation.

EXAMPLE 5   Define a linear transformation $T : \mathbb{R}^2 \to \mathbb{R}^2$ by

$$T(\mathbf{x}) = \begin{bmatrix} 0 & -1 \\ 1 & 0 \end{bmatrix}\begin{bmatrix} x_1 \\ x_2 \end{bmatrix} = \begin{bmatrix} -x_2 \\ x_1 \end{bmatrix}$$

Find the images under T of $\mathbf{u} = \begin{bmatrix} 4 \\ 1 \end{bmatrix}$, $\mathbf{v} = \begin{bmatrix} 2 \\ 3 \end{bmatrix}$, and $\mathbf{u} + \mathbf{v} = \begin{bmatrix} 6 \\ 4 \end{bmatrix}$.

SOLUTION

$$T(\mathbf{u}) = \begin{bmatrix} 0 & -1 \\ 1 & 0 \end{bmatrix}\begin{bmatrix} 4 \\ 1 \end{bmatrix} = \begin{bmatrix} -1 \\ 4 \end{bmatrix}, \qquad T(\mathbf{v}) = \begin{bmatrix} 0 & -1 \\ 1 & 0 \end{bmatrix}\begin{bmatrix} 2 \\ 3 \end{bmatrix} = \begin{bmatrix} -3 \\ 2 \end{bmatrix},$$

$$T(\mathbf{u} + \mathbf{v}) = \begin{bmatrix} 0 & -1 \\ 1 & 0 \end{bmatrix}\begin{bmatrix} 6 \\ 4 \end{bmatrix} = \begin{bmatrix} -4 \\ 6 \end{bmatrix}$$

Note that $T(\mathbf{u} + \mathbf{v})$ is obviously equal to $T(\mathbf{u}) + T(\mathbf{v})$. It appears from Figure 6 that T rotates $\mathbf{u}$, $\mathbf{v}$, and $\mathbf{u} + \mathbf{v}$ counterclockwise about the origin through $90°$. In fact, T transforms the entire parallelogram determined by $\mathbf{u}$ and $\mathbf{v}$ into the one determined by $T(\mathbf{u})$ and $T(\mathbf{v})$. (See Exercise 36.) ■

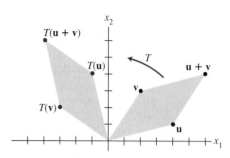

FIGURE 6 A rotation transformation.

The final example is not geometrical; instead, it shows how a linear mapping can transform one type of data into another.

EXAMPLE 6 A company manufactures two products, B and C. Using data from Example 7 in Section 1.3, we construct a "unit cost" matrix, $U = [\mathbf{b} \ \mathbf{c}]$, whose columns describe the "costs per dollar of output" for the products:

$$
\begin{array}{cc}
 & \text{Product} \\
 & \begin{array}{cc} \text{B} & \text{C} \end{array}
\end{array}
$$

$$
U = \begin{bmatrix} .45 & .40 \\ .25 & .30 \\ .15 & .15 \end{bmatrix} \begin{array}{l} \text{Materials} \\ \text{Labor} \\ \text{Overhead} \end{array}
$$

Let $\mathbf{x} = (x_1, x_2)$ be a "production" vector, corresponding to x_1 dollars of product B and x_2 dollars of product C, and define $T : \mathbb{R}^2 \to \mathbb{R}^3$ by

$$
T(\mathbf{x}) = U\mathbf{x} = x_1 \begin{bmatrix} .45 \\ .25 \\ .15 \end{bmatrix} + x_2 \begin{bmatrix} .40 \\ .30 \\ .15 \end{bmatrix} = \begin{bmatrix} \text{Total cost of materials} \\ \text{Total cost of labor} \\ \text{Total cost of overhead} \end{bmatrix}
$$

The mapping T transforms a list of production quantities (measured in dollars) into a list of total costs. The linearity of this mapping is reflected in two ways:

1. If production is increased by a factor of, say, 4, from $\mathbf{x}$ to $4\mathbf{x}$, then the costs will increase by the same factor, from $T(\mathbf{x})$ to $4T(\mathbf{x})$.

2. If $\mathbf{x}$ and $\mathbf{y}$ are production vectors, then the total cost vector associated with the combined production $\mathbf{x} + \mathbf{y}$ is precisely the sum of the cost vectors $T(\mathbf{x})$ and $T(\mathbf{y})$. ∎

Practice Problems

1. Suppose $T : \mathbb{R}^5 \to \mathbb{R}^2$ and $T(\mathbf{x}) = A\mathbf{x}$ for some matrix A and for each $\mathbf{x}$ in $\mathbb{R}^5$. How many rows and columns does A have?

2. Let $A = \begin{bmatrix} 1 & 0 \\ 0 & -1 \end{bmatrix}$. Give a geometric description of the transformation $\mathbf{x} \mapsto A\mathbf{x}$.

3. The line segment from $\mathbf{0}$ to a vector $\mathbf{u}$ is the set of points of the form $t\mathbf{u}$, where $0 \leq t \leq 1$. Show that a linear transformation T maps this segment into the segment between $\mathbf{0}$ and $T(\mathbf{u})$.

1.8 Exercises

1. Let $A = \begin{bmatrix} 2 & 0 \\ 0 & 2 \end{bmatrix}$, and define $T : \mathbb{R}^2 \to \mathbb{R}^2$ by $T(\mathbf{x}) = A\mathbf{x}$. Find the images under T of $\mathbf{u} = \begin{bmatrix} 1 \\ -3 \end{bmatrix}$ and $\mathbf{v} = \begin{bmatrix} a \\ b \end{bmatrix}$.

2. Let $A = \begin{bmatrix} .5 & 0 & 0 \\ 0 & .5 & 0 \\ 0 & 0 & .5 \end{bmatrix}$, $\mathbf{u} = \begin{bmatrix} 1 \\ 0 \\ -4 \end{bmatrix}$, and $\mathbf{v} = \begin{bmatrix} a \\ b \\ c \end{bmatrix}$. Define $T : \mathbb{R}^3 \to \mathbb{R}^3$ by $T(\mathbf{x}) = A\mathbf{x}$. Find $T(\mathbf{u})$ and $T(\mathbf{v})$.

In Exercises 3–6, with T defined by $T(\mathbf{x}) = A\mathbf{x}$, find a vector $\mathbf{x}$ whose image under T is $\mathbf{b}$, and determine whether $\mathbf{x}$ is unique.

3. $A = \begin{bmatrix} 1 & 0 & -2 \\ -2 & 1 & 6 \\ 3 & -2 & -5 \end{bmatrix}$, $\mathbf{b} = \begin{bmatrix} -1 \\ 7 \\ -3 \end{bmatrix}$

4. $A = \begin{bmatrix} 1 & -3 & 2 \\ 0 & 1 & -4 \\ 3 & -5 & -9 \end{bmatrix}$, $\mathbf{b} = \begin{bmatrix} 6 \\ -7 \\ -9 \end{bmatrix}$

5. $A = \begin{bmatrix} 1 & -5 & -7 \\ -3 & 7 & 5 \end{bmatrix}$, $\mathbf{b} = \begin{bmatrix} -2 \\ -2 \end{bmatrix}$

6. $A = \begin{bmatrix} 1 & -2 & 1 \\ 3 & -4 & 5 \\ 0 & 1 & 1 \\ -3 & 5 & -4 \end{bmatrix}$, $\mathbf{b} = \begin{bmatrix} 1 \\ 9 \\ 3 \\ -6 \end{bmatrix}$

7. Let A be a 6×5 matrix. What must a and b be in order to define $T : \mathbb{R}^a \to \mathbb{R}^b$ by $T(\mathbf{x}) = A\mathbf{x}$?

8. How many rows and columns must a matrix A have in order to define a mapping from $\mathbb{R}^4$ into $\mathbb{R}^5$ by the rule $T(\mathbf{x}) = A\mathbf{x}$?

For Exercises 9 and 10, find all $\mathbf{x}$ in $\mathbb{R}^4$ that are mapped into the zero vector by the transformation $\mathbf{x} \mapsto A\mathbf{x}$ for the given matrix A.

9. $A = \begin{bmatrix} 1 & -4 & 7 & -5 \\ 0 & 1 & -4 & 3 \\ 2 & -6 & 6 & -4 \end{bmatrix}$

10. $A = \begin{bmatrix} 1 & 3 & 9 & 2 \\ 1 & 0 & 3 & -4 \\ 0 & 1 & 2 & 3 \\ -2 & 3 & 0 & 5 \end{bmatrix}$

11. Let $\mathbf{b} = \begin{bmatrix} -1 \\ 1 \\ 0 \end{bmatrix}$, and let A be the matrix in Exercise 9. Is $\mathbf{b}$ in the range of the linear transformation $\mathbf{x} \mapsto A\mathbf{x}$? Why or why not?

12. Let $\mathbf{b} = \begin{bmatrix} -1 \\ 3 \\ -1 \\ 4 \end{bmatrix}$, and let A be the matrix in Exercise 10. Is $\mathbf{b}$ in the range of the linear transformation $\mathbf{x} \mapsto A\mathbf{x}$? Why or why not?

In Exercises 13–16, use a rectangular coordinate system to plot $\mathbf{u} = \begin{bmatrix} 5 \\ 2 \end{bmatrix}$, $\mathbf{v} = \begin{bmatrix} -2 \\ 4 \end{bmatrix}$, and their images under the given transformation T. (Make a separate and reasonably large sketch for each exercise.) Describe geometrically what T does to each vector $\mathbf{x}$ in $\mathbb{R}^2$.

13. $T(\mathbf{x}) = \begin{bmatrix} -1 & 0 \\ 0 & -1 \end{bmatrix} \begin{bmatrix} x_1 \\ x_2 \end{bmatrix}$

14. $T(\mathbf{x}) = \begin{bmatrix} .5 & 0 \\ 0 & .5 \end{bmatrix} \begin{bmatrix} x_1 \\ x_2 \end{bmatrix}$

15. $T(\mathbf{x}) = \begin{bmatrix} 0 & 0 \\ 0 & 1 \end{bmatrix} \begin{bmatrix} x_1 \\ x_2 \end{bmatrix}$

16. $T(\mathbf{x}) = \begin{bmatrix} 0 & 1 \\ 1 & 0 \end{bmatrix} \begin{bmatrix} x_1 \\ x_2 \end{bmatrix}$

17. Let $T : \mathbb{R}^2 \to \mathbb{R}^2$ be a linear transformation that maps $\mathbf{u} = \begin{bmatrix} 5 \\ 2 \end{bmatrix}$ into $\begin{bmatrix} 2 \\ 1 \end{bmatrix}$ and maps $\mathbf{v} = \begin{bmatrix} 1 \\ 3 \end{bmatrix}$ into $\begin{bmatrix} -1 \\ 3 \end{bmatrix}$. Use the fact that T is linear to find the images under T of $3\mathbf{u}$, $2\mathbf{v}$, and $3\mathbf{u} + 2\mathbf{v}$.

18. The figure shows vectors $\mathbf{u}$, $\mathbf{v}$, and $\mathbf{w}$, along with the images $T(\mathbf{u})$ and $T(\mathbf{v})$ under the action of a linear transformation $T : \mathbb{R}^2 \to \mathbb{R}^2$. Copy this figure carefully, and draw the image $T(\mathbf{w})$ as accurately as possible. [*Hint:* First, write $\mathbf{w}$ as a linear combination of $\mathbf{u}$ and $\mathbf{v}$.]

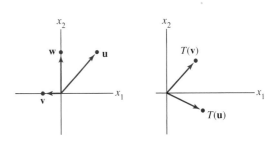

19. Let $\mathbf{e}_1 = \begin{bmatrix} 1 \\ 0 \end{bmatrix}$, $\mathbf{e}_2 = \begin{bmatrix} 0 \\ 1 \end{bmatrix}$, $\mathbf{y}_1 = \begin{bmatrix} 2 \\ 5 \end{bmatrix}$, and $\mathbf{y}_2 = \begin{bmatrix} -1 \\ 6 \end{bmatrix}$, and let $T : \mathbb{R}^2 \to \mathbb{R}^2$ be a linear transformation that maps $\mathbf{e}_1$ into $\mathbf{y}_1$ and maps $\mathbf{e}_2$ into $\mathbf{y}_2$. Find the images of $\begin{bmatrix} 5 \\ -3 \end{bmatrix}$ and $\begin{bmatrix} x_1 \\ x_2 \end{bmatrix}$.

20. Let $\mathbf{x} = \begin{bmatrix} x_1 \\ x_2 \end{bmatrix}$, $\mathbf{v}_1 = \begin{bmatrix} -2 \\ 5 \end{bmatrix}$, and $\mathbf{v}_2 = \begin{bmatrix} 7 \\ -3 \end{bmatrix}$, and let $T : \mathbb{R}^2 \to \mathbb{R}^2$ be a linear transformation that maps $\mathbf{x}$ into $x_1 \mathbf{v}_1 + x_2 \mathbf{v}_2$. Find a matrix A such that $T(\mathbf{x})$ is $A\mathbf{x}$ for each $\mathbf{x}$.

In Exercises 21–30, mark each statement True or False (**T/F**). Justify each answer.

21. (**T/F**) A linear transformation is a special type of function.

22. (**T/F**) Every matrix transformation is a linear transformation.

23. (**T/F**) If A is a 3×5 matrix and T is a transformation defined by $T(\mathbf{x}) = A\mathbf{x}$, then the domain of T is $\mathbb{R}^3$.

24. (**T/F**) The codomain of the transformation $\mathbf{x} \mapsto A\mathbf{x}$ is the set of all linear combinations of the columns of A.

25. (**T/F**) If A is an $m \times n$ matrix, then the range of the transformation $\mathbf{x} \mapsto A\mathbf{x}$ is $\mathbb{R}^m$.

26. (**T/F**) If $T : \mathbb{R}^n \to \mathbb{R}^m$ is a linear transformation and if $\mathbf{c}$ is in $\mathbb{R}^m$, then a uniqueness question is "Is $\mathbf{c}$ in the range of T?"

27. (**T/F**) Every linear transformation is a matrix transformation.

28. (**T/F**) A linear transformation preserves the operations of vector addition and scalar multiplication.

29. (**T/F**) A transformation T is linear if and only if $T(c_1\mathbf{v}_1 + c_2\mathbf{v}_2) = c_1 T(\mathbf{v}_1) + c_2 T(\mathbf{v}_2)$ for all $\mathbf{v}_1$ and $\mathbf{v}_2$ in the domain of T and for all scalars c_1 and c_2.

30. (**T/F**) The superposition principle is a physical description of a linear transformation.

31. Let $T : \mathbb{R}^2 \to \mathbb{R}^2$ be the linear transformation that reflects each point through the x_1-axis. (See Practice Problem 2.) Make two sketches similar to Figure 6 that illustrate properties (i) and (ii) of a linear transformation.

32. Suppose vectors $\mathbf{v}_1, \ldots, \mathbf{v}_p$ span $\mathbb{R}^n$, and let $T : \mathbb{R}^n \to \mathbb{R}^n$ be a linear transformation. Suppose $T(\mathbf{v}_i) = \mathbf{0}$ for $i = 1, \ldots, p$. Show that T is the zero transformation. That is, show that if $\mathbf{x}$ is any vector in $\mathbb{R}^n$, then $T(\mathbf{x}) = \mathbf{0}$.

33. Given $\mathbf{v} \neq \mathbf{0}$ and $\mathbf{p}$ in $\mathbb{R}^n$, the line through $\mathbf{p}$ in the direction of $\mathbf{v}$ has the parametric equation $\mathbf{x} = \mathbf{p} + t\mathbf{v}$. Show that a linear transformation $T : \mathbb{R}^n \to \mathbb{R}^n$ maps this line onto another line or onto a single point (a *degenerate line*).

34. Let $\mathbf{u}$ and $\mathbf{v}$ be linearly independent vectors in $\mathbb{R}^3$, and let P be the plane through $\mathbf{u}$, $\mathbf{v}$, and $\mathbf{0}$. The parametric equation of P is $\mathbf{x} = s\mathbf{u} + t\mathbf{v}$ (with s, t in $\mathbb{R}$). Show that a linear transformation $T : \mathbb{R}^3 \to \mathbb{R}^3$ maps P onto a plane through $\mathbf{0}$, or onto a line through $\mathbf{0}$, or onto just the origin in $\mathbb{R}^3$. What must be true about $T(\mathbf{u})$ and $T(\mathbf{v})$ in order for the image of the plane P to be a plane?

35. a. Show that the line through vectors $\mathbf{p}$ and $\mathbf{q}$ in $\mathbb{R}^n$ may be written in the parametric form $\mathbf{x} = (1 - t)\mathbf{p} + t\mathbf{q}$. (Refer to the figure with Exercises 25 and 26 in Section 1.5.)

    b. The line segment from $\mathbf{p}$ to $\mathbf{q}$ is the set of points of the form $(1 - t)\mathbf{p} + t\mathbf{q}$ for $0 \leq t \leq 1$ (as shown in the figure

below). Show that a linear transformation T maps this line segment onto a line segment or onto a single point.

36. Let $\mathbf{u}$ and $\mathbf{v}$ be vectors in $\mathbb{R}^n$. It can be shown that the set P of all points in the parallelogram determined by $\mathbf{u}$ and $\mathbf{v}$ has the form $a\mathbf{u} + b\mathbf{v}$, for $0 \leq a \leq 1, 0 \leq b \leq 1$. Let $T : \mathbb{R}^n \to \mathbb{R}^m$ be a linear transformation. Explain why the image of a point in P under the transformation T lies in the parallelogram determined by $T(\mathbf{u})$ and $T(\mathbf{v})$.

37. Define $f : \mathbb{R} \to \mathbb{R}$ by $f(x) = mx + b$.

    a. Show that f is a linear transformation when $b = 0$.

    b. Find a property of a linear transformation that is violated when $b \neq 0$.

    c. Why is f called a linear function?

38. An *affine transformation* $T : \mathbb{R}^n \to \mathbb{R}^m$ has the form $T(x) = A\mathbf{x} + \mathbf{b}$, with A an $m \times n$ matrix and $\mathbf{b}$ in $\mathbb{R}^m$. Show that T is *not* a linear transformation when $\mathbf{b} \neq \mathbf{0}$. (Affine transformations are important in computer graphics.)

39. Let $T : \mathbb{R}^n \to \mathbb{R}^m$ be a linear transformation, and let $\{\mathbf{v}_1, \mathbf{v}_2, \mathbf{v}_3\}$ be a linearly dependent set in $\mathbb{R}^n$. Explain why the set $\{T(\mathbf{v}_1), T(\mathbf{v}_2), T(\mathbf{v}_3)\}$ is linearly dependent.

In Exercises 40–44, column vectors are written as rows, such as $\mathbf{x} = (x_1, x_2)$, and $T(\mathbf{x})$ is written as $T(x_1, x_2)$.

40. Show that the transformation T defined by $T(x_1, x_2) = (4x_1 - 2x_2, 3|x_2|)$ is not linear.

41. Show that the transformation T defined by $T(x_1, x_2) = (2x_1 - 3x_2, x_1 + 4, 5x_2)$ is not linear.

42. Let $T : \mathbb{R}^n \to \mathbb{R}^m$ be a linear transformation. Show that if T maps two linearly independent vectors onto a linearly dependent set, then the equation $T(\mathbf{x}) = \mathbf{0}$ has a nontrivial solution. [*Hint:* Suppose $\mathbf{u}$ and $\mathbf{v}$ in $\mathbb{R}^n$ are linearly independent and yet $T(\mathbf{u})$ and $T(\mathbf{v})$ are linearly dependent. Then $c_1 T(\mathbf{u}) + c_2 T(\mathbf{v}) = \mathbf{0}$ for some weights c_1 and c_2, not both zero. Use this equation.]

43. Let $T : \mathbb{R}^3 \to \mathbb{R}^3$ be the transformation that reflects each vector $\mathbf{x} = (x_1, x_2, x_3)$ through the plane $x_3 = 0$ onto $T(\mathbf{x}) = (x_1, x_2, -x_3)$. Show that T is a linear transformation. [See Example 4 for ideas.]

44. Let $T : \mathbb{R}^3 \to \mathbb{R}^3$ be the transformation that projects each vector $\mathbf{x} = (x_1, x_2, x_3)$ onto the plane $x_2 = 0$, so $T(\mathbf{x}) = (x_1, 0, x_3)$. Show that T is a linear transformation.

T In Exercises 45 and 46, the given matrix determines a linear transformation T. Find all **x** such that $T(\mathbf{x}) = \mathbf{0}$.

45. $\begin{bmatrix} 4 & -2 & 5 & -5 \\ -9 & 7 & -8 & 0 \\ -6 & 4 & 5 & 3 \\ 5 & -3 & 8 & -4 \end{bmatrix}$

46. $\begin{bmatrix} -9 & -4 & -9 & 4 \\ 5 & -8 & -7 & 6 \\ 7 & 11 & 16 & -9 \\ 9 & -7 & -4 & 5 \end{bmatrix}$

T 47. Let $\mathbf{b} = \begin{bmatrix} 7 \\ 5 \\ 9 \\ 7 \end{bmatrix}$ and let A be the matrix in Exercise 45. Is **b** in the range of the transformation $\mathbf{x} \mapsto A\mathbf{x}$? If so, find an **x** whose image under the transformation is **b**.

T 48. Let $\mathbf{b} = \begin{bmatrix} -7 \\ -7 \\ 13 \\ -5 \end{bmatrix}$ and let A be the matrix in Exercise 46. Is **b** in the range of the transformation $\mathbf{x} \mapsto A\mathbf{x}$? If so, find an **x** whose image under the transformation is **b**.

STUDY GUIDE offers additional resources for mastering linear transformations.

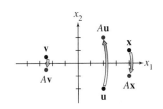

The transformation $\mathbf{x} \mapsto A\mathbf{x}$.

Solutions to Practice Problems

1. A must have five columns for $A\mathbf{x}$ to be defined. A must have two rows for the codomain of T to be $\mathbb{R}^2$.

2. Plot some random points (vectors) on graph paper to see what happens. A point such as $(4, 1)$ maps into $(4, -1)$. The transformation $\mathbf{x} \mapsto A\mathbf{x}$ reflects points through the x-axis (or x_1-axis).

3. Let $\mathbf{x} = t\mathbf{u}$ for some t such that $0 \le t \le 1$. Since T is linear, $T(t\mathbf{u}) = t\,T(\mathbf{u})$, which is a point on the line segment between $\mathbf{0}$ and $T(\mathbf{u})$.

1.9 The Matrix of a Linear Transformation

Whenever a linear transformation T arises geometrically or is described in words, we usually want a "formula" for $T(\mathbf{x})$. The discussion that follows shows that every linear transformation from $\mathbb{R}^n$ to $\mathbb{R}^m$ is actually a matrix transformation $\mathbf{x} \mapsto A\mathbf{x}$ and that important properties of T are intimately related to familiar properties of A. The key to finding A is to observe that T is completely determined by what it does to the columns of the $n \times n$ identity matrix I_n.

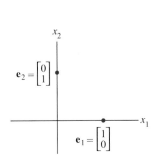

EXAMPLE 1   The columns of $I_2 = \begin{bmatrix} 1 & 0 \\ 0 & 1 \end{bmatrix}$ are $\mathbf{e}_1 = \begin{bmatrix} 1 \\ 0 \end{bmatrix}$ and $\mathbf{e}_2 = \begin{bmatrix} 0 \\ 1 \end{bmatrix}$.

Suppose T is a linear transformation from $\mathbb{R}^2$ into $\mathbb{R}^3$ such that

$$T(\mathbf{e}_1) = \begin{bmatrix} 5 \\ -7 \\ 2 \end{bmatrix} \quad \text{and} \quad T(\mathbf{e}_2) = \begin{bmatrix} -3 \\ 8 \\ 0 \end{bmatrix}$$

With no additional information, find a formula for the image of an arbitrary **x** in $\mathbb{R}^2$.

SOLUTION   Write

$$\mathbf{x} = \begin{bmatrix} x_1 \\ x_2 \end{bmatrix} = x_1 \begin{bmatrix} 1 \\ 0 \end{bmatrix} + x_2 \begin{bmatrix} 0 \\ 1 \end{bmatrix} = x_1 \mathbf{e}_1 + x_2 \mathbf{e}_2 \tag{1}$$

Since T is a *linear* transformation,

$$T(\mathbf{x}) = x_1 T(\mathbf{e}_1) + x_2 T(\mathbf{e}_2) \tag{2}$$

$$= x_1 \begin{bmatrix} 5 \\ -7 \\ 2 \end{bmatrix} + x_2 \begin{bmatrix} -3 \\ 8 \\ 0 \end{bmatrix} = \begin{bmatrix} 5x_1 - 3x_2 \\ -7x_1 + 8x_2 \\ 2x_1 + 0 \end{bmatrix}$$ ∎

The step from equation (1) to equation (2) explains why knowledge of $T(\mathbf{e}_1)$ and $T(\mathbf{e}_2)$ is sufficient to determine $T(\mathbf{x})$ for any $\mathbf{x}$. Moreover, since (2) expresses $T(\mathbf{x})$ as a linear combination of vectors, we can put these vectors into the columns of a matrix A and write (2) as

$$T(\mathbf{x}) = \begin{bmatrix} T(\mathbf{e}_1) & T(\mathbf{e}_2) \end{bmatrix} \begin{bmatrix} x_1 \\ x_2 \end{bmatrix} = A\mathbf{x}$$

THEOREM 10

Let $T : \mathbb{R}^n \rightarrow \mathbb{R}^m$ be a linear transformation. Then there exists a unique matrix A such that

$$T(\mathbf{x}) = A\mathbf{x} \quad \text{for all } \mathbf{x} \text{ in } \mathbb{R}^n$$

In fact, A is the $m \times n$ matrix whose jth column is the vector $T(\mathbf{e}_j)$, where $\mathbf{e}_j$ is the jth column of the identity matrix in $\mathbb{R}^n$:

$$A = \begin{bmatrix} T(\mathbf{e}_1) & \cdots & T(\mathbf{e}_n) \end{bmatrix} \tag{3}$$

PROOF   Write $\mathbf{x} = I_n\mathbf{x} = \begin{bmatrix} \mathbf{e}_1 & \cdots & \mathbf{e}_n \end{bmatrix}\mathbf{x} = x_1\mathbf{e}_1 + \cdots + x_n\mathbf{e}_n$, and use the linearity of T to compute

$$T(\mathbf{x}) = T(x_1\mathbf{e}_1 + \cdots + x_n\mathbf{e}_n) = x_1 T(\mathbf{e}_1) + \cdots + x_n T(\mathbf{e}_n)$$

$$= \begin{bmatrix} T(\mathbf{e}_1) & \cdots & T(\mathbf{e}_n) \end{bmatrix} \begin{bmatrix} x_1 \\ \vdots \\ x_n \end{bmatrix} = A\mathbf{x}$$

The uniqueness of A is treated in Exercise 41.   ∎

The matrix A in (3) is called the **standard matrix for the linear transformation** T.

We know now that every linear transformation from $\mathbb{R}^n$ to $\mathbb{R}^m$ can be viewed as a matrix transformation, and vice versa. The term *linear transformation* focuses on a property of a mapping, while *matrix transformation* describes how such a mapping is implemented, as Examples 2 and 3 illustrate.

EXAMPLE 2   Find the standard matrix A for the dilation transformation $T(\mathbf{x}) = 3\mathbf{x}$, for $\mathbf{x}$ in $\mathbb{R}^2$.

SOLUTION   Write

$$T(\mathbf{e}_1) = 3\mathbf{e}_1 = \begin{bmatrix} 3 \\ 0 \end{bmatrix} \quad \text{and} \quad T(\mathbf{e}_2) = 3\mathbf{e}_2 = \begin{bmatrix} 0 \\ 3 \end{bmatrix}$$

$$A = \begin{bmatrix} 3 & 0 \\ 0 & 3 \end{bmatrix}$$

∎

EXAMPLE 3   Let $T : \mathbb{R}^2 \rightarrow \mathbb{R}^2$ be the transformation that rotates each point in $\mathbb{R}^2$ about the origin through an angle φ, with counterclockwise rotation for a positive angle. We could show geometrically that such a transformation is linear. (See Figure 6 in Section 1.8.) Find the standard matrix A of this transformation.

SOLUTION $\begin{bmatrix} 1 \\ 0 \end{bmatrix}$ rotates into $\begin{bmatrix} \cos\varphi \\ \sin\varphi \end{bmatrix}$, and $\begin{bmatrix} 0 \\ 1 \end{bmatrix}$ rotates into $\begin{bmatrix} -\sin\varphi \\ \cos\varphi \end{bmatrix}$. See Figure 1. By Theorem 10,

$$A = \begin{bmatrix} \cos\varphi & -\sin\varphi \\ \sin\varphi & \cos\varphi \end{bmatrix}$$

Example 5 in Section 1.8 is a special case of this transformation, with $\varphi = \pi/2$.   ∎

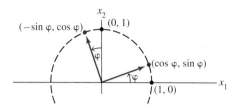

FIGURE 1 A rotation transformation.

Geometric Linear Transformations of $\mathbb{R}^2$

Examples 2 and 3 illustrate linear transformations that are described geometrically. Tables 1–4 illustrate other common geometric linear transformations of the plane. Because the transformations are linear, they are determined completely by what they do to the columns of I_2. Instead of showing only the images of $\mathbf{e}_1$ and $\mathbf{e}_2$, the tables show what a transformation does to the unit square (Figure 2).

Other transformations can be constructed from those listed in Tables 1–4 by applying one transformation after another. For instance, a horizontal shear could be followed by a reflection in the x_2-axis. Section 2.1 will show that such a *composition* of linear transformations is linear. (Also, see Exercise 44.)

FIGURE 2
The unit square.

Existence and Uniqueness Questions

The concept of a linear transformation provides a new way to understand the existence and uniqueness questions asked earlier. The next two definitions give the appropriate terminology for transformations.

DEFINITION

> A mapping $T : \mathbb{R}^n \to \mathbb{R}^m$ is said to be **onto** $\mathbb{R}^m$ if each $\mathbf{b}$ in $\mathbb{R}^m$ is the image of *at least one* $\mathbf{x}$ in $\mathbb{R}^n$.

Equivalently, T is onto $\mathbb{R}^m$ when the range of T is all of the codomain $\mathbb{R}^m$. That is, T maps $\mathbb{R}^n$ onto $\mathbb{R}^m$ if, for each $\mathbf{b}$ in the codomain $\mathbb{R}^m$, there exists at least one solution of $T(\mathbf{x}) = \mathbf{b}$. "Does T map $\mathbb{R}^n$ onto $\mathbb{R}^m$?" is an existence question. The mapping T is *not* onto when there is some $\mathbf{b}$ in $\mathbb{R}^m$ for which the equation $T(\mathbf{x}) = \mathbf{b}$ has no solution. See Figure 3.

T is *not* onto $\mathbb{R}^m$

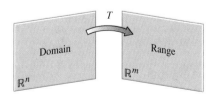

T is onto $\mathbb{R}^m$

FIGURE 3 Is the range of T all of $\mathbb{R}^m$?

TABLE I Reflections

Transformation	Image of the Unit Square	Standard Matrix
Reflection through the x_1-axis		$\begin{bmatrix} 1 & 0 \\ 0 & -1 \end{bmatrix}$
Reflection through the x_2-axis		$\begin{bmatrix} -1 & 0 \\ 0 & 1 \end{bmatrix}$
Reflection through the line $x_2 = x_1$		$\begin{bmatrix} 0 & 1 \\ 1 & 0 \end{bmatrix}$
Reflection through the line $x_2 = -x_1$		$\begin{bmatrix} 0 & -1 \\ -1 & 0 \end{bmatrix}$
Reflection through the origin		$\begin{bmatrix} -1 & 0 \\ 0 & -1 \end{bmatrix}$

TABLE 2   Contractions and Expansions

Transformation	Image of the Unit Square		Standard Matrix
Horizontal contraction and expansion	$0 < k < 1$	$k > 1$	$\begin{bmatrix} k & 0 \\ 0 & 1 \end{bmatrix}$
Vertical contraction and expansion	$0 < k < 1$	$k > 1$	$\begin{bmatrix} 1 & 0 \\ 0 & k \end{bmatrix}$

TABLE 3   Shears

Transformation	Image of the Unit Square		Standard Matrix
Horizontal shear	$k < 0$	$k > 0$	$\begin{bmatrix} 1 & k \\ 0 & 1 \end{bmatrix}$
Vertical shear	$k < 0$	$k > 0$	$\begin{bmatrix} 1 & 0 \\ k & 1 \end{bmatrix}$

TABLE 4 Projections

Transformation	Image of the Unit Square	Standard Matrix
Projection onto the x_1-axis		$\begin{bmatrix} 1 & 0 \\ 0 & 0 \end{bmatrix}$
Projection onto the x_2-axis		$\begin{bmatrix} 0 & 0 \\ 0 & 1 \end{bmatrix}$

DEFINITION

A mapping $T : \mathbb{R}^n \rightarrow \mathbb{R}^m$ is said to be **one-to-one** if each $\mathbf{b}$ in $\mathbb{R}^m$ is the image of *at most one* $\mathbf{x}$ in $\mathbb{R}^n$.

Equivalently, T is one-to-one if, for each $\mathbf{b}$ in $\mathbb{R}^m$, the equation $T(\mathbf{x}) = \mathbf{b}$ has either a unique solution or none at all. "Is T one-to-one?" is a uniqueness question. The mapping T is *not* one-to-one when some $\mathbf{b}$ in $\mathbb{R}^m$ is the image of more than one vector in $\mathbb{R}^n$. If there is no such $\mathbf{b}$, then T is one-to-one. See Figure 4.

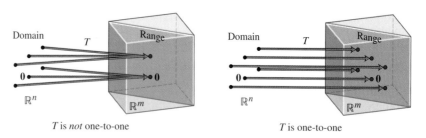

FIGURE 4 Is every $\mathbf{b}$ the image of at most one vector?

The projection transformations shown in Table 4 are *not* one-to-one and do *not* map $\mathbb{R}^2$ onto $\mathbb{R}^2$. The transformations in Tables 1, 2, and 3 are one-to-one *and* do map $\mathbb{R}^2$ onto $\mathbb{R}^2$. Other possibilities are shown in the two examples below.

Example 4 and the theorems that follow show how the function properties of being one-to-one and mapping onto are related to important concepts studied earlier in this chapter.

EXAMPLE 4   Let T be the linear transformation whose standard matrix is

$$A = \begin{bmatrix} 1 & -4 & 8 & 1 \\ 0 & 2 & -1 & 3 \\ 0 & 0 & 0 & 5 \end{bmatrix}$$

Does T map $\mathbb{R}^4$ onto $\mathbb{R}^3$? Is T a one-to-one mapping?

SOLUTION   Since A happens to be in echelon form, we can see at once that A has a pivot position in each row. By Theorem 4 in Section 1.4, for each $\mathbf{b}$ in $\mathbb{R}^3$, the equation $A\mathbf{x} = \mathbf{b}$ is consistent. In other words, the linear transformation T maps $\mathbb{R}^4$ (its domain) onto $\mathbb{R}^3$. However, since the equation $A\mathbf{x} = \mathbf{b}$ has a free variable (because there are four variables and only three basic variables), each $\mathbf{b}$ is the image of more than one $\mathbf{x}$. That is, T is *not* one-to-one.   ∎

THEOREM 11

Let $T : \mathbb{R}^n \to \mathbb{R}^m$ be a linear transformation. Then T is one-to-one if and only if the equation $T(\mathbf{x}) = \mathbf{0}$ has only the trivial solution.

Remark: To prove a theorem that says "statement P is true if and only if statement Q is true," one must establish two things: (1) If P is true, then Q is true and (2) If Q is true, then P is true. The second requirement can also be established by showing (2a): If P is false, then Q is false. (This is called contrapositive reasoning.) This proof uses (1) and (2a) to show that P and Q are either both true or both false.

PROOF   Since T is linear, $T(\mathbf{0}) = \mathbf{0}$. If T is one-to-one, then the equation $T(\mathbf{x}) = \mathbf{0}$ has at most one solution and hence only the trivial solution. If T is not one-to-one, then there is a $\mathbf{b}$ that is the image of at least two different vectors in $\mathbb{R}^n$—say, $\mathbf{u}$ and $\mathbf{v}$. That is, $T(\mathbf{u}) = \mathbf{b}$ and $T(\mathbf{v}) = \mathbf{b}$. But then, since T is linear,

$$T(\mathbf{u} - \mathbf{v}) = T(\mathbf{u}) - T(\mathbf{v}) = \mathbf{b} - \mathbf{b} = \mathbf{0}$$

The vector $\mathbf{u} - \mathbf{v}$ is not zero, since $\mathbf{u} \neq \mathbf{v}$. Hence the equation $T(\mathbf{x}) = \mathbf{0}$ has more than one solution. So, either the two conditions in the theorem are both true or they are both false.   ∎

THEOREM 12

Let $T : \mathbb{R}^n \to \mathbb{R}^m$ be a linear transformation, and let A be the standard matrix for T. Then:

a. T maps $\mathbb{R}^n$ onto $\mathbb{R}^m$ if and only if the columns of A span $\mathbb{R}^m$;

b. T is one-to-one if and only if the columns of A are linearly independent.

Remark: "If and only if" statements can be linked together. For example if "P if and only if Q" is known and "Q if and only if R" is known, then one can conclude "P if and only if R." This strategy is used repeatedly in this proof.

PROOF

a. By Theorem 4 in Section 1.4, the columns of A span $\mathbb{R}^m$ if and only if for each $\mathbf{b}$ in $\mathbb{R}^m$ the equation $A\mathbf{x} = \mathbf{b}$ is consistent—in other words, if and only if for every $\mathbf{b}$, the equation $T(\mathbf{x}) = \mathbf{b}$ has at least one solution. This is true if and only if T maps $\mathbb{R}^n$ onto $\mathbb{R}^m$.

b. The equations $T(\mathbf{x}) = \mathbf{0}$ and $A\mathbf{x} = \mathbf{0}$ are the same except for notation. So, by Theorem 11, T is one-to-one if and only if $A\mathbf{x} = \mathbf{0}$ has only the trivial solution. This happens if and only if the columns of A are linearly independent, as was already noted in the boxed statement (3) in Section 1.7. ∎

Statement (a) in Theorem 12 is equivalent to the statement "T maps $\mathbb{R}^n$ onto $\mathbb{R}^m$ if and only if every vector in $\mathbb{R}^m$ is a linear combination of the columns of A." See Theorem 4 in Section 1.4.

In the next example and in some exercises that follow, column vectors are written in rows, such as $\mathbf{x} = (x_1, x_2)$, and $T(\mathbf{x})$ is written as $T(x_1, x_2)$ instead of the more formal $T((x_1, x_2))$.

The transformation T is not onto $\mathbb{R}^3$.

EXAMPLE 5   Let $T(x_1, x_2) = (3x_1 + x_2, 5x_1 + 7x_2, x_1 + 3x_2)$. Show that T is a one-to-one linear transformation. Does T map $\mathbb{R}^2$ onto $\mathbb{R}^3$?

SOLUTION   When $\mathbf{x}$ and $T(\mathbf{x})$ are written as column vectors, you can determine the standard matrix of T by inspection, visualizing the row–vector computation of each entry in $A\mathbf{x}$.

$$T(\mathbf{x}) = \begin{bmatrix} 3x_1 + x_2 \\ 5x_1 + 7x_2 \\ x_1 + 3x_2 \end{bmatrix} = \begin{bmatrix} ? & ? \\ ? & ? \\ ? & ? \end{bmatrix} \underset{A}{\begin{bmatrix} x_1 \\ x_2 \end{bmatrix}} = \begin{bmatrix} 3 & 1 \\ 5 & 7 \\ 1 & 3 \end{bmatrix} \begin{bmatrix} x_1 \\ x_2 \end{bmatrix} \tag{4}$$

So T is indeed a linear transformation, with its standard matrix A shown in (4). The columns of A are linearly independent because they are not multiples. By Theorem 12(b), T is one-to-one. To decide if T is onto $\mathbb{R}^3$, examine the span of the columns of A. Since A is 3×2, the columns of A span $\mathbb{R}^3$ if and only if A has 3 pivot positions, by Theorem 4. This is impossible, since A has only 2 columns. So the columns of A do not span $\mathbb{R}^3$, and the associated linear transformation is not onto $\mathbb{R}^3$. ∎

Practice Problems

1. Let $T : \mathbb{R}^2 \to \mathbb{R}^2$ be the transformation that first performs a horizontal shear that maps $\mathbf{e}_2$ into $\mathbf{e}_2 - .5\mathbf{e}_1$ (but leaves $\mathbf{e}_1$ unchanged) and then reflects the result through the x_2-axis. Assuming that T is linear, find its standard matrix. [*Hint:* Determine the final location of the images of $\mathbf{e}_1$ and $\mathbf{e}_2$.]

2. Suppose A is a 7×5 matrix with 5 pivots. Let $T(\mathbf{x}) = A\mathbf{x}$ be a linear transformation from $\mathbb{R}^5$ into $\mathbb{R}^7$. Is T a one-to-one linear transformation? Is T onto $\mathbb{R}^7$?

1.9 Exercises

In Exercises 1–10, assume that T is a linear transformation. Find the standard matrix of T.

1. $T : \mathbb{R}^2 \to \mathbb{R}^4$, $T(\mathbf{e}_1) = (2, 1, 2, 1)$ and $T(\mathbf{e}_2) = (-5, 2, 0, 0)$, where $\mathbf{e}_1 = (1, 0)$ and $\mathbf{e}_2 = (0, 1)$.

2. $T : \mathbb{R}^3 \to \mathbb{R}^2$, $T(\mathbf{e}_1) = (1, 3)$, $T(\mathbf{e}_2) = (4, 2)$, and $T(\mathbf{e}_3) = (-5, 4)$, where $\mathbf{e}_1, \mathbf{e}_2, \mathbf{e}_3$ are the columns of the 3×3 identity matrix.

3. $T : \mathbb{R}^2 \to \mathbb{R}^2$ rotates points (about the origin) through $3\pi/2$ radians (in the counterclockwise direction).

4. $T : \mathbb{R}^2 \to \mathbb{R}^2$ rotates points (about the origin) through $-\pi/4$ radians (since the number is negative, the actual rotation is clockwise). [*Hint:* $T(\mathbf{e}_1) = (1/\sqrt{2}, -1/\sqrt{2})$.]

5. $T : \mathbb{R}^2 \to \mathbb{R}^2$ is a vertical shear transformation that maps $\mathbf{e}_1$ into $\mathbf{e}_1 - 2\mathbf{e}_2$ but leaves the vector $\mathbf{e}_2$ unchanged.

6. $T : \mathbb{R}^2 \to \mathbb{R}^2$ is a horizontal shear transformation that leaves $\mathbf{e}_1$ unchanged and maps $\mathbf{e}_2$ into $\mathbf{e}_2 + 3\mathbf{e}_1$.

7. $T : \mathbb{R}^2 \to \mathbb{R}^2$ first rotates points through $-3\pi/4$ radians (since the number is negative, the actual rotation is clockwise) and then reflects points through the horizontal x_1-axis. [*Hint:* $T(\mathbf{e}_1) = (-1/\sqrt{2}, 1/\sqrt{2})$.]

8. $T : \mathbb{R}^2 \to \mathbb{R}^2$ first reflects points through the horizontal x_1-axis and then reflects points through the line $x_2 = x_1$.

9. $T : \mathbb{R}^2 \to \mathbb{R}^2$ first performs a horizontal shear that transforms $\mathbf{e}_2$ into $\mathbf{e}_2 - 3\mathbf{e}_1$ (leaving $\mathbf{e}_1$ unchanged) and then reflects points through the line $x_2 = -x_1$.

10. $T : \mathbb{R}^2 \to \mathbb{R}^2$ first reflects points through the vertical x_2-axis and then rotates points $3\pi/2$ radians.

11. A linear transformation $T : \mathbb{R}^2 \to \mathbb{R}^2$ first reflects points through the x_1-axis and then reflects points through the x_2-axis. Show that T can also be described as a linear transformation that rotates points about the origin. What is the angle of that rotation?

12. Show that the transformation in Exercise 8 is merely a rotation about the origin. What is the angle of the rotation?

13. Let $T : \mathbb{R}^2 \to \mathbb{R}^2$ be the linear transformation such that $T(\mathbf{e}_1)$ and $T(\mathbf{e}_2)$ are the vectors shown in the figure. Using the figure, sketch the vector $T(2, 1)$.

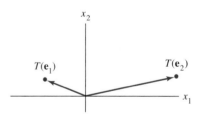

14. Let $T : \mathbb{R}^2 \to \mathbb{R}^2$ be a linear transformation with standard matrix $A = [\mathbf{a}_1 \quad \mathbf{a}_2]$, where $\mathbf{a}_1$ and $\mathbf{a}_2$ are shown in the figure. Using the figure, draw the image of $\begin{bmatrix} -1 \\ 3 \end{bmatrix}$ under the transformation T.

In Exercises 15 and 16, fill in the missing entries of the matrix, assuming that the equation holds for all values of the variables.

15. $\begin{bmatrix} ? & ? & ? \\ ? & ? & ? \\ ? & ? & ? \end{bmatrix} \begin{bmatrix} x_1 \\ x_2 \\ x_3 \end{bmatrix} = \begin{bmatrix} 2x_1 - 3x_3 \\ 4x_1 \\ x_1 - x_2 + x_3 \end{bmatrix}$

16. $\begin{bmatrix} ? & ? \\ ? & ? \\ ? & ? \end{bmatrix} \begin{bmatrix} x_1 \\ x_2 \end{bmatrix} = \begin{bmatrix} x_1 - 3x_2 \\ -2x_1 + x_2 \\ x_1 \end{bmatrix}$

In Exercises 17–20, show that T is a linear transformation by finding a matrix that implements the mapping. Note that $x_1, x_2, \ldots$ are not vectors but are entries in vectors.

17. $T(x_1, x_2, x_3, x_4) = (0, x_1 + x_2, x_2 + x_3, x_3 + x_4)$

18. $T(x_1, x_2) = (2x_2 - 3x_1, x_1 - 4x_2, 0, x_2)$

19. $T(x_1, x_2, x_3) = (x_1 - 5x_2 + 4x_3, x_2 - 6x_3)$

20. $T(x_1, x_2, x_3, x_4) = 2x_1 + 3x_3 - 4x_4 \qquad (T : \mathbb{R}^4 \to \mathbb{R})$

21. Let $T : \mathbb{R}^2 \to \mathbb{R}^2$ be a linear transformation such that $T(x_1, x_2) = (x_1 + x_2, 4x_1 + 5x_2)$. Find $\mathbf{x}$ such that $T(\mathbf{x}) = (3, 8)$.

22. Let $T : \mathbb{R}^2 \to \mathbb{R}^3$ be a linear transformation such that $T(x_1, x_2) = (x_1 - 2x_2, -x_1 + 3x_2, 3x_1 - 2x_2)$. Find $\mathbf{x}$ such that $T(\mathbf{x}) = (-1, 4, 9)$.

In Exercises 23–32, mark each statement True or False (**T/F**). Justify each answer.

23. (**T/F**) A linear transformation $T : \mathbb{R}^n \to \mathbb{R}^m$ is completely determined by its effect on the columns of the $n \times n$ identity matrix.

24. (**T/F**) A mapping $T : \mathbb{R}^n \to \mathbb{R}^m$ is one-to-one if each vector in $\mathbb{R}^n$ maps onto a unique vector in $\mathbb{R}^m$.

25. (**T/F**) If $T : \mathbb{R}^2 \to \mathbb{R}^2$ rotates vectors about the origin through an angle ϕ, then T is a linear transformation.

26. (**T/F**) The columns of the standard matrix for a linear transformation from $\mathbb{R}^n$ to $\mathbb{R}^m$ are the images of the columns of the $n \times n$ identity matrix.

27. (**T/F**) When two linear transformations are performed one after another, the combined effect may not always be a linear transformation.

28. (**T/F**) Not every linear transformation from $\mathbb{R}^n$ to $\mathbb{R}^m$ is a matrix transformation.

29. (**T/F**) A mapping $T : \mathbb{R}^n \to \mathbb{R}^m$ is onto $\mathbb{R}^m$ if every vector $\mathbf{x}$ in $\mathbb{R}^n$ maps onto some vector in $\mathbb{R}^m$.

30. (**T/F**) The standard matrix of a linear transformation from $\mathbb{R}^2$ to $\mathbb{R}^2$ that reflects points through the horizontal axis, the vertical axis, or the origin has the form $\begin{bmatrix} a & 0 \\ 0 & d \end{bmatrix}$, where a and d are ± 1.

31. (T/F) A is a 3×2 matrix, then the transformation $\mathbf{x} \mapsto A\mathbf{x}$ cannot be one-to-one.

32. (T/F) A is a 3×2 matrix, then the transformation $\mathbf{x} \mapsto A\mathbf{x}$ cannot map $\mathbb{R}^2$ onto $\mathbb{R}^3$.

In Exercises 33–36, determine if the specified linear transformation is (a) one-to-one and (b) onto. Justify each answer.

33. The transformation in Exercise 17

34. The transformation in Exercise 2

35. The transformation in Exercise 19

36. The transformation in Exercise 14

In Exercises 37 and 38, describe the possible echelon forms of the standard matrix for a linear transformation T. Use the notation of Example 1 in Section 1.2.

37. $T : \mathbb{R}^3 \rightarrow \mathbb{R}^4$ is one-to-one.

38. $T : \mathbb{R}^4 \rightarrow \mathbb{R}^3$ is onto.

39. Let $T : \mathbb{R}^n \rightarrow \mathbb{R}^m$ be a linear transformation, with A its standard matrix. Complete the following statement to make it true: "T is one-to-one if and only if A has _____ pivot columns." Explain why the statement is true. [*Hint:* Look in the exercises for Section 1.7.]

40. Let $T : \mathbb{R}^n \rightarrow \mathbb{R}^m$ be a linear transformation, with A its standard matrix. Complete the following statement to make it true: "T maps $\mathbb{R}^n$ onto $\mathbb{R}^m$ if and only if A has _____ pivot columns." Find some theorems that explain why the statement is true.

41. Verify the uniqueness of A in Theorem 10. Let $T : \mathbb{R}^n \rightarrow \mathbb{R}^m$ be a linear transformation such that $T(\mathbf{x}) = B\mathbf{x}$ for some

$m \times n$ matrix B. Show that if A is the standard matrix for T, then $A = B$. [*Hint:* Show that A and B have the same columns.]

42. Why is the question "Is the linear transformation T onto?" an existence question?

43. If a linear transformation $T : \mathbb{R}^n \rightarrow \mathbb{R}^m$ maps $\mathbb{R}^n$ *onto* $\mathbb{R}^m$, can you give a relation between m and n? If T is one-to-one, what can you say about m and n?

44. Let $S : \mathbb{R}^p \rightarrow \mathbb{R}^n$ and $T : \mathbb{R}^n \rightarrow \mathbb{R}^m$ be linear transformations. Show that the mapping $\mathbf{x} \mapsto T(S(\mathbf{x}))$ is a linear transformation (from $\mathbb{R}^p$ to $\mathbb{R}^m$). [*Hint:* Compute $T(S(c\mathbf{u} + d\mathbf{v}))$ for $\mathbf{u}, \mathbf{v}$ in $\mathbb{R}^p$ and scalars c and d. Justify each step of the computation, and explain why this computation gives the desired conclusion.]

▮ In Exercises 45–48, let T be the linear transformation whose standard matrix is given. In Exercises 45 and 46, decide if T is a one-to-one mapping. In Exercises 47 and 48, decide if T maps $\mathbb{R}^5$ onto $\mathbb{R}^5$. Justify your answers.

45. $\begin{bmatrix} -5 & 10 & -5 & 4 \\ 8 & 3 & -4 & 7 \\ 4 & -9 & 5 & -3 \\ -3 & -2 & 5 & 4 \end{bmatrix}$ **46.** $\begin{bmatrix} 7 & 5 & 4 & -9 \\ 10 & 6 & 16 & -4 \\ 12 & 8 & 12 & 7 \\ -8 & -6 & -2 & 5 \end{bmatrix}$

47. $\begin{bmatrix} 4 & -7 & 3 & 7 & 5 \\ 6 & -8 & 5 & 12 & -8 \\ -7 & 10 & -8 & -9 & 14 \\ 3 & -5 & 4 & 2 & -6 \\ -5 & 6 & -6 & -7 & 3 \end{bmatrix}$

48. $\begin{bmatrix} 9 & 13 & 5 & 6 & -1 \\ 14 & 15 & -7 & -6 & 4 \\ -8 & -9 & 12 & -5 & -9 \\ -5 & -6 & -8 & 9 & 8 \\ 13 & 14 & 15 & 2 & 11 \end{bmatrix}$

STUDY GUIDE offers additional resources for mastering existence and uniqueness.

Solution to Practice Problems

1. Follow what happens to $\mathbf{e}_1$ and $\mathbf{e}_2$. See Figure 5. First, $\mathbf{e}_1$ is unaffected by the shear and then is reflected into $-\mathbf{e}_1$. So $T(\mathbf{e}_1) = -\mathbf{e}_1$. Second, $\mathbf{e}_2$ goes to $\mathbf{e}_2 - .5\mathbf{e}_1$ by the shear transformation. Since reflection through the x_2-axis changes $\mathbf{e}_1$ into $-\mathbf{e}_1$ and leaves $\mathbf{e}_2$ unchanged, the vector $\mathbf{e}_2 - .5\mathbf{e}_1$ goes to $\mathbf{e}_2 + .5\mathbf{e}_1$. So $T(\mathbf{e}_2) = \mathbf{e}_2 + .5\mathbf{e}_1$.

Shear transformation Reflection through the x_2-axis

FIGURE 5 The composition of two transformations.

Thus the standard matrix of T is

$$\begin{bmatrix} T(\mathbf{e}_1) & T(\mathbf{e}_2) \end{bmatrix} = \begin{bmatrix} -\mathbf{e}_1 & \mathbf{e}_2 + .5\mathbf{e}_1 \end{bmatrix} = \begin{bmatrix} -1 & .5 \\ 0 & 1 \end{bmatrix}$$

2. The standard matrix representation of T is the matrix A. Since A has 5 columns and 5 pivots, there is a pivot in every column so the columns are linearly independent. By Theorem 12, T is one-to-one. Since A has 7 rows and only 5 pivots, there is not a pivot in every row hence the columns of A do not span $\mathbb{R}^7$. By Theorem 12, and T is not onto.

1.10 Linear Models in Business, Science, and Engineering

The mathematical models in this section are all *linear*; that is, each describes a problem by means of a linear equation, usually in vector or matrix form. The first model concerns nutrition but actually is representative of a general technique in linear programming problems. The second model comes from electrical engineering. The third model introduces the concept of a *linear difference equation*, a powerful mathematical tool for studying dynamic processes in a wide variety of fields such as engineering, ecology, economics, telecommunications, and the management sciences. Linear models are important because natural phenomena are often linear or nearly linear when the variables involved are held within reasonable bounds. Also, linear models are more easily adapted for computer calculation than are complex nonlinear models.

As you read about each model, pay attention to how its linearity reflects some property of the system being modeled.

Constructing a Nutritious Weight-Loss Diet

The formula for the Cambridge Diet, a popular diet in the 1980s, was based on years of research. A team of scientists headed by Dr. Alan H. Howard developed this diet at Cambridge University after more than eight years of clinical work with obese patients.[1] The very low-calorie powdered formula diet combines a precise balance of carbohydrate, high-quality protein, and fat, together with vitamins, minerals, trace elements, and electrolytes. Millions of persons have used the diet to achieve rapid and substantial weight loss.

To achieve the desired amounts and proportions of nutrients, Dr. Howard had to incorporate a large variety of foodstuffs in the diet. Each foodstuff supplied several of the required ingredients, but not in the correct proportions. For instance, nonfat milk was a major source of protein but contained too much calcium. So soy flour was used for part of the protein because soy flour contains little calcium. However, soy flour contains proportionally too much fat, so whey was added since it supplies less fat in relation to calcium. Unfortunately, whey contains too much carbohydrate. . . .

The following example illustrates the problem on a small scale. Listed in Table 1 are three of the ingredients in the diet, together with the amounts of certain nutrients supplied by 100 grams (g) of each ingredient.[2]

---

[1] The first announcement of this rapid weight-loss regimen was given in the *International Journal of Obesity* (1978) **2**, 321–332.

[2] Ingredients in the diet as of 1984; nutrient data for ingredients adapted from USDA Agricultural Handbooks No. 8-1 and 8-6, 1976.

TABLE I The Cambridge Diet

Amounts (g) Supplied per 100 g of Ingredient				Amounts (g) Supplied by Cambridge Diet in One Day
Nutrient	**Nonfat milk**	**Soy flour**	**Whey**	
Protein	36	51	13	33
Carbohydrate	52	34	74	45
Fat	0	7	1.1	3

EXAMPLE 1 If possible, find some combination of nonfat milk, soy flour, and whey to provide the exact amounts of protein, carbohydrate, and fat supplied by the diet in one day (Table 1).

SOLUTION Let x_1, x_2, and x_3, respectively, denote the number of units (100 g) of these foodstuffs. One approach to the problem is to derive equations for each nutrient separately. For instance, the product

$$\left\{ \begin{matrix} x_1 \text{ units of} \\ \text{nonfat milk} \end{matrix} \right\} \cdot \left\{ \begin{matrix} \text{protein per unit} \\ \text{of nonfat milk} \end{matrix} \right\}$$

gives the amount of protein supplied by x_1 units of nonfat milk. To this amount, we would then add similar products for soy flour and whey and set the resulting sum equal to the amount of protein we need. Analogous calculations would have to be made for each nutrient.

A more efficient method, and one that is conceptually simpler, is to consider a "nutrient vector" for each foodstuff and build just one vector equation. The amount of nutrients supplied by x_1 units of nonfat milk is the scalar multiple

$$\overset{\text{Scalar}}{\left\{ \begin{matrix} x_1 \text{ units of} \\ \text{nonfat milk} \end{matrix} \right\}} \cdot \overset{\text{Vector}}{\left\{ \begin{matrix} \text{nutrients per unit} \\ \text{of nonfat milk} \end{matrix} \right\}} = x_1 \mathbf{a}_1 \tag{1}$$

where $\mathbf{a}_1$ is the first column in Table 1. Let $\mathbf{a}_2$ and $\mathbf{a}_3$ be the corresponding vectors for soy flour and whey, respectively, and let $\mathbf{b}$ be the vector that lists the total nutrients required (the last column of the table). Then $x_2 \mathbf{a}_2$ and $x_3 \mathbf{a}_3$ give the nutrients supplied by x_2 units of soy flour and x_3 units of whey, respectively. So the relevant equation is

$$x_1 \mathbf{a}_1 + x_2 \mathbf{a}_2 + x_3 \mathbf{a}_3 = \mathbf{b} \tag{2}$$

Row reduction of the augmented matrix for the corresponding system of equations shows that

$$\begin{bmatrix} 36 & 51 & 13 & 33 \\ 52 & 34 & 74 & 45 \\ 0 & 7 & 1.1 & 3 \end{bmatrix} \sim \cdots \sim \begin{bmatrix} 1 & 0 & 0 & .277 \\ 0 & 1 & 0 & .392 \\ 0 & 0 & 1 & .233 \end{bmatrix}$$

To three significant digits, the diet requires .277 units of nonfat milk, .392 units of soy flour, and .233 units of whey in order to provide the desired amounts of protein, carbohydrate, and fat. ∎

It is important that the values of x_1, x_2, and x_3 found above are nonnegative. This is necessary for the solution to be physically feasible. (How could you use $-.233$ units of whey, for instance?) With a large number of nutrient requirements, it may be necessary to use a larger number of foodstuffs in order to produce a system of equations with a "nonnegative" solution. Thus many, many different combinations of foodstuffs may need to be examined in order to find a system of equations with such a solution. In fact, the manufacturer of the Cambridge Diet was able to supply 31 nutrients in precise amounts using only 33 ingredients.

The diet construction problem leads to the *linear* equation (2) because the amount of nutrients supplied by each foodstuff can be written as a scalar multiple of a vector, as in (1). That is, the nutrients supplied by a foodstuff are *proportional* to the amount of the foodstuff added to the diet mixture. Also, each nutrient in the mixture is the *sum* of the amounts from the various foodstuffs.

Problems of formulating specialized diets for humans and livestock occur frequently. Usually they are treated by linear programming techniques. Our method of constructing vector equations often simplifies the task of formulating such problems.

Linear Equations and Electrical Networks

Current flow in a simple electrical network can be described by a system of linear equations. A voltage source such as a battery forces a current of electrons to flow through the network. When the current passes through a resistor (such as a lightbulb or motor), some of the voltage is "used up"; by Ohm's law, this "voltage drop" across a resistor is given by

$$V = RI$$

where the voltage V is measured in *volts*, the resistance R in *ohms* (denoted by Ω), and the current flow I in *amperes* (*amps*, for short).

The network in Figure 1 contains three closed loops. The currents flowing in loops 1, 2, and 3 are denoted by I_1, I_2, and I_3, respectively. The designated directions of such *loop currents* are arbitrary. If a current turns out to be negative, then the actual direction of current flow is opposite to that chosen in the figure. If the current direction shown is away from the positive (longer) side of a battery (-|⊢) around to the negative (shorter) side, the voltage is positive; otherwise, the voltage is negative.

Current flow in a loop is governed by the following rule.

KIRCHHOFF'S VOLTAGE LAW

The algebraic sum of the RI voltage drops in one direction around a loop equals the algebraic sum of the voltage sources in the same direction around the loop.

EXAMPLE 2   Determine the loop currents in the network in Figure 1.

SOLUTION   For loop 1, the current I_1 flows through three resistors, and the sum of the RI voltage drops is

$$4I_1 + 4I_1 + 3I_1 = (4 + 4 + 3)I_1 = 11I_1$$

Current from loop 2 also flows in part of loop 1, through the short *branch* between A and B. The associated RI drop there is $3I_2$ volts. However, the current direction for the branch AB in loop 1 is opposite to that chosen for the flow in loop 2, so the algebraic sum of all RI drops for loop 1 is $11I_1 - 3I_2$. Since the voltage in loop 1 is $+30$ volts, Kirchhoff's voltage law implies that

$$11I_1 - 3I_2 = 30$$

FIGURE 1

The equation for loop 2 is

$$-3I_1 + 6I_2 - I_3 = 5$$

The term $-3I_1$ comes from the flow of the loop 1 current through the branch AB (with a negative voltage drop because the current flow there is opposite to the flow in loop 2). The term $6I_2$ is the sum of all resistances in loop 2, multiplied by the loop current. The term $-I_3 = -1 \cdot I_3$ comes from the loop 3 current flowing through the 1-ohm resistor in branch CD, in the direction opposite to the flow in loop 2. The loop 3 equation is

$$-I_2 + 3I_3 = -25$$

Note that the 5-volt battery in branch CD is counted as part of both loop 2 and loop 3, but it is -5 volts for loop 3 because of the direction chosen for the current in loop 3. The 20-volt battery is negative for the same reason.

The loop currents are found by solving the system

$$\begin{aligned} 11I_1 - 3I_2 &= 30 \\ -3I_1 + 6I_2 - I_3 &= 5 \\ -I_2 + 3I_3 &= -25 \end{aligned} \tag{3}$$

Row operations on the augmented matrix lead to the solution: $I_1 = 3$ amps, $I_2 = 1$ amp, and $I_3 = -8$ amps. The negative value of I_3 indicates that the actual current in loop 3 flows in the direction opposite to that shown in Figure 1. ∎

It is instructive to look at system (3) as a vector equation:

$$I_1 \begin{bmatrix} 11 \\ -3 \\ 0 \end{bmatrix} + I_2 \begin{bmatrix} -3 \\ 6 \\ -1 \end{bmatrix} + I_3 \begin{bmatrix} 0 \\ -1 \\ 3 \end{bmatrix} = \begin{bmatrix} 30 \\ 5 \\ -25 \end{bmatrix} \tag{4}$$
$$\quad\;\; \uparrow \qquad\quad \uparrow \qquad\quad \uparrow \qquad\quad \uparrow$$
$$\quad\;\; \mathbf{r}_1 \qquad\quad \mathbf{r}_2 \qquad\quad \mathbf{r}_3 \qquad\quad \mathbf{v}$$

The first entry of each vector concerns the first loop, and similarly for the second and third entries. The first resistor vector $\mathbf{r}_1$ lists the resistance in the various loops through which current I_1 flows. A resistance is written negatively when I_1 flows against the flow direction in another loop. Examine Figure 1 and see how to compute the entries in $\mathbf{r}_1$; then do the same for $\mathbf{r}_2$ and $\mathbf{r}_3$. The matrix form of equation (4),

$$R\mathbf{i} = \mathbf{v}, \quad \text{where} \quad R = [\,\mathbf{r}_1 \quad \mathbf{r}_2 \quad \mathbf{r}_3\,] \quad \text{and} \quad \mathbf{i} = \begin{bmatrix} I_1 \\ I_2 \\ I_3 \end{bmatrix}$$

provides a matrix version of Ohm's law. If all loop currents are chosen in the same direction (say, counterclockwise), then all entries off the main diagonal of R will be negative.

The matrix equation $R\mathbf{i} = \mathbf{v}$ makes the linearity of this model easy to see at a glance. For instance, if the voltage vector is doubled, then the current vector must double. Also, a *superposition principle* holds. That is, the solution of equation (4) is the sum of the solutions of the equations

$$R\mathbf{i} = \begin{bmatrix} 30 \\ 0 \\ 0 \end{bmatrix}, \quad R\mathbf{i} = \begin{bmatrix} 0 \\ 5 \\ 0 \end{bmatrix}, \quad \text{and} \quad R\mathbf{i} = \begin{bmatrix} 0 \\ 0 \\ -25 \end{bmatrix}$$

Each equation here corresponds to the circuit with only one voltage source (the other sources being replaced by wires that close each loop). The model for current flow is *linear* precisely because Ohm's law and Kirchhoff's law are linear: The voltage drop across a resistor is *proportional* to the current flowing through it (Ohm), and the *sum* of the voltage drops in a loop equals the sum of the voltage sources in the loop (Kirchhoff).

Loop currents in a network can be used to determine the current in any branch of the network. If only one loop current passes through a branch, such as from B to D in Figure 1, the branch current equals the loop current. If more than one loop current passes through a branch, such as from A to B, the branch current is the algebraic sum of the loop currents in the branch (*Kirchhoff's current law*). For instance, the current in branch AB is $I_1 - I_2 = 3 - 1 = 2$ amps, in the direction of I_1. The current in branch CD is $I_2 - I_3 = 9$ amps.

Difference Equations

In many fields, such as ecology, economics, and engineering, a need arises to model mathematically a dynamic system that changes over time. Several features of the system are each measured at discrete time intervals, producing a sequence of vectors $\mathbf{x}_0$, $\mathbf{x}_1$, $\mathbf{x}_2, \ldots$. The entries in $\mathbf{x}_k$ provide information about the *state* of the system at the time of the kth measurement.

If there is a matrix A such that $\mathbf{x}_1 = A\mathbf{x}_0$, $\mathbf{x}_2 = A\mathbf{x}_1$, and, in general,

$$\mathbf{x}_{k+1} = A\mathbf{x}_k \quad \text{for } k = 0, 1, 2, \ldots \tag{5}$$

then (5) is called a **linear difference equation** (or **recurrence relation**). Given such an equation, one can compute $\mathbf{x}_1$, $\mathbf{x}_2$, and so on, provided $\mathbf{x}_0$ is known. Sections 4.8 and several sections in Chapter 5 will develop formulas for $\mathbf{x}_k$ and describe what can happen to $\mathbf{x}_k$ as k increases indefinitely. The discussion below illustrates how a difference equation might arise.

A subject of interest to demographers is the movement of populations or groups of people from one region to another. The simple model here considers the changes in the population of a certain city and its surrounding suburbs over a period of years.

Fix an initial year—say, 2020—and denote the populations of the city and suburbs that year by r_0 and s_0, respectively. Let $\mathbf{x}_0$ be the population vector

$$\mathbf{x}_0 = \begin{bmatrix} r_0 \\ s_0 \end{bmatrix} \quad \begin{array}{l} \text{City population, 2020} \\ \text{Suburban population, 2020} \end{array}$$

For 2021 and subsequent years, denote the populations of the city and suburbs by the vectors

$$\mathbf{x}_1 = \begin{bmatrix} r_1 \\ s_1 \end{bmatrix}, \qquad \mathbf{x}_2 = \begin{bmatrix} r_2 \\ s_2 \end{bmatrix}, \qquad \mathbf{x}_3 = \begin{bmatrix} r_3 \\ s_3 \end{bmatrix}, \ldots$$

Our goal is to describe mathematically how these vectors might be related.

Suppose demographic studies show that each year about 5% of the city's population moves to the suburbs (and 95% remains in the city), while 3% of the suburban population moves to the city (and 97% remains in the suburbs). See Figure 2.

After 1 year, the original r_0 persons in the city are now distributed between city and suburbs as

$$\begin{bmatrix} .95r_0 \\ .05r_0 \end{bmatrix} = r_0 \begin{bmatrix} .95 \\ .05 \end{bmatrix} \quad \begin{array}{l} \text{Remain in city} \\ \text{Move to suburbs} \end{array} \tag{6}$$

The s_0 persons in the suburbs in 2020 are distributed 1 year later as

$$s_0 \begin{bmatrix} .03 \\ .97 \end{bmatrix} \quad \begin{array}{l} \text{Move to city} \\ \text{Remain in suburbs} \end{array} \tag{7}$$

FIGURE 2 Annual percentage migration between city and suburbs.

The vectors in (6) and (7) account for all of the population in 2021.[3] Thus

$$\begin{bmatrix} r_1 \\ s_1 \end{bmatrix} = r_0 \begin{bmatrix} .95 \\ .05 \end{bmatrix} + s_0 \begin{bmatrix} .03 \\ .97 \end{bmatrix} = \begin{bmatrix} .95 & .03 \\ .05 & .97 \end{bmatrix} \begin{bmatrix} r_0 \\ s_0 \end{bmatrix}$$

That is,

$$\mathbf{x}_1 = M\mathbf{x}_0 \tag{8}$$

where M is the **migration matrix** determined by the following table:

From:
City Suburbs To:
$$\begin{bmatrix} .95 & .03 \\ .05 & .97 \end{bmatrix}$$ City
Suburbs

Equation (8) describes how the population changes from 2020 to 2021. If the migration percentages remain constant, then the change from 2021 to 2022 is given by

$$\mathbf{x}_2 = M\mathbf{x}_1$$

and similarly for 2022 to 2023 and subsequent years. In general,

$$\mathbf{x}_{k+1} = M\mathbf{x}_k \quad \text{for } k = 0, 1, 2, \ldots \tag{9}$$

The sequence of vectors $\{\mathbf{x}_0, \mathbf{x}_1, \mathbf{x}_2, \ldots\}$ describes the population of the city/suburban region over a period of years.

EXAMPLE 3 Compute the population of the region just described for the years 2021 and 2022, given that the population in 2020 was 600,000 in the city and 400,000 in the suburbs.

SOLUTION The initial population in 2020 is $\mathbf{x}_0 = \begin{bmatrix} 600,000 \\ 400,000 \end{bmatrix}$. For 2021,

$$\mathbf{x}_1 = \begin{bmatrix} .95 & .03 \\ .05 & .97 \end{bmatrix} \begin{bmatrix} 600,000 \\ 400,000 \end{bmatrix} = \begin{bmatrix} 582,000 \\ 418,000 \end{bmatrix}$$

For 2022,

$$\mathbf{x}_2 = M\mathbf{x}_1 = \begin{bmatrix} .95 & .03 \\ .05 & .97 \end{bmatrix} \begin{bmatrix} 582,000 \\ 418,000 \end{bmatrix} = \begin{bmatrix} 565,440 \\ 434,560 \end{bmatrix} \qquad ■$$

---

[3] For simplicity, we ignore other influences on the population such as births, deaths, and migration into and out of the city/suburban region.

The model for population movement in (9) is *linear* because the correspondence $\mathbf{x}_k \mapsto \mathbf{x}_{k+1}$ is a linear transformation. The linearity depends on two facts: the number of people who chose to move from one area to another is *proportional* to the number of people in that area, as shown in (6) and (7), and the cumulative effect of these choices is found by *adding* the movement of people from the different areas.

Practice Problem

Find a matrix A and vectors $\mathbf{x}$ and $\mathbf{b}$ such that the problem in Example 1 amounts to solving the equation $A\mathbf{x} = \mathbf{b}$.

1.10 Exercises

1. The container of a breakfast cereal usually lists the number of calories and the amounts of protein, carbohydrate, and fat contained in one serving of the cereal. The amounts for two common cereals are given below. Suppose a mixture of these two cereals is to be prepared that contains exactly 295 calories, 9 g of protein, 48 g of carbohydrate, and 8 g of fat.

   a. Set up a vector equation for this problem. Include a statement of what the variables in your equation represent.

   b. Write an equivalent matrix equation, and then determine if the desired mixture of the two cereals can be prepared.

   Nutrition Information per Serving

Nutrient	General Mills Cheerios®	Quaker® 100% Natural Cereal
Calories	110	130
Protein (g)	4	3
Carbohydrate (g)	20	18
Fat (g)	2	5

2. One serving of Post Shredded Wheat® supplies 160 calories, 5 g of protein, 6 g of fiber, and 1 g of fat. One serving of Crispix® supplies 110 calories, 2 g of protein, .1 g of fiber, and .4 g of fat.

   a. Set up a matrix B and a vector $\mathbf{u}$ such that $B\mathbf{u}$ gives the amounts of calories, protein, fiber, and fat contained in a mixture of three servings of Shredded Wheat and two servings of Crispix.

   T b. Suppose that you want a cereal with more fiber than Crispix but fewer calories than Shredded Wheat. Is it possible for a mixture of the two cereals to supply 130 calories, 3.20 g of protein, 2.46 g of fiber, and .64 g of fat? If so, what is the mixture?

3. After taking a nutrition class, a big Annie's® Mac and Cheese fan decides to improve the levels of protein and fiber in her favorite lunch by adding broccoli and canned chicken. The nutritional information for the foods referred to in this are given in the table.

   Nutrition Information per Serving

Nutrient	Mac and Cheese	Broccoli	Chicken	Shells
Calories	270	51	70	260
Protein (g)	10	5.4	15	9
Fiber (g)	2	5.2	0	5

   T a. If she wants to limit her lunch to 400 calories but get 30 g of protein and 10 g of fiber, what proportions of servings of Mac and Cheese, broccoli, and chicken should she use?

   T b. She found that there was too much broccoli in the proportions from part (a), so she decided to switch from classical Mac and Cheese to Annie's® Whole Wheat Shells and White Cheddar. What proportions of servings of each food should she use to meet the same goals as in part (a)?

4. The Cambridge Diet supplies .8 g of calcium per day, in addition to the nutrients listed in Table 1 for Example 1. The amounts of calcium per unit (100 g) supplied by the three ingredients in the Cambridge Diet are as follows: 1.26 g from nonfat milk, .19 g from soy flour, and .8 g from whey. Another ingredient in the diet mixture is isolated soy protein, which provides the following nutrients in each unit: 80 g of protein, 0 g of carbohydrate, 3.4 g of fat, and .18 g of calcium.

   a. Set up a matrix equation whose solution determines the amounts of nonfat milk, soy flour, whey, and isolated soy protein necessary to supply the precise amounts of protein, carbohydrate, fat, and calcium in the Cambridge Diet. State what the variables in the equation represent.

   T b. Solve the equation in (a) and discuss your answer.

T In Exercises 5–8, write a matrix equation that determines the loop currents. If MATLAB or another matrix program is available, solve the system for the loop currents.

5.

6.

7.

8.

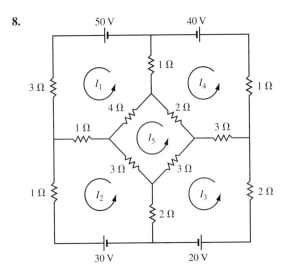

9. In a certain region, about 7% of a city's population moves to the surrounding suburbs each year, and about 5% of the suburban population moves into the city. In 2020, there were 800,000 residents in the city and 500,000 in the suburbs. Set up a difference equation that describes this situation, where x_0 is the initial population in 2020. Then estimate

the populations in the city and in the suburbs two years later, in 2022. (Ignore other factors that might influence the population sizes.)

10. In a certain region, about 6% of a city's population moves to the surrounding suburbs each year, and about 4% of the suburban population moves into the city. In 2020, there were 10,000,000 residents in the city and 800,000 in the suburbs. Set up a difference equation that describes this situation, where x_0 is the initial population in 2020. Then estimate the populations in the city and in the suburbs two years later, in 2022.

T 11. College Moving Truck Rental has a fleet of 20, 100, and 200 trucks in Pullman, Spokane, and Seattle, respectively. A truck rented at one location may be returned to any of the three locations. The various fractions of trucks returned to the three locations each month are shown in the matrix below. What will be the approximate distribution of the trucks after three months?

Trucks Rented From:

Pullman	Spokane	Seattle	Returned To:
.30	.15	.05	Airport
.30	.70	.05	East
.40	.15	.90	West

T 12. Budget® Rent a Car in Wichita, Kansas, has a fleet of about 500 cars, at three locations. A car rented at one location may be returned to any of the three locations. The various fractions of cars returned to the three locations are shown in the matrix below. Suppose that on Monday there are 295 cars at the airport (or rented from there), 55 cars at the east side office, and 150 cars at the west side office. What will be the approximate distribution of cars on Wednesday?

Cars Rented From:

Airport	East	West	Returned To:
.97	.05	.10	Airport
.00	.90	.05	East
.03	.05	.85	West

T 13. Let M and x_0 be as in Example 3.
   a. Compute the population vectors x_k for $k = 1, \ldots, 20$. Discuss what you find.
   b. Repeat part (a) with an initial population of 350,000 in the city and 650,000 in the suburbs. What do you find?

T 14. Study how changes in boundary temperatures on a steel plate affect the temperatures at interior points on the plate.
   a. Begin by estimating the temperatures T_1, T_2, T_3, T_4 at each of the sets of four points on the steel plate shown in the figure. In each case, the value of T_k is approximated by the average of the temperatures at the four closest points. See Exercises 43 and 44 in Section 1.1, where the values

(in degrees) turn out to be (20, 27.5, 30, 22.5). How is this list of values related to your results for the points in set (a) and set (b)?

b. Without making any computations, guess the interior temperatures in (a) when the boundary temperatures are all multiplied by 3. Check your guess.

c. Finally, make a general conjecture about the correspondence from the list of eight boundary temperatures to the list of four interior temperatures.

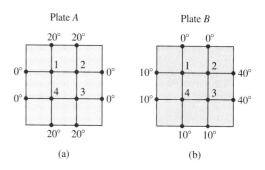

(a)   (b)

Solution to Practice Problem

$$A = \begin{bmatrix} 36 & 51 & 13 \\ 52 & 34 & 74 \\ 0 & 7 & 1.1 \end{bmatrix}, \quad \mathbf{x} = \begin{bmatrix} x_1 \\ x_2 \\ x_3 \end{bmatrix}, \quad \mathbf{b} = \begin{bmatrix} 33 \\ 45 \\ 3 \end{bmatrix}$$

CHAPTER 1 PROJECTS

Chapter 1 projects are available online at `bit.ly/30IM8gT`.

A. *Interpolating Polynomials*: This project shows how to use a system of linear equations to fit a polynomial through a set of points.

B. *Splines*: This project also shows how to use a system of linear equations to fit a piecewise polynomial curve through a set of points.

C. *Network Flows*: The purpose of this project is to show how systems of linear equations may be used to model flow through a network.

D. *The Art of Linear Transformations*: In this project, it is illustrated how to graph a polygon and then use linear transformations to change its shape and create a design.

E. *Loop Currents*: The purpose of this project is to provide more and larger examples of loop currents.

F. *Diet*: The purpose of this project is to provide examples of vector equations that result from balancing nutrients in a diet.

CHAPTER 1 SUPPLEMENTARY EXERCISES

Mark each statement True or False (**T/ F**). Justify each answer. (If true, cite appropriate facts or theorems. If false, explain why or give a counterexample that shows why the statement is not true in every case.

1. (**T/F**) Every matrix is row equivalent to a unique matrix in echelon form.

2. (**T/F**) Any system of n linear equations in n variables has at most n solutions.

3. (**T/F**) If a system of linear equations has two different solutions, it must have infinitely many solutions.

4. (**T/F**) If a system of linear equations has no free variables, then it has a unique solution.

5. (**T/F**) If an augmented matrix $[\,A \quad \mathbf{b}\,]$ is transformed into $[\,C \quad \mathbf{d}\,]$ by elementary row operations, then the equations $A\mathbf{x} = \mathbf{b}$ and $C\mathbf{x} = \mathbf{d}$ have exactly the same solution sets.

6. (**T/F**) If a system $A\mathbf{x} = \mathbf{b}$ has more than one solution, then so does the system $A\mathbf{x} = \mathbf{0}$.

7. (**T/F**) If A is an $m \times n$ matrix and the equation $A\mathbf{x} = \mathbf{b}$ is consistent for some $\mathbf{b}$, then the columns of A span $\mathbb{R}^m$.

8. (**T/F**) If an augmented matrix $[\,A \quad \mathbf{b}\,]$ can be transformed by elementary row operations into reduced echelon form, then the equation $A\mathbf{x} = \mathbf{b}$ is consistent.

9. (**T/F**) If matrices A and B are row equivalent, they have the same reduced echelon form.

10. (**T/F**) The equation $A\mathbf{x} = \mathbf{0}$ has the trivial solution if and only if there are no free variables.

11. (**T/F**) If A is an $m \times n$ matrix and the equation $A\mathbf{x} = \mathbf{b}$ is consistent for every $\mathbf{b}$ in $\mathbb{R}^m$, then A has m pivot columns.

12. (T/F) If an $m \times n$ matrix A has a pivot position in every row, then the equation $A\mathbf{x} = \mathbf{b}$ has a unique solution for each $\mathbf{b}$ in $\mathbb{R}^m$.

13. (T/F) If an $n \times n$ matrix A has n pivot positions, then the reduced echelon form of A is the $n \times n$ identity matrix.

14. (T/F) If 3×3 matrices A and B each have three pivot positions, then A can be transformed into B by elementary row operations.

15. (T/F) If A is an $m \times n$ matrix, if the equation $A\mathbf{x} = \mathbf{b}$ has at least two different solutions, and if the equation $A\mathbf{x} = \mathbf{c}$ is consistent, then the equation $A\mathbf{x} = \mathbf{c}$ has many solutions.

16. (T/F) If A and B are row equivalent $m \times n$ matrices and if the columns of A span $\mathbb{R}^m$, then so do the columns of B.

17. (T/F) If none of the vectors in the set $S = \{\mathbf{v}_1, \mathbf{v}_2, \mathbf{v}_3\}$ in $\mathbb{R}^3$ is a multiple of one of the other vectors, then S is linearly independent.

18. (T/F) If $\{\mathbf{u}, \mathbf{v}, \mathbf{w}\}$ is linearly independent, then $\mathbf{u}, \mathbf{v}$, and $\mathbf{w}$ are not in $\mathbb{R}^2$.

19. (T/F) In some cases, it is possible for four vectors to span $\mathbb{R}^5$.

20. (T/F) If $\mathbf{u}$ and $\mathbf{v}$ are in $\mathbb{R}^m$, then $-\mathbf{u}$ is in Span$\{\mathbf{u}, \mathbf{v}\}$.

21. (T/F) If $\mathbf{u}, \mathbf{v}$, and $\mathbf{w}$ are nonzero vectors in $\mathbb{R}^2$, then $\mathbf{w}$ is a linear combination of $\mathbf{u}$ and $\mathbf{v}$.

22. (T/F) If $\mathbf{w}$ is a linear combination of $\mathbf{u}$ and $\mathbf{v}$ in $\mathbb{R}^n$, then $\mathbf{u}$ is a linear combination of $\mathbf{v}$ and $\mathbf{w}$.

23. (T/F) Suppose that $\mathbf{v}_1, \mathbf{v}_2$, and $\mathbf{v}_3$ are in $\mathbb{R}^5$, $\mathbf{v}_2$ is not a multiple of $\mathbf{v}_1$, and $\mathbf{v}_3$ is not a linear combination of $\mathbf{v}_1$ and $\mathbf{v}_2$. Then $\{\mathbf{v}_1, \mathbf{v}_2, \mathbf{v}_3\}$ is linearly independent.

24. (T/F) A linear transformation is a function.

25. (T/F) If A is a 6×5 matrix, the linear transformation $\mathbf{x} \mapsto A\mathbf{x}$ cannot map $\mathbb{R}^5$ onto $\mathbb{R}^6$.

26. Let a and b represent real numbers. Describe the possible solution sets of the (linear) equation $ax = b$. [*Hint:* The number of solutions depends upon a and b.]

27. The solutions (x, y, z) of a single linear equation

$$ax + by + cz = d$$

form a plane in $\mathbb{R}^3$ when a, b, and c are not all zero. Construct sets of three linear equations whose graphs (a) intersect in a single line, (b) intersect in a single point, and (c) have

no points in common. Typical graphs are illustrated in the figure.

Three planes intersecting in a line
(a)

Three planes intersecting in a point
(b)

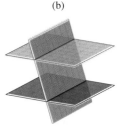

Three planes with no intersection
(c)

Three planes with no intersection
(c')

28. Suppose the coefficient matrix of a linear system of three equations in three variables has a pivot position in each column. Explain why the system has a unique solution.

29. Determine h and k such that the solution set of the system (i) is empty, (ii) contains a unique solution, and (iii) contains infinitely many solutions.

a. $\quad x_1 + 3x_2 = k$ b. $\quad -2x_1 + hx_2 = 1$
$\quad\quad 4x_1 + hx_2 = 8$ $6x_1 + kx_2 = -2$

30. Consider the problem of determining whether the following system of equations is consistent:

$$4x_1 - 2x_2 + 7x_3 = -5$$
$$8x_1 - 3x_2 + 10x_3 = -3$$

a. Define appropriate vectors, and restate the problem in terms of linear combinations. Then solve that problem.

b. Define an appropriate matrix, and restate the problem using the phrase "columns of A."

c. Define an appropriate linear transformation T using the matrix in (b), and restate the problem in terms of T.

31. Consider the problem of determining whether the following system of equations is consistent for all b_1, b_2, b_3:

$$2x_1 - 4x_2 - 2x_3 = b_1$$
$$-5x_1 + x_2 + x_3 = b_2$$
$$7x_1 - 5x_2 - 3x_3 = b_3$$

a. Define appropriate vectors, and restate the problem in terms of Span $\{\mathbf{v}_1, \mathbf{v}_2, \mathbf{v}_3\}$. Then solve that problem.

b. Define an appropriate matrix, and restate the problem using the phrase "columns of A."

c. Define an appropriate linear transformation T using the matrix in (b), and restate the problem in terms of T.

32. Describe the possible echelon forms of the matrix A. Use the notation of Example 1 in Section 1.2.

a. A is a 2×3 matrix whose columns span $\mathbb{R}^2$.

b. A is a 3×3 matrix whose columns span $\mathbb{R}^3$.

33. Write the vector $\begin{bmatrix} 5 \\ 6 \end{bmatrix}$ as the sum of two vectors, one on the line $\{(x, y) : y = 2x\}$ and one on the line $\{(x, y) : y = x/2\}$.

34. Let $\mathbf{a}_1$, $\mathbf{a}_2$, and $\mathbf{b}$ be the vectors in $\mathbb{R}^2$ shown in the figure, and let $A = [\mathbf{a}_1 \quad \mathbf{a}_2]$. Does the equation $A\mathbf{x} = \mathbf{b}$ have a solution? If so, is the solution unique? Explain.

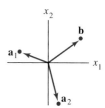

35. Construct a 2×3 matrix A, not in echelon form, such that the solution of $A\mathbf{x} = \mathbf{0}$ is a line in $\mathbb{R}^3$.

36. Construct a 2×3 matrix A, not in echelon form, such that the solution of $A\mathbf{x} = \mathbf{0}$ is a plane in $\mathbb{R}^3$.

37. Write the *reduced* echelon form of a 3×3 matrix A such that the first two columns of A are pivot columns and
$$A \begin{bmatrix} 3 \\ -2 \\ 1 \end{bmatrix} = \begin{bmatrix} 0 \\ 0 \\ 0 \end{bmatrix}.$$

38. Determine the value(s) of a such that $\left\{ \begin{bmatrix} 1 \\ a \end{bmatrix}, \begin{bmatrix} a \\ a+2 \end{bmatrix} \right\}$ is linearly independent.

39. In (a) and (b), suppose the vectors are linearly independent. What can you say about the numbers $a, \ldots, f$? Justify your answers. [*Hint:* Use a theorem for (b).]

a. $\begin{bmatrix} a \\ 0 \\ 0 \end{bmatrix}, \begin{bmatrix} b \\ c \\ 0 \end{bmatrix}, \begin{bmatrix} d \\ e \\ f \end{bmatrix}$ b. $\begin{bmatrix} a \\ 1 \\ 0 \\ 0 \end{bmatrix}, \begin{bmatrix} b \\ c \\ 1 \\ 0 \end{bmatrix}, \begin{bmatrix} d \\ e \\ f \\ 1 \end{bmatrix}$

40. Use Theorem 7 in Section 1.7 to explain why the columns of the matrix A are linearly independent.
$$A = \begin{bmatrix} 1 & 0 & 0 & 0 \\ 2 & 5 & 0 & 0 \\ 3 & 6 & 8 & 0 \\ 4 & 7 & 9 & 10 \end{bmatrix}$$

41. Explain why a set $\{\mathbf{v}_1, \mathbf{v}_2, \mathbf{v}_3, \mathbf{v}_4\}$ in $\mathbb{R}^5$ must be linearly independent when $\{\mathbf{v}_1, \mathbf{v}_2, \mathbf{v}_3\}$ is linearly independent and $\mathbf{v}_4$ is *not* in Span $\{\mathbf{v}_1, \mathbf{v}_2, \mathbf{v}_3\}$.

42. Suppose $\{\mathbf{v}_1, \mathbf{v}_2\}$ is a linearly independent set in $\mathbb{R}^n$. Show that $\{\mathbf{v}_1, \mathbf{v}_1 + \mathbf{v}_2\}$ is also linearly independent.

43. Suppose $\mathbf{v}_1, \mathbf{v}_2, \mathbf{v}_3$ are distinct points on one line in $\mathbb{R}^3$. The line need not pass through the origin. Show that $\{\mathbf{v}_1, \mathbf{v}_2, \mathbf{v}_3\}$ is linearly dependent.

44. Let $T : \mathbb{R}^n \to \mathbb{R}^m$ be a linear transformation, and suppose $T(\mathbf{u}) = \mathbf{v}$. Show that $T(-\mathbf{u}) = -\mathbf{v}$.

45. Let $T : \mathbb{R}^3 \to \mathbb{R}^3$ be the linear transformation that reflects each vector through the plane $x_2 = 0$. That is, $T(x_1, x_2, x_3) = (x_1, -x_2, x_3)$. Find the standard matrix of T.

46. Let A be a 3×3 matrix with the property that the linear transformation $\mathbf{x} \mapsto A\mathbf{x}$ maps $\mathbb{R}^3$ onto $\mathbb{R}^3$. Explain why the transformation must be one-to-one.

47. A *Givens rotation* is a linear transformation from $\mathbb{R}^n$ to $\mathbb{R}^n$ used in computer programs to create a zero entry in a vector (usually a column of a matrix). The standard matrix of a Givens rotation in $\mathbb{R}^2$ has the form
$$\begin{bmatrix} a & -b \\ b & a \end{bmatrix}, \qquad a^2 + b^2 = 1$$

Find a and b such that $\begin{bmatrix} 4 \\ 3 \end{bmatrix}$ is rotated into $\begin{bmatrix} 5 \\ 0 \end{bmatrix}$.

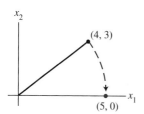

A Givens rotation in $\mathbb{R}^2$.

48. The following equation describes a Givens rotation in $\mathbb{R}^3$. Find a and b.
$$\begin{bmatrix} a & 0 & -b \\ 0 & 1 & 0 \\ b & 0 & a \end{bmatrix} \begin{bmatrix} 2 \\ 3 \\ 4 \end{bmatrix} = \begin{bmatrix} 2\sqrt{5} \\ 3 \\ 0 \end{bmatrix}, \qquad a^2 + b^2 = 1$$

49. A large apartment building is to be built using modular construction techniques. The arrangement of apartments on any particular floor is to be chosen from one of three basic floor plans. Plan A has 18 apartments on one floor, including 3 three-bedroom units, 7 two-bedroom units, and 8 one-bedroom units. Each floor of plan B includes 4 three-bedroom units, 4 two-bedroom units, and 8 one-bedroom units. Each floor of plan C includes 5 three-bedroom units,

3 two-bedroom units, and 9 one-bedroom units. Suppose the building contains a total of x_1 floors of plan A, x_2 floors of plan B, and x_3 floors of plan C.

a. What interpretation can be given to the vector $x_1 \begin{bmatrix} 3 \\ 7 \\ 8 \end{bmatrix}$?

b. Write a formal linear combination of vectors that expresses the total numbers of three-, two-, and one-bedroom apartments contained in the building.

T c. Is it possible to design the building with exactly 66 three-bedroom units, 74 two-bedroom units, and 136 one-bedroom units? If so, is there more than one way to do it? Explain your answer.

2 Matrix Algebra

COMPUTER MODELS IN AIRCRAFT DESIGN

To design the next generation of commercial and military aircraft, engineers at Boeing's Phantom Works use 3D modeling and computational fluid dynamics (CFD). They study the airflow around a virtual airplane to answer important design questions before physical models are created. This has drastically reduced design cycle times and cost—and linear algebra plays a crucial role in the process.

The virtual airplane begins as a mathematical "wire-frame" model that exists only in computer memory and on graphics display terminals. (Model of a Boeing 747 is shown.) This mathematical model organizes and influences each step of the design and manufacture of the airplane—both the exterior and interior. The CFD analysis concerns the exterior surface.

Although the finished skin of a plane may seem smooth, the geometry of the surface is complicated. In addition to wings and a fuselage, an aircraft has nacelles, stabilizers, slats, flaps, and ailerons. The way air flows around these structures determines how the plane moves through the sky. Equations that describe the airflow are complicated, and they must account for engine intake, engine exhaust, and the wakes left by the wings of the plane. To study the airflow, engineers need a highly refined description of the plane's surface.

A computer creates a model of the surface by first superimposing a three-dimensional grid of "boxes" on the original wire-frame model. Boxes in this grid lie either completely inside or completely outside the plane, or they intersect the surface of the plane. The computer selects the boxes that intersect the surface and subdivides them, retaining only the smaller boxes that still intersect the surface. The subdividing process is repeated until the grid is extremely fine. A typical grid can include more than 400,000 boxes.

The process for finding the airflow around the plane involves repeatedly solving a system of linear equations $A\mathbf{x} = \mathbf{b}$ that may involve up to 2 million equations and variables. The vector $\mathbf{b}$ changes each time, based on data from the grid and solutions of previous equations. Using the fastest computers available commercially, a Phantom Works team can spend from a few hours to several days setting up and solving a single airflow problem. After the team analyzes the solution, they may make small changes to the airplane surface and begin the whole process again. Thousands of CFD runs may be required.

This chapter presents two important concepts that assist in the solution of such massive systems of equations:

- *Partitioned matrices:* A typical CFD system of equations has a "sparse" coefficient matrix with mostly zero entries. Grouping the variables correctly leads to a partitioned matrix with many zero blocks. Section 2.4 introduces such matrices and describes some of their applications.

- *Matrix factorizations:* Even when written with partitioned matrices, the system of equations is complicated. To further simplify the computations, the CFD software at Boeing uses what is called an LU factorization of the coefficient matrix. Section 2.5 discusses LU and other useful matrix factorizations. Further details about factorizations appear at several points later in the text.

To analyze a solution of an airflow system, engineers want to visualize the airflow over the surface of the plane. They use computer graphics, and linear algebra provides the engine for the graphics. The wire-frame model of the plane's surface is stored as data in many matrices. Once the image has been rendered on a computer screen, engineers can change its scale, zoom in or out of small regions, and rotate the image to see parts that may be hidden from view.

TU-Delft and Air France-KLM are investigating a flying V aircraft design because of its potential for significantly better fuel economy.

Each of these operations is accomplished by appropriate matrix multiplications. Section 2.7 explains the basic ideas.

Our ability to analyze and solve equations will be greatly enhanced when we can perform algebraic operations with matrices. Furthermore, the definitions and theorems in this chapter provide some basic tools for handling the many applications of linear algebra that involve two or more matrices. For $n \times n$ matrices, the Invertible Matrix Theorem in Section 2.3 ties together most of the concepts treated earlier in the text. Sections 2.4 and 2.5 examine partitioned matrices and matrix factorizations, which appear in most modern uses of linear algebra. Sections 2.6 and 2.7 describe two interesting applications of matrix algebra: to economics and to computer graphics. Sections 2.8 and 2.9 provide readers enough information about subspaces to move directly into Chapters 5, 6, and 7, without covering Chapter 4. You may want to omit these two sections if you plan to cover Chapter 4 before moving to Chapter 5.

2.1 Matrix Operations

If A is an $m \times n$ matrix—that is, a matrix with m rows and n columns—then the scalar entry in the ith row and jth column of A is denoted by a_{ij} and is called the (i, j)-entry of A. See Figure 1. For instance, the $(3, 2)$-entry is the number a_{32} in the third row, second column. Each column of A is a list of m real numbers, which identifies a vector in $\mathbb{R}^m$. Often, these columns are denoted by $\mathbf{a}_1, \ldots, \mathbf{a}_n$, and the matrix A is written as

$$A = \begin{bmatrix} \mathbf{a}_1 & \mathbf{a}_2 & \cdots & \mathbf{a}_n \end{bmatrix}$$

Observe that the number a_{ij} is the ith entry (from the top) of the jth column vector $\mathbf{a}_j$.

The **diagonal entries** in an $m \times n$ matrix $A = [a_{ij}]$ are $a_{11}, a_{22}, a_{33}, \ldots$, and they form the **main diagonal** of A. A **diagonal matrix** is a square $n \times n$ matrix whose nondiagonal entries are zero. An example is the $n \times n$ identity matrix, I_n. An $m \times n$ matrix whose entries are all zero is a **zero matrix** and is written as 0. The size of a zero matrix is usually clear from the context.

$$A = \begin{bmatrix} a_{11} & \cdots & a_{1j} & \cdots & a_{1n} \\ \vdots & & \vdots & & \vdots \\ a_{i1} & \cdots & a_{ij} & \cdots & a_{in} \\ \vdots & & \vdots & & \vdots \\ a_{m1} & \cdots & a_{mj} & \cdots & a_{mn} \end{bmatrix} = A$$

FIGURE 1 Matrix notation.

Sums and Scalar Multiples

The arithmetic for vectors described earlier has a natural extension to matrices. We say that two matrices are **equal** if they have the same size (i.e., the same number of rows and the same number of columns) and if their corresponding columns are equal, which amounts to saying that their corresponding entries are equal. If A and B are $m \times n$ matrices, then the **sum** $A + B$ is the $m \times n$ matrix whose columns are the sums of the corresponding columns in A and B. Since vector addition of the columns is done entrywise, each entry in $A + B$ is the sum of the corresponding entries in A and B. The sum $A + B$ is defined only when A and B are the same size.

EXAMPLE 1   Let

$$A = \begin{bmatrix} 4 & 0 & 5 \\ -1 & 3 & 2 \end{bmatrix}, \qquad B = \begin{bmatrix} 1 & 1 & 1 \\ 3 & 5 & 7 \end{bmatrix}, \qquad C = \begin{bmatrix} 2 & -3 \\ 0 & 1 \end{bmatrix}$$

Then

$$A + B = \begin{bmatrix} 5 & 1 & 6 \\ 2 & 8 & 9 \end{bmatrix}$$

but $A + C$ is not defined because A and C have different sizes.   ■

If r is a scalar and A is a matrix, then the **scalar multiple** rA is the matrix whose columns are r times the corresponding columns in A. As with vectors, $-A$ stands for $(-1)A$, and $A - B$ is the same as $A + (-1)B$.

EXAMPLE 2   If A and B are the matrices in Example 1, then

$$2B = 2\begin{bmatrix} 1 & 1 & 1 \\ 3 & 5 & 7 \end{bmatrix} = \begin{bmatrix} 2 & 2 & 2 \\ 6 & 10 & 14 \end{bmatrix}$$

$$A - 2B = \begin{bmatrix} 4 & 0 & 5 \\ -1 & 3 & 2 \end{bmatrix} - \begin{bmatrix} 2 & 2 & 2 \\ 6 & 10 & 14 \end{bmatrix} = \begin{bmatrix} 2 & -2 & 3 \\ -7 & -7 & -12 \end{bmatrix}$$ ■

It was unnecessary in Example 2 to compute $A - 2B$ as $A + (-1)2B$ because the usual rules of algebra apply to sums and scalar multiples of matrices, as the following theorem shows.

THEOREM 1

Let A, B, and C be matrices of the same size, and let r and s be scalars.

a. $A + B = B + A$
b. $(A + B) + C = A + (B + C)$
c. $A + 0 = A$
d. $r(A + B) = rA + rB$
e. $(r + s)A = rA + sA$
f. $r(sA) = (rs)A$

Each equality in Theorem 1 is verified by showing that the matrix on the left side has the same size as the matrix on the right and that corresponding columns are equal. Size is no problem because A, B, and C are equal in size. The equality of columns follows immediately from analogous properties of vectors. For instance, if the jth columns of A, B, and C are $\mathbf{a}_j$, $\mathbf{b}_j$, and $\mathbf{c}_j$, respectively, then the jth columns of $(A + B) + C$ and $A + (B + C)$ are

$$(\mathbf{a}_j + \mathbf{b}_j) + \mathbf{c}_j \quad \text{and} \quad \mathbf{a}_j + (\mathbf{b}_j + \mathbf{c}_j)$$

respectively. Since these two vector sums are equal for each j, property (b) is verified.

Because of the associative property of addition, we can simply write $A + B + C$ for the sum, which can be computed either as $(A + B) + C$ or as $A + (B + C)$. The same applies to sums of four or more matrices.

Matrix Multiplication

When a matrix B multiplies a vector $\mathbf{x}$, it transforms $\mathbf{x}$ into the vector $B\mathbf{x}$. If this vector is then multiplied in turn by a matrix A, the resulting vector is $A(B\mathbf{x})$. See Figure 2.

FIGURE 2 Multiplication by B and then A.

Thus $A(B\mathbf{x})$ is produced from $\mathbf{x}$ by a *composition* of mappings—the linear transformations studied in Section 1.8. Our goal is to represent this composite mapping as multiplication by a single matrix, denoted by AB, so that

$$A(B\mathbf{x}) = (AB)\mathbf{x} \tag{1}$$

See Figure 3.

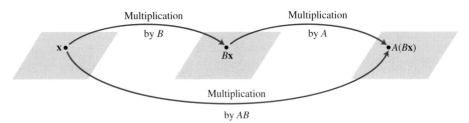

FIGURE 3 Multiplication by AB.

If A is $m \times n$, B is $n \times p$, and $\mathbf{x}$ is in $\mathbb{R}^p$, denote the columns of B by $\mathbf{b}_1, \ldots, \mathbf{b}_p$ and the entries in $\mathbf{x}$ by $x_1, \ldots, x_p$. Then

$$B\mathbf{x} = x_1\mathbf{b}_1 + \cdots + x_p\mathbf{b}_p$$

By the linearity of multiplication by A,

$$A(B\mathbf{x}) = A(x_1\mathbf{b}_1) + \cdots + A(x_p\mathbf{b}_p)$$
$$= x_1 A\mathbf{b}_1 + \cdots + x_p A\mathbf{b}_p$$

The vector $A(B\mathbf{x})$ is a linear combination of the vectors $A\mathbf{b}_1, \ldots, A\mathbf{b}_p$, using the entries in $\mathbf{x}$ as weights. In matrix notation, this linear combination is written as

$$A(B\mathbf{x}) = [\, A\mathbf{b}_1 \quad A\mathbf{b}_2 \quad \cdots \quad A\mathbf{b}_p \,]\mathbf{x}$$

Thus multiplication by $[\, A\mathbf{b}_1 \quad A\mathbf{b}_2 \quad \cdots \quad A\mathbf{b}_p \,]$ transforms $\mathbf{x}$ into $A(B\mathbf{x})$. We have found the matrix we sought!

DEFINITION

> If A is an $m \times n$ matrix, and if B is an $n \times p$ matrix with columns $\mathbf{b}_1, \ldots, \mathbf{b}_p$, then the product AB is the $m \times p$ matrix whose columns are $A\mathbf{b}_1, \ldots, A\mathbf{b}_p$. That is,
>
> $$AB = A\begin{bmatrix} \mathbf{b}_1 & \mathbf{b}_2 & \cdots & \mathbf{b}_p \end{bmatrix} = \begin{bmatrix} A\mathbf{b}_1 & A\mathbf{b}_2 & \cdots & A\mathbf{b}_p \end{bmatrix}$$

This definition makes equation (1) true for all $\mathbf{x}$ in $\mathbb{R}^p$. Equation (1) proves that the composite mapping in Figure 3 is a linear transformation and that its standard matrix is AB. *Multiplication of matrices corresponds to composition of linear transformations.*

EXAMPLE 3 Compute AB, where $A = \begin{bmatrix} 2 & 3 \\ 1 & -5 \end{bmatrix}$ and $B = \begin{bmatrix} 4 & 3 & 6 \\ 1 & -2 & 3 \end{bmatrix}$.

SOLUTION Write $B = [\, \mathbf{b}_1 \quad \mathbf{b}_2 \quad \mathbf{b}_3 \,]$, and compute:

$$A\mathbf{b}_1 = \begin{bmatrix} 2 & 3 \\ 1 & -5 \end{bmatrix}\begin{bmatrix} 4 \\ 1 \end{bmatrix}, \quad A\mathbf{b}_2 = \begin{bmatrix} 2 & 3 \\ 1 & -5 \end{bmatrix}\begin{bmatrix} 3 \\ -2 \end{bmatrix}, \quad A\mathbf{b}_3 = \begin{bmatrix} 2 & 3 \\ 1 & -5 \end{bmatrix}\begin{bmatrix} 6 \\ 3 \end{bmatrix}$$

$$= \begin{bmatrix} 11 \\ -1 \end{bmatrix} \qquad\qquad = \begin{bmatrix} 0 \\ 13 \end{bmatrix} \qquad\qquad = \begin{bmatrix} 21 \\ -9 \end{bmatrix}$$

Then

$$AB = A[\, \mathbf{b}_1 \quad \mathbf{b}_2 \quad \mathbf{b}_3 \,] = \begin{bmatrix} 11 & 0 & 21 \\ -1 & 13 & -9 \end{bmatrix}$$

$$\qquad\qquad\qquad\qquad\quad A\mathbf{b}_1 \quad A\mathbf{b}_2 \quad A\mathbf{b}_3$$

■

Notice that since the first column of AB is $A\mathbf{b}_1$, this column is a linear combination of the columns of A using the entries in $\mathbf{b}_1$ as weights. A similar statement is true for each column of AB.

> Each column of AB is a linear combination of the columns of A using weights from the corresponding column of B.

Obviously, the number of columns of A must match the number of rows in B in order for a linear combination such as $A\mathbf{b}_1$ to be defined. Also, the definition of AB shows that *AB has the same number of rows as A and the same number of columns as B.*

EXAMPLE 4 If A is a 3×5 matrix and B is a 5×2 matrix, what are the sizes of AB and BA, if they are defined?

SOLUTION Since A has 5 columns and B has 5 rows, the product AB is defined and is a 3×2 matrix:

$$
\begin{array}{ccc}
A & B & AB \\
\begin{bmatrix} * & * & * & * & * \\ * & * & * & * & * \\ * & * & * & * & * \end{bmatrix}
\begin{bmatrix} * & * \\ * & * \\ * & * \\ * & * \\ * & * \end{bmatrix}
& = &
\begin{bmatrix} * & * \\ * & * \\ * & * \end{bmatrix} \\
3 \times 5 \qquad\qquad 5 \times 2 & & 3 \times 2
\end{array}
$$

Match

Size of AB

The product BA is *not* defined because the 2 columns of B do not match the 3 rows of A. ∎

The definition of AB is important for theoretical work and applications, but the following rule provides a more efficient method for calculating the individual entries in AB when working small problems by hand.

ROW–COLUMN RULE FOR COMPUTING AB

If the product AB is defined, then the entry in row i and column j of AB is the sum of the products of corresponding entries from row i of A and column j of B. If $(AB)_{ij}$ denotes the (i, j)-entry in AB, and if A is an $m \times n$ matrix, then

$$(AB)_{ij} = a_{i1}b_{1j} + a_{i2}b_{2j} + \cdots + a_{in}b_{nj}$$

To verify this rule, let $B = [\, \mathbf{b}_1 \ \cdots \ \mathbf{b}_p \,]$. Column j of AB is $A\mathbf{b}_j$, and we can compute $A\mathbf{b}_j$ by the row–vector rule for computing $A\mathbf{x}$ from Section 1.4. The ith entry in $A\mathbf{b}_j$ is the sum of the products of corresponding entries from row i of A and the vector $\mathbf{b}_j$, which is precisely the computation described in the rule for computing the (i, j)-entry of AB.

EXAMPLE 5 Use the row–column rule to compute two of the entries in AB for the matrices in Example 3. An inspection of the numbers involved will make it clear how the two methods for calculating AB produce the same matrix.

SOLUTION To find the entry in row 1 and column 3 of AB, consider row 1 of A and column 3 of B. Multiply corresponding entries and add the results, as shown below:

$$
AB = \longrightarrow \begin{bmatrix} 2 & 3 \\ 1 & -5 \end{bmatrix}
\begin{bmatrix} 4 & 3 & 6 \\ 1 & -2 & 3 \end{bmatrix}
= \begin{bmatrix} \square & \square & 2(6) + 3(3) \\ \square & \square & \square \end{bmatrix}
= \begin{bmatrix} \square & \square & 21 \\ \square & \square & \square \end{bmatrix}
$$

For the entry in row 2 and column 2 of AB, use row 2 of A and column 2 of B:

$$
\longrightarrow \begin{bmatrix} 2 & 3 \\ 1 & -5 \end{bmatrix}
\begin{bmatrix} 4 & 3 & 6 \\ 1 & -2 & 3 \end{bmatrix}
= \begin{bmatrix} \square & \square & 21 \\ \square & 1(3) + -5(-2) & \square \end{bmatrix}
= \begin{bmatrix} \square & \square & 21 \\ \square & 13 & \square \end{bmatrix}
$$
∎

EXAMPLE 6 Find the entries in the second row of AB, where

$$A = \begin{bmatrix} 2 & -5 & 0 \\ -1 & 3 & -4 \\ 6 & -8 & -7 \\ -3 & 0 & 9 \end{bmatrix}, \qquad B = \begin{bmatrix} 4 & -6 \\ 7 & 1 \\ 3 & 2 \end{bmatrix}$$

SOLUTION By the row–column rule, the entries of the second row of AB come from row 2 of A (and the columns of B):

$$\rightarrow \begin{bmatrix} 2 & -5 & 0 \\ -1 & 3 & -4 \\ 6 & -8 & -7 \\ -3 & 0 & 9 \end{bmatrix} \begin{bmatrix} 4 & -6 \\ 7 & 1 \\ 3 & 2 \end{bmatrix}$$

$$= \begin{bmatrix} \square & \square \\ -4+21-12 & 6+3-8 \\ \square & \square \\ \square & \square \end{bmatrix} = \begin{bmatrix} \square & \square \\ 5 & 1 \\ \square & \square \\ \square & \square \end{bmatrix} \qquad \blacksquare$$

Notice that since Example 6 requested only the second row of AB, we could have written just the second row of A to the left of B and computed

$$\begin{bmatrix} -1 & 3 & -4 \end{bmatrix} \begin{bmatrix} 4 & -6 \\ 7 & 1 \\ 3 & 2 \end{bmatrix} = \begin{bmatrix} 5 & 1 \end{bmatrix}$$

This observation about rows of AB is true in general and follows from the row–column rule. Let $\text{row}_i(A)$ denote the ith row of a matrix A. Then

$$\text{row}_i(AB) = \text{row}_i(A) \cdot B \tag{2}$$

Properties of Matrix Multiplication

The following theorem lists the standard properties of matrix multiplication. Recall that I_m represents the $m \times m$ identity matrix and $I_m\mathbf{x} = \mathbf{x}$ for all $\mathbf{x}$ in $\mathbb{R}^m$.

THEOREM 2

Let A be an $m \times n$ matrix, and let B and C have sizes for which the indicated sums and products are defined.

a. $A(BC) = (AB)C$ (associative law of multiplication)
b. $A(B + C) = AB + AC$ (left distributive law)
c. $(B + C)A = BA + CA$ (right distributive law)
d. $r(AB) = (rA)B = A(rB)$
   for any scalar r
e. $I_m A = A = AI_n$ (identity for matrix multiplication)

PROOF Properties (b)–(e) are considered in the exercises. Property (a) follows from the fact that matrix multiplication corresponds to composition of linear transformations (which are functions), and it is known (or easy to check) that the composition of functions is associative. Here is another proof of (a) that rests on the "column definition" of the product of two matrices. Let

$$C = [\, \mathbf{c}_1 \; \cdots \; \mathbf{c}_p \,]$$

By the definition of matrix multiplication,

$$BC = [\, B\mathbf{c}_1 \; \cdots \; B\mathbf{c}_p \,]$$
$$A(BC) = [\, A(B\mathbf{c}_1) \; \cdots \; A(B\mathbf{c}_p) \,]$$

Recall from equation (1) that the definition of AB makes $A(B\mathbf{x}) = (AB)\mathbf{x}$ for all $\mathbf{x}$, so

$$A(BC) = [\, (AB)\mathbf{c}_1 \; \cdots \; (AB)\mathbf{c}_p \,] = (AB)C \qquad \blacksquare$$

The associative and distributive laws in Theorems 1 and 2 say essentially that pairs of parentheses in matrix expressions can be inserted and deleted in the same way as in the algebra of real numbers. In particular, we can write ABC for the product, which can be computed either as $A(BC)$ or as $(AB)C$.[1] Similarly, a product $ABCD$ of four matrices can be computed as $A(BCD)$ or $(ABC)D$ or $A(BC)D$, and so on. It does not matter how we group the matrices when computing the product, so long as the left-to-right order of the matrices is preserved.

The left-to-right order in products is critical because AB and BA are usually not the same. This is not surprising, because the columns of AB are linear combinations of the columns of A, whereas the columns of BA are constructed from the columns of B. The position of the factors in the product AB is emphasized by saying that A is *right-multiplied* by B or that B is *left-multiplied* by A. If $AB = BA$, we say that A and B **commute** with one another.

EXAMPLE 7 Let $A = \begin{bmatrix} 5 & 1 \\ 3 & -2 \end{bmatrix}$ and $B = \begin{bmatrix} 2 & 0 \\ 4 & 3 \end{bmatrix}$. Show that these matrices do not commute. That is, verify that $AB \neq BA$.

SOLUTION

$$AB = \begin{bmatrix} 5 & 1 \\ 3 & -2 \end{bmatrix}\begin{bmatrix} 2 & 0 \\ 4 & 3 \end{bmatrix} = \begin{bmatrix} 14 & 3 \\ -2 & -6 \end{bmatrix}$$
$$BA = \begin{bmatrix} 2 & 0 \\ 4 & 3 \end{bmatrix}\begin{bmatrix} 5 & 1 \\ 3 & -2 \end{bmatrix} = \begin{bmatrix} 10 & 2 \\ 29 & -2 \end{bmatrix} \qquad \blacksquare$$

Example 7 illustrates the first of the following list of important differences between matrix algebra and the ordinary algebra of real numbers. See Exercises 9–12 for examples of these situations.

***Warnings*:**

1. In general, $AB \neq BA$.

2. The cancellation laws do *not* hold for matrix multiplication. That is, if $AB = AC$, then it is *not* true in general that $B = C$. (See Exercise 10.)

3. If a product AB is the zero matrix, you *cannot* conclude in general that either $A = 0$ or $B = 0$. (See Exercise 12.)

---

[1] When B is square and C has fewer columns than A has rows, it is more efficient to compute $A(BC)$ than $(AB)C$.

Powers of a Matrix

If A is an $n \times n$ matrix and if k is a positive integer, then A^k denotes the product of k copies of A:

$$A^k = \underbrace{A \cdots A}_{k}$$

If A is nonzero and if $\mathbf{x}$ is in $\mathbb{R}^n$, then $A^k\mathbf{x}$ is the result of left-multiplying $\mathbf{x}$ by A repeatedly k times. If $k = 0$, then $A^0\mathbf{x}$ should be $\mathbf{x}$ itself. Thus A^0 is interpreted as the identity matrix. Matrix powers are useful in both theory and applications (Sections 2.6, 5.9, and later in the text).

The Transpose of a Matrix

Given an $m \times n$ matrix A, the **transpose** of A is the $n \times m$ matrix, denoted by A^T, whose columns are formed from the corresponding rows of A.

EXAMPLE 8 Let

$$A = \begin{bmatrix} a & b \\ c & d \end{bmatrix}, \quad B = \begin{bmatrix} -5 & 2 \\ 1 & -3 \\ 0 & 4 \end{bmatrix}, \quad C = \begin{bmatrix} 1 & 1 & 1 & 1 \\ -3 & 5 & -2 & 7 \end{bmatrix}$$

Then

$$A^T = \begin{bmatrix} a & c \\ b & d \end{bmatrix}, \quad B^T = \begin{bmatrix} -5 & 1 & 0 \\ 2 & -3 & 4 \end{bmatrix}, \quad C^T = \begin{bmatrix} 1 & -3 \\ 1 & 5 \\ 1 & -2 \\ 1 & 7 \end{bmatrix} \quad \blacksquare$$

THEOREM 3

Let A and B denote matrices whose sizes are appropriate for the following sums and products.

a. $(A^T)^T = A$
b. $(A + B)^T = A^T + B^T$
c. For any scalar r, $(rA)^T = rA^T$
d. $(AB)^T = B^TA^T$

Proofs of (a)–(c) are straightforward and are omitted. For (d), see Exercise 41. Usually, $(AB)^T$ is not equal to A^TB^T, even when A and B have sizes such that the product A^TB^T is defined.

The generalization of Theorem 3(d) to products of more than two factors can be stated in words as follows:

The transpose of a product of matrices equals the product of their transposes in the *reverse* order.

The exercises contain numerical examples that illustrate properties of transposes.

Artificial intelligence (AI) involves having a computer learn to recognize important information about anything that can be presented in a digitized format. One important area of AI is identifying whether the object in a picture matches a chosen object such as a number, fingerprint, or face.

In the next example, matrix transposition and matrix multiplication are used to tell whether or not a 2×2 block of colored squares matches the chosen checkerboard pattern in Figure 4.

EXAMPLE 9 In order to feed a 2×2 colored block into the computer, it first gets converted into a 4×1 vector by assigning a 1 to each block that is blue and a 0 to each block that is white. Then, the computer converts the block of numbers into a vector by placing the numbers in each column below the numbers in the column to its left.

FIGURE 4

Let $M = \begin{bmatrix} 1 & 0 & 0 & -1 \\ 0 & 1 & 0 & 0 \\ 0 & 0 & 1 & 0 \\ -1 & 0 & 0 & 1 \end{bmatrix}$.

Notice that $\mathbf{v}^T M \mathbf{v} = \begin{bmatrix} 1 & 0 & 0 & 1 \end{bmatrix} \begin{bmatrix} 1 & 0 & 0 & -1 \\ 0 & 1 & 0 & 0 \\ 0 & 0 & 1 & 0 \\ -1 & 0 & 0 & 1 \end{bmatrix} \begin{bmatrix} 1 \\ 0 \\ 0 \\ 1 \end{bmatrix} = 0,$

and $\mathbf{w}^T M \mathbf{w} = \begin{bmatrix} 0 & 0 & 0 & 0 \end{bmatrix} \begin{bmatrix} 1 & 0 & 0 & -1 \\ 0 & 1 & 0 & 0 \\ 0 & 0 & 1 & 0 \\ -1 & 0 & 0 & 1 \end{bmatrix} \begin{bmatrix} 0 \\ 0 \\ 0 \\ 0 \end{bmatrix} = 0,$ where $\mathbf{w}$ is the

vector generated by a 2×2 block of all white squares. It can be verified that for any other vector $\mathbf{x}$ generated from a 2×2 block of white and blue squares, if $\mathbf{x}$ is not $\mathbf{v}$ or $\mathbf{w}$, then the product $\mathbf{x}^T M \mathbf{x}$ is nonzero. Thus, if a computer checks the value of $\mathbf{x}^T M \mathbf{x}$ and finds it is nonzero, the computer knows that the pattern corresponding to $\mathbf{x}$ is not the checkerboard with a blue square in the top left corner.

This pattern is not the checkerboard pattern since $\mathbf{x}^T M \mathbf{x} \neq 0$.

This pattern is the checkerboard pattern since $\mathbf{x}^T M \mathbf{x} = 0$, but $\mathbf{x}^T \mathbf{x} \neq 0$.

FIGURE 5

However, if the computer finds that $\mathbf{x}^T M \mathbf{x} = 0$, then $\mathbf{x}$ could be either $\mathbf{v}$ or $\mathbf{w}$. To distinguish between the two, the computer can calculate the product $\mathbf{x}^T \mathbf{x}$, for $\mathbf{x}^T \mathbf{x}$ is zero if and only if $\mathbf{x}$ is $\mathbf{w}$.[2] Thus, to conclude that $\mathbf{x}$ is equal to $\mathbf{v}$, the computer must have $\mathbf{x}^T M \mathbf{x} = 0$ and $\mathbf{x}^T \mathbf{x} \neq 0$. ■

Example 5 of Section 6.3 illustrates one way to choose a matrix M so that matrix multiplication and transposition can be used to identify a particular pattern of colored squares.

Another important aspect of AI starts even before the data is fed to the machine. In Section 1.9, it is illustrated how matrix multiplication can be used to move vectors around in space. In the next example, matrix multiplication is used to *scrub* data and prepare it for processing.

EXAMPLE 10 The dates of ground crew accidents for January and February of 2020 are listed in the columns of matrix T for Toronto Pearson Airport and matrix C for Chicago O'Hare Airport:

Toronto: $T = \begin{bmatrix} 1 & 12 & 14 & 15 & 21 & 22 & 23 & 1 & 2 & 3 & 12 & 15 & 17 & 19 & 26 \\ 1 & 1 & 1 & 1 & 1 & 1 & 1 & 2 & 2 & 2 & 2 & 2 & 2 & 2 & 2 \end{bmatrix}$

Chicago: $C = \begin{bmatrix} 1 & 1 & 1 & 1 & 1 & 2 & 2 & 2 & 2 & 2 \\ 1 & 11 & 22 & 23 & 24 & 1 & 2 & 5 & 20 & 21 \end{bmatrix}$

Clearly the data is listed differently in the two matrices. Canada and the United states have different traditions for whether the month or day comes first when writing a date. For matrix T, the day is listed in the first row and the month is listed in the second row. For matrix C, the month is listed in the first row and the day is listed in the second row. In order to use this data, the first and second rows need to be swapped in one of the matrices. Reviewing the effects of matrix multiplication in Table 1 of Section 1.9, notice that the matrix $A = \begin{bmatrix} 0 & 1 \\ 1 & 0 \end{bmatrix}$ switches the x_1 and x_2 coordinates of any vector $x = \begin{bmatrix} x_1 \\ x_2 \end{bmatrix}$ it is applied to and indeed

$AT = \begin{bmatrix} 1 & 1 & 1 & 1 & 1 & 1 & 1 & 2 & 2 & 2 & 2 & 2 & 2 & 2 & 2 \\ 1 & 12 & 14 & 15 & 21 & 22 & 23 & 1 & 2 & 3 & 12 & 15 & 17 & 19 & 26 \end{bmatrix}$

has the data listed in the same order as it is listed in matrix C. The matrices AT and C can now be fed into the same machine. ■

In Exercises 51 and 52 you will be asked to *scrub* further data for this project.[3]

---

[2] To see why $\mathbf{x}^T \mathbf{x}$ is zero if and only if $\mathbf{x}$ is $\mathbf{w}$, let $\mathbf{x}^T = [x_1\ x_2\ x_3\ x_4]$. Then $\mathbf{x}^T \mathbf{x} = x_1^2 + x_2^2 + x_3^2 + x_4^2$ and this sum is zero if and only if the coordinates of $\mathbf{x}$ are all zero. That is, if and only if $\mathbf{x} = \mathbf{w}$.

[3] Although the data in this example and the corresponding exercises are fictitious, Data Analytics students at Washington State University identified scrubbing the data they received as an important first step in their actual analysis of ground crew accidents at three major airports in the United States.

> **Numerical Notes**
>
> 1. The fastest way to obtain AB on a computer depends on the way in which the computer stores matrices in its memory. The standard high-performance algorithms, such as in LAPACK, calculate AB by columns, as in our definition of the product. (A version of LAPACK written in C++ calculates AB by rows.)
> 2. The definition of AB lends itself well to parallel processing on a computer. The columns of B are assigned individually or in groups to different processors, which independently and hence simultaneously compute the corresponding columns of AB.

Practice Problems

1. Since vectors in $\mathbb{R}^n$ may be regarded as $n \times 1$ matrices, the properties of transposes in Theorem 3 apply to vectors, too. Let

$$A = \begin{bmatrix} 1 & -3 \\ -2 & 4 \end{bmatrix} \quad \text{and} \quad \mathbf{x} = \begin{bmatrix} 5 \\ 3 \end{bmatrix}$$

Compute $(A\mathbf{x})^T$, $\mathbf{x}^T A^T$, $\mathbf{x}\mathbf{x}^T$, and $\mathbf{x}^T\mathbf{x}$. Is $A^T\mathbf{x}^T$ defined?

2. Let A be a 4×4 matrix and let $\mathbf{x}$ be a vector in $\mathbb{R}^4$. What is the fastest way to compute $A^2\mathbf{x}$? Count the multiplications.

3. Suppose A is an $m \times n$ matrix, all of whose rows are identical. Suppose B is an $n \times p$ matrix, all of whose columns are identical. What can be said about the entries in AB?

2.1 Exercises

In Exercises 1 and 2, compute each matrix sum or product if it is defined. If an expression is undefined, explain why. Let

$$A = \begin{bmatrix} 2 & 0 & -1 \\ 4 & -3 & 2 \end{bmatrix}, \quad B = \begin{bmatrix} 7 & -5 & 1 \\ 1 & -4 & -3 \end{bmatrix},$$

$$C = \begin{bmatrix} 1 & 2 \\ -2 & 1 \end{bmatrix}, \quad D = \begin{bmatrix} 3 & 5 \\ -1 & 4 \end{bmatrix}, \quad E = \begin{bmatrix} -5 \\ 3 \end{bmatrix}$$

1. $-2A$, $B - 2A$, AC, CD

2. $A + 2B$, $3C - E$, CB, EB

In the rest of this exercise set and in those to follow, you should assume that each matrix expression is defined. That is, the sizes of the matrices (and vectors) involved "match" appropriately.

3. Let $A = \begin{bmatrix} 4 & -1 \\ 5 & -2 \end{bmatrix}$. Compute $3I_2 - A$ and $(3I_2)A$.

4. Compute $A - 5I_3$ and $(5I_3)A$, when

$$A = \begin{bmatrix} 9 & -1 & 3 \\ -8 & 7 & -3 \\ -4 & 1 & 8 \end{bmatrix}.$$

In Exercises 5 and 6, compute the product AB in two ways: (a) by the definition, where $A\mathbf{b}_1$ and $A\mathbf{b}_2$ are computed separately, and (b) by the row–column rule for computing AB.

5. $A = \begin{bmatrix} -1 & 2 \\ 5 & 4 \\ 2 & -3 \end{bmatrix}$, $B = \begin{bmatrix} 3 & -4 \\ -2 & 1 \end{bmatrix}$

6. $A = \begin{bmatrix} 4 & -2 \\ -3 & 0 \\ 3 & 5 \end{bmatrix}$, $B = \begin{bmatrix} 1 & 3 \\ 4 & -1 \end{bmatrix}$

7. If a matrix A is 5×3 and the product AB is 5×7, what is the size of B?

8. How many rows does B have if BC is a 3×4 matrix?

9. Let $A = \begin{bmatrix} 2 & 5 \\ -3 & 1 \end{bmatrix}$ and $B = \begin{bmatrix} 4 & -5 \\ 3 & k \end{bmatrix}$. What value(s) of k, if any, will make $AB = BA$?

10. Let $A = \begin{bmatrix} 2 & -3 \\ -4 & 6 \end{bmatrix}$, $B = \begin{bmatrix} 8 & 4 \\ 5 & 5 \end{bmatrix}$, and $C = \begin{bmatrix} 5 & -2 \\ 3 & 1 \end{bmatrix}$. Verify that $AB = AC$ and yet $B \neq C$.

11. Let $A = \begin{bmatrix} 1 & 1 & 1 \\ 1 & 2 & 3 \\ 1 & 4 & 5 \end{bmatrix}$ and $D = \begin{bmatrix} 2 & 0 & 0 \\ 0 & 3 & 0 \\ 0 & 0 & 5 \end{bmatrix}$. Compute AD and DA. Explain how the columns or rows of A change when A is multiplied by D on the right or on the left. Find a 3×3 matrix B, not the identity matrix or the zero matrix, such that $AB = BA$.

12. Let $A = \begin{bmatrix} 3 & -6 \\ -1 & 2 \end{bmatrix}$. Construct a 2×2 matrix B such that AB is the zero matrix. Use two different nonzero columns for B.

13. Let $\mathbf{r}_1, \ldots, \mathbf{r}_p$ be vectors in $\mathbb{R}^n$, and let Q be an $m \times n$ matrix. Write the matrix $[\, Q\mathbf{r}_1 \cdots Q\mathbf{r}_p \,]$ as a *product* of two matrices (neither of which is an identity matrix).

14. Let U be the 3×2 cost matrix described in Example 6 of Section 1.8. The first column of U lists the costs per dollar of output for manufacturing product B, and the second column lists the costs per dollar of output for product C. (The costs are categorized as materials, labor, and overhead.) Let $\mathbf{q}_1$ be a vector in $\mathbb{R}^2$ that lists the output (measured in dollars) of products B and C manufactured during the first quarter of the year, and let $\mathbf{q}_2, \mathbf{q}_3$, and $\mathbf{q}_4$ be the analogous vectors that list the amounts of products B and C manufactured in the second, third, and fourth quarters, respectively. Give an economic description of the data in the matrix UQ, where $Q = [\mathbf{q}_1 \quad \mathbf{q}_2 \quad \mathbf{q}_3 \quad \mathbf{q}_4]$.

Exercises 15–24 concern arbitrary matrices A, B, and C for which the indicated sums and products are defined. Mark each statement True or False (**T/F**). Justify each answer.

15. (**T/F**) If A and B are 2×2 with columns $\mathbf{a}_1, \mathbf{a}_2$, and $\mathbf{b}_1, \mathbf{b}_2$, respectively, then $AB = [\mathbf{a}_1\mathbf{b}_1 \quad \mathbf{a}_2\mathbf{b}_2]$.

16. (**T/F**) If A and B are 3×3 and $B = [\mathbf{b}_1 \quad \mathbf{b}_2 \quad \mathbf{b}_3]$, then $AB = [A\mathbf{b}_1 + A\mathbf{b}_2 + A\mathbf{b}_3]$.

17. (**T/F**) Each column of AB is a linear combination of the columns of B using weights from the corresponding column of A.

18. (**T/F**) The second row of AB is the second row of A multiplied on the right by B.

19. (**T/F**) $AB + AC = A(B + C)$

20. (**T/F**) $A^T + B^T = (A + B)^T$

21. (**T/F**) $(AB)C = (AC)B$

22. (**T/F**) $(AB)^T = A^T B^T$

23. (**T/F**) The transpose of a product of matrices equals the product of their transposes in the same order.

24. (**T/F**) The transpose of a sum of matrices equals the sum of their transposes.

25. If $A = \begin{bmatrix} 1 & -2 \\ -2 & 5 \end{bmatrix}$ and $AB = \begin{bmatrix} -1 & 2 & -1 \\ 6 & -9 & 3 \end{bmatrix}$, determine the first and second columns of B.

26. Suppose the first two columns, $\mathbf{b}_1$ and $\mathbf{b}_2$, of B are equal. What can you say about the columns of AB (if AB is defined)? Why?

27. Suppose the third column of B is the sum of the first two columns. What can you say about the third column of AB? Why?

28. Suppose the second column of B is all zeros. What can you say about the second column of AB?

29. Suppose the last column of AB is all zeros, but B itself has no column of zeros. What can you say about the columns of A?

30. Show that if the columns of B are linearly dependent, then so are the columns of AB.

31. Suppose $CA = I_n$ (the $n \times n$ identity matrix). Show that the equation $A\mathbf{x} = \mathbf{0}$ has only the trivial solution. Explain why A cannot have more columns than rows.

32. Suppose $AD = I_m$ (the $m \times m$ identity matrix). Show that for any $\mathbf{b}$ in $\mathbb{R}^m$, the equation $A\mathbf{x} = \mathbf{b}$ has a solution. [*Hint:* Think about the equation $AD\mathbf{b} = \mathbf{b}$.] Explain why A cannot have more rows than columns.

33. Suppose A is an $m \times n$ matrix and there exist $n \times m$ matrices C and D such that $CA = I_n$ and $AD = I_m$. Prove that $m = n$ and $C = D$. [*Hint:* Think about the product CAD.]

34. Suppose A is a $3 \times n$ matrix whose columns span $\mathbb{R}^3$. Explain how to construct an $n \times 3$ matrix D such that $AD = I_3$.

In Exercises 35 and 36, view vectors in $\mathbb{R}^n$ as $n \times 1$ matrices. For $\mathbf{u}$ and $\mathbf{v}$ in $\mathbb{R}^n$, the matrix product $\mathbf{u}^T\mathbf{v}$ is a 1×1 matrix, called the **scalar product**, or **inner product**, of $\mathbf{u}$ and $\mathbf{v}$. It is usually written as a single real number without brackets. The matrix product $\mathbf{u}\mathbf{v}^T$ is an $n \times n$ matrix, called the **outer product** of $\mathbf{u}$ and $\mathbf{v}$. The products $\mathbf{u}^T\mathbf{v}$ and $\mathbf{u}\mathbf{v}^T$ will appear later in the text.

35. Let $\mathbf{u} = \begin{bmatrix} -2 \\ 3 \\ -4 \end{bmatrix}$ and $\mathbf{v} = \begin{bmatrix} a \\ b \\ c \end{bmatrix}$. Compute $\mathbf{u}^T\mathbf{v}, \mathbf{v}^T\mathbf{u}, \mathbf{u}\mathbf{v}^T$, and $\mathbf{v}\mathbf{u}^T$.

36. If $\mathbf{u}$ and $\mathbf{v}$ are in $\mathbb{R}^n$, how are $\mathbf{u}^T\mathbf{v}$ and $\mathbf{v}^T\mathbf{u}$ related? How are $\mathbf{u}\mathbf{v}^T$ and $\mathbf{v}\mathbf{u}^T$ related?

37. Prove Theorem 2(b) and 2(c). Use the row–column rule. The (i, j)-entry in $A(B + C)$ can be written as

$$a_{i1}(b_{1j} + c_{1j}) + \cdots + a_{in}(b_{nj} + c_{nj}) \text{ or } \sum_{k=1}^{n} a_{ik}(b_{kj} + c_{kj})$$

38. Prove Theorem 2(d). [*Hint:* The (i, j)-entry in $(rA)B$ is $(ra_{i1})b_{1j} + \cdots + (ra_{in})b_{nj}$.]

39. Show that $I_m A = A$ when A is an $m \times n$ matrix. You can assume $I_m\mathbf{x} = \mathbf{x}$ for all $\mathbf{x}$ in $\mathbb{R}^m$.

40. Show that $AI_n = A$ when A is an $m \times n$ matrix. [*Hint:* Use the (column) definition of AI_n.]

41. Prove Theorem 3(d). [*Hint:* Consider the jth row of $(AB)^T$.]

42. Give a formula for $(AB\mathbf{x})^T$, where $\mathbf{x}$ is a vector and A and B are matrices of appropriate sizes.

T 43. Use a web search engine such as Google to find documentation for your matrix program, and write the commands that will produce the following matrices (without keying in each entry of the matrix).

a. A 5×6 matrix of zeros

b. A 3×5 matrix of ones

c. The 6×6 identity matrix

d. A 5×5 diagonal matrix, with diagonal entries 3, 5, 7, 2, 4

A useful way to test new ideas in matrix algebra, or to make conjectures, is to make calculations with matrices selected at random. Checking a property for a few matrices does not prove that the property holds in general, but it makes the property more believable. Also, if the property is actually false, you may discover this when you make a few calculations.

T 44. Write the command(s) that will create a 6×4 matrix with random entries. In what range of numbers do the entries lie? Tell how to create a 3×3 matrix with random integer entries between -9 and 9. [*Hint:* If x is a random number such that $0 < x < 1$, then $-9.5 < 19(x - .5) < 9.5$.]

T 45. Construct a random 4×4 matrix A and test whether $(A + I)(A - I) = A^2 - I$. The best way to do this is to compute $(A + I)(A - I) - (A^2 - I)$ and verify that this difference is the zero matrix. Do this for three random matrices. Then test $(A + B)(A - B) = A^2 - B^2$ the same way for three pairs of random 4×4 matrices. Report your conclusions.

T 46. Use at least three pairs of random 4×4 matrices A and B to test the equalities $(A + B)^T = A^T + B^T$ and $(AB)^T = A^T B^T$. (See Exercise 45.) Report your conclusions. [*Note:* Most matrix programs use A' for A^T.]

T 47. Let
$$S = \begin{bmatrix} 0 & 1 & 0 & 0 & 0 \\ 0 & 0 & 1 & 0 & 0 \\ 0 & 0 & 0 & 1 & 0 \\ 0 & 0 & 0 & 0 & 1 \\ 0 & 0 & 0 & 0 & 0 \end{bmatrix}$$
Compute S^k for $k = 2, \ldots, 6$.

T 48. Describe in words what happens when you compute A^5, A^{10}, A^{20}, and A^{30} for
$$A = \begin{bmatrix} 1/6 & 1/2 & 1/3 \\ 1/2 & 1/4 & 1/4 \\ 1/3 & 1/4 & 5/12 \end{bmatrix}$$

T 49. The matrix M can detect a particular 2×2 colored pattern like in Example 9. Create a nonzero 4×1 vector $\mathbf{x}$ by choosing each entry to be a zero or one. Test to see if $\mathbf{x}$ corresponds

to the right pattern by calculating $\mathbf{x}^T M \mathbf{x}$. If $\mathbf{x}^T M \mathbf{x} = 0$, then $\mathbf{x}$ is the pattern identified by M. If $\mathbf{x}^T M \mathbf{x} \neq 0$, try a different nonzero vector of zeros and ones. You may want to be systematic in the way that you choose each $\mathbf{x}$ in order to avoid testing the same vector twice. You are using "guess and check" to determine which pattern of 2×2 colored squares the matrix M detects.

$$M = \begin{bmatrix} 1 & 0 & -1 & 0 \\ 0 & 1 & 0 & 0 \\ -1 & 0 & 1 & 0 \\ 0 & 0 & 0 & 1 \end{bmatrix}$$

T 50. Repeat Exercise 49 with the matrix

$$M = \begin{bmatrix} 1 & 0 & 0 & -1 \\ 0 & 1 & 0 & -1 \\ 0 & 0 & 1 & 0 \\ -1 & -1 & 0 & 2 \end{bmatrix}$$

T 51. Use the matrix $A = \begin{bmatrix} 0 & 1 \\ 1 & 0 \end{bmatrix}$ to switch the first and second rows of the matrix M containing dates of accidents at the Montreal Trudeau Airport.

Montreal:
$$M = \begin{bmatrix} 2 & 3 & 16 & 24 & 25 & 26 & 6 & 7 & 19 & 26 \\ 1 & 1 & 1 & 1 & 1 & 1 & 2 & 2 & 2 & 2 \end{bmatrix}$$

This data in matrix M has been scrubbed in matrix AM and can be fed into the same machine as the other data from Example 10.

T 52. Use the matrix $B = \begin{bmatrix} 1 & 0 & 0 \\ 0 & 1 & 0 \end{bmatrix}$ to remove the last row from the matrix N containing dates of accidents at the New York JFK Airport.

New York:
$$N = \begin{bmatrix} 1 & 1 & 1 & 1 & 2 & 2 & 2 \\ 1 & 12 & 21 & 22 & 3 & 20 & 21 \\ 2020 & 2020 & 2020 & 2020 & 2020 & 2020 & 2020 \end{bmatrix}$$

The data in matrix N has been scrubbed in matrix BN and can be fed into the same machine as the other data from Example 10.

Solutions to Practice Problems

1. $A\mathbf{x} = \begin{bmatrix} 1 & -3 \\ -2 & 4 \end{bmatrix} \begin{bmatrix} 5 \\ 3 \end{bmatrix} = \begin{bmatrix} -4 \\ 2 \end{bmatrix}$. So $(A\mathbf{x})^T = \begin{bmatrix} -4 & 2 \end{bmatrix}$. Also,

$$\mathbf{x}^T A^T = \begin{bmatrix} 5 & 3 \end{bmatrix} \begin{bmatrix} 1 & -2 \\ -3 & 4 \end{bmatrix} = \begin{bmatrix} -4 & 2 \end{bmatrix}.$$

The quantities $(A\mathbf{x})^T$ and $\mathbf{x}^T A^T$ are equal, by Theorem 3(d). Next,

$$\mathbf{x}\mathbf{x}^T = \begin{bmatrix} 5 \\ 3 \end{bmatrix} \begin{bmatrix} 5 & 3 \end{bmatrix} = \begin{bmatrix} 25 & 15 \\ 15 & 9 \end{bmatrix}$$

$$\mathbf{x}^T\mathbf{x} = \begin{bmatrix} 5 & 3 \end{bmatrix} \begin{bmatrix} 5 \\ 3 \end{bmatrix} = [25 + 9] = 34$$

A 1×1 matrix such as $\mathbf{x}^T\mathbf{x}$ is usually written without the brackets. Finally, $A^T\mathbf{x}^T$ is not defined, because $\mathbf{x}^T$ does not have two rows to match the two columns of A^T.

2. The fastest way to compute $A^2\mathbf{x}$ is to compute $A(A\mathbf{x})$. The product $A\mathbf{x}$ requires 16 multiplications, 4 for each entry, and $A(A\mathbf{x})$ requires 16 more. In contrast, the product A^2 requires 64 multiplications, 4 for each of the 16 entries in A^2. After that, $A^2\mathbf{x}$ takes 16 more multiplications, for a total of 80.

3. First observe that by the definition of matrix multiplication,

$$AB = [A\mathbf{b}_1 \quad A\mathbf{b}_2 \quad \cdots \quad A\mathbf{b}_n] = [A\mathbf{b}_1 \quad A\mathbf{b}_1 \quad \cdots \quad A\mathbf{b}_1],$$

so the columns of AB are identical. Next, recall that $\text{row}_i(AB) = \text{row}_i(A) \cdot B$. Since all the rows of A are identical, all the rows of AB are identical. Putting this information about the rows and columns together, it follows that all the entries in AB are the same.

2.2 The Inverse of a Matrix

Matrix algebra provides tools for manipulating matrix equations and creating various useful formulas in ways similar to doing ordinary algebra with real numbers. This section investigates the matrix analogue of the reciprocal, or multiplicative inverse, of a nonzero number.

Recall that the multiplicative inverse of a number such as 5 is 1/5 or 5^{-1}. This inverse satisfies the equations

$$5^{-1}(5) = 1 \quad \text{and} \quad 5(5^{-1}) = 1$$

The matrix generalization requires *both* equations and avoids the slanted-line notation (for division) because matrix multiplication is not commutative. Furthermore, a full generalization is possible only if the matrices involved are square.[1]

An $n \times n$ matrix A is said to be **invertible** if there is an $n \times n$ matrix C such that

$$CA = I \quad \text{and} \quad AC = I$$

where $I = I_n$, the $n \times n$ identity matrix. In this case, C is an **inverse** of A. In fact, C is uniquely determined by A, because if B were another inverse of A, then $B = BI = B(AC) = (BA)C = IC = C$. This unique inverse is denoted by A^{-1}, so that

$$A^{-1}A = I \quad \text{and} \quad AA^{-1} = I$$

A matrix that is *not* invertible is sometimes called a **singular matrix**, and an invertible matrix is called a **nonsingular matrix**.

---

[1] One could say that an $m \times n$ matrix A is invertible if there exist $n \times m$ matrices C and D such that $CA = I_n$ and $AD = I_m$. However, these equations imply that A is square and $C = D$. Thus, A is invertible as defined above. See Exercises 31–33 in Section 2.1.

EXAMPLE 1 If $A = \begin{bmatrix} 2 & 5 \\ -3 & -7 \end{bmatrix}$ and $C = \begin{bmatrix} -7 & -5 \\ 3 & 2 \end{bmatrix}$, then

$$AC = \begin{bmatrix} 2 & 5 \\ -3 & -7 \end{bmatrix}\begin{bmatrix} -7 & -5 \\ 3 & 2 \end{bmatrix} = \begin{bmatrix} 1 & 0 \\ 0 & 1 \end{bmatrix} \quad \text{and}$$

$$CA = \begin{bmatrix} -7 & -5 \\ 3 & 2 \end{bmatrix}\begin{bmatrix} 2 & 5 \\ -3 & -7 \end{bmatrix} = \begin{bmatrix} 1 & 0 \\ 0 & 1 \end{bmatrix}$$

Thus $C = A^{-1}$. ∎

Here is a simple formula for the inverse of a 2×2 matrix, along with a test to tell if the inverse exists.

THEOREM 4

Let $A = \begin{bmatrix} a & b \\ c & d \end{bmatrix}$. If $ad - bc \neq 0$, then A is invertible and

$$A^{-1} = \frac{1}{ad - bc}\begin{bmatrix} d & -b \\ -c & a \end{bmatrix}$$

If $ad - bc = 0$, then A is not invertible.

The simple proof of Theorem 4 is outlined in Exercises 35 and 36. The quantity $ad - bc$ is called the **determinant** of A, and we write

$$\det A = ad - bc$$

Theorem 4 says that a 2×2 matrix A is invertible if and only if $\det A \neq 0$.

EXAMPLE 2 Find the inverse of $A = \begin{bmatrix} 3 & 4 \\ 5 & 6 \end{bmatrix}$.

SOLUTION Since $\det A = 3(6) - 4(5) = -2 \neq 0$, A is invertible, and

$$A^{-1} = \frac{1}{-2}\begin{bmatrix} 6 & -4 \\ -5 & 3 \end{bmatrix} = \begin{bmatrix} 6/(-2) & -4/(-2) \\ -5/(-2) & 3/(-2) \end{bmatrix} = \begin{bmatrix} -3 & 2 \\ 5/2 & -3/2 \end{bmatrix} \quad ∎$$

Invertible matrices are indispensable in linear algebra—mainly for algebraic calculations and formula derivations, as in the next theorem. There are also occasions when an inverse matrix provides insight into a mathematical model of a real-life situation, as in Example 3.

THEOREM 5

If A is an invertible $n \times n$ matrix, then for each $\mathbf{b}$ in $\mathbb{R}^n$, the equation $A\mathbf{x} = \mathbf{b}$ has the unique solution $\mathbf{x} = A^{-1}\mathbf{b}$.

PROOF Take any $\mathbf{b}$ in $\mathbb{R}^n$. A solution exists because if $A^{-1}\mathbf{b}$ is substituted for $\mathbf{x}$, then $A\mathbf{x} = A(A^{-1}\mathbf{b}) = (AA^{-1})\mathbf{b} = I\mathbf{b} = \mathbf{b}$. So $A^{-1}\mathbf{b}$ is a solution. To prove that the solution is unique, show that if $\mathbf{u}$ is any solution, then $\mathbf{u}$, in fact, must be $A^{-1}\mathbf{b}$. Indeed, if $A\mathbf{u} = \mathbf{b}$, we can multiply both sides by A^{-1} and obtain

$$A^{-1}A\mathbf{u} = A^{-1}\mathbf{b}, \quad I\mathbf{u} = A^{-1}\mathbf{b}, \quad \text{and} \quad \mathbf{u} = A^{-1}\mathbf{b} \quad ∎$$

EXAMPLE 3   A horizontal elastic beam is supported at each end and is subjected to forces at points 1, 2, and 3, as shown in Figure 1. Let $\mathbf{f}$ in $\mathbb{R}^3$ list the forces at these points, and let $\mathbf{y}$ in $\mathbb{R}^3$ list the amounts of deflection (that is, movement) of the beam at the three points. Using Hooke's law from physics, it can be shown that

$$\mathbf{y} = D\mathbf{f}$$

where D is a *flexibility matrix*. Its inverse is called the *stiffness matrix*. Describe the physical significance of the columns of D and D^{-1}.

FIGURE 1   Deflection of an elastic beam.

SOLUTION   Write $I_3 = [\,\mathbf{e}_1 \quad \mathbf{e}_2 \quad \mathbf{e}_3\,]$ and observe that

$$D = DI_3 = [\,D\mathbf{e}_1 \quad D\mathbf{e}_2 \quad D\mathbf{e}_3\,]$$

Interpret the vector $\mathbf{e}_1 = (1, 0, 0)$ as a unit force applied downward at point 1 on the beam (with zero force at the other two points). Then $D\mathbf{e}_1$, the first column of D, lists the beam deflections due to a unit force at point 1. Similar descriptions apply to the second and third columns of D.

To study the stiffness matrix D^{-1}, observe that the equation $\mathbf{f} = D^{-1}\mathbf{y}$ computes a force vector $\mathbf{f}$ when a deflection vector $\mathbf{y}$ is given. Write

$$D^{-1} = D^{-1}I_3 = [\,D^{-1}\mathbf{e}_1 \quad D^{-1}\mathbf{e}_2 \quad D^{-1}\mathbf{e}_3\,]$$

Now interpret $\mathbf{e}_1$ as a deflection vector. Then $D^{-1}\mathbf{e}_1$ lists the forces that create the deflection. That is, the first column of D^{-1} lists the forces that must be applied at the three points to produce a unit deflection at point 1 and zero deflections at the other points. Similarly, columns 2 and 3 of D^{-1} list the forces required to produce unit deflections at points 2 and 3, respectively. In each column, one or two of the forces must be negative (point upward) to produce a unit deflection at the desired point and zero deflections at the other two points. If the flexibility is measured, for example, in inches of deflection per pound of load, then the stiffness matrix entries are given in pounds of load per inch of deflection.   ∎

The formula in Theorem 5 is seldom used to solve an equation $A\mathbf{x} = \mathbf{b}$ numerically because row reduction of $[\,A \quad \mathbf{b}\,]$ is nearly always faster. (Row reduction is usually more accurate, too, when computations involve rounding off numbers.) One possible exception is the 2×2 case. In this case, mental computations to solve $A\mathbf{x} = \mathbf{b}$ are sometimes easier using the formula for A^{-1}, as in the next example.

EXAMPLE 4   Use the inverse of the matrix A in Example 2 to solve the system

$$3x_1 + 4x_2 = 3$$
$$5x_1 + 6x_2 = 7$$

SOLUTION   This system is equivalent to $A\mathbf{x} = \mathbf{b}$, so

$$\mathbf{x} = A^{-1}\mathbf{b} = \begin{bmatrix} -3 & 2 \\ 5/2 & -3/2 \end{bmatrix} \begin{bmatrix} 3 \\ 7 \end{bmatrix} = \begin{bmatrix} 5 \\ -3 \end{bmatrix}$$ ∎

The next theorem provides three useful facts about invertible matrices.

THEOREM 6

a. If A is an invertible matrix, then A^{-1} is invertible and

$$(A^{-1})^{-1} = A$$

b. If A and B are $n \times n$ invertible matrices, then so is AB, and the inverse of AB is the product of the inverses of A and B in the reverse order. That is,

$$(AB)^{-1} = B^{-1}A^{-1}$$

c. If A is an invertible matrix, then so is A^T, and the inverse of A^T is the transpose of A^{-1}. That is,

$$(A^T)^{-1} = (A^{-1})^T$$

PROOF To verify statement (a), find a matrix C such that

$$A^{-1}C = I \quad \text{and} \quad CA^{-1} = I$$

In fact, these equations are satisfied with A in place of C. Hence A^{-1} is invertible, and A is its inverse. Next, to prove statement (b), compute:

$$(AB)(B^{-1}A^{-1}) = A(BB^{-1})A^{-1} = AIA^{-1} = AA^{-1} = I$$

A similar calculation shows that $(B^{-1}A^{-1})(AB) = I$. For statement (c), use Theorem 3(d), read from right to left, $(A^{-1})^T A^T = (AA^{-1})^T = I^T = I$. Similarly, $A^T(A^{-1})^T = I^T = I$. Hence A^T is invertible, and its inverse is $(A^{-1})^T$. ∎

Remark: Part (b) illustrates the important role that definitions play in proofs. The theorem claims that $B^{-1}A^{-1}$ is the inverse of AB. The proof establishes this by showing that $B^{-1}A^{-1}$ satisfies the definition of what it means to be the inverse of AB. Now, the inverse of AB is a matrix that when multiplied on the left (or right) by AB, the product is the identity matrix I. So the proof consists of showing that $B^{-1}A^{-1}$ has this property.

The following generalization of Theorem 6(b) is needed later.

The product of $n \times n$ invertible matrices is invertible, and the inverse is the product of their inverses in the reverse order.

There is an important connection between invertible matrices and row operations that leads to a method for computing inverses. As we shall see, an invertible matrix A is row equivalent to an identity matrix, and we can find A^{-1} by *watching the row reduction of A to I*.

Elementary Matrices

An **elementary matrix** is one that is obtained by performing a single elementary row operation on an identity matrix. The next example illustrates the three kinds of elementary matrices.

EXAMPLE 5 Let

$$E_1 = \begin{bmatrix} 1 & 0 & 0 \\ 0 & 1 & 0 \\ -4 & 0 & 1 \end{bmatrix}, \quad E_2 = \begin{bmatrix} 0 & 1 & 0 \\ 1 & 0 & 0 \\ 0 & 0 & 1 \end{bmatrix}, \quad E_3 = \begin{bmatrix} 1 & 0 & 0 \\ 0 & 1 & 0 \\ 0 & 0 & 5 \end{bmatrix},$$

$$A = \begin{bmatrix} a & b & c \\ d & e & f \\ g & h & i \end{bmatrix}$$

Compute E_1A, E_2A, and E_3A, and describe how these products can be obtained by elementary row operations on A.

SOLUTION   Verify that

$$E_1A = \begin{bmatrix} a & b & c \\ d & e & f \\ g-4a & h-4b & i-4c \end{bmatrix}, \quad E_2A = \begin{bmatrix} d & e & f \\ a & b & c \\ g & h & i \end{bmatrix},$$

$$E_3A = \begin{bmatrix} a & b & c \\ d & e & f \\ 5g & 5h & 5i \end{bmatrix}.$$

Addition of -4 times row 1 of A to row 3 produces E_1A. (This is a row replacement operation.) An interchange of rows 1 and 2 of A produces E_2A, and multiplication of row 3 of A by 5 produces E_3A.   ∎

Left-multiplication (that is, multiplication on the left) by E_1 in Example 5 has the same effect on any $3 \times n$ matrix. It adds -4 times row 1 to row 3. In particular, since $E_1 \cdot I = E_1$, we see that E_1 *itself* is produced by this same row operation on the identity. Thus Example 5 illustrates the following general fact about elementary matrices. See Exercises 37 and 38.

> If an elementary row operation is performed on an $m \times n$ matrix A, the resulting matrix can be written as EA, where the $m \times m$ matrix E is created by performing the same row operation on I_m.

Since row operations are reversible, as shown in Section 1.1, elementary matrices are invertible, for if E is produced by a row operation on I, then there is another row operation of the same type that changes E back into I. Hence there is an elementary matrix F such that $FE = I$. Since E and F correspond to reverse operations, $EF = I$, too.

> Each elementary matrix E is invertible. The inverse of E is the elementary matrix of the same type that transforms E back into I.

EXAMPLE 6 Find the inverse of $E_1 = \begin{bmatrix} 1 & 0 & 0 \\ 0 & 1 & 0 \\ -4 & 0 & 1 \end{bmatrix}$.

SOLUTION   To transform E_1 into I, add $+4$ times row 1 to row 3. The elementary matrix that does this is

$$E_1^{-1} = \begin{bmatrix} 1 & 0 & 0 \\ 0 & 1 & 0 \\ +4 & 0 & 1 \end{bmatrix}$$ ∎

The following theorem provides the best way to "visualize" an invertible matrix, and the theorem leads immediately to a method for finding the inverse of a matrix.

THEOREM 7

An $n \times n$ matrix A is invertible if and only if A is row equivalent to I_n, and in this case, any sequence of elementary row operations that reduces A to I_n also transforms I_n into A^{-1}.

Remark: The comment on the proof of Theorem 11 in Chapter 1 noted that "P if and only if Q" is equivalent to two statements: (1) "If P then Q" and (2) "If Q then P." The second statement is called the *converse* of the first and explains the use of the word *conversely* in the second paragraph of this proof.

PROOF Suppose that A is invertible. Then, since the equation $A\mathbf{x} = \mathbf{b}$ has a solution for each $\mathbf{b}$ (Theorem 5), A has a pivot position in every row (Theorem 4 in Section 1.4). Because A is square, the n pivot positions must be on the diagonal, which implies that the reduced echelon form of A is I_n. That is, $A \sim I_n$.

Now suppose, conversely, that $A \sim I_n$. Then, since each step of the row reduction of A corresponds to left-multiplication by an elementary matrix, there exist elementary matrices $E_1, \ldots, E_p$ such that

$$A \sim E_1 A \sim E_2(E_1 A) \sim \cdots \sim E_p(E_{p-1} \cdots E_1 A) = I_n$$

That is,

$$E_p \cdots E_1 A = I_n \tag{1}$$

Since the product $E_p \cdots E_1$ of invertible matrices is invertible, (1) leads to

$$(E_p \cdots E_1)^{-1}(E_p \cdots E_1)A = (E_p \cdots E_1)^{-1} I_n$$
$$A = (E_p \cdots E_1)^{-1}$$

Thus A is invertible, as it is the inverse of an invertible matrix (Theorem 6). Also,

$$A^{-1} = \left[(E_p \cdots E_1)^{-1} \right]^{-1} = E_p \cdots E_1$$

Then $A^{-1} = E_p \cdots E_1 I_n$, which says that A^{-1} results from applying $E_1, \ldots, E_p$ successively to I_n. This is the same sequence in (1) that reduced A to I_n. ∎

An Algorithm for Finding A^{-1}

If we place A and I side by side to form an augmented matrix $[\, A \quad I \,]$, then row operations on this matrix produce identical operations on A and on I. By Theorem 7, either there are row operations that transform A to I_n and I_n to A^{-1} or else A is not invertible.

ALGORITHM FOR FINDING A^{-1}

Row reduce the augmented matrix $[\, A \quad I \,]$. If A is row equivalent to I, then $[\, A \quad I \,]$ is row equivalent to $[\, I \quad A^{-1} \,]$. Otherwise, A does not have an inverse.

EXAMPLE 7   Find the inverse of the matrix $A = \begin{bmatrix} 0 & 1 & 2 \\ 1 & 0 & 3 \\ 4 & -3 & 8 \end{bmatrix}$, if it exists.

SOLUTION

$$[A \quad I] = \begin{bmatrix} 0 & 1 & 2 & 1 & 0 & 0 \\ 1 & 0 & 3 & 0 & 1 & 0 \\ 4 & -3 & 8 & 0 & 0 & 1 \end{bmatrix} \sim \begin{bmatrix} 1 & 0 & 3 & 0 & 1 & 0 \\ 0 & 1 & 2 & 1 & 0 & 0 \\ 4 & -3 & 8 & 0 & 0 & 1 \end{bmatrix}$$

$$\sim \begin{bmatrix} 1 & 0 & 3 & 0 & 1 & 0 \\ 0 & 1 & 2 & 1 & 0 & 0 \\ 0 & -3 & -4 & 0 & -4 & 1 \end{bmatrix} \sim \begin{bmatrix} 1 & 0 & 3 & 0 & 1 & 0 \\ 0 & 1 & 2 & 1 & 0 & 0 \\ 0 & 0 & 2 & 3 & -4 & 1 \end{bmatrix}$$

$$\sim \begin{bmatrix} 1 & 0 & 3 & 0 & 1 & 0 \\ 0 & 1 & 2 & 1 & 0 & 0 \\ 0 & 0 & 1 & 3/2 & -2 & 1/2 \end{bmatrix}$$

$$\sim \begin{bmatrix} 1 & 0 & 0 & -9/2 & 7 & -3/2 \\ 0 & 1 & 0 & -2 & 4 & -1 \\ 0 & 0 & 1 & 3/2 & -2 & 1/2 \end{bmatrix}$$

Theorem 7 shows, since $A \sim I$, that A is invertible, and

$$A^{-1} = \begin{bmatrix} -9/2 & 7 & -3/2 \\ -2 & 4 & -1 \\ 3/2 & -2 & 1/2 \end{bmatrix}$$

∎

Reasonable Answers

Once you have found a candidate for the inverse of a matrix, you can check that your answer is correct by finding the product of A with A^{-1}. For the inverse found for matrix A in Example 7, notice

$$AA^{-1} = \begin{bmatrix} 0 & 1 & 2 \\ 1 & 0 & 3 \\ 4 & -3 & 8 \end{bmatrix} \begin{bmatrix} -9/2 & 7 & -3/2 \\ -2 & 4 & -1 \\ 3/2 & -2 & 1/2 \end{bmatrix} = \begin{bmatrix} 1 & 0 & 0 \\ 0 & 1 & 0 \\ 0 & 0 & 1 \end{bmatrix}$$

confirming that answer is correct. It is not necessary to check that $A^{-1}A = I$ since A is invertible.

Another View of Matrix Inversion

Denote the columns of I_n by $\mathbf{e}_1, \dots, \mathbf{e}_n$. Then row reduction of $[\, A \quad I \,]$ to $[\, I \quad A^{-1} \,]$ can be viewed as the simultaneous solution of the n systems

$$A\mathbf{x} = \mathbf{e}_1, \quad A\mathbf{x} = \mathbf{e}_2, \quad \dots, \quad A\mathbf{x} = \mathbf{e}_n \tag{2}$$

where the "augmented columns" of these systems have all been placed next to A to form $[\, A \quad \mathbf{e}_1 \quad \mathbf{e}_2 \quad \cdots \quad \mathbf{e}_n \,] = [\, A \quad I \,]$. The equation $AA^{-1} = I$ and the definition of matrix multiplication show that the columns of A^{-1} are precisely the solutions of the systems in (2). This observation is useful because some applied problems may require finding only one or two columns of A^{-1}. In this case, only the corresponding systems in (2) need to be solved.

> Numerical Note
>
> In practical work, A^{-1} is seldom computed, unless the entries of A^{-1} are needed. Computing both A^{-1} and $A^{-1}\mathbf{b}$ takes about three times as many arithmetic operations as solving $A\mathbf{x} = \mathbf{b}$ by row reduction, and row reduction may be more accurate.

Practice Problems

1. Use determinants to determine which of the following matrices are invertible.

    a. $\begin{bmatrix} 3 & -9 \\ 2 & 6 \end{bmatrix}$
    b. $\begin{bmatrix} 4 & -9 \\ 0 & 5 \end{bmatrix}$
    c. $\begin{bmatrix} 6 & -9 \\ -4 & 6 \end{bmatrix}$

2. Find the inverse of the matrix $A = \begin{bmatrix} 1 & -2 & -1 \\ -1 & 5 & 6 \\ 5 & -4 & 5 \end{bmatrix}$, if it exists.

3. If A is an invertible matrix, prove that $5A$ is an invertible matrix.

2.2 Exercises

Find the inverses of the matrices in Exercises 1–4.

1. $\begin{bmatrix} 8 & 3 \\ 5 & 2 \end{bmatrix}$
2. $\begin{bmatrix} 3 & 1 \\ 7 & 2 \end{bmatrix}$

3. $\begin{bmatrix} 8 & 3 \\ -7 & -3 \end{bmatrix}$
4. $\begin{bmatrix} 3 & -2 \\ 7 & -4 \end{bmatrix}$

5. Verify that the inverse you found in Exercise 1 is correct.

6. Verify that the inverse you found in Exercise 2 is correct.

7. Use the inverse found in Exercise 1 to solve the system

$$\begin{aligned} 8x_1 + 3x_2 &= 2 \\ 5x_1 + 2x_2 &= -1 \end{aligned}$$

8. Use the inverse found in Exercise 2 to solve the system

$$\begin{aligned} 3x_1 + x_2 &= -2 \\ 7x_1 + 2x_2 &= 3 \end{aligned}$$

9. Let $A = \begin{bmatrix} 1 & 2 \\ 5 & 12 \end{bmatrix}$, $\mathbf{b}_1 = \begin{bmatrix} -1 \\ 3 \end{bmatrix}$, $\mathbf{b}_2 = \begin{bmatrix} 1 \\ -5 \end{bmatrix}$, $\mathbf{b}_3 = \begin{bmatrix} 2 \\ 6 \end{bmatrix}$, and $\mathbf{b}_4 = \begin{bmatrix} 3 \\ 5 \end{bmatrix}$.

    a. Find A^{-1}, and use it to solve the four equations $A\mathbf{x} = \mathbf{b}_1, A\mathbf{x} = \mathbf{b}_2, A\mathbf{x} = \mathbf{b}_3, A\mathbf{x} = \mathbf{b}_4$

    b. The four equations in part (a) can be solved by the same set of row operations, since the coefficient matrix is the same in each case. Solve the four equations in part (a) by row reducing the augmented matrix $\begin{bmatrix} A & \mathbf{b}_1 & \mathbf{b}_2 & \mathbf{b}_3 & \mathbf{b}_4 \end{bmatrix}$

10. Use matrix algebra to show that if A is invertible and D satisfies $AD = I$, then $D = A^{-1}$.

In Exercises 11–20, mark each statement True or False (T/F). Justify each answer.

11. **(T/F)** In order for a matrix B to be the inverse of A, both equations $AB = I$ and $BA = I$ must be true.

12. **(T/F)** A product of invertible $n \times n$ matrices is invertible, and the inverse of the product is the product of their inverses in the same order.

13. **(T/F)** If A and B are $n \times n$ and invertible, then $A^{-1}B^{-1}$ is the inverse of AB.

14. **(T/F)** If A is invertible, then the inverse of A^{-1} is A itself.

15. **(T/F)** If $A = \begin{bmatrix} a & b \\ c & d \end{bmatrix}$ and $ab - cd \neq 0$, then A is invertible.

16. **(T/F)** If $A = \begin{bmatrix} a & b \\ c & d \end{bmatrix}$ and $ad = bc$, then A is not invertible.

17. **(T/F)** If A is an invertible $n \times n$ matrix, then the equation $A\mathbf{x} = \mathbf{b}$ is consistent for *each* $\mathbf{b}$ in $\mathbb{R}^n$.

18. **(T/F)** If A can be row reduced to the identity matrix, then A must be invertible.

19. **(T/F)** Each elementary matrix is invertible.

20. **(T/F)** If A is invertible, then the elementary row operations that reduce A to the identity I_n also reduce A^{-1} to I_n.

21. Let A be an invertible $n \times n$ matrix, and let B be an $n \times p$ matrix. Show that the equation $AX = B$ has a unique solution $A^{-1}B$.

22. Let A be an invertible $n \times n$ matrix, and let B be an $n \times p$ matrix. Explain why $A^{-1}B$ can be computed by row reduction:

If $[A \ \ B] \sim \cdots \sim [I \ \ X]$, then $X = A^{-1}B$.

If A is larger than 2×2, then row reduction of $[A \ \ B]$ is much faster than computing both A^{-1} and $A^{-1}B$.

23. Suppose $AB = AC$, where B and C are $n \times p$ matrices and A is invertible. Show that $B = C$. Is this true, in general, when A is not invertible?

24. Suppose $(B - C)D = 0$, where B and C are $m \times n$ matrices and D is invertible. Show that $B = C$.

25. Suppose A, B, and C are invertible $n \times n$ matrices. Show that ABC is also invertible by producing a matrix D such that $(ABC)D = I$ and $D(ABC) = I$.

26. Suppose A and B are $n \times n$, B is invertible, and AB is invertible. Show that A is invertible. [*Hint:* Let $C = AB$, and solve this equation for A.]

27. Solve the equation $AB = BC$ for A, assuming that A, B, and C are square and B is invertible.

28. Suppose P is invertible and $A = PBP^{-1}$. Solve for B in terms of A.

29. If A, B, and C are $n \times n$ invertible matrices, does the equation $C^{-1}(A + X)B^{-1} = I_n$ have a solution, X? If so, find it.

30. Suppose A, B, and X are $n \times n$ matrices with A, X, and $A - AX$ invertible, and suppose

$$(A - AX)^{-1} = X^{-1}B \quad (3)$$

a. Explain why B is invertible.

b. Solve (3) for X. If you need to invert a matrix, explain why that matrix is invertible.

31. Explain why the columns of an $n \times n$ matrix A are linearly independent when A is invertible.

32. Explain why the columns of an $n \times n$ matrix A span $\mathbb{R}^n$ when A is invertible. [*Hint:* Review Theorem 4 in Section 1.4.]

33. Suppose A is $n \times n$ and the equation $A\mathbf{x} = \mathbf{0}$ has only the trivial solution. Explain why A has n pivot columns and A is row equivalent to I_n. By Theorem 7, this shows that A must be invertible. (This exercise and Exercise 34 will be cited in Section 2.3.)

34. Suppose A is $n \times n$ and the equation $A\mathbf{x} = \mathbf{b}$ has a solution for each $\mathbf{b}$ in $\mathbb{R}^n$. Explain why A must be invertible. [*Hint:* Is A row equivalent to I_n?]

Exercises 35 and 36 prove Theorem 4 for $A = \begin{bmatrix} a & b \\ c & d \end{bmatrix}$.

35. Show that if $ad - bc = 0$, then the equation $A\mathbf{x} = \mathbf{0}$ has more than one solution. Why does this imply that A is not invertible? [*Hint:* First, consider $a = b = 0$. Then, if a and b are not both zero, consider the vector $\mathbf{x} = \begin{bmatrix} -b \\ a \end{bmatrix}$.]

36. Show that if $ad - bc \neq 0$, the formula for A^{-1} works.

Exercises 37 and 38 prove special cases of the facts about elementary matrices stated in the box following Example 5. Here A is a 3×3 matrix and $I = I_3$. (A general proof would require slightly more notation.)

37. a. Use equation (1) from Section 2.1 to show that $\text{row}_i(A) = \text{row}_i(I) \cdot A$, for $i = 1, 2, 3$.

b. Show that if rows 1 and 2 of A are interchanged, then the result may be written as EA, where E is an elementary matrix formed by interchanging rows 1 and 2 of I.

c. Show that if row 3 of A is multiplied by 5, then the result may be written as EA, where E is formed by multiplying row 3 of I by 5.

38. Show that if row 3 of A is replaced by $\text{row}_3(A) - 4\text{row}_1(A)$, the result is EA, where E is formed from I by replacing $\text{row}_3(I)$ by $\text{row}_3(I) - 4\text{row}_1(I)$.

Find the inverses of the matrices in Exercises 39–42, if they exist. Use the algorithm introduced in this section.

39. $\begin{bmatrix} 1 & 2 \\ 4 & 7 \end{bmatrix}$ **40.** $\begin{bmatrix} 5 & 10 \\ 4 & 7 \end{bmatrix}$

41. $\begin{bmatrix} 1 & 0 & -2 \\ -3 & 1 & 4 \\ 2 & -3 & 4 \end{bmatrix}$ **42.** $\begin{bmatrix} 1 & -2 & 1 \\ 4 & -7 & 3 \\ -2 & 6 & -4 \end{bmatrix}$

43. Use the algorithm from this section to find the inverses of

$$\begin{bmatrix} 1 & 0 & 0 \\ 1 & 1 & 0 \\ 1 & 1 & 1 \end{bmatrix} \text{ and } \begin{bmatrix} 1 & 0 & 0 & 0 \\ 1 & 1 & 0 & 0 \\ 1 & 1 & 1 & 0 \\ 1 & 1 & 1 & 1 \end{bmatrix}.$$

Let A be the corresponding $n \times n$ matrix, and let B be its inverse. Guess the form of B, and then prove that $AB = I$ and $BA = I$.

44. Repeat the strategy of Exercise 43 to guess the inverse of

$$A = \begin{bmatrix} 1 & 0 & 0 & \cdots & 0 \\ 1 & 2 & 0 & & 0 \\ 1 & 2 & 3 & & 0 \\ \vdots & & & \ddots & \vdots \\ 1 & 2 & 3 & \cdots & n \end{bmatrix}. \text{ Prove that your guess is correct.}$$

45. Let $A = \begin{bmatrix} -2 & -7 & -9 \\ 2 & 5 & 6 \\ 1 & 3 & 4 \end{bmatrix}$. Find the third column of A^{-1} without computing the other columns.

46. Let $A = \begin{bmatrix} -25 & -9 & -27 \\ 546 & 180 & 537 \\ 154 & 50 & 149 \end{bmatrix}$. Find the second and third columns of A^{-1} without computing the first column.

47. Let $A = \begin{bmatrix} 1 & 2 \\ 1 & 3 \\ 1 & 5 \end{bmatrix}$. Construct a 2×3 matrix C (by trial and error) using only 1, -1, and 0 as entries, such that $CA = I_2$. Compute AC and note that $AC \neq I_3$.

48. Let $A = \begin{bmatrix} 1 & 1 & 1 & 0 \\ 0 & 1 & 1 & 1 \end{bmatrix}$. Construct a 4×2 matrix D using only 1 and 0 as entries, such that $AD = I_2$. Is it possible that $CA = I_4$ for some 4×2 matrix C? Why or why not?

49. Let $D = \begin{bmatrix} .005 & .002 & .001 \\ .002 & .004 & .002 \\ .001 & .002 & .005 \end{bmatrix}$ be a flexibility matrix, with flexibility measured in inches per pound. Suppose that forces of 30, 50, and 20 lb are applied at points 1, 2, and 3, respectively, in Figure 1 of Example 3. Find the corresponding deflections.

T 50. Compute the stiffness matrix D^{-1} for D in Exercise 49. List the forces needed to produce a deflection of .04 in. at point 3, with zero deflections at the other points.

T 51. Let $D = \begin{bmatrix} .0040 & .0030 & .0010 & .0005 \\ .0030 & .0050 & .0030 & .0010 \\ .0010 & .0030 & .0050 & .0030 \\ .0005 & .0010 & .0030 & .0040 \end{bmatrix}$ be a flexibility matrix for an elastic beam with four points at which force is applied. Units are centimeters per newton of force. Measurements at the four points show deflections of .08, .12, .16, and .12 cm. Determine the forces at the four points.

Deflection of elastic beam in Exercises 51 and 52.

T 52. With D as in Exercise 51, determine the forces that produce a deflection of .24 cm at the second point on the beam, with zero deflections at the other three points. How is the answer related to the entries in D^{-1}? [*Hint:* First answer the question when the deflection is 1 cm at the second point.]

Solutions to Practice Problems

1. a. $\det \begin{bmatrix} 3 & -9 \\ 2 & 6 \end{bmatrix} = 3 \cdot 6 - (-9) \cdot 2 = 18 + 18 = 36$. The determinant is nonzero, so the matrix is invertible.

   b. $\det \begin{bmatrix} 4 & -9 \\ 0 & 5 \end{bmatrix} = 4 \cdot 5 - (-9) \cdot 0 = 20 \neq 0$. The matrix is invertible.

   c. $\det \begin{bmatrix} 6 & -9 \\ -4 & 6 \end{bmatrix} = 6 \cdot 6 - (-9)(-4) = 36 - 36 = 0$. The matrix is not invertible.

2. $[\,A \quad I\,] \sim \begin{bmatrix} 1 & -2 & -1 & 1 & 0 & 0 \\ -1 & 5 & 6 & 0 & 1 & 0 \\ 5 & -4 & 5 & 0 & 0 & 1 \end{bmatrix}$

   $\sim \begin{bmatrix} 1 & -2 & -1 & 1 & 0 & 0 \\ 0 & 3 & 5 & 1 & 1 & 0 \\ 0 & 6 & 10 & -5 & 0 & 1 \end{bmatrix}$

   $\sim \begin{bmatrix} 1 & -2 & -1 & 1 & 0 & 0 \\ 0 & 3 & 5 & 1 & 1 & 0 \\ 0 & 0 & 0 & -7 & -2 & 1 \end{bmatrix}$

   So $[\,A \quad I\,]$ is row equivalent to a matrix of the form $[\,B \quad D\,]$, where B is square and has a row of zeros. Further row operations will not transform B into I, so we stop. A does not have an inverse.

3. Since A is an invertible matrix, there exists a matrix C such that $AC = I = CA$. The goal is to find a matrix D so that $(5A)D = I = D(5A)$. Set $D = 1/5\,C$. Applying Theorem 2 from Section 2.1 establishes that $(5A)(1/5\,C) = (5)(1/5)(AC) = 1\,I = I$, and $(1/5\,C)(5A) = (1/5)(5)(CA) = 1\,I = I$. Thus $1/5\,C$ is indeed the inverse of A, proving that A is invertible.

2.3 Characterizations of Invertible Matrices

This section provides a review of most of the concepts introduced in Chapter 1, in relation to systems of n linear equations in n unknowns and to *square* matrices. The main result is Theorem 8.

THEOREM 8

The Invertible Matrix Theorem

Let A be a square $n \times n$ matrix. Then the following statements are equivalent. That is, for a given A, the statements are either all true or all false.

a. A is an invertible matrix.

b. A is row equivalent to the $n \times n$ identity matrix.

c. A has n pivot positions.

d. The equation $A\mathbf{x} = \mathbf{0}$ has only the trivial solution.

e. The columns of A form a linearly independent set.

f. The linear transformation $\mathbf{x} \mapsto A\mathbf{x}$ is one-to-one.

g. The equation $A\mathbf{x} = \mathbf{b}$ has at least one solution for each $\mathbf{b}$ in $\mathbb{R}^n$.

h. The columns of A span $\mathbb{R}^n$.

i. The linear transformation $\mathbf{x} \mapsto A\mathbf{x}$ maps $\mathbb{R}^n$ onto $\mathbb{R}^n$.

j. There is an $n \times n$ matrix C such that $CA = I$.

k. There is an $n \times n$ matrix D such that $AD = I$.

l. A^T is an invertible matrix.

FIGURE 1

First, we need some notation. If the truth of statement (a) always implies that statement (j) is true, we say that (a) *implies* (j) and write (a) $\Rightarrow$ (j). The proof will establish the "circle" of implications shown in Figure 1. If any one of these five statements is true, then so are the others. Finally, the proof will link the remaining statements of the theorem to the statements in this circle.

PROOF If statement (a) is true, then A^{-1} works for C in (j), so (a) $\Rightarrow$ (j). Next, (j) $\Rightarrow$ (d) by Exercise 31 in Section 2.1. (Turn back and read the exercise.) Also, (d) $\Rightarrow$ (c) by Exercise 33 in Section 2.2. If A is square and has n pivot positions, then the pivots must lie on the main diagonal, in which case the reduced echelon form of A is I_n. Thus (c) $\Rightarrow$ (b). Also, (b) $\Rightarrow$ (a) by Theorem 7 in Section 2.2. This completes the circle in Figure 1.

Next, (a) $\Rightarrow$ (k) because A^{-1} works for D. Also, (k) $\Rightarrow$ (g) by Exercise 32 in Section 2.1, and (g) $\Rightarrow$ (a) by Exercise 34 in Section 2.2. So (k) and (g) are linked to the circle. Further, (g), (h), and (i) are equivalent for any matrix, by Theorem 4 in Section 1.4 and Theorem 12(a) in Section 1.9. Thus, (h) and (i) are linked through (g) to the circle.

Since (d) is linked to the circle, so are (e) and (f), because (d), (e), and (f) are all equivalent for *any* matrix A. (See Section 1.7 and Theorem 12(b) in Section 1.9.) Finally, (a) $\Rightarrow$ (l) by Theorem 6(c) in Section 2.2, and (l) $\Rightarrow$ (a) by the same theorem with A and A^T interchanged. This completes the proof. ∎

Because of Theorem 5 in Section 2.2, statement (g) in Theorem 8 could also be written as "The equation $A\mathbf{x} = \mathbf{b}$ has a *unique* solution for each $\mathbf{b}$ in $\mathbb{R}^n$." This statement certainly implies (b) and hence implies that A is invertible.

The next fact follows from Theorem 8 and Exercise 10 in Section 2.2.

> Let A and B be square matrices. If $AB = I$, then A and B are both invertible, with $B = A^{-1}$ and $A = B^{-1}$.

The Invertible Matrix Theorem divides the set of all $n \times n$ matrices into two disjoint classes: the invertible (nonsingular) matrices, and the noninvertible (singular) matrices. Each statement in the theorem describes a property of every $n \times n$ invertible matrix. The *negation* of a statement in the theorem describes a property of every $n \times n$ singular matrix. For instance, an $n \times n$ singular matrix is *not* row equivalent to I_n, does *not* have n pivot positions, and has linearly *dependent* columns. Negations of other statements are considered in the exercises.

EXAMPLE 1 Use the Invertible Matrix Theorem to decide if A is invertible:

$$A = \begin{bmatrix} 1 & 0 & -2 \\ 3 & 1 & -2 \\ -5 & -1 & 9 \end{bmatrix}$$

SOLUTION

$$A \sim \begin{bmatrix} 1 & 0 & -2 \\ 0 & 1 & 4 \\ 0 & -1 & -1 \end{bmatrix} \sim \begin{bmatrix} 1 & 0 & -2 \\ 0 & 1 & 4 \\ 0 & 0 & 3 \end{bmatrix}$$

So A has three pivot positions and hence is invertible, by the Invertible Matrix Theorem, statement (c). ■

STUDY GUIDE offers an expanded table for the Invertible Matrix Theorem.

The power of the Invertible Matrix Theorem lies in the connections it provides among so many important concepts, such as linear independence of columns of a matrix A and the existence of solutions to equations of the form $A\mathbf{x} = \mathbf{b}$. It should be emphasized, however, that the Invertible Matrix Theorem *applies only to square matrices*. For example, if the columns of a 4×3 matrix are linearly independent, we cannot use the Invertible Matrix Theorem to conclude anything about the existence or nonexistence of solutions to equations of the form $A\mathbf{x} = \mathbf{b}$.

Invertible Linear Transformations

Recall from Section 2.1 that matrix multiplication corresponds to composition of linear transformations. When a matrix A is invertible, the equation $A^{-1}A\mathbf{x} = \mathbf{x}$ can be viewed as a statement about linear transformations. See Figure 2.

FIGURE 2 A^{-1} transforms $A\mathbf{x}$ back to $\mathbf{x}$.

A linear transformation $T : \mathbb{R}^n \to \mathbb{R}^n$ is said to be **invertible** if there exists a function $S : \mathbb{R}^n \to \mathbb{R}^n$ such that

$$S(T(\mathbf{x})) = \mathbf{x} \quad \text{for all } \mathbf{x} \text{ in } \mathbb{R}^n \tag{1}$$

$$T(S(\mathbf{x})) = \mathbf{x} \quad \text{for all } \mathbf{x} \text{ in } \mathbb{R}^n \tag{2}$$

The next theorem shows that if such an S exists, it is unique and must be a linear transformation. We call S the **inverse** of T and write it as T^{-1}.

THEOREM 9

Let $T : \mathbb{R}^n \to \mathbb{R}^n$ be a linear transformation and let A be the standard matrix for T. Then T is invertible if and only if A is an invertible matrix. In that case, the linear transformation S given by $S(\mathbf{x}) = A^{-1}\mathbf{x}$ is the unique function satisfying equations (1) and (2).

Remark: See the comment on the proof of Theorem 7.

PROOF Suppose that T is invertible. Then (2) shows that T is onto $\mathbb{R}^n$, for if $\mathbf{b}$ is in $\mathbb{R}^n$ and $\mathbf{x} = S(\mathbf{b})$, then $T(\mathbf{x}) = T(S(\mathbf{b})) = \mathbf{b}$, so each $\mathbf{b}$ is in the range of T. Thus A is invertible, by the Invertible Matrix Theorem, statement (i).

Conversely, suppose that A is invertible, and let $S(\mathbf{x}) = A^{-1}\mathbf{x}$. Then, S is a linear transformation, and S obviously satisfies (1) and (2). For instance,

$$S(T(\mathbf{x})) = S(A\mathbf{x}) = A^{-1}(A\mathbf{x}) = \mathbf{x}$$

Thus T is invertible. The proof that S is unique is outlined in Exercise 47. ∎

EXAMPLE 2 What can you say about a one-to-one linear transformation T from $\mathbb{R}^n$ into $\mathbb{R}^n$?

SOLUTION The columns of the standard matrix A of T are linearly independent (by Theorem 12 in Section 1.9). So A is invertible, by the Invertible Matrix Theorem, and T maps $\mathbb{R}^n$ *onto* $\mathbb{R}^n$. Also, T is invertible, by Theorem 9. ∎

Numerical Notes

In practical work, you might occasionally encounter a "nearly singular" or **ill-conditioned** matrix—an invertible matrix that can become singular if some of its entries are changed ever so slightly. In this case, row reduction may produce fewer than n pivot positions, as a result of roundoff error. Also, roundoff error can sometimes make a singular matrix appear to be invertible.

Some matrix programs will compute a **condition number** for a square matrix. The larger the condition number, the closer the matrix is to being singular. The condition number of the identity matrix is 1. A singular matrix has an infinite condition number. In extreme cases, a matrix program may not be able to distinguish between a singular matrix and an ill-conditioned matrix.

Exercises 49–53 show that matrix computations can produce substantial error when a condition number is large.

Practice Problems

1. Determine if $A = \begin{bmatrix} 2 & 3 & 4 \\ 2 & 3 & 4 \\ 2 & 3 & 4 \end{bmatrix}$ is invertible.

2. Suppose that for a certain $n \times n$ matrix A, statement (g) of the Invertible Matrix Theorem is *not* true. What can you say about equations of the form $A\mathbf{x} = \mathbf{b}$?

3. Suppose that A and B are $n \times n$ matrices and the equation $AB\mathbf{x} = \mathbf{0}$ has a nontrivial solution. What can you say about the matrix AB?

2.3 Exercises

Unless otherwise specified, assume that all matrices in these exercises are $n \times n$. Determine which of the matrices in Exercises 1–10 are invertible. Use as few calculations as possible. Justify your answers.

1. $\begin{bmatrix} 5 & 7 \\ -3 & -6 \end{bmatrix}$

2. $\begin{bmatrix} -4 & 6 \\ 6 & -9 \end{bmatrix}$

3. $\begin{bmatrix} 5 & 0 & 0 \\ -3 & -7 & 0 \\ 8 & 5 & -1 \end{bmatrix}$

4. $\begin{bmatrix} -7 & 0 & 4 \\ 3 & 0 & -1 \\ 2 & 0 & 9 \end{bmatrix}$

5. $\begin{bmatrix} 0 & 3 & -5 \\ 1 & 0 & 2 \\ -4 & -9 & 7 \end{bmatrix}$

6. $\begin{bmatrix} 1 & -5 & -4 \\ 0 & 3 & 4 \\ -3 & 6 & 0 \end{bmatrix}$

7. $\begin{bmatrix} -1 & -3 & 0 & 1 \\ 3 & 5 & 8 & -3 \\ -2 & -6 & 3 & 2 \\ 0 & -1 & 2 & 1 \end{bmatrix}$

8. $\begin{bmatrix} 1 & 3 & 7 & 4 \\ 0 & 5 & 9 & 6 \\ 0 & 0 & 2 & 8 \\ 0 & 0 & 0 & 10 \end{bmatrix}$

T 9. $\begin{bmatrix} 4 & 0 & -7 & -7 \\ -6 & 1 & 11 & 9 \\ 7 & -5 & 10 & 19 \\ -1 & 2 & 3 & -1 \end{bmatrix}$

T 10. $\begin{bmatrix} 5 & 3 & 1 & 7 & 9 \\ 6 & 4 & 2 & 8 & -8 \\ 7 & 5 & 3 & 10 & 9 \\ 9 & 6 & 4 & -9 & -5 \\ 8 & 5 & 2 & 11 & 4 \end{bmatrix}$

In Exercises 11–20, the matrices are all $n \times n$. Each part of the exercises is an *implication* of the form "If 'statement 1', then 'statement 2'." Mark an implication as True if the truth of "statement 2" *always* follows whenever "statement 1" happens to be true. An implication is False if there is an instance in which "statement 2" is false but "statement 1" is true. Justify each answer.

11. **(T/F)** If the equation $A\mathbf{x} = \mathbf{0}$ has only the trivial solution, then A is row equivalent to the $n \times n$ identity matrix.

12. **(T/F)** If there is an $n \times n$ matrix D such that $AD = I$, then there is also an $n \times n$ matrix C such that $CA = I$.

13. **(T/F)** If the columns of A span $\mathbb{R}^n$, then the columns are linearly independent.

14. **(T/F)** If the columns of A are linearly independent, then the columns of A span $\mathbb{R}^n$.

15. **(T/F)** If A is an $n \times n$ matrix, then the equation $A\mathbf{x} = \mathbf{b}$ has at least one solution for each $\mathbf{b}$ in $\mathbb{R}^n$.

16. **(T/F)** If the equation $A\mathbf{x} = \mathbf{b}$ has at least one solution for each $\mathbf{b}$ in $\mathbb{R}^n$, then the solution is unique for each $\mathbf{b}$.

17. **(T/F)** If the equation $A\mathbf{x} = \mathbf{0}$ has a nontrivial solution, then A has fewer than n pivot positions.

18. **(T/F)** If the linear transformation $\mathbf{x} \mapsto A\mathbf{x}$ maps $\mathbb{R}^n$ into $\mathbb{R}^n$, then A has n pivot positions.

19. **(T/F)** If A^T is not invertible, then A is not invertible.

20. **(T/F)** If there is a $\mathbf{b}$ in $\mathbb{R}^n$ such that the equation $A\mathbf{x} = \mathbf{b}$ is inconsistent, then the transformation $\mathbf{x} \mapsto A\mathbf{x}$ is not one-to-one.

21. An $m \times n$ **upper triangular matrix** is one whose entries *below* the main diagonal are 0's (as in Exercise 8). When is a square upper triangular matrix invertible? Justify your answer.

22. An $m \times n$ **lower triangular matrix** is one whose entries *above* the main diagonal are 0's (as in Exercise 3). When is a square lower triangular matrix invertible? Justify your answer.

23. Can a square matrix with two identical columns be invertible? Why or why not?

24. Is it possible for a 5×5 matrix to be invertible when its columns do not span $\mathbb{R}^5$? Why or why not?

25. If A is invertible, then the columns of A^{-1} are linearly independent. Explain why.

26. If C is 6×6 and the equation $C\mathbf{x} = \mathbf{v}$ is consistent for every $\mathbf{v}$ in $\mathbb{R}^6$, is it possible that for some $\mathbf{v}$, the equation $C\mathbf{x} = \mathbf{v}$ has more than one solution? Why or why not?

27. If the columns of a 7×7 matrix D are linearly independent, what can you say about solutions of $D\mathbf{x} = \mathbf{b}$? Why?

28. If $n \times n$ matrices E and F have the property that $EF = I$, then E and F commute. Explain why.

29. If the equation $G\mathbf{x} = \mathbf{y}$ has more than one solution for some $\mathbf{y}$ in $\mathbb{R}^n$, can the columns of G span $\mathbb{R}^n$? Why or why not?

30. If the equation $H\mathbf{x} = \mathbf{c}$ is inconsistent for some $\mathbf{c}$ in $\mathbb{R}^n$, what can you say about the equation $H\mathbf{x} = \mathbf{0}$? Why?

31. If an $n \times n$ matrix K cannot be row reduced to I_n, what can you say about the columns of K? Why?

32. If L is $n \times n$ and the equation $L\mathbf{x} = \mathbf{0}$ has the trivial solution, do the columns of L span $\mathbb{R}^n$? Why?

33. Verify the boxed statement preceding Example 1.

34. Explain why the columns of A^2 span $\mathbb{R}^n$ whenever the columns of A are linearly independent.

35. Show that if AB is invertible, so is A. You cannot use Theorem 6(b), because you cannot *assume* that A and B are invertible. [*Hint:* There is a matrix W such that $ABW = I$. Why?]

36. Show that if AB is invertible, so is B.

37. If A is an $n \times n$ matrix and the equation $A\mathbf{x} = \mathbf{b}$ has more than one solution for some $\mathbf{b}$, then the transformation $\mathbf{x} \mapsto A\mathbf{x}$ is

not one-to-one. What else can you say about this transformation? Justify your answer.

38. If A is an $n \times n$ matrix and the transformation $\mathbf{x} \mapsto A\mathbf{x}$ is one-to-one, what else can you say about this transformation? Justify your answer.

39. Suppose A is an $n \times n$ matrix with the property that the equation $A\mathbf{x} = \mathbf{b}$ has at least one solution for each $\mathbf{b}$ in $\mathbb{R}^n$. Without using Theorems 5 or 8, explain why each equation $A\mathbf{x} = \mathbf{b}$ has in fact exactly one solution.

40. Suppose A is an $n \times n$ matrix with the property that the equation $A\mathbf{x} = \mathbf{0}$ has only the trivial solution. Without using the Invertible Matrix Theorem, explain directly why the equation $A\mathbf{x} = \mathbf{b}$ must have a solution for each $\mathbf{b}$ in $\mathbb{R}^n$.

In Exercises 41 and 42, T is a linear transformation from $\mathbb{R}^2$ into $\mathbb{R}^2$. Show that T is invertible and find a formula for T^{-1}.

41. $T(x_1, x_2) = (-5x_1 + 9x_2, 4x_1 - 7x_2)$

42. $T(x_1, x_2) = (6x_1 - 8x_2, -5x_1 + 7x_2)$

43. Let $T : \mathbb{R}^n \to \mathbb{R}^n$ be an invertible linear transformation. Explain why T is both one-to-one and onto $\mathbb{R}^n$. Use equations (1) and (2). Then give a second explanation using one or more theorems.

44. Let T be a linear transformation that maps $\mathbb{R}^n$ onto $\mathbb{R}^n$. Show that T^{-1} exists and maps $\mathbb{R}^n$ onto $\mathbb{R}^n$. Is T^{-1} also one-to-one?

45. Suppose T and U are linear transformations from $\mathbb{R}^n$ to $\mathbb{R}^n$ such that $T(U\mathbf{x}) = \mathbf{x}$ for all $\mathbf{x}$ in $\mathbb{R}^n$. Is it true that $U(T\mathbf{x}) = \mathbf{x}$ for all $\mathbf{x}$ in $\mathbb{R}^n$? Why or why not?

46. Suppose a linear transformation $T : \mathbb{R}^n \to \mathbb{R}^n$ has the property that $T(\mathbf{u}) = T(\mathbf{v})$ for some pair of distinct vectors $\mathbf{u}$ and $\mathbf{v}$ in $\mathbb{R}^n$. Can T map $\mathbb{R}^n$ onto $\mathbb{R}^n$? Why or why not?

47. Let $T : \mathbb{R}^n \to \mathbb{R}^n$ be an invertible linear transformation, and let S and U be functions from $\mathbb{R}^n$ into $\mathbb{R}^n$ such that $S(T(\mathbf{x})) = \mathbf{x}$ and $U(T(\mathbf{x})) = \mathbf{x}$ for all $\mathbf{x}$ in $\mathbb{R}^n$. Show that $U(\mathbf{v}) = S(\mathbf{v})$ for all $\mathbf{v}$ in $\mathbb{R}^n$. This will show that T has a unique inverse, as asserted in Theorem 9. [*Hint:* Given any $\mathbf{v}$ in $\mathbb{R}^n$, we can write $\mathbf{v} = T(\mathbf{x})$ for some $\mathbf{x}$. Why? Compute $S(\mathbf{v})$ and $U(\mathbf{v})$.]

48. Suppose T and S satisfy the invertibility equations (1) and (2), where T is a linear transformation. Show directly that S is a linear transformation. [*Hint:* Given $\mathbf{u}$, $\mathbf{v}$ in $\mathbb{R}^n$, let $\mathbf{x} = S(\mathbf{u})$, $\mathbf{y} = S(\mathbf{v})$. Then $T(\mathbf{x}) = \mathbf{u}$, $T(\mathbf{y}) = \mathbf{v}$. Why? Apply S to both sides of the equation $T(\mathbf{x}) + T(\mathbf{y}) = T(\mathbf{x} + \mathbf{y})$. Also, consider $T(c\mathbf{x}) = cT(\mathbf{x})$.]

T 49. Suppose an experiment leads to the following system of equations:

$$4.5x_1 + 3.1x_2 = 19.249 \quad (3)$$
$$1.6x_1 + 1.1x_2 = 6.843$$

a. Solve system (3), and then solve system (4), below, in which the data on the right have been rounded to two decimal places. In each case, find the *exact* solution.

$$4.5x_1 + 3.1x_2 = 19.25 \quad (4)$$
$$1.6x_1 + 1.1x_2 = 6.84$$

b. The entries in (4) differ from those in (3) by less than .05%. Find the percentage error when using the solution of (4) as an approximation for the solution of (3).

c. Use your matrix program to produce the condition number of the coefficient matrix in (3).

Exercises 50–52 show how to use the condition number of a matrix A to estimate the accuracy of a computed solution of $A\mathbf{x} = \mathbf{b}$. If the entries of A and $\mathbf{b}$ are accurate to about r significant digits and if the condition number of A is approximately 10^k (with k a positive integer), then the computed solution of $A\mathbf{x} = \mathbf{b}$ should usually be accurate to at least $r - k$ significant digits.

T 50. Find the condition number of the matrix A in Exercise 9. Construct a random vector $\mathbf{x}$ in $\mathbb{R}^4$ and compute $\mathbf{b} = A\mathbf{x}$. Then use your matrix program to compute the solution $\mathbf{x}_1$ of $A\mathbf{x} = \mathbf{b}$. To how many digits do $\mathbf{x}$ and $\mathbf{x}_1$ agree? Find out the number of digits your matrix program stores accurately, and report how many digits of accuracy are lost when $\mathbf{x}_1$ is used in place of the exact solution $\mathbf{x}$.

T 51. Repeat Exercise 50 for the matrix in Exercise 10.

T 52. Solve an equation $A\mathbf{x} = \mathbf{b}$ for a suitable $\mathbf{b}$ to find the last column of the inverse of the *fifth-order Hilbert matrix*

$$A = \begin{bmatrix} 1 & 1/2 & 1/3 & 1/4 & 1/5 \\ 1/2 & 1/3 & 1/4 & 1/5 & 1/6 \\ 1/3 & 1/4 & 1/5 & 1/6 & 1/7 \\ 1/4 & 1/5 & 1/6 & 1/7 & 1/8 \\ 1/5 & 1/6 & 1/7 & 1/8 & 1/9 \end{bmatrix}$$

How many digits in each entry of $\mathbf{x}$ do you expect to be correct? Explain. [*Note:* The exact solution is $(630, -12600, 56700, -88200, 44100)$.]

T 53. Some matrix programs, such as MATLAB, have a command to create Hilbert matrices of various sizes. If possible, use an inverse command to compute the inverse of a twelfth-order or larger Hilbert matrix, A. Compute AA^{-1}. Report what you find.

Solutions to Practice Problems

1. The columns of A are obviously linearly dependent because columns 2 and 3 are multiples of column 1. Hence, A cannot be invertible (by the Invertible Matrix Theorem).

> **Solutions to Practice Problems (Continued)**
>
> **2.** If statement (g) is *not* true, then the equation $A\mathbf{x} = \mathbf{b}$ is inconsistent for at least one $\mathbf{b}$ in $\mathbb{R}^n$.
>
> **3.** Apply the Invertible Matrix Theorem to the matrix AB in place of A. Then statement (d) becomes: $AB\mathbf{x} = \mathbf{0}$ has only the trivial solution. This is not true. So AB is not invertible.

2.4 Partitioned Matrices

A key feature of our work with matrices has been the ability to regard a matrix A as a list of column vectors rather than just a rectangular array of numbers. This point of view has been so useful that we wish to consider other **partitions** of A, indicated by horizontal and vertical dividing rules, as in Example 1 below. Partitioned matrices appear in most modern applications of linear algebra because the notation highlights essential structures in matrix analysis, as in the chapter introductory example on aircraft design. This section provides an opportunity to review matrix algebra and use the Invertible Matrix Theorem.

EXAMPLE 1 The matrix

$$
A = \left[\begin{array}{rrr|rr|r}
3 & 0 & -1 & 5 & 9 & -2 \\
-5 & 2 & 4 & 0 & -3 & 1 \\
\hline
-8 & -6 & 3 & 1 & 7 & -4
\end{array}\right]
$$

can also be written as the 2×3 **partitioned** (or **block**) **matrix**

$$
A = \begin{bmatrix} A_{11} & A_{12} & A_{13} \\ A_{21} & A_{22} & A_{23} \end{bmatrix}
$$

whose entries are the *blocks* (or *submatrices*)

$$
A_{11} = \begin{bmatrix} 3 & 0 & -1 \\ -5 & 2 & 4 \end{bmatrix}, \quad A_{12} = \begin{bmatrix} 5 & 9 \\ 0 & -3 \end{bmatrix}, \quad A_{13} = \begin{bmatrix} -2 \\ 1 \end{bmatrix}
$$

$$
A_{21} = \begin{bmatrix} -8 & -6 & 3 \end{bmatrix}, \quad A_{22} = \begin{bmatrix} 1 & 7 \end{bmatrix}, \quad A_{23} = \begin{bmatrix} -4 \end{bmatrix} \qquad \blacksquare
$$

EXAMPLE 2 When a matrix A appears in a mathematical model of a physical system such as an electrical network, a transportation system, or a large corporation, it may be natural to regard A as a partitioned matrix. For instance, if a microcomputer circuit board consists mainly of three VLSI (very large-scale integrated) microchips, then the matrix for the circuit board might have the general form

$$
A = \left[\begin{array}{c|c|c}
A_{11} & A_{12} & A_{13} \\
\hline
A_{21} & A_{22} & A_{23} \\
\hline
A_{31} & A_{32} & A_{33}
\end{array}\right]
$$

The submatrices on the "diagonal" of A—namely A_{11}, A_{22}, and A_{33}—concern the three VLSI chips, while the other submatrices depend on the interconnections among those microchips. $\qquad \blacksquare$

Addition and Scalar Multiplication

If matrices A and B are the same size and are partitioned in exactly the same way, then it is natural to make the same partition of the ordinary matrix sum $A + B$. In this case, each block of $A + B$ is the (matrix) sum of the corresponding blocks of A and B. Multiplication of a partitioned matrix by a scalar is also computed block by block.

Multiplication of Partitioned Matrices

Partitioned matrices can be multiplied by the usual row–column rule as if the block entries were scalars, provided that for a product AB, the column partition of A matches the row partition of B.

EXAMPLE 3 Let

$$A = \begin{bmatrix} 2 & -3 & 1 & 0 & -4 \\ 1 & 5 & -2 & 3 & -1 \\ 0 & -4 & -2 & 7 & -1 \end{bmatrix} = \begin{bmatrix} A_{11} & A_{12} \\ A_{21} & A_{22} \end{bmatrix}, \quad B = \begin{bmatrix} 6 & 4 \\ -2 & 1 \\ -3 & 7 \\ -1 & 3 \\ 5 & 2 \end{bmatrix} = \begin{bmatrix} B_1 \\ B_2 \end{bmatrix}$$

The 5 columns of A are partitioned into a set of 3 columns and then a set of 2 columns. The 5 rows of B are partitioned in the same way—into a set of 3 rows and then a set of 2 rows. We say that the partitions of A and B are **conformable** for **block multiplication**. It can be shown that the ordinary product AB can be written as

$$AB = \begin{bmatrix} A_{11} & A_{12} \\ A_{21} & A_{22} \end{bmatrix} \begin{bmatrix} B_1 \\ B_2 \end{bmatrix} = \begin{bmatrix} A_{11}B_1 + A_{12}B_2 \\ A_{21}B_1 + A_{22}B_2 \end{bmatrix} = \begin{bmatrix} -5 & 4 \\ -6 & 2 \\ 2 & 1 \end{bmatrix}$$

It is important for each smaller product in the expression for AB to be written with the submatrix from A on the left, since matrix multiplication is not commutative. For instance,

$$A_{11}B_1 = \begin{bmatrix} 2 & -3 & 1 \\ 1 & 5 & -2 \end{bmatrix} \begin{bmatrix} 6 & 4 \\ -2 & 1 \\ -3 & 7 \end{bmatrix} = \begin{bmatrix} 15 & 12 \\ 2 & -5 \end{bmatrix}$$

$$A_{12}B_2 = \begin{bmatrix} 0 & -4 \\ 3 & -1 \end{bmatrix} \begin{bmatrix} -1 & 3 \\ 5 & 2 \end{bmatrix} = \begin{bmatrix} -20 & -8 \\ -8 & 7 \end{bmatrix}$$

Hence the top block in AB is

$$A_{11}B_1 + A_{12}B_2 = \begin{bmatrix} 15 & 12 \\ 2 & -5 \end{bmatrix} + \begin{bmatrix} -20 & -8 \\ -8 & 7 \end{bmatrix} = \begin{bmatrix} -5 & 4 \\ -6 & 2 \end{bmatrix} \qquad \blacksquare$$

The row–column rule for multiplication of block matrices provides the most general way to regard the product of two matrices. Each of the following views of a product has already been described using simple partitions of matrices: (1) the definition of $A\mathbf{x}$ using the columns of A, (2) the column definition of AB, (3) the row–column rule for computing AB, and (4) the rows of AB as products of the rows of A and the matrix B. A fifth view of AB, again using partitions, follows in Theorem 10.

The calculations in the next example prepare the way for Theorem 10. Here $\mathrm{col}_k(A)$ is the kth column of A, and $\mathrm{row}_k(B)$ is the kth row of B.

EXAMPLE 4 Let $A = \begin{bmatrix} -3 & 1 & 2 \\ 1 & -4 & 5 \end{bmatrix}$ and $B = \begin{bmatrix} a & b \\ c & d \\ e & f \end{bmatrix}$. Verify that

$$AB = \mathrm{col}_1(A)\,\mathrm{row}_1(B) + \mathrm{col}_2(A)\,\mathrm{row}_2(B) + \mathrm{col}_3(A)\,\mathrm{row}_3(B)$$

SOLUTION Each term in the preceding equation is an *outer product*. (See Exercises 35 and 36 in Section 2.1.) By the row–column rule for computing a matrix product,

$$\operatorname{col}_1(A)\operatorname{row}_1(B) = \begin{bmatrix} -3 \\ 1 \end{bmatrix}\begin{bmatrix} a & b \end{bmatrix} = \begin{bmatrix} -3a & -3b \\ a & b \end{bmatrix}$$

$$\operatorname{col}_2(A)\operatorname{row}_2(B) = \begin{bmatrix} 1 \\ -4 \end{bmatrix}\begin{bmatrix} c & d \end{bmatrix} = \begin{bmatrix} c & d \\ -4c & -4d \end{bmatrix}$$

$$\operatorname{col}_3(A)\operatorname{row}_3(B) = \begin{bmatrix} 2 \\ 5 \end{bmatrix}\begin{bmatrix} e & f \end{bmatrix} = \begin{bmatrix} 2e & 2f \\ 5e & 5f \end{bmatrix}$$

Thus

$$\sum_{k=1}^{3} \operatorname{col}_k(A)\operatorname{row}_k(B) = \begin{bmatrix} -3a + c + 2e & -3b + d + 2f \\ a - 4c + 5e & b - 4d + 5f \end{bmatrix}$$

This matrix is obviously AB. Notice that the $(1,1)$-entry in AB is the sum of the $(1,1)$-entries in the three outer products, the $(1,2)$-entry in AB is the sum of the $(1,2)$-entries in the three outer products, and so on. ∎

THEOREM 10

Column–Row Expansion of AB

If A is $m \times n$ and B is $n \times p$, then

$$AB = \begin{bmatrix} \operatorname{col}_1(A) & \operatorname{col}_2(A) & \cdots & \operatorname{col}_n(A) \end{bmatrix} \begin{bmatrix} \operatorname{row}_1(B) \\ \operatorname{row}_2(B) \\ \vdots \\ \operatorname{row}_n(B) \end{bmatrix} \tag{1}$$

$$= \operatorname{col}_1(A)\operatorname{row}_1(B) + \cdots + \operatorname{col}_n(A)\operatorname{row}_n(B)$$

PROOF For each row index i and column index j, the (i, j)-entry in $\operatorname{col}_k(A)\operatorname{row}_k(B)$ is the product of a_{ik} from $\operatorname{col}_k(A)$ and b_{kj} from $\operatorname{row}_k(B)$. Hence the (i, j)-entry in the sum shown in equation (1) is

$$\underset{(k=1)}{a_{i1}b_{1j}} \quad + \quad \underset{(k=2)}{a_{i2}b_{2j}} \quad + \quad \cdots \quad + \quad \underset{(k=n)}{a_{in}b_{nj}}$$

This sum is also the (i, j)-entry in AB, by the row–column rule. ∎

Inverses of Partitioned Matrices

The next example illustrates calculations involving inverses and partitioned matrices.

EXAMPLE 5 A matrix of the form

$$A = \begin{bmatrix} A_{11} & A_{12} \\ 0 & A_{22} \end{bmatrix}$$

is said to be *block upper triangular*. Assume that A_{11} is $p \times p$, A_{22} is $q \times q$, and A is invertible. Find a formula for A^{-1}.

SOLUTION Denote A^{-1} by B and partition B so that

$$\begin{bmatrix} A_{11} & A_{12} \\ 0 & A_{22} \end{bmatrix}\begin{bmatrix} B_{11} & B_{12} \\ B_{21} & B_{22} \end{bmatrix} = \begin{bmatrix} I_p & 0 \\ 0 & I_q \end{bmatrix} \tag{2}$$

This matrix equation provides four equations that will lead to the unknown blocks $B_{11}, \ldots, B_{22}$. Compute the product on the left side of equation (2), and equate each entry with the corresponding block in the identity matrix on the right. That is, set

$$A_{11}B_{11} + A_{12}B_{21} = I_p \tag{3}$$
$$A_{11}B_{12} + A_{12}B_{22} = 0 \tag{4}$$
$$A_{22}B_{21} = 0 \tag{5}$$
$$A_{22}B_{22} = I_q \tag{6}$$

By itself, equation (6) does not show that A_{22} is invertible. However, since A_{22} is square, the Invertible Matrix Theorem and (6) together show that A_{22} is invertible and $B_{22} = A_{22}^{-1}$. Next, left-multiply both sides of (5) by A_{22}^{-1} and obtain

$$B_{21} = A_{22}^{-1}0 = 0$$

so that (3) simplifies to

$$A_{11}B_{11} + 0 = I_p$$

Since A_{11} is square, this shows that A_{11} is invertible and $B_{11} = A_{11}^{-1}$. Finally, use these results with (4) to find that

$$A_{11}B_{12} = -A_{12}B_{22} = -A_{12}A_{22}^{-1} \quad \text{and} \quad B_{12} = -A_{11}^{-1}A_{12}A_{22}^{-1}$$

Thus

$$A^{-1} = \begin{bmatrix} A_{11} & A_{12} \\ 0 & A_{22} \end{bmatrix}^{-1} = \begin{bmatrix} A_{11}^{-1} & -A_{11}^{-1}A_{12}A_{22}^{-1} \\ 0 & A_{22}^{-1} \end{bmatrix} \qquad \blacksquare$$

A **block diagonal matrix** is a partitioned matrix with zero blocks off the main diagonal (of blocks). Such a matrix is invertible if and only if each block on the diagonal is invertible. See Exercises 15 and 16.

Numerical Notes

1. When matrices are too large to fit in a computer's high-speed memory, partitioning permits the computer to work with only two or three submatrices at a time. For instance, one linear programming research team simplified a problem by partitioning the matrix into 837 rows and 51 columns. The problem's solution took about 4 minutes on a Cray supercomputer.[1]

2. Some high-speed computers, particularly those with vector pipeline architecture, perform matrix calculations more efficiently when the algorithms use partitioned matrices.[2]

3. Professional software for high-performance numerical linear algebra, such as LAPACK, makes intensive use of partitioned matrix calculations.

---

[1] The solution time doesn't sound too impressive until you learn that each of the 51 block columns contained about 250,000 individual columns. The original problem had 837 equations and more than 12,750,000 variables! Nearly 100 million of the more than 10 billion entries in the matrix were nonzero. See Robert E. Bixby et al., "Very Large-Scale Linear Programming: A Case Study in Combining Interior Point and Simplex Methods," *Operations Research*, 40, no. 5 (1992): 885–897.

[2] The importance of block matrix algorithms for computer calculations is described in *Matrix Computations*, 3rd ed., by Gene H. Golub and Charles F. van Loan (Baltimore: Johns Hopkins University Press, 1996).

The exercises that follow give practice with matrix algebra and illustrate typical calculations found in applications.

Practice Problems

1. Show that $\begin{bmatrix} I & 0 \\ A & I \end{bmatrix}$ is invertible and find its inverse.

2. Compute $X^T X$, where X is partitioned as $\begin{bmatrix} X_1 & X_2 \end{bmatrix}$.

2.4 Exercises

In Exercises 1–9, assume that the matrices are partitioned conformably for block multiplication. Compute the products shown in Exercises 1–4.

1. $\begin{bmatrix} I & 0 \\ E & I \end{bmatrix}\begin{bmatrix} A & B \\ C & D \end{bmatrix}$

2. $\begin{bmatrix} E & 0 \\ 0 & F \end{bmatrix}\begin{bmatrix} A & B \\ C & D \end{bmatrix}$

3. $\begin{bmatrix} 0 & I \\ I & 0 \end{bmatrix}\begin{bmatrix} W & X \\ Y & Z \end{bmatrix}$

4. $\begin{bmatrix} I & 0 \\ -X & I \end{bmatrix}\begin{bmatrix} A & B \\ C & D \end{bmatrix}$

In Exercises 5–8, find formulas for X, Y, and Z in terms of A, B, and C, and justify your calculations. In some cases, you may need to make assumptions about the size of a matrix in order to produce a formula. [*Hint:* Compute the product on the left, and set it equal to the right side.]

5. $\begin{bmatrix} A & B \\ C & 0 \end{bmatrix}\begin{bmatrix} I & 0 \\ X & Y \end{bmatrix} = \begin{bmatrix} 0 & I \\ Z & 0 \end{bmatrix}$

6. $\begin{bmatrix} X & 0 \\ Y & Z \end{bmatrix}\begin{bmatrix} A & 0 \\ B & C \end{bmatrix} = \begin{bmatrix} I & 0 \\ 0 & I \end{bmatrix}$

7. $\begin{bmatrix} X & 0 & 0 \\ Y & 0 & I \end{bmatrix}\begin{bmatrix} A & Z \\ 0 & 0 \\ B & I \end{bmatrix} = \begin{bmatrix} I & 0 \\ 0 & I \end{bmatrix}$

8. $\begin{bmatrix} A & B \\ 0 & I \end{bmatrix}\begin{bmatrix} X & Y & Z \\ 0 & 0 & I \end{bmatrix} = \begin{bmatrix} I & 0 & 0 \\ 0 & 0 & I \end{bmatrix}$

9. Suppose A_{11} is an invertible matrix. Find matrices X and Y such that the product below has the form indicated. Also, compute B_{22}. [*Hint:* Compute the product on the left, and set it equal to the right side.]

$$\begin{bmatrix} I & 0 & 0 \\ X & I & 0 \\ Y & 0 & I \end{bmatrix}\begin{bmatrix} A_{11} & A_{12} \\ A_{21} & A_{22} \\ A_{31} & A_{32} \end{bmatrix} = \begin{bmatrix} B_{11} & B_{12} \\ 0 & B_{22} \\ 0 & B_{32} \end{bmatrix}$$

10. The inverse of $\begin{bmatrix} I & 0 & 0 \\ C & I & 0 \\ A & B & I \end{bmatrix}$ is $\begin{bmatrix} I & 0 & 0 \\ Z & I & 0 \\ X & Y & I \end{bmatrix}$.

    Find X, Y, and Z.

In Exercises 11–14, mark each statement True or False (**T/F**). Justify each answer.

11. **(T/F)** If $A = [A_1 \quad A_2]$ and $B = [B_1 \quad B_2]$, with A_1 and A_2 the same sizes as B_1 and B_2, respectively, then $A + B = [A_1 + B_1 \quad A_2 + B_2]$.

12. **(T/F)** The definition of the matrix–vector product $A\mathbf{x}$ is a special case of block multiplication.

13. **(T/F)** If $A = \begin{bmatrix} A_{11} & A_{12} \\ A_{21} & A_{22} \end{bmatrix}$ and $B = \begin{bmatrix} B_1 \\ B_2 \end{bmatrix}$, then the partitions of A and B are conformable for block multiplication.

14. **(T/F)** If A_1, A_2, B_1, and B_2 are $n \times n$ matrices, $A = \begin{bmatrix} A_1 \\ A_2 \end{bmatrix}$, and $B = \begin{bmatrix} B_1 & B_2 \end{bmatrix}$, then the product BA is defined, but AB is not.

15. Let $A = \begin{bmatrix} B & 0 \\ 0 & C \end{bmatrix}$, where B and C are square. Show that A is invertible if and only if both B and C are invertible.

16. Show that the block upper triangular matrix A in Example 5 is invertible if and only if both A_{11} and A_{22} are invertible. [*Hint:* If A_{11} and A_{22} are invertible, the formula for A^{-1} given in Example 5 actually works as the inverse of A.] This fact about A is an important part of several computer algorithms that estimate eigenvalues of matrices. Eigenvalues are discussed in Chapter 5.

17. Suppose A_{11} is invertible. Find X and Y such that

$$\begin{bmatrix} A_{11} & A_{12} \\ A_{21} & A_{22} \end{bmatrix} = \begin{bmatrix} I & 0 \\ X & I \end{bmatrix}\begin{bmatrix} A_{11} & 0 \\ 0 & S \end{bmatrix}\begin{bmatrix} I & Y \\ 0 & I \end{bmatrix} \qquad (7)$$

where $S = A_{22} - A_{21}A_{11}^{-1}A_{12}$. The matrix S is called the **Schur complement** of A_{11}. Likewise, if A_{22} is invertible, the matrix $A_{11} - A_{12}A_{22}^{-1}A_{21}$ is called the Schur complement of A_{22}. Such expressions occur frequently in the theory of systems engineering, and elsewhere.

18. Suppose the block matrix A on the left side of (7) is invertible and A_{11} is invertible. Show that the Schur complement S of A_{11} is invertible. [*Hint:* The outside factors on the right side of (7) are always invertible. Verify this.] When A and A_{11} are both invertible, (7) leads to a formula for A^{-1}, using S^{-1}, A_{11}^{-1}, and the other entries in A.

19. When a deep space probe is launched, corrections may be necessary to place the probe on a precisely calculated trajectory. Radio telemetry provides a stream of vectors, $\mathbf{x}_1, \ldots, \mathbf{x}_k$, giving information at different times about how the probe's position compares with its planned trajectory. Let X_k be the matrix $[\mathbf{x}_1 \cdots \mathbf{x}_k]$. The matrix $G_k = X_k X_k^T$ is computed as the radar data are analyzed. When $\mathbf{x}_{k+1}$ arrives, a new G_{k+1} must be computed. Since the data vectors arrive at high speed, the computational burden could be severe. But partitioned matrix multiplication helps tremendously. Compute the column–row expansions of G_k and G_{k+1}, and describe what must be computed in order to *update* G_k to form G_{k+1}.

The probe Galileo was launched October 18, 1989, and arrived near Jupiter in early December 1995.

20. Let X be an $m \times n$ data matrix such that $X^T X$ is invertible, and let $M = I_m - X(X^T X)^{-1} X^T$. Add a column $\mathbf{x}_0$ to the data and form

$$W = [X \quad \mathbf{x}_0]$$

Compute $W^T W$. The (1, 1)-entry is $X^T X$. Show that the Schur complement (Exercise 17) of $X^T X$ can be written in the form $\mathbf{x}_0^T M \mathbf{x}_0$. It can be shown that the quantity $(\mathbf{x}_0^T M \mathbf{x}_0)^{-1}$ is the (2, 2)-entry in $(W^T W)^{-1}$. This entry has a useful statistical interpretation, under appropriate hypotheses.

In the study of engineering control of physical systems, a standard set of differential equations is transformed by Laplace transforms into the following system of linear equations:

$$\begin{bmatrix} A - sI_n & B \\ C & I_m \end{bmatrix} \begin{bmatrix} \mathbf{x} \\ \mathbf{u} \end{bmatrix} = \begin{bmatrix} \mathbf{0} \\ \mathbf{y} \end{bmatrix} \tag{8}$$

where A is $n \times n$, B is $n \times m$, C is $m \times n$, and s is a variable. The vector $\mathbf{u}$ in $\mathbb{R}^m$ is the "input" to the system, $\mathbf{y}$ in $\mathbb{R}^m$ is the "output," and $\mathbf{x}$ in $\mathbb{R}^n$ is the "state" vector. (Actually, the vectors $\mathbf{x}$, $\mathbf{u}$, and $\mathbf{y}$ are functions of s, but we suppress this fact because it does not affect the algebraic calculations in Exercises 21 and 22.)

21. Assume $A - sI_n$ is invertible and view (8) as a system of two matrix equations. Solve the top equation for $\mathbf{x}$ and substitute

into the bottom equation. The result is an equation of the form $W(s)\mathbf{u} = \mathbf{y}$, where $W(s)$ is a matrix that depends on s. $W(s)$ is called the *transfer function* of the system because it transforms the input $\mathbf{u}$ into the output $\mathbf{y}$. Find $W(s)$ and describe how it is related to the partitioned *system matrix* on the left side of (8). See Exercise 17.

22. Suppose the transfer function $W(s)$ in Exercise 21 is invertible for some s. It can be shown that the inverse transfer function $W(s)^{-1}$, which transforms outputs into inputs, is the Schur complement of $A - BC - sI_n$ for the matrix below. Find this Schur complement. See Exercise 17.

$$\begin{bmatrix} A - BC - sI_n & B \\ -C & I_m \end{bmatrix}$$

23. a. Verify that $A^2 = I$ when $A = \begin{bmatrix} 1 & 0 \\ 3 & -1 \end{bmatrix}$.

b. Use partitioned matrices to show that $M^2 = I$ when

$$M = \begin{bmatrix} 1 & 0 & 0 & 0 \\ 3 & -1 & 0 & 0 \\ 1 & 0 & -1 & 0 \\ 0 & 1 & -3 & 1 \end{bmatrix}$$

24. Generalize the idea of Exercise 23(a) [not 23(b)] by constructing a 5×5 matrix $M = \begin{bmatrix} A & 0 \\ C & D \end{bmatrix}$ such that $M^2 = I$. Make C a nonzero 2×3 matrix. Show that your construction works.

25. Use partitioned matrices to prove by induction that the product of two lower triangular matrices is also lower triangular. [*Hint:* A $(k + 1) \times (k + 1)$ matrix A_1 can be written in the form below, where a is a scalar, $\mathbf{v}$ is in $\mathbb{R}^k$, and A is a $k \times k$ lower triangular matrix. See the *Study Guide* for help with induction.]

$$A_1 = \begin{bmatrix} a & \mathbf{0}^T \\ \mathbf{v} & A \end{bmatrix}$$

26. Use partitioned matrices to prove by induction that for $n = 2, 3, \ldots$, the $n \times n$ matrix A shown below is invertible and B is its inverse.

$$A = \begin{bmatrix} 1 & 0 & 0 & \cdots & 0 \\ 1 & 1 & 0 & & 0 \\ 1 & 1 & 1 & & 0 \\ \vdots & & & \ddots & \\ 1 & 1 & 1 & \cdots & 1 \end{bmatrix},$$

$$B = \begin{bmatrix} 1 & 0 & 0 & \cdots & 0 \\ -1 & 1 & 0 & & 0 \\ 0 & -1 & 1 & & 0 \\ \vdots & & \ddots & \ddots & \\ 0 & & \cdots & -1 & 1 \end{bmatrix}$$

For the induction step, assume A and B are $(k + 1) \times (k + 1)$ matrices, and partition A and B in a form similar to that displayed in Exercise 25.

27. Without using row reduction, find the inverse of

$$A = \begin{bmatrix} 1 & 2 & 0 & 0 & 0 \\ 3 & 5 & 0 & 0 & 0 \\ 0 & 0 & 2 & 0 & 0 \\ 0 & 0 & 0 & 7 & 8 \\ 0 & 0 & 0 & 5 & 6 \end{bmatrix}$$

T 28. For block operations, it may be necessary to access or enter submatrices of a large matrix. Describe the functions or commands of your matrix program that accomplish the following tasks. Suppose A is a 20×30 matrix.

a. Display the submatrix of A from rows 15 to 20 and columns 5 to 10.

b. Insert a 5×10 matrix B into A, beginning at row 10 and column 20.

c. Create a 50×50 matrix of the form $B = \begin{bmatrix} A & 0 \\ 0 & A^T \end{bmatrix}$.

[*Note:* It may not be necessary to specify the zero blocks in B.]

T 29. Suppose memory or size restrictions prevent your matrix program from working with matrices having more than 32 rows and 32 columns, and suppose some project involves 50×50 matrices A and B. Describe the commands or operations of your matrix program that accomplish the following tasks.

a. Compute $A + B$.

b. Compute AB.

c. Solve $Ax = \mathbf{b}$ for some vector $\mathbf{b}$ in $\mathbb{R}^{50}$, assuming that A can be partitioned into a 2×2 block matrix $[A_{ij}]$, with A_{11} an invertible 20×20 matrix, A_{22} an invertible 30×30 matrix, and A_{12} a zero matrix. [*Hint:* Describe appropriate smaller systems to solve, without using any matrix inverses.]

Solutions to Practice Problems

1. If $\begin{bmatrix} I & 0 \\ A & I \end{bmatrix}$ is invertible, its inverse has the form $\begin{bmatrix} W & X \\ Y & Z \end{bmatrix}$. Verify that

$$\begin{bmatrix} I & 0 \\ A & I \end{bmatrix}\begin{bmatrix} W & X \\ Y & Z \end{bmatrix} = \begin{bmatrix} W & X \\ AW + Y & AX + Z \end{bmatrix}$$

So W, X, Y, and Z must satisfy $W = I, X = 0, AW + Y = 0$, and $AX + Z = I$. It follows that $Y = -A$ and $Z = I$. Hence

$$\begin{bmatrix} I & 0 \\ A & I \end{bmatrix}\begin{bmatrix} I & 0 \\ -A & I \end{bmatrix} = \begin{bmatrix} I & 0 \\ 0 & I \end{bmatrix}$$

The product in the reverse order is also the identity, so the block matrix is invertible, and its inverse is $\begin{bmatrix} I & 0 \\ -A & I \end{bmatrix}$. (You could also appeal to the Invertible Matrix Theorem.)

2. $X^T X = \begin{bmatrix} X_1^T \\ X_2^T \end{bmatrix}\begin{bmatrix} X_1 & X_2 \end{bmatrix} = \begin{bmatrix} X_1^T X_1 & X_1^T X_2 \\ X_2^T X_1 & X_2^T X_2 \end{bmatrix}$. The partitions of X^T and X are automatically conformable for block multiplication because the columns of X^T are the rows of X. This partition of $X^T X$ is used in several computer algorithms for matrix computations.

2.5 Matrix Factorizations

A *factorization* of a matrix A is an equation that expresses A as a product of two or more matrices. Whereas matrix multiplication involves a *synthesis* of data (combining the effects of two or more linear transformations into a single matrix), matrix factorization is an *analysis* of data. In the language of computer science, the expression of A as a product amounts to a *preprocessing* of the data in A, organizing that data into two or more parts whose structures are more useful in some way, perhaps more accessible for computation.

Matrix factorizations and, later, factorizations of linear transformations will appear at a number of key points throughout the text. This section focuses on a factorization that lies at the heart of several important computer programs widely used in applications, such as the airflow problem described in the chapter introduction. Several other factorizations, to be studied later, are introduced in the exercises.

The LU Factorization

The LU factorization, described below, is motivated by the fairly common industrial and business problem of solving a sequence of equations, all with the same coefficient matrix:

$$A\mathbf{x} = \mathbf{b}_1, \quad A\mathbf{x} = \mathbf{b}_2, \quad \ldots, \quad A\mathbf{x} = \mathbf{b}_p \tag{1}$$

See Exercise 32, for example. Also see Section 5.8, where the inverse power method is used to estimate eigenvalues of a matrix by solving equations like those in sequence (1), one at a time.

When A is invertible, one could compute A^{-1} and then compute $A^{-1}\mathbf{b}_1$, $A^{-1}\mathbf{b}_2$, and so on. However, it is more efficient to solve the first equation in sequence (1) by row reduction and obtain an LU factorization of A at the same time. Thereafter, the remaining equations in sequence (1) are solved with the LU factorization.

At first, assume that A is an $m \times n$ matrix that can be row reduced to echelon form, *without row interchanges*. (Later, we will treat the general case.) Then A can be written in the form $A = LU$, where L is an $m \times m$ lower triangular matrix with 1's on the diagonal and U is an $m \times n$ echelon form of A. For instance, see Figure 1. Such a factorization is called an **LU factorization** of A. The matrix L is invertible and is called a *unit* lower triangular matrix.

$$A = \begin{bmatrix} 1 & 0 & 0 & 0 \\ * & 1 & 0 & 0 \\ * & * & 1 & 0 \\ * & * & * & 1 \end{bmatrix} \begin{bmatrix} \blacksquare & * & * & * & * \\ 0 & \blacksquare & * & * & * \\ 0 & 0 & 0 & \blacksquare & * \\ 0 & 0 & 0 & 0 & 0 \end{bmatrix}$$
$$\qquad\qquad L \qquad\qquad\qquad\qquad U$$

FIGURE 1 An LU factorization.

Before studying how to construct L and U, we should look at why they are so useful. When $A = LU$, the equation $A\mathbf{x} = \mathbf{b}$ can be written as $L(U\mathbf{x}) = \mathbf{b}$. Writing $\mathbf{y}$ for $U\mathbf{x}$, we can find $\mathbf{x}$ by solving the *pair* of equations

$$\boxed{\begin{array}{l} L\mathbf{y} = \mathbf{b} \\ U\mathbf{x} = \mathbf{y} \end{array}} \tag{2}$$

First solve $L\mathbf{y} = \mathbf{b}$ for $\mathbf{y}$, and then solve $U\mathbf{x} = \mathbf{y}$ for $\mathbf{x}$. See Figure 2. Each equation is easy to solve because L and U are triangular.

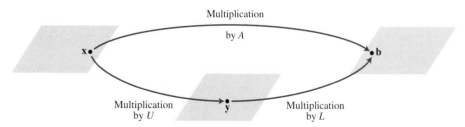

FIGURE 2 Factorization of the mapping $\mathbf{x} \mapsto A\mathbf{x}$.

EXAMPLE 1 It can be verified that

$$A = \begin{bmatrix} 3 & -7 & -2 & 2 \\ -3 & 5 & 1 & 0 \\ 6 & -4 & 0 & -5 \\ -9 & 5 & -5 & 12 \end{bmatrix} = \begin{bmatrix} 1 & 0 & 0 & 0 \\ -1 & 1 & 0 & 0 \\ 2 & -5 & 1 & 0 \\ -3 & 8 & 3 & 1 \end{bmatrix} \begin{bmatrix} 3 & -7 & -2 & 2 \\ 0 & -2 & -1 & 2 \\ 0 & 0 & -1 & 1 \\ 0 & 0 & 0 & -1 \end{bmatrix} = LU$$

Use this LU factorization of A to solve $A\mathbf{x} = \mathbf{b}$, where $\mathbf{b} = \begin{bmatrix} -9 \\ 5 \\ 7 \\ 11 \end{bmatrix}$.

SOLUTION The solution of $L\mathbf{y} = \mathbf{b}$ needs only 6 multiplications and 6 additions, because the arithmetic takes place only in column 5. (The zeros below each pivot in L are created automatically by the choice of row operations.)

$$[\, L \quad \mathbf{b} \,] = \begin{bmatrix} 1 & 0 & 0 & 0 & -9 \\ -1 & 1 & 0 & 0 & 5 \\ 2 & -5 & 1 & 0 & 7 \\ -3 & 8 & 3 & 1 & 11 \end{bmatrix} \sim \begin{bmatrix} 1 & 0 & 0 & 0 & -9 \\ 0 & 1 & 0 & 0 & -4 \\ 0 & 0 & 1 & 0 & 5 \\ 0 & 0 & 0 & 1 & 1 \end{bmatrix} = [\, I \quad \mathbf{y} \,]$$

Then, for $U\mathbf{x} = \mathbf{y}$, the "backward" phase of row reduction requires 4 divisions, 6 multiplications, and 6 additions. (For instance, creating the zeros in column 4 of $[\, U \quad \mathbf{y} \,]$ requires 1 division in row 4 and 3 multiplication–addition pairs to add multiples of row 4 to the rows above.)

$$[\, U \quad \mathbf{y} \,] = \begin{bmatrix} 3 & -7 & -2 & 2 & -9 \\ 0 & -2 & -1 & 2 & -4 \\ 0 & 0 & -1 & 1 & 5 \\ 0 & 0 & 0 & -1 & 1 \end{bmatrix} \sim \begin{bmatrix} 1 & 0 & 0 & 0 & 3 \\ 0 & 1 & 0 & 0 & 4 \\ 0 & 0 & 1 & 0 & -6 \\ 0 & 0 & 0 & 1 & -1 \end{bmatrix}, \quad \mathbf{x} = \begin{bmatrix} 3 \\ 4 \\ -6 \\ -1 \end{bmatrix}$$

To find $\mathbf{x}$ requires 28 arithmetic operations, or "flops" (floating point operations), excluding the cost of finding L and U. In contrast, row reduction of $[\, A \quad \mathbf{b} \,]$ to $[\, I \quad \mathbf{x} \,]$ takes 62 operations. ∎

The computational efficiency of the LU factorization depends on knowing L and U. The next algorithm shows that the row reduction of A to an echelon form U amounts to an LU factorization because it produces L with essentially no extra work. After the first row reduction, L and U are available for solving additional equations whose coefficient matrix is A.

An LU Factorization Algorithm

Suppose A can be reduced to an echelon form U using only row replacements that add a multiple of one row to another row *below it*. In this case, there exist unit lower triangular elementary matrices $E_1, \ldots, E_p$ such that

$$E_p \cdots E_1 A = U \tag{3}$$

Then

$$A = (E_p \cdots E_1)^{-1} U = LU$$

where

$$L = (E_p \cdots E_1)^{-1} \tag{4}$$

It can be shown that products and inverses of unit lower triangular matrices are also unit lower triangular. (For instance, see Exercise 19.) Thus L is unit lower triangular.

Note that the row operations in equation (3), which reduce A to U, also reduce the L in equation (4) to I, because $E_p \cdots E_1 L = (E_p \cdots E_1)(E_p \cdots E_1)^{-1} = I$. This observation is the key to *constructing L*.

ALGORITHM FOR AN LU FACTORIZATION

1. Reduce A to an echelon form U by a sequence of row replacement operations, if possible.

2. Place entries in L such that the *same sequence of row operations* reduces L to I.

Step 1 is not always possible, but when it is, the argument above shows that an LU factorization exists. Example 2 will show how to implement step 2. By construction, L will satisfy

$$(E_p \cdots E_1)L = I$$

using the same $E_1, \ldots, E_p$ as in equation (3). Thus L will be invertible, by the Invertible Matrix Theorem, with $(E_p \cdots E_1) = L^{-1}$. From (3), $L^{-1}A = U$, and $A = LU$. So step 2 will produce an acceptable L.

EXAMPLE 2   Find an LU factorization of

$$A = \begin{bmatrix} 2 & 4 & -1 & 5 & -2 \\ -4 & -5 & 3 & -8 & 1 \\ 2 & -5 & -4 & 1 & 8 \\ -6 & 0 & 7 & -3 & 1 \end{bmatrix}$$

SOLUTION   Since A has four rows, L should be 4×4. The first column of L is the first column of A divided by the top pivot entry:

$$L = \begin{bmatrix} 1 & 0 & 0 & 0 \\ -2 & 1 & 0 & 0 \\ 1 & & 1 & 0 \\ -3 & & & 1 \end{bmatrix}$$

Compare the first columns of A and L. *The row operations that create zeros in the first column of A will also create zeros in the first column of L.* To make this same correspondence of row operations on A hold for the rest of L, watch a row reduction of A to an echelon form U. That is, *highlight the entries* in each matrix that are used to determine the sequence of row operations that transform A into U. [See the highlighted entries in equation (5).]

$$A = \begin{bmatrix} 2 & 4 & -1 & 5 & -2 \\ -4 & -5 & 3 & -8 & 1 \\ 2 & -5 & -4 & 1 & 8 \\ -6 & 0 & 7 & -3 & 1 \end{bmatrix} \sim \begin{bmatrix} 2 & 4 & -1 & 5 & -2 \\ 0 & 3 & 1 & 2 & -3 \\ 0 & -9 & -3 & -4 & 10 \\ 0 & 12 & 4 & 12 & -5 \end{bmatrix} = A_1 \qquad (5)$$

$$\sim A_2 = \begin{bmatrix} 2 & 4 & -1 & 5 & -2 \\ 0 & 3 & 1 & 2 & -3 \\ 0 & 0 & 0 & 2 & 1 \\ 0 & 0 & 0 & 4 & 7 \end{bmatrix} \sim \begin{bmatrix} 2 & 4 & -1 & 5 & -2 \\ 0 & 3 & 1 & 2 & -3 \\ 0 & 0 & 0 & 2 & 1 \\ 0 & 0 & 0 & 0 & 5 \end{bmatrix} = U$$

These highlighted entries determine the row reduction of A to U. At each pivot column, divide the highlighted entries by the pivot and place the result into L:

$$
\begin{bmatrix} 2 \\ -4 \\ 2 \\ -6 \end{bmatrix}
\begin{bmatrix} 3 \\ -9 \\ 12 \end{bmatrix}
\begin{bmatrix} 2 \\ 4 \end{bmatrix}
\begin{bmatrix} 5 \end{bmatrix}
$$

$$
\begin{array}{cccc}
\div 2 & \div 3 & \div 2 & \div 5 \\
\downarrow & \downarrow & \downarrow & \downarrow
\end{array}
$$

$$
\begin{bmatrix}
1 & & & \\
-2 & 1 & & \\
1 & -3 & 1 & \\
-3 & 4 & 2 & 1
\end{bmatrix}, \quad \text{and} \quad
L = \begin{bmatrix}
1 & 0 & 0 & 0 \\
-2 & 1 & 0 & 0 \\
1 & -3 & 1 & 0 \\
-3 & 4 & 2 & 1
\end{bmatrix}
$$

An easy calculation verifies that this L and U satisfy $LU = A$. ∎

In practical work, row interchanges are nearly always needed, because partial pivoting is used for high accuracy. (Recall that this procedure selects, among the possible choices for a pivot, an entry in the column having the largest absolute value.) To handle row interchanges, the LU factorization above can be modified easily to produce an L that is *permuted lower triangular*, in the sense that a rearrangement (called a permutation) of the rows of L can make L (unit) lower triangular. The resulting *permuted LU factorization* solves $A\mathbf{x} = \mathbf{b}$ in the same way as before, except that the reduction of $[\,L \quad \mathbf{b}\,]$ to $[\,I \quad \mathbf{y}\,]$ follows the order of the pivots in L from left to right, starting with the pivot in the first column. A reference to an "LU factorization" usually includes the possibility that L might be permuted lower triangular. For details, see the *Study Guide*.

STUDY GUIDE offers information about permuted LU factorizations.

Numerical Notes

The following operation counts apply to an $n \times n$ dense matrix A (with most entries nonzero) for n moderately large, say, $n \geq 30$.[1]

1. Computing an LU factorization of A takes about $2n^3/3$ flops (about the same as row reducing $[\,A \quad \mathbf{b}\,]$), whereas finding A^{-1} requires about $2n^3$ flops.

2. Solving $L\mathbf{y} = \mathbf{b}$ and $U\mathbf{x} = \mathbf{y}$ requires about $2n^2$ flops, because any $n \times n$ triangular system can be solved in about n^2 flops.

3. Multiplication of $\mathbf{b}$ by A^{-1} also requires about $2n^2$ flops, but the result may not be as accurate as that obtained from L and U (because of roundoff error when computing both A^{-1} and $A^{-1}\mathbf{b}$).

4. If A is sparse (with mostly zero entries), then L and U may be sparse, too, whereas A^{-1} is likely to be dense. In this case, a solution of $A\mathbf{x} = \mathbf{b}$ with an LU factorization is *much* faster than using A^{-1}. See Exercise 31.

A Matrix Factorization in Electrical Engineering

Matrix factorization is intimately related to the problem of constructing an electrical network with specified properties. The following discussion gives just a glimpse of the connection between factorization and circuit design.

---

[1] See Section 3.8 in *Applied Linear Algebra*, 3rd ed., by Ben Noble and James W. Daniel (Englewood Cliffs, NJ: Prentice-Hall, 1988). Recall that for our purposes, a *flop* is $+$, $-$, $\times$, or $\div$.

Suppose the box in Figure 3 represents some sort of electric circuit, with an input and output. Record the input voltage and current by $\begin{bmatrix} v_1 \\ i_1 \end{bmatrix}$ (with voltage v in volts and current i in amps), and record the output voltage and current by $\begin{bmatrix} v_2 \\ i_2 \end{bmatrix}$. Frequently, the transformation $\begin{bmatrix} v_1 \\ i_1 \end{bmatrix} \mapsto \begin{bmatrix} v_2 \\ i_2 \end{bmatrix}$ is linear. That is, there is a matrix A, called the *transfer matrix*, such that

$$\begin{bmatrix} v_2 \\ i_2 \end{bmatrix} = A \begin{bmatrix} v_1 \\ i_1 \end{bmatrix}$$

FIGURE 3 A circuit with input and output terminals.

Figure 4 shows a *ladder network*, where two circuits (there could be more) are connected in series, so that the output of one circuit becomes the input of the next circuit. The left circuit in Figure 4 is called a *series circuit*, with resistance R_1 (in ohms).

FIGURE 4 A ladder network.

The right circuit in Figure 4 is a *shunt circuit*, with resistance R_2. Using Ohm's law and Kirchhoff's laws, one can show that the transfer matrices of the series and shunt circuits, respectively, are

$$\begin{bmatrix} 1 & -R_1 \\ 0 & 1 \end{bmatrix} \quad \text{and} \quad \begin{bmatrix} 1 & 0 \\ -1/R_2 & 1 \end{bmatrix}$$

Transfer matrix of series circuit   Transfer matrix of shunt circuit

EXAMPLE 3

a. Compute the transfer matrix of the ladder network in Figure 4.

b. Design a ladder network whose transfer matrix is $\begin{bmatrix} 1 & -8 \\ -.5 & 5 \end{bmatrix}$.

SOLUTION

a. Let A_1 and A_2 be the transfer matrices of the series and shunt circuits, respectively. Then an input vector $\mathbf{x}$ is transformed first into $A_1\mathbf{x}$ and then into $A_2(A_1\mathbf{x})$. The series connection of the circuits corresponds to composition of linear transformations, and the transfer matrix of the ladder network is (note the order)

$$A_2 A_1 = \begin{bmatrix} 1 & 0 \\ -1/R_2 & 1 \end{bmatrix} \begin{bmatrix} 1 & -R_1 \\ 0 & 1 \end{bmatrix} = \begin{bmatrix} 1 & -R_1 \\ -1/R_2 & 1 + R_1/R_2 \end{bmatrix} \quad (6)$$

b. To factor the matrix $\begin{bmatrix} 1 & -8 \\ -.5 & 5 \end{bmatrix}$ into the product of transfer matrices, as in equation (6), look for R_1 and R_2 in Figure 4 to satisfy

$$\begin{bmatrix} 1 & -R_1 \\ -1/R_2 & 1 + R_1/R_2 \end{bmatrix} = \begin{bmatrix} 1 & -8 \\ -.5 & 5 \end{bmatrix}$$

From the $(1, 2)$-entries, $R_1 = 8$ ohms, and from the $(2, 1)$-entries, $1/R_2 = .5$ ohm and $R_2 = 1/.5 = 2$ ohms. With these values, the network in Figure 4 has the desired transfer matrix. ∎

A network transfer matrix summarizes the input–output behavior (the design specifications) of the network without reference to the interior circuits. To physically build a network with specified properties, an engineer first determines if such a network can be constructed (or *realized*). Then the engineer tries to factor the transfer matrix into matrices corresponding to smaller circuits that perhaps are already manufactured and ready for assembly. In the common case of alternating current, the entries in the transfer matrix are usually rational complex-valued functions. (See Exercises 21 and 22 in Section 2.4.) A standard problem is to find a *minimal realization* that uses the smallest number of electrical components.

Practice Problem

Find an LU factorization of $A = \begin{bmatrix} 2 & -4 & -2 & 3 \\ 6 & -9 & -5 & 8 \\ 2 & -7 & -3 & 9 \\ 4 & -2 & -2 & -1 \\ -6 & 3 & 3 & 4 \end{bmatrix}$. [*Note:* It will turn out that A has only three pivot columns, so the method of Example 2 will produce only the first three columns of L. The remaining two columns of L come from I_5.]

2.5 Exercises

In Exercises 1–6, solve the equation $A\mathbf{x} = \mathbf{b}$ by using the LU factorization given for A. In Exercises 1 and 2, also solve $A\mathbf{x} = \mathbf{b}$ by ordinary row reduction.

1. $A = \begin{bmatrix} 3 & -7 & -2 \\ -3 & 5 & 1 \\ 6 & -4 & 0 \end{bmatrix}$, $\mathbf{b} = \begin{bmatrix} -7 \\ 5 \\ 2 \end{bmatrix}$

$A = \begin{bmatrix} 1 & 0 & 0 \\ -1 & 1 & 0 \\ 2 & -5 & 1 \end{bmatrix} \begin{bmatrix} 3 & -7 & -2 \\ 0 & -2 & -1 \\ 0 & 0 & -1 \end{bmatrix}$

2. $A = \begin{bmatrix} 4 & 3 & -5 \\ -4 & -5 & 7 \\ 8 & 6 & -8 \end{bmatrix}$, $\mathbf{b} = \begin{bmatrix} 2 \\ -4 \\ 6 \end{bmatrix}$

$A = \begin{bmatrix} 1 & 0 & 0 \\ -1 & 1 & 0 \\ 2 & 0 & 1 \end{bmatrix} \begin{bmatrix} 4 & 3 & -5 \\ 0 & -2 & 2 \\ 0 & 0 & 2 \end{bmatrix}$

3. $A = \begin{bmatrix} 2 & -1 & 2 \\ -6 & 0 & -2 \\ 8 & -1 & 5 \end{bmatrix}$, $\mathbf{b} = \begin{bmatrix} 1 \\ 0 \\ 4 \end{bmatrix}$

$A = \begin{bmatrix} 1 & 0 & 0 \\ -3 & 1 & 0 \\ 4 & -1 & 1 \end{bmatrix} \begin{bmatrix} 2 & -1 & 2 \\ 0 & -3 & 4 \\ 0 & 0 & 1 \end{bmatrix}$

4. $A = \begin{bmatrix} 2 & -2 & 4 \\ 1 & -3 & 1 \\ 3 & 7 & 5 \end{bmatrix}$, $\mathbf{b} = \begin{bmatrix} 0 \\ -5 \\ 7 \end{bmatrix}$

$A = \begin{bmatrix} 1 & 0 & 0 \\ 1/2 & 1 & 0 \\ 3/2 & -5 & 1 \end{bmatrix} \begin{bmatrix} 2 & -2 & 4 \\ 0 & -2 & -1 \\ 0 & 0 & -6 \end{bmatrix}$

5. $A = \begin{bmatrix} 1 & -2 & -4 & -3 \\ 2 & -7 & -7 & -6 \\ -1 & 2 & 6 & 4 \\ -4 & -1 & 9 & 8 \end{bmatrix}$, $\mathbf{b} = \begin{bmatrix} 1 \\ 7 \\ 0 \\ 3 \end{bmatrix}$

$A = \begin{bmatrix} 1 & 0 & 0 & 0 \\ 2 & 1 & 0 & 0 \\ -1 & 0 & 1 & 0 \\ -4 & 3 & -5 & 1 \end{bmatrix} \begin{bmatrix} 1 & -2 & -4 & -3 \\ 0 & -3 & 1 & 0 \\ 0 & 0 & 2 & 1 \\ 0 & 0 & 0 & 1 \end{bmatrix}$

6. $A = \begin{bmatrix} 1 & 3 & 4 & 0 \\ -3 & -6 & -7 & 2 \\ 3 & 3 & 0 & -4 \\ -5 & -3 & 2 & 9 \end{bmatrix}, \mathbf{b} = \begin{bmatrix} 1 \\ -2 \\ -1 \\ 2 \end{bmatrix}$

$A = \begin{bmatrix} 1 & 0 & 0 & 0 \\ -3 & 1 & 0 & 0 \\ 3 & -2 & 1 & 0 \\ -5 & 4 & -1 & 1 \end{bmatrix} \begin{bmatrix} 1 & 3 & 4 & 0 \\ 0 & 3 & 5 & 2 \\ 0 & 0 & -2 & 0 \\ 0 & 0 & 0 & 1 \end{bmatrix}$

Find an LU factorization of the matrices in Exercises 7–16 (with L unit lower triangular). Note that MATLAB will usually produce a permuted LU factorization because it uses partial pivoting for numerical accuracy.

7. $\begin{bmatrix} 2 & 5 \\ -3 & -4 \end{bmatrix}$

8. $\begin{bmatrix} 6 & 9 \\ 4 & 5 \end{bmatrix}$

9. $\begin{bmatrix} 3 & -1 & 2 \\ -3 & -2 & 10 \\ 9 & -5 & 6 \end{bmatrix}$

10. $\begin{bmatrix} -5 & 3 & 4 \\ 10 & -8 & -9 \\ 15 & 1 & 2 \end{bmatrix}$

11. $\begin{bmatrix} 3 & -6 & 3 \\ 6 & -7 & 2 \\ -1 & 7 & 0 \end{bmatrix}$

12. $\begin{bmatrix} 2 & -4 & 2 \\ 1 & 5 & -4 \\ -6 & -2 & 4 \end{bmatrix}$

13. $\begin{bmatrix} 1 & 3 & -5 & -3 \\ -1 & -5 & 8 & 4 \\ 4 & 2 & -5 & -7 \\ -2 & -4 & 7 & 5 \end{bmatrix}$

14. $\begin{bmatrix} 1 & 4 & -1 & 5 \\ 3 & 7 & -2 & 9 \\ -2 & -3 & 1 & -4 \\ -1 & 6 & -1 & 7 \end{bmatrix}$

15. $\begin{bmatrix} 2 & -4 & 4 & -2 \\ 6 & -9 & 7 & -3 \\ -1 & -4 & 8 & 0 \end{bmatrix}$

16. $\begin{bmatrix} 2 & -6 & 6 \\ -4 & 5 & -7 \\ 3 & 5 & -1 \\ -6 & 4 & -8 \\ 8 & -3 & 9 \end{bmatrix}$

17. When A is invertible, MATLAB finds A^{-1} by factoring $A = LU$ (where L may be permuted lower triangular), inverting L and U, and then computing $U^{-1}L^{-1}$. Use this method to compute the inverse of A in Exercise 2. (Apply the algorithm of Section 2.2 to L and to U.)

18. Find A^{-1} as in Exercise 17, using A from Exercise 3.

19. Let A be a lower triangular $n \times n$ matrix with nonzero entries on the diagonal. Show that A is invertible and A^{-1} is lower triangular. [*Hint:* Explain why A can be changed into I using only row replacements and scaling. (Where are the pivots?) Also, explain why the row operations that reduce A to I change I into a lower triangular matrix.]

20. Let $A = LU$ be an LU factorization. Explain why A can be row reduced to U using only replacement operations. (This fact is the converse of what was proved in the text.)

21. Suppose $A = BC$, where B is invertible. Show that any sequence of row operations that reduces B to I also reduces A to C. The converse is not true, since the zero matrix may be factored as $0 = B(0)$.

Exercises 22–26 provide a glimpse of some widely used matrix factorizations, some of which are discussed later in the text.

22. (*Reduced LU Factorization*) With A as in the Practice Problem, find a 5×3 matrix B and a 3×4 matrix C such that $A = BC$. Generalize this idea to the case where A is $m \times n$, $A = LU$, and U has only three nonzero rows.

23. (*Rank Factorization*) Suppose an $m \times n$ matrix A admits a factorization $A = CD$ where C is $m \times 4$ and D is $4 \times n$.
   a. Show that A is the sum of four outer products. (See Section 2.4.)
   b. Let $m = 400$ and $n = 100$. Explain why a computer programmer might prefer to store the data from A in the form of two matrices C and D.

24. (*QR Factorization*) Suppose $A = QR$, where Q and R are $n \times n$, R is invertible and upper triangular, and Q has the property that $Q^T Q = I$. Show that for each $\mathbf{b}$ in $\mathbb{R}^n$, the equation $A\mathbf{x} = \mathbf{b}$ has a unique solution. What computations with Q and R will produce the solution?

25. (*Singular Value Decomposition*) Suppose $A = UDV^T$, where U and V are $n \times n$ matrices with the property that $U^T U = I$ and $V^T V = I$, and where D is a diagonal matrix with positive numbers $\sigma_1, \ldots, \sigma_n$ on the diagonal. Show that A is invertible, and find a formula for A^{-1}.

26. (*Spectral Factorization*) Suppose a 3×3 matrix A admits a factorization as $A = PDP^{-1}$, where P is some invertible 3×3 matrix and D is the diagonal matrix

$$D = \begin{bmatrix} 1 & 0 & 0 \\ 0 & 1/2 & 0 \\ 0 & 0 & 1/3 \end{bmatrix}$$

Show that this factorization is useful when computing high powers of A. Find fairly simple formulas for A^2, A^3, and A^k (k a positive integer), using P and the entries in D.

27. Design two different ladder networks that each output 9 volts and 4 amps when the input is 12 volts and 6 amps.

28. Show that if three shunt circuits (with resistances R_1, R_2, R_3) are connected in series, the resulting network has the same transfer matrix as a single shunt circuit. Find a formula for the resistance in that circuit.

29. a. Compute the transfer matrix of the network in the figure.
   b. Let $A = \begin{bmatrix} 4/3 & -12 \\ -1/4 & 3 \end{bmatrix}$. Design a ladder network whose transfer matrix is A by finding a suitable matrix factorization of A.

30. Find a different factorization of the A in Exercise 29, and thereby design a different ladder network whose transfer matrix is A.

T 31. The solution to the steady-state heat flow problem for the plate in the figure is approximated by the solution to the equation $A\mathbf{x} = \mathbf{b}$, where $\mathbf{b} = (5, 15, 0, 10, 0, 10, 20, 30)$ and

$$A = \begin{bmatrix} 4 & -1 & -1 & & & & & \\ -1 & 4 & 0 & -1 & & & & \\ -1 & 0 & 4 & -1 & -1 & & & \\ & -1 & -1 & 4 & 0 & -1 & & \\ & & -1 & 0 & 4 & -1 & -1 & \\ & & & -1 & -1 & 4 & 0 & -1 \\ & & & & -1 & 0 & 4 & -1 \\ & & & & & -1 & -1 & 4 \end{bmatrix}$$

(Refer to Exercise 43 of Section 1.1.) The missing entries in A are zeros. The nonzero entries of A lie within a band along the main diagonal. Such *band matrices* occur in a variety of applications and often are extremely large (with thousands of rows and columns but relatively narrow bands).

a. Use the method of Example 2 to construct an LU factorization of A, and note that both factors are band matrices (with two nonzero diagonals below or above the main diagonal). Compute $LU - A$ to check your work.

b. Use the LU factorization to solve $A\mathbf{x} = \mathbf{b}$.

c. Obtain A^{-1} and note that A^{-1} is a dense matrix with no band structure. When A is large, L and U can be stored in much less space than A^{-1}. This fact is another reason for preferring the LU factorization of A to A^{-1} itself.

T 32. The band matrix A shown below can be used to estimate the unsteady conduction of heat in a rod when the temperatures at points $p_1, \ldots, p_5$ on the rod change with time.[2]

$$\overset{\Delta x \quad \Delta x}{\underset{0 \quad\quad p_1 \quad p_2 \quad p_3 \quad p_4 \quad p_5}{\rule{5cm}{0.4pt}}}$$

The constant C in the matrix depends on the physical nature of the rod, the distance Δx between the points on the rod, and the length of time Δt between successive temperature measurements. Suppose that for $k = 0, 1, 2, \ldots$, a vector $\mathbf{t}_k$ in $\mathbb{R}^5$ lists the temperatures at time $k\Delta t$. If the two ends of the rod are maintained at $0°$, then the temperature vectors satisfy the equation $A\mathbf{t}_{k+1} = \mathbf{t}_k$ ($k = 0, 1, \ldots$), where

$$A = \begin{bmatrix} (1+2C) & -C & & & \\ -C & (1+2C) & -C & & \\ & -C & (1+2C) & -C & \\ & & -C & (1+2C) & -C \\ & & & -C & (1+2C) \end{bmatrix}$$

a. Find the LU factorization of A when $C = 1$. A matrix such as A with three nonzero diagonals is called a *tridiagonal matrix*. The L and U factors are *bidiagonal matrices*.

b. Suppose $C = 1$ and $\mathbf{t}_0 = (10, 12, 12, 12, 10)$. Use the LU factorization of A to find the temperature distributions $\mathbf{t}_1, \mathbf{t}_2, \mathbf{t}_3,$ and $\mathbf{t}_4$.

---

[2] See Biswa N. Datta, *Numerical Linear Algebra and Applications* (Pacific Grove, CA: Brooks/Cole, 1994), pp. 200–201.

Solution to Practice Problem

$$A = \begin{bmatrix} 2 & -4 & -2 & 3 \\ 6 & -9 & -5 & 8 \\ 2 & -7 & -3 & 9 \\ 4 & -2 & -2 & -1 \\ -6 & 3 & 3 & 4 \end{bmatrix} \sim \begin{bmatrix} 2 & -4 & -2 & 3 \\ 0 & 3 & 1 & -1 \\ 0 & -3 & -1 & 6 \\ 0 & 6 & 2 & -7 \\ 0 & -9 & -3 & 13 \end{bmatrix}$$

$$\sim \begin{bmatrix} 2 & -4 & -2 & 3 \\ 0 & 3 & 1 & -1 \\ 0 & 0 & 0 & 5 \\ 0 & 0 & 0 & -5 \\ 0 & 0 & 0 & 10 \end{bmatrix} \sim \begin{bmatrix} 2 & -4 & -2 & 3 \\ 0 & 3 & 1 & -1 \\ 0 & 0 & 0 & 5 \\ 0 & 0 & 0 & 0 \\ 0 & 0 & 0 & 0 \end{bmatrix} = U$$

Divide the entries in each highlighted column by the pivot at the top. The resulting columns form the first three columns in the lower half of L. This suffices to make row reduction of L to I correspond to reduction of A to U. Use the last two columns of I_5

to make L unit lower triangular.

$$
\begin{bmatrix} 2 \\ 6 \\ 2 \\ 4 \\ -6 \end{bmatrix}
\begin{bmatrix} 3 \\ -3 \\ 6 \\ -9 \end{bmatrix}
\begin{bmatrix} 5 \\ -5 \\ 10 \end{bmatrix}
$$

$$
\begin{array}{ccc}
\div 2 & \div 3 & \div 5 \\
\downarrow & \downarrow & \downarrow
\end{array}
$$

$$
\begin{bmatrix}
1 & & & \\
3 & 1 & & \\
1 & -1 & 1 & \cdots \\
2 & 2 & -1 & \\
-3 & -3 & 2 &
\end{bmatrix}, \quad
L =
\begin{bmatrix}
1 & 0 & 0 & 0 & 0 \\
3 & 1 & 0 & 0 & 0 \\
1 & -1 & 1 & 0 & 0 \\
2 & 2 & -1 & 1 & 0 \\
-3 & -3 & 2 & 0 & 1
\end{bmatrix}
$$

2.6 The Leontief Input–Output Model

Linear algebra played an essential role in the Nobel prize–winning work of Wassily Leontief, as mentioned at the beginning of Chapter 1. The economic model described in this section is the basis for more elaborate models used in many parts of the world.

Suppose a nation's economy is divided into n sectors that produce goods or services, and let **x** be a **production vector** in $\mathbb{R}^n$ that lists the output of each sector for one year. Also, suppose another part of the economy (called the *open sector*) does not produce goods or services but only consumes them, and let **d** be a **final demand vector** (or **bill of final demands**) that lists the values of the goods and services demanded from the various sectors by the nonproductive part of the economy. The vector **d** can represent consumer demand, government consumption, surplus production, exports, or other external demands.

As the various sectors produce goods to meet consumer demand, the producers themselves create additional **intermediate demand** for goods they need as inputs for their own production. The interrelations between the sectors are very complex, and the connection between the final demand and the production is unclear. Leontief asked if there is a production level **x** such that the amounts produced (or "supplied") will exactly balance the total demand for that production, so that

$$
\left\{ \begin{array}{c} \text{amount} \\ \text{produced} \\ \mathbf{x} \end{array} \right\}
=
\left\{ \begin{array}{c} \text{intermediate} \\ \text{demand} \end{array} \right\}
+
\left\{ \begin{array}{c} \text{final} \\ \text{demand} \\ \mathbf{d} \end{array} \right\}
\tag{1}
$$

The basic assumption of Leontief's input–output model is that for each sector, there is a **unit consumption vector** in $\mathbb{R}^n$ that lists the inputs needed *per unit of output* of the sector. All input and output units are measured in millions of dollars, rather than in quantities such as tons or bushels. (Prices of goods and services are held constant.)

As a simple example, suppose the economy consists of three sectors—manufacturing, agriculture, and services—with unit consumption vectors $\mathbf{c}_1$, $\mathbf{c}_2$, and $\mathbf{c}_3$, as shown in the table that follows.

Purchased from	Inputs Consumed per Unit of Output		
	Manufacturing	Agriculture	Services
Manufacturing	.50	.40	.20
Agriculture	.20	.30	.10
Services	.10	.10	.30
	↑	↑	↑
	$\mathbf{c}_1$	$\mathbf{c}_2$	$\mathbf{c}_3$

EXAMPLE 1   What amounts will be consumed by the manufacturing sector if it decides to produce 100 units?

SOLUTION   Compute

$$100\mathbf{c}_1 = 100 \begin{bmatrix} .50 \\ .20 \\ .10 \end{bmatrix} = \begin{bmatrix} 50 \\ 20 \\ 10 \end{bmatrix}$$

To produce 100 units, manufacturing will order (i.e., "demand") and consume 50 units from other parts of the manufacturing sector, 20 units from agriculture, and 10 units from services. ∎

If manufacturing decides to produce x_1 units of output, then $x_1\mathbf{c}_1$ represents the *intermediate demands* of manufacturing, because the amounts in $x_1\mathbf{c}_1$ will be consumed in the process of creating the x_1 units of output. Likewise, if x_2 and x_3 denote the planned outputs of the agriculture and services sectors, $x_2\mathbf{c}_2$ and $x_3\mathbf{c}_3$ list their corresponding intermediate demands. The total intermediate demand from all three sectors is given by

$$\{\text{intermediate demand}\} = x_1\mathbf{c}_1 + x_2\mathbf{c}_2 + x_3\mathbf{c}_3$$
$$= C\mathbf{x} \tag{2}$$

where C is the **consumption matrix** $[\,\mathbf{c}_1 \quad \mathbf{c}_2 \quad \mathbf{c}_3\,]$, namely

$$C = \begin{bmatrix} .50 & .40 & .20 \\ .20 & .30 & .10 \\ .10 & .10 & .30 \end{bmatrix} \tag{3}$$

Equations (1) and (2) yield Leontief's model.

THE LEONTIEF INPUT–OUTPUT MODEL, OR PRODUCTION EQUATION

$$\underset{\substack{\text{Amount} \\ \text{produced}}}{\mathbf{x}} \;=\; \underset{\substack{\text{Intermediate} \\ \text{demand}}}{C\mathbf{x}} \;+\; \underset{\substack{\text{Final} \\ \text{demand}}}{\mathbf{d}} \tag{4}$$

Equation (4) may also be written as $I\mathbf{x} - C\mathbf{x} = \mathbf{d}$, or

$$(I - C)\mathbf{x} = \mathbf{d} \tag{5}$$

EXAMPLE 2   Consider the economy whose consumption matrix is given by (3). Suppose the final demand is 50 units for manufacturing, 30 units for agriculture, and 20 units for services. Find the production level $\mathbf{x}$ that will satisfy this demand.

SOLUTION   The coefficient matrix in (5) is

$$I - C = \begin{bmatrix} 1 & 0 & 0 \\ 0 & 1 & 0 \\ 0 & 0 & 1 \end{bmatrix} - \begin{bmatrix} .5 & .4 & .2 \\ .2 & .3 & .1 \\ .1 & .1 & .3 \end{bmatrix} = \begin{bmatrix} .5 & -.4 & -.2 \\ -.2 & .7 & -.1 \\ -.1 & -.1 & .7 \end{bmatrix}$$

To solve (5), row reduce the augmented matrix

$$\begin{bmatrix} .5 & -.4 & -.2 & 50 \\ -.2 & .7 & -.1 & 30 \\ -.1 & -.1 & .7 & 20 \end{bmatrix} \sim \begin{bmatrix} 5 & -4 & -2 & 500 \\ -2 & 7 & -1 & 300 \\ -1 & -1 & 7 & 200 \end{bmatrix} \sim \cdots \sim \begin{bmatrix} 1 & 0 & 0 & 226 \\ 0 & 1 & 0 & 119 \\ 0 & 0 & 1 & 78 \end{bmatrix}$$

The last column is rounded to the nearest whole unit. Manufacturing must produce approximately 226 units, agriculture 119 units, and services only 78 units.   ∎

If the matrix $I - C$ is invertible, then we can apply Theorem 5 in Section 2.2, with A replaced by $(I - C)$, and from the equation $(I - C)\mathbf{x} = \mathbf{d}$ obtain $\mathbf{x} = (I - C)^{-1}\mathbf{d}$. The theorem below shows that in most practical cases, $I - C$ *is* invertible and the production vector $\mathbf{x}$ is economically feasible, in the sense that the entries in $\mathbf{x}$ are nonnegative.

In the theorem, the term **column sum** denotes the sum of the entries in a column of a matrix. Under ordinary circumstances, the column sums of a consumption matrix are less than 1 because a sector should require less than one unit's worth of inputs to produce one unit of output.

THEOREM 11

Let C be the consumption matrix for an economy, and let $\mathbf{d}$ be the final demand. If C and $\mathbf{d}$ have nonnegative entries and if each column sum of C is less than 1, then $(I - C)^{-1}$ exists and the production vector

$$\mathbf{x} = (I - C)^{-1}\mathbf{d}$$

has nonnegative entries and is the unique solution of

$$\mathbf{x} = C\mathbf{x} + \mathbf{d}$$

The following discussion will suggest why the theorem is true and will lead to a new way to compute $(I - C)^{-1}$.

A Formula for $(I - C)^{-1}$

Imagine that the demand represented by $\mathbf{d}$ is presented to the various industries at the beginning of the year, and the industries respond by setting their production levels at $\mathbf{x} = \mathbf{d}$, which will exactly meet the final demand. As the industries prepare to produce $\mathbf{d}$, they send out orders for their raw materials and other inputs. This creates an intermediate demand of $C\mathbf{d}$ for inputs.

To meet the additional demand of $C\mathbf{d}$, the industries will need as additional inputs the amounts in $C(C\mathbf{d}) = C^2\mathbf{d}$. Of course, this creates a second round of intermediate demand, and when the industries decide to produce even more to meet this new demand, they create a third round of demand, namely $C(C^2\mathbf{d}) = C^3\mathbf{d}$. And so it goes.

Theoretically, this process could continue indefinitely, although in real life it would not take place in such a rigid sequence of events. We can diagram this hypothetical situation as follows:

	Demand That Must Be Met	Inputs Needed to Meet This Demand
Final demand	$\mathbf{d}$	$C\mathbf{d}$
Intermediate demand		
1st round	$C\mathbf{d}$	$C(C\mathbf{d}) = C^2\mathbf{d}$
2nd round	$C^2\mathbf{d}$	$C(C^2\mathbf{d}) = C^3\mathbf{d}$
3rd round	$C^3\mathbf{d}$	$C(C^3\mathbf{d}) = C^4\mathbf{d}$
$\vdots$		$\vdots$

The production level $\mathbf{x}$ that will meet all of this demand is

$$\mathbf{x} = \mathbf{d} + C\mathbf{d} + C^2\mathbf{d} + C^3\mathbf{d} + \cdots$$
$$= (I + C + C^2 + C^3 + \cdots)\mathbf{d} \tag{6}$$

To make sense of equation (6), consider the following algebraic identity:

$$(I - C)(I + C + C^2 + \cdots + C^m) = I - C^{m+1} \tag{7}$$

It can be shown that if the column sums in C are all strictly less than 1, then $I - C$ is invertible, C^m approaches the zero matrix as m gets arbitrarily large, and $I - C^{m+1} \to I$. (This fact is analogous to the fact that if a positive number t is less than 1, then $t^m \to 0$ as m increases.) Using equation (7), write

$$(I - C)^{-1} \approx I + C + C^2 + C^3 + \cdots + C^m$$
when the column sums of C are less than 1. $\tag{8}$

The approximation in (8) means that the right side can be made as close to $(I - C)^{-1}$ as desired by taking m sufficiently large.

In actual input–output models, powers of the consumption matrix approach the zero matrix rather quickly. So (8) really provides a practical way to compute $(I - C)^{-1}$. Likewise, for any $\mathbf{d}$, the vectors $C^m\mathbf{d}$ approach the zero vector quickly, and (6) is a practical way to solve $(I - C)\mathbf{x} = \mathbf{d}$. If the entries in C and $\mathbf{d}$ are nonnegative, then (6) shows that the entries in $\mathbf{x}$ are nonnegative, too.

The Economic Importance of Entries in $(I - C)^{-1}$

The entries in $(I - C)^{-1}$ are significant because they can be used to predict how the production $\mathbf{x}$ will have to change when the final demand $\mathbf{d}$ changes. In fact, the entries in column j of $(I - C)^{-1}$ are the *increased* amounts the various sectors will have to produce in order to satisfy *an increase of 1 unit* in the final demand for output from sector j. See Exercise 8.

Numerical Note

In any applied problem (not just in economics), an equation $A\mathbf{x} = \mathbf{b}$ can always be written as $(I - C)\mathbf{x} = \mathbf{b}$, with $C = I - A$. If the system is large and *sparse* (with mostly zero entries), it can happen that the column sums of the absolute values in C are less than 1. In this case, $C^m \to 0$. If C^m approaches zero quickly enough, (6) and (8) will provide practical formulas for solving $A\mathbf{x} = \mathbf{b}$ and finding A^{-1}.

Practice Problem

Suppose an economy has two sectors: goods and services. One unit of output from goods requires inputs of .2 unit from goods and .5 unit from services. One unit of output from services requires inputs of .4 unit from goods and .3 unit from services. There is a final demand of 20 units of goods and 30 units of services. Set up the Leontief input–output model for this situation.

2.6 Exercises

Exercises 1–4 refer to an economy that is divided into three sectors—manufacturing, agriculture, and services. For each unit of output, manufacturing requires .10 unit from other companies in that sector, .30 unit from agriculture, and .30 unit from services. For each unit of output, agriculture uses .20 unit of its own output, .60 unit from manufacturing, and .10 unit from services. For each unit of output, the services sector consumes .10 unit from services, .60 unit from manufacturing, but no agricultural products.

1. Construct the consumption matrix for this economy, and determine what intermediate demands are created if agriculture plans to produce 100 units.

2. Determine the production levels needed to satisfy a final demand of 18 units for agriculture, with no final demand for the other sectors. (Do not compute an inverse matrix.)

3. Determine the production levels needed to satisfy a final demand of 18 units for manufacturing, with no final demand for the other sectors. (Do not compute an inverse matrix.)

4. Determine the production levels needed to satisfy a final demand of 18 units for manufacturing, 18 units for agriculture, and 0 units for services.

5. Consider the production model $\mathbf{x} = C\mathbf{x} + \mathbf{d}$ for an economy with two sectors, where

$$C = \begin{bmatrix} .0 & .5 \\ .6 & .2 \end{bmatrix}, \qquad \mathbf{d} = \begin{bmatrix} 50 \\ 30 \end{bmatrix}$$

Use an inverse matrix to determine the production level necessary to satisfy the final demand.

6. Repeat Exercise 5 with $C = \begin{bmatrix} .1 & .6 \\ .5 & .2 \end{bmatrix}$, and $\mathbf{d} = \begin{bmatrix} 18 \\ 11 \end{bmatrix}$.

7. Let C and $\mathbf{d}$ be as in Exercise 5.

   a. Determine the production level necessary to satisfy a final demand for 1 unit of output from sector 1.

b. Use an inverse matrix to determine the production level necessary to satisfy a final demand of $\begin{bmatrix} 51 \\ 30 \end{bmatrix}$.

c. Use the fact that $\begin{bmatrix} 51 \\ 30 \end{bmatrix} = \begin{bmatrix} 50 \\ 30 \end{bmatrix} + \begin{bmatrix} 1 \\ 0 \end{bmatrix}$ to explain how and why the answers to parts (a) and (b) and to Exercise 5 are related.

8. Let C be an $n \times n$ consumption matrix whose column sums are less than 1. Let $\mathbf{x}$ be the production vector that satisfies a final demand $\mathbf{d}$, and let $\Delta\mathbf{x}$ be a production vector that satisfies a different final demand $\Delta\mathbf{d}$.

a. Show that if the final demand changes from $\mathbf{d}$ to $\mathbf{d} + \Delta\mathbf{d}$, then the new production level must be $\mathbf{x} + \Delta\mathbf{x}$. Thus $\Delta\mathbf{x}$ gives the amounts by which production must *change* in order to accommodate the *change* $\Delta\mathbf{d}$ in demand.

b. Let $\Delta\mathbf{d}$ be the vector in $\mathbb{R}^n$ with 1 as the first entry and 0's elsewhere. Explain why the corresponding production $\Delta\mathbf{x}$ is the first column of $(I - C)^{-1}$. This shows that the first column of $(I - C)^{-1}$ gives the amounts the various sectors must produce to satisfy an increase of 1 unit in the final demand for output from sector 1.

9. Solve the Leontief production equation for an economy with three sectors, given that

$$C = \begin{bmatrix} .2 & .2 & .0 \\ .3 & .1 & .3 \\ .1 & .0 & .2 \end{bmatrix} \quad \text{and} \quad \mathbf{d} = \begin{bmatrix} 40 \\ 60 \\ 80 \end{bmatrix}$$

10. The consumption matrix C for the U.S. economy in 1972 has the property that *every entry* in the matrix $(I - C)^{-1}$ is nonzero (and positive).[1] What does that say about the effect of raising the demand for the output of just one sector of the economy?

11. The Leontief production equation, $\mathbf{x} = C\mathbf{x} + \mathbf{d}$, is usually accompanied by a dual **price equation**,

$$\mathbf{p} = C^T\mathbf{p} + \mathbf{v}$$

where $\mathbf{p}$ is a **price vector** whose entries list the price per unit for each sector's output, and $\mathbf{v}$ is a **value added vector** whose entries list the value added per unit of output. (Value added includes wages, profit, depreciation, etc.) An important fact in economics is that the gross domestic product (GDP) can be expressed in two ways:

$$\{\text{gross domestic product}\} = \mathbf{p}^T\mathbf{d} = \mathbf{v}^T\mathbf{x}$$

Verify the second equality. [*Hint:* Compute $\mathbf{p}^T\mathbf{x}$ in two ways.]

---

[1] Wassily W. Leontief, "The World Economy of the Year 2000," *Scientific American*, September 1980, pp. 206–231.

12. Let C be a consumption matrix such that $C^m \to 0$ as $m \to \infty$, and for $m = 1, 2, \ldots$, let $D_m = I + C + \cdots + C^m$. Find a difference equation that relates D_m and D_{m+1} and thereby obtain an iterative procedure for computing formula (8) for $(I - C)^{-1}$.

T 13. The consumption matrix C below is based on input–output data for the U.S. economy in 1958, with data for 81 sectors grouped into 7 larger sectors: (1) nonmetal household and personal products, (2) final metal products (such as motor vehicles), (3) basic metal products and mining, (4) basic nonmetal products and agriculture, (5) energy, (6) services, and (7) entertainment and miscellaneous products.[2] Find the production levels needed to satisfy the final demand $\mathbf{d}$. (Units are in millions of dollars.)

$$\begin{bmatrix} .1588 & .0064 & .0025 & .0304 & .0014 & .0083 & .1594 \\ .0057 & .2645 & .0436 & .0099 & .0083 & .0201 & .3413 \\ .0264 & .1506 & .3557 & .0139 & .0142 & .0070 & .0236 \\ .3299 & .0565 & .0495 & .3636 & .0204 & .0483 & .0649 \\ .0089 & .0081 & .0333 & .0295 & .3412 & .0237 & .0020 \\ .1190 & .0901 & .0996 & .1260 & .1722 & .2368 & .3369 \\ .0063 & .0126 & .0196 & .0098 & .0064 & .0132 & .0012 \end{bmatrix},$$

$$\mathbf{d} = \begin{bmatrix} 74{,}000 \\ 56{,}000 \\ 10{,}500 \\ 25{,}000 \\ 17{,}500 \\ 196{,}000 \\ 5{,}000 \end{bmatrix}$$

T 14. The demand vector in Exercise 13 is reasonable for 1958 data, but Leontief's discussion of the economy in the reference cited there used a demand vector closer to 1964 data:

$$\mathbf{d} = (99640, 75548, 14444, 33501, 23527, 263985, 6526)$$

Find the production levels needed to satisfy this demand.

T 15. Use equation (6) to solve the problem in Exercise 13. Set $\mathbf{x}^{(0)} = \mathbf{d}$, and for $k = 1, 2, \ldots$, compute $\mathbf{x}^{(k)} = \mathbf{d} + C\mathbf{x}^{(k-1)}$. How many steps are needed to obtain the answer in Exercise 13 to four significant figures?

---

[2] Wassily W. Leontief, "The Structure of the U.S. Economy," *Scientific American*, April 1965, pp. 30–32.

Solution to Practice Problem

The following data are given:

Purchased from	Inputs Needed per Unit of Output		External Demand
	Goods	Services	
Goods	.2	.4	20
Services	.5	.3	30

The Leontief input–output model is $\mathbf{x} = C\mathbf{x} + \mathbf{d}$, where

$$C = \begin{bmatrix} .2 & .4 \\ .5 & .3 \end{bmatrix}, \qquad \mathbf{d} = \begin{bmatrix} 20 \\ 30 \end{bmatrix}$$

2.7 Applications to Computer Graphics

Computer graphics are images displayed or animated on a computer screen. Applications of computer graphics are widespread and growing rapidly. For instance, computer-aided design (CAD) is an integral part of many engineering processes such as the aircraft design process described in the chapter introduction. The entertainment industry has made the most spectacular use of computer graphics—from the special effects in *Amazing Spider-Man 2* to PlayStation 4 and Xbox One.

Most interactive computer software for business and industry makes use of computer graphics in the screen displays and for other functions, such as graphical display of data, desktop publishing, and slide production for commercial and educational presentations. Consequently, anyone studying a computer language invariably spends time learning how to use at least two-dimensional (2D) graphics.

This section examines some of the basic mathematics used to manipulate and display graphical images such as a wire-frame model of an airplane. Such an image (or picture) consists of a number of points, connecting lines or curves, and information about how to fill in closed regions bounded by the lines and curves. Often, curved lines are approximated by short straight-line segments, and a figure is defined mathematically by a list of points.

Among the simplest 2D graphics symbols are letters used for labels on the screen. Some letters are stored as wire-frame objects; others that have curved portions are stored with additional mathematical formulas for the curves.

EXAMPLE 1   The capital letter N in Figure 1 is determined by eight points, or *vertices*. The coordinates of the points can be stored in a data matrix, D.

$$\begin{array}{c} \text{Vertex:} \\ \begin{array}{cccccccc} 1 & 2 & 3 & 4 & 5 & 6 & 7 & 8 \end{array} \\ \begin{array}{c} x\text{-coordinate} \\ y\text{-coordinate} \end{array} \begin{bmatrix} 0 & .5 & .5 & 6 & 6 & 5.5 & 5.5 & 0 \\ 0 & 0 & 6.42 & 0 & 8 & 8 & 1.58 & 8 \end{bmatrix} = D \end{array}$$

In addition to D, it is necessary to specify which vertices are connected by lines, but we omit this detail.   ∎

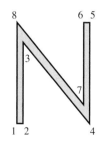

FIGURE 1
Regular *N*.

The main reason graphical objects are described by collections of straight-line segments is that the standard transformations in computer graphics map line segments onto other line segments. (For instance, see Exercise 35 in Section 1.8.) Once the vertices that describe an object have been transformed, their images can be connected with the appropriate straight lines to produce the complete image of the original object.

EXAMPLE 2   Given $A = \begin{bmatrix} 1 & .25 \\ 0 & 1 \end{bmatrix}$, describe the effect of the shear transformation $\mathbf{x} \mapsto A\mathbf{x}$ on the letter N in Example 1.

SOLUTION   By definition of matrix multiplication, the columns of the product AD contain the images of the vertices of the letter N.

$$AD = \begin{matrix} 1 & 2 & 3 & 4 & 5 & 6 & 7 & 8 \\ \begin{bmatrix} 0 & .5 & 2.105 & 6 & 8 & 7.5 & 5.895 & 2 \\ 0 & 0 & 6.420 & 0 & 8 & 8 & 1.580 & 8 \end{bmatrix} \end{matrix}$$

The transformed vertices are plotted in Figure 2, along with connecting line segments that correspond to those in the original figure.   ∎

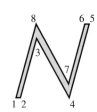

FIGURE 2
Slanted N.

The italic N in Figure 2 looks a bit too wide. To compensate, shrink the width by a scale transformation that affects the x-coordinates of the points.

EXAMPLE 3   Compute the matrix of the transformation that performs a shear transformation, as in Example 2, and then scales all x-coordinates by a factor of .75.

SOLUTION   The matrix that multiplies the x-coordinate of a point by .75 is

$$S = \begin{bmatrix} .75 & 0 \\ 0 & 1 \end{bmatrix}$$

So the matrix of the composite transformation is

$$SA = \begin{bmatrix} .75 & 0 \\ 0 & 1 \end{bmatrix} \begin{bmatrix} 1 & .25 \\ 0 & 1 \end{bmatrix}$$
$$= \begin{bmatrix} .75 & .1875 \\ 0 & 1 \end{bmatrix}$$

FIGURE 3
Composite transformation of N.

The result of this composite transformation is shown in Figure 3.   ∎

The mathematics of computer graphics is intimately connected with matrix multiplication. Unfortunately, translating an object on a screen does not correspond directly to matrix multiplication because translation is not a linear transformation. The standard way to avoid this difficulty is to introduce what are called *homogeneous coordinates*.

Homogeneous Coordinates

Each point (x, y) in $\mathbb{R}^2$ can be identified with the point $(x, y, 1)$ on the plane in $\mathbb{R}^3$ that lies one unit above the xy-plane. We say that (x, y) has *homogeneous coordinates* $(x, y, 1)$. For instance, the point $(0, 0)$ has homogeneous coordinates $(0, 0, 1)$. Homogeneous coordinates for points are not added or multiplied by scalars, but they can be transformed via multiplication by 3×3 matrices.

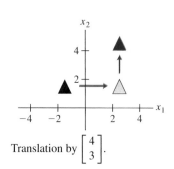

Translation by $\begin{bmatrix} 4 \\ 3 \end{bmatrix}$.

EXAMPLE 4   A translation of the form $(x, y) \mapsto (x + h, y + k)$ is written in homogeneous coordinates as $(x, y, 1) \mapsto (x + h, y + k, 1)$. This transformation can be computed via matrix multiplication:

$$\begin{bmatrix} 1 & 0 & h \\ 0 & 1 & k \\ 0 & 0 & 1 \end{bmatrix} \begin{bmatrix} x \\ y \\ 1 \end{bmatrix} = \begin{bmatrix} x + h \\ y + k \\ 1 \end{bmatrix}$$ ∎

EXAMPLE 5 Any linear transformation on $\mathbb{R}^2$ is represented with respect to homogeneous coordinates by a partitioned matrix of the form $\begin{bmatrix} A & 0 \\ 0 & 1 \end{bmatrix}$, where A is a 2×2 matrix. Typical examples are

$$\begin{bmatrix} \cos\varphi & -\sin\varphi & 0 \\ \sin\varphi & \cos\varphi & 0 \\ 0 & 0 & 1 \end{bmatrix}, \quad \begin{bmatrix} 0 & 1 & 0 \\ 1 & 0 & 0 \\ 0 & 0 & 1 \end{bmatrix}, \quad \begin{bmatrix} s & 0 & 0 \\ 0 & t & 0 \\ 0 & 0 & 1 \end{bmatrix}$$

Counterclockwise rotation about the origin, angle φ Reflection through $y = x$ Scale x by s and y by t ∎

Composite Transformations

The movement of a figure on a computer screen often requires two or more basic transformations. The composition of such transformations corresponds to matrix multiplication when homogeneous coordinates are used.

EXAMPLE 6 Find the 3×3 matrix that corresponds to the composite transformation of a scaling by .3, a rotation of 90° about the origin, and finally a translation that adds $(-.5, 2)$ to each point of a figure.

SOLUTION If $\varphi = \pi/2$, then $\sin\varphi = 1$ and $\cos\varphi = 0$. From Examples 4 and 5, we have

Original Figure

$$\begin{bmatrix} x \\ y \\ 1 \end{bmatrix} \xrightarrow{\text{Scale}} \begin{bmatrix} .3 & 0 & 0 \\ 0 & .3 & 0 \\ 0 & 0 & 1 \end{bmatrix}\begin{bmatrix} x \\ y \\ 1 \end{bmatrix}$$

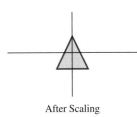

After Scaling

$$\xrightarrow{\text{Rotate}} \begin{bmatrix} 0 & -1 & 0 \\ 1 & 0 & 0 \\ 0 & 0 & 1 \end{bmatrix}\begin{bmatrix} .3 & 0 & 0 \\ 0 & .3 & 0 \\ 0 & 0 & 1 \end{bmatrix}\begin{bmatrix} x \\ y \\ 1 \end{bmatrix}$$

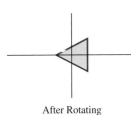

After Rotating

$$\xrightarrow{\text{Translate}} \begin{bmatrix} 1 & 0 & -.5 \\ 0 & 1 & 2 \\ 0 & 0 & 1 \end{bmatrix}\begin{bmatrix} 0 & -1 & 0 \\ 1 & 0 & 0 \\ 0 & 0 & 1 \end{bmatrix}\begin{bmatrix} .3 & 0 & 0 \\ 0 & .3 & 0 \\ 0 & 0 & 1 \end{bmatrix}\begin{bmatrix} x \\ y \\ 1 \end{bmatrix}$$

The matrix for the composite transformation is

$$\begin{bmatrix} 1 & 0 & -.5 \\ 0 & 1 & 2 \\ 0 & 0 & 1 \end{bmatrix}\begin{bmatrix} 0 & -1 & 0 \\ 1 & 0 & 0 \\ 0 & 0 & 1 \end{bmatrix}\begin{bmatrix} .3 & 0 & 0 \\ 0 & .3 & 0 \\ 0 & 0 & 1 \end{bmatrix}$$

$$= \begin{bmatrix} 0 & -1 & -.5 \\ 1 & 0 & 2 \\ 0 & 0 & 1 \end{bmatrix}\begin{bmatrix} .3 & 0 & 0 \\ 0 & .3 & 0 \\ 0 & 0 & 1 \end{bmatrix} = \begin{bmatrix} 0 & -.3 & -.5 \\ .3 & 0 & 2 \\ 0 & 0 & 1 \end{bmatrix}$$ ∎

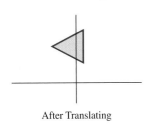

After Translating

3D Computer Graphics

Some of the newest and most exciting work in computer graphics is connected with molecular modeling. With 3D (three-dimensional) graphics, a biologist can examine a simulated protein molecule and search for active sites that might accept a drug molecule. The biologist can rotate and translate an experimental drug and attempt to attach it to the protein. This ability to *visualize* potential chemical reactions is vital to modern drug and cancer research. In fact, advances in drug design depend to some extent upon progress in the ability of computer graphics to construct realistic simulations of molecules and their interactions.[1]

Current research in molecular modeling is focused on *virtual reality*, an environment in which a researcher can see and *feel* the drug molecule slide into the protein. In Figure 4, such tactile feedback is provided by a force-displaying remote manipulator.

FIGURE 4 Molecular modeling in virtual reality.

Another design for virtual reality involves a helmet and glove that detect head, hand, and finger movements. The helmet contains two tiny computer screens, one for each eye. Making this virtual environment more realistic is a challenge to engineers, scientists, and mathematicians. The mathematics we examine here barely opens the door to this interesting field of research.

Homogeneous 3D Coordinates

By analogy with the 2D case, we say that $(x, y, z, 1)$ are homogeneous coordinates for the point (x, y, z) in $\mathbb{R}^3$. In general, (X, Y, Z, H) are **homogeneous coordinates** for (x, y, z) if $H \neq 0$ and

$$x = \frac{X}{H}, \qquad y = \frac{Y}{H}, \quad \text{and} \quad z = \frac{Z}{H} \qquad (1)$$

Each nonzero scalar multiple of $(x, y, z, 1)$ gives a set of homogeneous coordinates for (x, y, z). For instance, both $(10, -6, 14, 2)$ and $(-15, 9, -21, -3)$ are homogeneous coordinates for $(5, -3, 7)$.

The next example illustrates the transformations used in molecular modeling to move a drug into a protein molecule.

EXAMPLE 7 Give 4×4 matrices for the following transformations:

a. Rotation about the y-axis through an angle of $30°$. (By convention, a positive angle is the counterclockwise direction when looking toward the origin from the positive half of the axis of rotation—in this case, the y-axis.)

b. Translation by the vector $\mathbf{p} = (-6, 4, 5)$.

SOLUTION

a. First, construct the 3×3 matrix for the rotation. The vector $\mathbf{e}_1$ rotates down toward the negative z-axis, stopping at $(\cos 30°, 0, -\sin 30°) = (\sqrt{3}/2, 0, -.5)$. The vector $\mathbf{e}_2$ on the y-axis does not move, but $\mathbf{e}_3$ on the z-axis rotates down toward the positive

---

[1] Robert Pool, "Computing in Science," *Science* **256**, 3 April 1992, p. 45.

x-axis, stopping at $(\sin 30°, 0, \cos 30°) = (.5, 0, \sqrt{3}/2)$. See Figure 5. From Section 1.9, the standard matrix for this rotation is

$$\begin{bmatrix} \sqrt{3}/2 & 0 & .5 \\ 0 & 1 & 0 \\ -.5 & 0 & \sqrt{3}/2 \end{bmatrix}$$

So the rotation matrix for homogeneous coordinates is

$$A = \begin{bmatrix} \sqrt{3}/2 & 0 & .5 & 0 \\ 0 & 1 & 0 & 0 \\ -.5 & 0 & \sqrt{3}/2 & 0 \\ 0 & 0 & 0 & 1 \end{bmatrix}$$

FIGURE 5

b. We want $(x, y, z, 1)$ to map to $(x - 6, y + 4, z + 5, 1)$. The matrix that does this is

$$\begin{bmatrix} 1 & 0 & 0 & -6 \\ 0 & 1 & 0 & 4 \\ 0 & 0 & 1 & 5 \\ 0 & 0 & 0 & 1 \end{bmatrix}$$ ∎

Perspective Projections

A three-dimensional object is represented on the two-dimensional computer screen by projecting the object onto a *viewing plane*. (We ignore other important steps, such as selecting the portion of the viewing plane to display on the screen.) For simplicity, let the xy-plane represent the computer screen, and imagine that the eye of a viewer is along the positive z-axis, at a point $(0, 0, d)$. A *perspective projection* maps each point (x, y, z) onto an image point $(x^*, y^*, 0)$ so that the two points and the eye position, called the *center of projection*, are on a line. See Figure 6(a).

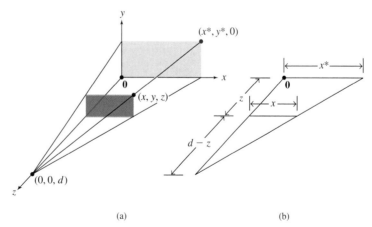

FIGURE 6 Perspective projection of (x, y, z) onto $(x^*, y^*, 0)$.

The triangle in the xz-plane in Figure 6(a) is redrawn in part (b) showing the lengths of line segments. Similar triangles show that

$$\frac{x^*}{d} = \frac{x}{d - z} \quad \text{and} \quad x^* = \frac{dx}{d - z} = \frac{x}{1 - z/d}$$

Similarly,

$$y^* = \frac{y}{1 - z/d}$$

Using homogeneous coordinates, we can represent the perspective projection by a matrix, say, P. We want $(x, y, z, 1)$ to map into $\left(\frac{x}{1 - z/d}, \frac{y}{1 - z/d}, 0, 1 \right)$. Scaling these coordinates by $1 - z/d$, we can also use $(x, y, 0, 1 - z/d)$ as homogeneous coordinates for the image. Now it is easy to display P. In fact,

$$P \begin{bmatrix} x \\ y \\ z \\ 1 \end{bmatrix} = \begin{bmatrix} 1 & 0 & 0 & 0 \\ 0 & 1 & 0 & 0 \\ 0 & 0 & 0 & 0 \\ 0 & 0 & -1/d & 1 \end{bmatrix} \begin{bmatrix} x \\ y \\ z \\ 1 \end{bmatrix} = \begin{bmatrix} x \\ y \\ 0 \\ 1 - z/d \end{bmatrix}$$

EXAMPLE 8 Let S be the box with vertices $(3, 1, 5)$, $(5, 1, 5)$, $(5, 0, 5)$, $(3, 0, 5)$, $(3, 1, 4)$, $(5, 1, 4)$, $(5, 0, 4)$, and $(3, 0, 4)$. Find the image of S under the perspective projection with center of projection at $(0, 0, 10)$.

SOLUTION Let P be the projection matrix, and let D be the data matrix for S using homogeneous coordinates. The data matrix for the image of S is

Vertex:

$$PD = \begin{bmatrix} 1 & 0 & 0 & 0 \\ 0 & 1 & 0 & 0 \\ 0 & 0 & 0 & 0 \\ 0 & 0 & -1/10 & 1 \end{bmatrix} \begin{matrix} \scriptstyle 1 & \scriptstyle 2 & \scriptstyle 3 & \scriptstyle 4 & \scriptstyle 5 & \scriptstyle 6 & \scriptstyle 7 & \scriptstyle 8 \\ \begin{bmatrix} 3 & 5 & 5 & 3 & 3 & 5 & 5 & 3 \\ 1 & 1 & 0 & 0 & 1 & 1 & 0 & 0 \\ 5 & 5 & 5 & 5 & 4 & 4 & 4 & 4 \\ 1 & 1 & 1 & 1 & 1 & 1 & 1 & 1 \end{bmatrix} \end{matrix}$$

$$= \begin{bmatrix} 3 & 5 & 5 & 3 & 3 & 5 & 5 & 3 \\ 1 & 1 & 0 & 0 & 1 & 1 & 0 & 0 \\ 0 & 0 & 0 & 0 & 0 & 0 & 0 & 0 \\ .5 & .5 & .5 & .5 & .6 & .6 & .6 & .6 \end{bmatrix}$$

S under the perspective transformation.

To obtain $\mathbb{R}^3$ coordinates, use equation (1) before Example 7, and divide the top three entries in each column by the corresponding entry in the fourth row:

Vertex:

$$\begin{matrix} \scriptstyle 1 & \scriptstyle 2 & \scriptstyle 3 & \scriptstyle 4 & \scriptstyle 5 & \scriptstyle 6 & \scriptstyle 7 & \scriptstyle 8 \\ \begin{bmatrix} 6 & 10 & 10 & 6 & 5 & 8.3 & 8.3 & 5 \\ 2 & 2 & 0 & 0 & 1.7 & 1.7 & 0 & 0 \\ 0 & 0 & 0 & 0 & 0 & 0 & 0 & 0 \end{bmatrix} \end{matrix} \blacksquare$$

This text's web site has some interesting applications of computer graphics, including a further discussion of perspective projections. One of the chapter projects involves simple animation.

Numerical Note

Continuous movement of graphical 3D objects requires intensive computation with 4×4 matrices, particularly when the surfaces are *rendered* to appear realistic, with texture and appropriate lighting. *High-end computer graphics boards* have 4×4 matrix operations and graphics algorithms embedded in their microchips and circuitry. Such boards can perform the billions of matrix multiplications per second needed for realistic color animation in 3D gaming programs.[2]

Further Reading

James D. Foley, Andries van Dam, Steven K. Feiner, and John F. Hughes, *Computer Graphics: Principles and Practice*, 3rd ed. (Boston, MA: Addison-Wesley, 2002), Chapters 5 and 6.

Practice Problem

Rotation of a figure about a point $\mathbf{p}$ in $\mathbb{R}^2$ is accomplished by first translating the figure by $-\mathbf{p}$, rotating about the origin, and then translating back by $\mathbf{p}$. See Figure 7. Construct the 3×3 matrix that rotates points $-30°$ about the point $(-2, 6)$, using homogeneous coordinates.

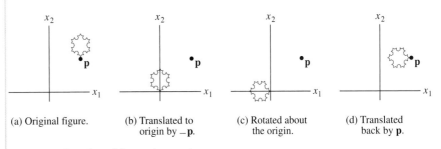

(a) Original figure.   (b) Translated to origin by $-\mathbf{p}$.   (c) Rotated about the origin.   (d) Translated back by $\mathbf{p}$.

FIGURE 7 Rotation of figure about point $\mathbf{p}$.

2.7 Exercises

1. What 3×3 matrix will have the same effect on homogeneous coordinates for $\mathbb{R}^2$ that the shear matrix A has in Example 2?

2. Use matrix multiplication to find the image of the triangle with data matrix $D = \begin{bmatrix} 5 & 2 & 4 \\ 0 & 2 & 3 \end{bmatrix}$ under the transformation that reflects points through the y-axis. Sketch both the original triangle and its image.

In Exercises 3–8, find the 3×3 matrices that produce the described composite 2D transformations, using homogeneous coordinates.

3. Translate by $(3, 1)$, and then rotate $45°$ about the origin.

4. Translate by $(-2, 3)$, and then scale the x-coordinate by .8 and the y-coordinated by 1.2.

5. Reflect points through the x-axis, and then rotate $30°$ about the origin.

6. Rotate points $30°$, and then reflect through the x-axis.

7. Rotate points through $60°$ about the point $(6, 8)$.

8. Rotate points through $45°$ about the point $(3, 7)$.

9. A 2×200 data matrix D contains the coordinates of 200 points. Compute the number of multiplications required to transform these points using two arbitrary 2×2 matrices A and B. Consider the two possibilities $A(BD)$ and $(AB)D$. Discuss the implications of your results for computer graphics calculations.

10. Consider the following geometric 2D transformations: D, a dilation (in which x-coordinates and y-coordinates are scaled by the same factor); R, a rotation; and T, a translation. Does D *commute* with R? That is, is $D(R(\mathbf{x})) = R(D(\mathbf{x}))$ for all $\mathbf{x}$ in $\mathbb{R}^2$? Does D commute with T? Does R commute with T?

---

[2] See Jan Ozer, "High-Performance Graphics Boards," *PC Magazine* **19**, September 1, 2000, pp. 187–200. Also, "The Ultimate Upgrade Guide: Moving On Up," *PC Magazine* **21**, January 29, 2002, pp. 82–91.

11. A rotation on a computer screen is sometimes implemented as the product of two shear-and-scale transformations, which can speed up calculations that determine how a graphic image actually appears in terms of screen pixels. (The screen consists of rows and columns of small dots, called *pixels*.) The first transformation A_1 shears vertically and then compresses each column of pixels; the second transformation A_2 shears horizontally and then stretches each row of pixels. Let

$$A_1 = \begin{bmatrix} 1 & 0 & 0 \\ \sin\varphi & \cos\varphi & 0 \\ 0 & 0 & 1 \end{bmatrix},$$

$$A_2 = \begin{bmatrix} \sec\varphi & -\tan\varphi & 0 \\ 0 & 1 & 0 \\ 0 & 0 & 1 \end{bmatrix}$$

Show that the composition of the two transformations is a rotation in $\mathbb{R}^2$.

12. A rotation in $\mathbb{R}^2$ usually requires four multiplications. Compute the product below, and show that the matrix for a rotation can be factored into three shear transformations (each of which requires only one multiplication).

$$\begin{bmatrix} 1 & -\tan\varphi/2 & 0 \\ 0 & 1 & 0 \\ 0 & 0 & 1 \end{bmatrix} \begin{bmatrix} 1 & 0 & 0 \\ \sin\varphi & 1 & 0 \\ 0 & 0 & 1 \end{bmatrix}$$
$$\begin{bmatrix} 1 & -\tan\varphi/2 & 0 \\ 0 & 1 & 0 \\ 0 & 0 & 1 \end{bmatrix}$$

13. The usual transformations on homogeneous coordinates for 2D computer graphics involve 3×3 matrices of the form $\begin{bmatrix} A & \mathbf{p} \\ \mathbf{0}^T & 1 \end{bmatrix}$ where A is a 2×2 matrix and $\mathbf{p}$ is in $\mathbb{R}^2$. Show that such a transformation amounts to a linear transformation on $\mathbb{R}^2$ followed by a translation. [*Hint:* Find an appropriate matrix factorization involving partitioned matrices.]

14. Show that the transformation in Exercise 7 is equivalent to a rotation about the origin followed by a translation by $\mathbf{p}$. Find $\mathbf{p}$.

15. What vector in $\mathbb{R}^3$ has homogeneous coordinates $\left(\frac{1}{2}, -\frac{1}{4}, \frac{1}{8}, \frac{1}{24}\right)$?

16. Are $(1, -2, 3, 4)$ and $(10, -20, 30, 40)$ homogeneous coordinates for the same point in $\mathbb{R}^3$? Why or why not?

17. Give the 4×4 matrix that rotates points in $\mathbb{R}^3$ about the x-axis through an angle of $60°$. (See the figure.)

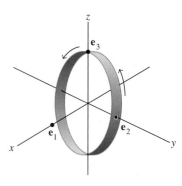

18. Give the 4×4 matrix that rotates points in $\mathbb{R}^3$ about the z-axis through an angle of $-30°$, and then translates by $\mathbf{p} = (5, -2, 1)$.

19. Let S be the triangle with vertices $(4.2, 1.2, 4), (6, 4, 2), (2, 2, 6)$. Find the image of S under the perspective projection with center of projection at $(0, 0, 10)$.

20. Let S be the triangle with vertices $(9, 3, -5), (12, 8, 2), (1.8, 2.7, 1)$. Find the image of S under the perspective projection with center of projection at $(0, 0, 10)$.

Exercises 21 and 22 concern the way in which color is specified for display in computer graphics. A color on a computer screen is encoded by three numbers (R, G, B) that list the amount of energy an electron gun must transmit to red, green, and blue phosphor dots on the computer screen. (A fourth number specifies the luminance or intensity of the color.)

T 21. The actual color a viewer sees on a screen is influenced by the specific type and amount of phosphors on the screen. So each computer screen manufacturer must convert between the (R, G, B) data and an international CIE standard for color, which uses three primary colors, called X, Y, and Z. A typical conversion for short-persistence phosphors is

$$\begin{bmatrix} .61 & .29 & .150 \\ .35 & .59 & .063 \\ .04 & .12 & .787 \end{bmatrix} \begin{bmatrix} R \\ G \\ B \end{bmatrix} = \begin{bmatrix} X \\ Y \\ Z \end{bmatrix}$$

A computer program will send a stream of color information to the screen, using standard CIE data (X, Y, Z). Find the equation that converts these data to the (R, G, B) data needed for the screen's electron gun.

T 22. The signal broadcast by commercial television describes each color by a vector (Y, I, Q). If the screen is black and white, only the Y-coordinate is used. (This gives a better monochrome picture than using CIE data for colors.) The correspondence between YIQ and a "standard" RGB color is given by

$$\begin{bmatrix} Y \\ I \\ Q \end{bmatrix} = \begin{bmatrix} .299 & .587 & .114 \\ .596 & -.275 & -.321 \\ .212 & -.528 & .311 \end{bmatrix} \begin{bmatrix} R \\ G \\ B \end{bmatrix}$$

(A screen manufacturer would change the matrix entries to work for its RGB screens.) Find the equation that converts the YIQ data transmitted by the television station to the RGB data needed for the television screen.

Solution to Practice Problem

Assemble the matrices right-to-left for the three operations. Using $\mathbf{p} = (-2, 6)$, $\cos(-30°) = \sqrt{3}/2$, and $\sin(-30°) = -.5$, we have

$$
\underset{\substack{\text{Translate} \\ \text{back by } p}}{\begin{bmatrix} 1 & 0 & -2 \\ 0 & 1 & 6 \\ 0 & 0 & 1 \end{bmatrix}}
\underset{\substack{\text{Rotate around} \\ \text{the origin}}}{\begin{bmatrix} \sqrt{3}/2 & 1/2 & 0 \\ -1/2 & \sqrt{3}/2 & 0 \\ 0 & 0 & 1 \end{bmatrix}}
\underset{\substack{\text{Translate} \\ \text{by } -p}}{\begin{bmatrix} 1 & 0 & 2 \\ 0 & 1 & -6 \\ 0 & 0 & 1 \end{bmatrix}}
$$

$$
= \begin{bmatrix} \sqrt{3}/2 & 1/2 & \sqrt{3} - 5 \\ -1/2 & \sqrt{3}/2 & -3\sqrt{3} + 5 \\ 0 & 0 & 1 \end{bmatrix}
$$

2.8 Subspaces of $\mathbb{R}^n$

This section focuses on important sets of vectors in $\mathbb{R}^n$ called *subspaces*. Often subspaces arise in connection with some matrix A, and they provide useful information about the equation $A\mathbf{x} = \mathbf{b}$. The concepts and terminology in this section will be used repeatedly throughout the rest of the book.[1]

DEFINITION

A **subspace** of $\mathbb{R}^n$ is any set H in $\mathbb{R}^n$ that has three properties:

a. The zero vector is in H.

b. For each $\mathbf{u}$ and $\mathbf{v}$ in H, the sum $\mathbf{u} + \mathbf{v}$ is in H.

c. For each $\mathbf{u}$ in H and each scalar c, the vector $c\mathbf{u}$ is in H.

In words, a subspace is *closed* under addition and scalar multiplication. As you will see in the next few examples, most sets of vectors discussed in Chapter 1 are subspaces. For instance, a plane through the origin is the standard way to visualize the subspace in Example 1. See Figure 1.

FIGURE 1

Span $\{\mathbf{v}_1, \mathbf{v}_2\}$ as a plane through the origin.

EXAMPLE 1 If $\mathbf{v}_1$ and $\mathbf{v}_2$ are in $\mathbb{R}^n$ and $H = \text{Span}\{\mathbf{v}_1, \mathbf{v}_2\}$, then H is a subspace of $\mathbb{R}^n$. To verify this statement, note that the zero vector is in H (because $0\mathbf{v}_1 + 0\mathbf{v}_2$ is a linear combination of $\mathbf{v}_1$ and $\mathbf{v}_2$). Now take two arbitrary vectors in H, say,

$$\mathbf{u} = s_1\mathbf{v}_1 + s_2\mathbf{v}_2 \quad \text{and} \quad \mathbf{v} = t_1\mathbf{v}_1 + t_2\mathbf{v}_2$$

Then

$$\mathbf{u} + \mathbf{v} = (s_1 + t_1)\mathbf{v}_1 + (s_2 + t_2)\mathbf{v}_2$$

which shows that $\mathbf{u} + \mathbf{v}$ is a linear combination of $\mathbf{v}_1$ and $\mathbf{v}_2$ and hence is in H. Also, for any scalar c, the vector $c\mathbf{u}$ is in H, because $c\mathbf{u} = c(s_1\mathbf{v}_1 + s_2\mathbf{v}_2) = (cs_1)\mathbf{v}_1 + (cs_2)\mathbf{v}_2$. ∎

If $\mathbf{v}_1$ is not zero and if $\mathbf{v}_2$ is a multiple of $\mathbf{v}_1$, then $\mathbf{v}_1$ and $\mathbf{v}_2$ simply span a *line* through the origin. So a line through the origin is another example of a subspace.

---

[1] Sections 2.8 and 2.9 are included here to permit readers to postpone the study of most or all of the next two chapters and to skip directly to Chapter 5, if so desired. *Omit* these two sections if you plan to work through Chapter 4 before beginning Chapter 5.

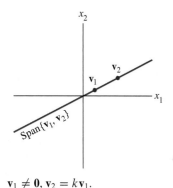

$\mathbf{v}_1 \neq \mathbf{0}, \mathbf{v}_2 = k\mathbf{v}_1.$

EXAMPLE 2 A line L *not* through the origin is *not* a subspace, because it does not contain the origin, as required. Also, Figure 2 shows that L is not closed under addition or scalar multiplication. ∎

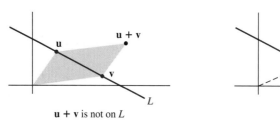

$\mathbf{u} + \mathbf{v}$ is not on L $2\mathbf{w}$ is not on L

FIGURE 2

EXAMPLE 3 For $\mathbf{v}_1, \ldots, \mathbf{v}_p$ in $\mathbb{R}^n$, the set of all linear combinations of $\mathbf{v}_1, \ldots, \mathbf{v}_p$ is a subspace of $\mathbb{R}^n$. The verification of this statement is similar to the argument given in Example 1. We shall now refer to Span $\{\mathbf{v}_1, \ldots, \mathbf{v}_p\}$ as **the subspace spanned** (or **generated**) by $\mathbf{v}_1, \ldots, \mathbf{v}_p$. ∎

Note that $\mathbb{R}^n$ is a subspace of itself because it has the three properties required for a subspace. Another special subspace is the set consisting of only the zero vector in $\mathbb{R}^n$. This set, called the **zero subspace**, also satisfies the conditions for a subspace.

Column Space and Null Space of a Matrix

Subspaces of $\mathbb{R}^n$ usually occur in applications and theory in one of two ways. In both cases, the subspace can be related to a matrix.

DEFINITION

The **column space** of a matrix A is the set Col A of all linear combinations of the columns of A.

If $A = [\mathbf{a}_1 \ \cdots \ \mathbf{a}_n]$, with the columns in $\mathbb{R}^m$, then Col A is the same as Span $\{\mathbf{a}_1, \ldots, \mathbf{a}_n\}$. Example 4 shows that the **column space of an $m \times n$ matrix is a subspace of $\mathbb{R}^m$**. Note that Col A equals $\mathbb{R}^m$ only when the columns of A span $\mathbb{R}^m$. Otherwise, Col A is only part of $\mathbb{R}^m$.

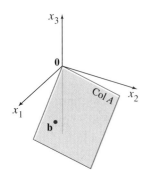

EXAMPLE 4 Let $A = \begin{bmatrix} 1 & -3 & -4 \\ -4 & 6 & -2 \\ -3 & 7 & 6 \end{bmatrix}$ and $\mathbf{b} = \begin{bmatrix} 3 \\ 3 \\ -4 \end{bmatrix}$. Determine whether $\mathbf{b}$ is in the column space of A.

SOLUTION The vector $\mathbf{b}$ is a linear combination of the columns of A if and only if $\mathbf{b}$ can be written as $A\mathbf{x}$ for some $\mathbf{x}$, that is, if and only if the equation $A\mathbf{x} = \mathbf{b}$ has a solution. Row reducing the augmented matrix $[\, A \ \ \mathbf{b}\,]$,

$$\begin{bmatrix} 1 & -3 & -4 & 3 \\ -4 & 6 & -2 & 3 \\ -3 & 7 & 6 & -4 \end{bmatrix} \sim \begin{bmatrix} 1 & -3 & -4 & 3 \\ 0 & -6 & -18 & 15 \\ 0 & -2 & -6 & 5 \end{bmatrix} \sim \begin{bmatrix} 1 & -3 & -4 & 3 \\ 0 & -6 & -18 & 15 \\ 0 & 0 & 0 & 0 \end{bmatrix}$$

we conclude that $A\mathbf{x} = \mathbf{b}$ is consistent and $\mathbf{b}$ is in Col A. ∎

The solution of Example 4 shows that when a system of linear equations is written in the form $A\mathbf{x} = \mathbf{b}$, the column space of A is the set of all $\mathbf{b}$ for which the system has a solution.

DEFINITION

The **null space** of a matrix A is the set Nul A of all solutions of the homogeneous equation $A\mathbf{x} = \mathbf{0}$.

When A has n columns, the solutions of $A\mathbf{x} = \mathbf{0}$ belong to $\mathbb{R}^n$, and the null space of A is a subset of $\mathbb{R}^n$. In fact, Nul A has the properties of a *subspace* of $\mathbb{R}^n$.

THEOREM 12

The null space of an $m \times n$ matrix A is a subspace of $\mathbb{R}^n$. Equivalently, the set of all solutions of a system $A\mathbf{x} = \mathbf{0}$ of m homogeneous linear equations in n unknowns is a subspace of $\mathbb{R}^n$.

PROOF   The zero vector is in Nul A (because $A\mathbf{0} = \mathbf{0}$). To show that Nul A satisfies the other two properties required for a subspace, take any $\mathbf{u}$ and $\mathbf{v}$ in Nul A. That is, suppose $A\mathbf{u} = \mathbf{0}$ and $A\mathbf{v} = \mathbf{0}$. Then, by a property of matrix multiplication,

$$A(\mathbf{u} + \mathbf{v}) = A\mathbf{u} + A\mathbf{v} = \mathbf{0} + \mathbf{0} = \mathbf{0}$$

Thus $\mathbf{u} + \mathbf{v}$ satisfies $A\mathbf{x} = \mathbf{0}$, and so $\mathbf{u} + \mathbf{v}$ is in Nul A. Also, for any scalar c, $A(c\mathbf{u}) = c(A\mathbf{u}) = c(\mathbf{0}) = \mathbf{0}$, which shows that $c\mathbf{u}$ is in Nul A.   ∎

To test whether a given vector $\mathbf{v}$ is in Nul A, just compute $A\mathbf{v}$ to see whether $A\mathbf{v}$ is the zero vector. Because Nul A is described by a condition that must be checked for each vector, we say that the null space is defined *implicitly*. In contrast, the column space is defined *explicitly*, because vectors in Col A can be constructed (by linear combinations) from the columns of A. To create an explicit description of Nul A, solve the equation $A\mathbf{x} = \mathbf{0}$ and write the solution in parametric vector form. (See Example 6.)[2]

Basis for a Subspace

Because a subspace typically contains an infinite number of vectors, some problems involving a subspace are handled best by working with a small finite set of vectors that span the subspace. The smaller the set, the better. It can be shown that the smallest possible spanning set must be linearly independent.

DEFINITION

A **basis** for a subspace H of $\mathbb{R}^n$ is a linearly independent set in H that spans H.

EXAMPLE 5   The columns of an invertible $n \times n$ matrix form a basis for all of $\mathbb{R}^n$ because they are linearly independent and span $\mathbb{R}^n$, by the Invertible Matrix Theorem. One such matrix is the $n \times n$ identity matrix. Its columns are denoted by $\mathbf{e}_1, \dots, \mathbf{e}_n$:

$$\mathbf{e}_1 = \begin{bmatrix} 1 \\ 0 \\ \vdots \\ 0 \end{bmatrix}, \quad \mathbf{e}_2 = \begin{bmatrix} 0 \\ 1 \\ \vdots \\ 0 \end{bmatrix}, \quad \dots, \quad \mathbf{e}_n = \begin{bmatrix} 0 \\ \vdots \\ 0 \\ 1 \end{bmatrix}$$

The set $\{\mathbf{e}_1, \dots, \mathbf{e}_n\}$ is called the **standard basis** for $\mathbb{R}^n$. See Figure 3.   ∎

FIGURE 3
The standard basis for $\mathbb{R}^3$.

[2] The contrast between Nul A and Col A is discussed further in Section 4.2.

The next example shows that the standard procedure for writing the solution set of $A\mathbf{x} = \mathbf{0}$ in parametric vector form actually identifies a basis for Nul A. This fact will be used throughout Chapter 5.

EXAMPLE 6 Find a basis for the null space of the matrix

$$A = \begin{bmatrix} -3 & 6 & -1 & 1 & -7 \\ 1 & -2 & 2 & 3 & -1 \\ 2 & -4 & 5 & 8 & -4 \end{bmatrix}$$

SOLUTION First, write the solution of $A\mathbf{x} = \mathbf{0}$ in parametric vector form:

$$\begin{bmatrix} A & \mathbf{0} \end{bmatrix} \sim \begin{bmatrix} 1 & -2 & 0 & -1 & 3 & 0 \\ 0 & 0 & 1 & 2 & -2 & 0 \\ 0 & 0 & 0 & 0 & 0 & 0 \end{bmatrix}, \quad \begin{matrix} x_1 - 2x_2 \quad - x_4 + 3x_5 = 0 \\ x_3 + 2x_4 - 2x_5 = 0 \\ 0 = 0 \end{matrix}$$

The general solution is $x_1 = 2x_2 + x_4 - 3x_5$, $x_3 = -2x_4 + 2x_5$, with x_2, x_4, and x_5 free.

$$\begin{bmatrix} x_1 \\ x_2 \\ x_3 \\ x_4 \\ x_5 \end{bmatrix} = \begin{bmatrix} 2x_2 + x_4 - 3x_5 \\ x_2 \\ -2x_4 + 2x_5 \\ x_4 \\ x_5 \end{bmatrix} = x_2 \underset{\mathbf{u}}{\begin{bmatrix} 2 \\ 1 \\ 0 \\ 0 \\ 0 \end{bmatrix}} + x_4 \underset{\mathbf{v}}{\begin{bmatrix} 1 \\ 0 \\ -2 \\ 1 \\ 0 \end{bmatrix}} + x_5 \underset{\mathbf{w}}{\begin{bmatrix} -3 \\ 0 \\ 2 \\ 0 \\ 1 \end{bmatrix}}$$

$$= x_2\mathbf{u} + x_4\mathbf{v} + x_5\mathbf{w} \tag{1}$$

Equation (1) shows that Nul A coincides with the set of all linear combinations of $\mathbf{u}$, $\mathbf{v}$, and $\mathbf{w}$. That is, $\{\mathbf{u}, \mathbf{v}, \mathbf{w}\}$ generates Nul A. In fact, this construction of $\mathbf{u}$, $\mathbf{v}$, and $\mathbf{w}$ automatically makes them linearly independent, because equation (1) shows that $\mathbf{0} = x_2\mathbf{u} + x_4\mathbf{v} + x_5\mathbf{w}$ only if the weights x_2, x_4, and x_5 are all zero. (Examine entries 2, 4, and 5 in the vector $x_2\mathbf{u} + x_4\mathbf{v} + x_5\mathbf{w}$.) So $\{\mathbf{u}, \mathbf{v}, \mathbf{w}\}$ is a *basis* for Nul A. ■

Finding a basis for the column space of a matrix is actually less work than finding a basis for the null space. However, the method requires some explanation. Let's begin with a simple case.

EXAMPLE 7 Find a basis for the column space of the matrix

$$B = \begin{bmatrix} 1 & 0 & -3 & 5 & 0 \\ 0 & 1 & 2 & -1 & 0 \\ 0 & 0 & 0 & 0 & 1 \\ 0 & 0 & 0 & 0 & 0 \end{bmatrix}$$

SOLUTION Denote the columns of B by $\mathbf{b}_1, \ldots, \mathbf{b}_5$ and note that $\mathbf{b}_3 = -3\mathbf{b}_1 + 2\mathbf{b}_2$ and $\mathbf{b}_4 = 5\mathbf{b}_1 - \mathbf{b}_2$. The fact that $\mathbf{b}_3$ and $\mathbf{b}_4$ are combinations of the pivot columns means that any combination of $\mathbf{b}_1, \ldots, \mathbf{b}_5$ is actually just a combination of $\mathbf{b}_1$, $\mathbf{b}_2$, and $\mathbf{b}_5$. Indeed, if $\mathbf{v}$ is any vector in Col B, say,

$$\mathbf{v} = c_1\mathbf{b}_1 + c_2\mathbf{b}_2 + c_3\mathbf{b}_3 + c_4\mathbf{b}_4 + c_5\mathbf{b}_5$$

then, substituting for $\mathbf{b}_3$ and $\mathbf{b}_4$, we can write $\mathbf{v}$ in the form

$$\mathbf{v} = c_1\mathbf{b}_1 + c_2\mathbf{b}_2 + c_3(-3\mathbf{b}_1 + 2\mathbf{b}_2) + c_4(5\mathbf{b}_1 - \mathbf{b}_2) + c_5\mathbf{b}_5$$

which is a linear combination of $\mathbf{b}_1$, $\mathbf{b}_2$, and $\mathbf{b}_5$. So $\{\mathbf{b}_1, \mathbf{b}_2, \mathbf{b}_5\}$ spans Col B. Also, $\mathbf{b}_1$, $\mathbf{b}_2$, and $\mathbf{b}_5$ are linearly independent, because they are columns from an identity matrix. So the pivot columns of B form a basis for Col B. ■

The matrix B in Example 7 is in reduced echelon form. To handle a general matrix A, recall that linear dependence relations among the columns of A can be expressed in the form $A\mathbf{x} = \mathbf{0}$ for some $\mathbf{x}$. (If some columns are not involved in a particular dependence relation, then the corresponding entries in $\mathbf{x}$ are zero.) When A is row reduced to echelon form B, the columns are drastically changed, but the equations $A\mathbf{x} = \mathbf{0}$ and $B\mathbf{x} = \mathbf{0}$ have the same set of solutions. That is, the columns of A have *exactly the same linear dependence relationships* as the columns of B.

EXAMPLE 8 It can be verified that the matrix

$$A = [\,\mathbf{a}_1 \quad \mathbf{a}_2 \quad \cdots \quad \mathbf{a}_5\,] = \begin{bmatrix} 1 & 3 & 3 & 2 & -9 \\ -2 & -2 & 2 & -8 & 2 \\ 2 & 3 & 0 & 7 & 1 \\ 3 & 4 & -1 & 11 & -8 \end{bmatrix}$$

is row equivalent to the matrix B in Example 7. Find a basis for Col A.

SOLUTION From Example 7, the pivot columns of A are columns 1, 2, and 5. Also, $\mathbf{b}_3 = -3\mathbf{b}_1 + 2\mathbf{b}_2$ and $\mathbf{b}_4 = 5\mathbf{b}_1 - \mathbf{b}_2$. Since row operations do not affect linear dependence relations among the columns of the matrix, we should have

$$\mathbf{a}_3 = -3\mathbf{a}_1 + 2\mathbf{a}_2 \quad \text{and} \quad \mathbf{a}_4 = 5\mathbf{a}_1 - \mathbf{a}_2$$

Check that this is true! By the argument in Example 7, $\mathbf{a}_3$ and $\mathbf{a}_4$ are not needed to generate the column space of A. Also, $\{\mathbf{a}_1, \mathbf{a}_2, \mathbf{a}_5\}$ must be linearly independent, because any dependence relation among $\mathbf{a}_1$, $\mathbf{a}_2$, and $\mathbf{a}_5$ would imply the same dependence relation among $\mathbf{b}_1$, $\mathbf{b}_2$, and $\mathbf{b}_5$. Since $\{\mathbf{b}_1, \mathbf{b}_2, \mathbf{b}_5\}$ is linearly independent, $\{\mathbf{a}_1, \mathbf{a}_2, \mathbf{a}_5\}$ is also linearly independent and hence is a basis for Col A. ■

The argument in Example 8 can be adapted to prove the following theorem.

THEOREM 13

The pivot columns of a matrix A form a basis for the column space of A.

Warning: Be careful to use *pivot columns of A itself* for the basis of Col A. The columns of an echelon form B are often not in the column space of A. (For instance, in Examples 7 and 8, the columns of B all have zeros in their last entries and cannot generate the columns of A.)

Practice Problems

1. Let $A = \begin{bmatrix} 1 & -1 & 5 \\ 2 & 0 & 7 \\ -3 & -5 & -3 \end{bmatrix}$ and $\mathbf{u} = \begin{bmatrix} -7 \\ 3 \\ 2 \end{bmatrix}$. Is $\mathbf{u}$ in Nul A? Is $\mathbf{u}$ in Col A? Justify each answer.

2. Given $A = \begin{bmatrix} 0 & 1 & 0 \\ 0 & 0 & 1 \\ 0 & 0 & 0 \end{bmatrix}$, find a vector in Nul A and a vector in Col A.

3. Suppose an $n \times n$ matrix A is invertible. What can you say about Col A? About Nul A?

2.8 Exercises

Exercises 1–4 display sets in $\mathbb{R}^2$. Assume the sets include the bounding lines. In each case, give a specific reason why the set H is *not* a subspace of $\mathbb{R}^2$. (For instance, find two vectors in H whose sum is *not* in H, or find a vector in H with a scalar multiple that is not in H. Draw a picture.)

1.

2.

3.

4.

5. Let $\mathbf{v}_1 = \begin{bmatrix} 2 \\ 3 \\ -5 \end{bmatrix}$, $\mathbf{v}_2 = \begin{bmatrix} -4 \\ -5 \\ 8 \end{bmatrix}$, and $\mathbf{w} = \begin{bmatrix} 8 \\ 2 \\ -9 \end{bmatrix}$. Determine if $\mathbf{w}$ is in the subspace of $\mathbb{R}^3$ generated by $\mathbf{v}_1$ and $\mathbf{v}_2$.

6. Let $\mathbf{v}_1 = \begin{bmatrix} 1 \\ -2 \\ 4 \\ 3 \end{bmatrix}$, $\mathbf{v}_2 = \begin{bmatrix} 4 \\ -7 \\ 9 \\ 7 \end{bmatrix}$, $\mathbf{v}_3 = \begin{bmatrix} 5 \\ -8 \\ 6 \\ 5 \end{bmatrix}$, and $\mathbf{u} = \begin{bmatrix} -4 \\ 10 \\ -7 \\ -5 \end{bmatrix}$. Determine if $\mathbf{u}$ is in the subspace of $\mathbb{R}^4$ generated by $\{\mathbf{v}_1, \mathbf{v}_2, \mathbf{v}_3\}$.

7. Let $\mathbf{v}_1 = \begin{bmatrix} 2 \\ -8 \\ 6 \end{bmatrix}$, $\mathbf{v}_2 = \begin{bmatrix} -3 \\ 8 \\ -7 \end{bmatrix}$, $\mathbf{v}_3 = \begin{bmatrix} -4 \\ 6 \\ -7 \end{bmatrix}$,

$\mathbf{p} = \begin{bmatrix} 6 \\ -10 \\ 11 \end{bmatrix}$, and $A = [\mathbf{v}_1 \ \mathbf{v}_2 \ \mathbf{v}_3]$.

a. How many vectors are in $\{\mathbf{v}_1, \mathbf{v}_2, \mathbf{v}_3\}$?

b. How many vectors are in Col A?

c. Is $\mathbf{p}$ in Col A? Why or why not?

8. Let $\mathbf{v}_1 = \begin{bmatrix} -3 \\ 0 \\ 6 \end{bmatrix}$, $\mathbf{v}_2 = \begin{bmatrix} -2 \\ 2 \\ 3 \end{bmatrix}$, $\mathbf{v}_3 = \begin{bmatrix} 0 \\ -6 \\ 3 \end{bmatrix}$, and $\mathbf{p} = \begin{bmatrix} 1 \\ 14 \\ -9 \end{bmatrix}$. Determine if $\mathbf{p}$ is in Col A, where $A = [\mathbf{v}_1 \ \mathbf{v}_2 \ \mathbf{v}_3]$.

9. With A and $\mathbf{p}$ as in Exercise 7, determine if $\mathbf{p}$ is in Nul A.

10. With $\mathbf{u} = (-2, 3, 1)$ and A as in Exercise 8, determine if $\mathbf{u}$ is in Nul A.

In Exercises 11 and 12, give integers p and q such that Nul A is a subspace of $\mathbb{R}^p$ and Col A is a subspace of $\mathbb{R}^q$.

11. $A = \begin{bmatrix} 3 & 2 & 1 & -5 \\ -9 & -4 & 1 & 7 \\ 9 & 2 & -5 & 1 \end{bmatrix}$

12. $A = \begin{bmatrix} 1 & 2 & 3 \\ 4 & 5 & 7 \\ -5 & -1 & 0 \\ 2 & 7 & 11 \end{bmatrix}$

13. For A as in Exercise 11, find a nonzero vector in Nul A and a nonzero vector in Col A.

14. For A as in Exercise 12, find a nonzero vector in Nul A and a nonzero vector in Col A.

Determine which sets in Exercises 15–20 are bases for $\mathbb{R}^2$ or $\mathbb{R}^3$. Justify each answer.

15. $\begin{bmatrix} 5 \\ -2 \end{bmatrix}, \begin{bmatrix} 10 \\ -3 \end{bmatrix}$ **16.** $\begin{bmatrix} -4 \\ 6 \end{bmatrix}, \begin{bmatrix} 2 \\ -3 \end{bmatrix}$

17. $\begin{bmatrix} 0 \\ 1 \\ -2 \end{bmatrix}, \begin{bmatrix} 5 \\ -7 \\ 4 \end{bmatrix}, \begin{bmatrix} 6 \\ 3 \\ 5 \end{bmatrix}$ **18.** $\begin{bmatrix} 1 \\ 1 \\ -2 \end{bmatrix}, \begin{bmatrix} -5 \\ -1 \\ 2 \end{bmatrix}, \begin{bmatrix} 7 \\ 0 \\ -5 \end{bmatrix}$

19. $\begin{bmatrix} 3 \\ -8 \\ 1 \end{bmatrix}, \begin{bmatrix} 6 \\ 2 \\ -5 \end{bmatrix}$

20. $\begin{bmatrix} 1 \\ -6 \\ -7 \end{bmatrix}, \begin{bmatrix} 3 \\ -4 \\ 7 \end{bmatrix}, \begin{bmatrix} -2 \\ 7 \\ 5 \end{bmatrix}, \begin{bmatrix} 0 \\ 8 \\ 9 \end{bmatrix}$

In Exercises 21–30, mark each statement Ture or False (T/F). Justify each answer.

21. (T/F) A subspace of $\mathbb{R}^n$ is any set H such that (i) the zero vector is in H, (ii) $\mathbf{u}$, $\mathbf{v}$, and $\mathbf{u} + \mathbf{v}$ are in H, and (iii) c is a scalar and $c\mathbf{u}$ is in H.

22. (T/F) A subset H of $\mathbb{R}^n$ is a subspace if the zero vector is in H.

23. (T/F) If $\mathbf{v}_1, \ldots, \mathbf{v}_p$ are in $\mathbb{R}^n$, then Span $\{\mathbf{v}_1, \ldots, \mathbf{v}_p\}$ is the same as the column space of the matrix $[\mathbf{v}_1 \ \ldots \ \mathbf{v}_p]$.

24. (T/F) Given vectors $\mathbf{v}_1, \ldots, \mathbf{v}_p$ in $\mathbb{R}^n$, the set of all linear combinations of these vectors is a subspace of $\mathbb{R}^n$.

25. (T/F) The set of all solutions of a system of m homogeneous equations in n unknowns is a subspace of $\mathbb{R}^m$.

26. (T/F) The null space of an $m \times n$ matrix is a subspace of $\mathbb{R}^n$.

27. (T/F) The columns of an invertible $n \times n$ matrix form a basis for $\mathbb{R}^n$.

28. (T/F) The column space of a matrix A is the set of solutions of $A\mathbf{x} = \mathbf{b}$.

29. (T/F) Row operations do not affect linear dependence relations among the columns of a matrix.

30. (T/F) If B is an echelon form of a matrix A, then the pivot columns of B form a basis for Col A.

Exercises 31–34 display a matrix A and an echelon form of A. Find a basis for Col A and a basis for Nul A.

31. $A = \begin{bmatrix} 4 & 5 & 9 & -2 \\ 6 & 5 & 1 & 12 \\ 3 & 4 & 8 & -3 \end{bmatrix} \sim \begin{bmatrix} 1 & 2 & 6 & -5 \\ 0 & 1 & 5 & -6 \\ 0 & 0 & 0 & 0 \end{bmatrix}$

32. $A = \begin{bmatrix} -3 & 9 & -2 & -7 \\ 2 & -6 & 4 & 8 \\ 3 & -9 & -2 & 2 \end{bmatrix} \sim \begin{bmatrix} 1 & -3 & 6 & 9 \\ 0 & 0 & 4 & 5 \\ 0 & 0 & 0 & 0 \end{bmatrix}$

33. $A = \begin{bmatrix} 1 & 4 & 8 & -3 & -7 \\ -1 & 2 & 7 & 3 & 4 \\ -2 & 2 & 9 & 5 & 5 \\ 3 & 6 & 9 & -5 & -2 \end{bmatrix}$

$\sim \begin{bmatrix} 1 & 4 & 8 & 0 & 5 \\ 0 & 2 & 5 & 0 & -1 \\ 0 & 0 & 0 & 1 & 4 \\ 0 & 0 & 0 & 0 & 0 \end{bmatrix}$

34. $A = \begin{bmatrix} 3 & -1 & 7 & 3 & 9 \\ -2 & 2 & -2 & 7 & 5 \\ -5 & 9 & 3 & 3 & 4 \\ -2 & 6 & 6 & 3 & 7 \end{bmatrix}$

$\sim \begin{bmatrix} 3 & -1 & 7 & 0 & 6 \\ 0 & 2 & 4 & 0 & 3 \\ 0 & 0 & 0 & 1 & 1 \\ 0 & 0 & 0 & 0 & 0 \end{bmatrix}$

35. Construct a nonzero 3×3 matrix A and a nonzero vector $\mathbf{b}$ such that $\mathbf{b}$ is in Col A, but $\mathbf{b}$ is not the same as any one of the columns of A.

36. Construct a nonzero 3×3 matrix A and a vector $\mathbf{b}$ such that $\mathbf{b}$ is *not* in Col A.

37. Construct a nonzero 3×3 matrix A and a nonzero vector $\mathbf{b}$ such that $\mathbf{b}$ is in Nul A.

38. Suppose the columns of a matrix $A = [\mathbf{a}_1 \ \cdots \ \mathbf{a}_p]$ are linearly independent. Explain why $\{\mathbf{a}_1, \ldots, \mathbf{a}_p\}$ is a basis for Col A.

In Exercises 39–44, respond as comprehensively as possible, and justify your answer.

39. Suppose F is a 5×5 matrix whose column space is not equal to $\mathbb{R}^5$. What can you say about Nul F?

40. If R is a 6×6 matrix and Nul R is *not* the zero subspace, what can you say about Col R?

41. If Q is a 4×4 matrix and Col $Q = \mathbb{R}^4$, what can you say about solutions of equations of the form $Q\mathbf{x} = \mathbf{b}$ for $\mathbf{b}$ in $\mathbb{R}^4$?

42. If P is a 5×5 matrix and Nul P is the zero subspace, what can you say about solutions of equations of the form $P\mathbf{x} = \mathbf{b}$ for $\mathbf{b}$ in $\mathbb{R}^5$?

43. What can you say about Nul B when B is a 5×4 matrix with linearly independent columns?

44. What can you say about the shape of an $m \times n$ matrix A when the columns of A form a basis for $\mathbb{R}^m$?

In Exercises 45 and 46, construct bases for the column space and the null space of the given matrix A. Justify your work.

T 45. $A = \begin{bmatrix} 3 & -5 & 0 & -1 & 3 \\ -7 & 9 & -4 & 9 & -11 \\ -5 & 7 & -2 & 5 & -7 \\ 3 & -7 & -3 & 4 & 0 \end{bmatrix}$

T 46. $A = \begin{bmatrix} 5 & 2 & 0 & -8 & -8 \\ 4 & 1 & 2 & -8 & -9 \\ 5 & 1 & 3 & 5 & 19 \\ -8 & -5 & 6 & 8 & 5 \end{bmatrix}$

Solutions to Practice Problems

1. To determine whether $\mathbf{u}$ is in Nul A, simply compute

$$A\mathbf{u} = \begin{bmatrix} 1 & -1 & 5 \\ 2 & 0 & 7 \\ -3 & -5 & -3 \end{bmatrix} \begin{bmatrix} -7 \\ 3 \\ 2 \end{bmatrix} = \begin{bmatrix} 0 \\ 0 \\ 0 \end{bmatrix}$$

The result shows that $\mathbf{u}$ is in Nul A. Deciding whether $\mathbf{u}$ is in Col A requires more work. Reduce the augmented matrix $[A \quad \mathbf{u}]$ to echelon form to determine whether the equation $A\mathbf{x} = \mathbf{u}$ is consistent:

$$\begin{bmatrix} 1 & -1 & 5 & -7 \\ 2 & 0 & 7 & 3 \\ -3 & -5 & -3 & 2 \end{bmatrix} \sim \begin{bmatrix} 1 & -1 & 5 & -7 \\ 0 & 2 & -3 & 17 \\ 0 & -8 & 12 & -19 \end{bmatrix} \sim \begin{bmatrix} 1 & -1 & 5 & -7 \\ 0 & 2 & -3 & 17 \\ 0 & 0 & 0 & 49 \end{bmatrix}$$

The equation $A\mathbf{x} = \mathbf{u}$ has no solution, so $\mathbf{u}$ is not in Col A.

2. In contrast to Practice Problem 1, finding a vector in Nul A requires more work than testing whether a specified vector is in Nul A. However, since A is already in reduced echelon form, the equation $A\mathbf{x} = \mathbf{0}$ shows that if $\mathbf{x} = (x_1, x_2, x_3)$, then $x_2 = 0$, $x_3 = 0$, and x_1 is a free variable. Thus, a basis for Nul A is $\mathbf{v} = (1, 0, 0)$. Finding just one vector in Col A is trivial, since each column of A is in Col A. In this particular case, the same vector $\mathbf{v}$ is in both Nul A *and* Col A. For most $n \times n$ matrices, the zero vector of $\mathbb{R}^n$ is the only vector in both Nul A and Col A.

3. If A is invertible, then the columns of A span $\mathbb{R}^n$, by the Invertible Matrix Theorem. By definition, the columns of any matrix always span the column space, so in this case Col A is all of $\mathbb{R}^n$. In symbols, Col $A = \mathbb{R}^n$. Also, since A is invertible, the equation $A\mathbf{x} = \mathbf{0}$ has only the trivial solution. This means that Nul A is the zero subspace. In symbols, Nul $A = \{\mathbf{0}\}$.

2.9 Dimension and Rank

This section continues the discussion of subspaces and bases for subspaces, beginning with the concept of a coordinate system. The definition and example below should make a useful new term, *dimension*, seem quite natural, at least for subspaces of $\mathbb{R}^3$.

Coordinate Systems

The main reason for selecting a basis for a subspace H, instead of merely a spanning set, is that each vector in H can be written in only one way as a linear combination of the basis vectors. To see why, suppose $\mathcal{B} = \{\mathbf{b}_1, \ldots, \mathbf{b}_p\}$ is a basis for H, and suppose a vector $\mathbf{x}$ in H can be generated in two ways, say,

$$\mathbf{x} = c_1\mathbf{b}_1 + \cdots + c_p\mathbf{b}_p \quad \text{and} \quad \mathbf{x} = d_1\mathbf{b}_1 + \cdots + d_p\mathbf{b}_p \tag{1}$$

Then, subtracting gives

$$\mathbf{0} = \mathbf{x} - \mathbf{x} = (c_1 - d_1)\mathbf{b}_1 + \cdots + (c_p - d_p)\mathbf{b}_p \tag{2}$$

Since $\mathcal{B}$ is linearly independent, the weights in (2) must all be zero. That is, $c_j = d_j$ for $1 \leq j \leq p$, which shows that the two representations in (1) are actually the same.

DEFINITION

Suppose the set $\mathcal{B} = \{\mathbf{b}_1, \ldots, \mathbf{b}_p\}$ is a basis for a subspace H. For each $\mathbf{x}$ in H, the **coordinates of x relative to the basis** $\mathcal{B}$ are the weights $c_1, \ldots, c_p$ such that $\mathbf{x} = c_1\mathbf{b}_1 + \cdots + c_p\mathbf{b}_p$, and the vector in $\mathbb{R}^p$

$$[\mathbf{x}]_\mathcal{B} = \begin{bmatrix} c_1 \\ \vdots \\ c_p \end{bmatrix}$$

is called the **coordinate vector of x (relative to** $\mathcal{B}$**)** or the $\mathcal{B}$**-coordinate vector of x.**[1]

EXAMPLE 1   Let $\mathbf{v}_1 = \begin{bmatrix} 3 \\ 6 \\ 2 \end{bmatrix}$, $\mathbf{v}_2 = \begin{bmatrix} -1 \\ 0 \\ 1 \end{bmatrix}$, $\mathbf{x} = \begin{bmatrix} 3 \\ 12 \\ 7 \end{bmatrix}$, and $\mathcal{B} = \{\mathbf{v}_1, \mathbf{v}_2\}$. Then $\mathcal{B}$ is a basis for $H = \text{Span}\{\mathbf{v}_1, \mathbf{v}_2\}$ because $\mathbf{v}_1$ and $\mathbf{v}_2$ are linearly independent. Determine if $\mathbf{x}$ is in H, and if it is, find the coordinate vector of $\mathbf{x}$ relative to $\mathcal{B}$.

SOLUTION   If $\mathbf{x}$ is in H, then the following vector equation is consistent:

$$c_1\begin{bmatrix} 3 \\ 6 \\ 2 \end{bmatrix} + c_2\begin{bmatrix} -1 \\ 0 \\ 1 \end{bmatrix} = \begin{bmatrix} 3 \\ 12 \\ 7 \end{bmatrix}$$

The scalars c_1 and c_2, if they exist, are the $\mathcal{B}$-coordinates of $\mathbf{x}$. Row operations show that

$$\begin{bmatrix} 3 & -1 & 3 \\ 6 & 0 & 12 \\ 2 & 1 & 7 \end{bmatrix} \sim \begin{bmatrix} 1 & 0 & 2 \\ 0 & 1 & 3 \\ 0 & 0 & 0 \end{bmatrix}$$

Thus $c_1 = 2$, $c_2 = 3$, and $[\mathbf{x}]_\mathcal{B} = \begin{bmatrix} 2 \\ 3 \end{bmatrix}$. The basis $\mathcal{B}$ determines a "coordinate system" on H, which can be visualized by the grid shown in Figure 1.   ∎

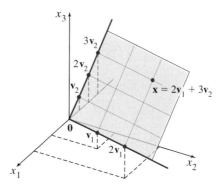

FIGURE 1 A coordinate system on a plane H in $\mathbb{R}^3$.

Notice that although points in H are also in $\mathbb{R}^3$, they are completely determined by their coordinate vectors, which belong to $\mathbb{R}^2$. The grid on the plane in Figure 1

---

[1] It is important that the elements of $\mathcal{B}$ are numbered because the entries in $[\mathbf{x}]_\mathcal{B}$ depend on the order of the vectors in $\mathcal{B}$.

makes H "look" like $\mathbb{R}^2$. The correspondence $\mathbf{x} \mapsto [\mathbf{x}]_{\mathcal{B}}$ is a one-to-one correspondence between H and $\mathbb{R}^2$ that preserves linear combinations. We call such a correspondence an *isomorphism*, and we say that H is *isomorphic* to $\mathbb{R}^2$.

In general, if $\mathcal{B} = \{\mathbf{b}_1, \ldots, \mathbf{b}_p\}$ is a basis for H, then the mapping $\mathbf{x} \mapsto [\mathbf{x}]_{\mathcal{B}}$ is a one-to-one correspondence that makes H look and act the same as $\mathbb{R}^p$ (even though the vectors in H themselves may have more than p entries). (Section 4.4 has more details.)

The Dimension of a Subspace

It can be shown that if a subspace H has a basis of p vectors, then every basis of H must consist of exactly p vectors. (See Exercises 35 and 36.) Thus the following definition makes sense.

DEFINITION

> The **dimension** of a nonzero subspace H, denoted by dim H, is the number of vectors in any basis for H. The dimension of the zero subspace $\{\mathbf{0}\}$ is defined to be zero.[2]

The space $\mathbb{R}^n$ has dimension n. Every basis for $\mathbb{R}^n$ consists of n vectors. A plane through $\mathbf{0}$ in $\mathbb{R}^3$ is two-dimensional, and a line through $\mathbf{0}$ is one-dimensional.

EXAMPLE 2 Recall that the null space of the matrix A in Example 6 in Section 2.8 had a basis of 3 vectors. So the dimension of Nul A in this case is 3. Observe how each basis vector corresponds to a free variable in the equation $A\mathbf{x} = \mathbf{0}$. Our construction always produces a basis in this way. So, to find the dimension of Nul A, simply identify and count the number of free variables in $A\mathbf{x} = \mathbf{0}$. ∎

DEFINITION

> The **rank** of a matrix A, denoted by rank A, is the dimension of the column space of A.

Since the pivot columns of A form a basis for Col A, the rank of A is just the number of pivot columns in A.

EXAMPLE 3 Determine the rank of the matrix

$$A = \begin{bmatrix} 2 & 5 & -3 & -4 & 8 \\ 4 & 7 & -4 & -3 & 9 \\ 6 & 9 & -5 & 2 & 4 \\ 0 & -9 & 6 & 5 & -6 \end{bmatrix}$$

SOLUTION Reduce A to echelon form:

$$A \sim \begin{bmatrix} 2 & 5 & -3 & -4 & 8 \\ 0 & -3 & 2 & 5 & -7 \\ 0 & -6 & 4 & 14 & -20 \\ 0 & -9 & 6 & 5 & -6 \end{bmatrix} \sim \cdots \sim \begin{bmatrix} 2 & 5 & -3 & -4 & 8 \\ 0 & -3 & 2 & 5 & -7 \\ 0 & 0 & 0 & 4 & -6 \\ 0 & 0 & 0 & 0 & 0 \end{bmatrix}$$

Pivot columns ⬆ ⬆ ⬆

The matrix A has 3 pivot columns, so rank $A = 3$. ∎

---

[2] The zero subspace has *no* basis (because the zero vector by itself forms a linearly dependent set).

The row reduction in Example 3 reveals that there are two free variables in $A\mathbf{x} = \mathbf{0}$, because two of the five columns of A are *not* pivot columns. (The nonpivot columns correspond to the free variables in $A\mathbf{x} = \mathbf{0}$.) Since the number of pivot columns plus the number of nonpivot columns is exactly the number of columns, the dimensions of Col A and Nul A have the following useful connection. (See the Rank Theorem in Section 4.6 for additional details.)

THEOREM 14

The Rank Theorem

If a matrix A has n columns, then rank A + dim Nul $A = n$.

The following theorem is important for applications and will be needed in Chapters 5 and 6. The theorem (proved in Section 4.5) is certainly plausible, if you think of a p-dimensional subspace as isomorphic to $\mathbb{R}^p$. The Invertible Matrix Theorem shows that p vectors in $\mathbb{R}^p$ are linearly independent if and only if they also span $\mathbb{R}^p$.

THEOREM 15

The Basis Theorem

Let H be a p-dimensional subspace of $\mathbb{R}^n$. Any linearly independent set of exactly p elements in H is automatically a basis for H. Also, any set of p elements of H that spans H is automatically a basis for H.

Rank and the Invertible Matrix Theorem

The various vector space concepts associated with a matrix provide several more statements for the Invertible Matrix Theorem. They are presented below to follow the statements in the original theorem in Section 2.3.

THEOREM

The Invertible Matrix Theorem (continued)

Let A be an $n \times n$ matrix. Then the following statements are each equivalent to the statement that A is an invertible matrix.

m. The columns of A form a basis of $\mathbb{R}^n$.

n. Col $A = \mathbb{R}^n$

o. rank $A = n$

p. dim Nul $A = 0$

q. Nul $A = \{\mathbf{0}\}$

PROOF Statement (m) is logically equivalent to statements (e) and (h) regarding linear independence and spanning. The other four statements are linked to the earlier ones of the theorem by the following chain of almost trivial implications:

$$(\text{g}) \Rightarrow (\text{n}) \Rightarrow (\text{o}) \Rightarrow (\text{p}) \Rightarrow (\text{q}) \Rightarrow (\text{d})$$

Statement (g), which says that the equation $A\mathbf{x} = \mathbf{b}$ has at least one solution for each $\mathbf{b}$ in $\mathbb{R}^n$, implies statement (n), because Col A is precisely the set of all $\mathbf{b}$ such that the equation $A\mathbf{x} = \mathbf{b}$ is consistent. The implications (n) $\Rightarrow$ (o) $\Rightarrow$ (p) follow from the definitions of *dimension* and *rank*. If the rank of A is n, the number of columns of A, then dim Nul $A = 0$, by the Rank Theorem, and so Nul $A = \{0\}$. Thus (p) $\Rightarrow$ (q). Also, statement (q) implies that the equation $A\mathbf{x} = 0$ has only the trivial solution, which is statement (d). Since statements (d) and (g) are already known to be equivalent to the statement that A is invertible, the proof is complete. ∎

STUDY GUIDE offers an expanded Invertible Matrix Theorem Table.

Numerical Notes

Many algorithms discussed in this text are useful for understanding concepts and making simple computations by hand. However, the algorithms are often unsuitable for large-scale problems in real life.

Rank determination is a good example. It would seem easy to reduce a matrix to echelon form and count the pivots. But unless exact arithmetic is performed on a matrix whose entries are specified exactly, row operations can change the apparent rank of a matrix. For instance, if the value of x in the matrix $\begin{bmatrix} 5 & 7 \\ 5 & x \end{bmatrix}$ is not stored exactly as 7 in a computer, then the rank may be 1 or 2, depending on whether the computer treats $x - 7$ as zero.

In practical applications, the effective rank of a matrix A is often determined from the singular value decomposition of A, to be discussed in Section 7.4.

Practice Problems

1. Determine the dimension of the subspace H of $\mathbb{R}^3$ spanned by the vectors $\mathbf{v}_1$, $\mathbf{v}_2$, and $\mathbf{v}_3$. (First, find a basis for H.)

$$\mathbf{v}_1 = \begin{bmatrix} 2 \\ -8 \\ 6 \end{bmatrix}, \quad \mathbf{v}_2 = \begin{bmatrix} 3 \\ -7 \\ -1 \end{bmatrix}, \quad \mathbf{v}_3 = \begin{bmatrix} -1 \\ 6 \\ -7 \end{bmatrix}$$

2. Consider the basis

$$\mathcal{B} = \left\{ \begin{bmatrix} 1 \\ .2 \end{bmatrix}, \begin{bmatrix} .2 \\ 1 \end{bmatrix} \right\}$$

for $\mathbb{R}^2$. If $[\,\mathbf{x}\,]_{\mathcal{B}} = \begin{bmatrix} 3 \\ 2 \end{bmatrix}$, what is $\mathbf{x}$?

3. Could $\mathbb{R}^3$ possibly contain a four-dimensional subspace? Explain.

2.9 Exercises

In Exercises 1 and 2, find the vector $\mathbf{x}$ determined by the given coordinate vector $[\mathbf{x}]_{\mathcal{B}}$ and the given basis $\mathcal{B}$. Illustrate your answer with a figure, as in the solution of Practice Problem 2.

1. $\mathcal{B} = \left\{ \begin{bmatrix} 1 \\ 1 \end{bmatrix}, \begin{bmatrix} 2 \\ -1 \end{bmatrix} \right\}, [\mathbf{x}]_{\mathcal{B}} = \begin{bmatrix} 3 \\ 2 \end{bmatrix}$

2. $\mathcal{B} = \left\{ \begin{bmatrix} -2 \\ 1 \end{bmatrix}, \begin{bmatrix} 3 \\ 1 \end{bmatrix} \right\}, [\mathbf{x}]_{\mathcal{B}} = \begin{bmatrix} -1 \\ 3 \end{bmatrix}$

In Exercises 3–6, the vector $\mathbf{x}$ is in a subspace H with a basis $\mathcal{B} = \{\mathbf{b}_1, \mathbf{b}_2\}$. Find the $\mathcal{B}$-coordinate vector of $\mathbf{x}$.

3. $\mathbf{b}_1 = \begin{bmatrix} 1 \\ -4 \end{bmatrix}, \mathbf{b}_2 = \begin{bmatrix} -2 \\ 7 \end{bmatrix}, \mathbf{x} = \begin{bmatrix} -3 \\ 7 \end{bmatrix}$

4. $\mathbf{b}_1 = \begin{bmatrix} 1 \\ -3 \end{bmatrix}, \mathbf{b}_2 = \begin{bmatrix} -3 \\ 5 \end{bmatrix}, \mathbf{x} = \begin{bmatrix} -7 \\ 5 \end{bmatrix}$

5. $\mathbf{b}_1 = \begin{bmatrix} 1 \\ 5 \\ -3 \end{bmatrix}, \mathbf{b}_2 = \begin{bmatrix} -3 \\ -7 \\ 5 \end{bmatrix}, \mathbf{x} = \begin{bmatrix} 4 \\ 10 \\ -7 \end{bmatrix}$

6. $\mathbf{b}_1 = \begin{bmatrix} -3 \\ 1 \\ -4 \end{bmatrix}, \mathbf{b}_2 = \begin{bmatrix} 7 \\ 5 \\ -6 \end{bmatrix}, \mathbf{x} = \begin{bmatrix} 11 \\ 0 \\ 7 \end{bmatrix}$

7. Let $\mathbf{b}_1 = \begin{bmatrix} 3 \\ 0 \end{bmatrix}$, $\mathbf{b}_2 = \begin{bmatrix} -1 \\ 2 \end{bmatrix}$, $\mathbf{w} = \begin{bmatrix} 7 \\ -2 \end{bmatrix}$, $\mathbf{x} = \begin{bmatrix} 4 \\ 1 \end{bmatrix}$, and $\mathcal{B} = \{\mathbf{b}_1, \mathbf{b}_2\}$. Use the figure to estimate $[\mathbf{w}]_{\mathcal{B}}$ and $[\mathbf{x}]_{\mathcal{B}}$. Confirm your estimate of $[\mathbf{x}]_{\mathcal{B}}$ by using it and $\{\mathbf{b}_1, \mathbf{b}_2\}$ to compute $\mathbf{x}$.

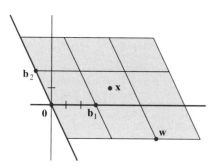

8. Let $\mathbf{b}_1 = \begin{bmatrix} 0 \\ 2 \end{bmatrix}$, $\mathbf{b}_2 = \begin{bmatrix} 2 \\ 1 \end{bmatrix}$, $\mathbf{x} = \begin{bmatrix} -2 \\ 3 \end{bmatrix}$, $\mathbf{y} = \begin{bmatrix} 2 \\ 4 \end{bmatrix}$,

$\mathbf{z} = \begin{bmatrix} -1 \\ -2.5 \end{bmatrix}$, and $\mathcal{B} = \{\mathbf{b}_1, \mathbf{b}_2\}$. Use the figure to estimate $[\mathbf{x}]_{\mathcal{B}}$, $[\mathbf{y}]_{\mathcal{B}}$, and $[\mathbf{z}]_{\mathcal{B}}$. Confirm your estimates of $[\mathbf{y}]_{\mathcal{B}}$ and $[\mathbf{z}]_{\mathcal{B}}$ by using them and $\{\mathbf{b}_1, \mathbf{b}_2\}$ to compute $\mathbf{y}$ and $\mathbf{z}$.

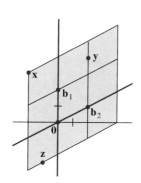

Exercises 9–12 display a matrix A and an echelon form of A. Find bases for Col A and Nul A, and then state the dimensions of these subspaces.

9. $A = \begin{bmatrix} 1 & -3 & 2 & -4 \\ -3 & 9 & -1 & 5 \\ 2 & -6 & 4 & -3 \\ -4 & 12 & 2 & 7 \end{bmatrix} \sim \begin{bmatrix} 1 & -3 & 2 & -4 \\ 0 & 0 & 5 & -7 \\ 0 & 0 & 0 & 5 \\ 0 & 0 & 0 & 0 \end{bmatrix}$

10. $A = \begin{bmatrix} 1 & -2 & 9 & 5 & 4 \\ 1 & -1 & 6 & 5 & -3 \\ -2 & 0 & -6 & 1 & -2 \\ 4 & 1 & 9 & 1 & -9 \end{bmatrix}$

$\sim \begin{bmatrix} 1 & -2 & 9 & 5 & 4 \\ 0 & 1 & -3 & 0 & -7 \\ 0 & 0 & 0 & 1 & -2 \\ 0 & 0 & 0 & 0 & 0 \end{bmatrix}$

11. $A = \begin{bmatrix} 1 & 2 & -5 & 0 & -1 \\ 2 & 5 & -8 & 4 & 3 \\ -3 & -9 & 9 & -7 & -2 \\ 3 & 10 & -7 & 11 & 7 \end{bmatrix}$

$\sim \begin{bmatrix} 1 & 2 & -5 & 0 & -1 \\ 0 & 1 & 2 & 4 & 5 \\ 0 & 0 & 0 & 1 & 2 \\ 0 & 0 & 0 & 0 & 0 \end{bmatrix}$

12. $A = \begin{bmatrix} 1 & 2 & -4 & 3 & 3 \\ 5 & 10 & -9 & -7 & 8 \\ 4 & 8 & -9 & -2 & 7 \\ -2 & -4 & 5 & 0 & -6 \end{bmatrix}$

$\sim \begin{bmatrix} 1 & 2 & -4 & 3 & 3 \\ 0 & 0 & 1 & -2 & 0 \\ 0 & 0 & 0 & 0 & -5 \\ 0 & 0 & 0 & 0 & 0 \end{bmatrix}$

In Exercises 13 and 14, find a basis for the subspace spanned by the given vectors. What is the dimension of the subspace?

13. $\begin{bmatrix} 1 \\ -3 \\ 2 \\ -4 \end{bmatrix}$, $\begin{bmatrix} -3 \\ 9 \\ -6 \\ 12 \end{bmatrix}$, $\begin{bmatrix} 2 \\ -1 \\ 4 \\ 2 \end{bmatrix}$, $\begin{bmatrix} -4 \\ 5 \\ -3 \\ 7 \end{bmatrix}$

14. $\begin{bmatrix} 1 \\ -1 \\ -2 \\ 5 \end{bmatrix}$, $\begin{bmatrix} 2 \\ -3 \\ -1 \\ 6 \end{bmatrix}$, $\begin{bmatrix} 0 \\ 2 \\ -6 \\ 8 \end{bmatrix}$, $\begin{bmatrix} -1 \\ 4 \\ -7 \\ 7 \end{bmatrix}$, $\begin{bmatrix} 3 \\ -8 \\ 9 \\ -5 \end{bmatrix}$

15. Suppose a 3×5 matrix A has three pivot columns. Is Col $A = \mathbb{R}^3$? Is Nul $A = \mathbb{R}^2$? Explain your answers.

16. Suppose a 4×7 matrix A has three pivot columns. Is Col $A = \mathbb{R}^3$? What is the dimension of Nul A? Explain your answers.

In Exercises 17–26, mark each statement True or False **(T/F)**. Justify each answer. Here A is an $m \times n$ matrix.

17. (T/F) If $\mathcal{B} = \{\mathbf{v}_1, \dots, \mathbf{v}_p\}$ is a basis for a subspace H and if $\mathbf{x} = c_1 \mathbf{v}_1 + \cdots + c_p \mathbf{v}_p$, then $c_1, \dots, c_p$ are the coordinates of $\mathbf{x}$ relative to the basis $\mathcal{B}$.

18. (T/F) If $\mathcal{B}$ is a basis for a subspace H, then each vector in H can be written in only one way as a linear combination of the vectors in $\mathcal{B}$.

19. (T/F) Each line in $\mathbb{R}^n$ is a one-dimensional subspace of $\mathbb{R}^n$.

20. (T/F) If $\mathcal{B} = \{\mathbf{v}_1, \dots, \mathbf{v}_p\}$ is a basis for a subspace H of $\mathbb{R}^n$, then the correspondence $\mathbf{x} \mapsto [\mathbf{x}]_{\mathcal{B}}$ makes H look and act the same as $\mathbb{R}^p$.

21. (T/F) The dimension of Col A is the number of pivot columns of A.

22. (T/F) The dimension of Nul A is the number of variables in the equation $A\mathbf{x} = \mathbf{0}$.

23. (T/F) The dimensions of Col A and Nul A add up to the number of columns of A.

24. (T/F) The dimension of the column space of A is rank A.

25. (T/F) If a set of p vectors spans a p-dimensional subspace H of $\mathbb{R}^n$, then these vectors form a basis for H.

26. (T/F) If H is a p-dimensional subspace of $\mathbb{R}^n$, then a linearly independent set of p vectors in H is a basis for H.

In Exercises 27–32, justify each answer or construction.

27. If the subspace of all solutions of $A\mathbf{x} = \mathbf{0}$ has a basis consisting of three vectors and if A is a 5×7 matrix, what is the rank of A?

28. What is the rank of a 4×5 matrix whose null space is three-dimensional?

29. If the rank of a 7×6 matrix A is 4, what is the dimension of the solution space of $A\mathbf{x} = \mathbf{0}$?

30. Show that a set of vectors $\{\mathbf{v}_1, \mathbf{v}_2, ..., \mathbf{v}_5\}$ in $\mathbb{R}^n$ is linearly dependent when dim Span $\{\mathbf{v}_1, \mathbf{v}_2, ..., \mathbf{v}_5\} = 4$.

31. If possible, construct a 3×4 matrix A such that dim Nul $A = 2$ and dim Col $A = 2$.

32. Construct a 4×3 matrix with rank 1.

33. Let A be an $n \times p$ matrix whose column space is p-dimensional. Explain why the columns of A must be linearly independent.

34. Suppose columns 1, 3, 5, and 6 of a matrix A are linearly independent (but are not necessarily pivot columns) and the rank of A is 4. Explain why the four columns mentioned must be a basis for the column space of A.

35. Suppose vectors $\mathbf{b}_1, ..., \mathbf{b}_p$ span a subspace W, and let $\{\mathbf{a}_1, ..., \mathbf{a}_q\}$ be any set in W containing more than p vectors. Fill in the details of the following argument to show that $\{\mathbf{a}_1, ..., \mathbf{a}_q\}$ must be linearly dependent. First, let $B = [\mathbf{b}_1 \ \cdots \ \mathbf{b}_p]$ and $A = [\mathbf{a}_1 \ \cdots \ \mathbf{a}_q]$.

a. Explain why for each vector $\mathbf{a}_j$, there exists a vector $\mathbf{c}_j$ in $\mathbb{R}^p$ such that $\mathbf{a}_j = B\mathbf{c}_j$.

b. Let $C = [\mathbf{c}_1 \ \cdots \ \mathbf{c}_q]$. Explain why there is a nonzero vector $\mathbf{u}$ such that $C\mathbf{u} = \mathbf{0}$.

c. Use B and C to show that $A\mathbf{u} = \mathbf{0}$. This shows that the columns of A are linearly dependent.

36. Use Exercise 35 to show that if $\mathcal{A}$ and $\mathcal{B}$ are bases for a subspace W of $\mathbb{R}^n$, then $\mathcal{A}$ cannot contain more vectors than $\mathcal{B}$, and, conversely, $\mathcal{B}$ cannot contain more vectors than $\mathcal{A}$.

37. Let $H = \text{Span} \{\mathbf{v}_1, \mathbf{v}_2\}$ and $\mathcal{B} = \{\mathbf{v}_1, \mathbf{v}_2\}$. Show that $\mathbf{x}$ is in H, and find the $\mathcal{B}$-coordinate vector of $\mathbf{x}$, when

$$\mathbf{v}_1 = \begin{bmatrix} 11 \\ -5 \\ 10 \\ 7 \end{bmatrix}, \mathbf{v}_2 = \begin{bmatrix} 14 \\ -8 \\ 13 \\ 10 \end{bmatrix}, \mathbf{x} = \begin{bmatrix} 19 \\ -13 \\ 18 \\ 15 \end{bmatrix}$$

38. Let $H = \text{Span}\{\mathbf{v}_1, \mathbf{v}_2, \mathbf{v}_3\}$ and $\mathcal{B} = \{\mathbf{v}_1, \mathbf{v}_2, \mathbf{v}_3\}$. Show that $\mathcal{B}$ is a basis for H and $\mathbf{x}$ is in H, and find the $\mathcal{B}$-coordinate vector of $\mathbf{x}$, when

$$\mathbf{v}_1 = \begin{bmatrix} -6 \\ 4 \\ -9 \\ 4 \end{bmatrix}, \mathbf{v}_2 = \begin{bmatrix} 8 \\ -3 \\ 7 \\ -3 \end{bmatrix}, \mathbf{v}_3 = \begin{bmatrix} -9 \\ 5 \\ -8 \\ 3 \end{bmatrix}, \mathbf{x} = \begin{bmatrix} 4 \\ 7 \\ -8 \\ 3 \end{bmatrix}$$

STUDY GUIDE
offers additional resources for mastering the concepts of dimension and rank.

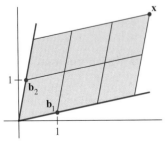

Solutions to Practice Problems

1. Construct $A = [\mathbf{v}_1 \ \ \mathbf{v}_2 \ \ \mathbf{v}_3]$ so that the subspace spanned by $\mathbf{v}_1, \mathbf{v}_2, \mathbf{v}_3$ is the column space of A. A basis for this space is provided by the pivot columns of A.

$$A = \begin{bmatrix} 2 & 3 & -1 \\ -8 & -7 & 6 \\ 6 & -1 & -7 \end{bmatrix} \sim \begin{bmatrix} 2 & 3 & -1 \\ 0 & 5 & 2 \\ 0 & -10 & -4 \end{bmatrix} \sim \begin{bmatrix} 2 & 3 & -1 \\ 0 & 5 & 2 \\ 0 & 0 & 0 \end{bmatrix}$$

The first two columns of A are pivot columns and form a basis for H. Thus dim $H = 2$.

2. If $[\mathbf{x}]_\mathcal{B} = \begin{bmatrix} 3 \\ 2 \end{bmatrix}$, then $\mathbf{x}$ is formed from a linear combination of the basis vectors using weights 3 and 2:

$$\mathbf{x} = 3\mathbf{b}_1 + 2\mathbf{b}_2 = 3\begin{bmatrix} 1 \\ .2 \end{bmatrix} + 2\begin{bmatrix} .2 \\ 1 \end{bmatrix} = \begin{bmatrix} 3.4 \\ 2.6 \end{bmatrix}$$

The basis $\{\mathbf{b}_1, \mathbf{b}_2\}$ determines a *coordinate system* for $\mathbb{R}^2$, illustrated by the grid in the figure. Note how $\mathbf{x}$ is 3 units in the $\mathbf{b}_1$-direction and 2 units in the $\mathbf{b}_2$-direction.

3. A four-dimensional subspace would contain a basis of four linearly independent vectors. This is impossible inside $\mathbb{R}^3$. Since any linearly independent set in $\mathbb{R}^3$ has no more than three vectors, any subspace of $\mathbb{R}^3$ has dimension no more than 3. The space $\mathbb{R}^3$ itself is the only three-dimensional subspace of $\mathbb{R}^3$. Other subspaces of $\mathbb{R}^3$ have dimension 2, 1, or 0.

CHAPTER 2 PROJECTS

Chapter 2 projects are available online at `bit.ly/30IM8gT`.

A. *Other Matrix Products*:   This project introduces two new operations on square matrices called the Jordan product and the commutator product and their properties are explored.

B. *Adjacency Matrices*:   The purpose of this project is to show how powers of a matrix may be used to investigate graphs.

C. *Dominance Matrices*:   The purpose of this project is to apply matrices and their powers to questions concerning various forms of competition between individuals and groups.

D. *Condition Numbers*:   The purpose of this project is to show how a condition number of a matrix A may be defined, and how its value affects the accuracy of solutions to systems of equations $A\mathbf{x} = \mathbf{b}$.

E. *Equilibrium Temperature Distributions*:   The purpose of this project is to discuss a physical situation in which solving a system of linear equations becomes necessary: that of determining the equilibrium temperature of a thin plate.

F. *The LU and QR Factorizations*:   The purpose of this project is to explore a relationship between two matrix factorizations: the LU factorization and the QR factorization.

G. *The Leontief Input–Output Model*:   The purpose of this project is to provide three more examples of the Leontief input–output model in action.

H. *The Art of Linear Transformations*:   This project illustrates how to graph a polygon and then use linear transformations to move it around in the plane.

CHAPTER 2 SUPPLEMENTARY EXERCISES

Assume that the matrices mentioned in Exercises 1–15 below have appropriate sizes. Mark each statement True or False **(T/F)**. Justify each answer.

1. **(T/F)** If A and B are $m \times n$, then both AB^T and A^TB are defined.

2. **(T/F)** If $AB = C$ and C has 2 columns, then A has 2 columns.

3. **(T/F)** Left-multiplying a matrix B by a diagonal matrix A, with nonzero entries on the diagonal, scales the rows of B.

4. **(T/F)** If $BC = BD$, then $C = D$.

5. **(T/F)** If $AC = 0$, then either $A = 0$ or $C = 0$.

6. **(T/F)** If A and B are $n \times n$, then $(A + B)(A - B) = A^2 - B^2$.

7. **(T/F)** An elementary $n \times n$ matrix has either n or $n + 1$ nonzero entries.

8. **(T/F)** The transpose of an elementary matrix is an elementary matrix.

9. **(T/F)** An elementary matrix must be square.

10. **(T/F)** Every square matrix is a product of elementary matrices.

11. **(T/F)** If A is a 3×3 matrix with three pivot positions, there exist elementary matrices $E_1, \ldots, E_p$ such that $E_p \cdots E_1 A = I$.

12. **(T/F)** If $AB = I$, then A is invertible.

13. **(T/F)** If A and B are square and invertible, then AB is invertible, and $(AB)^{-1} = A^{-1}B^{-1}$.

14. **(T/F)** If $AB = BA$ and if A is invertible, then $A^{-1}B = BA^{-1}$.

15. **(T/F)** If A is invertible and if $r \neq 0$, then $(rA)^{-1} = rA^{-1}$.

16. Find the matrix C whose inverse is $C^{-1} = \begin{bmatrix} 4 & 5 \\ 6 & 7 \end{bmatrix}$.

17. Let $A = \begin{bmatrix} 0 & 0 & 0 \\ 1 & 0 & 0 \\ 0 & 1 & 0 \end{bmatrix}$. Show that $A^3 = 0$. Use matrix algebra to compute the product $(I - A)(I + A + A^2)$.

18. Suppose $A^n = 0$ for some $n > 1$. Find an inverse for $I - A$.

19. Suppose an $n \times n$ matrix A satisfies the equation $A^2 - 2A + I = 0$. Show that $A^3 = 3A - 2I$ and $A^4 = 4A - 3I$.

20. Let $A = \begin{bmatrix} 1 & 0 \\ 0 & -1 \end{bmatrix}$, $B = \begin{bmatrix} 0 & 1 \\ 1 & 0 \end{bmatrix}$. These are *Pauli spin matrices* used in the study of electron spin in quantum mechanics. Show that $A^2 = I$, $B^2 = I$, and $AB = -BA$. Matrices such that $AB = -BA$ are said to *anticommute*.

21. Let $A = \begin{bmatrix} 1 & 3 & 8 \\ 2 & 4 & 11 \\ 1 & 2 & 5 \end{bmatrix}$ and $B = \begin{bmatrix} -3 & 5 \\ 1 & 5 \\ 3 & 4 \end{bmatrix}$. Compute $A^{-1}B$ without computing A^{-1}. [*Hint:* $A^{-1}B$ is the solution of the equation $AX = B$.]

22. Find a matrix A such that the transformation $\mathbf{x} \mapsto A\mathbf{x}$ maps $\begin{bmatrix} 1 \\ 3 \end{bmatrix}$ and $\begin{bmatrix} 2 \\ 7 \end{bmatrix}$ into $\begin{bmatrix} 1 \\ 1 \end{bmatrix}$ and $\begin{bmatrix} 3 \\ 1 \end{bmatrix}$, respectively. [*Hint:* Write a matrix equation involving A, and solve for A.]

23. Suppose $AB = \begin{bmatrix} 5 & 4 \\ -2 & 3 \end{bmatrix}$ and $B = \begin{bmatrix} 7 & 3 \\ 2 & 1 \end{bmatrix}$. Find A.

24. Suppose A is invertible. Explain why A^TA is also invertible. Then show that $A^{-1} = (A^TA)^{-1}A^T$.

25. Let $x_1, \ldots, x_n$ be fixed numbers. The matrix below, called a *Vandermonde matrix*, occurs in applications such as signal processing, error-correcting codes, and polynomial interpolation.

$$V = \begin{bmatrix} 1 & x_1 & x_1^2 & \cdots & x_1^{n-1} \\ 1 & x_2 & x_2^2 & \cdots & x_2^{n-1} \\ \vdots & \vdots & \vdots & & \vdots \\ 1 & x_n & x_n^2 & \cdots & x_n^{n-1} \end{bmatrix}$$

Given $\mathbf{y} = (y_1, \ldots, y_n)$ in $\mathbb{R}^n$, suppose $\mathbf{c} = (c_0, \ldots, c_{n-1})$ in $\mathbb{R}^n$ satisfies $V\mathbf{c} = \mathbf{y}$, and define the polynomial

$$p(t) = c_0 + c_1 t + c_2 t^2 + \cdots + c_{n-1}t^{n-1}.$$

a. Show that $p(x_1) = y_1, \ldots, p(x_n) = y_n$. We call $p(t)$ an *interpolating polynomial for the points* $(x_1, y_1), \ldots, (x_n, y_n)$ because the graph of $p(t)$ passes through the points.

b. Suppose $x_1, \ldots, x_n$ are distinct numbers. Show that the columns of V are linearly independent. [*Hint:* How many zeros can a polynomial of degree $n - 1$ have?]

c. Prove: "If $x_1, \ldots, x_n$ are distinct numbers, and $y_1, \ldots, y_n$ are arbitrary numbers, then there is an interpolating polynomial of degree $\leq n - 1$ for $(x_1, y_1), \ldots, (x_n, y_n)$."

26. Let $A = LU$, where L is an invertible lower triangular matrix and U is upper triangular. Explain why the first column of A is a multiple of the first column of L. How is the second column of A related to the columns of L?

27. Given $\mathbf{u}$ in $\mathbb{R}^n$ with $\mathbf{u}^T\mathbf{u} = 1$, let $P = \mathbf{u}\mathbf{u}^T$ (an outer product) and $Q = I - 2P$. Justify statements (a), (b), and (c).

a. $P^2 = P$ b. $P^T = P$ c. $Q^2 = I$

The transformation $\mathbf{x} \mapsto P\mathbf{x}$ is called a *projection*, and $\mathbf{x} \mapsto Q\mathbf{x}$ is called a *Householder reflection*. Such reflections are used in computer programs to create multiple zeros in a vector (usually a column of a matrix).

28. Let $\mathbf{u} = \begin{bmatrix} 0 \\ 0 \\ 1 \end{bmatrix}$ and $\mathbf{x} = \begin{bmatrix} 1 \\ 5 \\ 3 \end{bmatrix}$. Determine P and Q as in Exercise 27, and compute $P\mathbf{x}$ and $Q\mathbf{x}$. The figure shows that $Q\mathbf{x}$ is the reflection of $\mathbf{x}$ through the x_1x_2-plane.

A Householder reflection through the plane $x_3 = 0$.

29. Suppose $C = E_3E_2E_1B$, where E_1, E_2, and E_3 are elementary matrices. Explain why C is row equivalent to B.

30. Let A be an $n \times n$ singular matrix. Describe how to construct an $n \times n$ nonzero matrix B such that $AB = 0$.

31. Let A be a 6×4 matrix and B a 4×6 matrix. Show that the 6×6 matrix AB cannot be invertible.

32. Suppose A is a 5×3 matrix and there exists a 3×5 matrix C such that $CA = I_3$. Suppose further that for some given $\mathbf{b}$ in $\mathbb{R}^5$, the equation $A\mathbf{x} = \mathbf{b}$ has at least one solution. Show that this solution is unique.

T 33. Certain dynamical systems can be studied by examining powers of a matrix, such as those below. Determine what happens to A^k and B^k as k increases (for example, try $k = 2, \ldots, 16$). Try to identify what is special about A and B. Investigate large powers of other matrices of this type, and make a conjecture about such matrices.

$$A = \begin{bmatrix} .4 & .2 & .3 \\ .3 & .6 & .3 \\ .3 & .2 & .4 \end{bmatrix}, \quad B = \begin{bmatrix} 0 & .2 & .3 \\ .1 & .6 & .3 \\ .9 & .2 & .4 \end{bmatrix}$$

T 34. Let A_n be the $n \times n$ matrix with 0's on the main diagonal and 1's elsewhere. Compute A_n^{-1} for $n = 4, 5$, and 6, and make a conjecture about the general form of A_n^{-1} for larger values of n.

3 Determinants

Introductory Example

WEIGHING DIAMONDS

How is the value of a diamond determined? Jewelers use the four cs: cut, clarity, color, and carats; a carat is a unit of mass equal to 0.2 grams. When a jeweler receives a supply of diamonds, it is vital that they be weighed accurately as part of determining their value. The difference of half a carat can have a large impact on a diamond's value.

When weighing small objects, such as diamonds or other gemstones, one strategy is to weigh the objects individually, but there are more accurate strategies that involve weighing the objects in groups and then deducing the individual weights from the results.

Suppose there are n small objects to be weighed, labeled $s_1, s_2, \cdots, s_n$. One method of determining the weight of each small object uses a two-pan balance. A *weighing* consists of placing some of the small objects in the left pan and the rest in the right pan. The balance records the difference between the weights in the pans.

The jeweler (or other individual weighing small light objects) plans her strategy in advance by creating a design matrix D with entries determined by the following

scheme: If gemstone s_j is placed in the left pan during the ith weighing, then $d_{ij} = -1$ and if gemstone s_j is placed in the right pan during the ith weighing the $d_{ij} = 1$. Each row of the matrix D corresponds to a particular weighing. The jth column of D tells you where to put s_j at each weighing. Thus D is an $m \times n$ matrix, where m corresponds to the number of weighings and n corresponds to the number of objects. It has been shown that the accuracy of a weighing design is highest when a design matrix that maximizes the value of the *determinant* of $D^T D$ is chosen.

For example, consider the design matrix $D =$

$$\begin{bmatrix} 1 & 1 & 1 & 1 \\ 1 & -1 & 1 & 1 \\ 1 & 1 & -1 & 1 \\ 1 & 1 & 1 & -1 \end{bmatrix}$$ for weighing the gems

s_1, s_2, s_3, s_4. For this design, the first weighing has all four gems in the right tray (the first row of D consists of all ones). For the second weighing, gem s_2 is in the left tray and the rest of the gems are in the right tray (the second row of D has a -1 in the second column). For the third weighing, gem s_3 is in the left tray and the rest of the gems are in the right tray (the third row of D has a -1 in the third column). In the last weighing, gem s_4 is in the left tray and the remaining gems are in the right tray (the fourth row of D has a -1 in the fourth column). The determinant of $D^T D$ is 64.

However, this is not the best design for using four weighings to determine the weight of four objects.

171

If $D = \begin{bmatrix} 1 & 1 & 1 & 1 \\ 1 & 1 & -1 & -1 \\ 1 & -1 & 1 & -1 \\ -1 & 1 & 1 & -1 \end{bmatrix}$, then the determinant

of $D^T D = 256$, and hence this is a better design. Notice that the first weighing of this design is the same as the previous one, but then the remaining weighings each have two objects in each pan.

Calculating determinants of matrices and understanding their properties is the theme of this chapter. As you learn more about determinants, you may also come up with strategies for good and bad choices for a weighing design.

Another important use of the determinant is to calculate the area of a parallelogram or the volume of a parallelepiped. In Section 1.9, we saw that matrix multiplication can be used to change the shape of a box or other object. The determinant of the matrix used determines how much the area changes when it is multiplied by a matrix, just as a fish story can transform the size of the fish caught.

Indeed, the determinant has so many uses that a summary of the applications known in the early 1900s filled a four-volume treatise by Thomas Muir. With changes in emphasis and the greatly increased sizes of the matrices used in modem applications, many uses that were important then are no longer critical today. Nevertheless, the determinant still plays many important theoretical and practical roles.

Beyond introducing the determinant in Section 3.1, this chapter presents two important ideas. Section 3.2 derives an invertibility criterion for a square matrix that plays a pivotal role in Chapter 5. Section 3.3 shows how the determinant measures the amount by which a linear transformation changes the area of a figure. When applied locally, this technique answers the question of a map's expansion rate near the poles. This idea plays a critical role in multivariable calculus in the form of the Jacobian.

3.1 Introduction to Determinants

Recall from Section 2.2 that a 2×2 matrix is invertible if and only if its determinant is nonzero. To extend this useful fact to larger matrices, we need a definition for the determinant of an $n \times n$ matrix. We can discover the definition for the 3×3 case by watching what happens when an invertible 3×3 matrix A is row reduced.

Consider $A = [a_{ij}]$ with $a_{11} \neq 0$. If we multiply the second and third rows of A by a_{11} and then subtract appropriate multiples of the first row from the other two rows, we find that A is row equivalent to the following two matrices:

$$\begin{bmatrix} a_{11} & a_{12} & a_{13} \\ a_{11}a_{21} & a_{11}a_{22} & a_{11}a_{23} \\ a_{11}a_{31} & a_{11}a_{32} & a_{11}a_{33} \end{bmatrix} \sim \begin{bmatrix} a_{11} & a_{12} & a_{13} \\ 0 & a_{11}a_{22} - a_{12}a_{21} & a_{11}a_{23} - a_{13}a_{21} \\ 0 & a_{11}a_{32} - a_{12}a_{31} & a_{11}a_{33} - a_{13}a_{31} \end{bmatrix} \quad (1)$$

Since A is invertible, either the $(2, 2)$-entry or the $(3, 2)$-entry on the right in (1) is nonzero. Let us suppose that the $(2, 2)$-entry is nonzero. (Otherwise, we can make a row interchange before proceeding.) Multiply row 3 by $a_{11}a_{22} - a_{12}a_{21}$, and then to the new row 3 add $-(a_{11}a_{32} - a_{12}a_{31})$ times row 2. This will show that

$$A \sim \begin{bmatrix} a_{11} & a_{12} & a_{13} \\ 0 & a_{11}a_{22} - a_{12}a_{21} & a_{11}a_{23} - a_{13}a_{21} \\ 0 & 0 & a_{11}\Delta \end{bmatrix}$$

where

$$\Delta = a_{11}a_{22}a_{33} + a_{12}a_{23}a_{31} + a_{13}a_{21}a_{32} - a_{11}a_{23}a_{32} - a_{12}a_{21}a_{33} - a_{13}a_{22}a_{31} \quad (2)$$

Since A is invertible, Δ must be nonzero. The converse is true, too, as we will see in Section 3.2. We call Δ in (2) the **determinant** of the 3×3 matrix A.

Recall that the determinant of a 2×2 matrix, $A = [a_{ij}]$, is the number

$$\det A = a_{11}a_{22} - a_{12}a_{21}$$

For a 1×1 matrix—say, $A = [a_{11}]$—we define $\det A = a_{11}$. To generalize the definition of the determinant to larger matrices, we'll use 2×2 determinants to rewrite the 3×3 determinant Δ described above. Since the terms in Δ can be grouped as $(a_{11}a_{22}a_{33} - a_{11}a_{23}a_{32}) - (a_{12}a_{21}a_{33} - a_{12}a_{23}a_{31}) + (a_{13}a_{21}a_{32} - a_{13}a_{22}a_{31})$,

$$\Delta = a_{11} \det \begin{bmatrix} a_{22} & a_{23} \\ a_{32} & a_{33} \end{bmatrix} - a_{12} \det \begin{bmatrix} a_{21} & a_{23} \\ a_{31} & a_{33} \end{bmatrix} + a_{13} \det \begin{bmatrix} a_{21} & a_{22} \\ a_{31} & a_{32} \end{bmatrix}$$

For brevity, write

$$\Delta = a_{11} \det A_{11} - a_{12} \det A_{12} + a_{13} \det A_{13} \quad (3)$$

where A_{11}, A_{12}, and A_{13} are obtained from A by deleting the first row and one of the three columns. For any square matrix A, let A_{ij} denote the submatrix formed by deleting the ith row and jth column of A. For instance, if

$$A = \begin{bmatrix} 1 & -2 & 5 & 0 \\ 2 & 0 & 4 & -1 \\ 3 & 1 & 0 & 7 \\ 0 & 4 & -2 & 0 \end{bmatrix}$$

then A_{32} is obtained by crossing out row 3 and column 2,

$$\begin{bmatrix} 1 & -2 & 5 & 0 \\ 2 & 0 & 4 & -1 \\ 3 & 1 & 0 & 7 \\ 0 & 4 & -2 & 0 \end{bmatrix}$$

so that

$$A_{32} = \begin{bmatrix} 1 & 5 & 0 \\ 2 & 4 & -1 \\ 0 & -2 & 0 \end{bmatrix}$$

We can now give a *recursive* definition of a determinant. When $n = 3$, $\det A$ is defined using determinants of the 2×2 submatrices A_{1j}, as in (3) above. When $n = 4$, $\det A$ uses determinants of the 3×3 submatrices A_{1j}. In general, an $n \times n$ determinant is defined by determinants of $(n - 1) \times (n - 1)$ submatrices.

DEFINITION

For $n \geq 2$, the **determinant** of an $n \times n$ matrix $A = [a_{ij}]$ is the sum of n terms of the form $\pm a_{1j} \det A_{1j}$, with plus and minus signs alternating, where the entries $a_{11}, a_{12}, \ldots, a_{1n}$ are from the first row of A. In symbols,

$$\det A = a_{11} \det A_{11} - a_{12} \det A_{12} + \cdots + (-1)^{1+n} a_{1n} \det A_{1n}$$

$$= \sum_{j=1}^{n} (-1)^{1+j} a_{1j} \det A_{1j}$$

EXAMPLE 1 Compute the determinant of

$$A = \begin{bmatrix} 1 & 5 & 0 \\ 2 & 4 & -1 \\ 0 & -2 & 0 \end{bmatrix}$$

SOLUTION Compute $\det A = a_{11} \det A_{11} - a_{12} \det A_{12} + a_{13} \det A_{13}$:

$$\det A = 1 \det \begin{bmatrix} 4 & -1 \\ -2 & 0 \end{bmatrix} - 5 \det \begin{bmatrix} 2 & -1 \\ 0 & 0 \end{bmatrix} + 0 \det \begin{bmatrix} 2 & 4 \\ 0 & -2 \end{bmatrix}$$

$$= 1(0 - 2) - 5(0 - 0) + 0(-4 - 0) = -2 \qquad \blacksquare$$

Another common notation for the determinant of a matrix uses a pair of vertical lines in place of brackets. Thus the calculation in Example 1 can be written as

$$\det A = 1 \begin{vmatrix} 4 & -1 \\ -2 & 0 \end{vmatrix} - 5 \begin{vmatrix} 2 & -1 \\ 0 & 0 \end{vmatrix} + 0 \begin{vmatrix} 2 & 4 \\ 0 & -2 \end{vmatrix} = \cdots = -2$$

To state the next theorem, it is convenient to write the definition of $\det A$ in a slightly different form. Given $A = [a_{ij}]$, the **(i, j)-cofactor** of A is the number C_{ij} given by

$$C_{ij} = (-1)^{i+j} \det A_{ij} \qquad (4)$$

Then

$$\det A = a_{11} C_{11} + a_{12} C_{12} + \cdots + a_{1n} C_{1n}$$

This formula is called a **cofactor expansion across the first row** of A. We omit the proof of the following fundamental theorem to avoid a lengthy digression.

THEOREM 1

> The determinant of an $n \times n$ matrix A can be computed by a cofactor expansion across any row or down any column. The expansion across the ith row using the cofactors in (4) is
>
> $$\det A = a_{i1} C_{i1} + a_{i2} C_{i2} + \cdots + a_{in} C_{in}$$
>
> The cofactor expansion down the jth column is
>
> $$\det A = a_{1j} C_{1j} + a_{2j} C_{2j} + \cdots + a_{nj} C_{nj}$$

The plus or minus sign in the (i, j)-cofactor depends on the position of a_{ij} in the matrix, regardless of the sign of a_{ij} itself. The factor $(-1)^{i+j}$ determines the following checkerboard pattern of signs:

$$\begin{bmatrix} + & - & + & \cdots \\ - & + & - & \\ + & - & + & \\ \vdots & & & \ddots \end{bmatrix}$$

EXAMPLE 2 Use a cofactor expansion across the third row to compute $\det A$, where

$$A = \begin{bmatrix} 1 & 5 & 0 \\ 2 & 4 & -1 \\ 0 & -2 & 0 \end{bmatrix}$$

SOLUTION   Compute

$$\det A = a_{31}C_{31} + a_{32}C_{32} + a_{33}C_{33}$$
$$= (-1)^{3+1}a_{31} \det A_{31} + (-1)^{3+2}a_{32} \det A_{32} + (-1)^{3+3}a_{33} \det A_{33}$$
$$= 0 \begin{vmatrix} 5 & 0 \\ 4 & -1 \end{vmatrix} - (-2) \begin{vmatrix} 1 & 0 \\ 2 & -1 \end{vmatrix} + 0 \begin{vmatrix} 1 & 5 \\ 2 & 4 \end{vmatrix}$$
$$= 0 + 2(-1) + 0 = -2 \qquad \blacksquare$$

Theorem 1 is helpful for computing the determinant of a matrix that contains many zeros. For example, if a row is mostly zeros, then the cofactor expansion across that row has many terms that are zero, and the cofactors in those terms need not be calculated. The same approach works with a column that contains many zeros.

EXAMPLE 3   Compute $\det A$, where

$$A = \begin{bmatrix} 3 & -7 & 8 & 9 & -6 \\ 0 & 2 & -5 & 7 & 3 \\ 0 & 0 & 1 & 5 & 0 \\ 0 & 0 & 2 & 4 & -1 \\ 0 & 0 & 0 & -2 & 0 \end{bmatrix}$$

SOLUTION   The cofactor expansion down the first column of A has all terms equal to zero except the first. Thus

$$\det A = 3 \begin{vmatrix} 2 & -5 & 7 & 3 \\ 0 & 1 & 5 & 0 \\ 0 & 2 & 4 & -1 \\ 0 & 0 & -2 & 0 \end{vmatrix} + 0 C_{21} + 0 C_{31} + 0 C_{41} + 0 C_{51}$$

Henceforth we will omit the zero terms in the cofactor expansion. Next, expand this 4×4 determinant down the first column to take advantage of the zeros there. We have

$$\det A = 3(2) \begin{vmatrix} 1 & 5 & 0 \\ 2 & 4 & -1 \\ 0 & -2 & 0 \end{vmatrix}$$

This 3×3 determinant was computed in Example 1 and found to equal -2. Hence $\det A = 3(2)(-2) = -12$. $\blacksquare$

The matrix in Example 3 was nearly triangular. The method in that example is easily adapted to prove the following theorem.

THEOREM 2

If A is a triangular matrix, then $\det A$ is the product of the entries on the main diagonal of A.

The strategy in Example 3 of looking for zeros works extremely well when an entire row or column consists of zeros. In such a case, the cofactor expansion along such a row or column is a sum of zeros! So the determinant is zero. Unfortunately, most cofactor expansions are not so quickly evaluated.

Reasonable Answers

How big can a determinant be? Let A be an $n \times n$ matrix. Notice that taking the determinant of A consists of adding and subtracting terms with n products each. If p is the product of the n largest elements in absolute value (the same number may be repeated if it occurs more than once as a matrix entry), then the determinant must be between $-np$ and np. For example, consider $A = \begin{bmatrix} 6 & 5 \\ -7 & 9 \end{bmatrix}$ and $B = \begin{bmatrix} 7 & 6 \\ 7 & -9 \end{bmatrix}$. The largest number in absolute value of each matrix is 9, and the second largest number is 7. In these two cases, $p = 7(9) = 63$ and $np = 126$. The determinant of each of these matrices should be a number between -126 and 126. Notice that $\det A = 6(9) - 5(-7) = 54 + 35 = 89$, $\det B = 7(-9) - 6(7) = -63 - 42 = -105$, illustrating that because the products are added and subtracted, any number in the range between -126 and 126 could turn out to be the determinant.

Next, consider $C = \begin{bmatrix} 7 & 9 \\ 7 & 9 \end{bmatrix}$ and $D = \begin{bmatrix} -9 & 9 \\ 9 & 9 \end{bmatrix}$. In matrices C and D, the number 9 appears twice and so should be selected twice. In this case, $p = 9(9) = 81$ and $np = 162$, so the determinants of C and D should be numbers between -162 and 162. Indeed, $\det C = (7)(9) - (7)(9) = 0$ and $\det D = (-9)(9) - (9)(9) = -162$. Notice that it is important to choose 9 twice as the two largest numbers in matrix D in order to get the correct bounds for the determinant of D.

Numerical Note

By today's standards, a 25×25 matrix is small. Yet it would be impossible to calculate a 25×25 determinant by cofactor expansion. In general, a cofactor expansion requires more than $n!$ multiplications, and 25! is approximately 1.55×10^{25}.

If a computer performs one trillion multiplications per second, it would have to run for almost 500,000 years to compute a 25×25 determinant by this method. Fortunately, there are faster methods, as we'll soon discover.

Exercises 19–38 explore important properties of determinants, mostly for the 2×2 case. The results from Exercises 33–36 will be used in the next section to derive the analogous properties for $n \times n$ matrices.

Practice Problem

Compute $\begin{vmatrix} 5 & -7 & 2 & 2 \\ 0 & 3 & 0 & -4 \\ -5 & -8 & 0 & 3 \\ 0 & 5 & 0 & -6 \end{vmatrix}$.

3.1 Exercises

Compute the determinants in Exercises 1–8 using a cofactor expansion across the first row. In Exercises 1–4, also compute the determinant by a cofactor expansion down the second column.

1. $\begin{vmatrix} 3 & 0 & 4 \\ 2 & 3 & 2 \\ 0 & 5 & -1 \end{vmatrix}$ 2. $\begin{vmatrix} 0 & 4 & 1 \\ 5 & -3 & 0 \\ 2 & 4 & 1 \end{vmatrix}$

3. $\begin{vmatrix} 2 & -2 & 3 \\ 3 & 1 & 2 \\ 1 & 3 & -1 \end{vmatrix}$ 4. $\begin{vmatrix} 1 & 2 & 4 \\ 3 & 1 & 1 \\ 2 & 4 & 2 \end{vmatrix}$

5. $\begin{vmatrix} 2 & 3 & -3 \\ 4 & 0 & 3 \\ 6 & 1 & 5 \end{vmatrix}$ 6. $\begin{vmatrix} 5 & -2 & 3 \\ 0 & 3 & -3 \\ 2 & -4 & 7 \end{vmatrix}$

7. $\begin{vmatrix} 4 & 3 & 0 \\ 6 & 5 & 2 \\ 9 & 7 & 3 \end{vmatrix}$ 8. $\begin{vmatrix} 4 & 1 & 2 \\ 4 & 0 & 3 \\ 3 & -2 & 5 \end{vmatrix}$

Compute the determinants in Exercises 9–14 by cofactor expansions. At each step, choose a row or column that involves the least amount of computation.

9. $\begin{vmatrix} 4 & 0 & 0 & 5 \\ 1 & 7 & 2 & -5 \\ 3 & 0 & 0 & 0 \\ 8 & 3 & 1 & 7 \end{vmatrix}$ 10. $\begin{vmatrix} 1 & -2 & 4 & 2 \\ 0 & 0 & 3 & 0 \\ 2 & -4 & -3 & 5 \\ 2 & 0 & 3 & 5 \end{vmatrix}$

11. $\begin{vmatrix} 3 & 5 & -6 & 4 \\ 0 & -2 & 3 & -3 \\ 0 & 0 & 1 & 5 \\ 0 & 0 & 0 & 3 \end{vmatrix}$ 12. $\begin{vmatrix} 3 & 0 & 0 & 0 \\ 7 & -2 & 0 & 0 \\ 2 & 6 & 3 & 0 \\ 3 & -8 & 4 & -3 \end{vmatrix}$

13. $\begin{vmatrix} 4 & 0 & -7 & 3 & -5 \\ 0 & 0 & 2 & 0 & 0 \\ 7 & 3 & -6 & 4 & -8 \\ 5 & 0 & 5 & 2 & -3 \\ 0 & 0 & 9 & -1 & 2 \end{vmatrix}$

14. $\begin{vmatrix} 6 & 0 & 2 & 4 & 0 \\ 9 & 0 & -4 & 1 & 0 \\ 8 & -5 & 6 & 7 & 1 \\ 2 & 0 & 0 & 0 & 0 \\ 4 & 2 & 3 & 2 & 0 \end{vmatrix}$

The expansion of a 3×3 determinant can be remembered by the following device. Write a second copy of the first two columns to the right of the matrix, and compute the determinant by multiplying entries on six diagonals:

Add the downward diagonal products and subtract the upward products. Use this method to compute the determinants in Exercises 15–18. *Warning: This trick does not generalize in any reasonable way to 4×4 or larger matrices.*

15. $\begin{vmatrix} 1 & 0 & 4 \\ 2 & 3 & 2 \\ 0 & 5 & -2 \end{vmatrix}$ 16. $\begin{vmatrix} 0 & 3 & 1 \\ 4 & -5 & 0 \\ 3 & 4 & 1 \end{vmatrix}$

17. $\begin{vmatrix} 2 & -3 & 3 \\ 3 & 2 & 2 \\ 1 & 3 & -1 \end{vmatrix}$ 18. $\begin{vmatrix} 1 & 3 & 4 \\ 2 & 3 & 2 \\ 3 & 3 & 2 \end{vmatrix}$

In Exercises 19–24, explore the effect of an elementary row operation on the determinant of a matrix. In each case, state the row operation and describe how it affects the determinant.

19. $\begin{bmatrix} a & b \\ c & d \end{bmatrix}, \begin{bmatrix} c & d \\ a & b \end{bmatrix}$

20. $\begin{bmatrix} a & b \\ c & d \end{bmatrix}, \begin{bmatrix} a & b \\ kc & kd \end{bmatrix}$

21. $\begin{bmatrix} 3 & 2 \\ 5 & 4 \end{bmatrix}, \begin{bmatrix} 3 & 2 \\ 5+3k & 4+2k \end{bmatrix}$

22. $\begin{bmatrix} a & b \\ c & d \end{bmatrix}, \begin{bmatrix} a+kc & b+kd \\ c & d \end{bmatrix}$

23. $\begin{bmatrix} a & b & c \\ 3 & 2 & 1 \\ 4 & 5 & 6 \end{bmatrix}, \begin{bmatrix} 3 & 2 & 1 \\ a & b & c \\ 4 & 5 & 6 \end{bmatrix}$

24. $\begin{bmatrix} 1 & 0 & 1 \\ -3 & 4 & -4 \\ 2 & -3 & 1 \end{bmatrix}, \begin{bmatrix} k & 0 & k \\ -3 & 4 & -4 \\ 2 & -3 & 1 \end{bmatrix}$

Compute the determinants of the elementary matrices given in Exercises 25–30. (See Section 2.2, Examples 5 and 6.)

25. $\begin{bmatrix} 1 & 0 & 0 \\ 0 & 1 & 0 \\ 0 & k & 1 \end{bmatrix}$ 26. $\begin{bmatrix} 0 & 1 & 0 \\ 1 & 0 & 0 \\ 0 & 0 & 1 \end{bmatrix}$

27. $\begin{bmatrix} 1 & 0 & 0 \\ 0 & 1 & 0 \\ k & 0 & 1 \end{bmatrix}$ 28. $\begin{bmatrix} 0 & 0 & 1 \\ 0 & 1 & 0 \\ 1 & 0 & 0 \end{bmatrix}$

29. $\begin{bmatrix} 1 & 0 & 0 \\ 0 & k & 0 \\ 0 & 0 & 1 \end{bmatrix}$ 30. $\begin{bmatrix} k & 0 & 0 \\ 0 & 1 & 0 \\ 0 & 0 & 1 \end{bmatrix}$

Use Exercises 25–30 to answer the questions in Exercises 31 and 32. Give reasons for your answers.

31. What is the determinant of an elementary row replacement matrix?

32. What is the determinant of an elementary scaling matrix with k on the diagonal?

In Exercises 33–36, verify that $\det EA = (\det E)(\det A)$, where E is the elementary matrix shown and $A = \begin{bmatrix} a & b \\ c & d \end{bmatrix}$.

33. $\begin{bmatrix} 1 & k \\ 0 & 1 \end{bmatrix}$ **34.** $\begin{bmatrix} 1 & 0 \\ k & 1 \end{bmatrix}$

35. $\begin{bmatrix} 0 & 1 \\ 1 & 0 \end{bmatrix}$ **36.** $\begin{bmatrix} 1 & 0 \\ 0 & k \end{bmatrix}$

37. Let $A = \begin{bmatrix} 3 & 1 \\ 4 & 2 \end{bmatrix}$. Write $5A$. Is $\det 5A = 5 \det A$?

38. Let $A = \begin{bmatrix} a & b \\ c & d \end{bmatrix}$ and let k be a scalar. Find a formula that relates $\det kA$ to k and $\det A$.

In Exercises 39 through 42, A is an $n \times n$ matrix. Mark each statement True or False **(T/F)**. Justify each answer.

39. (T/F) An $n \times n$ determinant is defined by determinants of $(n-1) \times (n-1)$ submatrices.

40. (T/F) The (i, j)-cofactor of a matrix A is the matrix A_{ij} obtained by deleting from A its ith row and jth column.

41. (T/F) The cofactor expansion of $\det A$ down a column is equal to the cofactor expansion along a row.

42. (T/F) The determinant of a triangular matrix is the sum of the entries on the main diagonal.

43. Let $\mathbf{u} = \begin{bmatrix} 3 \\ 0 \end{bmatrix}$ and $\mathbf{v} = \begin{bmatrix} 1 \\ 2 \end{bmatrix}$. Compute the area of the parallelogram determined by $\mathbf{u}$, $\mathbf{v}$, $\mathbf{u} + \mathbf{v}$, and $\mathbf{0}$, and compute the determinant of $[\mathbf{u} \ \ \mathbf{v}]$. How do they compare? Replace the first entry of $\mathbf{v}$ by an arbitrary number x, and repeat the problem. Draw a picture and explain what you find.

44. Let $\mathbf{u} = \begin{bmatrix} a \\ b \end{bmatrix}$ and $\mathbf{v} = \begin{bmatrix} c \\ 0 \end{bmatrix}$, where a, b, and c are positive (for simplicity). Compute the area of the parallelogram determined by $\mathbf{u}$, $\mathbf{v}$, $\mathbf{u} + \mathbf{v}$, and $\mathbf{0}$, and compute the determinants of the matrices $[\mathbf{u} \ \ \mathbf{v}]$ and $[\mathbf{v} \ \ \mathbf{u}]$. Draw a picture and explain what you find.

45. Let A be a 2×2 matrix all of whose entries are numbers that are greater than or equal to -10 and less than or equal to 10. Decide if each of the following is a reasonable answer for $\det A$.

a. 0

b. 202

c. -110

d. 555

46. Let A be a 3×3 matrix all of whose entries are numbers that are greater than or equal to -5 and less than or equal to 5. Decide if each of the following is a reasonable answer for $\det A$.

a. 300

b. -220

c. 1000

d. 10

T 47. Construct a random 4×4 matrix A with integer entries between -9 and 9. How is $\det A^{-1}$ related to $\det A$? Experiment with random $n \times n$ integer matrices for $n = 4$, 5, and 6, and make a conjecture. *Note:* In the unlikely event that you encounter a matrix with a zero determinant, reduce it to echelon form and discuss what you find.

T 48. Is it true that $\det AB = (\det A)(\det B)$? To find out, generate random 5×5 matrices A and B, and compute $\det AB - (\det A \det B)$. Repeat the calculations for three other pairs of $n \times n$ matrices, for various values of n. Report your results.

T 49. Is it true that $\det(A + B) = \det A + \det B$? Experiment with four pairs of random matrices as in Exercise 48, and make a conjecture.

T 50. Construct a random 4×4 matrix A with integer entries between -9 and 9, and compare $\det A$ with $\det A^T$, $\det(-A)$, $\det(2A)$, and $\det(10A)$. Repeat with two other random 4×4 integer matrices, and make conjectures about how these determinants are related. (Refer to Exercise 44 in Section 2.1.) Then check your conjectures with several random 5×5 and 6×6 integer matrices. Modify your conjectures, if necessary, and report your results.

T 51. Recall from the introductory section that the larger the determinant of $D^T D$, where D is the design matrix, the better will be the accuracy of the calculated weights for small light objects. Which of the following matrices corresponds to the best design for four weighings of four objects? Describe which of the objects s_1, s_2, s_3, and s_4 you would put in the left and right pans for each weighing corresponding to the best design matrix.

a. $D = \begin{bmatrix} 1 & 1 & 1 & 1 \\ 1 & -1 & 1 & 1 \\ -1 & -1 & 1 & -1 \\ 1 & 1 & 1 & -1 \end{bmatrix}$

b. $D = \begin{bmatrix} -1 & -1 & -1 & -1 \\ 1 & -1 & -1 & 1 \\ -1 & 1 & -1 & 1 \\ -1 & -1 & 1 & 1 \end{bmatrix}$

c. $D = \begin{bmatrix} 1 & 1 & -1 & -1 \\ 1 & -1 & -1 & 1 \\ -1 & 1 & 1 & -1 \\ 1 & -1 & 1 & -1 \end{bmatrix}$

T 52. Repeat Exercise 51 for the case of five weighings of four objects and the following design matrices.

a. $D = \begin{bmatrix} 1 & 1 & 1 & 1 \\ 1 & -1 & 1 & 1 \\ -1 & -1 & 1 & -1 \\ 1 & 1 & 1 & -1 \\ -1 & -1 & -1 & 1 \end{bmatrix}$

c. $D = \begin{bmatrix} 1 & 1 & -1 & -1 \\ 1 & -1 & -1 & 1 \\ -1 & 1 & 1 & -1 \\ 1 & -1 & 1 & -1 \\ -1 & -1 & -1 & -1 \end{bmatrix}$

b. $D = \begin{bmatrix} -1 & -1 & -1 & -1 \\ 1 & -1 & -1 & 1 \\ -1 & 1 & 1 & 1 \\ -1 & -1 & 1 & 1 \\ 1 & 1 & 1 & -1 \end{bmatrix}$

Solution to Practice Problem

Take advantage of the zeros. Begin with a cofactor expansion down the third column to obtain a 3×3 matrix, which may be evaluated by an expansion down its first column.

$$\begin{vmatrix} 5 & -7 & 2 & 2 \\ 0 & 3 & 0 & -4 \\ -5 & -8 & 0 & 3 \\ 0 & 5 & 0 & -6 \end{vmatrix} = (-1)^{1+3}(2)\begin{vmatrix} 0 & 3 & -4 \\ -5 & -8 & 3 \\ 0 & 5 & -6 \end{vmatrix}$$

$$= 2(-1)^{2+1}(-5)\begin{vmatrix} 3 & -4 \\ 5 & -6 \end{vmatrix} = 20$$

The $(-1)^{2+1}$ in the next-to-last calculation came from the $(2,1)$-position of the -5 in the 3×3 determinant.

3.2 Properties of Determinants

The secret of determinants lies in how they change when row operations are performed. The following theorem generalizes the results of Exercises 19–24 in Section 3.1. The proof is at the end of this section.

THEOREM 3

Row Operations

Let A be a square matrix.

a. If a multiple of one row of A is added to another row to produce a matrix B, then $\det B = \det A$.

b. If two rows of A are interchanged to produce B, then $\det B = -\det A$.

c. If one row of A is multiplied by k to produce B, then $\det B = k \det A$.

The following examples show how to use Theorem 3 to find determinants efficiently.

EXAMPLE 1   Compute $\det A$, where $A = \begin{bmatrix} 1 & -4 & 2 \\ -2 & 8 & -9 \\ -1 & 7 & 0 \end{bmatrix}$.

SOLUTION   The strategy is to reduce A to echelon form and then to use the fact that the determinant of a triangular matrix is the product of the diagonal entries. The first two row replacements in column 1 do not change the determinant:

$$\det A = \begin{vmatrix} 1 & -4 & 2 \\ -2 & 8 & -9 \\ -1 & 7 & 0 \end{vmatrix} = \begin{vmatrix} 1 & -4 & 2 \\ 0 & 0 & -5 \\ -1 & 7 & 0 \end{vmatrix} = \begin{vmatrix} 1 & -4 & 2 \\ 0 & 0 & -5 \\ 0 & 3 & 2 \end{vmatrix}$$

An interchange of rows 2 and 3 reverses the sign of the determinant, so

$$\det A = -\begin{vmatrix} 1 & -4 & 2 \\ 0 & 3 & 2 \\ 0 & 0 & -5 \end{vmatrix} = -(1)(3)(-5) = 15 \qquad \blacksquare$$

A common use of Theorem 3(c) in hand calculations is to *factor out a common multiple of one row* of a matrix. For instance,

$$\begin{vmatrix} * & * & * \\ 5k & -2k & 3k \\ * & * & * \end{vmatrix} = k \begin{vmatrix} * & * & * \\ 5 & -2 & 3 \\ * & * & * \end{vmatrix}$$

where the starred entries are unchanged. We use this step in the next example.

EXAMPLE 2   Compute $\det A$, where $A = \begin{bmatrix} 2 & -8 & 6 & 8 \\ 3 & -9 & 5 & 10 \\ -3 & 0 & 1 & -2 \\ 1 & -4 & 0 & 6 \end{bmatrix}$.

SOLUTION   To simplify the arithmetic, we want a 1 in the upper-left corner. We could interchange rows 1 and 4. Instead, we factor out 2 from the top row, and then proceed with row replacements in the first column:

$$\det A = 2 \begin{vmatrix} 1 & -4 & 3 & 4 \\ 3 & -9 & 5 & 10 \\ -3 & 0 & 1 & -2 \\ 1 & -4 & 0 & 6 \end{vmatrix} = 2 \begin{vmatrix} 1 & -4 & 3 & 4 \\ 0 & 3 & -4 & -2 \\ 0 & -12 & 10 & 10 \\ 0 & 0 & -3 & 2 \end{vmatrix}$$

Next, we could factor out another 2 from row 3 or use the 3 in the second column as a pivot. We choose the latter operation, adding 4 times row 2 to row 3:

$$\det A = 2 \begin{vmatrix} 1 & -4 & 3 & 4 \\ 0 & 3 & -4 & -2 \\ 0 & 0 & -6 & 2 \\ 0 & 0 & -3 & 2 \end{vmatrix}$$

Finally, adding $-1/2$ times row 3 to row 4, and computing the "triangular" determinant, we find that

$$\det A = 2 \begin{vmatrix} 1 & -4 & 3 & 4 \\ 0 & 3 & -4 & -2 \\ 0 & 0 & -6 & 2 \\ 0 & 0 & 0 & 1 \end{vmatrix} = 2\,(1)(3)(-6)(1) = -36 \qquad \blacksquare$$

$$U = \begin{bmatrix} \blacksquare & * & * & * \\ 0 & \blacksquare & * & * \\ 0 & 0 & \blacksquare & * \\ 0 & 0 & 0 & \blacksquare \end{bmatrix}$$

$\det U \neq 0$

$$U = \begin{bmatrix} \blacksquare & * & * & * \\ 0 & \blacksquare & * & * \\ 0 & 0 & 0 & \blacksquare \\ 0 & 0 & 0 & 0 \end{bmatrix}$$

$\det U = 0$

FIGURE 1

Typical echelon forms of square matrices.

Suppose a square matrix A has been reduced to an echelon form U by row replacements and row interchanges. (This is always possible. See the row reduction algorithm in Section 1.2.) If there are r interchanges, then Theorem 3 shows that

$$\det A = (-1)^r \det U$$

Since U is in echelon form, it is triangular, and so $\det U$ is the product of the diagonal entries $u_{11}, \ldots, u_{nn}$. If A is invertible, the entries u_{ii} are all pivots (because $A \sim I_n$ and the u_{ii} have not been scaled to 1's). Otherwise, at least u_{nn} is zero, and the product $u_{11} \cdots u_{nn}$ is zero. See Figure 1. Thus

$$\det A = \begin{cases} (-1)^r \begin{pmatrix} \text{product of} \\ \text{pivots in } U \end{pmatrix} & \text{when } A \text{ is invertible} \\ 0 & \text{when } A \text{ is not invertible} \end{cases} \tag{1}$$

It is interesting to note that although the echelon form U described above is not unique (because it is not completely row reduced), and the pivots are not unique, the *product* of the pivots *is* unique, except for a possible minus sign.

Formula (1) not only gives a concrete interpretation of $\det A$ but also proves the main theorem of this section:

THEOREM 4   A square matrix A is invertible if and only if $\det A \neq 0$.

Theorem 4 adds the statement "$\det A \neq 0$" to the Invertible Matrix Theorem. A useful corollary is that $\det A = 0$ when the columns of A are linearly dependent. Also, $\det A = 0$ when the *rows* of A are linearly dependent. (Rows of A are columns of A^T, and linearly dependent columns of A^T make A^T singular. When A^T is singular, so is A, by the Invertible Matrix Theorem.) In practice, linear dependence is obvious when two columns or two rows are the same or a column or a row is zero.

EXAMPLE 3   Compute $\det A$, where $A = \begin{bmatrix} 3 & -1 & 2 & -5 \\ 0 & 5 & -3 & -6 \\ -6 & 7 & -7 & 4 \\ -5 & -8 & 0 & 9 \end{bmatrix}$.

SOLUTION   Add 2 times row 1 to row 3 to obtain

$$\det A = \det \begin{bmatrix} 3 & -1 & 2 & -5 \\ 0 & 5 & -3 & -6 \\ 0 & 5 & -3 & -6 \\ -5 & -8 & 0 & 9 \end{bmatrix} = 0$$

because the second and third rows of the second matrix are equal.   ∎

Numerical Notes

1. Most computer programs that compute $\det A$ for a general matrix A use the method of formula (1) above.

2. It can be shown that evaluation of an $n \times n$ determinant using row operations requires about $2n^3/3$ arithmetic operations. Any modern microcomputer can calculate a 25×25 determinant in a fraction of a second, since only about 10,000 operations are required.

Computers can also handle large "sparse" matrices, with special routines that take advantage of the presence of many zeros. Of course, zero entries can speed hand computations, too. The calculations in the next example combine the power of row operations with the strategy from Section 3.1 of using zero entries in cofactor expansions.

EXAMPLE 4 Compute det A, where $A = \begin{bmatrix} 0 & 1 & 2 & -1 \\ 2 & 5 & -7 & 3 \\ 0 & 3 & 6 & 2 \\ -2 & -5 & 4 & -2 \end{bmatrix}$.

SOLUTION A good way to begin is to use the 2 in column 1 as a pivot, eliminating the -2 below it. Then use a cofactor expansion to reduce the size of the determinant, followed by another row replacement operation. Thus

$$\det A = \begin{vmatrix} 0 & 1 & 2 & -1 \\ 2 & 5 & -7 & 3 \\ 0 & 3 & 6 & 2 \\ 0 & 0 & -3 & 1 \end{vmatrix} = -2\begin{vmatrix} 1 & 2 & -1 \\ 3 & 6 & 2 \\ 0 & -3 & 1 \end{vmatrix} = -2\begin{vmatrix} 1 & 2 & -1 \\ 0 & 0 & 5 \\ 0 & -3 & 1 \end{vmatrix}$$

An interchange of rows 2 and 3 would produce a "triangular determinant." Another approach is to make a cofactor expansion down the first column:

$$\det A = (-2)(1)\begin{vmatrix} 0 & 5 \\ -3 & 1 \end{vmatrix} = -2\,(15) = -30 \qquad \blacksquare$$

Column Operations

We can perform operations on the columns of a matrix in a way that is analogous to the row operations we have considered. The next theorem shows that column operations have the same effects on determinants as row operations.

Remark: The Principle of Mathematical Induction says the following: Let $P(n)$ be a statement that is either true or false for each natural number n. Then $P(n)$ is true for all $n \geq 1$ provided that $P(1)$ is true, and for each natural number k, if $P(k)$ is true, then $P(k+1)$ is true. The Principle of Mathematical Induction is used to prove the next theorem.

THEOREM 5   If A is an $n \times n$ matrix, then det $A^T = $ det A.

PROOF The theorem is obvious for $n = 1$. Suppose the theorem is true for $k \times k$ determinants and let $n = k + 1$. Then the cofactor of a_{1j} in A equals the cofactor of a_{j1} in A^T, because the cofactors involve $k \times k$ determinants. Hence the cofactor expansion of det A along the first *row* equals the cofactor expansion of det A^T down the first *column*. That is, A and A^T have equal determinants. The theorem is true for $n = 1$, and the truth of the theorem for one value of n implies its truth for the next value of n. By the Principle of Mathematical Induction, the theorem is true for all $n \geq 1$. $\blacksquare$

Because of Theorem 5, each statement in Theorem 3 is true when the word *row* is replaced everywhere by *column*. To verify this property, one merely applies the original Theorem 3 to A^T. A row operation on A^T amounts to a column operation on A.

Column operations are useful for both theoretical purposes and hand computations. However, for simplicity we'll perform only row operations in numerical calculations.

Determinants and Matrix Products

The proof of the following useful theorem is at the end of the section. Applications are in the exercises.

THEOREM 6

Multiplicative Property

If A and B are $n \times n$ matrices, then $\det AB = (\det A)(\det B)$.

EXAMPLE 5   Verify Theorem 6 for $A = \begin{bmatrix} 6 & 1 \\ 3 & 2 \end{bmatrix}$ and $B = \begin{bmatrix} 4 & 3 \\ 1 & 2 \end{bmatrix}$.

SOLUTION

$$AB = \begin{bmatrix} 6 & 1 \\ 3 & 2 \end{bmatrix}\begin{bmatrix} 4 & 3 \\ 1 & 2 \end{bmatrix} = \begin{bmatrix} 25 & 20 \\ 14 & 13 \end{bmatrix}$$

and

$$\det AB = 25(13) - 20(14) = 325 - 280 = 45$$

Since $\det A = 9$ and $\det B = 5$,

$$(\det A)(\det B) = 9(5) = 45 = \det AB \qquad \blacksquare$$

Warning: A common misconception is that Theorem 6 has an analogue for *sums* of matrices. However, $\det(A + B)$ is *not* equal to $\det A + \det B$, in general.

A Linearity Property of the Determinant Function

For an $n \times n$ matrix A, we can consider $\det A$ as a function of the n column vectors in A. We will show that if all columns except one are held fixed, then $\det A$ is a *linear function* of that one (vector) variable.

Suppose that the jth column of A is allowed to vary, and write

$$A = \begin{bmatrix} \mathbf{a}_1 & \cdots & \mathbf{a}_{j-1} & \mathbf{x} & \mathbf{a}_{j+1} & \cdots & \mathbf{a}_n \end{bmatrix}$$

Define a transformation T from $\mathbb{R}^n$ to $\mathbb{R}$ by

$$T(\mathbf{x}) = \det \begin{bmatrix} \mathbf{a}_1 & \cdots & \mathbf{a}_{j-1} & \mathbf{x} & \mathbf{a}_{j+1} & \cdots & \mathbf{a}_n \end{bmatrix}$$

Then,

$$T(c\mathbf{x}) = cT(\mathbf{x}) \quad \text{for all scalars } c \text{ and all } \mathbf{x} \text{ in } \mathbb{R}^n \tag{2}$$

$$T(\mathbf{u} + \mathbf{v}) = T(\mathbf{u}) + T(\mathbf{v}) \quad \text{for all } \mathbf{u}, \mathbf{v} \text{ in } \mathbb{R}^n \tag{3}$$

Property (2) is Theorem 3(c) applied to the columns of A. A proof of property (3) follows from a cofactor expansion of det A down the jth column. (See Exercise 49.) This (multi-) linearity property of the determinant turns out to have many useful consequences that are studied in more advanced courses.

Proofs of Theorems 3 and 6

It is convenient to prove Theorem 3 when it is stated in terms of the elementary matrices discussed in Section 2.2. We call an elementary matrix E a *row replacement* (*matrix*) if E is obtained from the identity I by adding a multiple of one row to another row; E is an *interchange* if E is obtained by interchanging two rows of I; and E is *a scale by r* if E is obtained by multiplying a row of I by a nonzero scalar r. With this terminology, Theorem 3 can be reformulated as follows:

If A is an $n \times n$ matrix and E is an $n \times n$ elementary matrix, then

$$\det EA = (\det E)(\det A)$$

where

$$\det E = \begin{cases} 1 & \text{if } E \text{ is a row replacement} \\ -1 & \text{if } E \text{ is an interchange} \\ r & \text{if } E \text{ is a scale by } r \end{cases}$$

PROOF OF THEOREM 3 The proof is by induction on the size of A. The case of a 2×2 matrix was verified in Exercises 33–36 of Section 3.1. Suppose the theorem has been verified for determinants of $k \times k$ matrices with $k \geq 2$, let $n = k + 1$, and let A be $n \times n$. The action of E on A involves either two rows or only one row. So we can expand det EA across a row that is unchanged by the action of E, say, row i. Let A_{ij} (respectively, B_{ij}) be the matrix obtained by deleting row i and column j from A (respectively, EA). Then the rows of B_{ij} are obtained from the rows of A_{ij} by the same type of elementary row operation that E performs on A. Since these submatrices are only $k \times k$, the induction assumption implies that

$$\det B_{ij} = \alpha \det A_{ij}$$

where $\alpha = 1, -1$, or r, depending on the nature of E. The cofactor expansion across row i is

$$\begin{aligned} \det EA &= a_{i1}(-1)^{i+1} \det B_{i1} + \cdots + a_{in}(-1)^{i+n} \det B_{in} \\ &= \alpha a_{i1}(-1)^{i+1} \det A_{i1} + \cdots + \alpha a_{in}(-1)^{i+n} \det A_{in} \\ &= \alpha \det A \end{aligned}$$

In particular, taking $A = I_n$, we see that det $E = 1, -1$, or r, depending on the nature of E. Thus the theorem is true for $n = 2$, and the truth of the theorem for one value of n implies its truth for the next value of n. By the principle of induction, the theorem must be true for $n \geq 2$. The theorem is trivially true for $n = 1$. ∎

PROOF OF THEOREM 6 If A is not invertible, then neither is AB, by Exercise 35 in Section 2.3. In this case, $\det AB = (\det A)(\det B)$, because both sides are zero, by Theorem 4. If A is invertible, then A and the identity matrix I_n are row equivalent by the Invertible Matrix Theorem. So there exist elementary matrices $E_1, \ldots, E_p$ such that

$$A = E_p E_{p-1} \cdots E_1 I_n = E_p E_{p-1} \cdots E_1$$

For brevity, write $|A|$ for det A. Then repeated application of Theorem 3, as rephrased above, shows that

$$|AB| = |E_p \cdots E_1 B| = |E_p||E_{p-1} \cdots E_1 B| = \cdots$$
$$= |E_p| \cdots |E_1||B| = \cdots = |E_p \cdots E_1||B|$$
$$= |A||B|$$ ∎

Practice Problems

1. Compute $\begin{vmatrix} 1 & -3 & 1 & -2 \\ 2 & -5 & -1 & -2 \\ 0 & -4 & 5 & 1 \\ -3 & 10 & -6 & 8 \end{vmatrix}$ in as few steps as possible.

2. Use a determinant to decide if $\mathbf{v}_1$, $\mathbf{v}_2$, and $\mathbf{v}_3$ are linearly independent, when

$$\mathbf{v}_1 = \begin{bmatrix} 5 \\ -7 \\ 9 \end{bmatrix}, \qquad \mathbf{v}_2 = \begin{bmatrix} -3 \\ 3 \\ -5 \end{bmatrix}, \qquad \mathbf{v}_3 = \begin{bmatrix} 2 \\ -7 \\ 5 \end{bmatrix}$$

3. Let A be an $n \times n$ matrix such that $A^2 = I$. Show that det $A = \pm 1$.

3.2 Exercises

Each equation in Exercises 1–4 illustrates a property of determinants. State the property.

1. $\begin{vmatrix} 0 & 5 & -2 \\ 1 & -3 & 6 \\ 4 & -1 & 8 \end{vmatrix} = - \begin{vmatrix} 1 & -3 & 6 \\ 0 & 5 & -2 \\ 4 & -1 & 8 \end{vmatrix}$

2. $\begin{vmatrix} 3 & -6 & 9 \\ 3 & 5 & -5 \\ 1 & 3 & 3 \end{vmatrix} = 3 \begin{vmatrix} 1 & -2 & 3 \\ 3 & 5 & -5 \\ 1 & 3 & 3 \end{vmatrix}$

3. $\begin{vmatrix} 1 & 2 & 2 \\ 0 & 3 & -4 \\ 2 & 7 & 4 \end{vmatrix} = \begin{vmatrix} 1 & 2 & 2 \\ 0 & 3 & -4 \\ 0 & 3 & 0 \end{vmatrix}$

4. $\begin{vmatrix} 1 & 3 & -4 \\ 2 & 0 & -3 \\ 3 & -5 & 2 \end{vmatrix} = \begin{vmatrix} 1 & 3 & -4 \\ 0 & -6 & 5 \\ 3 & -5 & 2 \end{vmatrix}$

Find the determinants in Exercises 5–10 by row reduction to echelon form.

5. $\begin{vmatrix} 1 & 5 & -4 \\ -1 & -4 & 5 \\ -2 & -8 & 7 \end{vmatrix}$

6. $\begin{vmatrix} 2 & 2 & -2 \\ 3 & 4 & -4 \\ 2 & -3 & -5 \end{vmatrix}$

7. $\begin{vmatrix} 1 & 3 & 0 & 2 \\ -2 & -5 & 7 & 4 \\ 3 & 5 & 2 & 1 \\ 1 & -1 & 2 & -3 \end{vmatrix}$

8. $\begin{vmatrix} 1 & 3 & 2 & -4 \\ 0 & 1 & 2 & -5 \\ 2 & 7 & 6 & -3 \\ -3 & -10 & -7 & 2 \end{vmatrix}$

9. $\begin{vmatrix} 1 & -1 & -3 & 0 \\ 0 & 1 & 5 & 4 \\ -1 & 0 & 5 & 3 \\ 3 & -3 & -2 & 3 \end{vmatrix}$

10. $\begin{vmatrix} 1 & 3 & -1 & 0 & -2 \\ 0 & 1 & -2 & -1 & -3 \\ -2 & -6 & 2 & 3 & 10 \\ 1 & 5 & -6 & 2 & -3 \\ 0 & 2 & -4 & 5 & 9 \end{vmatrix}$

Combine the methods of row reduction and cofactor expansion to compute the determinants in Exercises 11–14.

11. $\begin{vmatrix} 3 & 4 & -3 & -1 \\ 3 & 0 & 1 & -3 \\ -6 & 0 & -4 & 3 \\ 6 & 8 & -4 & -1 \end{vmatrix}$

12. $\begin{vmatrix} -1 & 2 & 3 & 0 \\ 3 & 4 & 3 & 0 \\ 11 & 4 & 6 & 6 \\ 4 & 2 & 4 & 3 \end{vmatrix}$

13. $\begin{vmatrix} 2 & 5 & 4 & 1 \\ 4 & 7 & 6 & 2 \\ 6 & -2 & -4 & 0 \\ -6 & 7 & 7 & 0 \end{vmatrix}$

14. $\begin{vmatrix} 1 & 5 & 4 & 1 \\ 0 & -3 & -6 & 0 \\ 3 & 5 & 4 & 1 \\ -6 & 5 & 5 & 0 \end{vmatrix}$

Find the determinants in Exercises 15–20, where

$$\begin{vmatrix} a & b & c \\ d & e & f \\ g & h & i \end{vmatrix} = 7.$$

15. $\begin{vmatrix} a & b & c \\ d & e & f \\ 3g & 3h & 3i \end{vmatrix}$ **16.** $\begin{vmatrix} a & b & c \\ d + 3g & e + 3h & f + 3i \\ g & h & i \end{vmatrix}$

17. $\begin{vmatrix} a + d & b + e & c + f \\ d & e & f \\ g & h & i \end{vmatrix}$

18. $\begin{vmatrix} a & b & c \\ 5d & 5e & 5f \\ g & h & i \end{vmatrix}$

19. $\begin{vmatrix} a & b & c \\ 2d + a & 2e + b & 2f + c \\ g & h & i \end{vmatrix}$

20. $\begin{vmatrix} d & e & f \\ a & b & c \\ g & h & i \end{vmatrix}$

In Exercises 21–23, use determinants to find out if the matrix is invertible.

21. $\begin{bmatrix} 2 & 6 & 0 \\ 1 & 3 & 2 \\ 3 & 9 & 2 \end{bmatrix}$ **22.** $\begin{bmatrix} 5 & 1 & -1 \\ 1 & -3 & -2 \\ 0 & 5 & 3 \end{bmatrix}$

23. $\begin{bmatrix} 2 & 0 & 0 & 6 \\ 1 & -7 & -5 & 0 \\ 3 & 8 & 6 & 0 \\ 0 & 7 & 5 & 4 \end{bmatrix}$

In Exercises 24–26, use determinants to decide if the set of vectors is linearly independent.

24. $\begin{bmatrix} 4 \\ 6 \\ 2 \end{bmatrix}, \begin{bmatrix} -6 \\ 0 \\ 6 \end{bmatrix}, \begin{bmatrix} -3 \\ -5 \\ -2 \end{bmatrix}$

25. $\begin{bmatrix} 7 \\ -4 \\ -6 \end{bmatrix}, \begin{bmatrix} -8 \\ 5 \\ 7 \end{bmatrix}, \begin{bmatrix} 7 \\ 0 \\ -5 \end{bmatrix}$

26. $\begin{bmatrix} 3 \\ 5 \\ -6 \\ 4 \end{bmatrix}, \begin{bmatrix} 2 \\ -6 \\ 0 \\ 7 \end{bmatrix}, \begin{bmatrix} -2 \\ -1 \\ 3 \\ 0 \end{bmatrix}, \begin{bmatrix} 0 \\ 0 \\ 0 \\ -2 \end{bmatrix}$

In Exercises 27–34, A and B are $n \times n$ matrices. Mark each statement True or False **(T/F)**. Justify each answer.

27. **(T/F)** A row replacement operation does not affect the determinant of a matrix.

28. **(T/F)** If det A is zero, then two rows or two columns are the same, or a row or a column is zero.

29. **(T/F)** If the columns of A are linearly dependent, then det $A = 0$.

30. **(T/F)** The determinant of A is the product of the diagonal entries in A.

31. **(T/F)** If three row interchanges are made in succession, then the new determinant equals the old determinant.

32. **(T/F)** The determinant of A is the product of the pivots in any echelon form U of A, multiplied by $(-1)^r$, where r is the number of row interchanges made during row reduction from A to U.

33. **(T/F)** $\det(A + B) = \det A + \det B$.

34. **(T/F)** $\det A^{-1} = (-1) \det A$.

35. Compute det B^4, where $B = \begin{bmatrix} 1 & 0 & 1 \\ 1 & 1 & 2 \\ 1 & 2 & 1 \end{bmatrix}$.

36. Use Theorem 3 (but not Theorem 4) to show that if two rows of a square matrix A are equal, then det $A = 0$. The same is true for two columns. Why?

In Exercises 37–42, mention an appropriate theorem in your explanation.

37. Show that if A is invertible, then $\det A^{-1} = \dfrac{1}{\det A}$.

38. Suppose that A is a square matrix such that $\det A^3 = 0$. Explain why A cannot be invertible.

39. Let A and B be square matrices. Show that even though AB and BA may not be equal, it is always true that $\det AB = \det BA$.

40. Let A and P be square matrices, with P invertible. Show that $\det(PAP^{-1}) = \det A$.

41. Let U be a square matrix such that $U^T U = I$. Show that $\det U = \pm 1$.

42. Find a formula for $\det(rA)$ when A is an $n \times n$ matrix.

Verify that $\det AB = (\det A)(\det B)$ for the matrices in Exercises 43 and 44. (Do not use Theorem 6.)

43. $A = \begin{bmatrix} 3 & 0 \\ 6 & 1 \end{bmatrix}, B = \begin{bmatrix} 2 & 0 \\ 5 & 4 \end{bmatrix}$

44. $A = \begin{bmatrix} 3 & 6 \\ -1 & -2 \end{bmatrix}, B = \begin{bmatrix} 4 & 3 \\ -1 & -3 \end{bmatrix}$

45. Let A and B be 3×3 matrices, with det $A = -2$ and det $B = 3$. Use properties of determinants (in the text and in the preceding exercises) to compute:
  a. det AB b. det $5A$ c. det B^T
  d. det A^{-1} e. det A^3

46. Let A and B be 4×4 matrices, with det $A = 4$ and det $B = -3$. Compute:
  a. det AB b. det B^5 c. det $2A$
  d. det $A^T BA$ e. det $B^{-1}AB$

47. Verify that $\det A = \det B + \det C$, where
$$A = \begin{bmatrix} a+e & b+f \\ c & d \end{bmatrix}, \ B = \begin{bmatrix} a & b \\ c & d \end{bmatrix}, \ C = \begin{bmatrix} e & f \\ c & d \end{bmatrix}$$

48. Let $A = \begin{bmatrix} 1 & 0 \\ 0 & 1 \end{bmatrix}$ and $B = \begin{bmatrix} a & b \\ c & d \end{bmatrix}$. Show that $\det(A + B) = \det A + \det B$ if and only if $a + d = 0$.

49. Verify that $\det A = \det B + \det C$, where
$$A = \begin{bmatrix} a_{11} & a_{12} & u_1 + v_1 \\ a_{21} & a_{22} & u_2 + v_2 \\ a_{31} & a_{32} & u_3 + v_3 \end{bmatrix},$$

$$B = \begin{bmatrix} a_{11} & a_{12} & u_1 \\ a_{21} & a_{22} & u_2 \\ a_{31} & a_{32} & u_3 \end{bmatrix}, \ C = \begin{bmatrix} a_{11} & a_{12} & v_1 \\ a_{21} & a_{22} & v_2 \\ a_{31} & a_{32} & v_3 \end{bmatrix}$$

Note, however, that A is *not* the same as $B + C$.

50. Right-multiplication by an elementary matrix E affects the *columns* of A in the same way that left-multiplication affects the *rows*. Use Theorems 5 and 3 and the obvious fact that E^T is another elementary matrix to show that
$$\det AE = (\det E)(\det A)$$
Do not use Theorem 6.

51. Suppose A is an $n \times n$ matrix and a computer suggests that $\det A = 5$ and $\det(A^{-1}) = 1$. Should you trust these answers? Why or why not?

52. Suppose A and B are $n \times n$ matrices and a computer suggests that $\det A = 5$, $\det B = 2$ and $\det AB = 7$. Should you trust these answers? Why or why not?

⊤ 53. Compute $\det A^T A$ and $\det A A^T$ for several random 4×5 matrices and several random 5×6 matrices. What can you say about $A^T A$ and $A A^T$ when A has more columns than rows?

⊤ 54. If $\det A$ is close to zero, is the matrix A nearly singular? Experiment with the nearly singular 4×4 matrix
$$A = \begin{bmatrix} 4 & 0 & -7 & -7 \\ -6 & 1 & 11 & 9 \\ 7 & -5 & 10 & 19 \\ -1 & 2 & 3 & -1 \end{bmatrix}$$

Compute the determinants of A, $10A$, and $0.1A$. In contrast, compute the condition numbers of these matrices. Repeat these calculations when A is the 4×4 identity matrix. Discuss your results.

Solutions to Practice Problems

1. Perform row replacements to create zeros in the first column, and then create a row of zeros.

$$\begin{vmatrix} 1 & -3 & 1 & -2 \\ 2 & -5 & -1 & -2 \\ 0 & -4 & 5 & 1 \\ -3 & 10 & -6 & 8 \end{vmatrix} = \begin{vmatrix} 1 & -3 & 1 & -2 \\ 0 & 1 & -3 & 2 \\ 0 & -4 & 5 & 1 \\ 0 & 1 & -3 & 2 \end{vmatrix} = \begin{vmatrix} 1 & -3 & 1 & -2 \\ 0 & 1 & -3 & 2 \\ 0 & -4 & 5 & 1 \\ 0 & 0 & 0 & 0 \end{vmatrix} = 0$$

2. $\det [\, \mathbf{v}_1 \ \ \mathbf{v}_2 \ \ \mathbf{v}_3 \,] = \begin{vmatrix} 5 & -3 & 2 \\ -7 & 3 & -7 \\ 9 & -5 & 5 \end{vmatrix} = \begin{vmatrix} 5 & -3 & 2 \\ -2 & 0 & -5 \\ 9 & -5 & 5 \end{vmatrix}$ Row 1 added to row 2

$$= -(-3) \begin{vmatrix} -2 & -5 \\ 9 & 5 \end{vmatrix} - (-5) \begin{vmatrix} 5 & 2 \\ -2 & -5 \end{vmatrix}$$ Cofactors of column 2

$$= 3(35) + 5(-21) = 0$$

By Theorem 4, the matrix $[\, \mathbf{v}_1 \ \ \mathbf{v}_2 \ \ \mathbf{v}_3 \,]$ is not invertible. The columns are linearly dependent, by the Invertible Matrix Theorem.

3. Recall that $\det I = 1$. By Theorem 6, $\det(AA) = (\det A)(\det A)$. Putting these two observations together results in
$$1 = \det I = \det A^2 = \det(AA) = (\det A)(\det A) = (\det A)^2$$

Taking the square root of both sides establishes that $\det A = \pm 1$.

3.3 Cramer's Rule, Volume, and Linear Transformations

This section applies the theory of the preceding sections to obtain important theoretical formulas and a geometric interpretation of the determinant.

Cramer's Rule

Cramer's rule is needed in a variety of theoretical calculations. For instance, it can be used to study how the solution of $A\mathbf{x} = \mathbf{b}$ is affected by changes in the entries of $\mathbf{b}$. However, the formula is inefficient for hand calculations, except for 2×2 or perhaps 3×3 matrices.

For any $n \times n$ matrix A and any $\mathbf{b}$ in $\mathbb{R}^n$, let $A_i(\mathbf{b})$ be the matrix obtained from A by replacing column i by the vector $\mathbf{b}$.

$$A_i(\mathbf{b}) = [\mathbf{a}_1 \quad \cdots \quad \underset{\underset{\text{col } i}{\uparrow}}{\mathbf{b}} \quad \cdots \quad \mathbf{a}_n]$$

THEOREM 7

Cramer's Rule

Let A be an invertible $n \times n$ matrix. For any $\mathbf{b}$ in $\mathbb{R}^n$, the unique solution $\mathbf{x}$ of $A\mathbf{x} = \mathbf{b}$ has entries given by

$$x_i = \frac{\det A_i(\mathbf{b})}{\det A}, \qquad i = 1, 2, \ldots, n \tag{1}$$

PROOF Denote the columns of A by $\mathbf{a}_1, \ldots, \mathbf{a}_n$ and the columns of the $n \times n$ identity matrix I by $\mathbf{e}_1, \ldots, \mathbf{e}_n$. If $A\mathbf{x} = \mathbf{b}$, the definition of matrix multiplication shows that

$$\begin{aligned} A(I_i(\mathbf{x})) &= A[\mathbf{e}_1 \quad \cdots \quad \mathbf{x} \quad \cdots \quad \mathbf{e}_n] = [A\mathbf{e}_1 \quad \cdots \quad A\mathbf{x} \quad \cdots \quad A\mathbf{e}_n] \\ &= [\mathbf{a}_1 \quad \cdots \quad \mathbf{b} \quad \cdots \quad \mathbf{a}_n] = A_i(\mathbf{b}) \end{aligned}$$

By the multiplicative property of determinants,

$$(\det A)(\det I_i(\mathbf{x})) = \det A_i(\mathbf{b})$$

The second determinant on the left is simply x_i. (Make a cofactor expansion along the ith row.) Hence $(\det A)\, x_i = \det A_i(\mathbf{b})$. This proves (1) because A is invertible and $\det A \neq 0$. ∎

EXAMPLE 1 Use Cramer's rule to solve the system

$$\begin{aligned} 3x_1 - 2x_2 &= 6 \\ -5x_1 + 4x_2 &= 8 \end{aligned}$$

SOLUTION View the system as $A\mathbf{x} = \mathbf{b}$. Using the notation introduced above,

$$A = \begin{bmatrix} 3 & -2 \\ -5 & 4 \end{bmatrix}, \qquad A_1(\mathbf{b}) = \begin{bmatrix} 6 & -2 \\ 8 & 4 \end{bmatrix}, \qquad A_2(\mathbf{b}) = \begin{bmatrix} 3 & 6 \\ -5 & 8 \end{bmatrix}$$

Since $\det A = 2$, the system has a unique solution. By Cramer's rule,

$$x_1 = \frac{\det A_1(\mathbf{b})}{\det A} = \frac{24 + 16}{2} = 20$$

$$x_2 = \frac{\det A_2(\mathbf{b})}{\det A} = \frac{24 + 30}{2} = 27 \qquad \blacksquare$$

Application to Engineering

A number of important engineering problems, particularly in electrical engineering and control theory, can be analyzed by *Laplace transforms*. This approach converts an appropriate system of linear differential equations into a system of linear algebraic equations whose coefficients involve a parameter. The next example illustrates the type of algebraic system that may arise.

EXAMPLE 2 Consider the following system in which s is an unspecified parameter. Determine the values of s for which the system has a unique solution, and use Cramer's rule to describe the solution.

$$3sx_1 - 2x_2 = 4$$
$$-6x_1 + sx_2 = 1$$

SOLUTION View the system as $A\mathbf{x} = \mathbf{b}$. Then

$$A = \begin{bmatrix} 3s & -2 \\ -6 & s \end{bmatrix}, \quad A_1(\mathbf{b}) = \begin{bmatrix} 4 & -2 \\ 1 & s \end{bmatrix}, \quad A_2(\mathbf{b}) = \begin{bmatrix} 3s & 4 \\ -6 & 1 \end{bmatrix}$$

Since

$$\det A = 3s^2 - 12 = 3(s + 2)(s - 2)$$

the system has a unique solution precisely when $s \neq \pm 2$. For such an s, the solution is (x_1, x_2), where

$$x_1 = \frac{\det A_1(\mathbf{b})}{\det A} = \frac{4s + 2}{3(s + 2)(s - 2)}$$

$$x_2 = \frac{\det A_2(\mathbf{b})}{\det A} = \frac{3s + 24}{3(s + 2)(s - 2)} = \frac{s + 8}{(s + 2)(s - 2)} \qquad \blacksquare$$

A Formula for A^{-1}

Cramer's rule leads easily to a general formula for the inverse of an $n \times n$ matrix A. The jth column of A^{-1} is a vector $\mathbf{x}$ that satisfies

$$A\mathbf{x} = \mathbf{e}_j$$

where $\mathbf{e}_j$ is the jth column of the identity matrix, and the ith entry of $\mathbf{x}$ is the (i, j)-entry of A^{-1}. By Cramer's rule,

$$\{(i, j)\text{-entry of } A^{-1}\} = x_i = \frac{\det A_i(\mathbf{e}_j)}{\det A} \qquad (2)$$

Recall that A_{ji} denotes the submatrix of A formed by deleting row j and column i. A cofactor expansion down column i of $A_i(\mathbf{e}_j)$ shows that

$$\det A_i(\mathbf{e}_j) = (-1)^{i+j} \det A_{ji} = C_{ji} \qquad (3)$$

where C_{ji} is a cofactor of A. By (2), the (i, j)-entry of A^{-1} is the cofactor C_{ji} divided by det A. [Note that the subscripts on C_{ji} are the reverse of (i, j).] Thus

$$A^{-1} = \frac{1}{\det A} \begin{bmatrix} C_{11} & C_{21} & \cdots & C_{n1} \\ C_{12} & C_{22} & \cdots & C_{n2} \\ \vdots & \vdots & & \vdots \\ C_{1n} & C_{2n} & \cdots & C_{nn} \end{bmatrix} \tag{4}$$

The matrix of cofactors on the right side of (4) is called the **adjugate** (or **classical adjoint**) of A, denoted by adj A. (The term *adjoint* also has another meaning in advanced texts on linear transformations.) The next theorem simply restates (4).

THEOREM 8

An Inverse Formula

Let A be an invertible $n \times n$ matrix. Then

$$A^{-1} = \frac{1}{\det A} \operatorname{adj} A$$

EXAMPLE 3 Find the inverse of the matrix $A = \begin{bmatrix} 2 & 1 & 3 \\ 1 & -1 & 1 \\ 1 & 4 & -2 \end{bmatrix}$.

SOLUTION The nine cofactors are

$$C_{11} = + \begin{vmatrix} -1 & 1 \\ 4 & -2 \end{vmatrix} = -2, \quad C_{12} = - \begin{vmatrix} 1 & 1 \\ 1 & -2 \end{vmatrix} = 3, \quad C_{13} = + \begin{vmatrix} 1 & -1 \\ 1 & 4 \end{vmatrix} = 5$$

$$C_{21} = - \begin{vmatrix} 1 & 3 \\ 4 & -2 \end{vmatrix} = 14, \quad C_{22} = + \begin{vmatrix} 2 & 3 \\ 1 & -2 \end{vmatrix} = -7, \quad C_{23} = - \begin{vmatrix} 2 & 1 \\ 1 & 4 \end{vmatrix} = -7$$

$$C_{31} = + \begin{vmatrix} 1 & 3 \\ -1 & 1 \end{vmatrix} = 4, \quad C_{32} = - \begin{vmatrix} 2 & 3 \\ 1 & 1 \end{vmatrix} = 1, \quad C_{33} = + \begin{vmatrix} 2 & 1 \\ 1 & -1 \end{vmatrix} = -3$$

The adjugate matrix is the *transpose* of the matrix of cofactors. [For instance, C_{12} goes in the $(2, 1)$ position.] Thus

$$\operatorname{adj} A = \begin{bmatrix} -2 & 14 & 4 \\ 3 & -7 & 1 \\ 5 & -7 & -3 \end{bmatrix}$$

We could compute det A directly, but the following computation provides a check on the calculations for adj A *and* produces det A:

$$(\operatorname{adj} A) A = \begin{bmatrix} -2 & 14 & 4 \\ 3 & -7 & 1 \\ 5 & -7 & -3 \end{bmatrix} \begin{bmatrix} 2 & 1 & 3 \\ 1 & -1 & 1 \\ 1 & 4 & -2 \end{bmatrix} = \begin{bmatrix} 14 & 0 & 0 \\ 0 & 14 & 0 \\ 0 & 0 & 14 \end{bmatrix} = 14I$$

Since $(\operatorname{adj} A)A = 14I$, Theorem 8 shows that det $A = 14$ and

$$A^{-1} = \frac{1}{14} \begin{bmatrix} -2 & 14 & 4 \\ 3 & -7 & 1 \\ 5 & -7 & -3 \end{bmatrix} = \begin{bmatrix} -1/7 & 1 & 2/7 \\ 3/14 & -1/2 & 1/14 \\ 5/14 & -1/2 & -3/14 \end{bmatrix}$$

■

Numerical Notes

Theorem 8 is useful mainly for theoretical calculations. The formula for A^{-1} permits one to deduce properties of the inverse without actually calculating it. Except for special cases, the algorithm in Section 2.2 gives a much better way to compute A^{-1}, if the inverse is really needed.

Cramer's rule is also a theoretical tool. It can be used to study how sensitive the solution of $A\mathbf{x} = \mathbf{b}$ is to changes in an entry in $\mathbf{b}$ or in A (perhaps due to experimental error when acquiring the entries for $\mathbf{b}$ or A). When A is a 3×3 matrix with *complex* entries, Cramer's rule is sometimes selected for hand computation because row reduction of $[\,A \quad \mathbf{b}\,]$ with complex arithmetic can be messy, and the determinants are fairly easy to compute. For a larger $n \times n$ matrix (real or complex), Cramer's rule is hopelessly inefficient. Computing just *one* determinant takes about as much work as solving $A\mathbf{x} = \mathbf{b}$ by row reduction.

Determinants as Area or Volume

In the next application, we verify the geometric interpretation of determinants described in the chapter introduction. Although a general discussion of length and distance in $\mathbb{R}^n$ will not be given until Chapter 6, we assume here that the usual Euclidean concepts of length, area, and volume are already understood for $\mathbb{R}^2$ and $\mathbb{R}^3$.

THEOREM 9

If A is a 2×2 matrix, the area of the parallelogram determined by the columns of A is $|\det A|$. If A is a 3×3 matrix, the volume of the parallelepiped determined by the columns of A is $|\det A|$.

STUDY GUIDE provides a geometric proof of the determinant as area.

FIGURE 1
Area $= |ad|$.

PROOF   The theorem is obviously true for any 2×2 diagonal matrix:

$$\left| \det \begin{bmatrix} a & 0 \\ 0 & d \end{bmatrix} \right| = |ad| = \left\{ \begin{array}{c} \text{area of} \\ \text{rectangle} \end{array} \right\}$$

See Figure 1. It will suffice to show that any 2×2 matrix $A = [\,\mathbf{a}_1 \quad \mathbf{a}_2\,]$ can be transformed into a diagonal matrix in a way that changes neither the area of the associated parallelogram nor $|\det A|$. From Section 3.2, we know that the absolute value of the determinant is unchanged when two columns are interchanged or a multiple of one column is added to another. And it is easy to see that such operations suffice to transform A into a diagonal matrix. Column interchanges do not change the parallelogram at all. So it suffices to prove the following simple geometric observation that applies to vectors in $\mathbb{R}^2$ or $\mathbb{R}^3$:

Let $\mathbf{a}_1$ and $\mathbf{a}_2$ be nonzero vectors. Then for any scalar c, the area of the parallelogram determined by $\mathbf{a}_1$ and $\mathbf{a}_2$ equals the area of the parallelogram determined by $\mathbf{a}_1$ and $\mathbf{a}_2 + c\mathbf{a}_1$.

To prove this statement, we may assume that $\mathbf{a}_2$ is not a multiple of $\mathbf{a}_1$, for otherwise the two parallelograms would be degenerate and have zero area. If L is the line through $\mathbf{0}$ and $\mathbf{a}_1$, then $\mathbf{a}_2 + L$ is the line through $\mathbf{a}_2$ parallel to L, and $\mathbf{a}_2 + c\mathbf{a}_1$ is on this line. See Figure 2. The points $\mathbf{a}_2$ and $\mathbf{a}_2 + c\mathbf{a}_1$ have the same perpendicular distance to L. Hence the two parallelograms in Figure 2 have the same area, since they share the base from $\mathbf{0}$ to $\mathbf{a}_1$. This completes the proof for $\mathbb{R}^2$.

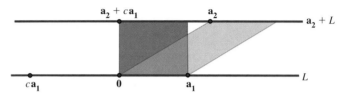

FIGURE 2 Two parallelograms of equal area.

The proof for $\mathbb{R}^3$ is similar. The theorem is obviously true for a 3×3 diagonal matrix. See Figure 3. And any 3×3 matrix A can be transformed into a diagonal matrix using column operations that do not change $|\det A|$. (Think about doing row operations on A^T.) So it suffices to show that these operations do not affect the volume of the parallelepiped determined by the columns of A.

A parallelepiped is shown in Figure 4 as a shaded box with two sloping sides. Its volume is the area of the base in the plane Span $\{\mathbf{a}_1, \mathbf{a}_3\}$ times the altitude of $\mathbf{a}_2$ above Span $\{\mathbf{a}_1, \mathbf{a}_3\}$. Any vector $\mathbf{a}_2 + c\mathbf{a}_1$ has the same altitude because $\mathbf{a}_2 + c\mathbf{a}_1$ lies in the plane $\mathbf{a}_2 + $ Span $\{\mathbf{a}_1, \mathbf{a}_3\}$, which is parallel to Span $\{\mathbf{a}_1, \mathbf{a}_3\}$. Hence the volume of the parallelepiped is unchanged when $[\, \mathbf{a}_1 \quad \mathbf{a}_2 \quad \mathbf{a}_3 \,]$ is changed to $[\, \mathbf{a}_1 \quad \mathbf{a}_2 + c\mathbf{a}_1 \quad \mathbf{a}_3 \,]$. Thus a column replacement operation does not affect the volume of the parallelepiped. Since column interchanges have no effect on the volume, the proof is complete. ∎

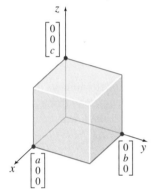

FIGURE 3
Volume $= |abc|$.

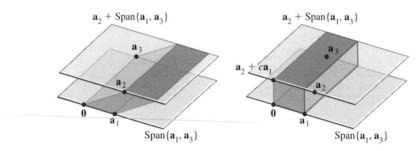

FIGURE 4 Two parallelepipeds of equal volume.

EXAMPLE 4 Calculate the area of the parallelogram determined by the points $(-2, -2)$, $(0, 3)$, $(4, -1)$, and $(6, 4)$. See Figure 5(a).

SOLUTION First translate the parallelogram to one having the origin as a vertex. For example, subtract the vertex $(-2, -2)$ from each of the four vertices. The new parallelogram has the same area, and its vertices are $(0, 0)$, $(2, 5)$, $(6, 1)$, and $(8, 6)$. See Figure 5(b). This parallelogram is determined by the columns of

$$A = \begin{bmatrix} 2 & 6 \\ 5 & 1 \end{bmatrix}$$

Since $|\det A| = |-28|$, the area of the parallelogram is 28. ∎

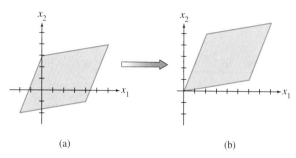

(a)   (b)

FIGURE 5 Translating a parallelogram does not change its area.

Linear Transformations

Determinants can be used to describe an important geometric property of linear transformations in the plane and in $\mathbb{R}^3$. If T is a linear transformation and S is a set in the domain of T, let $T(S)$ denote the set of images of points in S. We are interested in how the area (or volume) of $T(S)$ compares with the area (or volume) of the original set S. For convenience, when S is a region bounded by a parallelogram, we also refer to S as a parallelogram.

THEOREM 10

Let $T : \mathbb{R}^2 \to \mathbb{R}^2$ be the linear transformation determined by a 2×2 matrix A. If S is a parallelogram in $\mathbb{R}^2$, then

$$\{\text{area of } T(S)\} = |\det A| \cdot \{\text{area of } S\} \tag{5}$$

If T is determined by a 3×3 matrix A, and if S is a parallelepiped in $\mathbb{R}^3$, then

$$\{\text{volume of } T(S)\} = |\det A| \cdot \{\text{volume of } S\} \tag{6}$$

PROOF   Consider the 2×2 case, with $A = [\, \mathbf{a}_1 \quad \mathbf{a}_2 \,]$. A parallelogram at the origin in $\mathbb{R}^2$ determined by vectors $\mathbf{b}_1$ and $\mathbf{b}_2$ has the form

$$S = \{s_1\mathbf{b}_1 + s_2\mathbf{b}_2 : 0 \le s_1 \le 1, \, 0 \le s_2 \le 1\}$$

The image of S under T consists of points of the form

$$\begin{aligned} T(s_1\mathbf{b}_1 + s_2\mathbf{b}_2) &= s_1 T(\mathbf{b}_1) + s_2 T(\mathbf{b}_2) \\ &= s_1 A\mathbf{b}_1 + s_2 A\mathbf{b}_2 \end{aligned}$$

where $0 \le s_1 \le 1, \, 0 \le s_2 \le 1$. It follows that $T(S)$ is the parallelogram determined by the columns of the matrix $[\, A\mathbf{b}_1 \quad A\mathbf{b}_2 \,]$. This matrix can be written as AB, where $B = [\, \mathbf{b}_1 \quad \mathbf{b}_2 \,]$. By Theorem 9 and the product theorem for determinants,

$$\begin{aligned} \{\text{area of } T(S)\} &= |\det AB| = |\det A| \, |\det B| \\ &= |\det A| \cdot \{\text{area of } S\} \end{aligned} \tag{7}$$

An arbitrary parallelogram has the form $\mathbf{p} + S$, where $\mathbf{p}$ is a vector and S is a parallelogram at the origin, as seen previously. It is easy to see that T transforms $\mathbf{p} + S$

into $T(\mathbf{p}) + T(S)$. (See Exercise 26.) Since translation does not affect the area of a set,

$$
\begin{aligned}
\{\text{area of } T(\mathbf{p} + S)\} &= \{\text{area of } T(\mathbf{p}) + T(S)\} \\
&= \{\text{area of } T(S)\} && \text{Translation} \\
&= |\det A| \cdot \{\text{area of } S\} && \text{By equation (7)} \\
&= |\det A| \cdot \{\text{area of } (\mathbf{p} + S)\} && \text{Translation}
\end{aligned}
$$

This shows that (5) holds for all parallelograms in $\mathbb{R}^2$. The proof of (6) for the 3×3 case is analogous. ∎

When we attempt to generalize Theorem 10 to a region in $\mathbb{R}^2$ or $\mathbb{R}^3$ that is not bounded by straight lines or planes, we must face the problem of how to define and compute its area or volume. This is a question studied in calculus, and we shall only outline the basic idea for $\mathbb{R}^2$. If R is a planar region that has a finite area, then R can be approximated by a grid of small squares that lie inside R. By making the squares sufficiently small, the area of R may be approximated as closely as desired by the sum of the areas of the small squares. See Figure 6.

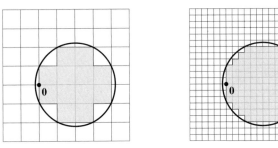

FIGURE 6 Approximating a planar region by a union of squares. The approximation improves as the grid becomes finer.

If T is a linear transformation associated with a 2×2 matrix A, then the image of a planar region R under T is approximated by the images of the small squares inside R. The proof of Theorem 10 shows that each such image is a parallelogram whose area is $|\det A|$ times the area of the square. If R' is the union of the squares inside R, then the area of $T(R')$ is $|\det A|$ times the area of R'. See Figure 7. Also, the area of $T(R')$ is close to the area of $T(R)$. An argument involving a limiting process may be given to justify the following generalization of Theorem 10.

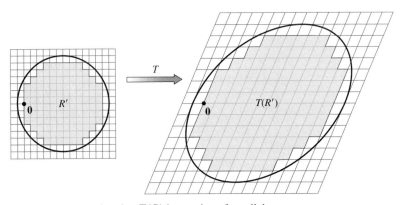

FIGURE 7 Approximating $T(R)$ by a union of parallelograms.

The conclusions of Theorem 10 hold whenever S is a region in $\mathbb{R}^2$ with finite area or a region in $\mathbb{R}^3$ with finite volume.

EXAMPLE 5 Let a and b be positive numbers. Find the area of the region E bounded by the ellipse whose equation is

$$\frac{x_1^2}{a^2} + \frac{x_2^2}{b^2} = 1$$

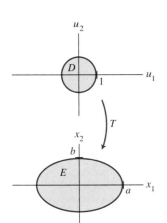

SOLUTION We claim that E is the image of the unit disk D under the linear transformation T determined by the matrix $A = \begin{bmatrix} a & 0 \\ 0 & b \end{bmatrix}$, because if $\mathbf{u} = \begin{bmatrix} u_1 \\ u_2 \end{bmatrix}$, $\mathbf{x} = \begin{bmatrix} x_1 \\ x_2 \end{bmatrix}$, and $\mathbf{x} = A\mathbf{u}$, then

$$u_1 = \frac{x_1}{a} \quad \text{and} \quad u_2 = \frac{x_2}{b}$$

It follows that $\mathbf{u}$ is in the unit disk, with $u_1^2 + u_2^2 \le 1$, if and only if $\mathbf{x}$ is in E, with $(x_1/a)^2 + (x_2/b)^2 \le 1$. By the generalization of Theorem 10,

$$\begin{aligned}
\{\text{area of ellipse}\} &= \{\text{area of } T(D)\} \\
&= |\det A| \cdot \{\text{area of } D\} \\
&= ab\pi(1)^2 = \pi ab
\end{aligned}$$
∎

Practice Problem

Let S be the parallelogram determined by the vectors $\mathbf{b}_1 = \begin{bmatrix} 1 \\ 3 \end{bmatrix}$ and $\mathbf{b}_2 = \begin{bmatrix} 5 \\ 1 \end{bmatrix}$, and let $A = \begin{bmatrix} 1 & -.1 \\ 0 & 2 \end{bmatrix}$. Compute the area of the image of S under the mapping $\mathbf{x} \mapsto A\mathbf{x}$.

3.3 Exercises

Use Cramer's rule to compute the solutions of the systems in Exercises 1–6.

1. $5x_1 + 7x_2 = 3$
$2x_1 + 4x_2 = 1$

2. $4x_1 + x_2 = 6$
$3x_1 + 2x_2 = 5$

3. $3x_1 - 2x_2 = 3$
$-4x_1 + 6x_2 = -5$

4. $-5x_1 + 2x_2 = 9$
$3x_1 - x_2 = -4$

5. $x_1 + x_2 = 3$
$-3x_1 + 2x_3 = 0$
$x_2 - 2x_3 = 2$

6. $x_1 + 3x_2 + x_3 = 8$
$-x_1 + 2x_3 = 4$
$3x_1 + x_2 = 4$

In Exercises 7–10, determine the values of the parameter s for which the system has a unique solution, and describe the solution.

7. $6sx_1 + 4x_2 = 5$
$9x_1 + 2sx_2 = -2$

8. $3sx_1 + 5x_2 = 3$
$12x_1 + 5sx_2 = 2$

9. $sx_1 + 2sx_2 = -1$
$3x_1 + 6sx_2 = 4$

10. $sx_1 - 2x_2 = 1$
$4sx_1 + 4sx_2 = 2$

In Exercises 11–16, compute the adjugate of the given matrix, and then use Theorem 8 to give the inverse of the matrix.

11. $\begin{bmatrix} 0 & -2 & -1 \\ 5 & 0 & 0 \\ -1 & 1 & 1 \end{bmatrix}$

12. $\begin{bmatrix} 1 & 1 & 3 \\ -2 & 2 & 1 \\ 0 & 1 & 1 \end{bmatrix}$

13. $\begin{bmatrix} 3 & 5 & 4 \\ 1 & 0 & 1 \\ 2 & 1 & 1 \end{bmatrix}$

14. $\begin{bmatrix} 1 & -1 & 2 \\ 0 & 2 & 1 \\ 3 & 0 & 6 \end{bmatrix}$

15. $\begin{bmatrix} 5 & 0 & 0 \\ -1 & 1 & 0 \\ -2 & 3 & -1 \end{bmatrix}$

16. $\begin{bmatrix} 1 & 2 & 4 \\ 0 & -3 & 1 \\ 0 & 0 & -2 \end{bmatrix}$

17. Show that if A is 2×2, then Theorem 8 gives the same formula for A^{-1} as that given by Theorem 4 in Section 2.2.

18. Suppose that all the entries in A are integers and $\det A = 1$. Explain why all the entries in A^{-1} are integers.

In Exercises 19–22, find the area of the parallelogram whose vertices are listed.

19. $(0, 0), (5, 2), (6, 4), (11, 6)$

20. $(0, 0), (-2, 4), (6, -5), (4, -1)$

21. $(-2, 0), (0, 3), (1, 3), (-1, 0)$

22. $(0, -2), (5, -2), (-3, 1), (2, 1)$

23. Find the volume of the parallelepiped with one vertex at the origin and adjacent vertices at $(1, 0, -3)$, $(1, 2, 4)$, and $(5, 1, 0)$.

24. Find the volume of the parallelepiped with one vertex at the origin and adjacent vertices at $(1, 3, 0)$, $(-2, 0, 2)$, and $(-1, 3, -1)$.

25. Use the concept of volume to explain why the determinant of a 3×3 matrix A is zero if and only if A is not invertible. Do not appeal to Theorem 4 in Section 3.2. [*Hint:* Think about the columns of A.]

26. Let $T : \mathbb{R}^m \to \mathbb{R}^n$ be a linear transformation, and let $\mathbf{p}$ be a vector and S a set in $\mathbb{R}^m$. Show that the image of $\mathbf{p} + S$ under T is the translated set $T(\mathbf{p}) + T(S)$ in $\mathbb{R}^n$.

27. Let S be the parallelogram determined by the vectors $\mathbf{b}_1 = \begin{bmatrix} -2 \\ 3 \end{bmatrix}$ and $\mathbf{b}_2 = \begin{bmatrix} -2 \\ 5 \end{bmatrix}$, and let $A = \begin{bmatrix} 6 & -3 \\ -3 & 2 \end{bmatrix}$. Compute the area of the image of S under the mapping $\mathbf{x} \mapsto A\mathbf{x}$.

28. Repeat Exercise 27 with $\mathbf{b}_1 = \begin{bmatrix} 4 \\ -7 \end{bmatrix}$, $\mathbf{b}_2 = \begin{bmatrix} 0 \\ 1 \end{bmatrix}$, and $A = \begin{bmatrix} 5 & 2 \\ 1 & 1 \end{bmatrix}$.

29. Find a formula for the area of the triangle whose vertices are $\mathbf{0}$, $\mathbf{v}_1$, and $\mathbf{v}_2$ in $\mathbb{R}^2$.

30. Let R be the triangle with vertices at (x_1, y_1), (x_2, y_2), and (x_3, y_3). Show that

$$\{\text{area of triangle}\} = \frac{1}{2} \det \begin{bmatrix} x_1 & y_1 & 1 \\ x_2 & y_2 & 1 \\ x_3 & y_3 & 1 \end{bmatrix}$$

[*Hint:* Translate R to the origin by subtracting one of the vertices, and use Exercise 29.]

31. Let $T : \mathbb{R}^3 \to \mathbb{R}^3$ be the linear transformation determined by the matrix $A = \begin{bmatrix} a & 0 & 0 \\ 0 & b & 0 \\ 0 & 0 & c \end{bmatrix}$, where a, b, and c are positive numbers. Let S be the unit ball, whose bounding surface has the equation $x_1^2 + x_2^2 + x_3^2 = 1$.

a. Show that $T(S)$ is bounded by the ellipsoid with the equation $\dfrac{x_1^2}{a^2} + \dfrac{x_2^2}{b^2} + \dfrac{x_3^2}{c^2} = 1$.

b. Use the fact that the volume of the unit ball is $4\pi/3$ to determine the volume of the region bounded by the ellipsoid in part (a).

32. Let S be the tetrahedron in $\mathbb{R}^3$ with vertices at the vectors $\mathbf{0}$, $\mathbf{e}_1$, $\mathbf{e}_2$, and $\mathbf{e}_3$, and let S' be the tetrahedron with vertices at vectors $\mathbf{0}$, $\mathbf{v}_1$, $\mathbf{v}_2$, and $\mathbf{v}_3$. See the figure.

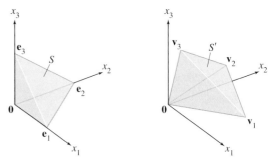

a. Describe a linear transformation that maps S onto S'.

b. Find a formula for the volume of the tetrahedron S' using the fact that

$$\{\text{volume of } S\} = (1/3) \cdot \{\text{area of base}\} \cdot \{\text{height}\}$$

33. Let A be an $n \times n$ matrix. If $A^{-1} = \dfrac{1}{\det A} \operatorname{adj} A$ is computed, what should AA^{-1} be equal to in order to confirm that A^{-1} has been found correctly?

34. If a parallelogram fits inside a circle radius 1 and $\det A = 2$, where A is the matrix whose columns correspond to the edges of the parallelogram, does it seem like A and its determinant have been calculated correctly to correspond to the area of this parallelogram? Explain why or why not.

In Exercises 35–38, mark each statement as True or False (**T/F**). Justify each answer.

35. (**T/F**) Two parallelograms with the same base and height have the same area.

36. (**T/F**) Applying a linear transformation to a region does not change its area.

37. (**T/F**) If A is an invertible $n \times n$ matrix, then $A^{-1} = \operatorname{adj} A$.

38. (**T/F**) Cramer's rule can only be used for invertible matrices.

T 39. Test the inverse formula of Theorem 8 for a random 4×4 matrix A. Use your matrix program to compute the cofactors of the 3×3 submatrices, construct the adjugate, and set $B = (\text{adj } A)/(\det A)$. Then compute $B - \text{inv}(A)$, where $\text{inv}(A)$ is the inverse of A as computed by the matrix program. Use floating point arithmetic with the maximum possible number of decimal places. Report your results.

T 40. Test Cramer's rule for a random 4×4 matrix A and a random 4×1 vector **b**. Compute each entry in the solution of $A\mathbf{x} = \mathbf{b}$, and compare these entries with the entries in $A^{-1}\mathbf{b}$. Write the command (or keystrokes) for your matrix program that uses Cramer's rule to produce the second entry of **x**.

T 41. If your version of MATLAB has the `flops` command, use it to count the number of floating point operations to compute A^{-1} for a random 30×30 matrix. Compare this number with the number of flops needed to form $(\text{adj } A)/(\det A)$.

Solution to Practice Problem

The area of S is $\left| \det \begin{bmatrix} 1 & 5 \\ 3 & 1 \end{bmatrix} \right| = 14$, and $\det A = 2$. By Theorem 10, the area of the image of S under the mapping $\mathbf{x} \mapsto A\mathbf{x}$ is

$$|\det A| \cdot \{\text{area of } S\} = 2 \cdot 14 = 28$$

CHAPTER 3 PROJECTS

Chapter 3 projects are available online at `bit.ly/30IM8gT`.

A. *Weighing Design*: This project develops the concept of weighing design and their corresponding matrices for use in weighing a few small, light objects.

B. *Jacobians*: This set of exercises examines how a particular determinant called the Jacobian may be used to allow us to change variables in double and triple integrals.

CHAPTER 3 SUPPLEMENTARY EXERCISES

In Exercises 1-15, mark each statement True or False (**T/F**). Justify each answer. Assume that all matrices here are square.

1. (T/F) If A is a 2×2 matrix with a zero determinant, then one column of A is a multiple of the other.

2. (T/F) If two rows of a 3×3 matrix A are the same, then $\det A = 0$.

3. (T/F) If A is a 3×3 matrix, then $\det 5A = 5 \det A$.

4. (T/F) If A and B are $n \times n$ matrices, with $\det A = 2$ and $\det B = 3$, then $\det(A + B) = 5$.

5. (T/F) If A is $n \times n$ and $\det A = 2$, then $\det A^3 = 6$.

6. (T/F) If B is produced by interchanging two rows of A, then $\det B = \det A$.

7. (T/F) If B is produced by multiplying row 3 of A by 5, then $\det B = 5 \det A$.

8. (T/F) If B is formed by adding to one row of A a linear combination of the other rows, then $\det B = \det A$.

9. (T/F) $\det A^T = -\det A$.

10. (T/F) $\det(-A) = -\det A$.

11. (T/F) $\det A^T A \geq 0$.

12. (T/F) Any system of n linear equations in n variables can be solved by Cramer's rule.

13. (T/F) If **u** and **v** are in $\mathbb{R}^2$ and $\det [\, \mathbf{u} \quad \mathbf{v} \,] = 10$, then the area of the triangle in the plane with vertices at **0**, **u**, and **v** is 10.

14. (T/F) If $A^3 = 0$, then $\det A = 0$.

15. (T/F) If A is invertible, then $\det A^{-1} = \det A$.

Use row operations to show that the determinants in Exercises 16–18 are all zero.

16. $\begin{vmatrix} 12 & 13 & 14 \\ 15 & 16 & 17 \\ 18 & 19 & 20 \end{vmatrix}$ **17.** $\begin{vmatrix} 1 & a & b+c \\ 1 & b & a+c \\ 1 & c & a+b \end{vmatrix}$

18. $\begin{vmatrix} a & b & c \\ a+x & b+x & c+x \\ a+y & b+y & c+y \end{vmatrix}$

Compute the determinants in Exercises 19 and 20.

19. $\begin{vmatrix} 9 & 1 & 9 & 9 & 9 \\ 9 & 0 & 9 & 9 & 2 \\ 4 & 0 & 0 & 5 & 0 \\ 9 & 0 & 3 & 9 & 0 \\ 6 & 0 & 0 & 7 & 0 \end{vmatrix}$

20. $\begin{vmatrix} 4 & 8 & 8 & 8 & 5 \\ 0 & 1 & 0 & 0 & 0 \\ 6 & 8 & 8 & 8 & 7 \\ 0 & 8 & 8 & 3 & 0 \\ 0 & 8 & 2 & 0 & 0 \end{vmatrix}$

21. Show that the equation of the line in $\mathbb{R}^2$ through distinct points (x_1, y_1) and (x_2, y_2) can be written as

$$\det \begin{bmatrix} 1 & x & y \\ 1 & x_1 & y_1 \\ 1 & x_2 & y_2 \end{bmatrix} = 0$$

22. Find a 3×3 determinant equation similar to that in Exercise 21 that describes the equation of the line through (x_1, y_1) with slope m.

Exercises 23 and 24 concern determinants of the following *Vandermonde matrices*.

$$T = \begin{bmatrix} 1 & a & a^2 \\ 1 & b & b^2 \\ 1 & c & c^2 \end{bmatrix}, \quad V(t) = \begin{bmatrix} 1 & t & t^2 & t^3 \\ 1 & x_1 & x_1^2 & x_1^3 \\ 1 & x_2 & x_2^2 & x_2^3 \\ 1 & x_3 & x_3^2 & x_3^3 \end{bmatrix}$$

23. Use row operations to show that

$$\det T = (b - a)(c - a)(c - b)$$

24. Let $f(t) = \det V$, with x_1, x_2, and x_3 all distinct. Explain why $f(t)$ is a cubic polynomial, show that the coefficient of t^3 is nonzero, and find three points on the graph of f.

25. Find the area of the parallelogram determined by the points $(1, 4)$, $(-1, 5)$, $(3, 9)$, and $(5, 8)$. How can you tell that the quadrilateral determined by the points is actually a parallelogram?

26. Use the concept of area of a parallelogram to write a statement about a 2×2 matrix A that is true if and only if A is invertible.

27. Show that if A is invertible, then adj A is invertible, and

$$(\text{adj } A)^{-1} = \frac{1}{\det A} A$$

[*Hint:* Given matrices B and C, what calculation(s) would show that C is the inverse of B?]

28. Let A, B, C, D, and I be $n \times n$ matrices. Use the definition or properties of a determinant to justify the following formulas. Part (c) is useful in applications of eigenvalues (Chapter 5).

a. $\det \begin{bmatrix} A & 0 \\ 0 & I \end{bmatrix} = \det A$ b. $\det \begin{bmatrix} I & 0 \\ C & D \end{bmatrix} = \det D$

c. $\det \begin{bmatrix} A & 0 \\ C & D \end{bmatrix} = (\det A)(\det D) = \det \begin{bmatrix} A & B \\ 0 & D \end{bmatrix}$

29. Let A, B, C, and D be $n \times n$ matrices with A invertible.

a. Find matrices X and Y to produce the block LU factorization

$$\begin{bmatrix} A & B \\ C & D \end{bmatrix} = \begin{bmatrix} I & 0 \\ X & I \end{bmatrix} \begin{bmatrix} A & B \\ 0 & Y \end{bmatrix}$$

and then show that

$$\det \begin{bmatrix} A & B \\ C & D \end{bmatrix} = (\det A) \cdot \det(D - CA^{-1}B)$$

b. Show that if $AC = CA$, then

$$\det \begin{bmatrix} A & B \\ C & D \end{bmatrix} = \det(AD - CB)$$

30. Let J be the $n \times n$ matrix of all 1's, and consider $A = (a - b)I + bJ$; that is,

$$A = \begin{bmatrix} a & b & b & \cdots & b \\ b & a & b & \cdots & b \\ b & b & a & \cdots & b \\ \vdots & \vdots & \vdots & \ddots & \vdots \\ b & b & b & \cdots & a \end{bmatrix}$$

Confirm that $\det A = (a - b)^{n-1}[a + (n - 1)b]$ as follows:

a. Subtract row 2 from row 1, row 3 from row 2, and so on, and explain why this does not change the determinant of the matrix.

b. With the resulting matrix from part (a), add column 1 to column 2, then add this new column 2 to column 3, and so on, and explain why this does not change the determinant.

c. Find the determinant of the resulting matrix from (b).

31. Let A be the original matrix given in Exercise 30, and let

$$B = \begin{bmatrix} a-b & b & b & \cdots & b \\ 0 & a & b & \cdots & b \\ 0 & b & a & \cdots & b \\ \vdots & \vdots & \vdots & \ddots & \vdots \\ 0 & b & b & \cdots & a \end{bmatrix},$$

$$C = \begin{bmatrix} b & b & b & \cdots & b \\ b & a & b & \cdots & b \\ b & b & a & \cdots & b \\ \vdots & \vdots & \vdots & \ddots & \vdots \\ b & b & b & \cdots & a \end{bmatrix}$$

Notice that A, B, and C are nearly the same except that the first column of A equals the sum of the first columns of B and C. A *linearity property* of the determinant function, discussed in Section 3.2, says that $\det A = \det B + \det C$. Use this fact to prove the formula in Exercise 30 by induction on the size of matrix A.

32. Apply the result of Exercise 30 to find the determinants of the following matrices, and confirm your answers using a matrix program.

$$\begin{bmatrix} 3 & 8 & 8 & 8 \\ 8 & 3 & 8 & 8 \\ 8 & 8 & 3 & 8 \\ 8 & 8 & 8 & 3 \end{bmatrix} \quad \begin{bmatrix} 8 & 3 & 3 & 3 & 3 \\ 3 & 8 & 3 & 3 & 3 \\ 3 & 3 & 8 & 3 & 3 \\ 3 & 3 & 3 & 8 & 3 \\ 3 & 3 & 3 & 3 & 8 \end{bmatrix}$$

T 33. Use a matrix program to compute the determinants of the following matrices.

$$\begin{bmatrix} 1 & 1 & 1 \\ 1 & 2 & 2 \\ 1 & 2 & 3 \end{bmatrix} \quad \begin{bmatrix} 1 & 1 & 1 & 1 \\ 1 & 2 & 2 & 2 \\ 1 & 2 & 3 & 3 \\ 1 & 2 & 3 & 4 \end{bmatrix}$$

$$\begin{bmatrix} 1 & 1 & 1 & 1 & 1 \\ 1 & 2 & 2 & 2 & 2 \\ 1 & 2 & 3 & 3 & 3 \\ 1 & 2 & 3 & 4 & 4 \\ 1 & 2 & 3 & 4 & 5 \end{bmatrix}$$

Use the results to guess the determinant of the matrix M, and confirm your guess by using row operations to evaluate that determinant.

$$M = \begin{bmatrix} 1 & 1 & 1 & \cdots & 1 \\ 1 & 2 & 2 & \cdots & 2 \\ 1 & 2 & 3 & \cdots & 3 \\ \vdots & \vdots & \vdots & \ddots & \vdots \\ 1 & 2 & 3 & \cdots & n \end{bmatrix}$$

T 34. Use the method of Exercise 33 to guess the determinant of

$$\begin{bmatrix} 1 & 1 & 1 & \cdots & 1 \\ 1 & 3 & 3 & \cdots & 3 \\ 1 & 3 & 6 & \cdots & 6 \\ \vdots & \vdots & \vdots & \ddots & \vdots \\ 1 & 3 & 6 & \cdots & 3(n-1) \end{bmatrix}$$

Justify your conjecture. [*Hint:* Use Exercise 28(c) and the result of Exercise 33.]

4 Vector Spaces

DISCRETE-TIME SIGNALS AND DIGITAL SIGNAL PROCESSING

What is digital signal processing? Just ask Alexa, who uses signal processing to record your question and deliver the answer. In 2500 BC, the Egyptians created the first recorded discrete-time signal by carving information about the flooding of the Nile into a Palermo Stone. Despite the early beginnings of discrete-time signals, it was not until the 1940s that Claude Shannon set off the digital revolution with the ideas articulated in his paper "A Mathematical Theory of Communication."

When a person speaks into a digital processor like Alexa, it converts the sounds made by the voice into a discrete-time signal—basically a sequence of numbers $\{y_k\}$, where k represents the time at which the value y_k was recorded. Then using linear time invariant (LTI) transformations, the signal is processed to filter out unwanted noise, such as the sound of a fan running in the background. The processed signal is then compared to the signals produced by recordings of the individual sounds that make up the language of the speaker. Figure 1 shows a recording of the word "yes" and of the word "no" illustrating that the signals produced are quite distinct. Once the sounds spoken in the question are identified, machine learning is used to make a best guess at the intended question for a digital processor like Alexa. The digital processor then searches through digitized data to find the most appropriate response. Finally, the signal is processed further to produce the virtual sounds that replicate a spoken answer.

Digital signal processing (DSP) is the branch of engineering that, in the span of just a few decades, has revolutionized interpersonal communication and the entertainment industry. By reworking the principles of electronics, telecommunication, and computer science into a unifying paradigm, DSP is at the heart of the digital revolution. A smartphone fits easily in the palm of your hand, replacing numerous other devices such as cameras, video recorders, CD players, day planners, and calculators, and taking the fantasy component out of Borges's imagined Library of Babel.

The usefulness of discrete-time signals and DSP goes well beyond systems engineering. Technical analysis is employed in the investment sector. Trading opportunities are identified by applying DSP to the discrete-time signals created when the price or volume traded of a stock is recorded over time. In Example 11 of Section 4.2, price data is smoothed using a linear transformation. In the entertainment industry, audio and video are produced

FIGURE 1

virtually and synthesized using DSP. In Example 3 of Section 4.7, we see how signal processing can be used to add richness to virtual sounds.

Discrete-time signals and DSP have become significant tools in many industries and areas of research. Mathematically speaking, discrete-time signals can be viewed as vectors that are processed using linear transformations. The operations of adding, scaling, and applying linear transformations to signals is completely analogous to the same operations for vectors in $\mathbb{R}^n$. For this reason, the set of all possible signals, $\mathbb{S}$, is treated as a *vector space*. In Sections 4.7 and 4.8, we look at the vector space of discrete-time signals in more detail.

The focus of Chapter 4 is to extend the theory of vectors in $\mathbb{R}^n$ to include signals and other mathematical structures that behave like the vectors you are already familiar with. Later on in the text, you will see how other vector spaces and their corresponding linear transformations arise in engineering, physics, biology, and statistics.

The mathematical seeds planted in Chapters 1 and 2 germinate and begin to blossom in this chapter. The beauty and power of linear algebra will be seen more clearly when you view $\mathbb{R}^n$ as only one of a variety of vector spaces that arise naturally in applied problems.

Beginning with basic definitions in Section 4.1, the general vector space framework develops gradually throughout the chapter. A goal of Sections 4.5 and 4.6 is to demonstrate how closely other vector spaces resemble $\mathbb{R}^n$. Sections 4.7 and 4.8 apply the theory of this chapter to discrete-time signals, DSP, and difference equations—the mathematics underlying the digital revolution.

4.1 Vector Spaces and Subspaces

Much of the theory in Chapters 1 and 2 rested on certain simple and obvious algebraic properties of $\mathbb{R}^n$, listed in Section 1.3. In fact, many other mathematical systems have the same properties. The specific properties of interest are listed in the following definition.

DEFINITION

A **vector space** is a nonempty set V of objects, called *vectors*, on which are defined two operations, called *addition* and *multiplication by scalars* (real numbers), subject to the ten axioms (or rules) listed below.[1] The axioms must hold for all vectors $\mathbf{u}$, $\mathbf{v}$, and $\mathbf{w}$ in V and for all scalars c and d.

1. The sum of $\mathbf{u}$ and $\mathbf{v}$, denoted by $\mathbf{u} + \mathbf{v}$, is in V.
2. $\mathbf{u} + \mathbf{v} = \mathbf{v} + \mathbf{u}$.
3. $(\mathbf{u} + \mathbf{v}) + \mathbf{w} = \mathbf{u} + (\mathbf{v} + \mathbf{w})$.
4. There is a **zero** vector $\mathbf{0}$ in V such that $\mathbf{u} + \mathbf{0} = \mathbf{u}$.
5. For each $\mathbf{u}$ in V, there is a vector $-\mathbf{u}$ in V such that $\mathbf{u} + (-\mathbf{u}) = \mathbf{0}$.

---

[1] Technically, V is a *real vector space*. All of the theory in this chapter also holds for a *complex vector space* in which the scalars and matrix entries are complex numbers. We will look at this briefly in Chapter 5. Until then, all scalars and matrix entries are assumed to be real.

6. The scalar multiple of $\mathbf{u}$ by c, denoted by $c\mathbf{u}$, is in V.

7. $c(\mathbf{u} + \mathbf{v}) = c\mathbf{u} + c\mathbf{v}$.

8. $(c + d)\mathbf{u} = c\mathbf{u} + d\mathbf{u}$.

9. $c(d\mathbf{u}) = (cd)\mathbf{u}$.

10. $1\mathbf{u} = \mathbf{u}$.

Using only these axioms, one can show that the zero vector in Axiom 4 is unique, and the vector $-\mathbf{u}$, called the **negative** of $\mathbf{u}$, in Axiom 5 is unique for each $\mathbf{u}$ in V. See Exercises 33 and 34. Proofs of the following simple facts are also outlined in the exercises:

For each $\mathbf{u}$ in V and scalar c,

$$0\mathbf{u} = \mathbf{0} \tag{1}$$

$$c\mathbf{0} = \mathbf{0} \tag{2}$$

$$-\mathbf{u} = (-1)\mathbf{u} \tag{3}$$

EXAMPLE 1 The spaces $\mathbb{R}^n$, where $n \geq 1$, are the premier examples of vector spaces. The geometric intuition developed for $\mathbb{R}^3$ will help you understand and visualize many concepts throughout the chapter. ∎

EXAMPLE 2 Let V be the set of all arrows (directed line segments) in three-dimensional space, with two arrows regarded as equal if they have the same length and point in the same direction. Define addition by the parallelogram rule (from Section 1.3), and for each $\mathbf{v}$ in V, define $c\mathbf{v}$ to be the arrow whose length is $|c|$ times the length of $\mathbf{v}$, pointing in the same direction as $\mathbf{v}$ if $c \geq 0$ and otherwise pointing in the opposite direction. (See Figure 1.) Show that V is a vector space. This space is a common model in physical problems for various forces.

FIGURE 1

SOLUTION The definition of V is geometric, using concepts of length and direction. No xyz-coordinate system is involved. An arrow of zero length is a single point and represents the zero vector. The negative of $\mathbf{v}$ is $(-1)\mathbf{v}$. So Axioms 1, 4, 5, 6, and 10 are evident. The rest are verified by geometry. For instance, see Figures 2 and 3. ∎

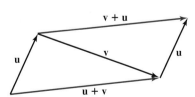

FIGURE 2 $\mathbf{u} + \mathbf{v} = \mathbf{v} + \mathbf{u}$.

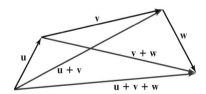

FIGURE 3 $(\mathbf{u} + \mathbf{v}) + \mathbf{w} = \mathbf{u} + (\mathbf{v} + \mathbf{w})$.

EXAMPLE 3 Let $\mathbb{S}$ be the space of all doubly infinite sequences of numbers (usually written in a row rather than a column):

$$\{y_k\} = (\ldots, y_{-2}, y_{-1}, y_0, y_1, y_2, \ldots)$$

If $\{z_k\}$ is another element of $\mathbb{S}$, then the sum $\{y_k\} + \{z_k\}$ is the sequence $\{y_k + z_k\}$ formed by adding corresponding terms of $\{y_k\}$ and $\{z_k\}$. The scalar multiple $c\{y_k\}$ is the sequence $\{cy_k\}$. The vector space axioms are verified in the same way as for $\mathbb{R}^n$.

Elements of $\mathbb{S}$ arise in engineering, for example, whenever a signal is measured (or sampled) at discrete times. A signal might be electrical, mechanical, optical, biological, audio, and so on. The digital signal processors mentioned in the chapter introduction use discrete (or digital) signals. For convenience, we will call $\mathbb{S}$ the space of (discrete-time) **signals**. A signal may be visualized by a graph as in Figure 4.

FIGURE 4 A discrete-time signal. ∎

EXAMPLE 4 For $n \geq 0$, the set $\mathbb{P}_n$ of polynomials of degree at most n consists of all polynomials of the form

$$\mathbf{p}(t) = a_0 + a_1 t + a_2 t^2 + \cdots + a_n t^n \tag{4}$$

where the coefficients $a_0, \ldots, a_n$ and the variable t are real numbers. The *degree* of $\mathbf{p}$ is the highest power of t in (4) whose coefficient is not zero. If $\mathbf{p}(t) = a_0 \neq 0$, the degree of $\mathbf{p}$ is zero. If all the coefficients are zero, $\mathbf{p}$ is called the *zero polynomial*. The zero polynomial is included in $\mathbb{P}_n$ even though its degree, for technical reasons, is not defined.

If $\mathbf{p}$ is given by (4) and if $\mathbf{q}(t) = b_0 + b_1 t + \cdots + b_n t^n$, then the sum $\mathbf{p} + \mathbf{q}$ is defined by

$$\begin{aligned}(\mathbf{p} + \mathbf{q})(t) &= \mathbf{p}(t) + \mathbf{q}(t) \\ &= (a_0 + b_0) + (a_1 + b_1)t + \cdots + (a_n + b_n)t^n\end{aligned}$$

The scalar multiple $c\mathbf{p}$ is the polynomial defined by

$$(c\mathbf{p})(t) = c\mathbf{p}(t) = ca_0 + (ca_1)t + \cdots + (ca_n)t^n$$

These definitions satisfy Axioms 1 and 6 because $\mathbf{p} + \mathbf{q}$ and $c\mathbf{p}$ are polynomials of degree less than or equal to n. Axioms 2, 3, and 7–10 follow from properties of the real numbers. Clearly, the zero polynomial acts as the zero vector in Axiom 4. Finally, $(-1)\mathbf{p}$ acts as the negative of $\mathbf{p}$, so Axiom 5 is satisfied. Thus $\mathbb{P}_n$ is a vector space.

The vector spaces $\mathbb{P}_n$ for various n are used, for instance, in statistical trend analysis of data, discussed in Section 6.8. ∎

EXAMPLE 5 Let V be the set of all real-valued functions defined on a set $\mathbb{D}$. (Typically, $\mathbb{D}$ is the set of real numbers or some interval on the real line.) Functions are added in the usual way: $\mathbf{f} + \mathbf{g}$ is the function whose value at t in the domain $\mathbb{D}$ is $\mathbf{f}(t) + \mathbf{g}(t)$. Likewise, for a scalar c and an $\mathbf{f}$ in V, the scalar multiple $c\mathbf{f}$ is the function whose value at t is $c\mathbf{f}(t)$. For instance, if $\mathbb{D} = \mathbb{R}$, $\mathbf{f}(t) = 1 + \sin 2t$, and $\mathbf{g}(t) = 2 + .5t$, then

$$(\mathbf{f} + \mathbf{g})(t) = 3 + \sin 2t + .5t \quad \text{and} \quad (2\mathbf{g})(t) = 4 + t$$

FIGURE 5

The sum of two vectors (functions).

Two functions in V are equal if and only if their values are equal for every t in $\mathbb{D}$. Hence the zero vector in V is the function that is identically zero, $\mathbf{f}(t) = 0$ for all t, and the negative of $\mathbf{f}$ is $(-1)\mathbf{f}$. Axioms 1 and 6 are obviously true, and the other axioms follow from properties of the real numbers, so V is a vector space. ∎

It is important to think of each function in the vector space V of Example 5 as a single object, as just one "point" or vector in the vector space. The sum of two vectors $\mathbf{f}$ and $\mathbf{g}$ (functions in V, or elements of *any* vector space) can be visualized as in Figure 5, because this can help you carry over to a general vector space the geometric intuition you have developed while working with the vector space $\mathbb{R}^n$. See the *Study Guide* for help as you learn to adopt this more general point of view.

Subspaces

In many problems, a vector space consists of an appropriate subset of vectors from some larger vector space. In this case, only three of the ten vector space axioms need to be checked; the rest are automatically satisfied.

DEFINITION

A **subspace** of a vector space V is a subset H of V that has three properties:

a. The zero vector of V is in H.[2]

b. H is closed under vector addition. That is, for each $\mathbf{u}$ and $\mathbf{v}$ in H, the sum $\mathbf{u} + \mathbf{v}$ is in H.

c. H is closed under multiplication by scalars. That is, for each $\mathbf{u}$ in H and each scalar c, the vector $c\mathbf{u}$ is in H.

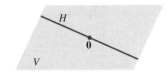

FIGURE 6

A subspace of V.

Properties (a), (b), and (c) guarantee that a subspace H of V is itself a *vector space*, under the vector space operations already defined in V. To verify this, note that properties (a), (b), and (c) are Axioms 1, 4, and 6. Axioms 2, 3, and 7–10 are automatically true in H because they apply to all elements of V, including those in H. Axiom 5 is also true in H, because if $\mathbf{u}$ is in H, then $(-1)\mathbf{u}$ is in H by property (c), and we know from equation (3) earlier in this section that $(-1)\mathbf{u}$ is the vector $-\mathbf{u}$ in Axiom 5.

So every subspace is a vector space. Conversely, every vector space is a subspace (of itself and possibly of other larger spaces). The term *subspace* is used when at least two vector spaces are in mind, with one inside the other, and the phrase *subspace of V* identifies V as the larger space. (See Figure 6.)

EXAMPLE 6   The set consisting of only the zero vector in a vector space V is a subspace of V, called the **zero subspace** and written as $\{\mathbf{0}\}$. ∎

EXAMPLE 7   Let $\mathbb{P}$ be the set of all polynomials with real coefficients, with operations in $\mathbb{P}$ defined as for functions. Then $\mathbb{P}$ is a subspace of the space of all real-valued functions defined on $\mathbb{R}$. Also, for each $n \geq 0$, $\mathbb{P}_n$ is a subspace of $\mathbb{P}$, because $\mathbb{P}_n$ is a subset of $\mathbb{P}$ that contains the zero polynomial, the sum of two polynomials in $\mathbb{P}_n$ is also in $\mathbb{P}_n$, and a scalar multiple of a polynomial in $\mathbb{P}_n$ is also in $\mathbb{P}_n$. ∎

---

[2] Some texts replace property (a) in this definition by the assumption that H is nonempty. Then (a) could be deduced from (c) and the fact that $0\mathbf{u} = \mathbf{0}$. But the best way to test for a subspace is to look first for the zero vector. If $\mathbf{0}$ is in H, then properties (b) and (c) must be checked. If $\mathbf{0}$ is *not* in H, then H cannot be a subspace and the other properties need not be checked.

EXAMPLE 8 The set of finitely supported signals $\mathbb{S}_f$ consists of the signals $\{y_k\}$, where only finitely many of the y_k are nonzero. Since the zero signal $\mathbf{0} = (\ldots, 0, 0, 0, \ldots)$ has no nonzero entries, it is clearly an element of $\mathbb{S}_f$. If two signals with finitely many nonzeros are added, the resulting signal will have finitely many nonzeros. Similarly if a signal with finitely many nonzeros is scaled, the result will still have finitely many nonzeros. Thus $\mathbb{S}_f$ is a subspace of $\mathbb{S}$, the discrete-time signals. See Figure 7.

FIGURE 7 ∎

FIGURE 8

The $x_1 x_2$-plane as a subspace of $\mathbb{R}^3$.

EXAMPLE 9 The vector space $\mathbb{R}^2$ is *not* a subspace of $\mathbb{R}^3$ because $\mathbb{R}^2$ is not even a subset of $\mathbb{R}^3$. (The vectors in $\mathbb{R}^3$ all have three entries, whereas the vectors in $\mathbb{R}^2$ have only two.) The set

$$H = \left\{ \begin{bmatrix} s \\ t \\ 0 \end{bmatrix} : s \text{ and } t \text{ are real} \right\}$$

is a subset of $\mathbb{R}^3$ that "looks" and "acts" like $\mathbb{R}^2$, although it is logically distinct from $\mathbb{R}^2$. See Figure 8. Show that H is a subspace of $\mathbb{R}^3$.

SOLUTION The zero vector is in H, and H is closed under vector addition and scalar multiplication because these operations on vectors in H always produce vectors whose third entries are zero (and so belong to H). Thus H is a subspace of $\mathbb{R}^3$. ∎

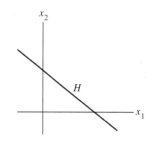

FIGURE 9

A line that is not a vector space.

EXAMPLE 10 A plane in $\mathbb{R}^3$ *not* through the origin is not a subspace of $\mathbb{R}^3$, because the plane does not contain the zero vector of $\mathbb{R}^3$. Similarly, a line in $\mathbb{R}^2$ *not* through the origin, such as in Figure 9, is *not* a subspace of $\mathbb{R}^2$. ∎

A Subspace Spanned by a Set

The next example illustrates one of the most common ways of describing a subspace. As in Chapter 1, the term **linear combination** refers to any sum of scalar multiples of vectors, and Span $\{\mathbf{v}_1, \ldots, \mathbf{v}_p\}$ denotes the set of all vectors that can be written as linear combinations of $\mathbf{v}_1, \ldots, \mathbf{v}_p$.

EXAMPLE 11 Given $\mathbf{v}_1$ and $\mathbf{v}_2$ in a vector space V, let $H = \text{Span}\{\mathbf{v}_1, \mathbf{v}_2\}$. Show that H is a subspace of V.

SOLUTION The zero vector is in H, since $\mathbf{0} = 0\mathbf{v}_1 + 0\mathbf{v}_2$. To show that H is closed under vector addition, take two arbitrary vectors in H, say,

$$\mathbf{u} = s_1\mathbf{v}_1 + s_2\mathbf{v}_2 \quad \text{and} \quad \mathbf{w} = t_1\mathbf{v}_1 + t_2\mathbf{v}_2$$

By Axioms 2, 3, and 8 for the vector space V,

$$\mathbf{u} + \mathbf{w} = (s_1\mathbf{v}_1 + s_2\mathbf{v}_2) + (t_1\mathbf{v}_1 + t_2\mathbf{v}_2)$$
$$= (s_1 + t_1)\mathbf{v}_1 + (s_2 + t_2)\mathbf{v}_2$$

So $\mathbf{u} + \mathbf{w}$ is in H. Furthermore, if c is any scalar, then by Axioms 7 and 9,

$$c\mathbf{u} = c(s_1\mathbf{v}_1 + s_2\mathbf{v}_2) = (cs_1)\mathbf{v}_1 + (cs_2)\mathbf{v}_2$$

which shows that $c\mathbf{u}$ is in H and H is closed under scalar multiplication. Thus H is a subspace of V. ∎

FIGURE 10

An example of a subspace.

In Section 4.5, you will see that every nonzero subspace of $\mathbb{R}^3$, other than $\mathbb{R}^3$ itself, is either Span $\{\mathbf{v}_1, \mathbf{v}_2\}$ for some linearly independent $\mathbf{v}_1$ and $\mathbf{v}_2$ or Span $\{\mathbf{v}\}$ for $\mathbf{v} \neq \mathbf{0}$. In the first case, the subspace is a plane through the origin; in the second case, it is a line through the origin. (See Figure 10.) It is helpful to keep these geometric pictures in mind, even for an abstract vector space.

The argument in Example 11 can easily be generalized to prove the following theorem.

THEOREM 1

> If $\mathbf{v}_1, \ldots, \mathbf{v}_p$ are in a vector space V, then Span $\{\mathbf{v}_1, \ldots, \mathbf{v}_p\}$ is a subspace of V.

We call Span $\{\mathbf{v}_1, \ldots, \mathbf{v}_p\}$ **the subspace spanned** (or **generated**) by $\{\mathbf{v}_1, \ldots, \mathbf{v}_p\}$. Given any subspace H of V, a **spanning** (or **generating**) **set** for H is a set $\{\mathbf{v}_1, \ldots, \mathbf{v}_p\}$ in H such that $H = $ Span $\{\mathbf{v}_1, \ldots, \mathbf{v}_p\}$.

The next example shows how to use Theorem 1.

EXAMPLE 12   Let H be the set of all vectors of the form $(a - 3b, b - a, a, b)$, where a and b are arbitrary scalars. That is, let $H = \{(a - 3b, b - a, a, b) : a$ and b in $\mathbb{R}\}$. Show that H is a subspace of $\mathbb{R}^4$.

SOLUTION   Write the vectors in H as column vectors. Then an arbitrary vector in H has the form

$$\begin{bmatrix} a - 3b \\ b - a \\ a \\ b \end{bmatrix} = a \underset{\mathbf{v}_1}{\begin{bmatrix} 1 \\ -1 \\ 1 \\ 0 \end{bmatrix}} + b \underset{\mathbf{v}_2}{\begin{bmatrix} -3 \\ 1 \\ 0 \\ 1 \end{bmatrix}}$$

This calculation shows that $H = $ Span $\{\mathbf{v}_1, \mathbf{v}_2\}$, where $\mathbf{v}_1$ and $\mathbf{v}_2$ are the vectors indicated above. Thus H is a subspace of $\mathbb{R}^4$ by Theorem 1. ∎

Example 12 illustrates a useful technique of expressing a subspace H as the set of linear combinations of some small collection of vectors. If $H = $ Span $\{\mathbf{v}_1, \ldots, \mathbf{v}_p\}$, we can think of the vectors $\mathbf{v}_1, \ldots, \mathbf{v}_p$ in the spanning set as "handles" that allow us to hold on to the subspace H. Calculations with the infinitely many vectors in H are often reduced to operations with the finite number of vectors in the spanning set.

EXAMPLE 13   For what value(s) of h will $\mathbf{y}$ be in the subspace of $\mathbb{R}^3$ spanned by $\mathbf{v}_1, \mathbf{v}_2, \mathbf{v}_3$, if

$$\mathbf{v}_1 = \begin{bmatrix} 1 \\ -1 \\ -2 \end{bmatrix}, \qquad \mathbf{v}_2 = \begin{bmatrix} 5 \\ -4 \\ -7 \end{bmatrix}, \qquad \mathbf{v}_3 = \begin{bmatrix} -3 \\ 1 \\ 0 \end{bmatrix}, \quad \text{and} \quad \mathbf{y} = \begin{bmatrix} -4 \\ 3 \\ h \end{bmatrix}$$

SOLUTION This question is Practice Problem 2 in Section 1.3, written here with the term *subspace* rather than Span $\{v_1, v_2, v_3\}$. The solution there shows that y is in Span $\{v_1, v_2, v_3\}$ if and only if $h = 5$. That solution is worth reviewing now, along with Exercises 11–16 and 19–21 in Section 1.3. ∎

Although many vector spaces in this chapter will be subspaces of $\mathbb{R}^n$, it is important to keep in mind that the abstract theory applies to other vector spaces as well. Vector spaces of functions arise in many applications, and they will receive more attention later.

Practice Problems

1. Show that the set H of all points in $\mathbb{R}^2$ of the form $(3s, 2 + 5s)$ is not a vector space, by showing that it is not closed under scalar multiplication. (Find a specific vector $\mathbf{u}$ in H and a scalar c such that $c\mathbf{u}$ is not in H.)

2. Let $W = \text{Span}\{v_1, \ldots, v_p\}$, where $v_1, \ldots, v_p$ are in a vector space V. Show that v_k is in W for $1 \le k \le p$. [*Hint:* First write an equation that shows that v_1 is in W. Then adjust your notation for the general case.]

3. An $n \times n$ matrix A is said to be symmetric if $A^T = A$. Let S be the set of all 3×3 symmetric matrices. Show that S is a subspace of $M_{3\times3}$, the vector space of 3×3 matrices.

4.1 Exercises

1. Let V be the first quadrant in the xy-plane; that is, let
$$V = \left\{ \begin{bmatrix} x \\ y \end{bmatrix} : x \ge 0, y \ge 0 \right\}$$

   a. If $\mathbf{u}$ and $\mathbf{v}$ are in V, is $\mathbf{u} + \mathbf{v}$ in V? Why?

   b. Find a specific vector $\mathbf{u}$ in V and a specific scalar c such that $c\mathbf{u}$ is *not* in V. (This is enough to show that V is *not* a vector space.)

2. Let W be the union of the first and third quadrants in the xy-plane. That is, let $W = \left\{ \begin{bmatrix} x \\ y \end{bmatrix} : xy \ge 0 \right\}$.

   a. If $\mathbf{u}$ is in W and c is any scalar, is $c\mathbf{u}$ in W? Why?

   b. Find specific vectors $\mathbf{u}$ and $\mathbf{v}$ in W such that $\mathbf{u} + \mathbf{v}$ is not in W. (This is enough to show that W is *not* a vector space.)

3. Let H be the set of points inside and on the unit circle in the xy-plane. That is, let $H = \left\{ \begin{bmatrix} x \\ y \end{bmatrix} : x^2 + y^2 \le 1 \right\}$. Find a specific example—two vectors or a vector and a scalar—to show that H is not a subspace of $\mathbb{R}^2$.

4. Construct a geometric figure that illustrates why a line in $\mathbb{R}^2$ *not* through the origin is not closed under vector addition.

In Exercises 5–8, determine if the given set is a subspace of $\mathbb{P}_n$ for an appropriate value of n. Justify your answers.

5. All polynomials of the form $\mathbf{p}(t) = at^2$, where a is in $\mathbb{R}$.

6. All polynomials of the form $\mathbf{p}(t) = a + t^2$, where a is in $\mathbb{R}$.

7. All polynomials of degree at most 3, with integers as coefficients.

8. All polynomials in $\mathbb{P}_n$ such that $\mathbf{p}(0) = 0$.

9. Let H be the set of all vectors of the form $\begin{bmatrix} s \\ 3s \\ 2s \end{bmatrix}$. Find a vector $\mathbf{v}$ in $\mathbb{R}^3$ such that $H = \text{Span}\{\mathbf{v}\}$. Why does this show that H is a subspace of $\mathbb{R}^3$?

10. Let H be the set of all vectors of the form $\begin{bmatrix} 2t \\ 0 \\ -t \end{bmatrix}$. Show that H is a subspace of $\mathbb{R}^3$. (Use the method of Exercise 9.)

11. Let W be the set of all vectors of the form $\begin{bmatrix} 5b + 2c \\ b \\ c \end{bmatrix}$, where b and c are arbitrary. Find vectors $\mathbf{u}$ and $\mathbf{v}$ such that $W = \text{Span}\{\mathbf{u}, \mathbf{v}\}$. Why does this show that W is a subspace of $\mathbb{R}^3$?

12. Let W be the set of all vectors of the form $\begin{bmatrix} s + 3t \\ s - t \\ 2s - t \\ 4t \end{bmatrix}$. Show that W is a subspace of $\mathbb{R}^4$. (Use the method of Exercise 11.)

13. Let $v_1 = \begin{bmatrix} 1 \\ 0 \\ -1 \end{bmatrix}$, $v_2 = \begin{bmatrix} 2 \\ 1 \\ 3 \end{bmatrix}$, $v_3 = \begin{bmatrix} 4 \\ 2 \\ 6 \end{bmatrix}$, and $\mathbf{w} = \begin{bmatrix} 3 \\ 1 \\ 2 \end{bmatrix}$.

a. Is **w** in $\{\mathbf{v}_1, \mathbf{v}_2, \mathbf{v}_3\}$? How many vectors are in $\{\mathbf{v}_1, \mathbf{v}_2, \mathbf{v}_3\}$?

b. How many vectors are in Span $\{\mathbf{v}_1, \mathbf{v}_2, \mathbf{v}_3\}$?

c. Is **w** in the subspace spanned by $\{\mathbf{v}_1, \mathbf{v}_2, \mathbf{v}_3\}$? Why?

14. Let $\mathbf{v}_1, \mathbf{v}_2, \mathbf{v}_3$ be as in Exercise 13, and let $\mathbf{w} = \begin{bmatrix} 8 \\ 4 \\ 7 \end{bmatrix}$. Is **w** in the subspace spanned by $\{\mathbf{v}_1, \mathbf{v}_2, \mathbf{v}_3\}$? Why?

In Exercises 15–18, let W be the set of all vectors of the form shown, where a, b, and c represent arbitrary real numbers. In each case, either find a set S of vectors that spans W or give an example to show that W is *not* a vector space.

15. $\begin{bmatrix} 3a + b \\ 4 \\ a - 5b \end{bmatrix}$ **16.** $\begin{bmatrix} -a + 1 \\ a - 6b \\ 2b + a \end{bmatrix}$

17. $\begin{bmatrix} a - b \\ b - c \\ c - a \\ b \end{bmatrix}$ **18.** $\begin{bmatrix} 4a + 3b \\ 0 \\ a + b + c \\ c - 2a \end{bmatrix}$

19. If a mass m is placed at the end of a spring, and if the mass is pulled downward and released, the mass–spring system will begin to oscillate. The displacement y of the mass from its resting position is given by a function of the form

$$y(t) = c_1 \cos \omega t + c_2 \sin \omega t \qquad (5)$$

where ω is a constant that depends on the spring and the mass. (See the figure below.) Show that the set of all functions described in (5) (with ω fixed and c_1, c_2 arbitrary) is a vector space.

20. The set of all continuous real-valued functions defined on a closed interval $[a, b]$ in $\mathbb{R}$ is denoted by $C[a, b]$. This set is a subspace of the vector space of all real-valued functions defined on $[a, b]$.

a. What facts about continuous functions should be proved in order to demonstrate that $C[a, b]$ is indeed a subspace as claimed? (These facts are usually discussed in a calculus class.)

b. Show that $\{\mathbf{f} \text{ in } C[a, b] : \mathbf{f}(a) = \mathbf{f}(b)\}$ is a subspace of $C[a, b]$.

For fixed positive integers m and n, the set $M_{m \times n}$ of all $m \times n$ matrices is a vector space, under the usual operations of addition of matrices and multiplication by real scalars.

21. Determine if the set H of all matrices of the form $\begin{bmatrix} a & b \\ 0 & d \end{bmatrix}$ is a subspace of $M_{2 \times 2}$.

22. Let F be a fixed 3×2 matrix, and let H be the set of all matrices A in $M_{2 \times 4}$ with the property that $FA = 0$ (the zero matrix in $M_{3 \times 4}$). Determine if H is a subspace of $M_{2 \times 4}$.

In Exercises 23–32, mark each statement True or False **(T/F)**. Justify each answer.

23. **(T/F)** If **f** is a function in the vector space V of all real-valued functions on $\mathbb{R}$ and if $\mathbf{f}(t) = 0$ for some t, then **f** is the zero vector in V.

24. **(T/F)** A vector is any element of a vector space.

25. **(T/F)** An arrow in three-dimensional space can be considered to be a vector.

26. **(T/F)** If **u** is a vector in a vector space V, then $(-1)\mathbf{u}$ is the same as the negative of **u**.

27. **(T/F)** A subset H of a vector space V is a subspace of V if the zero vector is in H.

28. **(T/F)** A vector space is also a subspace.

29. **(T/F)** A subspace is also a vector space.

30. **(T/F)** $\mathbb{R}^2$ is a subspace of $\mathbb{R}^3$.

31. **(T/F)** The polynomials of degree two or less are a subspace of the polynomials of degree three or less.

32. **(T/F)** A subset H of a vector space V is a subspace of V if the following conditions are satisfied: (i) the zero vector of V is in H, (ii) **u**, **v**, and $\mathbf{u} + \mathbf{v}$ are in H, and (iii) c is a scalar and $c\mathbf{u}$ is in H.

Exercises 33–36 show how the axioms for a vector space V can be used to prove the elementary properties described after the definition of a vector space. Fill in the blanks with the appropriate axiom numbers. Because of Axiom 2, Axioms 4 and 5 imply, respectively, that $\mathbf{0} + \mathbf{u} = \mathbf{u}$ and $-\mathbf{u} + \mathbf{u} = \mathbf{0}$ for all **u**.

33. Complete the following proof that the zero vector is unique. Suppose that **w** in V has the property that $\mathbf{u} + \mathbf{w} = \mathbf{w} + \mathbf{u} = \mathbf{u}$ for all **u** in V. In particular, $\mathbf{0} + \mathbf{w} = \mathbf{0}$. But $\mathbf{0} + \mathbf{w} = \mathbf{w}$, by Axiom _____. Hence $\mathbf{w} = \mathbf{0} + \mathbf{w} = \mathbf{0}$.

34. Complete the following proof that $-\mathbf{u}$ is the *unique vector* in V such that $\mathbf{u} + (-\mathbf{u}) = \mathbf{0}$. Suppose that **w** satisfies $\mathbf{u} + \mathbf{w} = \mathbf{0}$. Adding $-\mathbf{u}$ to both sides, we have

$$(-\mathbf{u}) + [\mathbf{u} + \mathbf{w}] = (-\mathbf{u}) + \mathbf{0}$$
$$[(-\mathbf{u}) + \mathbf{u}] + \mathbf{w} = (-\mathbf{u}) + \mathbf{0} \qquad \text{by Axiom } \_\_\_\_ \text{ (a)}$$
$$\mathbf{0} + \mathbf{w} = (-\mathbf{u}) + \mathbf{0} \qquad \text{by Axiom } \_\_\_\_ \text{ (b)}$$
$$\mathbf{w} = -\mathbf{u} \qquad \text{by Axiom } \_\_\_\_ \text{ (c)}$$

35. Fill in the missing axiom numbers in the following proof that $0\mathbf{u} = \mathbf{0}$ for every **u** in V.

$$0\mathbf{u} = (0 + 0)\mathbf{u} = 0\mathbf{u} + 0\mathbf{u} \qquad \text{by Axiom } \_\_\_\_ \text{ (a)}$$

Add the negative of $0\mathbf{u}$ to both sides:

$0\mathbf{u} + (-0\mathbf{u}) = [0\mathbf{u} + 0\mathbf{u}] + (-0\mathbf{u})$

$0\mathbf{u} + (-0\mathbf{u}) = 0\mathbf{u} + [0\mathbf{u} + (-0\mathbf{u})]$ by Axiom _____ (b)

$\mathbf{0} = 0\mathbf{u} + \mathbf{0}$ by Axiom _____ (c)

$\mathbf{0} = 0\mathbf{u}$ by Axiom _____ (d)

36. Fill in the missing axiom numbers in the following proof that $c\mathbf{0} = \mathbf{0}$ for every scalar c.

$c\mathbf{0} = c(\mathbf{0} + \mathbf{0})$ by Axiom _____ (a)

$\quad = c\mathbf{0} + c\mathbf{0}$ by Axiom _____ (b)

Add the negative of $c\mathbf{0}$ to both sides:

$c\mathbf{0} + (-c\mathbf{0}) = [c\mathbf{0} + c\mathbf{0}] + (-c\mathbf{0})$

$c\mathbf{0} + (-c\mathbf{0}) = c\mathbf{0} + [c\mathbf{0} + (-c\mathbf{0})]$ by Axiom _____ (c)

$\mathbf{0} = c\mathbf{0} + \mathbf{0}$ by Axiom _____ (d)

$\mathbf{0} = c\mathbf{0}$ by Axiom _____ (e)

37. Prove that $(-1)\mathbf{u} = -\mathbf{u}$. [*Hint:* Show that $\mathbf{u} + (-1)\mathbf{u} = \mathbf{0}$. Use some axioms and the results of Exercises 34 and 35.]

38. Suppose $c\mathbf{u} = \mathbf{0}$ for some nonzero scalar c. Show that $\mathbf{u} = \mathbf{0}$. Mention the axioms or properties you use.

39. Let $\mathbf{u}$ and $\mathbf{v}$ be vectors in a vector space V, and let H be any subspace of V that contains both $\mathbf{u}$ and $\mathbf{v}$. Explain why H also contains Span $\{\mathbf{u}, \mathbf{v}\}$. This shows that Span $\{\mathbf{u}, \mathbf{v}\}$ is the smallest subspace of V that contains both $\mathbf{u}$ and $\mathbf{v}$.

40. Let H and K be subspaces of a vector space V. The **intersection** of H and K, written as $H \cap K$, is the set of $\mathbf{v}$ in V that belong to both H and K. Show that $H \cap K$ is a subspace of V. (See the figure.) Give an example in $\mathbb{R}^2$ to show that the union of two subspaces is not, in general, a subspace.

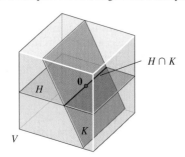

41. Given subspaces H and K of a vector space V, the **sum** of H and K, written as $H + K$, is the set of all vectors in V that

can be written as the sum of two vectors, one in H and the other in K; that is,

$H + K = \{\mathbf{w} : \mathbf{w} = \mathbf{u} + \mathbf{v}$ for some $\mathbf{u}$ in H

and some $\mathbf{v}$ in $K\}$

a. Show that $H + K$ is a subspace of V.

b. Show that H is a subspace of $H + K$ and K is a subspace of $H + K$.

42. Suppose $\mathbf{u}_1, \ldots, \mathbf{u}_p$ and $\mathbf{v}_1, \ldots, \mathbf{v}_q$ are vectors in a vector space V, and let

$H = \text{Span}\{\mathbf{u}_1, \ldots, \mathbf{u}_p\}$ and $K = \text{Span}\{\mathbf{v}_1, \ldots, \mathbf{v}_q\}$

Show that $H + K = \text{Span}\{\mathbf{u}_1, \ldots, \mathbf{u}_p, \mathbf{v}_1, \ldots, \mathbf{v}_q\}$.

T 43. Show that $\mathbf{w}$ is in the subspace of $\mathbb{R}^4$ spanned by $\mathbf{v}_1, \mathbf{v}_2, \mathbf{v}_3$, where

$$\mathbf{w} = \begin{bmatrix} 9 \\ -4 \\ -4 \\ 7 \end{bmatrix}, \mathbf{v}_1 = \begin{bmatrix} 8 \\ -4 \\ -3 \\ 9 \end{bmatrix}, \mathbf{v}_2 = \begin{bmatrix} -4 \\ 3 \\ -2 \\ -8 \end{bmatrix}, \mathbf{v}_3 = \begin{bmatrix} -7 \\ 6 \\ -5 \\ -18 \end{bmatrix}$$

T 44. Determine if $\mathbf{y}$ is in the subspace of $\mathbb{R}^4$ spanned by the columns of A, where

$$\mathbf{y} = \begin{bmatrix} -4 \\ -8 \\ 6 \\ -5 \end{bmatrix}, \quad A = \begin{bmatrix} 3 & -5 & -9 \\ 8 & 7 & -6 \\ -5 & -8 & 3 \\ 2 & -2 & -9 \end{bmatrix}$$

T 45. The vector space $H = \text{Span}\{1, \cos^2 t, \cos^4 t, \cos^6 t\}$ contains at least two interesting functions that will be used in a later exercise:

$\mathbf{f}(t) = 1 - 8\cos^2 t + 8\cos^4 t$

$\mathbf{g}(t) = -1 + 18\cos^2 t - 48\cos^4 t + 32\cos^6 t$

Study the graph of $\mathbf{f}$ for $0 \le t \le 2\pi$, and guess a simple formula for $\mathbf{f}(t)$. Verify your conjecture by graphing the difference between $1 + \mathbf{f}(t)$ and your formula for $\mathbf{f}(t)$. (Hopefully, you will see the constant function 1.) Repeat for $\mathbf{g}$.

T 46. Repeat Exercise 45 for the functions

$\mathbf{f}(t) = 3\sin t - 4\sin^3 t$

$\mathbf{g}(t) = 1 - 8\sin^2 t + 8\sin^4 t$

$\mathbf{h}(t) = 5\sin t - 20\sin^3 t + 16\sin^5 t$

in the vector space Span $\{1, \sin t, \sin^2 t, \ldots, \sin^5 t\}$.

Solutions to Practice Problems

1. Take any $\mathbf{u}$ in H—say, $\mathbf{u} = \begin{bmatrix} 3 \\ 7 \end{bmatrix}$—and take any $c \ne 1$—say, $c = 2$. Then $c\mathbf{u} = \begin{bmatrix} 6 \\ 14 \end{bmatrix}$. If this is in H, then there is some s such that

$$\begin{bmatrix} 3s \\ 2 + 5s \end{bmatrix} = \begin{bmatrix} 6 \\ 14 \end{bmatrix}$$

That is, $s = 2$ and $s = 12/5$, which is impossible. So $2\mathbf{u}$ is not in H and H is not a vector space.

2. $\mathbf{v}_1 = 1\mathbf{v}_1 + 0\mathbf{v}_2 + \cdots + 0\mathbf{v}_p$. This expresses $\mathbf{v}_1$ as a linear combination of $\mathbf{v}_1, \ldots, \mathbf{v}_p$, so $\mathbf{v}_1$ is in W. In general, $\mathbf{v}_k$ is in W because

$$\mathbf{v}_k = 0\mathbf{v}_1 + \cdots + 0\mathbf{v}_{k-1} + 1\mathbf{v}_k + 0\mathbf{v}_{k+1} + \cdots + 0\mathbf{v}_p$$

3. The subset S is a subspace of $M_{3\times3}$ since it satisfies all three of the requirements listed in the definition of a subspace:

   a. Observe that the $\mathbf{0}$ in $M_{3\times3}$ is the 3×3 zero matrix and since $\mathbf{0}^T = \mathbf{0}$, the matrix $\mathbf{0}$ is symmetric and hence $\mathbf{0}$ is in S.

   b. Let A and B in S. Notice that A and B are 3×3 symmetric matrices so $A^T = A$ and $B^T = B$. By the properties of transposes of matrices, $(A + B)^T = A^T + B^T = A + B$. Thus $A + B$ is symmetric and hence $A + B$ is in S.

   c. Let A be in S and let c be a scalar. Since A is symmetric, by the properties of symmetric matrices, $(cA)^T = c(A^T) = cA$. Thus cA is also a symmetric matrix and hence cA is in S.

4.2 Null Spaces, Column Spaces, Row Spaces, and Linear Transformations

In applications of linear algebra, subspaces of $\mathbb{R}^n$ usually arise in one of two ways: (1) as the set of all solutions to a system of homogeneous linear equations or (2) as the set of all linear combinations of certain specified vectors. In this section, we compare and contrast these two descriptions of subspaces, allowing us to practice using the concept of a subspace. Actually, as you will soon discover, we have been working with subspaces ever since Section 1.3. The main new feature here is the terminology. The section concludes with a discussion of the kernel and range of a linear transformation.

The Null Space of a Matrix

Consider the following system of homogeneous equations:

$$\begin{aligned} x_1 - 3x_2 - 2x_3 &= 0 \\ -5x_1 + 9x_2 + x_3 &= 0 \end{aligned} \tag{1}$$

In matrix form, this system is written as $A\mathbf{x} = \mathbf{0}$, where

$$A = \begin{bmatrix} 1 & -3 & -2 \\ -5 & 9 & 1 \end{bmatrix} \tag{2}$$

Recall that the set of all $\mathbf{x}$ that satisfy (1) is called the **solution set** of the system (1). Often it is convenient to relate this set directly to the matrix A and the equation $A\mathbf{x} = \mathbf{0}$. We call the set of $\mathbf{x}$ that satisfy $A\mathbf{x} = \mathbf{0}$ the **null space** of the matrix A.

DEFINITION

The **null space** of an $m \times n$ matrix A, written as Nul A, is the set of all solutions of the homogeneous equation $A\mathbf{x} = \mathbf{0}$. In set notation,

$$\text{Nul } A = \{\mathbf{x} : \mathbf{x} \text{ is in } \mathbb{R}^n \text{ and } A\mathbf{x} = \mathbf{0}\}$$

A more dynamic description of Nul A is the set of all $\mathbf{x}$ in $\mathbb{R}^n$ that are mapped into the zero vector of $\mathbb{R}^m$ via the linear transformation $\mathbf{x} \mapsto A\mathbf{x}$. See Figure 1.

FIGURE 1

EXAMPLE 1 Let A be the matrix in (2), and let $\mathbf{u} = \begin{bmatrix} 5 \\ 3 \\ -2 \end{bmatrix}$. Determine if $\mathbf{u}$ belongs to the null space of A.

SOLUTION To test if $\mathbf{u}$ satisfies $A\mathbf{u} = \mathbf{0}$, simply compute

$$A\mathbf{u} = \begin{bmatrix} 1 & -3 & -2 \\ -5 & 9 & 1 \end{bmatrix} \begin{bmatrix} 5 \\ 3 \\ -2 \end{bmatrix} = \begin{bmatrix} 5 - 9 + 4 \\ -25 + 27 - 2 \end{bmatrix} = \begin{bmatrix} 0 \\ 0 \end{bmatrix}$$

Thus $\mathbf{u}$ is in Nul A. ■

The term *space* in *null space* is appropriate because the null space of a matrix is a vector space, as shown in the next theorem.

THEOREM 2

The null space of an $m \times n$ matrix A is a subspace of $\mathbb{R}^n$. Equivalently, the set of all solutions to a system $A\mathbf{x} = \mathbf{0}$ of m homogeneous linear equations in n unknowns is a subspace of $\mathbb{R}^n$.

PROOF Certainly Nul A is a subset of $\mathbb{R}^n$ because A has n columns. We must show that Nul A satisfies the three properties of a subspace. Of course, $\mathbf{0}$ is in Nul A. Next, let $\mathbf{u}$ and $\mathbf{v}$ represent any two vectors in Nul A. Then

$$A\mathbf{u} = \mathbf{0} \quad \text{and} \quad A\mathbf{v} = \mathbf{0}$$

To show that $\mathbf{u} + \mathbf{v}$ is in Nul A, we must show that $A(\mathbf{u} + \mathbf{v}) = \mathbf{0}$. Using a property of matrix multiplication, compute

$$A(\mathbf{u} + \mathbf{v}) = A\mathbf{u} + A\mathbf{v} = \mathbf{0} + \mathbf{0} = \mathbf{0}$$

Thus $\mathbf{u} + \mathbf{v}$ is in Nul A, and Nul A is closed under vector addition. Finally, if c is any scalar, then

$$A(c\mathbf{u}) = c(A\mathbf{u}) = c(\mathbf{0}) = \mathbf{0}$$

which shows that $c\mathbf{u}$ is in Nul A. Thus Nul A is a subspace of $\mathbb{R}^n$. ■

EXAMPLE 2 Let H be the set of all vectors in $\mathbb{R}^4$ whose coordinates a, b, c, d satisfy the equations $a - 2b + 5c = d$ and $c - a = b$. Show that H is a subspace of $\mathbb{R}^4$.

SOLUTION Rearrange the equations that describe the elements of H, and note that H is the set of all solutions of the following system of homogeneous linear equations:

$$
\begin{aligned}
a - 2b + 5c - d &= 0 \\
-a - b + c &= 0
\end{aligned}
$$

By Theorem 2, H is a subspace of $\mathbb{R}^4$. ∎

It is important that the linear equations defining the set H are homogeneous. Otherwise, the set of solutions will definitely *not* be a subspace (because the zero vector is not a solution of a nonhomogeneous system). Also, in some cases, the set of solutions could be empty.

An Explicit Description of Nul A

There is no obvious relation between vectors in Nul A and the entries in A. We say that Nul A is defined *implicitly*, because it is defined by a condition that must be checked. No explicit list or description of the elements in Nul A is given. However, *solving the equation $A\mathbf{x} = \mathbf{0}$ amounts to producing an explicit description of Nul A*. The next example reviews the procedure from Section 1.5.

EXAMPLE 3 Find a spanning set for the null space of the matrix

$$
A = \begin{bmatrix} -3 & 6 & -1 & 1 & -7 \\ 1 & -2 & 2 & 3 & -1 \\ 2 & -4 & 5 & 8 & -4 \end{bmatrix}
$$

SOLUTION The first step is to find the general solution of $A\mathbf{x} = \mathbf{0}$ in terms of free variables. Row reduce the augmented matrix $[\, A \quad \mathbf{0} \,]$ to *reduced* echelon form in order to write the basic variables in terms of the free variables:

$$
\begin{bmatrix} 1 & -2 & 0 & -1 & 3 & 0 \\ 0 & 0 & 1 & 2 & -2 & 0 \\ 0 & 0 & 0 & 0 & 0 & 0 \end{bmatrix},
\qquad
\begin{aligned}
x_1 - 2x_2 \quad - x_4 + 3x_5 &= 0 \\
x_3 + 2x_4 - 2x_5 &= 0 \\
0 &= 0
\end{aligned}
$$

The general solution is $x_1 = 2x_2 + x_4 - 3x_5$, $x_3 = -2x_4 + 2x_5$, with x_2, x_4, and x_5 free. Next, decompose the vector giving the general solution into a linear combination of vectors where *the weights are the free variables*. That is,

$$
\begin{bmatrix} x_1 \\ x_2 \\ x_3 \\ x_4 \\ x_5 \end{bmatrix} = \begin{bmatrix} 2x_2 + x_4 - 3x_5 \\ x_2 \\ -2x_4 + 2x_5 \\ x_4 \\ x_5 \end{bmatrix} = x_2 \underset{\mathbf{u}}{\begin{bmatrix} 2 \\ 1 \\ 0 \\ 0 \\ 0 \end{bmatrix}} + x_4 \underset{\mathbf{v}}{\begin{bmatrix} 1 \\ 0 \\ -2 \\ 1 \\ 0 \end{bmatrix}} + x_5 \underset{\mathbf{w}}{\begin{bmatrix} -3 \\ 0 \\ 2 \\ 0 \\ 1 \end{bmatrix}}
$$

$$
= x_2\mathbf{u} + x_4\mathbf{v} + x_5\mathbf{w} \tag{3}
$$

Every linear combination of $\mathbf{u}$, $\mathbf{v}$, and $\mathbf{w}$ is an element of Nul A and vice versa. Thus $\{\mathbf{u}, \mathbf{v}, \mathbf{w}\}$ is a spanning set for Nul A. ∎

Two points should be made about the solution of Example 3 that apply to all problems of this type where Nul A contains nonzero vectors. We will use these facts later.

1. The spanning set produced by the method in Example 3 is automatically linearly independent because the free variables are the weights on the spanning vectors. For instance, look at the 2nd, 4th, and 5th entries in the solution vector in (3) and note that $x_2\mathbf{u} + x_4\mathbf{v} + x_5\mathbf{w}$ can be $\mathbf{0}$ only if the weights x_2, x_4, and x_5 are all zero.

2. When Nul A contains nonzero vectors, the number of vectors in the spanning set for Nul A equals the number of free variables in the equation $A\mathbf{x} = \mathbf{0}$.

The Column Space of a Matrix

Another important subspace associated with a matrix is its column space. Unlike the null space, the column space is defined explicitly via linear combinations.

DEFINITION

The **column space** of an $m \times n$ matrix A, written as Col A, is the set of all linear combinations of the columns of A. If $A = [\,\mathbf{a}_1 \ \cdots \ \mathbf{a}_n\,]$, then

$$\text{Col } A = \text{Span}\{\mathbf{a}_1, \ldots, \mathbf{a}_n\}$$

Since Span$\{\mathbf{a}_1, \ldots, \mathbf{a}_n\}$ is a subspace, by Theorem 1, the next theorem follows from the definition of Col A and the fact that the columns of A are in $\mathbb{R}^m$.

THEOREM 3

The column space of an $m \times n$ matrix A is a subspace of $\mathbb{R}^m$.

Note that a typical vector in Col A can be written as $A\mathbf{x}$ for some $\mathbf{x}$ because the notation $A\mathbf{x}$ stands for a linear combination of the columns of A. That is,

$$\boxed{\text{Col } A = \{\mathbf{b} : \mathbf{b} = A\mathbf{x} \text{ for some } \mathbf{x} \text{ in } \mathbb{R}^n\}}$$

The notation $A\mathbf{x}$ for vectors in Col A also shows that Col A is the *range* of the linear transformation $\mathbf{x} \mapsto A\mathbf{x}$. We will return to this point of view at the end of the section.

EXAMPLE 4 Find a matrix A such that $W = \text{Col } A$.

$$W = \left\{ \begin{bmatrix} 6a - b \\ a + b \\ -7a \end{bmatrix} : a, b \text{ in } \mathbb{R} \right\}$$

SOLUTION First, write W as a set of linear combinations.

$$W = \left\{ a \begin{bmatrix} 6 \\ 1 \\ -7 \end{bmatrix} + b \begin{bmatrix} -1 \\ 1 \\ 0 \end{bmatrix} : a, b \text{ in } \mathbb{R} \right\} = \text{Span} \left\{ \begin{bmatrix} 6 \\ 1 \\ -7 \end{bmatrix}, \begin{bmatrix} -1 \\ 1 \\ 0 \end{bmatrix} \right\}$$

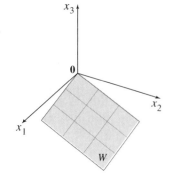

Second, use the vectors in the spanning set as the columns of A. Let $A = \begin{bmatrix} 6 & -1 \\ 1 & 1 \\ -7 & 0 \end{bmatrix}$.

Then $W = \text{Col } A$, as desired. ∎

Recall from Theorem 4 in Section 1.4 that the columns of A span $\mathbb{R}^m$ if and only if the equation $A\mathbf{x} = \mathbf{b}$ has a solution for each $\mathbf{b}$. We can restate this fact as follows:

> The column space of an $m \times n$ matrix A is all of $\mathbb{R}^m$ if and only if the equation $A\mathbf{x} = \mathbf{b}$ has a solution for each $\mathbf{b}$ in $\mathbb{R}^m$.

The Row Space

If A is an $m \times n$ matrix, each row of A has n entries and thus can be identified with a vector in $\mathbb{R}^n$. The set of all linear combinations of the row vectors is called the **row space** of A and is denoted by Row A. Each row has n entries, so Row A is a subspace of $\mathbb{R}^n$. Since the rows of A are identified with the columns of A^T, we could also write Col A^T in place of Row A.

EXAMPLE 5 Let

$$A = \begin{bmatrix} -2 & -5 & 8 & 0 & -17 \\ 1 & 3 & -5 & 1 & 5 \\ 3 & 11 & -19 & 7 & 1 \\ 1 & 7 & -13 & 5 & -3 \end{bmatrix} \quad \text{and} \quad \begin{aligned} \mathbf{r}_1 &= (-2, -5, 8, 0, -17) \\ \mathbf{r}_2 &= (1, 3, -5, 1, 5) \\ \mathbf{r}_3 &= (3, 11, -19, 7, 1) \\ \mathbf{r}_4 &= (1, 7, -13, 5, -3) \end{aligned}$$

The row space of A is the subspace of $\mathbb{R}^5$ spanned by $\{\mathbf{r}_1, \mathbf{r}_2, \mathbf{r}_3, \mathbf{r}_4\}$. That is, Row $A =$ Span $\{\mathbf{r}_1, \mathbf{r}_2, \mathbf{r}_3, \mathbf{r}_4\}$. It is natural to write row vectors horizontally; however, they may also be written as column vectors if that is more convenient. ∎

The Contrast Between Nul A and Col A

It is natural to wonder how the null space and column space of a matrix are related. In fact, the two spaces are quite dissimilar, as Examples 6–8 will show. Nevertheless, a surprising connection between the null space and column space will emerge in Section 4.5, after more theory is available.

EXAMPLE 6 Let

$$A = \begin{bmatrix} 2 & 4 & -2 & 1 \\ -2 & -5 & 7 & 3 \\ 3 & 7 & -8 & 6 \end{bmatrix}$$

a. If the column space of A is a subspace of $\mathbb{R}^k$, what is k?

b. If the null space of A is a subspace of $\mathbb{R}^k$, what is k?

SOLUTION

a. The columns of A each have three entries, so Col A is a subspace of $\mathbb{R}^k$, where $k = 3$.

b. A vector $\mathbf{x}$ such that $A\mathbf{x}$ is defined must have four entries, so Nul A is a subspace of $\mathbb{R}^k$, where $k = 4$. ∎

When a matrix is not square, as in Example 6, the vectors in Nul A and Col A live in entirely different "universes." For example, no linear combination of vectors in $\mathbb{R}^3$ can produce a vector in $\mathbb{R}^4$. When A is square, Nul A and Col A do have the zero vector in common, and in special cases it is possible that some nonzero vectors belong to both Nul A and Col A.

EXAMPLE 7 With A as in Example 6, find a nonzero vector in Col A and a nonzero vector in Nul A.

SOLUTION It is easy to find a vector in Col A. Any column of A will do, say, $\begin{bmatrix} 2 \\ -2 \\ 3 \end{bmatrix}$.

To find a nonzero vector in Nul A, row reduce the augmented matrix $[\,A \quad \mathbf{0}\,]$ and obtain

$$[\,A \quad \mathbf{0}\,] \sim \begin{bmatrix} 1 & 0 & 9 & 0 & 0 \\ 0 & 1 & -5 & 0 & 0 \\ 0 & 0 & 0 & 1 & 0 \end{bmatrix}$$

Thus, if $\mathbf{x}$ satisfies $A\mathbf{x} = \mathbf{0}$, then $x_1 = -9x_3$, $x_2 = 5x_3$, $x_4 = 0$, and x_3 is free. Assigning a nonzero value to x_3—say, $x_3 = 1$—we obtain a vector in Nul A, namely, $\mathbf{x} = (-9, 5, 1, 0)$. ∎

EXAMPLE 8 With A as in Example 6, let $\mathbf{u} = \begin{bmatrix} 3 \\ -2 \\ -1 \\ 0 \end{bmatrix}$ and $\mathbf{v} = \begin{bmatrix} 3 \\ -1 \\ 3 \end{bmatrix}$.

a. Determine if $\mathbf{u}$ is in Nul A. Could $\mathbf{u}$ be in Col A?

b. Determine if $\mathbf{v}$ is in Col A. Could $\mathbf{v}$ be in Nul A?

SOLUTION

a. An explicit description of Nul A is not needed here. Simply compute the product $A\mathbf{u}$.

$$A\mathbf{u} = \begin{bmatrix} 2 & 4 & -2 & 1 \\ -2 & -5 & 7 & 3 \\ 3 & 7 & -8 & 6 \end{bmatrix} \begin{bmatrix} 3 \\ -2 \\ -1 \\ 0 \end{bmatrix} = \begin{bmatrix} 0 \\ -3 \\ 3 \end{bmatrix} \neq \begin{bmatrix} 0 \\ 0 \\ 0 \end{bmatrix}$$

Obviously, $\mathbf{u}$ is *not* a solution of $A\mathbf{x} = \mathbf{0}$, so $\mathbf{u}$ is not in Nul A. Also, with four entries, $\mathbf{u}$ could not possibly be in Col A, since Col A is a subspace of $\mathbb{R}^3$.

b. Reduce $[\,A \quad \mathbf{v}\,]$ to an echelon form.

$$[\,A \quad \mathbf{v}\,] = \begin{bmatrix} 2 & 4 & -2 & 1 & 3 \\ -2 & -5 & 7 & 3 & -1 \\ 3 & 7 & -8 & 6 & 3 \end{bmatrix} \sim \begin{bmatrix} 2 & 4 & -2 & 1 & 3 \\ 0 & 1 & -5 & -4 & -2 \\ 0 & 0 & 0 & 17 & 1 \end{bmatrix}$$

At this point, it is clear that the equation $A\mathbf{x} = \mathbf{v}$ is consistent, so $\mathbf{v}$ is in Col A. With only three entries, $\mathbf{v}$ could not possibly be in Nul A, since Nul A is a subspace of $\mathbb{R}^4$. ∎

The table on page 217 summarizes what we have learned about Nul A and Col A. Item 8 is a restatement of Theorems 11 and 12(a) in Section 1.9.

Kernel and Range of a Linear Transformation

Subspaces of vector spaces other than $\mathbb{R}^n$ are often described in terms of a linear transformation instead of a matrix. To make this precise, we generalize the definition given in Section 1.8.

Contrast Between Nul *A* and Col *A* for an *m* × *n* Matrix *A*

Nul *A*	Col *A*
1. Nul *A* is a subspace of $\mathbb{R}^n$.	1. Col *A* is a subspace of $\mathbb{R}^m$.
2. Nul *A* is implicitly defined; that is, you are given only a condition ($A\mathbf{x} = \mathbf{0}$) that vectors in Nul *A* must satisfy.	2. Col *A* is explicitly defined; that is, you are told how to build vectors in Col *A*.
3. It takes time to find vectors in Nul *A*. Row operations on [*A* **0**] are required.	3. It is easy to find vectors in Col *A*. The columns of *A* are displayed; others are formed from them.
4. There is no obvious relation between Nul *A* and the entries in *A*.	4. There is an obvious relation between Col *A* and the entries in *A*, since each column of *A* is in Col *A*.
5. A typical vector **v** in Nul *A* has the property that $A\mathbf{v} = \mathbf{0}$.	5. A typical vector **v** in Col *A* has the property that the equation $A\mathbf{x} = \mathbf{v}$ is consistent.
6. Given a specific vector **v**, it is easy to tell if **v** is in Nul *A*. Just compute $A\mathbf{v}$.	6. Given a specific vector **v**, it may take time to tell if **v** is in Col *A*. Row operations on [*A* **v**] are required.
7. Nul $A = \{\mathbf{0}\}$ if and only if the equation $A\mathbf{x} = \mathbf{0}$ has only the trivial solution.	7. Col $A = \mathbb{R}^m$ if and only if the equation $A\mathbf{x} = \mathbf{b}$ has a solution for every **b** in $\mathbb{R}^m$.
8. Nul $A = \{\mathbf{0}\}$ if and only if the linear transformation $\mathbf{x} \mapsto A\mathbf{x}$ is one-to-one.	8. Col $A = \mathbb{R}^m$ if and only if the linear transformation $\mathbf{x} \mapsto A\mathbf{x}$ maps $\mathbb{R}^n$ *onto* $\mathbb{R}^m$.

DEFINITION

A **linear transformation** T from a vector space V into a vector space W is a rule that assigns to each vector **x** in V a unique vector $T(\mathbf{x})$ in W, such that

(i) $T(\mathbf{u} + \mathbf{v}) = T(\mathbf{u}) + T(\mathbf{v})$ for all **u**, **v** in V, and

(ii) $T(c\mathbf{u}) = cT(\mathbf{u})$ for all **u** in V and all scalars c.

The **kernel** (or **null space**) of such a T is the set of all **u** in V such that $T(\mathbf{u}) = \mathbf{0}$ (the zero vector in W). The **range** of T is the set of all vectors in W of the form $T(\mathbf{x})$ for some **x** in V. If T happens to arise as a matrix transformation—say, $T(\mathbf{x}) = A\mathbf{x}$ for some matrix A—then the kernel and the range of T are just the null space and the column space of A, as defined earlier.

It is not difficult to show that the kernel of T is a subspace of V. The proof is essentially the same as the one for Theorem 2. Also, the range of T is a subspace of W. See Figure 2 and Exercise 42.

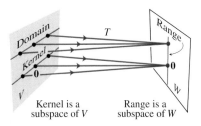

FIGURE 2 Subspaces associated with a linear transformation.

In applications, a subspace usually arises as either the kernel or the range of an appropriate linear transformation. For instance, the set of all solutions of a homogeneous linear differential equation turns out to be the kernel of a linear transformation. Typically, such a linear transformation is described in terms of one or more derivatives of a function. To explain this in any detail would take us too far afield at this point. So we consider only two examples. The first explains why the operation of differentiation is a linear transformation.

EXAMPLE 9 (Calculus required) Let V be the vector space of all real-valued functions f defined on an interval $[a, b]$ with the property that they are differentiable and their derivatives are continuous functions on $[a, b]$. Let W be the vector space $C[a, b]$ of all continuous functions on $[a, b]$, and let $D : V \to W$ be the transformation that changes f in V into its derivative f'. In calculus, two simple differentiation rules are

$$D(f + g) = D(f) + D(g) \quad \text{and} \quad D(cf) = cD(f)$$

That is, D is a linear transformation. It can be shown that the kernel of D is the set of constant functions on $[a, b]$ and the range of D is the set W of all continuous functions on $[a, b]$. ∎

EXAMPLE 10 (Calculus required) The differential equation

$$y'' + \omega^2 y = 0 \tag{4}$$

where ω is a constant, is used to describe a variety of physical systems, such as the vibration of a weighted spring, the movement of a pendulum, and the voltage in an inductance-capacitance electrical circuit. The set of solutions of (4) is precisely the kernel of the linear transformation that maps a function $y = f(t)$ into the function $f''(t) + \omega^2 f(t)$. Finding an explicit description of this vector space is a problem in differential equations. The solution set turns out to be the space described in Exercise 19 in Section 4.1. ∎

A common technique used in the stock market is technical analysis. Statistical trends gathered from stock-trading activity, such as price movement and volume, are analyzed. Technical analysts focus on patterns of stock-price movements, trading signals, and various other analytical charting tools to evaluate a security's strength or weakness. A moving average is a commonly used indicator in technical analysis. It smooths out price action by filtering out the effects from random price fluctuations. In the final example for this section, we examine the linear transformation that creates the two-day moving average from a "signal" of daily prices. We will look at moving average transformations that average over a longer period of time in Section 4.7.

EXAMPLE 11 Let $\{p_k\}$ in $\mathbb{S}$ represent the price of a stock that has been recorded daily over an extended period of time. Note that we can assume that $p_k = 0$ for k outside the time period under study. To create a two-day moving average, the mapping $M_2 : \mathbb{S} \to \mathbb{S}$ defined by $M_2(\{p_k\}) = \left\{ \dfrac{p_k + p_{k-1}}{2} \right\}$ is applied to the data. Show that M_2 is a linear transformation and find its kernel.

SOLUTION To see that M_2 is a linear transformation, observe that for two signals $\{p_k\}$ and $\{q_k\}$ in $\mathbb{S}$ and any scalar c,

$$M_2(\{p_k\} + \{q_k\}) = M_2(\{p_k + q_k\}) = \left\{ \frac{p_k + q_k + p_{k-1} + q_{k-1}}{2} \right\}$$

$$= \left\{ \frac{p_k + p_{k-1}}{2} \right\} + \left\{ \frac{q_k + q_{k-1}}{2} \right\}$$

$$= M_2(\{p_k\}) + M_2(\{q_k\})$$

and

$$M_2(c\{p_k\}) = M_2(\{cp_k\}) = \left\{ \frac{cp_k + cp_{k-1}}{2} \right\} = c \left\{ \frac{p_k + p_{k-1}}{2} \right\} = cM_2(\{p_k\})$$

thus M_2 is a linear transformation.

To find the kernel of M_2, notice that $\{p_k\}$ is in the kernel if and only if $\frac{p_k + p_{k-1}}{2} = 0$ for all k, and hence $p_k = -p_{k-1}$. Since this relationship is true for all integers k, it can be applied recursively resulting in $p_k = -p_{k-1} = (-1)^2 p_{k-2} = (-1)^3 p_{k-3} \ldots$. Working out from $k = 0$, any signal in the kernel can be written as $p_k = p_0(-1)^k$, a multiple of the alternating signal described by $\{(-1)^k\}$. Since the kernel of the two-day moving average function consists of all multiples of the alternating sequence, it smooths out daily fluctuations, without leveling out overall trends. (See Figure 3.)

- Original signal
- Moving average

FIGURE 3

Practice Problems

1. Let $W = \left\{ \begin{bmatrix} a \\ b \\ c \end{bmatrix} : a - 3b - c = 0 \right\}$. Show in two different ways that W is a subspace of $\mathbb{R}^3$. (Use two theorems.)

2. Let $A = \begin{bmatrix} 7 & -3 & 5 \\ -4 & 1 & -5 \\ -5 & 2 & -4 \end{bmatrix}$, $\mathbf{v} = \begin{bmatrix} 2 \\ 1 \\ -1 \end{bmatrix}$, and $\mathbf{w} = \begin{bmatrix} 7 \\ 6 \\ -3 \end{bmatrix}$. Suppose you know that the equations $A\mathbf{x} = \mathbf{v}$ and $A\mathbf{x} = \mathbf{w}$ are both consistent. What can you say about the equation $A\mathbf{x} = \mathbf{v} + \mathbf{w}$?

3. Let A be an $n \times n$ matrix. If Col $A = $ Nul A, show that Nul $A^2 = \mathbb{R}^n$.

4.2 Exercises

1. Determine if $\mathbf{w} = \begin{bmatrix} 1 \\ 3 \\ -4 \end{bmatrix}$ is in Nul A, where

$$A = \begin{bmatrix} 3 & -5 & -3 \\ 6 & -2 & 0 \\ -8 & 4 & 1 \end{bmatrix}.$$

2. Determine if $\mathbf{w} = \begin{bmatrix} 5 \\ -3 \\ 2 \end{bmatrix}$ is in Nul A, where

$$A = \begin{bmatrix} 5 & 21 & 19 \\ 13 & 23 & 2 \\ 8 & 14 & 1 \end{bmatrix}.$$

In Exercises 3–6, find an explicit description of Nul A by listing vectors that span the null space.

3. $A = \begin{bmatrix} 1 & 3 & 5 & 0 \\ 0 & 1 & 4 & -2 \end{bmatrix}$

4. $A = \begin{bmatrix} 1 & -6 & 4 & 0 \\ 0 & 0 & 2 & 0 \end{bmatrix}$

5. $A = \begin{bmatrix} 1 & -2 & 0 & 4 & 0 \\ 0 & 0 & 1 & -9 & 0 \\ 0 & 0 & 0 & 0 & 1 \end{bmatrix}$

6. $A = \begin{bmatrix} 1 & 5 & -4 & -3 & 1 \\ 0 & 1 & -2 & 1 & 0 \\ 0 & 0 & 0 & 0 & 0 \end{bmatrix}$

In Exercises 7–14, either use an appropriate theorem to show that the given set, W, is a vector space, or find a specific example to the contrary.

7. $\left\{ \begin{bmatrix} a \\ b \\ c \end{bmatrix} : a + b + c = 2 \right\}$ **8.** $\left\{ \begin{bmatrix} r \\ s \\ t \end{bmatrix} : 5r - 1 = s + 2t \right\}$

9. $\left\{ \begin{bmatrix} a \\ b \\ c \\ d \end{bmatrix} : \begin{array}{l} a - 2b = 4c \\ 2a = c + 3d \end{array} \right\}$ **10.** $\left\{ \begin{bmatrix} a \\ b \\ c \\ d \end{bmatrix} : \begin{array}{l} a + 3b = c \\ b + c + a = d \end{array} \right\}$

11. $\left\{ \begin{bmatrix} b - 2d \\ 5 + d \\ b + 3d \\ d \end{bmatrix} : b, d \text{ real} \right\}$ **12.** $\left\{ \begin{bmatrix} b - 5d \\ 2b \\ 2d + 1 \\ d \end{bmatrix} : b, d \text{ real} \right\}$

13. $\left\{ \begin{bmatrix} c - 6d \\ d \\ c \end{bmatrix} : c, d \text{ real} \right\}$ **14.** $\left\{ \begin{bmatrix} -a + 2b \\ a - 2b \\ 3a - 6b \end{bmatrix} : a, b \text{ real} \right\}$

In Exercises 15 and 16, find A such that the given set is Col A.

15. $\left\{ \begin{bmatrix} 2s + 3t \\ r + s - 2t \\ 4r + s \\ 3r - s - t \end{bmatrix} : r, s, t \text{ real} \right\}$

16. $\left\{ \begin{bmatrix} b - c \\ 2b + c + d \\ 5c - 4d \\ d \end{bmatrix} : b, c, d \text{ real} \right\}$

For the matrices in Exercises 17–20, (a) find k such that Nul A is a subspace of $\mathbb{R}^k$, and (b) find k such that Col A is a subspace of $\mathbb{R}^k$.

17. $A = \begin{bmatrix} 2 & -6 \\ -1 & 3 \\ -4 & 12 \\ 3 & -9 \end{bmatrix}$ **18.** $A = \begin{bmatrix} 7 & -2 & 0 \\ -2 & 0 & -5 \\ 0 & -5 & 7 \\ -5 & 7 & -2 \end{bmatrix}$

19. $A = \begin{bmatrix} 4 & 5 & -2 & 6 & 0 \\ 1 & 1 & 0 & 1 & 0 \end{bmatrix}$

20. $A = \begin{bmatrix} 1 & -3 & 9 & 0 & -5 \end{bmatrix}$

21. With A as in Exercise 17, find a nonzero vector in Nul A, a nonzero vector in Col A, and a nonzero vector in Row A.

22. With A as in Exercise 3, find a nonzero vector in Nul A, a nonzero vector in Col A, and a nonzero vector in Row A.

23. Let $A = \begin{bmatrix} -6 & 12 \\ -3 & 6 \end{bmatrix}$ and $\mathbf{w} = \begin{bmatrix} 2 \\ 1 \end{bmatrix}$. Determine if $\mathbf{w}$ is in Col A. Is $\mathbf{w}$ in Nul A?

24. Let $A = \begin{bmatrix} -8 & -2 & -9 \\ 6 & 4 & 8 \\ 4 & 0 & 4 \end{bmatrix}$ and $\mathbf{w} = \begin{bmatrix} 2 \\ 1 \\ -2 \end{bmatrix}$. Determine if $\mathbf{w}$ is in Col A. Is $\mathbf{w}$ in Nul A?

In Exercises 25–38, A denotes an $m \times n$ matrix. Mark each statement True or False (**T/F**). Justify each answer.

25. (**T/F**) The null space of A is the solution set of the equation $A\mathbf{x} = \mathbf{0}$.

26. (**T/F**) A null space is a vector space.

27. (**T/F**) The null space of an $m \times n$ matrix is in $\mathbb{R}^m$.

28. (**T/F**) The column space of an $m \times n$ matrix is in $\mathbb{R}^m$.

29. (**T/F**) The column space of A is the range of the mapping $\mathbf{x} \mapsto A\mathbf{x}$.

30. (**T/F**) Col A is the set of all solutions of $A\mathbf{x} = \mathbf{b}$.

31. (**T/F**) If the equation $A\mathbf{x} = \mathbf{b}$ is consistent, then Col $A = \mathbb{R}^m$.

32. (**T/F**) Nul A is the kernel of the mapping $\mathbf{x} \mapsto A\mathbf{x}$.

33. (**T/F**) The kernel of a linear transformation is a vector space.

34. (**T/F**) The range of a linear transformation is a vector space.

35. (**T/F**) Col A is the set of all vectors that can be written as $A\mathbf{x}$ for some $\mathbf{x}$.

36. (**T/F**) The set of all solutions of a homogeneous linear differential equation is the kernel of a linear transformation.

37. (**T/F**) The row space of A is the same as the column space of A^T.

38. (**T/F**) The null space of A is the same as the row space of A^T.

39. It can be shown that a solution of the system below is $x_1 = 3$, $x_2 = 2$, and $x_3 = -1$. Use this fact and the theory from this section to explain why another solution is $x_1 = 30$, $x_2 = 20$, and $x_3 = -10$. (Observe how the solutions are related, but make no other calculations.)

$$\begin{array}{r} x_1 - 3x_2 - 3x_3 = 0 \\ -2x_1 + 4x_2 + 2x_3 = 0 \\ -x_1 + 5x_2 + 7x_3 = 0 \end{array}$$

40. Consider the following two systems of equations:

$$5x_1 + x_2 - 3x_3 = 0 \qquad 5x_1 + x_2 - 3x_3 = 0$$
$$-9x_1 + 2x_2 + 5x_3 = 1 \qquad -9x_1 + 2x_2 + 5x_3 = 5$$
$$4x_1 + x_2 - 6x_3 = 9 \qquad 4x_1 + x_2 - 6x_3 = 45$$

It can be shown that the first system has a solution. Use this fact and the theory from this section to explain why the second system must also have a solution. (Make no row operations.)

41. Prove Theorem 3 as follows: Given an $m \times n$ matrix A, an element in Col A has the form $A\mathbf{x}$ for some $\mathbf{x}$ in $\mathbb{R}^n$. Let $A\mathbf{x}$ and $A\mathbf{w}$ represent any two vectors in Col A.

a. Explain why the zero vector is in Col A.

b. Show that the vector $A\mathbf{x} + A\mathbf{w}$ is in Col A.

c. Given a scalar c, show that $c(A\mathbf{x})$ is in Col A.

42. Let $T : V \to W$ be a linear transformation from a vector space V into a vector space W. Prove that the range of T is a subspace of W. [*Hint:* Typical elements of the range have the form $T(\mathbf{x})$ and $T(\mathbf{w})$ for some $\mathbf{x}, \mathbf{w}$ in V.]

43. Define $T : \mathbb{P}_2 \to \mathbb{R}^2$ by $T(\mathbf{p}) = \begin{bmatrix} \mathbf{p}(0) \\ \mathbf{p}(1) \end{bmatrix}$. For instance, if $\mathbf{p}(t) = 3 + 5t + 7t^2$, then $T(\mathbf{p}) = \begin{bmatrix} 3 \\ 15 \end{bmatrix}$.

a. Show that T is a linear transformation. [*Hint:* For arbitrary polynomials $\mathbf{p}, \mathbf{q}$ in $\mathbb{P}_2$, compute $T(\mathbf{p} + \mathbf{q})$ and $T(c\mathbf{p})$.]

b. Find a polynomial $\mathbf{p}$ in $\mathbb{P}_2$ that spans the kernel of T, and describe the range of T.

44. Define a linear transformation $T : \mathbb{P}_2 \to \mathbb{R}^2$ by $T(\mathbf{p}) = \begin{bmatrix} \mathbf{p}(0) \\ \mathbf{p}(0) \end{bmatrix}$. Find polynomials $\mathbf{p}_1$ and $\mathbf{p}_2$ in $\mathbb{P}_2$ that span the kernel of T, and describe the range of T.

45. Let $M_{2\times2}$ be the vector space of all 2×2 matrices, and define $T : M_{2\times2} \to M_{2\times2}$ by $T(A) = A + A^T$, where $A = \begin{bmatrix} a & b \\ c & d \end{bmatrix}$.

a. Show that T is a linear transformation.

b. Let B be any element of $M_{2\times2}$ such that $B^T = B$. Find an A in $M_{2\times2}$ such that $T(A) = B$.

c. Show that the range of T is the set of B in $M_{2\times2}$ with the property that $B^T = B$.

d. Describe the kernel of T.

46. (*Calculus required*) Define $T : C[0, 1] \to C[0, 1]$ as follows: For $\mathbf{f}$ in $C[0, 1]$, let $T(\mathbf{f})$ be the antiderivative $\mathbf{F}$ of $\mathbf{f}$ such that $\mathbf{F}(0) = 0$. Show that T is a linear transformation, and describe the kernel of T. (See the notation in Exercise 20 of Section 4.1.)

47. Let V and W be vector spaces, and let $T : V \to W$ be a linear transformation. Given a subspace U of V, let $T(U)$ denote the set of all images of the form $T(\mathbf{x})$, where $\mathbf{x}$ is in U. Show that $T(U)$ is a subspace of W.

48. Given $T : V \to W$ as in Exercise 47, and given a subspace Z of W, let U be the set of all $\mathbf{x}$ in V such that $T(\mathbf{x})$ is in Z. Show that U is a subspace of V.

T 49. Determine whether $\mathbf{w}$ is in the column space of A, the null space of A, or both, where

$$\mathbf{w} = \begin{bmatrix} 1 \\ 1 \\ -1 \\ -3 \end{bmatrix}, \quad A = \begin{bmatrix} 7 & 6 & -4 & 1 \\ -5 & -1 & 0 & -2 \\ 9 & -11 & 7 & -3 \\ 19 & -9 & 7 & 1 \end{bmatrix}$$

T 50. Determine whether $\mathbf{w}$ is in the column space of A, the null space of A, or both, where

$$\mathbf{w} = \begin{bmatrix} 1 \\ 2 \\ 1 \\ 0 \end{bmatrix}, \quad A = \begin{bmatrix} -8 & 5 & -2 & 0 \\ -5 & 2 & 1 & -2 \\ 10 & -8 & 6 & -3 \\ 3 & -2 & 1 & 0 \end{bmatrix}$$

T 51. Let $\mathbf{a}_1, \ldots, \mathbf{a}_5$ denote the columns of the matrix A, where

$$A = \begin{bmatrix} 5 & 1 & 2 & 2 & 0 \\ 3 & 3 & 2 & -1 & -12 \\ 8 & 4 & 4 & -5 & 12 \\ 2 & 1 & 1 & 0 & -2 \end{bmatrix}, \quad B = [\, \mathbf{a}_1 \; \mathbf{a}_2 \; \mathbf{a}_4 \,]$$

a. Explain why $\mathbf{a}_3$ and $\mathbf{a}_5$ are in the column space of B.

b. Find a set of vectors that spans Nul A.

c. Let $T : \mathbb{R}^5 \to \mathbb{R}^4$ be defined by $T(\mathbf{x}) = A\mathbf{x}$. Explain why T is neither one-to-one nor onto.

T 52. Let $H = \text{Span}\{\mathbf{v}_1, \mathbf{v}_2\}$ and $K = \text{Span}\{\mathbf{v}_3, \mathbf{v}_4\}$, where

$$\mathbf{v}_1 = \begin{bmatrix} 5 \\ 3 \\ 8 \end{bmatrix}, \mathbf{v}_2 = \begin{bmatrix} 1 \\ 3 \\ 4 \end{bmatrix}, \mathbf{v}_3 = \begin{bmatrix} 2 \\ -1 \\ 5 \end{bmatrix}, \mathbf{v}_4 = \begin{bmatrix} 0 \\ -12 \\ -28 \end{bmatrix}.$$

Then H and K are subspaces of $\mathbb{R}^3$. In fact, H and K are planes in $\mathbb{R}^3$ through the origin, and they intersect in a line through $\mathbf{0}$. Find a nonzero vector $\mathbf{w}$ that generates that line. [*Hint:* $\mathbf{w}$ can be written as $c_1\mathbf{v}_1 + c_2\mathbf{v}_2$ and also as $c_3\mathbf{v}_3 + c_4\mathbf{v}_4$. To build $\mathbf{w}$, solve the equation $c_1\mathbf{v}_1 + c_2\mathbf{v}_2 = c_3\mathbf{v}_3 + c_4\mathbf{v}_4$ for the unknown c_j's.]

STUDY GUIDE offers additional resources for mastering vector spaces, subspaces, and column row, and null spaces.

Solutions to Practice Problems

1. *First method:* W is a subspace of $\mathbb{R}^3$ by Theorem 2 because W is the set of all solutions to a system of homogeneous linear equations (where the system has only one equation). Equivalently, W is the null space of the 1×3 matrix $A = [\,1 \quad -3 \quad -1\,]$.

   Second method: Solve the equation $a - 3b - c = 0$ for the leading variable a in terms of the free variables b and c. Any solution has the form $\begin{bmatrix} 3b + c \\ b \\ c \end{bmatrix}$, where b and c are arbitrary, and

   $$\begin{bmatrix} 3b + c \\ b \\ c \end{bmatrix} = b \underset{\underset{\mathbf{v}_1}{\uparrow}}{\begin{bmatrix} 3 \\ 1 \\ 0 \end{bmatrix}} + c \underset{\underset{\mathbf{v}_2}{\uparrow}}{\begin{bmatrix} 1 \\ 0 \\ 1 \end{bmatrix}}$$

   This calculation shows that $W = \text{Span}\,\{\mathbf{v}_1, \mathbf{v}_2\}$. Thus W is a subspace of $\mathbb{R}^3$ by Theorem 1. We could also solve the equation $a - 3b - c = 0$ for b or c and get alternative descriptions of W as a set of linear combinations of two vectors.

2. Both $\mathbf{v}$ and $\mathbf{w}$ are in Col A. Since Col A is a vector space, $\mathbf{v} + \mathbf{w}$ must be in Col A. That is, the equation $A\mathbf{x} = \mathbf{v} + \mathbf{w}$ is consistent.

3. Let $\mathbf{x}$ be any vector in $\mathbb{R}^n$. Notice $A\mathbf{x}$ is in Col A, since it is a linear combination of the columns of A. Since Col $A = $ Nul A, the vector $A\mathbf{x}$ is also in Nul A. Hence $A^2\mathbf{x} = A(A\mathbf{x}) = \mathbf{0}$ establishing that every vector $\mathbf{x}$ from $\mathbb{R}^n$ is in Nul A^2.

4.3 Linearly Independent Sets; Bases

In this section we identify and study the subsets that span a vector space V or a subspace H as "efficiently" as possible. The key idea is that of linear independence, defined as in $\mathbb{R}^n$.

An indexed set of vectors $\{\mathbf{v}_1, \ldots, \mathbf{v}_p\}$ in V is said to be **linearly independent** if the vector equation

$$c_1\mathbf{v}_1 + c_2\mathbf{v}_2 + \cdots + c_p\mathbf{v}_p = \mathbf{0} \tag{1}$$

has *only* the trivial solution, $c_1 = 0, \ldots, c_p = 0.$[1]

The set $\{\mathbf{v}_1, \ldots, \mathbf{v}_p\}$ is said to be **linearly dependent** if (1) has a nontrivial solution, that is, if there are some weights, $c_1, \ldots, c_p$, *not all zero*, such that (1) holds. In such a case, (1) is called a **linear dependence relation** among $\mathbf{v}_1, \ldots, \mathbf{v}_p$.

Just as in $\mathbb{R}^n$, a set containing a single vector $\mathbf{v}$ is linearly independent if and only if $\mathbf{v} \neq \mathbf{0}$. Also, a set of two vectors is linearly dependent if and only if one of the vectors is a multiple of the other. And any set containing the zero vector is linearly dependent. The following theorem has the same proof as Theorem 7 in Section 1.7.

---

[1] It is convenient to use $c_1, \ldots, c_p$ in (1) for the scalars instead of $x_1, \ldots, x_p$, as we did previously.

THEOREM 4   An indexed set $\{\mathbf{v}_1, \ldots, \mathbf{v}_p\}$ of two or more vectors, with $\mathbf{v}_1 \neq \mathbf{0}$, is linearly dependent if and only if some $\mathbf{v}_j$ (with $j > 1$) is a linear combination of the preceding vectors, $\mathbf{v}_1, \ldots, \mathbf{v}_{j-1}$.

The main difference between linear dependence in $\mathbb{R}^n$ and in a general vector space is that when the vectors are not n-tuples, the homogeneous equation (1) usually cannot be written as a system of n linear equations. That is, the vectors cannot be made into the columns of a matrix A in order to study the equation $A\mathbf{x} = \mathbf{0}$. We must rely instead on the definition of linear dependence and on Theorem 4.

EXAMPLE 1   Let $\mathbf{p}_1(t) = 1$, $\mathbf{p}_2(t) = t$, and $\mathbf{p}_3(t) = 4 - t$. Then $\{\mathbf{p}_1, \mathbf{p}_2, \mathbf{p}_3\}$ is linearly dependent in $\mathbb{P}$ because $\mathbf{p}_3 = 4\mathbf{p}_1 - \mathbf{p}_2$.   ∎

EXAMPLE 2   The set $\{\sin t, \cos t\}$ is linearly independent in $C[0, 1]$, the space of all continuous functions on $0 \leq t \leq 1$, because $\sin t$ and $\cos t$ are not multiples of one another *as vectors in* $C[0, 1]$. That is, there is no scalar c such that $\cos t = c \cdot \sin t$ for all t in $[0, 1]$. (Look at the graphs of $\sin t$ and $\cos t$.) However, $\{\sin t \cos t, \sin 2t\}$ is linearly dependent because of the identity $\sin 2t = 2 \sin t \cos t$, for all t.   ∎

DEFINITION   Let H be a subspace of a vector space V. A set of vectors $\mathcal{B}$ in V is a **basis** for H if

(i) $\mathcal{B}$ is a linearly independent set, and
(ii) the subspace spanned by $\mathcal{B}$ coincides with H; that is,

$$H = \text{Span } \mathcal{B}$$

The definition of a basis applies to the case when $H = V$, because any vector space is a subspace of itself. Thus a basis of V is a linearly independent set that spans V. Observe that when $H \neq V$, condition (ii) includes the requirement that each of the vectors $\mathbf{b}$ in $\mathcal{B}$ must belong to H, because Span $\mathcal{B}$ contains every element in $\mathcal{B}$, as shown in Section 4.1.

EXAMPLE 3   Let A be an invertible $n \times n$ matrix—say, $A = [\,\mathbf{a}_1 \ \cdots \ \mathbf{a}_n\,]$. Then the columns of A form a basis for $\mathbb{R}^n$ because they are linearly independent and they span $\mathbb{R}^n$, by the Invertible Matrix Theorem.   ∎

EXAMPLE 4   Let $\mathbf{e}_1, \ldots, \mathbf{e}_n$ be the columns of the $n \times n$ identity matrix, I_n. That is,

$$\mathbf{e}_1 = \begin{bmatrix} 1 \\ 0 \\ \vdots \\ 0 \end{bmatrix}, \quad \mathbf{e}_2 = \begin{bmatrix} 0 \\ 1 \\ \vdots \\ 0 \end{bmatrix}, \quad \ldots, \quad \mathbf{e}_n = \begin{bmatrix} 0 \\ \vdots \\ 0 \\ 1 \end{bmatrix}$$

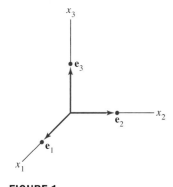

FIGURE 1
The standard basis for $\mathbb{R}^3$.

The set $\{\mathbf{e}_1, \ldots, \mathbf{e}_n\}$ is called the **standard basis** for $\mathbb{R}^n$ (Figure 1).   ∎

EXAMPLE 5 Let $\mathbf{v}_1 = \begin{bmatrix} 3 \\ 0 \\ -6 \end{bmatrix}$, $\mathbf{v}_2 = \begin{bmatrix} -4 \\ 1 \\ 7 \end{bmatrix}$, and $\mathbf{v}_3 = \begin{bmatrix} -2 \\ 1 \\ 5 \end{bmatrix}$. Determine if $\{\mathbf{v}_1, \mathbf{v}_2, \mathbf{v}_3\}$ is a basis for $\mathbb{R}^3$.

SOLUTION Since there are exactly three vectors here in $\mathbb{R}^3$, we can use any of several methods to determine if the matrix $A = [\,\mathbf{v}_1 \quad \mathbf{v}_2 \quad \mathbf{v}_3\,]$ is invertible. For instance, two row replacements reveal that A has three pivot positions. Thus A is invertible. As in Example 3, the columns of A form a basis for $\mathbb{R}^3$. ■

EXAMPLE 6 Let $S = \{1, t, t^2, \ldots, t^n\}$. Verify that S is a basis for $\mathbb{P}_n$. This basis is called the **standard basis** for $\mathbb{P}_n$.

SOLUTION Certainly S spans $\mathbb{P}_n$. To show that S is linearly independent, suppose that $c_0, \ldots, c_n$ satisfy

$$c_0 1 + c_1 t + c_2 t^2 + \cdots + c_n t^n = \mathbf{0}(t) \tag{2}$$

This equality means that the polynomial on the left has the same values as the zero polynomial on the right. A fundamental theorem in algebra says that the only polynomial in $\mathbb{P}_n$ with more than n zeros is the zero polynomial. That is, equation (2) holds for all t only if $c_0 = \cdots = c_n = 0$. This proves that S is linearly independent and hence is a basis for $\mathbb{P}_n$. See Figure 2. ■

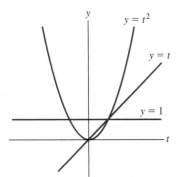

FIGURE 2
The standard basis for $\mathbb{P}_2$.

Problems involving linear independence and spanning in $\mathbb{P}_n$ are handled best by a technique to be discussed in Section 4.4.

The Spanning Set Theorem

As we will see, a basis is an "efficient" spanning set that contains no unnecessary vectors. In fact, a basis can be constructed from a spanning set by discarding unneeded vectors.

EXAMPLE 7 Let

$$\mathbf{v}_1 = \begin{bmatrix} 0 \\ 2 \\ -1 \end{bmatrix}, \quad \mathbf{v}_2 = \begin{bmatrix} 2 \\ 2 \\ 0 \end{bmatrix}, \quad \mathbf{v}_3 = \begin{bmatrix} 6 \\ 16 \\ -5 \end{bmatrix}, \quad \text{and} \quad H = \operatorname{Span}\{\mathbf{v}_1, \mathbf{v}_2, \mathbf{v}_3\}.$$

Note that $\mathbf{v}_3 = 5\mathbf{v}_1 + 3\mathbf{v}_2$, and show that $\operatorname{Span}\{\mathbf{v}_1, \mathbf{v}_2, \mathbf{v}_3\} = \operatorname{Span}\{\mathbf{v}_1, \mathbf{v}_2\}$. Then find a basis for the subspace H.

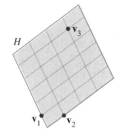

SOLUTION Every vector in $\operatorname{Span}\{\mathbf{v}_1, \mathbf{v}_2\}$ belongs to H because

$$c_1 \mathbf{v}_1 + c_2 \mathbf{v}_2 = c_1 \mathbf{v}_1 + c_2 \mathbf{v}_2 + 0\mathbf{v}_3$$

Now let $\mathbf{x}$ be any vector in H—say, $\mathbf{x} = c_1 \mathbf{v}_1 + c_2 \mathbf{v}_2 + c_3 \mathbf{v}_3$. Since $\mathbf{v}_3 = 5\mathbf{v}_1 + 3\mathbf{v}_2$, we may substitute

$$\mathbf{x} = c_1 \mathbf{v}_1 + c_2 \mathbf{v}_2 + c_3 (5\mathbf{v}_1 + 3\mathbf{v}_2)$$
$$= (c_1 + 5c_3)\mathbf{v}_1 + (c_2 + 3c_3)\mathbf{v}_2$$

Thus $\mathbf{x}$ is in $\operatorname{Span}\{\mathbf{v}_1, \mathbf{v}_2\}$, so every vector in H already belongs to $\operatorname{Span}\{\mathbf{v}_1, \mathbf{v}_2\}$. We conclude that H and $\operatorname{Span}\{\mathbf{v}_1, \mathbf{v}_2\}$ are actually the same set of vectors. It follows that $\{\mathbf{v}_1, \mathbf{v}_2\}$ is a basis of H since $\{\mathbf{v}_1, \mathbf{v}_2\}$ is obviously linearly independent. ■

The next theorem generalizes Example 7.

THEOREM 5 | **The Spanning Set Theorem**

Let $S = \{\mathbf{v}_1, \ldots, \mathbf{v}_p\}$ be a set in a vector space V, and let $H = \text{Span}\{\mathbf{v}_1, \ldots, \mathbf{v}_p\}$.

a. If one of the vectors in S—say, $\mathbf{v}_k$—is a linear combination of the remaining vectors in S, then the set formed from S by removing $\mathbf{v}_k$ still spans H.

b. If $H \neq \{\mathbf{0}\}$, some subset of S is a basis for H.

PROOF

a. By rearranging the list of vectors in S, if necessary, we may suppose that $\mathbf{v}_p$ is a linear combination of $\mathbf{v}_1, \ldots, \mathbf{v}_{p-1}$—say,

$$\mathbf{v}_p = a_1\mathbf{v}_1 + \cdots + a_{p-1}\mathbf{v}_{p-1} \tag{3}$$

Given any $\mathbf{x}$ in H, we may write

$$\mathbf{x} = c_1\mathbf{v}_1 + \cdots + c_{p-1}\mathbf{v}_{p-1} + c_p\mathbf{v}_p \tag{4}$$

for suitable scalars $c_1, \ldots, c_p$. Substituting the expression for $\mathbf{v}_p$ from (3) into (4), it is easy to see that $\mathbf{x}$ is a linear combination of $\mathbf{v}_1, \ldots, \mathbf{v}_{p-1}$. Thus $\{\mathbf{v}_1, \ldots, \mathbf{v}_{p-1}\}$ spans H, because $\mathbf{x}$ was an arbitrary element of H.

b. If the original spanning set S is linearly independent, then it is already a basis for H. Otherwise, one of the vectors in S depends on the others and can be deleted, by part (a). So long as there are two or more vectors in the spanning set, we can repeat this process until the spanning set is linearly independent and hence is a basis for H. If the spanning set is eventually reduced to one vector, that vector will be nonzero (and hence linearly independent) because $H \neq \{\mathbf{0}\}$. ∎

Bases for Nul A, Col A, and Row A

We already know how to find vectors that span the null space of a matrix A. The discussion in Section 4.2 pointed out that our method always produces a linearly independent set when Nul A contains nonzero vectors. So, in this case, that method produces a *basis* for Nul A.

The next two examples describe a simple algorithm for finding a basis for the column space.

EXAMPLE 8   Find a basis for Col B, where

$$B = \begin{bmatrix} \mathbf{b}_1 & \mathbf{b}_2 & \cdots & \mathbf{b}_5 \end{bmatrix} = \begin{bmatrix} 1 & 4 & 0 & 2 & 0 \\ 0 & 0 & 1 & -1 & 0 \\ 0 & 0 & 0 & 0 & 1 \\ 0 & 0 & 0 & 0 & 0 \end{bmatrix}$$

SOLUTION   Each nonpivot column of B is a linear combination of the pivot columns. In fact, $\mathbf{b}_2 = 4\mathbf{b}_1$ and $\mathbf{b}_4 = 2\mathbf{b}_1 - \mathbf{b}_3$. By the Spanning Set Theorem, we may discard $\mathbf{b}_2$ and $\mathbf{b}_4$, and $\{\mathbf{b}_1, \mathbf{b}_3, \mathbf{b}_5\}$ will still span Col B. Let

$$S = \{\mathbf{b}_1, \mathbf{b}_3, \mathbf{b}_5\} = \left\{ \begin{bmatrix} 1 \\ 0 \\ 0 \\ 0 \end{bmatrix}, \begin{bmatrix} 0 \\ 1 \\ 0 \\ 0 \end{bmatrix}, \begin{bmatrix} 0 \\ 0 \\ 1 \\ 0 \end{bmatrix} \right\}$$

Since $\mathbf{b}_1 \neq 0$ and no vector in S is a linear combination of the vectors that precede it, S is linearly independent (Theorem 4). Thus S is a basis for Col B. ∎

What about a matrix A that is *not* in reduced echelon form? Recall that any linear dependence relationship among the columns of A can be expressed in the form $A\mathbf{x} = \mathbf{0}$, where $\mathbf{x}$ is a column of weights. (If some columns are not involved in a particular dependence relation, then their weights are zero.) When A is row reduced to a matrix B, the columns of B are often totally different from the columns of A. However, the equations $A\mathbf{x} = \mathbf{0}$ and $B\mathbf{x} = \mathbf{0}$ have exactly the same set of solutions. If $A = [\, \mathbf{a}_1 \ \cdots \ \mathbf{a}_n \,]$ and $B = [\, \mathbf{b}_1 \ \cdots \ \mathbf{b}_n \,]$, then the vector equations

$$x_1\mathbf{a}_1 + \cdots + x_n\mathbf{a}_n = \mathbf{0} \quad \text{and} \quad x_1\mathbf{b}_1 + \cdots + x_n\mathbf{b}_n = \mathbf{0}$$

also have the same set of solutions. That is, the columns of A have *exactly the same linear dependence relationships* as the columns of B.

EXAMPLE 9 It can be shown that the matrix

$$A = [\, \mathbf{a}_1 \quad \mathbf{a}_2 \quad \cdots \quad \mathbf{a}_5 \,] = \begin{bmatrix} 1 & 4 & 0 & 2 & -1 \\ 3 & 12 & 1 & 5 & 5 \\ 2 & 8 & 1 & 3 & 2 \\ 5 & 20 & 2 & 8 & 8 \end{bmatrix}$$

is row equivalent to the matrix B in Example 8. Find a basis for Col A.

SOLUTION In Example 8 we saw that

$$\mathbf{b}_2 = 4\mathbf{b}_1 \quad \text{and} \quad \mathbf{b}_4 = 2\mathbf{b}_1 - \mathbf{b}_3$$

so we can expect that

$$\mathbf{a}_2 = 4\mathbf{a}_1 \quad \text{and} \quad \mathbf{a}_4 = 2\mathbf{a}_1 - \mathbf{a}_3$$

Check that this is indeed the case! Thus we may discard $\mathbf{a}_2$ and $\mathbf{a}_4$ when selecting a minimal spanning set for Col A. In fact, $\{\mathbf{a}_1, \mathbf{a}_3, \mathbf{a}_5\}$ must be linearly independent because any linear dependence relationship among $\mathbf{a}_1, \mathbf{a}_3, \mathbf{a}_5$ would imply a linear dependence relationship among $\mathbf{b}_1, \mathbf{b}_3, \mathbf{b}_5$. But we know that $\{\mathbf{b}_1, \mathbf{b}_3, \mathbf{b}_5\}$ is a linearly independent set. Thus $\{\mathbf{a}_1, \mathbf{a}_3, \mathbf{a}_5\}$ is a basis for Col A. The columns we have used for this basis are the pivot columns of A. ∎

Examples 8 and 9 illustrate the following useful fact.

THEOREM 6   The pivot columns of a matrix A form a basis for Col A.

PROOF The general proof uses the arguments discussed above. Let B be the reduced echelon form of A. The set of pivot columns of B is linearly independent, for no vector in the set is a linear combination of the vectors that precede it. Since A is row equivalent to B, the pivot columns of A are linearly independent as well, because any linear dependence relation among the columns of A corresponds to a linear dependence relation among the columns of B. For this same reason, every nonpivot column of A is a linear combination of the pivot columns of A. Thus the nonpivot columns of A may be discarded from the spanning set for Col A, by the Spanning Set Theorem. This leaves the pivot columns of A as a basis for Col A. ∎

Warning: The pivot columns of a matrix A are evident when A has been reduced only to *echelon* form. But, be careful to use the *pivot columns of A itself* for the basis of Col A. Row operations can change the column space of a matrix. The columns of an echelon form B of A are often not in the column space of A. For instance, the columns of matrix B in Example 8 all have zeros in their last entries, so they cannot span the column space of matrix A in Example 9.

In contrast, the following theorem establishes that row reduction does not change the row space of a matrix.

THEOREM 7

If two matrices A and B are row equivalent, then their row spaces are the same. If B is in echelon form, the nonzero rows of B form a basis for the row space of A as well as for that of B.

PROOF   If B is obtained from A by row operations, the rows of B are linear combinations of the rows of A. It follows that any linear combination of the rows of B is automatically a linear combination of the rows of A. Thus the row space of B is contained in the row space of A. Since row operations are reversible, the same argument shows that the row space of A is a subset of the row space of B. So the two row spaces are the same. If B is in echelon form, its nonzero rows are linearly independent because no nonzero row is a linear combination of the nonzero rows below it. (Apply Theorem 4 to the nonzero rows of B in reverse order, with the first row last.) Thus the nonzero rows of B form a basis of the (common) row space of B and A. ∎

EXAMPLE 10   Find a basis for the row space of the matrix A from Example 9.

SOLUTION   To find a basis for the row space, recall that matrix A from Example 9 is row equivalent to matrix B from Example 8:

$$A = \begin{bmatrix} 1 & 4 & 0 & 2 & -1 \\ 3 & 12 & 1 & 5 & 5 \\ 2 & 8 & 1 & 3 & 2 \\ 5 & 20 & 2 & 8 & 8 \end{bmatrix} \sim B = \begin{bmatrix} 1 & 4 & 0 & 2 & 0 \\ 0 & 0 & 1 & -1 & 0 \\ 0 & 0 & 0 & 0 & 1 \\ 0 & 0 & 0 & 0 & 0 \end{bmatrix}$$

By Theorem 7, the first three rows of B form a basis for the row space of A (as well as for the row space of B). Thus

$$\text{Basis for Row } A : \{(1,4,0,2,0), (0,0,1,-1,0), (0,0,0,0,1)\}$$

Observe that, unlike the basis for Col A, the bases for Row A and Nul A have no simple connection with the entries in A itself.[2]

Two Views of a Basis

When the Spanning Set Theorem is used, the deletion of vectors from a spanning set must stop when the set becomes linearly independent. If an additional vector is deleted,

[2] It is possible to find a basis for the row space Row A that uses rows of A. First form A^T, and then row reduce until the pivot columns of A^T are found. These pivot columns of A^T are rows of A, and they form a basis for the row space of A.

it will not be a linear combination of the remaining vectors, and hence the smaller set will no longer span V. Thus a basis is a spanning set that is as small as possible.

A basis is also a linearly independent set that is as large as possible. If S is a basis for V, and if S is enlarged by one vector—say, $\mathbf{w}$—from V, then the new set cannot be linearly independent, because S spans V, and $\mathbf{w}$ is therefore a linear combination of the elements in S.

EXAMPLE 11 The following three sets in $\mathbb{R}^3$ show how a linearly independent set can be enlarged to a basis and how further enlargement destroys the linear independence of the set. Also, a spanning set can be shrunk to a basis, but further shrinking destroys the spanning property.

$$\left\{\begin{bmatrix} 1 \\ 0 \\ 0 \end{bmatrix}, \begin{bmatrix} 2 \\ 3 \\ 0 \end{bmatrix}\right\} \qquad \left\{\begin{bmatrix} 1 \\ 0 \\ 0 \end{bmatrix}, \begin{bmatrix} 2 \\ 3 \\ 0 \end{bmatrix}, \begin{bmatrix} 4 \\ 5 \\ 6 \end{bmatrix}\right\} \qquad \left\{\begin{bmatrix} 1 \\ 0 \\ 0 \end{bmatrix}, \begin{bmatrix} 2 \\ 3 \\ 0 \end{bmatrix}, \begin{bmatrix} 4 \\ 5 \\ 6 \end{bmatrix}, \begin{bmatrix} 7 \\ 8 \\ 9 \end{bmatrix}\right\}$$

Linearly independent   A basis   Spans $\mathbb{R}^3$ but is
but does not span $\mathbb{R}^3$ for $\mathbb{R}^3$ linearly dependent ∎

Practice Problems

1. Let $\mathbf{v}_1 = \begin{bmatrix} 1 \\ -2 \\ 3 \end{bmatrix}$ and $\mathbf{v}_2 = \begin{bmatrix} -2 \\ 7 \\ -9 \end{bmatrix}$. Determine if $\{\mathbf{v}_1, \mathbf{v}_2\}$ is a basis for $\mathbb{R}^3$. Is $\{\mathbf{v}_1, \mathbf{v}_2\}$ a basis for $\mathbb{R}^2$?

2. Let $\mathbf{v}_1 = \begin{bmatrix} 1 \\ -3 \\ 4 \end{bmatrix}$, $\mathbf{v}_2 = \begin{bmatrix} 6 \\ 2 \\ -1 \end{bmatrix}$, $\mathbf{v}_3 = \begin{bmatrix} 2 \\ -2 \\ 3 \end{bmatrix}$, and $\mathbf{v}_4 = \begin{bmatrix} -4 \\ -8 \\ 9 \end{bmatrix}$. Find a basis for the subspace W spanned by $\{\mathbf{v}_1, \mathbf{v}_2, \mathbf{v}_3, \mathbf{v}_4\}$.

3. Let $\mathbf{v}_1 = \begin{bmatrix} 1 \\ 0 \\ 0 \end{bmatrix}$, $\mathbf{v}_2 = \begin{bmatrix} 0 \\ 1 \\ 0 \end{bmatrix}$, and $H = \left\{\begin{bmatrix} s \\ s \\ 0 \end{bmatrix} : s \text{ in } \mathbb{R}\right\}$. Then every vector in H is a linear combination of $\mathbf{v}_1$ and $\mathbf{v}_2$ because

$$\begin{bmatrix} s \\ s \\ 0 \end{bmatrix} = s \begin{bmatrix} 1 \\ 0 \\ 0 \end{bmatrix} + s \begin{bmatrix} 0 \\ 1 \\ 0 \end{bmatrix}$$

Is $\{\mathbf{v}_1, \mathbf{v}_2\}$ a basis for H?

4. Let V and W be vector spaces, let $T : V \to W$ and $U : V \to W$ be linear transformations, and let $\{\mathbf{v}_1, \ldots, \mathbf{v}_p\}$ be a basis for V. If $T(\mathbf{v}_j) = U(\mathbf{v}_j)$ for every value of j between 1 and p, show that $T(\mathbf{x}) = U(\mathbf{x})$ for every vector $\mathbf{x}$ in V.

STUDY GUIDE offers additional resources for mastering the concept of basis.

4.3 Exercises

Determine which sets in Exercises 1–8 are bases for $\mathbb{R}^3$. Of the sets that are *not* bases, determine which ones are linearly independent and which ones span $\mathbb{R}^3$. Justify your answers.

1. $\begin{bmatrix} 1 \\ 0 \\ 0 \end{bmatrix}, \begin{bmatrix} 1 \\ 1 \\ 0 \end{bmatrix}, \begin{bmatrix} 1 \\ 1 \\ 1 \end{bmatrix}$

2. $\begin{bmatrix} 1 \\ 0 \\ 1 \end{bmatrix}, \begin{bmatrix} 0 \\ 0 \\ 0 \end{bmatrix}, \begin{bmatrix} 0 \\ 1 \\ 0 \end{bmatrix}$

3. $\begin{bmatrix} 1 \\ 0 \\ -2 \end{bmatrix}, \begin{bmatrix} 3 \\ 2 \\ -4 \end{bmatrix}, \begin{bmatrix} -3 \\ -5 \\ 1 \end{bmatrix}$

4. $\begin{bmatrix} 2 \\ -2 \\ 1 \end{bmatrix}, \begin{bmatrix} 1 \\ -3 \\ 2 \end{bmatrix}, \begin{bmatrix} -7 \\ 5 \\ 4 \end{bmatrix}$

5. $\begin{bmatrix} 1 \\ -3 \\ 0 \end{bmatrix}, \begin{bmatrix} -2 \\ 9 \\ 0 \end{bmatrix}, \begin{bmatrix} 0 \\ 0 \\ 0 \end{bmatrix}, \begin{bmatrix} 0 \\ -3 \\ 5 \end{bmatrix}$

6. $\begin{bmatrix} 1 \\ 2 \\ -3 \end{bmatrix}, \begin{bmatrix} -4 \\ -5 \\ 6 \end{bmatrix}$

7. $\begin{bmatrix} -2 \\ 3 \\ 0 \end{bmatrix}, \begin{bmatrix} 6 \\ -1 \\ 5 \end{bmatrix}$ **8.** $\begin{bmatrix} 1 \\ -4 \\ 3 \end{bmatrix}, \begin{bmatrix} 0 \\ 3 \\ -1 \end{bmatrix}, \begin{bmatrix} 3 \\ -5 \\ 4 \end{bmatrix}, \begin{bmatrix} 0 \\ 2 \\ -2 \end{bmatrix}$

Find bases for the null spaces of the matrices given in Exercises 9 and 10. Refer to the remarks that follow Example 3 in Section 4.2.

9. $\begin{bmatrix} 1 & 0 & -3 & 2 \\ 0 & 1 & -5 & 4 \\ 3 & -2 & 1 & -2 \end{bmatrix}$ **10.** $\begin{bmatrix} 1 & 0 & -5 & 1 & 4 \\ -2 & 1 & 6 & -2 & -2 \\ 0 & 2 & -8 & 1 & 9 \end{bmatrix}$

11. Find a basis for the set of vectors in $\mathbb{R}^3$ in the plane $x + 2y + z = 0$. [*Hint:* Think of the equation as a "system" of homogeneous equations.]

12. Find a basis for the set of vectors in $\mathbb{R}^2$ on the line $y = 5x$.

In Exercises 13 and 14, assume that A is row equivalent to B. Find bases for Nul A, Col A, and Row A.

13. $A = \begin{bmatrix} -2 & 4 & -2 & -4 \\ 2 & -6 & -3 & 1 \\ -3 & 8 & 2 & -3 \end{bmatrix}$, $B = \begin{bmatrix} 1 & 0 & 6 & 5 \\ 0 & 2 & 5 & 3 \\ 0 & 0 & 0 & 0 \end{bmatrix}$

14. $A = \begin{bmatrix} 1 & 2 & -5 & 11 & -3 \\ 2 & 4 & -5 & 15 & 2 \\ 1 & 2 & 0 & 4 & 5 \\ 3 & 6 & -5 & 19 & -2 \end{bmatrix}$,

$B = \begin{bmatrix} 1 & 2 & 0 & 4 & 5 \\ 0 & 0 & 5 & -7 & 8 \\ 0 & 0 & 0 & 0 & -9 \\ 0 & 0 & 0 & 0 & 0 \end{bmatrix}$

In Exercises 15–18, find a basis for the space spanned by the given vectors, $\mathbf{v}_1, \ldots, \mathbf{v}_5$.

15. $\begin{bmatrix} 1 \\ 0 \\ -3 \\ 2 \end{bmatrix}, \begin{bmatrix} 0 \\ 1 \\ 2 \\ -3 \end{bmatrix}, \begin{bmatrix} -3 \\ -4 \\ 1 \\ 6 \end{bmatrix}, \begin{bmatrix} 1 \\ -3 \\ -8 \\ 7 \end{bmatrix}, \begin{bmatrix} 2 \\ 1 \\ -6 \\ 9 \end{bmatrix}$

16. $\begin{bmatrix} 1 \\ 0 \\ 0 \\ 1 \end{bmatrix}, \begin{bmatrix} -2 \\ 1 \\ -1 \\ 1 \end{bmatrix}, \begin{bmatrix} 6 \\ -1 \\ 2 \\ -1 \end{bmatrix}, \begin{bmatrix} 5 \\ -3 \\ 3 \\ -4 \end{bmatrix}, \begin{bmatrix} 0 \\ 3 \\ -1 \\ 1 \end{bmatrix}$

T 17. $\begin{bmatrix} 8 \\ 9 \\ -3 \\ -6 \\ 0 \end{bmatrix}, \begin{bmatrix} 4 \\ 5 \\ 1 \\ -4 \\ 4 \end{bmatrix}, \begin{bmatrix} -1 \\ -4 \\ -9 \\ 6 \\ -7 \end{bmatrix}, \begin{bmatrix} 6 \\ 8 \\ 4 \\ -7 \\ 10 \end{bmatrix}, \begin{bmatrix} -1 \\ 4 \\ 11 \\ -8 \\ -7 \end{bmatrix}$

T 18. $\begin{bmatrix} -8 \\ 7 \\ 6 \\ 5 \\ -7 \end{bmatrix}, \begin{bmatrix} 8 \\ -7 \\ -9 \\ -5 \\ 7 \end{bmatrix}, \begin{bmatrix} -8 \\ 7 \\ 4 \\ 5 \\ -7 \end{bmatrix}, \begin{bmatrix} 1 \\ 4 \\ 9 \\ 6 \\ -7 \end{bmatrix}, \begin{bmatrix} -9 \\ 3 \\ -4 \\ -1 \\ 0 \end{bmatrix}$

19. Let $\mathbf{v}_1 = \begin{bmatrix} 4 \\ -3 \\ 7 \end{bmatrix}$, $\mathbf{v}_2 = \begin{bmatrix} 1 \\ 9 \\ -2 \end{bmatrix}$, $\mathbf{v}_3 = \begin{bmatrix} 7 \\ 11 \\ 6 \end{bmatrix}$, and $H = $ Span $\{\mathbf{v}_1, \mathbf{v}_2, \mathbf{v}_3\}$. It can be verified that $4\mathbf{v}_1 + 5\mathbf{v}_2 - 3\mathbf{v}_3 = \mathbf{0}$. Use this information to find a basis for H. There is more than one answer.

20. Let $\mathbf{v}_1 = \begin{bmatrix} 7 \\ 4 \\ -9 \\ -5 \end{bmatrix}$, $\mathbf{v}_2 = \begin{bmatrix} 4 \\ -7 \\ 2 \\ 5 \end{bmatrix}$, $\mathbf{v}_3 = \begin{bmatrix} 1 \\ -5 \\ 3 \\ 4 \end{bmatrix}$. It can be verified that $\mathbf{v}_1 - 3\mathbf{v}_2 + 5\mathbf{v}_3 = \mathbf{0}$. Use this information to find a basis for $H = $ Span $\{\mathbf{v}_1, \mathbf{v}_2, \mathbf{v}_3\}$.

In Exercises 21–32, mark each statement True or False (**T/F**). Justify each answer.

21. (**T/F**) A single vector by itself is linearly dependent.

22. (**T/F**) A linearly independent set in a subspace H is a basis for H.

23. (**T/F**) If $H = $ Span $\{\mathbf{b}_1, \ldots, \mathbf{b}_p\}$, then $\{\mathbf{b}_1, \ldots, \mathbf{b}_p\}$ is a basis for H.

24. (**T/F**) If a finite set S of nonzero vectors spans a vector space V, then some subset of S is a basis for V.

25. (**T/F**) The columns of an invertible $n \times n$ matrix form a basis for $\mathbb{R}^n$.

26. (**T/F**) A basis is a linearly independent set that is as large as possible.

27. (**T/F**) A basis is a spanning set that is as large as possible.

28. (**T/F**) The standard method for producing a spanning set for Nul A, described in Section 4.2, sometimes fails to produce a basis for Nul A.

29. (**T/F**) In some cases, the linear dependence relations among the columns of a matrix can be affected by certain elementary row operations on the matrix.

30. (**T/F**) If B is an echelon form of a matrix A, then the pivot columns of B form a basis for Col A.

31. (**T/F**) Row operations preserve the linear dependence relations among the rows of A.

32. (**T/F**) If A and B are row equivalent, then their row spaces are the same.

33. Suppose $\mathbb{R}^4 = $ Span $\{\mathbf{v}_1, \ldots, \mathbf{v}_4\}$. Explain why $\{\mathbf{v}_1, \ldots, \mathbf{v}_4\}$ is a basis for $\mathbb{R}^4$.

34. Let $\mathcal{B} = \{\mathbf{v}_1, \ldots, \mathbf{v}_n\}$ be a linearly independent set in $\mathbb{R}^n$. Explain why $\mathcal{B}$ must be a basis for $\mathbb{R}^n$.

35. Let $\mathbf{v}_1 = \begin{bmatrix} 1 \\ 0 \\ 1 \end{bmatrix}$, $\mathbf{v}_2 = \begin{bmatrix} 0 \\ 1 \\ 1 \end{bmatrix}$, $\mathbf{v}_3 = \begin{bmatrix} 0 \\ 1 \\ 0 \end{bmatrix}$, and let H be the set of vectors in $\mathbb{R}^3$ whose second and third entries are equal. Then every vector in H has a unique expansion as a linear combination of $\mathbf{v}_1, \mathbf{v}_2, \mathbf{v}_3$, because

$$\begin{bmatrix} s \\ t \\ t \end{bmatrix} = s \begin{bmatrix} 1 \\ 0 \\ 1 \end{bmatrix} + (t-s) \begin{bmatrix} 0 \\ 1 \\ 1 \end{bmatrix} + s \begin{bmatrix} 0 \\ 1 \\ 0 \end{bmatrix}$$

for any s and t. Is $\{\mathbf{v}_1, \mathbf{v}_2, \mathbf{v}_3\}$ a basis for H? Why or why not?

36. In the vector space of all real-valued functions, find a basis for the subspace spanned by $\{\sin t, \sin 2t, \sin t \cos t\}$.

37. Let V be the vector space of functions that describe the vibration of a mass–spring system. (Refer to Exercise 19 in Section 4.1.) Find a basis for V.

38. (*RLC circuit*) The circuit in the figure consists of a resistor (R ohms), an inductor (L henrys), a capacitor (C farads), and an initial voltage source. Let $b = R/(2L)$, and suppose R, L, and C have been selected so that b also equals $1/\sqrt{LC}$. (This is done, for instance, when the circuit is used in a voltmeter.) Let $v(t)$ be the voltage (in volts) at time t, measured across the capacitor. It can be shown that v is in the null space H of the linear transformation that maps $v(t)$ into $Lv''(t) + Rv'(t) + (1/C)v(t)$, and H consists of all functions of the form $v(t) = e^{-bt}(c_1 + c_2t)$. Find a basis for H.

Exercises 39 and 40 show that every basis for $\mathbb{R}^n$ must contain exactly n vectors.

39. Let $S = \{\mathbf{v}_1, \ldots, \mathbf{v}_k\}$ be a set of k vectors in $\mathbb{R}^n$, with $k < n$. Use a theorem from Section 1.4 to explain why S cannot be a basis for $\mathbb{R}^n$.

40. Let $S = \{\mathbf{v}_1, \ldots, \mathbf{v}_k\}$ be a set of k vectors in $\mathbb{R}^n$, with $k > n$. Use a theorem from Chapter 1 to explain why S cannot be a basis for $\mathbb{R}^n$.

Exercises 41 and 42 reveal an important connection between linear independence and linear transformations and provide practice using the definition of linear dependence. Let V and W be vector spaces, let $T : V \to W$ be a linear transformation, and let $\{\mathbf{v}_1, \ldots, \mathbf{v}_p\}$ be a subset of V.

41. Show that if $\{\mathbf{v}_1, \ldots, \mathbf{v}_p\}$ is linearly dependent in V, then the set of images, $\{T(\mathbf{v}_1), \ldots, T(\mathbf{v}_p)\}$, is linearly dependent in W. This fact shows that if a linear transformation maps a set $\{\mathbf{v}_1, \ldots, \mathbf{v}_p\}$ onto a linearly *independent* set $\{T(\mathbf{v}_1), \ldots, T(\mathbf{v}_p)\}$, then the original set is linearly independent, too (because it cannot be linearly dependent).

42. Suppose that T is a one-to-one transformation, so that an equation $T(\mathbf{u}) = T(\mathbf{v})$ always implies $\mathbf{u} = \mathbf{v}$. Show that if the set of images $\{T(\mathbf{v}_1), \ldots, T(\mathbf{v}_p)\}$ is linearly dependent, then $\{\mathbf{v}_1, \ldots, \mathbf{v}_p\}$ is linearly dependent. This fact shows that *a one-to-one linear transformation maps a linearly independent set onto a linearly independent set* (because in this case the set of images cannot be linearly dependent).

43. Consider the polynomials $\mathbf{p}_1(t) = 1 + t^2$ and $\mathbf{p}_2(t) = 1 - t^2$. Is $\{\mathbf{p}_1, \mathbf{p}_2\}$ a linearly independent set in $\mathbb{P}_3$? Why or why not?

44. Consider the polynomials $\mathbf{p}_1(t) = 1 + t$, $\mathbf{p}_2(t) = 1 - t$, and $\mathbf{p}_3(t) = 2$ (for all t). By inspection, write a linear dependence relation among $\mathbf{p}_1$, $\mathbf{p}_2$, and $\mathbf{p}_3$. Then find a basis for Span $\{\mathbf{p}_1, \mathbf{p}_2, \mathbf{p}_3\}$.

45. Let V be a vector space that contains a linearly independent set $\{\mathbf{u}_1, \mathbf{u}_2, \mathbf{u}_3, \mathbf{u}_4\}$. Describe how to construct a set of vectors $\{\mathbf{v}_1, \mathbf{v}_2, \mathbf{v}_3, \mathbf{v}_4\}$ in V such that $\{\mathbf{v}_1, \mathbf{v}_3\}$ is a basis for Span $\{\mathbf{v}_1, \mathbf{v}_2, \mathbf{v}_3, \mathbf{v}_4\}$.

T 46. Let $H = \text{Span}\,\{\mathbf{u}_1, \mathbf{u}_2, \mathbf{u}_3\}$ and $K = \text{Span}\,\{\mathbf{v}_1, \mathbf{v}_2, \mathbf{v}_3\}$, where

$$\mathbf{u}_1 = \begin{bmatrix} 1 \\ 2 \\ 0 \\ -1 \end{bmatrix}, \quad \mathbf{u}_2 = \begin{bmatrix} 0 \\ 2 \\ -1 \\ 1 \end{bmatrix}, \quad \mathbf{u}_3 = \begin{bmatrix} 3 \\ 4 \\ 1 \\ -4 \end{bmatrix},$$

$$\mathbf{v}_1 = \begin{bmatrix} -2 \\ -2 \\ -1 \\ 3 \end{bmatrix}, \quad \mathbf{v}_2 = \begin{bmatrix} 2 \\ 3 \\ 2 \\ -6 \end{bmatrix}, \quad \mathbf{v}_3 = \begin{bmatrix} -1 \\ 4 \\ 6 \\ -2 \end{bmatrix}.$$

Find bases for H, K, and $H + K$. (See Exercises 41 and 42 in Section 4.1.)

T 47. Show that $\{t, \sin t, \cos 2t, \sin t \cos t\}$ is a linearly independent set of functions defined on $\mathbb{R}$. Start by assuming that

$$c_1 t + c_2 \sin t + c_3 \cos 2t + c_4 \sin t \cos t = 0 \qquad (5)$$

Equation (5) must hold for all real t, so choose several specific values of t (say, $t = 0$, .1, .2) until you get a system of enough equations to determine that all the c_j must be zero.

T 48. Show that $\{1, \cos t, \cos^2 t, \ldots, \cos^6 t\}$ is a linearly independent set of functions defined on $\mathbb{R}$. Use the method of Exercise 47. (This result will be needed in Exercise 54 in Section 4.5.)

Solutions to Practice Problems

1. Let $A = [\mathbf{v}_1 \quad \mathbf{v}_2]$. Row operations show that

$$A = \begin{bmatrix} 1 & -2 \\ -2 & 7 \\ 3 & -9 \end{bmatrix} \sim \begin{bmatrix} 1 & -2 \\ 0 & 3 \\ 0 & 0 \end{bmatrix}$$

Not every row of A contains a pivot position. So the columns of A do not span $\mathbb{R}^3$, by Theorem 4 in Section 1.4. Hence $\{\mathbf{v}_1, \mathbf{v}_2\}$ is not a basis for $\mathbb{R}^3$. Since $\mathbf{v}_1$ and $\mathbf{v}_2$ are not in $\mathbb{R}^2$, they cannot possibly be a basis for $\mathbb{R}^2$. However, since $\mathbf{v}_1$ and $\mathbf{v}_2$ are obviously linearly independent, they are a basis for a subspace of $\mathbb{R}^3$, namely Span $\{\mathbf{v}_1, \mathbf{v}_2\}$.

2. Set up a matrix A whose column space is the space spanned by $\{\mathbf{v}_1, \mathbf{v}_2, \mathbf{v}_3, \mathbf{v}_4\}$, and then row reduce A to find its pivot columns.

$$A = \begin{bmatrix} 1 & 6 & 2 & -4 \\ -3 & 2 & -2 & -8 \\ 4 & -1 & 3 & 9 \end{bmatrix} \sim \begin{bmatrix} 1 & 6 & 2 & -4 \\ 0 & 20 & 4 & -20 \\ 0 & -25 & -5 & 25 \end{bmatrix} \sim \begin{bmatrix} 1 & 6 & 2 & -4 \\ 0 & 5 & 1 & -5 \\ 0 & 0 & 0 & 0 \end{bmatrix}$$

The first two columns of A are the pivot columns and hence form a basis of Col $A = W$. Hence $\{\mathbf{v}_1, \mathbf{v}_2\}$ is a basis for W. Note that the reduced echelon form of A is not needed in order to locate the pivot columns.

3. Neither $\mathbf{v}_1$ nor $\mathbf{v}_2$ is in H, so $\{\mathbf{v}_1, \mathbf{v}_2\}$ cannot be a basis for H. In fact, $\{\mathbf{v}_1, \mathbf{v}_2\}$ is a basis for the *plane* of all vectors of the form $(c_1, c_2, 0)$, but H is only a *line*.

4. Since $\{\mathbf{v}_1, \dots, \mathbf{v}_p\}$ is a basis for V, for any vector $\mathbf{x}$ in V, there exist scalars $c_1, \dots, c_p$ such that $\mathbf{x} = c_1\mathbf{v}_1 + \cdots + c_p\mathbf{v}_p$. Then since T and U are linear transformations

$$\begin{aligned} T(\mathbf{x}) &= T(c_1\mathbf{v}_1 + \cdots + c_p\mathbf{v}_p) = c_1 T(\mathbf{v}_1) + \cdots + c_p T(\mathbf{v}_p) \\ &= c_1 U(\mathbf{v}_1) + \cdots + c_p U(\mathbf{v}_p) = U(c_1\mathbf{v}_1 + \cdots + c_p\mathbf{v}_p) \\ &= U(\mathbf{x}) \end{aligned}$$

4.4 Coordinate Systems

An important reason for specifying a basis $\mathcal{B}$ for a vector space V is to impose a "coordinate system" on V. This section will show that if $\mathcal{B}$ contains n vectors, then the coordinate system will make V act like $\mathbb{R}^n$. If V is already $\mathbb{R}^n$ itself, then $\mathcal{B}$ will determine a coordinate system that gives a new "view" of V.

The existence of coordinate systems rests on the following fundamental result.

THEOREM 8

The Unique Representation Theorem

Let $\mathcal{B} = \{\mathbf{b}_1, \dots, \mathbf{b}_n\}$ be a basis for a vector space V. Then for each $\mathbf{x}$ in V, there exists a unique set of scalars $c_1, \dots, c_n$ such that

$$\mathbf{x} = c_1\mathbf{b}_1 + \cdots + c_n\mathbf{b}_n \tag{1}$$

PROOF   Since $\mathcal{B}$ spans V, there exist scalars such that (1) holds. Suppose $\mathbf{x}$ also has the representation

$$\mathbf{x} = d_1\mathbf{b}_1 + \cdots + d_n\mathbf{b}_n$$

for scalars $d_1, \ldots, d_n$. Then, subtracting, we have

$$\mathbf{0} = \mathbf{x} - \mathbf{x} = (c_1 - d_1)\mathbf{b}_1 + \cdots + (c_n - d_n)\mathbf{b}_n \tag{2}$$

Since $\mathcal{B}$ is linearly independent, the weights in (2) must all be zero. That is, $c_j = d_j$ for $1 \le j \le n$.   ∎

DEFINITION

> Suppose $\mathcal{B} = \{\mathbf{b}_1, \ldots, \mathbf{b}_n\}$ is a basis for a vector space V and $\mathbf{x}$ is in V. The **coordinates of $\mathbf{x}$ relative to the basis $\mathcal{B}$** (or the **$\mathcal{B}$-coordinates of $\mathbf{x}$**) are the weights $c_1, \ldots, c_n$ such that $\mathbf{x} = c_1\mathbf{b}_1 + \cdots + c_n\mathbf{b}_n$.

If $c_1, \ldots, c_n$ are the $\mathcal{B}$-coordinates of $\mathbf{x}$, then the vector in $\mathbb{R}^n$

$$\left[\,\mathbf{x}\,\right]_{\mathcal{B}} = \begin{bmatrix} c_1 \\ \vdots \\ c_n \end{bmatrix}$$

is the **coordinate vector of $\mathbf{x}$ (relative to $\mathcal{B}$)**, or the **$\mathcal{B}$-coordinate vector of $\mathbf{x}$**. The mapping $\mathbf{x} \mapsto \left[\,\mathbf{x}\,\right]_{\mathcal{B}}$ is the **coordinate mapping (determined by $\mathcal{B}$)**.[1]

EXAMPLE 1   Consider a basis $\mathcal{B} = \{\mathbf{b}_1, \mathbf{b}_2\}$ for $\mathbb{R}^2$, where $\mathbf{b}_1 = \begin{bmatrix} 1 \\ 0 \end{bmatrix}$ and $\mathbf{b}_2 = \begin{bmatrix} 1 \\ 2 \end{bmatrix}$. Suppose an $\mathbf{x}$ in $\mathbb{R}^2$ has the coordinate vector $\left[\,\mathbf{x}\,\right]_{\mathcal{B}} = \begin{bmatrix} -2 \\ 3 \end{bmatrix}$. Find $\mathbf{x}$.

SOLUTION   The $\mathcal{B}$-coordinates of $\mathbf{x}$ tell how to build $\mathbf{x}$ from the vectors in $\mathcal{B}$. That is,

$$\mathbf{x} = (-2)\mathbf{b}_1 + 3\mathbf{b}_2 = (-2)\begin{bmatrix} 1 \\ 0 \end{bmatrix} + 3\begin{bmatrix} 1 \\ 2 \end{bmatrix} = \begin{bmatrix} 1 \\ 6 \end{bmatrix}$$ ∎

EXAMPLE 2   The entries in the vector $\mathbf{x} = \begin{bmatrix} 1 \\ 6 \end{bmatrix}$ are the coordinates of $\mathbf{x}$ relative to the *standard basis* $\mathcal{E} = \{\mathbf{e}_1, \mathbf{e}_2\}$, since

$$\begin{bmatrix} 1 \\ 6 \end{bmatrix} = 1\begin{bmatrix} 1 \\ 0 \end{bmatrix} + 6\begin{bmatrix} 0 \\ 1 \end{bmatrix} = 1\mathbf{e}_1 + 6\mathbf{e}_2$$

If $\mathcal{E} = \{\mathbf{e}_1, \mathbf{e}_2\}$, then $\left[\,\mathbf{x}\,\right]_{\mathcal{E}} = \mathbf{x}$.   ∎

A Graphical Interpretation of Coordinates

A coordinate system on a set consists of a one-to-one mapping of the points in the set into $\mathbb{R}^n$. For example, ordinary graph paper provides a coordinate system for the plane

---

[1] The concept of a coordinate mapping assumes that the basis $\mathcal{B}$ is an indexed set whose vectors are listed in some fixed preassigned order. This property makes the definition of $\left[\,\mathbf{x}\,\right]_{\mathcal{B}}$ unambiguous.

when one selects perpendicular axes and a unit of measurement on each axis. Figure 1 shows the standard basis $\{\mathbf{e}_1, \mathbf{e}_2\}$, the vectors $\mathbf{b}_1 (= \mathbf{e}_1)$ and $\mathbf{b}_2$ from Example 1, and the vector $\mathbf{x} = \begin{bmatrix} 1 \\ 6 \end{bmatrix}$. The coordinates 1 and 6 give the location of $\mathbf{x}$ relative to the standard basis: 1 unit in the $\mathbf{e}_1$ direction and 6 units in the $\mathbf{e}_2$ direction.

Figure 2 shows the vectors $\mathbf{b}_1$, $\mathbf{b}_2$, and $\mathbf{x}$ from Figure 1. (Geometrically, the three vectors lie on a vertical line in both figures.) However, the standard coordinate grid was erased and replaced by a grid especially adapted to the basis $\mathcal{B}$ in Example 1. The coordinate vector $[\mathbf{x}]_\mathcal{B} = \begin{bmatrix} -2 \\ 3 \end{bmatrix}$ gives the location of $\mathbf{x}$ on this new coordinate system: -2 units in the $\mathbf{b}_1$ direction and 3 units in the $\mathbf{b}_2$ direction.

FIGURE 1 Standard graph paper.

FIGURE 2 $\mathcal{B}$-graph paper.

EXAMPLE 3 In crystallography, the description of a crystal lattice is aided by choosing a basis $\{\mathbf{u}, \mathbf{v}, \mathbf{w}\}$ for $\mathbb{R}^3$ that corresponds to three adjacent edges of one "unit cell" of the crystal. An entire lattice is constructed by stacking together many copies of one cell. There are fourteen basic types of unit cells; three are displayed in Figure 3.[2]

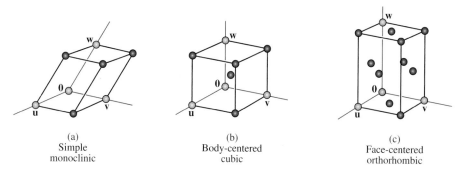

(a)
Simple
monoclinic

(b)
Body-centered
cubic

(c)
Face-centered
orthorhombic

FIGURE 3 Examples of unit cells.

The coordinates of atoms within the crystal are given relative to the basis for the lattice. For instance,

$$\begin{bmatrix} 1/2 \\ 1/2 \\ 1 \end{bmatrix}$$

identifies the top face-centered atom in the cell in Figure 3(c). ∎

---

[2] Adapted from *The Science and Engineering of Materials*, 4th Ed., by Donald R. Askeland (Boston: Prindle, Weber & Schmidt, © 2002), p. 36.

Coordinates in $\mathbb{R}^n$

When a basis $\mathcal{B}$ for $\mathbb{R}^n$ is fixed, the $\mathcal{B}$-coordinate vector of a specified $\mathbf{x}$ is easily found, as in the next example.

EXAMPLE 4 Let $\mathbf{b}_1 = \begin{bmatrix} 2 \\ 1 \end{bmatrix}$, $\mathbf{b}_2 = \begin{bmatrix} -1 \\ 1 \end{bmatrix}$, $\mathbf{x} = \begin{bmatrix} 4 \\ 5 \end{bmatrix}$, and $\mathcal{B} = \{\mathbf{b}_1, \mathbf{b}_2\}$. Find the coordinate vector $[\mathbf{x}]_{\mathcal{B}}$ of $\mathbf{x}$ relative to $\mathcal{B}$.

SOLUTION The $\mathcal{B}$-coordinates c_1, c_2 of $\mathbf{x}$ satisfy

$$c_1 \underbrace{\begin{bmatrix} 2 \\ 1 \end{bmatrix}}_{\mathbf{b}_1} + c_2 \underbrace{\begin{bmatrix} -1 \\ 1 \end{bmatrix}}_{\mathbf{b}_2} = \underbrace{\begin{bmatrix} 4 \\ 5 \end{bmatrix}}_{\mathbf{x}}$$

or

$$\underbrace{\begin{bmatrix} 2 & -1 \\ 1 & 1 \end{bmatrix}}_{\mathbf{b}_1 \quad \mathbf{b}_2} \begin{bmatrix} c_1 \\ c_2 \end{bmatrix} = \underbrace{\begin{bmatrix} 4 \\ 5 \end{bmatrix}}_{\mathbf{x}} \tag{3}$$

This equation can be solved by row operations on an augmented matrix or by multiplying the vector $\mathbf{x}$ by the inverse of the matrix. In any case, the solution is $c_1 = 3$, $c_2 = 2$. Thus $\mathbf{x} = 3\mathbf{b}_1 + 2\mathbf{b}_2$, and

$$[\mathbf{x}]_{\mathcal{B}} = \begin{bmatrix} c_1 \\ c_2 \end{bmatrix} = \begin{bmatrix} 3 \\ 2 \end{bmatrix}$$

See Figure 4. ∎

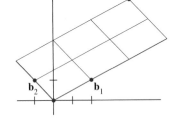

FIGURE 4
The $\mathcal{B}$-coordinate vector of $\mathbf{x}$ is $(3, 2)$.

The matrix in (3) changes the $\mathcal{B}$-coordinates of a vector $\mathbf{x}$ into the standard coordinates for $\mathbf{x}$. An analogous change of coordinates can be carried out in $\mathbb{R}^n$ for a basis $\mathcal{B} = \{\mathbf{b}_1, \ldots, \mathbf{b}_n\}$. Let

$$P_{\mathcal{B}} = \begin{bmatrix} \mathbf{b}_1 & \mathbf{b}_2 & \cdots & \mathbf{b}_n \end{bmatrix}$$

Then the vector equation

$$\mathbf{x} = c_1\mathbf{b}_1 + c_2\mathbf{b}_2 + \cdots + c_n\mathbf{b}_n$$

is equivalent to

$$\boxed{\mathbf{x} = P_{\mathcal{B}}[\mathbf{x}]_{\mathcal{B}}} \tag{4}$$

We call $P_{\mathcal{B}}$ the **change-of-coordinates matrix** from $\mathcal{B}$ to the standard basis in $\mathbb{R}^n$. Left-multiplication by $P_{\mathcal{B}}$ transforms the coordinate vector $[\mathbf{x}]_{\mathcal{B}}$ into $\mathbf{x}$. The change-of-coordinates equation (4) is important and will be needed at several points in Chapters 5 and 7.

Since the columns of $P_{\mathcal{B}}$ form a basis for $\mathbb{R}^n$, $P_{\mathcal{B}}$ is invertible (by the Invertible Matrix Theorem). Left-multiplication by $P_{\mathcal{B}}^{-1}$ converts $\mathbf{x}$ into its $\mathcal{B}$-coordinate vector:

$$P_{\mathcal{B}}^{-1}\mathbf{x} = [\mathbf{x}]_{\mathcal{B}}$$

The correspondence $\mathbf{x} \mapsto [\mathbf{x}]_{\mathcal{B}}$, produced here by $P_{\mathcal{B}}^{-1}$, is the coordinate mapping mentioned earlier. Since $P_{\mathcal{B}}^{-1}$ is an invertible matrix, the coordinate mapping is a one-to-one linear transformation from $\mathbb{R}^n$ onto $\mathbb{R}^n$, by the Invertible Matrix Theorem. (See also Theorem 12 in Section 1.9.) This property of the coordinate mapping is also true in a general vector space that has a basis, as we shall see.

The Coordinate Mapping

Choosing a basis $\mathcal{B} = \{\mathbf{b}_1, \ldots, \mathbf{b}_n\}$ for a vector space V introduces a coordinate system in V. The coordinate mapping $\mathbf{x} \mapsto [\,\mathbf{x}\,]_{\mathcal{B}}$ connects the possibly unfamiliar space V to the familiar space $\mathbb{R}^n$. See Figure 5. Points in V can now be identified by their new "names."

FIGURE 5 The coordinate mapping from V onto $\mathbb{R}^n$.

THEOREM 9

Let $\mathcal{B} = \{\mathbf{b}_1, \ldots, \mathbf{b}_n\}$ be a basis for a vector space V. Then the coordinate mapping $\mathbf{x} \mapsto [\,\mathbf{x}\,]_{\mathcal{B}}$ is a one-to-one linear transformation from V onto $\mathbb{R}^n$.

PROOF   Take two typical vectors in V, say,

$$\mathbf{u} = c_1 \mathbf{b}_1 + \cdots + c_n \mathbf{b}_n$$
$$\mathbf{w} = d_1 \mathbf{b}_1 + \cdots + d_n \mathbf{b}_n$$

Then, using vector operations,

$$\mathbf{u} + \mathbf{w} = (c_1 + d_1)\mathbf{b}_1 + \cdots + (c_n + d_n)\mathbf{b}_n$$

It follows that

$$[\,\mathbf{u} + \mathbf{w}\,]_{\mathcal{B}} = \begin{bmatrix} c_1 + d_1 \\ \vdots \\ c_n + d_n \end{bmatrix} = \begin{bmatrix} c_1 \\ \vdots \\ c_n \end{bmatrix} + \begin{bmatrix} d_1 \\ \vdots \\ d_n \end{bmatrix} = [\,\mathbf{u}\,]_{\mathcal{B}} + [\,\mathbf{w}\,]_{\mathcal{B}}$$

So the coordinate mapping preserves addition. If r is any scalar, then

$$r\mathbf{u} = r(c_1 \mathbf{b}_1 + \cdots + c_n \mathbf{b}_n) = (rc_1)\mathbf{b}_1 + \cdots + (rc_n)\mathbf{b}_n$$

So

$$[\,r\mathbf{u}\,]_{\mathcal{B}} = \begin{bmatrix} rc_1 \\ \vdots \\ rc_n \end{bmatrix} = r \begin{bmatrix} c_1 \\ \vdots \\ c_n \end{bmatrix} = r[\,\mathbf{u}\,]_{\mathcal{B}}$$

Thus the coordinate mapping also preserves scalar multiplication and hence is a linear transformation. See Exercises 27 and 28 for verification that the coordinate mapping is one-to-one and maps V onto $\mathbb{R}^n$.   ■

The linearity of the coordinate mapping extends to linear combinations, just as in Section 1.8. If $\mathbf{u}_1, \ldots, \mathbf{u}_p$ are in V and if $c_1, \ldots, c_p$ are scalars, then

$$[\,c_1 \mathbf{u}_1 + \cdots + c_p \mathbf{u}_p\,]_{\mathcal{B}} = c_1[\,\mathbf{u}_1\,]_{\mathcal{B}} + \cdots + c_p[\,\mathbf{u}_p\,]_{\mathcal{B}} \tag{5}$$

In words, (5) says that the $\mathcal{B}$-coordinate vector of a linear combination of $\mathbf{u}_1, \ldots, \mathbf{u}_p$ is the *same* linear combination of their coordinate vectors.

The coordinate mapping in Theorem 9 is an important example of an *isomorphism* from V onto $\mathbb{R}^n$. In general, a one-to-one linear transformation from a vector space V onto a vector space W is called an **isomorphism** from V onto W (*iso* from the Greek for "the same," and *morph* from the Greek for "form" or "structure"). The notation and terminology for V and W may differ, but the two spaces are indistinguishable as vector spaces. *Every vector space calculation in V is accurately reproduced in W, and vice versa.* In particular, any real vector space with a basis of n vectors is indistinguishable from $\mathbb{R}^n$. See Exercises 29 and 30.

STUDY GUIDE offers additional resources about isomorphic vector spaces.

EXAMPLE 5 Let $\mathcal{B}$ be the standard basis of the space $\mathbb{P}_3$ of polynomials; that is, let $\mathcal{B} = \{1, t, t^2, t^3\}$. A typical element $\mathbf{p}$ of $\mathbb{P}_3$ has the form

$$\mathbf{p}(t) = a_0 + a_1 t + a_2 t^2 + a_3 t^3$$

Since $\mathbf{p}$ is already displayed as a linear combination of the standard basis vectors, we conclude that

$$[\mathbf{p}]_{\mathcal{B}} = \begin{bmatrix} a_0 \\ a_1 \\ a_2 \\ a_3 \end{bmatrix}$$

Thus the coordinate mapping $\mathbf{p} \mapsto [\mathbf{p}]_{\mathcal{B}}$ is an isomorphism from $\mathbb{P}_3$ onto $\mathbb{R}^4$. All vector space operations in $\mathbb{P}_3$ correspond to operations in $\mathbb{R}^4$. ∎

If we think of $\mathbb{P}_3$ and $\mathbb{R}^4$ as displays on two computer screens that are connected via the coordinate mapping, then every vector space operation in $\mathbb{P}_3$ on one screen is exactly duplicated by a corresponding vector operation in $\mathbb{R}^4$ on the other screen. The vectors on the $\mathbb{P}_3$ screen look different from those on the $\mathbb{R}^4$ screen, but they "act" as vectors in exactly the same way. See Figure 6.

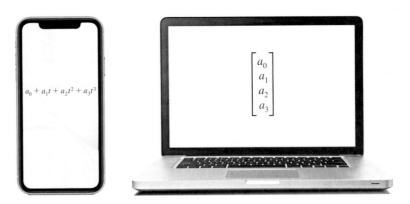

FIGURE 6 The space $\mathbb{P}_3$ is isomorphic to $\mathbb{R}^4$.

EXAMPLE 6 Use coordinate vectors to verify that the polynomials $1 + 2t^2$, $4 + t + 5t^2$, and $3 + 2t$ are linearly dependent in $\mathbb{P}_2$.

SOLUTION The coordinate mapping from Example 5 produces the coordinate vectors $(1, 0, 2)$, $(4, 1, 5)$, and $(3, 2, 0)$, respectively. Writing these vectors as the *columns* of a matrix A, we can determine their independence by row reducing the augmented matrix for $A\mathbf{x} = \mathbf{0}$:

$$\begin{bmatrix} 1 & 4 & 3 & 0 \\ 0 & 1 & 2 & 0 \\ 2 & 5 & 0 & 0 \end{bmatrix} \sim \begin{bmatrix} 1 & 4 & 3 & 0 \\ 0 & 1 & 2 & 0 \\ 0 & 0 & 0 & 0 \end{bmatrix}$$

The columns of A are linearly dependent, so the corresponding polynomials are linearly dependent. In fact, it is easy to check that column 3 of A is 2 times column 2 minus 5 times column 1. The corresponding relation for the polynomials is

$$3 + 2t = 2(4 + t + 5t^2) - 5(1 + 2t^2) \qquad\blacksquare$$

The final example concerns a plane in $\mathbb{R}^3$ that is isomorphic to $\mathbb{R}^2$.

EXAMPLE 7 Let

$$\mathbf{v}_1 = \begin{bmatrix} 3 \\ 6 \\ 2 \end{bmatrix}, \quad \mathbf{v}_2 = \begin{bmatrix} -1 \\ 0 \\ 1 \end{bmatrix}, \quad \mathbf{x} = \begin{bmatrix} 3 \\ 12 \\ 7 \end{bmatrix},$$

and $\mathcal{B} = \{\mathbf{v}_1, \mathbf{v}_2\}$. Then $\mathcal{B}$ is a basis for $H = \text{Span}\,\{\mathbf{v}_1, \mathbf{v}_2\}$. Determine if $\mathbf{x}$ is in H, and if it is, find the coordinate vector of $\mathbf{x}$ relative to $\mathcal{B}$.

SOLUTION If $\mathbf{x}$ is in H, then the following vector equation is consistent:

$$c_1 \begin{bmatrix} 3 \\ 6 \\ 2 \end{bmatrix} + c_2 \begin{bmatrix} -1 \\ 0 \\ 1 \end{bmatrix} = \begin{bmatrix} 3 \\ 12 \\ 7 \end{bmatrix}$$

The scalars c_1 and c_2, if they exist, are the $\mathcal{B}$-coordinates of $\mathbf{x}$. Using row operations, we obtain

$$\begin{bmatrix} 3 & -1 & 3 \\ 6 & 0 & 12 \\ 2 & 1 & 7 \end{bmatrix} \sim \begin{bmatrix} 1 & 0 & 2 \\ 0 & 1 & 3 \\ 0 & 0 & 0 \end{bmatrix}$$

Thus $c_1 = 2$, $c_2 = 3$, and $[\,\mathbf{x}\,]_{\mathcal{B}} = \begin{bmatrix} 2 \\ 3 \end{bmatrix}$. The coordinate system on H determined by $\mathcal{B}$ is shown in Figure 7. $\qquad\blacksquare$

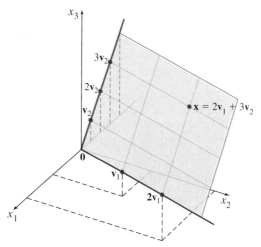

FIGURE 7 A coordinate system on a plane H in $\mathbb{R}^3$.

If a different basis for H were chosen, would the associated coordinate system also make H isomorphic to $\mathbb{R}^2$? Surely, this must be true. We shall prove it in the next section.

Practice Problems

1. Let $\mathbf{b}_1 = \begin{bmatrix} 1 \\ 0 \\ 0 \end{bmatrix}$, $\mathbf{b}_2 = \begin{bmatrix} -3 \\ 4 \\ 0 \end{bmatrix}$, $\mathbf{b}_3 = \begin{bmatrix} 3 \\ -6 \\ 3 \end{bmatrix}$, and $\mathbf{x} = \begin{bmatrix} -8 \\ 2 \\ 3 \end{bmatrix}$.

   a. Show that the set $\mathcal{B} = \{\mathbf{b}_1, \mathbf{b}_2, \mathbf{b}_3\}$ is a basis of $\mathbb{R}^3$.

   b. Find the change-of-coordinates matrix from $\mathcal{B}$ to the standard basis.

   c. Write the equation that relates $\mathbf{x}$ in $\mathbb{R}^3$ to $[\mathbf{x}]_\mathcal{B}$.

   d. Find $[\mathbf{x}]_\mathcal{B}$, for the $\mathbf{x}$ given above.

2. The set $\mathcal{B} = \{1 + t, 1 + t^2, t + t^2\}$ is a basis for $\mathbb{P}_2$. Find the coordinate vector of $\mathbf{p}(t) = 6 + 3t - t^2$ relative to $\mathcal{B}$.

4.4 Exercises

In Exercises 1–4, find the vector $\mathbf{x}$ determined by the given coordinate vector $[\mathbf{x}]_\mathcal{B}$ and the given basis $\mathcal{B}$.

1. $\mathcal{B} = \left\{ \begin{bmatrix} 3 \\ -5 \end{bmatrix}, \begin{bmatrix} -4 \\ 6 \end{bmatrix} \right\}$, $[\mathbf{x}]_\mathcal{B} = \begin{bmatrix} 5 \\ 3 \end{bmatrix}$

2. $\mathcal{B} = \left\{ \begin{bmatrix} 4 \\ 5 \end{bmatrix}, \begin{bmatrix} 6 \\ 7 \end{bmatrix} \right\}$, $[\mathbf{x}]_\mathcal{B} = \begin{bmatrix} 8 \\ -5 \end{bmatrix}$

3. $\mathcal{B} = \left\{ \begin{bmatrix} 1 \\ -4 \\ 3 \end{bmatrix}, \begin{bmatrix} 5 \\ 2 \\ -2 \end{bmatrix}, \begin{bmatrix} 4 \\ -7 \\ 0 \end{bmatrix} \right\}$, $[\mathbf{x}]_\mathcal{B} = \begin{bmatrix} 3 \\ 0 \\ -1 \end{bmatrix}$

4. $\mathcal{B} = \left\{ \begin{bmatrix} -1 \\ 2 \\ 0 \end{bmatrix}, \begin{bmatrix} 3 \\ -5 \\ 2 \end{bmatrix}, \begin{bmatrix} 4 \\ -7 \\ 3 \end{bmatrix} \right\}$, $[\mathbf{x}]_\mathcal{B} = \begin{bmatrix} -4 \\ 8 \\ -7 \end{bmatrix}$

In Exercises 5–8, find the coordinate vector $[\mathbf{x}]_\mathcal{B}$ of $\mathbf{x}$ relative to the given basis $\mathcal{B} = \{\mathbf{b}_1, \ldots, \mathbf{b}_n\}$.

5. $\mathbf{b}_1 = \begin{bmatrix} 1 \\ -3 \end{bmatrix}$, $\mathbf{b}_2 = \begin{bmatrix} 2 \\ -5 \end{bmatrix}$, $\mathbf{x} = \begin{bmatrix} -2 \\ 1 \end{bmatrix}$

6. $\mathbf{b}_1 = \begin{bmatrix} 1 \\ -2 \end{bmatrix}$, $\mathbf{b}_2 = \begin{bmatrix} 5 \\ -6 \end{bmatrix}$, $\mathbf{x} = \begin{bmatrix} 4 \\ 0 \end{bmatrix}$

7. $\mathbf{b}_1 = \begin{bmatrix} 1 \\ -1 \\ -3 \end{bmatrix}$, $\mathbf{b}_2 = \begin{bmatrix} -3 \\ 4 \\ 9 \end{bmatrix}$, $\mathbf{b}_3 = \begin{bmatrix} 2 \\ -2 \\ 4 \end{bmatrix}$, $\mathbf{x} = \begin{bmatrix} 8 \\ -9 \\ 6 \end{bmatrix}$

8. $\mathbf{b}_1 = \begin{bmatrix} 1 \\ 0 \\ 3 \end{bmatrix}$, $\mathbf{b}_2 = \begin{bmatrix} 2 \\ 1 \\ 8 \end{bmatrix}$, $\mathbf{b}_3 = \begin{bmatrix} 1 \\ -1 \\ 2 \end{bmatrix}$, $\mathbf{x} = \begin{bmatrix} 3 \\ -5 \\ 4 \end{bmatrix}$

In Exercises 9 and 10, find the change-of-coordinates matrix from $\mathcal{B}$ to the standard basis in $\mathbb{R}^n$.

9. $\mathcal{B} = \left\{ \begin{bmatrix} 2 \\ -9 \end{bmatrix}, \begin{bmatrix} 1 \\ 8 \end{bmatrix} \right\}$

10. $\mathcal{B} = \left\{ \begin{bmatrix} 3 \\ -1 \\ 4 \end{bmatrix}, \begin{bmatrix} 2 \\ 0 \\ -5 \end{bmatrix}, \begin{bmatrix} 8 \\ -2 \\ 7 \end{bmatrix} \right\}$

In Exercises 11 and 12, use an inverse matrix to find $[\mathbf{x}]_\mathcal{B}$ for the given $\mathbf{x}$ and $\mathcal{B}$.

11. $\mathcal{B} = \left\{ \begin{bmatrix} 3 \\ -5 \end{bmatrix}, \begin{bmatrix} -4 \\ 6 \end{bmatrix} \right\}$, $\mathbf{x} = \begin{bmatrix} 2 \\ -6 \end{bmatrix}$

12. $\mathcal{B} = \left\{ \begin{bmatrix} 4 \\ 5 \end{bmatrix}, \begin{bmatrix} 6 \\ 7 \end{bmatrix} \right\}$, $\mathbf{x} = \begin{bmatrix} 2 \\ 0 \end{bmatrix}$

13. The set $\mathcal{B} = \{1 + t^2, t + t^2, 1 + 2t + t^2\}$ is a basis for $\mathbb{P}_2$. Find the coordinate vector of $\mathbf{p}(t) = 1 + 4t + 7t^2$ relative to $\mathcal{B}$.

14. The set $\mathcal{B} = \{1 - t^2, t - t^2, 2 - 2t + t^2\}$ is a basis for $\mathbb{P}_2$. Find the coordinate vector of $\mathbf{p}(t) = 3 + t - 6t^2$ relative to $\mathcal{B}$.

In Exercises 15–20, mark each statement True or False (**T/F**). Justify each answer. Unless stated otherwise, $\mathcal{B}$ is a basis for a vector space V.

15. (**T/F**) If $\mathbf{x}$ is in V and if $\mathcal{B}$ contains n vectors, then the $\mathcal{B}$-coordinate vector of $\mathbf{x}$ is in $\mathbb{R}^n$.

16. (**T/F**) If $\mathcal{B}$ is the standard basis for $\mathbb{R}^n$, then the $\mathcal{B}$-coordinate vector of an $\mathbf{x}$ in $\mathbb{R}^n$ is $\mathbf{x}$ itself.

17. (**T/F**) If $P_\mathcal{B}$ is the change-of-coordinates matrix, then $[\mathbf{x}]_\mathcal{B} = P_\mathcal{B}\mathbf{x}$, for $\mathbf{x}$ in V.

18. (**T/F**) The correspondence $[\mathbf{x}]_\mathcal{B} \mapsto \mathbf{x}$ is called the coordinate mapping.

19. (**T/F**) The vector spaces $\mathbb{P}_3$ and $\mathbb{R}^3$ are isomorphic.

20. (**T/F**) In some cases, a plane in $\mathbb{R}^3$ can be isomorphic to $\mathbb{R}^2$.

21. The vectors $\mathbf{v}_1 = \begin{bmatrix} 1 \\ -3 \end{bmatrix}, \mathbf{v}_2 = \begin{bmatrix} 2 \\ -8 \end{bmatrix}, \mathbf{v}_3 = \begin{bmatrix} -3 \\ 7 \end{bmatrix}$ span $\mathbb{R}^2$ but do not form a basis. Find two different ways to express $\begin{bmatrix} 1 \\ 1 \end{bmatrix}$ as a linear combination of $\mathbf{v}_1, \mathbf{v}_2, \mathbf{v}_3$.

22. Let $\mathcal{B} = \{\mathbf{b}_1, \ldots, \mathbf{b}_n\}$ be a basis for a vector space V. Explain why the $\mathcal{B}$-coordinate vectors of $\mathbf{b}_1, \ldots, \mathbf{b}_n$ are the columns $\mathbf{e}_1, \ldots, \mathbf{e}_n$ of the $n \times n$ identity matrix.

23. Let S be a finite set in a vector space V with the property that every $\mathbf{x}$ in V has a unique representation as a linear combination of elements of S. Show that S is a basis of V.

24. Suppose $\{\mathbf{v}_1, \ldots, \mathbf{v}_4\}$ is a linearly dependent spanning set for a vector space V. Show that each $\mathbf{w}$ in V can be expressed in more than one way as a linear combination of $\mathbf{v}_1, \ldots, \mathbf{v}_4$. [*Hint:* Let $\mathbf{w} = k_1\mathbf{v}_1 + \cdots + k_4\mathbf{v}_4$ be an arbitrary vector in V. Use the linear dependence of $\{\mathbf{v}_1, \ldots, \mathbf{v}_4\}$ to produce another representation of $\mathbf{w}$ as a linear combination of $\mathbf{v}_1, \ldots, \mathbf{v}_4$.]

25. Let $\mathcal{B} = \left\{ \begin{bmatrix} 1 \\ -4 \end{bmatrix}, \begin{bmatrix} -2 \\ 9 \end{bmatrix} \right\}$. Since the coordinate mapping determined by $\mathcal{B}$ is a linear transformation from $\mathbb{R}^2$ into $\mathbb{R}^2$, this mapping must be implemented by some 2×2 matrix A. Find it. [*Hint:* Multiplication by A should transform a vector $\mathbf{x}$ into its coordinate vector $[\mathbf{x}]_{\mathcal{B}}$.]

26. Let $\mathcal{B} = \{\mathbf{b}_1, \ldots, \mathbf{b}_n\}$ be a basis for $\mathbb{R}^n$. Produce a description of an $n \times n$ matrix A that implements the coordinate mapping $\mathbf{x} \mapsto [\mathbf{x}]_{\mathcal{B}}$. (See Exercise 25.)

Exercises 27–30 concern a vector space V, a basis $\mathcal{B} = \{\mathbf{b}_1, \ldots, \mathbf{b}_n\}$, and the coordinate mapping $\mathbf{x} \mapsto [\mathbf{x}]_{\mathcal{B}}$.

27. Show that the coordinate mapping is one-to-one. [*Hint:* Suppose $[\mathbf{u}]_{\mathcal{B}} = [\mathbf{w}]_{\mathcal{B}}$ for some $\mathbf{u}$ and $\mathbf{w}$ in V, and show that $\mathbf{u} = \mathbf{w}$.]

28. Show that the coordinate mapping is *onto* $\mathbb{R}^n$. That is, given any $\mathbf{y}$ in $\mathbb{R}^n$, with entries $y_1, \ldots, y_n$, produce $\mathbf{u}$ in V such that $[\mathbf{u}]_{\mathcal{B}} = \mathbf{y}$.

29. Show that a subset $\{\mathbf{u}_1, \ldots, \mathbf{u}_p\}$ in V is linearly independent if and only if the set of coordinate vectors $\{[\mathbf{u}_1]_{\mathcal{B}}, \ldots, [\mathbf{u}_p]_{\mathcal{B}}\}$ is linearly independent in $\mathbb{R}^n$. [*Hint:* Since the coordinate mapping is one-to-one, the following equations have the same solutions, $c_1, \ldots, c_p$.]

$$c_1\mathbf{u}_1 + \cdots + c_p\mathbf{u}_p = \mathbf{0} \qquad \text{The zero vector in } V$$
$$[c_1\mathbf{u}_1 + \cdots + c_p\mathbf{u}_p]_{\mathcal{B}} = [\mathbf{0}]_{\mathcal{B}} \qquad \text{The zero vector in } \mathbb{R}^n$$

30. Given vectors $\mathbf{u}_1, \ldots, \mathbf{u}_p$, and $\mathbf{w}$ in V, show that $\mathbf{w}$ is a linear combination of $\mathbf{u}_1, \ldots, \mathbf{u}_p$ if and only if $[\mathbf{w}]_{\mathcal{B}}$ is a linear combination of the coordinate vectors $[\mathbf{u}_1]_{\mathcal{B}}, \ldots, [\mathbf{u}_p]_{\mathcal{B}}$.

In Exercises 31–34, use coordinate vectors to test the linear independence of the sets of polynomials. Explain your work.

31. $\{1 + 2t^3, 2 + t - 3t^2, -t + 2t^2 - t^3\}$

32. $\{1 - 2t^2 - t^3, t + 2t^3, 1 + t - 2t^2\}$

33. $\{(1 - t)^2, t - 2t^2 + t^3, (1 - t)^3\}$

34. $\{(2 - t)^3, (3 - t)^2, 1 + 6t - 5t^2 + t^3\}$

35. Use coordinate vectors to test whether the following sets of polynomials span $\mathbb{P}_2$. Justify your conclusions.
  a. $\{1 - 3t + 5t^2, -3 + 5t - 7t^2, -4 + 5t - 6t^2, 1 - t^2\}$
  b. $\{5t + t^2, 1 - 8t - 2t^2, -3 + 4t + 2t^2, 2 - 3t\}$

36. Let $\mathbf{p}_1(t) = 1 + t^2, \mathbf{p}_2(t) = t - 3t^2, \mathbf{p}_3(t) = 1 + t - 3t^2$.
  a. Use coordinate vectors to show that these polynomials form a basis for $\mathbb{P}_2$.
  b. Consider the basis $\mathcal{B} = \{\mathbf{p}_1, \mathbf{p}_2, \mathbf{p}_3\}$ for $\mathbb{P}_2$. Find $\mathbf{q}$ in $\mathbb{P}_2$, given that $[\mathbf{q}]_{\mathcal{B}} = \begin{bmatrix} -1 \\ 1 \\ 2 \end{bmatrix}$.

In Exercises 37 and 38, determine whether the sets of polynomials form a basis for $\mathbb{P}_3$. Justify your conclusions.

T 37. $3 + 7t, 5 + t - 2t^3, t - 2t^2, 1 + 16t - 6t^2 + 2t^3$

T 38. $5 - 3t + 4t^2 + 2t^3, 9 + t + 8t^2 - 6t^3, 6 - 2t + 5t^2, t^3$

T 39. Let $H = \text{Span}\{\mathbf{v}_1, \mathbf{v}_2\}$ and $\mathcal{B} = \{\mathbf{v}_1, \mathbf{v}_2\}$. Show that $\mathbf{x}$ is in H and find the $\mathcal{B}$-coordinate vector of $\mathbf{x}$, for

$$\mathbf{v}_1 = \begin{bmatrix} 11 \\ -5 \\ 10 \\ 7 \end{bmatrix}, \mathbf{v}_2 = \begin{bmatrix} 14 \\ -8 \\ 13 \\ 10 \end{bmatrix}, \mathbf{x} = \begin{bmatrix} 19 \\ -13 \\ 18 \\ 15 \end{bmatrix}$$

T 40. Let $H = \text{Span}\{\mathbf{v}_1, \mathbf{v}_2, \mathbf{v}_3\}$ and $\mathcal{B} = \{\mathbf{v}_1, \mathbf{v}_2, \mathbf{v}_3\}$. Show that $\mathcal{B}$ is a basis for H and $\mathbf{x}$ is in H, and find the $\mathcal{B}$-coordinate vector of $\mathbf{x}$, for

$$\mathbf{v}_1 = \begin{bmatrix} -6 \\ 4 \\ -9 \\ 4 \end{bmatrix}, \mathbf{v}_2 = \begin{bmatrix} 8 \\ -3 \\ 7 \\ -3 \end{bmatrix}, \mathbf{v}_3 = \begin{bmatrix} -9 \\ 5 \\ -8 \\ 3 \end{bmatrix}, \mathbf{x} = \begin{bmatrix} 4 \\ 7 \\ -8 \\ 3 \end{bmatrix}$$

Exercises 41 and 42 concern the crystal lattice for titanium, which has the hexagonal structure shown on the left in the accompanying figure. The vectors $\begin{bmatrix} 2.6 \\ -1.5 \\ 0 \end{bmatrix}, \begin{bmatrix} 0 \\ 3 \\ 0 \end{bmatrix}, \begin{bmatrix} 0 \\ 0 \\ 4.8 \end{bmatrix}$ in $\mathbb{R}^3$ form a basis for the unit cell shown on the right. The numbers here are Ångstrom units ($1 \text{ Å} = 10^{-8}$ cm). In alloys of titanium, some additional atoms may be in the unit cell at the *octahedral* and *tetrahedral* sites (so named because of the geometric objects formed by atoms at these locations).

The hexagonal close-packed lattice and its unit cell.

41. One of the octahedral sites is $\begin{bmatrix} 1/2 \\ 1/4 \\ 1/6 \end{bmatrix}$, relative to the lattice basis. Determine the coordinates of this site relative to the standard basis of $\mathbb{R}^3$.

42. One of the tetrahedral sites is $\begin{bmatrix} 1/2 \\ 1/2 \\ 1/3 \end{bmatrix}$. Determine the coordinates of this site relative to the standard basis of $\mathbb{R}^3$.

Solutions to Practice Problems

1. a. It is evident that the matrix $P_{\mathcal{B}} = [\, \mathbf{b}_1 \quad \mathbf{b}_2 \quad \mathbf{b}_3 \,]$ is row-equivalent to the identity matrix. By the Invertible Matrix Theorem, $P_{\mathcal{B}}$ is invertible and its columns form a basis for $\mathbb{R}^3$.

   b. From part (a), the change-of-coordinates matrix is $P_{\mathcal{B}} = \begin{bmatrix} 1 & -3 & 3 \\ 0 & 4 & -6 \\ 0 & 0 & 3 \end{bmatrix}$.

   c. $\mathbf{x} = P_{\mathcal{B}}[\, \mathbf{x} \,]_{\mathcal{B}}$

   d. To solve the equation in (c), it is probably easier to row reduce an augmented matrix than to compute $P_{\mathcal{B}}^{-1}$:

$$
\underbrace{\begin{bmatrix} 1 & -3 & 3 \\ 0 & 4 & -6 \\ 0 & 0 & 3 \end{bmatrix}}_{P_{\mathcal{B}}} \underbrace{\begin{matrix} -8 \\ 2 \\ 3 \end{matrix}}_{\mathbf{x}} \sim \underbrace{\begin{bmatrix} 1 & 0 & 0 \\ 0 & 1 & 0 \\ 0 & 0 & 1 \end{bmatrix}}_{I} \underbrace{\begin{matrix} -5 \\ 2 \\ 1 \end{matrix}}_{[\, \mathbf{x} \,]_{\mathcal{B}}}
$$

   Hence

$$
[\, \mathbf{x} \,]_{\mathcal{B}} = \begin{bmatrix} -5 \\ 2 \\ 1 \end{bmatrix}
$$

2. The coordinates of $\mathbf{p}(t) = 6 + 3t - t^2$ with respect to $\mathcal{B}$ satisfy

$$
c_1(1 + t) + c_2(1 + t^2) + c_3(t + t^2) = 6 + 3t - t^2
$$

Equating coefficients of like powers of t, we have

$$
\begin{array}{rcl}
c_1 + c_2 & = & 6 \\
c_1 \quad + c_3 & = & 3 \\
c_2 + c_3 & = & -1
\end{array}
$$

Solving, we find that $c_1 = 5$, $c_2 = 1$, $c_3 = -2$, and $[\, \mathbf{p} \,]_{\mathcal{B}} = \begin{bmatrix} 5 \\ 1 \\ -2 \end{bmatrix}$.

4.5 The Dimension of a Vector Space

Theorem 9 in Section 4.4 implies that a vector space V with a basis $\mathcal{B}$ containing n vectors is isomorphic to $\mathbb{R}^n$. This section shows that this number n is an intrinsic property (called the dimension) of the space V that does not depend on the particular choice of basis. The discussion of dimension will give additional insight into properties of bases.

The first theorem generalizes a well-known result about the vector space $\mathbb{R}^n$.

THEOREM 10

> If a vector space V has a basis $\mathcal{B} = \{\mathbf{b}_1, \ldots, \mathbf{b}_n\}$, then any set in V containing more than n vectors must be linearly dependent.

PROOF   Let $\{\mathbf{u}_1, \ldots, \mathbf{u}_p\}$ be a set in V with more than n vectors. The coordinate vectors $[\mathbf{u}_1]_{\mathcal{B}}, \ldots, [\mathbf{u}_p]_{\mathcal{B}}$ form a linearly dependent set in $\mathbb{R}^n$, because there are more vectors (p) than entries (n) in each vector. So there exist scalars $c_1, \ldots, c_p$, not all zero, such that

$$c_1[\mathbf{u}_1]_{\mathcal{B}} + \cdots + c_p[\mathbf{u}_p]_{\mathcal{B}} = \begin{bmatrix} 0 \\ \vdots \\ 0 \end{bmatrix} \qquad \text{The zero vector in } \mathbb{R}^n$$

Since the coordinate mapping is a linear transformation,

$$\left[c_1\mathbf{u}_1 + \cdots + c_p\mathbf{u}_p \right]_{\mathcal{B}} = \begin{bmatrix} 0 \\ \vdots \\ 0 \end{bmatrix}$$

The zero vector on the right displays the n weights needed to build the vector $c_1\mathbf{u}_1 + \cdots + c_p\mathbf{u}_p$ from the basis vectors in $\mathcal{B}$. That is, $c_1\mathbf{u}_1 + \cdots + c_p\mathbf{u}_p = 0\mathbf{b}_1 + \cdots + 0\mathbf{b}_n = \mathbf{0}$. Since the c_i are not all zero, $\{\mathbf{u}_1, \ldots, \mathbf{u}_p\}$ is linearly dependent.[1]   ∎

Theorem 10 implies that if a vector space V has a basis $\mathcal{B} = \{\mathbf{b}_1, \ldots, \mathbf{b}_n\}$, then each linearly independent set in V has no more than n vectors.

THEOREM 11

> If a vector space V has a basis of n vectors, then every basis of V must consist of exactly n vectors.

PROOF   Let $\mathcal{B}_1$ be a basis of n vectors and $\mathcal{B}_2$ be any other basis (of V). Since $\mathcal{B}_1$ is a basis and $\mathcal{B}_2$ is linearly independent, $\mathcal{B}_2$ has no more than n vectors, by Theorem 10. Also, since $\mathcal{B}_2$ is a basis and $\mathcal{B}_1$ is linearly independent, $\mathcal{B}_2$ has at least n vectors. Thus $\mathcal{B}_2$ consists of exactly n vectors.   ∎

---

[1] Theorem 10 also applies to infinite sets in V. An infinite set is said to be linearly dependent if some finite subset is linearly dependent; otherwise, the set is linearly independent. If S is an infinite set in V, take any subset $\{\mathbf{u}_1, \ldots, \mathbf{u}_p\}$ of S, with $p > n$. The proof above shows that this subset is linearly dependent and hence so is S.

If a nonzero vector space V is spanned by a finite set S, then a subset of S is a basis for V, by the Spanning Set Theorem. In this case, Theorem 11 ensures that the following definition makes sense.

DEFINITION

If a vector space V is spanned by a finite set, then V is said to be **finite-dimensional**, and the **dimension** of V, written as dim V, is the number of vectors in a basis for V. The dimension of the zero vector space $\{\mathbf{0}\}$ is defined to be zero. If V is not spanned by a finite set, then V is said to be **infinite-dimensional**.

EXAMPLE 1 The standard basis for $\mathbb{R}^n$ contains n vectors, so dim $\mathbb{R}^n = n$. The standard polynomial basis $\{1, t, t^2\}$ shows that dim $\mathbb{P}_2 = 3$. In general, dim $\mathbb{P}_n = n + 1$. The space $\mathbb{P}$ of all polynomials is infinite-dimensional. ∎

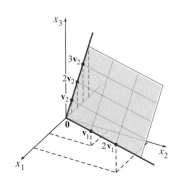

EXAMPLE 2 Let $H = \text{Span}\{\mathbf{v}_1, \mathbf{v}_2\}$, where $\mathbf{v}_1 = \begin{bmatrix} 3 \\ 6 \\ 2 \end{bmatrix}$ and $\mathbf{v}_2 = \begin{bmatrix} -1 \\ 0 \\ 1 \end{bmatrix}$. Then H is the plane studied in Example 7 in Section 4.4. A basis for H is $\{\mathbf{v}_1, \mathbf{v}_2\}$, since $\mathbf{v}_1$ and $\mathbf{v}_2$ are not multiples and hence are linearly independent. Thus dim $H = 2$. ∎

EXAMPLE 3 Find the dimension of the subspace

$$H = \left\{ \begin{bmatrix} a - 3b + 6c \\ 5a + 4d \\ b - 2c - d \\ 5d \end{bmatrix} : a, b, c, d \text{ in } \mathbb{R} \right\}$$

SOLUTION It is easy to see that H is the set of all linear combinations of the vectors

$$\mathbf{v}_1 = \begin{bmatrix} 1 \\ 5 \\ 0 \\ 0 \end{bmatrix}, \quad \mathbf{v}_2 = \begin{bmatrix} -3 \\ 0 \\ 1 \\ 0 \end{bmatrix}, \quad \mathbf{v}_3 = \begin{bmatrix} 6 \\ 0 \\ -2 \\ 0 \end{bmatrix}, \quad \mathbf{v}_4 = \begin{bmatrix} 0 \\ 4 \\ -1 \\ 5 \end{bmatrix}$$

Clearly, $\mathbf{v}_1 \neq \mathbf{0}$, $\mathbf{v}_2$ is not a multiple of $\mathbf{v}_1$, but $\mathbf{v}_3$ is a multiple of $\mathbf{v}_2$. By the Spanning Set Theorem, we may discard $\mathbf{v}_3$ and still have a set that spans H. Finally, $\mathbf{v}_4$ is not a linear combination of $\mathbf{v}_1$ and $\mathbf{v}_2$. So $\{\mathbf{v}_1, \mathbf{v}_2, \mathbf{v}_4\}$ is linearly independent (by Theorem 4 in Section 4.3) and hence is a basis for H. Thus dim $H = 3$. ∎

EXAMPLE 4 The subspaces of $\mathbb{R}^3$ can be classified by dimension. See Figure 1.

0-dimensional subspaces. Only the zero subspace.

1-dimensional subspaces. Any subspace spanned by a single nonzero vector. Such subspaces are lines through the origin.

2-dimensional subspaces. Any subspace spanned by two linearly independent vectors. Such subspaces are planes through the origin.

3-dimensional subspaces. Only $\mathbb{R}^3$ itself. Any three linearly independent vectors in $\mathbb{R}^3$ span all of $\mathbb{R}^3$, by the Invertible Matrix Theorem. ∎

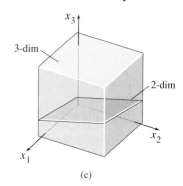

FIGURE 1 Sample subspaces of $\mathbb{R}^3$.

Subspaces of a Finite-Dimensional Space

The next theorem is a natural counterpart to the Spanning Set Theorem.

THEOREM 12

Let H be a subspace of a finite-dimensional vector space V. Any linearly independent set in H can be expanded, if necessary, to a basis for H. Also, H is finite-dimensional and

$$\dim H \leq \dim V$$

PROOF If $H = \{\mathbf{0}\}$, then certainly $\dim H = 0 \leq \dim V$. Otherwise, let $S = \{\mathbf{u}_1, \ldots, \mathbf{u}_k\}$ be any linearly independent set in H. If S spans H, then S is a basis for H. Otherwise, there is some $\mathbf{u}_{k+1}$ in H that is not in Span S. But then $\{\mathbf{u}_1, \ldots, \mathbf{u}_k, \mathbf{u}_{k+1}\}$ will be linearly independent, because no vector in the set can be a linear combination of vectors that precede it (by Theorem 4).

So long as the new set does not span H, we can continue this process of expanding S to a larger linearly independent set in H. But the number of vectors in a linearly independent expansion of S can never exceed the dimension of V, by Theorem 10. So eventually the expansion of S will span H and hence will be a basis for H, and $\dim H \leq \dim V$.   ■

When the dimension of a vector space or subspace is known, the search for a basis is simplified by the next theorem. It says that if a set has the right number of elements, then one has only to show either that the set is linearly independent or that it spans the space. The theorem is of critical importance in numerous applied problems (involving differential equations or difference equations, for example) where linear independence is much easier to verify than spanning.

THEOREM 13

The Basis Theorem

Let V be a p-dimensional vector space, $p \geq 1$. Any linearly independent set of exactly p elements in V is automatically a basis for V. Any set of exactly p elements that spans V is automatically a basis for V.

PROOF By Theorem 12, a linearly independent set S of p elements can be extended to a basis for V. But that basis must contain exactly p elements, since $\dim V = p$. So S must already be a basis for V. Now suppose that S has p elements and spans V. Since V is nonzero, the Spanning Set Theorem implies that a subset S' of S is a basis of V. Since $\dim V = p$, S' must contain p vectors. Hence $S = S'$.   ■

The Dimensions of Nul A, Col A, and Row A

Since the dimensions of the null space and column space of an $m \times n$ matrix are referred to frequently, they have specific names:

DEFINITION

> The **rank** of an $m \times n$ matrix A is the dimension of the column space and the **nullity** of A is the dimension of the null space.

The pivot columns of a matrix A form a basis for Col A, so the rank of A is just the number of pivot columns. Since a basis for Row A can be found by taking the pivot rows from the row reduced echelon form of A, the dimension of Row A is also equal to the rank of A.

The nullity of A might seem to require more work, since finding a basis for Nul A usually takes more time than finding a basis for Col A. There is a shortcut: Let A be an $m \times n$ matrix, and suppose the equation $A\mathbf{x} = \mathbf{0}$ has k free variables. From Section 4.2, we know that the standard method of finding a spanning set for Nul A will produce exactly k linearly independent vectors—say, $\mathbf{u}_1, \ldots, \mathbf{u}_k$ – one for each free variable. So $\mathbf{u}_1, \ldots, \mathbf{u}_k$ is a basis for Nul A, and the number of free variables determines the size of the basis.

To summarize these facts for future reference:

> The rank of an $m \times n$ matrix A is the number of pivot columns and the nullity of A is the number of free variables. Since the dimension of the row space is the number of pivot rows, it is also equal to the rank of A.

Putting these observations together results in the rank theorem.

THEOREM 14

> **The Rank Theorem**
>
> The dimensions of the column space and the null space of an $m \times n$ matrix A satisfy the equation
>
> $$\text{rank } A + \text{nullity } A = \text{number of columns in } A$$

PROOF By Theorem 6 in Section 4.3, rank A is the number of pivot columns in A. The nullity of A equals the number of free variables in the equation $A\mathbf{x} = \mathbf{0}$. Expressed another way, the nullity of A is the number of columns of A that are *not* pivot columns. (It is the number of these columns, not the columns themselves, that is related to Nul A.) Obviously,

$$\left\{ \begin{array}{c} \text{number of} \\ \text{pivot columns} \end{array} \right\} + \left\{ \begin{array}{c} \text{number of} \\ \text{nonpivot columns} \end{array} \right\} = \left\{ \begin{array}{c} \text{number of} \\ \text{columns} \end{array} \right\}$$

This proves the theorem. ∎

EXAMPLE 5   Find the nullity and rank of

$$A = \begin{bmatrix} -3 & 6 & -1 & 1 & -7 \\ 1 & -2 & 2 & 3 & -1 \\ 2 & -4 & 5 & 8 & -4 \end{bmatrix}$$

SOLUTION   Row reduce the augmented matrix $[A \ \mathbf{0}]$ to echelon form:

$$B = \begin{bmatrix} 1 & -2 & 2 & 3 & -1 & 0 \\ 0 & 0 & 1 & 2 & -2 & 0 \\ 0 & 0 & 0 & 0 & 0 & 0 \end{bmatrix}$$

There are three free variables: x_2, x_4, and x_5. Hence the nullity of A is 3. Also, the rank of A is 2 because A has two pivot columns.   ∎

The ideas behind Theorem 14 are visible in the calculations in Example 5. The two pivot positions in B, an echelon form of A, determine the basic variables and identify the basis vectors for Col A and those for Row A.

EXAMPLE 6

a.   If A is a 7×9 matrix with nullity 2, what is the rank of A?

b.   Could a 6×9 matrix have nullity 2?

SOLUTION

a.   Since A has 9 columns, (rank A) $+ 2 = 9$, and hence rank $A = 7$.

b.   No. If a 6×9 matrix, call it B, had a two-dimensional null space, it would have to have rank 7, by the Rank Theorem. But the columns of B are vectors in $\mathbb{R}^6$, and so the dimension of Col B cannot exceed 6; that is, rank B cannot exceed 6.   ∎

The next example provides a nice way to visualize the subspaces we have been studying. In Chapter 6, we will learn that Row A and Nul A have only the zero vector in common and are actually perpendicular to each other. The same fact applies to Row A^T ($= $ Col$A$) and Nul A^T. So Figure 2, which accompanies Example 7, creates a good mental image for the general case.

EXAMPLE 7   Let $A = \begin{bmatrix} 3 & 0 & -1 \\ 3 & 0 & -1 \\ 4 & 0 & 5 \end{bmatrix}$. It is readily checked that Nul A is the

x_2-axis, Row A is the $x_1 x_3$-plane, Col A is the plane whose equation is $x_1 - x_2 = 0$, and Nul A^T is the set of all multiples of $(1, -1, 0)$. Figure 2 shows Nul A and Row A in the domain of the linear transformation $\mathbf{x} \mapsto A\mathbf{x}$; the range of this mapping, Col A, is shown in a separate copy of $\mathbb{R}^3$, along with Nul A^T.   ∎

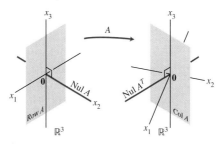

FIGURE 2 Subspaces determined by a matrix A.

Applications to Systems of Equations

The Rank Theorem is a powerful tool for processing information about systems of linear equations. The next example simulates the way a real-life problem using linear equations might be stated, without explicit mention of linear algebra terms such as matrix, subspace, and dimension.

EXAMPLE 8　A scientist has found two solutions to a homogeneous system of 40 equations in 42 variables. The two solutions are not multiples, and all other solutions can be constructed by adding together appropriate multiples of these two solutions. Can the scientist be *certain* that an associated nonhomogeneous system (with the same coefficients) has a solution?

SOLUTION　Yes. Let A be the 40×42 coefficient matrix of the system. The given information implies that the two solutions are linearly independent and span Nul A. So nullity $A = 2$. By the Rank Theorem, rank $A = 42 - 2 = 40$. Since $\mathbb{R}^{40}$ is the only subspace of $\mathbb{R}^{40}$ whose dimension is 40, Col A must be all of $\mathbb{R}^{40}$. This means that every nonhomogeneous equation $A\mathbf{x} = \mathbf{b}$ has a solution. ∎

Rank and the Invertible Matrix Theorem

The various vector space concepts associated with a matrix provide several more statements for the Invertible Matrix Theorem. The new statements listed here follow those in the original Invertible Matrix Theorem in Section 2.3 and other theorems in the text where statements have been added to it.

THEOREM

> **The Invertible Matrix Theorem (continued)**
>
> Let A be an $n \times n$ matrix. Then the following statements are each equivalent to the statement that A is an invertible matrix.
>
> m. The columns of A form a basis of $\mathbb{R}^n$.
>
> n. Col $A = \mathbb{R}^n$
>
> o. rank $A = n$
>
> p. nullity $A = 0$
>
> q. Nul $A = \{\mathbf{0}\}$

PROOF　Statement (m) is logically equivalent to statements (e) and (h) regarding linear independence and spanning. The other five statements are linked to the earlier ones of the theorem by the following chain of almost trivial implications:

$$(g) \Rightarrow (n) \Rightarrow (o) \Rightarrow (p) \Rightarrow (q) \Rightarrow (d)$$

Statement (g), which says that the equation $A\mathbf{x} = \mathbf{b}$ has at least one solution for each $\mathbf{b}$ in $\mathbb{R}^n$, implies (n), because Col A is precisely the set of all $\mathbf{b}$ such that the equation $A\mathbf{x} = \mathbf{b}$ is consistent. The implication (n) $\Rightarrow$ (o) follows from the definitions of dimension and rank. If the rank of A is n, the number of columns of A, then nullity $A = 0$, by the Rank Theorem, and so Nul $A = \{\mathbf{0}\}$. Thus (o) $\Rightarrow$ (p) $\Rightarrow$ (q). Also, (q) implies that the equation $A\mathbf{x} = \mathbf{0}$ has only the trivial solution, which is statement (d). Since statements (d) and (g) are already known to be equivalent to the statement that A is invertible, the proof is complete. ∎

We have refrained from adding to the Invertible Matrix Theorem obvious statements about the row space of A, because the row space is the column space of A^T. Recall from statement (1) of the Invertible Matrix Theorem that A is invertible if and only if A^T is invertible. Hence every statement in the Invertible Matrix Theorem can also be stated for A^T. To do so would double the length of the theorem and produce a list of more than 30 statements!

Numerical Notes

Many algorithms discussed in this text are useful for understanding concepts and making simple computations by hand. However, the algorithms are often unsuitable for large-scale problems in real life.

Rank determination is a good example. It would seem easy to reduce a matrix to echelon form and count the pivots. But unless exact arithmetic is performed on a matrix whose entries are specified exactly, row operations can change the apparent rank of a matrix. For instance, if the value of x in the matrix $\begin{bmatrix} 5 & 7 \\ 5 & x \end{bmatrix}$ is not stored exactly as 7 in a computer, then the rank may be 1 or 2, depending on whether the computer treats $x - 7$ as zero.

In practical applications, the effective rank of a matrix A is often determined from the singular value decomposition of A, to be discussed in Section 7.4. This decomposition is also a reliable source of bases for Col A, Row A, Nul A, and Nul A^T.

Practice Problems

1. Decide whether each statement is True or False, and give a reason for each answer. Here V is a nonzero finite-dimensional vector space.

   a. If dim $V = p$ and if S is a linearly dependent subset of V, then S contains more than p vectors.

   b. If S spans V and if T is a subset of V that contains more vectors than S, then T is linearly dependent.

2. Let H and K be subspaces of a vector space V. In Section 4.1, Exercise 40, it is established that $H \cap K$ is also a subspace of V. Prove dim $(H \cap K) \le$ dim H.

4.5 Exercises

For each subspace in Exercises 1–8, (a) find a basis, and (b) state the dimension.

1. $\left\{ \begin{bmatrix} s - 2t \\ s + t \\ 3t \end{bmatrix} : s, t \text{ in } \mathbb{R} \right\}$ 2. $\left\{ \begin{bmatrix} 4s \\ -3s \\ -t \end{bmatrix} : s, t \text{ in } \mathbb{R} \right\}$

3. $\left\{ \begin{bmatrix} 2c \\ a - b \\ b - 3c \\ a + 2b \end{bmatrix} : a, b, c \text{ in } \mathbb{R} \right\}$ 4. $\left\{ \begin{bmatrix} a + b \\ 2a \\ 3a - b \\ -b \end{bmatrix} : a, b \text{ in } \mathbb{R} \right\}$

5. $\left\{ \begin{bmatrix} a - 4b - 2c \\ 2a + 5b - 4c \\ -a + 2c \\ -3a + 7b + 6c \end{bmatrix} : a, b, c \text{ in } \mathbb{R} \right\}$

6. $\left\{ \begin{bmatrix} 3a + 6b - c \\ 6a - 2b - 2c \\ -9a + 5b + 3c \\ -3a + b + c \end{bmatrix} : a, b, c \text{ in } \mathbb{R} \right\}$

7. $\{(a, b, c) : a - 3b + c = 0, b - 2c = 0, 2b - c = 0\}$

8. $\{(a, b, c, d) : a - 3b + c = 0\}$

In Exercises 9 and 10, find the dimension of the subspace spanned by the given vectors.

9. $\begin{bmatrix} 1 \\ 0 \\ 2 \end{bmatrix}, \begin{bmatrix} 3 \\ 1 \\ 1 \end{bmatrix}, \begin{bmatrix} 9 \\ 4 \\ -2 \end{bmatrix}, \begin{bmatrix} -7 \\ -3 \\ 1 \end{bmatrix}$

10. $\begin{bmatrix} 1 \\ -2 \\ 0 \end{bmatrix}, \begin{bmatrix} -3 \\ 4 \\ 1 \end{bmatrix}, \begin{bmatrix} -8 \\ 6 \\ 5 \end{bmatrix}, \begin{bmatrix} -3 \\ 0 \\ 7 \end{bmatrix}$

Determine the dimensions of Nul A, Col A, and Row A for the matrices shown in Exercises 11–16.

11. $A = \begin{bmatrix} 1 & -6 & 9 & 0 & -2 \\ 0 & 1 & 2 & -4 & 5 \\ 0 & 0 & 0 & 5 & 1 \\ 0 & 0 & 0 & 0 & 0 \end{bmatrix}$

12. $A = \begin{bmatrix} 1 & 3 & -4 & 2 & -1 & 6 \\ 0 & 0 & 1 & -3 & 7 & 0 \\ 0 & 0 & 0 & 1 & 4 & -3 \\ 0 & 0 & 0 & 0 & 0 & 0 \end{bmatrix}$

13. $A = \begin{bmatrix} 1 & 0 & 9 & 5 \\ 0 & 0 & 1 & -4 \end{bmatrix}$

14. $A = \begin{bmatrix} 3 & 4 \\ -6 & 10 \end{bmatrix}$

15. $A = \begin{bmatrix} 1 & -1 & 0 \\ 0 & 4 & 7 \\ 0 & 0 & 5 \end{bmatrix}$ **16.** $A = \begin{bmatrix} 1 & 4 & -1 \\ 0 & 7 & 0 \\ 0 & 0 & 0 \end{bmatrix}$

In Exercises 17–26, V is a vector space and A is an $m \times n$ matrix. Mark each statement True or False (**T/F**). Justify each answer.

17. (**T/F**) The number of pivot columns of a matrix equals the dimension of its column space.

18. (**T/F**) The number of variables in the equation $A\mathbf{x} = \mathbf{0}$ equals the nullity A.

19. (**T/F**) A plane in $\mathbb{R}^3$ is a two-dimensional subspace of $\mathbb{R}^3$.

20. (**T/F**) The dimension of the vector space $\mathbb{P}_4$ is 4.

21. (**T/F**) The dimension of the vector space of signals, $\mathbb{S}$, is 10.

22. (**T/F**) The dimensions of the row space and the column space of A are the same, even if A is not square.

23. (**T/F**) If B is any echelon form of A, then the pivot columns of B form a basis for the column space of A.

24. (**T/F**) The nullity of A is the number of columns of A that are not pivot columns.

25. (**T/F**) If a set $\{\mathbf{v}_1, \ldots, \mathbf{v}_p\}$ spans a finite-dimensional vector space V and if T is a set of more than p vectors in V, then T is linearly dependent.

26. (**T/F**) A vector space is infinite-dimensional if it is spanned by an infinite set.

27. The first four Hermite polynomials are $1, 2t, -2 + 4t^2$, and $-12t + 8t^3$. These polynomials arise naturally in the study of certain important differential equations in mathematical

physics.[2] Show that the first four Hermite polynomials form a basis of $\mathbb{P}_3$.

28. The first four Laguerre polynomials are $1, 1 - t, 2 - 4t + t^2$, and $6 - 18t + 9t^2 - t^3$. Show that these polynomials form a basis of $\mathbb{P}_3$.

29. Let $\mathcal{B}$ be the basis of $\mathbb{P}_3$ consisting of the Hermite polynomials in Exercise 27, and let $\mathbf{p}(t) = 7 - 12t - 8t^2 + 12t^3$. Find the coordinate vector of $\mathbf{p}$ relative to $\mathcal{B}$.

30. Let $\mathcal{B}$ be the basis of $\mathbb{P}_2$ consisting of the first three Laguerre polynomials listed in Exercise 28, and let $\mathbf{p}(t) = 7 - 8t + 3t^2$. Find the coordinate vector of $\mathbf{p}$ relative to $\mathcal{B}$.

31. Let S be a subset of an n-dimensional vector space V, and suppose S contains fewer than n vectors. Explain why S cannot span V.

32. Let H be an n-dimensional subspace of an n-dimensional vector space V. Show that $H = V$.

33. If a 3×8 matrix A has rank 3, find nullity A, rank A, and rank A^T.

34. If a 6×3 matrix A has rank 3, find nullity A, rank A, and rank A^T.

35. Suppose a 4×7 matrix A has four pivot columns. Is Col $A = \mathbb{R}^4$? Is Nul $A = \mathbb{R}^3$? Explain your answers.

36. Suppose a 5×6 matrix A has four pivot columns. What is nullity A? Is Col $A = \mathbb{R}^4$? Why or why not?

37. If the nullity of a 5×6 matrix A is 4, what are the dimensions of the column and row spaces of A?

38. If the nullity of a 7×6 matrix A is 5, what are the dimensions of the column and row spaces of A?

39. If A is a 7×5 matrix, what is the largest possible rank of A? If A is a 5×7 matrix, what is the largest possible rank of A? Explain your answers.

40. If A is a 4×3 matrix, what is the largest possible dimension of the row space of A? If A is a 3×4 matrix, what is the largest possible dimension of the row space of A? Explain.

41. Explain why the space $\mathbb{P}$ of all polynomials is an infinite-dimensional space.

42. Show that the space $C(\mathbb{R})$ of all continuous functions defined on the real line is an infinite-dimensional space.

In Exercises 43–48, V is a nonzero finite-dimensional vector space, and the vectors listed belong to V. Mark each statement True or False (**T/F**). Justify each answer. (These questions are more difficult than those in Exercises 17–26.)

---

[2] See *Introduction to Functional Analysis*, 2nd ed., by A. E. Taylor and David C. Lay (New York: John Wiley & Sons, 1980), pp. 92–93. Other sets of polynomials are discussed there, too.

43. (T/F) If there exists a set $\{\mathbf{v}_1, \ldots, \mathbf{v}_p\}$ that spans V, then $\dim V \le p$.

44. (T/F) If there exists a linearly dependent set $\{\mathbf{v}_1, \ldots, \mathbf{v}_p\}$ in V, then $\dim V \le p$.

45. (T/F) If there exists a linearly independent set $\{\mathbf{v}_1, \ldots, \mathbf{v}_p\}$ in V, then $\dim V \ge p$.

46. (T/F) If $\dim V = p$, then there exists a spanning set of $p + 1$ vectors in V.

47. (T/F) If every set of p elements in V fails to span V, then $\dim V > p$.

48. (T/F) If $p \ge 2$ and $\dim V = p$, then every set of $p - 1$ nonzero vectors is linearly independent.

49. Justify the following equality: $\dim \operatorname{Row} A + \operatorname{nullity} A = n$, the number of columns of A

50. Justify the following equality: $\dim \operatorname{Row} A + \operatorname{nullity} A^T = m$, the number of rows of A

Exercises 51 and 52 concern finite-dimensional vector spaces V and W and a linear transformation $T : V \to W$.

51. Let H be a nonzero subspace of V, and let $T(H)$ be the set of images of vectors in H. Then $T(H)$ is a subspace of W, by Exercise 47 in Section 4.2. Prove that $\dim T(H) \le \dim H$.

52. Let H be a nonzero subspace of V, and suppose T is a one-to-one (linear) mapping of V into W. Prove that $\dim T(H) = \dim H$. If T happens to be a one-to-one mapping of V *onto* W, then $\dim V = \dim W$. Isomorphic finite-dimensional vector spaces have the same dimension.

T 53. According to Theorem 12, a linearly independent set $\{\mathbf{v}_1, \ldots, \mathbf{v}_k\}$ in $\mathbb{R}^n$ can be expanded to a basis for $\mathbb{R}^n$. One way to do this is to create $A = [\, \mathbf{v}_1 \ \cdots \ \mathbf{v}_k \ \mathbf{e}_1 \ \cdots \ \mathbf{e}_n \,]$, with $\mathbf{e}_1, \ldots, \mathbf{e}_n$ the columns of the identity matrix; the pivot columns of A form a basis for $\mathbb{R}^n$.

a. Use the method described to extend the following vectors to a basis for $\mathbb{R}^5$:

$$\mathbf{v}_1 = \begin{bmatrix} -9 \\ -7 \\ 8 \\ -5 \\ 7 \end{bmatrix}, \quad \mathbf{v}_2 = \begin{bmatrix} 9 \\ 4 \\ 1 \\ 6 \\ -7 \end{bmatrix}, \quad \mathbf{v}_3 = \begin{bmatrix} 6 \\ 7 \\ -8 \\ 5 \\ -7 \end{bmatrix}$$

b. Explain why the method works in general: Why are the original vectors $\mathbf{v}_1, \ldots, \mathbf{v}_k$ included in the basis found for $\operatorname{Col} A$? Why is $\operatorname{Col} A = \mathbb{R}^n$?

T 54. Let $\mathcal{B} = \{1, \cos t, \cos^2 t, \ldots, \cos^6 t\}$ and $\mathcal{C} = \{1, \cos t, \cos 2t, \ldots, \cos 6t\}$. Assume the following trigonometric identities (see Exercise 45 in Section 4.1).

$$\cos 2t = -1 + 2\cos^2 t$$
$$\cos 3t = -3\cos t + 4\cos^3 t$$
$$\cos 4t = 1 - 8\cos^2 t + 8\cos^4 t$$
$$\cos 5t = 5\cos t - 20\cos^3 t + 16\cos^5 t$$
$$\cos 6t = -1 + 18\cos^2 t - 48\cos^4 t + 32\cos^6 t$$

Let H be the subspace of functions spanned by the functions in $\mathcal{B}$. Then $\mathcal{B}$ is a basis for H, by Exercise 48 in Section 4.3.

a. Write the $\mathcal{B}$-coordinate vectors of the vectors in $\mathcal{C}$, and use them to show that $\mathcal{C}$ is a linearly independent set in H.

b. Explain why $\mathcal{C}$ is a basis for H.

Solutions to Practice Problems

1. a. False. Consider the set $\{\mathbf{0}\}$.

   b. True. By the Spanning Set Theorem, S contains a basis for V; call that basis S'. Then T will contain more vectors than S'. By Theorem 10, T is linearly dependent.

2. Let $\{\mathbf{v}_1, \ldots, \mathbf{v}_p\}$ be a basis for $H \cap K$. Notice $\{\mathbf{v}_1, \ldots, \mathbf{v}_p\}$ is a linearly independent subset of H, hence by Theorem 12, $\{\mathbf{v}_1, \ldots, \mathbf{v}_p\}$ can be expanded, if necessary, to a basis for H. Since the dimension of a subspace is just the number of vectors in a basis, it follows that $\dim (H \cap K) = p \le \dim H$.

4.6 Change of Basis

When a basis $\mathcal{B}$ is chosen for an n-dimensional vector space V, the associated coordinate mapping onto $\mathbb{R}^n$ provides a coordinate system for V. Each $\mathbf{x}$ in V is identified uniquely by its $\mathcal{B}$-coordinate vector $[\,\mathbf{x}\,]_{\mathcal{B}}$.[1]

---
[1] Think of $[\,\mathbf{x}\,]_{\mathcal{B}}$ as a name for $\mathbf{x}$ that lists the weights used to build $\mathbf{x}$ as a linear combination of the basis vectors in $\mathcal{B}$.

In some applications, a problem is described initially using a basis $\mathcal{B}$, but the problem's solution is aided by changing $\mathcal{B}$ to a new basis $\mathcal{C}$. (Examples will be given in Chapters 5 and 7.) Each vector is assigned a new $\mathcal{C}$-coordinate vector. In this section, we study how $[\mathbf{x}]_{\mathcal{C}}$ and $[\mathbf{x}]_{\mathcal{B}}$ are related for each $\mathbf{x}$ in V.

To visualize the problem, consider the two coordinate systems in Figure 1. In Figure 1(a), $\mathbf{x} = 3\mathbf{b}_1 + \mathbf{b}_2$, while in Figure 1(b), the same $\mathbf{x}$ is shown as $\mathbf{x} = 6\mathbf{c}_1 + 4\mathbf{c}_2$. That is,

$$[\mathbf{x}]_{\mathcal{B}} = \begin{bmatrix} 3 \\ 1 \end{bmatrix} \quad \text{and} \quad [\mathbf{x}]_{\mathcal{C}} = \begin{bmatrix} 6 \\ 4 \end{bmatrix}$$

Our problem is to find the connection between the two coordinate vectors. Example 1 shows how to do this, provided we know how $\mathbf{b}_1$ and $\mathbf{b}_2$ are formed from $\mathbf{c}_1$ and $\mathbf{c}_2$.

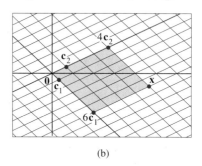

(a)   (b)

FIGURE 1 Two coordinate systems for the same vector space.

EXAMPLE 1 Consider two bases $\mathcal{B} = \{\mathbf{b}_1, \mathbf{b}_2\}$ and $\mathcal{C} = \{\mathbf{c}_1, \mathbf{c}_2\}$ for a vector space V, such that

$$\mathbf{b}_1 = 4\mathbf{c}_1 + \mathbf{c}_2 \quad \text{and} \quad \mathbf{b}_2 = -6\mathbf{c}_1 + \mathbf{c}_2 \tag{1}$$

Suppose

$$\mathbf{x} = 3\mathbf{b}_1 + \mathbf{b}_2 \tag{2}$$

That is, suppose $[\mathbf{x}]_{\mathcal{B}} = \begin{bmatrix} 3 \\ 1 \end{bmatrix}$. Find $[\mathbf{x}]_{\mathcal{C}}$.

SOLUTION Apply the coordinate mapping determined by $\mathcal{C}$ to $\mathbf{x}$ in (2). Since the coordinate mapping is a linear transformation,

$$[\mathbf{x}]_{\mathcal{C}} = [3\mathbf{b}_1 + \mathbf{b}_2]_{\mathcal{C}}$$
$$= 3[\mathbf{b}_1]_{\mathcal{C}} + [\mathbf{b}_2]_{\mathcal{C}}$$

We can write this vector equation as a matrix equation, using the vectors in the linear combination as the columns of a matrix:

$$[\mathbf{x}]_{\mathcal{C}} = \begin{bmatrix} [\mathbf{b}_1]_{\mathcal{C}} & [\mathbf{b}_2]_{\mathcal{C}} \end{bmatrix} \begin{bmatrix} 3 \\ 1 \end{bmatrix} \tag{3}$$

This formula gives $[\mathbf{x}]_{\mathcal{C}}$, once we know the columns of the matrix. From (1),

$$[\mathbf{b}_1]_{\mathcal{C}} = \begin{bmatrix} 4 \\ 1 \end{bmatrix} \quad \text{and} \quad [\mathbf{b}_2]_{\mathcal{C}} = \begin{bmatrix} -6 \\ 1 \end{bmatrix}$$

Thus (3) provides the solution:

$$[\mathbf{x}]_{\mathcal{C}} = \begin{bmatrix} 4 & -6 \\ 1 & 1 \end{bmatrix} \begin{bmatrix} 3 \\ 1 \end{bmatrix} = \begin{bmatrix} 6 \\ 4 \end{bmatrix}$$

The $\mathcal{C}$-coordinates of $\mathbf{x}$ match those of the $\mathbf{x}$ in Figure 1. ∎

The argument used to derive formula (3) can be generalized to yield the following result. (See Exercises 17 and 18.)

THEOREM 15

Let $\mathcal{B} = \{\mathbf{b}_1, \ldots, \mathbf{b}_n\}$ and $\mathcal{C} = \{\mathbf{c}_1, \ldots, \mathbf{c}_n\}$ be bases of a vector space V. Then there is a unique $n \times n$ matrix $\underset{\mathcal{C} \leftarrow \mathcal{B}}{P}$ such that

$$[\mathbf{x}]_{\mathcal{C}} = \underset{\mathcal{C} \leftarrow \mathcal{B}}{P} [\mathbf{x}]_{\mathcal{B}} \tag{4}$$

The columns of $\underset{\mathcal{C} \leftarrow \mathcal{B}}{P}$ are the $\mathcal{C}$-coordinate vectors of the vectors in the basis $\mathcal{B}$. That is,

$$\underset{\mathcal{C} \leftarrow \mathcal{B}}{P} = \begin{bmatrix} [\mathbf{b}_1]_{\mathcal{C}} & [\mathbf{b}_2]_{\mathcal{C}} & \cdots & [\mathbf{b}_n]_{\mathcal{C}} \end{bmatrix} \tag{5}$$

The matrix $\underset{\mathcal{C} \leftarrow \mathcal{B}}{P}$ in Theorem 15 is called the **change-of-coordinates matrix from** $\mathcal{B}$ **to** $\mathcal{C}$. Multiplication by $\underset{\mathcal{C} \leftarrow \mathcal{B}}{P}$ converts $\mathcal{B}$-coordinates into $\mathcal{C}$-coordinates.[2] Figure 2 illustrates the change-of-coordinates equation (4).

FIGURE 2 Two coordinate systems for V.

The columns of $\underset{\mathcal{C} \leftarrow \mathcal{B}}{P}$ are linearly independent because they are the coordinate vectors of the linearly independent set $\mathcal{B}$. (See Exercise 29 in Section 4.4.) Since $\underset{\mathcal{C} \leftarrow \mathcal{B}}{P}$ is square, it must be invertible, by the Invertible Matrix Theorem. Left-multiplying both sides of equation (4) by $(\underset{\mathcal{C} \leftarrow \mathcal{B}}{P})^{-1}$ yields

$$(\underset{\mathcal{C} \leftarrow \mathcal{B}}{P})^{-1} [\mathbf{x}]_{\mathcal{C}} = [\mathbf{x}]_{\mathcal{B}}$$

Thus $(\underset{\mathcal{C} \leftarrow \mathcal{B}}{P})^{-1}$ is the matrix that converts $\mathcal{C}$-coordinates into $\mathcal{B}$-coordinates. That is,

$$(\underset{\mathcal{C} \leftarrow \mathcal{B}}{P})^{-1} = \underset{\mathcal{B} \leftarrow \mathcal{C}}{P} \tag{6}$$

Change of Basis in $\mathbb{R}^n$

If $\mathcal{B} = \{\mathbf{b}_1, \ldots, \mathbf{b}_n\}$ and $\mathcal{E}$ is the *standard basis* $\{\mathbf{e}_1, \ldots, \mathbf{e}_n\}$ in $\mathbb{R}^n$, then $[\mathbf{b}_1]_{\mathcal{E}} = \mathbf{b}_1$, and likewise for the other vectors in $\mathcal{B}$. In this case, $\underset{\mathcal{E} \leftarrow \mathcal{B}}{P}$ is the same as the change-of-coordinates matrix $P_{\mathcal{B}}$ introduced in Section 4.4, namely

$$P_{\mathcal{B}} = \begin{bmatrix} \mathbf{b}_1 & \mathbf{b}_2 & \cdots & \mathbf{b}_n \end{bmatrix}$$

---

[2] To remember how to construct the matrix, think of $\underset{\mathcal{C} \leftarrow \mathcal{B}}{P} [\mathbf{x}]_{\mathcal{B}}$ as a linear combination of the columns of $\underset{\mathcal{C} \leftarrow \mathcal{B}}{P}$. The matrix-vector product is a $\mathcal{C}$-coordinate vector, so the columns of $\underset{\mathcal{C} \leftarrow \mathcal{B}}{P}$ should be $\mathcal{C}$-coordinate vectors, too.

To change coordinates between two nonstandard bases in $\mathbb{R}^n$, we need Theorem 15. The theorem shows that to solve the change-of-basis problem, we need the coordinate vectors of the old basis relative to the new basis.

EXAMPLE 2 Let $\mathbf{b}_1 = \begin{bmatrix} -9 \\ 1 \end{bmatrix}$, $\mathbf{b}_2 = \begin{bmatrix} -5 \\ -1 \end{bmatrix}$, $\mathbf{c}_1 = \begin{bmatrix} 1 \\ -4 \end{bmatrix}$, $\mathbf{c}_2 = \begin{bmatrix} 3 \\ -5 \end{bmatrix}$, and consider the bases for $\mathbb{R}^2$ given by $\mathcal{B} = \{\mathbf{b}_1, \mathbf{b}_2\}$ and $\mathcal{C} = \{\mathbf{c}_1, \mathbf{c}_2\}$. Find the change-of-coordinates matrix from $\mathcal{B}$ to $\mathcal{C}$.

SOLUTION The matrix $\underset{\mathcal{C}\leftarrow\mathcal{B}}{P}$ involves the $\mathcal{C}$-coordinate vectors of $\mathbf{b}_1$ and $\mathbf{b}_2$. Let $[\mathbf{b}_1]_{\mathcal{C}} = \begin{bmatrix} x_1 \\ x_2 \end{bmatrix}$ and $[\mathbf{b}_2]_{\mathcal{C}} = \begin{bmatrix} y_1 \\ y_2 \end{bmatrix}$. Then, by definition,

$$\begin{bmatrix} \mathbf{c}_1 & \mathbf{c}_2 \end{bmatrix} \begin{bmatrix} x_1 \\ x_2 \end{bmatrix} = \mathbf{b}_1 \quad \text{and} \quad \begin{bmatrix} \mathbf{c}_1 & \mathbf{c}_2 \end{bmatrix} \begin{bmatrix} y_1 \\ y_2 \end{bmatrix} = \mathbf{b}_2$$

To solve both systems simultaneously, augment the coefficient matrix with $\mathbf{b}_1$ *and* $\mathbf{b}_2$, and row reduce:

$$\begin{bmatrix} \mathbf{c}_1 & \mathbf{c}_2 & \vdots & \mathbf{b}_1 & \mathbf{b}_2 \end{bmatrix} = \begin{bmatrix} 1 & 3 & \vdots & -9 & -5 \\ -4 & -5 & \vdots & 1 & -1 \end{bmatrix} \sim \begin{bmatrix} 1 & 0 & \vdots & 6 & 4 \\ 0 & 1 & \vdots & -5 & -3 \end{bmatrix} \quad (7)$$

Thus

$$[\mathbf{b}_1]_{\mathcal{C}} = \begin{bmatrix} 6 \\ -5 \end{bmatrix} \quad \text{and} \quad [\mathbf{b}_2]_{\mathcal{C}} = \begin{bmatrix} 4 \\ -3 \end{bmatrix}$$

The desired change-of-coordinates matrix is therefore

$$\underset{\mathcal{C}\leftarrow\mathcal{B}}{P} = \begin{bmatrix} [\mathbf{b}_1]_{\mathcal{C}} & [\mathbf{b}_2]_{\mathcal{C}} \end{bmatrix} = \begin{bmatrix} 6 & 4 \\ -5 & -3 \end{bmatrix}$$

∎

Observe that the matrix $\underset{\mathcal{C}\leftarrow\mathcal{B}}{P}$ in Example 2 already appeared in (7). This is not surprising because the first column of $\underset{\mathcal{C}\leftarrow\mathcal{B}}{P}$ results from row reducing $\begin{bmatrix} \mathbf{c}_1 & \mathbf{c}_2 & \vdots & \mathbf{b}_1 \end{bmatrix}$ to $\begin{bmatrix} I & \vdots & [\mathbf{b}_1]_{\mathcal{C}} \end{bmatrix}$, and similarly for the second column of $\underset{\mathcal{C}\leftarrow\mathcal{B}}{P}$. Thus

$$\boxed{\begin{bmatrix} \mathbf{c}_1 & \mathbf{c}_2 & \vdots & \mathbf{b}_1 & \mathbf{b}_2 \end{bmatrix} \sim \begin{bmatrix} I & \vdots & \underset{\mathcal{C}\leftarrow\mathcal{B}}{P} \end{bmatrix}}$$

An analogous procedure works for finding the change-of-coordinates matrix between any two bases in $\mathbb{R}^n$.

EXAMPLE 3 Let $\mathbf{b}_1 = \begin{bmatrix} 1 \\ -3 \end{bmatrix}$, $\mathbf{b}_2 = \begin{bmatrix} -2 \\ 4 \end{bmatrix}$, $\mathbf{c}_1 = \begin{bmatrix} -7 \\ 9 \end{bmatrix}$, $\mathbf{c}_2 = \begin{bmatrix} -5 \\ 7 \end{bmatrix}$, and consider the bases for $\mathbb{R}^2$ given by $\mathcal{B} = \{\mathbf{b}_1, \mathbf{b}_2\}$ and $\mathcal{C} = \{\mathbf{c}_1, \mathbf{c}_2\}$.

a. Find the change-of-coordinates matrix from $\mathcal{C}$ to $\mathcal{B}$.

b. Find the change-of-coordinates matrix from $\mathcal{B}$ to $\mathcal{C}$.

SOLUTION

a. Notice that $\underset{\mathcal{B}\leftarrow\mathcal{C}}{P}$ is needed rather than $\underset{\mathcal{C}\leftarrow\mathcal{B}}{P}$, and compute

$$\begin{bmatrix} \mathbf{b}_1 & \mathbf{b}_2 & \vdots & \mathbf{c}_1 & \mathbf{c}_2 \end{bmatrix} = \begin{bmatrix} 1 & -2 & \vdots & -7 & -5 \\ -3 & 4 & \vdots & 9 & 7 \end{bmatrix} \sim \begin{bmatrix} 1 & 0 & \vdots & 5 & 3 \\ 0 & 1 & \vdots & 6 & 4 \end{bmatrix}$$

So

$$\underset{\mathcal{B}\leftarrow\mathcal{C}}{P} = \begin{bmatrix} 5 & 3 \\ 6 & 4 \end{bmatrix}$$

b. By part (a) and property (6) (with $\mathcal{B}$ and $\mathcal{C}$ interchanged),

$$\underset{\mathcal{C}\leftarrow\mathcal{B}}{P} = (\underset{\mathcal{B}\leftarrow\mathcal{C}}{P})^{-1} = \frac{1}{2}\begin{bmatrix} 4 & -3 \\ -6 & 5 \end{bmatrix} = \begin{bmatrix} 2 & -3/2 \\ -3 & 5/2 \end{bmatrix} \quad\blacksquare$$

Another description of the change-of-coordinates matrix $\underset{\mathcal{C}\leftarrow\mathcal{B}}{P}$ uses the change-of-coordinate matrices $P_{\mathcal{B}}$ and $P_{\mathcal{C}}$ that convert $\mathcal{B}$-coordinates and $\mathcal{C}$-coordinates, respectively, into standard coordinates. Recall that for each $\mathbf{x}$ in $\mathbb{R}^n$,

$$P_{\mathcal{B}}[\mathbf{x}]_{\mathcal{B}} = \mathbf{x}, \quad P_{\mathcal{C}}[\mathbf{x}]_{\mathcal{C}} = \mathbf{x}, \quad \text{and} \quad [\mathbf{x}]_{\mathcal{C}} = P_{\mathcal{C}}^{-1}\mathbf{x}$$

Thus

$$[\mathbf{x}]_{\mathcal{C}} = P_{\mathcal{C}}^{-1}\mathbf{x} = P_{\mathcal{C}}^{-1}P_{\mathcal{B}}[\mathbf{x}]_{\mathcal{B}}$$

In $\mathbb{R}^n$, the change-of-coordinates matrix $\underset{\mathcal{C}\leftarrow\mathcal{B}}{P}$ may be computed as $P_{\mathcal{C}}^{-1}P_{\mathcal{B}}$. Actually, for matrices larger than 2×2, an algorithm analogous to the one in Example 3 is faster than computing $P_{\mathcal{C}}^{-1}$ and then $P_{\mathcal{C}}^{-1}P_{\mathcal{B}}$. See Exercise 22 in Section 2.2.

Practice Problems

1. Let $\mathcal{F} = \{\mathbf{f}_1, \mathbf{f}_2\}$ and $\mathcal{G} = \{\mathbf{g}_1, \mathbf{g}_2\}$ be bases for a vector space V, and let P be a matrix whose columns are $[\mathbf{f}_1]_{\mathcal{G}}$ and $[\mathbf{f}_2]_{\mathcal{G}}$. Which of the following equations is satisfied by P for all $\mathbf{v}$ in V?

   (i) $[\mathbf{v}]_{\mathcal{F}} = P[\mathbf{v}]_{\mathcal{G}}$ (ii) $[\mathbf{v}]_{\mathcal{G}} = P[\mathbf{v}]_{\mathcal{F}}$

2. Let $\mathcal{B}$ and $\mathcal{C}$ be as in Example 1. Use the results of that example to find the change-of-coordinates matrix from $\mathcal{C}$ to $\mathcal{B}$.

4.6 Exercises

1. Let $\mathcal{B} = \{\mathbf{b}_1, \mathbf{b}_2\}$ and $\mathcal{C} = \{\mathbf{c}_1, \mathbf{c}_2\}$ be bases for a vector space V, and suppose $\mathbf{b}_1 = 6\mathbf{c}_1 - 2\mathbf{c}_2$ and $\mathbf{b}_2 = 9\mathbf{c}_1 - 4\mathbf{c}_2$.

   a. Find the change-of-coordinates matrix from $\mathcal{B}$ to $\mathcal{C}$.

   b. Find $[\mathbf{x}]_{\mathcal{C}}$ for $\mathbf{x} = -3\mathbf{b}_1 + 2\mathbf{b}_2$. Use part (a).

2. Let $\mathcal{B} = \{\mathbf{b}_1, \mathbf{b}_2\}$ and $\mathcal{C} = \{\mathbf{c}_1, \mathbf{c}_2\}$ be bases for a vector space V, and suppose $\mathbf{b}_1 = -\mathbf{c}_1 + 4\mathbf{c}_2$ and $\mathbf{b}_2 = 5\mathbf{c}_1 - 3\mathbf{c}_2$.

   a. Find the change-of-coordinates matrix from $\mathcal{B}$ to $\mathcal{C}$.

   b. Find $[\mathbf{x}]_{\mathcal{C}}$ for $\mathbf{x} = 5\mathbf{b}_1 + 3\mathbf{b}_2$.

3. Let $\mathcal{U} = \{\mathbf{u}_1, \mathbf{u}_2\}$ and $\mathcal{W} = \{\mathbf{w}_1, \mathbf{w}_2\}$ be bases for V, and let P be a matrix whose columns are $[\mathbf{u}_1]_{\mathcal{W}}$ and $[\mathbf{u}_2]_{\mathcal{W}}$. Which of the following equations is satisfied by P for all $\mathbf{x}$ in V?

   (i) $[\mathbf{x}]_{\mathcal{U}} = P[\mathbf{x}]_{\mathcal{W}}$ (ii) $[\mathbf{x}]_{\mathcal{W}} = P[\mathbf{x}]_{\mathcal{U}}$

4. Let $\mathcal{A} = \{\mathbf{a}_1, \mathbf{a}_2, \mathbf{a}_3\}$ and $\mathcal{D} = \{\mathbf{d}_1, \mathbf{d}_2, \mathbf{d}_3\}$ be bases for V, and let $P = [\,[\mathbf{d}_1]_{\mathcal{A}} \;\; [\mathbf{d}_2]_{\mathcal{A}} \;\; [\mathbf{d}_3]_{\mathcal{A}}\,]$. Which of the following equations is satisfied by P for all $\mathbf{x}$ in V?

   (i) $[\mathbf{x}]_{\mathcal{A}} = P[\mathbf{x}]_{\mathcal{D}}$ (ii) $[\mathbf{x}]_{\mathcal{D}} = P[\mathbf{x}]_{\mathcal{A}}$

5. Let $\mathcal{A} = \{\mathbf{a}_1, \mathbf{a}_2, \mathbf{a}_3\}$ and $\mathcal{B} = \{\mathbf{b}_1, \mathbf{b}_2, \mathbf{b}_3\}$ be bases for a vector space V, and suppose $\mathbf{a}_1 = 4\mathbf{b}_1 - \mathbf{b}_2$, $\mathbf{a}_2 = -\mathbf{b}_1 + \mathbf{b}_2 + \mathbf{b}_3$, and $\mathbf{a}_3 = \mathbf{b}_2 - 2\mathbf{b}_3$.

   a. Find the change-of-coordinates matrix from $\mathcal{A}$ to $\mathcal{B}$.

   b. Find $[\mathbf{x}]_{\mathcal{B}}$ for $\mathbf{x} = 3\mathbf{a}_1 + 4\mathbf{a}_2 + \mathbf{a}_3$.

6. Let $\mathcal{D} = \{\mathbf{d}_1, \mathbf{d}_2, \mathbf{d}_3\}$ and $\mathcal{F} = \{\mathbf{f}_1, \mathbf{f}_2, \mathbf{f}_3\}$ be bases for a vector space V, and suppose $\mathbf{f}_1 = 2\mathbf{d}_1 - \mathbf{d}_2 + \mathbf{d}_3$, $\mathbf{f}_2 = 3\mathbf{d}_2 + \mathbf{d}_3$, and $\mathbf{f}_3 = -3\mathbf{d}_1 + 2\mathbf{d}_3$.

   a. Find the change-of-coordinates matrix from $\mathcal{F}$ to $\mathcal{D}$.

   b. Find $[\mathbf{x}]_{\mathcal{D}}$ for $\mathbf{x} = \mathbf{f}_1 - 2\mathbf{f}_2 + 2\mathbf{f}_3$.

In Exercises 7–10, let $\mathcal{B} = \{\mathbf{b}_1, \mathbf{b}_2\}$ and $\mathcal{C} = \{\mathbf{c}_1, \mathbf{c}_2\}$ be bases for $\mathbb{R}^2$. In each exercise, find the change-of-coordinates matrix from $\mathcal{B}$ to $\mathcal{C}$ and the change-of-coordinates matrix from $\mathcal{C}$ to $\mathcal{B}$.

7. $\mathbf{b}_1 = \begin{bmatrix} 7 \\ 5 \end{bmatrix}$, $\mathbf{b}_2 = \begin{bmatrix} -3 \\ -1 \end{bmatrix}$, $\mathbf{c}_1 = \begin{bmatrix} 1 \\ -5 \end{bmatrix}$, $\mathbf{c}_2 = \begin{bmatrix} -2 \\ 2 \end{bmatrix}$

8. $\mathbf{b}_1 = \begin{bmatrix} -1 \\ 8 \end{bmatrix}$, $\mathbf{b}_2 = \begin{bmatrix} 1 \\ -5 \end{bmatrix}$, $\mathbf{c}_1 = \begin{bmatrix} 1 \\ 4 \end{bmatrix}$, $\mathbf{c}_2 = \begin{bmatrix} 1 \\ 1 \end{bmatrix}$

9. $\mathbf{b}_1 = \begin{bmatrix} -6 \\ -1 \end{bmatrix}$, $\mathbf{b}_2 = \begin{bmatrix} 2 \\ 0 \end{bmatrix}$, $\mathbf{c}_1 = \begin{bmatrix} 2 \\ -1 \end{bmatrix}$, $\mathbf{c}_2 = \begin{bmatrix} 6 \\ -2 \end{bmatrix}$

10. $\mathbf{b}_1 = \begin{bmatrix} 7 \\ -2 \end{bmatrix}$, $\mathbf{b}_2 = \begin{bmatrix} 2 \\ -1 \end{bmatrix}$, $\mathbf{c}_1 = \begin{bmatrix} 4 \\ 1 \end{bmatrix}$, $\mathbf{c}_2 = \begin{bmatrix} 5 \\ 2 \end{bmatrix}$

In Exercises 11–14, $\mathcal{B}$ and $\mathcal{C}$ are bases for a vector space V. Mark each statement True or False **(T/F)**. Justify each answer.

11. **(T/F)** The columns of the change-of-coordinates matrix $\underset{\mathcal{C}\leftarrow\mathcal{B}}{P}$ are $\mathcal{B}$-coordinate vectors of the vectors in $\mathcal{C}$.

12. **(T/F)** The columns of $\underset{\mathcal{C}\leftarrow\mathcal{B}}{P}$ are linearly independent.

13. **(T/F)** If $V = \mathbb{R}^n$ and $\mathcal{C}$ is the *standard* basis for V, then $\underset{\mathcal{C}\leftarrow\mathcal{B}}{P}$ is the same as the change-of-coordinates matrix $P_{\mathcal{B}}$ introduced in Section 4.4.

14. **(T/F)** If $V = \mathbb{R}^2$, $\mathcal{B} = \{\mathbf{b}_1, \mathbf{b}_2\}$, and $\mathcal{C} = \{\mathbf{c}_1, \mathbf{c}_2\}$, then row reduction of $[\,\mathbf{c}_1 \quad \mathbf{c}_2 \quad \mathbf{b}_1 \quad \mathbf{b}_2\,]$ to $[\,I \quad P\,]$ produces a matrix P that satisfies $[\,\mathbf{x}\,]_{\mathcal{B}} = P[\,\mathbf{x}\,]_{\mathcal{C}}$ for all $\mathbf{x}$ in V.

15. In $\mathbb{P}_2$, find the change-of-coordinates matrix from the basis $\mathcal{B} = \{1 - 2t + t^2, 3 - 5t + 4t^2, 2t + 3t^2\}$ to the standard basis $\mathcal{C} = \{1, t, t^2\}$. Then find the $\mathcal{B}$-coordinate vector for $-1 + 2t$.

16. In $\mathbb{P}_2$, find the change-of-coordinates matrix from the basis $\mathcal{B} = \{1 - 3t^2, 2 + t - 5t^2, 1 + 2t\}$ to the standard basis. Then write t^2 as a linear combination of the polynomials in $\mathcal{B}$.

Exercises 17 and 18 provide a proof of Theorem 15. Fill in a justification for each step.

17. Given $\mathbf{v}$ in V, there exist scalars $x_1, \ldots, x_n$, such that

$$\mathbf{v} = x_1\mathbf{b}_1 + x_2\mathbf{b}_2 + \cdots + x_n\mathbf{b}_n$$

because (a) _____. Apply the coordinate mapping determined by the basis $\mathcal{C}$, and obtain

$$[\mathbf{v}]_{\mathcal{C}} = x_1[\mathbf{b}_1]_{\mathcal{C}} + x_2[\mathbf{b}_2]_{\mathcal{C}} + \cdots + x_n[\mathbf{b}_n]_{\mathcal{C}}$$

because (b) _____. This equation may be written in the form

$$[\,\mathbf{v}\,]_{\mathcal{C}} = \begin{bmatrix} [\,\mathbf{b}_1\,]_{\mathcal{C}} & [\,\mathbf{b}_2\,]_{\mathcal{C}} & \cdots & [\,\mathbf{b}_n\,]_{\mathcal{C}} \end{bmatrix} \begin{bmatrix} x_1 \\ \vdots \\ x_n \end{bmatrix} \qquad (8)$$

by the definition of (c) _____. This shows that the matrix $\underset{\mathcal{C}\leftarrow\mathcal{B}}{P}$ shown in (5) satisfies $[\mathbf{v}]_{\mathcal{C}} = \underset{\mathcal{C}\leftarrow\mathcal{B}}{P}[\mathbf{v}]_{\mathcal{B}}$ for each $\mathbf{v}$ in V, because the vector on the right side of (8) is (d) _____.

18. Suppose Q is any matrix such that

$$[\mathbf{v}]_{\mathcal{C}} = Q[\mathbf{v}]_{\mathcal{B}} \quad \text{for each } \mathbf{v} \text{ in } V \qquad (9)$$

Set $\mathbf{v} = \mathbf{b}_1$ in (9). Then (9) shows that $[\mathbf{b}_1]_{\mathcal{C}}$ is the first column of Q because (a) _____. Similarly, for $k = 2, \ldots, n$, the kth column of Q is (b) _____ because (c) _____. This shows

that the matrix $\underset{\mathcal{C}\leftarrow\mathcal{B}}{P}$ defined by (5) in Theorem 15 is the only matrix that satisfies condition (4).

T 19. Let $\mathcal{B} = \{\mathbf{x}_0, \ldots, \mathbf{x}_6\}$ and $\mathcal{C} = \{\mathbf{y}_0, \ldots, \mathbf{y}_6\}$, where $\mathbf{x}_k$ is the function $\cos^k t$ and $\mathbf{y}_k$ is the function $\cos kt$. Exercise 54 in Section 4.5 showed that both $\mathcal{B}$ and $\mathcal{C}$ are bases for the vector space $H = \text{Span}\{\mathbf{x}_0, \ldots, \mathbf{x}_6\}$.

a. Set $P = \begin{bmatrix} [\,\mathbf{y}_0\,]_{\mathcal{B}} & \cdots & [\,\mathbf{y}_6\,]_{\mathcal{B}} \end{bmatrix}$, and calculate P^{-1}.

b. Explain why the columns of P^{-1} are the $\mathcal{C}$-coordinate vectors of $\mathbf{x}_0, \ldots, \mathbf{x}_6$. Then use these coordinate vectors to write trigonometric identities that express powers of $\cos t$ in terms of the functions in $\mathcal{C}$.

See the *Study Guide*.

T 20. (*Calculus required*)[3] Recall from calculus that integrals such as

$$\int (5\cos^3 t - 6\cos^4 t + 5\cos^5 t - 12\cos^6 t)\,dt \qquad (10)$$

are tedious to compute. (The usual method is to apply integration by parts repeatedly and use the half-angle formula.) Use the matrix P or P^{-1} from Exercise 19 to transform (10); then compute the integral.

T 21. Let

$$P = \begin{bmatrix} 1 & 2 & -1 \\ -3 & -5 & 0 \\ 4 & 6 & 1 \end{bmatrix},$$

$$\mathbf{v}_1 = \begin{bmatrix} -2 \\ 2 \\ 3 \end{bmatrix}, \quad \mathbf{v}_2 = \begin{bmatrix} -8 \\ 5 \\ 2 \end{bmatrix}, \quad \mathbf{v}_3 = \begin{bmatrix} -7 \\ 2 \\ 6 \end{bmatrix}$$

a. Find a basis $\{\mathbf{u}_1, \mathbf{u}_2, \mathbf{u}_3\}$ for $\mathbb{R}^3$ such that P is the change-of-coordinates matrix from $\{\mathbf{u}_1, \mathbf{u}_2, \mathbf{u}_3\}$ to the basis $\{\mathbf{v}_1, \mathbf{v}_2, \mathbf{v}_3\}$. [*Hint:* What do the columns of $\underset{\mathcal{C}\leftarrow\mathcal{B}}{P}$ represent?]

b. Find a basis $\{\mathbf{w}_1, \mathbf{w}_2, \mathbf{w}_3\}$ for $\mathbb{R}^3$ such that P is the change-of-coordinates matrix from $\{\mathbf{v}_1, \mathbf{v}_2, \mathbf{v}_3\}$ to $\{\mathbf{w}_1, \mathbf{w}_2, \mathbf{w}_3\}$.

T 22. Let $\mathcal{B} = \{\mathbf{b}_1, \mathbf{b}_2\}$, $\mathcal{C} = \{\mathbf{c}_1, \mathbf{c}_2\}$, and $\mathcal{D} = \{\mathbf{d}_1, \mathbf{d}_2\}$ be bases for a two-dimensional vector space.

a. Write an equation that relates the matrices $\underset{\mathcal{C}\leftarrow\mathcal{B}}{P}$, $\underset{\mathcal{D}\leftarrow\mathcal{C}}{P}$, and $\underset{\mathcal{D}\leftarrow\mathcal{B}}{P}$. Justify your result.

b. Use a matrix program either to help you find the equation or to check the equation you write. Work with three bases for $\mathbb{R}^2$. (See Exercises 7–10.)

---

[3] The idea for Exercises 19 and 20 and five related exercises in earlier sections came from a paper by Jack W. Rogers, Jr., of Auburn University, presented at a meeting of the International Linear Algebra Society, August 1995. See "Applications of Linear Algebra in Calculus," *American Mathematical Monthly* **104** (1), 1997.

Solutions to Practice Problems

1. Since the columns of P are $\mathcal{G}$-coordinate vectors, a vector of the form $P\mathbf{x}$ must be a $\mathcal{G}$-coordinate vector. Thus P satisfies equation (ii).

2. The coordinate vectors found in Example 1 show that

$$\underset{\mathcal{C}\leftarrow\mathcal{B}}{P} = \begin{bmatrix} [\,\mathbf{b}_1\,]_\mathcal{C} & [\,\mathbf{b}_2\,]_\mathcal{C} \end{bmatrix} = \begin{bmatrix} 4 & -6 \\ 1 & 1 \end{bmatrix}$$

Hence

$$\underset{\mathcal{B}\leftarrow\mathcal{C}}{P} = (\underset{\mathcal{C}\leftarrow\mathcal{B}}{P})^{-1} = \frac{1}{10}\begin{bmatrix} 1 & 6 \\ -1 & 4 \end{bmatrix} = \begin{bmatrix} .1 & .6 \\ -.1 & .4 \end{bmatrix}$$

4.7 Digital Signal Processing

Introduction

In the space of just a few decades, digital signal processing (DSP) has led to a dramatic shift in how data is collected, processed, and synthesized. DSP models unify the approach to dealing with data that was previously viewed as unrelated. From stock market analysis to telecommunications and computer science, the data collected over time can be viewed as discrete-time signals and DSP used to store and process the data for more efficient and effective use. Not only do digital signals arise in electrical and control systems engineering, but discrete-data sequences are also generated in biology, physics, economics, demography, and many other areas, wherever a process is measured, or *sampled*, at discrete time intervals. In this section, we will explore the properties of the discrete-time signal space, $\mathbb{S}$, and some of its subspaces, as well as how linear transformations can be used to process, filter, and synthesize the data contained in signals.

Discrete-Time Signals

The vector space $\mathbb{S}$ of discrete-time signals was introduced in Section 4.1. A **signal** in $\mathbb{S}$ is an infinite sequence of numbers, $\{y_k\}$, where the subscripts k range over all integers. Table 1 shows several examples of signals.

TABLE I Examples of Signals

Signals

Name	Symbol	Vector	Formal Description
delta	δ	$(\dots,0,0,0,1,0,0,0,\dots)$	$\{d_k\}$, where $d_k = \begin{cases} 1 & \text{if } k = 0 \\ 0 & \text{if } k \neq 0 \end{cases}$
unit step	υ	$(\dots,0,0,0,1,1,1,1,\dots)$	$\{u_k\}$, where $u_k = \begin{cases} 1 & \text{if } k \geq 0 \\ 0 & \text{if } k < 0 \end{cases}$
constant	χ	$(\dots,1,1,1,1,1,1,1,\dots)$	$\{c_k\}$, where $c_k = 1$
alternating	α	$(\dots,-1,1,-1,1,-1,1,-1,\dots)$	$\{a_k\}$, where $a_k = (-1)^k$
Fibonacci	F	$(\dots,2,-1,1,0,1,1,2,\dots)$	$\{f_k\}$, where $f_k = \begin{cases} 0 & \text{if } k = 0 \\ 1 & \text{if } k = 1 \\ f_{k-1} + f_{k-2} & \text{if } k > 1 \\ f_{k+2} - f_{k+1} & \text{if } k < 0 \end{cases}$
exponential	ϵ_c	$(\dots,c^{-2},c^{-1},c^0,c^1,c^2,\dots)$ $\underset{k=0}{\uparrow}$	$\{e_k\}$, where $e_k = c^k$

- $\left\{\cos\left(\frac{1}{6}\pi k + \frac{1}{4}\pi\right)\right\}$

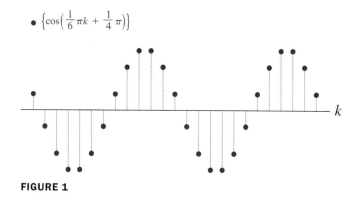

FIGURE 1

Another set of commonly used signals are the **periodic signals**—specifically signals $\{p_k\}$ for which there exists a positive integer q such that $p_k = p_{k+q}$ for all integers k. In particular, the sinusoidal signals, described by $\sigma_{f,\theta} = \{\cos(fk\pi + \theta\pi)\}$ where f and θ are fixed rational numbers, are periodic functions. (See Figure 1.)

Linear Time Invariant Transformations

Linear time invariant (LTI) transformations are used to process signals. One type of processing is to create signals as they are needed, rather than using valuable storage space to store the signals themselves.

To describe the standard basis for $\mathbb{R}^n$ given in Example 4 of Section 4.3, n vectors, $e_1, e_2, \ldots, e_n$, are listed where e_j has a value of 1 in the j-th position and zeros elsewhere. In Example 1 that follows, the analogous signal to each e_j can be created by repeatedly applying a shift LTI transformation to just one signal, that of δ in Table 1.

EXAMPLE 1 Let S be the transformation that shifts each element in a signal to the right, specifically $S(\{x_k\}) = \{y_k\}$, where $y_k = x_{k-1}$. For ease of notation, write $S(\{x_k\}) = \{x_{k-1}\}$. To shift a signal to the left, consider $S^{-1}(\{x_k\}) = \{x_{k+1}\}$. Notice $S^{-1}S(\{x_k\}) = S^{-1}(\{x_{k-1}\}) = \{x_{(k-1)+1}\} = \{x_k\}$. It is easy to verify that $S^{-1}S = SS^{-1} = S^0 = I$, the identity transformation, and hence S is an example of an invertible transformation. Table 2 illustrates the effect of repeatedly applying S and S^{-1} to delta, and the resulting signals can be visualized using Figure 2.

TABLE 2 Applying a Shift Signal

$\vdots$	$\vdots$	$\vdots$
$S^{-2}(\delta)$	$(\ldots, 1, 0, 0, 0, 0, \ldots)$	$\{w_k\}$, where $w_k = \begin{cases} 1 & \text{if } k = -2 \\ 0 & \text{if } k \neq -2 \end{cases}$
$S^{-1}(\delta)$	$(\ldots, 0, 1, 0, 0, 0, \ldots)$	$\{x_k\}$, where $x_k = \begin{cases} 1 & \text{if } k = -1 \\ 0 & \text{if } k \neq -1 \end{cases}$
δ	$(\ldots, 0, 0, 1, 0, 0, \ldots)$	$\{d_k\}$, where $d_k = \begin{cases} 1 & \text{if } k = 0 \\ 0 & \text{if } k \neq 0 \end{cases}$
$S^1(\delta)$	$(\ldots, 0, 0, 0, 1, 0, \ldots)$	$\{y_k\}$, where $y_k = \begin{cases} 1 & \text{if } k = 1 \\ 0 & \text{if } k \neq 1 \end{cases}$
$S^2(\delta)$	$(\ldots, 0, 0, 0, 0, 1, \ldots)$	$\{z_k\}$, where $z_k = \begin{cases} 1 & \text{if } k = 2 \\ 0 & \text{if } k \neq 2 \end{cases}$
$\vdots$	$\underset{k=0}{\uparrow}$	$\vdots$

FIGURE 2

Notice that S satisfies the properties of a linear transformation. Specifically, for any scalar c and signals $\{x_k\}$ and $\{y_k\}$, applying S results in $S(\{x_k\} + \{y_k\}) = \{x_{k-1} + y_{k-1}\} = \{x_{k-1}\} + \{y_{k-1}\} = S(\{x_k\}) + S(\{y_k\})$ and $S(c\{x_k\}) = \{cx_{k-1}\} = cS(\{x_k\})$. The mapping S has an additional property. Notice that for any integer q, $S(\{x_{k+q}\}) = \{x_{k-1+q}\}$. One can think of this last property as the *time invariance* property. Transformations with the same properties as S are referred to as linear time invariant (LTI).

DEFINITION

Linear Time Invariant (LTI) Transformations

A transformation $T : \mathbb{S} \to \mathbb{S}$ is **linear time invariant** provided

(i) $T(\{x_k + y_k\}) = T(\{x_k\}) + T(\{y_k\})$ for all signals $\{x_k\}$ and $\{y_k\}$;

(ii) $T(c\{x_k\}) = cT(\{x_k\})$ for all scalars c and all signals $\{x_k\}$;

(iii) If $T(\{x_k\}) = \{y_k\}$, then $T(\{x_{k+q}\}) = \{y_{k+q}\}$ for all integers q and all signals $\{x_k\}$.

The first two properties in the definition of LTI transformations are the same as the two properties listed in the definition of a linear transformation resulting in the following theorem:

THEOREM 16

LTI Transformations are Linear Transformations

A linear time invariant transformation on the signal space $\mathbb{S}$ is a special type of linear transformation.

Digital Signal Processing

LTI transformations, like the shift transformation, can be used to create new signals from signals that are already stored in a system. Another type of LTI transformation is used for *smoothing* or *filtering* data. In Example 11 of Section 4.2, a two-day moving average LTI transformation is used to smooth out stock price fluctuations. In Example 2, this mapping is extended to encompass longer time periods. Smoothing out a signal can make it easier to spot trends in data. Filtering will be discussed in more detail in Section 4.8.

EXAMPLE 2 For any positive integer m, the **moving average** LTI transformation with time period m is given by

$$M_m(\{x_k\}) = \{y_k\} \text{ where } y_k = \frac{1}{m} \sum_{j=k-m+1}^{k} x_k$$

Figure 3 illustrates how M_3 smooths out a signal. Section 4.2, Figure 3 illustrates the smoothing that occurred when M_2 was applied to the same data. As m is increased, applying M_m smooths the signal even more.

FIGURE 3

The kernel of M_2 is calculated in Example 11 of Section 4.2. It is the span of the alternating sequence α listed in Table 1. The kernel of the LTI transformation describes what is smoothed out of the original signal. Exercises 10, 12, and 14 explore properties of M_3 further. ■

Another type of DSP does the opposite of smoothing or filtering - it combines signals to increase their complexity. **Auralization** is a process used in the entertainment industry to give a more acoustic quality to virtually generated sounds. In Example 3, we illustrate how combining signals enhances the sound generated by the signal $\{\cos(440\pi k)\}$.

EXAMPLE 3 Combining several signals can be used to produce more realistic virtual sounds. In Figure 4, notice that the original cosine wave contains very little variation, whereas by enhancing the equation used, the waves created contain more variation by introducing echos or allowing a sound to fade out. ■

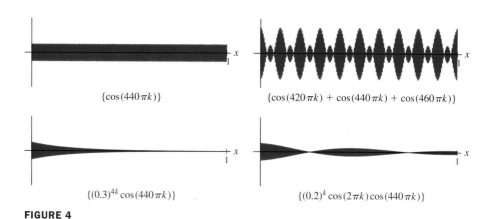

FIGURE 4

Generating Bases for Subspaces of $\mathbb{S}$

If several sets of data are being sampled over the same n time periods, it may be advantageous to view the signals created as part of $\mathbb{S}_n$. The set of **signals of length n**, $\mathbb{S}_n$, is defined to be the set of all signals $\{y_k\}$ such that $y_k = 0$ whenever $k < 0$ or $k > n$. Theorem 17 establishes that $\mathbb{S}_n$ is isomorphic to $\mathbb{R}^{n+1}$. A basis for $\mathbb{S}_n$ can be generated using the shift LTI transformation S from Example 1 and the signal δ as illustrated in Table 2.

THEOREM 17

The set $\mathbb{S}_n$ is a subspace of $\mathbb{S}$ isomorphic to $\mathbb{R}^{n+1}$, and the set of signals $\mathcal{B}_n = \{\delta, S(\delta), S^2(\delta), \ldots, S^n(\delta)\}$ forms a basis for $\mathbb{S}_n$.

PROOF Since the zero signal is in $\mathbb{S}_n$, and adding or scaling signals cannot create nonzeros in the positions that must contain zeros, the set $\mathbb{S}_n$ is a subspace of $\mathbb{S}$. Let $\{y_k\}$ be any signal in $\mathbb{S}_n$. Notice

$$\{y_k\} = \sum_{j=0}^{n} y_j S^j(\delta),$$

so $\mathcal{B}_n$ is a spanning set for $\mathbb{S}_n$. Conversely, if $c_0, \ldots, c_n$ are scalars such that

$$c_0 \delta + c_1 S(\delta) + \ldots + c_n S^n(\delta) = \{0\},$$

specifically

$$(\ldots\, 0,\, 0,\, c_0,\, c_1,\, \ldots,\, c_n,\, 0,\, 0,\, \ldots) = (\ldots,\, 0,\, 0,\, 0,\, 0,\, \ldots,\, 0,\, 0,\, 0,\, \ldots),$$

then $c_0 = c_1 = \cdots = c_n = 0$, and thus the vectors in $\mathcal{B}_n$ form a linearly independent set. This establishes that $\mathcal{B}_n$ is a basis for $\mathbb{S}_n$ and hence it is an $n + 1$ dimensional vector space isomorphic to $\mathbb{R}^{n+1}$.

Since $\mathbb{S}_n$ has a finite basis, any vector in $\mathbb{S}_n$ can be represented as a vector in $\mathbb{R}^{n+1}$. ∎

EXAMPLE 4 Using the basis $\mathcal{B}_2 = \{\delta, S(\delta), S^2(\delta)\}$ for $\mathbb{S}_2$, represent the signal $\{y_k\}$, where

$$y_k = \begin{cases} 0 & \text{if } k < 0 \text{ or } k > 3 \\ 2 & \text{if } k = 0 \\ 3 & \text{if } k = 1 \\ -1 & \text{if } k = 2 \end{cases}$$

as a vector in $\mathbb{R}^3$.

SOLUTION First write $\{y_k\}$ as a linear combination of the basis vectors in $\mathcal{B}_2$.

$$\{y_k\} = 2\delta + 3S(\delta) + (-1)S^2(\delta)$$

The coefficients of this linear combination are precisely the entries in the coordinate vector. Thus $[\{y_k\}]_{\mathcal{B}_2} = \begin{bmatrix} 2 \\ 3 \\ -1 \end{bmatrix}$ ∎

The set of **finitely supported signals**, $\mathbb{S}_f$, is the set of signals $\{y_k\}$, where only finitely many of the entries are nonzero. In Example 8 of Section 4.1, it is established that $\mathbb{S}_f$ is a subspace of $\mathbb{S}$. The signals created by recording the daily price of a stock increase in length each day, but remain finitely supported, and hence these signals belong to $\mathbb{S}_f$, but not to any particular $\mathbb{S}_n$. Conversely, if a signal is in $\mathbb{S}_n$ for some positive integer n, then it is also in $\mathbb{S}_f$. In Theorem 18, we see that $\mathbb{S}_f$ is an infinite dimensional subspace and so it is not isomorphic to $\mathbb{R}^n$ for any n.

THEOREM 18   The set $\mathcal{B}_f = \{S^j(\delta) : \text{where } j \in \mathbb{Z}\}$ is a basis for the infinite dimensional vector space $\mathbb{S}_f$.

PROOF   Let $\{y_k\}$ be any signal in $\mathbb{S}_f$. Since only finitely many entries in $\{y_k\}$ are nonzero, there exist integers p and q such that $y_k = 0$ for all $k < p$ and and $k > q$. Thus

$$\{y_k\} = \sum_{j=p}^{q} y_j S^j(\delta),$$

so $\mathcal{B}_f$ is a spanning set for $\mathbb{S}_n$. Moreover, if a linear combination of signals with scalars $c_p, c_{p+1}, \ldots, c_q$ add to zero,

$$\sum_{j=p}^{q} c_j S^j(\delta), = \{0\},$$

then $c_p = c_{p+1} = \cdots = c_q = 0$, and thus the vectors in $\mathcal{B}_f$ form a linearly independent set. This establishes that $\mathcal{B}_f$ is a basis for $\mathbb{S}_f$. Since $\mathcal{B}_f$ contains infinitely many signals, $\mathbb{S}_f$ is an infinite dimensional vector space. ∎

The creative power of the shift LTI transformation falls short of being able to create a basis for $\mathbb{S}$ itself. The definition of linear combination requires that only finitely many vectors and scalars are used in a sum. Consider the unit step signal, υ, from Table 1. Although $\upsilon = \sum_{j=0}^{\infty} S^j(\delta)$, this is an infinite sum of vectors and hence not technically considered a *linear combination* of the basis elements from $\mathcal{B}_f$.

In calculus, sums with infinitely many terms are studied in detail. Although it can be shown that every vector space has a basis (using a finite number of terms in each linear combination), the proof relies on the Axiom of Choice and hence establishing that $\mathbb{S}$ has a basis is a topic you may see in higher level math classes. The sinusoidal and exponential signals, which have infinite support, are explored in detail in Section 4.8

Practice Problems

1. Find $\upsilon + \chi$ from Table 1. Express the answer as a vector and give its formal description.
2. Show that $T(\{x_k\}) = \{3x_k - 2x_{k-1}\}$ is a linear time invariant transformation.
3. Find a nonzero vector in the kernel of T for the linear time invariant transformation given in Practice Problem 2.

4.7 Exercises

For Exercises 1–4, find the indicated sums of the signals in Table 1.

1. $\chi + \alpha$
2. $\chi - \alpha$
3. $\upsilon + 2\alpha$
4. $\upsilon - 3\alpha$

For Exercises 5–8, recall that $I(\{x_k\}) = \{x_k\}$ and $S(\{x_k\}) = \{x_{k-1}\}$.

5. Which signals from Table 1 are in the kernel of $I + S$?
6. Which signals from Table 1 are in the kernel of $I - S$?

7. Which signals from Table 1 are in the kernel of $I - cS$ for a fixed nonzero scalar $c \neq 1$?

8. Which signals from Table 1 are in the kernel of $I - S - S^2$?

9. Show that $T(\{x_k\}) = \{x_k - x_{k-1}\}$ is a linear time invariant transformation.

10. Show that $M_3(\{x_k\}) = \left\{ \frac{1}{3}(x_{k-2} + x_{k-1} + x_k) \right\}$ is a linear time invariant transformation.

11. Find a nonzero signal in the kernel of T from Exercise 9.

12. Find a nonzero signal in the kernel of M_3 from Exercise 10.

13. Find a nonzero signal in the range of T from Exercise 9.

14. Find a nonzero signal in the range of M_3 from Exercise 10.

In Exercises 15–22, V is a vector space and A is an $m \times n$ matrix. Mark each statement True or False (**T/F**). Justify each answer.

15. (**T/F**) The set of signals of length n, $\mathbb{S}_n$, has a basis with $n + 1$ signals.

16. (**T/F**) The set of signals, $\mathbb{S}$, has a finite basis.

17. (**T/F**) Every subspace of the set of signals $\mathbb{S}$ is infinite dimensional.

18. (**T/F**) The vector space $\mathbb{R}^{n+1}$ is a subspace of $\mathbb{S}$.

19. (**T/F**) Every linear time invariant transformation is a linear transformation.

20. (**T/F**) The moving average function is a linear time invariant transformation.

21. (**T/F**) If you scale a signal by a fixed constant, the result is not a signal.

22. (**T/F**) If you scale a linear time invariant transformation by a fixed constant, the result is no longer a linear transformation.

Guess and check or working backwards through the solution to Practice Problem 3 are two good ways to find solutions to Exercises 23 and 24.

23. Construct a linear time invariant transformation that has the signal $\{x_k\} = \left\{ \left(\frac{3}{4} \right)^k \right\}$ in its kernel.

24. Construct a linear time invariant transformation that has the signal $\{x_k\} = \left\{ \left(\frac{-2}{3} \right)^k \right\}$ in its kernel.

25. Let $W = \left\{ \{x_k\} \mid x_k = \begin{cases} 0 & \text{if } k \text{ is a multiple of } 2 \\ r & \text{if } k \text{ is not a multiple of } 2 \end{cases} \right.$
where r can be any real number$\left. \right\}$. A typical signal in W looks like

$$(\ldots, r, 0, r, 0, r, 0, r, \ldots)$$
$$\uparrow$$
$$k = 0$$

Show that W is a subspace of $\mathbb{S}$.

26. Let $W = \left\{ \{x_k\} \mid x_k = \begin{cases} 0 & \text{if } k < 0 \\ r & \text{if } k \geq 0 \end{cases} \right.$ where r can be any real number.$\left. \right\}$ A typical signal in W looks like

$$(\ldots, 0, 0, 0, r, r, r, r, \ldots)$$
$$\uparrow$$
$$k = 0$$

Show that W is a subspace of $\mathbb{S}$.

27. Find a basis for the subspace W in Exercise 25. What is the dimension of this subspace?

28. Find a basis for the subspace W in Exercise 26. What is the dimension of this subspace?

29. Let $W = \left\{ \{x_k\} \mid x_k = \begin{cases} 0 & \text{if } k \text{ is a multiple of } 2 \\ r_k & \text{if } k \text{ is not a multiple of } 2 \end{cases} \right.$
where each r_k can be any real number$\left. \right\}$. A typical signal in W looks like

$$(\ldots, r_{-3}, 0, r_{-1}, 0, r_1, 0, r_3, \ldots)$$
$$\uparrow$$
$$k = 0$$

Show that W is a subspace of $\mathbb{S}$.

30. Let $W = \left\{ \{x_k\} \mid x_k = \begin{cases} 0 & \text{if } k < 0 \\ r_k & \text{if } k \geq 0 \end{cases} \right.$
where each r_k can be any real number$\left. \right\}$. A typical signal in W looks like

$$(\ldots, 0, 0, 0, r_0, r_1, r_2, r_3, \ldots)$$
$$\uparrow$$
$$k = 0$$

Show that W is a subspace of $\mathbb{S}$.

31. Describe an infinite linearly independent subset of the subspace W in Exercise 29. Does this establish that W is infinite dimensional? Justify your answer.

32. Describe an infinite linearly independent subset of the subspace W in Exercise 30. Does this establish that W is infinite dimensional? Justify your.

Solutions to Practice Problems

1. First add $v + \chi$ in vector form:

$$
\begin{array}{rl}
 & (\dots, 0, 0, 0, 1, 1, 1, 1, \dots) \\
+ & (\dots, 1, 1, 1, 1, 1, 1, 1, \dots) \\
= & (\dots, 1, 1, 1, 2, 2, 2, 2, \dots) \\
 & \qquad\quad \uparrow \\
 & \qquad\quad k = 0
\end{array}
$$

Then add the terms in the formal description to get a new formal description:

$$
v + \chi = \{z_k\}, \text{ where } z_k = u_k + c_k = \begin{cases} 1+1 & \text{if } k \geq 0 \\ 0+1 & \text{if } k < 0 \end{cases} = \begin{cases} 2 & \text{if } k \geq 0 \\ 1 & \text{if } k < 0 \end{cases}
$$

2. Verify that the three conditions for a linear time invariant transformation hold. Specifically, for any two signals $\{x_k\}$ and $\{y_k\}$, and scalar c, observe that

   a. $T(\{x_k + y_k\}) = \{3(x_k + y_k) - 2(x_{k-1} + y_{k-1})\} = \{3x_k - 2x_{k-1}\} + \{3y_k - 2y_{k-1}\} = T(\{x_k\}) + T(\{y_k\})$

   b. $T(c\{x_k\}) = \{3cx_k - 2cx_{k-1}\} = c\{3x_k - 2x_{k-1}\} = cT(\{x_k\})$

   c. $T(\{x_k\}) = \{3x_k - 2x_{k-1}\}$ and $T(\{x_{k+q}\}) = \{3x_{k+q} - 2x_{k+q-1}\} = \{3x_{k+q} - 2x_{k-1+q}\}$ for all integers q.

   Thus T is a linear time invariant transformation.

3. To find a vector in the kernel of T, set $T(\{x_k\}) = \{3x_k - 2x_{k-1}\} = \{0\}$. Then for each k, notice $3x_k - 2x_{k-1} = 0$ and hence $x_k = \frac{2}{3}x_{k-1}$. Picking a nonzero value for x_0, say $x_0 = 1$, then $x_1 = \frac{2}{3}$, $x_2 = \left(\frac{2}{3}\right)^2$, and in general, $x_k = \left(\frac{2}{3}\right)^k$. To verify that this signal is indeed in the kernel of T observe that $T\left(\left\{\left(\frac{2}{3}\right)^k\right\}\right) = \left\{3\left(\frac{2}{3}\right)^k - 2\left(\frac{2}{3}\right)^{k-1}\right\} = \left\{\left(\frac{2}{3}\right)^{k-1}\left(3\left(\frac{2}{3}\right) - 2\right)\right\} = \{0\}$. Notice that $\left\{\left(\frac{2}{3}\right)^k\right\}$ is the exponential signal with $c = \frac{2}{3}$.

4.8 Applications to Difference Equations

Continuing our study of discrete-time signals, in this section we explore difference equations, a valuable tool used to filter the data contained in signals. Even when a differential equation is used to model a continuous process, a numerical solution is often produced from a related difference equation. This section highlights some fundamental properties of linear difference equations that are explained using linear algebra.

Linear Independence in the Space $\mathbb{S}$ of Signals

To simplify notation, we consider a set of only three signals in $\mathbb{S}$, say, $\{u_k\}$, $\{v_k\}$, and $\{w_k\}$. They are linearly independent precisely when the equation

$$
c_1 u_k + c_2 v_k + c_3 w_k = 0 \quad \text{for all } k \tag{1}
$$

implies that $c_1 = c_2 = c_3 = 0$. The phrase "for all k" means for all integers—positive, negative, and zero. One could also consider signals that start with $k = 0$, for example, in which case, "for all k" would mean for all integers $k \geq 0$.

Suppose c_1, c_2, c_3 satisfy (1). Then equation (1) holds for any three consecutive values of k, say, k, $k + 1$, and $k + 2$. Thus (1) implies that

$$c_1 u_{k+1} + c_2 v_{k+1} + c_3 w_{k+1} = 0 \quad \text{for all } k$$

and

$$c_1 u_{k+2} + c_2 v_{k+2} + c_3 w_{k+2} = 0 \quad \text{for all } k$$

Hence c_1, c_2, c_3 satisfy

$$\begin{bmatrix} u_k & v_k & w_k \\ u_{k+1} & v_{k+1} & w_{k+1} \\ u_{k+2} & v_{k+2} & w_{k+2} \end{bmatrix} \begin{bmatrix} c_1 \\ c_2 \\ c_3 \end{bmatrix} = \begin{bmatrix} 0 \\ 0 \\ 0 \end{bmatrix} \quad \text{for all } k \tag{2}$$

STUDY GUIDE offers additional resources on the Casorati Test.

The coefficient matrix in this system is called the **Casorati matrix**, $C(k)$, of the signals, and the determinant of the matrix is called the **Casoratian** of $\{u_k\}$, $\{v_k\}$, and $\{w_k\}$. If the Casorati matrix is invertible for at least one value of k, then (2) will imply that $c_1 = c_2 = c_3 = 0$, which will prove that the three signals are linearly independent.

EXAMPLE 1 Verify that $\{1^k\}$, $\{(-2)^k\}$, and $\{3^k\}$ are linearly independent signals.

SOLUTION   The Casorati matrix is

$$\begin{bmatrix} 1^k & (-2)^k & 3^k \\ 1^{k+1} & (-2)^{k+1} & 3^{k+1} \\ 1^{k+2} & (-2)^{k+2} & 3^{k+2} \end{bmatrix}$$

Row operations can show fairly easily that this matrix is always invertible. However, it is faster to substitute a value for k—say, $k = 0$—and row reduce the numerical matrix:

$$\begin{bmatrix} 1 & 1 & 1 \\ 1 & -2 & 3 \\ 1 & 4 & 9 \end{bmatrix} \sim \begin{bmatrix} 1 & 1 & 1 \\ 0 & -3 & 2 \\ 0 & 3 & 8 \end{bmatrix} \sim \begin{bmatrix} 1 & 1 & 1 \\ 0 & -3 & 2 \\ 0 & 0 & 10 \end{bmatrix}$$

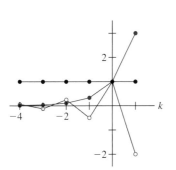

The signals 1^k, $(-2)^k$, and 3^k.

The Casorati matrix is invertible for $k = 0$. So $\{1^k\}$, $\{(-2)^k\}$, and $\{3^k\}$ are linearly independent. ∎

If a Casorati matrix is not invertible, the associated signals being tested may or may not be linearly dependent. (See Exercise 35.) However, it can be shown that if the signals are all solutions of the *same* homogeneous difference equation (described below), then either the Casorati matrix is invertible for all k and the signals are linearly independent, or else the Casorati matrix is not invertible for all k and the signals are linearly dependent. A nice proof using linear transformations is in the *Study Guide*.

Linear Difference Equations

Given scalars $a_0, \ldots, a_n$, with a_0 and a_n nonzero, and given a signal $\{z_k\}$, the equation

$$a_0 y_{k+n} + a_1 y_{k+n-1} + \cdots + a_{n-1} y_{k+1} + a_n y_k = z_k \quad \text{for all } k \tag{3}$$

is called a **linear difference equation** (or **linear recurrence relation**) **of order n**. For simplicity, a_0 is often taken equal to 1. If $\{z_k\}$ is the zero sequence, the equation is **homogeneous**; otherwise, the equation is **nonhomogeneous**.

In digital signal processing (DSP), a difference equation such as (3) describes a **linear time invariant (LTI) filter**, and $a_0, \ldots, a_n$ are called the **filter coefficients**. The shift LTI transformations $S(\{y_k\}) = \{y_{k-1}\}$ and $S^{-1}(\{y_k\}) = \{y_{k+1}\}$ were introduced in Example 1 of Section 4.7 and are used here to describe the LTI filter associated with a linear difference equation. Define

$$T = a_0 S^{-n} + a_1 S^{-n+1} + \cdots + a_{n-1} S^{-1} + a_n S^0.$$

Notice if $\{z_k\} = T(\{y_k\})$, then for any k, Equation (3) describes the relationship between terms in the two signals.

EXAMPLE 2 Let us feed two different signals into the filter

$$.35y_{k+2} + .5y_{k+1} + .35y_k = z_k$$

Here $.35$ is an abbreviation for $\sqrt{2}/4$. The first signal is created by sampling the continuous signal $y = \cos(\pi t/4)$ at integer values of t, as in Figure 1(a). The discrete signal is

$$\{y_k\} = (\ldots, \cos(0), \cos(\pi/4), \cos(2\pi/4), \cos(3\pi/4), \ldots)$$

For simplicity, write $\pm.7$ in place of $\pm\sqrt{2}/2$, so that

$$\{y_k\} = (\ldots, 1, \ .7, \ 0, \ -.7, \ -1, \ -.7, \ 0, \ .7, \ 1, \ .7, \ 0, \ldots)$$
$$\uparrow$$
$$k = 0$$

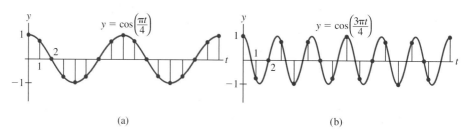

FIGURE 1 Discrete signals with different frequencies.

Table 1 shows a calculation of the output sequence $\{z_k\}$, where $.35(.7)$ is an abbreviation for $(\sqrt{2}/4)(\sqrt{2}/2) = .25$. The output is $\{y_k\}$, shifted by one term.

TABLE 1 Computing the Output of a Filter

k	y_k	y_{k+1}	y_{k+2}	$.35y_k$	$+ .5y_{k+1}$	$+ .35y_{k+2}$	$= z_k$
0	1	.7	0	$.35(1)$	$+ .5(.7)$	$+ .35(0)$	$= .7$
1	.7	0	$-.7$	$.35(.7)$	$+ .5(0)$	$+ .35(-.7) =$	0
2	0	$-.7$	-1	$.35(0)$	$+ .5(-.7)$	$+ .35(-1) =$	$-.7$
3	$-.7$	-1	$-.7$	$.35(-.7)$	$+ .5(-1)$	$+ .35(-.7) =$	-1
4	-1	$-.7$	0	$.35(-1)$	$+ .5(-.7)$	$+ .35(0)$	$= -.7$
5	$-.7$	0	.7	$.35(-.7)$	$+ .5(0)$	$+ .35(.7)$	$= 0$
$\vdots$	$\vdots$						$\vdots$

A different input signal is produced from the higher frequency signal $y = \cos(3\pi t/4)$, shown in Figure 1(b). Sampling at the same rate as before produces a new input sequence:

$$\{w_k\} = (\ldots, 1, \ -.7, \ 0, \ .7, \ -1, \ .7, \ 0, \ -.7, \ 1, \ -.7, \ 0, \ldots)$$
$$\uparrow$$
$$k = 0$$

When $\{w_k\}$ is fed into the filter, the output is the zero sequence. The filter, called a *low-pass filter*, lets $\{y_k\}$ pass through, but stops the higher frequency $\{w_k\}$. ∎

In many applications, a sequence $\{z_k\}$ is specified for the right side of a difference equation (3), and a $\{y_k\}$ that satisfies (3) is called a **solution** of the equation. The next example shows how to find solutions for a homogeneous equation.

EXAMPLE 3 Solutions of a homogeneous difference equation often have the form $\{y_k\} = \{r^k\}$ for some r. Find some solutions of the equation

$$y_{k+3} - 2y_{k+2} - 5y_{k+1} + 6y_k = 0 \quad \text{for all } k \tag{4}$$

SOLUTION Substitute r^k for y_k in the equation and factor the left side:

$$r^{k+3} - 2r^{k+2} - 5r^{k+1} + 6r^k = 0 \tag{5}$$

$$r^k(r^3 - 2r^2 - 5r + 6) = 0$$

$$r^k(r - 1)(r + 2)(r - 3) = 0 \tag{6}$$

Since (5) is equivalent to (6), $\{r^k\}$ satisfies the difference equation (4) if and only if r satisfies (6). Thus $\{1^k\}$, $\{(-2)^k\}$, and $\{3^k\}$ are all solutions of (4). For instance, to verify that $\{3^k\}$ is a solution of (4), compute

$$3^{k+3} - 2 \cdot 3^{k+2} - 5 \cdot 3^{k+1} + 6 \cdot 3^k$$

$$= 3^k(27 - 18 - 15 + 6) = 0 \quad \text{for all } k \quad ∎$$

In general, a nonzero signal $\{r^k\}$ satisfies the homogeneous difference equation

$$y_{k+n} + a_1 y_{k+n-1} + \cdots + a_{n-1} y_{k+1} + a_n y_k = 0 \quad \text{for all } k$$

if and only if r is a root of the **auxiliary equation**

$$r^n + a_1 r^{n-1} + \cdots + a_{n-1} r + a_n = 0$$

We will not consider the case in which r is a repeated root of the auxiliary equation. When the auxiliary equation has a *complex root*, the difference equation has solutions of the form $\{s^k \cos k\omega\}$ and $\{s^k \sin k\omega\}$, for constants s and ω. This happened in Example 2.

Solution Sets of Linear Difference Equations

Given $a_1, \ldots, a_n$, recall that the LTI transformation $T : \mathbb{S} \to \mathbb{S}$ given by

$$T = a_0 S^{-n} + a_1 S^{-n+1} + \cdots + a_{n-1} S^{-1} + a_n S^0$$

transforms a signal $\{y_k\}$ into the signal $\{w_k\}$ given by

$$w_k = y_{k+n} + a_1 y_{k+n-1} + \cdots + a_{n-1} y_{k+1} + a_n y_k \text{ for all } k$$

This implies that the solution set of the homogeneous equation

$$y_{k+n} + a_1 y_{k+n-1} + \cdots + a_{n-1} y_{k+1} + a_n y_k = 0 \text{ for all } k$$

is the kernel of T and describes the signals that are *filtered out* or transformed into the zero signal. Since the kernel of any linear transformation with domain $\mathbb{S}$ is a *subspace* of $\mathbb{S}$, so is the solution set of a homogeneous equation. Any linear combination of solutions is again a solution.

The next theorem, a simple but basic result, will lead to more information about the solution sets of difference equations.

THEOREM 19

If $a_n \neq 0$ and if $\{z_k\}$ is given, the equation

$$y_{k+n} + a_1 y_{k+n-1} + \cdots + a_{n-1} y_{k+1} + a_n y_k = z_k \quad \text{for all } k \qquad (7)$$

has a unique solution whenever $y_0, \ldots, y_{n-1}$ are specified.

PROOF If $y_0, \ldots, y_{n-1}$ are specified, use (7) to *define*

$$y_n = z_0 - [a_1 y_{n-1} + \cdots + a_{n-1} y_1 + a_n y_0]$$

And now that $y_1, \ldots, y_n$ are specified, use (7) to define y_{n+1}. In general, use the recurrence relation

$$y_{n+k} = z_k - [a_1 y_{k+n-1} + \cdots + a_n y_k] \qquad (8)$$

to define y_{n+k} for $k \geq 0$. To define y_k for $k < 0$, use the recurrence relation

$$y_k = \frac{1}{a_n} z_k - \frac{1}{a_n} [y_{k+n} + a_1 y_{k+n-1} + \cdots + a_{n-1} y_{k+1}] \qquad (9)$$

This produces a signal that satisfies (7). Conversely, any signal that satisfies (7) for all k certainly satisfies (8) and (9), so the solution of (7) is unique. ∎

THEOREM 20

The set H of all solutions of the nth-order homogeneous linear difference equation

$$y_{k+n} + a_1 y_{k+n-1} + \cdots + a_{n-1} y_{k+1} + a_n y_k = 0 \quad \text{for all } k \qquad (10)$$

is an n-dimensional vector space.

PROOF As was pointed out earlier, H is a subspace of $\mathbb{S}$ because H is the kernel of a linear transformation. For $\{y_k\}$ in H, let $F\{y_k\}$ be the vector in $\mathbb{R}^n$ given by $(y_0, y_1, \ldots, y_{n-1})$. It is readily verified that $F : H \to \mathbb{R}^n$ is a linear transformation. Given any vector $(y_0, y_1, \ldots, y_{n-1})$ in $\mathbb{R}^n$, Theorem 19 says that there is a unique signal $\{y_k\}$ in H such that $F\{y_k\} = (y_0, y_1, \ldots, y_{n-1})$. This means that F is a one-to-one linear transformation of H onto $\mathbb{R}^n$; that is, F is an isomorphism. Thus $\dim H = \dim \mathbb{R}^n = n$. (See Exercise 52 in Section 4.5.) ∎

EXAMPLE 4 Find a basis for the set of all solutions to the difference equation

$$y_{k+3} - 2y_{k+2} - 5y_{k+1} + 6y_k = 0 \quad \text{for all } k$$

SOLUTION Our work in linear algebra really pays off now! We know from Examples 1 and 3 that $\{1^k\}$, $\{(-2)^k\}$, and $\{3^k\}$ are linearly independent solutions. In general, it can be difficult to verify directly that a set of signals *spans* the solution space. But that is no problem here because of two key theorems—Theorem 20, which shows that the solution space is exactly three-dimensional, and the Basis Theorem in Section 4.5, which says that a linearly independent set of n vectors in an n-dimensional space is automatically a basis. So $\{1^k\}$, $\{(-2)^k\}$, and $\{3^k\}$ form a basis for the solution space. ∎

The standard way to describe the "general solution" of the difference equation (10) is to exhibit a basis for the subspace of all solutions. Such a basis is usually called a

fundamental set of solutions of (10). In practice, if you can find n linearly independent signals that satisfy (10), they will automatically span the n-dimensional solution space, as explained in Example 4.

Nonhomogeneous Equations

The general solution of the nonhomogeneous difference equation

$$y_{k+n} + a_1 y_{k+n-1} + \cdots + a_{n-1} y_{k+1} + a_n y_k = z_k \quad \text{for all } k \tag{11}$$

can be written as one particular solution of (11) plus an arbitrary linear combination of a fundamental set of solutions of the corresponding homogeneous equation (10). This fact is analogous to the result in Section 1.5 showing that the solution sets of $A\mathbf{x} = \mathbf{b}$ and $A\mathbf{x} = \mathbf{0}$ are parallel. Both results have the same explanation: The mapping $\mathbf{x} \mapsto A\mathbf{x}$ is linear, and the mapping that transforms the signal $\{y_k\}$ into the signal $\{z_k\}$ in (11) is linear.

EXAMPLE 5   Verify that the signal $\{y_k\} = \{k^2\}$ satisfies the difference equation

$$y_{k+2} - 4y_{k+1} + 3y_k = -4k \quad \text{for all } k \tag{12}$$

Then find a description of all solutions of this equation.

SOLUTION   Substitute k^2 for y_k on the left side of (12):

$$(k+2)^2 - 4(k+1)^2 + 3k^2$$
$$= (k^2 + 4k + 4) - 4(k^2 + 2k + 1) + 3k^2$$
$$= -4k$$

So k^2 is indeed a solution of (12). The next step is to solve the homogeneous equation

$$y_{k+2} - 4y_{k+1} + 3y_k = 0 \quad \text{for all } k \tag{13}$$

The auxiliary equation is

$$r^2 - 4r + 3 = (r-1)(r-3) = 0$$

The roots are $r = 1, 3$. So two solutions of the homogeneous difference equation are $\{1^k\}$ and $\{3^k\}$. They are obviously not multiples of each other, so they are linearly independent signals. By Theorem 20, the solution space is two-dimensional, so $\{1^k\}$ and $\{3^k\}$ form a basis for the set of solutions of equation (13). Translating that set by a particular solution of the nonhomogeneous equation (12), we obtain the general solution of (12):

$$\{k^2\} + c_1\{1^k\} + c_2\{3^k\}, \quad \text{or} \quad \{k^2 + c_1 + c_2 3^k\}$$

Figure 2 gives a geometric visualization of the two solution sets. Each point in the figure corresponds to one signal in $\mathbb{S}$.   ∎

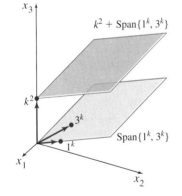

FIGURE 2

Solution sets of difference equations (12) and (13).

Reduction to Systems of First-Order Equations

A modern way to study a homogeneous nth-order linear difference equation is to replace it by an equivalent system of first-order difference equations, written in the form

$$\mathbf{x}_{k+1} = A\mathbf{x}_k \quad \text{for all } k$$

where the vectors $\mathbf{x}_k$ are in $\mathbb{R}^n$ and A is an $n \times n$ matrix.

A simple example of such a (vector-valued) difference equation was already studied in Section 1.10. Further examples will be covered in Sections 5.6 and 5.9.

EXAMPLE 6 Write the following difference equation as a first-order system:

$$y_{k+3} - 2y_{k+2} - 5y_{k+1} + 6y_k = 0 \quad \text{for all } k$$

SOLUTION For each k, set

$$\mathbf{x}_k = \begin{bmatrix} y_k \\ y_{k+1} \\ y_{k+2} \end{bmatrix}$$

The difference equation says that $y_{k+3} = -6y_k + 5y_{k+1} + 2y_{k+2}$, so

$$\mathbf{x}_{k+1} = \begin{bmatrix} y_{k+1} \\ y_{k+2} \\ y_{k+3} \end{bmatrix} = \begin{bmatrix} 0 & + & y_{k+1} & + 0 \\ 0 & + 0 & & + y_{k+2} \\ -6y_k & + 5y_{k+1} & + 2y_{k+2} \end{bmatrix} = \begin{bmatrix} 0 & 1 & 0 \\ 0 & 0 & 1 \\ -6 & 5 & 2 \end{bmatrix} \begin{bmatrix} y_k \\ y_{k+1} \\ y_{k+2} \end{bmatrix}$$

That is,

$$\mathbf{x}_{k+1} = A\mathbf{x}_k \quad \text{for all } k, \quad \text{where} \quad A = \begin{bmatrix} 0 & 1 & 0 \\ 0 & 0 & 1 \\ -6 & 5 & 2 \end{bmatrix} \quad ■$$

In general, the equation

$$y_{k+n} + a_1 y_{k+n-1} + \cdots + a_{n-1} y_{k+1} + a_n y_k = 0 \quad \text{for all } k$$

can be rewritten as $\mathbf{x}_{k+1} = A\mathbf{x}_k$ for all k, where

$$\mathbf{x}_k = \begin{bmatrix} y_k \\ y_{k+1} \\ \vdots \\ y_{k+n-1} \end{bmatrix}, \quad A = \begin{bmatrix} 0 & 1 & 0 & \cdots & 0 \\ 0 & 0 & 1 & & 0 \\ \vdots & & & \ddots & \vdots \\ 0 & 0 & 0 & & 1 \\ -a_n & -a_{n-1} & -a_{n-2} & \cdots & -a_1 \end{bmatrix}$$

Practice Problem

It can be shown that the signals 2^k, $3^k \sin \frac{k\pi}{2}$, and $3^k \cos \frac{k\pi}{2}$ are solutions of

$$y_{k+3} - 2y_{k+2} + 9y_{k+1} - 18y_k = 0$$

Show that these signals form a basis for the set of all solutions of the difference equation.

4.8 Exercises

Verify that the signals in Exercises 1 and 2 are solutions of the accompanying difference equation.

1. $2^k, (-4)^k$; $y_{k+2} + 2y_{k+1} - 8y_k = 0$

2. $3^k, (-3)^k$; $y_{k+2} - 9y_k = 0$

Show that the signals in Exercises 3–6 form a basis for the solution set of the accompanying difference equation.

3. The signals and equation in Exercise 1

4. The signals and equation in Exercise 2

5. $(-3)^k, k(-3)^k$; $y_{k+2} + 6y_{k+1} + 9y_k = 0$

6. $5^k \cos \frac{k\pi}{2}, 5^k \sin \frac{k\pi}{2}$; $y_{k+2} + 25y_k = 0$

In Exercises 7–12, assume the signals listed are solutions of the given difference equation. Determine if the signals form a basis for the solution space of the equation. Justify your answers using appropriate theorems.

7. $1^k, 2^k, (-2)^k$; $y_{k+3} - y_{k+2} - 4y_{k+1} + 4y_k = 0$

8. $2^k, 4^k, (-5)^k$; $y_{k+3} - y_{k+2} - 22y_{k+1} + 40y_k = 0$

9. $1^k, 3^k \cos \frac{k\pi}{2}, 3^k \sin \frac{k\pi}{2}$; $y_{k+3} - y_{k+2} + 9y_{k+1} - 9y_k = 0$

10. $(-1)^k, k(-1)^k, 5^k$; $y_{k+3} - 3y_{k+2} - 9y_{k+1} - 5y_k = 0$

11. $(-1)^k, 3^k$; $y_{k+3} + y_{k+2} - 9y_{k+1} - 9y_k = 0$

12. $1^k, (-1)^k$; $y_{k+4} - 2y_{k+2} + y_k = 0$

In Exercises 13–16, find a basis for the solution space of the difference equation. Prove that the solutions you find span the solution set.

13. $y_{k+2} - y_{k+1} + \frac{2}{9}y_k = 0$ **14.** $y_{k+2} - 7y_{k+1} + 12y_k = 0$

15. $y_{k+2} - 25y_k = 0$ **16.** $16y_{k+2} + 8y_{k+1} - 3y_k = 0$

17. The Fibonacci Sequence is listed in Table 1 of Section 4.7. It can be viewed as the sequence of numbers where each number is the sum of the two numbers before it. It can be described as the homogeneous difference equation

$$y_{k+2} - y_{k+1} - y_k = 0$$

with the initial conditions $y_0 = 0$ and $y_1 = 1$. Find the general solution of the Fibonacci sequence.

18. If the initial conditions are changed to $y_0 = 1$ and $y_1 = 2$ for the Fibonacci sequence in Exercise 17, list the terms of the sequence for $k = 2, 3, 4$ and 5. Find the solution to the difference equation from 17 with these new initial conditions.

Exercises 19 and 20 concern a simple model of the national economy described by the difference equation

$$Y_{k+2} - a(1+b)Y_{k+1} + abY_k = 1 \tag{14}$$

Here Y_k is the total national income during year k, a is a constant less than 1, called the *marginal propensity to consume*, and b is a positive *constant of adjustment* that describes how changes in consumer spending affect the annual rate of private investment.[1]

19. Find the general solution of equation (14) when $a = .9$ and $b = \frac{4}{9}$. What happens to Y_k as k increases? [*Hint:* First find a particular solution of the form $Y_k = T$, where T is a constant, called the equilibrium level of national income.]

20. Find the general solution of equation (14) when $a = .9$ and $b = .5$.

A lightweight cantilevered beam is supported at N points spaced 10 ft apart, and a weight of 500 lb is placed at the end of the beam, 10 ft from the first support, as in the figure. Let y_k be the bending moment at the kth support. Then $y_1 = 5000$ ft-lb. Suppose the beam is rigidly attached at the Nth support and the bending moment there is zero. In between, the moments satisfy the *three-moment equation*

$$y_{k+2} + 4y_{k+1} + y_k = 0 \quad \text{for } k = 1, 2, \ldots, N-2 \tag{15}$$

[1] For example, see *Discrete Dynamical Systems*, by James T. Sandefur (Oxford: Clarendon Press, 1990), pp. 267–276. The original *accelerator-multiplier model* is attributed to the economist P. A. Samuelson.

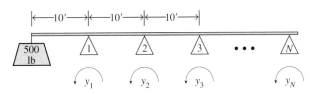

Bending moments on a cantilevered beam.

21. Find the general solution of difference equation (15). Justify your answer.

22. Find the particular solution of (15) that satisfies the *boundary conditions* $y_1 = 5000$ and $y_N = 0$. (The answer involves N.)

23. When a signal is produced from a sequence of measurements made on a process (a chemical reaction, a flow of heat through a tube, a moving robot arm, etc.), the signal usually contains random *noise* produced by measurement errors. A standard method of preprocessing the data to reduce the noise is to smooth or filter the data. One simple filter is a *moving average* that replaces each y_k by its average with the two adjacent values:

$$\tfrac{1}{3}y_{k+1} + \tfrac{1}{3}y_k + \tfrac{1}{3}y_{k-1} = z_k \quad \text{for } k = 1, 2, \ldots$$

Suppose a signal y_k, for $k = 0, \ldots, 14$, is

9, 5, 7, 3, 2, 4, 6, 5, 7, 6, 8, 10, 9, 5, 7

Use the filter to compute $z_1, \ldots, z_{13}$. Make a broken-line graph that superimposes the original signal and the smoothed signal.

24. Let $\{y_k\}$ be the sequence produced by sampling the continuous signal $2\cos\frac{\pi t}{4} + \cos\frac{3\pi t}{4}$ at $t = 0, 1, 2, \ldots$, as shown in the figure. The values of y_k, beginning with $k = 0$, are

3, .7, 0, −.7, −3, −.7, 0, .7, 3, .7, 0, ...

where .7 is an abbreviation for $\sqrt{2}/2$.

a. Compute the output signal $\{z_k\}$ when $\{y_k\}$ is fed into the filter in Example 2.

b. Explain how and why the output in part (a) is related to the calculations in Example 2.

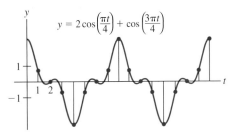

Sampled data from $2\cos\frac{\pi t}{4} + \cos\frac{3\pi t}{4}$.

Exercises 25 and 26 refer to a difference equation of the form $y_{k+1} - ay_k = b$, for suitable constants a and b.

25. A loan of $10,000 has an interest rate of 1% per month and a monthly payment of $450. The loan is made at month $k = 0$, and the first payment is made one month later, at $k = 1$. For

$k = 0, 1, 2, \ldots,$ let y_k be the unpaid balance of the loan just after the kth monthly payment. Thus

$$
\begin{array}{llll}
y_1 & = & 10{,}000 & + \; (.01)10{,}000 - 450 \\
& \text{New} & \text{Balance} & \text{Interest} \quad \text{Payment} \\
& \text{balance} & \text{due} & \text{added}
\end{array}
$$

a. Write a difference equation satisfied by $\{y_k\}$.

T b. Create a table showing k and the balance y_k at month k. List the program or the keystrokes you used to create the table.

T c. What will k be when the last payment is made? How much will the last payment be? How much money did the borrower pay in total?

26. At time $k = 0$, an initial investment of \$1000 is made into a savings account that pays 6% interest per year compounded monthly. (The interest rate per month is .005.) Each month after the initial investment, an additional \$200 is added to the account. For $k = 0, 1, 2, \ldots,$ let y_k be the amount in the account at time k, just after a deposit has been made.

a. Write a difference equation satisfied by $\{y_k\}$.

T b. Create a table showing k and the total amount in the savings account at month k, for $k = 0$ through 60. List your program or the keystrokes you used to create the table.

T c. How much will be in the account after two years (that is, 24 months), four years, and five years? How much of the five-year total is interest?

In Exercises 27–30, show that the given signal is a solution of the difference equation. Then find the general solution of that difference equation.

27. $y_k = k^2$; $y_{k+2} + 3y_{k+1} - 4y_k = 7 + 10k$

28. $y_k = 1 + k$; $y_{k+2} - 8y_{k+1} + 15y_k = 2 + 8k$

29. $y_k = 2 - 2k$; $y_{k+2} - \frac{9}{2}y_{k+1} + 2y_k = 2 + 3k$

30. $y_k = 2k - 4$; $y_{k+2} + \frac{3}{2}y_{k+1} - y_k = 1 + 3k$

Write the difference equations in Exercises 31 and 32 as first-order systems, $\mathbf{x}_{k+1} = A\mathbf{x}_k$, for all k.

31. $y_{k+4} - 6y_{k+3} + 8y_{k+2} + 6y_{k+1} - 9y_k = 0$

32. $y_{k+3} - \frac{3}{4}y_{k+2} + \frac{1}{16}y_k = 0$

33. Is the following difference equation of order 3? Explain.

$$y_{k+3} + 5y_{k+2} + 6y_{k+1} = 0$$

34. What is the order of the following difference equation? Explain your answer.

$$y_{k+3} + a_1 y_{k+2} + a_2 y_{k+1} + a_3 y_k = 0$$

35. Let $y_k = k^2$ and $z_k = 2k|k|$. Are the signals $\{y_k\}$ and $\{z_k\}$ linearly independent? Evaluate the associated Casorati matrix $C(k)$ for $k = 0$, $k = -1$, and $k = -2$, and discuss your results.

36. Let f, g, and h be linearly independent functions defined for all real numbers, and construct three signals by sampling the values of the functions at the integers:

$$u_k = f(k), \qquad v_k = g(k), \qquad w_k = h(k)$$

Must the signals be linearly independent in $\mathbb{S}$? Discuss.

Solution to Practice Problem

Examine the Casorati matrix:

$$
C(k) = \begin{bmatrix}
2^k & 3^k \sin \frac{k\pi}{2} & 3^k \cos \frac{k\pi}{2} \\
2^{k+1} & 3^{k+1} \sin \frac{(k+1)\pi}{2} & 3^{k+1} \cos \frac{(k+1)\pi}{2} \\
2^{k+2} & 3^{k+2} \sin \frac{(k+2)\pi}{2} & 3^{k+2} \cos \frac{(k+2)\pi}{2}
\end{bmatrix}
$$

Set $k = 0$ and row reduce the matrix to verify that it has three pivot positions and hence is invertible:

$$
C(0) = \begin{bmatrix}
1 & 0 & 1 \\
2 & 3 & 0 \\
4 & 0 & -9
\end{bmatrix}
\sim
\begin{bmatrix}
1 & 0 & 1 \\
0 & 3 & -2 \\
0 & 0 & -13
\end{bmatrix}
$$

The Casorati matrix is invertible at $k = 0$, so the signals are linearly independent. Since there are three signals, and the solution space H of the difference equation has dimension 3 (Theorem 20), the signals form a basis for H, by the Basis Theorem.

CHAPTER 4 PROJECTS

Chapter 4 projects are available online at `bit.ly/30IM8gT`.

A. *Exploring Subspaces*: This project explores subspaces with a more hands-on approach.

B. *Hill Substitution Ciphers*: This project shows how to use matrices to encode and decode messages.

C. *Error Detecting and Error Correcting*: In this project, a method detecting and correcting errors made in the

transmission of encoded messages is constructed. It will turn out that abstract vector spaces and the concepts of null space, rank, and dimension are needed for this construction.

D. *Signal Processing*: This project examines signal processing in more detail.

E. *Fibonacci Sequences*: The purpose of this project is to investigate further the Fibonacci sequence, which arises in number theory, applied mathematics, and biology.

CHAPTER 4 SUPPLEMENTARY EXERCISES

In Exercises 1–19, mark each statement True or False (**T/F**). Justify each answer. (If true, cite appropriate facts or theorems. If false, explain why or give a counterexample that shows why the statement is not true in every case.) In Exercises 1–6, $\mathbf{v}_1, \ldots, \mathbf{v}_p$ are vectors in a nonzero finite-dimensional vector space V, and $S = \{\mathbf{v}_1, \ldots, \mathbf{v}_p\}$.

1. **(T/F)** The set of all linear combinations of $\mathbf{v}_1, \ldots, \mathbf{v}_p$ is a vector space.

2. **(T/F)** If $\{\mathbf{v}_1, \ldots, \mathbf{v}_{p-1}\}$ spans V, then S spans V.

3. **(T/F)** If $\{\mathbf{v}_1, \ldots, \mathbf{v}_{p-1}\}$ is linearly independent, then so is S.

4. **(T/F)** If S is linearly independent, then S is a basis for V.

5. **(T/F)** If Span $S = V$, then some subset of S is a basis for V.

6. **(T/F)** If $\dim V = p$ and Span $S = V$, then S cannot be linearly dependent.

7. **(T/F)** A plane in $\mathbb{R}^3$ is a two-dimensional subspace.

8. **(T/F)** The nonpivot columns of a matrix are always linearly dependent.

9. **(T/F)** Row operations on a matrix A can change the linear dependence relations among the rows of A.

10. **(T/F)** Row operations on a matrix can change the null space.

11. **(T/F)** The rank of a matrix equals the number of nonzero rows.

12. **(T/F)** If an $m \times n$ matrix A is row equivalent to an echelon matrix U and if U has k nonzero rows, then the dimension of the solution space of $A\mathbf{x} = \mathbf{0}$ is $m - k$.

13. **(T/F)** If B is obtained from a matrix A by several elementary row operations, then rank B = rank A.

14. **(T/F)** The nonzero rows of a matrix A form a basis for Row A.

15. **(T/F)** If matrices A and B have the same reduced echelon form, then Row A = Row B.

16. **(T/F)** If H is a subspace of $\mathbb{R}^3$, then there is a 3×3 matrix A such that $H = \text{Col } A$.

17. **(T/F)** If A is $m \times n$ and rank $A = m$, then the linear transformation $\mathbf{x} \mapsto A\mathbf{x}$ is one-to-one.

18. **(T/F)** If A is $m \times n$ and the linear transformation $\mathbf{x} \mapsto A\mathbf{x}$ is onto, then rank $A = m$.

19. **(T/F)** A change-of-coordinates matrix is always invertible.

20. Find a basis for the set of all vectors of the form
$$\begin{bmatrix} a - 2b + 5c \\ 2a + 5b - 8c \\ -a - 4b + 7c \\ 3a + b + c \end{bmatrix}. \quad \text{(Be careful.)}$$

21. Let $\quad \mathbf{u}_1 = \begin{bmatrix} -2 \\ 4 \\ -6 \end{bmatrix}, \quad \mathbf{u}_2 = \begin{bmatrix} 1 \\ 2 \\ -5 \end{bmatrix}, \quad \mathbf{b} = \begin{bmatrix} b_1 \\ b_2 \\ b_3 \end{bmatrix}, \quad$ and $W = \text{Span}\{\mathbf{u}_1, \mathbf{u}_2\}$. Find an *implicit* description of W; that is, find a set of one or more homogeneous equations that characterize the points of W. [*Hint:* When is $\mathbf{b}$ in W?]

22. Explain what is wrong with the following discussion: Let $\mathbf{f}(t) = 3 + t$ and $\mathbf{g}(t) = 3t + t^2$, and note that $\mathbf{g}(t) = t\mathbf{f}(t)$. Then $\{\mathbf{f}, \mathbf{g}\}$ is linearly dependent because $\mathbf{g}$ is a multiple of $\mathbf{f}$.

23. Consider the polynomials $\mathbf{p}_1(t) = 1 + t$, $\mathbf{p}_2(t) = 1 - t$, $\mathbf{p}_3(t) = 4$, $\mathbf{p}_4(t) = t + t^2$, and $\mathbf{p}_5(t) = 1 + 2t + t^2$, and let H be the subspace of $\mathbb{P}_5$ spanned by the set $S = \{\mathbf{p}_1, \mathbf{p}_2, \mathbf{p}_3, \mathbf{p}_4, \mathbf{p}_5\}$. Use the method described in the proof of the Spanning Set Theorem (Section 4.3) to produce a basis for H. (Explain how to select appropriate members of S.)

24. Suppose $\mathbf{p}_1, \mathbf{p}_2, \mathbf{p}_3$, and $\mathbf{p}_4$ are specific polynomials that span a two-dimensional subspace H of $\mathbb{P}_5$. Describe how one can find a basis for H by examining the four polynomials and making almost no computations.

25. What would you have to know about the solution set of a homogeneous system of 18 linear equations in 20 variables in order to know that every associated nonhomogeneous equation has a solution? Discuss.

26. Let H be an n-dimensional subspace of an n-dimensional vector space V. Explain why $H = V$.

27. Let $T : \mathbb{R}^n \rightarrow \mathbb{R}^m$ be a linear transformation.

   a. What is the dimension of the range of T if T is a one-to-one mapping? Explain.

   b. What is the dimension of the kernel of T (see Section 4.2) if T maps $\mathbb{R}^n$ onto $\mathbb{R}^m$? Explain.

28. Let S be a maximal linearly independent subset of a vector space V. That is, S has the property that if a vector not in S is adjoined to S, then the new set will no longer be linearly independent. Prove that S must be a basis for V. [*Hint:* What if S were linearly independent but not a basis of V?]

29. Let S be a finite minimal spanning set of a vector space V. That is, S has the property that if a vector is removed from S, then the new set will no longer span V. Prove that S must be a basis for V.

Exercises 30–35 develop properties of rank that are sometimes needed in applications. Assume the matrix A is $m \times n$.

30. Show from parts (a) and (b) that rank AB cannot exceed the rank of A or the rank of B. (In general, the rank of a product of matrices cannot exceed the rank of any factor in the product.)

   a. Show that if B is $n \times p$, then rank $AB \leq$ rank A. [*Hint:* Explain why every vector in the column space of AB is in the column space of A.]

   b. Show that if B is $n \times p$, then rank $AB \leq$ rank B. [*Hint:* Use part (a) to study rank$(AB)^T$.]

31. Show that if P is an invertible $m \times m$ matrix, then rank $PA =$ rank A. [*Hint:* Apply Exercise 30 to PA and $P^{-1}(PA)$.]

32. Show that if Q is invertible, then rank $AQ =$ rank A. [*Hint:* Use Exercise 31 to study rank$(AQ)^T$.]

33. Let A be an $m \times n$ matrix, and let B be an $n \times p$ matrix such that $AB = 0$. Show that rank $A +$ rank $B \leq n$. [*Hint:* One of the four subspaces Nul A, Col A, Nul B, and Col B is contained in one of the other three subspaces.]

34. If A is an $m \times n$ matrix of rank r, then a *rank factorization* of A is an equation of the form $A = CR$, where C is an $m \times r$ matrix of rank r and R is an $r \times n$ matrix of rank r. Show that such a factorization always exists. Then show that given any two $m \times n$ matrices A and B.

rank$(A + B) \leq$ rank $A +$ rank B

[*Hint:* Write $A + B$ as the product of two partitioned matrices.]

35. A **submatrix** of a matrix A is any matrix that results from deleting some (or no) rows and/or columns of A. It can be shown that A has rank r if and only if A contains an invertible $r \times r$ submatrix and no larger square submatrix is invertible. Demonstrate part of this statement by explaining (a) why an $m \times n$ matrix A of rank r has an $m \times r$ submatrix A_1 of rank r, and (b) why A_1 has an invertible $r \times r$ submatrix A_2.

The concept of rank plays an important role in the design of engineering control systems. A *state-space model* of a control system includes a difference equation of the form

$$\mathbf{x}_{k+1} = A\mathbf{x}_k + B\mathbf{u}_k \quad \text{for } k = 0, 1, \ldots \qquad (1)$$

where A is $n \times n$, B is $n \times m$, $\{\mathbf{x}_k\}$ is a sequence of "state vectors" in $\mathbb{R}^n$ that describe the state of the system at discrete times, and $\{\mathbf{u}_k\}$ is a *control*, or *input*, sequence. The pair (A, B) is said to be **controllable** if

$$\text{rank}\begin{bmatrix} B & AB & A^2B & \cdots & A^{n-1}B \end{bmatrix} = n \qquad (2)$$

The matrix that appears in (2) is called the **controllability matrix** for the system. If (A, B) is controllable, then the system can be controlled, or driven from the state $\mathbf{0}$ to any specified state $\mathbf{v}$ (in $\mathbb{R}^n$) in at most n steps, simply by choosing an appropriate control sequence in $\mathbb{R}^m$. This fact is illustrated in Exercise 36 for $n = 4$ and $m = 2$.

36. Suppose A is a 4×4 matrix and B is a 4×2 matrix, and let $\mathbf{u}_0, \ldots, \mathbf{u}_3$ represent a sequence of input vectors in $\mathbb{R}^2$.

   a. Set $\mathbf{x}_0 = \mathbf{0}$, compute $\mathbf{x}_1, \ldots, \mathbf{x}_4$ from equation (1), and write a formula for $\mathbf{x}_4$ involving the controllability matrix M appearing in equation (2). (*Note:* The matrix M is constructed as a partitioned matrix. Its overall size here is 4×8.)

   b. Suppose (A, B) is controllable and $\mathbf{v}$ is any vector in $\mathbb{R}^4$. Explain why there exists a control sequence $\mathbf{u}_0, \ldots, \mathbf{u}_3$ in $\mathbb{R}^2$ such that $\mathbf{x}_4 = \mathbf{v}$.

Determine if the matrix pairs in Exercises 37–40 are controllable.

37. $A = \begin{bmatrix} .9 & 1 & 0 \\ 0 & -.9 & 0 \\ 0 & 0 & .5 \end{bmatrix}, B = \begin{bmatrix} 0 \\ 1 \\ 1 \end{bmatrix}$

38. $A = \begin{bmatrix} .8 & -.3 & 0 \\ .2 & .5 & 1 \\ 0 & 0 & -.5 \end{bmatrix}, B = \begin{bmatrix} 1 \\ 1 \\ 0 \end{bmatrix}$

T 39. $A = \begin{bmatrix} 0 & 1 & 0 & 0 \\ 0 & 0 & 1 & 0 \\ 0 & 0 & 0 & 1 \\ -2 & -4.2 & -4.8 & -3.6 \end{bmatrix}, B = \begin{bmatrix} 1 \\ 0 \\ 0 \\ -1 \end{bmatrix}$

T 40. $A = \begin{bmatrix} 0 & 1 & 0 & 0 \\ 0 & 0 & 1 & 0 \\ 0 & 0 & 0 & 1 \\ -1 & -13 & -12.2 & -1.5 \end{bmatrix}, B = \begin{bmatrix} 1 \\ 0 \\ 0 \\ -1 \end{bmatrix}$

5 Eigenvalues and Eigenvectors

DYNAMICAL SYSTEMS AND SPOTTED OWLS

In 1990, the northern spotted owl became the center of a nationwide controversy over the use and misuse of the majestic forests in the Pacific Northwest. Environmentalists convinced the federal government that the owl was threatened with extinction if logging continued in the old-growth forests (with trees more than 200 years old), where the owls prefer to live. The timber industry, anticipating the loss of 30,000–100,000 jobs as a result of new government restrictions on logging, argued that the owl should not be classified as a "threatened species" and cited a number of published scientific reports to support its case.[1]

The population of spotted owls continues to decline, and it remains a species caught in the crossfire between economic opportunities and conservation efforts. Mathematical ecologists help to analyze the effects on the spotted owl population of factors such as logging techniques, wildfires, and competition for habitat with the invasive barred owl. The life cycle of a spotted owl divides naturally into three stages: juvenile (up to 1 year old), subadult (1–2 years), and adult (older than 2 years). The owls mate for life during the subadult and adult stages, begin to breed as adults, and live for up to 20 years. Each owl pair requires about 1000 hectares (4 square miles) for its own home territory. A critical time in the life cycle is when the juveniles leave the nest. To survive and become a subadult, a juvenile must successfully find a new home range (and usually a mate).

A first step in studying the population dynamics is to model the population at yearly intervals, at times denoted by $k = 0, 1, 2, \ldots$. Usually, one assumes that there is a 1:1 ratio of males to females in each life stage and counts only the females. The population at year k can be described by a vector $\mathbf{x}_k = (j_k, s_k, a_k)$, where j_k, s_k, and a_k are the numbers of females in the juvenile, subadult, and adult stages, respectively.

Using actual field data from demographic studies, R. Lamberson and coworkers considered the following *stage-matrix model*:[2]

$$\begin{bmatrix} j_{k+1} \\ s_{k+1} \\ a_{k+1} \end{bmatrix} = \begin{bmatrix} 0 & 0 & .33 \\ .18 & 0 & 0 \\ 0 & .71 & .94 \end{bmatrix} \begin{bmatrix} j_k \\ s_k \\ a_k \end{bmatrix}$$

Here the number of new juvenile females in year $k + 1$ is .33 times the number of adult females in year k (based on the average birth rate per owl pair). Also, 18% of the juveniles survive to become subadults, and 71% of the subadults and 94% of the adults survive to be counted as adults.

The stage-matrix model is a difference equation of the form $\mathbf{x}_{k+1} = A\mathbf{x}_k$. Such an equation is often called a **dynamical system** (or a **discrete linear dynamical**

---

[1] "The Great Spotted Owl War," *Reader's Digest*, November 1992, pp. 91–95.

[2] R. H. Lamberson, R. McKelvey, B. R. Noon, and C. Voss, "A Dynamic Analysis of the Viability of the Northern Spotted Owl in a Fragmented Forest Environment," *Conservation Biology* **6** (1992), 505–512. Also, a private communication from Professor Lamberson, 1993.

system) because it describes the changes in a system as time passes.

The 18% juvenile survival rate in the Lamberson stage matrix is the entry affected most by the amount of old-growth forest available. Actually, 60% of the juveniles normally survive to leave the nest, but in the Willow Creek region of California studied by Lamberson and his colleagues, only 30% of the juveniles that left the nest were able to find new home ranges. The rest perished during the search process.

A significant reason for the failure of owls to find new home ranges is the increasing fragmentation of old-growth timber stands due to clear-cutting of scattered areas on the old-growth land. When an owl leaves the protective canopy of the forest and crosses a clear-cut area, the risk of attack by predators increases dramatically. Section 5.6 will show that the model described in the chapter introduction predicts the eventual demise of the spotted owl, but that if 50% of the juveniles who survive to leave the nest also find new home ranges, then the owl population will thrive.

The goal of this chapter is to dissect the action of a linear transformation $\mathbf{x} \mapsto A\mathbf{x}$ into elements that are easily visualized. All matrices in the chapter are square. The main applications described here are to discrete dynamical systems, differential equations, and Markov chains. However, the basic concepts—eigenvectors and eigenvalues—are useful throughout pure and applied mathematics, and they appear in settings far more general than we consider here. Eigenvalues are also used to study differential equations and *continuous* dynamical systems, they provide critical information in engineering design, and they arise naturally in fields such as physics and chemistry.

5.1 Eigenvectors and Eigenvalues

Although a transformation $\mathbf{x} \mapsto A\mathbf{x}$ may move vectors in a variety of directions, it often happens that there are special vectors on which the action of A is quite simple.

EXAMPLE 1 Let $A = \begin{bmatrix} 3 & -2 \\ 1 & 0 \end{bmatrix}$, $\mathbf{u} = \begin{bmatrix} -1 \\ 1 \end{bmatrix}$, and $\mathbf{v} = \begin{bmatrix} 2 \\ 1 \end{bmatrix}$. The images of $\mathbf{u}$ and $\mathbf{v}$ under multiplication by A are shown in Figure 1. In fact, $A\mathbf{v}$ is just $2\mathbf{v}$. So A only "stretches" or dilates $\mathbf{v}$. ∎

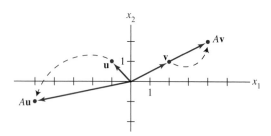

FIGURE 1 Effects of multiplication by A.

This section studies equations such as

$$A\mathbf{x} = 2\mathbf{x} \quad \text{or} \quad A\mathbf{x} = -4\mathbf{x}$$

where special vectors are transformed by A into scalar multiples of themselves.

DEFINITION

An **eigenvector** of an $n \times n$ matrix A is a nonzero vector $\mathbf{x}$ such that $A\mathbf{x} = \lambda\mathbf{x}$ for some scalar λ. A scalar λ is called an **eigenvalue** of A if there is a nontrivial solution $\mathbf{x}$ of $A\mathbf{x} = \lambda\mathbf{x}$; such an $\mathbf{x}$ is called an *eigenvector corresponding to* λ.[1]

It is easy to determine if a given vector is an eigenvector of a matrix. See Example 2. It is also easy to decide if a specified scalar is an eigenvalue. See Example 3.

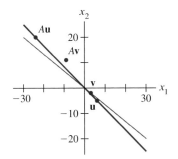

$A\mathbf{u} = -4\mathbf{u}$, but $A\mathbf{v} \neq \lambda\mathbf{v}$.

EXAMPLE 2 Let $A = \begin{bmatrix} 1 & 6 \\ 5 & 2 \end{bmatrix}$, $\mathbf{u} = \begin{bmatrix} 6 \\ -5 \end{bmatrix}$, and $\mathbf{v} = \begin{bmatrix} 3 \\ -2 \end{bmatrix}$. Are $\mathbf{u}$ and $\mathbf{v}$ eigenvectors of A?

SOLUTION

$$A\mathbf{u} = \begin{bmatrix} 1 & 6 \\ 5 & 2 \end{bmatrix}\begin{bmatrix} 6 \\ -5 \end{bmatrix} = \begin{bmatrix} -24 \\ 20 \end{bmatrix} = -4\begin{bmatrix} 6 \\ -5 \end{bmatrix} = -4\mathbf{u}$$

$$A\mathbf{v} = \begin{bmatrix} 1 & 6 \\ 5 & 2 \end{bmatrix}\begin{bmatrix} 3 \\ -2 \end{bmatrix} = \begin{bmatrix} -9 \\ 11 \end{bmatrix} \neq \lambda\begin{bmatrix} 3 \\ -2 \end{bmatrix}$$

Thus $\mathbf{u}$ is an eigenvector corresponding to an eigenvalue (-4), but $\mathbf{v}$ is not an eigenvector of A, because $A\mathbf{v}$ is not a multiple of $\mathbf{v}$. ∎

EXAMPLE 3 Show that 7 is an eigenvalue of matrix A in Example 2, and find the corresponding eigenvectors.

SOLUTION The scalar 7 is an eigenvalue of A if and only if the equation

$$A\mathbf{x} = 7\mathbf{x} \tag{1}$$

has a nontrivial solution. But (1) is equivalent to $A\mathbf{x} - 7\mathbf{x} = \mathbf{0}$, or

$$(A - 7I)\mathbf{x} = \mathbf{0} \tag{2}$$

To solve this homogeneous equation, form the matrix

$$A - 7I = \begin{bmatrix} 1 & 6 \\ 5 & 2 \end{bmatrix} - \begin{bmatrix} 7 & 0 \\ 0 & 7 \end{bmatrix} = \begin{bmatrix} -6 & 6 \\ 5 & -5 \end{bmatrix}$$

The columns of $A - 7I$ are obviously linearly dependent, so (2) has nontrivial solutions. Thus 7 *is* an eigenvalue of A. To find the corresponding eigenvectors, use row operations:

$$\begin{bmatrix} -6 & 6 & 0 \\ 5 & -5 & 0 \end{bmatrix} \sim \begin{bmatrix} 1 & -1 & 0 \\ 0 & 0 & 0 \end{bmatrix}$$

The general solution has the form $x_2\begin{bmatrix} 1 \\ 1 \end{bmatrix}$. Each vector of this form with $x_2 \neq 0$ is an eigenvector corresponding to $\lambda = 7$. ∎

---

[1] Note that an eigenvector must be *nonzero*, by definition, but an eigenvalue may be zero. The case in which the number 0 is an eigenvalue is discussed after Example 5.

Warning: Although row reduction was used in Example 3 to find eigen*vectors*, it cannot be used to find eigen*values*. An echelon form of a matrix A usually does *not* display the eigenvalues of A.

The equivalence of equations (1) and (2) obviously holds for any λ in place of $\lambda = 7$. Thus λ is an eigenvalue of an $n \times n$ matrix A if and only if the equation

$$(A - \lambda I)\mathbf{x} = \mathbf{0} \tag{3}$$

has a nontrivial solution. The set of *all* solutions of (3) is just the null space of the matrix $A - \lambda I$. So this set is a *subspace* of $\mathbb{R}^n$ and is called the **eigenspace** of A corresponding to λ. The eigenspace consists of the zero vector and all the eigenvectors corresponding to λ.

Example 3 shows that for matrix A in Example 2, the eigenspace corresponding to $\lambda = 7$ consists of *all* multiples of $(1, 1)$, which is the line through $(1, 1)$ and the origin. From Example 2, you can check that the eigenspace corresponding to $\lambda = -4$ is the line through $(6, -5)$. These eigenspaces are shown in Figure 2, along with eigenvectors $(1, 1)$ and $(3/2, -5/4)$ and the geometric action of the transformation $\mathbf{x} \mapsto A\mathbf{x}$ on each eigenspace.

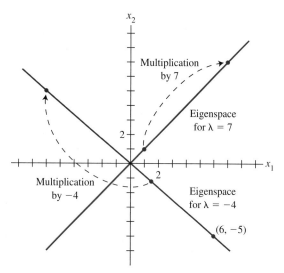

FIGURE 2   Eigenspaces for $\lambda = -4$ and $\lambda = 7$.

EXAMPLE 4   Let $A = \begin{bmatrix} 4 & -1 & 6 \\ 2 & 1 & 6 \\ 2 & -1 & 8 \end{bmatrix}$. An eigenvalue of A is 2. Find a basis for the corresponding eigenspace.

SOLUTION   Form

$$A - 2I = \begin{bmatrix} 4 & -1 & 6 \\ 2 & 1 & 6 \\ 2 & -1 & 8 \end{bmatrix} - \begin{bmatrix} 2 & 0 & 0 \\ 0 & 2 & 0 \\ 0 & 0 & 2 \end{bmatrix} = \begin{bmatrix} 2 & -1 & 6 \\ 2 & -1 & 6 \\ 2 & -1 & 6 \end{bmatrix}$$

and row reduce the augmented matrix for $(A - 2I)\mathbf{x} = \mathbf{0}$:

$$\begin{bmatrix} 2 & -1 & 6 & 0 \\ 2 & -1 & 6 & 0 \\ 2 & -1 & 6 & 0 \end{bmatrix} \sim \begin{bmatrix} 2 & -1 & 6 & 0 \\ 0 & 0 & 0 & 0 \\ 0 & 0 & 0 & 0 \end{bmatrix}$$

At this point, it is clear that 2 is indeed an eigenvalue of A because the equation $(A - 2I)\mathbf{x} = \mathbf{0}$ has free variables. The general solution is

$$\begin{bmatrix} x_1 \\ x_2 \\ x_3 \end{bmatrix} = x_2 \begin{bmatrix} 1/2 \\ 1 \\ 0 \end{bmatrix} + x_3 \begin{bmatrix} -3 \\ 0 \\ 1 \end{bmatrix}, \quad x_2 \text{ and } x_3 \text{ free}$$

The eigenspace, shown in Figure 3, is a two-dimensional subspace of $\mathbb{R}^3$. A basis is

$$\left\{ \begin{bmatrix} 1 \\ 2 \\ 0 \end{bmatrix}, \begin{bmatrix} -3 \\ 0 \\ 1 \end{bmatrix} \right\}$$
∎

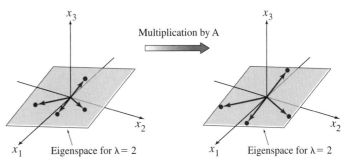

FIGURE 3 A acts as a dilation on the eigenspace.

Reasonable Answers

Remember that once you find a potential eigenvector $\mathbf{v}$, you can easily check your answer: just find $A\mathbf{v}$ and see if it is a multiple of $\mathbf{v}$. For example, to check whether $\mathbf{v} = \begin{bmatrix} 1 \\ 1 \end{bmatrix}$ is an eigenvector of $A = \begin{bmatrix} 1 & 2 \\ -1 & -2 \end{bmatrix}$, notice $A\mathbf{v} = \begin{bmatrix} 3 \\ -3 \end{bmatrix}$, which is not a multiple of $\mathbf{v} = \begin{bmatrix} 1 \\ 1 \end{bmatrix}$, establishing that $\mathbf{v}$ is not an eigenvector. It turns out we had a sign error. The vector $\mathbf{u} = \begin{bmatrix} 1 \\ -1 \end{bmatrix}$ is a correct eigenvector for A since $A\mathbf{u} = \begin{bmatrix} -1 \\ 1 \end{bmatrix} = -1 \begin{bmatrix} 1 \\ -1 \end{bmatrix} = -1\,\mathbf{u}$.

Numerical Notes

Example 4 shows a good method for manual computation of eigenvectors in simple cases when an eigenvalue is known. Using a matrix program and row reduction to find an eigenspace (for a specified eigenvalue) usually works, too, but this is not entirely reliable. Roundoff error can lead occasionally to a reduced echelon form with the wrong number of pivots. The best computer programs compute approximations for eigenvalues and eigenvectors simultaneously, to any desired degree of accuracy, for matrices that are not too large. The size of matrices that can be analyzed increases each year as computing power and software improve.

The following theorem describes one of the few special cases in which eigenvalues can be found precisely. Calculation of eigenvalues will also be discussed in Section 5.2.

THEOREM 1 The eigenvalues of a triangular matrix are the entries on its main diagonal.

PROOF For simplicity, consider the 3×3 case. If A is upper triangular, then $A - \lambda I$ has the form

$$A - \lambda I = \begin{bmatrix} a_{11} & a_{12} & a_{13} \\ 0 & a_{22} & a_{23} \\ 0 & 0 & a_{33} \end{bmatrix} - \begin{bmatrix} \lambda & 0 & 0 \\ 0 & \lambda & 0 \\ 0 & 0 & \lambda \end{bmatrix}$$

$$= \begin{bmatrix} a_{11} - \lambda & a_{12} & a_{13} \\ 0 & a_{22} - \lambda & a_{23} \\ 0 & 0 & a_{33} - \lambda \end{bmatrix}$$

The scalar λ is an eigenvalue of A if and only if the equation $(A - \lambda I)\mathbf{x} = \mathbf{0}$ has a nontrivial solution, that is, if and only if the equation has a free variable. Because of the zero entries in $A - \lambda I$, it is easy to see that $(A - \lambda I)\mathbf{x} = \mathbf{0}$ has a free variable if and only if at least one of the entries on the diagonal of $A - \lambda I$ is zero. This happens if and only if λ equals one of the entries a_{11}, a_{22}, a_{33} in A. For the case in which A is lower triangular, see Exercise 36. ∎

EXAMPLE 5 Let $A = \begin{bmatrix} 3 & 6 & -8 \\ 0 & 0 & 6 \\ 0 & 0 & 2 \end{bmatrix}$ and $B = \begin{bmatrix} 4 & 0 & 0 \\ -2 & 1 & 0 \\ 5 & 3 & 4 \end{bmatrix}$. The eigenvalues of A are 3, 0, and 2. The eigenvalues of B are 4 and 1. ∎

What does it mean for a matrix A to have an eigenvalue of 0, such as in Example 5? This happens if and only if the equation

$$A\mathbf{x} = 0\mathbf{x} \tag{4}$$

has a nontrivial solution. But (4) is equivalent to $A\mathbf{x} = \mathbf{0}$, which has a nontrivial solution if and only if A is not invertible. Thus 0 *is an eigenvalue of A if and only if A is not invertible*. This fact will be added to the Invertible Matrix Theorem in Section 5.2.

The following important theorem will be needed later. Its proof illustrates a typical calculation with eigenvectors. One way to prove the statement "If P then Q" is to show that P and the negation of Q leads to a contradiction. This strategy is used in the proof of the theorem.

THEOREM 2 If $\mathbf{v}_1, \ldots, \mathbf{v}_r$ are eigenvectors that correspond to distinct eigenvalues $\lambda_1, \ldots, \lambda_r$ of an $n \times n$ matrix A, then the set $\{\mathbf{v}_1, \ldots, \mathbf{v}_r\}$ is linearly independent.

PROOF Suppose $\{\mathbf{v}_1, \ldots, \mathbf{v}_r\}$ is linearly dependent. Since $\mathbf{v}_1$ is nonzero, Theorem 7 in Section 1.7 says that one of the vectors in the set is a linear combination of the preceding vectors. Let p be the least index such that $\mathbf{v}_{p+1}$ is a linear combination of the preceding (linearly independent) vectors. Then there exist scalars $c_1, \ldots, c_p$ such that

$$c_1\mathbf{v}_1 + \cdots + c_p\mathbf{v}_p = \mathbf{v}_{p+1} \tag{5}$$

Multiplying both sides of (5) by A and using the fact that $A\mathbf{v}_k = \lambda_k \mathbf{v}_k$ for each k, we obtain

$$c_1 A\mathbf{v}_1 + \cdots + c_p A\mathbf{v}_p = A\mathbf{v}_{p+1}$$
$$c_1 \lambda_1 \mathbf{v}_1 + \cdots + c_p \lambda_p \mathbf{v}_p = \lambda_{p+1} \mathbf{v}_{p+1} \tag{6}$$

Multiplying both sides of (5) by λ_{p+1} and subtracting the result from (6), we have

$$c_1(\lambda_1 - \lambda_{p+1})\mathbf{v}_1 + \cdots + c_p(\lambda_p - \lambda_{p+1})\mathbf{v}_p = \mathbf{0} \tag{7}$$

Since $\{\mathbf{v}_1, \ldots, \mathbf{v}_p\}$ is linearly independent, the weights in (7) are all zero. But none of the factors $\lambda_i - \lambda_{p+1}$ are zero, because the eigenvalues are distinct. Hence $c_i = 0$ for $i = 1, \ldots, p$. But then (5) says that $\mathbf{v}_{p+1} = \mathbf{0}$, which is impossible. Hence $\{\mathbf{v}_1, \ldots, \mathbf{v}_r\}$ cannot be linearly dependent and therefore must be linearly independent. ∎

Eigenvectors and Difference Equations

This section concludes by showing how to construct solutions of the first-order difference equation discussed in the chapter introductory example:

$$\mathbf{x}_{k+1} = A\mathbf{x}_k \quad (k = 0, 1, 2, \ldots) \tag{8}$$

If A is an $n \times n$ matrix, then (8) is a *recursive* description of a sequence $\{\mathbf{x}_k\}$ in $\mathbb{R}^n$. A **solution** of (8) is an explicit description of $\{\mathbf{x}_k\}$ whose formula for each $\mathbf{x}_k$ does not depend directly on A or on the preceding terms in the sequence other than the initial term $\mathbf{x}_0$.

The simplest way to build a solution of (8) is to take an eigenvector $\mathbf{x}_0$ and its corresponding eigenvalue λ and let

$$\mathbf{x}_k = \lambda^k \mathbf{x}_0 \quad (k = 1, 2, \ldots) \tag{9}$$

This sequence is a solution because

$$A\mathbf{x}_k = A(\lambda^k \mathbf{x}_0) = \lambda^k (A\mathbf{x}_0) = \lambda^k (\lambda \mathbf{x}_0) = \lambda^{k+1} \mathbf{x}_0 = \mathbf{x}_{k+1}$$

Linear combinations of solutions in the form of equation (9) are solutions, too! See Exercise 41.

Practice Problems

1. Is 5 an eigenvalue of $A = \begin{bmatrix} 6 & -3 & 1 \\ 3 & 0 & 5 \\ 2 & 2 & 6 \end{bmatrix}$?

2. If $\mathbf{x}$ is an eigenvector of A corresponding to λ, what is $A^3 \mathbf{x}$?

3. Suppose that $\mathbf{b}_1$ and $\mathbf{b}_2$ are eigenvectors corresponding to distinct eigenvalues λ_1 and λ_2, respectively, and suppose that $\mathbf{b}_3$ and $\mathbf{b}_4$ are linearly independent eigenvectors corresponding to a third distinct eigenvalue λ_3. Does it necessarily follow that $\{\mathbf{b}_1, \mathbf{b}_2, \mathbf{b}_3, \mathbf{b}_4\}$ is a linearly independent set? [*Hint:* Consider the equation $c_1\mathbf{b}_1 + c_2\mathbf{b}_2 + (c_3\mathbf{b}_3 + c_4\mathbf{b}_4) = \mathbf{0}$.]

4. If A is an $n \times n$ matrix and λ is an eigenvalue of A, show that 2λ is an eigenvalue of $2A$.

5.1 Exercises

1. Is $\lambda = 2$ an eigenvalue of $\begin{bmatrix} 3 & 2 \\ 3 & 8 \end{bmatrix}$? Why or why not?

2. Is $\lambda = -2$ an eigenvalue of $\begin{bmatrix} 7 & 3 \\ 3 & -1 \end{bmatrix}$? Why or why not?

3. Is $\begin{bmatrix} 1 \\ 4 \end{bmatrix}$ an eigenvector of $\begin{bmatrix} -3 & 1 \\ -3 & 8 \end{bmatrix}$? If so, find the eigenvalue.

4. Is $\begin{bmatrix} -1 \\ 1 \end{bmatrix}$ an eigenvector of $\begin{bmatrix} 4 & 2 \\ 2 & 4 \end{bmatrix}$? If so, find the eigenvalue.

5. Is $\begin{bmatrix} 4 \\ -3 \\ 1 \end{bmatrix}$ an eigenvector of $\begin{bmatrix} 3 & 7 & 9 \\ -4 & -5 & 1 \\ 2 & 4 & 4 \end{bmatrix}$? If so, find the eigenvalue.

6. Is $\begin{bmatrix} 1 \\ -2 \\ 1 \end{bmatrix}$ an eigenvector of $\begin{bmatrix} 2 & 6 & 7 \\ 3 & 2 & 7 \\ 5 & 6 & 4 \end{bmatrix}$? If so, find the eigenvalue.

7. Is $\lambda = 4$ an eigenvalue of $\begin{bmatrix} 3 & 0 & -1 \\ 2 & 3 & 1 \\ -3 & 4 & 5 \end{bmatrix}$? If so, find one corresponding eigenvector.

8. Is $\lambda = 3$ an eigenvalue of $\begin{bmatrix} 1 & 2 & 2 \\ 3 & -2 & 1 \\ 0 & 1 & 1 \end{bmatrix}$? If so, find one corresponding eigenvector.

In Exercises 9–16, find a basis for the eigenspace corresponding to each listed eigenvalue.

9. $A = \begin{bmatrix} 5 & 0 \\ 2 & 1 \end{bmatrix}, \lambda = 1, 5$

10. $A = \begin{bmatrix} 10 & -9 \\ 4 & -2 \end{bmatrix}, \lambda = 4$

11. $A = \begin{bmatrix} 4 & -2 \\ -3 & 9 \end{bmatrix}, \lambda = 10$

12. $A = \begin{bmatrix} 1 & 4 \\ 3 & 2 \end{bmatrix}, \lambda = -2, 5$

13. $A = \begin{bmatrix} 4 & 0 & 1 \\ -2 & 1 & 0 \\ -2 & 0 & 1 \end{bmatrix}, \lambda = 1, 2, 3$

14. $A = \begin{bmatrix} 3 & -1 & 3 \\ -1 & 3 & 3 \\ 6 & 6 & 2 \end{bmatrix}, \lambda = -4$

15. $A = \begin{bmatrix} 4 & 2 & 3 \\ -1 & 1 & -3 \\ 2 & 4 & 9 \end{bmatrix}, \lambda = 3$

16. $A = \begin{bmatrix} 3 & 0 & 2 & 0 \\ 1 & 3 & 1 & 0 \\ 0 & 1 & 1 & 0 \\ 0 & 0 & 0 & 4 \end{bmatrix}, \lambda = 4$

Find the eigenvalues of the matrices in Exercises 17 and 18.

17. $\begin{bmatrix} 0 & 0 & 0 \\ 0 & 2 & 5 \\ 0 & 0 & -1 \end{bmatrix}$ **18.** $\begin{bmatrix} 4 & 0 & 0 \\ 0 & 0 & 0 \\ 1 & 0 & -3 \end{bmatrix}$

19. For $A = \begin{bmatrix} 1 & 2 & 3 \\ 1 & 2 & 3 \\ 1 & 2 & 3 \end{bmatrix}$, find one eigenvalue, with no calculation. Justify your answer.

20. Without calculation, find one eigenvalue and two linearly independent eigenvectors of $A = \begin{bmatrix} 5 & -5 & 5 \\ 5 & -5 & 5 \\ 5 & -5 & 5 \end{bmatrix}$. Justify your answer.

In Exercises 21–30, A is an $n \times n$ matrix. Mark each statement True or False (**T/F**). Justify each answer.

21. (**T/F**) If $A\mathbf{x} = \lambda\mathbf{x}$ for some vector $\mathbf{x}$, then λ is an eigenvalue of A.

22. (**T/F**) If $A\mathbf{x} = \lambda\mathbf{x}$ for some scalar λ, then $\mathbf{x}$ is an eigenvector of A.

23. (**T/F**) A matrix A is invertible if and only if 0 is an eigenvalue of A.

24. (**T/F**) A number c is an eigenvalue of A if and only if the equation $(A - cI)\mathbf{x} = 0$ has a nontrivial solution.

25. (**T/F**) Finding an eigenvector of A may be difficult, but checking whether a given vector is in fact an eigenvector is easy.

26. (**T/F**) To find the eigenvalues of A, reduce A to echelon form.

27. (**T/F**) If $\mathbf{v}_1$ and $\mathbf{v}_2$ are linearly independent eigenvectors, then they correspond to distinct eigenvalues.

28. (**T/F**) The eigenvalues of a matrix are on its main diagonal.

29. (**T/F**) If $\mathbf{v}$ is an eigenvector with eigenvalue 2, then $2\mathbf{v}$ is an eigenvector with eigenvalue 4.

30. (**T/F**) An eigenspace of A is a null space of a certain matrix.

31. Explain why a 2×2 matrix can have at most two distinct eigenvalues. Explain why an $n \times n$ matrix can have at most n distinct eigenvalues.

32. Construct an example of a 2×2 matrix with only one distinct eigenvalue.

33. Let λ be an eigenvalue of an invertible matrix A. Show that λ^{-1} is an eigenvalue of A^{-1}. [*Hint:* Suppose a nonzero $\mathbf{x}$ satisfies $A\mathbf{x} = \lambda\mathbf{x}$.]

34. Show that if A^2 is the zero matrix, then the only eigenvalue of A is 0.

35. Show that λ is an eigenvalue of A if and only if λ is an eigenvalue of A^T. [*Hint:* Find out how $A - \lambda I$ and $A^T - \lambda I$ are related.]

36. Use Exercise 35 to complete the proof of Theorem 1 for the case when A is lower triangular.

37. Consider an $n \times n$ matrix A with the property that the row sums all equal the same number s. Show that s is an eigenvalue of A. [*Hint:* Find an eigenvector.]

38. Consider an $n \times n$ matrix A with the property that the column sums all equal the same number s. Show that s is an eigenvalue of A. [*Hint:* Use Exercises 35 and 37.]

In Exercises 39 and 40, let A be the matrix of the linear transformation T. Without writing A, find an eigenvalue of A and describe the eigenspace.

39. T is the transformation on $\mathbb{R}^2$ that reflects points across some line through the origin.

40. T is the transformation on $\mathbb{R}^3$ that rotates points about some line through the origin.

41. Let $\mathbf{u}$ and $\mathbf{v}$ be eigenvectors of a matrix A, with corresponding eigenvalues λ and μ, and let c_1 and c_2 be scalars. Define
$$\mathbf{x}_k = c_1\lambda^k\mathbf{u} + c_2\mu^k\mathbf{v} \quad (k = 0, 1, 2, \ldots)$$
a. What is $\mathbf{x}_{k+1}$, by definition?
b. Compute $A\mathbf{x}_k$ from the formula for $\mathbf{x}_k$, and show that $A\mathbf{x}_k = \mathbf{x}_{k+1}$. This calculation will prove that the sequence $\{\mathbf{x}_k\}$ defined above satisfies the difference equation $\mathbf{x}_{k+1} = A\mathbf{x}_k \ (k = 0, 1, 2, \ldots)$.

42. Describe how you might try to build a solution of a difference equation $\mathbf{x}_{k+1} = A\mathbf{x}_k \ (k = 0, 1, 2, \ldots)$ if you were given the initial $\mathbf{x}_0$ and this vector did not happen to be an eigenvector of A. [*Hint:* How might you relate $\mathbf{x}_0$ to eigenvectors of A?]

43. Let $\mathbf{u}$ and $\mathbf{v}$ be the vectors shown in the figure, and suppose $\mathbf{u}$ and $\mathbf{v}$ are eigenvectors of a 2×2 matrix A that correspond to eigenvalues 2 and 3, respectively. Let $T : \mathbb{R}^2 \to \mathbb{R}^2$ be the linear transformation given by $T(\mathbf{x}) = A\mathbf{x}$ for each $\mathbf{x}$ in $\mathbb{R}^2$, and let $\mathbf{w} = \mathbf{u} + \mathbf{v}$. Make a copy of the figure, and on the same coordinate system, carefully plot the vectors $T(\mathbf{u})$, $T(\mathbf{v})$, and $T(\mathbf{w})$.

44. Repeat Exercise 43, assuming $\mathbf{u}$ and $\mathbf{v}$ are eigenvectors of A that correspond to eigenvalues -1 and 3, respectively.

T In Exercises 45–48, use a matrix program to find the eigenvalues of the matrix. Then use the method of Example 4 with a row reduction routine to produce a basis for each eigenspace.

45. $\begin{bmatrix} 8 & -10 & -5 \\ 2 & 17 & 2 \\ -9 & -18 & 4 \end{bmatrix}$

46. $\begin{bmatrix} 9 & -4 & -2 & -4 \\ -56 & 32 & -28 & 44 \\ -14 & -14 & 6 & -14 \\ 42 & -33 & 21 & -45 \end{bmatrix}$

47. $\begin{bmatrix} 4 & -9 & -7 & 8 & 2 \\ -7 & -9 & 0 & 7 & 14 \\ 5 & 10 & 5 & -5 & -10 \\ -2 & 3 & 7 & 0 & 4 \\ -3 & -13 & -7 & 10 & 11 \end{bmatrix}$

48. $\begin{bmatrix} -4 & -4 & 20 & -8 & -1 \\ 14 & 12 & 46 & 18 & 2 \\ 6 & 4 & -18 & 8 & 1 \\ 11 & 7 & -37 & 17 & 2 \\ 18 & 12 & -60 & 24 & 5 \end{bmatrix}$

Solutions to Practice Problems

1. The number 5 is an eigenvalue of A if and only if the equation $(A - 5I)\mathbf{x} = \mathbf{0}$ has a nontrivial solution. Form
$$A - 5I = \begin{bmatrix} 6 & -3 & 1 \\ 3 & 0 & 5 \\ 2 & 2 & 6 \end{bmatrix} - \begin{bmatrix} 5 & 0 & 0 \\ 0 & 5 & 0 \\ 0 & 0 & 5 \end{bmatrix} = \begin{bmatrix} 1 & -3 & 1 \\ 3 & -5 & 5 \\ 2 & 2 & 1 \end{bmatrix}$$

and row reduce the augmented matrix:
$$\begin{bmatrix} 1 & -3 & 1 & 0 \\ 3 & -5 & 5 & 0 \\ 2 & 2 & 1 & 0 \end{bmatrix} \sim \begin{bmatrix} 1 & -3 & 1 & 0 \\ 0 & 4 & 2 & 0 \\ 0 & 8 & -1 & 0 \end{bmatrix} \sim \begin{bmatrix} 1 & -3 & 1 & 0 \\ 0 & 4 & 2 & 0 \\ 0 & 0 & -5 & 0 \end{bmatrix}$$

Solutions to Practice Problems (Continued)

At this point, it is clear that the homogeneous system has no free variables. Thus $A - 5I$ is an invertible matrix, which means that 5 is *not* an eigenvalue of A.

2. If $\mathbf{x}$ is an eigenvector of A corresponding to λ, then $A\mathbf{x} = \lambda\mathbf{x}$ and so

$$A^2\mathbf{x} = A(\lambda\mathbf{x}) = \lambda A\mathbf{x} = \lambda^2\mathbf{x}$$

Again, $A^3\mathbf{x} = A(A^2\mathbf{x}) = A(\lambda^2\mathbf{x}) = \lambda^2 A\mathbf{x} = \lambda^3\mathbf{x}$. The general pattern, $A^k\mathbf{x} = \lambda^k\mathbf{x}$, is proved by induction.

3. Yes. Suppose $c_1\mathbf{b}_1 + c_2\mathbf{b}_2 + (c_3\mathbf{b}_3 + c_4\mathbf{b}_4) = \mathbf{0}$. Since any linear combination of eigenvectors corresponding to the same eigenvalue is in the eigenspace for that eigenvalue, $c_3\mathbf{b}_3 + c_4\mathbf{b}_4$ is either $\mathbf{0}$ or an eigenvector for λ_3. If $c_3\mathbf{b}_3 + c_4\mathbf{b}_4$ were an eigenvector for λ_3, then by Theorem 2, $\{\mathbf{b}_1, \mathbf{b}_2, c_3\mathbf{b}_3 + c_4\mathbf{b}_4\}$ would be a linearly independent set, which would force $c_1 = c_2 = 0$ and $c_3\mathbf{b}_3 + c_4\mathbf{b}_4 = \mathbf{0}$, contradicting that $c_3\mathbf{b}_3 + c_4\mathbf{b}_4$ is an eigenvector. Thus $c_3\mathbf{b}_3 + c_4\mathbf{b}_4$ must be $\mathbf{0}$, implying that $c_1\mathbf{b}_1 + c_2\mathbf{b}_2 = \mathbf{0}$ also. By Theorem 2, $\{\mathbf{b}_1, \mathbf{b}_2\}$ is a linearly independent set so $c_1 = c_2 = 0$. Moreover, $\{\mathbf{b}_3, \mathbf{b}_4\}$ is a linearly independent set so $c_3 = c_4 = 0$. Since all of the coefficients c_1, c_2, c_3, and c_4 must be zero, it follows that $\{\mathbf{b}_1, \mathbf{b}_2, \mathbf{b}_3, \mathbf{b}_4\}$ is a linearly independent set.

4. Since λ is an eigenvalue of A, there is a nonzero vector $\mathbf{x}$ in $\mathbb{R}^n$ such that $A\mathbf{x} = \lambda\mathbf{x}$. Multiplying both sides of this equation by 2 results in the equation $2(A\mathbf{x}) = 2(\lambda\mathbf{x})$. Thus $(2A)\mathbf{x} = (2\lambda)\mathbf{x}$ and hence 2λ is an eigenvalue of $2A$.

5.2 The Characteristic Equation

Useful information about the eigenvalues of a square matrix A is encoded in a special scalar equation called the characteristic equation of A. A simple example will lead to the general case.

EXAMPLE 1 Find the eigenvalues of $A = \begin{bmatrix} 2 & 3 \\ 3 & -6 \end{bmatrix}$.

SOLUTION We must find all scalars λ such that the matrix equation

$$(A - \lambda I)\mathbf{x} = \mathbf{0}$$

has a nontrivial solution. By the Invertible Matrix Theorem in Section 2.3, this problem is equivalent to finding all λ such that the matrix $A - \lambda I$ is *not* invertible, where

$$A - \lambda I = \begin{bmatrix} 2 & 3 \\ 3 & -6 \end{bmatrix} - \begin{bmatrix} \lambda & 0 \\ 0 & \lambda \end{bmatrix} = \begin{bmatrix} 2 - \lambda & 3 \\ 3 & -6 - \lambda \end{bmatrix}$$

By Theorem 4 in Section 2.2, this matrix fails to be invertible precisely when its determinant is zero. So the eigenvalues of A are the solutions of the equation

$$\det(A - \lambda I) = \det\begin{bmatrix} 2 - \lambda & 3 \\ 3 & -6 - \lambda \end{bmatrix} = 0$$

Recall that

$$\det\begin{bmatrix} a & b \\ c & d \end{bmatrix} = ad - bc$$

So

$$\begin{aligned}
\det(A - \lambda I) &= (2 - \lambda)(-6 - \lambda) - (3)(3) \\
&= -12 + 6\lambda - 2\lambda + \lambda^2 - 9 \\
&= \lambda^2 + 4\lambda - 21 \\
&= (\lambda - 3)(\lambda + 7)
\end{aligned}$$

If $\det(A - \lambda I) = 0$, then $\lambda = 3$ or $\lambda = -7$. So the eigenvalues of A are 3 and -7.   ∎

Determinants

The determinant in Example 1 transformed the matrix equation $(A - \lambda I)\mathbf{x} = \mathbf{0}$, which involves *two* unknowns λ and $\mathbf{x}$, into the scalar equation $\lambda^2 + 4\lambda - 21 = 0$, which involves only *one* unknown. The same idea works for $n \times n$ matrices.

Before turning to larger matrices, recall from Section 3.1 that the matrix A_{ij} is obtained from A by deleting the ith row and jth column. The determinant of an $n \times n$ matrix A can be computed by an expansion across any row or down any column. The expansion across the ith row is given by

$$\det A = (-1)^{i+1}a_{i1} \det A_{i1} + (-1)^{i+2}a_{i2} \det A_{i2} + \cdots + (-1)^{i+n}a_{in} \det A_{in}$$

The expansion down the jth column is given by

$$\det A = (-1)^{1+j}a_{1j} \det A_{1j} + (-1)^{2+j}a_{2j} \det A_{2j} + \cdots + (-1)^{n+j}a_{nj} \det A_{nj}$$

EXAMPLE 2   Compute the determinant of

$$A = \begin{bmatrix} 2 & 3 & 1 \\ 4 & 0 & -1 \\ 0 & 2 & 1 \end{bmatrix}$$

SOLUTION   Any row or column can be chosen for the expansion. For example, expanding down the first column of A results in

$$\begin{aligned}
\det A &= a_{11} \det A_{11} - a_{21} \det A_{21} + a_{31} \det A_{31} \\
&= 2 \det \begin{bmatrix} 0 & -1 \\ 2 & 1 \end{bmatrix} - 4 \det \begin{bmatrix} 3 & 1 \\ 2 & 1 \end{bmatrix} + 0 \det \begin{bmatrix} 3 & 1 \\ 0 & -1 \end{bmatrix} \\
&= 2(0 - (-2)) - 4(3 - 2) + 0(-3 - 0) = 0
\end{aligned}$$

The next theorem lists facts from Sections 3.1 and 3.2 and is included here for convenient reference.   ∎

THEOREM 3

Properties of Determinants

Let A and B be $n \times n$ matrices.

a. A is invertible if and only if $\det A \neq 0$.

b. $\det AB = (\det A)(\det B)$.

c. $\det A^T = \det A$.

d. If A is triangular, then $\det A$ is the product of the entries on the main diagonal of A.

e. A row replacement operation on A does not change the determinant. A row interchange changes the sign of the determinant. A row scaling also scales the determinant by the same scalar factor.

Recall that A is invertible if and only if the equation $A\mathbf{x} = \mathbf{0}$ has only the trivial solution. Notice that the number 0 is an eigenvalue of A if and only if there *is* a nonzero vector $\mathbf{x}$ such that $A\mathbf{x} = 0\mathbf{x} = \mathbf{0}$, which happens if and only if $0 = \det(A - 0I) = \det A$. Hence A is invertible if and only if 0 is *not* an eigenvalue.

THEOREM

The Invertible Matrix Theorem (continued)

Let A be an $n \times n$ matrix. Then A is invertible if and only if

r. The number 0 is *not* an eigenvalue of A.

The Characteristic Equation

Theorem 3(a) shows how to determine when a matrix of the form $A - \lambda I$ is *not* invertible. The scalar equation $\det(A - \lambda I) = 0$ is called the **characteristic equation** of A, and the argument in Example 1 justifies the following fact.

A scalar λ is an eigenvalue of an $n \times n$ matrix A if and only if λ satisfies the characteristic equation

$$\det(A - \lambda I) = 0$$

EXAMPLE 3   Find the characteristic equation of

$$A = \begin{bmatrix} 5 & -2 & 6 & -1 \\ 0 & 3 & -8 & 0 \\ 0 & 0 & 5 & 4 \\ 0 & 0 & 0 & 1 \end{bmatrix}$$

SOLUTION   Form $A - \lambda I$, and use Theorem 3(d):

$$\det(A - \lambda I) = \det \begin{bmatrix} 5 - \lambda & -2 & 6 & -1 \\ 0 & 3 - \lambda & -8 & 0 \\ 0 & 0 & 5 - \lambda & 4 \\ 0 & 0 & 0 & 1 - \lambda \end{bmatrix}$$

$$= (5 - \lambda)(3 - \lambda)(5 - \lambda)(1 - \lambda)$$

The characteristic equation is

$$(5 - \lambda)^2(3 - \lambda)(1 - \lambda) = 0$$

or

$$(\lambda - 5)^2(\lambda - 3)(\lambda - 1) = 0$$

Expanding the product, we can also write

$$\lambda^4 - 14\lambda^3 + 68\lambda^2 - 130\lambda + 75 = 0 \qquad \blacksquare$$

Reasonable Answers

If you want to verify λ is an eigenvalue of A, row reduce $A - \lambda I$. If you get a pivot in every column, something is amiss—the scalar λ is not an eigenvalue of A. Looking back at Example 3, notice that $A - 5I$, $A - 3I$, and $A - I$ all have at least one column without a pivot; however, if λ is chosen to be any number other than 5, 3, or 1, the matrix $A - \lambda I$ has a pivot in every column.

In Examples 1 and 3, $\det(A - \lambda I)$ is a polynomial in λ. It can be shown that if A is an $n \times n$ matrix, then $\det(A - \lambda I)$ is a polynomial of degree n called the **characteristic polynomial** of A.

The eigenvalue 5 in Example 3 is said to have *multiplicity* 2 because $(\lambda - 5)$ occurs two times as a factor of the characteristic polynomial. In general, the (**algebraic**) **multiplicity** of an eigenvalue λ is its multiplicity as a root of the characteristic equation.

EXAMPLE 4 The characteristic polynomial of a 6×6 matrix is $\lambda^6 - 4\lambda^5 - 12\lambda^4$. Find the eigenvalues and their multiplicities.

SOLUTION Factor the polynomial

$$\lambda^6 - 4\lambda^5 - 12\lambda^4 = \lambda^4(\lambda^2 - 4\lambda - 12) = \lambda^4(\lambda - 6)(\lambda + 2)$$

The eigenvalues are 0 (multiplicity 4), 6 (multiplicity 1), and -2 (multiplicity 1). $\blacksquare$

We could also list the eigenvalues in Example 4 as $0, 0, 0, 0, 6$, and -2, so that the eigenvalues are repeated according to their multiplicities.

Because the characteristic equation for an $n \times n$ matrix involves an nth-degree polynomial, the equation has exactly n roots, counting multiplicities, provided complex roots are allowed. Such complex roots, called *complex eigenvalues*, will be discussed in Section 5.5. Until then, we consider only real eigenvalues, and scalars will continue to be real numbers.

The characteristic equation is important for theoretical purposes. In practical work, however, eigenvalues of any matrix larger than 2×2 should be found by a computer, unless the matrix is triangular or has other special properties. Although a 3×3 characteristic polynomial is easy to compute by hand, factoring it can be difficult (unless the matrix is carefully chosen). See the Numerical Notes at the end of this section.

STUDY GUIDE has advice on how to factor a polynomial.

Similarity

The next theorem illustrates one use of the characteristic polynomial, and it provides the foundation for several iterative methods that *approximate* eigenvalues. If A and B are $n \times n$ matrices, then A **is similar to** B if there is an invertible matrix P such that $P^{-1}AP = B$, or, equivalently, $A = PBP^{-1}$. Writing Q for P^{-1}, we have $Q^{-1}BQ = A$. So B is also similar to A, and we say simply that A and B **are similar**. Changing A into $P^{-1}AP$ is called a **similarity transformation**.

THEOREM 4 If $n \times n$ matrices A and B are similar, then they have the same characteristic polynomial and hence the same eigenvalues (with the same multiplicities).

PROOF If $B = P^{-1}AP$, then

$$B - \lambda I = P^{-1}AP - \lambda P^{-1}P = P^{-1}(AP - \lambda P) = P^{-1}(A - \lambda I)P$$

Using the multiplicative property (b) in Theorem 3, we compute

$$\det(B - \lambda I) = \det[P^{-1}(A - \lambda I)P]$$
$$= \det(P^{-1}) \cdot \det(A - \lambda I) \cdot \det(P) \qquad (1)$$

Since $\det(P^{-1}) \cdot \det(P) = \det(P^{-1}P) = \det I = 1$, we see from equation (1) that $\det(B - \lambda I) = \det(A - \lambda I)$. ∎

Warnings:

1. The matrices

$$\begin{bmatrix} 2 & 1 \\ 0 & 2 \end{bmatrix} \quad \text{and} \quad \begin{bmatrix} 2 & 0 \\ 0 & 2 \end{bmatrix}$$

   are not similar even though they have the same eigenvalues.

2. Similarity is not the same as row equivalence. (If A is row equivalent to B, then $B = EA$ for some invertible matrix E.) Row operations on a matrix usually change its eigenvalues.

Application to Dynamical Systems

Eigenvalues and eigenvectors hold the key to the discrete evolution of a dynamical system, as mentioned in the chapter introduction.

EXAMPLE 5 Let $A = \begin{bmatrix} .95 & .03 \\ .05 & .97 \end{bmatrix}$. Analyze the long-term behavior (as k increases) of the dynamical system defined by $\mathbf{x}_{k+1} = A\mathbf{x}_k$ ($k = 0, 1, 2, \ldots$), with $\mathbf{x}_0 = \begin{bmatrix} .6 \\ .4 \end{bmatrix}$.

SOLUTION The first step is to find the eigenvalues of A and a basis for each eigenspace. The characteristic equation for A is

$$0 = \det \begin{bmatrix} .95 - \lambda & .03 \\ .05 & .97 - \lambda \end{bmatrix} = (.95 - \lambda)(.97 - \lambda) - (.03)(.05)$$
$$= \lambda^2 - 1.92\lambda + .92$$

By the quadratic formula

$$\lambda = \frac{1.92 \pm \sqrt{(-1.92)^2 - 4(.92)}}{2} = \frac{1.92 \pm \sqrt{.0064}}{2}$$
$$= \frac{1.92 \pm .08}{2} = 1 \quad \text{or} \quad .92$$

It is readily checked that eigenvectors corresponding to $\lambda = 1$ and $\lambda = .92$ are multiples of

$$\mathbf{v}_1 = \begin{bmatrix} 3 \\ 5 \end{bmatrix} \quad \text{and} \quad \mathbf{v}_2 = \begin{bmatrix} 1 \\ -1 \end{bmatrix}$$

respectively.

The next step is to write the given $\mathbf{x}_0$ in terms of $\mathbf{v}_1$ and $\mathbf{v}_2$. This can be done because $\{\mathbf{v}_1, \mathbf{v}_2\}$ is obviously a basis for $\mathbb{R}^2$. (Why?) So there exist weights c_1 and c_2 such that

$$\mathbf{x}_0 = c_1\mathbf{v}_1 + c_2\mathbf{v}_2 = [\, \mathbf{v}_1 \;\; \mathbf{v}_2 \,]\begin{bmatrix} c_1 \\ c_2 \end{bmatrix} \tag{2}$$

In fact,

$$\begin{bmatrix} c_1 \\ c_2 \end{bmatrix} = [\, \mathbf{v}_1 \;\; \mathbf{v}_2 \,]^{-1}\mathbf{x}_0 = \begin{bmatrix} 3 & 1 \\ 5 & -1 \end{bmatrix}^{-1}\begin{bmatrix} .60 \\ .40 \end{bmatrix}$$

$$= \frac{1}{-8}\begin{bmatrix} -1 & -1 \\ -5 & 3 \end{bmatrix}\begin{bmatrix} .60 \\ .40 \end{bmatrix} = \begin{bmatrix} .125 \\ .225 \end{bmatrix} \tag{3}$$

Because $\mathbf{v}_1$ and $\mathbf{v}_2$ in (3) are eigenvectors of A, with $A\mathbf{v}_1 = \mathbf{v}_1$ and $A\mathbf{v}_2 = .92\mathbf{v}_2$, we easily compute each $\mathbf{x}_k$:

$$\begin{aligned} \mathbf{x}_1 = A\mathbf{x}_0 &= c_1 A\mathbf{v}_1 + c_2 A\mathbf{v}_2 & \text{Using linearity of } \mathbf{x} \mapsto A\mathbf{x} \\ &= c_1\mathbf{v}_1 + c_2(.92)\mathbf{v}_2 & \mathbf{v}_1 \text{ and } \mathbf{v}_2 \text{ are eigenvectors.} \\ \mathbf{x}_2 = A\mathbf{x}_1 &= c_1 A\mathbf{v}_1 + c_2(.92)A\mathbf{v}_2 \\ &= c_1\mathbf{v}_1 + c_2(.92)^2\mathbf{v}_2 \end{aligned}$$

and so on. In general,

$$\mathbf{x}_k = c_1\mathbf{v}_1 + c_2(.92)^k\mathbf{v}_2 \quad (k = 0, 1, 2, \ldots)$$

Using c_1 and c_2 from (4),

$$\mathbf{x}_k = .125\begin{bmatrix} 3 \\ 5 \end{bmatrix} + .225(.92)^k\begin{bmatrix} 1 \\ -1 \end{bmatrix} \quad (k = 0, 1, 2, \ldots) \tag{4}$$

This explicit formula for $\mathbf{x}_k$ gives the solution of the difference equation $\mathbf{x}_{k+1} = A\mathbf{x}_k$. As $k \to \infty$, $(.92)^k$ tends to zero and $\mathbf{x}_k$ tends to $\begin{bmatrix} .375 \\ .625 \end{bmatrix} = .125\mathbf{v}_1$.   ∎

The calculations in Example 5 have an interesting application to a Markov chain discussed in Section 5.9. Those who read that section may recognize that matrix A in Example 5 above is the same as the migration matrix M in Section 5.9, $\mathbf{x}_0$ is the initial population distribution between city and suburbs, and $\mathbf{x}_k$ represents the population distribution after k years.

Numerical Notes

1. Computer software such as Mathematica and Maple can use symbolic calculations to find the characteristic polynomial of a moderate-sized matrix. But there is no formula or finite algorithm to solve the characteristic equation of a general $n \times n$ matrix for $n \geq 5$.

Numerical Notes (Continued)

2. The best numerical methods for finding eigenvalues avoid the characteristic polynomial entirely. In fact, MATLAB finds the characteristic polynomial of a matrix A by first computing the eigenvalues $\lambda_1, \ldots, \lambda_n$ of A and then expanding the product $(\lambda - \lambda_1)(\lambda - \lambda_2)\cdots(\lambda - \lambda_n)$.

3. Several common algorithms for estimating the eigenvalues of a matrix A are based on Theorem 4. The powerful *QR algorithm* is discussed in the exercises. Another technique, called *Jacobi's method*, works when $A = A^T$ and computes a sequence of matrices of the form

$$A_1 = A \quad \text{and} \quad A_{k+1} = P_k^{-1} A_k P_k \quad (k = 1, 2, \ldots)$$

Each matrix in the sequence is similar to A and so has the same eigenvalues as A. The nondiagonal entries of A_{k+1} tend to zero as k increases, and the diagonal entries tend to approach the eigenvalues of A.

4. Other methods of estimating eigenvalues are discussed in Section 5.8.

Practice Problem

Find the characteristic equation and eigenvalues of $A = \begin{bmatrix} 1 & -4 \\ 4 & 2 \end{bmatrix}$.

5.2 Exercises

Find the characteristic polynomial and the eigenvalues of the matrices in Exercises 1–8.

1. $\begin{bmatrix} 2 & 7 \\ 7 & 2 \end{bmatrix}$

2. $\begin{bmatrix} 5 & 3 \\ 3 & 5 \end{bmatrix}$

3. $\begin{bmatrix} 3 & -2 \\ 1 & -1 \end{bmatrix}$

4. $\begin{bmatrix} 4 & -3 \\ -4 & 2 \end{bmatrix}$

5. $\begin{bmatrix} 2 & 1 \\ -1 & 4 \end{bmatrix}$

6. $\begin{bmatrix} 1 & -4 \\ 4 & 6 \end{bmatrix}$

7. $\begin{bmatrix} 5 & 3 \\ -4 & 4 \end{bmatrix}$

8. $\begin{bmatrix} 7 & -2 \\ 2 & 3 \end{bmatrix}$

Exercises 9–14 require techniques from Section 3.1. Find the characteristic polynomial of each matrix using expansion across a row or down a column. [*Note:* Finding the characteristic polynomial of a 3×3 matrix is not easy to do with just row operations, because the variable λ is involved.]

9. $\begin{bmatrix} 1 & 0 & -1 \\ 2 & 3 & -1 \\ 0 & 6 & 0 \end{bmatrix}$

10. $\begin{bmatrix} 0 & 3 & 1 \\ 3 & 0 & 2 \\ 1 & 2 & 0 \end{bmatrix}$

11. $\begin{bmatrix} 4 & 0 & 0 \\ 5 & 3 & 2 \\ -2 & 0 & 2 \end{bmatrix}$

12. $\begin{bmatrix} 1 & 0 & 1 \\ -3 & 6 & 1 \\ 0 & 0 & 4 \end{bmatrix}$

13. $\begin{bmatrix} 6 & -2 & 0 \\ -2 & 9 & 0 \\ 5 & 8 & 3 \end{bmatrix}$

14. $\begin{bmatrix} 3 & -2 & 3 \\ 0 & -1 & 0 \\ 6 & 7 & -4 \end{bmatrix}$

For the matrices in Exercises 15–17, list the eigenvalues, repeated according to their multiplicities.

15. $\begin{bmatrix} 4 & -7 & 0 & 2 \\ 0 & 3 & -4 & 6 \\ 0 & 0 & 3 & -8 \\ 0 & 0 & 0 & 1 \end{bmatrix}$

16. $\begin{bmatrix} 5 & 0 & 0 & 0 \\ 8 & -4 & 0 & 0 \\ 0 & 7 & 1 & 0 \\ 1 & -5 & 2 & 1 \end{bmatrix}$

17. $\begin{bmatrix} 3 & 0 & 0 & 0 & 0 \\ -5 & 1 & 0 & 0 & 0 \\ 3 & 8 & 0 & 0 & 0 \\ 0 & -7 & 2 & 1 & 0 \\ -4 & 1 & 9 & -2 & 3 \end{bmatrix}$

18. It can be shown that the algebraic multiplicity of an eigenvalue λ is always greater than or equal to the dimension of the eigenspace corresponding to λ. Find h in the matrix A below such that the eigenspace for $\lambda = 5$ is two-dimensional:

$$A = \begin{bmatrix} 5 & -2 & 6 & -1 \\ 0 & 3 & h & 0 \\ 0 & 0 & 5 & 4 \\ 0 & 0 & 0 & 1 \end{bmatrix}$$

19. Let A be an $n \times n$ matrix, and suppose A has n real eigenvalues, $\lambda_1, \ldots, \lambda_n$, repeated according to multiplicities, so that $\det(A - \lambda I) = (\lambda_1 - \lambda)(\lambda_2 - \lambda)\cdots(\lambda_n - \lambda)$

Explain why $\det A$ is the product of the n eigenvalues of A. (This result is true for any square matrix when complex eigenvalues are considered.)

20. Use a property of determinants to show that A and A^T have the same characteristic polynomial.

In Exercises 21–30, A and B are $n \times n$ matrices. Mark each statement True or False (**T/F**). Justify each answer.

21. (**T/F**) If 0 is an eigenvalue of A, then A is invertible.

22. (**T/F**) The zero vector is in the eigenspace of A associated with an eigenvalue λ.

23. (**T/F**) The matrix A and its transpose, A^T, have different sets of eigenvalues.

24. (**T/F**) The matrices A and $B^{-1}AB$ have the same sets of eigenvalues for every invertible matrix B.

25. (**T/F**) If 2 is an eigenvalue of A, then $A - 2I$ is not invertible.

26. (**T/F**) If two matrices have the same set of eigenvalues, then they are similar.

27. (**T/F**) If $\lambda + 5$ is a factor of the characteristic polynomial of A, then 5 is an eigenvalue of A.

28. (**T/F**) The multiplicity of a root r of the characteristic equation of A is called the algebraic multiplicity of r as an eigenvalue of A.

29. (**T/F**) The eigenvalue of the $n \times n$ identity matrix is 1 with algebraic multiplicity n.

30. (**T/F**) The matrix A can have more than n eigenvalues.

A widely used method for estimating eigenvalues of a general matrix A is the *QR algorithm*. Under suitable conditions, this algorithm produces a sequence of matrices, all similar to A, that become almost upper triangular, with diagonal entries that approach the eigenvalues of A. The main idea is to factor A (or another matrix similar to A) in the form $A = Q_1 R_1$, where $Q_1^T = Q_1^{-1}$ and R_1 is upper triangular. The factors are interchanged to form $A_1 = R_1 Q_1$, which is again factored as $A_1 = Q_2 R_2$; then to form $A_2 = R_2 Q_2$, and so on. The similarity of $A, A_1, \dots$ follows from the more general result in Exercise 31.

31. Show that if $A = QR$ with Q invertible, then A is similar to $A_1 = RQ$.

32. Show that if A and B are similar, then $\det A = \det B$.

T 33. Construct a random integer-valued 4×4 matrix A, and verify that A and A^T have the same characteristic polynomial (the same eigenvalues with the same multiplicities). Do A and A^T have the same eigenvectors? Make the same analysis of a 5×5 matrix. Report the matrices and your conclusions.

T 34. Construct a random integer-valued 4×4 matrix A.

a. Reduce A to echelon form U with no row scaling, and compute $\det A$. (If A happens to be singular, start over with a new random matrix.)

b. Compute the eigenvalues of A and the product of these eigenvalues (as accurately as possible).

c. List the matrix A, and, to four decimal places, list the pivots in U and the eigenvalues of A. Compute $\det A$ with your matrix program, and compare it with the products you found in (a) and (b).

T 35. Let $A = \begin{bmatrix} -6 & 28 & 21 \\ 4 & -15 & -12 \\ -8 & a & 25 \end{bmatrix}$. For each value of a in the set $\{32, 31.9, 31.8, 32.1, 32.2\}$, compute the characteristic polynomial of A and the eigenvalues. In each case, create a graph of the characteristic polynomial $p(t) = \det(A - tI)$ for $0 \le t \le 3$. If possible, construct all graphs on one coordinate system. Describe how the graphs reveal the changes in the eigenvalues as a changes.

Solution to Practice Problem

The characteristic equation is

$$0 = \det(A - \lambda I) = \det \begin{bmatrix} 1 - \lambda & -4 \\ 4 & 2 - \lambda \end{bmatrix}$$

$$= (1 - \lambda)(2 - \lambda) - (-4)(4) = \lambda^2 - 3\lambda + 18$$

From the quadratic formula,

$$\lambda = \frac{3 \pm \sqrt{(-3)^2 - 4(18)}}{2} = \frac{3 \pm \sqrt{-63}}{2}$$

It is clear that the characteristic equation has no real solutions, so A has no real eigenvalues. The matrix A is acting on the real vector space $\mathbb{R}^2$, and there is no nonzero vector $\mathbf{v}$ in $\mathbb{R}^2$ such that $A\mathbf{v} = \lambda\mathbf{v}$ for some scalar λ.

5.3 Diagonalization

In many cases, the eigenvalue–eigenvector information contained within a matrix A can be displayed in a useful factorization of the form $A = PDP^{-1}$ where D is a diagonal matrix. In this section, the factorization enables us to compute A^k quickly for large values of k, a fundamental idea in several applications of linear algebra. Later, in Sections 5.6 and 5.7, the factorization will be used to analyze (and *decouple*) dynamical systems.

The following example illustrates that powers of a diagonal matrix are easy to compute.

EXAMPLE 1 If $D = \begin{bmatrix} 5 & 0 \\ 0 & 3 \end{bmatrix}$, then $D^2 = \begin{bmatrix} 5 & 0 \\ 0 & 3 \end{bmatrix}\begin{bmatrix} 5 & 0 \\ 0 & 3 \end{bmatrix} = \begin{bmatrix} 5^2 & 0 \\ 0 & 3^2 \end{bmatrix}$

and

$$D^3 = DD^2 = \begin{bmatrix} 5 & 0 \\ 0 & 3 \end{bmatrix}\begin{bmatrix} 5^2 & 0 \\ 0 & 3^2 \end{bmatrix} = \begin{bmatrix} 5^3 & 0 \\ 0 & 3^3 \end{bmatrix}$$

In general,

$$D^k = \begin{bmatrix} 5^k & 0 \\ 0 & 3^k \end{bmatrix} \quad \text{for } k \geq 1 \qquad \blacksquare$$

If $A = PDP^{-1}$ for some invertible P and diagonal D, then A^k is also easy to compute, as the next example shows.

EXAMPLE 2 Let $A = \begin{bmatrix} 7 & 2 \\ -4 & 1 \end{bmatrix}$. Find a formula for A^k, given that $A = PDP^{-1}$, where

$$P = \begin{bmatrix} 1 & 1 \\ -1 & -2 \end{bmatrix} \quad \text{and} \quad D = \begin{bmatrix} 5 & 0 \\ 0 & 3 \end{bmatrix}$$

SOLUTION The standard formula for the inverse of a 2×2 matrix yields

$$P^{-1} = \begin{bmatrix} 2 & 1 \\ -1 & -1 \end{bmatrix}$$

Then, by associativity of matrix multiplication,

$$A^2 = (PDP^{-1})(PDP^{-1}) = PD\underbrace{(P^{-1}P)}_{I}DP^{-1} = PDDP^{-1}$$

$$= PD^2P^{-1} = \begin{bmatrix} 1 & 1 \\ -1 & -2 \end{bmatrix}\begin{bmatrix} 5^2 & 0 \\ 0 & 3^2 \end{bmatrix}\begin{bmatrix} 2 & 1 \\ -1 & -1 \end{bmatrix}$$

Again,

$$A^3 = (PDP^{-1})A^2 = (PDP^{-1})PD^2P^{-1} = PD\underbrace{D^2}_{I}P^{-1} = PD^3P^{-1}$$

In general, for $k \geq 1$,

$$A^k = PD^kP^{-1} = \begin{bmatrix} 1 & 1 \\ -1 & -2 \end{bmatrix}\begin{bmatrix} 5^k & 0 \\ 0 & 3^k \end{bmatrix}\begin{bmatrix} 2 & 1 \\ -1 & -1 \end{bmatrix}$$

$$= \begin{bmatrix} 2 \cdot 5^k - 3^k & 5^k - 3^k \\ 2 \cdot 3^k - 2 \cdot 5^k & 2 \cdot 3^k - 5^k \end{bmatrix} \qquad \blacksquare$$

A square matrix A is said to be **diagonalizable** if A is similar to a diagonal matrix, that is, if $A = PDP^{-1}$ for some invertible matrix P and some diagonal matrix D. The next theorem gives a characterization of diagonalizable matrices and tells how to construct a suitable factorization.

THEOREM 5

The Diagonalization Theorem

An $n \times n$ matrix A is diagonalizable if and only if A has n linearly independent eigenvectors.

In fact, $A = PDP^{-1}$, with D a diagonal matrix, if and only if the columns of P are n linearly independent eigenvectors of A. In this case, the diagonal entries of D are eigenvalues of A that correspond, respectively, to the eigenvectors in P.

In other words, A is diagonalizable if and only if there are enough eigenvectors to form a basis of $\mathbb{R}^n$. We call such a basis an **eigenvector basis** of $\mathbb{R}^n$.

PROOF First, observe that if P is any $n \times n$ matrix with columns $\mathbf{v}_1, \ldots, \mathbf{v}_n$, and if D is any diagonal matrix with diagonal entries $\lambda_1, \ldots, \lambda_n$, then

$$AP = A[\, \mathbf{v}_1 \quad \mathbf{v}_2 \quad \cdots \quad \mathbf{v}_n \,] = [\, A\mathbf{v}_1 \quad A\mathbf{v}_2 \quad \cdots \quad A\mathbf{v}_n \,] \tag{1}$$

while

$$PD = P \begin{bmatrix} \lambda_1 & 0 & \cdots & 0 \\ 0 & \lambda_2 & \cdots & 0 \\ \vdots & \vdots & & \vdots \\ 0 & 0 & \cdots & \lambda_n \end{bmatrix} = [\, \lambda_1 \mathbf{v}_1 \quad \lambda_2 \mathbf{v}_2 \quad \cdots \quad \lambda_n \mathbf{v}_n \,] \tag{2}$$

Now suppose A is diagonalizable and $A = PDP^{-1}$. Then right-multiplying this relation by P, we have $AP = PD$. In this case, equations (1) and (2) imply that

$$[\, A\mathbf{v}_1 \quad A\mathbf{v}_2 \quad \cdots \quad A\mathbf{v}_n \,] = [\, \lambda_1 \mathbf{v}_1 \quad \lambda_2 \mathbf{v}_2 \quad \cdots \quad \lambda_n \mathbf{v}_n \,] \tag{3}$$

Equating columns, we find that

$$A\mathbf{v}_1 = \lambda_1 \mathbf{v}_1, \quad A\mathbf{v}_2 = \lambda_2 \mathbf{v}_2, \quad \ldots, \quad A\mathbf{v}_n = \lambda_n \mathbf{v}_n \tag{4}$$

Since P is invertible, its columns $\mathbf{v}_1, \ldots, \mathbf{v}_n$ must be linearly independent. Also, since these columns are nonzero, the equations in (4) show that $\lambda_1, \ldots, \lambda_n$ are eigenvalues and $\mathbf{v}_1, \ldots, \mathbf{v}_n$ are corresponding eigenvectors. This argument proves the "only if" parts of the first and second statements, along with the third statement, of the theorem.

Finally, given any n eigenvectors $\mathbf{v}_1, \ldots, \mathbf{v}_n$, use them to construct the columns of P and use corresponding eigenvalues $\lambda_1, \ldots, \lambda_n$ to construct D. By equations (1)–(3), $AP = PD$. This is true without any condition on the eigenvectors. If, in fact, the eigenvectors are linearly independent, then P is invertible (by the Invertible Matrix Theorem), and $AP = PD$ implies that $A = PDP^{-1}$. ∎

Diagonalizing Matrices

EXAMPLE 3 Diagonalize the following matrix, if possible.

$$A = \begin{bmatrix} 1 & 3 & 3 \\ -3 & -5 & -3 \\ 3 & 3 & 1 \end{bmatrix}$$

That is, find an invertible matrix P and a diagonal matrix D such that $A = PDP^{-1}$.

SOLUTION There are four steps to implement the description in Theorem 5.

Step 1. Find the eigenvalues of A. As mentioned in Section 5.2, the mechanics of this step are appropriate for a computer when the matrix is larger than 2×2. To avoid unnecessary distractions, the text will usually supply information needed for this step. In the present case, the characteristic equation turns out to involve a cubic polynomial that can be factored:

$$
\begin{aligned}
0 = \det (A - \lambda I) &= -\lambda^3 - 3\lambda^2 + 4 \\
&= -(\lambda - 1)(\lambda + 2)^2
\end{aligned}
$$

The eigenvalues are $\lambda = 1$ and $\lambda = -2$.

Step 2. Find three linearly independent eigenvectors of A. *Three* vectors are needed because A is a 3×3 matrix. This is the critical step. If it fails, then Theorem 5 says that A cannot be diagonalized. The method in Section 5.1 produces a basis for each eigenspace:

$$
\text{Basis for } \lambda = 1: \quad \mathbf{v}_1 = \begin{bmatrix} 1 \\ -1 \\ 1 \end{bmatrix}
$$

$$
\text{Basis for } \lambda = -2: \quad \mathbf{v}_2 = \begin{bmatrix} -1 \\ 1 \\ 0 \end{bmatrix} \quad \text{and} \quad \mathbf{v}_3 = \begin{bmatrix} -1 \\ 0 \\ 1 \end{bmatrix}
$$

You can check that $\{\mathbf{v}_1, \mathbf{v}_2, \mathbf{v}_3\}$ is a linearly independent set.

Step 3. Construct P from the vectors in step 2. The vectors may be listed in any order. Using the order chosen in step 2, form

$$
P = \begin{bmatrix} \mathbf{v}_1 & \mathbf{v}_2 & \mathbf{v}_3 \end{bmatrix} = \begin{bmatrix} 1 & -1 & -1 \\ -1 & 1 & 0 \\ 1 & 0 & 1 \end{bmatrix}
$$

Step 4. Construct D from the corresponding eigenvalues. In this step, it is essential that the order of the eigenvalues matches the order chosen for the columns of P. Use the eigenvalue $\lambda = -2$ twice, once for each of the eigenvectors corresponding to $\lambda = -2$:

$$
D = \begin{bmatrix} 1 & 0 & 0 \\ 0 & -2 & 0 \\ 0 & 0 & -2 \end{bmatrix}
$$

It is a good idea to check that P and D really work. To avoid computing P^{-1}, simply verify that $AP = PD$. This is equivalent to $A = PDP^{-1}$ when P is invertible. (However, be sure that P is invertible!) Compute

$$
AP = \begin{bmatrix} 1 & 3 & 3 \\ -3 & -5 & -3 \\ 3 & 3 & 1 \end{bmatrix} \begin{bmatrix} 1 & -1 & -1 \\ -1 & 1 & 0 \\ 1 & 0 & 1 \end{bmatrix} = \begin{bmatrix} 1 & 2 & 2 \\ -1 & -2 & 0 \\ 1 & 0 & -2 \end{bmatrix}
$$

$$
PD = \begin{bmatrix} 1 & -1 & -1 \\ -1 & 1 & 0 \\ 1 & 0 & 1 \end{bmatrix} \begin{bmatrix} 1 & 0 & 0 \\ 0 & -2 & 0 \\ 0 & 0 & -2 \end{bmatrix} = \begin{bmatrix} 1 & 2 & 2 \\ -1 & -2 & 0 \\ 1 & 0 & -2 \end{bmatrix} \quad \blacksquare
$$

EXAMPLE 4   Diagonalize the following matrix, if possible.

$$A = \begin{bmatrix} 2 & 4 & 3 \\ -4 & -6 & -3 \\ 3 & 3 & 1 \end{bmatrix}$$

SOLUTION   The characteristic equation of A turns out to be exactly the same as that in Example 3:

$$0 = \det(A - \lambda I) = -\lambda^3 - 3\lambda^2 + 4 = -(\lambda - 1)(\lambda + 2)^2$$

The eigenvalues are $\lambda = 1$ and $\lambda = -2$. However, it is easy to verify that each eigenspace is only one-dimensional:

$$\text{Basis for } \lambda = 1: \qquad \mathbf{v}_1 = \begin{bmatrix} 1 \\ -1 \\ 1 \end{bmatrix}$$

$$\text{Basis for } \lambda = -2: \qquad \mathbf{v}_2 = \begin{bmatrix} -1 \\ 1 \\ 0 \end{bmatrix}$$

There are no other eigenvalues, and every eigenvector of A is a multiple of either $\mathbf{v}_1$ or $\mathbf{v}_2$. Hence it is impossible to construct a basis of $\mathbb{R}^3$ using eigenvectors of A. By Theorem 5, A is *not* diagonalizable. ■

The following theorem provides a *sufficient* condition for a matrix to be diagonalizable.

THEOREM 6   An $n \times n$ matrix with n distinct eigenvalues is diagonalizable.

PROOF   Let $\mathbf{v}_1, \ldots, \mathbf{v}_n$ be eigenvectors corresponding to the n distinct eigenvalues of a matrix A. Then $\{\mathbf{v}_1, \ldots, \mathbf{v}_n\}$ is linearly independent, by Theorem 2 in Section 5.1. Hence A is diagonalizable, by Theorem 5. ■

It is not *necessary* for an $n \times n$ matrix to have n distinct eigenvalues in order to be diagonalizable. The 3×3 matrix in Example 3 is diagonalizable even though it has only two distinct eigenvalues.

EXAMPLE 5   Determine if the following matrix is diagonalizable.

$$A = \begin{bmatrix} 5 & -8 & 1 \\ 0 & 0 & 7 \\ 0 & 0 & -2 \end{bmatrix}$$

SOLUTION   This is easy! Since the matrix is triangular, its eigenvalues are obviously 5, 0, and -2. Since A is a 3×3 matrix with three distinct eigenvalues, A is diagonalizable. ■

Matrices Whose Eigenvalues Are Not Distinct

If an $n \times n$ matrix A has n distinct eigenvalues, with corresponding eigenvectors $\mathbf{v}_1, \ldots,$ $\mathbf{v}_n$, and if $P = [\, \mathbf{v}_1 \quad \cdots \quad \mathbf{v}_n \,]$, then P is automatically invertible because its columns are linearly independent, by Theorem 2. When A is diagonalizable but has fewer than n distinct eigenvalues, it is still possible to build P in a way that makes P automatically invertible, as the next theorem shows.[1]

THEOREM 7   Let A be an $n \times n$ matrix whose distinct eigenvalues are $\lambda_1, \ldots, \lambda_p$.

a. For $1 \le k \le p$, the dimension of the eigenspace for λ_k is less than or equal to the multiplicity of the eigenvalue λ_k.

b. The matrix A is diagonalizable if and only if the sum of the dimensions of the eigenspaces equals n, and this happens if and only if (*i*) the characteristic polynomial factors completely into linear factors and (*ii*) the dimension of the eigenspace for each λ_k equals the multiplicity of λ_k.

c. If A is diagonalizable and $\mathcal{B}_k$ is a basis for the eigenspace corresponding to λ_k for each k, then the total collection of vectors in the sets $\mathcal{B}_1, \ldots, \mathcal{B}_p$ forms an eigenvector basis for $\mathbb{R}^n$.

EXAMPLE 6   Diagonalize the following matrix, if possible.

$$A = \begin{bmatrix} 5 & 0 & 0 & 0 \\ 0 & 5 & 0 & 0 \\ 1 & 4 & -3 & 0 \\ -1 & -2 & 0 & -3 \end{bmatrix}$$

SOLUTION   Since A is a triangular matrix, the eigenvalues are 5 and -3, each with multiplicity 2. Using the method in Section 5.1, we find a basis for each eigenspace.

$$\text{Basis for } \lambda = 5: \quad \mathbf{v}_1 = \begin{bmatrix} -8 \\ 4 \\ 1 \\ 0 \end{bmatrix} \quad \text{and} \quad \mathbf{v}_2 = \begin{bmatrix} -16 \\ 4 \\ 0 \\ 1 \end{bmatrix}$$

$$\text{Basis for } \lambda = -3: \quad \mathbf{v}_3 = \begin{bmatrix} 0 \\ 0 \\ 1 \\ 0 \end{bmatrix} \quad \text{and} \quad \mathbf{v}_4 = \begin{bmatrix} 0 \\ 0 \\ 0 \\ 1 \end{bmatrix}$$

The set $\{\mathbf{v}_1, \ldots, \mathbf{v}_4\}$ is linearly independent, by Theorem 7. So the matrix $P = [\mathbf{v}_1 \ \cdots \ \mathbf{v}_4]$ is invertible, and $A = PDP^{-1}$, where

$$P = \begin{bmatrix} -8 & -16 & 0 & 0 \\ 4 & 4 & 0 & 0 \\ 1 & 0 & 1 & 0 \\ 0 & 1 & 0 & 1 \end{bmatrix} \quad \text{and} \quad D = \begin{bmatrix} 5 & 0 & 0 & 0 \\ 0 & 5 & 0 & 0 \\ 0 & 0 & -3 & 0 \\ 0 & 0 & 0 & -3 \end{bmatrix} \quad \blacksquare$$

---

[1] The proof of Theorem 7 is somewhat lengthy but not difficult. For instance, see S. Friedberg, A. Insel, and L. Spence, *Linear Algebra*, 4th ed. (Englewood Cliffs, NJ: Prentice-Hall, 2002), Section 5.2.

Practice Problems

1. Compute A^8, where $A = \begin{bmatrix} 4 & -3 \\ 2 & -1 \end{bmatrix}$.

2. Let $A = \begin{bmatrix} -3 & 12 \\ -2 & 7 \end{bmatrix}$, $\mathbf{v}_1 = \begin{bmatrix} 3 \\ 1 \end{bmatrix}$, and $\mathbf{v}_2 = \begin{bmatrix} 2 \\ 1 \end{bmatrix}$. Suppose you are told that $\mathbf{v}_1$ and $\mathbf{v}_2$ are eigenvectors of A. Use this information to diagonalize A.

3. Let A be a 4×4 matrix with eigenvalues 5, 3, and -2, and suppose you know that the eigenspace for $\lambda = 3$ is two-dimensional. Do you have enough information to determine if A is diagonalizable?

5.3 Exercises

In Exercises 1 and 2, let $A = PDP^{-1}$ and compute A^4.

1. $P = \begin{bmatrix} 5 & 7 \\ 2 & 3 \end{bmatrix}$, $D = \begin{bmatrix} 2 & 0 \\ 0 & 1 \end{bmatrix}$

2. $P = \begin{bmatrix} 2 & -3 \\ -3 & 5 \end{bmatrix}$, $D = \begin{bmatrix} 1 & 0 \\ 0 & -1 \end{bmatrix}$

In Exercises 3 and 4, use the factorization $A = PDP^{-1}$ to compute A^k, where k represents an arbitrary positive integer.

3. $\begin{bmatrix} a & 0 \\ 3(a-b) & b \end{bmatrix} = \begin{bmatrix} 1 & 0 \\ 3 & 1 \end{bmatrix} \begin{bmatrix} a & 0 \\ 0 & b \end{bmatrix} \begin{bmatrix} 1 & 0 \\ -3 & 1 \end{bmatrix}$

4. $\begin{bmatrix} -6 & 8 \\ -4 & 6 \end{bmatrix} = \begin{bmatrix} 1 & 2 \\ 1 & 1 \end{bmatrix} \begin{bmatrix} 2 & 0 \\ 0 & -2 \end{bmatrix} \begin{bmatrix} -1 & 2 \\ 1 & -1 \end{bmatrix}$

In Exercises 5 and 6, the matrix A is factored in the form PDP^{-1}. Use the Diagonalization Theorem to find the eigenvalues of A and a basis for each eigenspace.

5. $\begin{bmatrix} 2 & 2 & 1 \\ 1 & 3 & 1 \\ 1 & 2 & 2 \end{bmatrix} = $
$\begin{bmatrix} 1 & 1 & 2 \\ 1 & 0 & -1 \\ 1 & -1 & 0 \end{bmatrix} \begin{bmatrix} 5 & 0 & 0 \\ 0 & 1 & 0 \\ 0 & 0 & 1 \end{bmatrix} \begin{bmatrix} 1/4 & 1/2 & 1/4 \\ 1/4 & 1/2 & -3/4 \\ 1/4 & -1/2 & 1/4 \end{bmatrix}$

6. $\begin{bmatrix} 5 & -2 & -2 \\ 1 & 2 & -1 \\ 0 & 0 & 3 \end{bmatrix} = $
$\begin{bmatrix} 2 & -1 & -2 \\ 1 & -1 & -1 \\ 1 & 0 & 0 \end{bmatrix} \begin{bmatrix} 3 & 0 & 0 \\ 0 & 3 & 0 \\ 0 & 0 & 4 \end{bmatrix} \begin{bmatrix} 0 & 0 & 1 \\ 1 & -2 & 0 \\ -1 & 1 & 1 \end{bmatrix}$

Diagonalize the matrices in Exercises 7–20, if possible. The eigenvalues for Exercises 11–16 are as follows: (11) $\lambda = 1, 2, 3$; (12) $\lambda = 1, 4$; (13) $\lambda = 5, 1$; (14) $\lambda = 3, 4$; (15) $\lambda = 3, 1$; (16) $\lambda = 2, 1$. For Exercise 18, one eigenvalue is $\lambda = 5$ and one eigenvector is $(-2, 1, 2)$.

7. $\begin{bmatrix} 1 & 0 \\ 6 & -1 \end{bmatrix}$

8. $\begin{bmatrix} 5 & 1 \\ 0 & 5 \end{bmatrix}$

9. $\begin{bmatrix} 3 & -1 \\ 1 & 5 \end{bmatrix}$

10. $\begin{bmatrix} 2 & 3 \\ 4 & 1 \end{bmatrix}$

11. $\begin{bmatrix} -1 & 4 & -2 \\ -3 & 4 & 0 \\ -3 & 1 & 3 \end{bmatrix}$

12. $\begin{bmatrix} 3 & -1 & -1 \\ -1 & 3 & -1 \\ -1 & -1 & 3 \end{bmatrix}$

13. $\begin{bmatrix} 2 & 2 & -1 \\ 1 & 3 & -1 \\ -1 & -2 & 2 \end{bmatrix}$

14. $\begin{bmatrix} 4 & 0 & 2 \\ 2 & 3 & 4 \\ 0 & 0 & 3 \end{bmatrix}$

15. $\begin{bmatrix} 7 & 4 & 16 \\ 2 & 5 & 8 \\ -2 & -2 & -5 \end{bmatrix}$

16. $\begin{bmatrix} 0 & -4 & -6 \\ -1 & 0 & -3 \\ 1 & 2 & 5 \end{bmatrix}$

17. $\begin{bmatrix} 4 & 0 & 0 \\ 1 & 4 & 0 \\ 0 & 0 & 5 \end{bmatrix}$

18. $\begin{bmatrix} -7 & -16 & 4 \\ 6 & 13 & -2 \\ 12 & 16 & 1 \end{bmatrix}$

19. $\begin{bmatrix} 5 & -3 & 0 & 9 \\ 0 & 3 & 1 & -2 \\ 0 & 0 & 2 & 0 \\ 0 & 0 & 0 & 2 \end{bmatrix}$

20. $\begin{bmatrix} 2 & 0 & 0 & 0 \\ 0 & 2 & 0 & 0 \\ 0 & 0 & 2 & 0 \\ 1 & 0 & 0 & 2 \end{bmatrix}$

In Exercises 21–28, A, P, and D are $n \times n$ matrices. Mark each statement True or False **(T/F)**. Justify each answer. (Study Theorems 5 and 6 and the examples in this section carefully before you try these exercises.)

21. **(T/F)** A is diagonalizable if $A = PDP^{-1}$ for some matrix D and some invertible matrix P.

22. **(T/F)** If $\mathbb{R}^n$ has a basis of eigenvectors of A, then A is diagonalizable.

23. **(T/F)** A is diagonalizable if and only if A has n eigenvalues, counting multiplicities.

24. **(T/F)** If A is diagonalizable, then A is invertible.

25. **(T/F)** A is diagonalizable if A has n eigenvectors.

26. **(T/F)** If A is diagonalizable, then A has n distinct eigenvalues.

27. **(T/F)** If $AP = PD$, with D diagonal, then the nonzero columns of P must be eigenvectors of A.

28. **(T/F)** If A is invertible, then A is diagonalizable.

29. A is a 5×5 matrix with two eigenvalues. One eigenspace is three-dimensional, and the other eigenspace is two-dimensional. Is A diagonalizable? Why?

30. A is a 3×3 matrix with two eigenvalues. Each eigenspace is one-dimensional. Is A diagonalizable? Why?

31. A is a 4×4 matrix with three eigenvalues. One eigenspace is one-dimensional, and one of the other eigenspaces is two-dimensional. Is it possible that A is *not* diagonalizable? Justify your answer.

32. A is a 7×7 matrix with three eigenvalues. One eigenspace is two-dimensional, and one of the other eigenspaces is three-dimensional. Is it possible that A is *not* diagonalizable? Justify your answer.

33. Show that if A is both diagonalizable and invertible, then so is A^{-1}.

34. Show that if A has n linearly independent eigenvectors, then so does A^T. [*Hint:* Use the Diagonalization Theorem.]

35. A factorization $A = PDP^{-1}$ is not unique. Demonstrate this for the matrix A in Example 2. With $D_1 = \begin{bmatrix} 3 & 0 \\ 0 & 5 \end{bmatrix}$, use the information in Example 2 to find a matrix P_1 such that $A = P_1 D_1 P_1^{-1}$.

36. With A and D as in Example 2, find an invertible P_2 unequal to the P in Example 2, such that $A = P_2 D P_2^{-1}$.

37. Construct a nonzero 2×2 matrix that is invertible but not diagonalizable.

38. Construct a nondiagonal 2×2 matrix that is diagonalizable but not invertible.

Diagonalize the matrices in Exercises 39–42. Use your matrix program's eigenvalue command to find the eigenvalues, and then compute bases for the eigenspaces as in Section 5.1.

T 39. $\begin{bmatrix} -6 & 4 & 0 & 9 \\ -3 & 0 & 1 & 6 \\ -1 & -2 & 1 & 0 \\ -4 & 4 & 0 & 7 \end{bmatrix}$
T 40. $\begin{bmatrix} 0 & 13 & 8 & 4 \\ 4 & 9 & 8 & 4 \\ 8 & 6 & 12 & 8 \\ 0 & 5 & 0 & -4 \end{bmatrix}$

T 41. $\begin{bmatrix} 11 & -6 & 4 & -10 & -4 \\ -3 & 5 & -2 & 4 & 1 \\ -8 & 12 & -3 & 12 & 4 \\ 1 & 6 & -2 & 3 & -1 \\ 8 & -18 & 8 & -14 & -1 \end{bmatrix}$

T 42. $\begin{bmatrix} 4 & 4 & 2 & 3 & -2 \\ 0 & 1 & -2 & -2 & 2 \\ 6 & 12 & 11 & 2 & -4 \\ 9 & 20 & 10 & 10 & -6 \\ 15 & 28 & 14 & 5 & -3 \end{bmatrix}$

Solutions to Practice Problems

1. $\det (A - \lambda I) = \lambda^2 - 3\lambda + 2 = (\lambda - 2)(\lambda - 1)$. The eigenvalues are 2 and 1, and the corresponding eigenvectors are $\mathbf{v}_1 = \begin{bmatrix} 3 \\ 2 \end{bmatrix}$ and $\mathbf{v}_2 = \begin{bmatrix} 1 \\ 1 \end{bmatrix}$. Next, form

$$P = \begin{bmatrix} 3 & 1 \\ 2 & 1 \end{bmatrix}, \qquad D = \begin{bmatrix} 2 & 0 \\ 0 & 1 \end{bmatrix}, \quad \text{and} \quad P^{-1} = \begin{bmatrix} 1 & -1 \\ -2 & 3 \end{bmatrix}$$

Since $A = PDP^{-1}$,

$$A^8 = PD^8P^{-1} = \begin{bmatrix} 3 & 1 \\ 2 & 1 \end{bmatrix} \begin{bmatrix} 2^8 & 0 \\ 0 & 1^8 \end{bmatrix} \begin{bmatrix} 1 & -1 \\ -2 & 3 \end{bmatrix}$$

$$= \begin{bmatrix} 3 & 1 \\ 2 & 1 \end{bmatrix} \begin{bmatrix} 256 & 0 \\ 0 & 1 \end{bmatrix} \begin{bmatrix} 1 & -1 \\ -2 & 3 \end{bmatrix}$$

$$= \begin{bmatrix} 766 & -765 \\ 510 & -509 \end{bmatrix}$$

2. Compute $A\mathbf{v}_1 = \begin{bmatrix} -3 & 12 \\ -2 & 7 \end{bmatrix} \begin{bmatrix} 3 \\ 1 \end{bmatrix} = \begin{bmatrix} 3 \\ 1 \end{bmatrix} = 1 \cdot \mathbf{v}_1$, and

$$A\mathbf{v}_2 = \begin{bmatrix} -3 & 12 \\ -2 & 7 \end{bmatrix} \begin{bmatrix} 2 \\ 1 \end{bmatrix} = \begin{bmatrix} 6 \\ 3 \end{bmatrix} = 3 \cdot \mathbf{v}_2$$

So, $\mathbf{v}_1$ and $\mathbf{v}_2$ are eigenvectors for the eigenvalues 1 and 3, respectively. Thus

$$A = PDP^{-1}, \quad \text{where} \quad P = \begin{bmatrix} 3 & 2 \\ 1 & 1 \end{bmatrix} \quad \text{and} \quad D = \begin{bmatrix} 1 & 0 \\ 0 & 3 \end{bmatrix}$$

3. Yes, A is diagonalizable. There is a basis $\{\mathbf{v}_1, \mathbf{v}_2\}$ for the eigenspace corresponding to $\lambda = 3$. In addition, there will be at least one eigenvector for $\lambda = 5$ and one for $\lambda = -2$. Call them $\mathbf{v}_3$ and $\mathbf{v}_4$. Then $\{\mathbf{v}_1, \mathbf{v}_2, \mathbf{v}_3, \mathbf{v}_4\}$ is linearly independent by Theorem 2 and Practice Problem 3 in Section 5.1. There can be no additional eigenvectors that are linearly independent from $\mathbf{v}_1, \mathbf{v}_2, \mathbf{v}_3, \mathbf{v}_4$, because the vectors are all in $\mathbb{R}^4$. Hence the eigenspaces for $\lambda = 5$ and $\lambda = -2$ are both one-dimensional. It follows that A is diagonalizable by Theorem 7(b).

STUDY GUIDE has advice on mastering eigenvalues and eigenspaces.

5.4 Eigenvectors and Linear Transformations

In this section, we will look at eigenvalues and eigenvectors of linear transformations $T : V \to V$, where V is any vector space. In the case where V is a finite dimensional vector space and there is a basis for V consisting of eigenvectors of T, we will see how to represent the transformation T as left multiplication by a diagonal matrix.

Eigenvectors of Linear Transformations

Previously, we looked at a variety of vector spaces including the discrete-time signal space, $\mathbb{S}$, and the set of polynomials, $\mathbb{P}$. Eigenvalues and eigenvectors can be defined for linear transformations from any vector space to itself.

DEFINITION

Let V be a vector space. An **eigenvector** of a linear transformation $T : V \to V$ is a nonzero vector $\mathbf{x}$ in V such that $T(\mathbf{x}) = \lambda \mathbf{x}$ for some scalar λ. A scalar λ is called an **eigenvalue** of T if there is a nontrivial solution $\mathbf{x}$ of $T(\mathbf{x}) = \lambda \mathbf{x}$; such an $\mathbf{x}$ is called an **eigenvector** corresponding to λ.

EXAMPLE 1 The sinusoidal signals were studied in detail in Sections 4.7 and 4.8. Consider the signal defined by $\{s_k\} = \left\{ \cos\left(\dfrac{k\pi}{2} \right) \right\}$, where k ranges over all integers. The left double-shift linear transformation D is defined by $D(\{x_k\}) = \{x_{k+2}\}$. Show that $\{s_k\}$ is an eigenvector of D and determine the associated eigenvalue.

SOLUTION The trigonometric formula $\cos(\theta + \pi) = -\cos(\theta)$ is useful here. Set $\{y_k\} = D(\{s_k\})$ and observe that

$$y_k = s_{k+2} = \cos\left(\frac{(k+2)\pi}{2} \right) = \cos\left(\frac{k\pi}{2} + \pi \right) = -\cos\left(\frac{k\pi}{2} \right) = -s_k$$

and so $D(\{s_k\}) = \{-s_k\} = -\{s_k\}$. This establishes that $\{s_k\}$ is an eigenvector of D with eigenvalue -1. ∎

In Figure 1, different values for the frequency, f, are chosen to graph a section of the sinusoidal signals $\left\{ \cos\left(\dfrac{fk\pi}{4} \right) \right\}$ and $D\left(\left\{ \cos\left(\dfrac{fk\pi}{4} \right) \right\} \right)$. Setting $f = 2$ illustrates the eigenvector for D established in Example 1. What is the relationship in the patterns of the dots that signifies an eigenvector relationship between the original signal and the transformed signal? Which other choices of the frequency, f, create a signal that is an eigenvector for D? What are the associated eigenvalues? In Figure 1, the graph on the left illustrates the sinusoidal signal with $f = 1$ and the graph on the right illustrates the sinusoidal signal with $f = 2$.

FIGURE 1

The Matrix of a Linear Transformation

There are branches of linear algebra that use infinite dimensional matrices to transform infinite dimensional vector spaces; however, in the remainder of this chapter we will restrict our study to linear transformations and matrices associated with finite dimensional vector spaces.

Let V be an n-dimensional vector space and let T be any linear transformation from V to V. To associate a matrix with T, choose any basis $\mathcal{B}$ for V. Given any $\mathbf{x}$ in V, the coordinate vector $[\mathbf{x}]_\mathcal{B}$ is in $\mathbb{R}^n$, as is the coordinate vector of its image, $[T(\mathbf{x})]_\mathcal{B}$.

The connection between $[\mathbf{x}]_\mathcal{B}$ and $[T(\mathbf{x})]_\mathcal{B}$ is easy to find. Let $\{\mathbf{b}_1, \dots, \mathbf{b}_n\}$ be the basis $\mathcal{B}$ for V. If $\mathbf{x} = r_1 \mathbf{b}_1 + \cdots + r_n \mathbf{b}_n$, then

$$[\mathbf{x}]_\mathcal{B} = \begin{bmatrix} r_1 \\ \vdots \\ r_n \end{bmatrix}$$

and

$$T(\mathbf{x}) = T(r_1 \mathbf{b}_1 + \cdots + r_n \mathbf{b}_n) = r_1 T(\mathbf{b}_1) + \cdots + r_n T(\mathbf{b}_n) \tag{1}$$

because T is linear. Now, since the coordinate mapping from V to $\mathbb{R}^n$ is linear (Theorem 8 in Section 4.4), equation (1) leads to

$$[T(\mathbf{x})]_\mathcal{B} = r_1 [T(\mathbf{b}_1)]_\mathcal{B} + \cdots + r_n [T(\mathbf{b}_n)]_\mathcal{B} \tag{2}$$

Since $\mathcal{B}$-coordinate vectors are in $\mathbb{R}^n$, the vector equation (2) can be written as a matrix equation, namely

$$[T(\mathbf{x})]_\mathcal{B} = M [\mathbf{x}]_\mathcal{B} \tag{3}$$

where

$$M = [[T(\mathbf{b}_1)]_\mathcal{B} \; [T(\mathbf{b}_2)]_\mathcal{B} \; \cdots \; [T(\mathbf{b}_n)]_\mathcal{B}] \tag{4}$$

The matrix M is a matrix representation of T, called the **matrix for T relative to the basis $\mathcal{B}$** and denoted by $[T]_\mathcal{B}$. See Figure 2.

Equation (3) says that, so far as coordinate vectors are concerned, the action of T on $\mathbf{x}$ may be viewed as left-multiplication by M.

$$\begin{array}{ccc} \mathbf{x} & \xrightarrow{\;\;T\;\;} & T(\mathbf{x}) \\ \downarrow & & \downarrow \\ [\mathbf{x}]_\mathcal{B} & \xrightarrow[\text{by } M=[T]_\mathcal{B}]{\text{Multiplication}} & [T(\mathbf{x})]_\mathcal{B} \end{array}$$

FIGURE 2

EXAMPLE 2 Suppose $\mathcal{B} = \{\mathbf{b}_1, \mathbf{b}_2\}$ is a basis for V. Let $T : V \to V$ be a linear transformation with the property that

$$T(\mathbf{b}_1) = 3\mathbf{b}_1 - 2\mathbf{b}_2 \quad \text{and} \quad T(\mathbf{b}_2) = 4\mathbf{b}_1 + 7\mathbf{b}_2$$

Find the matrix M for T relative to $\mathcal{B}$.

SOLUTION The $\mathcal{B}$-coordinate vectors of the *images* of $\mathbf{b}_1$ and $\mathbf{b}_2$ are

$$[T(\mathbf{b}_1)]_{\mathcal{B}} = \begin{bmatrix} 3 \\ -2 \end{bmatrix} \quad \text{and} \quad [T(\mathbf{b}_2)]_{\mathcal{B}} = \begin{bmatrix} 4 \\ 7 \end{bmatrix}$$

Hence

$$M = \begin{bmatrix} 3 & 4 \\ -2 & 7 \end{bmatrix}$$

■

EXAMPLE 3 The mapping $T : \mathbb{P}_2 \to \mathbb{P}_2$ defined by

$$T\left(a_0 + a_1 t + a_2 t^2\right) = a_1 + 2a_2 t$$

is a linear transformation. (Calculus students will recognize T as the differentiation operator.)

a. Find the $\mathcal{B}$-matrix for T, when $\mathcal{B}$ is the basis $\{1, t, t^2\}$.
b. Verify that $[T(\mathbf{p})]_{\mathcal{B}} = [T]_{\mathcal{B}}[\mathbf{p}]_{\mathcal{B}}$ for each $\mathbf{p}$ in $\mathbb{P}_2$.

SOLUTION

a. Compute the images of the basis vectors:

$$T(1) = 0 \quad \text{The zero polynomial}$$
$$T(t) = 1 \quad \text{The polynomial whose value is always 1}$$
$$T\left(t^2\right) = 2t$$

Then write the $\mathcal{B}$-coordinate vectors of $T(1)$, $T(t)$, and $T\left(t^2\right)$ (which are found by inspection in this example) and place them together as the $\mathcal{B}$-matrix for T:

$$[T(1)]_{\mathcal{B}} = \begin{bmatrix} 0 \\ 0 \\ 0 \end{bmatrix}, [T(t)]_{\mathcal{B}} = \begin{bmatrix} 1 \\ 0 \\ 0 \end{bmatrix}, \left[T\left(t^2\right)\right]_{\mathcal{B}} = \begin{bmatrix} 0 \\ 2 \\ 0 \end{bmatrix}$$

$$[T]_{\mathcal{B}} = \begin{bmatrix} 0 & 1 & 0 \\ 0 & 0 & 2 \\ 0 & 0 & 0 \end{bmatrix}$$

b. For a general $\mathbf{p}(t) = a_0 + a_1 t + a_2 t^2$

$$[T(\mathbf{p})]_{\mathcal{B}} = [a_1 + 2a_2 t]_{\mathcal{B}} = \begin{bmatrix} a_1 \\ 2a_2 \\ 0 \end{bmatrix}$$

$$= \begin{bmatrix} 0 & 1 & 0 \\ 0 & 0 & 2 \\ 0 & 0 & 0 \end{bmatrix} \begin{bmatrix} a_0 \\ a_1 \\ a_2 \end{bmatrix} = [T]_{\mathcal{B}}[\mathbf{p}]_{\mathcal{B}}$$

See Figure 3.

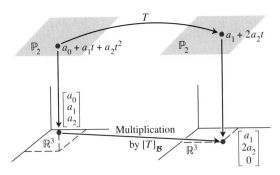

FIGURE 3 Matrix representation of a linear transformation.

Linear Transformations on $\mathbb{R}^n$

In an applied problem involving $\mathbb{R}^n$, a linear transformation T usually appears first as a matrix transformation, $\mathbf{x} \mapsto A\mathbf{x}$. If A is diagonalizable, then there is a basis $\mathcal{B}$ for $\mathbb{R}^n$ consisting of eigenvectors of A. Theorem 8 below shows that, in this case, the $\mathcal{B}$-matrix for T is diagonal. Diagonalizing A amounts to finding a diagonal matrix representation of $\mathbf{x} \mapsto A\mathbf{x}$.

THEOREM 8

Diagonal Matrix Representation

Suppose $A = PDP^{-1}$, where D is a diagonal $n \times n$ matrix. If $\mathcal{B}$ is the basis for $\mathbb{R}^n$ formed from the columns of P, then D is the $\mathcal{B}$-matrix for the transformation $\mathbf{x} \mapsto A\mathbf{x}$.

PROOF Denote the columns of P by $\mathbf{b}_1, \ldots, \mathbf{b}_n$, so that $\mathcal{B} = \{\mathbf{b}_1, \ldots, \mathbf{b}_n\}$ and $P = [\mathbf{b}_1 \ \cdots \ \mathbf{b}_n]$. In this case, P is the change-of-coordinates matrix $P_{\mathcal{B}}$ discussed in Section 4.4, where

$$P[\mathbf{x}]_{\mathcal{B}} = \mathbf{x} \quad \text{and} \quad [\mathbf{x}]_{\mathcal{B}} = P^{-1}\mathbf{x}$$

If $T(\mathbf{x}) = A\mathbf{x}$ for $\mathbf{x}$ in $\mathbb{R}^n$, then

$$
\begin{aligned}
[T]_{\mathcal{B}} &= \left[\, [T(\mathbf{b}_1)]_{\mathcal{B}} \quad \cdots \quad [T(\mathbf{b}_n)]_{\mathcal{B}} \,\right] && \text{Definition of } [T]_{\mathcal{B}} \\
&= \left[\, [A\mathbf{b}_1]_{\mathcal{B}} \quad \cdots \quad [A\mathbf{b}_n]_{\mathcal{B}} \,\right] && \text{Since } T(\mathbf{x}) = A\mathbf{x} \\
&= [\, P^{-1}A\mathbf{b}_1 \quad \cdots \quad P^{-1}A\mathbf{b}_n \,] && \text{Change of coordinates} \\
&= P^{-1}A[\, \mathbf{b}_1 \quad \cdots \quad \mathbf{b}_n \,] && \text{Matrix multiplication} \\
&= P^{-1}AP && (6)
\end{aligned}
$$

Since $A = PDP^{-1}$, we have $[T]_{\mathcal{B}} = P^{-1}AP = D$. ∎

EXAMPLE 4 Define $T : \mathbb{R}^2 \to \mathbb{R}^2$ by $T(\mathbf{x}) = A\mathbf{x}$, where $A = \begin{bmatrix} 7 & 2 \\ -4 & 1 \end{bmatrix}$. Find a basis $\mathcal{B}$ for $\mathbb{R}^2$ with the property that the $\mathcal{B}$-matrix for T is a diagonal matrix.

SOLUTION From Example 2 in Section 5.3, we know that $A = PDP^{-1}$, where

$$P = \begin{bmatrix} 1 & 1 \\ -1 & -2 \end{bmatrix} \quad \text{and} \quad D = \begin{bmatrix} 5 & 0 \\ 0 & 3 \end{bmatrix}$$

The columns of P, call them $\mathbf{b}_1$ and $\mathbf{b}_2$, are eigenvectors of A. By Theorem 8, D is the $\mathcal{B}$-matrix for T when $\mathcal{B} = \{\mathbf{b}_1, \mathbf{b}_2\}$. The mappings $\mathbf{x} \mapsto A\mathbf{x}$ and $\mathbf{u} \mapsto D\mathbf{u}$ describe the same linear transformation, relative to different bases. ∎

Similarity of Matrix Representations

The proof of Theorem 8 did not use the information that D was diagonal. Hence, if A is similar to a matrix C, with $A = PCP^{-1}$, then C is the $\mathcal{B}$-matrix for the transformation $\mathbf{x} \mapsto A\mathbf{x}$ when the basis $\mathcal{B}$ is formed from the columns of P. The factorization $A = PCP^{-1}$ is shown in Figure 4.

FIGURE 4 Similarity of two matrix representations: $A = PCP^{-1}$.

Conversely, if $T : \mathbb{R}^n \to \mathbb{R}^n$ is defined by $T(\mathbf{x}) = A\mathbf{x}$, and if $\mathcal{B}$ is any basis for $\mathbb{R}^n$, then the $\mathcal{B}$-matrix for T is similar to A. In fact, the calculations in the proof of Theorem 8 show that if P is the matrix whose columns come from the vectors in $\mathcal{B}$, then $[T]_{\mathcal{B}} = P^{-1}AP$. This important connection between the matrix of a linear transformation and similar matrices is highlighted here.

> The set of all matrices similar to a matrix A coincides with the set of all matrix representations of the transformation $\mathbf{x} \mapsto A\mathbf{x}$.

EXAMPLE 5 Let $A = \begin{bmatrix} 4 & -9 \\ 4 & -8 \end{bmatrix}$, $\mathbf{b}_1 = \begin{bmatrix} 3 \\ 2 \end{bmatrix}$, and $\mathbf{b}_2 = \begin{bmatrix} 2 \\ 1 \end{bmatrix}$. The characteristic polynomial of A is $(\lambda + 2)^2$, but the eigenspace for the eigenvalue -2 is only one-dimensional; so A is not diagonalizable. However, the basis $\mathcal{B} = \{\mathbf{b}_1, \mathbf{b}_2\}$ has the property that the $\mathcal{B}$-matrix for the transformation $\mathbf{x} \mapsto A\mathbf{x}$ is a triangular matrix called the *Jordan form* of A.[1] Find this $\mathcal{B}$-matrix.

SOLUTION If $P = [\,\mathbf{b}_1 \quad \mathbf{b}_2\,]$, then the $\mathcal{B}$-matrix is $P^{-1}AP$. Compute

$$AP = \begin{bmatrix} 4 & -9 \\ 4 & -8 \end{bmatrix} \begin{bmatrix} 3 & 2 \\ 2 & 1 \end{bmatrix} = \begin{bmatrix} -6 & -1 \\ -4 & 0 \end{bmatrix}$$

$$P^{-1}AP = \begin{bmatrix} -1 & 2 \\ 2 & -3 \end{bmatrix} \begin{bmatrix} -6 & -1 \\ -4 & 0 \end{bmatrix} = \begin{bmatrix} -2 & 1 \\ 0 & -2 \end{bmatrix}$$

Notice that the eigenvalue of A is on the diagonal. ∎

---

[1] Every square matrix A is similar to a matrix in Jordan form. The basis used to produce a Jordan form consists of eigenvectors and so-called "generalized eigenvectors" of A. See Chapter 9 of *Applied Linear Algebra*, 3rd ed. (Englewood Cliffs, NJ: Prentice-Hall, 1988), by B. Noble and J. W. Daniel.

> ### Numerical Notes
>
> An efficient way to compute a $\mathcal{B}$-matrix $P^{-1}AP$ is to compute AP and then to row reduce the augmented matrix $[\,P \quad AP\,]$ to $[\,I \quad P^{-1}AP\,]$. A separate computation of P^{-1} is unnecessary. See Exercise 22 in Section 2.2.

Practice Problems

1. Find $T(a_0 + a_1 t + a_2 t^2)$, if T is the linear transformation from $\mathbb{P}_2$ to $\mathbb{P}_2$ whose matrix relative to $\mathcal{B} = \{1, t, t^2\}$ is

$$[T]_{\mathcal{B}} = \begin{bmatrix} 3 & 4 & 0 \\ 0 & 5 & -1 \\ 1 & -2 & 7 \end{bmatrix}$$

2. Let A, B, and C be $n \times n$ matrices. The text has shown that if A is similar to B, then B is similar to A. This property, together with the statements below, shows that "similar to" is an *equivalence relation*. (Row equivalence is another example of an equivalence relation.) Verify parts (a) and (b).

   a. A is similar to A.

   b. If A is similar to B and B is similar to C, then A is similar to C.

5.4 Exercises

1. Let $\mathcal{B} = \{\mathbf{b}_1, \mathbf{b}_2, \mathbf{b}_3\}$ be a basis for the vector space V. Let $T : V \to V$ be a linear transformation with the property that

   $$T(\mathbf{b}_1) = 3\mathbf{b}_1 - 5\mathbf{b}_2, \quad T(\mathbf{b}_2) = -\mathbf{b}_1 + 6\mathbf{b}_2, \quad T(\mathbf{b}_3) = 4\mathbf{b}_2$$

   Find $[T]_{\mathcal{B}}$, the matrix for T relative to $\mathcal{B}$.

2. Let $\mathcal{B} = \{\mathbf{b}_1, \mathbf{b}_2\}$ be a basis for vector space V. Let $T : V \to V$ be a linear transformation with the property that

   $$T(\mathbf{b}_1) = 2\mathbf{b}_1 - 3\mathbf{b}_2, \quad T(\mathbf{b}_2) = -4\mathbf{b}_1 + 5\mathbf{b}_2$$

   Find $[T]_{\mathcal{B}}$, the matrix for T relative to $\mathcal{B}$.

3. Assume the mapping $T : \mathbb{P}_2 \to \mathbb{P}_2$ defined by

   $$T\left(a_0 + a_1 t + a_2 t^2\right) = 3a_0 + (5a_0 - 2a_1)t + (4a_1 + a_2)t^2$$

   is linear. Find the matrix representation of T relative to the basis $\mathcal{B} = \{1, t, t^2\}$.

4. Define $T : \mathbb{P}_2 \to \mathbb{P}_2$ by $T(\mathbf{p}) = \mathbf{p}(0) - \mathbf{p}(1)t + \mathbf{p}(2)t^2$.

   a. Show that T is a linear transformation.

   b. Find $T(\mathbf{p})$ when $\mathbf{p}(t) = -2 + t$. Is $\mathbf{p}$ an eigenvector of T?

   c. Find the matrix for T relative to the basis $\{1, t, t^2\}$ for $\mathbb{P}_2$.

5. Let $\mathcal{B} = \{\mathbf{b}_1, \mathbf{b}_2, \mathbf{b}_3\}$ be a basis for a vector space V. Find $T(3\mathbf{b}_1 - 4\mathbf{b}_2)$ when T is a linear transformation from V to V whose matrix relative to $\mathcal{B}$ is

$$[T]_{\mathcal{B}} = \begin{bmatrix} 0 & -6 & 1 \\ 0 & 5 & -1 \\ 1 & -2 & 7 \end{bmatrix}$$

6. Let $\mathcal{B} = \{\mathbf{b}_1, \mathbf{b}_2, \mathbf{b}_3\}$ be a basis for a vector space V. Find $T(2\mathbf{b}_1 - \mathbf{b}_2 + 4\mathbf{b}_3)$ when T is a linear transformation from V to V whose matrix relative to $\mathcal{B}$ is

$$[T]_{\mathcal{B}} = \begin{bmatrix} 0 & -6 & 1 \\ 0 & 5 & -1 \\ 1 & -2 & 7 \end{bmatrix}$$

In Exercises 7 and 8, find the $\mathcal{B}$-matrix for the transformation $\mathbf{x} \mapsto A\mathbf{x}$, when $\mathcal{B} = \{\mathbf{b}_1, \mathbf{b}_2\}$.

7. $A = \begin{bmatrix} 3 & 4 \\ -1 & -1 \end{bmatrix}$, $\mathbf{b}_1 = \begin{bmatrix} 2 \\ -1 \end{bmatrix}$, $\mathbf{b}_2 = \begin{bmatrix} 1 \\ 2 \end{bmatrix}$

8. $A = \begin{bmatrix} -1 & 4 \\ -2 & 3 \end{bmatrix}$, $\mathbf{b}_1 = \begin{bmatrix} 3 \\ 2 \end{bmatrix}$, $\mathbf{b}_2 = \begin{bmatrix} -1 \\ 1 \end{bmatrix}$

In Exercises 9–12, define $T : \mathbb{R}^2 \to \mathbb{R}^2$ by $T(\mathbf{x}) = A\mathbf{x}$. Find a basis $\mathcal{B}$ for $\mathbb{R}^2$ with the property that $[T]_{\mathcal{B}}$ is diagonal.

9. $A = \begin{bmatrix} 0 & 1 \\ -3 & 4 \end{bmatrix}$

10. $A = \begin{bmatrix} 5 & -3 \\ -7 & 1 \end{bmatrix}$

11. $A = \begin{bmatrix} 4 & -2 \\ -1 & 3 \end{bmatrix}$

12. $A = \begin{bmatrix} 2 & -6 \\ -1 & 3 \end{bmatrix}$

13. Let $A = \begin{bmatrix} 1 & 1 \\ -1 & 3 \end{bmatrix}$ and $\mathcal{B} = \{\mathbf{b}_1, \mathbf{b}_2\}$, for $\mathbf{b}_1 = \begin{bmatrix} 1 \\ 1 \end{bmatrix}$, $\mathbf{b}_2 = \begin{bmatrix} 5 \\ 4 \end{bmatrix}$. Define $T : \mathbb{R}^2 \to \mathbb{R}^2$ by $T(\mathbf{x}) = A\mathbf{x}$.

a. Verify that $\mathbf{b}_1$ is an eigenvector of A but A is not diagonalizable.

b. Find the $\mathcal{B}$ matrix for T.

14. Define $T : \mathbb{R}^3 \to \mathbb{R}^3$ by $T(\mathbf{x}) = A\mathbf{x}$, where A is a 3×3 matrix with eigenvalues 5 and -2. Does there exist a basis $\mathcal{B}$ for $\mathbb{R}^3$ such that the $\mathcal{B}$-matrix for T is a diagonal matrix? Discuss.

15. Define $T : \mathbb{P}_2 \to \mathbb{P}_2$ by $T(\mathbf{p}) = \mathbf{p}(1) + \mathbf{p}(1)t + \mathbf{p}(1)t^2$.

a. Find $T(\mathbf{p})$ when $\mathbf{p}(t) = 1 + t + t^2$. Is $\mathbf{p}$ an eigenvector of T? If $\mathbf{p}$ is an eigenvector, what is its eigenvalue?

b. Find $T(\mathbf{p})$ when $\mathbf{p}(t) = -2 + t$. Is $\mathbf{p}$ an eigenvector of T? If $\mathbf{p}$ is an eigenvector, what is its eigenvalue?

16. Define $T : \mathbb{P}_3 \to \mathbb{P}_3$ by $T(\mathbf{p}) = \mathbf{p}(0) - \mathbf{p}(1)t - \mathbf{p}(1)t^2 + \mathbf{p}(0)t^3$.

a. Find $T(\mathbf{p})$ when $\mathbf{p}(t) = 1 + t + t^2 + t^3$. Is $\mathbf{p}$ an eigenvector of T? If $\mathbf{p}$ is an eigenvector, what is its eigenvalue?

b. Find $T(\mathbf{p})$ when $\mathbf{p}(t) = t + t^2$. Is $\mathbf{p}$ an eigenvector of T? If $\mathbf{p}$ is an eigenvector, what is its eigenvalue?

In Exercises 17 through 20, mark each statement True or False (**T/F**). Justify each answer.

17. (**T/F**) Similar matrices have the same eigenvalues.

18. (**T/F**) Similar matrices have the same eigenvectors.

19. (**T/F**) Only linear transformations on finite vectors spaces have eigenvectors.

20. (**T/F**) If there is a nonzero vector in the kernel of a linear transformation T, then 0 is an eigenvalue of T.

Verify the statements in Exercises 21–28 by providing justification for each statement. In each case, the matrices are square.

21. If A is invertible and similar to B, then B is invertible and A^{-1} is similar to B^{-1}. [*Hint:* $P^{-1}AP = B$ for some invertible P. Explain why B is invertible. Then find an invertible Q such that $Q^{-1}A^{-1}Q = B^{-1}$.]

22. If A is similar to B, then A^2 is similar to B^2.

23. If B is similar to A and C is similar to A, then B is similar to C.

24. If A is diagonalizable and B is similar to A, then B is also diagonalizable.

25. If $B = P^{-1}AP$ and $\mathbf{x}$ is an eigenvector of A corresponding to an eigenvalue λ, then $P^{-1}\mathbf{x}$ is an eigenvector of B corresponding also to λ.

26. If A and B are similar, then they have the same rank. [*Hint:* Refer to Supplementary Exercises 31 and 32 for Chapter 4.]

27. The *trace* of a square matrix A is the sum of the diagonal entries in A and is denoted by tr A. It can be verified that $\text{tr}(FG) = \text{tr}(GF)$ for any two $n \times n$ matrices F and G. Show that if A and B are similar, then tr $A = $ tr B.

28. It can be shown that the trace of a matrix A equals the sum of the eigenvalues of A. Verify this statement for the case when A is diagonalizable.

Exercises 29–32 refer to the vector space of signals, $\mathbb{S}$, from Section 4.7. The shift transformation, $S(\{y_k\}) = \{y_{k-1}\}$, shifts each entry in the signal one position to the right. The moving average transformation, $M_2(\{y_k\}) = \left\{\dfrac{y_k + y_{k-1}}{2}\right\}$, creates a new signal by averaging two consecutive terms in the given signal. The constant signal of all ones is given by $\chi = \{1^k\}$ and the alternating signal by $\alpha = \{(-1)^k\}$.

29. Show that χ is an eigenvector of the shift transformation S. What is the associated eigenvalue?

30. Show that α is an eigenvector of the shift transformation S. What is the associated eigenvalue?

31. Show that α is an eigenvector of the moving average transformation M_2. What is the associated eigenvalue?

32. Show that χ is an eigenvector of the moving average transformation M_2. What is the associated eigenvalue?

In Exercises 33 and 34, find the $\mathcal{B}$-matrix for the transformation $\mathbf{x} \mapsto A\mathbf{x}$ when $\mathcal{B} = \{\mathbf{b}_1, \mathbf{b}_2, \mathbf{b}_3\}$.

T 33. $A = \begin{bmatrix} -14 & 4 & -14 \\ -33 & 9 & -31 \\ 11 & -4 & 11 \end{bmatrix}$, $\mathbf{b}_1 = \begin{bmatrix} -1 \\ -2 \\ 1 \end{bmatrix}$,

$\mathbf{b}_2 = \begin{bmatrix} -1 \\ -1 \\ 1 \end{bmatrix}$, $\mathbf{b}_3 = \begin{bmatrix} -1 \\ -2 \\ 0 \end{bmatrix}$

T 34. $A = \begin{bmatrix} -7 & -48 & -16 \\ 1 & 14 & 6 \\ -3 & -45 & -19 \end{bmatrix}$, $\mathbf{b}_1 = \begin{bmatrix} -3 \\ 1 \\ -3 \end{bmatrix}$,

$\mathbf{b}_2 = \begin{bmatrix} -2 \\ 1 \\ -3 \end{bmatrix}$, $\mathbf{b}_3 = \begin{bmatrix} 3 \\ -1 \\ 0 \end{bmatrix}$

T 35. Let T be the transformation whose standard matrix is given below. Find a basis for $\mathbb{R}^4$ with the property that $[T]_{\mathcal{B}}$ is diagonal.

$$A = \begin{bmatrix} 15 & -66 & -44 & -33 \\ 0 & 13 & 21 & -15 \\ 1 & -15 & -21 & 12 \\ 2 & -18 & -22 & 8 \end{bmatrix}$$

Solutions to Practice Problems

1. Let $\mathbf{p}(t) = a_0 + a_1 t + a_2 t^2$ and compute

$$[T(\mathbf{p})]_{\mathcal{B}} = [T]_{\mathcal{B}}[\mathbf{p}]_{\mathcal{B}} = \begin{bmatrix} 3 & 4 & 0 \\ 0 & 5 & -1 \\ 1 & -2 & 7 \end{bmatrix} \begin{bmatrix} a_0 \\ a_1 \\ a_2 \end{bmatrix} = \begin{bmatrix} 3a_0 + 4a_1 \\ 5a_1 - a_2 \\ a_0 - 2a_1 + 7a_2 \end{bmatrix}$$

So $T(\mathbf{p}) = (3a_0 + 4a_1) + (5a_1 - a_2)t + (a_0 - 2a_1 + 7a_2)t^2$.

2. a. $A = (I)^{-1}AI$, so A is similar to A.

   b. By hypothesis, there exist invertible matrices P and Q with the property that $B = P^{-1}AP$ and $C = Q^{-1}BQ$. Substitute the formula for B into the formula for C, and use a fact about the inverse of a product:

$$C = Q^{-1}BQ = Q^{-1}(P^{-1}AP)Q = (PQ)^{-1}A(PQ)$$

   This equation has the proper form to show that A is similar to C.

5.5 Complex Eigenvalues

Since the characteristic equation of an $n \times n$ matrix involves a polynomial of degree n, the equation always has exactly n roots, counting multiplicities, *provided that possibly complex roots are included*. This section shows that if the characteristic equation of a real matrix A has some complex roots, then these roots provide critical information about A. The key is to let A act on the space $\mathbb{C}^n$ of n-tuples of complex numbers.[1]

Our interest in $\mathbb{C}^n$ does not arise from a desire to "generalize" the results of the earlier chapters, although that would in fact open up significant new applications of linear algebra.[2] Rather, this study of complex eigenvalues is essential in order to uncover "hidden" information about certain matrices with real entries that arise in a variety of real-life problems. Such problems include many real dynamical systems that involve periodic motion, vibration, or some type of rotation in space.

The matrix eigenvalue–eigenvector theory already developed for $\mathbb{R}^n$ applies equally well to $\mathbb{C}^n$. So a complex scalar λ satisfies $\det(A - \lambda I) = 0$ if and only if there is a nonzero vector $\mathbf{x}$ in $\mathbb{C}^n$ such that $A\mathbf{x} = \lambda\mathbf{x}$. We call λ a (**complex**) **eigenvalue** and $\mathbf{x}$ a (**complex**) **eigenvector** corresponding to λ.

EXAMPLE 1 If $A = \begin{bmatrix} 0 & -1 \\ 1 & 0 \end{bmatrix}$, then the linear transformation $\mathbf{x} \mapsto A\mathbf{x}$ on $\mathbb{R}^2$ rotates the plane counterclockwise through a quarter-turn. The action of A is periodic, since after four quarter-turns, a vector is back where it started. Obviously, no nonzero vector is mapped into a multiple of itself, so A has no eigenvectors in $\mathbb{R}^2$ and hence no real eigenvalues. In fact, the characteristic equation of A is

$$\lambda^2 + 1 = 0$$

---

[1] Refer to Appendix B for a brief discussion of complex numbers. Matrix algebra and concepts about real vector spaces carry over to the case with complex entries and scalars. In particular, $A(c\mathbf{x} + d\mathbf{y}) = cA\mathbf{x} + dA\mathbf{y}$, for A an $m \times n$ matrix with complex entries, $\mathbf{x}, \mathbf{y}$ in $\mathbb{C}^n$, and c, d in $\mathbb{C}$.

[2] A second course in linear algebra often discusses such topics. They are of particular importance in electrical engineering.

The only roots are complex: $\lambda = i$ and $\lambda = -i$. However, if we permit A to act on $\mathbb{C}^2$, then

$$\begin{bmatrix} 0 & -1 \\ 1 & 0 \end{bmatrix}\begin{bmatrix} 1 \\ -i \end{bmatrix} = \begin{bmatrix} i \\ 1 \end{bmatrix} = i\begin{bmatrix} 1 \\ -i \end{bmatrix}$$

$$\begin{bmatrix} 0 & -1 \\ 1 & 0 \end{bmatrix}\begin{bmatrix} 1 \\ i \end{bmatrix} = \begin{bmatrix} -i \\ 1 \end{bmatrix} = -i\begin{bmatrix} 1 \\ i \end{bmatrix}$$

Thus i and $-i$ are eigenvalues, with $\begin{bmatrix} 1 \\ -i \end{bmatrix}$ and $\begin{bmatrix} 1 \\ i \end{bmatrix}$ as corresponding eigenvectors. (A method for *finding* complex eigenvectors is discussed in Example 2.)   ∎

The main focus of this section will be on the matrix in the next example.

EXAMPLE 2   Let $A = \begin{bmatrix} .5 & -.6 \\ .75 & 1.1 \end{bmatrix}$. Find the eigenvalues of A, and find a basis for each eigenspace.

SOLUTION   The characteristic equation of A is

$$0 = \det\begin{bmatrix} .5 - \lambda & -.6 \\ .75 & 1.1 - \lambda \end{bmatrix} = (.5 - \lambda)(1.1 - \lambda) - (-.6)(.75)$$
$$= \lambda^2 - 1.6\lambda + 1$$

From the quadratic formula, $\lambda = \frac{1}{2}[1.6 \pm \sqrt{(-1.6)^2 - 4}] = .8 \pm .6i$. For the eigenvalue $\lambda = .8 - .6i$, construct

$$A - (.8 - .6i)I = \begin{bmatrix} .5 & -.6 \\ .75 & 1.1 \end{bmatrix} - \begin{bmatrix} .8 - .6i & 0 \\ 0 & .8 - .6i \end{bmatrix}$$
$$= \begin{bmatrix} -.3 + .6i & -.6 \\ .75 & .3 + .6i \end{bmatrix} \tag{1}$$

Row reduction of the usual augmented matrix is quite unpleasant by hand because of the complex arithmetic. However, here is a nice observation that really simplifies matters: Since $.8 - .6i$ is an eigenvalue, the system

$$\begin{aligned} (-.3 + .6i)x_1 - \quad\quad .6x_2 &= 0 \\ .75x_1 + (.3 + .6i)x_2 &= 0 \end{aligned} \tag{2}$$

has a nontrivial solution (with x_1 and x_2 possibly complex numbers). Therefore, *both equations in (2) determine the same relationship between x_1 and x_2, and either equation can be used to express one variable in terms of the other.*[3]
   The second equation in (2) leads to

$$.75x_1 = (-.3 - .6i)x_2$$
$$x_1 = (-.4 - .8i)x_2$$

Choose $x_2 = 5$ to eliminate the decimals, and obtain $x_1 = -2 - 4i$. A basis for the eigenspace corresponding to $\lambda = .8 - .6i$ is

$$\mathbf{v}_1 = \begin{bmatrix} -2 - 4i \\ 5 \end{bmatrix}$$

---

[3] Another way to see this is to realize that the matrix in equation (1) is not invertible, so its rows are linearly dependent (as vectors in $\mathbb{C}^2$), and hence one row is a (complex) multiple of the other.

Analogous calculations for $\lambda = .8 + .6i$ produce the eigenvector

$$\mathbf{v}_2 = \begin{bmatrix} -2 + 4i \\ 5 \end{bmatrix}$$

As a check on the work, compute

$$A\mathbf{v}_2 = \begin{bmatrix} .5 & -.6 \\ .75 & 1.1 \end{bmatrix} \begin{bmatrix} -2 + 4i \\ 5 \end{bmatrix} = \begin{bmatrix} -4 + 2i \\ 4 + 3i \end{bmatrix} = (.8 + .6i)\mathbf{v}_2 \qquad \blacksquare$$

Surprisingly, the matrix A in Example 2 determines a transformation $\mathbf{x} \mapsto A\mathbf{x}$ that is essentially a rotation. This fact becomes evident when appropriate points are plotted, as illustrated in Figure 1.

EXAMPLE 3   One way to see how multiplication by the matrix A in Example 2 affects points is to plot an arbitrary initial point—say, $\mathbf{x}_0 = (2, 0)$—and then to plot successive images of this point under repeated multiplications by A. That is, plot

$$\mathbf{x}_1 = A\mathbf{x}_0 = \begin{bmatrix} .5 & -.6 \\ .75 & 1.1 \end{bmatrix} \begin{bmatrix} 2 \\ 0 \end{bmatrix} = \begin{bmatrix} 1.0 \\ 1.5 \end{bmatrix}$$

$$\mathbf{x}_2 = A\mathbf{x}_1 = \begin{bmatrix} .5 & -.6 \\ .75 & 1.1 \end{bmatrix} \begin{bmatrix} 1.0 \\ 1.5 \end{bmatrix} = \begin{bmatrix} -.4 \\ 2.4 \end{bmatrix}$$

$$\mathbf{x}_3 = A\mathbf{x}_2, \ldots$$

Figure 1 shows $\mathbf{x}_0, \ldots, \mathbf{x}_8$ as larger dots. The smaller dots are the locations of $\mathbf{x}_9, \ldots,$ $\mathbf{x}_{100}$. The sequence lies along an elliptical orbit. $\qquad \blacksquare$

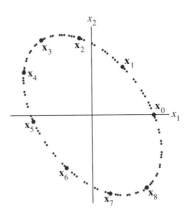

FIGURE 1 Iterates of a point $\mathbf{x}_0$ under the action of a matrix with a complex eigenvalue.

Of course, Figure 1 does not explain *why* the rotation occurs. The secret to the rotation is hidden in the real and imaginary parts of a complex eigenvector.

Real and Imaginary Parts of Vectors

The complex conjugate of a complex vector $\mathbf{x}$ in $\mathbb{C}^n$ is the vector $\bar{\mathbf{x}}$ in $\mathbb{C}^n$ whose entries are the complex conjugates of the entries in $\mathbf{x}$. The **real** and **imaginary parts** of a complex vector **x** are the vectors Re **x** and Im **x** in $\mathbb{R}^n$ formed from the real and imaginary parts of the entries of **x**. Thus,

$$\mathbf{x} = \text{Re } \mathbf{x} + i\text{Im } \mathbf{x} \tag{3}$$

EXAMPLE 4   If $\mathbf{x} = \begin{bmatrix} 3-i \\ i \\ 2+5i \end{bmatrix} = \begin{bmatrix} 3 \\ 0 \\ 2 \end{bmatrix} + i \begin{bmatrix} -1 \\ 1 \\ 5 \end{bmatrix}$, then

$$\text{Re } \mathbf{x} = \begin{bmatrix} 3 \\ 0 \\ 2 \end{bmatrix}, \quad \text{Im } \mathbf{x} = \begin{bmatrix} -1 \\ 1 \\ 5 \end{bmatrix}, \quad \text{and} \quad \overline{\mathbf{x}} = \begin{bmatrix} 3 \\ 0 \\ 2 \end{bmatrix} - i \begin{bmatrix} -1 \\ 1 \\ 5 \end{bmatrix} = \begin{bmatrix} 3+i \\ -i \\ 2-5i \end{bmatrix} \quad \blacksquare$$

If B is an $m \times n$ matrix with possibly complex entries, then $\overline{B}$ denotes the matrix whose entries are the complex conjugates of the entries in B. Let r be a complex number and **x** any vector. Properties of conjugates for complex numbers carry over to complex matrix algebra:

$$\overline{r\mathbf{x}} = \overline{r}\,\overline{\mathbf{x}}, \quad \overline{B\mathbf{x}} = \overline{B}\,\overline{\mathbf{x}}, \quad \overline{BC} = \overline{B}\,\overline{C}, \quad \text{and} \quad \overline{rB} = \overline{r}\,\overline{B}$$

Eigenvalues and Eigenvectors of a Real Matrix That Acts on $\mathbb{C}^n$

Let A be an $n \times n$ matrix whose entries are real. Then $\overline{A\mathbf{x}} = \overline{A}\overline{\mathbf{x}} = A\overline{\mathbf{x}}$. If λ is an eigenvalue of A and **x** is a corresponding eigenvector in $\mathbb{C}^n$, then

$$A\overline{\mathbf{x}} = \overline{A\mathbf{x}} = \overline{\lambda\mathbf{x}} = \overline{\lambda}\,\overline{\mathbf{x}}$$

Hence $\overline{\lambda}$ is also an eigenvalue of A, with $\overline{\mathbf{x}}$ a corresponding eigenvector. This shows that *when A is real, its complex eigenvalues occur in conjugate pairs*. (Here and elsewhere, we use the term *complex eigenvalue* to refer to an eigenvalue $\lambda = a + bi$, with $b \neq 0$.)

EXAMPLE 5   The eigenvalues of the real matrix in Example 2 are complex conjugates, namely $.8 - .6i$ and $.8 + .6i$. The corresponding eigenvectors found in Example 2 are also conjugates:

$$\mathbf{v}_1 = \begin{bmatrix} -2-4i \\ 5 \end{bmatrix} \quad \text{and} \quad \mathbf{v}_2 = \begin{bmatrix} -2+4i \\ 5 \end{bmatrix} = \overline{\mathbf{v}}_1 \quad \blacksquare$$

The next example provides the basic "building block" for all real 2×2 matrices with complex eigenvalues.

EXAMPLE 6   If $C = \begin{bmatrix} a & -b \\ b & a \end{bmatrix}$, where a and b are real and not both zero, then the eigenvalues of C are $\lambda = a \pm bi$. (See the Practice Problem at the end of this section.) Also, if $r = |\lambda| = \sqrt{a^2 + b^2}$, then

$$C = r\begin{bmatrix} a/r & -b/r \\ b/r & a/r \end{bmatrix} = \begin{bmatrix} r & 0 \\ 0 & r \end{bmatrix}\begin{bmatrix} \cos\varphi & -\sin\varphi \\ \sin\varphi & \cos\varphi \end{bmatrix}$$

where φ is the angle between the positive x-axis and the ray from $(0,0)$ through (a, b). See Figure 2 and Appendix B. The angle φ is called the *argument* of $\lambda = a + bi$. Thus

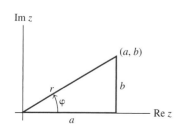

FIGURE 2

the transformation $\mathbf{x} \mapsto C\mathbf{x}$ may be viewed as the composition of a rotation through the angle φ and a scaling by $|\lambda|$ (see Figure 3). ■

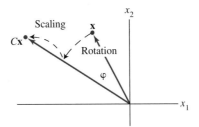

FIGURE 3 A rotation followed by a scaling.

Finally, we are ready to uncover the rotation that is hidden within a real matrix having a complex eigenvalue.

EXAMPLE 7 Let $A = \begin{bmatrix} .5 & -.6 \\ .75 & 1.1 \end{bmatrix}$, $\lambda = .8 - .6i$, and $\mathbf{v}_1 = \begin{bmatrix} -2 - 4i \\ 5 \end{bmatrix}$, as in Example 2. Also, let P be the 2×2 real matrix, described in Theorem 9,

$$P = \begin{bmatrix} \operatorname{Re} \mathbf{v}_1 & \operatorname{Im} \mathbf{v}_1 \end{bmatrix} = \begin{bmatrix} -2 & -4 \\ 5 & 0 \end{bmatrix}$$

and let

$$C = P^{-1}AP = \frac{1}{20} \begin{bmatrix} 0 & 4 \\ -5 & -2 \end{bmatrix} \begin{bmatrix} .5 & -.6 \\ .75 & 1.1 \end{bmatrix} \begin{bmatrix} -2 & -4 \\ 5 & 0 \end{bmatrix} = \begin{bmatrix} .8 & -.6 \\ .6 & .8 \end{bmatrix}$$

By Example 6, C is a pure rotation because $|\lambda|^2 = (.8)^2 + (.6)^2 = 1$. From $C = P^{-1}AP$, we obtain

$$A = PCP^{-1} = P \begin{bmatrix} .8 & -.6 \\ .6 & .8 \end{bmatrix} P^{-1}$$

Here is the rotation "inside" A! The matrix P provides a change of variable, say, $\mathbf{x} = P\mathbf{u}$. The action of A amounts to a change of variable from $\mathbf{x}$ to $\mathbf{u}$, followed by a rotation, and then a return to the original variable. See Figure 4. The rotation produces an ellipse, as in Figure 1, instead of a circle, because the coordinate system determined by the columns of P is not rectangular and does not have equal unit lengths on the two axes. ■

FIGURE 4 Rotation due to a complex eigenvalue.

The next theorem shows that the calculations in Example 7 can be carried out for any 2×2 real matrix A having a complex eigenvalue λ. The proof uses the fact that if the entries in A are real, then $A(\operatorname{Re} \mathbf{x}) = \operatorname{Re}(A\mathbf{x})$ and $A(\operatorname{Im} \mathbf{x}) = \operatorname{Im}(A\mathbf{x})$, and if $\mathbf{x}$ is an eigenvector for a complex eigenvalue, then $\operatorname{Re} \mathbf{x}$ and $\operatorname{Im} \mathbf{x}$ are linearly independent in $\mathbb{R}^2$. (See Exercises 29 and 30.) The details are omitted.

THEOREM 9

Let A be a real 2×2 matrix with a complex eigenvalue $\lambda = a - bi$ ($b \neq 0$) and an associated eigenvector $\mathbf{v}$ in $\mathbb{C}^2$. Then

$$A = PCP^{-1}, \quad \text{where} \quad P = [\,\text{Re}\,\mathbf{v} \quad \text{Im}\,\mathbf{v}\,] \quad \text{and} \quad C = \begin{bmatrix} a & -b \\ b & a \end{bmatrix}$$

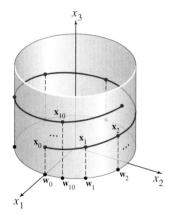

FIGURE 5

Iterates of two points under the action of a 3×3 matrix with a complex eigenvalue.

The phenomenon displayed in Example 7 persists in higher dimensions. For instance, if A is a 3×3 matrix with a complex eigenvalue, then there is a plane in $\mathbb{R}^3$ on which A acts as a rotation (possibly combined with scaling). Every vector in that plane is rotated into another point on the same plane. We say that the plane is **invariant** under A.

EXAMPLE 8 The matrix $A = \begin{bmatrix} .8 & -.6 & 0 \\ .6 & .8 & 0 \\ 0 & 0 & 1.07 \end{bmatrix}$ has eigenvalues $.8 \pm .6i$ and 1.07. Any vector $\mathbf{w}_0$ in the x_1x_2-plane (with third coordinate 0) is rotated by A into another point in the plane. Any vector $\mathbf{x}_0$ not in the plane has its x_3-coordinate multiplied by 1.07. The iterates of the points $\mathbf{w}_0 = (2, 0, 0)$ and $\mathbf{x}_0 = (2, 0, 1)$ under multiplication by A are shown in Figure 5. ∎

FIGURE 6

EXAMPLE 9 Many robots work by rotating at various joints, just as matrices with complex eigenvalues rotate points in space. Figure 6 illustrates a robot arm made using linear transformations, each with a pair of complex eigenvalues. In Project C, at the end of the chapter, you will be asked to find videos of robots on the web that use rotations as a key element of their functioning.

Practice Problem

Show that if a and b are real, then the eigenvalues of $A = \begin{bmatrix} a & -b \\ b & a \end{bmatrix}$ are $a \pm bi$, with corresponding eigenvectors $\begin{bmatrix} 1 \\ -i \end{bmatrix}$ and $\begin{bmatrix} 1 \\ i \end{bmatrix}$.

5.5 Exercises

Let each matrix in Exercises 1–6 act on $\mathbb{C}^2$. Find the eigenvalues and a basis for each eigenspace in $\mathbb{C}^2$.

1. $\begin{bmatrix} 1 & -2 \\ 1 & 3 \end{bmatrix}$

2. $\begin{bmatrix} -1 & -1 \\ 5 & -5 \end{bmatrix}$

3. $\begin{bmatrix} 1 & 5 \\ -2 & 3 \end{bmatrix}$

4. $\begin{bmatrix} -3 & -1 \\ 2 & -5 \end{bmatrix}$

5. $\begin{bmatrix} 0 & 1 \\ -8 & 4 \end{bmatrix}$

6. $\begin{bmatrix} 4 & 3 \\ -3 & 4 \end{bmatrix}$

In Exercises 7–12, use Example 6 to list the eigenvalues of A. In each case, the transformation $\mathbf{x} \mapsto A\mathbf{x}$ is the composition of a rotation and a scaling. Give the angle φ of the rotation, where $-\pi < \varphi \leq \pi$, and give the scale factor r.

7. $\begin{bmatrix} \sqrt{3} & -1 \\ 1 & \sqrt{3} \end{bmatrix}$

8. $\begin{bmatrix} \sqrt{3} & 3 \\ -3 & \sqrt{3} \end{bmatrix}$

9. $\begin{bmatrix} -\sqrt{3}/2 & 1/2 \\ -1/2 & -\sqrt{3}/2 \end{bmatrix}$

10. $\begin{bmatrix} 3 & 3 \\ -3 & 3 \end{bmatrix}$

11. $\begin{bmatrix} .1 & .1 \\ -.1 & .1 \end{bmatrix}$

12. $\begin{bmatrix} 0 & 4 \\ -4 & 0 \end{bmatrix}$

In Exercises 13–20, find an invertible matrix P and a matrix C of the form $\begin{bmatrix} a & -b \\ b & a \end{bmatrix}$ such that the given matrix has the form $A = PCP^{-1}$. For Exercises 13–16, use information from Exercises 1–4.

13. $\begin{bmatrix} 1 & -2 \\ 1 & 3 \end{bmatrix}$

14. $\begin{bmatrix} -1 & -1 \\ 5 & -5 \end{bmatrix}$

15. $\begin{bmatrix} 1 & 5 \\ -2 & 3 \end{bmatrix}$

16. $\begin{bmatrix} -3 & -1 \\ 2 & -5 \end{bmatrix}$

17. $\begin{bmatrix} 1 & -.8 \\ 4 & -2.2 \end{bmatrix}$

18. $\begin{bmatrix} 1 & -1 \\ .4 & .6 \end{bmatrix}$

19. $\begin{bmatrix} 1.52 & -.7 \\ .56 & .4 \end{bmatrix}$

20. $\begin{bmatrix} -1.64 & -2.4 \\ 1.92 & 2.2 \end{bmatrix}$

21. In Example 2, solve the first equation in (2) for x_2 in terms of x_1, and from that produce the eigenvector $\mathbf{y} = \begin{bmatrix} 2 \\ -1 + 2i \end{bmatrix}$ for the matrix A. Show that this $\mathbf{y}$ is a (complex) multiple of the vector $\mathbf{v}_1$ used in Example 2.

22. Let A be a complex (or real) $n \times n$ matrix, and let $\mathbf{x}$ in $\mathbb{C}^n$ be an eigenvector corresponding to an eigenvalue λ in $\mathbb{C}$. Show that for each nonzero complex scalar μ, the vector $\mu\mathbf{x}$ is an eigenvector of A.

In Exercises 23–26, A is a 2×2 matrix with real entries, and $\mathbf{x}$ is a vector in $\mathbb{R}^2$. Mark each statement True or False (**T/F**). Justify each answer.

23. (T/F) The matrix A can have one real and one complex eigenvalue.

24. (T/F) The points $A\mathbf{x}, A^2\mathbf{x}, A^3\mathbf{x}, \ldots$ always lie on the same circle.

25. (T/F) The matrix A always has two eigenvalues, but sometimes they have algebraic multiplicity 2 or are complex numbers.

26. (T/F) If the matrix A has two complex eigenvalues, then it also has two linearly independent real eigenvectors.

Chapter 7 will focus on matrices A with the property that $A^T = A$. Exercises 27 and 28 show that every eigenvalue of such a matrix is necessarily real.

27. Let A be an $n \times n$ real matrix with the property that $A^T = A$, let $\mathbf{x}$ be any vector in $\mathbb{C}^n$, and let $q = \overline{\mathbf{x}}^T A \mathbf{x}$. The equalities below show that q is a real number by verifying that $\overline{q} = q$. Give a reason for each step.

$$\overline{q} = \overline{\overline{\mathbf{x}}^T A \mathbf{x}} = \mathbf{x}^T \overline{A\mathbf{x}} = \mathbf{x}^T A\overline{\mathbf{x}} = (\mathbf{x}^T A\overline{\mathbf{x}})^T = \overline{\mathbf{x}}^T A^T \mathbf{x} = q$$
$$\quad\ \text{(a)}\qquad \text{(b)}\qquad \text{(c)}\qquad \text{(d)}\qquad \text{(e)}$$

28. Let A be an $n \times n$ real matrix with the property that $A^T = A$. Show that if $A\mathbf{x} = \lambda\mathbf{x}$ for some nonzero vector $\mathbf{x}$ in $\mathbb{C}^n$, then, in fact, λ is real and the real part of $\mathbf{x}$ is an eigenvector of A. [*Hint:* Compute $\overline{\mathbf{x}}^T A\mathbf{x}$, and use Exercise 27. Also, examine the real and imaginary parts of $A\mathbf{x}$.]

29. Let A be a real $n \times n$ matrix, and let $\mathbf{x}$ be a vector in $\mathbb{C}^n$. Show that $\text{Re}(A\mathbf{x}) = A(\text{Re}\,\mathbf{x})$ and $\text{Im}(A\mathbf{x}) = A(\text{Im}\,\mathbf{x})$.

30. Let A be a real 2×2 matrix with a complex eigenvalue $\lambda = a - bi$ $(b \neq 0)$ and an associated eigenvector $\mathbf{v}$ in $\mathbb{C}^2$.

a. Show that $A(\text{Re}\,\mathbf{v}) = a\,\text{Re}\,\mathbf{v} + b\,\text{Im}\,\mathbf{v}$ and $A(\text{Im}\,\mathbf{v}) = -b\,\text{Re}\,\mathbf{v} + a\,\text{Im}\,\mathbf{v}$. [*Hint:* Write $\mathbf{v} = \text{Re}\,\mathbf{v} + i\,\text{Im}\,\mathbf{v}$, and compute $A\mathbf{v}$.]

b. Verify that if P and C are given as in Theorem 9, then $AP = PC$.

T In Exercises 31 and 32, find a factorization of the given matrix A in the form $A = PCP^{-1}$, where C is a block-diagonal matrix with 2×2 blocks of the form shown in Example 6. (For each conjugate pair of eigenvalues, use the real and imaginary parts of one eigenvector in $\mathbb{C}^4$ to create two columns of P.)

31. $\begin{bmatrix} .7 & 1.1 & 2.0 & 1.7 \\ -2.0 & -4.0 & -8.6 & -7.4 \\ 0 & -.5 & -1.0 & -1.0 \\ 1.0 & 2.8 & 6.0 & 5.3 \end{bmatrix}$

32. $\begin{bmatrix} -1.4 & -2.0 & -2.0 & -2.0 \\ -1.3 & -.8 & -.1 & -.6 \\ .3 & -1.9 & -1.6 & -1.4 \\ 2.0 & 3.3 & 2.3 & 2.6 \end{bmatrix}$

Solution to Practice Problem

Remember that it is easy to test whether a vector is an eigenvector. There is no need to examine the characteristic equation. Compute

$$A\mathbf{x} = \begin{bmatrix} a & -b \\ b & a \end{bmatrix} \begin{bmatrix} 1 \\ -i \end{bmatrix} = \begin{bmatrix} a + bi \\ b - ai \end{bmatrix} = (a + bi) \begin{bmatrix} 1 \\ -i \end{bmatrix}$$

Thus $\begin{bmatrix} 1 \\ -i \end{bmatrix}$ is an eigenvector corresponding to $\lambda = a + bi$. From the discussion in this section, $\begin{bmatrix} 1 \\ i \end{bmatrix}$ must be an eigenvector corresponding to $\overline{\lambda} = a - bi$.

5.6 Discrete Dynamical Systems

Eigenvalues and eigenvectors provide the key to understanding the long-term behavior, or *evolution*, of a dynamical system described by a difference equation $\mathbf{x}_{k+1} = A\mathbf{x}_k$. Such an equation was used to model population movement in Section 1.10, and will be used in various Markov chains in Section 5.9 and the spotted owl population in the introductory example for this chapter. The vectors $\mathbf{x}_k$ give information about the system as time (denoted by k) passes, where k is a nonnegative integer. In the spotted owl example, for instance, $\mathbf{x}_k$ listed the numbers of owls in three age classes at time k.

The applications in this section focus on ecological problems because they are easier to state and explain than, say, problems in physics or engineering. However, dynamical systems arise in many scientific fields. For instance, standard undergraduate courses in control systems discuss several aspects of dynamical systems. The modern *state-space* design method in such courses relies heavily on matrix algebra.[1] The *steady-state response* of a control system is the engineering equivalent of what we call here the "long-term behavior" of the dynamical system $\mathbf{x}_{k+1} = A\mathbf{x}_k$.

Until Example 6, we assume that A is diagonalizable, with n linearly independent eigenvectors, $\mathbf{v}_1, \ldots, \mathbf{v}_n$, and corresponding eigenvalues, $\lambda_1, \ldots, \lambda_n$. For convenience, assume the eigenvectors are arranged so that $|\lambda_1| \geq |\lambda_2| \geq \cdots \geq |\lambda_n|$. Since $\{\mathbf{v}_1, \ldots, \mathbf{v}_n\}$ is a basis for $\mathbb{R}^n$, any initial vector $\mathbf{x}_0$ can be written uniquely as

$$\mathbf{x}_0 = c_1 \mathbf{v}_1 + \cdots + c_n \mathbf{v}_n \tag{1}$$

This *eigenvector decomposition* of $\mathbf{x}_0$ determines what happens to the sequence $\{\mathbf{x}_k\}$. The next calculation generalizes the simple case examined in Example 5 of Section 5.2. Since the $\mathbf{v}_i$ are eigenvectors,

$$\mathbf{x}_1 = A\mathbf{x}_0 = c_1 A\mathbf{v}_1 + \cdots + c_n A\mathbf{v}_n$$
$$= c_1 \lambda_1 \mathbf{v}_1 + \cdots + c_n \lambda_n \mathbf{v}_n$$

and

$$\mathbf{x}_2 = A\mathbf{x}_1 = c_1 \lambda_1 A\mathbf{v}_1 + \cdots + c_n \lambda_n A\mathbf{v}_n$$
$$= c_1 (\lambda_1)^2 \mathbf{v}_1 + \cdots + c_n (\lambda_n)^2 \mathbf{v}_n$$

---

[1] See G. F. Franklin, J. D. Powell, and A. Emami-Naeimi, *Feedback Control of Dynamic Systems*, 5th ed. (Upper Saddle River, NJ: Prentice-Hall, 2006). This undergraduate text has a nice introduction to dynamic models (Chapter 2). State-space design is covered in Chapters 7 and 8.

In general,

$$\mathbf{x}_k = c_1(\lambda_1)^k \mathbf{v}_1 + \cdots + c_n(\lambda_n)^k \mathbf{v}_n \qquad (k = 0, 1, 2, \ldots) \tag{2}$$

The examples that follow illustrate what can happen in (2) as $k \to \infty$.

A Predator–Prey System

Deep in the redwood forests of California, dusky-footed wood rats provide up to 80% of the diet for the spotted owl, the main predator of the wood rat. Example 1 uses a linear dynamical system to model the physical system of the owls and the rats. (Admittedly, the model is unrealistic in several respects, but it can provide a starting point for the study of more complicated nonlinear models used by environmental scientists.)

EXAMPLE 1 Denote the owl and wood rat populations at time k by $\mathbf{x}_k = \begin{bmatrix} O_k \\ R_k \end{bmatrix}$, where k is the time in months, O_k is the number of owls in the region studied, and R_k is the number of rats (measured in thousands). Suppose

$$\begin{aligned} O_{k+1} &= (.5)O_k + (.4)R_k \\ R_{k+1} &= -p \cdot O_k + (1.1)R_k \end{aligned} \tag{3}$$

where p is a positive parameter to be specified. The $(.5)O_k$ in the first equation says that with no wood rats for food, only half of the owls will survive each month, while the $(1.1)R_k$ in the second equation says that with no owls as predators, the rat population will grow by 10% per month. If rats are plentiful, the $(.4)R_k$ will tend to make the owl population rise, while the negative term $-p \cdot O_k$ measures the deaths of rats due to predation by owls. (In fact, $1000p$ is the average number of rats eaten by one owl in one month.) Determine the evolution of this system when the predation parameter p is .104.

SOLUTION When $p = .104$, the eigenvalues of the coefficient matrix $A = \begin{bmatrix} .5 & .4 \\ -p & 1.1 \end{bmatrix}$ for the equations in (3) turn out to be $\lambda_1 = 1.02$ and $\lambda_2 = .58$. Corresponding eigenvectors are

$$\mathbf{v}_1 = \begin{bmatrix} 10 \\ 13 \end{bmatrix}, \qquad \mathbf{v}_2 = \begin{bmatrix} 5 \\ 1 \end{bmatrix}$$

An initial $\mathbf{x}_0$ can be written as $\mathbf{x}_0 = c_1\mathbf{v}_1 + c_2\mathbf{v}_2$. Then, for $k \geq 0$,

$$\begin{aligned} \mathbf{x}_k &= c_1(1.02)^k \mathbf{v}_1 + c_2(.58)^k \mathbf{v}_2 \\ &= c_1(1.02)^k \begin{bmatrix} 10 \\ 13 \end{bmatrix} + c_2(.58)^k \begin{bmatrix} 5 \\ 1 \end{bmatrix} \end{aligned}$$

As $k \to \infty$, $(.58)^k$ rapidly approaches zero. Assume $c_1 > 0$. Then, for all sufficiently large k, $\mathbf{x}_k$ is approximately the same as $c_1(1.02)^k \mathbf{v}_1$, and we write

$$\mathbf{x}_k \approx c_1(1.02)^k \begin{bmatrix} 10 \\ 13 \end{bmatrix} \tag{4}$$

The approximation in (4) improves as k increases, and so for large k,

$$\mathbf{x}_{k+1} \approx c_1(1.02)^{k+1} \begin{bmatrix} 10 \\ 13 \end{bmatrix} = (1.02)c_1(1.02)^k \begin{bmatrix} 10 \\ 13 \end{bmatrix} \approx 1.02\mathbf{x}_k \tag{5}$$

The approximation in (5) says that eventually both entries of $\mathbf{x}_k$ (the numbers of owls and rats) grow by a factor of almost 1.02 each month, a 2% monthly growth rate. By (4), $\mathbf{x}_k$ is approximately a multiple of (10, 13), so the entries in $\mathbf{x}_k$ are nearly in the same ratio as 10 to 13. That is, for every 10 owls there are about 13 thousand rats. ∎

Example 1 illustrates two general facts about a dynamical system $\mathbf{x}_{k+1} = A\mathbf{x}_k$ in which A is $n \times n$, its eigenvalues satisfy $|\lambda_1| \geq 1$ and $1 > |\lambda_j|$ for $j = 2, \ldots, n$, and $\mathbf{v}_1$ is an eigenvector corresponding to λ_1. If $\mathbf{x}_0$ is given by equation (1), with $c_1 \neq 0$, then for all sufficiently large k,

$$\mathbf{x}_{k+1} \approx \lambda_1 \mathbf{x}_k \tag{6}$$

and

$$\mathbf{x}_k \approx c_1 (\lambda_1)^k \mathbf{v}_1 \tag{7}$$

The approximations in (6) and (7) can be made as close as desired by taking k sufficiently large. By (6), the $\mathbf{x}_k$ eventually grow almost by a factor of λ_1 each time, so λ_1 determines the eventual growth rate of the system. Also, by (7), the ratio of any two entries in $\mathbf{x}_k$ (for large k) is nearly the same as the ratio of the corresponding entries in $\mathbf{v}_1$. The case in which $\lambda_1 = 1$ is illustrated in Example 5 in Section 5.2.

Graphical Description of Solutions

When A is 2×2, algebraic calculations can be supplemented by a geometric description of a system's evolution. We can view the equation $\mathbf{x}_{k+1} = A\mathbf{x}_k$ as a description of what happens to an initial point $\mathbf{x}_0$ in $\mathbb{R}^2$ as it is transformed repeatedly by the mapping $\mathbf{x} \mapsto A\mathbf{x}$. The graph of $\mathbf{x}_0, \mathbf{x}_1, \ldots$ is called a **trajectory** of the dynamical system.

EXAMPLE 2 Plot several trajectories of the dynamical system $\mathbf{x}_{k+1} = A\mathbf{x}_k$, when

$$A = \begin{bmatrix} .80 & 0 \\ 0 & .64 \end{bmatrix}$$

SOLUTION The eigenvalues of A are .8 and .64, with eigenvectors $\mathbf{v}_1 = \begin{bmatrix} 1 \\ 0 \end{bmatrix}$ and $\mathbf{v}_2 = \begin{bmatrix} 0 \\ 1 \end{bmatrix}$. If $\mathbf{x}_0 = c_1 \mathbf{v}_1 + c_2 \mathbf{v}_2$, then

$$\mathbf{x}_k = c_1 (.8)^k \begin{bmatrix} 1 \\ 0 \end{bmatrix} + c_2 (.64)^k \begin{bmatrix} 0 \\ 1 \end{bmatrix}$$

Of course, $\mathbf{x}_k$ tends to $\mathbf{0}$ because $(.8)^k$ and $(.64)^k$ both approach 0 as $k \to \infty$. But *the way* $\mathbf{x}_k$ goes toward $\mathbf{0}$ is interesting. Figure 1 shows the first few terms of several trajectories

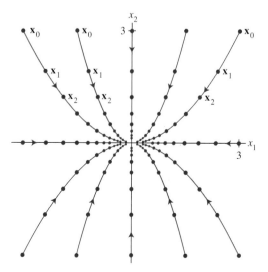

FIGURE 1 The origin as an attractor.

that begin at points on the boundary of the box with corners at $(\pm 3, \pm 3)$. The points on each trajectory are connected by a thin curve, to make the trajectory easier to see. ∎

In Example 2, the origin is called an **attractor** of the dynamical system because all trajectories tend toward $\mathbf{0}$. This occurs whenever both eigenvalues are less than 1 in magnitude. The direction of greatest attraction is along the line through $\mathbf{0}$ and the eigenvector $\mathbf{v}_2$ for the eigenvalue of smaller magnitude.

In the next example, both eigenvalues of A are larger than 1 in magnitude, and $\mathbf{0}$ is called a **repeller** of the dynamical system. All solutions of $\mathbf{x}_{k+1} = A\mathbf{x}_k$ except the (constant) zero solution are unbounded and tend away from the origin.[2]

EXAMPLE 3 Plot several typical solutions of the equation $\mathbf{x}_{k+1} = A\mathbf{x}_k$, where

$$A = \begin{bmatrix} 1.44 & 0 \\ 0 & 1.2 \end{bmatrix}$$

SOLUTION The eigenvalues of A are 1.44 and 1.2. If $\mathbf{x}_0 = \begin{bmatrix} c_1 \\ c_2 \end{bmatrix}$, then

$$\mathbf{x}_k = c_1 (1.44)^k \begin{bmatrix} 1 \\ 0 \end{bmatrix} + c_2 (1.2)^k \begin{bmatrix} 0 \\ 1 \end{bmatrix}$$

Both terms grow in size, but the first term grows faster. So the direction of greatest repulsion is the line through $\mathbf{0}$ and the eigenvector for the eigenvalue of larger magnitude. Figure 2 shows several trajectories that begin at points quite close to $\mathbf{0}$. ∎

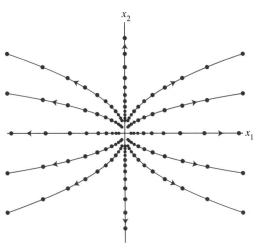

FIGURE 2 The origin as a repeller.

In the next example, $\mathbf{0}$ is called a **saddle point** because the origin attracts solutions from some directions and repels them in other directions. This occurs whenever one eigenvalue is greater than 1 in magnitude and the other is less than 1 in magnitude. The direction of greatest attraction is determined by an eigenvector for the eigenvalue of smaller magnitude. The direction of greatest repulsion is determined by an eigenvector for the eigenvalue of greater magnitude.

---

[2] The origin is the only possible attractor or repeller in a *linear* dynamical system, but there can be multiple attractors and repellers in a more general dynamical system for which the mapping $\mathbf{x}_k \mapsto \mathbf{x}_{k+1}$ is not linear. In such a system, attractors and repellers are defined in terms of the eigenvalues of a special matrix (with variable entries) called the *Jacobian matrix* of the system.

EXAMPLE 4   Plot several typical solutions of the equation $\mathbf{y}_{k+1} = D\mathbf{y}_k$, where

$$D = \begin{bmatrix} 2.0 & 0 \\ 0 & 0.5 \end{bmatrix}$$

(We write D and $\mathbf{y}$ here instead of A and $\mathbf{x}$ because this example will be used later.) Show that a solution $\{\mathbf{y}_k\}$ is unbounded if its initial point is not on the x_2-axis.

SOLUTION   The eigenvalues of D are 2 and .5. If $\mathbf{y}_0 = \begin{bmatrix} c_1 \\ c_2 \end{bmatrix}$, then

$$\mathbf{y}_k = c_1 2^k \begin{bmatrix} 1 \\ 0 \end{bmatrix} + c_2 (.5)^k \begin{bmatrix} 0 \\ 1 \end{bmatrix} \tag{8}$$

If $\mathbf{y}_0$ is on the x_2-axis, then $c_1 = 0$ and $\mathbf{y}_k \to \mathbf{0}$ as $k \to \infty$. But if $\mathbf{y}_0$ is not on the x_2-axis, then the first term in the sum for $\mathbf{y}_k$ becomes arbitrarily large, and so $\{\mathbf{y}_k\}$ is unbounded. Figure 3 shows ten trajectories that begin near or on the x_2-axis. ∎

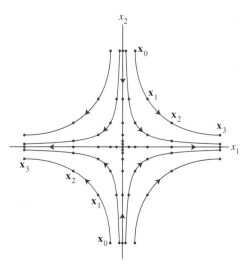

FIGURE 3   The origin as a saddle point.

Change of Variable

The preceding three examples involved diagonal matrices. To handle the nondiagonal case, we return for a moment to the $n \times n$ case in which eigenvectors of A form a basis $\{\mathbf{v}_1, \dots, \mathbf{v}_n\}$ for $\mathbb{R}^n$. Let $P = [\,\mathbf{v}_1 \;\; \cdots \;\; \mathbf{v}_n\,]$, and let D be the diagonal matrix with the corresponding eigenvalues on the diagonal. Given a sequence $\{\mathbf{x}_k\}$ satisfying $\mathbf{x}_{k+1} = A\mathbf{x}_k$, define a new sequence $\{\mathbf{y}_k\}$ by

$$\mathbf{y}_k = P^{-1}\mathbf{x}_k, \quad \text{or equivalently,} \quad \mathbf{x}_k = P\mathbf{y}_k$$

Substituting these relations into the equation $\mathbf{x}_{k+1} = A\mathbf{x}_k$ and using the fact that $A = PDP^{-1}$, we find that

$$P\mathbf{y}_{k+1} = AP\mathbf{y}_k = (PDP^{-1})P\mathbf{y}_k = PD\mathbf{y}_k$$

Left-multiplying both sides by P^{-1}, we obtain

$$\mathbf{y}_{k+1} = D\mathbf{y}_k$$

If we write $\mathbf{y}_k$ as $\mathbf{y}(k)$ and denote the entries in $\mathbf{y}(k)$ by $y_1(k), \ldots, y_n(k)$, then

$$
\begin{bmatrix} y_1(k+1) \\ y_2(k+1) \\ \vdots \\ y_n(k+1) \end{bmatrix} = \begin{bmatrix} \lambda_1 & 0 & \cdots & 0 \\ 0 & \lambda_2 & & \vdots \\ \vdots & & \ddots & 0 \\ 0 & \cdots & 0 & \lambda_n \end{bmatrix} \begin{bmatrix} y_1(k) \\ y_2(k) \\ \vdots \\ y_n(k) \end{bmatrix}
$$

The change of variable from $\mathbf{x}_k$ to $\mathbf{y}_k$ has *decoupled* the system of difference equations. The evolution of $y_1(k)$, for example, is unaffected by what happens to $y_2(k), \ldots, y_n(k)$, because $y_1(k+1) = \lambda_1 \cdot y_1(k)$ for each k.

The equation $\mathbf{x}_k = P\mathbf{y}_k$ says that $\mathbf{y}_k$ is the coordinate vector of $\mathbf{x}_k$ with respect to the eigenvector basis $\{\mathbf{v}_1, \ldots, \mathbf{v}_n\}$. We can decouple the system $\mathbf{x}_{k+1} = A\mathbf{x}_k$ by making calculations in the new eigenvector coordinate system. When $n = 2$, this amounts to using graph paper with axes in the directions of the two eigenvectors.

EXAMPLE 5   Show that the origin is a saddle point for solutions of $\mathbf{x}_{k+1} = A\mathbf{x}_k$, where

$$
A = \begin{bmatrix} 1.25 & -.75 \\ -.75 & 1.25 \end{bmatrix}
$$

Find the directions of greatest attraction and greatest repulsion.

SOLUTION   Using standard techniques, we find that A has eigenvalues 2 and .5, with corresponding eigenvectors $\mathbf{v}_1 = \begin{bmatrix} 1 \\ -1 \end{bmatrix}$ and $\mathbf{v}_2 = \begin{bmatrix} 1 \\ 1 \end{bmatrix}$, respectively. Since $|2| > 1$ and $|.5| < 1$, the origin is a saddle point of the dynamical system. If $\mathbf{x}_0 = c_1\mathbf{v}_1 + c_2\mathbf{v}_2$, then

$$
\mathbf{x}_k = c_1 2^k \mathbf{v}_1 + c_2 (.5)^k \mathbf{v}_2 \tag{9}
$$

This equation looks just like equation (8) in Example 4, with $\mathbf{v}_1$ and $\mathbf{v}_2$ in place of the standard basis.

On graph paper, draw axes through $\mathbf{0}$ and the eigenvectors $\mathbf{v}_1$ and $\mathbf{v}_2$. See Figure 4. Movement along these axes corresponds to movement along the standard axes in Figure 3. In Figure 4, the direction of greatest *repulsion* is the line through $\mathbf{0}$ and the

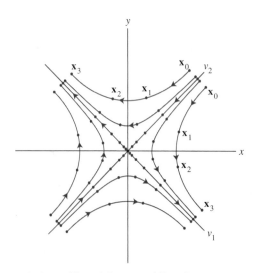

FIGURE 4   The origin as a saddle point.

eigenvector $\mathbf{v}_1$ whose eigenvalue is greater than 1 in magnitude. If $\mathbf{x}_0$ is on this line, the c_2 in (9) is zero and $\mathbf{x}_k$ moves quickly away from $\mathbf{0}$. The direction of greatest *attraction* is determined by the eigenvector $\mathbf{v}_2$ whose eigenvalue is less than 1 in magnitude.

A number of trajectories are shown in Figure 4. When this graph is viewed in terms of the eigenvector axes, the picture "looks" essentially the same as the picture in Figure 3. ∎

Complex Eigenvalues

When a real 2×2 matrix A has complex eigenvalues, A is not diagonalizable (when acting on $\mathbb{R}^2$), but the dynamical system $\mathbf{x}_{k+1} = A\mathbf{x}_k$ is easy to describe. Example 3 of Section 5.5 illustrated the case in which the eigenvalues have absolute value 1. The iterates of a point $\mathbf{x}_0$ spiral around the origin along an elliptical trajectory.

If A has two complex eigenvalues whose absolute value is greater than 1, then $\mathbf{0}$ is a repeller and iterates of $\mathbf{x}_0$ will spiral outward around the origin. If the absolute values of the complex eigenvalues are less than 1, then the origin is an attractor and the iterates of $\mathbf{x}_0$ spiral inward toward the origin, as in the following example.

EXAMPLE 6 It can be verified that the matrix

$$A = \begin{bmatrix} .8 & .5 \\ -.1 & 1.0 \end{bmatrix}$$

has eigenvalues $.9 \pm .2i$, with eigenvectors $\begin{bmatrix} 1 \mp 2i \\ 1 \end{bmatrix}$. Figure 5 shows three trajectories of the system $\mathbf{x}_{k+1} = A\mathbf{x}_k$, with initial vectors $\begin{bmatrix} 0 \\ 2.5 \end{bmatrix}$, $\begin{bmatrix} 3 \\ 0 \end{bmatrix}$, and $\begin{bmatrix} 0 \\ -2.5 \end{bmatrix}$. ∎

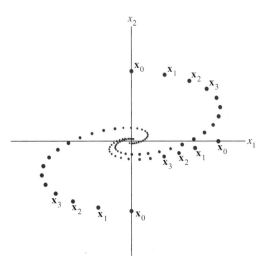

FIGURE 5 Rotation associated with complex eigenvalues.

Survival of the Spotted Owls

Recall from this chapter's introductory example that the spotted owl population in the Willow Creek area of California was modeled by a dynamical system $\mathbf{x}_{k+1} = A\mathbf{x}_k$ in which the entries in $\mathbf{x}_k = (j_k, s_k, a_k)$ listed the numbers of females (at time k) in the juvenile, subadult, and adult life stages, respectively, and A is the stage-matrix

$$A = \begin{bmatrix} 0 & 0 & .33 \\ .18 & 0 & 0 \\ 0 & .71 & .94 \end{bmatrix} \tag{10}$$

MATLAB shows that the eigenvalues of A are approximately $\lambda_1 = .98$, $\lambda_2 = -.02 + .21i$, and $\lambda_3 = -.02 - .21i$. Observe that all three eigenvalues are less than 1 in magnitude, because $|\lambda_2|^2 = |\lambda_3|^2 = (-.02)^2 + (.21)^2 = .0445$.

For the moment, let A act on the complex vector space $\mathbb{C}^3$. Then, because A has three distinct eigenvalues, the three corresponding eigenvectors are linearly independent and form a basis for $\mathbb{C}^3$. Denote the eigenvectors by $\mathbf{v}_1$, $\mathbf{v}_2$, and $\mathbf{v}_3$. Then the general solution of $\mathbf{x}_{k+1} = A\mathbf{x}_k$ (using vectors in $\mathbb{C}^3$) has the form

$$\mathbf{x}_k = c_1(\lambda_1)^k \mathbf{v}_1 + c_2(\lambda_2)^k \mathbf{v}_2 + c_3(\lambda_3)^k \mathbf{v}_3 \tag{11}$$

If $\mathbf{x}_0$ is a real initial vector, then $\mathbf{x}_1 = A\mathbf{x}_0$ is real because A is real. Similarly, the equation $\mathbf{x}_{k+1} = A\mathbf{x}_k$ shows that each $\mathbf{x}_k$ on the left side of (11) is real, even though it is expressed as a sum of complex vectors. However, each term on the right side of (11) is approaching the zero vector, because the eigenvalues are all less than 1 in magnitude. Therefore the real sequence $\mathbf{x}_k$ approaches the zero vector, too. Sadly, this model predicts that the spotted owls will eventually all perish.

Is there hope for the spotted owl? Recall from the introductory example that the 18% entry in the matrix A in (10) comes from the fact that although 60% of the juvenile owls live long enough to leave the nest and search for new home territories, only 30% of that group survive the search and find new home ranges. Search survival is strongly influenced by the number of clear-cut areas in the forest, which make the search more difficult and dangerous.

Some owl populations live in areas with few or no clear-cut areas. It may be that a larger percentage of the juvenile owls there survive and find new home ranges. Of course, the problem of the spotted owl is more complex than we have described, but the final example provides a happy ending to the story.

EXAMPLE 7 Suppose the search survival rate of the juvenile owls is 50%, so the $(2, 1)$-entry in the stage-matrix A in (10) is .3 instead of .18. What does the stage-matrix model predict about this spotted owl population?

SOLUTION Now the eigenvalues of A turn out to be approximately $\lambda_1 = 1.01$, $\lambda_2 = -.03 + .26i$, and $\lambda_3 = -.03 - .26i$. An eigenvector for λ_1 is approximately $\mathbf{v}_1 = (10, 3, 31)$. Let $\mathbf{v}_2$ and $\mathbf{v}_3$ be (complex) eigenvectors for λ_2 and λ_3. In this case, equation (11) becomes

$$\mathbf{x}_k = c_1(1.01)^k \mathbf{v}_1 + c_2(-.03 + .26i)^k \mathbf{v}_2 + c_3(-.03 - .26i)^k \mathbf{v}_3$$

As $k \to \infty$, the second two vectors tend to zero. So $\mathbf{x}_k$ becomes more and more like the (real) vector $c_1(1.01)^k \mathbf{v}_1$. The approximations in equations (6) and (7), following Example 1, apply here. Also, it can be shown that the constant c_1 in the initial

Further Reading: Franklin, G. F., J. D. Powell, and M. L. Workman. *Digital Control of Dynamic Systems*, 3rd ed. Reading, MA: Addison-Wesley, 1998; Sandefur, James T. *Discrete Dynamical Systems—Theory and Applications*. Oxford: Oxford University Press, 1990; Tuchinsky, Philip. *Management of a Buffalo Herd*, UMAP Module 207. Lexington, MA: COMAP, 1980.

decomposition of $\mathbf{x}_0$ is positive when the entries in $\mathbf{x}_0$ are nonnegative. Thus the owl population will grow slowly, with a long-term growth rate of 1.01. The eigenvector $\mathbf{v}_1$ describes the eventual distribution of the owls by life stages: for every 31 adults, there will be about 10 juveniles and 3 subadults. ∎

Practice Problems

1. The matrix A below has eigenvalues 1, $\frac{2}{3}$, and $\frac{1}{3}$, with corresponding eigenvectors $\mathbf{v}_1$, $\mathbf{v}_2$, and $\mathbf{v}_3$:

$$A = \frac{1}{9}\begin{bmatrix} 7 & -2 & 0 \\ -2 & 6 & 2 \\ 0 & 2 & 5 \end{bmatrix}, \quad \mathbf{v}_1 = \begin{bmatrix} -2 \\ 2 \\ 1 \end{bmatrix}, \quad \mathbf{v}_2 = \begin{bmatrix} 2 \\ 1 \\ 2 \end{bmatrix}, \quad \mathbf{v}_3 = \begin{bmatrix} 1 \\ 2 \\ -2 \end{bmatrix}$$

Find the general solution of the equation $\mathbf{x}_{k+1} = A\mathbf{x}_k$ if $\mathbf{x}_0 = \begin{bmatrix} 1 \\ 11 \\ -2 \end{bmatrix}$.

2. What happens to the sequence $\{\mathbf{x}_k\}$ in Practice Problem 1 as $k \to \infty$?

5.6 Exercises

1. Let A be a 2×2 matrix with eigenvalues 3 and $1/3$ and corresponding eigenvectors $\mathbf{v}_1 = \begin{bmatrix} 1 \\ 1 \end{bmatrix}$ and $\mathbf{v}_2 = \begin{bmatrix} -1 \\ 1 \end{bmatrix}$. Let $\{\mathbf{x}_k\}$ be a solution of the difference equation $\mathbf{x}_{k+1} = A\mathbf{x}_k$, $\mathbf{x}_0 = \begin{bmatrix} 9 \\ 1 \end{bmatrix}$.

   a. Compute $\mathbf{x}_1 = A\mathbf{x}_0$. [*Hint:* You do not need to know A itself.]

   b. Find a formula for $\mathbf{x}_k$ involving k and the eigenvectors $\mathbf{v}_1$ and $\mathbf{v}_2$.

2. Suppose the eigenvalues of a 3×3 matrix A are 3, $4/5$, and $3/5$, with corresponding eigenvectors $\begin{bmatrix} 1 \\ 0 \\ -3 \end{bmatrix}$, $\begin{bmatrix} 2 \\ 1 \\ -5 \end{bmatrix}$, and $\begin{bmatrix} -3 \\ -3 \\ 7 \end{bmatrix}$. Let $\mathbf{x}_0 = \begin{bmatrix} -2 \\ -5 \\ 3 \end{bmatrix}$. Find the solution of the equation $\mathbf{x}_{k+1} = A\mathbf{x}_k$ for the specified $\mathbf{x}_0$, and describe what happens as $k \to \infty$.

In Exercises 3–6, assume that any initial vector $\mathbf{x}_0$ has an eigenvector decomposition such that the coefficient c_1 in equation (1) of this section is positive.[3]

3. Determine the evolution of the dynamical system in Example 1 when the predation parameter p is .2 in equation (3). (Give a formula for $\mathbf{x}_k$.) Does the owl population grow or decline? What about the wood rat population?

---

[3] One of the limitations of the model in Example 1 is that there always exist initial population vectors $\mathbf{x}_0$ with positive entries such that the coefficient c_1 is negative. The approximation (7) is still valid, but the entries in $\mathbf{x}_k$ eventually become negative.

4. Determine the evolution of the dynamical system in Example 1 when the predation parameter p is .125. (Give a formula for $\mathbf{x}_k$.) As time passes, what happens to the sizes of the owl and wood rat populations? The system tends toward what is sometimes called an unstable equilibrium. What do you think might happen to the system if some aspect of the model (such as birth rates or the predation rate) were to change slightly?

5. In old-growth forests of Douglas fir, the spotted owl dines mainly on flying squirrels. Suppose the predator–prey matrix for these two populations is $A = \begin{bmatrix} .4 & .3 \\ -p & 1.2 \end{bmatrix}$. Show that if the predation parameter p is .325, both populations grow. Estimate the long-term growth rate and the eventual ratio of owls to flying squirrels.

6. Show that if the predation parameter p in Exercise 5 is .5, both the owls and the squirrels will eventually perish. Find a value of p for which populations of both owls and squirrels tend toward constant levels. What are the relative population sizes in this case?

7. Let A have the properties described in Exercise 1.

   a. Is the origin an attractor, a repeller, or a saddle point of the dynamical system $\mathbf{x}_{k+1} = A\mathbf{x}_k$?

   b. Find the directions of greatest attraction and/or repulsion for this dynamical system.

   c. Make a graphical description of the system, showing the directions of greatest attraction or repulsion. Include a rough sketch of several typical trajectories (without computing specific points).

8. Determine the nature of the origin (attractor, repeller, or saddle point) for the dynamical system $\mathbf{x}_{k+1} = A\mathbf{x}_k$ if A has

the properties described in Exercise 2. Find the directions of greatest attraction or repulsion.

In Exercises 9–14, classify the origin as an attractor, repeller, or saddle point of the dynamical system $\mathbf{x}_{k+1} = A\mathbf{x}_k$. Find the directions of greatest attraction and/or repulsion.

9. $A = \begin{bmatrix} 1.7 & -.3 \\ -1.2 & .8 \end{bmatrix}$

10. $A = \begin{bmatrix} .3 & .4 \\ -.3 & 1.1 \end{bmatrix}$

11. $A = \begin{bmatrix} .4 & .5 \\ -.4 & 1.3 \end{bmatrix}$

12. $A = \begin{bmatrix} .5 & .6 \\ -.3 & 1.4 \end{bmatrix}$

13. $A = \begin{bmatrix} .8 & .3 \\ -.4 & 1.5 \end{bmatrix}$

14. $A = \begin{bmatrix} 1.7 & .6 \\ -.4 & .7 \end{bmatrix}$

15. Let $A = \begin{bmatrix} .4 & 0 & .2 \\ .3 & .8 & .3 \\ .3 & .2 & .5 \end{bmatrix}$. The vector $\mathbf{v}_1 = \begin{bmatrix} .1 \\ .6 \\ .3 \end{bmatrix}$ is an eigenvector for A, and two eigenvalues are .5 and .2. Construct the solution of the dynamical system $\mathbf{x}_{k+1} = A\mathbf{x}_k$ that satisfies $\mathbf{x}_0 = (0, .3, .7)$. What happens to $\mathbf{x}_k$ as $k \to \infty$?

T 16. Produce the general solution of the dynamical system $\mathbf{x}_{k+1} = A\mathbf{x}_k$ when $A = \begin{bmatrix} .90 & .01 & .09 \\ .01 & .90 & .01 \\ .09 & .09 & .90 \end{bmatrix}$.

17. Construct a stage-matrix model for an animal species that has two life stages: juvenile (up to 1 year old) and adult. Suppose the female adults give birth each year to an average of 1.6 female juveniles. Each year, 30% of the juveniles survive to become adults and 80% of the adults survive. For $k \geq 0$,

let $\mathbf{x}_k = (j_k, a_k)$, where the entries in $\mathbf{x}_k$ are the numbers of female juveniles and female adults in year k.

a. Construct the stage-matrix A such that $\mathbf{x}_{k+1} = A\mathbf{x}_k$ for $k \geq 0$.

b. Show that the population is growing, compute the eventual growth rate of the population, and give the eventual ratio of juveniles to adults.

T c. Suppose that initially there are 15 juveniles and 10 adults in the population. Produce four graphs that show how the population changes over eight years: (a) the number of juveniles, (b) the number of adults, (c) the total population, and (d) the ratio of juveniles to adults (each year). When does the ratio in (d) seem to stabilize? Include a listing of the program or keystrokes used to produce the graphs for (c) and (d).

18. A herd of American buffalo (bison) can be modeled by a stage matrix similar to that for the spotted owls. The females can be divided into calves (up to 1 year old), yearlings (1 to 2 years), and adults. Suppose an average of 42 female calves are born each year per 100 adult females. (Only adults produce offspring.) Each year, about 60% of the calves survive, 75% of the yearlings survive, and 95% of the adults survive. For $k \geq 0$, let $\mathbf{x}_k = (c_k, y_k, a_k)$, where the entries in $\mathbf{x}_k$ are the numbers of females in each life stage at year k.

a. Construct the stage-matrix A for the buffalo herd, such that $\mathbf{x}_{k+1} = A\mathbf{x}_k$ for $k \geq 0$.

T b. Show that the buffalo herd is growing, determine the expected growth rate after many years, and give the expected numbers of calves and yearlings present per 100 adults.

Solutions to Practice Problems

1. The first step is to write $\mathbf{x}_0$ as a linear combination of $\mathbf{v}_1$, $\mathbf{v}_2$, and $\mathbf{v}_3$. Row reduction of $[\,\mathbf{v}_1 \quad \mathbf{v}_2 \quad \mathbf{v}_3 \quad \mathbf{x}_0\,]$ produces the weights $c_1 = 2$, $c_2 = 1$, and $c_3 = 3$, so that

$$\mathbf{x}_0 = 2\mathbf{v}_1 + 1\mathbf{v}_2 + 3\mathbf{v}_3$$

Since the eigenvalues are 1, $\frac{2}{3}$, and $\frac{1}{3}$, the general solution is

$$\mathbf{x}_k = 2 \cdot 1^k \mathbf{v}_1 + 1 \cdot \left(\frac{2}{3}\right)^k \mathbf{v}_2 + 3 \cdot \left(\frac{1}{3}\right)^k \mathbf{v}_3$$

$$= 2 \begin{bmatrix} -2 \\ 2 \\ 1 \end{bmatrix} + \left(\frac{2}{3}\right)^k \begin{bmatrix} 2 \\ 1 \\ 2 \end{bmatrix} + 3 \cdot \left(\frac{1}{3}\right)^k \begin{bmatrix} 1 \\ 2 \\ -2 \end{bmatrix} \tag{12}$$

2. As $k \to \infty$, the second and third terms in (12) tend to the zero vector, and

$$\mathbf{x}_k = 2\mathbf{v}_1 + \left(\frac{2}{3}\right)^k \mathbf{v}_2 + 3\left(\frac{1}{3}\right)^k \mathbf{v}_3 \to 2\mathbf{v}_1 = \begin{bmatrix} -4 \\ 4 \\ 2 \end{bmatrix}$$

5.7 Applications to Differential Equations

This section describes continuous analogues of the difference equations studied in Section 5.6. In many applied problems, several quantities are varying continuously in time, and they are related by a system of differential equations:

$$
\begin{aligned}
x_1' &= a_{11}x_1 + \cdots + a_{1n}x_n \\
x_2' &= a_{21}x_1 + \cdots + a_{2n}x_n \\
&\ \ \vdots \\
x_n' &= a_{n1}x_1 + \cdots + a_{nn}x_n
\end{aligned}
$$

Here $x_1, \ldots, x_n$ are differentiable functions of t, with derivatives $x_1', \ldots, x_n'$, and the a_{ij} are constants. The crucial feature of this system is that it is *linear*. To see this, write the system as a matrix differential equation

$$
\mathbf{x}'(t) = A\mathbf{x}(t) \tag{1}
$$

where

$$
\mathbf{x}(t) = \begin{bmatrix} x_1(t) \\ \vdots \\ x_n(t) \end{bmatrix}, \quad
\mathbf{x}'(t) = \begin{bmatrix} x_1'(t) \\ \vdots \\ x_n'(t) \end{bmatrix}, \quad \text{and} \quad
A = \begin{bmatrix} a_{11} & \cdots & a_{1n} \\ \vdots & & \vdots \\ a_{n1} & \cdots & a_{nn} \end{bmatrix}
$$

A **solution** of equation (1) is a vector-valued function that satisfies (1) for all t in some interval of real numbers, such as $t \geq 0$.

Equation (1) is *linear* because both differentiation of functions and multiplication of vectors by a matrix are linear transformations. Thus, if $\mathbf{u}$ and $\mathbf{v}$ are solutions of $\mathbf{x}' = A\mathbf{x}$, then $c\mathbf{u} + d\mathbf{v}$ is also a solution, because

$$
\begin{aligned}
(c\mathbf{u} + d\mathbf{v})' &= c\mathbf{u}' + d\mathbf{v}' \\
&= cA\mathbf{u} + dA\mathbf{v} = A(c\mathbf{u} + d\mathbf{v})
\end{aligned}
$$

(Engineers call this property *superposition* of solutions.) Also, the identically zero function is a (trivial) solution of (1). In the terminology of Chapter 4, the set of all solutions of (1) is a *subspace* of the set of all continuous functions with values in $\mathbb{R}^n$.

Standard texts on differential equations show that there always exists what is called a **fundamental set of solutions** to (1). If A is $n \times n$, then there are n linearly independent functions in a fundamental set, and each solution of (1) is a unique linear combination of these n functions. That is, a fundamental set of solutions is a *basis* for the set of all solutions of (1), and the solution set is an n-dimensional vector space of functions. If a vector $\mathbf{x}_0$ is specified, then the **initial value problem** is to construct the (unique) function $\mathbf{x}$ such that $\mathbf{x}' = A\mathbf{x}$ and $\mathbf{x}(0) = \mathbf{x}_0$.

When A is a diagonal matrix, the solutions of (1) can be produced by elementary calculus. For instance, consider

$$
\begin{bmatrix} x_1'(t) \\ x_2'(t) \end{bmatrix} = \begin{bmatrix} 3 & 0 \\ 0 & -5 \end{bmatrix} \begin{bmatrix} x_1(t) \\ x_2(t) \end{bmatrix} \tag{2}
$$

that is,

$$
\begin{aligned}
x_1'(t) &= 3x_1(t) \\
x_2'(t) &= -5x_2(t)
\end{aligned} \tag{3}
$$

The system (2) is said to be *decoupled* because each derivative of a function depends only on the function itself, not on some combination or "coupling" of both $x_1(t)$ and $x_2(t)$. From calculus, the solutions of (3) are $x_1(t) = c_1e^{3t}$ and $x_2(t) = c_2e^{-5t}$, for any constants c_1 and c_2. Each solution of equation (2) can be written in the form

$$\begin{bmatrix} x_1(t) \\ x_2(t) \end{bmatrix} = \begin{bmatrix} c_1e^{3t} \\ c_2e^{-5t} \end{bmatrix} = c_1\begin{bmatrix} 1 \\ 0 \end{bmatrix}e^{3t} + c_2\begin{bmatrix} 0 \\ 1 \end{bmatrix}e^{-5t}$$

This example suggests that for the general equation $\mathbf{x}' = A\mathbf{x}$, a solution might be a linear combination of functions of the form

$$\mathbf{x}(t) = \mathbf{v}e^{\lambda t} \tag{4}$$

for some scalar λ and some fixed nonzero vector $\mathbf{v}$. [If $\mathbf{v} = \mathbf{0}$, the function $\mathbf{x}(t)$ is identically zero and hence satisfies $\mathbf{x}' = A\mathbf{x}$.] Observe that

$$\mathbf{x}'(t) = \lambda\mathbf{v}e^{\lambda t} \quad \text{By calculus, since } \mathbf{v} \text{ is a constant vector}$$
$$A\mathbf{x}(t) = A\mathbf{v}e^{\lambda t} \quad \text{Multiplying both sides of (4) by } A$$

Since $e^{\lambda t}$ is never zero, $\mathbf{x}'(t)$ will equal $A\mathbf{x}(t)$ if and only if $\lambda\mathbf{v} = A\mathbf{v}$, that is, if and only if λ is an eigenvalue of A and $\mathbf{v}$ is a corresponding eigenvector. Thus each eigenvalue–eigenvector pair provides a solution (4) of $\mathbf{x}' = A\mathbf{x}$. Such solutions are sometimes called *eigenfunctions* of the differential equation. Eigenfunctions provide the key to solving systems of differential equations.

EXAMPLE 1   The circuit in Figure 1 can be described by the differential equation

$$\begin{bmatrix} x_1'(t) \\ x_2'(t) \end{bmatrix} = \begin{bmatrix} -(1/R_1 + 1/R_2)/C_1 & 1/(R_2C_1) \\ 1/(R_2C_2) & -1/(R_2C_2) \end{bmatrix}\begin{bmatrix} x_1(t) \\ x_2(t) \end{bmatrix}$$

where $x_1(t)$ and $x_2(t)$ are the voltages across the two capacitors at time t. Suppose resistor R_1 is 1 ohm, R_2 is 2 ohms, capacitor C_1 is 1 farad, and C_2 is .5 farad, and suppose there is an initial charge of 5 volts on capacitor C_1 and 4 volts on capacitor C_2. Find formulas for $x_1(t)$ and $x_2(t)$ that describe how the voltages change over time.

FIGURE 1

SOLUTION   Let A denote the matrix displayed above, and let $\mathbf{x}(t) = \begin{bmatrix} x_1(t) \\ x_2(t) \end{bmatrix}$. For the data given, $A = \begin{bmatrix} -1.5 & .5 \\ 1 & -1 \end{bmatrix}$, and $\mathbf{x}(0) = \begin{bmatrix} 5 \\ 4 \end{bmatrix}$. The eigenvalues of A are $\lambda_1 = -.5$ and $\lambda_2 = -2$, with corresponding eigenvectors

$$\mathbf{v}_1 = \begin{bmatrix} 1 \\ 2 \end{bmatrix} \quad \text{and} \quad \mathbf{v}_2 = \begin{bmatrix} -1 \\ 1 \end{bmatrix}$$

The eigenfunctions $\mathbf{x}_1(t) = \mathbf{v}_1e^{\lambda_1 t}$ and $\mathbf{x}_2(t) = \mathbf{v}_2e^{\lambda_2 t}$ both satisfy $\mathbf{x}' = A\mathbf{x}$, and so does any linear combination of $\mathbf{x}_1$ and $\mathbf{x}_2$. Set

$$\mathbf{x}(t) = c_1\mathbf{v}_1e^{\lambda_1 t} + c_2\mathbf{v}_2e^{\lambda_2 t} = c_1\begin{bmatrix} 1 \\ 2 \end{bmatrix}e^{-.5t} + c_2\begin{bmatrix} -1 \\ 1 \end{bmatrix}e^{-2t}$$

and note that $\mathbf{x}(0) = c_1\mathbf{v}_1 + c_2\mathbf{v}_2$. Since $\mathbf{v}_1$ and $\mathbf{v}_2$ are obviously linearly independent and hence span $\mathbb{R}^2$, c_1 and c_2 can be found to make $\mathbf{x}(0)$ equal to $\mathbf{x}_0$. In fact, the equation

$$c_1\underset{\underset{\mathbf{v}_1}{\uparrow}}{\begin{bmatrix} 1 \\ 2 \end{bmatrix}} + c_2\underset{\underset{\mathbf{v}_2}{\uparrow}}{\begin{bmatrix} -1 \\ 1 \end{bmatrix}} = \underset{\underset{\mathbf{x}_0}{\uparrow}}{\begin{bmatrix} 5 \\ 4 \end{bmatrix}}$$

leads easily to $c_1 = 3$ and $c_2 = -2$. Thus the desired solution of the differential equation $\mathbf{x}' = A\mathbf{x}$ is

$$\mathbf{x}(t) = 3\begin{bmatrix} 1 \\ 2 \end{bmatrix} e^{-.5t} - 2 \begin{bmatrix} -1 \\ 1 \end{bmatrix} e^{-2t}$$

or

$$\begin{bmatrix} x_1(t) \\ x_2(t) \end{bmatrix} = \begin{bmatrix} 3e^{-.5t} + 2e^{-2t} \\ 6e^{-.5t} - 2e^{-2t} \end{bmatrix}$$

Figure 2 shows the graph, or *trajectory*, of $\mathbf{x}(t)$, for $t \geq 0$, along with trajectories for some other initial points. The trajectories of the two eigenfunctions $\mathbf{x}_1$ and $\mathbf{x}_2$ lie in the eigenspaces of A.

The functions $\mathbf{x}_1$ and $\mathbf{x}_2$ both decay to zero as $t \to \infty$, but the values of $\mathbf{x}_2$ decay faster because its exponent is more negative. The entries in the corresponding eigenvector $\mathbf{v}_2$ show that the voltages across the capacitors will decay to zero as rapidly as possible if the initial voltages are equal in magnitude but opposite in sign. ∎

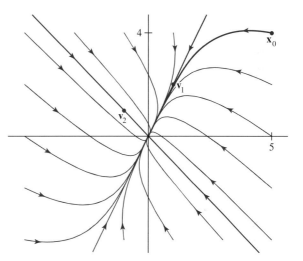

FIGURE 2 The origin as an attractor.

In Figure 2, the origin is called an **attractor**, or **sink**, of the dynamical system because all trajectories are drawn into the origin. The direction of greatest attraction is along the trajectory of the eigenfunction $\mathbf{x}_2$ (along the line through $\mathbf{0}$ and $\mathbf{v}_2$) corresponding to the more negative eigenvalue, $\lambda = -2$. Trajectories that begin at points not on this line become asymptotic to the line through $\mathbf{0}$ and $\mathbf{v}_1$ because their components in the $\mathbf{v}_2$ direction decay so rapidly.

If the eigenvalues in Example 1 were positive instead of negative, the corresponding trajectories would be similar in shape, but the trajectories would be traversed *away* from the origin. In such a case, the origin is called a **repeller**, or **source**, of the dynamical system, and the direction of greatest repulsion is the line containing the trajectory of the eigenfunction corresponding to the more positive eigenvalue.

EXAMPLE 2 Suppose a particle is moving in a planar force field and its position vector $\mathbf{x}$ satisfies $\mathbf{x}' = A\mathbf{x}$ and $\mathbf{x}(0) = \mathbf{x}_0$, where

$$A = \begin{bmatrix} 4 & -5 \\ -2 & 1 \end{bmatrix}, \qquad \mathbf{x}_0 = \begin{bmatrix} 2.9 \\ 2.6 \end{bmatrix}$$

Solve this initial value problem for $t \geq 0$, and sketch the trajectory of the particle.

SOLUTION The eigenvalues of A turn out to be $\lambda_1 = 6$ and $\lambda_2 = -1$, with corresponding eigenvectors $\mathbf{v}_1 = (-5, 2)$ and $\mathbf{v}_2 = (1, 1)$. For any constants c_1 and c_2, the function

$$\mathbf{x}(t) = c_1 \mathbf{v}_1 e^{\lambda_1 t} + c_2 \mathbf{v}_2 e^{\lambda_2 t} = c_1 \begin{bmatrix} -5 \\ 2 \end{bmatrix} e^{6t} + c_2 \begin{bmatrix} 1 \\ 1 \end{bmatrix} e^{-t}$$

is a solution of $\mathbf{x}' = A\mathbf{x}$. We want c_1 and c_2 to satisfy $\mathbf{x}(0) = \mathbf{x}_0$, that is,

$$c_1 \begin{bmatrix} -5 \\ 2 \end{bmatrix} + c_2 \begin{bmatrix} 1 \\ 1 \end{bmatrix} = \begin{bmatrix} 2.9 \\ 2.6 \end{bmatrix} \quad \text{or} \quad \begin{bmatrix} -5 & 1 \\ 2 & 1 \end{bmatrix} \begin{bmatrix} c_1 \\ c_2 \end{bmatrix} = \begin{bmatrix} 2.9 \\ 2.6 \end{bmatrix}$$

Calculations show that $c_1 = -3/70$ and $c_2 = 188/70$, and so the desired function is

$$\mathbf{x}(t) = \frac{-3}{70} \begin{bmatrix} -5 \\ 2 \end{bmatrix} e^{6t} + \frac{188}{70} \begin{bmatrix} 1 \\ 1 \end{bmatrix} e^{-t}$$

Trajectories of $\mathbf{x}$ and other solutions are shown in Figure 3. ∎

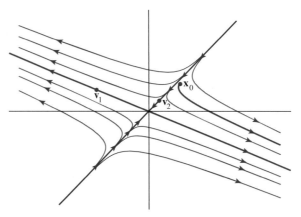

FIGURE 3 The origin as a saddle point.

In Figure 3, the origin is called a **saddle point** of the dynamical system because some trajectories approach the origin at first and then change direction and move away from the origin. A saddle point arises whenever the matrix A has both positive and negative eigenvalues. The direction of greatest repulsion is the line through $\mathbf{v}_1$ and $\mathbf{0}$, corresponding to the positive eigenvalue. The direction of greatest attraction is the line through $\mathbf{v}_2$ and $\mathbf{0}$, corresponding to the negative eigenvalue.

Decoupling a Dynamical System

The following discussion shows that the method of Examples 1 and 2 produces a fundamental set of solutions for any dynamical system described by $\mathbf{x}' = A\mathbf{x}$ when A is $n \times n$ and has n linearly independent eigenvectors, that is, when A is diagonalizable. Suppose the eigenfunctions for A are

$$\mathbf{v}_1 e^{\lambda_1 t}, \ldots, \mathbf{v}_n e^{\lambda_n t}$$

with $\mathbf{v}_1, \ldots, \mathbf{v}_n$ linearly independent eigenvectors. Let $P = [\, \mathbf{v}_1 \ \cdots \ \mathbf{v}_n \,]$, and let D be the diagonal matrix with entries $\lambda_1, \ldots, \lambda_n$, so that $A = PDP^{-1}$. Now make a *change of variable*, defining a new function $\mathbf{y}$ by

$$\mathbf{y}(t) = P^{-1}\mathbf{x}(t) \quad \text{or, equivalently,} \quad \mathbf{x}(t) = P\mathbf{y}(t)$$

The equation $\mathbf{x}(t) = P\mathbf{y}(t)$ says that $\mathbf{y}(t)$ is the coordinate vector of $\mathbf{x}(t)$ relative to the eigenvector basis. Substitution of $P\mathbf{y}$ for $\mathbf{x}$ in the equation $\mathbf{x}' = A\mathbf{x}$ gives

$$\frac{d}{dt}(P\mathbf{y}) = A(P\mathbf{y}) = (PDP^{-1})P\mathbf{y} = PD\mathbf{y} \tag{5}$$

Since P is a constant matrix, the left side of (5) is $P\mathbf{y}'$. Left-multiply both sides of (5) by P^{-1} and obtain $\mathbf{y}' = D\mathbf{y}$, or

$$\begin{bmatrix} y_1'(t) \\ y_2'(t) \\ \vdots \\ y_n'(t) \end{bmatrix} = \begin{bmatrix} \lambda_1 & 0 & \cdots & 0 \\ 0 & \lambda_2 & & \vdots \\ \vdots & & \ddots & 0 \\ 0 & \cdots & 0 & \lambda_n \end{bmatrix} \begin{bmatrix} y_1(t) \\ y_2(t) \\ \vdots \\ y_n(t) \end{bmatrix}$$

The change of variable from $\mathbf{x}$ to $\mathbf{y}$ has *decoupled* the system of differential equations, because the derivative of each scalar function y_k depends only on y_k. (Review the analogous change of variables in Section 5.6.) Since $y_1' = \lambda_1 y_1$, we have $y_1(t) = c_1 e^{\lambda_1 t}$, with similar formulas for $y_2, \ldots, y_n$. Thus

$$\mathbf{y}(t) = \begin{bmatrix} c_1 e^{\lambda_1 t} \\ \vdots \\ c_n e^{\lambda_n t} \end{bmatrix}, \quad \text{where} \quad \begin{bmatrix} c_1 \\ \vdots \\ c_n \end{bmatrix} = \mathbf{y}(0) = P^{-1}\mathbf{x}(0) = P^{-1}\mathbf{x}_0$$

To obtain the general solution $\mathbf{x}$ of the original system, compute

$$\mathbf{x}(t) = P\mathbf{y}(t) = [\, \mathbf{v}_1 \quad \cdots \quad \mathbf{v}_n \,]\mathbf{y}(t)$$
$$= c_1 \mathbf{v}_1 e^{\lambda_1 t} + \cdots + c_n \mathbf{v}_n e^{\lambda_n t}$$

This is the eigenfunction expansion constructed as in Example 1.

Complex Eigenvalues

In the next example, a real matrix A has a pair of complex eigenvalues λ and $\overline{\lambda}$, with associated complex eigenvectors $\mathbf{v}$ and $\overline{\mathbf{v}}$. (Recall from Section 5.5 that for a real matrix, complex eigenvalues and associated eigenvectors come in conjugate pairs.) So two solutions of $\mathbf{x}' = A\mathbf{x}$ are

$$\mathbf{x}_1(t) = \mathbf{v}e^{\lambda t} \quad \text{and} \quad \mathbf{x}_2(t) = \overline{\mathbf{v}}e^{\overline{\lambda} t} \tag{6}$$

It can be shown that $\mathbf{x}_2(t) = \overline{\mathbf{x}_1(t)}$ by using a power series representation for the complex exponential function. Although the complex eigenfunctions $\mathbf{x}_1$ and $\mathbf{x}_2$ are convenient for some calculations (particularly in electrical engineering), real functions are more appropriate for many purposes. Fortunately, the real and imaginary parts of $\mathbf{x}_1$ are (real) solutions of $\mathbf{x}' = A\mathbf{x}$, because they are linear combinations of the solutions in (6):

$$\text{Re}(\mathbf{v}e^{\lambda t}) = \frac{1}{2}[\mathbf{x}_1(t) + \overline{\mathbf{x}_1(t)}], \qquad \text{Im}(\mathbf{v}e^{\lambda t}) = \frac{1}{2i}[\mathbf{x}_1(t) - \overline{\mathbf{x}_1(t)}]$$

To understand the nature of $\text{Re}(\mathbf{v}e^{\lambda t})$, recall from calculus that for any number x, the exponential function e^x can be computed from the power series:

$$e^x = 1 + x + \frac{1}{2!}x^2 + \cdots + \frac{1}{n!}x^n + \cdots$$

This series can be used to define $e^{\lambda t}$ when λ is complex:

$$e^{\lambda t} = 1 + (\lambda t) + \frac{1}{2!}(\lambda t)^2 + \cdots + \frac{1}{n!}(\lambda t)^n + \cdots$$

By writing $\lambda = a + bi$ (with a and b real), and using similar power series for the cosine and sine functions, one can show that

$$e^{(a+bi)t} = e^{at}e^{ibt} = e^{at}(\cos bt + i \sin bt) \tag{7}$$

Hence

$$
\begin{aligned}
\mathbf{v}e^{\lambda t} &= (\operatorname{Re}\mathbf{v} + i \operatorname{Im}\mathbf{v})\,(e^{at})(\cos bt + i \sin bt) \\
&= [\,(\operatorname{Re}\mathbf{v})\cos bt - (\operatorname{Im}\mathbf{v})\sin bt\,]e^{at} \\
&\quad + i[\,(\operatorname{Re}\mathbf{v})\sin bt + (\operatorname{Im}\mathbf{v})\cos bt\,]e^{at}
\end{aligned}
$$

So two real solutions of $\mathbf{x}' = A\mathbf{x}$ are

$$
\begin{aligned}
\mathbf{y}_1(t) &= \operatorname{Re}\mathbf{x}_1(t) = [\,(\operatorname{Re}\mathbf{v})\cos bt - (\operatorname{Im}\mathbf{v})\sin bt\,]\,e^{at} \\
\mathbf{y}_2(t) &= \operatorname{Im}\mathbf{x}_1(t) = [\,(\operatorname{Re}\mathbf{v})\sin bt + (\operatorname{Im}\mathbf{v})\cos bt\,]\,e^{at}
\end{aligned}
$$

It can be shown that $\mathbf{y}_1$ and $\mathbf{y}_2$ are linearly independent functions (when $b \neq 0$).[1]

EXAMPLE 3 The circuit in Figure 4 can be described by the equation

$$
\begin{bmatrix} i_L' \\ v_C' \end{bmatrix} = \begin{bmatrix} -R_2/L & -1/L \\ 1/C & -1/(R_1 C) \end{bmatrix} \begin{bmatrix} i_L \\ v_C \end{bmatrix}
$$

where i_L is the current passing through the inductor L and v_C is the voltage drop across the capacitor C. Suppose R_1 is 5 ohms, R_2 is .8 ohm, C is .1 farad, and L is .4 henry. Find formulas for i_L and v_C, if the initial current through the inductor is 3 amperes and the initial voltage across the capacitor is 3 volts.

FIGURE 4

SOLUTION For the data given, $A = \begin{bmatrix} -2 & -2.5 \\ 10 & -2 \end{bmatrix}$ and $\mathbf{x}_0 = \begin{bmatrix} 3 \\ 3 \end{bmatrix}$. The method discussed in Section 5.5 produces the eigenvalue $\lambda = -2 + 5i$ and the corresponding eigenvector $\mathbf{v}_1 = \begin{bmatrix} i \\ 2 \end{bmatrix}$. The complex solutions of $\mathbf{x}' = A\mathbf{x}$ are complex linear combinations of

$$\mathbf{x}_1(t) = \begin{bmatrix} i \\ 2 \end{bmatrix} e^{(-2+5i)t} \quad \text{and} \quad \mathbf{x}_2(t) = \begin{bmatrix} -i \\ 2 \end{bmatrix} e^{(-2-5i)t}$$

Next, use equation (7) to write

$$\mathbf{x}_1(t) = \begin{bmatrix} i \\ 2 \end{bmatrix} e^{-2t}(\cos 5t + i \sin 5t)$$

The real and imaginary parts of $\mathbf{x}_1$ provide real solutions:

$$\mathbf{y}_1(t) = \begin{bmatrix} -\sin 5t \\ 2\cos 5t \end{bmatrix} e^{-2t}, \qquad \mathbf{y}_2(t) = \begin{bmatrix} \cos 5t \\ 2\sin 5t \end{bmatrix} e^{-2t}$$

---

[1] Since $\mathbf{x}_2(t)$ is the complex conjugate of $\mathbf{x}_1(t)$, the real and imaginary parts of $\mathbf{x}_2(t)$ are $\mathbf{y}_1(t)$ and $-\mathbf{y}_2(t)$, respectively. Thus one can use either $\mathbf{x}_1(t)$ or $\mathbf{x}_2(t)$, but not both, to produce two real linearly independent solutions of $\mathbf{x}' = A\mathbf{x}$.

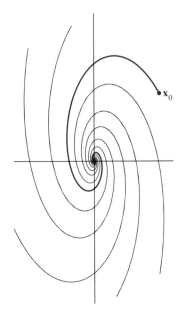

FIGURE 5
The origin as a spiral point.

Since $\mathbf{y}_1$ and $\mathbf{y}_2$ are linearly independent functions, they form a basis for the two-dimensional real vector space of solutions of $\mathbf{x}' = A\mathbf{x}$. Thus the general solution is

$$\mathbf{x}(t) = c_1 \begin{bmatrix} -\sin 5t \\ 2\cos 5t \end{bmatrix} e^{-2t} + c_2 \begin{bmatrix} \cos 5t \\ 2\sin 5t \end{bmatrix} e^{-2t}$$

To satisfy $\mathbf{x}(0) = \begin{bmatrix} 3 \\ 3 \end{bmatrix}$, we need $c_1 \begin{bmatrix} 0 \\ 2 \end{bmatrix} + c_2 \begin{bmatrix} 1 \\ 0 \end{bmatrix} = \begin{bmatrix} 3 \\ 3 \end{bmatrix}$, which leads to $c_1 = 1.5$ and $c_2 = 3$. Thus

$$\mathbf{x}(t) = 1.5 \begin{bmatrix} -\sin 5t \\ 2\cos 5t \end{bmatrix} e^{-2t} + 3 \begin{bmatrix} \cos 5t \\ 2\sin 5t \end{bmatrix} e^{-2t}$$

or

$$\begin{bmatrix} i_L(t) \\ v_C(t) \end{bmatrix} = \begin{bmatrix} -1.5\sin 5t + 3\cos 5t \\ 3\cos 5t + 6\sin 5t \end{bmatrix} e^{-2t}$$

See Figure 5. ∎

In Figure 5, the origin is called a **spiral point** of the dynamical system. The rotation is caused by the sine and cosine functions that arise from a complex eigenvalue. The trajectories spiral inward because the factor e^{-2t} tends to zero. Recall that -2 is the real part of the eigenvalue in Example 3. When A has a complex eigenvalue with positive real part, the trajectories spiral outward. If the real part of the eigenvalue is zero, the trajectories form ellipses around the origin.

Practice Problems

A real 3×3 matrix A has eigenvalues $-.5$, $.2 + .3i$, and $.2 - .3i$, with corresponding eigenvectors

$$\mathbf{v}_1 = \begin{bmatrix} 1 \\ -2 \\ 1 \end{bmatrix}, \quad \mathbf{v}_2 = \begin{bmatrix} 1 + 2i \\ 4i \\ 2 \end{bmatrix}, \quad \text{and} \quad \mathbf{v}_3 = \begin{bmatrix} 1 - 2i \\ -4i \\ 2 \end{bmatrix}$$

1. Is A diagonalizable as $A = PDP^{-1}$, using complex matrices?
2. Write the general solution of $\mathbf{x}' = A\mathbf{x}$ using complex eigenfunctions, and then find the general real solution.
3. Describe the shapes of typical trajectories.

5.7 Exercises

1. A particle moving in a planar force field has a position vector $\mathbf{x}$ that satisfies $\mathbf{x}' = A\mathbf{x}$. The 2×2 matrix A has eigenvalues 4 and 2, with corresponding eigenvectors $\mathbf{v}_1 = \begin{bmatrix} -3 \\ 1 \end{bmatrix}$ and $\mathbf{v}_2 = \begin{bmatrix} -1 \\ 1 \end{bmatrix}$. Find the position of the particle at time t, assuming that $\mathbf{x}(0) = \begin{bmatrix} -6 \\ 1 \end{bmatrix}$.

2. Let A be a 2×2 matrix with eigenvalues -3 and -1 and corresponding eigenvectors $\mathbf{v}_1 = \begin{bmatrix} -1 \\ 1 \end{bmatrix}$ and $\mathbf{v}_2 = \begin{bmatrix} 1 \\ 1 \end{bmatrix}$. Let $\mathbf{x}(t)$ be the position of a particle at time t. Solve the initial value problem $\mathbf{x}' = A\mathbf{x}$, $\mathbf{x}(0) = \begin{bmatrix} 2 \\ 3 \end{bmatrix}$.

In Exercises 3–6, solve the initial value problem $\mathbf{x}'(t) = A\mathbf{x}(t)$ for $t \geq 0$, with $\mathbf{x}(0) = (3, 2)$. Classify the nature of the origin as an attractor, repeller, or saddle point of the dynamical system described by $\mathbf{x}' = A\mathbf{x}$. Find the directions of greatest attraction and/or repulsion. When the origin is a saddle point, sketch typical trajectories.

3. $A = \begin{bmatrix} 2 & 3 \\ -1 & -2 \end{bmatrix}$

4. $A = \begin{bmatrix} -2 & -5 \\ 1 & 4 \end{bmatrix}$

5. $A = \begin{bmatrix} 7 & -1 \\ 3 & 3 \end{bmatrix}$

6. $A = \begin{bmatrix} 1 & -2 \\ 3 & -4 \end{bmatrix}$

In Exercises 7 and 8, make a change of variable that decouples the equation $\mathbf{x}' = A\mathbf{x}$. Write the equation $\mathbf{x}(t) = P\mathbf{y}(t)$ and show the calculation that leads to the uncoupled system $\mathbf{y}' = D\mathbf{y}$, specifying P and D.

7. A as in Exercise 5   **8.** A as in Exercise 6

In Exercises 9–18, construct the general solution of $\mathbf{x}' = A\mathbf{x}$ involving complex eigenfunctions and then obtain the general real solution. Describe the shapes of typical trajectories.

9. $A = \begin{bmatrix} -3 & 2 \\ -1 & -1 \end{bmatrix}$ **10.** $A = \begin{bmatrix} 3 & 1 \\ -2 & 1 \end{bmatrix}$

11. $A = \begin{bmatrix} -3 & -9 \\ 2 & 3 \end{bmatrix}$ **12.** $A = \begin{bmatrix} -7 & 10 \\ -4 & 5 \end{bmatrix}$

13. $A = \begin{bmatrix} 4 & -3 \\ 6 & -2 \end{bmatrix}$ **14.** $A = \begin{bmatrix} -2 & 1 \\ -8 & 2 \end{bmatrix}$

T 15. $A = \begin{bmatrix} -8 & -12 & -6 \\ 2 & 1 & 2 \\ 7 & 12 & 5 \end{bmatrix}$

T 16. $A = \begin{bmatrix} -6 & -11 & 16 \\ 2 & 5 & -4 \\ -4 & -5 & 10 \end{bmatrix}$

T 17. $A = \begin{bmatrix} 30 & 64 & 23 \\ -11 & -23 & -9 \\ 6 & 15 & 4 \end{bmatrix}$

T 18. $A = \begin{bmatrix} 53 & -30 & -2 \\ 90 & -52 & -3 \\ 20 & -10 & 2 \end{bmatrix}$

T 19. Find formulas for the voltages v_1 and v_2 (as functions of time t) for the circuit in Example 1, assuming that $R_1 = 1/5$ ohm, $R_2 = 1/3$ ohm, $C_1 = 4$ farads, $C_2 = 3$ farads, and the initial charge on each capacitor is 4 volts.

T 20. Find formulas for the voltages v_1 and v_2 for the circuit in Example 1, assuming that $R_1 = 1/15$ ohm, $R_2 = 1/3$ ohm, $C_1 = 9$ farads, $C_2 = 2$ farads, and the initial charge on each capacitor is 3 volts.

T 21. Find formulas for the current i_L and the voltage v_C for the circuit in Example 3, assuming that $R_1 = 1$ ohm, $R_2 = .125$ ohm, $C = .2$ farad, $L = .125$ henry, the initial current is 0 amp, and the initial voltage is 15 volts.

T 22. The circuit in the figure is described by the equation

$$\begin{bmatrix} i_L' \\ v_C' \end{bmatrix} = \begin{bmatrix} 0 & 1/L \\ -1/C & -1/(RC) \end{bmatrix} \begin{bmatrix} i_L \\ v_C \end{bmatrix}$$

where i_L is the current through the inductor L and v_C is the voltage drop across the capacitor C. Find formulas for i_L and v_C when $R = .5$ ohm, $C = 2.5$ farads, $L = .5$ henry, the initial current is 0 amp, and the initial voltage is 12 volts.

Solutions to Practice Problems

1. Yes, the 3×3 matrix is diagonalizable because it has three distinct eigenvalues. Theorem 2 in Section 5.1 and Theorem 6 in Section 5.3 are valid when complex scalars are used. (The proofs are essentially the same as for real scalars.)

2. The general solution has the form

$$\mathbf{x}(t) = c_1 \begin{bmatrix} 1 \\ -2 \\ 1 \end{bmatrix} e^{-.5t} + c_2 \begin{bmatrix} 1 + 2i \\ 4i \\ 2 \end{bmatrix} e^{(.2+.3i)t} + c_3 \begin{bmatrix} 1 - 2i \\ -4i \\ 2 \end{bmatrix} e^{(.2-.3i)t}$$

The scalars c_1, c_2, and c_3 here can be any complex numbers. The first term in $\mathbf{x}(t)$ is real, provided c_1 is real. Two more real solutions can be produced using the real and imaginary parts of the second term in $\mathbf{x}(t)$:

$$\begin{bmatrix} 1 + 2i \\ 4i \\ 2 \end{bmatrix} e^{.2t} (\cos .3t + i \sin .3t)$$

The general real solution has the following form, with *real* scalars c_1, c_2, and c_3:

$$c_1 \begin{bmatrix} 1 \\ -2 \\ 1 \end{bmatrix} e^{-.5t} + c_2 \begin{bmatrix} \cos .3t - 2\sin .3t \\ -4\sin .3t \\ 2\cos .3t \end{bmatrix} e^{.2t} + c_3 \begin{bmatrix} \sin .3t + 2\cos .3t \\ 4\cos .3t \\ 2\sin .3t \end{bmatrix} e^{.2t}$$

3. Any solution with $c_2 = c_3 = 0$ is attracted to the origin because of the negative exponential factor. Other solutions have components that grow without bound, and the trajectories spiral outward.

Be careful not to mistake this problem for one in Section 5.6. There the condition for attraction toward $\mathbf{0}$ was that an eigenvalue be less than 1 in magnitude, to make $|\lambda|^k \to 0$. Here the real part of the eigenvalue must be negative, to make $e^{\lambda t} \to 0$.

5.8 Iterative Estimates for Eigenvalues

In scientific applications of linear algebra, eigenvalues are seldom known precisely. Fortunately, a close numerical approximation is usually quite satisfactory. In fact, some applications require only a rough approximation to the largest eigenvalue. The first algorithm described below can work well for this case. Also, it provides a foundation for a more powerful method that can give fast estimates for other eigenvalues as well.

The Power Method

The power method applies to an $n \times n$ matrix A with a **strictly dominant eigenvalue** λ_1, which means that λ_1 must be larger in absolute value than all the other eigenvalues. In this case, the power method produces a scalar sequence that approaches λ_1 and a vector sequence that approaches a corresponding eigenvector. The background for the method rests on the eigenvector decomposition used at the beginning of Section 5.6.

Assume for simplicity that A is diagonalizable and $\mathbb{R}^n$ has a basis of eigenvectors $\mathbf{v}_1, \dots, \mathbf{v}_n$, arranged so their corresponding eigenvalues $\lambda_1, \dots, \lambda_n$ decrease in size, with the strictly dominant eigenvalue first. That is,

$$|\lambda_1| > |\lambda_2| \geq |\lambda_3| \geq \cdots \geq |\lambda_n| \tag{1}$$

$\llcorner$ Strictly larger

As we saw in equation (2) of Section 5.6, if $\mathbf{x}$ in $\mathbb{R}^n$ is written as $\mathbf{x} = c_1 \mathbf{v}_1 + \cdots + c_n \mathbf{v}_n$, then

$$A^k \mathbf{x} = c_1 (\lambda_1)^k \mathbf{v}_1 + c_2 (\lambda_2)^k \mathbf{v}_2 + \cdots + c_n (\lambda_n)^k \mathbf{v}_n \quad (k = 1, 2, \dots)$$

Assume $c_1 \neq 0$. Then, dividing by $(\lambda_1)^k$,

$$\frac{1}{(\lambda_1)^k} A^k \mathbf{x} = c_1 \mathbf{v}_1 + c_2 \left(\frac{\lambda_2}{\lambda_1} \right)^k \mathbf{v}_2 + \cdots + c_n \left(\frac{\lambda_n}{\lambda_1} \right)^k \mathbf{v}_n \quad (k = 1, 2, \dots) \tag{2}$$

From inequality (1), the fractions $\lambda_2/\lambda_1, \dots, \lambda_n/\lambda_1$ are all less than 1 in magnitude and so their powers go to zero. Hence

$$(\lambda_1)^{-k} A^k \mathbf{x} \to c_1 \mathbf{v}_1 \quad \text{as } k \to \infty \tag{3}$$

Thus, for large k, a scalar multiple of $A^k \mathbf{x}$ determines almost the same *direction* as the eigenvector $c_1 \mathbf{v}_1$. Since positive scalar multiples do not change the direction of a vector, $A^k \mathbf{x}$ itself points almost in the same direction as $\mathbf{v}_1$ or $-\mathbf{v}_1$, provided $c_1 \neq 0$.

EXAMPLE 1 Let $A = \begin{bmatrix} 1.8 & .8 \\ .2 & 1.2 \end{bmatrix}$, $\mathbf{v}_1 = \begin{bmatrix} 4 \\ 1 \end{bmatrix}$, and $\mathbf{x} = \begin{bmatrix} -.5 \\ 1 \end{bmatrix}$. Then A has eigenvalues 2 and 1, and the eigenspace for $\lambda_1 = 2$ is the line through $\mathbf{0}$ and $\mathbf{v}_1$. For $k = 0, \dots, 8$, compute $A^k \mathbf{x}$ and construct the line through $\mathbf{0}$ and $A^k \mathbf{x}$. What happens as k increases?

SOLUTION The first three calculations are

$$A\mathbf{x} = \begin{bmatrix} 1.8 & .8 \\ .2 & 1.2 \end{bmatrix} \begin{bmatrix} -.5 \\ 1 \end{bmatrix} = \begin{bmatrix} -.1 \\ 1.1 \end{bmatrix}$$

$$A^2\mathbf{x} = A(A\mathbf{x}) = \begin{bmatrix} 1.8 & .8 \\ .2 & 1.2 \end{bmatrix} \begin{bmatrix} -.1 \\ 1.1 \end{bmatrix} = \begin{bmatrix} .7 \\ 1.3 \end{bmatrix}$$

$$A^3\mathbf{x} = A(A^2\mathbf{x}) = \begin{bmatrix} 1.8 & .8 \\ .2 & 1.2 \end{bmatrix} \begin{bmatrix} .7 \\ 1.3 \end{bmatrix} = \begin{bmatrix} 2.3 \\ 1.7 \end{bmatrix}$$

Analogous calculations complete Table 1.

TABLE 1 Iterates of a Vector

k	0	1	2	3	4	5	6	7	8
$A^k\mathbf{x}$	$\begin{bmatrix} -.5 \\ 1 \end{bmatrix}$	$\begin{bmatrix} -.1 \\ 1.1 \end{bmatrix}$	$\begin{bmatrix} .7 \\ 1.3 \end{bmatrix}$	$\begin{bmatrix} 2.3 \\ 1.7 \end{bmatrix}$	$\begin{bmatrix} 5.5 \\ 2.5 \end{bmatrix}$	$\begin{bmatrix} 11.9 \\ 4.1 \end{bmatrix}$	$\begin{bmatrix} 24.7 \\ 7.3 \end{bmatrix}$	$\begin{bmatrix} 50.3 \\ 13.7 \end{bmatrix}$	$\begin{bmatrix} 101.5 \\ 26.5 \end{bmatrix}$

The vectors $\mathbf{x}, A\mathbf{x}, \ldots, A^4\mathbf{x}$ are shown in Figure 1. The other vectors are growing too long to display. However, line segments are drawn showing the directions of those vectors. In fact, the directions of the vectors are what we really want to see, not the vectors themselves. The lines seem to be approaching the line representing the eigenspace spanned by $\mathbf{v}_1$. More precisely, the angle between the line (subspace) determined by $A^k\mathbf{x}$ and the line (eigenspace) determined by $\mathbf{v}_1$ goes to zero as $k \to \infty$. ∎

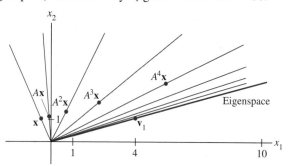

FIGURE 1 Directions determined by $\mathbf{x}, A\mathbf{x}, A^2\mathbf{x}, \ldots, A^7\mathbf{x}$.

The vectors $(\lambda_1)^{-k} A^k\mathbf{x}$ in (3) are scaled to make them converge to $c_1\mathbf{v}_1$, provided $c_1 \neq 0$. We cannot scale $A^k\mathbf{x}$ in this way because we do not know λ_1. But we can scale each $A^k\mathbf{x}$ to make its largest entry a 1. It turns out that the resulting sequence $\{\mathbf{x}_k\}$ will converge to a multiple of $\mathbf{v}_1$ whose largest entry is 1. Figure 2 shows the scaled sequence for Example 1. The eigenvalue λ_1 can be estimated from the sequence $\{\mathbf{x}_k\}$, too. When $\mathbf{x}_k$

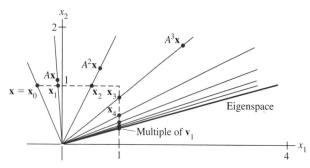

FIGURE 2 Scaled multiples of $\mathbf{x}, A\mathbf{x}, A^2\mathbf{x}, \ldots, A^7\mathbf{x}$.

is close to an eigenvector for λ_1, the vector $A\mathbf{x}_k$ is close to $\lambda_1\mathbf{x}_k$, with each entry in $A\mathbf{x}_k$ approximately λ_1 times the corresponding entry in $\mathbf{x}_k$. Because the largest entry in $\mathbf{x}_k$ is 1, the largest entry in $A\mathbf{x}_k$ is close to λ_1. (Careful proofs of these statements are omitted.)

THE POWER METHOD FOR ESTIMATING A STRICTLY DOMINANT EIGENVALUE

1. Select an initial vector $\mathbf{x}_0$ whose largest entry is 1.
2. For $k = 0, 1, \ldots$,
   a. Compute $A\mathbf{x}_k$.
   b. Let μ_k be an entry in $A\mathbf{x}_k$ whose absolute value is as large as possible.
   c. Compute $\mathbf{x}_{k+1} = (1/\mu_k)A\mathbf{x}_k$.
3. For almost all choices of $\mathbf{x}_0$, the sequence $\{\mu_k\}$ approaches the dominant eigenvalue, and the sequence $\{\mathbf{x}_k\}$ approaches a corresponding eigenvector.

EXAMPLE 2 Apply the power method to $A = \begin{bmatrix} 6 & 5 \\ 1 & 2 \end{bmatrix}$ with $\mathbf{x}_0 = \begin{bmatrix} 0 \\ 1 \end{bmatrix}$. Stop when $k = 5$, and estimate the dominant eigenvalue and a corresponding eigenvector of A.

SOLUTION Calculations in this example and the next were made with MATLAB, which computes with 16-digit accuracy, although we show only a few significant figures here. To begin, compute $A\mathbf{x}_0$ and identify the largest entry μ_0 in $A\mathbf{x}_0$:

$$A\mathbf{x}_0 = \begin{bmatrix} 6 & 5 \\ 1 & 2 \end{bmatrix} \begin{bmatrix} 0 \\ 1 \end{bmatrix} = \begin{bmatrix} 5 \\ 2 \end{bmatrix}, \quad \mu_0 = 5$$

Scale $A\mathbf{x}_0$ by $1/\mu_0$ to get $\mathbf{x}_1$, compute $A\mathbf{x}_1$, and identify the largest entry in $A\mathbf{x}_1$:

$$\mathbf{x}_1 = \frac{1}{\mu_0}A\mathbf{x}_0 = \frac{1}{5}\begin{bmatrix} 5 \\ 2 \end{bmatrix} = \begin{bmatrix} 1 \\ .4 \end{bmatrix}$$

$$A\mathbf{x}_1 = \begin{bmatrix} 6 & 5 \\ 1 & 2 \end{bmatrix} \begin{bmatrix} 1 \\ .4 \end{bmatrix} = \begin{bmatrix} 8 \\ 1.8 \end{bmatrix}, \quad \mu_1 = 8$$

Scale $A\mathbf{x}_1$ by $1/\mu_1$ to get $\mathbf{x}_2$, compute $A\mathbf{x}_2$, and identify the largest entry in $A\mathbf{x}_2$:

$$\mathbf{x}_2 = \frac{1}{\mu_1}A\mathbf{x}_1 = \frac{1}{8}\begin{bmatrix} 8 \\ 1.8 \end{bmatrix} = \begin{bmatrix} 1 \\ .225 \end{bmatrix}$$

$$A\mathbf{x}_2 = \begin{bmatrix} 6 & 5 \\ 1 & 2 \end{bmatrix} \begin{bmatrix} 1 \\ .225 \end{bmatrix} = \begin{bmatrix} 7.125 \\ 1.450 \end{bmatrix}, \quad \mu_2 = 7.125$$

Scale $A\mathbf{x}_2$ by $1/\mu_2$ to get $\mathbf{x}_3$, and so on. The results of MATLAB calculations for the first five iterations are arranged in Table 2.

TABLE 2 The Power Method for Example 2

k	0	1	2	3	4	5
$\mathbf{x}_k$	$\begin{bmatrix} 0 \\ 1 \end{bmatrix}$	$\begin{bmatrix} 1 \\ .4 \end{bmatrix}$	$\begin{bmatrix} 1 \\ .225 \end{bmatrix}$	$\begin{bmatrix} 1 \\ .2035 \end{bmatrix}$	$\begin{bmatrix} 1 \\ .2005 \end{bmatrix}$	$\begin{bmatrix} 1 \\ .20007 \end{bmatrix}$
$A\mathbf{x}_k$	$\begin{bmatrix} 5 \\ 2 \end{bmatrix}$	$\begin{bmatrix} 8 \\ 1.8 \end{bmatrix}$	$\begin{bmatrix} 7.125 \\ 1.450 \end{bmatrix}$	$\begin{bmatrix} 7.0175 \\ 1.4070 \end{bmatrix}$	$\begin{bmatrix} 7.0025 \\ 1.4010 \end{bmatrix}$	$\begin{bmatrix} 7.00036 \\ 1.40014 \end{bmatrix}$
μ_k	5	8	7.125	7.0175	7.0025	7.00036

The evidence from Table 2 strongly suggests that $\{\mathbf{x}_k\}$ approaches $(1, .2)$ and $\{\mu_k\}$ approaches 7. If so, then $(1, .2)$ is an eigenvector and 7 is the dominant eigenvalue. This is easily verified by computing

$$A\begin{bmatrix} 1 \\ .2 \end{bmatrix} = \begin{bmatrix} 6 & 5 \\ 1 & 2 \end{bmatrix}\begin{bmatrix} 1 \\ .2 \end{bmatrix} = \begin{bmatrix} 7 \\ 1.4 \end{bmatrix} = 7\begin{bmatrix} 1 \\ .2 \end{bmatrix}$$ ■

The sequence $\{\mu_k\}$ in Example 2 converged quickly to $\lambda_1 = 7$ because the second eigenvalue of A was much smaller. (In fact, $\lambda_2 = 1$.) In general, the rate of convergence depends on the ratio $|\lambda_2/\lambda_1|$, because the vector $c_2(\lambda_2/\lambda_1)^k\mathbf{v}_2$ in equation (2) is the main source of error when using a scaled version of $A^k\mathbf{x}$ as an estimate of $c_1\mathbf{v}_1$. (The other fractions λ_j/λ_1 are likely to be smaller.) If $|\lambda_2/\lambda_1|$ is close to 1, then $\{\mu_k\}$ and $\{\mathbf{x}_k\}$ can converge very slowly, and other approximation methods may be preferred.

With the power method, there is a slight chance that the chosen initial vector $\mathbf{x}$ will have no component in the $\mathbf{v}_1$ direction (when $c_1 = 0$). But computer rounding errors during the calculations of the $\mathbf{x}_k$ are likely to create a vector with at least a small component in the direction of $\mathbf{v}_1$. If that occurs, the $\mathbf{x}_k$ will start to converge to a multiple of $\mathbf{v}_1$.

The Inverse Power Method

This method provides an approximation for *any* eigenvalue, provided a good initial estimate α of the eigenvalue λ is known. In this case, we let $B = (A - \alpha I)^{-1}$ and apply the power method to B. It can be shown that if the eigenvalues of A are $\lambda_1, \ldots, \lambda_n$, then the eigenvalues of B are

$$\frac{1}{\lambda_1 - \alpha}, \quad \frac{1}{\lambda_2 - \alpha}, \quad \ldots, \quad \frac{1}{\lambda_n - \alpha}$$

and the corresponding eigenvectors are the same as those for A. (See Exercises 15 and 16.)

Suppose, for example, that α is closer to λ_2 than to the other eigenvalues of A. Then $1/(\lambda_2 - \alpha)$ will be a strictly dominant eigenvalue of B. If α is really close to λ_2, then $1/(\lambda_2 - \alpha)$ is *much* larger than the other eigenvalues of B, and the inverse power method produces a very rapid approximation to λ_2 for almost all choices of $\mathbf{x}_0$. The following algorithm gives the details.

THE INVERSE POWER METHOD FOR ESTIMATING AN EIGENVALUE λ OF A

1. Select an initial estimate α sufficiently close to λ.

2. Select an initial vector $\mathbf{x}_0$ whose largest entry is 1.

3. For $k = 0, 1, \ldots,$

   a. Solve $(A - \alpha I)\mathbf{y}_k = \mathbf{x}_k$ for $\mathbf{y}_k$.

   b. Let μ_k be an entry in $\mathbf{y}_k$ with the largest absolute value.

   c. Compute $\nu_k = \alpha + (1/\mu_k)$.

   d. Compute $\mathbf{x}_{k+1} = (1/\mu_k)\mathbf{y}_k$.

4. For almost all choices of $\mathbf{x}_0$, the sequence $\{\nu_k\}$ approaches the eigenvalue λ of A, and the sequence $\{\mathbf{x}_k\}$ approaches a corresponding eigenvector.

Notice that B, or rather $(A - \alpha I)^{-1}$, does not appear in the algorithm. Instead of computing $(A - \alpha I)^{-1}\mathbf{x}_k$ to get the next vector in the sequence, it is better to *solve* the equation $(A - \alpha I)\mathbf{y}_k = \mathbf{x}_k$ for $\mathbf{y}_k$ (and then scale $\mathbf{y}_k$ to produce $\mathbf{x}_{k+1}$). Since this equation for $\mathbf{y}_k$ must be solved for each k, an LU factorization of $A - \alpha I$ will speed up the process.

EXAMPLE 3 It is not uncommon in some applications to need to know the smallest eigenvalue of a matrix A and to have at hand rough estimates of the eigenvalues. Suppose 21, 3.3, and 1.9 are estimates for the eigenvalues of the matrix A below. Find the smallest eigenvalue, accurate to six decimal places.

$$A = \begin{bmatrix} 10 & -8 & -4 \\ -8 & 13 & 4 \\ -4 & 5 & 4 \end{bmatrix}$$

SOLUTION The two smallest eigenvalues seem close together, so we use the inverse power method for $A - 1.9I$. Results of a MATLAB calculation are shown in Table 3. Here $\mathbf{x}_0$ was chosen arbitrarily, $\mathbf{y}_k = (A - 1.9I)^{-1}\mathbf{x}_k$, μ_k is the largest entry in $\mathbf{y}_k$, $\nu_k = 1.9 + 1/\mu_k$, and $\mathbf{x}_{k+1} = (1/\mu_k)\mathbf{y}_k$. As it turns out, the initial eigenvalue estimate was fairly good, and the inverse power sequence converged quickly. The smallest eigenvalue is exactly 2. ∎

TABLE 3 The Inverse Power Method

k	**0**	**1**	**2**	**3**	**4**
$\mathbf{x}_k$	$\begin{bmatrix} 1 \\ 1 \\ 1 \end{bmatrix}$	$\begin{bmatrix} .5736 \\ .0646 \\ 1 \end{bmatrix}$	$\begin{bmatrix} .5054 \\ .0045 \\ 1 \end{bmatrix}$	$\begin{bmatrix} .5004 \\ .0003 \\ 1 \end{bmatrix}$	$\begin{bmatrix} .50003 \\ .00002 \\ 1 \end{bmatrix}$
$\mathbf{y}_k$	$\begin{bmatrix} 4.45 \\ .50 \\ 7.76 \end{bmatrix}$	$\begin{bmatrix} 5.0131 \\ .0442 \\ 9.9197 \end{bmatrix}$	$\begin{bmatrix} 5.0012 \\ .0031 \\ 9.9949 \end{bmatrix}$	$\begin{bmatrix} 5.0001 \\ .0002 \\ 9.9996 \end{bmatrix}$	$\begin{bmatrix} 5.000006 \\ .000015 \\ 9.999975 \end{bmatrix}$
μ_k	7.76	9.9197	9.9949	9.9996	9.999975
ν_k	2.03	2.0008	2.00005	2.000004	2.0000002

If an estimate for the smallest eigenvalue of a matrix is not available, one can simply take $\alpha = 0$ in the inverse power method. This choice of α works reasonably well if the smallest eigenvalue is much closer to zero than to the other eigenvalues.

The two algorithms presented in this section are practical tools for many simple situations, and they provide an introduction to the problem of eigenvalue estimation. A more robust and widely used iterative method is the QR algorithm. For instance, it is the heart of the MATLAB command `eig(A)`, which rapidly computes eigenvalues and eigenvectors of A. A brief description of the QR algorithm was given in the exercises for Section 5.2. Further details are presented in most modern numerical analysis texts.

Practice Problem

How can you tell if a given vector $\mathbf{x}$ is a good approximation to an eigenvector of a matrix A? If it is, how would you estimate the corresponding eigenvalue? Experiment with

$$A = \begin{bmatrix} 5 & 8 & 4 \\ 8 & 3 & -1 \\ 4 & -1 & 2 \end{bmatrix} \quad \text{and} \quad \mathbf{x} = \begin{bmatrix} 1.0 \\ -4.3 \\ 8.1 \end{bmatrix}$$

5.8 Exercises

In Exercises 1–4, the matrix A is followed by a sequence $\{\mathbf{x}_k\}$ produced by the power method. Use these data to estimate the largest eigenvalue of A, and give a corresponding eigenvector.

1. $A = \begin{bmatrix} 4 & 3 \\ 1 & 2 \end{bmatrix}$;

$\begin{bmatrix} 1 \\ 0 \end{bmatrix}, \begin{bmatrix} 1 \\ .25 \end{bmatrix}, \begin{bmatrix} 1 \\ .3158 \end{bmatrix}, \begin{bmatrix} 1 \\ .3298 \end{bmatrix}, \begin{bmatrix} 1 \\ .3326 \end{bmatrix}$

2. $A = \begin{bmatrix} 1.8 & -.8 \\ -3.2 & 4.2 \end{bmatrix}$;

$\begin{bmatrix} 1 \\ 0 \end{bmatrix}, \begin{bmatrix} -.5625 \\ 1 \end{bmatrix}, \begin{bmatrix} -.3021 \\ 1 \end{bmatrix}, \begin{bmatrix} -.2601 \\ 1 \end{bmatrix}, \begin{bmatrix} -.2520 \\ 1 \end{bmatrix}$

3. $A = \begin{bmatrix} .5 & .2 \\ .4 & .7 \end{bmatrix}$;

$\begin{bmatrix} 1 \\ 0 \end{bmatrix}, \begin{bmatrix} 1 \\ .8 \end{bmatrix}, \begin{bmatrix} .6875 \\ 1 \end{bmatrix}, \begin{bmatrix} .5577 \\ 1 \end{bmatrix}, \begin{bmatrix} .5188 \\ 1 \end{bmatrix}$

4. $A = \begin{bmatrix} 4.1 & -6 \\ 3 & -4.4 \end{bmatrix}$;

$\begin{bmatrix} 1 \\ 1 \end{bmatrix}, \begin{bmatrix} 1 \\ .7368 \end{bmatrix}, \begin{bmatrix} 1 \\ .7541 \end{bmatrix}, \begin{bmatrix} 1 \\ .7490 \end{bmatrix}, \begin{bmatrix} 1 \\ .7502 \end{bmatrix}$

5. Let $A = \begin{bmatrix} 15 & 16 \\ -20 & -21 \end{bmatrix}$. The vectors $\mathbf{x}, \ldots, A^5\mathbf{x}$ are $\begin{bmatrix} 1 \\ 1 \end{bmatrix}$,

$\begin{bmatrix} 31 \\ -41 \end{bmatrix}, \begin{bmatrix} -191 \\ 241 \end{bmatrix}, \begin{bmatrix} 991 \\ -1241 \end{bmatrix}, \begin{bmatrix} -4991 \\ 6241 \end{bmatrix}, \begin{bmatrix} 24991 \\ -31241 \end{bmatrix}$.

Find a vector with a 1 in the second entry that is close to an eigenvector of A. Use four decimal places. Check your estimate, and give an estimate for the dominant eigenvalue of A.

6. Let $A = \begin{bmatrix} -2 & -3 \\ 6 & 7 \end{bmatrix}$. Repeat Exercise 5, using the following sequence $\mathbf{x}, A\mathbf{x}, \ldots, A^5\mathbf{x}$.

$\begin{bmatrix} 1 \\ 1 \end{bmatrix}, \begin{bmatrix} -5 \\ 13 \end{bmatrix}, \begin{bmatrix} -29 \\ 61 \end{bmatrix}, \begin{bmatrix} -125 \\ 253 \end{bmatrix}, \begin{bmatrix} -509 \\ 1021 \end{bmatrix}, \begin{bmatrix} -2045 \\ 4093 \end{bmatrix}$

Exercises 7–12 require MATLAB or other computational aid. In Exercises 7 and 8, use the power method with the $\mathbf{x}_0$ given. List $\{\mathbf{x}_k\}$ and $\{\mu_k\}$ for $k = 1, \ldots, 5$. In Exercises 9 and 10, list μ_5 and μ_6.

T 7. $A = \begin{bmatrix} 6 & 7 \\ 8 & 5 \end{bmatrix}, \mathbf{x}_0 = \begin{bmatrix} 1 \\ 0 \end{bmatrix}$

T 8. $A = \begin{bmatrix} 2 & 1 \\ 4 & 5 \end{bmatrix}, \mathbf{x}_0 = \begin{bmatrix} 1 \\ 0 \end{bmatrix}$

T 9. $A = \begin{bmatrix} 8 & 0 & 12 \\ 1 & -2 & 1 \\ 0 & 3 & 0 \end{bmatrix}, \mathbf{x}_0 = \begin{bmatrix} 1 \\ 0 \\ 0 \end{bmatrix}$

T 10. $A = \begin{bmatrix} 1 & 2 & -2 \\ 1 & 1 & 9 \\ 0 & 1 & 9 \end{bmatrix}, \mathbf{x}_0 = \begin{bmatrix} 1 \\ 0 \\ 0 \end{bmatrix}$

Another estimate can be made for an eigenvalue when an approximate eigenvector is available. Observe that if $A\mathbf{x} = \lambda\mathbf{x}$, then $\mathbf{x}^T A\mathbf{x} = \mathbf{x}^T(\lambda\mathbf{x}) = \lambda(\mathbf{x}^T\mathbf{x})$, and the **Rayleigh quotient**

$$R(\mathbf{x}) = \frac{\mathbf{x}^T A\mathbf{x}}{\mathbf{x}^T\mathbf{x}}$$

equals λ. If $\mathbf{x}$ is close to an eigenvector for λ, then this quotient is close to λ. When A is a symmetric matrix ($A^T = A$), the Rayleigh quotient $R(\mathbf{x}_k) = (\mathbf{x}_k^T A\mathbf{x}_k)/(\mathbf{x}_k^T\mathbf{x}_k)$ will have roughly twice as many digits of accuracy as the scaling factor μ_k in the power method. Verify this increased accuracy in Exercises 11 and 12 by computing μ_k and $R(\mathbf{x}_k)$ for $k = 1, \ldots, 4$.

T 11. $A = \begin{bmatrix} 5 & 2 \\ 2 & 2 \end{bmatrix}, \mathbf{x}_0 = \begin{bmatrix} 1 \\ 0 \end{bmatrix}$

T 12. $A = \begin{bmatrix} -3 & 2 \\ 2 & 0 \end{bmatrix}, \mathbf{x}_0 = \begin{bmatrix} 1 \\ 0 \end{bmatrix}$

Exercises 13 and 14 apply to a 3×3 matrix A whose eigenvalues are estimated to be 4, −4, and 3.

13. If the eigenvalues close to 4 and −4 are known to have different absolute values, will the power method work? Is it likely to be useful?

14. Suppose the eigenvalues close to 4 and −4 are known to have exactly the same absolute value. Describe how one might obtain a sequence that estimates the eigenvalue close to 4.

15. Suppose $A\mathbf{x} = \lambda\mathbf{x}$ with $\mathbf{x} \neq \mathbf{0}$. Let α be a scalar different from the eigenvalues of A, and let $B = (A - \alpha I)^{-1}$. Subtract $\alpha\mathbf{x}$ from both sides of the equation $A\mathbf{x} = \lambda\mathbf{x}$, and use algebra to show that $1/(\lambda - \alpha)$ is an eigenvalue of B, with $\mathbf{x}$ a corresponding eigenvector.

16. Suppose μ is an eigenvalue of the B in Exercise 15, and that $\mathbf{x}$ is a corresponding eigenvector, so that $(A - \alpha I)^{-1}\mathbf{x} = \mu\mathbf{x}$. Use this equation to find an eigenvalue of A in terms of μ and α. [*Note:* $\mu \neq 0$ because B is invertible.]

T 17. Use the inverse power method to estimate the middle eigenvalue of the A in Example 3, with accuracy to four decimal places. Set $\mathbf{x}_0 = (1, 0, 0)$.

T 18. Let A be as in Exercise 9. Use the inverse power method with $\mathbf{x}_0 = (1, 0, 0)$ to estimate the eigenvalue of A near $\alpha = -1.4$, with an accuracy to four decimal places.

In Exercises 19 and 20, find (a) the largest eigenvalue and (b) the eigenvalue closest to zero. In each case, set $\mathbf{x}_0 = (1, 0, 0, 0)$ and carry out approximations until the approximating sequence seems accurate to four decimal places. Include the approximate eigenvector.

19. $A = \begin{bmatrix} 10 & 7 & 8 & 7 \\ 7 & 5 & 6 & 5 \\ 8 & 6 & 10 & 9 \\ 7 & 5 & 9 & 10 \end{bmatrix}$

20. $A = \begin{bmatrix} 1 & 2 & 3 & 2 \\ 2 & 12 & 13 & 11 \\ -2 & 3 & 0 & 2 \\ 4 & 5 & 7 & 2 \end{bmatrix}$

21. A common misconception is that if A has a strictly dominant eigenvalue, then, for any sufficiently large value of k, the

vector $A^k \mathbf{x}$ is approximately equal to an eigenvector of A. For the three matrices below, study what happens to $A^k \mathbf{x}$ when $\mathbf{x} = (.5, .5)$, and try to draw general conclusions (for a 2×2 matrix).

a. $A = \begin{bmatrix} .8 & 0 \\ 0 & .2 \end{bmatrix}$ 　　b. $A = \begin{bmatrix} 1 & 0 \\ 0 & .8 \end{bmatrix}$

c. $A = \begin{bmatrix} 8 & 0 \\ 0 & 2 \end{bmatrix}$

Solution to Practice Problem

For the given A and $\mathbf{x}$,

$$A\mathbf{x} = \begin{bmatrix} 5 & 8 & 4 \\ 8 & 3 & -1 \\ 4 & -1 & 2 \end{bmatrix} \begin{bmatrix} 1.00 \\ -4.30 \\ 8.10 \end{bmatrix} = \begin{bmatrix} 3.00 \\ -13.00 \\ 24.50 \end{bmatrix}$$

If $A\mathbf{x}$ is nearly a multiple of $\mathbf{x}$, then the ratios of corresponding entries in the two vectors should be nearly constant. So compute:

{entry in $A\mathbf{x}$}	÷ {entry in $\mathbf{x}$}	= {ratio}
3.00	1.00	3.000
−13.00	−4.30	3.023
24.50	8.10	3.025

Each entry in $A\mathbf{x}$ is about 3 times the corresponding entry in $\mathbf{x}$, so $\mathbf{x}$ is close to an eigenvector. Any of the ratios above is an estimate for the eigenvalue. (To five decimal places, the eigenvalue is 3.02409.)

5.9 Applications to Markov Chains

The Markov chains described in this section are used as mathematical models of a wide variety of situations in biology, business, chemistry, engineering, physics, and elsewhere. In each case, the model is used to describe an experiment or measurement that is performed many times in the same way, where the outcome of each trial of the experiment will be one of several specified possible outcomes, and where the outcome of one trial depends only on the immediately preceding trial.

For example, if the population of a city and its suburbs were measured each year, then a vector such as

$$\mathbf{x}_0 = \begin{bmatrix} .60 \\ .40 \end{bmatrix} \tag{1}$$

could indicate that 60% of the population lives in the city and 40% in the suburbs. The decimals in $\mathbf{x}_0$ add up to 1 because they account for the entire population of the region. Percentages are more convenient for our purposes here than population totals.

DEFINITION

A vector with nonnegative entries that add up to 1 is called a **probability vector**. A **stochastic matrix** is a square matrix whose columns are probability vectors.

A **Markov chain** is a sequence of probability vectors $\mathbf{x}_0, \mathbf{x}_1, \mathbf{x}_2, \ldots$, together with a stochastic matrix P, such that

$$\mathbf{x}_1 = P\mathbf{x}_0, \quad \mathbf{x}_2 = P\mathbf{x}_1, \quad \mathbf{x}_3 = P\mathbf{x}_2, \quad \ldots$$

Thus the Markov chain is described by the first-order difference equation

$$\mathbf{x}_{k+1} = P\mathbf{x}_k \quad \text{for } k = 0, 1, 2, \ldots$$

When a Markov chain of vectors in $\mathbb{R}^n$ describes a system or a sequence of experiments, the entries in $\mathbf{x}_k$ list, respectively, the probabilities that the system is in each of n possible states, or the probabilities that the outcome of the experiment is one of n possible outcomes. For this reason, $\mathbf{x}_k$ is often called a **state vector**.

EXAMPLE 1 Section 1.10 examined a model for population movement between a city and its suburbs. See Figure 1. The annual migration between these two parts of the metropolitan region was governed by the *migration matrix M*:

$$
\begin{array}{c}
\text{From} \\
\begin{array}{cc}
\text{City} & \text{Suburbs}
\end{array} \\
M = \begin{bmatrix} .95 & .03 \\ .05 & .97 \end{bmatrix} \begin{array}{l} \text{To} \\ \text{City} \\ \text{Suburbs} \end{array}
\end{array}
$$

That is, each year 5% of the city population moves to the suburbs, and 3% of the suburban population moves to the city. The columns of M are probability vectors, so M is a stochastic matrix. Suppose the 2020 population of the region is 600,000 in the city and 400,000 in the suburbs. Then the initial distribution of the population in the region is given previously by $\mathbf{x}_0$ in (1). What is the distribution of the population in 2021? In 2022?

FIGURE 1 Annual percentage migration between city and suburbs.

SOLUTION In Example 3 of Section 1.10, we saw that after one year, the population vector $\begin{bmatrix} 600,000 \\ 400,000 \end{bmatrix}$ changed to

$$\begin{bmatrix} .95 & .03 \\ .05 & .97 \end{bmatrix} \begin{bmatrix} 600,000 \\ 400,000 \end{bmatrix} = \begin{bmatrix} 582,000 \\ 418,000 \end{bmatrix}$$

If we divide both sides of this equation by the total population of 1 million, and use the fact that $kM\mathbf{x} = M(k\mathbf{x})$, we find that

$$\begin{bmatrix} .95 & .03 \\ .05 & .97 \end{bmatrix} \begin{bmatrix} .600 \\ .400 \end{bmatrix} = \begin{bmatrix} .582 \\ .418 \end{bmatrix}$$

The vector $\mathbf{x}_1 = \begin{bmatrix} .582 \\ .418 \end{bmatrix}$ gives the population distribution in 2021. That is, 58.2% of the region lived in the city and 41.8% lived in the suburbs. Similarly, the population distribution in 2022 is described by a vector $\mathbf{x}_2$, where

$$\mathbf{x}_2 = M\mathbf{x}_1 = \begin{bmatrix} .95 & .03 \\ .05 & .97 \end{bmatrix} \begin{bmatrix} .582 \\ .418 \end{bmatrix} = \begin{bmatrix} .565 \\ .435 \end{bmatrix} \qquad \blacksquare$$

EXAMPLE 2 Suppose the voting results of a congressional election at a certain voting precinct are represented by a vector $\mathbf{x}$ in $\mathbb{R}^3$:

$$\mathbf{x} = \begin{bmatrix} \% \text{ voting Democratic (D)} \\ \% \text{ voting Republican (R)} \\ \% \text{ voting Other (O)} \end{bmatrix}$$

Suppose we record the outcome of the congressional election every two years by a vector of this type and the outcome of one election depends only on the results of the preceding election. Then the sequence of vectors that describe the votes every two years may be a Markov chain. As an example of a stochastic matrix P for this chain, we take

$$\begin{array}{cccc} & \text{From} & & \\ & D \quad R \quad O & \text{To} \\ P = \begin{bmatrix} .70 & .10 & .30 \\ .20 & .80 & .30 \\ .10 & .10 & .40 \end{bmatrix} & \begin{array}{c} D \\ R \\ O \end{array} \end{array}$$

The entries in the first column, labeled D, describe what the persons voting Democratic in one election will do in the next election. Here we have supposed that 70% will vote D again in the next election, 20% will vote R, and 10% will vote O. Similar interpretations hold for the other columns of P. A diagram for this matrix is shown in Figure 2.

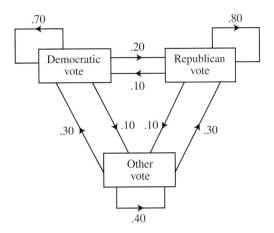

FIGURE 2 Voting changes from one election to the next.

If the "transition" percentages remain constant over many years from one election to the next, then the sequence of vectors that give the voting outcomes forms a Markov chain. Suppose the outcome of one election is given by

$$\mathbf{x}_0 = \begin{bmatrix} .55 \\ .40 \\ .05 \end{bmatrix}$$

Determine the likely outcome of the next election and the likely outcome of the election after that.

SOLUTION The outcome of the next election is described by the state vector $\mathbf{x}_1$ and that of the election after that by $\mathbf{x}_2$, where

$$\mathbf{x}_1 = P\mathbf{x}_0 = \begin{bmatrix} .70 & .10 & .30 \\ .20 & .80 & .30 \\ .10 & .10 & .40 \end{bmatrix} \begin{bmatrix} .55 \\ .40 \\ .05 \end{bmatrix} = \begin{bmatrix} .440 \\ .445 \\ .115 \end{bmatrix} \quad \begin{matrix} \text{44\% will vote D.} \\ \text{44.5\% will vote R.} \\ \text{11.5\% will vote O.} \end{matrix}$$

$$\mathbf{x}_2 = P\mathbf{x}_1 = \begin{bmatrix} .70 & .10 & .30 \\ .20 & .80 & .30 \\ .10 & .10 & .40 \end{bmatrix} \begin{bmatrix} .440 \\ .445 \\ .115 \end{bmatrix} = \begin{bmatrix} .3870 \\ .4785 \\ .1345 \end{bmatrix} \quad \begin{matrix} \text{38.7\% will vote D.} \\ \text{47.9\% will vote R.} \\ \text{13.5\% will vote O.} \end{matrix}$$

To understand why $\mathbf{x}_1$ does indeed give the outcome of the next election, suppose 1000 persons voted in the "first" election, with 550 voting D, 400 voting R, and 50 voting O. (See the percentages in $\mathbf{x}_0$.) In the next election, 70% of the 550 will vote D again, 10% of the 400 will switch from R to D, and 30% of the 50 will switch from O to D. Thus the total D vote will be

$$.70(550) + .10(400) + .30(50) = 385 + 40 + 15 = 440 \tag{2}$$

Thus 44% of the vote next time will be for the D candidate. The calculation in (2) is essentially the same as that used to compute the first entry in $\mathbf{x}_1$. Analogous calculations could be made for the other entries in $\mathbf{x}_1$, for the entries in $\mathbf{x}_2$, and so on. ∎

Predicting the Distant Future

The most interesting aspect of Markov chains is the study of a chain's long-term behavior. For instance, what can be said in Example 2 about the voting after many elections have passed (assuming that the given stochastic matrix continues to describe the transition percentages from one election to the next)? Or, what happens to the population distribution in Example 1 "in the long run"? Here our work on eigenvalues and eigenvectors becomes helpful.

THEOREM 10

Stochastic Matrices

If P is a stochastic matrix, then 1 is an eigenvalue of P.

PROOF Since the columns of P sum to 1, the rows of P^T will also sum to 1. Let $\mathbf{e}$ represent the vector for which every entry is 1. Notice that multiplying P^T by $\mathbf{e}$ has the effect of adding up the values in each row, hence $P^T\mathbf{e} = \mathbf{e}$, establishing that $\mathbf{e}$ is an eigenvector of P^T with eigenvalue 1. Since P and P^T have the same eigenvalues (Exercise 20 in Section 5.2), 1 is also an eigenvalue of P. ∎

In the next example, we see that the vectors generated in a Markov chain are almost the same as the vectors generated using the power method outlined in Section 5.8 – the only difference is that in a Markov chain, the vectors are not scaled at each step. Based on our experience from Section 5.8, as k increases we expect $\mathbf{x}_k \to \mathbf{q}$, where $\mathbf{q}$ is an eigenvector of P.

EXAMPLE 3   Let $P = \begin{bmatrix} .5 & .2 & .3 \\ .3 & .8 & .3 \\ .2 & 0 & .4 \end{bmatrix}$ and $\mathbf{x}_0 = \begin{bmatrix} 1 \\ 0 \\ 0 \end{bmatrix}$. Consider a system whose state is described by the Markov chain $\mathbf{x}_{k+1} = P\mathbf{x}_k$, for $k = 0, 1, \ldots$ What happens to the system as time passes? Compute the state vectors $\mathbf{x}_1, \ldots, \mathbf{x}_{15}$ to find out.

SOLUTION

$$\mathbf{x}_1 = P\mathbf{x}_0 = \begin{bmatrix} .5 & .2 & .3 \\ .3 & .8 & .3 \\ .2 & 0 & .4 \end{bmatrix} \begin{bmatrix} 1 \\ 0 \\ 0 \end{bmatrix} = \begin{bmatrix} .5 \\ .3 \\ .2 \end{bmatrix}$$

$$\mathbf{x}_2 = P\mathbf{x}_1 = \begin{bmatrix} .5 & .2 & .3 \\ .3 & .8 & .3 \\ .2 & 0 & .4 \end{bmatrix} \begin{bmatrix} .5 \\ .3 \\ .2 \end{bmatrix} = \begin{bmatrix} .37 \\ .45 \\ .18 \end{bmatrix}$$

$$\mathbf{x}_3 = P\mathbf{x}_2 = \begin{bmatrix} .5 & .2 & .3 \\ .3 & .8 & .3 \\ .2 & 0 & .4 \end{bmatrix} \begin{bmatrix} .37 \\ .45 \\ .18 \end{bmatrix} = \begin{bmatrix} .329 \\ .525 \\ .146 \end{bmatrix}$$

The results of further calculations are shown below, with entries rounded to four or five significant figures.

$$\mathbf{x}_4 = \begin{bmatrix} .3133 \\ .5625 \\ .1242 \end{bmatrix}, \quad \mathbf{x}_5 = \begin{bmatrix} .3064 \\ .5813 \\ .1123 \end{bmatrix}, \quad \mathbf{x}_6 = \begin{bmatrix} .3032 \\ .5906 \\ .1062 \end{bmatrix}, \quad \mathbf{x}_7 = \begin{bmatrix} .3016 \\ .5953 \\ .1031 \end{bmatrix}$$

$$\mathbf{x}_8 = \begin{bmatrix} .3008 \\ .5977 \\ .1016 \end{bmatrix}, \quad \mathbf{x}_9 = \begin{bmatrix} .3004 \\ .5988 \\ .1008 \end{bmatrix}, \quad \mathbf{x}_{10} = \begin{bmatrix} .3002 \\ .5994 \\ .1004 \end{bmatrix}, \quad \mathbf{x}_{11} = \begin{bmatrix} .3001 \\ .5997 \\ .1002 \end{bmatrix}$$

$$\mathbf{x}_{12} = \begin{bmatrix} .30005 \\ .59985 \\ .10010 \end{bmatrix}, \quad \mathbf{x}_{13} = \begin{bmatrix} .30002 \\ .59993 \\ .10005 \end{bmatrix}, \quad \mathbf{x}_{14} = \begin{bmatrix} .30001 \\ .59996 \\ .10002 \end{bmatrix}, \quad \mathbf{x}_{15} = \begin{bmatrix} .30001 \\ .59998 \\ .10001 \end{bmatrix}$$

These vectors seem to be approaching $\mathbf{q} = \begin{bmatrix} .3 \\ .6 \\ .1 \end{bmatrix}$. The probabilities are hardly changing from one value of k to the next. Observe that the following calculation is exact (with no rounding error):

$$P\mathbf{q} = \begin{bmatrix} .5 & .2 & .3 \\ .3 & .8 & .3 \\ .2 & 0 & .4 \end{bmatrix} \begin{bmatrix} .3 \\ .6 \\ .1 \end{bmatrix} = \begin{bmatrix} .15 + .12 + .03 \\ .09 + .48 + .03 \\ .06 + 0 + .04 \end{bmatrix} = \begin{bmatrix} .30 \\ .60 \\ .10 \end{bmatrix} = \mathbf{q}$$

When the system is in state $\mathbf{q}$, there is no change in the system from one measurement to the next. ∎

Steady-State Vectors

If P is a stochastic matrix, then a **steady-state vector** (or **equilibrium vector**) for P is a probability vector $\mathbf{q}$ such that

$$P\mathbf{q} = \mathbf{q}$$

In Theorem 10, it is established that 1 is an eigenvalue of any stochastic matrix. It can be shown that 1 is actually the largest eigenvalue of a stochastic matrix and the associated eigenvector can be chosen to be a steady-state vector. In Example 3, $\mathbf{q}$ is a steady-state vector for P.

EXAMPLE 4 The probability vector $\mathbf{q} = \begin{bmatrix} .375 \\ .625 \end{bmatrix}$ is a steady-state vector for the population migration matrix M in Example 1, because

$$M\mathbf{q} = \begin{bmatrix} .95 & .03 \\ .05 & .97 \end{bmatrix} \begin{bmatrix} .375 \\ .625 \end{bmatrix} = \begin{bmatrix} .35625 + .01875 \\ .01875 + .60625 \end{bmatrix} = \begin{bmatrix} .375 \\ .625 \end{bmatrix} = \mathbf{q} \qquad \blacksquare$$

If the total population of the metropolitan region in Example 1 is 1 million, then $\mathbf{q}$ from Example 4 would correspond to having 375,000 persons in the city and 625,000 in the suburbs. At the end of one year, the migration *out of* the city would be $(.05)(375,000) = 18,750$ persons, and the migration *into* the city from the suburbs would be $(.03)(625,000) = 18,750$ persons. As a result, the population in the city would remain the same. Similarly, the suburban population would be stable.

The next example shows how to *find* a steady-state vector. Notice that we are just finding an eigenvector associated with the eigenvalue 1 and then scaling it to create a probability vector.

EXAMPLE 5 Let $P = \begin{bmatrix} .6 & .3 \\ .4 & .7 \end{bmatrix}$. Find a steady-state vector for P.

SOLUTION First, solve the equation $P\mathbf{x} = \mathbf{x}$.

$$P\mathbf{x} - \mathbf{x} = \mathbf{0}$$
$$P\mathbf{x} - I\mathbf{x} = \mathbf{0} \qquad \text{Recall from Section 1.4 that } I\mathbf{x} = \mathbf{x}.$$
$$(P - I)\mathbf{x} = \mathbf{0}$$

For P as above,

$$P - I = \begin{bmatrix} .6 & .3 \\ .4 & .7 \end{bmatrix} - \begin{bmatrix} 1 & 0 \\ 0 & 1 \end{bmatrix} = \begin{bmatrix} -.4 & .3 \\ .4 & -.3 \end{bmatrix}$$

To find all solutions of $(P - I)\mathbf{x} = \mathbf{0}$, row reduce the augmented matrix:

$$\begin{bmatrix} -.4 & .3 & 0 \\ .4 & -.3 & 0 \end{bmatrix} \sim \begin{bmatrix} -.4 & .3 & 0 \\ 0 & 0 & 0 \end{bmatrix} \sim \begin{bmatrix} 1 & -3/4 & 0 \\ 0 & 0 & 0 \end{bmatrix}$$

Then $x_1 = \frac{3}{4}x_2$ and x_2 is free. The general solution is $x_2 \begin{bmatrix} 3/4 \\ 1 \end{bmatrix}$.

Next, choose a simple basis for the solution space. One obvious choice is $\begin{bmatrix} 3/4 \\ 1 \end{bmatrix}$ but a better choice with no fractions is $\mathbf{w} = \begin{bmatrix} 3 \\ 4 \end{bmatrix}$ (corresponding to $x_2 = 4$).

Finally, find a probability vector in the set of all solutions of $P\mathbf{x} = \mathbf{x}$. This process is easy, since every solution is a multiple of the solution $\mathbf{w}$. Divide $\mathbf{w}$ by the sum of its entries and obtain

$$\mathbf{q} = \begin{bmatrix} 3/7 \\ 4/7 \end{bmatrix}$$

As a check, compute

$$P\mathbf{q} = \begin{bmatrix} 6/10 & 3/10 \\ 4/10 & 7/10 \end{bmatrix} \begin{bmatrix} 3/7 \\ 4/7 \end{bmatrix} = \begin{bmatrix} 18/70 + 12/70 \\ 12/70 + 28/70 \end{bmatrix} = \begin{bmatrix} 30/70 \\ 40/70 \end{bmatrix} = \mathbf{q} \qquad \blacksquare$$

The next theorem shows that what happened in Example 3 is typical of many stochastic matrices. We say that a stochastic matrix is **regular** if some matrix power P^k contains only strictly positive entries. For P in Example 3,

$$P^2 = \begin{bmatrix} .37 & .26 & .33 \\ .45 & .70 & .45 \\ .18 & .04 & .22 \end{bmatrix}$$

Since every entry in P^2 is strictly positive, P is a regular stochastic matrix.

Also, we say that a sequence of vectors, $\mathbf{x}_k$ for $k = 1, 2, \ldots$, **converges** to a vector $\mathbf{q}$ as $k \to \infty$, if the entries in $\mathbf{x}_k$ can be made as close as desired to the corresponding entries in $\mathbf{q}$ by taking k sufficiently large.

THEOREM 11

> If P is an $n \times n$ regular stochastic matrix, then P has a unique steady-state vector $\mathbf{q}$. Further, if $\mathbf{x}_0$ is any initial state and $\mathbf{x}_{k+1} = P\mathbf{x}_k$ for $k = 0, 1, 2, \ldots$, then the Markov chain $\{\mathbf{x}_k\}$ converges to $\mathbf{q}$ as $k \to \infty$.

This theorem is proved in standard texts on Markov chains. The amazing part of the theorem is that the initial state has no effect on the long-term behavior of the Markov chain.

EXAMPLE 6 In Example 2, what percentage of the voters are likely to vote for the Republican candidate in some election many years from now, assuming that the election outcomes form a Markov chain?

SOLUTION If you want to compute the precise entries of the steady-state vector by hand, it is better to recognize that it is an eigenvector with eigenvalue 1 rather than to pick some initial vector $\mathbf{x}_0$ and compute $\mathbf{x}_1, \ldots, \mathbf{x}_k$ for some large value of k. You have no way of knowing how many vectors to compute, and you cannot be sure of the limiting values of the entries in $\mathbf{x}_k$.

A better approach is to compute the steady-state vector and then appeal to Theorem 11. Given P as in Example 2, form $P - I$ by subtracting 1 from each diagonal entry in P. Then row reduce the augmented matrix:

$$[(P - I) \quad \mathbf{0}] = \begin{bmatrix} -.3 & .1 & .3 & 0 \\ .2 & -.2 & .3 & 0 \\ .1 & .1 & -.6 & 0 \end{bmatrix}$$

Recall from earlier work with decimals that the arithmetic is simplified by multiplying each row by 10.[1]

$$\begin{bmatrix} -3 & 1 & 3 & 0 \\ 2 & -2 & 3 & 0 \\ 1 & 1 & -6 & 0 \end{bmatrix} \sim \begin{bmatrix} 1 & 0 & -9/4 & 0 \\ 0 & 1 & -15/4 & 0 \\ 0 & 0 & 0 & 0 \end{bmatrix}$$

The general solution of $(P - I)\mathbf{x} = \mathbf{0}$ is $x_1 = \frac{9}{4}x_3$, $x_2 = \frac{15}{4}x_3$, and x_3 is free. Choosing $x_3 = 4$, we obtain a basis for the solution space whose entries are integers, and from this we easily find the steady-state vector whose entries sum to 1:

$$\mathbf{w} = \begin{bmatrix} 9 \\ 15 \\ 4 \end{bmatrix}, \quad \text{and} \quad \mathbf{q} = \begin{bmatrix} 9/28 \\ 15/28 \\ 4/28 \end{bmatrix} \approx \begin{bmatrix} .32 \\ .54 \\ .14 \end{bmatrix}$$

The entries in $\mathbf{q}$ describe the distribution of votes at an election to be held many years from now (assuming the stochastic matrix continues to describe the changes from one election to the next). Thus, eventually, about 54% of the vote will be for the Republican candidate. ∎

Numerical Notes

You may have noticed that if $\mathbf{x}_{k+1} = P\mathbf{x}_k$ for $k = 0, 1, \ldots$, then

$$\mathbf{x}_2 = P\mathbf{x}_1 = P(P\mathbf{x}_0) = P^2\mathbf{x}_0,$$

and, in general,

$$\mathbf{x}_k = P^k\mathbf{x}_0 \quad \text{for } k = 0, 1, \ldots$$

To compute a specific vector such as $\mathbf{x}_3$, fewer arithmetic operations are needed to compute $\mathbf{x}_1$, $\mathbf{x}_2$, and $\mathbf{x}_3$, rather than P^3 and $P^3\mathbf{x}_0$. However, if P is small—say, 30×30—the machine computation time is insignificant for both methods, and a command to compute $P^3\mathbf{x}_0$ might be preferred because it requires fewer human keystrokes.

Practice Problems

1. Suppose the residents of a metropolitan region move according to the probabilities in the migration matrix M in Example 1 and a resident is chosen "at random." Then a state vector for a certain year may be interpreted as giving the probabilities that the person is a city resident or a suburban resident at that time.

   a. Suppose the person chosen is a city resident now, so that $\mathbf{x}_0 = \begin{bmatrix} 1 \\ 0 \end{bmatrix}$. What is the likelihood that the person will live in the suburbs next year?

   b. What is the likelihood that the person will be living in the suburbs in two years?

2. Let $P = \begin{bmatrix} .6 & .2 \\ .4 & .8 \end{bmatrix}$ and $\mathbf{q} = \begin{bmatrix} .3 \\ .7 \end{bmatrix}$. Is $\mathbf{q}$ a steady-state vector for P?

3. What percentage of the population in Example 1 will live in the suburbs after many years?

---

[1] **Warning:** Don't multiply only P by 10. Instead multiply the augmented matrix for equation $(P - I)\mathbf{x} = \mathbf{0}$ by 10.

5.9 Exercises

1. A small remote village receives radio broadcasts from two radio stations, a news station and a music station. Of the listeners who are tuned to the news station, 70% will remain listening to the news after the station break that occurs each half hour, while 30% will switch to the music station at the station break. Of the listeners who are tuned to the music station, 60% will switch to the news station at the station break, while 40% will remain listening to the music. Suppose everyone is listening to the news at 8:15 A.M.

   a. Give the stochastic matrix that describes how the radio listeners tend to change stations at each station break. Label the rows and columns.

   b. Give the initial state vector.

   c. What percentage of the listeners will be listening to the music station at 9:25 A.M. (after the station breaks at 8:30 and 9:00 A.M.)?

2. A laboratory animal may eat any one of three foods each day. Laboratory records show that if the animal chooses one food on one trial, it will choose the same food on the next trial with a probability of 50%, and it will choose the other foods on the next trial with equal probabilities of 25%.

   a. What is the stochastic matrix for this situation?

   b. If the animal chooses food #1 on an initial trial, what is the probability that it will choose food #2 on the second trial after the initial trial?

3. On any given day, a student is either healthy or ill. Of the students who are healthy today, 95% will be healthy tomorrow. Of the students who are ill today, 55% will still be ill tomorrow.

   a. What is the stochastic matrix for this situation?

   b. Suppose 20% of the students are ill on Monday. What fraction or percentage of the students are likely to be ill on Tuesday? On Wednesday?

   c. If a student is well today, what is the probability that he or she will be well two days from now?

4. The weather in Columbus is either good, indifferent, or bad on any given day. If the weather is good today, there is a 60% chance the weather will be good tomorrow, a 30% chance the weather will be indifferent, and a 10% chance the weather will be bad. If the weather is indifferent today, it will be good tomorrow with probability .40 and indifferent with probability .30. Finally, if the weather is bad today, it will be good tomorrow with probability .40 and indifferent with probability .50.

   a. What is the stochastic matrix for this situation?

   b. Suppose there is a 50% chance of good weather today and a 50% chance of indifferent weather. What are the chances of bad weather tomorrow?

   c. Suppose the predicted weather for Monday is 40% indifferent weather and 60% bad weather. What are the chances for good weather on Wednesday?

In Exercises 5–8, find the steady-state vector.

5. $\begin{bmatrix} .1 & .6 \\ .9 & .4 \end{bmatrix}$

6. $\begin{bmatrix} .8 & .5 \\ .2 & .5 \end{bmatrix}$

7. $\begin{bmatrix} .7 & .1 & .1 \\ .2 & .8 & .2 \\ .1 & .1 & .7 \end{bmatrix}$

8. $\begin{bmatrix} .7 & .2 & .2 \\ 0 & .2 & .4 \\ .3 & .6 & .4 \end{bmatrix}$

9. Determine if $P = \begin{bmatrix} .2 & 1 \\ .8 & 0 \end{bmatrix}$ is a regular stochastic matrix.

10. Determine if $P = \begin{bmatrix} 1 & .2 \\ 0 & .8 \end{bmatrix}$ is a regular stochastic matrix.

11. a. Find the steady-state vector for the Markov chain in Exercise 1.

    b. At some time late in the day, what fraction of the listeners will be listening to the news?

12. Refer to Exercise 2. Which food will the animal prefer after many trials?

13. a. Find the steady-state vector for the Markov chain in Exercise 3.

    b. What is the probability that after many days a specific student is ill? Does it matter if that person is ill today?

14. Refer to Exercise 4. In the long run, how likely is it for the weather in Columbus to be good on a given day?

In Exercises 15–20, P is an $n \times n$ stochastic matrix. Mark each statement True or False (**T/F**). Justify each answer.

15. (**T/F**) The steady state vector is an eigenvector of P.

16. (**T/F**) Every eigenvector of P is a steady state vector.

17. (**T/F**) The all ones vector is an eigenvector of P^T.

18. (**T/F**) The number 2 can be an eigenvalue of a stochastic matrix.

19. (**T/F**) The number 1/2 can be an eigenvalue of a stochastic matrix.

20. (**T/F**) All stochastic matrices are regular.

21. Is $\mathbf{q} = \begin{bmatrix} .6 \\ .8 \end{bmatrix}$ a steady state vector for $A = \begin{bmatrix} .2 & .6 \\ .8 & .4 \end{bmatrix}$? Justify your answer.

22. Is $\mathbf{q} = \begin{bmatrix} .4 \\ .4 \end{bmatrix}$ a steady state vector for $A = \begin{bmatrix} .2 & .8 \\ .8 & .2 \end{bmatrix}$? Justify your answer.

23. Is $\mathbf{q} = \begin{bmatrix} .6 \\ .4 \end{bmatrix}$ a steady state vector for $A = \begin{bmatrix} .4 & .6 \\ .6 & .4 \end{bmatrix}$? Justify your answer.

24. Is $\mathbf{q} = \begin{bmatrix} 3/7 \\ 4/7 \end{bmatrix}$ a steady state vector for $A = \begin{bmatrix} .2 & .6 \\ .8 & .4 \end{bmatrix}$? Justify your answer.

T 25. Suppose the following matrix describes the likelihood that an individual will switch between an iOS and an Android smartphone:

$$
\begin{array}{cc}
\text{From} & \text{To} \\
\text{iOS Android} & \\
\begin{bmatrix} .70 & .15 \\ .30 & .85 \end{bmatrix} & \begin{array}{c} \text{iOS} \\ \text{Android} \end{array}
\end{array}
$$

In the long run, what percentage of smartphone owners would you expect to have an Android operating system?

T 26. In Detroit, Hertz Rent A Car has a fleet of about 2000 cars. The pattern of rental and return locations is given by the fractions in the table below. On a typical day, about how many cars will be rented or ready to rent from the downtown location?

$$
\begin{array}{ccc}
\multicolumn{3}{c}{\text{Cars Rented from}} \\
\text{City} & \text{Down-} & \text{Metro} \\
\text{Airport} & \text{town} & \text{Airport} \\
\begin{bmatrix} .90 & .01 & .09 \\ .01 & .90 & .01 \\ .09 & .09 & .90 \end{bmatrix} & & \begin{array}{c} \text{Returned to} \\ \text{City Airport} \\ \text{Downtown} \\ \text{Metro Airport} \end{array}
\end{array}
$$

27. Let P be an $n \times n$ stochastic matrix. The following argument shows that the equation $P\mathbf{x} = \mathbf{x}$ has a nontrivial solution. (In fact, a steady-state solution exists with nonnegative entries. A proof is given in some advanced texts.) Justify each assertion below. (Mention a theorem when appropriate.)

a. If all the other rows of $P - I$ are added to the bottom row, the result is a row of zeros.

b. The rows of $P - I$ are linearly dependent.

c. The dimension of the row space of $P - I$ is less than n.

d. $P - I$ has a nontrivial null space.

28. Show that every 2×2 stochastic matrix has at least one steady-state vector. Any such matrix can be written in the form $P = \begin{bmatrix} 1 - \alpha & \beta \\ \alpha & 1 - \beta \end{bmatrix}$, where α and β are constants between 0 and 1. (There are two linearly independent steady-state vectors if $\alpha = \beta = 0$. Otherwise, there is only one.)

29. Let S be the $1 \times n$ row matrix with a 1 in each column,
$$S = [\, 1 \quad 1 \quad \cdots \quad 1 \,]$$

a. Explain why a vector $\mathbf{x}$ in $\mathbb{R}^n$ is a probability vector if and only if its entries are nonnegative and $S\mathbf{x} = 1$. (A 1×1 matrix such as the product $S\mathbf{x}$ is usually written without the matrix bracket symbols.)

b. Let P be an $n \times n$ stochastic matrix. Explain why $SP = S$.

c. Let P be an $n \times n$ stochastic matrix, and let $\mathbf{x}$ be a probability vector. Show that $P\mathbf{x}$ is also a probability vector.

30. Use Exercise 29 to show that if P is an $n \times n$ stochastic matrix, then so is P^2.

T 31. Examine powers of a regular stochastic matrix.

a. Compute P^k for $k = 2, 3, 4, 5$, when

$$
P = \begin{bmatrix} .3355 & .3682 & .3067 & .0389 \\ .2663 & .2723 & .3277 & .5451 \\ .1935 & .1502 & .1589 & .2395 \\ .2047 & .2093 & .2067 & .1765 \end{bmatrix}
$$

Display calculations to four decimal places. What happens to the columns of P^k as k increases? Compute the steady-state vector for P.

b. Compute Q^k for $k = 10, 20, \ldots, 80$, when

$$
Q = \begin{bmatrix} .97 & .05 & .10 \\ 0 & .90 & .05 \\ .03 & .05 & .85 \end{bmatrix}
$$

(Stability for Q^k to four decimal places may require $k = 116$ or more.) Compute the steady-state vector for Q. Conjecture what might be true for any regular stochastic matrix.

c. Use Theorem 11 to explain what you found in parts (a) and (b).

T 32. Compare two methods for finding the steady-state vector $\mathbf{q}$ of a regular stochastic matrix P: (1) computing $\mathbf{q}$ as in Example 5, or (2) computing P^k for some large value of k and using one of the columns of P^k as an approximation for $\mathbf{q}$. [The *Study Guide* describes a program *nulbasis* that almost automates method (1).]

Experiment with the largest random stochastic matrices your matrix program will allow, and use $k = 100$ or some other large value. For each method, describe the time *you* need to enter the keystrokes and run your program. (Some versions of MATLAB have commands flops and tic ... toc that record the number of floating point operations and the total elapsed time MATLAB uses.) Contrast the advantages of each method, and state which you prefer.

Solutions to Practice Problems

1. a. Since 5% of the city residents will move to the suburbs within one year, there is a 5% chance of choosing such a person. Without further knowledge about the person, we say that there is a 5% chance the person will move to the suburbs. This fact is contained in the second entry of the state vector $\mathbf{x}_1$, where

$$\mathbf{x}_1 = M\mathbf{x}_0 = \begin{bmatrix} .95 & .03 \\ .05 & .97 \end{bmatrix}\begin{bmatrix} 1 \\ 0 \end{bmatrix} = \begin{bmatrix} .95 \\ .05 \end{bmatrix}$$

b. The likelihood that the person will be living in the suburbs after two years is 9.6%, because

$$\mathbf{x}_2 = M\mathbf{x}_1 = \begin{bmatrix} .95 & .03 \\ .05 & .97 \end{bmatrix}\begin{bmatrix} .95 \\ .05 \end{bmatrix} = \begin{bmatrix} .904 \\ .096 \end{bmatrix}$$

2. The steady-state vector satisfies $P\mathbf{x} = \mathbf{x}$. Since

$$P\mathbf{q} = \begin{bmatrix} .6 & .2 \\ .4 & .8 \end{bmatrix}\begin{bmatrix} .3 \\ .7 \end{bmatrix} = \begin{bmatrix} .32 \\ .68 \end{bmatrix} \neq \mathbf{q}$$

we conclude that $\mathbf{q}$ is *not* the steady-state vector for P.

3. M in Example 1 is a regular stochastic matrix because its entries are all strictly positive. So we may use Theorem 11. We already know the steady-state vector from Example 4. Thus the population distribution vectors $\mathbf{x}_k$ converge to

$$\mathbf{q} = \begin{bmatrix} .375 \\ .625 \end{bmatrix}$$

Eventually 62.5% of the population will live in the suburbs.

CHAPTER 5 PROJECTS

Chapter 5 projects are available online at `bit.ly/3OIM8gT`.

A. *Power Method for Finding Eigenvalues*:   This project shows how to find the eigenvector associated with the eigenvalue corresponding to the largest eigenvalue.

B. *Integration by Parts*:   The purpose of this project is to show how the matrix of a linear transformation relative to a basis

B may be used to find antiderivatives usually found using integration by parts.

C. *Robotics*:   In this project, students are asked to find online examples of robots that use 3D rotations to function.

D. *Dynamical Systems and Markov Chains*:   This project applies the techniques of discrete dynamical systems to Markov chains.

CHAPTER 5 SUPPLEMENTARY EXERCISES

Throughout these supplementary exercises, A and B represent square matrices of appropriate sizes.

For Exercises 1–23, mark each statement as True or False (**T/F**). Justify each answer.

1. (**T/F**) If A is invertible and 1 is an eigenvalue for A, then 1 is also an eigenvalue of A^{-1}.

2. (**T/F**) If A is row equivalent to the identity matrix I, then A is diagonalizable.

3. (**T/F**) If A contains a row or column of zeros, then 0 is an eigenvalue of A.

4. (**T/F**) Each eigenvalue of A is also an eigenvalue of A^2.

5. (**T/F**) Each eigenvector of A is also an eigenvector of A^2.

6. (**T/F**) Each eigenvector of an invertible matrix A is also an eigenvector of A^{-1}.

7. (**T/F**) Eigenvalues must be nonzero scalars.

8. (T/F) Eigenvectors must be nonzero vectors.

9. (T/F) Two eigenvectors corresponding to the same eigenvalue are always linearly dependent.

10. (T/F) Similar matrices always have exactly the same eigenvalues.

11. (T/F) Similar matrices always have exactly the same eigenvectors.

12. (T/F) The sum of two eigenvectors of a matrix A is also an eigenvector of A.

13. (T/F) The eigenvalues of an upper triangular matrix A are exactly the nonzero entries on the diagonal of A.

14. (T/F) The matrices A and A^T have the same eigenvalues, counting multiplicities.

15. (T/F) If a 5×5 matrix A has fewer than 5 distinct eigenvalues, then A is not diagonalizable.

16. (T/F) There exists a 2×2 matrix that has no eigenvectors in $\mathbb{R}^2$.

17. (T/F) If A is diagonalizable, then the columns of A are linearly independent.

18. (T/F) A nonzero vector cannot correspond to two different eigenvalues of A.

19. (T/F) A (square) matrix A is invertible if and only if there is a coordinate system in which the transformation $\mathbf{x} \mapsto A\mathbf{x}$ is represented by a diagonal matrix.

20. (T/F) If each vector $\mathbf{e}_j$ in the standard basis for $\mathbb{R}^n$ is an eigenvector of A, then A is a diagonal matrix.

21. (T/F) If A is similar to a diagonalizable matrix B, then A is also diagonalizable.

22. (T/F) If A and B are invertible $n \times n$ matrices, then AB is similar to BA.

23. (T/F) An $n \times n$ matrix with n linearly independent eigenvectors is invertible.

24. Show that if $\mathbf{x}$ is an eigenvector of the matrix product AB and $B\mathbf{x} \neq \mathbf{0}$, then $B\mathbf{x}$ is an eigenvector of BA.

25. Suppose $\mathbf{x}$ is an eigenvector of A corresponding to an eigenvalue λ.

    a. Show that $\mathbf{x}$ is an eigenvector of $5I - A$. What is the corresponding eigenvalue?

    b. Show that $\mathbf{x}$ is an eigenvector of $5I - 3A + A^2$. What is the corresponding eigenvalue?

26. Use mathematical induction to show that if λ is an eigenvalue of an $n \times n$ matrix A, with $\mathbf{x}$ a corresponding eigenvector, then, for each positive integer m, λ^m is an eigenvalue of A^m, with $\mathbf{x}$ a corresponding eigenvector.

27. If $p(t) = c_0 + c_1 t + c_2 t^2 + \cdots + c_n t^n$, define $p(A)$ to be the matrix formed by replacing each power of t in $p(t)$ by the corresponding power of A (with $A^0 = I$). That is,

$$p(A) = c_0 I + c_1 A + c_2 A^2 + \cdots + c_n A^n$$

Show that if λ is an eigenvalue of A, then one eigenvalue of $p(A)$ is $p(\lambda)$.

28. Suppose $A = PDP^{-1}$, where P is 2×2 and $D = \begin{bmatrix} 2 & 0 \\ 0 & 7 \end{bmatrix}$.

    a. Let $B = 5I - 3A + A^2$. Show that B is diagonalizable by finding a suitable factorization of B.

    b. Given $p(t)$ and $p(A)$ as in Exercise 27, show that $p(A)$ is diagonalizable.

29. Suppose A is diagonalizable and $p(t)$ is the characteristic polynomial of A. Define $p(A)$ as in Exercise 27, and show that $p(A)$ is the zero matrix. This fact, which is also true for *any* square matrix, is called the *Cayley–Hamilton theorem*.

30. a. Let A be a diagonalizable $n \times n$ matrix. Show that if the multiplicity of an eigenvalue λ is n, then $A = \lambda I$.

    b. Use part (a) to show that the matrix $A = \begin{bmatrix} 3 & 1 \\ 0 & 3 \end{bmatrix}$ is not diagonalizable.

31. Show that $I - A$ is invertible when all the eigenvalues of A are less than 1 in magnitude. [*Hint:* What would be true if $I - A$ were not invertible?]

32. Show that if A is diagonalizable, with all eigenvalues less than 1 in magnitude, then A^k tends to the zero matrix as $k \to \infty$. [*Hint:* Consider $A^k \mathbf{x}$ where $\mathbf{x}$ represents any one of the columns of I.]

33. Let $\mathbf{u}$ be an eigenvector of A corresponding to an eigenvalue λ, and let H be the line in $\mathbb{R}^n$ through $\mathbf{u}$ and the origin.

    a. Explain why H is invariant under A in the sense that $A\mathbf{x}$ is in H whenever $\mathbf{x}$ is in H.

    b. Let K be a one-dimensional subspace of $\mathbb{R}^n$ that is invariant under A. Explain why K contains an eigenvector of A.

34. Let $G = \begin{bmatrix} A & X \\ 0 & B \end{bmatrix}$. Use formula for the determinant in Section 5.2 to explain why $\det G = (\det A)(\det B)$. From this, deduce that the characteristic polynomial of G is the product of the characteristic polynomials of A and B.

Use Exercise 34 to find the eigenvalues of the matrices in Exercises 35 and 36.

35. $A = \begin{bmatrix} 3 & -2 & 8 \\ 0 & 5 & -2 \\ 0 & -4 & 3 \end{bmatrix}$

36. $A = \begin{bmatrix} 1 & 5 & -6 & -7 \\ 2 & 4 & 5 & 2 \\ 0 & 0 & -7 & -4 \\ 0 & 0 & 3 & 1 \end{bmatrix}$

37. Let J be the $n \times n$ matrix of all 1's, and consider $A = (a - b)I + bJ$; that is,

$$A = \begin{bmatrix} a & b & b & \cdots & b \\ b & a & b & \cdots & b \\ b & b & a & \cdots & b \\ \vdots & \vdots & \vdots & \ddots & \vdots \\ b & b & b & \cdots & a \end{bmatrix}$$

Use the results of Exercise 30 in the Supplementary Exercises for Chapter 3 to show that the eigenvalues of A are $a - b$ and $a + (n - 1)b$. What are the multiplicities of these eigenvalues?

38. Apply the result of Exercise 37 to find the eigenvalues of the

matrices $\begin{bmatrix} 1 & 2 & 2 \\ 2 & 1 & 2 \\ 2 & 2 & 1 \end{bmatrix}$ and $\begin{bmatrix} 7 & 3 & 3 & 3 & 3 \\ 3 & 7 & 3 & 3 & 3 \\ 3 & 3 & 7 & 3 & 3 \\ 3 & 3 & 3 & 7 & 3 \\ 3 & 3 & 3 & 3 & 7 \end{bmatrix}$.

39. Let $A = \begin{bmatrix} a_{11} & a_{12} \\ a_{21} & a_{22} \end{bmatrix}$. The tr A (the trace of A) is the sum of the diagonal entries in A. Show that the characteristic polynomial of A is

$$\lambda^2 - (\text{tr } A)\lambda + \det A$$

Then show that the eigenvalues of a 2×2 matrix A are both real if and only if $\det A \le \left(\dfrac{\text{tr } A}{2}\right)^2$.

40. Let $A = \begin{bmatrix} .4 & -.3 \\ .4 & 1.2 \end{bmatrix}$. Explain why A^k approaches $\begin{bmatrix} -.5 & -.75 \\ 1.0 & 1.50 \end{bmatrix}$ as $k \to \infty$.

Exercises 41–45 concern the polynomial

$$p(t) = a_0 + a_1 t + \cdots + a_{n-1}t^{n-1} + t^n$$

and an $n \times n$ matrix C_p called the **companion matrix** of p:

$$C_p = \begin{bmatrix} 0 & 1 & 0 & \cdots & 0 \\ 0 & 0 & 1 & & 0 \\ \vdots & & & & \vdots \\ 0 & 0 & 0 & & 1 \\ -a_0 & -a_1 & -a_2 & \cdots & -a_{n-1} \end{bmatrix}$$

41. Write the companion matrix C_p for $p(t) = 6 - 5t + t^2$, and then find the characteristic polynomial of C_p.

42. Let $p(t) = (t - 2)(t - 3)(t - 4) = -24 + 26t - 9t^2 + t^3$. Write the companion matrix for $p(t)$, and use techniques from Chapter 3 to find its characteristic polynomial.

43. Use mathematical induction to prove that for $n \ge 2$,

$$\det(C_p - \lambda I) = (-1)^n(a_0 + a_1\lambda + \cdots + a_{n-1}\lambda^{n-1} + \lambda^n)$$
$$= (-1)^n p(\lambda)$$

[*Hint:* Expanding by cofactors down the first column, show that $\det(C_p - \lambda I)$ has the form $(-\lambda)B + (-1)^n a_0$, where B is a certain polynomial (by the induction assumption).]

44. Let $p(t) = a_0 + a_1 t + a_2 t^2 + t^3$, and let λ be a zero of p.
   a. Write the companion matrix for p.
   b. Explain why $\lambda^3 = -a_0 - a_1\lambda - a_2\lambda^2$, and show that $(1, \lambda, \lambda^2)$ is an eigenvector of the companion matrix for p.

45. Let p be the polynomial in Exercise 44, and suppose the equation $p(t) = 0$ has distinct roots $\lambda_1, \lambda_2, \lambda_3$. Let V be the Vandermonde matrix

$$V = \begin{bmatrix} 1 & 1 & 1 \\ \lambda_1 & \lambda_2 & \lambda_3 \\ \lambda_1^2 & \lambda_2^2 & \lambda_3^2 \end{bmatrix}$$

Use Exercise 44 and a theorem from this chapter to deduce that V is invertible (but do not compute V^{-1}). Then explain why $V^{-1}C_p V$ is a diagonal matrix.

T 46. The MATLAB command `roots (p)` computes the roots of the polynomial equation $p(t) = 0$. Read a MATLAB manual, and then describe the basic idea behind the algorithm for the `roots` command.

T 47. Use a matrix program to diagonalize

$$A = \begin{bmatrix} -3 & -2 & 0 \\ 14 & 7 & -1 \\ -6 & -3 & 1 \end{bmatrix}$$

if possible. Use the eigenvalue command to create the diagonal matrix D. If the program has a command that produces eigenvectors, use it to create an invertible matrix P. Then compute $AP - PD$ and PDP^{-1}. Discuss your results.

T 48. Repeat Exercise 47 for $A = \begin{bmatrix} -8 & 5 & -2 & 0 \\ -5 & 2 & 1 & -2 \\ 10 & -8 & 6 & -3 \\ 3 & -2 & 1 & 0 \end{bmatrix}$.

6

Orthogonality and Least Squares

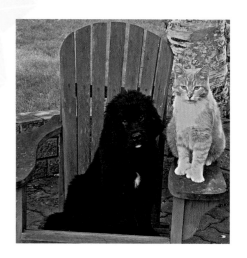

Introductory Example

ARTIFICIAL INTELLIGENCE AND MACHINE LEARNING

Can you tell a puppy from a kitten regardless of breed or coloring? Of course! Both may be furry bundles of joy, but one is clearly a feline and one is clearly a canine to the human eye. Simple. But, as blatantly obvious as it may seem to our trained eyes that are capable of interpreting meaning in the pixels of an image, this turns out to be a significant challenge for a machine. With the advancements in artificial intelligence (AI) and machine learning, computers are rapidly improving their capability to identify the creature on the left in the picture above as a puppy and the one on the right as a cat.

Many industries are now using AI technology to speed up the process of what once took hours of mindless work, such as post office scanners that can read bar codes and handwriting on envelopes to sort the mail with precision and speed. Nordstrom is using machine learning to design, display, organize, and recommend clothing ensembles to customers, exemplifying how even in creative aesthetic fields machine learning can be used to interpret color and shape patterns in pixels and organize visual possibilities into that which our eyes register as pleasing.

When calling a service desk number, one is often greeted by a machine that asks a series of questions and provides suggestions. Only persistence in interacting with this machine gets the caller through to a real person. More and more service calls are answered by machines, making it easier for customers' simple questions to be answered succinctly without waiting time and the obnoxious jingles of hold music. Google has designed an AI assistant that will handle making service calls for you too—booking a restaurant or hair appointment on your behalf.

AI and machine learning comprise developing systems that interpret external data correctly, learn from such data, and use that learning to achieve specific goals and tasks through flexibility and adaptation. Often, the driving engine behind these techniques is linear algebra. In Section 6.2, we see a simple way to design a matrix so that matrix multiplication can identify the correct pattern of blue and white squares. In Sections 6.5, 6.6, and 6.8, we explore techniques used in machine learning.

In order to find an approximate solution to an inconsistent system of equations that has no actual solution, a well-defined notion of nearness is needed. Section 6.1 introduces the concepts of distance and orthogonality in a vector space. Sections 6.2 and 6.3 show how orthogonality can be used to identify the point within a subspace W that is nearest to a point $\mathbf{y}$ lying outside of W. By taking W to be the column space of a matrix, Section 6.5 develops a method for producing approximate ("least-squares") solutions for inconsistent linear systems, an important technique in machine learning, which is discussed in Sections 6.6 and 6.8.

Section 6.4 provides another opportunity to see orthogonal projections at work, creating a matrix factorization widely used in numerical linear algebra. The remaining sections examine some of the many least-squares problems that arise in applications, including those in vector spaces more general than $\mathbb{R}^n$.

6.1 Inner Product, Length, and Orthogonality

Geometric concepts of length, distance, and perpendicularity, which are well known for $\mathbb{R}^2$ and $\mathbb{R}^3$, are defined here for $\mathbb{R}^n$. These concepts provide powerful geometric tools for solving many applied problems, including the least-squares problems mentioned above. All three notions are defined in terms of the inner product of two vectors.

The Inner Product

If $\mathbf{u}$ and $\mathbf{v}$ are vectors in $\mathbb{R}^n$, then we regard $\mathbf{u}$ and $\mathbf{v}$ as $n \times 1$ matrices. The transpose $\mathbf{u}^T$ is a $1 \times n$ matrix, and the matrix product $\mathbf{u}^T \mathbf{v}$ is a 1×1 matrix, which we write as a single real number (a scalar) without brackets. The number $\mathbf{u}^T \mathbf{v}$ is called the **inner product** of $\mathbf{u}$ and $\mathbf{v}$, and often it is written as $\mathbf{u} \cdot \mathbf{v}$. This inner product, mentioned in the exercises for Section 2.1, is also referred to as a **dot product**. If

$$\mathbf{u} = \begin{bmatrix} u_1 \\ u_2 \\ \vdots \\ u_n \end{bmatrix} \quad \text{and} \quad \mathbf{v} = \begin{bmatrix} v_1 \\ v_2 \\ \vdots \\ v_n \end{bmatrix}$$

then the inner product of $\mathbf{u}$ and $\mathbf{v}$ is

$$\begin{bmatrix} u_1 & u_2 & \cdots & u_n \end{bmatrix} \begin{bmatrix} v_1 \\ v_2 \\ \vdots \\ v_n \end{bmatrix} = u_1 v_1 + u_2 v_2 + \cdots + u_n v_n$$

EXAMPLE 1 Compute $\mathbf{u} \cdot \mathbf{v}$ and $\mathbf{v} \cdot \mathbf{u}$ for $\mathbf{u} = \begin{bmatrix} 2 \\ -5 \\ -1 \end{bmatrix}$ and $\mathbf{v} = \begin{bmatrix} 3 \\ 2 \\ -3 \end{bmatrix}$.

SOLUTION

$$\mathbf{u}\cdot\mathbf{v} = \mathbf{u}^T\mathbf{v} = \begin{bmatrix} 2 & -5 & -1 \end{bmatrix}\begin{bmatrix} 3 \\ 2 \\ -3 \end{bmatrix} = (2)(3) + (-5)(2) + (-1)(-3) = -1$$

$$\mathbf{v}\cdot\mathbf{u} = \mathbf{v}^T\mathbf{u} = \begin{bmatrix} 3 & 2 & -3 \end{bmatrix}\begin{bmatrix} 2 \\ -5 \\ -1 \end{bmatrix} = (3)(2) + (2)(-5) + (-3)(-1) = -1 \quad \blacksquare$$

It is clear from the calculations in Example 1 why $\mathbf{u}\cdot\mathbf{v} = \mathbf{v}\cdot\mathbf{u}$. This commutativity of the inner product holds in general. The following properties of the inner product are easily deduced from properties of the transpose operation in Section 2.1. (See Exercises 29 and 30 at the end of this section.)

THEOREM I

Let $\mathbf{u}$, $\mathbf{v}$, and $\mathbf{w}$ be vectors in $\mathbb{R}^n$, and let c be a scalar. Then

a. $\mathbf{u}\cdot\mathbf{v} = \mathbf{v}\cdot\mathbf{u}$
b. $(\mathbf{u} + \mathbf{v})\cdot\mathbf{w} = \mathbf{u}\cdot\mathbf{w} + \mathbf{v}\cdot\mathbf{w}$
c. $(c\mathbf{u})\cdot\mathbf{v} = c(\mathbf{u}\cdot\mathbf{v}) = \mathbf{u}\cdot(c\mathbf{v})$
d. $\mathbf{u}\cdot\mathbf{u} \geq 0,$ and $\mathbf{u}\cdot\mathbf{u} = 0$ if and only if $\mathbf{u} = \mathbf{0}$

Properties (b) and (c) can be combined several times to produce the following useful rule:

$$(c_1\mathbf{u}_1 + \cdots + c_p\mathbf{u}_p)\cdot\mathbf{w} = c_1(\mathbf{u}_1\cdot\mathbf{w}) + \cdots + c_p(\mathbf{u}_p\cdot\mathbf{w})$$

The Length of a Vector

If $\mathbf{v}$ is in $\mathbb{R}^n$, with entries $v_1, \ldots, v_n$, then the square root of $\mathbf{v}\cdot\mathbf{v}$ is defined because $\mathbf{v}\cdot\mathbf{v}$ is nonnegative.

DEFINITION

The **length** (or **norm**) of $\mathbf{v}$ is the nonnegative scalar $\|\mathbf{v}\|$ defined by

$$\|\mathbf{v}\| = \sqrt{\mathbf{v}\cdot\mathbf{v}} = \sqrt{v_1^2 + v_2^2 + \cdots + v_n^2}, \quad \text{and} \quad \|\mathbf{v}\|^2 = \mathbf{v}\cdot\mathbf{v}$$

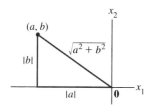

FIGURE 1
Interpretation of $\|\mathbf{v}\|$ as length.

Suppose $\mathbf{v}$ is in $\mathbb{R}^2$, say, $\mathbf{v} = \begin{bmatrix} a \\ b \end{bmatrix}$. If we identify $\mathbf{v}$ with a geometric point in the plane, as usual, then $\|\mathbf{v}\|$ coincides with the standard notion of the length of the line segment from the origin to $\mathbf{v}$. This follows from the Pythagorean Theorem applied to a triangle such as the one in Figure 1.

A similar calculation with the diagonal of a rectangular box shows that the definition of length of a vector $\mathbf{v}$ in $\mathbb{R}^3$ coincides with the usual notion of length.

For any scalar c, the length of $c\mathbf{v}$ is $|c|$ times the length of $\mathbf{v}$. That is,

$$\|c\mathbf{v}\| = |c|\|\mathbf{v}\|$$

(To see this, compute $\|c\mathbf{v}\|^2 = (c\mathbf{v})\cdot(c\mathbf{v}) = c^2\mathbf{v}\cdot\mathbf{v} = c^2\|\mathbf{v}\|^2$ and take square roots.)

A vector whose length is 1 is called a **unit vector**. If we *divide* a nonzero vector **v** by its length—that is, multiply by $1/\|\mathbf{v}\|$—we obtain a unit vector **u** because the length of **u** is $(1/\|\mathbf{v}\|)\|\mathbf{v}\|$. The process of creating **u** from **v** is sometimes called **normalizing v**, and we say that **u** is *in the same direction* as **v**.

Several examples that follow use the space-saving notation for (column) vectors.

EXAMPLE 2 Let $\mathbf{v} = (1, -2, 2, 0)$. Find a unit vector **u** in the same direction as **v**.

SOLUTION First, compute the length of **v**:

$$\|\mathbf{v}\|^2 = \mathbf{v}\cdot\mathbf{v} = (1)^2 + (-2)^2 + (2)^2 + (0)^2 = 9$$
$$\|\mathbf{v}\| = \sqrt{9} = 3$$

Then, multiply **v** by $1/\|\mathbf{v}\|$ to obtain

$$\mathbf{u} = \frac{1}{\|\mathbf{v}\|}\mathbf{v} = \frac{1}{3}\mathbf{v} = \frac{1}{3}\begin{bmatrix} 1 \\ -2 \\ 2 \\ 0 \end{bmatrix} = \begin{bmatrix} 1/3 \\ -2/3 \\ 2/3 \\ 0 \end{bmatrix}$$

To check that $\|\mathbf{u}\| = 1$, it suffices to show that $\|\mathbf{u}\|^2 = 1$.

$$\|\mathbf{u}\|^2 = \mathbf{u}\cdot\mathbf{u} = \left(\tfrac{1}{3}\right)^2 + \left(-\tfrac{2}{3}\right)^2 + \left(\tfrac{2}{3}\right)^2 + (0)^2$$
$$= \tfrac{1}{9} + \tfrac{4}{9} + \tfrac{4}{9} + 0 = 1 \qquad\blacksquare$$

EXAMPLE 3 Let W be the subspace of $\mathbb{R}^2$ spanned by $\mathbf{x} = (\tfrac{2}{3}, 1)$. Find a unit vector **z** that is a basis for W.

SOLUTION W consists of all multiples of **x**, as in Figure 2(a). Any nonzero vector in W is a basis for W. To simplify the calculation, "scale" **x** to eliminate fractions. That is, multiply **x** by 3 to get

$$\mathbf{y} = \begin{bmatrix} 2 \\ 3 \end{bmatrix}$$

Now compute $\|\mathbf{y}\|^2 = 2^2 + 3^2 = 13$, $\|\mathbf{y}\| = \sqrt{13}$, and normalize **y** to get

$$\mathbf{z} = \frac{1}{\sqrt{13}}\begin{bmatrix} 2 \\ 3 \end{bmatrix} = \begin{bmatrix} 2/\sqrt{13} \\ 3/\sqrt{13} \end{bmatrix}$$

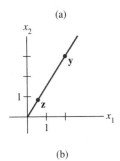

FIGURE 2

Normalizing a vector to produce a unit vector.

See Figure 2(b). Another unit vector is $(-2/\sqrt{13}, -3/\sqrt{13})$. $\qquad\blacksquare$

Distance in $\mathbb{R}^n$

We are ready now to describe how close one vector is to another. Recall that if a and b are real numbers, the distance on the number line between a and b is the number $|a - b|$. Two examples are shown in Figure 3. This definition of distance in $\mathbb{R}$ has a direct analogue in $\mathbb{R}^n$.

$$|2 - 8| = |-6| = 6 \quad\text{or}\quad |8 - 2| = |6| = 6$$

$$|(-3) - 4| = |-7| = 7 \quad\text{or}\quad |4 - (-3)| = |7| = 7$$

FIGURE 3 Distances in $\mathbb{R}$.

DEFINITION

For **u** and **v** in $\mathbb{R}^n$, the **distance between u and v**, written as dist $(\mathbf{u}, \mathbf{v})$, is the length of the vector $\mathbf{u} - \mathbf{v}$. That is,

$$\text{dist}(\mathbf{u}, \mathbf{v}) = \|\mathbf{u} - \mathbf{v}\|$$

In $\mathbb{R}^2$ and $\mathbb{R}^3$, this definition of distance coincides with the usual formulas for the Euclidean distance between two points, as the next two examples show.

EXAMPLE 4   Compute the distance between the vectors $\mathbf{u} = (7, 1)$ and $\mathbf{v} = (3, 2)$.

SOLUTION   Calculate

$$\mathbf{u} - \mathbf{v} = \begin{bmatrix} 7 \\ 1 \end{bmatrix} - \begin{bmatrix} 3 \\ 2 \end{bmatrix} = \begin{bmatrix} 4 \\ -1 \end{bmatrix}$$

$$\|\mathbf{u} - \mathbf{v}\| = \sqrt{4^2 + (-1)^2} = \sqrt{17}$$

The vectors **u**, **v**, and $\mathbf{u} - \mathbf{v}$ are shown in Figure 4. When the vector $\mathbf{u} - \mathbf{v}$ is added to **v**, the result is **u**. Notice that the parallelogram in Figure 4 shows that the distance from **u** to **v** is the same as the distance from $\mathbf{u} - \mathbf{v}$ to **0**.   ∎

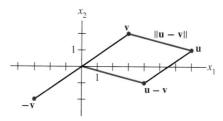

FIGURE 4   The distance between **u** and **v** is the length of $\mathbf{u} - \mathbf{v}$.

EXAMPLE 5   If $\mathbf{u} = (u_1, u_2, u_3)$ and $\mathbf{v} = (v_1, v_2, v_3)$, then

$$\text{dist}(\mathbf{u}, \mathbf{v}) = \|\mathbf{u} - \mathbf{v}\| = \sqrt{(\mathbf{u} - \mathbf{v}) \cdot (\mathbf{u} - \mathbf{v})}$$

$$= \sqrt{(u_1 - v_1)^2 + (u_2 - v_2)^2 + (u_3 - v_3)^2}$$ ∎

Orthogonal Vectors

The rest of this chapter depends on the fact that the concept of perpendicular lines in ordinary Euclidean geometry has an analogue in $\mathbb{R}^n$.

Consider $\mathbb{R}^2$ or $\mathbb{R}^3$ and two lines through the origin determined by vectors **u** and **v**. The two lines shown in Figure 5 are geometrically perpendicular if and only if the distance from **u** to **v** is the same as the distance from **u** to $-\mathbf{v}$. This is the same as requiring the squares of the distances to be the same. Now

$$[\,\text{dist}(\mathbf{u}, -\mathbf{v})\,]^2 = \|\mathbf{u} - (-\mathbf{v})\|^2 = \|\mathbf{u} + \mathbf{v}\|^2$$

$$= (\mathbf{u} + \mathbf{v}) \cdot (\mathbf{u} + \mathbf{v})$$

$$= \mathbf{u} \cdot (\mathbf{u} + \mathbf{v}) + \mathbf{v} \cdot (\mathbf{u} + \mathbf{v}) \qquad \text{Theorem 1(b)}$$

$$= \mathbf{u} \cdot \mathbf{u} + \mathbf{u} \cdot \mathbf{v} + \mathbf{v} \cdot \mathbf{u} + \mathbf{v} \cdot \mathbf{v} \qquad \text{Theorem 1(a), (b)}$$

$$= \|\mathbf{u}\|^2 + \|\mathbf{v}\|^2 + 2\mathbf{u} \cdot \mathbf{v} \qquad \text{Theorem 1(a)} \qquad (1)$$

FIGURE 5

The same calculations with $\mathbf{v}$ and $-\mathbf{v}$ interchanged show that

$$[\mathrm{dist}\,(\mathbf{u}, \mathbf{v})]^2 = \|\mathbf{u}\|^2 + \|-\mathbf{v}\|^2 + 2\mathbf{u}\cdot(-\mathbf{v})$$
$$= \|\mathbf{u}\|^2 + \|\mathbf{v}\|^2 - 2\mathbf{u}\cdot\mathbf{v}$$

The two squared distances are equal if and only if $2\mathbf{u}\cdot\mathbf{v} = -2\mathbf{u}\cdot\mathbf{v}$, which happens if and only if $\mathbf{u}\cdot\mathbf{v} = 0$.

This calculation shows that when vectors $\mathbf{u}$ and $\mathbf{v}$ are identified with geometric points, the corresponding lines through the points and the origin are perpendicular if and only if $\mathbf{u}\cdot\mathbf{v} = 0$. The following definition generalizes to $\mathbb{R}^n$ this notion of perpendicularity (or *orthogonality*, as it is commonly called in linear algebra).

DEFINITION

> Two vectors $\mathbf{u}$ and $\mathbf{v}$ in $\mathbb{R}^n$ are **orthogonal** (to each other) if $\mathbf{u}\cdot\mathbf{v} = 0$.

Observe that the zero vector is orthogonal to every vector in $\mathbb{R}^n$ because $\mathbf{0}^T\mathbf{v} = 0$ for all $\mathbf{v}$.

The next theorem provides a useful fact about orthogonal vectors. The proof follows immediately from the calculation in (1) and the definition of orthogonality. The right triangle shown in Figure 6 provides a visualization of the lengths that appear in the theorem.

THEOREM 2

> **The Pythagorean Theorem**
> Two vectors $\mathbf{u}$ and $\mathbf{v}$ are orthogonal if and only if $\|\mathbf{u} + \mathbf{v}\|^2 = \|\mathbf{u}\|^2 + \|\mathbf{v}\|^2$.

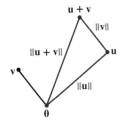

FIGURE 6

Orthogonal Complements

To provide practice using inner products, we introduce a concept here that will be of use in Section 6.3 and elsewhere in the chapter. If a vector $\mathbf{z}$ is orthogonal to every vector in a subspace W of $\mathbb{R}^n$, then $\mathbf{z}$ is said to be **orthogonal to** W. The set of all vectors $\mathbf{z}$ that are orthogonal to W is called the **orthogonal complement** of W and is denoted by $W^{\perp}$ (and read as "W perpendicular" or simply "W perp").

EXAMPLE 6 Let W be a plane through the origin in $\mathbb{R}^3$, and let L be the line through the origin and perpendicular to W. If $\mathbf{z}$ and $\mathbf{w}$ are nonzero, $\mathbf{z}$ is on L, and $\mathbf{w}$ is in W, then the line segment from $\mathbf{0}$ to $\mathbf{z}$ is perpendicular to the line segment from $\mathbf{0}$ to $\mathbf{w}$; that is, $\mathbf{z}\cdot\mathbf{w} = 0$. See Figure 7. So each vector on L is orthogonal to every $\mathbf{w}$ in W. In fact, L consists of *all* vectors that are orthogonal to the $\mathbf{w}$'s in W, and W consists of all vectors orthogonal to the $\mathbf{z}$'s in L. That is,

$$L = W^{\perp} \quad \text{and} \quad W = L^{\perp} \qquad\blacksquare$$

FIGURE 7

A plane and line through $\mathbf{0}$ as orthogonal complements.

The following two facts about $W^{\perp}$, with W a subspace of $\mathbb{R}^n$, are needed later in the chapter. Proofs are suggested in Exercises 37 and 38. Exercises 35–39 provide excellent practice using properties of the inner product.

1. A vector $\mathbf{x}$ is in $W^{\perp}$ if and only if $\mathbf{x}$ is orthogonal to every vector in a set that spans W.
2. $W^{\perp}$ is a subspace of $\mathbb{R}^n$.

The next theorem and Exercise 39 verify the claims made in Section 4.5 concerning the subspaces shown in Figure 8.

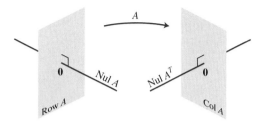

FIGURE 8 The fundamental subspaces determined by an $m \times n$ matrix A.

Remark: A common way to prove that two sets, say S and T, are equal is to show that S is a subset of T and T is a subset of S. The proof of the next theorem that Nul $A =$ (Row A)$^\perp$ is established by showing that Nul A is a subset of (Row A)$^\perp$ and (Row A)$^\perp$ is a subset of Nul A. That is, an arbitrary element **x** in Nul A is shown to be in (Row A)$^\perp$, and then an arbitrary element **x** in (Row A)$^\perp$ is shown to be in Nul A.

THEOREM 3

> Let A be an $m \times n$ matrix. The orthogonal complement of the row space of A is the null space of A, and the orthogonal complement of the column space of A is the null space of A^T:
>
> $$(\text{Row } A)^\perp = \text{Nul } A \quad \text{and} \quad (\text{Col } A)^\perp = \text{Nul } A^T$$

PROOF   The row–column rule for computing $A\mathbf{x}$ shows that if **x** is in Nul A, then **x** is orthogonal to each row of A (with the rows treated as vectors in $\mathbb{R}^n$). Since the rows of A span the row space, **x** is orthogonal to Row A. Conversely, if **x** is orthogonal to Row A, then **x** is certainly orthogonal to each row of A, and hence $A\mathbf{x} = \mathbf{0}$. This proves the first statement of the theorem. Since this statement is true for any matrix, it is true for A^T. That is, the orthogonal complement of the row space of A^T is the null space of A^T. This proves the second statement, because Row $A^T = \text{Col } A$.

Angles in $\mathbb{R}^2$ and $\mathbb{R}^3$ (Optional)

If **u** and **v** are nonzero vectors in either $\mathbb{R}^2$ or $\mathbb{R}^3$, then there is a nice connection between their inner product and the angle ϑ between the two line segments from the origin to the points identified with **u** and **v**. The formula is

$$\mathbf{u} \cdot \mathbf{v} = \|\mathbf{u}\| \, \|\mathbf{v}\| \cos \vartheta \tag{2}$$

To verify this formula for vectors in $\mathbb{R}^2$, consider the triangle shown in Figure 9, with sides of lengths $\|\mathbf{u}\|$, $\|\mathbf{v}\|$, and $\|\mathbf{u} - \mathbf{v}\|$. By the law of cosines,

$$\|\mathbf{u} - \mathbf{v}\|^2 = \|\mathbf{u}\|^2 + \|\mathbf{v}\|^2 - 2\|\mathbf{u}\| \, \|\mathbf{v}\| \cos \vartheta$$

FIGURE 9 The angle between two vectors.

which can be rearranged to produce

$$\|\mathbf{u}\| \|\mathbf{v}\| \cos \vartheta = \frac{1}{2} \left[\|\mathbf{u}\|^2 + \|\mathbf{v}\|^2 - \|\mathbf{u} - \mathbf{v}\|^2 \right]$$

$$= \frac{1}{2} \left[u_1^2 + u_2^2 + v_1^2 + v_2^2 - (u_1 - v_1)^2 - (u_2 - v_2)^2 \right]$$

$$= u_1 v_1 + u_2 v_2$$

$$= \mathbf{u} \cdot \mathbf{v}$$

The verification for $\mathbb{R}^3$ is similar. When $n > 3$, formula (2) may be used to *define* the angle between two vectors in $\mathbb{R}^n$. In statistics, for instance, the value of $\cos \vartheta$ defined by (2) for suitable vectors $\mathbf{u}$ and $\mathbf{v}$ is what statisticians call a *correlation coefficient*.

Practice Problems

1. Let $\mathbf{a} = \begin{bmatrix} -2 \\ 1 \end{bmatrix}$ and $\mathbf{b} = \begin{bmatrix} -3 \\ 1 \end{bmatrix}$. Compute $\dfrac{\mathbf{a} \cdot \mathbf{b}}{\mathbf{a} \cdot \mathbf{a}}$ and $\left(\dfrac{\mathbf{a} \cdot \mathbf{b}}{\mathbf{a} \cdot \mathbf{a}} \right) \mathbf{a}$.

2. Let $\mathbf{c} = \begin{bmatrix} 4/3 \\ -1 \\ 2/3 \end{bmatrix}$ and $\mathbf{d} = \begin{bmatrix} 5 \\ 6 \\ -1 \end{bmatrix}$.

   a. Find a unit vector $\mathbf{u}$ in the direction of $\mathbf{c}$.

   b. Show that $\mathbf{d}$ is orthogonal to $\mathbf{c}$.

   c. Use the results of (a) and (b) to explain why $\mathbf{d}$ must be orthogonal to the unit vector $\mathbf{u}$.

3. Let W be a subspace of $\mathbb{R}^n$. Exercise 38 establishes that $W^\perp$ is also a subspace of $\mathbb{R}^n$. Prove that $\dim W + \dim W^\perp = n$.

6.1 Exercises

Compute the quantities in Exercises 1–8 using the vectors

$$\mathbf{u} = \begin{bmatrix} -1 \\ 2 \end{bmatrix}, \quad \mathbf{v} = \begin{bmatrix} 2 \\ 3 \end{bmatrix}, \quad \mathbf{w} = \begin{bmatrix} 3 \\ -1 \\ -5 \end{bmatrix}, \quad \mathbf{x} = \begin{bmatrix} 6 \\ -2 \\ 3 \end{bmatrix}$$

1. $\mathbf{u} \cdot \mathbf{u}$, $\mathbf{v} \cdot \mathbf{u}$, and $\dfrac{\mathbf{v} \cdot \mathbf{u}}{\mathbf{u} \cdot \mathbf{u}}$

2. $\mathbf{w} \cdot \mathbf{w}$, $\mathbf{x} \cdot \mathbf{w}$, and $\dfrac{\mathbf{x} \cdot \mathbf{w}}{\mathbf{w} \cdot \mathbf{w}}$

3. $\dfrac{1}{\mathbf{w} \cdot \mathbf{w}} \mathbf{w}$

4. $\dfrac{1}{\mathbf{u} \cdot \mathbf{u}} \mathbf{u}$

5. $\left(\dfrac{\mathbf{u} \cdot \mathbf{v}}{\mathbf{v} \cdot \mathbf{v}} \right) \mathbf{v}$

6. $\left(\dfrac{\mathbf{x} \cdot \mathbf{w}}{\mathbf{x} \cdot \mathbf{x}} \right) \mathbf{x}$

7. $\|\mathbf{w}\|$

8. $\|\mathbf{x}\|$

In Exercises 9–12, find a unit vector in the direction of the given vector.

9. $\begin{bmatrix} -30 \\ 40 \end{bmatrix}$

10. $\begin{bmatrix} 3 \\ 6 \\ -3 \end{bmatrix}$

11. $\begin{bmatrix} 7/4 \\ 1/2 \\ 1 \end{bmatrix}$

12. $\begin{bmatrix} 8/3 \\ 1 \end{bmatrix}$

13. Find the distance between $\mathbf{x} = \begin{bmatrix} 10 \\ -3 \end{bmatrix}$ and $\mathbf{y} = \begin{bmatrix} -1 \\ -5 \end{bmatrix}$.

14. Find the distance between $\mathbf{u} = \begin{bmatrix} 0 \\ -5 \\ 2 \end{bmatrix}$ and $\mathbf{z} = \begin{bmatrix} -4 \\ -1 \\ 4 \end{bmatrix}$.

Determine which pairs of vectors in Exercises 15–18 are orthogonal.

15. $\mathbf{a} = \begin{bmatrix} 8 \\ -5 \end{bmatrix}$, $\mathbf{b} = \begin{bmatrix} -2 \\ -3 \end{bmatrix}$

16. $\mathbf{u} = \begin{bmatrix} 12 \\ 3 \\ -5 \end{bmatrix}$, $\mathbf{v} = \begin{bmatrix} 2 \\ -3 \\ 3 \end{bmatrix}$

17. $\mathbf{u} = \begin{bmatrix} 3 \\ 2 \\ -5 \\ 0 \end{bmatrix}$, $\mathbf{v} = \begin{bmatrix} -4 \\ 1 \\ -2 \\ 6 \end{bmatrix}$

18. $\mathbf{y} = \begin{bmatrix} -3 \\ 7 \\ 4 \\ 0 \end{bmatrix}$, $\mathbf{z} = \begin{bmatrix} 1 \\ -8 \\ 15 \\ -7 \end{bmatrix}$

In Exercises 19–28, all vectors are in $\mathbb{R}^n$. Mark each statement True or False (**T/F**). Justify each answer.

19. (**T/F**) $\mathbf{v} \cdot \mathbf{v} = \|\mathbf{v}\|^2$.

20. (**T/F**) $\mathbf{u} \cdot \mathbf{v} - \mathbf{v} \cdot \mathbf{u} = 0$.

21. (T/F) If the distance from **u** to **v** equals the distance from **u** to −**v**, then **u** and **v** are orthogonal.

22. (T/F) If $\|\mathbf{u}\|^2 + \|\mathbf{v}\|^2 = \|\mathbf{u} + \mathbf{v}\|^2$, then **u** and **v** are orthogonal.

23. (T/F) If vectors $\mathbf{v}_1, \ldots, \mathbf{v}_p$ span a subspace W and if **x** is orthogonal to each $\mathbf{v}_j$ for $j = 1, \ldots, p$, then **x** is in $W^\perp$.

24. (T/F) If **x** is orthogonal to every vector in a subspace W then **x** is in $W^\perp$.

25. (T/F) For any scalar c, $\|c\mathbf{v}\| = c\|\mathbf{v}\|$.

26. (T/F) For any scalar c, $\mathbf{u} \cdot (c\mathbf{v}) = c(\mathbf{u} \cdot \mathbf{v})$.

27. (T/F) For a square matrix A, vectors in Col A are orthogonal to vectors in Nul A.

28. (T/F) For an $m \times n$ matrix A, vectors in the null space of A are orthogonal to vectors in the row space of A.

29. Use the transpose definition of the inner product to verify parts (b) and (c) of Theorem 1. Mention the appropriate facts from Chapter 2.

30. Let $\mathbf{u} = (u_1, u_2, u_3)$. Explain why $\mathbf{u} \cdot \mathbf{u} \geq 0$. When is $\mathbf{u} \cdot \mathbf{u} = 0$?

31. Let $\mathbf{u} = \begin{bmatrix} 2 \\ -5 \\ -1 \end{bmatrix}$ and $\mathbf{v} = \begin{bmatrix} -7 \\ -4 \\ 6 \end{bmatrix}$. Compute and compare $\mathbf{u} \cdot \mathbf{v}$, $\|\mathbf{u}\|^2$, $\|\mathbf{v}\|^2$, and $\|\mathbf{u} + \mathbf{v}\|^2$. Do not use the Pythagorean Theorem.

32. Verify the *parallelogram law* for vectors **u** and **v** in $\mathbb{R}^n$:
$$\|\mathbf{u} + \mathbf{v}\|^2 + \|\mathbf{u} - \mathbf{v}\|^2 = 2\|\mathbf{u}\|^2 + 2\|\mathbf{v}\|^2$$

33. Let $\mathbf{v} = \begin{bmatrix} a \\ b \end{bmatrix}$. Describe the set H of vectors $\begin{bmatrix} x \\ y \end{bmatrix}$ that are orthogonal to **v**. [*Hint:* Consider $\mathbf{v} = \mathbf{0}$ and $\mathbf{v} \neq \mathbf{0}$.]

34. Let $\mathbf{u} = \begin{bmatrix} 5 \\ -6 \\ 7 \end{bmatrix}$, and let W be the set of all **x** in $\mathbb{R}^3$ such that $\mathbf{u} \cdot \mathbf{x} = 0$. What theorem in Chapter 4 can be used to show that W is a subspace of $\mathbb{R}^3$? Describe W in geometric language.

35. Suppose a vector **y** is orthogonal to vectors **u** and **v**. Show that **y** is orthogonal to the vector $\mathbf{u} + \mathbf{v}$.

36. Suppose **y** is orthogonal to **u** and **v**. Show that **y** is orthogonal to every **w** in Span $\{\mathbf{u}, \mathbf{v}\}$. [*Hint:* An arbitrary **w** in Span $\{\mathbf{u}, \mathbf{v}\}$ has the form $\mathbf{w} = c_1\mathbf{u} + c_2\mathbf{v}$. Show that **y** is orthogonal to such a vector **w**.]

Span{**u**, **v**}

37. Let $W = \text{Span} \{\mathbf{v}_1, \ldots, \mathbf{v}_p\}$. Show that if **x** is orthogonal to each $\mathbf{v}_j$, for $1 \leq j \leq p$, then **x** is orthogonal to every vector in W.

38. Let W be a subspace of $\mathbb{R}^n$, and let $W^\perp$ be the set of all vectors orthogonal to W. Show that $W^\perp$ is a subspace of $\mathbb{R}^n$ using the following steps.

a. Take **z** in $W^\perp$, and let **u** represent any element of W. Then $\mathbf{z} \cdot \mathbf{u} = 0$. Take any scalar c and show that $c\mathbf{z}$ is orthogonal to **u**. (Since **u** was an arbitrary element of W, this will show that $c\mathbf{z}$ is in $W^\perp$.)

b. Take $\mathbf{z}_1$ and $\mathbf{z}_2$ in $W^\perp$, and let **u** be any element of W. Show that $\mathbf{z}_1 + \mathbf{z}_2$ is orthogonal to **u**. What can you conclude about $\mathbf{z}_1 + \mathbf{z}_2$? Why?

c. Finish the proof that $W^\perp$ is a subspace of $\mathbb{R}^n$.

39. Show that if **x** is in both W and $W^\perp$, then $\mathbf{x} = \mathbf{0}$.

T 40. Construct a pair **u**, **v** of random vectors in $\mathbb{R}^4$, and let

$$A = \begin{bmatrix} .5 & .5 & .5 & .5 \\ .5 & .5 & -.5 & -.5 \\ .5 & -.5 & .5 & -.5 \\ .5 & -.5 & -.5 & .5 \end{bmatrix}$$

a. Denote the columns of A by $\mathbf{a}_1, \ldots, \mathbf{a}_4$. Compute the length of each column, and compute $\mathbf{a}_1 \cdot \mathbf{a}_2$, $\mathbf{a}_1 \cdot \mathbf{a}_3$, $\mathbf{a}_1 \cdot \mathbf{a}_4$, $\mathbf{a}_2 \cdot \mathbf{a}_3$, $\mathbf{a}_2 \cdot \mathbf{a}_4$, and $\mathbf{a}_3 \cdot \mathbf{a}_4$.

b. Compute and compare the lengths of **u**, $A\mathbf{u}$, **v**, and $A\mathbf{v}$.

c. Use equation (2) in this section to compute the cosine of the angle between **u** and **v**. Compare this with the cosine of the angle between $A\mathbf{u}$ and $A\mathbf{v}$.

d. Repeat parts (b) and (c) for two other pairs of random vectors. What do you conjecture about the effect of A on vectors?

T 41. Generate random vectors **x**, **y**, and **v** in $\mathbb{R}^4$ with integer entries (and $\mathbf{v} \neq \mathbf{0}$), and compute the quantities

$$\left(\frac{\mathbf{x} \cdot \mathbf{v}}{\mathbf{v} \cdot \mathbf{v}}\right) \mathbf{v}, \quad \left(\frac{\mathbf{y} \cdot \mathbf{v}}{\mathbf{v} \cdot \mathbf{v}}\right) \mathbf{v}, \quad \frac{(\mathbf{x} + \mathbf{y}) \cdot \mathbf{v}}{\mathbf{v} \cdot \mathbf{v}} \mathbf{v}, \quad \frac{(10\mathbf{x}) \cdot \mathbf{v}}{\mathbf{v} \cdot \mathbf{v}} \mathbf{v}$$

Repeat the computations with new random vectors **x** and **y**. What do you conjecture about the mapping $\mathbf{x} \mapsto T(\mathbf{x}) = \left(\frac{\mathbf{x} \cdot \mathbf{v}}{\mathbf{v} \cdot \mathbf{v}}\right) \mathbf{v}$ (for $\mathbf{v} \neq \mathbf{0}$)? Verify your conjecture algebraically.

T 42. Let $A = \begin{bmatrix} -6 & 3 & -27 & -33 & -13 \\ 6 & -5 & 25 & 28 & 14 \\ 8 & -6 & 34 & 38 & 18 \\ 12 & -10 & 50 & 41 & 23 \\ 14 & -21 & 49 & 29 & 33 \end{bmatrix}$. Construct a matrix N whose columns form a basis for Nul A, and construct a matrix R whose *rows* form a basis for Row A (see Section 4.6 for details). Perform a matrix computation with N and R that illustrates a fact from Theorem 3.

Solutions to Practice Problems

1. $\mathbf{a}\cdot\mathbf{b} = 7, \mathbf{a}\cdot\mathbf{a} = 5$. Hence $\dfrac{\mathbf{a}\cdot\mathbf{b}}{\mathbf{a}\cdot\mathbf{a}} = \dfrac{7}{5}$, and $\left(\dfrac{\mathbf{a}\cdot\mathbf{b}}{\mathbf{a}\cdot\mathbf{a}}\right)\mathbf{a} = \dfrac{7}{5}\mathbf{a} = \begin{bmatrix} -14/5 \\ 7/5 \end{bmatrix}$.

2. a. Scale $\mathbf{c}$, multiplying by 3 to get $\mathbf{y} = \begin{bmatrix} 4 \\ -3 \\ 2 \end{bmatrix}$. Compute $\|\mathbf{y}\|^2 = 29$ and $\|\mathbf{y}\| = \sqrt{29}$. The unit vector in the direction of both $\mathbf{c}$ and $\mathbf{y}$ is
$$\mathbf{u} = \frac{1}{\|\mathbf{y}\|}\mathbf{y} = \begin{bmatrix} 4/\sqrt{29} \\ -3/\sqrt{29} \\ 2/\sqrt{29} \end{bmatrix}.$$

   b. $\mathbf{d}$ is orthogonal to $\mathbf{c}$, because
$$\mathbf{d}\cdot\mathbf{c} = \begin{bmatrix} 5 \\ 6 \\ -1 \end{bmatrix}\cdot\begin{bmatrix} 4/3 \\ -1 \\ 2/3 \end{bmatrix} = \frac{20}{3} - 6 - \frac{2}{3} = 0$$

   c. $\mathbf{d}$ is orthogonal to $\mathbf{u}$, because $\mathbf{u}$ has the form $k\mathbf{c}$ for some k, and
$$\mathbf{d}\cdot\mathbf{u} = \mathbf{d}\cdot(k\mathbf{c}) = k(\mathbf{d}\cdot\mathbf{c}) = k(0) = 0$$

3. If $W \neq \{\mathbf{0}\}$, let $\{\mathbf{b}_1, \dots, \mathbf{b}_p\}$ be a basis for W, where $1 \leq p \leq n$. Let A be the $p \times n$ matrix having rows $\mathbf{b}_1^T, \dots, \mathbf{b}_p^T$. It follows that W is the row space of A. Theorem 3 implies that $W^{\perp} = (\text{Row } A)^{\perp} = \text{Nul } A$ and hence dim $W^{\perp} = $ dim Nul A. Thus, dim $W + $ dim $W^{\perp} = $ dim Row $A + $ dim Nul $A = $ rank $A + $ dim Nul $A = n$, by the Rank Theorem. If $W = \{\mathbf{0}\}$, then $W^{\perp} = \mathbb{R}^n$, and the result follows.

6.2 Orthogonal Sets

A set of vectors $\{\mathbf{u}_1, \dots, \mathbf{u}_p\}$ in $\mathbb{R}^n$ is said to be an **orthogonal set** if each pair of distinct vectors from the set is orthogonal, that is, if $\mathbf{u}_i\cdot\mathbf{u}_j = 0$ whenever $i \neq j$.

EXAMPLE 1 Show that $\{\mathbf{u}_1, \mathbf{u}_2, \mathbf{u}_3\}$ is an orthogonal set, where
$$\mathbf{u}_1 = \begin{bmatrix} 3 \\ 1 \\ 1 \end{bmatrix}, \quad \mathbf{u}_2 = \begin{bmatrix} -1 \\ 2 \\ 1 \end{bmatrix}, \quad \mathbf{u}_3 = \begin{bmatrix} -1/2 \\ -2 \\ 7/2 \end{bmatrix}$$

SOLUTION Consider the three possible pairs of distinct vectors, namely $\{\mathbf{u}_1, \mathbf{u}_2\}$, $\{\mathbf{u}_1, \mathbf{u}_3\}$, and $\{\mathbf{u}_2, \mathbf{u}_3\}$.
$$\mathbf{u}_1\cdot\mathbf{u}_2 = 3(-1) + 1(2) + 1(1) = 0$$
$$\mathbf{u}_1\cdot\mathbf{u}_3 = 3\left(-\tfrac{1}{2}\right) + 1(-2) + 1\left(\tfrac{7}{2}\right) = 0$$
$$\mathbf{u}_2\cdot\mathbf{u}_3 = -1\left(-\tfrac{1}{2}\right) + 2(-2) + 1\left(\tfrac{7}{2}\right) = 0$$

Each pair of distinct vectors is orthogonal, and so $\{\mathbf{u}_1, \mathbf{u}_2, \mathbf{u}_3\}$ is an orthogonal set. See Figure 1; the three line segments are mutually perpendicular. ∎

FIGURE 1

THEOREM 4   If $S = \{\mathbf{u}_1, \dots, \mathbf{u}_p\}$ is an orthogonal set of nonzero vectors in $\mathbb{R}^n$, then S is linearly independent and hence is a basis for the subspace spanned by S.

PROOF If $\mathbf{0} = c_1\mathbf{u}_1 + \cdots + c_p\mathbf{u}_p$ for some scalars $c_1, \ldots, c_p$, then

$$
\begin{aligned}
0 = \mathbf{0}\cdot\mathbf{u}_1 &= (c_1\mathbf{u}_1 + c_2\mathbf{u}_2 + \cdots + c_p\mathbf{u}_p)\cdot\mathbf{u}_1 \\
&= (c_1\mathbf{u}_1)\cdot\mathbf{u}_1 + (c_2\mathbf{u}_2)\cdot\mathbf{u}_1 + \cdots + (c_p\mathbf{u}_p)\cdot\mathbf{u}_1 \\
&= c_1(\mathbf{u}_1\cdot\mathbf{u}_1) + c_2(\mathbf{u}_2\cdot\mathbf{u}_1) + \cdots + c_p(\mathbf{u}_p\cdot\mathbf{u}_1) \\
&= c_1(\mathbf{u}_1\cdot\mathbf{u}_1)
\end{aligned}
$$

because $\mathbf{u}_1$ is orthogonal to $\mathbf{u}_2, \ldots, \mathbf{u}_p$. Since $\mathbf{u}_1$ is nonzero, $\mathbf{u}_1\cdot\mathbf{u}_1$ is not zero and so $c_1 = 0$. Similarly, $c_2, \ldots, c_p$ must be zero. Thus S is linearly independent. ∎

DEFINITION

> An **orthogonal basis** for a subspace W of $\mathbb{R}^n$ is a basis for W that is also an orthogonal set.

The next theorem suggests why an orthogonal basis is much nicer than other bases. The weights in a linear combination can be computed easily.

THEOREM 5

> Let $\{\mathbf{u}_1, \ldots, \mathbf{u}_p\}$ be an orthogonal basis for a subspace W of $\mathbb{R}^n$. For each $\mathbf{y}$ in W, the weights in the linear combination
>
> $$ \mathbf{y} = c_1\mathbf{u}_1 + \cdots + c_p\mathbf{u}_p $$
>
> are given by
>
> $$ c_j = \frac{\mathbf{y}\cdot\mathbf{u}_j}{\mathbf{u}_j\cdot\mathbf{u}_j} \qquad (j = 1, \ldots, p) $$

PROOF As in the preceding proof, the orthogonality of $\{\mathbf{u}_1, \ldots, \mathbf{u}_p\}$ shows that

$$ \mathbf{y}\cdot\mathbf{u}_1 = (c_1\mathbf{u}_1 + c_2\mathbf{u}_2 + \cdots + c_p\mathbf{u}_p)\cdot\mathbf{u}_1 = c_1(\mathbf{u}_1\cdot\mathbf{u}_1) $$

Since $\mathbf{u}_1\cdot\mathbf{u}_1$ is not zero, the equation above can be solved for c_1. To find c_j for $j = 2, \ldots, p$, compute $\mathbf{y}\cdot\mathbf{u}_j$ and solve for c_j. ∎

EXAMPLE 2 The set $S = \{\mathbf{u}_1, \mathbf{u}_2, \mathbf{u}_3\}$ in Example 1 is an orthogonal basis for $\mathbb{R}^3$. Express the vector $\mathbf{y} = \begin{bmatrix} 6 \\ 1 \\ -8 \end{bmatrix}$ as a linear combination of the vectors in S.

SOLUTION Compute

$$
\begin{aligned}
\mathbf{y}\cdot\mathbf{u}_1 &= 11, & \mathbf{y}\cdot\mathbf{u}_2 &= -12, & \mathbf{y}\cdot\mathbf{u}_3 &= -33 \\
\mathbf{u}_1\cdot\mathbf{u}_1 &= 11, & \mathbf{u}_2\cdot\mathbf{u}_2 &= 6, & \mathbf{u}_3\cdot\mathbf{u}_3 &= 33/2
\end{aligned}
$$

By Theorem 5,

$$
\begin{aligned}
\mathbf{y} &= \frac{\mathbf{y}\cdot\mathbf{u}_1}{\mathbf{u}_1\cdot\mathbf{u}_1}\mathbf{u}_1 + \frac{\mathbf{y}\cdot\mathbf{u}_2}{\mathbf{u}_2\cdot\mathbf{u}_2}\mathbf{u}_2 + \frac{\mathbf{y}\cdot\mathbf{u}_3}{\mathbf{u}_3\cdot\mathbf{u}_3}\mathbf{u}_3 \\
&= \frac{11}{11}\mathbf{u}_1 + \frac{-12}{6}\mathbf{u}_2 + \frac{-33}{33/2}\mathbf{u}_3 \\
&= \mathbf{u}_1 - 2\mathbf{u}_2 - 2\mathbf{u}_3
\end{aligned}
$$
∎

Notice how easy it is to compute the weights needed to build $\mathbf{y}$ from an orthogonal basis. If the basis were not orthogonal, it would be necessary to solve a system of linear equations in order to find the weights, as in Chapter 1.

We turn next to a construction that will become a key step in many calculations involving orthogonality, and it will lead to a geometric interpretation of Theorem 5.

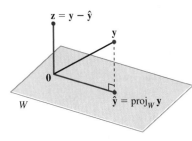

FIGURE 2
Finding α to make $\mathbf{y} - \hat{\mathbf{y}}$ orthogonal to $\mathbf{u}$.

An Orthogonal Projection

Given a nonzero vector $\mathbf{u}$ in $\mathbb{R}^n$, consider the problem of decomposing a vector $\mathbf{y}$ in $\mathbb{R}^n$ into the sum of two vectors, one a multiple of $\mathbf{u}$ and the other orthogonal to $\mathbf{u}$. We wish to write

$$\mathbf{y} = \hat{\mathbf{y}} + \mathbf{z} \tag{1}$$

where $\hat{\mathbf{y}} = \alpha\mathbf{u}$ for some scalar α and $\mathbf{z}$ is some vector orthogonal to $\mathbf{u}$. See Figure 2. Given any scalar α, let $\mathbf{z} = \mathbf{y} - \alpha\mathbf{u}$, so that (1) is satisfied. Then $\mathbf{y} - \hat{\mathbf{y}}$ is orthogonal to $\mathbf{u}$ if and only if

$$0 = (\mathbf{y} - \alpha\mathbf{u})\cdot\mathbf{u} = \mathbf{y}\cdot\mathbf{u} - (\alpha\mathbf{u})\cdot\mathbf{u} = \mathbf{y}\cdot\mathbf{u} - \alpha(\mathbf{u}\cdot\mathbf{u})$$

That is, (1) is satisfied with $\mathbf{z}$ orthogonal to $\mathbf{u}$ if and only if $\alpha = \dfrac{\mathbf{y}\cdot\mathbf{u}}{\mathbf{u}\cdot\mathbf{u}}$ and $\hat{\mathbf{y}} = \dfrac{\mathbf{y}\cdot\mathbf{u}}{\mathbf{u}\cdot\mathbf{u}}\mathbf{u}$. The vector $\hat{\mathbf{y}}$ is called the **orthogonal projection of y onto u**, and the vector $\mathbf{z}$ is called the **component of y orthogonal to u**.

If c is any nonzero scalar and if $\mathbf{u}$ is replaced by $c\mathbf{u}$ in the definition of $\hat{\mathbf{y}}$, then the orthogonal projection of $\mathbf{y}$ onto $c\mathbf{u}$ is exactly the same as the orthogonal projection of $\mathbf{y}$ onto $\mathbf{u}$ (Exercise 39). Hence this projection is determined by the *subspace L* spanned by $\mathbf{u}$ (the line through $\mathbf{u}$ and $\mathbf{0}$). Sometimes $\hat{\mathbf{y}}$ is denoted by $\text{proj}_L\,\mathbf{y}$ and is called the **orthogonal projection of y onto L**. That is,

$$\hat{\mathbf{y}} = \text{proj}_L\,\mathbf{y} = \frac{\mathbf{y}\cdot\mathbf{u}}{\mathbf{u}\cdot\mathbf{u}}\mathbf{u} \tag{2}$$

EXAMPLE 3 Let $\mathbf{y} = \begin{bmatrix} 7 \\ 6 \end{bmatrix}$ and $\mathbf{u} = \begin{bmatrix} 4 \\ 2 \end{bmatrix}$. Find the orthogonal projection of $\mathbf{y}$ onto $\mathbf{u}$. Then write $\mathbf{y}$ as the sum of two orthogonal vectors, one in Span $\{\mathbf{u}\}$ and one orthogonal to $\mathbf{u}$.

SOLUTION Compute

$$\mathbf{y}\cdot\mathbf{u} = \begin{bmatrix} 7 \\ 6 \end{bmatrix}\cdot\begin{bmatrix} 4 \\ 2 \end{bmatrix} = 40$$

$$\mathbf{u}\cdot\mathbf{u} = \begin{bmatrix} 4 \\ 2 \end{bmatrix}\cdot\begin{bmatrix} 4 \\ 2 \end{bmatrix} = 20$$

The orthogonal projection of $\mathbf{y}$ onto $\mathbf{u}$ is

$$\hat{\mathbf{y}} = \frac{\mathbf{y}\cdot\mathbf{u}}{\mathbf{u}\cdot\mathbf{u}}\mathbf{u} = \frac{40}{20}\mathbf{u} = 2\begin{bmatrix} 4 \\ 2 \end{bmatrix} = \begin{bmatrix} 8 \\ 4 \end{bmatrix}$$

and the component of $\mathbf{y}$ orthogonal to $\mathbf{u}$ is

$$\mathbf{y} - \hat{\mathbf{y}} = \begin{bmatrix} 7 \\ 6 \end{bmatrix} - \begin{bmatrix} 8 \\ 4 \end{bmatrix} = \begin{bmatrix} -1 \\ 2 \end{bmatrix}$$

The sum of these two vectors is $\mathbf{y}$. That is,

$$\underset{\underset{\mathbf{y}}{\uparrow}}{\begin{bmatrix} 7 \\ 6 \end{bmatrix}} = \underset{\underset{\hat{\mathbf{y}}}{\uparrow}}{\begin{bmatrix} 8 \\ 4 \end{bmatrix}} + \underset{\underset{(\mathbf{y}-\hat{\mathbf{y}})}{\uparrow}}{\begin{bmatrix} -1 \\ 2 \end{bmatrix}}$$

This decomposition of $\mathbf{y}$ is illustrated in Figure 3. *Note:* If the calculations above are correct, then $\{\hat{\mathbf{y}}, \mathbf{y} - \hat{\mathbf{y}}\}$ will be an orthogonal set. As a check, compute

$$\hat{\mathbf{y}}\cdot(\mathbf{y} - \hat{\mathbf{y}}) = \begin{bmatrix} 8 \\ 4 \end{bmatrix}\cdot\begin{bmatrix} -1 \\ 2 \end{bmatrix} = -8 + 8 = 0 \qquad \blacksquare$$

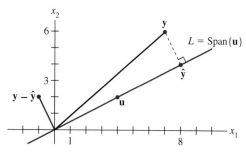

FIGURE 3 The orthogonal projection of **y** onto a line L through the origin.

Since the line segment in Figure 3 between **y** and $\hat{\mathbf{y}}$ is perpendicular to L, by construction of $\hat{\mathbf{y}}$, the point identified with $\hat{\mathbf{y}}$ is the closest point of L to **y**. (This can be proved from geometry. We will assume this for $\mathbb{R}^2$ now and prove it for $\mathbb{R}^n$ in Section 6.3.)

EXAMPLE 4 Find the distance in Figure 3 from **y** to L.

SOLUTION The distance from **y** to L is the length of the perpendicular line segment from **y** to the orthogonal projection $\hat{\mathbf{y}}$. This length equals the length of $\mathbf{y} - \hat{\mathbf{y}}$. Thus the distance is

$$\|\mathbf{y} - \hat{\mathbf{y}}\| = \sqrt{(-1)^2 + 2^2} = \sqrt{5}$$ ∎

A Geometric Interpretation of Theorem 5

The formula for the orthogonal projection $\hat{\mathbf{y}}$ in (2) has the same appearance as each of the terms in Theorem 5. Thus Theorem 5 decomposes a vector **y** into a sum of orthogonal projections onto one-dimensional subspaces.

It is easy to visualize the case in which $W = \mathbb{R}^2 = \text{Span}\{\mathbf{u}_1, \mathbf{u}_2\}$, with $\mathbf{u}_1$ and $\mathbf{u}_2$ orthogonal. Any **y** in $\mathbb{R}^2$ can be written in the form

$$\mathbf{y} = \frac{\mathbf{y} \cdot \mathbf{u}_1}{\mathbf{u}_1 \cdot \mathbf{u}_1} \mathbf{u}_1 + \frac{\mathbf{y} \cdot \mathbf{u}_2}{\mathbf{u}_2 \cdot \mathbf{u}_2} \mathbf{u}_2 \tag{3}$$

The first term in (3) is the projection of **y** onto the subspace spanned by $\mathbf{u}_1$ (the line through $\mathbf{u}_1$ and the origin), and the second term is the projection of **y** onto the subspace spanned by $\mathbf{u}_2$. Thus (3) expresses **y** as the sum of its projections onto the (orthogonal) axes determined by $\mathbf{u}_1$ and $\mathbf{u}_2$. See Figure 4.

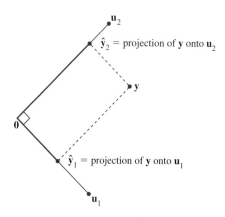

FIGURE 4 A vector decomposed into the sum of two projections.

Theorem 5 decomposes each **y** in Span $\{\mathbf{u}_1, \ldots, \mathbf{u}_p\}$ into the sum of p projections onto one-dimensional subspaces that are mutually orthogonal.

Decomposing a Force into Component Forces

The decomposition in Figure 4 can occur in physics when some sort of force is applied to an object. Choosing an appropriate coordinate system allows the force to be represented by a vector **y** in $\mathbb{R}^2$ or $\mathbb{R}^3$. Often the problem involves some particular direction of interest, which is represented by another vector **u**. For instance, if the object is moving in a straight line when the force is applied, the vector **u** might point in the direction of movement, as in Figure 5. A key step in the problem is to decompose the force into a component in the direction of **u** and a component orthogonal to **u**. The calculations would be analogous to those previously made in Example 3.

FIGURE 5

Orthonormal Sets

A set $\{\mathbf{u}_1, \ldots, \mathbf{u}_p\}$ is an **orthonormal set** if it is an orthogonal set of unit vectors. If W is the subspace spanned by such a set, then $\{\mathbf{u}_1, \ldots, \mathbf{u}_p\}$ is an **orthonormal basis** for W, since the set is automatically linearly independent, by Theorem 4.

The simplest example of an orthonormal set is the standard basis $\{\mathbf{e}_1, \ldots, \mathbf{e}_n\}$ for $\mathbb{R}^n$. Any nonempty subset of $\{\mathbf{e}_1, \ldots, \mathbf{e}_n\}$ is orthonormal, too. Here is a more complicated example.

EXAMPLE 5   Show that $\{\mathbf{v}_1, \mathbf{v}_2, \mathbf{v}_3\}$ is an orthonormal basis of $\mathbb{R}^3$, where

$$\mathbf{v}_1 = \begin{bmatrix} 3/\sqrt{11} \\ 1/\sqrt{11} \\ 1/\sqrt{11} \end{bmatrix}, \quad \mathbf{v}_2 = \begin{bmatrix} -1/\sqrt{6} \\ 2/\sqrt{6} \\ 1/\sqrt{6} \end{bmatrix}, \quad \mathbf{v}_3 = \begin{bmatrix} -1/\sqrt{66} \\ -4/\sqrt{66} \\ 7/\sqrt{66} \end{bmatrix}$$

SOLUTION   Compute

$$\mathbf{v}_1 \cdot \mathbf{v}_2 = -3/\sqrt{66} + 2/\sqrt{66} + 1/\sqrt{66} = 0$$
$$\mathbf{v}_1 \cdot \mathbf{v}_3 = -3/\sqrt{726} - 4/\sqrt{726} + 7/\sqrt{726} = 0$$
$$\mathbf{v}_2 \cdot \mathbf{v}_3 = 1/\sqrt{396} - 8/\sqrt{396} + 7/\sqrt{396} = 0$$

Thus $\{\mathbf{v}_1, \mathbf{v}_2, \mathbf{v}_3\}$ is an orthogonal set. Also,

$$\mathbf{v}_1 \cdot \mathbf{v}_1 = 9/11 + 1/11 + 1/11 = 1$$
$$\mathbf{v}_2 \cdot \mathbf{v}_2 = 1/6 + 4/6 + 1/6 = 1$$
$$\mathbf{v}_3 \cdot \mathbf{v}_3 = 1/66 + 16/66 + 49/66 = 1$$

which shows that $\mathbf{v}_1$, $\mathbf{v}_2$, and $\mathbf{v}_3$ are unit vectors. Thus $\{\mathbf{v}_1, \mathbf{v}_2, \mathbf{v}_3\}$ is an orthonormal set. Since the set is linearly independent, its three vectors form a basis for $\mathbb{R}^3$. See Figure 6.

FIGURE 6

When the vectors in an orthogonal set of nonzero vectors are *normalized* to have unit length, the new vectors will still be orthogonal, and hence the new set will be an orthonormal set. See Exercise 40. It is easy to check that the vectors in Figure 6 (Example 5) are simply the unit vectors in the directions of the vectors in Figure 1 (Example 1).

Matrices whose columns form an orthonormal set are important in applications and in computer algorithms for matrix computations. Their main properties are given in Theorems 6 and 7.

THEOREM 6   An $m \times n$ matrix U has orthonormal columns if and only if $U^T U = I$.

PROOF   To simplify notation, we suppose that U has only three columns, each a vector in $\mathbb{R}^m$. The proof of the general case is essentially the same. Let $U = [\, \mathbf{u}_1 \quad \mathbf{u}_2 \quad \mathbf{u}_3 \,]$ and compute

$$U^T U = \begin{bmatrix} \mathbf{u}_1^T \\ \mathbf{u}_2^T \\ \mathbf{u}_3^T \end{bmatrix} \begin{bmatrix} \mathbf{u}_1 & \mathbf{u}_2 & \mathbf{u}_3 \end{bmatrix} = \begin{bmatrix} \mathbf{u}_1^T \mathbf{u}_1 & \mathbf{u}_1^T \mathbf{u}_2 & \mathbf{u}_1^T \mathbf{u}_3 \\ \mathbf{u}_2^T \mathbf{u}_1 & \mathbf{u}_2^T \mathbf{u}_2 & \mathbf{u}_2^T \mathbf{u}_3 \\ \mathbf{u}_3^T \mathbf{u}_1 & \mathbf{u}_3^T \mathbf{u}_2 & \mathbf{u}_3^T \mathbf{u}_3 \end{bmatrix} \tag{4}$$

The entries in the matrix at the right are inner products, using transpose notation. The columns of U are orthogonal if and only if

$$\mathbf{u}_1^T \mathbf{u}_2 = \mathbf{u}_2^T \mathbf{u}_1 = 0, \quad \mathbf{u}_1^T \mathbf{u}_3 = \mathbf{u}_3^T \mathbf{u}_1 = 0, \quad \mathbf{u}_2^T \mathbf{u}_3 = \mathbf{u}_3^T \mathbf{u}_2 = 0 \tag{5}$$

The columns of U all have unit length if and only if

$$\mathbf{u}_1^T \mathbf{u}_1 = 1, \quad \mathbf{u}_2^T \mathbf{u}_2 = 1, \quad \mathbf{u}_3^T \mathbf{u}_3 = 1 \tag{6}$$

The theorem follows immediately from (4)–(6).   ∎

THEOREM 7   Let U be an $m \times n$ matrix with orthonormal columns, and let $\mathbf{x}$ and $\mathbf{y}$ be in $\mathbb{R}^n$. Then

a. $\|U\mathbf{x}\| = \|\mathbf{x}\|$
b. $(U\mathbf{x}) \cdot (U\mathbf{y}) = \mathbf{x} \cdot \mathbf{y}$
c. $(U\mathbf{x}) \cdot (U\mathbf{y}) = 0$ if and only if $\mathbf{x} \cdot \mathbf{y} = 0$

Properties (a) and (c) say that the linear mapping $\mathbf{x} \mapsto U\mathbf{x}$ preserves lengths and orthogonality. These properties are crucial for many computer algorithms. See Exercise 33 for the proof of Theorem 7.

EXAMPLE 6   Let $U = \begin{bmatrix} 1/\sqrt{2} & 2/3 \\ 1/\sqrt{2} & -2/3 \\ 0 & 1/3 \end{bmatrix}$ and $\mathbf{x} = \begin{bmatrix} \sqrt{2} \\ 3 \end{bmatrix}$. Notice that U has orthonormal columns and

$$U^T U = \begin{bmatrix} 1/\sqrt{2} & 1/\sqrt{2} & 0 \\ 2/3 & -2/3 & 1/3 \end{bmatrix} \begin{bmatrix} 1/\sqrt{2} & 2/3 \\ 1/\sqrt{2} & -2/3 \\ 0 & 1/3 \end{bmatrix} = \begin{bmatrix} 1 & 0 \\ 0 & 1 \end{bmatrix}$$

Verify that $\|U\mathbf{x}\| = \|\mathbf{x}\|$.

SOLUTION

$$Ux = \begin{bmatrix} 1/\sqrt{2} & 2/3 \\ 1/\sqrt{2} & -2/3 \\ 0 & 1/3 \end{bmatrix} \begin{bmatrix} \sqrt{2} \\ 3 \end{bmatrix} = \begin{bmatrix} 3 \\ -1 \\ 1 \end{bmatrix}$$

$$\|Ux\| = \sqrt{9+1+1} = \sqrt{11}$$

$$\|x\| = \sqrt{2+9} = \sqrt{11}$$ ∎

Theorems 6 and 7 are particularly useful when applied to *square* matrices. An **orthogonal matrix** is a square invertible matrix U such that $U^{-1} = U^T$. By Theorem 6, such a matrix has orthonormal columns.[1] It is easy to see that any *square* matrix with orthonormal columns is an orthogonal matrix. Surprisingly, such a matrix must have orthonormal *rows*, too. See Exercises 35 and 36. Orthogonal matrices will appear frequently in Chapter 7.

EXAMPLE 7 The matrix

$$U = \begin{bmatrix} 3/\sqrt{11} & -1/\sqrt{6} & -1/\sqrt{66} \\ 1/\sqrt{11} & 2/\sqrt{6} & -4/\sqrt{66} \\ 1/\sqrt{11} & 1/\sqrt{6} & 7/\sqrt{66} \end{bmatrix}$$

is an orthogonal matrix because it is square and because its columns are orthonormal, by Example 5. Verify that the rows are orthonormal, too! ∎

Practice Problems

1. Let $\mathbf{u}_1 = \begin{bmatrix} -1/\sqrt{5} \\ 2/\sqrt{5} \end{bmatrix}$ and $\mathbf{u}_2 = \begin{bmatrix} 2/\sqrt{5} \\ 1/\sqrt{5} \end{bmatrix}$. Show that $\{\mathbf{u}_1, \mathbf{u}_2\}$ is an orthonormal basis for $\mathbb{R}^2$.

2. Let $\mathbf{y}$ and L be as in Example 3 and Figure 3. Compute the orthogonal projection $\hat{\mathbf{y}}$ of $\mathbf{y}$ onto L using $\mathbf{u} = \begin{bmatrix} 2 \\ 1 \end{bmatrix}$ instead of the $\mathbf{u}$ in Example 3.

3. Let U and $\mathbf{x}$ be as in Example 6, and let $\mathbf{y} = \begin{bmatrix} -3\sqrt{2} \\ 6 \end{bmatrix}$. Verify that $U\mathbf{x} \cdot U\mathbf{y} = \mathbf{x} \cdot \mathbf{y}$.

4. Let U be an $n \times n$ matrix with orthonormal columns. Show that $\det U = \pm 1$.

6.2 Exercises

In Exercises 1–6, determine which sets of vectors are orthogonal.

1. $\begin{bmatrix} -1 \\ 4 \\ -3 \end{bmatrix}, \begin{bmatrix} 5 \\ 2 \\ 1 \end{bmatrix}, \begin{bmatrix} 3 \\ -4 \\ -7 \end{bmatrix}$ 2. $\begin{bmatrix} 1 \\ -2 \\ 1 \end{bmatrix}, \begin{bmatrix} 0 \\ 1 \\ 2 \end{bmatrix}, \begin{bmatrix} -5 \\ -2 \\ 1 \end{bmatrix}$

3. $\begin{bmatrix} 2 \\ -7 \\ -1 \end{bmatrix}, \begin{bmatrix} -6 \\ -3 \\ 9 \end{bmatrix}, \begin{bmatrix} 3 \\ 1 \\ -1 \end{bmatrix}$ 4. $\begin{bmatrix} 2 \\ -5 \\ -3 \end{bmatrix}, \begin{bmatrix} 0 \\ 0 \\ 0 \end{bmatrix}, \begin{bmatrix} 4 \\ 2 \\ 6 \end{bmatrix}$

5. $\begin{bmatrix} 3 \\ -2 \\ 1 \\ 3 \end{bmatrix}, \begin{bmatrix} -1 \\ 3 \\ -3 \\ 4 \end{bmatrix}, \begin{bmatrix} 3 \\ 8 \\ 7 \\ 0 \end{bmatrix}$ 6. $\begin{bmatrix} 5 \\ -4 \\ 0 \\ 3 \end{bmatrix}, \begin{bmatrix} -4 \\ 1 \\ -3 \\ 8 \end{bmatrix}, \begin{bmatrix} 3 \\ 3 \\ 5 \\ -1 \end{bmatrix}$

In Exercises 7–10, show that $\{\mathbf{u}_1, \mathbf{u}_2\}$ or $\{\mathbf{u}_1, \mathbf{u}_2, \mathbf{u}_3\}$ is an orthogonal basis for $\mathbb{R}^2$ or $\mathbb{R}^3$, respectively. Then express $\mathbf{x}$ as a linear combination of the $\mathbf{u}$'s.

7. $\mathbf{u}_1 = \begin{bmatrix} 2 \\ -3 \end{bmatrix}, \mathbf{u}_2 = \begin{bmatrix} 6 \\ 4 \end{bmatrix}$, and $\mathbf{x} = \begin{bmatrix} 9 \\ -7 \end{bmatrix}$

---

[1] A better name might be *orthonormal matrix*, and this term is found in some statistics texts. However, *orthogonal matrix* is the standard term in linear algebra.

8. $\mathbf{u}_1 = \begin{bmatrix} 3 \\ 1 \end{bmatrix}$, $\mathbf{u}_2 = \begin{bmatrix} -2 \\ 6 \end{bmatrix}$, and $\mathbf{x} = \begin{bmatrix} -4 \\ 3 \end{bmatrix}$

9. $\mathbf{u}_1 = \begin{bmatrix} 1 \\ 0 \\ 1 \end{bmatrix}$, $\mathbf{u}_2 = \begin{bmatrix} -1 \\ 4 \\ 1 \end{bmatrix}$, $\mathbf{u}_3 = \begin{bmatrix} 2 \\ 1 \\ -2 \end{bmatrix}$, and $\mathbf{x} = \begin{bmatrix} 8 \\ -4 \\ -3 \end{bmatrix}$

10. $\mathbf{u}_1 = \begin{bmatrix} 3 \\ -3 \\ 0 \end{bmatrix}$, $\mathbf{u}_2 = \begin{bmatrix} 2 \\ 2 \\ -1 \end{bmatrix}$, $\mathbf{u}_3 = \begin{bmatrix} 1 \\ 1 \\ 4 \end{bmatrix}$, and $\mathbf{x} = \begin{bmatrix} 5 \\ -3 \\ 1 \end{bmatrix}$

11. Compute the orthogonal projection of $\begin{bmatrix} 1 \\ 7 \end{bmatrix}$ onto the line through $\begin{bmatrix} -4 \\ 2 \end{bmatrix}$ and the origin.

12. Compute the orthogonal projection of $\begin{bmatrix} 1 \\ -1 \end{bmatrix}$ onto the line through $\begin{bmatrix} -1 \\ 2 \end{bmatrix}$ and the origin.

13. Let $\mathbf{y} = \begin{bmatrix} 2 \\ 3 \end{bmatrix}$ and $\mathbf{u} = \begin{bmatrix} 4 \\ -7 \end{bmatrix}$. Write $\mathbf{y}$ as the sum of two orthogonal vectors, one in Span $\{\mathbf{u}\}$ and one orthogonal to $\mathbf{u}$.

14. Let $\mathbf{y} = \begin{bmatrix} 2 \\ 6 \end{bmatrix}$ and $\mathbf{u} = \begin{bmatrix} 6 \\ 1 \end{bmatrix}$. Write $\mathbf{y}$ as the sum of a vector in Span $\{\mathbf{u}\}$ and a vector orthogonal to $\mathbf{u}$.

15. Let $\mathbf{y} = \begin{bmatrix} 3 \\ 1 \end{bmatrix}$ and $\mathbf{u} = \begin{bmatrix} 8 \\ 6 \end{bmatrix}$. Compute the distance from $\mathbf{y}$ to the line through $\mathbf{u}$ and the origin.

16. Let $\mathbf{y} = \begin{bmatrix} -3 \\ 9 \end{bmatrix}$ and $\mathbf{u} = \begin{bmatrix} 1 \\ 2 \end{bmatrix}$. Compute the distance from $\mathbf{y}$ to the line through $\mathbf{u}$ and the origin.

In Exercises 17–22, determine which sets of vectors are orthonormal. If a set is only orthogonal, normalize the vectors to produce an orthonormal set.

17. $\begin{bmatrix} 1/3 \\ 1/3 \\ 1/3 \end{bmatrix}$, $\begin{bmatrix} -1/2 \\ 0 \\ 1/2 \end{bmatrix}$ **18.** $\begin{bmatrix} 0 \\ 0 \\ 1 \end{bmatrix}$, $\begin{bmatrix} 0 \\ -1 \\ 0 \end{bmatrix}$

19. $\begin{bmatrix} -.6 \\ .8 \end{bmatrix}$, $\begin{bmatrix} .8 \\ .6 \end{bmatrix}$ **20.** $\begin{bmatrix} -2/3 \\ 1/3 \\ 2/3 \end{bmatrix}$, $\begin{bmatrix} 1/3 \\ 2/3 \\ 0 \end{bmatrix}$

21. $\begin{bmatrix} 1/\sqrt{10} \\ 3/\sqrt{20} \\ 3/\sqrt{20} \end{bmatrix}$, $\begin{bmatrix} 3/\sqrt{10} \\ -1/\sqrt{20} \\ -1/\sqrt{20} \end{bmatrix}$, $\begin{bmatrix} 0 \\ -1/\sqrt{2} \\ 1/\sqrt{2} \end{bmatrix}$

22. $\begin{bmatrix} 1/\sqrt{18} \\ 4/\sqrt{18} \\ 1/\sqrt{18} \end{bmatrix}$, $\begin{bmatrix} 1/\sqrt{2} \\ 0 \\ -1/\sqrt{2} \end{bmatrix}$, $\begin{bmatrix} -2/3 \\ 1/3 \\ -2/3 \end{bmatrix}$

In Exercises 23–32, all vectors are in $\mathbb{R}^n$. Mark each statement True or False **(T/F)**. Justify each answer.

23. **(T/F)** Not every linearly independent set in $\mathbb{R}^n$ is an orthogonal set.

24. **(T/F)** Not every orthogonal set in $\mathbb{R}^n$ is linearly independent.

25. **(T/F)** If $\mathbf{y}$ is a linear combination of nonzero vectors from an orthogonal set, then the weights in the linear combination can be computed without row operations on a matrix.

26. **(T/F)** If a set $S = \{\mathbf{u}_1, \ldots, \mathbf{u}_p\}$ has the property that $\mathbf{u}_i \cdot \mathbf{u}_j = 0$ whenever $i \neq j$, then S is an orthonormal set.

27. **(T/F)** If the vectors in an orthogonal set of nonzero vectors are normalized, then some of the new vectors may not be orthogonal.

28. **(T/F)** If the columns of an $m \times n$ matrix A are orthonormal, then the linear mapping $\mathbf{x} \mapsto A\mathbf{x}$ preserves lengths.

29. **(T/F)** A matrix with orthonormal columns is an orthogonal matrix.

30. **(T/F)** The orthogonal projection of $\mathbf{y}$ onto $\mathbf{v}$ is the same as the orthogonal projection of $\mathbf{y}$ onto $c\mathbf{v}$ whenever $c \neq 0$.

31. **(T/F)** If L is a line through $\mathbf{0}$ and if $\hat{\mathbf{y}}$ is the orthogonal projection of $\mathbf{y}$ onto L, then $\|\hat{\mathbf{y}}\|$ gives the distance from $\mathbf{y}$ to L.

32. **(T/F)** An orthogonal matrix is invertible.

33. Prove Theorem 7. [*Hint:* For (a), compute $\|U\mathbf{x}\|^2$, or prove (b) first.]

34. Suppose W is a subspace of $\mathbb{R}^n$ spanned by n nonzero orthogonal vectors. Explain why $W = \mathbb{R}^n$.

35. Let U be a square matrix with orthonormal columns. Explain why U is invertible. (Mention the theorems you use.)

36. Let U be an $n \times n$ orthogonal matrix. Show that the rows of U form an orthonormal basis of $\mathbb{R}^n$.

37. Let U and V be $n \times n$ orthogonal matrices. Explain why UV is an orthogonal matrix. [That is, explain why UV is invertible and its inverse is $(UV)^T$.]

38. Let U be an orthogonal matrix, and construct V by interchanging some of the columns of U. Explain why V is an orthogonal matrix.

39. Show that the orthogonal projection of a vector $\mathbf{y}$ onto a line L through the origin in $\mathbb{R}^2$ does not depend on the choice of the nonzero $\mathbf{u}$ in L used in the formula for $\hat{\mathbf{y}}$. To do this, suppose $\mathbf{y}$ and $\mathbf{u}$ are given and $\hat{\mathbf{y}}$ has been computed by formula (2) in this section. Replace $\mathbf{u}$ in that formula by $c\mathbf{u}$, where c is an unspecified nonzero scalar. Show that the new formula gives the same $\hat{\mathbf{y}}$.

40. Let $\{\mathbf{v}_1, \mathbf{v}_2\}$ be an orthogonal set of nonzero vectors, and let c_1, c_2 be any nonzero scalars. Show that $\{c_1\mathbf{v}_1, c_2\mathbf{v}_2\}$ is also an orthogonal set. Since orthogonality of a set is defined in terms of pairs of vectors, this shows that if the vectors in an orthogonal set are normalized, the new set will still be orthogonal.

41. Given $\mathbf{u} \neq \mathbf{0}$ in $\mathbb{R}^n$, let $L = \text{Span}\{\mathbf{u}\}$. Show that the mapping $\mathbf{x} \mapsto \text{proj}_L \mathbf{x}$ is a linear transformation.

42. Given $\mathbf{u} \neq \mathbf{0}$ in $\mathbb{R}^n$, let $L = \text{Span}\{\mathbf{u}\}$. For $\mathbf{y}$ in $\mathbb{R}^n$, the **reflection of y in L** is the point $\text{refl}_L \, \mathbf{y}$ defined by

$$\text{refl}_L \, \mathbf{y} = 2 \, \text{proj}_L \, \mathbf{y} - \mathbf{y}$$

See the figure, which shows that $\text{refl}_L \, \mathbf{y}$ is the sum of $\hat{\mathbf{y}} = \text{proj}_L \, \mathbf{y}$ and $\hat{\mathbf{y}} - \mathbf{y}$. Show that the mapping $\mathbf{y} \mapsto \text{refl}_L \, \mathbf{y}$ is a linear transformation.

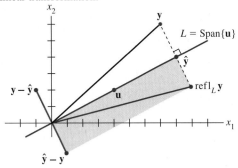

The reflection of **y** in a line through the origin.

T 43. Show that the columns of the matrix A are orthogonal by making an appropriate matrix calculation. State the calculation you use.

$$A = \begin{bmatrix} -6 & -3 & 6 & 1 \\ -1 & 2 & 1 & -6 \\ 3 & 6 & 3 & -2 \\ 6 & -3 & 6 & -1 \\ 2 & -1 & 2 & 3 \\ -3 & 6 & 3 & 2 \\ -2 & -1 & 2 & -3 \\ 1 & 2 & 1 & 6 \end{bmatrix}$$

T 44. In parts (a)–(d), let U be the matrix formed by normalizing each column of the matrix A in Exercise 43.

a. Compute $U^T U$ and $U U^T$. How do they differ?

b. Generate a random vector $\mathbf{y}$ in $\mathbb{R}^8$, and compute $\mathbf{p} = U U^T \mathbf{y}$ and $\mathbf{z} = \mathbf{y} - \mathbf{p}$. Explain why $\mathbf{p}$ is in Col A. Verify that $\mathbf{z}$ is orthogonal to $\mathbf{p}$.

c. Verify that $\mathbf{z}$ is orthogonal to each column of U.

d. Notice that $\mathbf{y} = \mathbf{p} + \mathbf{z}$, with $\mathbf{p}$ in Col A. Explain why $\mathbf{z}$ is in $(\text{Col } A)^\perp$. (The significance of this decomposition of $\mathbf{y}$ will be explained in the next section.)

Solutions to Practice Problems

1. The vectors are orthogonal because

$$\mathbf{u}_1 \cdot \mathbf{u}_2 = -2/5 + 2/5 = 0$$

They are unit vectors because

$$\|\mathbf{u}_1\|^2 = (-1/\sqrt{5})^2 + (2/\sqrt{5})^2 = 1/5 + 4/5 = 1$$
$$\|\mathbf{u}_2\|^2 = (2/\sqrt{5})^2 + (1/\sqrt{5})^2 = 4/5 + 1/5 = 1$$

In particular, the set $\{\mathbf{u}_1, \mathbf{u}_2\}$ is linearly independent, and hence is a basis for $\mathbb{R}^2$ since there are two vectors in the set.

2. When $\mathbf{y} = \begin{bmatrix} 7 \\ 6 \end{bmatrix}$ and $\mathbf{u} = \begin{bmatrix} 2 \\ 1 \end{bmatrix}$,

$$\hat{\mathbf{y}} = \frac{\mathbf{y} \cdot \mathbf{u}}{\mathbf{u} \cdot \mathbf{u}} \mathbf{u} = \frac{20}{5} \begin{bmatrix} 2 \\ 1 \end{bmatrix} = 4 \begin{bmatrix} 2 \\ 1 \end{bmatrix} = \begin{bmatrix} 8 \\ 4 \end{bmatrix}$$

This is the same $\hat{\mathbf{y}}$ found in Example 3. The orthogonal projection does not depend on the $\mathbf{u}$ chosen on the line. See Exercise 39.

3. $U\mathbf{y} = \begin{bmatrix} 1/\sqrt{2} & 2/3 \\ 1/\sqrt{2} & -2/3 \\ 0 & 1/3 \end{bmatrix} \begin{bmatrix} -3\sqrt{2} \\ 6 \end{bmatrix} = \begin{bmatrix} 1 \\ -7 \\ 2 \end{bmatrix}$

Also, from Example 6, $\mathbf{x} = \begin{bmatrix} \sqrt{2} \\ 3 \end{bmatrix}$ and $U\mathbf{x} = \begin{bmatrix} 3 \\ -1 \\ 1 \end{bmatrix}$. Hence

$$U\mathbf{x} \cdot U\mathbf{y} = 3 + 7 + 2 = 12, \quad \text{and} \quad \mathbf{x} \cdot \mathbf{y} = -6 + 18 = 12$$

4. Since U is an $n \times n$ matrix with orthonormal columns, by Theorem 6, $U^T U = I$. Taking the determinant of the left side of this equation, and applying Theorems 5 and 6 from Section 3.2 results in det $U^T U = (\det U^T)(\det U) = (\det U)(\det U) = (\det U)^2$. Recall det $I = 1$. Putting the two sides of the equation back together results in $(\det U)^2 = 1$ and hence det $U = \pm 1$.

6.3 Orthogonal Projections

The orthogonal projection of a point in $\mathbb{R}^2$ onto a line through the origin has an important analogue in $\mathbb{R}^n$. Given a vector $\mathbf{y}$ and a subspace W in $\mathbb{R}^n$, there is a vector $\hat{\mathbf{y}}$ in W such that (1) $\hat{\mathbf{y}}$ is the unique vector in W for which $\mathbf{y} - \hat{\mathbf{y}}$ is orthogonal to W, and (2) $\hat{\mathbf{y}}$ is the unique vector in W closest to $\mathbf{y}$. See Figure 1. These two properties of $\hat{\mathbf{y}}$ provide the key to finding least-squares solutions of linear systems.

To prepare for the first theorem, observe that whenever a vector $\mathbf{y}$ is written as a linear combination of vectors $\mathbf{u}_1, \ldots, \mathbf{u}_n$ in $\mathbb{R}^n$, the terms in the sum for $\mathbf{y}$ can be grouped into two parts so that $\mathbf{y}$ can be written as

$$\mathbf{y} = \mathbf{z}_1 + \mathbf{z}_2$$

where $\mathbf{z}_1$ is a linear combination of some of the $\mathbf{u}_i$ and $\mathbf{z}_2$ is a linear combination of the rest of the $\mathbf{u}_i$. This idea is particularly useful when $\{\mathbf{u}_1, \ldots, \mathbf{u}_n\}$ is an orthogonal basis. Recall from Section 6.1 that $W^\perp$ denotes the set of all vectors orthogonal to a subspace W.

FIGURE 1

EXAMPLE 1   Let $\{\mathbf{u}_1, \ldots, \mathbf{u}_5\}$ be an orthogonal basis for $\mathbb{R}^5$ and let

$$\mathbf{y} = c_1 \mathbf{u}_1 + \cdots + c_5 \mathbf{u}_5$$

Consider the subspace $W = \text{Span}\{\mathbf{u}_1, \mathbf{u}_2\}$, and write $\mathbf{y}$ as the sum of a vector $\mathbf{z}_1$ in W and a vector $\mathbf{z}_2$ in $W^\perp$.

SOLUTION   Write

$$\mathbf{y} = \underbrace{c_1 \mathbf{u}_1 + c_2 \mathbf{u}_2}_{\mathbf{z}_1} + \underbrace{c_3 \mathbf{u}_3 + c_4 \mathbf{u}_4 + c_5 \mathbf{u}_5}_{\mathbf{z}_2}$$

where $\mathbf{z}_1 = c_1 \mathbf{u}_1 + c_2 \mathbf{u}_2$ is in Span$\{\mathbf{u}_1, \mathbf{u}_2\}$

and $\mathbf{z}_2 = c_3 \mathbf{u}_3 + c_4 \mathbf{u}_4 + c_5 \mathbf{u}_5$ is in Span$\{\mathbf{u}_3, \mathbf{u}_4, \mathbf{u}_5\}$.

To show that $\mathbf{z}_2$ is in $W^\perp$, it suffices to show that $\mathbf{z}_2$ is orthogonal to the vectors in the basis $\{\mathbf{u}_1, \mathbf{u}_2\}$ for W. (See Section 6.1.) Using properties of the inner product, compute

$$\begin{aligned}
\mathbf{z}_2 \cdot \mathbf{u}_1 &= (c_3 \mathbf{u}_3 + c_4 \mathbf{u}_4 + c_5 \mathbf{u}_5) \cdot \mathbf{u}_1 \\
&= c_3 \mathbf{u}_3 \cdot \mathbf{u}_1 + c_4 \mathbf{u}_4 \cdot \mathbf{u}_1 + c_5 \mathbf{u}_5 \cdot \mathbf{u}_1 \\
&= 0
\end{aligned}$$

because $\mathbf{u}_1$ is orthogonal to $\mathbf{u}_3$, $\mathbf{u}_4$, and $\mathbf{u}_5$. A similar calculation shows that $\mathbf{z}_2 \cdot \mathbf{u}_2 = 0$. Thus $\mathbf{z}_2$ is in $W^\perp$.   ∎

The next theorem shows that the decomposition $\mathbf{y} = \mathbf{z}_1 + \mathbf{z}_2$ in Example 1 can be computed without having an orthogonal basis for $\mathbb{R}^n$. It is enough to have an orthogonal basis only for W.

THEOREM 8

The Orthogonal Decomposition Theorem

Let W be a subspace of $\mathbb{R}^n$. Then each $\mathbf{y}$ in $\mathbb{R}^n$ can be written uniquely in the form

$$\mathbf{y} = \hat{\mathbf{y}} + \mathbf{z} \tag{1}$$

where $\hat{\mathbf{y}}$ is in W and $\mathbf{z}$ is in $W^{\perp}$. In fact, if $\{\mathbf{u}_1, \ldots, \mathbf{u}_p\}$ is any orthogonal basis of W, then

$$\hat{\mathbf{y}} = \frac{\mathbf{y} \cdot \mathbf{u}_1}{\mathbf{u}_1 \cdot \mathbf{u}_1} \mathbf{u}_1 + \cdots + \frac{\mathbf{y} \cdot \mathbf{u}_p}{\mathbf{u}_p \cdot \mathbf{u}_p} \mathbf{u}_p \tag{2}$$

and $\mathbf{z} = \mathbf{y} - \hat{\mathbf{y}}$.

The vector $\hat{\mathbf{y}}$ in (2) is called the **orthogonal projection of y onto** W and often is written as $\text{proj}_W \mathbf{y}$. See Figure 2. When W is a one-dimensional subspace, the formula for $\hat{\mathbf{y}}$ matches the formula given in Section 6.2.

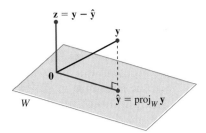

FIGURE 2 The orthogonal projection of $\mathbf{y}$ onto W.

PROOF Let $\{\mathbf{u}_1, \ldots, \mathbf{u}_p\}$ be any orthogonal basis for W, and define $\hat{\mathbf{y}}$ by (2).[1] Then $\hat{\mathbf{y}}$ is in W because $\hat{\mathbf{y}}$ is a linear combination of the basis $\mathbf{u}_1, \ldots, \mathbf{u}_p$. Let $\mathbf{z} = \mathbf{y} - \hat{\mathbf{y}}$. Since $\mathbf{u}_1$ is orthogonal to $\mathbf{u}_2, \ldots, \mathbf{u}_p$, it follows from (2) that

$$\mathbf{z} \cdot \mathbf{u}_1 = (\mathbf{y} - \hat{\mathbf{y}}) \cdot \mathbf{u}_1 = \mathbf{y} \cdot \mathbf{u}_1 - \left(\frac{\mathbf{y} \cdot \mathbf{u}_1}{\mathbf{u}_1 \cdot \mathbf{u}_1} \right) \mathbf{u}_1 \cdot \mathbf{u}_1 - 0 - \cdots - 0$$
$$= \mathbf{y} \cdot \mathbf{u}_1 - \mathbf{y} \cdot \mathbf{u}_1 = 0$$

Thus $\mathbf{z}$ is orthogonal to $\mathbf{u}_1$. Similarly, $\mathbf{z}$ is orthogonal to each $\mathbf{u}_j$ in the basis for W. Hence $\mathbf{z}$ is orthogonal to every vector in W. That is, $\mathbf{z}$ is in $W^{\perp}$.

To show that the decomposition in (1) is unique, suppose $\mathbf{y}$ can also be written as $\mathbf{y} = \hat{\mathbf{y}}_1 + \mathbf{z}_1$, with $\hat{\mathbf{y}}_1$ in W and $\mathbf{z}_1$ in $W^{\perp}$. Then $\hat{\mathbf{y}} + \mathbf{z} = \hat{\mathbf{y}}_1 + \mathbf{z}_1$ (since both sides equal $\mathbf{y}$), and so

$$\hat{\mathbf{y}} - \hat{\mathbf{y}}_1 = \mathbf{z}_1 - \mathbf{z}$$

This equality shows that the vector $\mathbf{v} = \hat{\mathbf{y}} - \hat{\mathbf{y}}_1$ is in W and in $W^{\perp}$ (because $\mathbf{z}_1$ and $\mathbf{z}$ are both in $W^{\perp}$, and $W^{\perp}$ is a subspace). Hence $\mathbf{v} \cdot \mathbf{v} = 0$, which shows that $\mathbf{v} = \mathbf{0}$. This proves that $\hat{\mathbf{y}} = \hat{\mathbf{y}}_1$ and also $\mathbf{z}_1 = \mathbf{z}$. ∎

The uniqueness of the decomposition (1) shows that the orthogonal projection $\hat{\mathbf{y}}$ depends only on W and not on the particular basis used in (2).

---

[1] We may assume that W is not the zero subspace, for otherwise $W^{\perp} = \mathbb{R}^n$ and (1) is simply $\mathbf{y} = \mathbf{0} + \mathbf{y}$. The next section will show that any nonzero subspace of $\mathbb{R}^n$ has an orthogonal basis.

EXAMPLE 2   Let $\mathbf{u}_1 = \begin{bmatrix} 2 \\ 5 \\ -1 \end{bmatrix}$, $\mathbf{u}_2 = \begin{bmatrix} -2 \\ 1 \\ 1 \end{bmatrix}$, and $\mathbf{y} = \begin{bmatrix} 1 \\ 2 \\ 3 \end{bmatrix}$. Observe that $\{\mathbf{u}_1, \mathbf{u}_2\}$ is an orthogonal basis for $W = \text{Span}\{\mathbf{u}_1, \mathbf{u}_2\}$. Write $\mathbf{y}$ as the sum of a vector in W and a vector orthogonal to W.

SOLUTION   The orthogonal projection of $\mathbf{y}$ onto W is

$$\hat{\mathbf{y}} = \frac{\mathbf{y} \cdot \mathbf{u}_1}{\mathbf{u}_1 \cdot \mathbf{u}_1} \mathbf{u}_1 + \frac{\mathbf{y} \cdot \mathbf{u}_2}{\mathbf{u}_2 \cdot \mathbf{u}_2} \mathbf{u}_2$$

$$= \frac{9}{30} \begin{bmatrix} 2 \\ 5 \\ -1 \end{bmatrix} + \frac{3}{6} \begin{bmatrix} -2 \\ 1 \\ 1 \end{bmatrix} = \frac{9}{30} \begin{bmatrix} 2 \\ 5 \\ -1 \end{bmatrix} + \frac{15}{30} \begin{bmatrix} -2 \\ 1 \\ 1 \end{bmatrix} = \begin{bmatrix} -2/5 \\ 2 \\ 1/5 \end{bmatrix}$$

Also

$$\mathbf{y} - \hat{\mathbf{y}} = \begin{bmatrix} 1 \\ 2 \\ 3 \end{bmatrix} - \begin{bmatrix} -2/5 \\ 2 \\ 1/5 \end{bmatrix} = \begin{bmatrix} 7/5 \\ 0 \\ 14/5 \end{bmatrix}$$

Theorem 8 ensures that $\mathbf{y} - \hat{\mathbf{y}}$ is in $W^{\perp}$. To check the calculations, however, it is a good idea to verify that $\mathbf{y} - \hat{\mathbf{y}}$ is orthogonal to both $\mathbf{u}_1$ and $\mathbf{u}_2$ and hence to all of W. The desired decomposition of $\mathbf{y}$ is

$$\mathbf{y} = \begin{bmatrix} 1 \\ 2 \\ 3 \end{bmatrix} = \begin{bmatrix} -2/5 \\ 2 \\ 1/5 \end{bmatrix} + \begin{bmatrix} 7/5 \\ 0 \\ 14/5 \end{bmatrix}$$ ∎

A Geometric Interpretation of the Orthogonal Projection

When W is a one-dimensional subspace, the formula (2) for $\text{proj}_W \mathbf{y}$ contains just one term. Thus, when dim $W > 1$, each term in (2) is itself an orthogonal projection of $\mathbf{y}$ onto a one-dimensional subspace spanned by one of the $\mathbf{u}$'s in the basis for W. Figure 3 illustrates this when W is a subspace of $\mathbb{R}^3$ spanned by $\mathbf{u}_1$ and $\mathbf{u}_2$. Here $\hat{\mathbf{y}}_1$ and $\hat{\mathbf{y}}_2$ denote the projections of $\mathbf{y}$ onto the lines spanned by $\mathbf{u}_1$ and $\mathbf{u}_2$, respectively. The orthogonal projection $\hat{\mathbf{y}}$ of $\mathbf{y}$ onto W is the sum of the projections of $\mathbf{y}$ onto one-dimensional subspaces that are orthogonal to each other. The vector $\hat{\mathbf{y}}$ in Figure 3 corresponds to the vector $\mathbf{y}$ in Figure 4 of Section 6.2, because now it is $\hat{\mathbf{y}}$ that is in W.

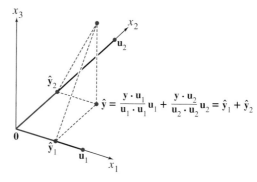

FIGURE 3 The orthogonal projection of $\mathbf{y}$ is the sum of its projections onto one-dimensional subspaces that are mutually orthogonal.

Properties of Orthogonal Projections

If $\{\mathbf{u}_1, \ldots, \mathbf{u}_p\}$ is an orthogonal basis for W and if $\mathbf{y}$ happens to be in W, then the formula for $\text{proj}_W \mathbf{y}$ is exactly the same as the representation of $\mathbf{y}$ given in Theorem 5 in Section 6.2. In this case, $\text{proj}_W \mathbf{y} = \mathbf{y}$.

> If $\mathbf{y}$ is in $W = \text{Span}\{\mathbf{u}_1, \ldots, \mathbf{u}_p\}$, then $\text{proj}_W \mathbf{y} = \mathbf{y}$.

This fact also follows from the next theorem.

THEOREM 9

The Best Approximation Theorem

Let W be a subspace of $\mathbb{R}^n$, let $\mathbf{y}$ be any vector in $\mathbb{R}^n$, and let $\hat{\mathbf{y}}$ be the orthogonal projection of $\mathbf{y}$ onto W. Then $\hat{\mathbf{y}}$ is the closest point in W to $\mathbf{y}$, in the sense that

$$\|\mathbf{y} - \hat{\mathbf{y}}\| < \|\mathbf{y} - \mathbf{v}\| \tag{3}$$

for all $\mathbf{v}$ in W distinct from $\hat{\mathbf{y}}$.

The vector $\hat{\mathbf{y}}$ in Theorem 9 is called **the best approximation to y by elements of** W. Later sections in the text will examine problems where a given $\mathbf{y}$ must be replaced, or *approximated*, by a vector $\mathbf{v}$ in some fixed subspace W. The distance from $\mathbf{y}$ to $\mathbf{v}$, given by $\|\mathbf{y} - \mathbf{v}\|$, can be regarded as the "error" of using $\mathbf{v}$ in place of $\mathbf{y}$. Theorem 9 says that this error is minimized when $\mathbf{v} = \hat{\mathbf{y}}$.

Inequality (3) leads to a new proof that $\hat{\mathbf{y}}$ does not depend on the particular orthogonal basis used to compute it. If a different orthogonal basis for W was used to construct an orthogonal projection of $\mathbf{y}$, then this projection would also be the closest point in W to $\mathbf{y}$, namely $\hat{\mathbf{y}}$.

PROOF Take $\mathbf{v}$ in W distinct from $\hat{\mathbf{y}}$. See Figure 4. Then $\hat{\mathbf{y}} - \mathbf{v}$ is in W. By the Orthogonal Decomposition Theorem, $\mathbf{y} - \hat{\mathbf{y}}$ is orthogonal to W. In particular, $\mathbf{y} - \hat{\mathbf{y}}$ is orthogonal to $\hat{\mathbf{y}} - \mathbf{v}$ (which is in W). Since

$$\mathbf{y} - \mathbf{v} = (\mathbf{y} - \hat{\mathbf{y}}) + (\hat{\mathbf{y}} - \mathbf{v})$$

the Pythagorean Theorem gives

$$\|\mathbf{y} - \mathbf{v}\|^2 = \|\mathbf{y} - \hat{\mathbf{y}}\|^2 + \|\hat{\mathbf{y}} - \mathbf{v}\|^2$$

(See the right triangle outlined in teal in Figure 4. The length of each side is labeled.) Now $\|\hat{\mathbf{y}} - \mathbf{v}\|^2 > 0$ because $\hat{\mathbf{y}} - \mathbf{v} \neq \mathbf{0}$, and so inequality (3) follows immediately. ∎

FIGURE 4 The orthogonal projection of $\mathbf{y}$ onto W is the closest point in W to $\mathbf{y}$.

EXAMPLE 3 If $\mathbf{u}_1 = \begin{bmatrix} 2 \\ 5 \\ -1 \end{bmatrix}$, $\mathbf{u}_2 = \begin{bmatrix} -2 \\ 1 \\ 1 \end{bmatrix}$, $\mathbf{y} = \begin{bmatrix} 1 \\ 2 \\ 3 \end{bmatrix}$, and $W = \text{Span}\{\mathbf{u}_1, \mathbf{u}_2\}$, as in Example 2, then the closest point in W to $\mathbf{y}$ is

$$\hat{\mathbf{y}} = \frac{\mathbf{y} \cdot \mathbf{u}_1}{\mathbf{u}_1 \cdot \mathbf{u}_1}\mathbf{u}_1 + \frac{\mathbf{y} \cdot \mathbf{u}_2}{\mathbf{u}_2 \cdot \mathbf{u}_2}\mathbf{u}_2 = \begin{bmatrix} -2/5 \\ 2 \\ 1/5 \end{bmatrix}$$

∎

EXAMPLE 4 The distance from a point $\mathbf{y}$ in $\mathbb{R}^n$ to a subspace W is defined as the distance from $\mathbf{y}$ to the nearest point in W. Find the distance from $\mathbf{y}$ to $W = \text{Span}\{\mathbf{u}_1, \mathbf{u}_2\}$, where

$$\mathbf{y} = \begin{bmatrix} -1 \\ -5 \\ 10 \end{bmatrix}, \quad \mathbf{u}_1 = \begin{bmatrix} 5 \\ -2 \\ 1 \end{bmatrix}, \quad \mathbf{u}_2 = \begin{bmatrix} 1 \\ 2 \\ -1 \end{bmatrix}$$

SOLUTION By the Best Approximation Theorem, the distance from $\mathbf{y}$ to W is $\|\mathbf{y} - \hat{\mathbf{y}}\|$, where $\hat{\mathbf{y}} = \text{proj}_W \mathbf{y}$. Since $\{\mathbf{u}_1, \mathbf{u}_2\}$ is an orthogonal basis for W,

$$\hat{\mathbf{y}} = \frac{15}{30}\mathbf{u}_1 + \frac{-21}{6}\mathbf{u}_2 = \frac{1}{2}\begin{bmatrix} 5 \\ -2 \\ 1 \end{bmatrix} - \frac{7}{2}\begin{bmatrix} 1 \\ 2 \\ -1 \end{bmatrix} = \begin{bmatrix} -1 \\ -8 \\ 4 \end{bmatrix}$$

$$\mathbf{y} - \hat{\mathbf{y}} = \begin{bmatrix} -1 \\ -5 \\ 10 \end{bmatrix} - \begin{bmatrix} -1 \\ -8 \\ 4 \end{bmatrix} = \begin{bmatrix} 0 \\ 3 \\ 6 \end{bmatrix}$$

$$\|\mathbf{y} - \hat{\mathbf{y}}\|^2 = 3^2 + 6^2 = 45$$

The distance from $\mathbf{y}$ to W is $\sqrt{45} = 3\sqrt{5}$.

∎

The final theorem in this section shows how formula (2) for $\text{proj}_W \mathbf{y}$ is simplified when the basis for W is an orthonormal set.

THEOREM 10

If $\{\mathbf{u}_1, \ldots, \mathbf{u}_p\}$ is an orthonormal basis for a subspace W of $\mathbb{R}^n$, then

$$\text{proj}_W \mathbf{y} = (\mathbf{y} \cdot \mathbf{u}_1)\mathbf{u}_1 + (\mathbf{y} \cdot \mathbf{u}_2)\mathbf{u}_2 + \cdots + (\mathbf{y} \cdot \mathbf{u}_p)\mathbf{u}_p \qquad (4)$$

If $U = [\,\mathbf{u}_1 \quad \mathbf{u}_2 \quad \cdots \quad \mathbf{u}_p\,]$, then

$$\text{proj}_W \mathbf{y} = UU^T\mathbf{y} \quad \text{for all } \mathbf{y} \text{ in } \mathbb{R}^n \qquad (5)$$

PROOF Formula (4) follows immediately from (2) in Theorem 8. Also, (4) shows that $\text{proj}_W \mathbf{y}$ is a linear combination of the columns of U using the weights $\mathbf{y} \cdot \mathbf{u}_1$, $\mathbf{y} \cdot \mathbf{u}_2, \ldots, \mathbf{y} \cdot \mathbf{u}_p$. The weights can be written as $\mathbf{u}_1^T\mathbf{y}, \mathbf{u}_2^T\mathbf{y}, \ldots, \mathbf{u}_p^T\mathbf{y}$, showing that they are the entries in $U^T\mathbf{y}$ and justifying (5).

∎

Suppose U is an $n \times p$ matrix with orthonormal columns, and let W be the column space of U. Then

$$U^TU\mathbf{x} = I_p\mathbf{x} = \mathbf{x} \quad \text{for all } \mathbf{x} \text{ in } \mathbb{R}^p \qquad \text{Theorem 6}$$

$$UU^T\mathbf{y} = \text{proj}_W \mathbf{y} \quad \text{for all } \mathbf{y} \text{ in } \mathbb{R}^n \qquad \text{Theorem 10}$$

If U is an $n \times n$ (square) matrix with orthonormal columns, then U is an *orthogonal* matrix, the column space W is all of $\mathbb{R}^n$, and $UU^T\mathbf{y} = I\mathbf{y} = \mathbf{y}$ for all $\mathbf{y}$ in $\mathbb{R}^n$.

Although formula (4) is important for theoretical purposes, in practice it usually involves calculations with square roots of numbers (in the entries of the $\mathbf{u}_i$). Formula (2) is recommended for hand calculations.

Example 9 of Section 2.1 illustrates how matrix multiplication and transposition are used to detect a specified pattern illustrated using blue and white squares. Now that we have more experience working with bases for W and $W^{\perp}$, we are ready to discuss how to set up the matrix M in Figure 6. Let $\mathbf{w}$ be the vector generated from a pattern of blue and white squares by turning each blue square into a 1 and each white square into a 0, and then lining up each column below the column before it. See Figure 5.

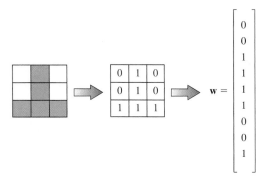

FIGURE 5 Creating a vector from colored squares.

Let $W = \text{span} \{\mathbf{w}\}$. Choose a basis $\{\mathbf{v}_1, \mathbf{v}_2, \ldots, \mathbf{v}_{n-1}\}$ for $W^{\perp}$. Create the matrix

$$B = \begin{bmatrix} \mathbf{v}_1^T \\ \mathbf{v}_2^T \\ \vdots \\ \mathbf{v}_{n-1}^T \end{bmatrix}.$$ Notice $B\mathbf{u} = \mathbf{0}$ if and only if $\mathbf{u}$ is orthogonal to a set of basis vectors

for $W^{\perp}$, which happens if and only if $\mathbf{u}$ is in W. Set $M = B^T B$. Then $\mathbf{u}^T M \mathbf{u} = \mathbf{u}^T B^T B \mathbf{u} = (B\mathbf{u})^T B\mathbf{u}$. By Theorem 1, $(B\mathbf{u})^T B\mathbf{u} = 0$ if and only if $B\mathbf{u} = 0$, and hence $\mathbf{u}^T M \mathbf{u} = 0$ if and only if $\mathbf{u} \in W$. But there are only two vectors in W consisting of zeros and ones: $1\mathbf{w} = \mathbf{w}$ and $0\mathbf{w} = \mathbf{0}$. Thus we can conclude that if $\mathbf{u}^T M \mathbf{u} = 0$, but $\mathbf{u}^T \mathbf{u} \neq 0$, then $\mathbf{u} = \mathbf{w}$. See Figure 6.

EXAMPLE 5 Find a matrix M that can be used in Figure 6 to identify the perp symbol.

SOLUTION First change the symbol into a vector. Set $\mathbf{w} = [0 \ 0 \ 1 \ 1 \ 1 \ 1 \ 0 \ 0 \ 1]^T$. Next set $W = \text{span} \{\mathbf{w}\}$ and find a basis for $W^{\perp}$: solving $\mathbf{x}^T \mathbf{w} = 0$ creates the homogeneous system of equations:

$$x_3 + x_4 + x_5 + x_6 + x_9 = 0$$

Treating x_3 as the basic variable and the remaining variables as free variables we get a basis for $W^{\perp}$. Transposing each vector in the basis and inserting it as a row of B we get

$$\mathbf{w} = \begin{bmatrix} 1 \\ 0 \\ 1 \\ 1 \\ 1 \\ 1 \\ 1 \\ 0 \\ 1 \end{bmatrix}$$

$\mathbf{w}^T M \mathbf{w} = 2$ and $\mathbf{w}^T \mathbf{w} = 7$

This pattern is not the perpendicular symbol since $\mathbf{w}^T M \mathbf{w} \neq 0$.

$$\mathbf{w} = \begin{bmatrix} 0 \\ 0 \\ 1 \\ 1 \\ 1 \\ 1 \\ 0 \\ 0 \\ 1 \end{bmatrix}$$

$\mathbf{w}^T M \mathbf{w} = 0$ and $\mathbf{w}^T \mathbf{w} = 5$

This pattern is the perpendicular symbol since $\mathbf{w}^T M \mathbf{w} = 0$, but $\mathbf{w}^T \mathbf{w} \neq 0$.

FIGURE 6 How AI detects the perp symbol.

$$B = \begin{bmatrix} 1 & 0 & 0 & 0 & 0 & 0 & 0 & 0 \\ 0 & 1 & 0 & 0 & 0 & 0 & 0 & 0 \\ 0 & 0 & -1 & 1 & 0 & 0 & 0 & 0 \\ 0 & 0 & -1 & 0 & 1 & 0 & 0 & 0 \\ 0 & 0 & -1 & 0 & 0 & 1 & 0 & 0 \\ 0 & 0 & 0 & 0 & 0 & 0 & 1 & 0 \\ 0 & 0 & 0 & 0 & 0 & 0 & 1 & 0 \\ 0 & 0 & -1 & 0 & 0 & 0 & 0 & 1 \end{bmatrix} \text{ and } M = B^T B = \begin{bmatrix} 1 & 0 & 0 & 0 & 0 & 0 & 0 & 0 & 0 \\ 0 & 1 & 0 & 0 & 0 & 0 & 0 & 0 & 0 \\ 0 & 0 & 4 & -1 & -1 & -1 & 0 & 0 & -1 \\ 0 & 0 & -1 & 1 & 0 & 0 & 0 & 0 & 0 \\ 0 & 0 & -1 & 0 & 1 & 0 & 0 & 0 & 0 \\ 0 & 0 & -1 & 0 & 0 & 1 & 0 & 0 & 0 \\ 0 & 0 & 0 & 0 & 0 & 0 & 1 & 0 & 0 \\ 0 & 0 & 0 & 0 & 0 & 0 & 0 & 1 & 0 \\ 0 & 0 & -1 & 0 & 0 & 0 & 0 & 0 & 1 \end{bmatrix}$$

Notice $\mathbf{w}^T M \mathbf{w} = 0$, but $\mathbf{w}^T \mathbf{w} \neq 0$. ∎

Practice Problems

1. Let $\mathbf{u}_1 = \begin{bmatrix} -7 \\ 1 \\ 4 \end{bmatrix}$, $\mathbf{u}_2 = \begin{bmatrix} -1 \\ 1 \\ -2 \end{bmatrix}$, $\mathbf{y} = \begin{bmatrix} -9 \\ 1 \\ 6 \end{bmatrix}$, and $W = \text{Span}\{\mathbf{u}_1, \mathbf{u}_2\}$. Use the fact that $\mathbf{u}_1$ and $\mathbf{u}_2$ are orthogonal to compute $\text{proj}_W \mathbf{y}$.

2. Let W be a subspace of $\mathbb{R}^n$. Let $\mathbf{x}$ and $\mathbf{y}$ be vectors in $\mathbb{R}^n$ and let $\mathbf{z} = \mathbf{x} + \mathbf{y}$. If $\mathbf{u}$ is the projection of $\mathbf{x}$ onto W and $\mathbf{v}$ is the projection of $\mathbf{y}$ onto W, show that $\mathbf{u} + \mathbf{v}$ is the projection of $\mathbf{z}$ onto W.

6.3 Exercises

In Exercises 1 and 2, you may assume that $\{\mathbf{u}_1, \dots, \mathbf{u}_4\}$ is an orthogonal basis for $\mathbb{R}^4$.

1. $\mathbf{u}_1 = \begin{bmatrix} 0 \\ 1 \\ -4 \\ -1 \end{bmatrix}$, $\mathbf{u}_2 = \begin{bmatrix} 3 \\ 5 \\ 1 \\ 1 \end{bmatrix}$, $\mathbf{u}_3 = \begin{bmatrix} 1 \\ 0 \\ 1 \\ -4 \end{bmatrix}$, $\mathbf{u}_4 = \begin{bmatrix} 5 \\ -3 \\ -1 \\ 1 \end{bmatrix}$,

$\mathbf{x} = \begin{bmatrix} 10 \\ -8 \\ 2 \\ 0 \end{bmatrix}$. Write $\mathbf{x}$ as the sum of two vectors, one in Span $\{\mathbf{u}_1, \mathbf{u}_2, \mathbf{u}_3\}$ and the other in Span $\{\mathbf{u}_4\}$.

2. $\mathbf{u}_1 = \begin{bmatrix} 1 \\ 2 \\ 1 \\ 1 \end{bmatrix}$, $\mathbf{u}_2 = \begin{bmatrix} -2 \\ 1 \\ -1 \\ 1 \end{bmatrix}$, $\mathbf{u}_3 = \begin{bmatrix} 1 \\ 1 \\ -2 \\ -1 \end{bmatrix}$, $\mathbf{u}_4 = \begin{bmatrix} -1 \\ 1 \\ 1 \\ -2 \end{bmatrix}$,

$\mathbf{v} = \begin{bmatrix} 4 \\ 5 \\ -2 \\ 2 \end{bmatrix}$. Write $\mathbf{v}$ as the sum of two vectors, one in Span $\{\mathbf{u}_1\}$ and the other in Span $\{\mathbf{u}_2, \mathbf{u}_3, \mathbf{u}_4\}$.

In Exercises 3–6, verify that $\{\mathbf{u}_1, \mathbf{u}_2\}$ is an orthogonal set, and then find the orthogonal projection of $\mathbf{y}$ onto Span $\{\mathbf{u}_1, \mathbf{u}_2\}$.

3. $\mathbf{y} = \begin{bmatrix} -1 \\ 4 \\ 3 \end{bmatrix}$, $\mathbf{u}_1 = \begin{bmatrix} 1 \\ 1 \\ 0 \end{bmatrix}$, $\mathbf{u}_2 = \begin{bmatrix} -1 \\ 1 \\ 0 \end{bmatrix}$

4. $\mathbf{y} = \begin{bmatrix} 4 \\ 3 \\ -2 \end{bmatrix}$, $\mathbf{u}_1 = \begin{bmatrix} 3 \\ 4 \\ 0 \end{bmatrix}$, $\mathbf{u}_2 = \begin{bmatrix} -4 \\ 3 \\ 0 \end{bmatrix}$

5. $\mathbf{y} = \begin{bmatrix} -1 \\ 2 \\ 6 \end{bmatrix}$, $\mathbf{u}_1 = \begin{bmatrix} 3 \\ -1 \\ 2 \end{bmatrix}$, $\mathbf{u}_2 = \begin{bmatrix} 1 \\ -1 \\ -2 \end{bmatrix}$

6. $\mathbf{y} = \begin{bmatrix} 4 \\ 4 \\ 1 \end{bmatrix}$, $\mathbf{u}_1 = \begin{bmatrix} -4 \\ -1 \\ 1 \end{bmatrix}$, $\mathbf{u}_2 = \begin{bmatrix} 0 \\ 1 \\ 1 \end{bmatrix}$

In Exercises 7–10, let W be the subspace spanned by the $\mathbf{u}$'s, and write $\mathbf{y}$ as the sum of a vector in W and a vector orthogonal to W.

7. $\mathbf{y} = \begin{bmatrix} 1 \\ 3 \\ 5 \end{bmatrix}$, $\mathbf{u}_1 = \begin{bmatrix} 1 \\ 3 \\ -2 \end{bmatrix}$, $\mathbf{u}_2 = \begin{bmatrix} 5 \\ 1 \\ 4 \end{bmatrix}$

8. $\mathbf{y} = \begin{bmatrix} -1 \\ 4 \\ 3 \end{bmatrix}$, $\mathbf{u}_1 = \begin{bmatrix} 1 \\ 1 \\ 1 \end{bmatrix}$, $\mathbf{u}_2 = \begin{bmatrix} -1 \\ 3 \\ -2 \end{bmatrix}$

9. $\mathbf{y} = \begin{bmatrix} 4 \\ 3 \\ 3 \\ -1 \end{bmatrix}$, $\mathbf{u}_1 = \begin{bmatrix} 1 \\ 1 \\ 0 \\ 1 \end{bmatrix}$, $\mathbf{u}_2 = \begin{bmatrix} -1 \\ 3 \\ 1 \\ -2 \end{bmatrix}$, $\mathbf{u}_3 = \begin{bmatrix} -1 \\ 0 \\ 1 \\ 1 \end{bmatrix}$

10. $\mathbf{y} = \begin{bmatrix} 3 \\ 4 \\ 5 \\ 4 \end{bmatrix}$, $\mathbf{u}_1 = \begin{bmatrix} 1 \\ 1 \\ 0 \\ -1 \end{bmatrix}$, $\mathbf{u}_2 = \begin{bmatrix} 1 \\ 0 \\ 1 \\ 1 \end{bmatrix}$, $\mathbf{u}_3 = \begin{bmatrix} 0 \\ -1 \\ 1 \\ -1 \end{bmatrix}$

In Exercises 11 and 12, find the closest point to $\mathbf{y}$ in the subspace W spanned by $\mathbf{v}_1$ and $\mathbf{v}_2$.

11. $\mathbf{y} = \begin{bmatrix} 3 \\ 1 \\ 5 \\ 1 \end{bmatrix}$, $\mathbf{v}_1 = \begin{bmatrix} 3 \\ 1 \\ -1 \\ 1 \end{bmatrix}$, $\mathbf{v}_2 = \begin{bmatrix} 1 \\ -1 \\ 1 \\ -1 \end{bmatrix}$

12. $\mathbf{y} = \begin{bmatrix} 3 \\ -1 \\ 1 \\ 13 \end{bmatrix}$, $\mathbf{v}_1 = \begin{bmatrix} 1 \\ -2 \\ -1 \\ 2 \end{bmatrix}$, $\mathbf{v}_2 = \begin{bmatrix} -4 \\ 1 \\ 0 \\ 3 \end{bmatrix}$

In Exercises 13 and 14, find the best approximation to $\mathbf{z}$ by vectors of the form $c_1\mathbf{v}_1 + c_2\mathbf{v}_2$.

13. $\mathbf{z} = \begin{bmatrix} 3 \\ -7 \\ 2 \\ 3 \end{bmatrix}$, $\mathbf{v}_1 = \begin{bmatrix} 2 \\ -1 \\ -3 \\ 1 \end{bmatrix}$, $\mathbf{v}_2 = \begin{bmatrix} 1 \\ 1 \\ 0 \\ -1 \end{bmatrix}$

14. $\mathbf{z} = \begin{bmatrix} 2 \\ 4 \\ 0 \\ -1 \end{bmatrix}$, $\mathbf{v}_1 = \begin{bmatrix} 2 \\ 0 \\ -1 \\ -3 \end{bmatrix}$, $\mathbf{v}_2 = \begin{bmatrix} 5 \\ -2 \\ 4 \\ 2 \end{bmatrix}$

15. Let $\mathbf{y} = \begin{bmatrix} 5 \\ -9 \\ 5 \end{bmatrix}$, $\mathbf{u}_1 = \begin{bmatrix} -3 \\ -5 \\ 1 \end{bmatrix}$, $\mathbf{u}_2 = \begin{bmatrix} -3 \\ 2 \\ 1 \end{bmatrix}$. Find the distance from $\mathbf{y}$ to the plane in $\mathbb{R}^3$ spanned by $\mathbf{u}_1$ and $\mathbf{u}_2$.

16. Let $\mathbf{y}$, $\mathbf{v}_1$, and $\mathbf{v}_2$ be as in Exercise 12. Find the distance from $\mathbf{y}$ to the subspace of $\mathbb{R}^4$ spanned by $\mathbf{v}_1$ and $\mathbf{v}_2$.

17. Let $\mathbf{y} = \begin{bmatrix} 4 \\ 8 \\ 1 \end{bmatrix}$, $\mathbf{u}_1 = \begin{bmatrix} 2/3 \\ 1/3 \\ 2/3 \end{bmatrix}$, $\mathbf{u}_2 = \begin{bmatrix} -2/3 \\ 2/3 \\ 1/3 \end{bmatrix}$, and $W = \text{Span}\,\{\mathbf{u}_1, \mathbf{u}_2\}$.

a. Let $U = [\,\mathbf{u}_1 \ \ \mathbf{u}_2\,]$. Compute $U^T U$ and $U U^T$.

b. Compute $\text{proj}_W \mathbf{y}$ and $(U U^T)\mathbf{y}$.

18. Let $\mathbf{y} = \begin{bmatrix} 7 \\ 9 \end{bmatrix}$, $\mathbf{u}_1 = \begin{bmatrix} 1/\sqrt{10} \\ -3/\sqrt{10} \end{bmatrix}$, and $W = \text{Span}\,\{\mathbf{u}_1\}$.

a. Let U be the 2×1 matrix whose only column is $\mathbf{u}_1$. Compute $U^T U$ and $U U^T$.

b. Compute $\text{proj}_W \mathbf{y}$ and $(U U^T)\mathbf{y}$.

19. Let $\mathbf{u}_1 = \begin{bmatrix} 1 \\ 1 \\ -2 \end{bmatrix}$, $\mathbf{u}_2 = \begin{bmatrix} 5 \\ -1 \\ 2 \end{bmatrix}$, and $\mathbf{u}_3 = \begin{bmatrix} 0 \\ 0 \\ 1 \end{bmatrix}$. Note that $\mathbf{u}_1$ and $\mathbf{u}_2$ are orthogonal but that $\mathbf{u}_3$ is not orthogonal to $\mathbf{u}_1$ or $\mathbf{u}_2$. It can be shown that $\mathbf{u}_3$ is not in the subspace W spanned by $\mathbf{u}_1$ and $\mathbf{u}_2$. Use this fact to construct a nonzero vector $\mathbf{v}$ in $\mathbb{R}^3$ that is orthogonal to $\mathbf{u}_1$ and $\mathbf{u}_2$.

20. Let $\mathbf{u}_1$ and $\mathbf{u}_2$ be as in Exercise 19, and let $\mathbf{u}_4 = \begin{bmatrix} 0 \\ 1 \\ 0 \end{bmatrix}$. It can be shown that $\mathbf{u}_4$ is not in the subspace W spanned by $\mathbf{u}_1$ and $\mathbf{u}_2$. Use this fact to construct a nonzero vector $\mathbf{v}$ in $\mathbb{R}^3$ that is orthogonal to $\mathbf{u}_1$ and $\mathbf{u}_2$.

In Exercises 21–30, all vectors and subspaces are in $\mathbb{R}^n$. Mark each statement True or False (**T/F**). Justify each answer.

21. (**T/F**) If $\mathbf{z}$ is orthogonal to $\mathbf{u}_1$ and to $\mathbf{u}_2$ and if $W =$ Span $\{\mathbf{u}_1, \mathbf{u}_2\}$, then $\mathbf{z}$ must be in $W^{\perp}$.

22. (**T/F**) For each $\mathbf{y}$ and each subspace W, the vector $\mathbf{y} - \mathrm{proj}_W \mathbf{y}$ is orthogonal to W.

23. (**T/F**) The orthogonal projection $\hat{\mathbf{y}}$ of $\mathbf{y}$ onto a subspace W can sometimes depend on the orthogonal basis for W used to compute $\hat{\mathbf{y}}$.

24. (**T/F**) If $\mathbf{y}$ is in a subspace W, then the orthogonal projection of $\mathbf{y}$ onto W is $\mathbf{y}$ itself.

25. (**T/F**) The best approximation to $\mathbf{y}$ by elements of a subspace W is given by the vector $\mathbf{y} - \mathrm{proj}_W \mathbf{y}$.

26. (**T/F**) If W is a subspace of $\mathbb{R}^n$ and if $\mathbf{v}$ is in both W and $W^{\perp}$, then $\mathbf{v}$ must be the zero vector.

27. (**T/F**) In the Orthogonal Decomposition Theorem, each term in formula (2) for $\hat{\mathbf{y}}$ is itself an orthogonal projection of $\mathbf{y}$ onto a subspace of W.

28. (**T/F**) If $\mathbf{y} = \mathbf{z}_1 + \mathbf{z}_2$, where $\mathbf{z}_1$ is in a subspace W and $\mathbf{z}_2$ is in $W^{\perp}$, then $\mathbf{z}_1$ must be the orthogonal projection of $\mathbf{y}$ onto W.

29. (**T/F**) If the columns of an $n \times p$ matrix U are orthonormal, then $UU^T\mathbf{y}$ is the orthogonal projection of $\mathbf{y}$ onto the column space of U.

30. (**T/F**) If an $n \times p$ matrix U has orthonormal columns, then $UU^T\mathbf{x} = \mathbf{x}$ for all $\mathbf{x}$ in $\mathbb{R}^n$.

31. Let A be an $m \times n$ matrix. Prove that every vector $\mathbf{x}$ in $\mathbb{R}^n$ can be written in the form $\mathbf{x} = \mathbf{p} + \mathbf{u}$, where $\mathbf{p}$ is in Row A and $\mathbf{u}$ is in Nul A. Also, show that if the equation $A\mathbf{x} = \mathbf{b}$ is consistent, then there is a unique $\mathbf{p}$ in Row A such that $A\mathbf{p} = \mathbf{b}$.

32. Let W be a subspace of $\mathbb{R}^n$ with an orthogonal basis $\{\mathbf{w}_1, \ldots, \mathbf{w}_p\}$, and let $\{\mathbf{v}_1, \ldots, \mathbf{v}_q\}$ be an orthogonal basis for $W^{\perp}$.

   a. Explain why $\{\mathbf{w}_1, \ldots, \mathbf{w}_p, \mathbf{v}_1, \ldots, \mathbf{v}_q\}$ is an orthogonal set.

   b. Explain why the set in part (a) spans $\mathbb{R}^n$.

   c. Show that $\dim W + \dim W^{\perp} = n$.

In Exercises 33–36, first change the given pattern into a vector $\mathbf{w}$ of zeros and ones and then use the method illustrated in Example 5 to find a matrix M so that $\mathbf{w}^T M \mathbf{w} = 0$, but $\mathbf{u}^T M \mathbf{u} \neq 0$ for all other nonzero vectors $\mathbf{u}$ of zeros and ones.

33.   34.

35.   36.

T 37. Let U be the 8×4 matrix in Exercise 43 in Section 6.2. Find the closest point to $\mathbf{y} = (1, 1, 1, 1, 1, 1, 1, 1)$ in Col U. Write the keystrokes or commands you use to solve this problem.

T 38. Let U be the matrix in Exercise 37. Find the distance from $\mathbf{b} = (1, 1, 1, 1, -1, -1, -1, -1)$ to Col U.

Solution to Practice Problems

1. Compute

$$\mathrm{proj}_W \mathbf{y} = \frac{\mathbf{y} \cdot \mathbf{u}_1}{\mathbf{u}_1 \cdot \mathbf{u}_1}\mathbf{u}_1 + \frac{\mathbf{y} \cdot \mathbf{u}_2}{\mathbf{u}_2 \cdot \mathbf{u}_2}\mathbf{u}_2 = \frac{88}{66}\mathbf{u}_1 + \frac{-2}{6}\mathbf{u}_2$$

$$= \frac{4}{3}\begin{bmatrix} -7 \\ 1 \\ 4 \end{bmatrix} - \frac{1}{3}\begin{bmatrix} -1 \\ 1 \\ -2 \end{bmatrix} = \begin{bmatrix} -9 \\ 1 \\ 6 \end{bmatrix} = \mathbf{y}$$

In this case, $\mathbf{y}$ happens to be a linear combination of $\mathbf{u}_1$ and $\mathbf{u}_2$, so $\mathbf{y}$ is in W. The closest point in W to $\mathbf{y}$ is $\mathbf{y}$ itself.

2. Using Theorem 10, let U be a matrix whose columns consist of an orthonormal basis for W. Then $\mathrm{proj}_W \mathbf{z} = UU^T\mathbf{z} = UU^T(\mathbf{x} + \mathbf{y}) = UU^T\mathbf{x} + UU^T\mathbf{y} = \mathrm{proj}_W\mathbf{x} + \mathrm{proj}_W\mathbf{y} = \mathbf{u} + \mathbf{v}$.

6.4 The Gram–Schmidt Process

The Gram–Schmidt process is a simple algorithm for producing an orthogonal or orthonormal basis for any nonzero subspace of $\mathbb{R}^n$. The first two examples of the process are aimed at hand calculation.

EXAMPLE 1 Let $W = \text{Span}\{\mathbf{x}_1, \mathbf{x}_2\}$, where $\mathbf{x}_1 = \begin{bmatrix} 3 \\ 6 \\ 0 \end{bmatrix}$ and $\mathbf{x}_2 = \begin{bmatrix} 1 \\ 2 \\ 2 \end{bmatrix}$. Construct an orthogonal basis $\{\mathbf{v}_1, \mathbf{v}_2\}$ for W.

SOLUTION The subspace W is shown in Figure 1, along with $\mathbf{x}_1, \mathbf{x}_2$, and the projection $\mathbf{p}$ of $\mathbf{x}_2$ onto $\mathbf{x}_1$. The component of $\mathbf{x}_2$ orthogonal to $\mathbf{x}_1$ is $\mathbf{x}_2 - \mathbf{p}$, which is in W because it is formed from $\mathbf{x}_2$ and a multiple of $\mathbf{x}_1$. Let $\mathbf{v}_1 = \mathbf{x}_1$ and

$$
\mathbf{v}_2 = \mathbf{x}_2 - \mathbf{p} = \mathbf{x}_2 - \frac{\mathbf{x}_2 \cdot \mathbf{x}_1}{\mathbf{x}_1 \cdot \mathbf{x}_1}\mathbf{x}_1 = \begin{bmatrix} 1 \\ 2 \\ 2 \end{bmatrix} - \frac{15}{45}\begin{bmatrix} 3 \\ 6 \\ 0 \end{bmatrix} = \begin{bmatrix} 0 \\ 0 \\ 2 \end{bmatrix}
$$

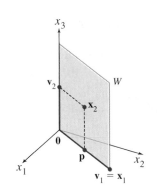

FIGURE 1
Construction of an orthogonal basis $\{\mathbf{v}_1, \mathbf{v}_2\}$.

Then $\{\mathbf{v}_1, \mathbf{v}_2\}$ is an orthogonal set of nonzero vectors in W. Since $\dim W = 2$, the set $\{\mathbf{v}_1, \mathbf{v}_2\}$ is a basis for W. ∎

The next example fully illustrates the Gram–Schmidt process. Study it carefully.

EXAMPLE 2 Let $\mathbf{x}_1 = \begin{bmatrix} 1 \\ 1 \\ 1 \\ 1 \end{bmatrix}$, $\mathbf{x}_2 = \begin{bmatrix} 0 \\ 1 \\ 1 \\ 1 \end{bmatrix}$, and $\mathbf{x}_3 = \begin{bmatrix} 0 \\ 0 \\ 1 \\ 1 \end{bmatrix}$. Then $\{\mathbf{x}_1, \mathbf{x}_2, \mathbf{x}_3\}$ is clearly linearly independent and thus is a basis for a subspace W of $\mathbb{R}^4$. Construct an orthogonal basis for W.

SOLUTION

Step 1. Let $\mathbf{v}_1 = \mathbf{x}_1$ and $W_1 = \text{Span}\{\mathbf{x}_1\} = \text{Span}\{\mathbf{v}_1\}$.

Step 2. Let $\mathbf{v}_2$ be the vector produced by subtracting from $\mathbf{x}_2$ its projection onto the subspace W_1. That is, let

$$
\begin{aligned}
\mathbf{v}_2 &= \mathbf{x}_2 - \text{proj}_{W_1}\mathbf{x}_2 \\
&= \mathbf{x}_2 - \frac{\mathbf{x}_2 \cdot \mathbf{v}_1}{\mathbf{v}_1 \cdot \mathbf{v}_1}\mathbf{v}_1 \qquad \text{Since } \mathbf{v}_1 = \mathbf{x}_1 \\
&= \begin{bmatrix} 0 \\ 1 \\ 1 \\ 1 \end{bmatrix} - \frac{3}{4}\begin{bmatrix} 1 \\ 1 \\ 1 \\ 1 \end{bmatrix} = \begin{bmatrix} -3/4 \\ 1/4 \\ 1/4 \\ 1/4 \end{bmatrix}
\end{aligned}
$$

As in Example 1, $\mathbf{v}_2$ is the component of $\mathbf{x}_2$ orthogonal to $\mathbf{x}_1$, and $\{\mathbf{v}_1, \mathbf{v}_2\}$ is an orthogonal basis for the subspace W_2 spanned by $\mathbf{x}_1$ and $\mathbf{x}_2$.

Step 2′ (optional). If appropriate, scale $\mathbf{v}_2$ to simplify later computations. Since $\mathbf{v}_2$ has fractional entries, it is convenient to scale it by a factor of 4 and replace $\{\mathbf{v}_1, \mathbf{v}_2\}$ by the orthogonal basis

$$
\mathbf{v}_1 = \begin{bmatrix} 1 \\ 1 \\ 1 \\ 1 \end{bmatrix}, \quad \mathbf{v}_2' = \begin{bmatrix} -3 \\ 1 \\ 1 \\ 1 \end{bmatrix}
$$

Step 3. Let $\mathbf{v}_3$ be the vector produced by subtracting from $\mathbf{x}_3$ its projection onto the subspace W_2. Use the orthogonal basis $\{\mathbf{v}_1, \mathbf{v}_2'\}$ to compute this projection onto W_2:

$$
\operatorname{proj}_{W_2} \mathbf{x}_3 = \underbrace{\frac{\mathbf{x}_3 \cdot \mathbf{v}_1}{\mathbf{v}_1 \cdot \mathbf{v}_1} \mathbf{v}_1}_{\substack{\text{Projection of} \\ \mathbf{x}_3 \text{ onto } \mathbf{v}_1}} + \underbrace{\frac{\mathbf{x}_3 \cdot \mathbf{v}_2'}{\mathbf{v}_2' \cdot \mathbf{v}_2'} \mathbf{v}_2'}_{\substack{\text{Projection of} \\ \mathbf{x}_3 \text{ onto } \mathbf{v}_2'}} = \frac{2}{4} \begin{bmatrix} 1 \\ 1 \\ 1 \\ 1 \end{bmatrix} + \frac{2}{12} \begin{bmatrix} -3 \\ 1 \\ 1 \\ 1 \end{bmatrix} = \begin{bmatrix} 0 \\ 2/3 \\ 2/3 \\ 2/3 \end{bmatrix}
$$

Then $\mathbf{v}_3$ is the component of $\mathbf{x}_3$ orthogonal to W_2, namely

$$
\mathbf{v}_3 = \mathbf{x}_3 - \operatorname{proj}_{W_2} \mathbf{x}_3 = \begin{bmatrix} 0 \\ 0 \\ 1 \\ 1 \end{bmatrix} - \begin{bmatrix} 0 \\ 2/3 \\ 2/3 \\ 2/3 \end{bmatrix} = \begin{bmatrix} 0 \\ -2/3 \\ 1/3 \\ 1/3 \end{bmatrix}
$$

See Figure 2 for a diagram of this construction. Observe that $\mathbf{v}_3$ is in W, because $\mathbf{x}_3$ and $\operatorname{proj}_{W_2}\mathbf{x}_3$ are both in W. Thus $\{\mathbf{v}_1, \mathbf{v}_2', \mathbf{v}_3\}$ is an orthogonal set of nonzero vectors and hence a linearly independent set in W. Note that W is three-dimensional since it was defined by a basis of three vectors. Hence, by the Basis Theorem in Section 4.5, $\{\mathbf{v}_1, \mathbf{v}_2', \mathbf{v}_3\}$ is an orthogonal basis for W. ∎

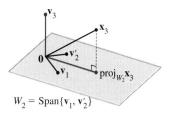

FIGURE 2 The construction of $\mathbf{v}_3$ from $\mathbf{x}_3$ and W_2.

The proof of the next theorem shows that this strategy really works. Scaling of vectors is not mentioned because that is used only to simplify hand calculations.

THEOREM 11

The Gram–Schmidt Process

Given a basis $\{\mathbf{x}_1, \ldots, \mathbf{x}_p\}$ for a nonzero subspace W of $\mathbb{R}^n$, define

$$
\mathbf{v}_1 = \mathbf{x}_1
$$
$$
\mathbf{v}_2 = \mathbf{x}_2 - \frac{\mathbf{x}_2 \cdot \mathbf{v}_1}{\mathbf{v}_1 \cdot \mathbf{v}_1} \mathbf{v}_1
$$
$$
\mathbf{v}_3 = \mathbf{x}_3 - \frac{\mathbf{x}_3 \cdot \mathbf{v}_1}{\mathbf{v}_1 \cdot \mathbf{v}_1} \mathbf{v}_1 - \frac{\mathbf{x}_3 \cdot \mathbf{v}_2}{\mathbf{v}_2 \cdot \mathbf{v}_2} \mathbf{v}_2
$$
$$
\vdots
$$
$$
\mathbf{v}_p = \mathbf{x}_p - \frac{\mathbf{x}_p \cdot \mathbf{v}_1}{\mathbf{v}_1 \cdot \mathbf{v}_1} \mathbf{v}_1 - \frac{\mathbf{x}_p \cdot \mathbf{v}_2}{\mathbf{v}_2 \cdot \mathbf{v}_2} \mathbf{v}_2 - \cdots - \frac{\mathbf{x}_p \cdot \mathbf{v}_{p-1}}{\mathbf{v}_{p-1} \cdot \mathbf{v}_{p-1}} \mathbf{v}_{p-1}
$$

Then $\{\mathbf{v}_1, \ldots, \mathbf{v}_p\}$ is an orthogonal basis for W. In addition

$$
\operatorname{Span}\{\mathbf{v}_1, \ldots, \mathbf{v}_k\} = \operatorname{Span}\{\mathbf{x}_1, \ldots, \mathbf{x}_k\} \qquad \text{for } 1 \le k \le p \tag{1}
$$

PROOF   For $1 \leq k \leq p$, let $W_k = \text{Span}\{\mathbf{x}_1, \ldots, \mathbf{x}_k\}$. Set $\mathbf{v}_1 = \mathbf{x}_1$, so that $\text{Span}\{\mathbf{v}_1\} = \text{Span}\{\mathbf{x}_1\}$. Suppose, for some $k < p$, we have constructed $\mathbf{v}_1, \ldots, \mathbf{v}_k$ so that $\{\mathbf{v}_1, \ldots, \mathbf{v}_k\}$ is an orthogonal basis for W_k. Define

$$\mathbf{v}_{k+1} = \mathbf{x}_{k+1} - \text{proj}_{W_k} \mathbf{x}_{k+1} \tag{2}$$

By the Orthogonal Decomposition Theorem, $\mathbf{v}_{k+1}$ is orthogonal to W_k. Note that $\text{proj}_{W_k} \mathbf{x}_{k+1}$ is in W_k and hence also in W_{k+1}. Since $\mathbf{x}_{k+1}$ is in W_{k+1}, so is $\mathbf{v}_{k+1}$ (because W_{k+1} is a subspace and is closed under subtraction). Furthermore, $\mathbf{v}_{k+1} \neq \mathbf{0}$ because $\mathbf{x}_{k+1}$ is not in $W_k = \text{Span}\{\mathbf{x}_1, \ldots, \mathbf{x}_k\}$. Hence $\{\mathbf{v}_1, \ldots, \mathbf{v}_{k+1}\}$ is an orthogonal set of nonzero vectors in the $(k + 1)$-dimensional space W_{k+1}. By the Basis Theorem in Section 4.5, this set is an orthogonal basis for W_{k+1}. Hence $W_{k+1} = \text{Span}\{\mathbf{v}_1, \ldots, \mathbf{v}_{k+1}\}$. When $k + 1 = p$, the process stops.   ■

Theorem 11 shows that any nonzero subspace W of $\mathbb{R}^n$ has an orthogonal basis, because an ordinary basis $\{\mathbf{x}_1, \ldots, \mathbf{x}_p\}$ is always available (by Theorem 12 in Section 4.5), and the Gram–Schmidt process depends only on the existence of orthogonal projections onto subspaces of W that already have orthogonal bases.

Orthonormal Bases

An orthonormal basis is constructed easily from an orthogonal basis $\{\mathbf{v}_1, \ldots, \mathbf{v}_p\}$: simply normalize (i.e., "scale") all the $\mathbf{v}_k$. When working problems by hand, this is easier than normalizing each $\mathbf{v}_k$ as soon as it is found (because it avoids unnecessary writing of square roots).

EXAMPLE 3   Example 1 constructed the orthogonal basis

$$\mathbf{v}_1 = \begin{bmatrix} 3 \\ 6 \\ 0 \end{bmatrix}, \quad \mathbf{v}_2 = \begin{bmatrix} 0 \\ 0 \\ 2 \end{bmatrix}$$

An orthonormal basis is

$$\mathbf{u}_1 = \frac{1}{\|\mathbf{v}_1\|} \mathbf{v}_1 = \frac{1}{\sqrt{45}} \begin{bmatrix} 3 \\ 6 \\ 0 \end{bmatrix} = \begin{bmatrix} 1/\sqrt{5} \\ 2/\sqrt{5} \\ 0 \end{bmatrix}$$

$$\mathbf{u}_2 = \frac{1}{\|\mathbf{v}_2\|} \mathbf{v}_2 = \begin{bmatrix} 0 \\ 0 \\ 1 \end{bmatrix}$$ ■

QR Factorization of Matrices

If an $m \times n$ matrix A has linearly independent columns $\mathbf{x}_1, \ldots, \mathbf{x}_n$, then applying the Gram–Schmidt process (with normalizations) to $\mathbf{x}_1, \ldots, \mathbf{x}_n$ amounts to *factoring A*, as described in the next theorem. This factorization is widely used in computer algorithms for various computations, such as solving equations (discussed in Section 6.5) and finding eigenvalues (mentioned in the exercises for Section 5.2).

THEOREM 12   **The QR Factorization**

If A is an $m \times n$ matrix with linearly independent columns, then A can be factored as $A = QR$, where Q is an $m \times n$ matrix whose columns form an orthonormal basis for Col A and R is an $n \times n$ upper triangular invertible matrix with positive entries on its diagonal.

PROOF The columns of A form a basis $\{\mathbf{x}_1, \ldots, \mathbf{x}_n\}$ for Col A. Construct an orthonormal basis $\{\mathbf{u}_1, \ldots, \mathbf{u}_n\}$ for $W = \text{Col } A$ with property (1) in Theorem 11. This basis may be constructed by the Gram–Schmidt process or some other means. Let

$$Q = [\, \mathbf{u}_1 \quad \mathbf{u}_2 \quad \cdots \quad \mathbf{u}_n \,]$$

For $k = 1, \ldots, n, \mathbf{x}_k$ is in Span $\{\mathbf{x}_1, \ldots, \mathbf{x}_k\} = \text{Span}\{\mathbf{u}_1, \ldots, \mathbf{u}_k\}$. So there are constants, $r_{1k}, \ldots, r_{kk}$, such that

$$\mathbf{x}_k = r_{1k}\mathbf{u}_1 + \cdots + r_{kk}\mathbf{u}_k + 0\,\mathbf{u}_{k+1} + \cdots + 0\,\mathbf{u}_n$$

We may assume that $r_{kk} \geq 0$. (If $r_{kk} < 0$, multiply both r_{kk} and $\mathbf{u}_k$ by -1.) This shows that $\mathbf{x}_k$ is a linear combination of the columns of Q using as weights the entries in the vector

$$\mathbf{r}_k = \begin{bmatrix} r_{1k} \\ \vdots \\ r_{kk} \\ 0 \\ \vdots \\ 0 \end{bmatrix}$$

That is, $\mathbf{x}_k = Q\mathbf{r}_k$ for $k = 1, \ldots, n$. Let $R = [\, \mathbf{r}_1 \quad \cdots \quad \mathbf{r}_n \,]$. Then

$$A = [\, \mathbf{x}_1 \quad \cdots \quad \mathbf{x}_n \,] = [\, Q\mathbf{r}_1 \quad \cdots \quad Q\mathbf{r}_n \,] = QR$$

The fact that R is invertible follows easily from the fact that the columns of A are linearly independent (Exercise 23). Since R is clearly upper triangular, its nonnegative diagonal entries must be positive. ∎

EXAMPLE 4 Find a QR factorization of $A = \begin{bmatrix} 1 & 0 & 0 \\ 1 & 1 & 0 \\ 1 & 1 & 1 \\ 1 & 1 & 1 \end{bmatrix}$.

SOLUTION The columns of A are the vectors $\mathbf{x}_1$, $\mathbf{x}_2$, and $\mathbf{x}_3$ in Example 2. An orthogonal basis for Col $A = \text{Span}\{\mathbf{x}_1, \mathbf{x}_2, \mathbf{x}_3\}$ was found in that example:

$$\mathbf{v}_1 = \begin{bmatrix} 1 \\ 1 \\ 1 \\ 1 \end{bmatrix}, \quad \mathbf{v}_2' = \begin{bmatrix} -3 \\ 1 \\ 1 \\ 1 \end{bmatrix}, \quad \mathbf{v}_3 = \begin{bmatrix} 0 \\ -2/3 \\ 1/3 \\ 1/3 \end{bmatrix}$$

To simplify the arithmetic that follows, scale $\mathbf{v}_3$ by letting $\mathbf{v}_3' = 3\mathbf{v}_3$. Then normalize the three vectors to obtain $\mathbf{u}_1$, $\mathbf{u}_2$, and $\mathbf{u}_3$, and use these vectors as the columns of Q:

$$Q = \begin{bmatrix} 1/2 & -3/\sqrt{12} & 0 \\ 1/2 & 1/\sqrt{12} & -2/\sqrt{6} \\ 1/2 & 1/\sqrt{12} & 1/\sqrt{6} \\ 1/2 & 1/\sqrt{12} & 1/\sqrt{6} \end{bmatrix}$$

By construction, the first k columns of Q are an orthonormal basis of Span $\{x_1, \ldots, x_k\}$. From the proof of Theorem 12, $A = QR$ for some R. To find R, observe that $Q^TQ = I$, because the columns of Q are orthonormal. Hence

$$Q^TA = Q^T(QR) = IR = R$$

and

$$R = \begin{bmatrix} 1/2 & 1/2 & 1/2 & 1/2 \\ -3/\sqrt{12} & 1/\sqrt{12} & 1/\sqrt{12} & 1/\sqrt{12} \\ 0 & -2/\sqrt{6} & 1/\sqrt{6} & 1/\sqrt{6} \end{bmatrix} \begin{bmatrix} 1 & 0 & 0 \\ 1 & 1 & 0 \\ 1 & 1 & 1 \\ 1 & 1 & 1 \end{bmatrix}$$

$$= \begin{bmatrix} 2 & 3/2 & 1 \\ 0 & 3/\sqrt{12} & 2/\sqrt{12} \\ 0 & 0 & 2/\sqrt{6} \end{bmatrix}$$

◾

Numerical Notes

1. When the Gram–Schmidt process is run on a computer, roundoff error can build up as the vectors u_k are calculated, one by one. For j and k large but unequal, the inner products $u_j^T u_k$ may not be sufficiently close to zero. This loss of orthogonality can be reduced substantially by rearranging the order of the calculations.[1] However, a different computer-based QR factorization is usually preferred to this modified Gram–Schmidt method because it yields a more accurate orthonormal basis, even though the factorization requires about twice as much arithmetic.

2. To produce a QR factorization of a matrix A, a computer program usually left-multiplies A by a sequence of orthogonal matrices until A is transformed into an upper triangular matrix. This construction is analogous to the left-multiplication by elementary matrices that produces an LU factorization of A.

Practice Problems

1. Let $W = $ Span $\{x_1, x_2\}$, where $x_1 = \begin{bmatrix} 1 \\ 1 \\ 1 \end{bmatrix}$ and $x_2 = \begin{bmatrix} 1/3 \\ 1/3 \\ -2/3 \end{bmatrix}$. Construct an orthonormal basis for W.

2. Suppose $A = QR$, where Q is an $m \times n$ matrix with orthogonal columns and R is an $n \times n$ matrix. Show that if the columns of A are linearly dependent, then R cannot be invertible.

6.4 Exercises

In Exercises 1–6, the given set is a basis for a subspace W. Use the Gram–Schmidt process to produce an orthogonal basis for W.

1. $\begin{bmatrix} 3 \\ 0 \\ -1 \end{bmatrix}, \begin{bmatrix} 8 \\ 5 \\ -6 \end{bmatrix}$

2. $\begin{bmatrix} 0 \\ 4 \\ 2 \end{bmatrix}, \begin{bmatrix} 5 \\ 6 \\ -7 \end{bmatrix}$

3. $\begin{bmatrix} 2 \\ -5 \\ 1 \end{bmatrix}, \begin{bmatrix} 4 \\ -1 \\ 2 \end{bmatrix}$

4. $\begin{bmatrix} 3 \\ -4 \\ 5 \end{bmatrix}, \begin{bmatrix} -3 \\ 14 \\ -7 \end{bmatrix}$

5. $\begin{bmatrix} 1 \\ -4 \\ 0 \\ 1 \end{bmatrix}, \begin{bmatrix} 7 \\ -7 \\ -4 \\ 1 \end{bmatrix}$

6. $\begin{bmatrix} 3 \\ -1 \\ 2 \\ -1 \end{bmatrix}, \begin{bmatrix} -5 \\ 9 \\ -9 \\ 3 \end{bmatrix}$

[1] See *Fundamentals of Matrix Computations*, by David S. Watkins (New York: John Wiley & Sons, 1991), pp. 167–180.

7. Find an orthonormal basis of the subspace spanned by the vectors in Exercise 3.

8. Find an orthonormal basis of the subspace spanned by the vectors in Exercise 4.

Find an orthogonal basis for the column space of each matrix in Exercises 9–12.

9. $\begin{bmatrix} 3 & -5 & 1 \\ 1 & 1 & 1 \\ -1 & 5 & -2 \\ 3 & -7 & 8 \end{bmatrix}$ **10.** $\begin{bmatrix} -1 & 6 & 6 \\ 3 & -8 & 3 \\ 1 & -2 & 6 \\ 1 & -4 & -3 \end{bmatrix}$

11. $\begin{bmatrix} 1 & 2 & 5 \\ -1 & 1 & -4 \\ -1 & 4 & -3 \\ 1 & -4 & 7 \\ 1 & 2 & 1 \end{bmatrix}$ **12.** $\begin{bmatrix} 1 & 3 & 5 \\ -1 & -3 & 1 \\ 0 & 2 & 3 \\ 1 & 5 & 2 \\ 1 & 5 & 8 \end{bmatrix}$

In Exercises 13 and 14, the columns of Q were obtained by applying the Gram–Schmidt process to the columns of A. Find an upper triangular matrix R such that $A = QR$. Check your work.

13. $A = \begin{bmatrix} 5 & 9 \\ 1 & 7 \\ -3 & -5 \\ 1 & 5 \end{bmatrix}$, $Q = \begin{bmatrix} 5/6 & -1/6 \\ 1/6 & 5/6 \\ -3/6 & 1/6 \\ 1/6 & 3/6 \end{bmatrix}$

14. $A = \begin{bmatrix} -2 & 3 \\ 5 & 7 \\ 2 & -2 \\ 4 & 6 \end{bmatrix}$, $Q = \begin{bmatrix} -2/7 & 5/7 \\ 5/7 & 2/7 \\ 2/7 & -4/7 \\ 4/7 & 2/7 \end{bmatrix}$

15. Find a QR factorization of the matrix in Exercise 11.

16. Find a QR factorization of the matrix in Exercise 12.

In Exercises 17–22, all vectors and subspaces are in $\mathbb{R}^n$. Mark each statement True or False (**T/F**). Justify each answer.

17. (**T/F**) If $\{\mathbf{v}_1, \mathbf{v}_2, \mathbf{v}_3\}$ is an orthogonal basis for W, then multiplying $\mathbf{v}_3$ by a scalar c gives a new orthogonal basis $\{\mathbf{v}_1, \mathbf{v}_2, c\mathbf{v}_3\}$.

18. (**T/F**) If $W = \text{Span}\{\mathbf{x}_1, \mathbf{x}_2, \mathbf{x}_3\}$ with $\{\mathbf{x}_1, \mathbf{x}_2, \mathbf{x}_3\}$ linearly independent, and if $\{\mathbf{v}_1, \mathbf{v}_2, \mathbf{v}_3\}$ is an orthogonal set in W, then $\{\mathbf{v}_1, \mathbf{v}_2, \mathbf{v}_3\}$ is a basis for W.

19. (**T/F**) The Gram–Schmidt process produces from a linearly independent set $\{\mathbf{x}_1, \dots, \mathbf{x}_p\}$ an orthogonal set $\{\mathbf{v}_1, \dots, \mathbf{v}_p\}$ with the property that for each k, the vectors $\mathbf{v}_1, \dots, \mathbf{v}_k$ span the same subspace as that spanned by $\mathbf{x}_1, \dots, \mathbf{x}_k$.

20. (**T/F**) If $\mathbf{x}$ is not in a subspace W, then $\mathbf{x} - \text{proj}_W \mathbf{x}$ is not zero.

21. (**T/F**) If $A = QR$, where Q has orthonormal columns, then $R = Q^T A$.

22. (**T/F**) In a QR factorization, say $A = QR$ (when A has linearly independent columns), the columns of Q form an orthonormal basis for the column space of A.

23. Suppose $A = QR$, where Q is $m \times n$ and R is $n \times n$. Show that if the columns of A are linearly independent, then R must be invertible. [*Hint:* Study the equation $R\mathbf{x} = \mathbf{0}$ and use the fact that $A = QR$.]

24. Suppose $A = QR$, where R is an invertible matrix. Show that A and Q have the same column space. [*Hint:* Given $\mathbf{y}$ in Col A, show that $\mathbf{y} = Q\mathbf{x}$ for some $\mathbf{x}$. Also, given $\mathbf{y}$ in Col Q, show that $\mathbf{y} = A\mathbf{x}$ for some $\mathbf{x}$.]

25. Given $A = QR$ as in Theorem 12, describe how to find an orthogonal $m \times m$ (square) matrix Q_1 and an invertible $n \times n$ upper triangular matrix R such that

$$A = Q_1 \begin{bmatrix} R \\ 0 \end{bmatrix}$$

The MATLAB `qr` command supplies this "full" QR factorization when rank $A = n$.

26. Let $\mathbf{u}_1, \dots, \mathbf{u}_p$ be an orthogonal basis for a subspace W of $\mathbb{R}^n$, and let $T : \mathbb{R}^n \to \mathbb{R}^n$ be defined by $T(\mathbf{x}) = \text{proj}_W \mathbf{x}$. Show that T is a linear transformation.

27. Suppose $A = QR$ is a QR factorization of an $m \times n$ matrix A (with linearly independent columns). Partition A as $[A_1 \quad A_2]$, where A_1 has p columns. Show how to obtain a QR factorization of A_1, and explain why your factorization has the appropriate properties.

T 28. Use the Gram–Schmidt process as in Example 2 to produce an orthogonal basis for the column space of

$$A = \begin{bmatrix} -10 & 13 & 7 & -11 \\ 2 & 1 & -5 & 3 \\ -6 & 3 & 13 & -3 \\ 16 & -16 & -2 & 5 \\ 2 & 1 & -5 & -7 \end{bmatrix}$$

T 29. Use the method in this section to produce a QR factorization of the matrix in Exercise 28.

T 30. For a matrix program, the Gram–Schmidt process works better with orthonormal vectors. Starting with $\mathbf{x}_1, \dots, \mathbf{x}_p$ as in Theorem 11, let $A = [\mathbf{x}_1 \quad \cdots \quad \mathbf{x}_p]$. Suppose Q is an $n \times k$ matrix whose columns form an orthonormal basis for the subspace W_k spanned by the first k columns of A. Then for $\mathbf{x}$ in $\mathbb{R}^n$, $QQ^T\mathbf{x}$ is the orthogonal projection of $\mathbf{x}$ onto W_k (Theorem 10 in Section 6.3). If $\mathbf{x}_{k+1}$ is the next column of A, then equation (2) in the proof of Theorem 11 becomes

$$\mathbf{v}_{k+1} = \mathbf{x}_{k+1} - Q(Q^T \mathbf{x}_{k+1})$$

(The parentheses above reduce the number of arithmetic operations.) Let $\mathbf{u}_{k+1} = \mathbf{v}_{k+1}/\|\mathbf{v}_{k+1}\|$. The new Q for the next step is $[Q \quad \mathbf{u}_{k+1}]$. Use this procedure to compute the QR factorization of the matrix in Exercise 28. Write the keystrokes or commands you use.

Solution to Practice Problems

1. Let $\mathbf{v}_1 = \mathbf{x}_1 = \begin{bmatrix} 1 \\ 1 \\ 1 \end{bmatrix}$ and $\mathbf{v}_2 = \mathbf{x}_2 - \dfrac{\mathbf{x}_2 \cdot \mathbf{v}_1}{\mathbf{v}_1 \cdot \mathbf{v}_1} \mathbf{v}_1 = \mathbf{x}_2 - 0\mathbf{v}_1 = \mathbf{x}_2$. So $\{\mathbf{x}_1, \mathbf{x}_2\}$ is already orthogonal. All that is needed is to normalize the vectors. Let

$$\mathbf{u}_1 = \frac{1}{\|\mathbf{v}_1\|} \mathbf{v}_1 = \frac{1}{\sqrt{3}} \begin{bmatrix} 1 \\ 1 \\ 1 \end{bmatrix} = \begin{bmatrix} 1/\sqrt{3} \\ 1/\sqrt{3} \\ 1/\sqrt{3} \end{bmatrix}$$

Instead of normalizing $\mathbf{v}_2$ directly, normalize $\mathbf{v}_2' = 3\mathbf{v}_2$ instead:

$$\mathbf{u}_2 = \frac{1}{\|\mathbf{v}_2'\|} \mathbf{v}_2' = \frac{1}{\sqrt{1^2 + 1^2 + (-2)^2}} \begin{bmatrix} 1 \\ 1 \\ -2 \end{bmatrix} = \begin{bmatrix} 1/\sqrt{6} \\ 1/\sqrt{6} \\ -2/\sqrt{6} \end{bmatrix}$$

Then $\{\mathbf{u}_1, \mathbf{u}_2\}$ is an orthonormal basis for W.

2. Since the columns of A are linearly dependent, there is a nontrivial vector $\mathbf{x}$ such that $A\mathbf{x} = \mathbf{0}$. But then $QR\mathbf{x} = \mathbf{0}$. Applying Theorem 7 from Section 6.2 results in $\|R\mathbf{x}\| = \|QR\mathbf{x}\| = \|\mathbf{0}\| = 0$. But $\|R\mathbf{x}\| = 0$ implies $R\mathbf{x} = \mathbf{0}$, by Theorem 1 from Section 6.1. Thus there is a nontrivial vector $\mathbf{x}$ such that $R\mathbf{x} = \mathbf{0}$ and hence, by the Invertible Matrix Theorem, R cannot be invertible.

6.5 Least-Squares Problems

Inconsistent systems arise often in applications. When a solution is demanded and none exists, the best one can do is to find an $\mathbf{x}$ that makes $A\mathbf{x}$ as close as possible to $\mathbf{b}$.

Think of $A\mathbf{x}$ as an *approximation* to $\mathbf{b}$. The smaller the distance between $\mathbf{b}$ and $A\mathbf{x}$, given by $\|\mathbf{b} - A\mathbf{x}\|$, the better the approximation. The **general least-squares problem** is to find an $\mathbf{x}$ that makes $\|\mathbf{b} - A\mathbf{x}\|$ as small as possible. The adjective "least-squares" arises from the fact that $\|\mathbf{b} - A\mathbf{x}\|$ is the square root of a sum of squares.

DEFINITION

If A is $m \times n$ and $\mathbf{b}$ is in $\mathbb{R}^m$, a **least-squares solution** of $A\mathbf{x} = \mathbf{b}$ is an $\hat{\mathbf{x}}$ in $\mathbb{R}^n$ such that

$$\|\mathbf{b} - A\hat{\mathbf{x}}\| \leq \|\mathbf{b} - A\mathbf{x}\|$$

for all $\mathbf{x}$ in $\mathbb{R}^n$.

The most important aspect of the least-squares problem is that no matter what $\mathbf{x}$ we select, the vector $A\mathbf{x}$ will necessarily be in the column space, Col A. So we seek an $\mathbf{x}$ that makes $A\mathbf{x}$ the closest point in Col A to $\mathbf{b}$. See Figure 1. (Of course, if $\mathbf{b}$ happens to be in Col A, then $\mathbf{b}$ *is* $A\mathbf{x}$ for some $\mathbf{x}$, and such an $\mathbf{x}$ is a "least-squares solution.")

Solution of the General Least-Squares Problem

Given A and $\mathbf{b}$ as above, apply the Best Approximation Theorem in Section 6.3 to the subspace Col A. Let

$$\hat{\mathbf{b}} = \text{proj}_{\text{Col } A} \mathbf{b}$$

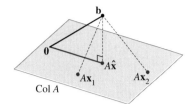

FIGURE 1 The vector **b** is closer to $A\hat{\mathbf{x}}$ than to $A\mathbf{x}$ for other **x**.

Because $\hat{\mathbf{b}}$ is in the column space of A, the equation $A\mathbf{x} = \hat{\mathbf{b}}$ *is* consistent, and there is an $\hat{\mathbf{x}}$ in $\mathbb{R}^n$ such that

$$A\hat{\mathbf{x}} = \hat{\mathbf{b}} \tag{1}$$

Since $\hat{\mathbf{b}}$ is the closest point in Col A to **b**, a vector $\hat{\mathbf{x}}$ is a least-squares solution of $A\mathbf{x} = \mathbf{b}$ if and only if $\hat{\mathbf{x}}$ satisfies (1). Such an $\hat{\mathbf{x}}$ in $\mathbb{R}^n$ is a list of weights that will build $\hat{\mathbf{b}}$ out of the columns of A. See Figure 2. [There are many solutions of (1) if the equation has free variables.]

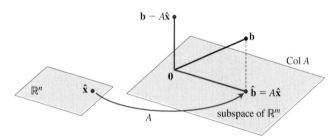

FIGURE 2 The least-squares solution $\hat{\mathbf{x}}$ is in $\mathbb{R}^n$.

Suppose $\hat{\mathbf{x}}$ satisfies $A\hat{\mathbf{x}} = \hat{\mathbf{b}}$. By the Orthogonal Decomposition Theorem in Section 6.3, the projection $\hat{\mathbf{b}}$ has the property that $\mathbf{b} - \hat{\mathbf{b}}$ is orthogonal to Col A, so $\mathbf{b} - A\hat{\mathbf{x}}$ is orthogonal to each column of A. If $\mathbf{a}_j$ is any column of A, then $\mathbf{a}_j \cdot (\mathbf{b} - A\hat{\mathbf{x}}) = 0$, and $\mathbf{a}_j^T (\mathbf{b} - A\hat{\mathbf{x}}) = 0$. Since each $\mathbf{a}_j^T$ is a row of A^T,

$$A^T (\mathbf{b} - A\hat{\mathbf{x}}) = \mathbf{0} \tag{2}$$

(This equation also follows from Theorem 3 in Section 6.1.) Thus

$$A^T \mathbf{b} - A^T A\hat{\mathbf{x}} = \mathbf{0}$$
$$A^T A\hat{\mathbf{x}} = A^T \mathbf{b}$$

These calculations show that each least-squares solution of $A\mathbf{x} = \mathbf{b}$ satisfies the equation

$$A^T A\mathbf{x} = A^T \mathbf{b} \tag{3}$$

The matrix equation (3) represents a system of equations called the **normal equations** for $A\mathbf{x} = \mathbf{b}$. A solution of (3) is often denoted by $\hat{\mathbf{x}}$.

THEOREM 13 The set of least-squares solutions of $A\mathbf{x} = \mathbf{b}$ coincides with the nonempty set of solutions of the normal equations $A^T A\mathbf{x} = A^T \mathbf{b}$.

PROOF As shown, the set of least-squares solutions is nonempty and each least-squares solution $\hat{\mathbf{x}}$ satisfies the normal equations. Conversely, suppose $\hat{\mathbf{x}}$ satisfies $A^T A \hat{\mathbf{x}} = A^T \mathbf{b}$. Then $\hat{\mathbf{x}}$ satisfies (2), which shows that $\mathbf{b} - A\hat{\mathbf{x}}$ is orthogonal to the rows of A^T and hence is orthogonal to the columns of A. Since the columns of A span Col A, the vector $\mathbf{b} - A\hat{\mathbf{x}}$ is orthogonal to all of Col A. Hence the equation

$$\mathbf{b} = A\hat{\mathbf{x}} + (\mathbf{b} - A\hat{\mathbf{x}})$$

is a decomposition of $\mathbf{b}$ into the sum of a vector in Col A and a vector orthogonal to Col A. By the uniqueness of the orthogonal decomposition, $A\hat{\mathbf{x}}$ must be the orthogonal projection of $\mathbf{b}$ onto Col A. That is, $A\hat{\mathbf{x}} = \hat{\mathbf{b}}$, and $\hat{\mathbf{x}}$ is a least-squares solution. ■

EXAMPLE 1 Find a least-squares solution of the inconsistent system $A\mathbf{x} = \mathbf{b}$ for

$$A = \begin{bmatrix} 4 & 0 \\ 0 & 2 \\ 1 & 1 \end{bmatrix}, \quad \mathbf{b} = \begin{bmatrix} 2 \\ 0 \\ 11 \end{bmatrix}$$

SOLUTION To use normal equations (3), compute:

$$A^T A = \begin{bmatrix} 4 & 0 & 1 \\ 0 & 2 & 1 \end{bmatrix} \begin{bmatrix} 4 & 0 \\ 0 & 2 \\ 1 & 1 \end{bmatrix} = \begin{bmatrix} 17 & 1 \\ 1 & 5 \end{bmatrix}$$

$$A^T \mathbf{b} = \begin{bmatrix} 4 & 0 & 1 \\ 0 & 2 & 1 \end{bmatrix} \begin{bmatrix} 2 \\ 0 \\ 11 \end{bmatrix} = \begin{bmatrix} 19 \\ 11 \end{bmatrix}$$

Then the equation $A^T A\mathbf{x} = A^T \mathbf{b}$ becomes

$$\begin{bmatrix} 17 & 1 \\ 1 & 5 \end{bmatrix} \begin{bmatrix} x_1 \\ x_2 \end{bmatrix} = \begin{bmatrix} 19 \\ 11 \end{bmatrix}$$

Row operations can be used to solve this system, but since $A^T A$ is invertible and 2×2, it is probably faster to compute

$$(A^T A)^{-1} = \frac{1}{84} \begin{bmatrix} 5 & -1 \\ -1 & 17 \end{bmatrix}$$

and then to solve $A^T A\mathbf{x} = A^T \mathbf{b}$ as

$$\hat{\mathbf{x}} = (A^T A)^{-1} A^T \mathbf{b}$$
$$= \frac{1}{84} \begin{bmatrix} 5 & -1 \\ -1 & 17 \end{bmatrix} \begin{bmatrix} 19 \\ 11 \end{bmatrix} = \frac{1}{84} \begin{bmatrix} 84 \\ 168 \end{bmatrix} = \begin{bmatrix} 1 \\ 2 \end{bmatrix}$$ ■

In many calculations, $A^T A$ is invertible, but this is not always the case. The next example involves a matrix of the sort that appears in what are called *analysis of variance* problems in statistics.

EXAMPLE 2 Find a least-squares solution of $A\mathbf{x} = \mathbf{b}$ for

$$A = \begin{bmatrix} 1 & 1 & 0 & 0 \\ 1 & 1 & 0 & 0 \\ 1 & 0 & 1 & 0 \\ 1 & 0 & 1 & 0 \\ 1 & 0 & 0 & 1 \\ 1 & 0 & 0 & 1 \end{bmatrix}, \quad \mathbf{b} = \begin{bmatrix} -3 \\ -1 \\ 0 \\ 2 \\ 5 \\ 1 \end{bmatrix}$$

SOLUTION   Compute

$$A^T A = \begin{bmatrix} 1 & 1 & 1 & 1 & 1 & 1 \\ 1 & 1 & 0 & 0 & 0 & 0 \\ 0 & 0 & 1 & 1 & 0 & 0 \\ 0 & 0 & 0 & 0 & 1 & 1 \end{bmatrix} \begin{bmatrix} 1 & 1 & 0 & 0 \\ 1 & 1 & 0 & 0 \\ 1 & 0 & 1 & 0 \\ 1 & 0 & 1 & 0 \\ 1 & 0 & 0 & 1 \\ 1 & 0 & 0 & 1 \end{bmatrix} = \begin{bmatrix} 6 & 2 & 2 & 2 \\ 2 & 2 & 0 & 0 \\ 2 & 0 & 2 & 0 \\ 2 & 0 & 0 & 2 \end{bmatrix}$$

$$A^T \mathbf{b} = \begin{bmatrix} 1 & 1 & 1 & 1 & 1 & 1 \\ 1 & 1 & 0 & 0 & 0 & 0 \\ 0 & 0 & 1 & 1 & 0 & 0 \\ 0 & 0 & 0 & 0 & 1 & 1 \end{bmatrix} \begin{bmatrix} -3 \\ -1 \\ 0 \\ 2 \\ 5 \\ 1 \end{bmatrix} = \begin{bmatrix} 4 \\ -4 \\ 2 \\ 6 \end{bmatrix}$$

The augmented matrix for $A^T A \mathbf{x} = A^T \mathbf{b}$ is

$$\begin{bmatrix} 6 & 2 & 2 & 2 & 4 \\ 2 & 2 & 0 & 0 & -4 \\ 2 & 0 & 2 & 0 & 2 \\ 2 & 0 & 0 & 2 & 6 \end{bmatrix} \sim \begin{bmatrix} 1 & 0 & 0 & 1 & 3 \\ 0 & 1 & 0 & -1 & -5 \\ 0 & 0 & 1 & -1 & -2 \\ 0 & 0 & 0 & 0 & 0 \end{bmatrix}$$

The general solution is $x_1 = 3 - x_4$, $x_2 = -5 + x_4$, $x_3 = -2 + x_4$, and x_4 is free. So the general least-squares solution of $A\mathbf{x} = \mathbf{b}$ has the form

$$\hat{\mathbf{x}} = \begin{bmatrix} 3 \\ -5 \\ -2 \\ 0 \end{bmatrix} + x_4 \begin{bmatrix} -1 \\ 1 \\ 1 \\ 1 \end{bmatrix}$$ ∎

The next theorem gives useful criteria for determining when there is only one least-squares solution of $A\mathbf{x} = \mathbf{b}$. (Of course, the orthogonal projection $\hat{\mathbf{b}}$ is always unique.)

THEOREM 14

Let A be an $m \times n$ matrix. The following statements are logically equivalent:

a. The equation $A\mathbf{x} = \mathbf{b}$ has a unique least-squares solution for each $\mathbf{b}$ in $\mathbb{R}^m$.

b. The columns of A are linearly independent.

c. The matrix $A^T A$ is invertible.

When these statements are true, the least-squares solution $\hat{\mathbf{x}}$ is given by

$$\hat{\mathbf{x}} = (A^T A)^{-1} A^T \mathbf{b} \tag{4}$$

The main elements of a proof of Theorem 14 are outlined in Exercises 27–29, which also review concepts from Chapter 4. Formula (4) for $\hat{\mathbf{x}}$ is useful mainly for theoretical purposes and for hand calculations when $A^T A$ is a 2×2 invertible matrix.

When a least-squares solution $\hat{\mathbf{x}}$ is used to produce $A\hat{\mathbf{x}}$ as an approximation to $\mathbf{b}$, the distance from $\mathbf{b}$ to $A\hat{\mathbf{x}}$ is called the **least-squares error** of this approximation.

EXAMPLE 3   Given A and $\mathbf{b}$ as in Example 1, determine the least-squares error in the least-squares solution of $A\mathbf{x} = \mathbf{b}$.

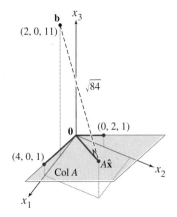

FIGURE 3

SOLUTION From Example 1,

$$\mathbf{b} = \begin{bmatrix} 2 \\ 0 \\ 11 \end{bmatrix} \quad \text{and} \quad A\hat{\mathbf{x}} = \begin{bmatrix} 4 & 0 \\ 0 & 2 \\ 1 & 1 \end{bmatrix} \begin{bmatrix} 1 \\ 2 \end{bmatrix} = \begin{bmatrix} 4 \\ 4 \\ 3 \end{bmatrix}$$

Hence

$$\mathbf{b} - A\hat{\mathbf{x}} = \begin{bmatrix} 2 \\ 0 \\ 11 \end{bmatrix} - \begin{bmatrix} 4 \\ 4 \\ 3 \end{bmatrix} = \begin{bmatrix} -2 \\ -4 \\ 8 \end{bmatrix}$$

and

$$\|\mathbf{b} - A\hat{\mathbf{x}}\| = \sqrt{(-2)^2 + (-4)^2 + 8^2} = \sqrt{84}$$

The least-squares error is $\sqrt{84}$. For any $\mathbf{x}$ in $\mathbb{R}^2$, the distance between $\mathbf{b}$ and the vector $A\mathbf{x}$ is at least $\sqrt{84}$. See Figure 3. Note that the least-squares solution $\hat{\mathbf{x}}$ itself does not appear in the figure. ∎

Alternative Calculations of Least-Squares Solutions

The next example shows how to find a least-squares solution of $A\mathbf{x} = \mathbf{b}$ when the columns of A are orthogonal. Such matrices often appear in linear regression problems, discussed in the next section.

EXAMPLE 4 Find a least-squares solution of $A\mathbf{x} = \mathbf{b}$ for

$$A = \begin{bmatrix} 1 & -6 \\ 1 & -2 \\ 1 & 1 \\ 1 & 7 \end{bmatrix}, \quad \mathbf{b} = \begin{bmatrix} -1 \\ 2 \\ 1 \\ 6 \end{bmatrix}$$

SOLUTION Because the columns $\mathbf{a}_1$ and $\mathbf{a}_2$ of A are orthogonal, the orthogonal projection of $\mathbf{b}$ onto Col A is given by

$$\hat{\mathbf{b}} = \frac{\mathbf{b} \cdot \mathbf{a}_1}{\mathbf{a}_1 \cdot \mathbf{a}_1} \mathbf{a}_1 + \frac{\mathbf{b} \cdot \mathbf{a}_2}{\mathbf{a}_2 \cdot \mathbf{a}_2} \mathbf{a}_2 = \frac{8}{4} \mathbf{a}_1 + \frac{45}{90} \mathbf{a}_2 \tag{5}$$

$$= \begin{bmatrix} 2 \\ 2 \\ 2 \\ 2 \end{bmatrix} + \begin{bmatrix} -3 \\ -1 \\ 1/2 \\ 7/2 \end{bmatrix} = \begin{bmatrix} -1 \\ 1 \\ 5/2 \\ 11/2 \end{bmatrix}$$

Now that $\hat{\mathbf{b}}$ is known, we can solve $A\hat{\mathbf{x}} = \hat{\mathbf{b}}$. But this is trivial, since we already know what weights to place on the columns of A to produce $\hat{\mathbf{b}}$. It is clear from (5) that

$$\hat{\mathbf{x}} = \begin{bmatrix} 8/4 \\ 45/90 \end{bmatrix} = \begin{bmatrix} 2 \\ 1/2 \end{bmatrix} \qquad ∎$$

In some cases, the normal equations for a least-squares problem can be *ill-conditioned*; that is, small errors in the calculations of the entries of $A^T A$ can sometimes cause relatively large errors in the solution $\hat{\mathbf{x}}$. If the columns of A are linearly independent, the least-squares solution can often be computed more reliably through a QR factorization of A (described in Section 6.4).[1]

---

[1] The QR method is compared with the standard normal equation method in G. Golub and C. Van Loan, *Matrix Computations*, 3rd ed. (Baltimore: Johns Hopkins Press, 1996), pp. 230–231.

THEOREM 15

Given an $m \times n$ matrix A with linearly independent columns, let $A = QR$ be a QR factorization of A as in Theorem 12. Then, for each $\mathbf{b}$ in $\mathbb{R}^m$, the equation $A\mathbf{x} = \mathbf{b}$ has a unique least-squares solution, given by

$$\hat{\mathbf{x}} = R^{-1}Q^T\mathbf{b} \tag{6}$$

PROOF   Let $\hat{\mathbf{x}} = R^{-1}Q^T\mathbf{b}$. Then

$$A\hat{\mathbf{x}} = QR\hat{\mathbf{x}} = QRR^{-1}Q^T\mathbf{b} = QQ^T\mathbf{b}$$

By Theorem 12, the columns of Q form an orthonormal basis for Col A. Hence, by Theorem 10, $QQ^T\mathbf{b}$ is the orthogonal projection $\hat{\mathbf{b}}$ of $\mathbf{b}$ onto Col A. Then $A\hat{\mathbf{x}} = \hat{\mathbf{b}}$, which shows that $\hat{\mathbf{x}}$ is a least-squares solution of $A\mathbf{x} = \mathbf{b}$. The uniqueness of $\hat{\mathbf{x}}$ follows from Theorem 14. ∎

> **Numerical Notes**
>
> Since R in Theorem 15 is upper triangular, $\hat{\mathbf{x}}$ should be calculated as the exact solution of the equation
>
> $$R\mathbf{x} = Q^T\mathbf{b} \tag{7}$$
>
> It is much faster to solve (7) by back-substitution or row operations than to compute R^{-1} and use (6).

EXAMPLE 5   Find the least-squares solution of $A\mathbf{x} = \mathbf{b}$ for

$$A = \begin{bmatrix} 1 & 3 & 5 \\ 1 & 1 & 0 \\ 1 & 1 & 2 \\ 1 & 3 & 3 \end{bmatrix}, \quad \mathbf{b} = \begin{bmatrix} 3 \\ 5 \\ 7 \\ -3 \end{bmatrix}$$

SOLUTION   The QR factorization of A can be obtained as in Section 6.4.

$$A = QR = \begin{bmatrix} 1/2 & 1/2 & 1/2 \\ 1/2 & -1/2 & -1/2 \\ 1/2 & -1/2 & 1/2 \\ 1/2 & 1/2 & -1/2 \end{bmatrix} \begin{bmatrix} 2 & 4 & 5 \\ 0 & 2 & 3 \\ 0 & 0 & 2 \end{bmatrix}$$

Then

$$Q^T\mathbf{b} = \begin{bmatrix} 1/2 & 1/2 & 1/2 & 1/2 \\ 1/2 & -1/2 & -1/2 & 1/2 \\ 1/2 & -1/2 & 1/2 & -1/2 \end{bmatrix} \begin{bmatrix} 3 \\ 5 \\ 7 \\ -3 \end{bmatrix} = \begin{bmatrix} 6 \\ -6 \\ 4 \end{bmatrix}$$

The least-squares solution $\hat{\mathbf{x}}$ satisfies $R\mathbf{x} = Q^T\mathbf{b}$; that is,

$$\begin{bmatrix} 2 & 4 & 5 \\ 0 & 2 & 3 \\ 0 & 0 & 2 \end{bmatrix} \begin{bmatrix} x_1 \\ x_2 \\ x_3 \end{bmatrix} = \begin{bmatrix} 6 \\ -6 \\ 4 \end{bmatrix}$$

This equation is solved easily and yields $\hat{\mathbf{x}} = \begin{bmatrix} 10 \\ -6 \\ 2 \end{bmatrix}$. ∎

Practice Problems

1. Let $A = \begin{bmatrix} 1 & -3 & -3 \\ 1 & 5 & 1 \\ 1 & 7 & 2 \end{bmatrix}$ and $\mathbf{b} = \begin{bmatrix} 5 \\ -3 \\ -5 \end{bmatrix}$. Find a least-squares solution of $A\mathbf{x} = \mathbf{b}$, and compute the associated least-squares error.

2. What can you say about the least-squares solution of $A\mathbf{x} = \mathbf{b}$ when $\mathbf{b}$ is orthogonal to the columns of A?

6.5 Exercises

In Exercises 1–4, find a least-squares solution of $A\mathbf{x} = \mathbf{b}$ by (a) constructing the normal equations for $\hat{\mathbf{x}}$ and (b) solving for $\hat{\mathbf{x}}$.

1. $A = \begin{bmatrix} -1 & 2 \\ 2 & -3 \\ -1 & 3 \end{bmatrix}, \mathbf{b} = \begin{bmatrix} 4 \\ 1 \\ 2 \end{bmatrix}$

2. $A = \begin{bmatrix} 2 & 1 \\ -2 & 0 \\ 2 & 3 \end{bmatrix}, \mathbf{b} = \begin{bmatrix} -5 \\ 8 \\ 1 \end{bmatrix}$

3. $A = \begin{bmatrix} 1 & -2 \\ -1 & 2 \\ 0 & 3 \\ 2 & 5 \end{bmatrix}, \mathbf{b} = \begin{bmatrix} 3 \\ 1 \\ -4 \\ 2 \end{bmatrix}$

4. $A = \begin{bmatrix} 1 & 3 \\ 1 & -1 \\ 1 & 1 \end{bmatrix}, \mathbf{b} = \begin{bmatrix} 5 \\ 1 \\ 0 \end{bmatrix}$

In Exercises 5 and 6, describe all least-squares solutions of the equation $A\mathbf{x} = \mathbf{b}$.

5. $A = \begin{bmatrix} 1 & 1 & 0 \\ 1 & 1 & 0 \\ 1 & 0 & 1 \\ 1 & 0 & 1 \end{bmatrix}, \mathbf{b} = \begin{bmatrix} 1 \\ 3 \\ 8 \\ 2 \end{bmatrix}$

6. $A = \begin{bmatrix} 1 & 1 & 0 \\ 1 & 1 & 0 \\ 1 & 1 & 0 \\ 1 & 0 & 1 \\ 1 & 0 & 1 \\ 1 & 0 & 1 \end{bmatrix}, \mathbf{b} = \begin{bmatrix} 7 \\ 2 \\ 3 \\ 6 \\ 5 \\ 4 \end{bmatrix}$

7. Compute the least-squares error associated with the least-squares solution found in Exercise 3.

8. Compute the least-squares error associated with the least-squares solution found in Exercise 4.

In Exercises 9–12, find (a) the orthogonal projection of $\mathbf{b}$ onto Col A and (b) a least-squares solution of $A\mathbf{x} = \mathbf{b}$.

9. $A = \begin{bmatrix} 1 & 5 \\ 3 & 1 \\ -2 & 4 \end{bmatrix}, \mathbf{b} = \begin{bmatrix} 4 \\ -2 \\ -3 \end{bmatrix}$

10. $A = \begin{bmatrix} 1 & 2 \\ -1 & 4 \\ 1 & 2 \end{bmatrix}, \mathbf{b} = \begin{bmatrix} 3 \\ -1 \\ 5 \end{bmatrix}$

11. $A = \begin{bmatrix} 4 & 0 & 1 \\ 1 & -5 & 1 \\ 6 & 1 & 0 \\ 1 & -1 & -5 \end{bmatrix}, \mathbf{b} = \begin{bmatrix} 9 \\ 0 \\ 0 \\ 0 \end{bmatrix}$

12. $A = \begin{bmatrix} 1 & 1 & 0 \\ 1 & 0 & -1 \\ 0 & 1 & 1 \\ -1 & 1 & -1 \end{bmatrix}, \mathbf{b} = \begin{bmatrix} 2 \\ 5 \\ 6 \\ 6 \end{bmatrix}$

13. Let $A = \begin{bmatrix} 3 & 4 \\ -2 & 1 \\ 3 & 4 \end{bmatrix}$, $\mathbf{b} = \begin{bmatrix} 11 \\ -9 \\ 5 \end{bmatrix}$, $\mathbf{u} = \begin{bmatrix} 5 \\ -1 \end{bmatrix}$, and $\mathbf{v} = \begin{bmatrix} 5 \\ -2 \end{bmatrix}$. Compute $A\mathbf{u}$ and $A\mathbf{v}$, and compare them with $\mathbf{b}$. Could $\mathbf{u}$ possibly be a least-squares solution of $A\mathbf{x} = \mathbf{b}$? (Answer this without computing a least-squares solution.)

14. Let $A = \begin{bmatrix} 2 & 1 \\ -3 & -4 \\ 3 & 2 \end{bmatrix}$, $\mathbf{b} = \begin{bmatrix} 5 \\ 4 \\ 4 \end{bmatrix}$, $\mathbf{u} = \begin{bmatrix} 4 \\ -5 \end{bmatrix}$, and $\mathbf{v} = \begin{bmatrix} 6 \\ -5 \end{bmatrix}$. Compute $A\mathbf{u}$ and $A\mathbf{v}$, and compare them with $\mathbf{b}$. Is it possible that at least one of $\mathbf{u}$ or $\mathbf{v}$ could be a least-squares solution of $A\mathbf{x} = \mathbf{b}$? (Answer this without computing a least-squares solution.)

In Exercises 15 and 16, use the factorization $A = QR$ to find the least-squares solution of $A\mathbf{x} = \mathbf{b}$.

15. $A = \begin{bmatrix} 2 & 3 \\ 2 & 4 \\ 1 & 1 \end{bmatrix} = \begin{bmatrix} 2/3 & -1/3 \\ 2/3 & 2/3 \\ 1/3 & -2/3 \end{bmatrix} \begin{bmatrix} 3 & 5 \\ 0 & 1 \end{bmatrix}, \mathbf{b} = \begin{bmatrix} 7 \\ 3 \\ 1 \end{bmatrix}$

16. $A = \begin{bmatrix} 1 & -1 \\ 1 & 4 \\ 1 & -1 \\ 1 & 4 \end{bmatrix} = \begin{bmatrix} 1/2 & -1/2 \\ 1/2 & 1/2 \\ 1/2 & -1/2 \\ 1/2 & 1/2 \end{bmatrix} \begin{bmatrix} 2 & 3 \\ 0 & 5 \end{bmatrix}, \mathbf{b} = \begin{bmatrix} -1 \\ 6 \\ 5 \\ 7 \end{bmatrix}$

In Exercises 17–26, A is an $m \times n$ matrix and $\mathbf{b}$ is in $\mathbb{R}^m$. Mark each statement True or False (**T/F**). Justify each answer.

17. (**T/F**) The general least-squares problem is to find an $\mathbf{x}$ that makes $A\mathbf{x}$ as close as possible to $\mathbf{b}$.

18. (T/F) If $\mathbf{b}$ is in the column space of A, then every solution of $A\mathbf{x} = \mathbf{b}$ is a least-squares solution.

19. (T/F) A least-squares solution of $A\mathbf{x} = \mathbf{b}$ is a vector $\hat{\mathbf{x}}$ that satisfies $A\hat{\mathbf{x}} = \hat{\mathbf{b}}$, where $\hat{\mathbf{b}}$ is the orthogonal projection of $\mathbf{b}$ onto Col A.

20. (T/F) A least-squares solution of $A\mathbf{x} = \mathbf{b}$ is a vector $\hat{\mathbf{x}}$ such that $\|\mathbf{b} - A\mathbf{x}\| \le \|\mathbf{b} - A\hat{\mathbf{x}}\|$ for all $\mathbf{x}$ in $\mathbb{R}^n$.

21. (T/F) Any solution of $A^T A\mathbf{x} = A^T \mathbf{b}$ is a least-squares solution of $A\mathbf{x} = \mathbf{b}$.

22. (T/F) If the columns of A are linearly independent, then the equation $A\mathbf{x} = \mathbf{b}$ has exactly one least-squares solution.

23. (T/F) The least-squares solution of $A\mathbf{x} = \mathbf{b}$ is the point in the column space of A closest to $\mathbf{b}$.

24. (T/F) A least-squares solution of $A\mathbf{x} = \mathbf{b}$ is a list of weights that, when applied to the columns of A, produces the orthogonal projection of $\mathbf{b}$ onto Col A.

25. (T/F) The normal equations always provide a reliable method for computing least-squares solutions.

26. (T/F) If A has a QR factorization, say $A = QR$, then the best way to find the least-squares solution of $A\mathbf{x} = \mathbf{b}$ is to compute $\hat{\mathbf{x}} = R^{-1}Q^T \mathbf{b}$.

27. Let A be an $m \times n$ matrix. Use the steps below to show that a vector $\mathbf{x}$ in $\mathbb{R}^n$ satisfies $A\mathbf{x} = \mathbf{0}$ if and only if $A^T A\mathbf{x} = \mathbf{0}$. This will show that Nul $A =$ Nul $A^T A$.
   a. Show that if $A\mathbf{x} = \mathbf{0}$, then $A^T A\mathbf{x} = \mathbf{0}$.
   b. Suppose $A^T A\mathbf{x} = \mathbf{0}$. Explain why $\mathbf{x}^T A^T A\mathbf{x} = \mathbf{0}$, and use this to show that $A\mathbf{x} = \mathbf{0}$.

28. Let A be an $m \times n$ matrix such that $A^T A$ is invertible. Show that the columns of A are linearly independent. [*Careful:* You may not assume that A is invertible; it may not even be square.]

29. Let A be an $m \times n$ matrix whose columns are linearly independent. [*Careful: A* need not be square.]
   a. Use Exercise 27 to show that $A^T A$ is an invertible matrix.
   b. Explain why A must have at least as many rows as columns.
   c. Determine the rank of A.

30. Use Exercise 27 to show that rank $A^T A =$ rank A. [*Hint:* How many columns does $A^T A$ have? How is this connected with the rank of $A^T A$?]

31. Suppose A is $m \times n$ with linearly independent columns and $\mathbf{b}$ is in $\mathbb{R}^m$. Use the normal equations to produce a formula for $\hat{\mathbf{b}}$, the projection of $\mathbf{b}$ onto Col A. [*Hint:* Find $\hat{\mathbf{x}}$ first. The formula does not require an orthogonal basis for Col A.]

32. Find a formula for the least-squares solution of $A\mathbf{x} = \mathbf{b}$ when the columns of A are orthonormal.

33. Describe all least-squares solutions of the system

$$x + y = 2$$
$$x + y = 4$$

T 34. Example 2 in Section 4.8 displayed a low-pass linear filter that changed a signal $\{y_k\}$ into $\{y_{k+1}\}$ and changed a higher-frequency signal $\{w_k\}$ into the zero signal, where $y_k = \cos(\pi k/4)$ and $w_k = \cos(3\pi k/4)$. The following calculations will design a filter with approximately those properties. The filter equation is

$$a_0 y_{k+2} + a_1 y_{k+1} + a_2 y_k = z_k \qquad \text{for all } k \qquad (8)$$

Because the signals are periodic, with period 8, it suffices to study equation (8) for $k = 0, \dots, 7$. The action on the two signals described above translates into two sets of eight equations, shown below:

$$
\begin{array}{c}
k=0 \\ k=1 \\ \vdots \\ \\ \\ \\ \\ k=7
\end{array}
\begin{array}{ccc}
y_{k+2} & y_{k+1} & y_k \\
\end{array}
\begin{bmatrix}
0 & .7 & 1 \\
-.7 & 0 & .7 \\
-1 & -.7 & 0 \\
-.7 & -1 & -.7 \\
0 & -.7 & -1 \\
.7 & 0 & -.7 \\
1 & .7 & 0 \\
.7 & 1 & .7
\end{bmatrix}
\begin{bmatrix} a_0 \\ a_1 \\ a_2 \end{bmatrix}
=
\begin{array}{c} y_{k+1} \\ \end{array}
\begin{bmatrix}
.7 \\ 0 \\ -.7 \\ -1 \\ -.7 \\ 0 \\ .7 \\ 1
\end{bmatrix}
$$

$$
\begin{array}{c}
k=0 \\ k=1 \\ \vdots \\ \\ \\ \\ \\ k=7
\end{array}
\begin{array}{ccc}
w_{k+2} & w_{k+1} & w_k \\
\end{array}
\begin{bmatrix}
0 & -.7 & 1 \\
.7 & 0 & -.7 \\
-1 & .7 & 0 \\
.7 & -1 & .7 \\
0 & .7 & -1 \\
-.7 & 0 & .7 \\
1 & -.7 & 0 \\
-.7 & 1 & -.7
\end{bmatrix}
\begin{bmatrix} a_0 \\ a_1 \\ a_2 \end{bmatrix}
=
\begin{bmatrix}
0 \\ 0 \\ 0 \\ 0 \\ 0 \\ 0 \\ 0 \\ 0
\end{bmatrix}
$$

Write an equation $A\mathbf{x} = \mathbf{b}$, where A is a 16×3 matrix formed from the two coefficient matrices above and where $\mathbf{b}$ in $\mathbb{R}^{16}$ is formed from the two right sides of the equations. Find a_0, a_1, and a_2 given by the least-squares solution of $A\mathbf{x} = \mathbf{b}$. (The .7 in the data above was used as an approximation for $\sqrt{2}/2$, to illustrate how a typical computation in an applied problem might proceed. If .707 were used instead, the resulting filter coefficients would agree to at least seven decimal places with $\sqrt{2}/4, 1/2$, and $\sqrt{2}/4$, the values produced by exact arithmetic calculations.)

Solutions to Practice Problems

1. First, compute

$$A^T A = \begin{bmatrix} 1 & 1 & 1 \\ -3 & 5 & 7 \\ -3 & 1 & 2 \end{bmatrix} \begin{bmatrix} 1 & -3 & -3 \\ 1 & 5 & 1 \\ 1 & 7 & 2 \end{bmatrix} = \begin{bmatrix} 3 & 9 & 0 \\ 9 & 83 & 28 \\ 0 & 28 & 14 \end{bmatrix}$$

$$A^T \mathbf{b} = \begin{bmatrix} 1 & 1 & 1 \\ -3 & 5 & 7 \\ -3 & 1 & 2 \end{bmatrix} \begin{bmatrix} 5 \\ -3 \\ -5 \end{bmatrix} = \begin{bmatrix} -3 \\ -65 \\ -28 \end{bmatrix}$$

Next, row reduce the augmented matrix for the normal equations, $A^T A\mathbf{x} = A^T \mathbf{b}$:

$$\begin{bmatrix} 3 & 9 & 0 & -3 \\ 9 & 83 & 28 & -65 \\ 0 & 28 & 14 & -28 \end{bmatrix} \sim \begin{bmatrix} 1 & 3 & 0 & -1 \\ 0 & 56 & 28 & -56 \\ 0 & 28 & 14 & -28 \end{bmatrix} \sim \cdots \sim \begin{bmatrix} 1 & 0 & -3/2 & 2 \\ 0 & 1 & 1/2 & -1 \\ 0 & 0 & 0 & 0 \end{bmatrix}$$

The general least-squares solution is $x_1 = 2 + \frac{3}{2}x_3$, $x_2 = -1 - \frac{1}{2}x_3$, with x_3 free. For one specific solution, take $x_3 = 0$ (for example), and get

$$\hat{\mathbf{x}} = \begin{bmatrix} 2 \\ -1 \\ 0 \end{bmatrix}$$

To find the least-squares error, compute

$$\hat{\mathbf{b}} = A\hat{\mathbf{x}} = \begin{bmatrix} 1 & -3 & -3 \\ 1 & 5 & 1 \\ 1 & 7 & 2 \end{bmatrix} \begin{bmatrix} 2 \\ -1 \\ 0 \end{bmatrix} = \begin{bmatrix} 5 \\ -3 \\ -5 \end{bmatrix}$$

It turns out that $\hat{\mathbf{b}} = \mathbf{b}$, so $\|\mathbf{b} - \hat{\mathbf{b}}\| = 0$. The least-squares error is zero because $\mathbf{b}$ happens to be in Col A.

2. If $\mathbf{b}$ is orthogonal to the columns of A, then the projection of $\mathbf{b}$ onto the column space of A is $\mathbf{0}$. In this case, a least-squares solution $\hat{\mathbf{x}}$ of $A\mathbf{x} = \mathbf{b}$ satisfies $A\hat{\mathbf{x}} = \mathbf{0}$.

6.6 Machine Learning and Linear Models

Machine Learning

Machine learning uses linear models in situations where the machine is being *trained* to predict the outcome (dependent variables) based on the values of the inputs (independent variables). The machine is given a set of training data where the values of the independent and dependent variables are known. The machine then *learns* the relationship between the independent variables and the dependent variables. One type of learning is to fit a curve, such as a least-squares line or parabola, to the data. Once the machine has learned the pattern from the training data, it can then estimate the value of the output based on a given value for the input.

Least-Squares Lines

A common task in science and engineering is to analyze and understand relationships among several quantities that vary. This section describes a variety of situations in which data are used to build or verify a formula that predicts the value of one variable as a function of other variables. In each case, the problem will amount to solving a least-squares problem.

For easy application of the discussion to real problems that you may encounter later in your career, we choose notation that is commonly used in the statistical analysis of scientific and engineering data. Instead of $A\mathbf{x} = \mathbf{b}$, we write $X\boldsymbol{\beta} = \mathbf{y}$ and refer to X as the **design matrix**, $\boldsymbol{\beta}$ as the **parameter vector**, and $\mathbf{y}$ as the **observation vector**.

The simplest relation between two variables x and y is the linear equation $y = \beta_0 + \beta_1 x$.[1] Experimental data often produce points $(x_1, y_1), \ldots, (x_n, y_n)$ that, when graphed, seem to lie close to a line. We want to determine the parameters β_0 and β_1 that make the line as "close" to the points as possible.

Suppose β_0 and β_1 are fixed, and consider the line $y = \beta_0 + \beta_1 x$ in Figure 1. Corresponding to each data point (x_j, y_j) there is a point $(x_j, \beta_0 + \beta_1 x_j)$ on the line with the same x-coordinate. We call y_j the *observed* value of y and $\beta_0 + \beta_1 x_j$ the *predicted* y-value (determined by the line). The difference between an observed y-value and a predicted y-value is called a *residual*.

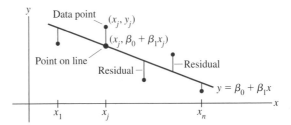

FIGURE 1 Fitting a line to experimental data.

There are several ways to measure how "close" the line is to the data. The usual choice (primarily because the mathematical calculations are simple) is to add the squares of the residuals. The **least-squares line** is the line $y = \beta_0 + \beta_1 x$ that minimizes the sum of the squares of the residuals. This line is also called a **line of regression of y on x**, because any errors in the data are assumed to be only in the y-coordinates. The coefficients β_0, β_1 of the line are called (linear) **regression coefficients**.[2]

If the data points were on the line, the parameters β_0 and β_1 would satisfy the equations

Predicted y-value		Observed y-value
$\beta_0 + \beta_1 x_1$	$=$	y_1
$\beta_0 + \beta_1 x_2$	$=$	y_2
$\vdots$		$\vdots$
$\beta_0 + \beta_1 x_n$	$=$	y_n

---

[1] This notation is commonly used for least-squares lines instead of $y = mx + b$.

[2] If the measurement errors are in x instead of y, simply interchange the coordinates of the data (x_j, y_j) before plotting the points and computing the regression line. If both coordinates are subject to possible error, then you might choose the line that minimizes the sum of the squares of the *orthogonal* (perpendicular) distances from the points to the line.

We can write this system as

$$X\boldsymbol{\beta} = \mathbf{y}, \quad \text{where } X = \begin{bmatrix} 1 & x_1 \\ 1 & x_2 \\ \vdots & \vdots \\ 1 & x_n \end{bmatrix}, \quad \boldsymbol{\beta} = \begin{bmatrix} \beta_0 \\ \beta_1 \end{bmatrix}, \quad \mathbf{y} = \begin{bmatrix} y_1 \\ y_2 \\ \vdots \\ y_n \end{bmatrix} \tag{1}$$

Of course, if the data points don't lie on a line, then there are no parameters β_0, β_1 for which the predicted y-values in $X\boldsymbol{\beta}$ equal the observed y-values in $\mathbf{y}$, and $X\boldsymbol{\beta} = \mathbf{y}$ has no solution. This is a least-squares problem, $A\mathbf{x} = \mathbf{b}$, with different notation!

The square of the distance between the vectors $X\boldsymbol{\beta}$ and $\mathbf{y}$ is precisely the sum of the squares of the residuals. The $\boldsymbol{\beta}$ that minimizes this sum also minimizes the distance between $X\boldsymbol{\beta}$ and $\mathbf{y}$. *Computing the least-squares solution of $X\boldsymbol{\beta} = \mathbf{y}$ is equivalent to finding the $\boldsymbol{\beta}$ that determines the least-squares line in Figure 1.*

EXAMPLE 1 Find the equation $y = \beta_0 + \beta_1 x$ of the least-squares line that best fits the data points $(2, 1)$, $(5, 2)$, $(7, 3)$, and $(8, 3)$.

SOLUTION Use the x-coordinates of the data to build the design matrix X in (1) and the y-coordinates to build the observation vector $\mathbf{y}$:

$$X = \begin{bmatrix} 1 & 2 \\ 1 & 5 \\ 1 & 7 \\ 1 & 8 \end{bmatrix}, \quad \mathbf{y} = \begin{bmatrix} 1 \\ 2 \\ 3 \\ 3 \end{bmatrix}$$

For the least-squares solution of $X\boldsymbol{\beta} = \mathbf{y}$, obtain the normal equations (with the new notation):

$$X^T X \boldsymbol{\beta} = X^T \mathbf{y}$$

That is, compute

$$X^T X = \begin{bmatrix} 1 & 1 & 1 & 1 \\ 2 & 5 & 7 & 8 \end{bmatrix} \begin{bmatrix} 1 & 2 \\ 1 & 5 \\ 1 & 7 \\ 1 & 8 \end{bmatrix} = \begin{bmatrix} 4 & 22 \\ 22 & 142 \end{bmatrix}$$

$$X^T \mathbf{y} = \begin{bmatrix} 1 & 1 & 1 & 1 \\ 2 & 5 & 7 & 8 \end{bmatrix} \begin{bmatrix} 1 \\ 2 \\ 3 \\ 3 \end{bmatrix} = \begin{bmatrix} 9 \\ 57 \end{bmatrix}$$

The normal equations are

$$\begin{bmatrix} 4 & 22 \\ 22 & 142 \end{bmatrix} \begin{bmatrix} \beta_0 \\ \beta_1 \end{bmatrix} = \begin{bmatrix} 9 \\ 57 \end{bmatrix}$$

Hence

$$\begin{bmatrix} \beta_0 \\ \beta_1 \end{bmatrix} = \begin{bmatrix} 4 & 22 \\ 22 & 142 \end{bmatrix}^{-1} \begin{bmatrix} 9 \\ 57 \end{bmatrix} = \frac{1}{84} \begin{bmatrix} 142 & -22 \\ -22 & 4 \end{bmatrix} \begin{bmatrix} 9 \\ 57 \end{bmatrix} = \frac{1}{84} \begin{bmatrix} 24 \\ 30 \end{bmatrix} = \begin{bmatrix} 2/7 \\ 5/14 \end{bmatrix}$$

Thus the least-squares line has the equation

$$y = \frac{2}{7} + \frac{5}{14} x$$

See Figure 2. ∎

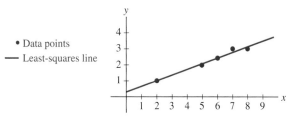

FIGURE 2 The least-squares line $y = \frac{2}{7} + \frac{5}{14}x$.

EXAMPLE 2   If a machine learns the data from Example 1 by creating a least-squares line, what outcome will it predict for the inputs 4 and 6?

SOLUTION   The machine would perform the same calculations as in Example 1 to arrive at the least-squares line

$$y = \frac{2}{7} + \frac{5}{14}x$$

as a reasonable pattern to use to predict the outcomes.

For the value $x = 4$, the machine will predict an output of $y = \frac{2}{7} + \frac{5}{14}(4) = \frac{12}{7}$.

For the value $x = 6$, the machine will predict an output of $y = \frac{2}{7} + \frac{5}{14}(6) = \frac{17}{7}$.
See Figure 3.

Machine Learned Output

FIGURE 3 Machine-learned output. ■

A common practice before computing a least-squares line is to compute the average $\overline{x}$ of the original x-values and form a new variable $x^* = x - \overline{x}$. The new x-data are said to be in **mean-deviation form**. In this case, the two columns of the design matrix will be orthogonal. Solution of the normal equations is simplified, just as in Example 4 in Section 6.5. See Exercises 23 and 24.

The General Linear Model

In some applications, it is necessary to fit data points with something other than a straight line. In the examples that follow, the matrix equation is still $X\boldsymbol{\beta} = \mathbf{y}$, but the specific form of X changes from one problem to the next. Statisticians usually introduce a **residual vector** $\boldsymbol{\epsilon}$, defined by $\boldsymbol{\epsilon} = \mathbf{y} - X\boldsymbol{\beta}$, and write

$$\mathbf{y} = X\boldsymbol{\beta} + \boldsymbol{\epsilon}$$

Any equation of this form is referred to as a **linear model**. Once X and $\mathbf{y}$ are determined, the goal is to minimize the length of $\boldsymbol{\epsilon}$, which amounts to finding a least-squares solution of $X\boldsymbol{\beta} = \mathbf{y}$. In each case, the least-squares solution $\hat{\boldsymbol{\beta}}$ is a solution of the normal equations

$$X^T X \boldsymbol{\beta} = X^T \mathbf{y}$$

Least-Squares Fitting of Other Curves

When data points $(x_1, y_1), \ldots, (x_n, y_n)$ on a scatter plot do not lie close to any line, it may be appropriate to postulate some other functional relationship between x and y.

The next two examples show how to fit data by curves that have the general form

$$y = \beta_0 f_0(x) + \beta_1 f_1(x) + \cdots + \beta_k f_k(x) \tag{2}$$

where $f_0, \ldots, f_k$ are known functions and $\beta_0, \ldots, \beta_k$ are parameters that must be determined. As we will see, equation (2) describes a linear model because it is linear in the unknown parameters.

For a particular value of x, (2) gives a predicted, or "fitted," value of y. The difference between the observed value and the predicted value is the residual. The parameters $\beta_0, \ldots, \beta_k$ must be determined so as to minimize the sum of the squares of the residuals.

FIGURE 4

Average cost curve.

EXAMPLE 3 Suppose data points $(x_1, y_1), \ldots, (x_n, y_n)$ appear to lie along some sort of parabola instead of a straight line. For instance, if the x-coordinate denotes the production level for a company, and y denotes the average cost per unit of operating at a level of x units per day, then a typical average cost curve looks like a parabola that opens upward (Figure 4). In ecology, a parabolic curve that opens downward is used to model the net primary production of nutrients in a plant, as a function of the surface area of the foliage (Figure 5). Suppose we wish to approximate the data by an equation of the form

$$y = \beta_0 + \beta_1 x + \beta_2 x^2 \tag{3}$$

Describe the linear model that produces a "least-squares fit" of the data by equation (3).

SOLUTION Equation (3) describes the ideal relationship. Suppose the actual values of the parameters are $\beta_0, \beta_1, \beta_2$. Then the coordinates of the first data point (x_1, y_1) satisfy an equation of the form

$$y_1 = \beta_0 + \beta_1 x_1 + \beta_2 x_1^2 + \epsilon_1$$

FIGURE 5

Production of nutrients.

where ϵ_1 is the residual error between the observed value y_1 and the predicted y-value $\beta_0 + \beta_1 x_1 + \beta_2 x_1^2$. Each data point determines a similar equation:

$$y_1 = \beta_0 + \beta_1 x_1 + \beta_2 x_1^2 + \epsilon_1$$
$$y_2 = \beta_0 + \beta_1 x_2 + \beta_2 x_2^2 + \epsilon_2$$
$$\vdots \qquad\qquad \vdots$$
$$y_n = \beta_0 + \beta_1 x_n + \beta_2 x_n^2 + \epsilon_n$$

It is a simple matter to write this system of equations in the form $\mathbf{y} = X\boldsymbol{\beta} + \boldsymbol{\epsilon}$. To find X, inspect the first few rows of the system and look for the pattern.

$$
\begin{bmatrix} y_1 \\ y_2 \\ \vdots \\ y_n \end{bmatrix} = \begin{bmatrix} 1 & x_1 & x_1^2 \\ 1 & x_2 & x_2^2 \\ \vdots & \vdots & \vdots \\ 1 & x_n & x_n^2 \end{bmatrix} \begin{bmatrix} \beta_0 \\ \beta_1 \\ \beta_2 \end{bmatrix} + \begin{bmatrix} \epsilon_1 \\ \epsilon_2 \\ \vdots \\ \epsilon_n \end{bmatrix}
$$
$$
\mathbf{y} \quad\quad = \quad\quad X \quad\quad \boldsymbol{\beta} \; + \; \boldsymbol{\epsilon}
$$

∎

EXAMPLE 4 If data points tend to follow a pattern such as in Figure 6, then an appropriate model might be an equation of the form

$$
y = \beta_0 + \beta_1 x + \beta_2 x^2 + \beta_3 x^3
$$

Such data, for instance, could come from a company's total costs, as a function of the level of production. Describe the linear model that gives a least-squares fit of this type to data $(x_1, y_1), \ldots, (x_n, y_n)$.

SOLUTION By an analysis similar to that in Example 2, we obtain

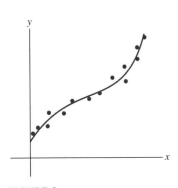

FIGURE 6

Data points along a cubic curve.

| Observation vector | Design matrix | Parameter vector | Residual vector |

$$
\mathbf{y} = \begin{bmatrix} y_1 \\ y_2 \\ \vdots \\ y_n \end{bmatrix}, \quad X = \begin{bmatrix} 1 & x_1 & x_1^2 & x_1^3 \\ 1 & x_2 & x_2^2 & x_2^3 \\ \vdots & \vdots & \vdots & \vdots \\ 1 & x_n & x_n^2 & x_n^3 \end{bmatrix}, \quad \boldsymbol{\beta} = \begin{bmatrix} \beta_0 \\ \beta_1 \\ \beta_2 \\ \beta_3 \end{bmatrix}, \quad \boldsymbol{\epsilon} = \begin{bmatrix} \epsilon_1 \\ \epsilon_2 \\ \vdots \\ \epsilon_n \end{bmatrix}
$$

∎

Multiple Regression

Suppose an experiment involves two independent variables—say, u and v—and one dependent variable, y. A simple equation for predicting y from u and v has the form

$$
y = \beta_0 + \beta_1 u + \beta_2 v \tag{4}
$$

A more general prediction equation might have the form

$$
y = \beta_0 + \beta_1 u + \beta_2 v + \beta_3 u^2 + \beta_4 uv + \beta_5 v^2 \tag{5}
$$

This equation is used in geology, for instance, to model erosion surfaces, glacial cirques, soil pH, and other quantities. In such cases, the least-squares fit is called a *trend surface*.

Equations (4) and (5) both lead to a linear model because they are linear in the unknown parameters (even though u and v are multiplied). In general, a linear model will arise whenever y is to be predicted by an equation of the form

$$
y = \beta_0 f_0(u, v) + \beta_1 f_1(u, v) + \cdots + \beta_k f_k(u, v)
$$

with $f_0, \ldots, f_k$ any sort of known functions and $\beta_0, \ldots, \beta_k$ unknown weights.

EXAMPLE 5 In geography, local models of terrain are constructed from data $(u_1, v_1, y_1), \ldots, (u_n, v_n, y_n)$, where u_j, v_j, and y_j are latitude, longitude, and altitude, respectively. Describe the linear model based on (4) that gives a least-squares fit to such data. The solution is called the *least-squares plane*. See Figure 7.

FIGURE 7 A least-squares plane.

SOLUTION We expect the data to satisfy the following equations:

$$y_1 = \beta_0 + \beta_1 u_1 + \beta_2 v_1 + \epsilon_1$$
$$y_2 = \beta_0 + \beta_1 u_2 + \beta_2 v_2 + \epsilon_2$$
$$\vdots \qquad\qquad \vdots$$
$$y_n = \beta_0 + \beta_1 u_n + \beta_2 v_n + \epsilon_n$$

This system has the matrix form $\mathbf{y} = X\boldsymbol{\beta} + \boldsymbol{\epsilon}$, where

$$
\underset{\substack{\text{Observation}\\\text{vector}}}{\mathbf{y} = \begin{bmatrix} y_1 \\ y_2 \\ \vdots \\ y_n \end{bmatrix}}, \quad
\underset{\substack{\text{Design}\\\text{matrix}}}{X = \begin{bmatrix} 1 & u_1 & v_1 \\ 1 & u_2 & v_2 \\ \vdots & \vdots & \vdots \\ 1 & u_n & v_n \end{bmatrix}}, \quad
\underset{\substack{\text{Parameter}\\\text{vector}}}{\boldsymbol{\beta} = \begin{bmatrix} \beta_0 \\ \beta_1 \\ \beta_2 \end{bmatrix}}, \quad
\underset{\substack{\text{Residual}\\\text{vector}}}{\boldsymbol{\epsilon} = \begin{bmatrix} \epsilon_1 \\ \epsilon_2 \\ \vdots \\ \epsilon_n \end{bmatrix}}
$$

∎

Example 5 shows that the linear model for multiple regression has the same abstract form as the model for the simple regression in the earlier examples. Linear algebra gives us the power to understand the general principle behind all the linear models. Once X is defined properly, the normal equations for $\boldsymbol{\beta}$ have the same matrix form, no matter how many variables are involved. Thus, for any linear model where $X^T X$ is invertible, the least-squares $\hat{\boldsymbol{\beta}}$ is given by $(X^T X)^{-1} X^T \mathbf{y}$.

STUDY GUIDE offers additional resources for understanding the geometry of a linear model.

Practice Problem

When the monthly sales of a product are subject to seasonal fluctuations, a curve that approximates the sales data might have the form

$$y = \beta_0 + \beta_1 x + \beta_2 \sin(2\pi x / 12)$$

where x is the time in months. The term $\beta_0 + \beta_1 x$ gives the basic sales trend, and the sine term reflects the seasonal changes in sales. Give the design matrix and the parameter vector for the linear model that leads to a least-squares fit of the equation above. Assume the data are $(x_1, y_1), \ldots, (x_n, y_n)$.

6.6 Exercises

In Exercises 1–4, find the equation $y = \beta_0 + \beta_1 x$ of the least-squares line that best fits the given data points.

1. $(0, 1), (1, 1), (2, 2), (3, 2)$

2. $(1, 0), (2, 1), (4, 2), (5, 3)$

3. $(-1, 0), (0, 1), (1, 2), (2, 4)$

4. $(2, 3), (3, 2), (5, 1), (6, 0)$

5. If a machine learns the least-squares line that best fits the data in Exercise 1, what will the machine pick for the value of y when $x = 4$?

6. If a machine learns the least-squares line that best fits the data in Exercise 2, what will the machine pick for the value of y when $x = 3$?

7. If a machine learns the least-squares line that best fits the data in Exercise 1, what will the machine pick for the value of y when $x = 3$? How closely does this match the data point at $x = 3$ fed into the machine?

8. If a machine learns the least-squares line that best fits the data in Exercise 2, what will the machine pick for the value of y when $x = 2$? How closely does this match the data point at $x = 2$ fed into the machine?

9. If you enter the data from Exercise 1 into a machine and it returns a y value of 20 when $x = 2.5$, should you trust the machine? Justify your answer.

10. If you enter the data from Exercise 2 into a machine and it returns a y value of -4 when $x = 2.5$, should you trust the machine? Justify your answer.

11. Let X be the design matrix used to find the least-squares line to fit data $(x_1, y_1), \ldots, (x_n, y_n)$. Use a theorem in Section 6.5 to show that the normal equations have a unique solution if and only if the data include at least two data points with different x-coordinates.

12. Let X be the design matrix in Example 2 corresponding to a least-squares fit of a parabola to data $(x_1, y_1), \ldots, (x_n, y_n)$. Suppose x_1, x_2, and x_3 are distinct. Explain why there is only one parabola that fits the data best, in a least-squares sense. (See Exercise 11.)

13. A certain experiment produces the data $(1, 1.8)$, $(2, 2.7)$, $(3, 3.4)$, $(4, 3.8)$, $(5, 3.9)$. Describe the model that produces a least-squares fit of these points by a function of the form

    $$y = \beta_1 x + \beta_2 x^2$$

    Such a function might arise, for example, as the revenue from the sale of x units of a product, when the amount offered for sale affects the price to be set for the product.

    a. Give the design matrix, the observation vector, and the unknown parameter vector.

    T b. Find the associated least-squares curve for the data.

c. If a machine learned the curve you found in (b), what output would it provide for an input of $x = 6$?

14. A simple curve that often makes a good model for the variable costs of a company, as a function of the sales level x, has the form $y = \beta_1 x + \beta_2 x^2 + \beta_3 x^3$. There is no constant term because fixed costs are not included.

    a. Give the design matrix and the parameter vector for the linear model that leads to a least-squares fit of the equation above, with data $(x_1, y_1), \ldots, (x_n, y_n)$.

    T b. Find the least-squares curve of the form above to fit the data $(4, 1.58)$, $(6, 2.08)$, $(8, 2.5)$, $(10, 2.8)$, $(12, 3.1)$, $(14, 3.4)$, $(16, 3.8)$, and $(18, 4.32)$, with values in thousands. If possible, produce a graph that shows the data points and the graph of the cubic approximation.

    c. If a machine learned the curve you found in (b), what output would it provide for an input of $x = 9$?

15. A certain experiment produces the data $(1, 7.9)$, $(2, 5.4)$, and $(3, -.9)$. Describe the model that produces a least-squares fit of these points by a function of the form

    $$y = A \cos x + B \sin x$$

16. Suppose radioactive substances A and B have decay constants of .02 and .07, respectively. If a mixture of these two substances at time $t = 0$ contains M_A grams of A and M_B grams of B, then a model for the total amount y of the mixture present at time t is

    $$y = M_A e^{-.02t} + M_B e^{-.07t} \qquad (6)$$

    Suppose the initial amounts M_A and M_B are unknown, but a scientist is able to measure the total amounts present at several times and records the following points (t_i, y_i): $(10, 21.34)$, $(11, 20.68)$, $(12, 20.05)$, $(14, 18.87)$, and $(15, 18.30)$.

    a. Describe a linear model that can be used to estimate M_A and M_B.

    T b. Find the least-squares curve based on (6).

Halley's Comet last appeared in 1986 and will reappear in 2061.

T 17. According to Kepler's first law, a comet should have an elliptic, parabolic, or hyperbolic orbit (with gravitational attractions from the planets ignored). In suitable polar coordinates, the position (r, ϑ) of a comet satisfies an equation of the form

$$r = \beta + e(r \cdot \cos \vartheta)$$

where β is a constant and e is the *eccentricity* of the orbit, with $0 \le e < 1$ for an ellipse, $e = 1$ for a parabola, and $e > 1$ for a hyperbola. Suppose observations of a newly discovered comet provide the data below. Determine the type of orbit, and predict where the comet will be when $\vartheta = 4.6$ (radians).[3]

ϑ	.88	1.10	1.42	1.77	2.14
r	3.00	2.30	1.65	1.25	1.01

T 18. A healthy child's systolic blood pressure p (in millimeters of mercury) and weight w (in pounds) are approximately related by the equation

$$\beta_0 + \beta_1 \ln w = p$$

Use the following experimental data to estimate the systolic blood pressure of a healthy child weighing 100 pounds.

w	44	61	81	113	131
$\ln w$	3.78	4.11	4.39	4.73	4.88
p	91	98	103	110	112

T 19. To measure the takeoff performance of an airplane, the horizontal position of the plane was measured every second, from $t = 0$ to $t = 12$. The positions (in feet) were: 0, 8.8, 29.9, 62.0, 104.7, 159.1, 222.0, 294.5, 380.4, 471.1, 571.7, 686.8, and 809.2.

   a. Find the least-squares cubic curve $y = \beta_0 + \beta_1 t + \beta_2 t^2 + \beta_3 t^3$ for these data.

   b. If a machine learned the curve given in part (a), what would it estimate the velocity of the plane to be when $t = 4.5$ seconds?

20. Let $\bar{x} = \dfrac{1}{n}(x_1 + \cdots + x_n)$ and $\bar{y} = \dfrac{1}{n}(y_1 + \cdots + y_n)$. Show that the least-squares line for the data $(x_1, y_1), \ldots, (x_n, y_n)$ must pass through $(\bar{x}, \bar{y})$. That is, show that $\bar{x}$ and $\bar{y}$ satisfy the linear equation $\bar{y} = \hat{\beta}_0 + \hat{\beta}_1 \bar{x}$. [*Hint:* Derive this equation from the vector equation $\mathbf{y} = X\hat{\boldsymbol{\beta}} + \boldsymbol{\epsilon}$. Denote the first column of X by $\mathbf{1}$. Use the fact that the residual vector $\boldsymbol{\epsilon}$ is orthogonal to the column space of X and hence is orthogonal to $\mathbf{1}$.]

Given data for a least-squares problem, $(x_1, y_1), \ldots, (x_n, y_n)$, the following abbreviations are helpful:

$$\sum x = \sum_{i=1}^n x_i, \quad \sum x^2 = \sum_{i=1}^n x_i^2,$$
$$\sum y = \sum_{i=1}^n y_i, \quad \sum xy = \sum_{i=1}^n x_i y_i$$

The normal equations for a least-squares line $y = \hat{\beta}_0 + \hat{\beta}_1 x$ may be written in the form

$$\begin{aligned} n\hat{\beta}_0 + \hat{\beta}_1 \sum x &= \sum y \\ \hat{\beta}_0 \sum x + \hat{\beta}_1 \sum x^2 &= \sum xy \end{aligned} \tag{7}$$

21. Derive the normal equations (7) from the matrix form given in this section.

22. Use a matrix inverse to solve the system of equations in (7) and thereby obtain formulas for $\hat{\beta}_0$ and $\hat{\beta}_1$ that appear in many statistics texts.

23. a. Rewrite the data in Example 1 with new x-coordinates in mean deviation form. Let X be the associated design matrix. Why are the columns of X orthogonal?

   b. Write the normal equations for the data in part (a), and solve them to find the least-squares line, $y = \beta_0 + \beta_1 x^*$, where $x^* = x - 5.5$.

24. Suppose the x-coordinates of the data $(x_1, y_1), \ldots, (x_n, y_n)$ are in mean deviation form, so that $\sum x_i = 0$. Show that if X is the design matrix for the least-squares line in this case, then $X^T X$ is a diagonal matrix.

Exercises 25 and 26 involve a design matrix X with two or more columns and a least-squares solution $\hat{\boldsymbol{\beta}}$ of $\mathbf{y} = X\boldsymbol{\beta}$. Consider the following numbers.

  (i) $\|X\hat{\boldsymbol{\beta}}\|^2$—the sum of the squares of the "regression term." Denote this number by SS(R).

  (ii) $\|\mathbf{y} - X\hat{\boldsymbol{\beta}}\|^2$—the sum of the squares for the error term. Denote this number by SS(E).

  (iii) $\|\mathbf{y}\|^2$—the "total" sum of the squares of the y-values. Denote this number by SS(T).

Every statistics text that discusses regression and the linear model $\mathbf{y} = X\boldsymbol{\beta} + \boldsymbol{\epsilon}$ introduces these numbers, though terminology and notation vary somewhat. To simplify matters, assume that the mean of the y-values is zero. In this case, SS(T) is proportional to what is called the *variance* of the set of y-values.

25. Justify the equation SS(T) = SS(R) + SS(E). [*Hint:* Use a theorem, and explain why the hypotheses of the theorem are satisfied.] This equation is extremely important in statistics, both in regression theory and in the analysis of variance.

26. Show that $\|X\hat{\boldsymbol{\beta}}\|^2 = \hat{\boldsymbol{\beta}}^T X^T \mathbf{y}$. [*Hint:* Rewrite the left side and use the fact that $\hat{\boldsymbol{\beta}}$ satisfies the normal equations.] This formula for SS(R) is used in statistics. From this and from Exercise 25, obtain the standard formula for SS(E):

$$\text{SS(E)} = \mathbf{y}^T \mathbf{y} - \hat{\boldsymbol{\beta}}^T X^T \mathbf{y}$$

---

[3] The basic idea of least-squares fitting of data is due to K. F. Gauss (and, independently, to A. Legendre), whose initial rise to fame occurred in 1801 when he used the method to determine the path of the asteroid *Ceres*. Forty days after the asteroid was discovered, it disappeared behind the sun. Gauss predicted it would appear ten months later and gave its location. The accuracy of the prediction astonished the European scientific community.

Sales trend with seasonal fluctuations.

Solution to Practice Problem

Construct X and $\boldsymbol{\beta}$ so that the kth row of $X\boldsymbol{\beta}$ is the predicted y-value that corresponds to the data point (x_k, y_k), namely

$$\beta_0 + \beta_1 x_k + \beta_2 \sin(2\pi x_k/12)$$

It should be clear that

$$X = \begin{bmatrix} 1 & x_1 & \sin(2\pi x_1/12) \\ \vdots & \vdots & \vdots \\ 1 & x_n & \sin(2\pi x_n/12) \end{bmatrix}, \quad \boldsymbol{\beta} = \begin{bmatrix} \beta_0 \\ \beta_1 \\ \beta_2 \end{bmatrix}$$

6.7 Inner Product Spaces

Notions of length, distance, and orthogonality are often important in applications involving a vector space. For $\mathbb{R}^n$, these concepts were based on the properties of the inner product listed in Theorem 1 of Section 6.1. For other spaces, we need analogues of the inner product with the same properties. The conclusions of Theorem 1 now become *axioms* in the following definition.

DEFINITION

An **inner product** on a vector space V is a function that, to each pair of vectors **u** and **v** in V, associates a real number $\langle \mathbf{u}, \mathbf{v} \rangle$ and satisfies the following axioms, for all **u**, **v**, and **w** in V and all scalars c:

1. $\langle \mathbf{u}, \mathbf{v} \rangle = \langle \mathbf{v}, \mathbf{u} \rangle$
2. $\langle \mathbf{u} + \mathbf{v}, \mathbf{w} \rangle = \langle \mathbf{u}, \mathbf{w} \rangle + \langle \mathbf{v}, \mathbf{w} \rangle$
3. $\langle c\mathbf{u}, \mathbf{v} \rangle = c \langle \mathbf{u}, \mathbf{v} \rangle$
4. $\langle \mathbf{u}, \mathbf{u} \rangle \geq 0$ and $\langle \mathbf{u}, \mathbf{u} \rangle = 0$ if and only if $\mathbf{u} = \mathbf{0}$

A vector space with an inner product is called an **inner product space**.

The vector space $\mathbb{R}^n$ with the standard inner product is an inner product space, and nearly everything discussed in this chapter for $\mathbb{R}^n$ carries over to inner product spaces. The examples in this section and the next lay the foundation for a variety of applications treated in courses in engineering, physics, mathematics, and statistics.

EXAMPLE 1 Fix any two positive numbers—say, 4 and 5—and for vectors $\mathbf{u} = (u_1, u_2)$ and $\mathbf{v} = (v_1, v_2)$ in $\mathbb{R}^2$, set

$$\langle \mathbf{u}, \mathbf{v} \rangle = 4u_1v_1 + 5u_2v_2 \tag{1}$$

Show that equation (1) defines an inner product.

SOLUTION Certainly Axiom 1 is satisfied, because $\langle \mathbf{u}, \mathbf{v} \rangle = 4u_1v_1 + 5u_2v_2 = 4v_1u_1 + 5v_2u_2 = \langle \mathbf{v}, \mathbf{u} \rangle$. If $\mathbf{w} = (w_1, w_2)$, then

$$\begin{aligned} \langle \mathbf{u} + \mathbf{v}, \mathbf{w} \rangle &= 4(u_1 + v_1)w_1 + 5(u_2 + v_2)w_2 \\ &= 4u_1w_1 + 5u_2w_2 + 4v_1w_1 + 5v_2w_2 \\ &= \langle \mathbf{u}, \mathbf{w} \rangle + \langle \mathbf{v}, \mathbf{w} \rangle \end{aligned}$$

This verifies Axiom 2. For Axiom 3, compute

$$\langle c\mathbf{u}, \mathbf{v} \rangle = 4(cu_1)v_1 + 5(cu_2)v_2 = c(4u_1v_1 + 5u_2v_2) = c\langle \mathbf{u}, \mathbf{v} \rangle$$

For Axiom 4, note that $\langle \mathbf{u}, \mathbf{u} \rangle = 4u_1^2 + 5u_2^2 \geq 0$, and $4u_1^2 + 5u_2^2 = 0$ only if $u_1 = u_2 = 0$, that is, if $\mathbf{u} = \mathbf{0}$. Also, $\langle \mathbf{0}, \mathbf{0} \rangle = 0$. So (1) defines an inner product on $\mathbb{R}^2$. ∎

Inner products similar to (1) can be defined on $\mathbb{R}^n$. They arise naturally in connection with "weighted least-squares" problems, in which weights are assigned to the various entries in the sum for the inner product in such a way that more importance is given to the more reliable measurements.

From now on, when an inner product space involves polynomials or other functions, we will write the functions in the familiar way, rather than use the boldface type for vectors. Nevertheless, it is important to remember that each function *is* a vector when it is treated as an element of a vector space.

EXAMPLE 2 Let $t_0, \ldots, t_n$ be distinct real numbers. For p and q in $\mathbb{P}_n$, define

$$\langle p, q \rangle = p(t_0)q(t_0) + p(t_1)q(t_1) + \cdots + p(t_n)q(t_n) \tag{2}$$

Inner product Axioms 1–3 are readily checked. For Axiom 4, note that

$$\langle p, p \rangle = [p(t_0)]^2 + [p(t_1)]^2 + \cdots + [p(t_n)]^2 \geq 0$$

Also, $\langle \mathbf{0}, \mathbf{0} \rangle = 0$. (The boldface zero here denotes the zero polynomial, the zero vector in $\mathbb{P}_n$.) If $\langle p, p \rangle = 0$, then p must vanish at $n + 1$ points: $t_0, \ldots, t_n$. This is possible only if p is the zero polynomial, because the degree of p is less than $n + 1$. Thus (2) defines an inner product on $\mathbb{P}_n$. ∎

EXAMPLE 3 Let V be $\mathbb{P}_2$, with the inner product from Example 2, where $t_0 = 0$, $t_1 = \frac{1}{2}$, and $t_2 = 1$. Let $p(t) = 12t^2$ and $q(t) = 2t - 1$. Compute $\langle p, q \rangle$ and $\langle q, q \rangle$.

SOLUTION

$$\begin{aligned} \langle p, q \rangle &= p(0)q(0) + p\left(\tfrac{1}{2}\right) q\left(\tfrac{1}{2}\right) + p(1)q(1) \\ &= (0)(-1) + (3)(0) + (12)(1) = 12 \\ \langle q, q \rangle &= [q(0)]^2 + [q\left(\tfrac{1}{2}\right)]^2 + [q(1)]^2 \\ &= (-1)^2 + (0)^2 + (1)^2 = 2 \end{aligned}$$

∎

Lengths, Distances, and Orthogonality

Let V be an inner product space, with the inner product denoted by $\langle \mathbf{u}, \mathbf{v} \rangle$. Just as in $\mathbb{R}^n$, we define the **length**, or **norm**, of a vector $\mathbf{v}$ to be the scalar

$$\|\mathbf{v}\| = \sqrt{\langle \mathbf{v}, \mathbf{v} \rangle}$$

Equivalently, $\|\mathbf{v}\|^2 = \langle \mathbf{v}, \mathbf{v} \rangle$. (This definition makes sense because $\langle \mathbf{v}, \mathbf{v} \rangle \geq 0$, but the definition *does not* say that $\langle \mathbf{v}, \mathbf{v} \rangle$ is a "sum of squares," because $\mathbf{v}$ need not be an element of $\mathbb{R}^n$.)

A **unit vector** is one whose length is 1. The **distance between u and v** is $\|\mathbf{u} - \mathbf{v}\|$. Vectors $\mathbf{u}$ and $\mathbf{v}$ are **orthogonal** if $\langle \mathbf{u}, \mathbf{v} \rangle = 0$.

EXAMPLE 4 Let $\mathbb{P}_2$ have the inner product (2) of Example 3. Compute the lengths of the vectors $p(t) = 12t^2$ and $q(t) = 2t - 1$.

SOLUTION

$$\|p\|^2 = \langle p, p \rangle = [p(0)]^2 + \left[p \left(\tfrac{1}{2} \right) \right]^2 + [p(1)]^2$$
$$= 0 + [3]^2 + [12]^2 = 153$$
$$\|p\| = \sqrt{153}$$

From Example 3, $\langle q, q \rangle = 2$. Hence $\|q\| = \sqrt{2}$. ∎

The Gram–Schmidt Process

The existence of orthogonal bases for finite-dimensional subspaces of an inner product space can be established by the Gram–Schmidt process, just as in $\mathbb{R}^n$. Certain orthogonal bases that arise frequently in applications can be constructed by this process.

The orthogonal projection of a vector onto a subspace W with an orthogonal basis can be constructed as usual. The projection does not depend on the choice of orthogonal basis, and it has the properties described in the Orthogonal Decomposition Theorem and the Best Approximation Theorem.

EXAMPLE 5 Let V be $\mathbb{P}_4$ with the inner product in Example 2, involving evaluation of polynomials at -2, -1, 0, 1, and 2, and view $\mathbb{P}_2$ as a subspace of V. Produce an orthogonal basis for $\mathbb{P}_2$ by applying the Gram–Schmidt process to the polynomials 1, t, and t^2.

SOLUTION The inner product depends only on the values of a polynomial at $-2, \ldots, 2$, so we list the values of each polynomial as a vector in $\mathbb{R}^5$, underneath the name of the polynomial:[1]

Polynomial:	1	t	t^2
Vector of values:	$\begin{bmatrix} 1 \\ 1 \\ 1 \\ 1 \\ 1 \end{bmatrix}$,	$\begin{bmatrix} -2 \\ -1 \\ 0 \\ 1 \\ 2 \end{bmatrix}$,	$\begin{bmatrix} 4 \\ 1 \\ 0 \\ 1 \\ 4 \end{bmatrix}$

The inner product of two polynomials in V equals the (standard) inner product of their corresponding vectors in $\mathbb{R}^5$. Observe that t is orthogonal to the constant function 1. So take $p_0(t) = 1$ and $p_1(t) = t$. For p_2, use the vectors in $\mathbb{R}^5$ to compute the projection of t^2 onto Span $\{p_0, p_1\}$:

$$\langle t^2, p_0 \rangle = \langle t^2, 1 \rangle = 4 + 1 + 0 + 1 + 4 = 10$$
$$\langle p_0, p_0 \rangle = 5$$
$$\langle t^2, p_1 \rangle = \langle t^2, t \rangle = -8 + (-1) + 0 + 1 + 8 = 0$$

The orthogonal projection of t^2 onto Span $\{1, t\}$ is $\frac{10}{5} p_0 + 0 p_1$. Thus

$$p_2(t) = t^2 - 2 p_0(t) = t^2 - 2$$

---

[1] Each polynomial in $\mathbb{P}_4$ is uniquely determined by its value at the five numbers $-2, \ldots, 2$. In fact, the correspondence between p and its vector of values is an isomorphism, that is, a one-to-one mapping onto $\mathbb{R}^5$ that preserves linear combinations.

An orthogonal basis for the subspace $\mathbb{P}_2$ of V is

$$
\begin{array}{cccc}
\text{Polynomial} & p_0 & p_1 & p_2 \\[4pt]
\text{Vector of values} &
\begin{bmatrix} 1 \\ 1 \\ 1 \\ 1 \\ 1 \end{bmatrix}, &
\begin{bmatrix} -2 \\ -1 \\ 0 \\ 1 \\ 2 \end{bmatrix}, &
\begin{bmatrix} 2 \\ -1 \\ -2 \\ -1 \\ 2 \end{bmatrix}
\end{array}
\tag{3}
$$

∎

Best Approximation in Inner Product Spaces

A common problem in applied mathematics involves a vector space V whose elements are functions. The problem is to approximate a function f in V by a function g from a specified subspace W of V. The "closeness" of the approximation of f depends on the way $\|f - g\|$ is defined. We will consider only the case in which the distance between f and g is determined by an inner product. In this case, the *best approximation to f by functions in W* is the orthogonal projection of f onto the subspace W.

EXAMPLE 6 Let V be $\mathbb{P}_4$ with the inner product in Example 5, and let p_0, p_1, and p_2 be the orthogonal basis found in Example 5 for the subspace $\mathbb{P}_2$. Find the best approximation to $p(t) = 5 - \frac{1}{2}t^4$ by polynomials in $\mathbb{P}_2$.

SOLUTION The values of p_0, p_1, and p_2 at the numbers $-2, -1, 0, 1$, and 2 are listed in $\mathbb{R}^5$ vectors in (3) above. The corresponding values for p are $-3, 9/2, 5, 9/2$, and -3. Compute

$$
\langle p, p_0 \rangle = 8, \qquad \langle p, p_1 \rangle = 0, \qquad \langle p, p_2 \rangle = -31
$$
$$
\langle p_0, p_0 \rangle = 5, \qquad\qquad\qquad\qquad \langle p_2, p_2 \rangle = 14
$$

Then the best approximation in V to p by polynomials in $\mathbb{P}_2$ is

$$
\hat{p} = \text{proj}_{\mathbb{P}_2} p = \frac{\langle p, p_0 \rangle}{\langle p_0, p_0 \rangle} p_0 + \frac{\langle p, p_1 \rangle}{\langle p_1, p_1 \rangle} p_1 + \frac{\langle p, p_2 \rangle}{\langle p_2, p_2 \rangle} p_2
$$
$$
= \tfrac{8}{5} p_0 + \tfrac{-31}{14} p_2 = \tfrac{8}{5} - \tfrac{31}{14}(t^2 - 2).
$$

This polynomial is the closest to p of all polynomials in $\mathbb{P}_2$, when the distance between polynomials is measured only at $-2, -1, 0, 1$, and 2. See Figure 1. ∎

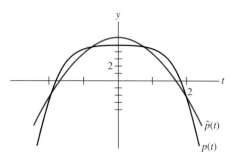

FIGURE 1

The polynomials p_0, p_1, and p_2 in Examples 5 and 6 belong to a class of polynomials that are referred to in statistics as *orthogonal polynomials*.[2] The orthogonality refers to the type of inner product described in Example 2.

Two Inequalities

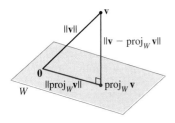

FIGURE 2
The hypotenuse is the longest side.

Given a vector $\mathbf{v}$ in an inner product space V and given a finite-dimensional subspace W, we may apply the Pythagorean Theorem to the orthogonal decomposition of $\mathbf{v}$ with respect to W and obtain

$$\|\mathbf{v}\|^2 = \|\operatorname{proj}_W \mathbf{v}\|^2 + \|\mathbf{v} - \operatorname{proj}_W \mathbf{v}\|^2$$

See Figure 2. In particular, this shows that the norm of the projection of $\mathbf{v}$ onto W does not exceed the norm of $\mathbf{v}$ itself. This simple observation leads to the following important inequality.

THEOREM 16

The Cauchy–Schwarz Inequality

For all $\mathbf{u}, \mathbf{v}$ in V,

$$|\langle \mathbf{u}, \mathbf{v} \rangle| \le \|\mathbf{u}\|\,\|\mathbf{v}\| \tag{4}$$

PROOF   If $\mathbf{u} = \mathbf{0}$, then both sides of (4) are zero, and hence the inequality is true in this case. (See Practice Problem 1.) If $\mathbf{u} \ne \mathbf{0}$, let W be the subspace spanned by $\mathbf{u}$. Recall that $\|c\mathbf{u}\| = |c|\,\|\mathbf{u}\|$ for any scalar c. Thus

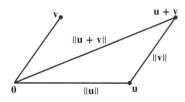

FIGURE 3
The lengths of the sides of a triangle.

$$\|\operatorname{proj}_W \mathbf{v}\| = \left\| \frac{\langle \mathbf{v}, \mathbf{u} \rangle}{\langle \mathbf{u}, \mathbf{u} \rangle} \mathbf{u} \right\| = \frac{|\langle \mathbf{v}, \mathbf{u} \rangle|}{|\langle \mathbf{u}, \mathbf{u} \rangle|}\|\mathbf{u}\| = \frac{|\langle \mathbf{v}, \mathbf{u} \rangle|}{\|\mathbf{u}\|^2}\|\mathbf{u}\| = \frac{|\langle \mathbf{u}, \mathbf{v} \rangle|}{\|\mathbf{u}\|}$$

Since $\|\operatorname{proj}_W \mathbf{v}\| \le \|\mathbf{v}\|$, we have $\dfrac{|\langle \mathbf{u}, \mathbf{v} \rangle|}{\|\mathbf{u}\|} \le \|\mathbf{v}\|$, which gives (4).   ∎

The Cauchy–Schwarz inequality is useful in many branches of mathematics. A few simple applications are presented in the exercises. Our main need for this inequality here is to prove another fundamental inequality involving norms of vectors. See Figure 3.

THEOREM 17

The Triangle Inequality

For all $\mathbf{u}, \mathbf{v}$ in V,

$$\|\mathbf{u} + \mathbf{v}\| \le \|\mathbf{u}\| + \|\mathbf{v}\|$$

PROOF

$$\begin{aligned}
\|\mathbf{u} + \mathbf{v}\|^2 &= \langle \mathbf{u} + \mathbf{v}, \mathbf{u} + \mathbf{v} \rangle = \langle \mathbf{u}, \mathbf{u} \rangle + 2\langle \mathbf{u}, \mathbf{v} \rangle + \langle \mathbf{v}, \mathbf{v} \rangle \\
&\le \|\mathbf{u}\|^2 + 2|\langle \mathbf{u}, \mathbf{v} \rangle| + \|\mathbf{v}\|^2 \\
&\le \|\mathbf{u}\|^2 + 2\|\mathbf{u}\|\,\|\mathbf{v}\| + \|\mathbf{v}\|^2 \qquad \text{Cauchy–Schwarz} \\
&= (\|\mathbf{u}\| + \|\mathbf{v}\|)^2
\end{aligned}$$

The triangle inequality follows immediately by taking square roots of both sides.   ∎

---

[2] See *Statistics and Experimental Design in Engineering and the Physical Sciences*, 2nd ed., by Norman L. Johnson and Fred C. Leone (New York: John Wiley & Sons, 1977). Tables there list "Orthogonal Polynomials," which are simply the values of the polynomial at numbers such as -2, -1, 0, 1, and 2.

An Inner Product for $C[a, b]$ (Calculus required)

Probably the most widely used inner product space for applications is the vector space $C[a, b]$ of all continuous functions on an interval $a \le t \le b$, with an inner product that we will describe.

We begin by considering a polynomial p and any integer n larger than or equal to the degree of p. Then p is in $\mathbb{P}_n$, and we may compute a "length" for p using the inner product of Example 2 involving evaluation at $n + 1$ points in $[a, b]$. However, this length of p captures the behavior at only those $n + 1$ points. Since p is in $\mathbb{P}_n$ for all large n, we could use a much larger n, with many more points for the "evaluation" inner product. See Figure 4.

FIGURE 4 Using different numbers of evaluation points in $[a, b]$ to compute $\|p\|^2$.

Let us partition $[a, b]$ into $n + 1$ subintervals of length $\Delta t = (b - a)/(n + 1)$, and let $t_0, \ldots, t_n$ be arbitrary points in these subintervals.

If n is large, the inner product on $\mathbb{P}_n$ determined by $t_0, \ldots, t_n$ will tend to give a large value to $\langle p, p \rangle$, so we scale it down and divide by $n + 1$. Observe that $1/(n + 1) = \Delta t/(b - a)$, and define

$$\langle p, q \rangle = \frac{1}{n+1} \sum_{j=0}^{n} p(t_j)q(t_j) = \frac{1}{b-a} \left[\sum_{j=0}^{n} p(t_j)q(t_j)\Delta t \right]$$

Now, let n increase without bound. Since polynomials p and q are continuous functions, the expression in brackets is a Riemann sum that approaches a definite integral, and we are led to consider the *average value of $p(t)q(t)$* on the interval $[a, b]$:

$$\frac{1}{b-a} \int_a^b p(t)q(t)\, dt$$

This quantity is defined for polynomials of any degree (in fact, for all continuous functions), and it has all the properties of an inner product, as the next example shows. The scale factor $1/(b - a)$ is inessential and is often omitted for simplicity.

EXAMPLE 7 For f, g in $C[a, b]$, set

$$\langle f, g \rangle = \int_a^b f(t)g(t)\, dt \tag{5}$$

Show that (5) defines an inner product on $C[a, b]$.

SOLUTION Inner product Axioms 1–3 follow from elementary properties of definite integrals. For Axiom 4, observe that

$$\langle f, f \rangle = \int_a^b [f(t)]^2 \, dt \geq 0$$

The function $[f(t)]^2$ is continuous and nonnegative on $[a, b]$. If the definite integral of $[f(t)]^2$ is zero, then $[f(t)]^2$ must be identically zero on $[a, b]$, by a theorem in advanced calculus, in which case f is the zero function. Thus $\langle f, f \rangle = 0$ implies that f is the zero function on $[a, b]$. So (5) defines an inner product on $C[a, b]$. ∎

EXAMPLE 8 Let V be the space $C[0, 1]$ with the inner product of Example 7, and let W be the subspace spanned by the polynomials $p_1(t) = 1$, $p_2(t) = 2t - 1$, and $p_3(t) = 12t^2$. Use the Gram–Schmidt process to find an orthogonal basis for W.

SOLUTION Let $q_1 = p_1$, and compute

$$\langle p_2, q_1 \rangle = \int_0^1 (2t - 1)(1) \, dt = (t^2 - t) \Big|_0^1 = 0$$

So p_2 is already orthogonal to q_1, and we can take $q_2 = p_2$. For the projection of p_3 onto $W_2 = \text{Span}\{q_1, q_2\}$, compute

$$\langle p_3, q_1 \rangle = \int_0^1 12t^2 \cdot 1 \, dt = 4t^3 \Big|_0^1 = 4$$

$$\langle q_1, q_1 \rangle = \int_0^1 1 \cdot 1 \, dt = t \Big|_0^1 = 1$$

$$\langle p_3, q_2 \rangle = \int_0^1 12t^2 (2t - 1) \, dt = \int_0^1 (24t^3 - 12t^2) \, dt = 2$$

$$\langle q_2, q_2 \rangle = \int_0^1 (2t - 1)^2 \, dt = \frac{1}{6}(2t - 1)^3 \Big|_0^1 = \frac{1}{3}$$

Then

$$\text{proj}_{W_2} p_3 = \frac{\langle p_3, q_1 \rangle}{\langle q_1, q_1 \rangle} q_1 + \frac{\langle p_3, q_2 \rangle}{\langle q_2, q_2 \rangle} q_2 = \frac{4}{1} q_1 + \frac{2}{1/3} q_2 = 4q_1 + 6q_2$$

and

$$q_3 = p_3 - \text{proj}_{W_2} p_3 = p_3 - 4q_1 - 6q_2$$

As a function, $q_3(t) = 12t^2 - 4 - 6(2t - 1) = 12t^2 - 12t + 2$. The orthogonal basis for the subspace W is $\{q_1, q_2, q_3\}$. ∎

Practice Problems

Use the inner product axioms to verify the following statements.

1. $\langle \mathbf{v}, \mathbf{0} \rangle = \langle \mathbf{0}, \mathbf{v} \rangle = 0$.

2. $\langle \mathbf{u}, \mathbf{v} + \mathbf{w} \rangle = \langle \mathbf{u}, \mathbf{v} \rangle + \langle \mathbf{u}, \mathbf{w} \rangle$.

6.7 Exercises

1. Let $\mathbb{R}^2$ have the inner product of Example 1, and let $\mathbf{x} = (1, 1)$ and $\mathbf{y} = (5, -1)$.
   a. Find $\|\mathbf{x}\|$, $\|\mathbf{y}\|$, and $|\langle \mathbf{x}, \mathbf{y} \rangle|^2$.
   b. Describe all vectors (z_1, z_2) that are orthogonal to $\mathbf{y}$.

2. Let $\mathbb{R}^2$ have the inner product of Example 1. Show that the Cauchy–Schwarz inequality holds for $\mathbf{x} = (3, -2)$ and $\mathbf{y} = (-2, 1)$. [*Suggestion:* Study $|\langle \mathbf{x}, \mathbf{y} \rangle|^2$.]

Exercises 3–8 refer to $\mathbb{P}_2$ with the inner product given by evaluation at -1, 0, and 1. (See Example 2.)

3. Compute $\langle p, q \rangle$, where $p(t) = 4 + t$, $q(t) = 5 - 4t^2$.

4. Compute $\langle p, q \rangle$, where $p(t) = 3t - t^2$, $q(t) = 3 + 2t^2$.

5. Compute $\|p\|$ and $\|q\|$, for p and q in Exercise 3.

6. Compute $\|p\|$ and $\|q\|$, for p and q in Exercise 4.

7. Compute the orthogonal projection of q onto the subspace spanned by p, for p and q in Exercise 3.

8. Compute the orthogonal projection of q onto the subspace spanned by p, for p and q in Exercise 4.

9. Let $\mathbb{P}_3$ have the inner product given by evaluation at -3, -1, 1, and 3. Let $p_0(t) = 1$, $p_1(t) = t$, and $p_2(t) = t^2$.
   a. Compute the orthogonal projection of p_2 onto the subspace spanned by p_0 and p_1.
   b. Find a polynomial q that is orthogonal to p_0 and p_1, such that $\{p_0, p_1, q\}$ is an orthogonal basis for Span $\{p_0, p_1, p_2\}$. Scale the polynomial q so that its vector of values at $(-3, -1, 1, 3)$ is $(1, -1, -1, 1)$.

10. Let $\mathbb{P}_3$ have the inner product as in Exercise 9, with p_0, p_1, and q the polynomials described there. Find the best approximation to $p(t) = t^3$ by polynomials in Span $\{p_0, p_1, q\}$.

11. Let p_0, p_1, and p_2 be the orthogonal polynomials described in Example 5, where the inner product on $\mathbb{P}_4$ is given by evaluation at -2, -1, 0, 1, and 2. Find the orthogonal projection of t^3 onto Span $\{p_0, p_1, p_2\}$.

12. Find a polynomial p_3 such that $\{p_0, p_1, p_2, p_3\}$ (see Exercise 11) is an orthogonal basis for the subspace $\mathbb{P}_3$ of $\mathbb{P}_4$. Scale the polynomial p_3 so that its vector of values is $(-1, 2, 0, -2, 1)$.

13. Let A be any invertible $n \times n$ matrix. Show that for $\mathbf{u}$, $\mathbf{v}$ in $\mathbb{R}^n$, the formula $\langle \mathbf{u}, \mathbf{v} \rangle = (A\mathbf{u}) \cdot (A\mathbf{v}) = (A\mathbf{u})^T (A\mathbf{v})$ defines an inner product on $\mathbb{R}^n$.

14. Let T be a one-to-one linear transformation from a vector space V into $\mathbb{R}^n$. Show that for $\mathbf{u}$, $\mathbf{v}$ in V, the formula $\langle \mathbf{u}, \mathbf{v} \rangle = T(\mathbf{u}) \cdot T(\mathbf{v})$ defines an inner product on V.

Use the inner product axioms and other results of this section to verify the statements in Exercises 15–18.

15. $\langle \mathbf{u}, c\mathbf{v} \rangle = c\langle \mathbf{u}, \mathbf{v} \rangle$ for all scalars c.

16. If $\{\mathbf{u}, \mathbf{v}\}$ is an orthonormal set in V, then $\|\mathbf{u} - \mathbf{v}\| = \sqrt{2}$.

17. $\langle \mathbf{u}, \mathbf{v} \rangle = \frac{1}{4}\|\mathbf{u} + \mathbf{v}\|^2 - \frac{1}{4}\|\mathbf{u} - \mathbf{v}\|^2$.

18. $\|\mathbf{u} + \mathbf{v}\|^2 + \|\mathbf{u} - \mathbf{v}\|^2 = 2\|\mathbf{u}\|^2 + 2\|\mathbf{v}\|^2$.

In Exercises 19–24, $\mathbf{u}$, $\mathbf{v}$, and $\mathbf{w}$ are vectors. Mark each statement True or False (**T/F**). Justify each answer.

19. (**T/F**) If $\langle \mathbf{u}, \mathbf{u} \rangle = 0$, then $\mathbf{u} = \mathbf{0}$.

20. (**T/F**) If $\langle \mathbf{u}, \mathbf{v} \rangle = 0$, then either $\mathbf{u} = \mathbf{0}$ or $\mathbf{v} = \mathbf{0}$.

21. (**T/F**) $\langle \mathbf{u} + \mathbf{v}, \mathbf{w} \rangle = \langle \mathbf{w}, \mathbf{u} \rangle + \langle \mathbf{w}, \mathbf{v} \rangle$.

22. (**T/F**) $\langle c\mathbf{u}, c\mathbf{v} \rangle = c\langle \mathbf{u}, \mathbf{v} \rangle$.

23. (**T/F**) $|\langle \mathbf{u}, \mathbf{u} \rangle| = \langle \mathbf{u}, \mathbf{u} \rangle$.

24. (**T/F**) $|\langle \mathbf{u}, \mathbf{v} \rangle| \leq \|\mathbf{u}\| \|\mathbf{v}\|$.

25. Given $a \geq 0$ and $b \geq 0$, let $\mathbf{u} = \begin{bmatrix} \sqrt{a} \\ \sqrt{b} \end{bmatrix}$ and $\mathbf{v} = \begin{bmatrix} \sqrt{b} \\ \sqrt{a} \end{bmatrix}$. Use the Cauchy–Schwarz inequality to compare the geometric mean $\sqrt{ab}$ with the arithmetic mean $(a + b)/2$.

26. Let $\mathbf{u} = \begin{bmatrix} a \\ b \end{bmatrix}$ and $\mathbf{v} = \begin{bmatrix} 1 \\ 1 \end{bmatrix}$. Use the Cauchy–Schwarz inequality to show that
$$\left(\frac{a + b}{2} \right)^2 \leq \frac{a^2 + b^2}{2}$$

Exercises 27–30 refer to $V = C[0, 1]$, with the inner product given by an integral, as in Example 7.

27. Compute $\langle f, g \rangle$, where $f(t) = 1 - 3t^2$ and $g(t) = t - t^3$.

28. Compute $\langle f, g \rangle$, where $f(t) = 5t - 3$ and $g(t) = t^3 - t^2$.

29. Compute $\|f\|$ for f in Exercise 27.

30. Compute $\|g\|$ for g in Exercise 28.

31. Let V be the space $C[-1, 1]$ with the inner product of Example 7. Find an orthogonal basis for the subspace spanned by the polynomials 1, t, and t^2. The polynomials in this basis are called *Legendre polynomials*.

32. Let V be the space $C[-2, 2]$ with the inner product of Example 7. Find an orthogonal basis for the subspace spanned by the polynomials 1, t, and t^2.

T 33. Let $\mathbb{P}_4$ have the inner product as in Example 5, and let p_0, p_1, p_2 be the orthogonal polynomials from that example. Using your matrix program, apply the Gram–Schmidt process to the set $\{p_0, p_1, p_2, t^3, t^4\}$ to create an orthogonal basis for $\mathbb{P}_4$.

T 34. Let V be the space $C[0, 2\pi]$ with the inner product of Example 7. Use the Gram–Schmidt process to create an orthogonal basis for the subspace spanned by $\{1, \cos t, \cos^2 t, \cos^3 t\}$. Use a matrix program or computational program to compute the appropriate definite integrals.

1. By Axiom 1, $\langle \mathbf{v}, \mathbf{0} \rangle = \langle \mathbf{0}, \mathbf{v} \rangle$. Then $\langle \mathbf{0}, \mathbf{v} \rangle = \langle 0\mathbf{v}, \mathbf{v} \rangle = 0\langle \mathbf{v}, \mathbf{v} \rangle$, by Axiom 3, so $\langle \mathbf{0}, \mathbf{v} \rangle = 0$.
2. By Axioms 1, 2, and then 1 again, $\langle \mathbf{u}, \mathbf{v} + \mathbf{w} \rangle = \langle \mathbf{v} + \mathbf{w}, \mathbf{u} \rangle = \langle \mathbf{v}, \mathbf{u} \rangle + \langle \mathbf{w}, \mathbf{u} \rangle = \langle \mathbf{u}, \mathbf{v} \rangle + \langle \mathbf{u}, \mathbf{w} \rangle$.

6.8 Applications of Inner Product Spaces

The examples in this section suggest how the inner product spaces defined in Section 6.7 arise in practical problems. Like in Section 6.6, important components of machine learning are analyzed.

Weighted Least-Squares

Let $\mathbf{y}$ be a vector of n observations, $y_1, \ldots, y_n$, and suppose we wish to approximate $\mathbf{y}$ by a vector $\hat{\mathbf{y}}$ that belongs to some specified subspace of $\mathbb{R}^n$. (In Section 6.5, $\hat{\mathbf{y}}$ was written as $A\mathbf{x}$ so that $\hat{\mathbf{y}}$ was in the column space of A.) Denote the entries in $\hat{\mathbf{y}}$ by $\hat{y}_1, \ldots, \hat{y}_n$. Then the *sum of the squares for error*, or SS(E), in approximating $\mathbf{y}$ by $\hat{\mathbf{y}}$ is

$$\text{SS(E)} = (y_1 - \hat{y}_1)^2 + \cdots + (y_n - \hat{y}_n)^2 \tag{1}$$

This is simply $\|\mathbf{y} - \hat{\mathbf{y}}\|^2$, using the standard length in $\mathbb{R}^n$.

Now suppose the measurements that produced the entries in $\mathbf{y}$ are not equally reliable. The entries in $\mathbf{y}$ might be computed from various samples of measurements, with unequal sample sizes. Then it becomes appropriate to weight the squared errors in (1) in such a way that more importance is assigned to the more reliable measurements.[1] If the weights are denoted by $w_1^2, \ldots, w_n^2$, then the weighted sum of the squares for error is

$$\text{Weighted SS(E)} = w_1^2(y_1 - \hat{y}_1)^2 + \cdots + w_n^2(y_n - \hat{y}_n)^2 \tag{2}$$

This is the square of the length of $\mathbf{y} - \hat{\mathbf{y}}$, where the length is derived from an inner product analogous to that in Example 1 in Section 6.7, namely

$$\langle \mathbf{x}, \mathbf{y} \rangle = w_1^2 x_1 y_1 + \cdots + w_n^2 x_n y_n$$

It is sometimes convenient to transform a weighted least-squares problem into an equivalent ordinary least-squares problem. Let W be the diagonal matrix with (positive) $w_1, \ldots, w_n$ on its diagonal, so that

$$W\mathbf{y} = \begin{bmatrix} w_1 & 0 & \cdots & 0 \\ 0 & w_2 & & \\ \vdots & & \ddots & \vdots \\ 0 & & \cdots & w_n \end{bmatrix} \begin{bmatrix} y_1 \\ y_2 \\ \vdots \\ y_n \end{bmatrix} = \begin{bmatrix} w_1 y_1 \\ w_2 y_2 \\ \vdots \\ w_n y_n \end{bmatrix}$$

with a similar expression for $W\hat{\mathbf{y}}$. Observe that the jth term in (2) can be written as

$$w_j^2(y_j - \hat{y}_j)^2 = (w_j y_j - w_j \hat{y}_j)^2$$

[1] Note for readers with a background in statistics: Suppose the errors in measuring the y_i are independent random variables with means equal to zero and variances of $\sigma_1^2, \ldots, \sigma_n^2$. Then the appropriate weights in (2) are $w_i^2 = 1/\sigma_i^2$. The larger the variance of the error, the smaller the weight.

It follows that the weighted SS(E) in (2) is the square of the ordinary length in $\mathbb{R}^n$ of $W\mathbf{y} - W\hat{\mathbf{y}}$, which we write as $\|W\mathbf{y} - W\hat{\mathbf{y}}\|^2$.

Now suppose the approximating vector $\hat{\mathbf{y}}$ is to be constructed from the columns of a matrix A. Then we seek an $\hat{\mathbf{x}}$ that makes $A\hat{\mathbf{x}} = \hat{\mathbf{y}}$ as close to $\mathbf{y}$ as possible. However, the measure of closeness is the weighted error,

$$\|W\mathbf{y} - W\hat{\mathbf{y}}\|^2 = \|W\mathbf{y} - WA\hat{\mathbf{x}}\|^2$$

Thus $\hat{\mathbf{x}}$ is the (ordinary) least-squares solution of the equation

$$WA\mathbf{x} = W\mathbf{y}$$

The normal equation for the least-squares solution is

$$(WA)^T WA\mathbf{x} = (WA)^T W\mathbf{y}$$

EXAMPLE 1   Find the least-squares line $y = \beta_0 + \beta_1 x$ that best fits the data $(-2, 3)$, $(-1, 5)$, $(0, 5)$, $(1, 4)$, and $(2, 3)$. Suppose the errors in measuring the y-values of the last two data points are greater than for the other points. Weight these data half as much as the rest of the data.

SOLUTION   As in Section 6.6, write X for the matrix A and $\boldsymbol{\beta}$ for the vector $\mathbf{x}$, and obtain

$$X = \begin{bmatrix} 1 & -2 \\ 1 & -1 \\ 1 & 0 \\ 1 & 1 \\ 1 & 2 \end{bmatrix}, \quad \boldsymbol{\beta} = \begin{bmatrix} \beta_0 \\ \beta_1 \end{bmatrix}, \quad \mathbf{y} = \begin{bmatrix} 3 \\ 5 \\ 5 \\ 4 \\ 3 \end{bmatrix}$$

For a weighting matrix, choose W with diagonal entries 2, 2, 2, 1, and 1. Left-multiplication by W scales the rows of X and $\mathbf{y}$:

$$WX = \begin{bmatrix} 2 & -4 \\ 2 & -2 \\ 2 & 0 \\ 1 & 1 \\ 1 & 2 \end{bmatrix}, \quad W\mathbf{y} = \begin{bmatrix} 6 \\ 10 \\ 10 \\ 4 \\ 3 \end{bmatrix}$$

For the normal equation, compute

$$(WX)^T WX = \begin{bmatrix} 14 & -9 \\ -9 & 25 \end{bmatrix} \quad \text{and} \quad (WX)^T W\mathbf{y} = \begin{bmatrix} 59 \\ -34 \end{bmatrix}$$

and solve

$$\begin{bmatrix} 14 & -9 \\ -9 & 25 \end{bmatrix} \begin{bmatrix} \beta_0 \\ \beta_1 \end{bmatrix} = \begin{bmatrix} 59 \\ -34 \end{bmatrix}$$

The solution of the normal equation is (to two significant digits) $\beta_0 = 4.3$ and $\beta_1 = .20$. The desired line is

$$y = 4.3 + .20x$$

In contrast, the ordinary least-squares line for these data is

$$y = 4.0 - .10x$$

Both lines are displayed in Figure 1.

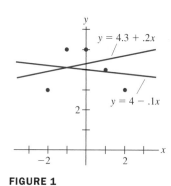

FIGURE 1

Weighted and ordinary least-squares lines.

Trend Analysis of Data

Let f represent an unknown function whose values are known (perhaps only approximately) at $t_0, \ldots, t_n$. If there is a "linear trend" in the data $f(t_0), \ldots, f(t_n)$, then we might expect to approximate the values of f by a function of the form $\beta_0 + \beta_1 t$. If there is a "quadratic trend" to the data, then we would try a function of the form $\beta_0 + \beta_1 t + \beta_2 t^2$. This was discussed in Section 6.6, from a different point of view.

In some statistical problems, it is important to be able to separate the linear trend from the quadratic trend (and possibly cubic or higher-order trends). For instance, suppose engineers are analyzing the performance of a new car, and $f(t)$ represents the distance between the car at time t and some reference point. If the car is traveling at constant velocity, then the graph of $f(t)$ should be a straight line whose slope is the car's velocity. If the gas pedal is suddenly pressed to the floor, the graph of $f(t)$ will change to include a quadratic term and possibly a cubic term (due to the acceleration). To analyze the ability of the car to pass another car, for example, engineers may want to separate the quadratic and cubic components from the linear term.

If the function is approximated by a curve of the form $y = \beta_0 + \beta_1 t + \beta_2 t^2$, the coefficient β_2 may not give the desired information about the quadratic trend in the data, because it may not be "independent" in a statistical sense from the other β_i. To make what is known as a **trend analysis** of the data, we introduce an inner product on the space $\mathbb{P}_n$ analogous to that given in Example 2 in Section 6.7. For p, q in $\mathbb{P}_n$, define

$$\langle p, q \rangle = p(t_0)q(t_0) + \cdots + p(t_n)q(t_n)$$

In practice, statisticians seldom need to consider trends in data of degree higher than cubic or quartic. So let p_0, p_1, p_2, p_3 denote an orthogonal basis of the subspace $\mathbb{P}_3$ of $\mathbb{P}_n$, obtained by applying the Gram–Schmidt process to the polynomials $1, t, t^2$, and t^3. There is a polynomial g in $\mathbb{P}_n$ whose values at $t_0, \ldots, t_n$ coincide with those of the unknown function f. Let $\hat{g}$ be the orthogonal projection (with respect to the given inner product) of g onto $\mathbb{P}_3$, say,

$$\hat{g} = c_0 p_0 + c_1 p_1 + c_2 p_2 + c_3 p_3$$

Then $\hat{g}$ is called a cubic **trend function**, and $c_0, \ldots, c_3$ are the **trend coefficients** of the data. The coefficient c_1 measures the linear trend, c_2 the quadratic trend, and c_3 the cubic trend. It turns out that if the data have certain properties, these coefficients are statistically independent.

Since $p_0, \ldots, p_3$ are orthogonal, the trend coefficients may be computed one at a time, independently of one another. (Recall that $c_i = \langle g, p_i \rangle / \langle p_i, p_i \rangle$.) We can ignore p_3 and c_3 if we want only the quadratic trend. And if, for example, we needed to determine the quartic trend, we would have to find (via Gram–Schmidt) only a polynomial p_4 in $\mathbb{P}_4$ that is orthogonal to $\mathbb{P}_3$ and compute $\langle g, p_4 \rangle / \langle p_4, p_4 \rangle$.

EXAMPLE 2   The simplest and most common use of trend analysis occurs when the points $t_0, \ldots, t_n$ can be adjusted so that they are evenly spaced and sum to zero. Fit a quadratic trend function to the data $(-2, 3)$, $(-1, 5)$, $(0, 5)$, $(1, 4)$, and $(2, 3)$.

SOLUTION The t-coordinates are suitably scaled to use the orthogonal polynomials found in Example 5 of Section 6.7:

Polynomial	p_0	p_1	p_2	Data: g
Vector of values	$\begin{bmatrix} 1 \\ 1 \\ 1 \\ 1 \\ 1 \end{bmatrix}$,	$\begin{bmatrix} -2 \\ -1 \\ 0 \\ 1 \\ 2 \end{bmatrix}$,	$\begin{bmatrix} 2 \\ -1 \\ -2 \\ -1 \\ 2 \end{bmatrix}$,	$\begin{bmatrix} 3 \\ 5 \\ 5 \\ 4 \\ 3 \end{bmatrix}$

The calculations involve only these vectors, not the specific formulas for the orthogonal polynomials. The best approximation to the data by polynomials in $\mathbb{P}_2$ is the orthogonal projection given by

$$\hat{p} = \frac{\langle g, p_0 \rangle}{\langle p_0, p_0 \rangle} p_0 + \frac{\langle g, p_1 \rangle}{\langle p_1, p_1 \rangle} p_1 + \frac{\langle g, p_2 \rangle}{\langle p_2, p_2 \rangle} p_2$$

$$= \frac{20}{5} p_0 - \frac{1}{10} p_1 - \frac{7}{14} p_2$$

and

$$\hat{p}(t) = 4 - .1t - .5(t^2 - 2) \tag{3}$$

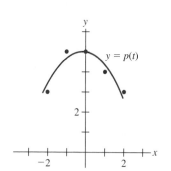

FIGURE 2

Approximation by a quadratic trend function.

Since the coefficient of p_2 is not extremely small, it would be reasonable to conclude that the trend is at least quadratic. This is confirmed by the graph in Figure 2. ∎

Fourier Series (Calculus required)

Continuous functions are often approximated by linear combinations of sine and cosine functions. For instance, a continuous function might represent a sound wave, an electric signal of some type, or the movement of a vibrating mechanical system.

For simplicity, we consider functions on $0 \le t \le 2\pi$. It turns out that any function in $C[0, 2\pi]$ can be approximated as closely as desired by a function of the form

$$\frac{a_0}{2} + a_1 \cos t + \cdots + a_n \cos nt + b_1 \sin t + \cdots + b_n \sin nt \tag{4}$$

for a sufficiently large value of n. The function (4) is called a **trigonometric polynomial**. If a_n and b_n are not both zero, the polynomial is said to be of **order n**. The connection between trigonometric polynomials and other functions in $C[0, 2\pi]$ depends on the fact that for any $n \ge 1$, the set

$$\{1, \cos t, \cos 2t, \ldots, \cos nt, \sin t, \sin 2t, \ldots, \sin nt\} \tag{5}$$

is orthogonal with respect to the inner product

$$\langle f, g \rangle = \int_0^{2\pi} f(t) g(t) \, dt \tag{6}$$

This orthogonality is verified as in the following example and in Exercises 5 and 6.

EXAMPLE 3 Let $C[0, 2\pi]$ have the inner product (6), and let m and n be unequal positive integers. Show that $\cos mt$ and $\cos nt$ are orthogonal.

SOLUTION Use a trigonometric identity. When $m \neq n$,

$$\langle \cos mt, \cos nt \rangle = \int_0^{2\pi} \cos mt \cos nt \, dt$$

$$= \frac{1}{2} \int_0^{2\pi} [\cos(mt + nt) + \cos(mt - nt)] \, dt$$

$$= \frac{1}{2} \left[\frac{\sin(mt + nt)}{m + n} + \frac{\sin(mt - nt)}{m - n} \right]\Big|_0^{2\pi} = 0 \qquad \blacksquare$$

Let W be the subspace of $C[0, 2\pi]$ spanned by the functions in (5). Given f in $C[0, 2\pi]$, the best approximation to f by functions in W is called the **nth-order Fourier approximation** to f on $[0, 2\pi]$. Since the functions in (5) are orthogonal, the best approximation is given by the orthogonal projection onto W. In this case, the coefficients a_k and b_k in (4) are called the **Fourier coefficients** of f. The standard formula for an orthogonal projection shows that

$$a_k = \frac{\langle f, \cos kt \rangle}{\langle \cos kt, \cos kt \rangle}, \quad b_k = \frac{\langle f, \sin kt \rangle}{\langle \sin kt, \sin kt \rangle}, \quad k \geq 1$$

Exercise 7 asks you to show that $\langle \cos kt, \cos kt \rangle = \pi$ and $\langle \sin kt, \sin kt \rangle = \pi$. Thus

$$a_k = \frac{1}{\pi} \int_0^{2\pi} f(t) \cos kt \, dt, \quad b_k = \frac{1}{\pi} \int_0^{2\pi} f(t) \sin kt \, dt \qquad (7)$$

The coefficient of the (constant) function 1 in the orthogonal projection is

$$\frac{\langle f, 1 \rangle}{\langle 1, 1 \rangle} = \frac{1}{2\pi} \int_0^{2\pi} f(t) \cdot 1 \, dt = \frac{1}{2} \left[\frac{1}{\pi} \int_0^{2\pi} f(t) \cos(0 \cdot t) \, dt \right] = \frac{a_0}{2}$$

where a_0 is defined by (7) for $k = 0$. This explains why the constant term in (4) is written as $a_0/2$.

EXAMPLE 4 Find the nth-order Fourier approximation to the function $f(t) = t$ on the interval $[0, 2\pi]$.

SOLUTION Compute

$$\frac{a_0}{2} = \frac{1}{2} \cdot \frac{1}{\pi} \int_0^{2\pi} t \, dt = \frac{1}{2\pi} \left[\frac{1}{2} t^2 \Big|_0^{2\pi} \right] = \pi$$

and for $k > 0$, using integration by parts,

$$a_k = \frac{1}{\pi} \int_0^{2\pi} t \cos kt \, dt = \frac{1}{\pi} \left[\frac{1}{k^2} \cos kt + \frac{t}{k} \sin kt \right]_0^{2\pi} = 0$$

$$b_k = \frac{1}{\pi} \int_0^{2\pi} t \sin kt \, dt = \frac{1}{\pi} \left[\frac{1}{k^2} \sin kt - \frac{t}{k} \cos kt \right]_0^{2\pi} = -\frac{2}{k}$$

Thus the nth-order Fourier approximation of $f(t) = t$ is

$$\pi - 2 \sin t - \sin 2t - \frac{2}{3} \sin 3t - \cdots - \frac{2}{n} \sin nt$$

Figure 3 shows the third- and fourth-order Fourier approximations of f. $\blacksquare$

(a) Third order   (b) Fourth order

FIGURE 3 Fourier approximations of the function $f(t) = t$.

The norm of the difference between f and a Fourier approximation is called the **mean square error** in the approximation. (The term *mean* refers to the fact that the norm is determined by an integral.) It can be shown that the mean square error approaches zero as the order of the Fourier approximation increases. For this reason, it is common to write

$$f(t) = \frac{a_0}{2} + \sum_{m=1}^{\infty} (a_m \cos mt + b_m \sin mt)$$

This expression for $f(t)$ is called the **Fourier series** for f on $[0, 2\pi]$. The term $a_m \cos mt$, for example, is the projection of f onto the one-dimensional subspace spanned by $\cos mt$.

Practice Problems

1. Let $q_1(t) = 1$, $q_2(t) = t$, and $q_3(t) = 3t^2 - 4$. Verify that $\{q_1, q_2, q_3\}$ is an orthogonal set in $C[-2, 2]$ with the inner product of Example 7 in Section 6.7 (integration from -2 to 2).

2. Find the first-order and third-order Fourier approximations to

$$f(t) = 3 - 2\sin t + 5\sin 2t - 6\cos 2t$$

6.8 Exercises

1. Find the least-squares line $y = \beta_0 + \beta_1 x$ that best fits the data $(-2, 0)$, $(-1, 0)$, $(0, 2)$, $(1, 4)$, and $(2, 4)$, assuming that the first and last data points are less reliable. Weight them half as much as the three interior points.

2. Suppose 5 out of 25 data points in a weighted least-squares problem have a y-measurement that is less reliable than the others, and they are to be weighted half as much as the other 20 points. One method is to weight the 20 points by a factor of 1 and the other 5 by a factor of $\frac{1}{2}$. A second method is to weight the 20 points by a factor of 2 and the other 5 by a factor of 1. Do the two methods produce different results? Explain.

3. Fit a cubic trend function to the data in Example 2. The orthogonal cubic polynomial is $p_3(t) = \frac{5}{6}t^3 - \frac{17}{6}t$.

4. To make a trend analysis of six evenly spaced data points, one can use orthogonal polynomials with respect to evaluation at the points $t = -5, -3, -1, 1, 3$, and 5.

   a. Show that the first three orthogonal polynomials are

   $$p_0(t) = 1, \quad p_1(t) = t, \quad \text{and} \quad p_2(t) = \tfrac{3}{8}t^2 - \tfrac{35}{8}$$

   (The polynomial p_2 has been scaled so that its values at the evaluation points are small integers.)

   b. Fit a quadratic trend function to the data

   $$(-5, 1), (-3, 1), (-1, 4), (1, 4), (3, 6), (5, 8)$$

In Exercises 5–14, the space is $C[0, 2\pi]$ with the inner product (6).

5. Show that $\sin mt$ and $\sin nt$ are orthogonal when $m \neq n$.

6. Show that $\sin mt$ and $\cos nt$ are orthogonal for all positive integers m and n.

7. Show that $\| \cos kt \|^2 = \pi$ and $\| \sin kt \|^2 = \pi$ for $k > 0$.

8. Find the third-order Fourier approximation to $f(t) = t - 1$.

9. Find the third-order Fourier approximation to $f(t) = 2\pi - t$.

10. Find the third-order Fourier approximation to the *square wave function* $f(t) = 1$ for $0 \le t < \pi$ and $f(t) = -1$ for $\pi \le t < 2\pi$.

11. Find the third-order Fourier approximation to $\sin^2 t$, without performing any integration calculations.

12. Find the third-order Fourier approximation to $\cos^3 t$, without performing any integration calculations.

13. Explain why a Fourier coefficient of the sum of two functions is the sum of the corresponding Fourier coefficients of the two functions.

14. Suppose the first few Fourier coefficients of some function f in $C[0, 2\pi]$ are a_0, a_1, a_2, and b_1, b_2, b_3. Which of the following trigonometric polynomials is closer to f? Defend your answer.

$$g(t) = \frac{a_0}{2} + a_1 \cos t + a_2 \cos 2t + b_1 \sin t$$

$$h(t) = \frac{a_0}{2} + a_1 \cos t + a_2 \cos 2t + b_1 \sin t + b_2 \sin 2t$$

T 15. Refer to the data in Exercise 19 in Section 6.6, concerning the takeoff performance of an airplane. Suppose the possible measurement errors become greater as the speed of the airplane increases, and let W be the diagonal weighting matrix whose diagonal entries are 1, 1, 1, .9, .9, .8, .7, .6, .5, .4, .3, .2, and .1. Find the cubic curve that fits the data with minimum weighted least-squares error, and use it to estimate the velocity of the plane when $t = 4.5$ seconds.

T 16. Let f_4 and f_5 be the fourth-order and fifth-order Fourier approximations in $C[0, 2\pi]$ to the square wave function in Exercise 10. Produce separate graphs of f_4 and f_5 on the interval $[0, 2\pi]$, and produce a graph of f_5 on $[-2\pi, 2\pi]$.

STUDY GUIDE offers additional resources for mastering orthogonal projections.

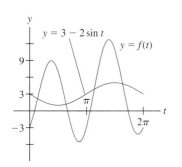

First- and third-order approximations to $f(t)$.

Solutions to Practice Problems

1. Compute

$$\langle q_1, q_2 \rangle = \int_{-2}^{2} 1 \cdot t \, dt = \frac{1}{2} t^2 \Big|_{-2}^{2} = 0$$

$$\langle q_1, q_3 \rangle = \int_{-2}^{2} 1 \cdot (3t^2 - 4) \, dt = (t^3 - 4t) \Big|_{-2}^{2} = 0$$

$$\langle q_2, q_3 \rangle = \int_{-2}^{2} t \cdot (3t^2 - 4) \, dt = \left(\frac{3}{4} t^4 - 2t^2 \right) \Big|_{-2}^{2} = 0$$

2. The third-order Fourier approximation to f is the best approximation in $C[0, 2\pi]$ to f by functions (vectors) in the subspace spanned by 1, $\cos t$, $\cos 2t$, $\cos 3t$, $\sin t$, $\sin 2t$, and $\sin 3t$. But f is obviously *in* this subspace, so f is its own best approximation:

$$f(t) = 3 - 2 \sin t + 5 \sin 2t - 6 \cos 2t$$

For the first-order approximation, the closest function to f in the subspace $W = $ Span$\{1, \cos t, \sin t\}$ is $3 - 2 \sin t$. The other two terms in the formula for $f(t)$ are orthogonal to the functions in W, so they contribute nothing to the integrals that give the Fourier coefficients for a first-order approximation.

CHAPTER 6 PROJECTS

Chapter 6 projects are available online at `bit.ly/30IM8gT`.

A. *The QR Method for Finding Eigenvalues*: This project shows how the QR factorization of a matrix may be used to calculate the eigenvalues of the matrix.

B. *Finding the Roots of a Polynomial with Eigenvalues*: This project shows how the real roots of a polynomial can be calculated by finding the eigenvalues of a particular matrix. These eigenvalues will be found by the QR method.

CHAPTER 6 SUPPLEMENTARY EXERCISES

The statements in Exercises 1–19 refer to vectors in $\mathbb{R}^n$ (or $\mathbb{R}^m$) with the standard inner product. Mark each statement True or False (T/F). Justify each answer.

1. **(T/F)** The length of every vector is a positive number.

2. **(T/F)** A vector $\mathbf{v}$ and its negative, $-\mathbf{v}$, have equal lengths.

3. **(T/F)** The distance between $\mathbf{u}$ and $\mathbf{v}$ is $\|\mathbf{u} - \mathbf{v}\|$.

4. **(T/F)** If r is any scalar, then $\|r\mathbf{v}\| = r\|\mathbf{v}\|$.

5. **(T/F)** If two vectors are orthogonal, they are linearly independent.

6. **(T/F)** If $\mathbf{x}$ is orthogonal to both $\mathbf{u}$ and $\mathbf{v}$, then $\mathbf{x}$ must be orthogonal to $\mathbf{u} - \mathbf{v}$.

7. **(T/F)** If $\|\mathbf{u} + \mathbf{v}\|^2 = \|\mathbf{u}\|^2 + \|\mathbf{v}\|^2$, then $\mathbf{u}$ and $\mathbf{v}$ are orthogonal.

8. **(T/F)** If $\|\mathbf{u} - \mathbf{v}\|^2 = \|\mathbf{u}\|^2 + \|\mathbf{v}\|^2$, then $\mathbf{u}$ and $\mathbf{v}$ are orthogonal.

9. **(T/F)** The orthogonal projection of $\mathbf{y}$ onto $\mathbf{u}$ is a scalar multiple of $\mathbf{y}$.

10. **(T/F)** If a vector $\mathbf{y}$ coincides with its orthogonal projection onto a subspace W, then $\mathbf{y}$ is in W.

11. **(T/F)** The set of all vectors in $\mathbb{R}^n$ orthogonal to one fixed vector is a subspace of $\mathbb{R}^n$.

12. **(T/F)** If W is a subspace of $\mathbb{R}^n$, then W and $W^\perp$ have no vectors in common.

13. **(T/F)** If $\{\mathbf{v}_1, \mathbf{v}_2, \mathbf{v}_3\}$ is an orthogonal set and if c_1, c_2, and c_3 are scalars, then $\{c_1\mathbf{v}_1, c_2\mathbf{v}_2, c_3\mathbf{v}_3\}$ is an orthogonal set.

14. **(T/F)** If a matrix U has orthonormal columns, then $UU^T = I$.

15. **(T/F)** A square matrix with orthogonal columns is an orthogonal matrix.

16. **(T/F)** If a square matrix has orthonormal columns, then it also has orthonormal rows.

17. **(T/F)** If W is a subspace, then $\|\operatorname{proj}_W \mathbf{v}\|^2 + \|\mathbf{v} - \operatorname{proj}_W \mathbf{v}\|^2 = \|\mathbf{v}\|^2$.

18. **(T/F)** A least-squares solution of $A\mathbf{x} = \mathbf{b}$ is the vector $A\hat{\mathbf{x}}$ in Col A closest to $\mathbf{b}$, so that $\|\mathbf{b} - A\hat{\mathbf{x}}\| \le \|\mathbf{b} - A\mathbf{x}\|$ for all $\mathbf{x}$.

19. **(T/F)** The normal equations for a least-squares solution of $A\mathbf{x} = \mathbf{b}$ are given by $\hat{\mathbf{x}} = (A^TA)^{-1}A^T\mathbf{b}$.

20. Let $\{\mathbf{v}_1, \ldots, \mathbf{v}_p\}$ be an orthonormal set. Verify the following equality by induction, beginning with $p = 2$. If $\mathbf{x} = c_1\mathbf{v}_1 + \cdots + c_p\mathbf{v}_p$, then $\|\mathbf{x}\|^2 = |c_1|^2 + \cdots + |c_p|^2$

21. Let $\{\mathbf{v}_1, \ldots, \mathbf{v}_p\}$ be an orthonormal set in $\mathbb{R}^n$. Verify the following inequality, called *Bessel's inequality*, which is true for each $\mathbf{x}$ in $\mathbb{R}^n$:
$$\|\mathbf{x}\|^2 \ge |\mathbf{x} \cdot \mathbf{v}_1|^2 + |\mathbf{x} \cdot \mathbf{v}_2|^2 + \cdots + |\mathbf{x} \cdot \mathbf{v}_p|^2$$

22. Let U be an $n \times n$ orthogonal matrix. Show that if $\{\mathbf{v}_1, \ldots, \mathbf{v}_n\}$ is an orthonormal basis for $\mathbb{R}^n$, then so is $\{U\mathbf{v}_1, \ldots, U\mathbf{v}_n\}$.

23. Show that if an $n \times n$ matrix U satisfies $(U\mathbf{x}) \cdot (U\mathbf{y}) = \mathbf{x} \cdot \mathbf{y}$ for all $\mathbf{x}$ and $\mathbf{y}$ in $\mathbb{R}^n$, then U is an orthogonal matrix.

24. Show that if U is an orthogonal matrix, then any real eigenvalue of U must be ± 1.

25. A *Householder matrix*, or an *elementary reflector*, has the form $Q = I - 2\mathbf{u}\mathbf{u}^T$ where $\mathbf{u}$ is a unit vector. Show that Q is an orthogonal matrix. (Elementary reflectors are often used in computer programs to produce a QR factorization of a matrix A. If A has linearly independent columns, then left-multiplication by a sequence of elementary reflectors can produce an upper triangular matrix.)

26. Let $T : \mathbb{R}^n \to \mathbb{R}^n$ be a linear transformation that preserves lengths; that is, $\|T(\mathbf{x})\| = \|\mathbf{x}\|$ for all $\mathbf{x}$ in $\mathbb{R}^n$.
    a. Show that T also preserves orthogonality; that is, $T(\mathbf{x}) \cdot T(\mathbf{y}) = 0$ whenever $\mathbf{x} \cdot \mathbf{y} = 0$.
    b. Show that the standard matrix of T is an orthogonal matrix.

27. Let $\mathbf{u}$ and $\mathbf{v}$ be linearly independent vectors in $\mathbb{R}^n$ that are *not* orthogonal. Describe how to find the best approximation to $\mathbf{z}$ in $\mathbb{R}^n$ by vectors of the form $x_1\mathbf{u} + x_2\mathbf{v}$ without first constructing an orthogonal basis for Span $\{\mathbf{u}, \mathbf{v}\}$.

28. Suppose the columns of A are linearly independent. Determine what happens to the least-squares solution $\hat{\mathbf{x}}$ of $A\mathbf{x} = \mathbf{b}$ when $\mathbf{b}$ is replaced by $c\mathbf{b}$ for some nonzero scalar c.

29. If a, b, and c are distinct numbers, then the following system is inconsistent because the graphs of the equations are parallel planes. Show that the set of all least-squares solutions of the system is precisely the plane whose equation is $x - 2y + 5z = (a + b + c)/3$.

$$x - 2y + 5z = a$$
$$x - 2y + 5z = b$$
$$x - 2y + 5z = c$$

30. Consider the problem of finding an eigenvalue of an $n \times n$ matrix A when an approximate eigenvector $\mathbf{v}$ is known. Since $\mathbf{v}$ is not exactly correct, the equation

$$A\mathbf{v} = \lambda\mathbf{v} \qquad (1)$$

will probably not have a solution. However, λ can be estimated by a least-squares solution when (1) is viewed properly. Think of $\mathbf{v}$ as an $n \times 1$ matrix V, think of λ as a vector in $\mathbb{R}^1$, and denote the vector $A\mathbf{v}$ by the symbol $\mathbf{b}$. Then (1) becomes $\mathbf{b} = \lambda V$, which may also be written as $V\lambda = \mathbf{b}$. Find the least-squares solution of this system of n equations in the one unknown λ, and write this solution using the original symbols. The resulting estimate for λ is called a *Rayleigh quotient*.

31. Use the steps below to prove the following relations among the four fundamental subspaces determined by an $m \times n$ matrix A.

$\text{Row } A = (\text{Nul } A)^\perp, \quad \text{Col } A = (\text{Nul } A^T)^\perp$

a. Show that Row A is contained in $(\text{Nul } A)^\perp$. (Show that if $\mathbf{x}$ is in Row A, then $\mathbf{x}$ is orthogonal to every $\mathbf{u}$ in Nul A.)

b. Suppose rank $A = r$. Find dim Nul A and dim $(\text{Nul } A)^\perp$, and then deduce from part (a) that Row $A = (\text{Nul } A)^\perp$. [*Hint:* Study the exercises for Section 6.3.]

c. Explain why Col $A = (\text{Nul } A^T)^\perp$.

32. Explain why an equation $A\mathbf{x} = \mathbf{b}$ has a solution if and only if $\mathbf{b}$ is orthogonal to all solutions of the equation $A^T\mathbf{x} = \mathbf{0}$.

Exercises 33 and 34 concern the (real) *Schur factorization* of an $n \times n$ matrix A in the form $A = URU^T$, where U is an orthogonal matrix and R is an $n \times n$ upper triangular matrix.[1]

33. Show that if A admits a (real) Schur factorization, $A = URU^T$, then A has n real eigenvalues, counting multiplicities.

34. Let A be an $n \times n$ matrix with n real eigenvalues, counting multiplicities, denoted by $\lambda_1, \ldots, \lambda_n$. It can be shown that A admits a (real) Schur factorization. Parts (a) and (b) show the key ideas in the proof. The rest of the proof amounts to repeating (a) and (b) for successively smaller matrices, and then piecing together the results.

a. Let $\mathbf{u}_1$ be a unit eigenvector corresponding to λ_1, let $\mathbf{u}_2, \ldots, \mathbf{u}_n$ be any other vectors such that $\{\mathbf{u}_1, \ldots, \mathbf{u}_n\}$ is an orthonormal basis for $\mathbb{R}^n$, and then let $U = [\,\mathbf{u}_1 \quad \mathbf{u}_2 \quad \cdots \quad \mathbf{u}_n\,]$. Show that the first column of $U^T AU$ is $\lambda_1\mathbf{e}_1$, where $\mathbf{e}_1$ is the first column of the $n \times n$ identity matrix.

---

[1] If complex numbers are allowed, *every $n \times n$ matrix A admits a (complex) Schur factorization*, $A = URU^{-1}$, where R is upper triangular and U^{-1} is the *conjugate* transpose of U. This very useful fact is discussed in *Matrix Analysis*, by Roger A. Horn and Charles R. Johnson (Cambridge: Cambridge University Press, 1985), pp. 79–100.

b. Part (a) implies that $U^T AU$ has the form shown below. Explain why the eigenvalues of A_1 are $\lambda_2, \ldots, \lambda_n$. [*Hint:* See the Supplementary Exercises for Chapter 5.]

$$U^T AU = \begin{bmatrix} \lambda_1 & * & * & * & * \\ 0 & & & & \\ \vdots & & A_1 & & \\ 0 & & & & \end{bmatrix}$$

T When the right side of an equation $A\mathbf{x} = \mathbf{b}$ is changed slightly—say, to $A\mathbf{x} = \mathbf{b} + \Delta\mathbf{b}$ for some vector $\Delta\mathbf{b}$—the solution changes from $\mathbf{x}$ to $\mathbf{x} + \Delta\mathbf{x}$, where $\Delta\mathbf{x}$ satisfies $A(\Delta\mathbf{x}) = \Delta\mathbf{b}$. The quotient $\|\Delta\mathbf{b}\|/\|\mathbf{b}\|$ is called the **relative change** in $\mathbf{b}$ (or the **relative error** in $\mathbf{b}$ when $\Delta\mathbf{b}$ represents possible error in the entries of $\mathbf{b}$). The relative change in the solution is $\|\Delta\mathbf{x}\|/\|\mathbf{x}\|$. When A is invertible, the **condition number** of A, written as $\text{cond}(A)$, produces a bound on how large the relative change in $\mathbf{x}$ can be:

$$\frac{\|\Delta\mathbf{x}\|}{\|\mathbf{x}\|} \le \text{cond}(A) \cdot \frac{\|\Delta\mathbf{b}\|}{\|\mathbf{b}\|} \tag{2}$$

In Exercises 35–38, solve $A\mathbf{x} = \mathbf{b}$ and $A(\Delta\mathbf{x}) = \Delta\mathbf{b}$, and show that the inequality (2) holds in each case. (See the discussion of *ill-conditioned* matrices in Exercises 49–51 in Section 2.3.)

T 35. $A = \begin{bmatrix} 4.5 & 3.1 \\ 1.6 & 1.1 \end{bmatrix}$, $\mathbf{b} = \begin{bmatrix} 19.249 \\ 6.843 \end{bmatrix}$, $\Delta\mathbf{b} = \begin{bmatrix} .001 \\ -.003 \end{bmatrix}$

T 36. $A = \begin{bmatrix} 4.5 & 3.1 \\ 1.6 & 1.1 \end{bmatrix}$, $\mathbf{b} = \begin{bmatrix} .500 \\ -1.407 \end{bmatrix}$, $\Delta\mathbf{b} = \begin{bmatrix} .001 \\ -.003 \end{bmatrix}$

T 37. $A = \begin{bmatrix} 7 & -6 & -4 & 1 \\ -5 & 1 & 0 & -2 \\ 10 & 11 & 7 & -3 \\ 19 & 9 & 7 & 1 \end{bmatrix}$, $\mathbf{b} = \begin{bmatrix} .100 \\ 2.888 \\ -1.404 \\ 1.462 \end{bmatrix}$,

$\Delta\mathbf{b} = 10^{-4}\begin{bmatrix} .49 \\ -1.28 \\ 5.78 \\ 8.04 \end{bmatrix}$

T 38. $A = \begin{bmatrix} 7 & -6 & -4 & 1 \\ -5 & 1 & 0 & -2 \\ 10 & 11 & 7 & -3 \\ 19 & 9 & 7 & 1 \end{bmatrix}$, $\mathbf{b} = \begin{bmatrix} 4.230 \\ -11.043 \\ 49.991 \\ 69.536 \end{bmatrix}$,

$\Delta\mathbf{b} = 10^{-4}\begin{bmatrix} .27 \\ 7.76 \\ -3.77 \\ 3.93 \end{bmatrix}$

7 Symmetric Matrices and Quadratic Forms

MULTICHANNEL IMAGE PROCESSING

Around the world in little more than 80 *minutes*, the two Landsat satellites streak silently across the sky in near polar orbits, recording images of terrain and coastline, in swaths 185 kilometers wide. Every 16 days, each satellite passes over almost every square kilometer of the earth's surface, so any location can be monitored every 8 days.

The Landsat images are useful for many purposes. Developers and urban planners use them to study the rate and direction of urban growth, industrial development, and other changes in land usage. Rural countries can analyze soil moisture, classify the vegetation in remote regions, and locate inland lakes and streams. Governments can detect and assess damage from natural disasters, such as forest fires, lava flows, floods, and hurricanes. Environmental agencies can identify pollution from smokestacks and measure water temperatures in lakes and rivers near power plants.

Sensors aboard the satellite acquire seven simultaneous images of any region on earth to be studied. The sensors record energy from separate wavelength bands—three in the visible light spectrum and four in infrared and thermal bands. Each image is digitized and stored as a rectangular array of numbers, each number indicating the signal intensity at a corresponding small point (or *pixel*) on the image. Each of the seven images is one channel of a *multichannel* or *multispectral image*.

The seven Landsat images of one fixed region typically contain much redundant information, since some features will appear in several images. Yet other features, because of their color or temperature, may reflect light that is recorded by only one or two sensors. One goal of multichannel image processing is to view the data in a way that extracts information better than studying each image separately.

Principal component analysis is an effective way to suppress redundant information and provide in only one or two composite images most of the information from the initial data. Roughly speaking, the goal is to find a special linear combination of the images, that is, a list of weights that at each pixel combine all seven corresponding image values into one new value. The weights are chosen in a way that makes the range of light intensities—the *scene variance*—in the composite image (called the *first principal component*) greater than that in any of the original images. Additional *component* images can also be constructed by criteria that will be explained in Section 7.5.

Principal component analysis is illustrated in the photos on the next page, taken over Railroad Valley, Nevada. Images from three Landsat spectral bands are shown in (a)–(c). The total information in the three bands is rearranged in the three principal component images in (d)–(f). The first component (d) displays (or "explains") 93.5% of the scene variance present in the initial data.

417

In this way, the three-channel initial data have been reduced to one-channel data, with a loss in some sense of only 6.5% of the scene variance.

Earth Satellite Corporation of Rockville, Maryland, which kindly supplied the photos shown here, is experimenting with images from 224 separate spectral bands. Principal component analysis, essential for such massive data sets, typically reduces the data to about 15 usable principal components.

(a) Spectral band 1: Visible blue.

(b) Spectral band 4: Near infrared.

(c) Spectral band 7: Mid-infrared.

(d) Principal component 1: 93.5%.

(e) Principal component 2: 5.3%.

(f) Principal component 3: 1.2%.

Symmetric matrices arise more often in applications, in one way or another, than any other major class of matrices. The theory is rich and beautiful, depending in an essential way on both diagonalization from Chapter 5 and orthogonality from Chapter 6. The diagonalization of a symmetric matrix, described in Section 7.1, is the foundation for the discussion in Sections 7.2 and 7.3 concerning quadratic forms. Section 7.3, in turn, is needed for the final two sections on the singular value decomposition and on the image processing described in the introductory example. Throughout the chapter, all vectors and matrices have real entries.

7.1 Diagonalization of Symmetric Matrices

A **symmetric** matrix is a matrix A such that $A^T = A$. Such a matrix is necessarily square. Its main diagonal entries are arbitrary, but its other entries occur in pairs—on opposite sides of the main diagonal.

EXAMPLE 1 Of the following matrices, only the first three are symmetric:

$$\text{Symmetric: } \begin{bmatrix} 1 & 0 \\ 0 & -3 \end{bmatrix}, \quad \begin{bmatrix} 0 & -1 & 0 \\ -1 & 5 & 8 \\ 0 & 8 & -7 \end{bmatrix}, \quad \begin{bmatrix} a & b & c \\ b & d & e \\ c & e & f \end{bmatrix}$$

$$\text{Nonsymmetric: } \begin{bmatrix} 1 & -3 \\ 3 & 0 \end{bmatrix}, \quad \begin{bmatrix} 1 & -4 & 0 \\ -6 & 1 & -4 \\ 0 & -6 & 1 \end{bmatrix}, \quad \begin{bmatrix} 5 & 4 & 3 & 2 \\ 4 & 3 & 2 & 1 \\ 3 & 2 & 1 & 0 \end{bmatrix} \qquad \blacksquare$$

To begin the study of symmetric matrices, it is helpful to review the diagonalization process of Section 5.3.

EXAMPLE 2 If possible, diagonalize the matrix $A = \begin{bmatrix} 6 & -2 & -1 \\ -2 & 6 & -1 \\ -1 & -1 & 5 \end{bmatrix}$.

SOLUTION The characteristic equation of A is

$$0 = -\lambda^3 + 17\lambda^2 - 90\lambda + 144 = -(\lambda - 8)(\lambda - 6)(\lambda - 3)$$

Standard calculations produce a basis for each eigenspace:

$$\lambda = 8: \mathbf{v}_1 = \begin{bmatrix} -1 \\ 1 \\ 0 \end{bmatrix}; \quad \lambda = 6: \mathbf{v}_2 = \begin{bmatrix} -1 \\ -1 \\ 2 \end{bmatrix}; \quad \lambda = 3: \mathbf{v}_3 = \begin{bmatrix} 1 \\ 1 \\ 1 \end{bmatrix}$$

These three vectors form a basis for $\mathbb{R}^3$. In fact, it is easy to check that $\{\mathbf{v}_1, \mathbf{v}_2, \mathbf{v}_3\}$ is an *orthogonal* basis for $\mathbb{R}^3$. Experience from Chapter 6 suggests that an *orthonormal* basis might be useful for calculations, so here are the normalized (unit) eigenvectors.

$$\mathbf{u}_1 = \begin{bmatrix} -1/\sqrt{2} \\ 1/\sqrt{2} \\ 0 \end{bmatrix}, \quad \mathbf{u}_2 = \begin{bmatrix} -1/\sqrt{6} \\ -1/\sqrt{6} \\ 2/\sqrt{6} \end{bmatrix}, \quad \mathbf{u}_3 = \begin{bmatrix} 1/\sqrt{3} \\ 1/\sqrt{3} \\ 1/\sqrt{3} \end{bmatrix}$$

Let

$$P = \begin{bmatrix} -1/\sqrt{2} & -1/\sqrt{6} & 1/\sqrt{3} \\ 1/\sqrt{2} & -1/\sqrt{6} & 1/\sqrt{3} \\ 0 & 2/\sqrt{6} & 1/\sqrt{3} \end{bmatrix}, \quad D = \begin{bmatrix} 8 & 0 & 0 \\ 0 & 6 & 0 \\ 0 & 0 & 3 \end{bmatrix}$$

Then $A = PDP^{-1}$, as usual. But this time, since P is square and has orthonormal columns, P is an *orthogonal* matrix, and P^{-1} is simply P^T. (See Section 6.2.) $\qquad \blacksquare$

Theorem 1 explains why the eigenvectors in Example 2 are orthogonal—they correspond to distinct eigenvalues.

THEOREM 1

> If A is symmetric, then any two eigenvectors from different eigenspaces are orthogonal.

PROOF Let $\mathbf{v}_1$ and $\mathbf{v}_2$ be eigenvectors that correspond to distinct eigenvalues, say, λ_1 and λ_2. To show that $\mathbf{v}_1 \cdot \mathbf{v}_2 = 0$, compute

$$
\begin{aligned}
\lambda_1 \mathbf{v}_1 \cdot \mathbf{v}_2 &= (\lambda_1 \mathbf{v}_1)^T \mathbf{v}_2 = (A\mathbf{v}_1)^T \mathbf{v}_2 && \text{Since } \mathbf{v}_1 \text{ is an eigenvector} \\
&= (\mathbf{v}_1^T A^T)\mathbf{v}_2 = \mathbf{v}_1^T (A\mathbf{v}_2) && \text{Since } A^T = A \\
&= \mathbf{v}_1^T (\lambda_2 \mathbf{v}_2) && \text{Since } \mathbf{v}_2 \text{ is an eigenvector} \\
&= \lambda_2 \mathbf{v}_1^T \mathbf{v}_2 = \lambda_2 \mathbf{v}_1 \cdot \mathbf{v}_2
\end{aligned}
$$

Hence $(\lambda_1 - \lambda_2)\mathbf{v}_1 \cdot \mathbf{v}_2 = 0$. But $\lambda_1 - \lambda_2 \neq 0$, so $\mathbf{v}_1 \cdot \mathbf{v}_2 = 0$. ∎

The special type of diagonalization in Example 2 is crucial for the theory of symmetric matrices. An $n \times n$ matrix A is said to be **orthogonally diagonalizable** if there are an orthogonal matrix P (with $P^{-1} = P^T$) and a diagonal matrix D such that

$$
A = PDP^T = PDP^{-1} \tag{1}
$$

Such a diagonalization requires n linearly independent and orthonormal eigenvectors. When is this possible? If A is orthogonally diagonalizable as in (1), then

$$
A^T = (PDP^T)^T = P^{TT} D^T P^T = PDP^T = A
$$

Thus A is symmetric! Theorem 2 below shows that, conversely, every symmetric matrix is orthogonally diagonalizable. The proof is much harder and is omitted; the main idea for a proof will be given after Theorem 3.

THEOREM 2

> An $n \times n$ matrix A is orthogonally diagonalizable if and only if A is a symmetric matrix.

This theorem is rather amazing, because the work in Chapter 5 would suggest that it is usually impossible to tell when a matrix is diagonalizable. But this is not the case for symmetric matrices.

The next example treats a matrix whose eigenvalues are not all distinct.

EXAMPLE 3 Orthogonally diagonalize the matrix $A = \begin{bmatrix} 3 & -2 & 4 \\ -2 & 6 & 2 \\ 4 & 2 & 3 \end{bmatrix}$, whose characteristic equation is

$$
0 = -\lambda^3 + 12\lambda^2 - 21\lambda - 98 = -(\lambda - 7)^2(\lambda + 2)
$$

SOLUTION The usual calculations produce bases for the eigenspaces:

$$
\lambda = 7: \mathbf{v}_1 = \begin{bmatrix} 1 \\ 0 \\ 1 \end{bmatrix}, \mathbf{v}_2 = \begin{bmatrix} -1/2 \\ 1 \\ 0 \end{bmatrix}; \qquad \lambda = -2: \mathbf{v}_3 = \begin{bmatrix} -1 \\ -1/2 \\ 1 \end{bmatrix}
$$

Although $\mathbf{v}_1$ and $\mathbf{v}_2$ are linearly independent, they are not orthogonal. Recall from Section 6.2 that the projection of $\mathbf{v}_2$ onto $\mathbf{v}_1$ is $\dfrac{\mathbf{v}_2 \cdot \mathbf{v}_1}{\mathbf{v}_1 \cdot \mathbf{v}_1}\mathbf{v}_1$, and the component of $\mathbf{v}_2$ orthogonal to $\mathbf{v}_1$ is

$$\mathbf{z}_2 = \mathbf{v}_2 - \frac{\mathbf{v}_2 \cdot \mathbf{v}_1}{\mathbf{v}_1 \cdot \mathbf{v}_1}\mathbf{v}_1 = \begin{bmatrix} -1/2 \\ 1 \\ 0 \end{bmatrix} - \frac{-1/2}{2}\begin{bmatrix} 1 \\ 0 \\ 1 \end{bmatrix} = \begin{bmatrix} -1/4 \\ 1 \\ 1/4 \end{bmatrix}$$

Then $\{\mathbf{v}_1, \mathbf{z}_2\}$ is an orthogonal set in the eigenspace for $\lambda = 7$. (Note that $\mathbf{z}_2$ is a linear combination of the eigenvectors $\mathbf{v}_1$ and $\mathbf{v}_2$, so $\mathbf{z}_2$ is in the eigenspace. This construction of $\mathbf{z}_2$ is just the Gram–Schmidt process of Section 6.4.) Since the eigenspace is two-dimensional (with basis $\mathbf{v}_1, \mathbf{v}_2$), the orthogonal set $\{\mathbf{v}_1, \mathbf{z}_2\}$ is an *orthogonal basis* for the eigenspace, by the Basis Theorem. (See Section 2.9 or 4.5.)

Normalize $\mathbf{v}_1$ and $\mathbf{z}_2$ to obtain the following orthonormal basis for the eigenspace for $\lambda = 7$:

$$\mathbf{u}_1 = \begin{bmatrix} 1/\sqrt{2} \\ 0 \\ 1/\sqrt{2} \end{bmatrix}, \quad \mathbf{u}_2 = \begin{bmatrix} -1/\sqrt{18} \\ 4/\sqrt{18} \\ 1/\sqrt{18} \end{bmatrix}$$

An orthonormal basis for the eigenspace for $\lambda = -2$ is

$$\mathbf{u}_3 = \frac{1}{\|2\mathbf{v}_3\|}2\mathbf{v}_3 = \frac{1}{3}\begin{bmatrix} -2 \\ -1 \\ 2 \end{bmatrix} = \begin{bmatrix} -2/3 \\ -1/3 \\ 2/3 \end{bmatrix}$$

By Theorem 1, $\mathbf{u}_3$ is orthogonal to the other eigenvectors $\mathbf{u}_1$ and $\mathbf{u}_2$. Hence $\{\mathbf{u}_1, \mathbf{u}_2, \mathbf{u}_3\}$ is an orthonormal set. Let

$$P = [\mathbf{u}_1 \ \ \mathbf{u}_2 \ \ \mathbf{u}_3] = \begin{bmatrix} 1/\sqrt{2} & -1/\sqrt{18} & -2/3 \\ 0 & 4/\sqrt{18} & -1/3 \\ 1/\sqrt{2} & 1/\sqrt{18} & 2/3 \end{bmatrix}, \quad D = \begin{bmatrix} 7 & 0 & 0 \\ 0 & 7 & 0 \\ 0 & 0 & -2 \end{bmatrix}$$

Then P orthogonally diagonalizes A, and $A = PDP^{-1}$. ∎

In Example 3, the eigenvalue 7 has multiplicity two and the eigenspace is two-dimensional. This fact is not accidental, as the next theorem shows.

The Spectral Theorem

The set of eigenvalues of a matrix A is sometimes called the *spectrum* of A, and the following description of the eigenvalues is called a *spectral theorem*.

THEOREM 3

The Spectral Theorem for Symmetric Matrices

An $n \times n$ symmetric matrix A has the following properties:

a. A has n real eigenvalues, counting multiplicities.

b. The dimension of the eigenspace for each eigenvalue λ equals the multiplicity of λ as a root of the characteristic equation.

c. The eigenspaces are mutually orthogonal, in the sense that eigenvectors corresponding to different eigenvalues are orthogonal.

d. A is orthogonally diagonalizable.

Part (a) follows from Exercise 28 in Section 5.5. Part (b) follows easily from part (d). (See Exercise 37.) Part (c) is Theorem 1. Because of (a), a proof of (d) can be given using Exercise 38 and the Schur factorization discussed in Supplementary Exercise 34 in Chapter 6. The details are omitted.

Spectral Decomposition

Suppose $A = PDP^{-1}$, where the columns of P are orthonormal eigenvectors $\mathbf{u}_1, \ldots, \mathbf{u}_n$ of A and the corresponding eigenvalues $\lambda_1, \ldots, \lambda_n$ are in the diagonal matrix D. Then, since $P^{-1} = P^T$,

$$A = PDP^T = \begin{bmatrix} \mathbf{u}_1 & \cdots & \mathbf{u}_n \end{bmatrix} \begin{bmatrix} \lambda_1 & & 0 \\ & \ddots & \\ 0 & & \lambda_n \end{bmatrix} \begin{bmatrix} \mathbf{u}_1^T \\ \vdots \\ \mathbf{u}_n^T \end{bmatrix}$$

$$= \begin{bmatrix} \lambda_1 \mathbf{u}_1 & \cdots & \lambda_n \mathbf{u}_n \end{bmatrix} \begin{bmatrix} \mathbf{u}_1^T \\ \vdots \\ \mathbf{u}_n^T \end{bmatrix}$$

Using the column–row expansion of a product (Theorem 10 in Section 2.4), we can write

$$A = \lambda_1 \mathbf{u}_1 \mathbf{u}_1^T + \lambda_2 \mathbf{u}_2 \mathbf{u}_2^T + \cdots + \lambda_n \mathbf{u}_n \mathbf{u}_n^T \tag{2}$$

This representation of A is called a **spectral decomposition** of A because it breaks up A into pieces determined by the spectrum (eigenvalues) of A. Each term in (2) is an $n \times n$ matrix of rank 1. For example, every column of $\lambda_1 \mathbf{u}_1 \mathbf{u}_1^T$ is a multiple of $\mathbf{u}_1$. Furthermore, each matrix $\mathbf{u}_j \mathbf{u}_j^T$ is a **projection matrix** in the sense that for each $\mathbf{x}$ in $\mathbb{R}^n$, the vector $(\mathbf{u}_j \mathbf{u}_j^T)\mathbf{x}$ is the orthogonal projection of $\mathbf{x}$ onto the subspace spanned by $\mathbf{u}_j$. (See Exercise 41.)

EXAMPLE 4 Construct a spectral decomposition of the matrix A that has the orthogonal diagonalization

$$A = \begin{bmatrix} 7 & 2 \\ 2 & 4 \end{bmatrix} = \begin{bmatrix} 2/\sqrt{5} & -1/\sqrt{5} \\ 1/\sqrt{5} & 2/\sqrt{5} \end{bmatrix} \begin{bmatrix} 8 & 0 \\ 0 & 3 \end{bmatrix} \begin{bmatrix} 2/\sqrt{5} & 1/\sqrt{5} \\ -1/\sqrt{5} & 2/\sqrt{5} \end{bmatrix}$$

SOLUTION Denote the columns of P by $\mathbf{u}_1$ and $\mathbf{u}_2$. Then

$$A = 8\mathbf{u}_1 \mathbf{u}_1^T + 3\mathbf{u}_2 \mathbf{u}_2^T$$

To verify this decomposition of A, compute

$$\mathbf{u}_1 \mathbf{u}_1^T = \begin{bmatrix} 2/\sqrt{5} \\ 1/\sqrt{5} \end{bmatrix} \begin{bmatrix} 2/\sqrt{5} & 1/\sqrt{5} \end{bmatrix} = \begin{bmatrix} 4/5 & 2/5 \\ 2/5 & 1/5 \end{bmatrix}$$

$$\mathbf{u}_2 \mathbf{u}_2^T = \begin{bmatrix} -1/\sqrt{5} \\ 2/\sqrt{5} \end{bmatrix} \begin{bmatrix} -1/\sqrt{5} & 2/\sqrt{5} \end{bmatrix} = \begin{bmatrix} 1/5 & -2/5 \\ -2/5 & 4/5 \end{bmatrix}$$

and

$$8\mathbf{u}_1 \mathbf{u}_1^T + 3\mathbf{u}_2 \mathbf{u}_2^T = \begin{bmatrix} 32/5 & 16/5 \\ 16/5 & 8/5 \end{bmatrix} + \begin{bmatrix} 3/5 & -6/5 \\ -6/5 & 12/5 \end{bmatrix} = \begin{bmatrix} 7 & 2 \\ 2 & 4 \end{bmatrix} = A \quad \blacksquare$$

Numerical Notes

When A is symmetric and not too large, modern high-performance computer algorithms calculate eigenvalues and eigenvectors with great precision. They apply a sequence of similarity transformations to A involving orthogonal matrices. The diagonal entries of the transformed matrices converge rapidly to the eigenvalues of A. (See the Numerical Notes in Section 5.2.) Using orthogonal matrices generally prevents numerical errors from accumulating during the process. When A is symmetric, the sequence of orthogonal matrices combines to form an orthogonal matrix whose columns are eigenvectors of A.

A nonsymmetric matrix cannot have a full set of orthogonal eigenvectors, but the algorithm still produces fairly accurate eigenvalues. After that, nonorthogonal techniques are needed to calculate eigenvectors.

Practice Problems

1. Show that if A is a symmetric matrix, then A^2 is symmetric.
2. Show that if A is orthogonally diagonalizable, then so is A^2.

7.1 Exercises

Determine which of the matrices in Exercises 1–6 are symmetric.

1. $\begin{bmatrix} 3 & 5 \\ 5 & -7 \end{bmatrix}$

2. $\begin{bmatrix} 3 & -5 \\ -5 & -3 \end{bmatrix}$

3. $\begin{bmatrix} 2 & 3 \\ 2 & 4 \end{bmatrix}$

4. $\begin{bmatrix} 0 & 8 & 3 \\ 8 & 0 & -4 \\ 3 & 2 & 0 \end{bmatrix}$

5. $\begin{bmatrix} -6 & 2 & 0 \\ 2 & -6 & 2 \\ 0 & 2 & -6 \end{bmatrix}$

6. $\begin{bmatrix} 1 & 2 & 2 & 1 \\ 2 & 2 & 2 & 1 \\ 2 & 2 & 1 & 2 \end{bmatrix}$

Determine which of the matrices in Exercises 7–12 are orthogonal. If orthogonal, find the inverse.

7. $\begin{bmatrix} .6 & .8 \\ .8 & -.6 \end{bmatrix}$

8. $\begin{bmatrix} 1 & 1 \\ 1 & -1 \end{bmatrix}$

9. $\begin{bmatrix} -4/5 & 3/5 \\ 3/5 & 4/5 \end{bmatrix}$

10. $\begin{bmatrix} 1/3 & 2/3 & 2/3 \\ 2/3 & 1/3 & -2/3 \\ 2/3 & -2/3 & 1/3 \end{bmatrix}$

11. $\begin{bmatrix} 2/3 & 2/3 & 1/3 \\ 0 & 1/3 & -2/3 \\ 5/3 & -4/3 & -2/3 \end{bmatrix}$

12. $\begin{bmatrix} .5 & .5 & -.5 & -.5 \\ .5 & .5 & .5 & .5 \\ .5 & -.5 & -.5 & .5 \\ .5 & -.5 & .5 & -.5 \end{bmatrix}$

Orthogonally diagonalize the matrices in Exercises 13–22, giving an orthogonal matrix P and a diagonal matrix D. To save you time, the eigenvalues in Exercises 17–22 are the following: (17) $-4, 4, 7$; (18) $-3, -6, 9$; (19) $-2, 7$; (20) $-3, 15$; (21) 1, 5, 9; (22) 3, 5.

13. $\begin{bmatrix} 3 & 1 \\ 1 & 3 \end{bmatrix}$

14. $\begin{bmatrix} 1 & -5 \\ -5 & 1 \end{bmatrix}$

15. $\begin{bmatrix} 3 & 4 \\ 4 & 9 \end{bmatrix}$

16. $\begin{bmatrix} 6 & -2 \\ -2 & 9 \end{bmatrix}$

17. $\begin{bmatrix} 1 & 1 & 5 \\ 1 & 5 & 1 \\ 5 & 1 & 1 \end{bmatrix}$

18. $\begin{bmatrix} 1 & -6 & 4 \\ -6 & 2 & -2 \\ 4 & -2 & -3 \end{bmatrix}$

19. $\begin{bmatrix} 3 & -2 & 4 \\ -2 & 6 & 2 \\ 4 & 2 & 3 \end{bmatrix}$

20. $\begin{bmatrix} 5 & 8 & -4 \\ 8 & 5 & -4 \\ -4 & -4 & -1 \end{bmatrix}$

21. $\begin{bmatrix} 4 & 3 & 1 & 1 \\ 3 & 4 & 1 & 1 \\ 1 & 1 & 4 & 3 \\ 1 & 1 & 3 & 4 \end{bmatrix}$

22. $\begin{bmatrix} 4 & 0 & 1 & 0 \\ 0 & 4 & 0 & 1 \\ 1 & 0 & 4 & 0 \\ 0 & 1 & 0 & 4 \end{bmatrix}$

23. Let $A = \begin{bmatrix} 4 & -1 & -1 \\ -1 & 4 & -1 \\ -1 & -1 & 4 \end{bmatrix}$ and $\mathbf{v} = \begin{bmatrix} 1 \\ 1 \\ 1 \end{bmatrix}$. Verify that 5 is an eigenvalue of A and $\mathbf{v}$ is an eigenvector. Then orthogonally diagonalize A.

24. Let $A = \begin{bmatrix} 2 & -1 & 1 \\ -1 & 2 & -1 \\ 1 & -1 & 2 \end{bmatrix}$, $\mathbf{v}_1 = \begin{bmatrix} -1 \\ 0 \\ 1 \end{bmatrix}$, and $\mathbf{v}_2 = \begin{bmatrix} 1 \\ -1 \\ 1 \end{bmatrix}$. Verify that $\mathbf{v}_1$ and $\mathbf{v}_2$ are eigenvectors of A. Then orthogonally diagonalize A.

In Exercises 25–32, mark each statement True or False (**T/F**). Justify each answer.

25. (**T/F**) An $n \times n$ matrix that is orthogonally diagonalizable must be symmetric.

26. (**T/F**) There are symmetric matrices that are not orthogonally diagonalizable.

27. (**T/F**) An orthogonal matrix is orthogonally diagonalizable.

28. (**T/F**) If $B = PDP^T$, where $P^T = P^{-1}$ and D is a diagonal matrix, then B is a symmetric matrix.

29. (**T/F**) For a nonzero $\mathbf{v}$ in $\mathbb{R}^n$, the matrix $\mathbf{v}\mathbf{v}^T$ is called a projection matrix.

30. (**T/F**) If $A^T = A$ and if vectors $\mathbf{u}$ and $\mathbf{v}$ satisfy $A\mathbf{u} = 3\mathbf{u}$ and $A\mathbf{v} = 4\mathbf{v}$, then $\mathbf{u} \cdot \mathbf{v} = 0$.

31. (**T/F**) An $n \times n$ symmetric matrix has n distinct real eigenvalues.

32. (**T/F**) The dimension of an eigenspace of a symmetric matrix is sometimes less than the multiplicity of the corresponding eigenvalue.

33. Show that if A is an $n \times n$ symmetric matrix, then $(A\mathbf{x})\cdot\mathbf{y} = \mathbf{x}\cdot(A\mathbf{y})$ for all $\mathbf{x}, \mathbf{y}$ in $\mathbb{R}^n$.

34. Suppose A is a symmetric $n \times n$ matrix and B is any $n \times m$ matrix. Show that B^TAB, B^TB, and BB^T are symmetric matrices.

35. Suppose A is invertible and orthogonally diagonalizable. Explain why A^{-1} is also orthogonally diagonalizable.

36. Suppose A and B are both orthogonally diagonalizable and $AB = BA$. Explain why AB is also orthogonally diagonalizable.

37. Let $A = PDP^{-1}$, where P is orthogonal and D is diagonal, and let λ be an eigenvalue of A of multiplicity k. Then λ appears k times on the diagonal of D. Explain why the dimension of the eigenspace for λ is k.

38. Suppose $A = PRP^{-1}$, where P is orthogonal and R is upper triangular. Show that if A is symmetric, then R is symmetric and hence is actually a diagonal matrix.

39. Construct a spectral decomposition of A from Example 2.

40. Construct a spectral decomposition of A from Example 3.

41. Let $\mathbf{u}$ be a unit vector in $\mathbb{R}^n$, and let $B = \mathbf{u}\mathbf{u}^T$.

a. Given any $\mathbf{x}$ in $\mathbb{R}^n$, compute $B\mathbf{x}$ and show that $B\mathbf{x}$ is the orthogonal projection of $\mathbf{x}$ onto $\mathbf{u}$, as described in Section 6.2.

b. Show that B is a symmetric matrix and $B^2 = B$.

c. Show that $\mathbf{u}$ is an eigenvector of B. What is the corresponding eigenvalue?

42. Let B be an $n \times n$ symmetric matrix such that $B^2 = B$. Any such matrix is called a **projection matrix** (or an **orthogonal projection matrix**). Given any $\mathbf{y}$ in $\mathbb{R}^n$, let $\hat{\mathbf{y}} = B\mathbf{y}$ and $\mathbf{z} = \mathbf{y} - \hat{\mathbf{y}}$.

a. Show that $\mathbf{z}$ is orthogonal to $\hat{\mathbf{y}}$.

b. Let W be the column space of B. Show that $\mathbf{y}$ is the sum of a vector in W and a vector in $W^\perp$. Why does this prove that $B\mathbf{y}$ is the orthogonal projection of $\mathbf{y}$ onto the column space of B?

Orthogonally diagonalize the matrices in Exercises 43–46. To practice the methods of this section, do not use an eigenvector routine from your matrix program. Instead, use the program to find the eigenvalues, and, for each eigenvalue λ, find an orthonormal basis for $\text{Nul}(A - \lambda I)$, as in Examples 2 and 3.

T 43. $\begin{bmatrix} 6 & 2 & 9 & -6 \\ 2 & 6 & -6 & 9 \\ 9 & -6 & 6 & 2 \\ -6 & 9 & 2 & 6 \end{bmatrix}$

T 44. $\begin{bmatrix} .63 & -.18 & -.06 & -.04 \\ -.18 & .84 & -.04 & .12 \\ -.06 & -.04 & .72 & -.12 \\ -.04 & .12 & -.12 & .66 \end{bmatrix}$

T 45. $\begin{bmatrix} .31 & .58 & .08 & .44 \\ .58 & -.56 & .44 & -.58 \\ .08 & .44 & .19 & -.08 \\ .44 & -.58 & -.08 & .31 \end{bmatrix}$

T 46. $\begin{bmatrix} 8 & 2 & 2 & -6 & 9 \\ 2 & 8 & 2 & -6 & 9 \\ 2 & 2 & 8 & -6 & 9 \\ -6 & -6 & -6 & 24 & 9 \\ 9 & 9 & 9 & 9 & -21 \end{bmatrix}$

Solutions to Practice Problems

1. $(A^2)^T = (AA)^T = A^TA^T$, by a property of transposes. By hypothesis, $A^T = A$. So $(A^2)^T = AA = A^2$, which shows that A^2 is symmetric.

2. If A is orthogonally diagonalizable, then A is symmetric, by Theorem 2. By Practice Problem 1, A^2 is symmetric and hence is orthogonally diagonalizable (Theorem 2).

7.2 Quadratic Forms

Until now, our attention in this text has focused on linear equations, except for the sums of squares encountered in Chapter 6 when computing $\mathbf{x}^T\mathbf{x}$. Such sums and more general expressions, called *quadratic forms*, occur frequently in applications of linear algebra to engineering (in design criteria and optimization) and signal processing (as output noise power). They also arise, for example, in physics (as potential and kinetic energy), differential geometry (as normal curvature of surfaces), economics (as utility functions), and statistics (in confidence ellipsoids). Some of the mathematical background for such applications flows easily from our work on symmetric matrices.

A **quadratic form** on $\mathbb{R}^n$ is a function Q defined on $\mathbb{R}^n$ whose value at a vector $\mathbf{x}$ in $\mathbb{R}^n$ can be computed by an expression of the form $Q(\mathbf{x}) = \mathbf{x}^TA\mathbf{x}$, where A is an $n \times n$ symmetric matrix. The matrix A is called the **matrix of the quadratic form**.

The simplest example of a nonzero quadratic form is $Q(\mathbf{x}) = \mathbf{x}^TI\mathbf{x} = \|\mathbf{x}\|^2$. Examples 1 and 2 show the connection between any symmetric matrix A and the quadratic form $\mathbf{x}^TA\mathbf{x}$.

EXAMPLE 1 Let $\mathbf{x} = \begin{bmatrix} x_1 \\ x_2 \end{bmatrix}$. Compute $\mathbf{x}^TA\mathbf{x}$ for the following matrices:

a. $A = \begin{bmatrix} 4 & 0 \\ 0 & 3 \end{bmatrix}$ b. $A = \begin{bmatrix} 3 & -2 \\ -2 & 7 \end{bmatrix}$

SOLUTION

a. $\mathbf{x}^TA\mathbf{x} = \begin{bmatrix} x_1 & x_2 \end{bmatrix} \begin{bmatrix} 4 & 0 \\ 0 & 3 \end{bmatrix} \begin{bmatrix} x_1 \\ x_2 \end{bmatrix} = \begin{bmatrix} x_1 & x_2 \end{bmatrix} \begin{bmatrix} 4x_1 \\ 3x_2 \end{bmatrix} = 4x_1^2 + 3x_2^2.$

b. There are two -2 entries in A. Watch how they enter the calculations. The $(1, 2)$-entry in A is in boldface type.

$$\mathbf{x}^TA\mathbf{x} = \begin{bmatrix} x_1 & x_2 \end{bmatrix} \begin{bmatrix} 3 & \mathbf{-2} \\ -2 & 7 \end{bmatrix} \begin{bmatrix} x_1 \\ x_2 \end{bmatrix} = \begin{bmatrix} x_1 & x_2 \end{bmatrix} \begin{bmatrix} 3x_1 - \mathbf{2}x_2 \\ -2x_1 + 7x_2 \end{bmatrix}$$

$$= x_1(3x_1 - \mathbf{2}x_2) + x_2(-2x_1 + 7x_2)$$

$$= 3x_1^2 - \mathbf{2}x_1x_2 - 2x_2x_1 + 7x_2^2$$

$$= 3x_1^2 - 4x_1x_2 + 7x_2^2 \qquad \blacksquare$$

The presence of $-4x_1x_2$ in the quadratic form in Example 1(b) is due to the -2 entries off the diagonal in the matrix A. In contrast, the quadratic form associated with the diagonal matrix A in Example 1(a) has no x_1x_2 *cross-product* term.

EXAMPLE 2 For $\mathbf{x}$ in $\mathbb{R}^3$, let $Q(\mathbf{x}) = 5x_1^2 + 3x_2^2 + 2x_3^2 - x_1x_2 + 8x_2x_3$. Write this quadratic form as $\mathbf{x}^TA\mathbf{x}$.

SOLUTION The coefficients of x_1^2, x_2^2, x_3^2 go on the diagonal of A. To make A symmetric, the coefficient of x_ix_j for $i \neq j$ must be split evenly between the (i, j)- and (j, i)-entries in A. The coefficient of x_1x_3 is 0. It is readily checked that

$$Q(\mathbf{x}) = \mathbf{x}^TA\mathbf{x} = \begin{bmatrix} x_1 & x_2 & x_3 \end{bmatrix} \begin{bmatrix} 5 & -1/2 & 0 \\ -1/2 & 3 & 4 \\ 0 & 4 & 2 \end{bmatrix} \begin{bmatrix} x_1 \\ x_2 \\ x_3 \end{bmatrix} \qquad \blacksquare$$

EXAMPLE 3 Let $Q(\mathbf{x}) = x_1^2 - 8x_1x_2 - 5x_2^2$. Compute the value of $Q(\mathbf{x})$ for $\mathbf{x} = \begin{bmatrix} -3 \\ 1 \end{bmatrix}, \begin{bmatrix} 2 \\ -2 \end{bmatrix}$, and $\begin{bmatrix} 1 \\ -3 \end{bmatrix}$.

SOLUTION

$$Q(-3, 1) = (-3)^2 - 8(-3)(1) - 5(1)^2 = 28$$
$$Q(2, -2) = (2)^2 - 8(2)(-2) - 5(-2)^2 = 16$$
$$Q(1, -3) = (1)^2 - 8(1)(-3) - 5(-3)^2 = -20$$

∎

In some cases, quadratic forms are easier to use when they have no cross-product terms—that is, when the matrix of the quadratic form is a diagonal matrix. Fortunately, the cross-product term can be eliminated by making a suitable change of variable.

Change of Variable in a Quadratic Form

If $\mathbf{x}$ represents a variable vector in $\mathbb{R}^n$, then a **change of variable** is an equation of the form

$$\mathbf{x} = P\mathbf{y}, \qquad \text{or equivalently,} \qquad \mathbf{y} = P^{-1}\mathbf{x} \tag{1}$$

where P is an invertible matrix and $\mathbf{y}$ is a new variable vector in $\mathbb{R}^n$. Here $\mathbf{y}$ is the coordinate vector of $\mathbf{x}$ relative to the basis of $\mathbb{R}^n$ determined by the columns of P. (See Section 4.4.)

If the change of variable (1) is made in a quadratic form $\mathbf{x}^TA\mathbf{x}$, then

$$\mathbf{x}^TA\mathbf{x} = (P\mathbf{y})^TA(P\mathbf{y}) = \mathbf{y}^TP^TAP\mathbf{y} = \mathbf{y}^T(P^TAP)\mathbf{y} \tag{2}$$

and the new matrix of the quadratic form is P^TAP. Since A is symmetric, Theorem 2 guarantees that there is an *orthogonal* matrix P such that P^TAP is a diagonal matrix D, and the quadratic form in (2) becomes $\mathbf{y}^TD\mathbf{y}$. This is the strategy of the next example.

EXAMPLE 4 Make a change of variable that transforms the quadratic form in Example 3 into a quadratic form with no cross-product term.

SOLUTION The matrix of the quadratic form in Example 3 is

$$A = \begin{bmatrix} 1 & -4 \\ -4 & -5 \end{bmatrix}$$

The first step is to orthogonally diagonalize A. Its eigenvalues turn out to be $\lambda = 3$ and $\lambda = -7$. Associated unit eigenvectors are

$$\lambda = 3: \begin{bmatrix} 2/\sqrt{5} \\ -1/\sqrt{5} \end{bmatrix}; \qquad \lambda = -7: \begin{bmatrix} 1/\sqrt{5} \\ 2/\sqrt{5} \end{bmatrix}$$

These vectors are automatically orthogonal (because they correspond to distinct eigenvalues) and so provide an orthonormal basis for $\mathbb{R}^2$. Let

$$P = \begin{bmatrix} 2/\sqrt{5} & 1/\sqrt{5} \\ -1/\sqrt{5} & 2/\sqrt{5} \end{bmatrix}, \qquad D = \begin{bmatrix} 3 & 0 \\ 0 & -7 \end{bmatrix}$$

Then $A = PDP^{-1}$ and $D = P^{-1}AP = P^TAP$, as pointed out earlier. A suitable change of variable is

$$\mathbf{x} = P\mathbf{y}, \qquad \text{where } \mathbf{x} = \begin{bmatrix} x_1 \\ x_2 \end{bmatrix} \text{ and } \mathbf{y} = \begin{bmatrix} y_1 \\ y_2 \end{bmatrix}$$

Then

$$x_1^2 - 8x_1x_2 - 5x_2^2 = \mathbf{x}^T A\mathbf{x} = (P\mathbf{y})^T A(P\mathbf{y})$$
$$= \mathbf{y}^T P^T A P\mathbf{y} = \mathbf{y}^T D\mathbf{y}$$
$$= 3y_1^2 - 7y_2^2 \qquad\blacksquare$$

To illustrate the meaning of the equality of quadratic forms in Example 4, we can compute $Q(\mathbf{x})$ for $\mathbf{x} = (2, -2)$ using the new quadratic form. First, since $\mathbf{x} = P\mathbf{y}$,

$$\mathbf{y} = P^{-1}\mathbf{x} = P^T\mathbf{x}$$

so

$$\mathbf{y} = \begin{bmatrix} 2/\sqrt{5} & -1/\sqrt{5} \\ 1/\sqrt{5} & 2/\sqrt{5} \end{bmatrix} \begin{bmatrix} 2 \\ -2 \end{bmatrix} = \begin{bmatrix} 6/\sqrt{5} \\ -2/\sqrt{5} \end{bmatrix}$$

Hence

$$3y_1^2 - 7y_2^2 = 3(6/\sqrt{5})^2 - 7(-2/\sqrt{5})^2 = 3(36/5) - 7(4/5)$$
$$= 80/5 = 16$$

This is the value of $Q(\mathbf{x})$ in Example 3 when $\mathbf{x} = (2, -2)$. See Figure 1.

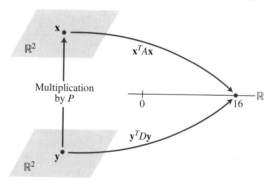

FIGURE 1 Change of variable in $\mathbf{x}^T A\mathbf{x}$.

Example 4 illustrates the following theorem. The proof of the theorem was essentially given before Example 4.

THEOREM 4

The Principal Axes Theorem

Let A be an $n \times n$ symmetric matrix. Then there is an orthogonal change of variable, $\mathbf{x} = P\mathbf{y}$, that transforms the quadratic form $\mathbf{x}^T A\mathbf{x}$ into a quadratic form $\mathbf{y}^T D\mathbf{y}$ with no cross-product term.

The columns of P in the theorem are called the **principal axes** of the quadratic form $\mathbf{x}^T A\mathbf{x}$. The vector $\mathbf{y}$ is the coordinate vector of $\mathbf{x}$ relative to the orthonormal basis of $\mathbb{R}^n$ given by these principal axes.

A Geometric View of Principal Axes

Suppose $Q(\mathbf{x}) = \mathbf{x}^T A\mathbf{x}$, where A is an invertible 2×2 symmetric matrix, and let c be a constant. It can be shown that the set of all $\mathbf{x}$ in $\mathbb{R}^2$ that satisfy

$$\mathbf{x}^T A\mathbf{x} = c \qquad (3)$$

either corresponds to an ellipse (or circle), a hyperbola, two intersecting lines, or a single point, or contains no points at all. If A is a diagonal matrix, the graph is in *standard position*, such as in Figure 2. If A is not a diagonal matrix, the graph of equation (3) is

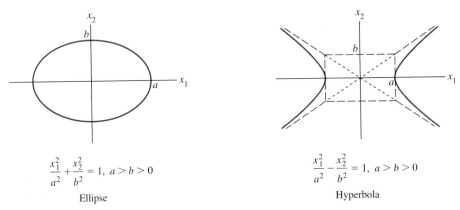

$$\frac{x_1^2}{a^2} + \frac{x_2^2}{b^2} = 1, \ a > b > 0$$

Ellipse

$$\frac{x_1^2}{a^2} - \frac{x_2^2}{b^2} = 1, \ a > b > 0$$

Hyperbola

FIGURE 2 An ellipse and a hyperbola in standard position.

rotated out of standard position, as in Figure 3. Finding the *principal axes* (determined by the eigenvectors of A) amounts to finding a new coordinate system with respect to which the graph is in standard position.

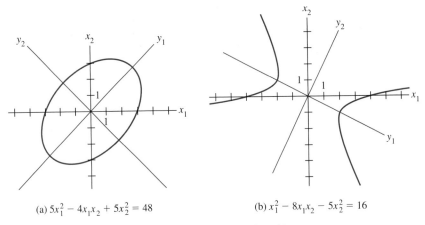

(a) $5x_1^2 - 4x_1x_2 + 5x_2^2 = 48$ (b) $x_1^2 - 8x_1x_2 - 5x_2^2 = 16$

FIGURE 3 An ellipse and a hyperbola *not* in standard position.

The hyperbola in Figure 3(b) is the graph of the equation $\mathbf{x}^T A \mathbf{x} = 16$, where A is the matrix in Example 4. The positive y_1-axis in Figure 3(b) is in the direction of the first column of the matrix P in Example 4, and the positive y_2-axis is in the direction of the second column of P.

EXAMPLE 5 The ellipse in Figure 3(a) is the graph of the equation $5x_1^2 - 4x_1x_2 + 5x_2^2 = 48$. Find a change of variable that removes the cross-product term from the equation.

SOLUTION The matrix of the quadratic form is $A = \begin{bmatrix} 5 & -2 \\ -2 & 5 \end{bmatrix}$. The eigenvalues of A turn out to be 3 and 7, with corresponding unit eigenvectors

$$\mathbf{u}_1 = \begin{bmatrix} 1/\sqrt{2} \\ 1/\sqrt{2} \end{bmatrix}, \quad \mathbf{u}_2 = \begin{bmatrix} -1/\sqrt{2} \\ 1/\sqrt{2} \end{bmatrix}$$

Let $P = [\,\mathbf{u}_1 \quad \mathbf{u}_2\,] = \begin{bmatrix} 1/\sqrt{2} & -1/\sqrt{2} \\ 1/\sqrt{2} & 1/\sqrt{2} \end{bmatrix}$. Then P orthogonally diagonalizes A, so the change of variable $\mathbf{x} = P\mathbf{y}$ produces the quadratic form $\mathbf{y}^T D\mathbf{y} = 3y_1^2 + 7y_2^2$. The new axes for this change of variable are shown in Figure 3(a).   ∎

Classifying Quadratic Forms

When A is an $n \times n$ matrix, the quadratic form $Q(\mathbf{x}) = \mathbf{x}^T A\mathbf{x}$ is a real-valued function with domain $\mathbb{R}^n$. Figure 4 displays the graphs of four quadratic forms with domain $\mathbb{R}^2$. For each point $\mathbf{x} = (x_1, x_2)$ in the domain of a quadratic form Q, the graph displays the point (x_1, x_2, z) where $z = Q(\mathbf{x})$. Notice that except at $\mathbf{x} = \mathbf{0}$, the values of $Q(\mathbf{x})$ are all positive in Figure 4(a) and all negative in Figure 4(d). The horizontal cross-sections of the graphs are ellipses in Figures 4(a) and 4(d) and hyperbolas in Figure 4(c).

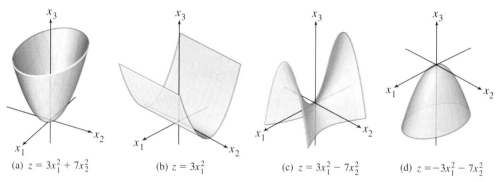

(a) $z = 3x_1^2 + 7x_2^2$ (b) $z = 3x_1^2$ (c) $z = 3x_1^2 - 7x_2^2$ (d) $z = -3x_1^2 - 7x_2^2$

FIGURE 4 Graphs of quadratic forms.

The simple 2×2 examples in Figure 4 illustrate the following definitions.

DEFINITION

A quadratic form Q is

a. **positive definite** if $Q(\mathbf{x}) > 0$ for all $\mathbf{x} \neq \mathbf{0}$,
b. **negative definite** if $Q(\mathbf{x}) < 0$ for all $\mathbf{x} \neq \mathbf{0}$,
c. **indefinite** if $Q(\mathbf{x})$ assumes both positive and negative values.

Also, Q is said to be **positive semidefinite** if $Q(\mathbf{x}) \geq 0$ for all $\mathbf{x}$, and to be **negative semidefinite** if $Q(\mathbf{x}) \leq 0$ for all $\mathbf{x}$. The quadratic forms in parts (a) and (b) of Figure 4 are both positive semidefinite, but the form in (a) is better described as positive definite. Theorem 5 characterizes some quadratic forms in terms of eigenvalues.

THEOREM 5

Quadratic Forms and Eigenvalues

Let A be an $n \times n$ symmetric matrix. Then a quadratic form $\mathbf{x}^T A\mathbf{x}$ is

a. positive definite if and only if the eigenvalues of A are all positive,
b. negative definite if and only if the eigenvalues of A are all negative, or
c. indefinite if and only if A has both positive and negative eigenvalues.

Positive definite

Negative definite

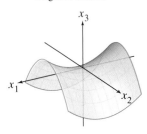

Indefinite

PROOF By the Principal Axes Theorem, there exists an orthogonal change of variable $\mathbf{x} = P\mathbf{y}$ such that

$$Q(\mathbf{x}) = \mathbf{x}^T A\mathbf{x} = \mathbf{y}^T D\mathbf{y} = \lambda_1 y_1^2 + \lambda_2 y_2^2 + \cdots + \lambda_n y_n^2 \tag{4}$$

where $\lambda_1, \ldots, \lambda_n$ are the eigenvalues of A. Since P is invertible, there is a one-to-one correspondence between all nonzero $\mathbf{x}$ and all nonzero $\mathbf{y}$. Thus the values of $Q(\mathbf{x})$ for $\mathbf{x} \neq \mathbf{0}$ coincide with the values of the expression on the right side of (4), which is obviously controlled by the signs of the eigenvalues $\lambda_1, \ldots, \lambda_n$, in the three ways described in the theorem. ∎

EXAMPLE 6 Is $Q(\mathbf{x}) = 3x_1^2 + 2x_2^2 + x_3^2 + 4x_1x_2 + 4x_2x_3$ positive definite?

SOLUTION Because of all the plus signs, this form "looks" positive definite. But the matrix of the form is

$$A = \begin{bmatrix} 3 & 2 & 0 \\ 2 & 2 & 2 \\ 0 & 2 & 1 \end{bmatrix}$$

and the eigenvalues of A turn out to be 5, 2, and -1. So Q is an indefinite quadratic form, not positive definite. ∎

The classification of a quadratic form is often carried over to the matrix of the form. Thus a **positive definite matrix** A is a *symmetric* matrix for which the quadratic form $\mathbf{x}^T A\mathbf{x}$ is positive definite. Other terms, such as **positive semidefinite matrix**, are defined analogously.

Numerical Notes

A fast way to determine whether a symmetric matrix A is positive definite is to attempt to factor A in the form $A = R^T R$, where R is upper triangular with positive diagonal entries. (A slightly modified algorithm for an LU factorization is one approach.) Such a *Cholesky factorization* is possible if and only if A is positive definite. See Supplementary Exercise 23 at the end of Chapter 7.

Practice Problem

Describe a positive semidefinite matrix A in terms of its eigenvalues.

7.2 Exercises

1. Compute the quadratic form $\mathbf{x}^T A\mathbf{x}$, when $A = \begin{bmatrix} 5 & 1/3 \\ 1/3 & 1 \end{bmatrix}$ and

   a. $\mathbf{x} = \begin{bmatrix} x_1 \\ x_2 \end{bmatrix}$ b. $\mathbf{x} = \begin{bmatrix} 6 \\ 1 \end{bmatrix}$ c. $\mathbf{x} = \begin{bmatrix} 1 \\ 3 \end{bmatrix}$

2. Compute the quadratic form $\mathbf{x}^T A\mathbf{x}$, for $A = \begin{bmatrix} 3 & 1 & 0 \\ 1 & 1 & 2 \\ 0 & 2 & 0 \end{bmatrix}$ and

   a. $\mathbf{x} = \begin{bmatrix} x_1 \\ x_2 \\ x_3 \end{bmatrix}$ b. $\mathbf{x} = \begin{bmatrix} -2 \\ -1 \\ 5 \end{bmatrix}$ c. $\mathbf{x} = \begin{bmatrix} 1/\sqrt{3} \\ 1/\sqrt{3} \\ 1/\sqrt{3} \end{bmatrix}$

3. Find the matrix of the quadratic form. Assume $\mathbf{x}$ is in $\mathbb{R}^2$.
   a. $3x_1^2 - 4x_1x_2 + 5x_2^2$ b. $3x_1^2 + 2x_1x_2$

4. Find the matrix of the quadratic form. Assume $\mathbf{x}$ is in $\mathbb{R}^2$.
   a. $5x_1^2 + 16x_1x_2 - 5x_2^2$ b. $2x_1x_2$

5. Find the matrix of the quadratic form. Assume $\mathbf{x}$ is in $\mathbb{R}^3$.
   a. $3x_1^2 + 2x_2^2 - 5x_3^2 - 6x_1x_2 + 8x_1x_3 - 4x_2x_3$
   b. $6x_1x_2 + 4x_1x_3 - 10x_2x_3$

6. Find the matrix of the quadratic form. Assume $\mathbf{x}$ is in $\mathbb{R}^3$.
   a. $3x_1^2 - 2x_2^2 + 5x_3^2 + 4x_1x_2 - 6x_1x_3$
   b. $4x_3^2 - 2x_1x_2 + 4x_2x_3$

7. Make a change of variable, $\mathbf{x} = P\mathbf{y}$, that transforms the quadratic form $x_1^2 + 10x_1x_2 + x_2^2$ into a quadratic form with no cross-product term. Give P and the new quadratic form.

8. Let A be the matrix of the quadratic form

$$9x_1^2 + 7x_2^2 + 11x_3^2 - 8x_1x_2 + 8x_1x_3$$

It can be shown that the eigenvalues of A are 3, 9, and 15. Find an orthogonal matrix P such that the change of variable $\mathbf{x} = P\mathbf{y}$ transforms $\mathbf{x}^T A\mathbf{x}$ into a quadratic form with no cross-product term. Give P and the new quadratic form.

Classify the quadratic forms in Exercises 9–18. Then make a change of variable, $\mathbf{x} = P\mathbf{y}$, that transforms the quadratic form into one with no cross-product term. Write the new quadratic form. Construct P using the methods of Section 7.1.

9. $4x_1^2 - 4x_1x_2 + 4x_2^2$

10. $2x_1^2 + 6x_1x_2 - 6x_2^2$

11. $2x_1^2 - 4x_1x_2 - x_2^2$

12. $-x_1^2 - 2x_1x_2 - x_2^2$

13. $x_1^2 - 6x_1x_2 + 9x_2^2$

14. $3x_1^2 + 4x_1x_2$

T 15. $-3x_1^2 - 7x_2^2 - 10x_3^2 - 10x_4^2 + 4x_1x_2 + 4x_1x_3 + 4x_1x_4 + 6x_3x_4$

T 16. $4x_1^2 + 4x_2^2 + 4x_3^2 + 4x_4^2 + 8x_1x_2 + 8x_3x_4 - 6x_1x_4 + 6x_2x_3$

T 17. $11x_1^2 + 11x_2^2 + 11x_3^2 + 11x_4^2 + 16x_1x_2 - 12x_1x_4 + 12x_2x_3 + 16x_3x_4$

T 18. $2x_1^2 + 2x_2^2 - 6x_1x_2 - 6x_1x_3 - 6x_1x_4 - 6x_2x_3 - 6x_2x_4 - 2x_3x_4$

19. What is the largest possible value of the quadratic form $5x_1^2 + 8x_2^2$ if $\mathbf{x} = (x_1, x_2)$ and $\mathbf{x}^T\mathbf{x} = 1$, that is, if $x_1^2 + x_2^2 = 1$? (Try some examples of $\mathbf{x}$.)

20. What is the largest value of the quadratic form $5x_1^2 - 3x_2^2$ if $\mathbf{x}^T\mathbf{x} = 1$?

In Exercises 21–30, matrices are $n \times n$ and vectors are in $\mathbb{R}^n$. Mark each statement True or False (**T/F**). Justify each answer.

21. (**T/F**) The matrix of a quadratic form is a symmetric matrix.

22. (**T/F**) The expression $\|\mathbf{x}\|^2$ is not a quadratic form.

23. (**T/F**) A quadratic form has no cross-product terms if and only if the matrix of the quadratic form is a diagonal matrix.

24. (**T/F**) If A is symmetric and P is an orthogonal matrix, then the change of variable $\mathbf{x} = P\mathbf{y}$ transforms $\mathbf{x}^T A\mathbf{x}$ into a quadratic form with no cross-product term.

25. (**T/F**) The principal axes of a quadratic form $\mathbf{x}^T A\mathbf{x}$ are eigenvectors of A.

26. (**T/F**) If the eigenvalues of a symmetric matrix A are all positive, then the quadratic form $\mathbf{x}^T A\mathbf{x}$ is positive definite.

27. (**T/F**) A positive definite quadratic form Q satisfies $Q(\mathbf{x}) > 0$ for all $\mathbf{x}$ in $\mathbb{R}^n$.

28. (**T/F**) An indefinite quadratic form is neither positive semidefinite nor negative semidefinite.

29. (**T/F**) A Cholesky factorization of a symmetric matrix A has the form $A = R^T R$, for an upper triangular matrix R with positive diagonal entries.

30. (**T/F**) If A is symmetric and the quadratic form $\mathbf{x}^T A\mathbf{x}$ has only negative values for $\mathbf{x} \neq \mathbf{0}$, then the eigenvalues of A are all positive.

Exercises 31 and 32 show how to classify a quadratic form $Q(\mathbf{x}) = \mathbf{x}^T A\mathbf{x}$, when $A = \begin{bmatrix} a & b \\ b & d \end{bmatrix}$ and $\det A \neq 0$, without finding the eigenvalues of A.

31. If λ_1 and λ_2 are the eigenvalues of A, then the characteristic polynomial of A can be written in two ways: $\det(A - \lambda I)$ and $(\lambda - \lambda_1)(\lambda - \lambda_2)$. Use this fact to show that $\lambda_1 + \lambda_2 = a + d$ (the diagonal entries of A) and $\lambda_1\lambda_2 = \det A$.

32. Verify the following statements:
   a. Q is positive definite if $\det A > 0$ and $a > 0$.
   b. Q is negative definite if $\det A > 0$ and $a < 0$.
   c. Q is indefinite if $\det A < 0$.

33. Show that if B is $m \times n$, then $B^T B$ is positive semidefinite; and if B is $n \times n$ and invertible, then $B^T B$ is positive definite.

34. Show that if an $n \times n$ matrix A is positive definite, then there exists a positive definite matrix B such that $A = B^T B$. [*Hint:* Write $A = PDP^T$, with $P^T = P^{-1}$. Produce a diagonal matrix C such that $D = C^T C$, and let $B = PCP^T$. Show that B works.]

35. Let A and B be symmetric $n \times n$ matrices whose eigenvalues are all positive. Show that the eigenvalues of $A + B$ are all positive. [*Hint:* Consider quadratic forms.]

36. Let A be an $n \times n$ invertible symmetric matrix. Show that if the quadratic form $\mathbf{x}^T A\mathbf{x}$ is positive definite, then so is the quadratic form $\mathbf{x}^T A^{-1}\mathbf{x}$. [*Hint:* Consider eigenvalues.]

STUDY GUIDE offers additional resources on diagonalization and quadratic forms.

x_3

x_2

x_1

Positive semidefinite

Solution to Practice Problem

Make an orthogonal change of variable $\mathbf{x} = P\mathbf{y}$, and write

$$\mathbf{x}^T A \mathbf{x} = \mathbf{y}^T D \mathbf{y} = \lambda_1 y_1^2 + \lambda_2 y_2^2 + \cdots + \lambda_n y_n^2$$

as in equation (4). If an eigenvalue—say, λ_i—were negative, then $\mathbf{x}^T A \mathbf{x}$ would be negative for the $\mathbf{x}$ corresponding to $\mathbf{y} = \mathbf{e}_i$ (the ith column of I_n). So the eigenvalues of a positive semidefinite quadratic form must all be nonnegative. Conversely, if the eigenvalues are nonnegative, the expansion above shows that $\mathbf{x}^T A \mathbf{x}$ must be positive semidefinite.

7.3 Constrained Optimization

Engineers, economists, scientists, and mathematicians often need to find the maximum or minimum value of a quadratic form $Q(\mathbf{x})$ for $\mathbf{x}$ in some specified set. Typically, the problem can be arranged so that $\mathbf{x}$ varies over the set of unit vectors. This *constrained optimization problem* has an interesting and elegant solution. Example 6 and the discussion in Section 7.5 will illustrate how such problems arise in practice.

The requirement that a vector $\mathbf{x}$ in $\mathbb{R}^n$ be a unit vector can be stated in several equivalent ways:

$$\|\mathbf{x}\| = 1, \qquad \|\mathbf{x}\|^2 = 1, \qquad \mathbf{x}^T \mathbf{x} = 1$$

and

$$x_1^2 + x_2^2 + \cdots + x_n^2 = 1 \tag{1}$$

The expanded version (1) of $\mathbf{x}^T \mathbf{x} = 1$ is commonly used in applications.

When a quadratic form Q has no cross-product terms, it is easy to find the maximum and minimum of $Q(\mathbf{x})$ for $\mathbf{x}^T \mathbf{x} = 1$.

EXAMPLE 1 Find the maximum and minimum values of $Q(\mathbf{x}) = 9x_1^2 + 4x_2^2 + 3x_3^2$ subject to the constraint $\mathbf{x}^T \mathbf{x} = 1$.

SOLUTION Since x_2^2 and x_3^2 are nonnegative, note that

$$4x_2^2 \leq 9x_2^2 \qquad \text{and} \qquad 3x_3^2 \leq 9x_3^2$$

and hence

$$\begin{aligned} Q(\mathbf{x}) &= 9x_1^2 + 4x_2^2 + 3x_3^2 \\ &\leq 9x_1^2 + 9x_2^2 + 9x_3^2 \\ &= 9(x_1^2 + x_2^2 + x_3^2) \\ &= 9 \end{aligned}$$

whenever $x_1^2 + x_2^2 + x_3^2 = 1$. So the maximum value of $Q(\mathbf{x})$ cannot exceed 9 when $\mathbf{x}$ is a unit vector. Furthermore, $Q(\mathbf{x}) = 9$ when $\mathbf{x} = (1, 0, 0)$. Thus 9 is the maximum value of $Q(\mathbf{x})$ for $\mathbf{x}^T \mathbf{x} = 1$.

To find the minimum value of $Q(\mathbf{x})$, observe that

$$9x_1^2 \geq 3x_1^2, \qquad 4x_2^2 \geq 3x_2^2$$

and hence

$$Q(\mathbf{x}) \geq 3x_1^2 + 3x_2^2 + 3x_3^2 = 3(x_1^2 + x_2^2 + x_3^2) = 3$$

whenever $x_1^2 + x_2^2 + x_3^2 = 1$. Also, $Q(\mathbf{x}) = 3$ when $x_1 = 0$, $x_2 = 0$, and $x_3 = 1$. So 3 is the minimum value of $Q(\mathbf{x})$ when $\mathbf{x}^T \mathbf{x} = 1$. ∎

It is easy to see in Example 1 that the matrix of the quadratic form Q has eigenvalues 9, 4, and 3 and that the greatest and least eigenvalues equal, respectively, the (constrained) maximum and minimum of $Q(\mathbf{x})$. The same holds true for any quadratic form, as we shall see.

EXAMPLE 2 Let $A = \begin{bmatrix} 3 & 0 \\ 0 & 7 \end{bmatrix}$, and let $Q(\mathbf{x}) = \mathbf{x}^T A \mathbf{x}$ for $\mathbf{x}$ in $\mathbb{R}^2$. Figure 1 displays the graph of Q. Figure 2 shows only the portion of the graph inside a cylinder; the intersection of the cylinder with the surface is the set of points (x_1, x_2, z) such that $z = Q(x_1, x_2)$ and $x_1^2 + x_2^2 = 1$. The "heights" of these points are the constrained values of $Q(\mathbf{x})$. Geometrically, the constrained optimization problem is to locate the highest and lowest points on the intersection curve.

The two highest points on the curve are 7 units above the $x_1 x_2$-plane, occurring where $x_1 = 0$ and $x_2 = \pm 1$. These points correspond to the eigenvalue 7 of A and the eigenvectors $\mathbf{x} = (0, 1)$ and $-\mathbf{x} = (0, -1)$. Similarly, the two lowest points on the curve are 3 units above the $x_1 x_2$-plane. They correspond to the eigenvalue 3 and the eigenvectors $(1, 0)$ and $(-1, 0)$. ∎

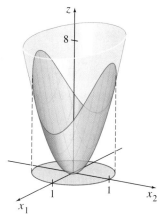

FIGURE 1 $z = 3x_1^2 + 7x_2^2$.

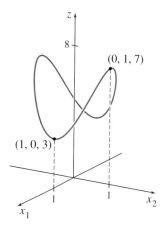

FIGURE 2 The intersection of $z = 3x_1^2 + 7x_2^2$ and the cylinder $x_1^2 + x_2^2 = 1$.

Every point on the intersection curve in Figure 2 has a z-coordinate between 3 and 7, and for any number t between 3 and 7, there is a unit vector $\mathbf{x}$ such that $Q(\mathbf{x}) = t$. In other words, the set of all possible values of $\mathbf{x}^T A \mathbf{x}$, for $\|\mathbf{x}\| = 1$, is the closed interval $3 \le t \le 7$.

It can be shown that for any symmetric matrix A, the set of all possible values of $\mathbf{x}^T A \mathbf{x}$, for $\|\mathbf{x}\| = 1$, is a closed interval on the real axis. (See Exercise 13.) Denote the left and right endpoints of this interval by m and M, respectively. That is, let

$$m = \min \{\mathbf{x}^T A \mathbf{x} : \|\mathbf{x}\| = 1\}, \quad M = \max \{\mathbf{x}^T A \mathbf{x} : \|\mathbf{x}\| = 1\} \tag{2}$$

Exercise 12 asks you to prove that if λ is an eigenvalue of A, then $m \le \lambda \le M$. The next theorem says that m and M are themselves eigenvalues of A, just as in Example 2.[1]

---

[1] The use of *minimum* and *maximum* in (2), and *least* and *greatest* in the theorem, refers to the natural ordering of the real numbers, not to magnitudes.

THEOREM 6

Let A be a symmetric matrix, and define m and M as in (2). Then M is the greatest eigenvalue λ_1 of A and m is the least eigenvalue of A. The value of $\mathbf{x}^T A \mathbf{x}$ is M when $\mathbf{x}$ is a unit eigenvector $\mathbf{u}_1$ corresponding to M. The value of $\mathbf{x}^T A \mathbf{x}$ is m when $\mathbf{x}$ is a unit eigenvector corresponding to m.

PROOF Orthogonally diagonalize A as PDP^{-1}. We know that

$$\mathbf{x}^T A \mathbf{x} = \mathbf{y}^T D \mathbf{y} \quad \text{when } \mathbf{x} = P\mathbf{y} \tag{3}$$

Also,

$$\|\mathbf{x}\| = \|P\mathbf{y}\| = \|\mathbf{y}\| \quad \text{for all } \mathbf{y}$$

because $P^T P = I$ and $\|P\mathbf{y}\|^2 = (P\mathbf{y})^T (P\mathbf{y}) = \mathbf{y}^T P^T P \mathbf{y} = \mathbf{y}^T \mathbf{y} = \|\mathbf{y}\|^2$. In particular, $\|\mathbf{y}\| = 1$ if and only if $\|\mathbf{x}\| = 1$. Thus $\mathbf{x}^T A \mathbf{x}$ and $\mathbf{y}^T D \mathbf{y}$ assume the same set of values as $\mathbf{x}$ and $\mathbf{y}$ range over the set of all unit vectors.

To simplify notation, suppose that A is a 3×3 matrix with eigenvalues $a \geq b \geq c$. Arrange the (eigenvector) columns of P so that $P = [\, \mathbf{u}_1 \quad \mathbf{u}_2 \quad \mathbf{u}_3 \,]$ and

$$D = \begin{bmatrix} a & 0 & 0 \\ 0 & b & 0 \\ 0 & 0 & c \end{bmatrix}$$

Given any unit vector $\mathbf{y}$ in $\mathbb{R}^3$ with coordinates y_1, y_2, y_3, observe that

$$ay_1^2 = ay_1^2$$
$$by_2^2 \leq ay_2^2$$
$$cy_3^2 \leq ay_3^2$$

and obtain these inequalities:

$$\begin{aligned} \mathbf{y}^T D \mathbf{y} &= ay_1^2 + by_2^2 + cy_3^2 \\ &\leq ay_1^2 + ay_2^2 + ay_3^2 \\ &= a(y_1^2 + y_2^2 + y_3^2) \\ &= a\|\mathbf{y}\|^2 = a \end{aligned}$$

Thus $M \leq a$, by definition of M. However, $\mathbf{y}^T D \mathbf{y} = a$ when $\mathbf{y} = \mathbf{e}_1 = (1, 0, 0)$, so in fact $M = a$. By (3), the $\mathbf{x}$ that corresponds to $\mathbf{y} = \mathbf{e}_1$ is the eigenvector $\mathbf{u}_1$ of A, because

$$\mathbf{x} = P\mathbf{e}_1 = \begin{bmatrix} \mathbf{u}_1 & \mathbf{u}_2 & \mathbf{u}_3 \end{bmatrix} \begin{bmatrix} 1 \\ 0 \\ 0 \end{bmatrix} = \mathbf{u}_1$$

Thus $M = a = \mathbf{e}_1^T D \mathbf{e}_1 = \mathbf{u}_1^T A \mathbf{u}_1$, which proves the statement about M. A similar argument shows that m is the least eigenvalue, c, and this value of $\mathbf{x}^T A \mathbf{x}$ is attained when $\mathbf{x} = P\mathbf{e}_3 = \mathbf{u}_3$. ∎

EXAMPLE 3 Let $A = \begin{bmatrix} 3 & 2 & 1 \\ 2 & 3 & 1 \\ 1 & 1 & 4 \end{bmatrix}$. Find the maximum value of the quadratic form $\mathbf{x}^T A \mathbf{x}$ subject to the constraint $\mathbf{x}^T \mathbf{x} = 1$, and find a unit vector at which this maximum value is attained.

SOLUTION By Theorem 6, the desired maximum value is the greatest eigenvalue of A. The characteristic equation turns out to be

$$0 = -\lambda^3 + 10\lambda^2 - 27\lambda + 18 = -(\lambda - 6)(\lambda - 3)(\lambda - 1)$$

The greatest eigenvalue is 6.

The constrained maximum of $\mathbf{x}^T A \mathbf{x}$ is attained when $\mathbf{x}$ is a unit eigenvector for $\lambda = 6$. Solve $(A - 6I)\mathbf{x} = 0$ and find an eigenvector $\begin{bmatrix} 1 \\ 1 \\ 1 \end{bmatrix}$. Set $\mathbf{u}_1 = \begin{bmatrix} 1/\sqrt{3} \\ 1/\sqrt{3} \\ 1/\sqrt{3} \end{bmatrix}$. ∎

In Theorem 7 and in later applications, the values of $\mathbf{x}^T A \mathbf{x}$ are computed with additional constraints on the unit vector $\mathbf{x}$.

THEOREM 7 Let A, λ_1, and $\mathbf{u}_1$ be as in Theorem 6. Then the maximum value of $\mathbf{x}^T A \mathbf{x}$ subject to the constraints

$$\mathbf{x}^T \mathbf{x} = 1, \quad \mathbf{x}^T \mathbf{u}_1 = 0$$

is the second greatest eigenvalue, λ_2, and this maximum is attained when $\mathbf{x}$ is an eigenvector $\mathbf{u}_2$ corresponding to λ_2.

Theorem 7 can be proved by an argument similar to the one above in which the theorem is reduced to the case where the matrix of the quadratic form is diagonal. The next example gives an idea of the proof for the case of a diagonal matrix.

EXAMPLE 4 Find the maximum value of $9x_1^2 + 4x_2^2 + 3x_3^2$ subject to the constraints $\mathbf{x}^T \mathbf{x} = 1$ and $\mathbf{x}^T \mathbf{u}_1 = 0$, where $\mathbf{u}_1 = (1, 0, 0)$. Note that $\mathbf{u}_1$ is a unit eigenvector corresponding to the greatest eigenvalue $\lambda = 9$ of the matrix of the quadratic form.

SOLUTION If the coordinates of $\mathbf{x}$ are x_1, x_2, x_3, then the constraint $\mathbf{x}^T \mathbf{u}_1 = 0$ means simply that $x_1 = 0$. For such a unit vector, $x_2^2 + x_3^2 = 1$, and

$$\begin{aligned} 9x_1^2 + 4x_2^2 + 3x_3^2 &= 4x_2^2 + 3x_3^2 \\ &\le 4x_2^2 + 4x_3^2 \\ &= 4(x_2^2 + x_3^2) \\ &= 4 \end{aligned}$$

Thus the constrained maximum of the quadratic form does not exceed 4. And this value is attained for $\mathbf{x} = (0, 1, 0)$, which is an eigenvector for the second greatest eigenvalue of the matrix of the quadratic form. ∎

EXAMPLE 5 Let A be the matrix in Example 3 and let $\mathbf{u}_1$ be a unit eigenvector corresponding to the greatest eigenvalue of A. Find the maximum value of $\mathbf{x}^T A \mathbf{x}$ subject to the conditions

$$\mathbf{x}^T \mathbf{x} = 1, \quad \mathbf{x}^T \mathbf{u}_1 = 0 \tag{4}$$

SOLUTION From Example 3, the second greatest eigenvalue of A is $\lambda = 3$. Solve $(A - 3I)\mathbf{x} = \mathbf{0}$ to find an eigenvector, and normalize it to obtain

$$\mathbf{u}_2 = \begin{bmatrix} 1/\sqrt{6} \\ 1/\sqrt{6} \\ -2/\sqrt{6} \end{bmatrix}$$

The vector $\mathbf{u}_2$ is automatically orthogonal to $\mathbf{u}_1$ because the vectors correspond to different eigenvalues. Thus the maximum of $\mathbf{x}^T A \mathbf{x}$ subject to the constraints in (4) is 3, attained when $\mathbf{x} = \mathbf{u}_2$. ∎

The next theorem generalizes Theorem 7 and, together with Theorem 6, gives a useful characterization of *all* the eigenvalues of A. The proof is omitted.

THEOREM 8

Let A be a symmetric $n \times n$ matrix with an orthogonal diagonalization $A = PDP^{-1}$, where the entries on the diagonal of D are arranged so that $\lambda_1 \geq \lambda_2 \geq \cdots \geq \lambda_n$ and where the columns of P are corresponding unit eigenvectors $\mathbf{u}_1, \dots, \mathbf{u}_n$. Then for $k = 2, \dots, n$, the maximum value of $\mathbf{x}^T A \mathbf{x}$ subject to the constraints

$$\mathbf{x}^T \mathbf{x} = 1, \quad \mathbf{x}^T \mathbf{u}_1 = 0, \quad \dots, \quad \mathbf{x}^T \mathbf{u}_{k-1} = 0$$

is the eigenvalue λ_k, and this maximum is attained at $\mathbf{x} = \mathbf{u}_k$.

Theorem 8 will be helpful in Sections 7.4 and 7.5. The following application requires only Theorem 6.

EXAMPLE 6 During the next year, a county government is planning to repair x hundred miles of public roads and bridges and to improve y hundred acres of parks and recreation areas. The county must decide how to allocate its resources (funds, equipment, labor, etc.) between these two projects. If it is more cost effective to work simultaneously on both projects rather than on only one, then x and y might satisfy a *constraint* such as

$$4x^2 + 9y^2 \leq 36$$

See Figure 3. Each point (x, y) in the shaded *feasible set* represents a possible public works schedule for the year. The points on the constraint curve, $4x^2 + 9y^2 = 36$, use the maximum amounts of resources available.

FIGURE 3 Public works schedules.

In choosing its public works schedule, the county wants to consider the opinions of the county residents. To measure the value, or *utility*, that the residents would assign to the various work schedules (x, y), economists sometimes use a function such as

$$q(x, y) = xy$$

The set of points (x, y) at which $q(x, y)$ is a constant is called an *indifference curve*. Three such curves are shown in Figure 4. Points along an indifference curve correspond to alternatives that county residents as a group would find equally valuable.[2] Find the public works schedule that maximizes the utility function q.

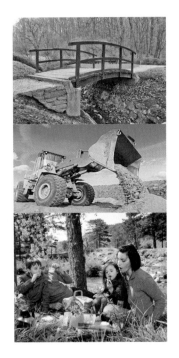

---

[2] Indifference curves are discussed in Michael D. Intriligator, Ronald G. Bodkin, and Cheng Hsiao, *Econometric Models, Techniques, and Applications* (Upper Saddle River, NJ: Prentice Hall, 1996).

FIGURE 4 The optimum public works schedule is $(2.1, 1.4)$.

SOLUTION   The constraint equation $4x^2 + 9y^2 = 36$ does not describe a set of unit vectors, but a change of variable can fix that problem. Rewrite the constraint in the form

$$\left(\frac{x}{3}\right)^2 + \left(\frac{y}{2}\right)^2 = 1$$

and define

$$x_1 = \frac{x}{3}, \quad x_2 = \frac{y}{2}, \quad \text{that is,} \quad x = 3x_1 \quad \text{and} \quad y = 2x_2$$

Then the constraint equation becomes

$$x_1^2 + x_2^2 = 1$$

and the utility function becomes $q(3x_1, 2x_2) = (3x_1)(2x_2) = 6x_1x_2$. Let $\mathbf{x} = \begin{bmatrix} x_1 \\ x_2 \end{bmatrix}$.
Then the problem is to maximize $Q(\mathbf{x}) = 6x_1x_2$ subject to $\mathbf{x}^T\mathbf{x} = 1$. Note that $Q(\mathbf{x}) = \mathbf{x}^T A \mathbf{x}$, where

$$A = \begin{bmatrix} 0 & 3 \\ 3 & 0 \end{bmatrix}$$

The eigenvalues of A are ± 3, with eigenvectors $\begin{bmatrix} 1/\sqrt{2} \\ 1/\sqrt{2} \end{bmatrix}$ for $\lambda = 3$ and $\begin{bmatrix} -1/\sqrt{2} \\ 1/\sqrt{2} \end{bmatrix}$ for
$\lambda = -3$. Thus the maximum value of $Q(\mathbf{x}) = q(x_1, x_2)$ is 3, attained when $x_1 = 1/\sqrt{2}$ and $x_2 = 1/\sqrt{2}$.

In terms of the original variables, the optimum public works schedule is $x = 3x_1 = 3/\sqrt{2} \approx 2.1$ hundred miles of roads and bridges and $y = 2x_2 = \sqrt{2} \approx 1.4$ hundred acres of parks and recreational areas. The optimum public works schedule is the point where the constraint curve and the indifference curve $q(x, y) = 3$ just meet. Points (x, y) with a higher utility lie on indifference curves that do not touch the constraint curve. See Figure 4.   ∎

Practice Problems

1. Let $Q(\mathbf{x}) = 3x_1^2 + 3x_2^2 + 2x_1x_2$. Find a change of variable that transforms Q into a quadratic form with no cross-product term, and give the new quadratic form.

2. With Q as in Problem 1, find the maximum value of $Q(\mathbf{x})$ subject to the constraint $\mathbf{x}^T\mathbf{x} = 1$, and find a unit vector at which the maximum is attained.

7.3 Exercises

In Exercises 1 and 2, find the change of variable $\mathbf{x} = P\mathbf{y}$ that transforms the quadratic form $\mathbf{x}^T A \mathbf{x}$ into $\mathbf{y}^T D \mathbf{y}$ as shown.

1. $5x_1^2 + 6x_2^2 + 7x_3^2 + 4x_1x_2 - 4x_2x_3 = 9y_1^2 + 6y_2^2 + 3y_3^2$

2. $3x_1^2 + 3x_2^2 + 5x_3^2 + 6x_1x_2 + 2x_1x_3 + 2x_2x_3 = 7y_1^2 + 4y_2^2$

Hint: $\mathbf{x}$ and $\mathbf{y}$ must have the same number of coordinates, so the quadratic form shown here must have a coefficient of zero for y_3^2. In Exercises 3–6, find (a) the maximum value of $Q(\mathbf{x})$ subject to the constraint $\mathbf{x}^T \mathbf{x} = 1$, (b) a unit vector $\mathbf{u}$ where this maximum is attained, and (c) the maximum of $Q(\mathbf{x})$ subject to the constraints $\mathbf{x}^T \mathbf{x} = 1$ and $\mathbf{x}^T \mathbf{u} = 0$.

3. $Q(\mathbf{x}) = 5x_1^2 + 6x_2^2 + 7x_3^2 + 4x_1x_2 - 4x_2x_3$
   (See Exercise 1.)

4. $Q(\mathbf{x}) = 3x_1^2 + 3x_2^2 + 5x_3^2 + 6x_1x_2 + 2x_1x_3 + 2x_2x_3$ (See Exercise 2.)

5. $Q(\mathbf{x}) = x_1^2 + x_2^2 - 10x_1x_2$

6. $Q(\mathbf{x}) = 3x_1^2 + 9x_2^2 + 8x_1x_2$

7. Let $Q(\mathbf{x}) = -2x_1^2 - x_2^2 + 4x_1x_2 + 4x_2x_3$. Find a unit vector $\mathbf{x}$ in $\mathbb{R}^3$ at which $Q(\mathbf{x})$ is maximized, subject to $\mathbf{x}^T \mathbf{x} = 1$. [*Hint:* The eigenvalues of the matrix of the quadratic form Q are 2, −1, and −4.]

8. Let $Q(\mathbf{x}) = 7x_1^2 + x_2^2 + 7x_3^2 - 8x_1x_2 - 4x_1x_3 - 8x_2x_3$. Find a unit vector $\mathbf{x}$ in $\mathbb{R}^3$ at which $Q(\mathbf{x})$ is maximized, subject to $\mathbf{x}^T \mathbf{x} = 1$. [*Hint:* The eigenvalues of the matrix of the quadratic form Q are 9 and −3.]

9. Find the maximum value of $Q(\mathbf{x}) = 7x_1^2 + 3x_2^2 - 2x_1x_2$, subject to the constraint $x_1^2 + x_2^2 = 1$. (Do not go on to find a vector where the maximum is attained.)

10. Find the maximum value of $Q(\mathbf{x}) = -3x_1^2 + 5x_2^2 - 2x_1x_2$, subject to the constraint $x_1^2 + x_2^2 = 1$. (Do not go on to find a vector where the maximum is attained.)

11. Suppose $\mathbf{x}$ is a unit eigenvector of a matrix A corresponding to an eigenvalue 3. What is the value of $\mathbf{x}^T A \mathbf{x}$?

12. Let λ be any eigenvalue of a symmetric matrix A. Justify the statement made in this section that $m \leq \lambda \leq M$, where m and M are defined as in (2). [*Hint:* Find an $\mathbf{x}$ such that $\lambda = \mathbf{x}^T A \mathbf{x}$.]

13. Let A be an $n \times n$ symmetric matrix, let M and m denote the maximum and minimum values of the quadratic form $\mathbf{x}^T A \mathbf{x}$, where $\mathbf{x}^T \mathbf{x} = 1$, and denote corresponding unit eigenvectors by $\mathbf{u}_1$ and $\mathbf{u}_n$. The following calculations show that given any number t between M and m, there is a unit vector $\mathbf{x}$ such that $t = \mathbf{x}^T A \mathbf{x}$. Verify that $t = (1 - \alpha)m + \alpha M$ for some number α between 0 and 1. Then let $\mathbf{x} = \sqrt{1 - \alpha}\,\mathbf{u}_n + \sqrt{\alpha}\,\mathbf{u}_1$, and show that $\mathbf{x}^T \mathbf{x} = 1$ and $\mathbf{x}^T A \mathbf{x} = t$.

In Exercises 14–17, follow the instructions given for Exercises 3–6.

T 14. $3x_1x_2 + 5x_1x_3 + 7x_1x_4 + 7x_2x_3 + 5x_2x_4 + 3x_3x_4$

T 15. $4x_1^2 - 6x_1x_2 - 10x_1x_3 - 10x_1x_4 - 6x_2x_3 - 6x_2x_4 - 2x_3x_4$

T 16. $-6x_1^2 - 10x_2^2 - 13x_3^2 - 13x_4^2 - 4x_1x_2 - 4x_1x_3 - 4x_1x_4 + 6x_3x_4$

T 17. $x_1x_2 + 3x_1x_3 + 30x_1x_4 + 30x_2x_3 + 3x_2x_4 + x_3x_4$

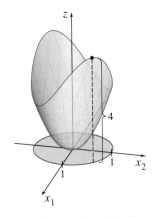

The maximum value of $Q(\mathbf{x})$ subject to $\mathbf{x}^T \mathbf{x} = 1$ is 4.

Solutions to Practice Problems

1. The matrix of the quadratic form is $A = \begin{bmatrix} 3 & 1 \\ 1 & 3 \end{bmatrix}$. It is easy to find the eigenvalues, 4 and 2, and corresponding unit eigenvectors, $\begin{bmatrix} 1/\sqrt{2} \\ 1/\sqrt{2} \end{bmatrix}$ and $\begin{bmatrix} -1/\sqrt{2} \\ 1/\sqrt{2} \end{bmatrix}$. So the desired change of variable is $\mathbf{x} = P\mathbf{y}$, where $P = \begin{bmatrix} 1/\sqrt{2} & -1/\sqrt{2} \\ 1/\sqrt{2} & 1/\sqrt{2} \end{bmatrix}$. (A common error here is to forget to normalize the eigenvectors.) The new quadratic form is $\mathbf{y}^T D \mathbf{y} = 4y_1^2 + 2y_2^2$.

2. The maximum of $Q(\mathbf{x})$, for a unit vector $\mathbf{x}$, is 4 and the maximum is attained at the unit eigenvector $\begin{bmatrix} 1/\sqrt{2} \\ 1/\sqrt{2} \end{bmatrix}$. [A common incorrect answer is $\begin{bmatrix} 1 \\ 0 \end{bmatrix}$. This vector maximizes the quadratic form $\mathbf{y}^T D \mathbf{y}$ instead of $Q(\mathbf{x})$.]

7.4 The Singular Value Decomposition

The diagonalization theorems in Sections 5.3 and 7.1 play a part in many interesting applications. Unfortunately, as we know, not all matrices can be factored as $A = PDP^{-1}$ with D diagonal. However, a factorization $A = QDP^{-1}$ *is* possible for any $m \times n$ matrix A! A special factorization of this type, called the *singular value decomposition*, is one of the most useful matrix factorizations in applied linear algebra.

The singular value decomposition is based on the following property of the ordinary diagonalization that can be imitated for rectangular matrices: The absolute values of the eigenvalues of a symmetric matrix A measure the amounts that A stretches or shrinks certain vectors (the eigenvectors). If $A\mathbf{x} = \lambda\mathbf{x}$ and $\|\mathbf{x}\| = 1$, then

$$\|A\mathbf{x}\| = \|\lambda\mathbf{x}\| = |\lambda| \, \|\mathbf{x}\| = |\lambda| \tag{1}$$

If λ_1 is the eigenvalue with the greatest magnitude, then a corresponding unit eigenvector $\mathbf{v}_1$ identifies a direction in which the stretching effect of A is greatest. That is, the length of $A\mathbf{x}$ is maximized when $\mathbf{x} = \mathbf{v}_1$, and $\|A\mathbf{v}_1\| = |\lambda_1|$, by (1). This description of $\mathbf{v}_1$ and $|\lambda_1|$ has an analogue for rectangular matrices that will lead to the singular value decomposition.

EXAMPLE 1 If $A = \begin{bmatrix} 4 & 11 & 14 \\ 8 & 7 & -2 \end{bmatrix}$, then the linear transformation $\mathbf{x} \mapsto A\mathbf{x}$ maps the unit sphere $\{\mathbf{x} : \|\mathbf{x}\| = 1\}$ in $\mathbb{R}^3$ onto an ellipse in $\mathbb{R}^2$, shown in Figure 1. Find a unit vector $\mathbf{x}$ at which the length $\|A\mathbf{x}\|$ is maximized, and compute this maximum length.

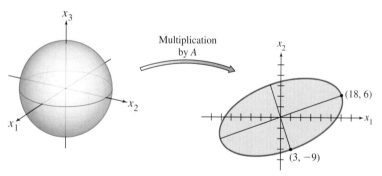

FIGURE 1 A transformation from $\mathbb{R}^3$ to $\mathbb{R}^2$.

SOLUTION The quantity $\|A\mathbf{x}\|^2$ is maximized at the same $\mathbf{x}$ that maximizes $\|A\mathbf{x}\|$, and $\|A\mathbf{x}\|^2$ is easier to study. Observe that

$$\|A\mathbf{x}\|^2 = (A\mathbf{x})^T (A\mathbf{x}) = \mathbf{x}^T A^T A\mathbf{x} = \mathbf{x}^T (A^T A)\mathbf{x}$$

Also, $A^T A$ is a symmetric matrix, since $(A^T A)^T = A^T A^{TT} = A^T A$. So the problem now is to maximize the quadratic form $\mathbf{x}^T (A^T A)\mathbf{x}$ subject to the constraint $\|\mathbf{x}\| = 1$. By Theorem 6 in Section 7.3, the maximum value is the greatest eigenvalue λ_1 of $A^T A$. Also, the maximum value is attained at a unit eigenvector of $A^T A$ corresponding to λ_1.

For the matrix A in this example,

$$A^T A = \begin{bmatrix} 4 & 8 \\ 11 & 7 \\ 14 & -2 \end{bmatrix} \begin{bmatrix} 4 & 11 & 14 \\ 8 & 7 & -2 \end{bmatrix} = \begin{bmatrix} 80 & 100 & 40 \\ 100 & 170 & 140 \\ 40 & 140 & 200 \end{bmatrix}$$

The eigenvalues of A^TA are $\lambda_1 = 360$, $\lambda_2 = 90$, and $\lambda_3 = 0$. Corresponding unit eigenvectors are, respectively,

$$\mathbf{v}_1 = \begin{bmatrix} 1/3 \\ 2/3 \\ 2/3 \end{bmatrix}, \quad \mathbf{v}_2 = \begin{bmatrix} -2/3 \\ -1/3 \\ 2/3 \end{bmatrix}, \quad \mathbf{v}_3 = \begin{bmatrix} 2/3 \\ -2/3 \\ 1/3 \end{bmatrix}$$

The maximum value of $\|A\mathbf{x}\|^2$ is 360, attained when $\mathbf{x}$ is the unit vector $\mathbf{v}_1$. The vector $A\mathbf{v}_1$ is a point on the ellipse in Figure 1 farthest from the origin, namely

$$A\mathbf{v}_1 = \begin{bmatrix} 4 & 11 & 14 \\ 8 & 7 & -2 \end{bmatrix} \begin{bmatrix} 1/3 \\ 2/3 \\ 2/3 \end{bmatrix} = \begin{bmatrix} 18 \\ 6 \end{bmatrix}$$

For $\|\mathbf{x}\| = 1$, the maximum value of $\|A\mathbf{x}\|$ is $\|A\mathbf{v}_1\| = \sqrt{360} = 6\sqrt{10}$. ∎

Example 1 suggests that the effect of A on the unit sphere in $\mathbb{R}^3$ is related to the quadratic form $\mathbf{x}^T(A^TA)\mathbf{x}$. In fact, the entire geometric behavior of the transformation $\mathbf{x} \mapsto A\mathbf{x}$ is captured by this quadratic form, as we shall see.

The Singular Values of an $m \times n$ Matrix

Let A be an $m \times n$ matrix. Then A^TA is symmetric and can be orthogonally diagonalized. Let $\{\mathbf{v}_1, \ldots, \mathbf{v}_n\}$ be an orthonormal basis for $\mathbb{R}^n$ consisting of eigenvectors of A^TA, and let $\lambda_1, \ldots, \lambda_n$ be the associated eigenvalues of A^TA. Then, for $1 \le i \le n$,

$$\begin{aligned} \|A\mathbf{v}_i\|^2 &= (A\mathbf{v}_i)^T A\mathbf{v}_i = \mathbf{v}_i^T A^T A\mathbf{v}_i \\ &= \mathbf{v}_i^T(\lambda_i \mathbf{v}_i) \qquad \text{Since } \mathbf{v}_i \text{ is an eigenvector of } A^TA \\ &= \lambda_i \qquad\qquad\;\; \text{Since } \mathbf{v}_i \text{ is a unit vector} \end{aligned} \tag{2}$$

So the eigenvalues of A^TA are all nonnegative. By renumbering, if necessary, we may assume that the eigenvalues are arranged so that

$$\lambda_1 \ge \lambda_2 \ge \cdots \ge \lambda_n \ge 0$$

The **singular values** of A are the square roots of the eigenvalues of A^TA, denoted by $\sigma_1, \ldots, \sigma_n$, and they are arranged in decreasing order. That is, $\sigma_i = \sqrt{\lambda_i}$ for $1 \le i \le n$. By equation (2), *the singular values of A are the lengths of the vectors $A\mathbf{v}_1, \ldots, A\mathbf{v}_n$.*

EXAMPLE 2 Let A be the matrix in Example 1. Since the eigenvalues of A^TA are 360, 90, and 0, the singular values of A are

$$\sigma_1 = \sqrt{360} = 6\sqrt{10}, \quad \sigma_2 = \sqrt{90} = 3\sqrt{10}, \quad \sigma_3 = 0$$

From Example 1, the first singular value of A is the maximum of $\|A\mathbf{x}\|$ over all unit vectors, and the maximum is attained at the unit eigenvector $\mathbf{v}_1$. Theorem 7 in Section 7.3 shows that the second singular value of A is the maximum of $\|A\mathbf{x}\|$ over all unit vectors that are *orthogonal to* $\mathbf{v}_1$, and this maximum is attained at the second unit eigenvector, $\mathbf{v}_2$ (Exercise 22). For the $\mathbf{v}_2$ in Example 1,

$$A\mathbf{v}_2 = \begin{bmatrix} 4 & 11 & 14 \\ 8 & 7 & -2 \end{bmatrix} \begin{bmatrix} -2/3 \\ -1/3 \\ 2/3 \end{bmatrix} = \begin{bmatrix} 3 \\ -9 \end{bmatrix}$$

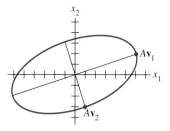

FIGURE 2

This point is on the minor axis of the ellipse in Figure 1, just as $A\mathbf{v}_1$ is on the major axis. (See Figure 2.) The first two singular values of A are the lengths of the major and minor semiaxes of the ellipse. ∎

The fact that $A\mathbf{v}_1$ and $A\mathbf{v}_2$ are orthogonal in Figure 2 is no accident, as the next theorem shows.

THEOREM 9

Suppose $\{\mathbf{v}_1, \ldots, \mathbf{v}_n\}$ is an orthonormal basis of $\mathbb{R}^n$ consisting of eigenvectors of A^TA, arranged so that the corresponding eigenvalues of A^TA satisfy $\lambda_1 \geq \cdots \geq \lambda_n$, and suppose A has r nonzero singular values. Then $\{A\mathbf{v}_1, \ldots, A\mathbf{v}_r\}$ is an orthogonal basis for Col A, and rank $A = r$.

PROOF Because $\mathbf{v}_i$ and $\lambda_j \mathbf{v}_j$ are orthogonal for $i \neq j$,

$$(A\mathbf{v}_i)^T (A\mathbf{v}_j) = \mathbf{v}_i^T A^T A \mathbf{v}_j = \mathbf{v}_i^T (\lambda_j \mathbf{v}_j) = 0$$

Thus $\{A\mathbf{v}_1, \ldots, A\mathbf{v}_n\}$ is an orthogonal set. Furthermore, since the lengths of the vectors $A\mathbf{v}_1, \ldots, A\mathbf{v}_n$ are the singular values of A, and since there are r nonzero singular values, $A\mathbf{v}_i \neq \mathbf{0}$ if and only if $1 \leq i \leq r$. So $A\mathbf{v}_1, \ldots, A\mathbf{v}_r$ are linearly independent vectors, and they are in Col A. Finally, for any $\mathbf{y}$ in Col A—say, $\mathbf{y} = A\mathbf{x}$—we can write $\mathbf{x} = c_1 \mathbf{v}_1 + \cdots + c_n \mathbf{v}_n$, and

$$\mathbf{y} = A\mathbf{x} = c_1 A\mathbf{v}_1 + \cdots + c_r A\mathbf{v}_r + c_{r+1} A\mathbf{v}_{r+1} + \cdots + c_n A\mathbf{v}_n$$
$$= c_1 A\mathbf{v}_1 + \cdots + c_r A\mathbf{v}_r + 0 + \cdots + 0$$

Thus $\mathbf{y}$ is in Span$\{A\mathbf{v}_1, \ldots, A\mathbf{v}_r\}$, which shows that $\{A\mathbf{v}_1, \ldots, A\mathbf{v}_r\}$ is an orthogonal basis for Col A. Hence rank $A = \dim \text{Col } A = r$. ∎

Numerical Notes

In some cases, the rank of A may be very sensitive to small changes in the entries of A. The obvious method of counting the number of pivot columns in A does not work well if A is row reduced by a computer. Roundoff error often creates an echelon form with full rank.

In practice, the most reliable way to estimate the rank of a large matrix A is to count the number of nonzero singular values. In this case, extremely small nonzero singular values are assumed to be zero for all practical purposes, and the *effective rank* of the matrix is the number obtained by counting the remaining nonzero singular values.[1]

---

[1] In general, rank estimation is not a simple problem. For a discussion of the subtle issues involved, see Philip E. Gill, Walter Murray, and Margaret H. Wright, *Numerical Linear Algebra and Optimization*, vol. 1 (Redwood City, CA: Addison-Wesley, 1991), Sec. 5.8.

The Singular Value Decomposition

The decomposition of A involves an $m \times n$ "diagonal" matrix Σ of the form

$$\Sigma = \begin{bmatrix} D & 0 \\ 0 & 0 \end{bmatrix} \;\begin{matrix} \leftarrow m - r \text{ rows} \\ \\ \end{matrix} \tag{3}$$
$$\begin{matrix} \uparrow \\ n - r \text{ columns} \end{matrix}$$

where D is an $r \times r$ diagonal matrix for some r not exceeding the smaller of m and n. (If r equals m or n or both, some or all of the zero matrices do not appear.)

THEOREM 10

The Singular Value Decomposition

Let A be an $m \times n$ matrix with rank r. Then there exists an $m \times n$ matrix Σ as in (3) for which the diagonal entries in D are the first r singular values of A, $\sigma_1 \geq \sigma_2 \geq \cdots \geq \sigma_r > 0$, and there exist an $m \times m$ orthogonal matrix U and an $n \times n$ orthogonal matrix V such that

$$A = U \Sigma V^T$$

Any factorization $A = U \Sigma V^T$, with U and V orthogonal, Σ as in (3), and positive diagonal entries in D, is called a **singular value decomposition** (or **SVD**) of A. The matrices U and V are not uniquely determined by A, but the diagonal entries of Σ are necessarily the singular values of A. See Exercise 19. The columns of U in such a decomposition are called **left singular vectors** of A, and the columns of V are called **right singular vectors** of A.

PROOF Let λ_i and $\mathbf{v}_i$ be as in Theorem 9, so that $\{A\mathbf{v}_1, \ldots, A\mathbf{v}_r\}$ is an orthogonal basis for Col A. Normalize each $A\mathbf{v}_i$ to obtain an orthonormal basis $\{\mathbf{u}_1, \ldots, \mathbf{u}_r\}$, where

$$\mathbf{u}_i = \frac{1}{\|A\mathbf{v}_i\|} A\mathbf{v}_i = \frac{1}{\sigma_i} A\mathbf{v}_i$$

and

$$A\mathbf{v}_i = \sigma_i \mathbf{u}_i \qquad (1 \leq i \leq r) \tag{4}$$

Now extend $\{\mathbf{u}_1, \ldots, \mathbf{u}_r\}$ to an orthonormal basis $\{\mathbf{u}_1, \ldots, \mathbf{u}_m\}$ of $\mathbb{R}^m$, and let

$$U = \begin{bmatrix} \mathbf{u}_1 & \mathbf{u}_2 & \cdots & \mathbf{u}_m \end{bmatrix} \quad \text{and} \quad V = \begin{bmatrix} \mathbf{v}_1 & \mathbf{v}_2 & \cdots & \mathbf{v}_n \end{bmatrix}$$

By construction, U and V are orthogonal matrices. Also, from (4),

$$AV = \begin{bmatrix} A\mathbf{v}_1 & \cdots & A\mathbf{v}_r & \mathbf{0} & \cdots & \mathbf{0} \end{bmatrix} = \begin{bmatrix} \sigma_1\mathbf{u}_1 & \cdots & \sigma_r\mathbf{u}_r & \mathbf{0} & \cdots & \mathbf{0} \end{bmatrix}$$

Let D be the diagonal matrix with diagonal entries $\sigma_1, \ldots, \sigma_r$, and let Σ be as in (3) above. Then

$$U\Sigma = [\begin{array}{cccc} \mathbf{u}_1 & \mathbf{u}_2 & \cdots & \mathbf{u}_m \end{array}] \begin{bmatrix} \sigma_1 & & & & 0 \\ & \sigma_2 & & & & 0 \\ & & \ddots & & \\ 0 & & & \sigma_r & \\ \hline & & 0 & & & 0 \end{bmatrix}$$

$$= [\begin{array}{ccccc} \sigma_1\mathbf{u}_1 & \cdots & \sigma_r\mathbf{u}_r & \mathbf{0} & \cdots & \mathbf{0} \end{array}]$$
$$= AV$$

Since V is an orthogonal matrix, $U\Sigma V^T = AVV^T = A$. ∎

The next two examples focus attention on the internal structure of a singular value decomposition. An efficient and numerically stable algorithm for this decomposition would use a different approach. See the Numerical Note at the end of the section.

EXAMPLE 3 Use the results of Examples 1 and 2 to construct a singular value decomposition of $A = \begin{bmatrix} 4 & 11 & 14 \\ 8 & 7 & -2 \end{bmatrix}$.

SOLUTION A construction can be divided into three steps.

STUDY GUIDE offers additional resources for learning to compute an SVD.

Step 1. ***Find an orthogonal diagonalization of A^TA.*** That is, find the eigenvalues of A^TA and a corresponding orthonormal set of eigenvectors. If A had only two columns, the calculations could be done by hand. Larger matrices usually require a matrix program. However, for the matrix A here, the eigendata for A^TA are provided in Example 1.

Step 2. ***Set up V and Σ.*** Arrange the eigenvalues of A^TA in decreasing order. In Example 1, the eigenvalues are already listed in decreasing order: 360, 90, and 0. The corresponding unit eigenvectors, $\mathbf{v}_1, \mathbf{v}_2$, and $\mathbf{v}_3$, are the right singular vectors of A. Using Example 1, construct

$$V = [\begin{array}{ccc} \mathbf{v}_1 & \mathbf{v}_2 & \mathbf{v}_3 \end{array}] = \begin{bmatrix} 1/3 & -2/3 & 2/3 \\ 2/3 & -1/3 & -2/3 \\ 2/3 & 2/3 & 1/3 \end{bmatrix}$$

The square roots of the eigenvalues are the singular values:

$$\sigma_1 = 6\sqrt{10}, \quad \sigma_2 = 3\sqrt{10}, \quad \sigma_3 = 0$$

The nonzero singular values are the diagonal entries of D. The matrix Σ is the same size as A, with D in its upper left corner and with 0's elsewhere.

$$D = \begin{bmatrix} 6\sqrt{10} & 0 \\ 0 & 3\sqrt{10} \end{bmatrix}, \quad \Sigma = [\begin{array}{cc} D & 0 \end{array}] = \begin{bmatrix} 6\sqrt{10} & 0 & 0 \\ 0 & 3\sqrt{10} & 0 \end{bmatrix}$$

Step 3. ***Construct U.*** When A has rank r, the first r columns of U are the normalized vectors obtained from $A\mathbf{v}_1, \ldots, A\mathbf{v}_r$. In this example, A has two nonzero singular values, so rank $A = 2$. Recall from equation (2) and the paragraph before Example 2 that $\|A\mathbf{v}_1\| = \sigma_1$ and $\|A\mathbf{v}_2\| = \sigma_2$. Thus

$$\mathbf{u}_1 = \frac{1}{\sigma_1}A\mathbf{v}_1 = \frac{1}{6\sqrt{10}}\begin{bmatrix} 18 \\ 6 \end{bmatrix} = \begin{bmatrix} 3/\sqrt{10} \\ 1/\sqrt{10} \end{bmatrix}$$

$$\mathbf{u}_2 = \frac{1}{\sigma_2}A\mathbf{v}_2 = \frac{1}{3\sqrt{10}}\begin{bmatrix} 3 \\ -9 \end{bmatrix} = \begin{bmatrix} 1/\sqrt{10} \\ -3/\sqrt{10} \end{bmatrix}$$

Note that $\{\mathbf{u}_1, \mathbf{u}_2\}$ is already a basis for $\mathbb{R}^2$. Thus no additional vectors are needed for U, and $U = [\,\mathbf{u}_1 \quad \mathbf{u}_2\,]$. The singular value decomposition of A is

$$A = \begin{bmatrix} 3/\sqrt{10} & 1/\sqrt{10} \\ 1/\sqrt{10} & -3/\sqrt{10} \end{bmatrix} \begin{bmatrix} 6\sqrt{10} & 0 & 0 \\ 0 & 3\sqrt{10} & 0 \end{bmatrix} \begin{bmatrix} 1/3 & 2/3 & 2/3 \\ -2/3 & -1/3 & 2/3 \\ 2/3 & -2/3 & 1/3 \end{bmatrix}$$

$$\underset{U}{\uparrow} \qquad\qquad \underset{\Sigma}{\uparrow} \qquad\qquad\qquad \underset{V^T}{\uparrow} \qquad\qquad \blacksquare$$

EXAMPLE 4 Find a singular value decomposition of $A = \begin{bmatrix} 1 & -1 \\ -2 & 2 \\ 2 & -2 \end{bmatrix}$.

SOLUTION First, compute $A^TA = \begin{bmatrix} 9 & -9 \\ -9 & 9 \end{bmatrix}$. The eigenvalues of A^TA are 18 and 0, with corresponding unit eigenvectors

$$\mathbf{v}_1 = \begin{bmatrix} 1/\sqrt{2} \\ -1/\sqrt{2} \end{bmatrix}, \quad \mathbf{v}_2 = \begin{bmatrix} 1/\sqrt{2} \\ 1/\sqrt{2} \end{bmatrix}$$

These unit vectors form the columns of V:

$$V = [\,\mathbf{v}_1 \quad \mathbf{v}_2\,] = \begin{bmatrix} 1/\sqrt{2} & 1/\sqrt{2} \\ -1/\sqrt{2} & 1/\sqrt{2} \end{bmatrix}$$

The singular values are $\sigma_1 = \sqrt{18} = 3\sqrt{2}$ and $\sigma_2 = 0$. Since there is only one nonzero singular value, the "matrix" D may be written as a single number. That is, $D = 3\sqrt{2}$. The matrix Σ is the same size as A, with D in its upper left corner:

$$\Sigma = \begin{bmatrix} D & 0 \\ 0 & 0 \\ 0 & 0 \end{bmatrix} = \begin{bmatrix} 3\sqrt{2} & 0 \\ 0 & 0 \\ 0 & 0 \end{bmatrix}$$

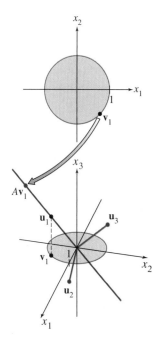

FIGURE 3

To construct U, first construct $A\mathbf{v}_1$ and $A\mathbf{v}_2$:

$$A\mathbf{v}_1 = \begin{bmatrix} 2/\sqrt{2} \\ -4/\sqrt{2} \\ 4/\sqrt{2} \end{bmatrix}, \quad A\mathbf{v}_2 = \begin{bmatrix} 0 \\ 0 \\ 0 \end{bmatrix}$$

As a check on the calculations, verify that $\|A\mathbf{v}_1\| = \sigma_1 = 3\sqrt{2}$. Of course, $A\mathbf{v}_2 = \mathbf{0}$ because $\|A\mathbf{v}_2\| = \sigma_2 = 0$. The only column found for U so far is

$$\mathbf{u}_1 = \frac{1}{3\sqrt{2}} A\mathbf{v}_1 = \begin{bmatrix} 1/3 \\ -2/3 \\ 2/3 \end{bmatrix}$$

The other columns of U are found by extending the set $\{\mathbf{u}_1\}$ to an orthonormal basis for $\mathbb{R}^3$. In this case, we need two orthogonal unit vectors $\mathbf{u}_2$ and $\mathbf{u}_3$ that are orthogonal to $\mathbf{u}_1$. (See Figure 3.) Each vector must satisfy $\mathbf{u}_1^T \mathbf{x} = 0$, which is equivalent to the equation $x_1 - 2x_2 + 2x_3 = 0$. A basis for the solution set of this equation is

$$\mathbf{w}_1 = \begin{bmatrix} 2 \\ 1 \\ 0 \end{bmatrix}, \quad \mathbf{w}_2 = \begin{bmatrix} -2 \\ 0 \\ 1 \end{bmatrix}$$

(Check that $\mathbf{w}_1$ and $\mathbf{w}_2$ are each orthogonal to $\mathbf{u}_1$.) Apply the Gram–Schmidt process (with normalizations) to $\{\mathbf{w}_1, \mathbf{w}_2\}$, and obtain

$$\mathbf{u}_2 = \begin{bmatrix} 2/\sqrt{5} \\ 1/\sqrt{5} \\ 0 \end{bmatrix}, \quad \mathbf{u}_3 = \begin{bmatrix} -2/\sqrt{45} \\ 4/\sqrt{45} \\ 5/\sqrt{45} \end{bmatrix}$$

Finally, set $U = [\,\mathbf{u}_1 \quad \mathbf{u}_2 \quad \mathbf{u}_3\,]$, take Σ and V^T from above, and write

$$A = \begin{bmatrix} 1 & -1 \\ -2 & 2 \\ 2 & -2 \end{bmatrix} = \begin{bmatrix} 1/3 & 2/\sqrt{5} & -2/\sqrt{45} \\ -2/3 & 1/\sqrt{5} & 4/\sqrt{45} \\ 2/3 & 0 & 5/\sqrt{45} \end{bmatrix} \begin{bmatrix} 3\sqrt{2} & 0 \\ 0 & 0 \\ 0 & 0 \end{bmatrix} \begin{bmatrix} 1/\sqrt{2} & -1/\sqrt{2} \\ 1/\sqrt{2} & 1/\sqrt{2} \end{bmatrix}$$

∎

Applications of the Singular Value Decomposition

The SVD is often used to estimate the rank of a matrix, as noted above. Several other numerical applications are described briefly below, and an application to image processing is presented in Section 7.5.

EXAMPLE 5 (The Condition Number) Most numerical calculations involving an equation $A\mathbf{x} = \mathbf{b}$ are as reliable as possible when the SVD of A is used. The two orthogonal matrices U and V do not affect lengths of vectors or angles between vectors (Theorem 7 in Section 6.2). Any possible instabilities in numerical calculations are identified in Σ. If the singular values of A are extremely large or small, roundoff errors are almost inevitable, but an error analysis is aided by knowing the entries in Σ and V.

If A is an invertible $n \times n$ matrix, then the ratio σ_1/σ_n of the largest and smallest singular values gives the **condition number** of A. Exercises 50–52 in Section 2.3 showed how the condition number affects the sensitivity of a solution of $A\mathbf{x} = \mathbf{b}$ to changes (or errors) in the entries of A. (Actually, a "condition number" of A can be computed in several ways, but the definition given here is widely used for studying $A\mathbf{x} = \mathbf{b}$.)   ∎

EXAMPLE 6 (Bases for Fundamental Subspaces) Given an SVD for an $m \times n$ matrix A, let $\mathbf{u}_1, \ldots, \mathbf{u}_m$ be the left singular vectors, $\mathbf{v}_1, \ldots, \mathbf{v}_n$ the right singular vectors, and $\sigma_1, \ldots, \sigma_n$ the singular values, and let r be the rank of A. By Theorem 9,

$$\{\mathbf{u}_1, \ldots, \mathbf{u}_r\} \tag{5}$$

is an orthonormal basis for Col A.

Recall from Theorem 3 in Section 6.1 that $(\text{Col } A)^{\perp} = \text{Nul } A^T$. Hence

$$\{\mathbf{u}_{r+1}, \ldots, \mathbf{u}_m\} \tag{6}$$

is an orthonormal basis for Nul A^T.

Since $\|A\mathbf{v}_i\| = \sigma_i$ for $1 \le i \le n$, and σ_i is 0 if and only if $i > r$, the vectors $\mathbf{v}_{r+1}, \ldots, \mathbf{v}_n$ span a subspace of Nul A of dimension $n - r$. By the Rank Theorem, $\dim \text{Nul } A = n - \text{rank } A$. It follows that

$$\{\mathbf{v}_{r+1}, \ldots, \mathbf{v}_n\} \tag{7}$$

is an orthonormal basis for Nul A, by the Basis Theorem (in Section 4.5).

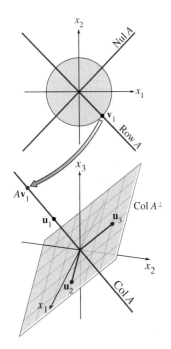

The fundamental subspaces in Example 4.

From (5) and (6), the orthogonal complement of Nul A^T is Col A. Interchanging A and A^T, note that $(\text{Nul } A)^\perp = \text{Col } A^T = \text{Row } A$. Hence, from (7),

$$\{\mathbf{v}_1, \ldots, \mathbf{v}_r\} \tag{8}$$

is an orthonormal basis for Row A.

Figure 4 summarizes (5)–(8), but shows the orthogonal basis $\{\sigma_1 \mathbf{u}_1, \ldots, \sigma_r \mathbf{u}_r\}$ for Col A instead of the normalized basis, to remind you that $A\mathbf{v}_i = \sigma_i \mathbf{u}_i$ for $1 \le i \le r$. Explicit orthonormal bases for the four fundamental subspaces determined by A are useful in some calculations, particularly in constrained optimization problems. ■

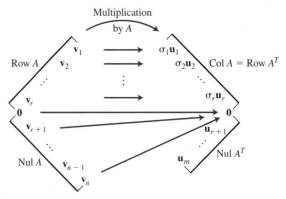

FIGURE 4 The four fundamental subspaces and the action of A.

The four fundamental subspaces and the concept of singular values provide the final statements of the Invertible Matrix Theorem. (Recall that statements about A^T have been omitted from the theorem to avoid nearly doubling the number of statements.) The other statements were given in Sections 2.3, 2.9, 3.2, 4.5, and 5.2.

THEOREM

The Invertible Matrix Theorem (concluded)

Let A be an $n \times n$ matrix. Then the following statements are each equivalent to the statement that A is an invertible matrix:

s. $(\text{Col } A)^\perp = \{\mathbf{0}\}$.

t. $(\text{Nul } A)^\perp = \mathbb{R}^n$.

u. Row $A = \mathbb{R}^n$.

v. A has n nonzero singular values.

EXAMPLE 7 (Reduced SVD and the Pseudoinverse of A) When Σ contains rows or columns of zeros, a more compact decomposition of A is possible. Using the notation established above, let $r = \text{rank } A$, and partition U and V into submatrices whose first blocks contain r columns:

$$U = [\, U_r \quad U_{m-r} \,], \quad \text{where } U_r = [\, \mathbf{u}_1 \quad \cdots \quad \mathbf{u}_r \,]$$
$$V = [\, V_r \quad V_{n-r} \,], \quad \text{where } V_r = [\, \mathbf{v}_1 \quad \cdots \quad \mathbf{v}_r \,]$$

Then U_r is $m \times r$ and V_r is $n \times r$. (To simplify notation, we consider U_{m-r} or V_{n-r} even though one of them may have no columns.) Then partitioned matrix multiplication shows that

$$A = [\, U_r \quad U_{m-r} \,] \begin{bmatrix} D & 0 \\ 0 & 0 \end{bmatrix} \begin{bmatrix} V_r^T \\ V_{n-r}^T \end{bmatrix} = U_r D V_r^T \tag{9}$$

This factorization of A is called a **reduced singular value decomposition** of A. Since the diagonal entries in D are nonzero, D is invertible. The following matrix is called the **pseudoinverse** (also, the **Moore–Penrose inverse**) of A:

$$A^+ = V_r D^{-1} U_r^T \tag{10}$$

Supplementary Exercises 28–30 at the end of the chapter explore some of the properties of the reduced singular value decomposition and the pseudoinverse. ■

EXAMPLE 8 (Least-Squares Solution) Given the equation $A\mathbf{x} = \mathbf{b}$, use the pseudoinverse of A in (10) to define

$$\hat{\mathbf{x}} = A^+ \mathbf{b} = V_r D^{-1} U_r^T \mathbf{b}$$

Then, from the SVD in (9),

$$\begin{aligned} A\hat{\mathbf{x}} &= (U_r D V_r^T)(V_r D^{-1} U_r^T \mathbf{b}) \\ &= U_r D D^{-1} U_r^T \mathbf{b} \qquad \text{Because } V_r^T V_r = I_r \\ &= U_r U_r^T \mathbf{b} \end{aligned}$$

It follows from (5) that $U_r U_r^T \mathbf{b}$ is the orthogonal projection $\hat{\mathbf{b}}$ of $\mathbf{b}$ onto Col A. (See Theorem 10 in Section 6.3.) Thus $\hat{\mathbf{x}}$ is a least-squares solution of $A\mathbf{x} = \mathbf{b}$. In fact, this $\hat{\mathbf{x}}$ has the smallest length among all least-squares solutions of $A\mathbf{x} = \mathbf{b}$. See Supplementary Exercise 30. ■

Numerical Notes

Examples 1–4 and the exercises illustrate the concept of singular values and suggest how to perform calculations by hand. In practice, the computation of $A^T A$ should be avoided, since any errors in the entries of A are squared in the entries of $A^T A$. There exist fast iterative methods that produce the singular values and singular vectors of A accurately to many decimal places.

Practice Problems

1. Given a singular value decomposition, $A = U \Sigma V^T$, find an SVD of A^T. How are the singular values of A and A^T related?

2. For any $n \times n$ matrix A, use the SVD to show that there is an $n \times n$ orthogonal matrix Q such that $A^T A = Q^T (A^T A) Q$.

Remark: Practice Problem 2 establishes that for any $n \times n$ matrix A, the matrices $A A^T$ and $A^T A$ are *orthogonally similar*.

7.4 Exercises

Find the singular values of the matrices in Exercises 1–4.

1. $\begin{bmatrix} 1 & 0 \\ 0 & -3 \end{bmatrix}$ **2.** $\begin{bmatrix} -3 & 0 \\ 0 & 0 \end{bmatrix}$

3. $\begin{bmatrix} 2 & 3 \\ 0 & 2 \end{bmatrix}$ **4.** $\begin{bmatrix} 3 & 0 \\ 8 & 3 \end{bmatrix}$

Find an SVD of each matrix in Exercises 5–12. [*Hint:* In Exercise 11, one choice for U is $\begin{bmatrix} -1/3 & 2/3 & 2/3 \\ 2/3 & -1/3 & 2/3 \\ 2/3 & 2/3 & -1/3 \end{bmatrix}$. In Exercise 12, one column of U can be $\begin{bmatrix} 1/\sqrt{6} \\ -2/\sqrt{6} \\ 1/\sqrt{6} \end{bmatrix}$.

5. $\begin{bmatrix} -2 & 0 \\ 0 & 0 \end{bmatrix}$ **6.** $\begin{bmatrix} -3 & 0 \\ 0 & -2 \end{bmatrix}$

7. $\begin{bmatrix} 2 & -1 \\ 2 & 2 \end{bmatrix}$ **8.** $\begin{bmatrix} 4 & 6 \\ 0 & 4 \end{bmatrix}$

9. $\begin{bmatrix} 3 & -3 \\ 0 & 0 \\ 1 & 1 \end{bmatrix}$ **10.** $\begin{bmatrix} 7 & 1 \\ 5 & 5 \\ 0 & 0 \end{bmatrix}$

11. $\begin{bmatrix} -3 & 1 \\ 6 & -2 \\ 6 & -2 \end{bmatrix}$ **12.** $\begin{bmatrix} 1 & 1 \\ 0 & 1 \\ -1 & 1 \end{bmatrix}$

13. Find the SVD of $A = \begin{bmatrix} 3 & 2 & 2 \\ 2 & 3 & -2 \end{bmatrix}$ [*Hint:* Work with A^T.]

14. In Exercise 7, find a unit vector $\mathbf{x}$ at which $A\mathbf{x}$ has maximum length.

15. Suppose the factorization below is an SVD of a matrix A, with the entries in U and V rounded to two decimal places.

$$A = \begin{bmatrix} .40 & -.78 & .47 \\ .37 & -.33 & -.87 \\ -.84 & -.52 & -.16 \end{bmatrix} \begin{bmatrix} 7.10 & 0 & 0 \\ 0 & 3.10 & 0 \\ 0 & 0 & 0 \end{bmatrix}$$
$$\times \begin{bmatrix} .30 & -.51 & -.81 \\ .76 & .64 & -.12 \\ .58 & -.58 & .58 \end{bmatrix}$$

a. What is the rank of A?

b. Use this decomposition of A, with no calculations, to write a basis for Col A and a basis for Nul A. [*Hint:* First write the columns of V.]

16. Repeat Exercise 15 for the following SVD of a 3×4 matrix A:

$$A = \begin{bmatrix} -.86 & -.11 & -.50 \\ .31 & .68 & -.67 \\ .41 & -.73 & -.55 \end{bmatrix} \begin{bmatrix} 12.48 & 0 & 0 & 0 \\ 0 & 6.34 & 0 & 0 \\ 0 & 0 & 0 & 0 \end{bmatrix}$$
$$\times \begin{bmatrix} .66 & -.03 & -.35 & .66 \\ -.13 & -.90 & -.39 & -.13 \\ .65 & .08 & -.16 & -.73 \\ -.34 & .42 & -.84 & -.08 \end{bmatrix}$$

In Exercises 17–24, A is an $m \times n$ matrix with a singular value decomposition $A = U\Sigma V^T$, where U is an $m \times m$ orthogonal matrix, Σ is an $m \times n$ "diagonal" matrix with r positive entries and no negative entries, and V is an $n \times n$ orthogonal matrix. Justify each answer.

17. Show that if A is square, then $|\det A|$ is the product of the singular values of A.

18. Suppose A is square and invertible. Find a singular value decomposition of A^{-1}.

19. Show that the columns of V are eigenvectors of A^TA, the columns of U are eigenvectors of AA^T, and the diagonal entries of Σ are the singular values of A. [*Hint:* Use the SVD to compute A^TA and AA^T.]

20. Show that if P is an orthogonal $m \times m$ matrix, then PA has the same singular values as A.

21. Justify the statement in Example 2 that the second singular value of a matrix A is the maximum of $\|A\mathbf{x}\|$ as $\mathbf{x}$ varies over all unit vectors orthogonal to $\mathbf{v}_1$, with $\mathbf{v}_1$ a right singular vector corresponding to the first singular value of A. [*Hint:* Use Theorem 7 in Section 7.3.]

22. Show that if A is an $n \times n$ positive definite matrix, then an orthogonal diagonalization $A = PDP^T$ is a singular value decomposition of A.

23. Let $U = [\mathbf{u}_1 \;\cdots\; \mathbf{u}_m]$ and $V = [\mathbf{v}_1 \;\cdots\; \mathbf{v}_n]$, where the $\mathbf{u}_i$ and $\mathbf{v}_i$ are as in Theorem 10. Show that

$$A = \sigma_1 \mathbf{u}_1 \mathbf{v}_1^T + \sigma_2 \mathbf{u}_2 \mathbf{v}_2^T + \cdots + \sigma_r \mathbf{u}_r \mathbf{v}_r^T.$$

24. Using the notation of Exercise 23, show that $A^T\mathbf{u}_j = \sigma_j \mathbf{v}_j$ for $1 \leq j \leq r = \text{rank } A$.

25. Let $T : \mathbb{R}^n \to \mathbb{R}^m$ be a linear transformation. Describe how to find a basis $\mathcal{B}$ for $\mathbb{R}^n$ and a basis $\mathcal{C}$ for $\mathbb{R}^m$ such that the matrix for T relative to $\mathcal{B}$ and $\mathcal{C}$ is an $m \times n$ "diagonal" matrix.

Compute an SVD of each matrix in Exercises 26 and 27. Report the final matrix entries accurate to two decimal places. Use the method of Examples 3 and 4.

26. $A = \begin{bmatrix} -18 & 13 & -4 & 4 \\ 2 & 19 & -4 & 12 \\ -14 & 11 & -12 & 8 \\ -2 & 21 & 4 & 8 \end{bmatrix}$

T 27. $A = \begin{bmatrix} 6 & -8 & -4 & 5 & -4 \\ 2 & 7 & -5 & -6 & 4 \\ 0 & -1 & -8 & 2 & 2 \\ -1 & -2 & 4 & 4 & -8 \end{bmatrix}$

T 28. Compute the singular values of the 4×4 matrix in Exercise 9 in Section 2.3, and compute the condition number σ_1/σ_4.

T 29. Compute the singular values of the 5×5 matrix in Exercise 10 in Section 2.3, and compute the condition number σ_1/σ_5.

Solutions to Practice Problems

1. If $A = U\Sigma V^T$, where Σ is $m \times n$, then $A^T = (V^T)^T \Sigma^T U^T = V\Sigma^T U^T$. This is an SVD of A^T because V and U are orthogonal matrices and Σ^T is an $n \times m$ "diagonal" matrix. Since Σ and Σ^T have the same nonzero diagonal entries, A and A^T have the same nonzero singular values. [*Note:* If A is $2 \times n$, then AA^T is only 2×2 and its eigenvalues may be easier to compute (by hand) than the eigenvalues of A^TA.]

2. Use the SVD to write $A = U\Sigma V^T$, where U and V are $n \times n$ orthogonal matrices and Σ is an $n \times n$ diagonal matrix. Notice that $U^TU = I = V^TV$ and $\Sigma^T = \Sigma$, since U and V are orthogonal matrices and Σ is a diagonal matrix. Substituting the SVD for A into AA^T and A^TA results in

$$AA^T = U\Sigma V^T(U\Sigma V^T)^T = U\Sigma V^T V\Sigma^T U^T = U\Sigma\Sigma^T U^T = U\Sigma^2 U^T,$$

and

$$A^TA = (U\Sigma V^T)^T U\Sigma V^T = V\Sigma^T U^T U\Sigma V^T = V\Sigma^T\Sigma V^T = V\Sigma^2 V^T.$$

Let $Q = VU^T$. Then

$$Q^T(A^TA)Q = (VU^T)^T(V\Sigma^2 V^T)(VU^T) = UV^TV\Sigma^2 V^TVU^T$$

$$= U\Sigma^2 U^T = AA^T.$$

7.5 Applications to Image Processing and Statistics

The satellite photographs in this chapter's introduction provide an example of multidimensional, or *multivariate*, data—information organized so that each datum in the data set is identified with a point (vector) in $\mathbb{R}^n$. The main goal of this section is to explain a technique, called *principal component analysis*, used to analyze such multivariate data. The calculations will illustrate the use of orthogonal diagonalization and the singular value decomposition.

Principal component analysis can be applied to any data that consist of lists of measurements made on a collection of objects or individuals. For instance, consider a chemical process that produces a plastic material. To monitor the process, 300 samples are taken of the material produced, and each sample is subjected to a battery of eight tests, such as melting point, density, ductility, tensile strength, and so on. The laboratory report for each sample is a vector in $\mathbb{R}^8$, and the set of such vectors forms an 8×300 matrix, called the **matrix of observations**.

Loosely speaking, we can say that the process control data are eight-dimensional. The next two examples describe data that can be visualized graphically.

EXAMPLE 1 An example of two-dimensional data is given by a set of weights and heights of N college students. Let $\mathbf{X}_j$ denote the **observation vector** in $\mathbb{R}^2$ that lists the weight and height of the jth student. If w denotes weight and h height, then the matrix of observations has the form

$$\begin{bmatrix} w_1 & w_2 & \cdots & w_N \\ h_1 & h_2 & \cdots & h_N \end{bmatrix}$$
$$\uparrow \quad \uparrow \qquad\quad \uparrow$$
$$\mathbf{X}_1 \quad \mathbf{X}_2 \qquad \mathbf{X}_N$$

The set of observation vectors can be visualized as a two-dimensional *scatter plot*. See Figure 1. ∎

FIGURE 1 A scatter plot of observation vectors $\mathbf{X}_1, \ldots, \mathbf{X}_N$.

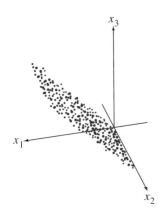

FIGURE 2

A scatter plot of spectral data for a satellite image.

EXAMPLE 2 The first three photographs of Railroad Valley, Nevada, shown in the chapter introduction can be viewed as *one* image of the region, with *three spectral components*, because simultaneous measurements of the region were made at three separate wavelengths. Each photograph gives different information about the same physical region. For instance, the first pixel in the upper-left corner of each photograph corresponds to the same place on the ground (about 30 meters by 30 meters). To each pixel there corresponds an observation vector in $\mathbb{R}^3$ that lists the signal intensities for that pixel in the three spectral bands.

Typically, the image is 2000×2000 pixels, so there are 4 million pixels in the image. The data for the image form a matrix with 3 rows and 4 million columns (with columns arranged in any convenient order). In this case, the "multidimensional" character of the data refers to the three *spectral* dimensions rather than the two *spatial* dimensions that naturally belong to any photograph. The data can be visualized as a cluster of 4 million points in $\mathbb{R}^3$, perhaps as in Figure 2. ∎

Mean and Covariance

To prepare for principal component analysis, let $[\, \mathbf{X}_1 \quad \cdots \quad \mathbf{X}_N \,]$ be a $p \times N$ matrix of observations, such as described above. The **sample mean**, $\mathbf{M}$, of the observation vectors $\mathbf{X}_1, \ldots, \mathbf{X}_N$ is given by

$$\mathbf{M} = \frac{1}{N}(\mathbf{X}_1 + \cdots + \mathbf{X}_N)$$

For the data in Figure 1, the sample mean is the point in the "center" of the scatter plot. For $k = 1, \ldots, N$, let

$$\hat{\mathbf{X}}_k = \mathbf{X}_k - \mathbf{M}$$

The columns of the $p \times N$ matrix

$$B = [\, \hat{\mathbf{X}}_1 \quad \hat{\mathbf{X}}_2 \quad \cdots \quad \hat{\mathbf{X}}_N \,]$$

FIGURE 3

Weight–height data in mean-deviation form.

have a zero sample mean, and B is said to be in **mean-deviation form**. When the sample mean is subtracted from the data in Figure 1, the resulting scatter plot has the form in Figure 3.

The (**sample**) **covariance matrix** is the $p \times p$ matrix S defined by

$$S = \frac{1}{N-1} BB^T$$

Since any matrix of the form BB^T is positive semidefinite, so is S. (See Exercise 33 in Section 7.2 with B and B^T interchanged.)

EXAMPLE 3 Three measurements are made on each of four individuals in a random sample from a population. The observation vectors are

$$\mathbf{X}_1 = \begin{bmatrix} 1 \\ 2 \\ 1 \end{bmatrix}, \quad \mathbf{X}_2 = \begin{bmatrix} 4 \\ 2 \\ 13 \end{bmatrix}, \quad \mathbf{X}_3 = \begin{bmatrix} 7 \\ 8 \\ 1 \end{bmatrix}, \quad \mathbf{X}_4 = \begin{bmatrix} 8 \\ 4 \\ 5 \end{bmatrix}$$

Compute the sample mean and the covariance matrix.

SOLUTION The sample mean is

$$\mathbf{M} = \frac{1}{4} \left(\begin{bmatrix} 1 \\ 2 \\ 1 \end{bmatrix} + \begin{bmatrix} 4 \\ 2 \\ 13 \end{bmatrix} + \begin{bmatrix} 7 \\ 8 \\ 1 \end{bmatrix} + \begin{bmatrix} 8 \\ 4 \\ 5 \end{bmatrix} \right) = \frac{1}{4} \begin{bmatrix} 20 \\ 16 \\ 20 \end{bmatrix} = \begin{bmatrix} 5 \\ 4 \\ 5 \end{bmatrix}$$

Subtract the sample mean from $\mathbf{X}_1, \ldots, \mathbf{X}_4$ to obtain

$$\hat{\mathbf{X}}_1 = \begin{bmatrix} -4 \\ -2 \\ -4 \end{bmatrix}, \quad \hat{\mathbf{X}}_2 = \begin{bmatrix} -1 \\ -2 \\ 8 \end{bmatrix}, \quad \hat{\mathbf{X}}_3 = \begin{bmatrix} 2 \\ 4 \\ -4 \end{bmatrix}, \quad \hat{\mathbf{X}}_4 = \begin{bmatrix} 3 \\ 0 \\ 0 \end{bmatrix}$$

and

$$B = \begin{bmatrix} -4 & -1 & 2 & 3 \\ -2 & -2 & 4 & 0 \\ -4 & 8 & -4 & 0 \end{bmatrix}$$

The sample covariance matrix is

$$S = \frac{1}{3} \begin{bmatrix} -4 & -1 & 2 & 3 \\ -2 & -2 & 4 & 0 \\ -4 & 8 & -4 & 0 \end{bmatrix} \begin{bmatrix} -4 & -2 & -4 \\ -1 & -2 & 8 \\ 2 & 4 & -4 \\ 3 & 0 & 0 \end{bmatrix}$$

$$= \frac{1}{3} \begin{bmatrix} 30 & 18 & 0 \\ 18 & 24 & -24 \\ 0 & -24 & 96 \end{bmatrix} = \begin{bmatrix} 10 & 6 & 0 \\ 6 & 8 & -8 \\ 0 & -8 & 32 \end{bmatrix} \quad\blacksquare$$

To discuss the entries in $S = [s_{ij}]$, let $\mathbf{X}$ represent a vector that varies over the set of observation vectors and denote the coordinates of $\mathbf{X}$ by $x_1, \ldots, x_p$. Then x_1, for example, is a scalar that varies over the set of first coordinates of $\mathbf{X}_1, \ldots, \mathbf{X}_N$. For $j = 1, \ldots, p$, the diagonal entry s_{jj} in S is called the **variance** of x_j.

The variance of x_j measures the spread of the values of x_j. (See Exercise 13.) In Example 3, the variance of x_1 is 10 and the variance of x_3 is 32. The fact that 32 is more than 10 indicates that the set of third entries in the response vectors contains a wider spread of values than the set of first entries.

The **total variance** of the data is the sum of the variances on the diagonal of S. In general, the sum of the diagonal entries of a square matrix S is called the **trace** of the matrix, written $\text{tr}(S)$. Thus

$$\{\text{total variance}\} = \text{tr}(S)$$

The entry s_{ij} in S for $i \neq j$ is called the **covariance** of x_i and x_j. Observe that in Example 3, the covariance between x_1 and x_3 is 0 because the $(1, 3)$-entry in S is 0. Statisticians say that x_1 and x_3 are **uncorrelated**. Analysis of the multivariate data in $\mathbf{X}_1, \ldots, \mathbf{X}_N$ is greatly simplified when most or all of the variables $x_1, \ldots, x_p$ are uncorrelated, that is, when the covariance matrix of $\mathbf{X}_1, \ldots, \mathbf{X}_N$ is diagonal or nearly diagonal.

Principal Component Analysis

For simplicity, assume that the matrix $[\,\mathbf{X}_1 \ \cdots \ \mathbf{X}_N\,]$ is already in mean-deviation form. The goal of principal component analysis is to find an orthogonal $p \times p$ matrix $P = [\,\mathbf{u}_1 \ \cdots \ \mathbf{u}_p\,]$ that determines a change of variable, $\mathbf{X} = P\mathbf{Y}$, or

$$
\begin{bmatrix} x_1 \\ x_2 \\ \vdots \\ x_p \end{bmatrix} = \begin{bmatrix} \mathbf{u}_1 & \mathbf{u}_2 & \cdots & \mathbf{u}_p \end{bmatrix} \begin{bmatrix} y_1 \\ y_2 \\ \vdots \\ y_p \end{bmatrix}
$$

with the property that the new variables $y_1, \ldots, y_p$ are uncorrelated and are arranged in order of decreasing variance.

The orthogonal change of variable $\mathbf{X} = P\mathbf{Y}$ means that each observation vector $\mathbf{X}_k$ receives a "new name," $\mathbf{Y}_k$, such that $\mathbf{X}_k = P\mathbf{Y}_k$. Notice that $\mathbf{Y}_k$ is the coordinate vector of $\mathbf{X}_k$ with respect to the columns of P, and $\mathbf{Y}_k = P^{-1}\mathbf{X}_k = P^T\mathbf{X}_k$ for $k = 1, \ldots, N$.

It is not difficult to verify that for any orthogonal P, the covariance matrix of $\mathbf{Y}_1, \ldots, \mathbf{Y}_N$ is P^TSP (Exercise 11). So the desired orthogonal matrix P is one that makes P^TSP diagonal. Let D be a diagonal matrix with the eigenvalues $\lambda_1, \ldots, \lambda_p$ of S on the diagonal, arranged so that $\lambda_1 \geq \lambda_2 \geq \cdots \geq \lambda_p \geq 0$, and let P be an orthogonal matrix whose columns are the corresponding unit eigenvectors $\mathbf{u}_1, \ldots, \mathbf{u}_p$. Then $S = PDP^T$ and $P^TSP = D$.

The unit eigenvectors $\mathbf{u}_1, \ldots, \mathbf{u}_p$ of the covariance matrix S are called the **principal components** of the data (in the matrix of observations). The **first principal component** is the eigenvector corresponding to the largest eigenvalue of S, the **second principal component** is the eigenvector corresponding to the second largest eigenvalue, and so on.

The first principal component $\mathbf{u}_1$ determines the new variable y_1 in the following way. Let $c_1, \ldots, c_p$ be the entries in $\mathbf{u}_1$. Since $\mathbf{u}_1^T$ is the first row of P^T, the equation $\mathbf{Y} = P^T\mathbf{X}$ shows that

$$y_1 = \mathbf{u}_1^T\mathbf{X} = c_1 x_1 + c_2 x_2 + \cdots + c_p x_p$$

Thus y_1 is a linear combination of the original variables $x_1, \ldots, x_p$, using the entries in the eigenvector $\mathbf{u}_1$ as weights. In a similar fashion, $\mathbf{u}_2$ determines the variable y_2, and so on.

EXAMPLE 4 The initial data for the multispectral image of Railroad Valley (Example 2) consisted of 4 million vectors in $\mathbb{R}^3$. The associated covariance matrix is[1]

$$
S = \begin{bmatrix} 2382.78 & 2611.84 & 2136.20 \\ 2611.84 & 3106.47 & 2553.90 \\ 2136.20 & 2553.90 & 2650.71 \end{bmatrix}
$$

---

[1] Data for Example 4 and Exercises 5 and 6 were provided by Earth Satellite Corporation, Rockville, Maryland.

Find the principal components of the data, and list the new variable determined by the first principal component.

SOLUTION The eigenvalues of S and the associated principal components (the unit eigenvectors) are

$$\lambda_1 = 7614.23 \qquad \lambda_2 = 427.63 \qquad \lambda_3 = 98.10$$

$$\mathbf{u}_1 = \begin{bmatrix} .5417 \\ .6295 \\ .5570 \end{bmatrix} \quad \mathbf{u}_2 = \begin{bmatrix} -.4894 \\ -.3026 \\ .8179 \end{bmatrix} \quad \mathbf{u}_3 = \begin{bmatrix} .6834 \\ -.7157 \\ .1441 \end{bmatrix}$$

Using two decimal places for simplicity, the variable for the first principal component is

$$y_1 = .54x_1 + .63x_2 + .56x_3$$

This equation was used to create photograph (d) in the chapter introduction. The variables x_1, x_2, and x_3 are the signal intensities in the three spectral bands. The values of x_1, converted to a gray scale between black and white, produced photograph (a). Similarly, the values of x_2 and x_3 produced photographs (b) and (c), respectively. At each pixel in photograph (d), the gray scale value is computed from y_1, a weighted linear combination of x_1, x_2, and x_3. In this sense, photograph (d) "displays" the first principal component of the data. ∎

In Example 4, the covariance matrix for the transformed data, using variables y_1, y_2, and y_3, is

$$D = \begin{bmatrix} 7614.23 & 0 & 0 \\ 0 & 427.63 & 0 \\ 0 & 0 & 98.10 \end{bmatrix}$$

Although D is obviously simpler than the original covariance matrix S, the merit of constructing the new variables is not yet apparent. However, the variances of the variables y_1, y_2, and y_3 appear on the diagonal of D, and obviously the first variance in D is much larger than the other two. As we shall see, this fact will permit us to view the data as essentially one-dimensional rather than three-dimensional.

Reducing the Dimension of Multivariate Data

Principal component analysis is potentially valuable for applications in which most of the variation, or dynamic range, in the data is due to variations in *only a few* of the new variables, $y_1, \ldots, y_p$.

It can be shown that an orthogonal change of variables, $\mathbf{X} = P\mathbf{Y}$, does not change the total variance of the data. (Roughly speaking, this is true because left-multiplication by P does not change the lengths of vectors or the angles between them. See Exercise 12.) This means that if $S = PDP^T$, then

$$\begin{Bmatrix} \text{total variance} \\ \text{of } x_1, \ldots, x_p \end{Bmatrix} = \begin{Bmatrix} \text{total variance} \\ \text{of } y_1, \ldots, y_p \end{Bmatrix} = \text{tr}(D) = \lambda_1 + \cdots + \lambda_p$$

The variance of y_j is λ_j, and the quotient $\lambda_j / \text{tr}(S)$ measures the fraction of the total variance that is "explained" or "captured" by y_j.

EXAMPLE 5   Compute the various percentages of variance of the Railroad Valley multispectral data that are displayed in the principal component photographs, (d)–(f), shown in the chapter introduction.

SOLUTION   The total variance of the data is

$$\text{tr}(D) = 7614.23 + 427.63 + 98.10 = 8139.96$$

[Verify that this number also equals $\text{tr}(S)$.] The percentages of the total variance explained by the principal components are

First component	Second component	Third component
$\dfrac{7614.23}{8139.96} = 93.5\%$	$\dfrac{427.63}{8139.96} = 5.3\%$	$\dfrac{98.10}{8139.96} = 1.2\%$

In a sense, 93.5% of the information collected by Landsat for the Railroad Valley region is displayed in photograph (d), with 5.3% in (e) and only 1.2% remaining for (f).   ∎

The calculations in Example 5 show that the data have practically no variance in the third (new) coordinate. The values of y_3 are all close to zero. Geometrically, the data points lie nearly in the plane $y_3 = 0$, and their locations can be determined fairly accurately by knowing only the values of y_1 and y_2. In fact, y_2 also has relatively small variance, which means that the points lie approximately along a line, and the data are essentially one-dimensional. See Figure 2, in which the data resemble a popsicle stick.

Characterizations of Principal Component Variables

If $y_1, \ldots, y_p$ arise from a principal component analysis of a $p \times N$ matrix of observations, then the variance of y_1 is as large as possible in the following sense: If $\mathbf{u}$ is any unit vector and if $y = \mathbf{u}^T \mathbf{X}$, then the variance of the values of y as $\mathbf{X}$ varies over the original data $\mathbf{X}_1, \ldots, \mathbf{X}_N$ turns out to be $\mathbf{u}^T S \mathbf{u}$. By Theorem 8 in Section 7.3, the maximum value of $\mathbf{u}^T S \mathbf{u}$, over all unit vectors $\mathbf{u}$, is the largest eigenvalue λ_1 of S, and this variance is attained when $\mathbf{u}$ is the corresponding eigenvector $\mathbf{u}_1$. In the same way, Theorem 8 shows that y_2 has maximum possible variance among all variables $y = \mathbf{u}^T \mathbf{X}$ that are *uncorrelated* with y_1. Likewise, y_3 has maximum possible variance among all variables uncorrelated with both y_1 and y_2, and so on.

Numerical Notes

The singular value decomposition is the main tool for performing principal component analysis in practical applications. If B is a $p \times N$ matrix of observations in mean-deviation form, and if $A = \left(1/\sqrt{N-1}\right)B^T$, then $A^T A$ is the covariance matrix, S. The squares of the singular values of A are the p eigenvalues of S, and the right singular vectors of A are the principal components of the data.

As mentioned in Section 7.4, iterative calculation of the SVD of A is faster and more accurate than an eigenvalue decomposition of S. This is particularly true, for instance, in the hyperspectral image processing (with $p = 224$) mentioned in the chapter introduction. Principal component analysis is completed in seconds on specialized workstations.

Practice Problems

The following table lists the weights and heights of five boys:

Boy	#1	#2	#3	#4	#5
Weight (lb)	120	125	125	135	145
Height (in.)	61	60	64	68	72

1. Find the covariance matrix for the data.

2. Make a principal component analysis of the data to find a single *size index* that explains most of the variation in the data.

7.5 Exercises

In Exercises 1 and 2, convert the matrix of observations to mean-deviation form, and construct the sample covariance matrix.

1. $\begin{bmatrix} 19 & 22 & 6 & 3 & 2 & 20 \\ 12 & 6 & 9 & 15 & 13 & 5 \end{bmatrix}$

2. $\begin{bmatrix} 1 & 5 & 2 & 6 & 7 & 3 \\ 3 & 11 & 6 & 8 & 15 & 11 \end{bmatrix}$

3. Find the principal components of the data for Exercise 1.

4. Find the principal components of the data for Exercise 2.

T 5. A Landsat image with three spectral components was made of Homestead Air Force Base in Florida (after the base was hit by Hurricane Andrew in 1992). The covariance matrix of the data is shown below. Find the first principal component of the data, and compute the percentage of the total variance that is contained in this component.

$$S = \begin{bmatrix} 164.12 & 32.73 & 81.04 \\ 32.73 & 539.44 & 249.13 \\ 81.04 & 249.13 & 189.11 \end{bmatrix}$$

T 6. The covariance matrix below was obtained from a Landsat image of the Columbia River in Washington, using data from three spectral bands. Let x_1, x_2, x_3 denote the spectral components of each pixel in the image. Find a new variable of the form $y_1 = c_1x_1 + c_2x_2 + c_3x_3$ that has maximum possible variance, subject to the constraint that $c_1^2 + c_2^2 + c_3^2 = 1$. What percentage of the total variance in the data is explained by y_1?

$$S = \begin{bmatrix} 29.64 & 18.38 & 5.00 \\ 18.38 & 20.82 & 14.06 \\ 5.00 & 14.06 & 29.21 \end{bmatrix}$$

7. Let x_1, x_2 denote the variables for the two-dimensional data in Exercise 1. Find a new variable y_1 of the form $y_1 = c_1x_1 + c_2x_2$, with $c_1^2 + c_2^2 = 1$, such that y_1 has maximum possible variance over the given data. How much of the variance in the data is explained by y_1?

8. Repeat Exercise 7 for the data in Exercise 2.

9. Suppose three tests are administered to a random sample of college students. Let $\mathbf{X}_1, \ldots, \mathbf{X}_N$ be observation vectors in $\mathbb{R}^3$ that list the three scores of each student, and for $j = 1, 2, 3$, let x_j denote a student's score on the jth exam. Suppose the covariance matrix of the data is

$$S = \begin{bmatrix} 5 & 2 & 0 \\ 2 & 6 & 2 \\ 0 & 2 & 7 \end{bmatrix}$$

Let y be an "index" of student performance, with $y = c_1x_1 + c_2x_2 + c_3x_3$ and $c_1^2 + c_2^2 + c_3^2 = 1$. Choose c_1, c_2, c_3 so that the variance of y over the data set is as large as possible. [*Hint:* The eigenvalues of the sample covariance matrix are $\lambda = 3, 6$, and 9.]

T 10. Repeat Exercise 9 with $S = \begin{bmatrix} 5 & 4 & 2 \\ 4 & 11 & 4 \\ 2 & 4 & 5 \end{bmatrix}$.

11. Given multivariate data $\mathbf{X}_1, \ldots, \mathbf{X}_N$ (in $\mathbb{R}^p$) in mean-deviation form, let P be a $p \times p$ matrix, and define $\mathbf{Y}_k = P^T\mathbf{X}_k$ for $k = 1, \ldots, N$.

   a. Show that $\mathbf{Y}_1, \ldots, \mathbf{Y}_N$ are in mean-deviation form. [*Hint:* Let $\mathbf{w}$ be the vector in $\mathbb{R}^N$ with a 1 in each entry. Then $\begin{bmatrix} \mathbf{X}_1 & \cdots & \mathbf{X}_N \end{bmatrix} \mathbf{w} = \mathbf{0}$ (the zero vector in $\mathbb{R}^p$).]

   b. Show that if the covariance matrix of $\mathbf{X}_1, \ldots, \mathbf{X}_N$ is S, then the covariance matrix of $\mathbf{Y}_1, \ldots, \mathbf{Y}_N$ is P^TSP.

12. Let $\mathbf{X}$ denote a vector that varies over the columns of a $p \times N$ matrix of observations, and let P be a $p \times p$ orthogonal matrix. Show that the change of variable $\mathbf{X} = P\mathbf{Y}$ does not change the total variance of the data. [*Hint:* By Exercise 11, it suffices to show that $\text{tr}(P^TSP) = \text{tr}(S)$. Use a property of the trace mentioned in Exercise 27 in Section 5.4.]

13. The sample covariance matrix is a generalization of a formula for the variance of a sample of N scalar measurements, say, $t_1, \ldots, t_N$. If m is the average of $t_1, \ldots, t_N$, then the *sample variance* is given by

$$\frac{1}{N-1} \sum_{k=1}^{n} (t_k - m)^2 \tag{1}$$

Show how the sample covariance matrix, S, defined prior to Example 3, may be written in a form similar to (1). [*Hint:* Use partitioned matrix multiplication to write S as $1/(N-1)$ times the sum of N matrices of size $p \times p$. For $1 \leq k \leq N$, write $\mathbf{X}_k - \mathbf{M}$ in place of $\hat{\mathbf{X}}_k$.]

Solutions to Practice Problems

1. First arrange the data in mean-deviation form. The sample mean vector is easily seen to be $\mathbf{M} = \begin{bmatrix} 130 \\ 65 \end{bmatrix}$. Subtract $\mathbf{M}$ from the observation vectors (the columns in the table) and obtain

$$B = \begin{bmatrix} -10 & -5 & -5 & 5 & 15 \\ -4 & -5 & -1 & 3 & 7 \end{bmatrix}$$

Then the sample covariance matrix is

$$S = \frac{1}{5-1} \begin{bmatrix} -10 & -5 & -5 & 5 & 15 \\ -4 & -5 & -1 & 3 & 7 \end{bmatrix} \begin{bmatrix} -10 & -4 \\ -5 & -5 \\ -5 & -1 \\ 5 & 3 \\ 15 & 7 \end{bmatrix}$$

$$= \frac{1}{4} \begin{bmatrix} 400 & 190 \\ 190 & 100 \end{bmatrix} = \begin{bmatrix} 100.0 & 47.5 \\ 47.5 & 25.0 \end{bmatrix}$$

2. The eigenvalues of S are (to two decimal places)

$$\lambda_1 = 123.02 \quad \text{and} \quad \lambda_2 = 1.98$$

The unit eigenvector corresponding to λ_1 is $\mathbf{u} = \begin{bmatrix} .900 \\ .436 \end{bmatrix}$. (Since S is 2×2, the computations can be done by hand if a matrix program is not available.) For the *size index*, set

$$y = .900\hat{w} + .436\hat{h}$$

where $\hat{w}$ and $\hat{h}$ are weight and height, respectively, in mean-deviation form. The variance of this index over the data set is 123.02. Because the total variance is $\text{tr}(S) = 100 + 25 = 125$, the size index accounts for practically all (98.4%) of the variance of the data.

The original data for Practice Problem 1 and the line determined by the first principal component $\mathbf{u}$ are shown in Figure 4. (In parametric vector form, the line

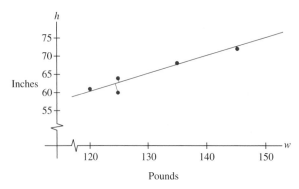

FIGURE 4 An orthogonal regression line determined by the first principal component of the data.

is $\mathbf{x} = \mathbf{M} + t\mathbf{u}$.) It can be shown that the line is the best approximation to the data, in the sense that the sum of the squares of the *orthogonal* distances to the line is minimized. In fact, principal component analysis is equivalent to what is termed *orthogonal regression*, but that is a story for another day.

CHAPTER 7 PROJECTS

Chapter 7 projects are available online at `bit.ly/30IM8gT`.

A. *Conic Sections and Quadric Surfaces*: This project shows how quadratic forms and the Principal Axes Theorem may be used to classify conic sections and quadric surfaces.

B. *Extrema for Functions of Several Variables*: This project shows how quadratic forms may be used to investigate maximum and minimum values of functions of several variables.

CHAPTER 7 SUPPLEMENTARY EXERCISES

Mark each statement True or False. Justify each answer. In each part, A represents an $n \times n$ matrix.

1. **(T/F)** If A is orthogonally diagonalizable, then A is symmetric.

2. **(T/F)** If A is an orthogonal matrix, then A is symmetric.

3. **(T/F)** If A is an orthogonal matrix, then $\|A\mathbf{x}\| = \|\mathbf{x}\|$ for all $\mathbf{x}$ in $\mathbb{R}^n$.

4. **(T/F)** The principal axes of a quadratic form $\mathbf{x}^T A\mathbf{x}$ can be the columns of any matrix P that diagonalizes A.

5. **(T/F)** If P is an $n \times n$ matrix with orthogonal columns, then $P^T = P^{-1}$.

6. **(T/F)** If every coefficient in a quadratic form is positive, then the quadratic form is positive definite.

7. **(T/F)** If $\mathbf{x}^T A\mathbf{x} > 0$ for some $\mathbf{x}$, then the quadratic form $\mathbf{x}^T A\mathbf{x}$ is positive definite.

8. **(T/F)** By a suitable change of variable, any quadratic form can be changed into one with no cross-product term.

9. **(T/F)** The largest value of a quadratic form $\mathbf{x}^T A\mathbf{x}$, for $\|\mathbf{x}\| = 1$, is the largest entry on the diagonal of A.

10. **(T/F)** The maximum value of a positive definite quadratic form $\mathbf{x}^T A\mathbf{x}$ is the greatest eigenvalue of A.

11. **(T/F)** A positive definite quadratic form can be changed into a negative definite form by a suitable change of variable $\mathbf{x} = P\mathbf{u}$, for some orthogonal matrix P.

12. **(T/F)** An indefinite quadratic form is one whose eigenvalues are not definite.

13. **(T/F)** If P is an $n \times n$ orthogonal matrix, then the change of variable $\mathbf{x} = P\mathbf{u}$ transforms $\mathbf{x}^T A\mathbf{x}$ into a quadratic form whose matrix is $P^{-1}AP$.

14. **(T/F)** If U is $m \times n$ with orthogonal columns, then $UU^T\mathbf{x}$ is the orthogonal projection of $\mathbf{x}$ onto Col U.

15. **(T/F)** If B is $m \times n$ and $\mathbf{x}$ is a unit vector in $\mathbb{R}^n$, then $\|B\mathbf{x}\| \leq \sigma_1$, where σ_1 is the first singular value of B.

16. **(T/F)** A singular value decomposition of an $m \times n$ matrix B can be written as $B = P\Sigma Q$, where P is an $m \times m$ orthogonal matrix, Q is an $n \times n$ orthogonal matrix, and Σ is an $m \times n$ "diagonal" matrix.

17. **(T/F)** If A is $n \times n$, then A and $A^T A$ have the same singular values.

18. Let $\{\mathbf{u}_1, \ldots, \mathbf{u}_n\}$ be an orthonormal basis for $\mathbb{R}^n$, and let $\lambda_1, \ldots, \lambda_n$ be any real scalars. Define

    $$A = \lambda_1 \mathbf{u}_1 \mathbf{u}_1^T + \cdots + \lambda_n \mathbf{u}_n \mathbf{u}_n^T$$

    a. Show that A is symmetric.

    b. Show that $\lambda_1, \ldots, \lambda_n$ are the eigenvalues of A.

19. Let A be an $n \times n$ symmetric matrix of rank r. Explain why the spectral decomposition of A represents A as the sum of r rank 1 matrices.

20. Let A be an $n \times n$ symmetric matrix.

    a. Show that $(\text{Col } A)^\perp = \text{Nul } A$. [*Hint:* See Section 6.1.]

    b. Show that each $\mathbf{y}$ in $\mathbb{R}^n$ can be written in the form $\mathbf{y} = \hat{\mathbf{y}} + \mathbf{z}$, with $\hat{\mathbf{y}}$ in Col A and $\mathbf{z}$ in Nul A.

21. Show that if $\mathbf{v}$ is an eigenvector of an $n \times n$ matrix A and $\mathbf{v}$ corresponds to a nonzero eigenvalue of A, then $\mathbf{v}$ is in Col A. [*Hint:* Use the definition of an eigenvector.]

22. Let A be an $n \times n$ symmetric matrix. Use Exercise 21 and an eigenvector basis for $\mathbb{R}^n$ to give a second proof of the decomposition in Exercise 20(b).

23. Prove that an $n \times n$ matrix A is positive definite if and only if A admits a *Cholesky factorization*, namely $A = R^T R$ for

some invertible upper triangular matrix R whose diagonal entries are all positive. [*Hint:* Use a QR factorization and Exercise 34 in Section 7.2.]

24. Use Exercise 23 to show that if A is positive definite, then A has an LU factorization, $A = LU$, where U has positive pivots on its diagonal. (The converse is true, too.)

If A is $m \times n$, then the matrix $G = A^T A$ is called the *Gram matrix* of A. In this case, the entries of G are the inner products of the columns of A. (See Exercises 25 and 26.)

25. Show that the Gram matrix of any matrix A is positive semidefinite, with the same rank as A. (See the Exercises in Section 6.5.)

26. Show that if an $n \times n$ matrix G is positive semidefinite and has rank r, then G is the Gram matrix of some $r \times n$ matrix A. This is called a *rank-revealing factorization* of G. [*Hint:* Consider the spectral decomposition of G, and first write G as BB^T for an $n \times r$ matrix B.]

27. Prove that any $n \times n$ matrix A admits a *polar decomposition* of the form $A = PQ$, where P is an $n \times n$ positive semidefinite matrix with the same rank as A and where Q is an $n \times n$ orthogonal matrix. [*Hint:* Use a singular value decomposition, $A = U \Sigma V^T$, and observe that $A = (U \Sigma U^T)(U V^T)$.] This decomposition is used, for instance, in mechanical engineering to model the deformation of a material. The matrix P describes the stretching or compression of the material (in the directions of the eigenvectors of P), and Q describes the rotation of the material in space.

Exercises 28–30 concern an $m \times n$ matrix A with a reduced singular value decomposition, $A = U_r D V_r^T$, and the pseudoinverse $A^+ = V_r D^{-1} U_r^T$.

28. Verify the properties of A^+:

a. For each $\mathbf{y}$ in $\mathbb{R}^m$, $AA^+\mathbf{y}$ is the orthogonal projection of $\mathbf{y}$ onto Col A.

b. For each $\mathbf{x}$ in $\mathbb{R}^n$, $A^+ A\mathbf{x}$ is the orthogonal projection of $\mathbf{x}$ onto Row A.

c. $AA^+ A = A$ and $A^+ AA^+ = A^+$.

29. Suppose the equation $A\mathbf{x} = \mathbf{b}$ is consistent, and let $\mathbf{x}^+ = A^+\mathbf{b}$. By Exercise 31 in Section 6.3, there is exactly one vector $\mathbf{p}$ in Row A such that $A\mathbf{p} = \mathbf{b}$. The following steps prove that $\mathbf{x}^+ = \mathbf{p}$ and $\mathbf{x}^+$ is the *minimum length solution* of $A\mathbf{x} = \mathbf{b}$.

a. Show that $\mathbf{x}^+$ is in Row A. [*Hint:* Write $\mathbf{b}$ as $A\mathbf{x}$ for some $\mathbf{x}$, and use Exercise 28.]

b. Show that $\mathbf{x}^+$ is a solution of $A\mathbf{x} = \mathbf{b}$.

c. Show that if $\mathbf{u}$ is any solution of $A\mathbf{x} = \mathbf{b}$, then $\|\mathbf{x}^+\| \le \|\mathbf{u}\|$, with equality only if $\mathbf{u} = \mathbf{x}^+$.

30. Given any $\mathbf{b}$ in $\mathbb{R}^m$, adapt Exercise 28 to show that $A^+\mathbf{b}$ is the *least-squares solution of minimum length*. [*Hint:* Consider the equation $A\mathbf{x} = \hat{\mathbf{b}}$, where $\hat{\mathbf{b}}$ is the orthogonal projection of $\mathbf{b}$ onto Col A.]

⊤ In Exercises 31 and 32, construct the pseudoinverse of A. Begin by using a matrix program to produce the SVD of A, or, if that is not available, begin with an orthogonal diagonalization of $A^T A$. Use the pseudoinverse to solve $A\mathbf{x} = \mathbf{b}$, for $\mathbf{b} = (6, -1, -4, 6)$, and let $\hat{\mathbf{x}}$ be the solution. Make a calculation to verify that $\hat{\mathbf{x}}$ is in Row A. Find a nonzero vector $\mathbf{u}$ in Nul A, and verify that $\|\hat{\mathbf{x}}\| < \|\hat{\mathbf{x}} + \mathbf{u}\|$, which must be true by Exercise 29(c).

⊤ **31.** $A = \begin{bmatrix} -3 & -3 & -6 & 6 & 1 \\ -1 & -1 & -1 & 1 & -2 \\ 0 & 0 & -1 & 1 & -1 \\ 0 & 0 & -1 & 1 & -1 \end{bmatrix}$

⊤ **32.** $A = \begin{bmatrix} 4 & 0 & -1 & -2 & 0 \\ -5 & 0 & 3 & 5 & 0 \\ 2 & 0 & -1 & -2 & 0 \\ 6 & 0 & -3 & -6 & 0 \end{bmatrix}$

8

The Geometry of Vector Spaces

THE PLATONIC SOLIDS

In the city of Athens in 387 B.C., the Greek philosopher Plato founded an Academy, sometimes referred to as the world's first university. While the curriculum included astronomy, biology, political theory, and philosophy, the subject closest to his heart was geometry. Indeed, inscribed over the doors of his academy were these words: *"Let no one destitute of geometry enter my doors."*

The Greeks were greatly impressed by geometric patterns such as the regular solids. A polyhedron is called regular if its faces are congruent regular polygons and all the angles at the vertices are equal. As early as 100 years before Plato, the Pythagoreans knew at least three of the regular solids: the tetrahedron (4 triangular faces), the cube (6 square faces), and the octahedron (8 triangular faces). (See Figure 1.) These shapes occur naturally as crystals of common minerals. There are only five such regular solids, the remaining two being the dodecahedron (12 pentagonal faces) and the icosahedron (20 triangular faces).

Plato discussed the basic theory of these five solids in the dialogue *Timaeus*, and since then they have carried his name: the Platonic solids.

For centuries there was no need to envision geometric objects in more than three dimensions. But nowadays mathematicians regularly deal with objects in vector spaces having four, five, or even hundreds of dimensions. It is not necessarily clear what geometrical properties one might ascribe to these objects in higher dimensions.

For example, what properties do lines have in 2-space and planes have in 3-space that would be useful in higher dimensions? How can one characterize such objects? Sections 8.1 and 8.4 provide some answers. The hyperplanes of Section 8.4 will be important for understanding the multidimensional nature of the linear programming problems in Chapter 9.

What would the analogue of a polyhedron "look like" in more than three dimensions? A partial answer is provided by two-dimensional projections of the four-dimensional object, created in a manner analogous to two-dimensional projections of a three-dimensional object. Section 8.5 illustrates this idea for the four-dimensional "cube" and the four-dimensional "simplex."

The study of geometry in higher dimensions not only provides new ways of visualizing abstract algebraic concepts, but also creates tools that may be applied in $\mathbb{R}^3$. For instance, Sections 8.2 and 8.6 include applications to computer graphics, and Section 8.5 outlines a proof (in Exercise 28) that there are only five regular polyhedra in $\mathbb{R}^3$.

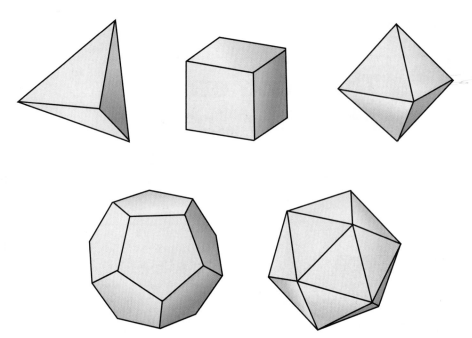

FIGURE 1 The five Platonic solids.

Most applications in earlier chapters involved algebraic calculations with subspaces and linear combinations of vectors. This chapter studies sets of vectors that can be visualized as geometric objects such as line segments, polygons, and solid objects. Individual vectors are viewed as points. The concepts introduced here are used in computer graphics, linear programming (in Chapter 9), and other areas of mathematics.[1]

Throughout the chapter, sets of vectors are described by linear combinations, but with various restrictions on the weights used in the combinations. For instance, in Section 8.1, the sum of the weights is 1, while in Section 8.3, the weights are positive and sum to 1. The visualizations are in $\mathbb{R}^2$ or $\mathbb{R}^3$, of course, but the concepts also apply to $\mathbb{R}^n$ and other vector spaces.

8.1 Affine Combinations

An affine combination of vectors is a special kind of linear combination. Given vectors (or "points") $\mathbf{v}_1, \mathbf{v}_2, \ldots, \mathbf{v}_p$ in $\mathbb{R}^n$ and scalars $c_1, \ldots, c_p$, an **affine combination** of $\mathbf{v}_1, \mathbf{v}_2, \ldots, \mathbf{v}_p$ is a linear combination

$$c_1 \mathbf{v}_1 + \cdots + c_p \mathbf{v}_p$$

such that the weights satisfy $c_1 + \cdots + c_p = 1$.

DEFINITION The set of all affine combinations of points in a set S is called the **affine hull** (or **affine span**) of S, denoted by aff S.

---

[1] See Foley, van Dam, Feiner, and Hughes, *Computer Graphics—Principles and Practice*, 2nd edition (Boston: Addison-Wesley, 1996), pp. 1083–1112. That material also discusses coordinate-free "affine spaces."

The affine hull of a single point $\mathbf{v}_1$ is just the set $\{\mathbf{v}_1\}$, since it has the form $c_1\mathbf{v}_1$ where $c_1 = 1$. The affine hull of two distinct points is often written in a special way. Suppose $\mathbf{y} = c_1\mathbf{v}_1 + c_2\mathbf{v}_2$ with $c_1 + c_2 = 1$. Write t in place of c_2, so that $c_1 = 1 - c_2 = 1 - t$. Then the affine hull of $\{\mathbf{v}_1, \mathbf{v}_2\}$ is the set

$$\mathbf{y} = (1 - t)\mathbf{v}_1 + t\mathbf{v}_2, \quad \text{with } t \text{ in } \mathbb{R} \tag{1}$$

This set of points includes $\mathbf{v}_1$ (when $t = 0$) and $\mathbf{v}_2$ (when $t = 1$). If $\mathbf{v}_2 = \mathbf{v}_1$, then (1) again describes just one point. Otherwise, (1) describes the *line* through $\mathbf{v}_1$ and $\mathbf{v}_2$. To see this, rewrite (1) in the form

$$\mathbf{y} = \mathbf{v}_1 + t(\mathbf{v}_2 - \mathbf{v}_1) = \mathbf{p} + t\mathbf{u}, \quad \text{with } t \text{ in } \mathbb{R}$$

where $\mathbf{p}$ is $\mathbf{v}_1$ and $\mathbf{u}$ is $\mathbf{v}_2 - \mathbf{v}_1$. The set of all multiples of $\mathbf{u}$ is Span $\{\mathbf{u}\}$, the line through $\mathbf{u}$ and the origin. Adding $\mathbf{p}$ to each point on this line translates Span $\{\mathbf{u}\}$ into the line through $\mathbf{p}$ parallel to the line through $\mathbf{u}$ and the origin. See Figure 1. (Compare this figure with Figure 5 in Section 1.5.)

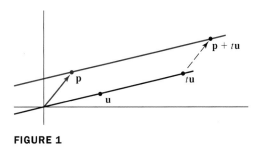

FIGURE 1

Figure 2 uses the original points $\mathbf{v}_1$ and $\mathbf{v}_2$, and displays aff $\{\mathbf{v}_1, \mathbf{v}_2\}$ as the line through $\mathbf{v}_1$ and $\mathbf{v}_2$.

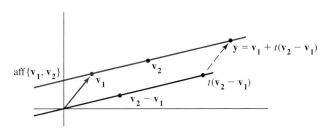

FIGURE 2

Notice that while the point $\mathbf{y}$ in Figure 2 is an affine combination of $\mathbf{v}_1$ and $\mathbf{v}_2$, the point $\mathbf{y} - \mathbf{v}_1$ equals $t(\mathbf{v}_2 - \mathbf{v}_1)$, which is a linear combination (in fact, a multiple) of $\mathbf{v}_2 - \mathbf{v}_1$. This relation between $\mathbf{y}$ and $\mathbf{y} - \mathbf{v}_1$ holds for any affine combination of points, as the following theorem shows.

THEOREM I

> A point $\mathbf{y}$ in $\mathbb{R}^n$ is an affine combination of $\mathbf{v}_1, \ldots, \mathbf{v}_p$ in $\mathbb{R}^n$ if and only if $\mathbf{y} - \mathbf{v}_1$ is a linear combination of the translated points $\mathbf{v}_2 - \mathbf{v}_1, \ldots, \mathbf{v}_p - \mathbf{v}_1$.

PROOF If $\mathbf{y} - \mathbf{v}_1$ is a linear combination of $\mathbf{v}_2 - \mathbf{v}_1, \ldots, \mathbf{v}_p - \mathbf{v}_1$, there exist weights $c_2, \ldots, c_p$ such that

$$\mathbf{y} - \mathbf{v}_1 = c_2(\mathbf{v}_2 - \mathbf{v}_1) + \cdots + c_p(\mathbf{v}_p - \mathbf{v}_1) \tag{2}$$

Then

$$\mathbf{y} = (1 - c_2 - \cdots - c_p)\mathbf{v}_1 + c_2\mathbf{v}_2 + \cdots + c_p\mathbf{v}_p \tag{3}$$

and the weights in this linear combination sum to 1. So $\mathbf{y}$ is an affine combination of $\mathbf{v}_1, \ldots, \mathbf{v}_p$. Conversely, suppose

$$\mathbf{y} = c_1\mathbf{v}_1 + c_2\mathbf{v}_2 + \cdots + c_p\mathbf{v}_p \tag{4}$$

where $c_1 + \cdots + c_p = 1$. Since $c_1 = 1 - c_2 - \cdots - c_p$, equation (4) may be written as in (3), and this leads to (2), which shows that $\mathbf{y} - \mathbf{v}_1$ is a linear combination of $\mathbf{v}_2 - \mathbf{v}_1, \ldots, \mathbf{v}_p - \mathbf{v}_1$. ∎

In the statement of Theorem 1, the point $\mathbf{v}_1$ could be replaced by any of the other points in the list $\mathbf{v}_1, \ldots, \mathbf{v}_p$. Only the notation in the proof would change.

EXAMPLE 1 Let $\mathbf{v}_1 = \begin{bmatrix} 1 \\ 2 \end{bmatrix}$, $\mathbf{v}_2 = \begin{bmatrix} 2 \\ 5 \end{bmatrix}$, $\mathbf{v}_3 = \begin{bmatrix} 1 \\ 3 \end{bmatrix}$, $\mathbf{v}_4 = \begin{bmatrix} -2 \\ 2 \end{bmatrix}$, and $\mathbf{y} = \begin{bmatrix} 4 \\ 1 \end{bmatrix}$.
If possible, write $\mathbf{y}$ as an affine combination of $\mathbf{v}_1, \mathbf{v}_2, \mathbf{v}_3$, and $\mathbf{v}_4$.

SOLUTION Compute the translated points

$$\mathbf{v}_2 - \mathbf{v}_1 = \begin{bmatrix} 1 \\ 3 \end{bmatrix}, \quad \mathbf{v}_3 - \mathbf{v}_1 = \begin{bmatrix} 0 \\ 1 \end{bmatrix}, \quad \mathbf{v}_4 - \mathbf{v}_1 = \begin{bmatrix} -3 \\ 0 \end{bmatrix}, \quad \mathbf{y} - \mathbf{v}_1 = \begin{bmatrix} 3 \\ -1 \end{bmatrix}$$

To find scalars c_2, c_3, and c_4 such that

$$c_2(\mathbf{v}_2 - \mathbf{v}_1) + c_3(\mathbf{v}_3 - \mathbf{v}_1) + c_4(\mathbf{v}_4 - \mathbf{v}_1) = \mathbf{y} - \mathbf{v}_1 \tag{5}$$

row reduce the augmented matrix having these points as columns:

$$\begin{bmatrix} 1 & 0 & -3 & 3 \\ 3 & 1 & 0 & -1 \end{bmatrix} \sim \begin{bmatrix} 1 & 0 & -3 & 3 \\ 0 & 1 & 9 & -10 \end{bmatrix}$$

This shows that equation (5) is consistent, and the general solution is $c_2 = 3c_4 + 3$, $c_3 = -9c_4 - 10$, with c_4 free. When $c_4 = 0$,

$$\mathbf{y} - \mathbf{v}_1 = 3(\mathbf{v}_2 - \mathbf{v}_1) - 10(\mathbf{v}_3 - \mathbf{v}_1) + 0(\mathbf{v}_4 - \mathbf{v}_1)$$

and

$$\mathbf{y} = 8\mathbf{v}_1 + 3\mathbf{v}_2 - 10\mathbf{v}_3$$

As another example, take $c_4 = 1$. Then $c_2 = 6$ and $c_3 = -19$, so

$$\mathbf{y} - \mathbf{v}_1 = 6(\mathbf{v}_2 - \mathbf{v}_1) - 19(\mathbf{v}_3 - \mathbf{v}_1) + 1(\mathbf{v}_4 - \mathbf{v}_1)$$

and

$$\mathbf{y} = 13\mathbf{v}_1 + 6\mathbf{v}_2 - 19\mathbf{v}_3 + \mathbf{v}_4$$ ∎

While the procedure in Example 1 works for arbitrary points $\mathbf{v}_1, \mathbf{v}_2, \ldots, \mathbf{v}_p$ in $\mathbb{R}^n$, the question can be answered more directly if the chosen points $\mathbf{v}_i$ are a basis for $\mathbb{R}^n$. For example, let $\mathcal{B} = \{\mathbf{b}_1, \ldots, \mathbf{b}_n\}$ be such a basis. Then any $\mathbf{y}$ in $\mathbb{R}^n$ is a unique *linear* combination of $\mathbf{b}_1, \ldots, \mathbf{b}_n$. This combination is an affine combination of the $\mathbf{b}$'s if and only if the weights sum to 1. (These weights are just the $\mathcal{B}$-coordinates of $\mathbf{y}$, as in Section 4.4.)

EXAMPLE 2 Let $\mathbf{b}_1 = \begin{bmatrix} 4 \\ 0 \\ 3 \end{bmatrix}$, $\mathbf{b}_2 = \begin{bmatrix} 0 \\ 4 \\ 2 \end{bmatrix}$, $\mathbf{b}_3 = \begin{bmatrix} 5 \\ 2 \\ 4 \end{bmatrix}$, $\mathbf{p}_1 = \begin{bmatrix} 2 \\ 0 \\ 0 \end{bmatrix}$, and $\mathbf{p}_2 = \begin{bmatrix} 1 \\ 2 \\ 2 \end{bmatrix}$.

The set $\mathcal{B} = \{\mathbf{b}_1, \mathbf{b}_2, \mathbf{b}_3\}$ is a basis for $\mathbb{R}^3$. Determine whether the points $\mathbf{p}_1$ and $\mathbf{p}_2$ are affine combinations of the points in $\mathcal{B}$.

SOLUTION Find the $\mathcal{B}$-coordinates of $\mathbf{p}_1$ and $\mathbf{p}_2$. These two calculations can be combined by row reducing the matrix $[\,\mathbf{b}_1 \ \ \mathbf{b}_2 \ \ \mathbf{b}_3 \ \ \mathbf{p}_1 \ \ \mathbf{p}_2\,]$, with two augmented columns:

$$
\begin{bmatrix} 4 & 0 & 5 & 2 & 1 \\ 0 & 4 & 2 & 0 & 2 \\ 3 & 2 & 4 & 0 & 2 \end{bmatrix} \sim \begin{bmatrix} 1 & 0 & 0 & -2 & \frac{2}{3} \\ 0 & 1 & 0 & -1 & \frac{2}{3} \\ 0 & 0 & 1 & 2 & -\frac{1}{3} \end{bmatrix}
$$

Read column 4 to build $\mathbf{p}_1$, and read column 5 to build $\mathbf{p}_2$:

$$
\mathbf{p}_1 = -2\mathbf{b}_1 - \mathbf{b}_2 + 2\mathbf{b}_3 \quad \text{and} \quad \mathbf{p}_2 = \tfrac{2}{3}\mathbf{b}_1 + \tfrac{2}{3}\mathbf{b}_2 - \tfrac{1}{3}\mathbf{b}_3
$$

The sum of the weights in the linear combination for $\mathbf{p}_1$ is -1, not 1, so $\mathbf{p}_1$ is *not* an affine combination of the $\mathbf{b}$'s. However, $\mathbf{p}_2$ *is* an affine combination of the $\mathbf{b}$'s, because the sum of the weights for $\mathbf{p}_2$ is 1. ∎

DEFINITION

A set S is **affine** if $\mathbf{p}, \mathbf{q} \in S$ implies that $(1-t)\mathbf{p} + t\mathbf{q} \in S$ for each real number t.

Geometrically, a set is affine if whenever two points are in the set, the entire line through these points is in the set. (If S contains only one point, $\mathbf{p}$, then the line through $\mathbf{p}$ and $\mathbf{p}$ is just a point, a "degenerate" line.) Algebraically, for a set S to be affine, the definition requires that every affine combination of two points of S belong to S. Remarkably, this is equivalent to requiring that S contain every affine combination of an arbitrary number of points of S.

THEOREM 2

A set S is affine if and only if every affine combination of points of S lies in S. That is, S is affine if and only if $S = \text{aff } S$.

Remark: See the remark prior to Theorem 5 in Chapter 3 regarding mathematical induction.

PROOF Suppose that S is affine and use induction on the number m of points of S occurring in an affine combination. When m is 1 or 2, an affine combination of m points of S lies in S, by the definition of an affine set. Now, assume that every affine combination of k or fewer points of S yields a point in S, and consider a combination of $k + 1$ points. Take $\mathbf{v}_i$ in S for $i = 1, \ldots, k + 1$, and let $\mathbf{y} = c_1\mathbf{v}_1 + \cdots + c_k\mathbf{v}_k + c_{k+1}\mathbf{v}_{k+1}$, where $c_1 + \cdots + c_{k+1} = 1$. Since the c_i's sum to 1, at least one of them must not be equal to 1. By reindexing the $\mathbf{v}_i$ and c_i, if necessary, we may assume that $c_{k+1} \neq 1$. Let $t = c_1 + \cdots + c_k$. Then $t = 1 - c_{k+1} \neq 0$, and

$$
\mathbf{y} = (1 - c_{k+1})\left(\frac{c_1}{t}\mathbf{v}_1 + \cdots + \frac{c_k}{t}\mathbf{v}_k\right) + c_{k+1}\mathbf{v}_{k+1} \tag{6}
$$

By the induction hypothesis, the point $\mathbf{z} = (c_1/t)\mathbf{v}_1 + \cdots + (c_k/t)\mathbf{v}_k$ is in S, since the coefficients sum to 1. Thus (6) displays $\mathbf{y}$ as an affine combination of two points in S, and so $\mathbf{y} \in S$. By the principle of induction, every affine combination of such points lies in S. That is, aff $S \subseteq S$. But the reverse inclusion, $S \subseteq$ aff S, always applies. Thus, when S is affine, $S =$ aff S. Conversely, if $S =$ aff S, then affine combinations of two (or more) points of S lie in S, so S is affine. ∎

The next definition provides terminology for affine sets that emphasizes their close connection with subspaces of $\mathbb{R}^n$.

DEFINITION

A translate of a set S in $\mathbb{R}^n$ by a vector $\mathbf{p}$ is the set $S + \mathbf{p} = \{\mathbf{s} + \mathbf{p} : \mathbf{s} \in S\}$.[2] A **flat** in $\mathbb{R}^n$ is a translate of a subspace of $\mathbb{R}^n$. Two flats are **parallel** if one is a translate of the other. The **dimension of a flat** is the dimension of the corresponding parallel subspace. The **dimension of a set** S, written as dim S, is the dimension of the smallest flat containing S. A **line** in $\mathbb{R}^n$ is a flat of dimension 1. A **hyperplane** in $\mathbb{R}^n$ is a flat of dimension $n - 1$.

In $\mathbb{R}^3$, the proper subspaces[3] consist of the origin $\mathbf{0}$, the set of all lines through $\mathbf{0}$, and the set of all planes through $\mathbf{0}$. Thus the proper flats in $\mathbb{R}^3$ are points (zero-dimensional), lines (one-dimensional), and planes (two-dimensional), which may or may not pass through the origin.

The next theorem shows that these geometric descriptions of lines and planes in $\mathbb{R}^3$ (as translates of subspaces) actually coincide with their earlier algebraic descriptions as sets of all affine combinations of two or three points, respectively.

THEOREM 3

A nonempty set S is affine if and only if it is a flat.

Remark: Notice the key role that definitions play in this proof. For example, the first part assumes that S is affine and seeks to show that S is a flat. By definition, a flat is a translate of a subspace. By choosing $\mathbf{p}$ in S and defining $W = S + (-\mathbf{p})$, the set S is translated to the origin and $S = W + \mathbf{p}$. It remains to show that W is a subspace, for then S will be a translate of a subspace and hence a flat.

PROOF Suppose that S is affine. Let $\mathbf{p}$ be any fixed point in S and let

$$W = S + (-\mathbf{p}), \text{ so that } S = W + \mathbf{p}$$

To show that S is a flat, it suffices to show that W is a subspace of $\mathbb{R}^n$. Since $\mathbf{p}$ is in S, the zero vector is in W. To show that W is closed under sums and scalar multiples, it suffices to show that if $\mathbf{u}_1$ and $\mathbf{u}_2$ are elements of W, then $\mathbf{u}_1 + t\mathbf{u}_2$ is in W for every real t. That is, we want to show that $\mathbf{u}_1 + t\mathbf{u}_2$ is in $S + (-\mathbf{p})$. Since $\mathbf{u}_1$ and $\mathbf{u}_2$ are in W, there exist $\mathbf{s}_1$ and $\mathbf{s}_2$ in S such that

$$\mathbf{u}_1 = \mathbf{s}_1 - \mathbf{p} \text{ and } \mathbf{u}_2 = \mathbf{s}_2 - \mathbf{p}$$

It follows that

$$\mathbf{u}_1 + t\mathbf{u}_2 = \mathbf{s}_1 - \mathbf{p} + t(\mathbf{s}_2 - \mathbf{p})$$
$$= \mathbf{s}_1 + t\mathbf{s}_2 - t\mathbf{p} - \mathbf{p}$$

---

[2] If $\mathbf{p} = \mathbf{0}$, then the translate is just S itself. See Figure 4 in Section 1.5.

[3] A subset A of a set B is called a **proper** subset of B if $A \neq B$. The same condition applies to proper subspaces and proper flats in $\mathbb{R}^n$: they are not equal to $\mathbb{R}^n$.

Regrouping the first three terms, we find that $\mathbf{s}_1 + t\mathbf{s}_2 - t\mathbf{p}$ is in S since the coefficients sum to 1 and S is affine. (See Theorem 2.) So $\mathbf{u}_1 + t\mathbf{u}_2$ is in $S - \mathbf{p} = W$. This shows that W is a subspace of $\mathbb{R}^n$. Thus S is a flat, because $S = W + \mathbf{p}$.

Conversely, suppose that S is a flat. That is, $S = W + \mathbf{p}$ for some $\mathbf{p}$ in $\mathbb{R}^n$ and some subspace W. To show that S is affine, it suffices to show that for any pair $\mathbf{s}_1$ and $\mathbf{s}_2$ of points in S, the line through $\mathbf{s}_1$ and $\mathbf{s}_2$ lies in S. By the definition of W, there exist $\mathbf{u}_1$ and $\mathbf{u}_2$ in W such that

$$\mathbf{s}_1 = \mathbf{u}_1 + \mathbf{p} \text{ and } \mathbf{s}_2 = \mathbf{u}_2 + \mathbf{p}$$

So, for each real t we have

$$
\begin{aligned}
(1-t)\mathbf{s}_1 + t\mathbf{s}_2 &= (1-t)(\mathbf{u}_1 + \mathbf{p}) + t(\mathbf{u}_2 + \mathbf{p}) \\
&= (1-t)\mathbf{u}_1 + (1-t)\mathbf{p} + t\mathbf{u}_2 + t\mathbf{p} \\
&= (1-t)\mathbf{u}_1 + t\mathbf{u}_2 + \mathbf{p}
\end{aligned}
$$

Since W is a subspace, $(1-t)\mathbf{u}_1 + t\mathbf{u}_2$ is in W and so $(1-t)\mathbf{s}_1 + t\mathbf{s}_2$ is in $W + \mathbf{p} = S$. Thus, S is affine. ∎

Theorem 3 provides a geometric way to view the affine hull of a set: it is the flat that consists of all the affine combinations of points in the set. For instance, Figure 3 shows the points studied in Example 2. Although the set of all *linear* combinations of $\mathbf{b}_1$, $\mathbf{b}_2$, and $\mathbf{b}_3$ is all of $\mathbb{R}^3$, the set of all *affine* combinations is only the plane through $\mathbf{b}_1$, $\mathbf{b}_2$, and $\mathbf{b}_3$. Note that $\mathbf{p}_2$ (from Example 2) is in the plane through $\mathbf{b}_1$, $\mathbf{b}_2$, and $\mathbf{b}_3$, while $\mathbf{p}_1$ is not in that plane. Also, see Exercise 22.

The next example takes a fresh look at a familiar set—the set of all solutions of a system $A\mathbf{x} = \mathbf{b}$.

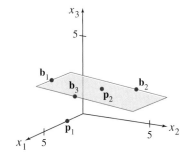

FIGURE 3

EXAMPLE 3   Suppose that the solutions of an equation $A\mathbf{x} = \mathbf{b}$ are all of the form $\mathbf{x} = x_3\mathbf{u} + \mathbf{p}$, where $\mathbf{u} = \begin{bmatrix} 2 \\ -3 \\ 1 \end{bmatrix}$ and $\mathbf{p} = \begin{bmatrix} 4 \\ 0 \\ -3 \end{bmatrix}$. Recall from Section 1.5 that this set is parallel to the solution set of $A\mathbf{x} = \mathbf{0}$, which consists of all points of the form $x_3\mathbf{u}$. Find points $\mathbf{v}_1$ and $\mathbf{v}_2$ such that the solution set of $A\mathbf{x} = \mathbf{b}$ is aff $\{\mathbf{v}_1, \mathbf{v}_2\}$.

SOLUTION   The solution set is a line through $\mathbf{p}$ in the direction of $\mathbf{u}$, as in Figure 1. Since aff $\{\mathbf{v}_1, \mathbf{v}_2\}$ is a line through $\mathbf{v}_1$ and $\mathbf{v}_2$, identify two points on the line $\mathbf{x} = x_3\mathbf{u} + \mathbf{p}$. Two simple choices appear when $x_3 = 0$ and $x_3 = 1$. That is, take $\mathbf{v}_1 = \mathbf{p}$ and $\mathbf{v}_2 = \mathbf{u} + \mathbf{p}$, so that

$$\mathbf{v}_2 = \mathbf{u} + \mathbf{p} = \begin{bmatrix} 2 \\ -3 \\ 1 \end{bmatrix} + \begin{bmatrix} 4 \\ 0 \\ -3 \end{bmatrix} = \begin{bmatrix} 6 \\ -3 \\ -2 \end{bmatrix}$$

In this case, the solution set is described as the set of all affine combinations of the form

$$\mathbf{x} = (1 - x_3)\begin{bmatrix} 4 \\ 0 \\ -3 \end{bmatrix} + x_3\begin{bmatrix} 6 \\ -3 \\ -2 \end{bmatrix} \qquad ∎$$

Earlier, Theorem 1 displayed an important connection between affine combinations and linear combinations. The next theorem provides another view of affine combinations, which for $\mathbb{R}^2$ and $\mathbb{R}^3$ is closely connected to applications in computer graphics, discussed in the next section (and in Section 2.7).

DEFINITION

For $\mathbf{v}$ in $\mathbb{R}^n$, the standard **homogeneous form** of $\mathbf{v}$ is the point $\tilde{\mathbf{v}} = \begin{bmatrix} \mathbf{v} \\ 1 \end{bmatrix}$ in $\mathbb{R}^{n+1}$.

THEOREM 4

A point $\mathbf{y}$ in $\mathbb{R}^n$ is an affine combination of $\mathbf{v}_1, \ldots, \mathbf{v}_p$ in $\mathbb{R}^n$ if and only if the homogeneous form of $\mathbf{y}$ is in Span $\{\tilde{\mathbf{v}}_1, \ldots, \tilde{\mathbf{v}}_p\}$. In fact, $\mathbf{y} = c_1\mathbf{v}_1 + \cdots + c_p\mathbf{v}_p$, with $c_1 + \cdots + c_p = 1$, if and only if $\tilde{\mathbf{y}} = c_1\tilde{\mathbf{v}}_1 + \cdots + c_p\tilde{\mathbf{v}}_p$.

PROOF A point $\mathbf{y}$ is in aff $\{\mathbf{v}_1, \ldots, \mathbf{v}_p\}$ if and only if there exist weights $c_1, \ldots, c_p$ such that

$$\begin{bmatrix} \mathbf{y} \\ 1 \end{bmatrix} = c_1 \begin{bmatrix} \mathbf{v}_1 \\ 1 \end{bmatrix} + c_2 \begin{bmatrix} \mathbf{v}_2 \\ 1 \end{bmatrix} + \cdots + c_p \begin{bmatrix} \mathbf{v}_p \\ 1 \end{bmatrix}$$

This happens if and only if $\tilde{\mathbf{y}}$ is in Span $\{\tilde{\mathbf{v}}_1, \tilde{\mathbf{v}}_2, \ldots, \tilde{\mathbf{v}}_p\}$. ∎

EXAMPLE 4 Let $\mathbf{v}_1 = \begin{bmatrix} 3 \\ 1 \\ 1 \end{bmatrix}$, $\mathbf{v}_2 = \begin{bmatrix} 1 \\ 2 \\ 2 \end{bmatrix}$, $\mathbf{v}_3 = \begin{bmatrix} 1 \\ 7 \\ 1 \end{bmatrix}$, and $\mathbf{p} = \begin{bmatrix} 4 \\ 3 \\ 0 \end{bmatrix}$. Use Theorem 4 to write $\mathbf{p}$ as an affine combination of $\mathbf{v}_1$, $\mathbf{v}_2$, and $\mathbf{v}_3$, if possible.

SOLUTION Row reduce the augmented matrix for the equation

$$x_1\tilde{\mathbf{v}}_1 + x_2\tilde{\mathbf{v}}_2 + x_3\tilde{\mathbf{v}}_3 = \tilde{\mathbf{p}}$$

To simplify the arithmetic, move the fourth row of 1's to the top (equivalent to three row interchanges). After this, the number of arithmetic operations here is basically the same as the number needed for the method using Theorem 1.

$$[\,\tilde{\mathbf{v}}_1 \ \ \tilde{\mathbf{v}}_2 \ \ \tilde{\mathbf{v}}_3 \ \ \tilde{\mathbf{p}}\,] \sim \begin{bmatrix} 1 & 1 & 1 & 1 \\ 3 & 1 & 1 & 4 \\ 1 & 2 & 7 & 3 \\ 1 & 2 & 1 & 0 \end{bmatrix} \sim \begin{bmatrix} 1 & 1 & 1 & 1 \\ 0 & -2 & -2 & 1 \\ 0 & 1 & 6 & 2 \\ 0 & 1 & 0 & -1 \end{bmatrix}$$

$$\sim \cdots \sim \begin{bmatrix} 1 & 0 & 0 & 1.5 \\ 0 & 1 & 0 & -1 \\ 0 & 0 & 1 & .5 \\ 0 & 0 & 0 & 0 \end{bmatrix}$$

By Theorem 4, $1.5\mathbf{v}_1 - \mathbf{v}_2 + .5\mathbf{v}_3 = \mathbf{p}$. See Figure 4, which shows the plane that contains $\mathbf{v}_1$, $\mathbf{v}_2$, $\mathbf{v}_3$, and $\mathbf{p}$ (together with points on the coordinate axes). ∎

Practice Problem

Plot the points $\mathbf{v}_1 = \begin{bmatrix} 1 \\ 0 \end{bmatrix}$, $\mathbf{v}_2 = \begin{bmatrix} -1 \\ 2 \end{bmatrix}$, $\mathbf{v}_3 = \begin{bmatrix} 3 \\ 1 \end{bmatrix}$, and $\mathbf{p} = \begin{bmatrix} 4 \\ 3 \end{bmatrix}$ on graph paper, and explain why $\mathbf{p}$ *must* be an affine combination of $\mathbf{v}_1$, $\mathbf{v}_2$, and $\mathbf{v}_3$. Then find the affine combination for $\mathbf{p}$. [*Hint:* What is the dimension of aff $\{\mathbf{v}_1, \mathbf{v}_2, \mathbf{v}_3\}$?]

FIGURE 4

8.1 Exercises

In Exercises 1–4, write **y** as an affine combination of the other points listed, if possible.

1. $\mathbf{v}_1 = \begin{bmatrix} 1 \\ 2 \end{bmatrix}$, $\mathbf{v}_2 = \begin{bmatrix} -2 \\ 2 \end{bmatrix}$, $\mathbf{v}_3 = \begin{bmatrix} 0 \\ 4 \end{bmatrix}$, $\mathbf{v}_4 = \begin{bmatrix} 3 \\ 7 \end{bmatrix}$, $\mathbf{y} = \begin{bmatrix} 5 \\ 3 \end{bmatrix}$

2. $\mathbf{v}_1 = \begin{bmatrix} 1 \\ 1 \end{bmatrix}$, $\mathbf{v}_2 = \begin{bmatrix} -1 \\ 2 \end{bmatrix}$, $\mathbf{v}_3 = \begin{bmatrix} 3 \\ 2 \end{bmatrix}$, $\mathbf{y} = \begin{bmatrix} 5 \\ 7 \end{bmatrix}$

3. $\mathbf{v}_1 = \begin{bmatrix} -3 \\ 1 \\ 1 \end{bmatrix}$, $\mathbf{v}_2 = \begin{bmatrix} 0 \\ 4 \\ -2 \end{bmatrix}$, $\mathbf{v}_3 = \begin{bmatrix} 4 \\ -2 \\ 6 \end{bmatrix}$, $\mathbf{y} = \begin{bmatrix} 17 \\ 1 \\ 5 \end{bmatrix}$

4. $\mathbf{v}_1 = \begin{bmatrix} 1 \\ 2 \\ 0 \end{bmatrix}$, $\mathbf{v}_2 = \begin{bmatrix} 2 \\ -6 \\ 7 \end{bmatrix}$, $\mathbf{v}_3 = \begin{bmatrix} 4 \\ 3 \\ 1 \end{bmatrix}$, $\mathbf{y} = \begin{bmatrix} -3 \\ 4 \\ -4 \end{bmatrix}$

In Exercises 5 and 6, let $\mathbf{b}_1 = \begin{bmatrix} 2 \\ 1 \\ 1 \end{bmatrix}$, $\mathbf{b}_2 = \begin{bmatrix} 1 \\ 0 \\ -2 \end{bmatrix}$, $\mathbf{b}_3 = \begin{bmatrix} 2 \\ -5 \\ 1 \end{bmatrix}$,

and $S = \{\mathbf{b}_1, \mathbf{b}_2, \mathbf{b}_3\}$. Note that S is an orthogonal basis for $\mathbb{R}^3$. Write each of the given points as an affine combination of the points in the set S, if possible. [*Hint:* Use Theorem 5 in Section 6.2 instead of row reduction to find the weights.]

5. a. $\mathbf{p}_1 = \begin{bmatrix} 3 \\ 8 \\ 4 \end{bmatrix}$ b. $\mathbf{p}_2 = \begin{bmatrix} 6 \\ -3 \\ 3 \end{bmatrix}$ c. $\mathbf{p}_3 = \begin{bmatrix} 0 \\ -1 \\ -5 \end{bmatrix}$

6. a. $\mathbf{p}_1 = \begin{bmatrix} 0 \\ -19 \\ -5 \end{bmatrix}$ b. $\mathbf{p}_2 = \begin{bmatrix} 1.5 \\ -1.3 \\ -.5 \end{bmatrix}$ c. $\mathbf{p}_3 = \begin{bmatrix} 5 \\ -4 \\ 0 \end{bmatrix}$

7. Let

$$\mathbf{v}_1 = \begin{bmatrix} 1 \\ 0 \\ 3 \\ 0 \end{bmatrix}, \quad \mathbf{v}_2 = \begin{bmatrix} 2 \\ -1 \\ 0 \\ 4 \end{bmatrix}, \quad \mathbf{v}_3 = \begin{bmatrix} -1 \\ 2 \\ 1 \\ 1 \end{bmatrix},$$

$$\mathbf{p}_1 = \begin{bmatrix} 5 \\ -3 \\ 5 \\ 3 \end{bmatrix}, \quad \mathbf{p}_2 = \begin{bmatrix} -9 \\ 10 \\ 9 \\ -13 \end{bmatrix}, \quad \mathbf{p}_3 = \begin{bmatrix} 4 \\ 2 \\ 8 \\ 5 \end{bmatrix},$$

and $S = \{\mathbf{v}_1, \mathbf{v}_2, \mathbf{v}_3\}$. It can be shown that S is linearly independent.

a. Is $\mathbf{p}_1$ in Span S? Is $\mathbf{p}_1$ in aff S?

b. Is $\mathbf{p}_2$ in Span S? Is $\mathbf{p}_2$ in aff S?

c. Is $\mathbf{p}_3$ in Span S? Is $\mathbf{p}_3$ in aff S?

8. Repeat Exercise 7 when

$$\mathbf{v}_1 = \begin{bmatrix} 1 \\ 0 \\ 3 \\ -2 \end{bmatrix}, \quad \mathbf{v}_2 = \begin{bmatrix} 2 \\ 1 \\ 6 \\ -5 \end{bmatrix}, \quad \mathbf{v}_3 = \begin{bmatrix} 3 \\ 0 \\ 12 \\ -6 \end{bmatrix},$$

$$\mathbf{p}_1 = \begin{bmatrix} 4 \\ -1 \\ 15 \\ -7 \end{bmatrix}, \quad \mathbf{p}_2 = \begin{bmatrix} -5 \\ 3 \\ -8 \\ 6 \end{bmatrix}, \quad \text{and} \quad \mathbf{p}_3 = \begin{bmatrix} 1 \\ 6 \\ -6 \\ -8 \end{bmatrix}.$$

9. Suppose that the solutions of an equation $A\mathbf{x} = \mathbf{b}$ are all of the form $\mathbf{x} = x_3\mathbf{u} + \mathbf{p}$, where $\mathbf{u} = \begin{bmatrix} 4 \\ -2 \end{bmatrix}$ and $\mathbf{p} = \begin{bmatrix} -3 \\ 0 \end{bmatrix}$. Find points $\mathbf{v}_1$ and $\mathbf{v}_2$ such that the solution set of $A\mathbf{x} = \mathbf{b}$ is aff $\{\mathbf{v}_1, \mathbf{v}_2\}$.

10. Suppose that the solutions of an equation $A\mathbf{x} = \mathbf{b}$ are all of the form $\mathbf{x} = x_3\mathbf{u} + \mathbf{p}$, where $\mathbf{u} = \begin{bmatrix} 5 \\ 1 \\ -2 \end{bmatrix}$ and $\mathbf{p} = \begin{bmatrix} 1 \\ -3 \\ 4 \end{bmatrix}$.

Find points $\mathbf{v}_1$ and $\mathbf{v}_2$ such that the solution set of $A\mathbf{x} = \mathbf{b}$ is aff $\{\mathbf{v}_1, \mathbf{v}_2\}$.

In Exercises 11–20, mark each statement True or False (**T/F**). Justify each answer.

11. (**T/F**) The set of all affine combinations of points in a set S is called the affine hull of S.

12. (**T/F**) If $S = \{\mathbf{x}\}$, then aff S is the empty set.

13. (**T/F**) If $\{\mathbf{b}_1, \ldots, \mathbf{b}_k\}$ is a linearly independent subset of $\mathbb{R}^n$ and if $\mathbf{p}$ is a linear combination of $\mathbf{b}_1, \ldots, \mathbf{b}_k$, then $\mathbf{p}$ is an affine combination of $\mathbf{b}_1, \ldots, \mathbf{b}_k$.

14. (**T/F**) A set is affine if and only if it contains its affine hull.

15. (**T/F**) The affine hull of two distinct points is called a line.

16. (**T/F**) A flat of dimension 1 is called a line.

17. (**T/F**) A flat is a subspace.

18. (**T/F**) A flat of dimension 2 is called a hyperplane.

19. (**T/F**) A plane in $\mathbb{R}^3$ is a hyperplane.

20. (**T/F**) A flat through the origin is a subspace.

21. Suppose $\{\mathbf{v}_1, \mathbf{v}_2, \mathbf{v}_3\}$ is a basis for $\mathbb{R}^3$. Show that Span $\{\mathbf{v}_2 - \mathbf{v}_1, \mathbf{v}_3 - \mathbf{v}_1\}$ is a plane in $\mathbb{R}^3$. [*Hint:* What can you say about $\mathbf{u}$ and $\mathbf{v}$ when Span $\{\mathbf{u}, \mathbf{v}\}$ is a plane?]

22. Show that if $\{\mathbf{v}_1, \mathbf{v}_2, \mathbf{v}_3\}$ is a basis for $\mathbb{R}^3$, then aff $\{\mathbf{v}_1, \mathbf{v}_2, \mathbf{v}_3\}$ is the plane through $\mathbf{v}_1, \mathbf{v}_2$, and $\mathbf{v}_3$.

23. Let A be an $m \times n$ matrix and, given $\mathbf{b}$ in $\mathbb{R}^m$, show that the set S of all solutions $A\mathbf{x} = \mathbf{b}$ is an affine subset of $\mathbb{R}^n$.

24. Let $\mathbf{v} \in \mathbb{R}^n$ and let $k \in \mathbb{R}$. Prove that $S = \{\mathbf{x} \in \mathbb{R}^n : \mathbf{x} \cdot \mathbf{v} = k\}$ is an affine subset of $\mathbb{R}^n$.

25. Choose a set S of three points such that aff S is the plane in $\mathbb{R}^3$ whose equation is $x_3 = 5$. Justify your work.

26. Choose a set S of four distinct points in $\mathbb{R}^3$ such that aff S is the plane $2x_1 + x_2 - 3x_3 = 12$. Justify your work.

27. Let S be an affine subset of $\mathbb{R}^n$, suppose $f : \mathbb{R}^n \to \mathbb{R}^m$ is a linear transformation, and let $f(S)$ denote the set of images $\{f(\mathbf{x}) : \mathbf{x} \in S\}$. Prove that $f(S)$ is an affine subset of $\mathbb{R}^m$.

28. Let $f : \mathbb{R}^n \to \mathbb{R}^m$ be a linear transformation, let T be an affine subset of $\mathbb{R}^m$, and let $S = \{\mathbf{x} \in \mathbb{R}^n : f(\mathbf{x}) \in T\}$. Show that S is an affine subset of $\mathbb{R}^n$.

In Exercises 29–34, prove the given statement about subsets A and B of $\mathbb{R}^n$, or provide the required example in $\mathbb{R}^2$. A proof for an exercise may use results from earlier exercises (as well as theorems already available in the text).

29. If $A \subseteq B$ and B is affine, then aff $A \subseteq B$.

30. If $A \subseteq B$, then aff $A \subseteq$ aff B.

31. $[(\text{aff } A) \cup (\text{aff } B)] \subseteq \text{aff } (A \cup B)$. [*Hint:* To show that $D \cup E \subseteq F$, show that $D \subseteq F$ and $E \subseteq F$.]

32. Find an example in $\mathbb{R}^2$ to show that equality need not hold in the statement of Exercise 31. [*Hint:* Consider sets A and B, each of which contains only one or two points.]

33. aff $(A \cap B) \subseteq (\text{aff } A \cap \text{aff } B)$.

34. Find an example in $\mathbb{R}^2$ to show that equality need not hold in the statement of Exercise 33.

Solution to Practice Problem

Since the points $\mathbf{v}_1$, $\mathbf{v}_2$, and $\mathbf{v}_3$ are not collinear (that is, not on a single line), aff $\{\mathbf{v}_1, \mathbf{v}_2, \mathbf{v}_3\}$ cannot be one-dimensional. Thus, aff $\{\mathbf{v}_1, \mathbf{v}_2, \mathbf{v}_3\}$ must equal $\mathbb{R}^2$. To find the actual weights used to express $\mathbf{p}$ as an affine combination of $\mathbf{v}_1$, $\mathbf{v}_2$, and $\mathbf{v}_3$, first compute

$$\mathbf{v}_2 - \mathbf{v}_1 = \begin{bmatrix} -2 \\ 2 \end{bmatrix}, \quad \mathbf{v}_3 - \mathbf{v}_1 = \begin{bmatrix} 2 \\ 1 \end{bmatrix}, \quad \text{and} \quad \mathbf{p} - \mathbf{v}_1 = \begin{bmatrix} 3 \\ 3 \end{bmatrix}$$

To write $\mathbf{p} - \mathbf{v}_1$ as a linear combination of $\mathbf{v}_2 - \mathbf{v}_1$ and $\mathbf{v}_3 - \mathbf{v}_1$, row reduce the matrix having these points as columns:

$$\begin{bmatrix} -2 & 2 & 3 \\ 2 & 1 & 3 \end{bmatrix} \sim \begin{bmatrix} 1 & 0 & \frac{1}{2} \\ 0 & 1 & 2 \end{bmatrix}$$

Thus $\mathbf{p} - \mathbf{v}_1 = \frac{1}{2}(\mathbf{v}_2 - \mathbf{v}_1) + 2(\mathbf{v}_3 - \mathbf{v}_1)$, which shows that

$$\mathbf{p} = \left(1 - \frac{1}{2} - 2\right)\mathbf{v}_1 + \frac{1}{2}\mathbf{v}_2 + 2\mathbf{v}_3 = -\frac{3}{2}\mathbf{v}_1 + \frac{1}{2}\mathbf{v}_2 + 2\mathbf{v}_3$$

This expresses $\mathbf{p}$ as an affine combination of $\mathbf{v}_1$, $\mathbf{v}_2$, and $\mathbf{v}_3$, because the coefficients sum to 1.

Alternatively, use the method of Example 4 and row reduce:

$$
\begin{bmatrix} \mathbf{v}_1 & \mathbf{v}_2 & \mathbf{v}_3 & \mathbf{p} \\ 1 & 1 & 1 & 1 \end{bmatrix} \sim \begin{bmatrix} 1 & 1 & 1 & 1 \\ 1 & -1 & 3 & 4 \\ 0 & 2 & 1 & 3 \end{bmatrix} \sim \begin{bmatrix} 1 & 0 & 0 & -\frac{3}{2} \\ 0 & 1 & 0 & \frac{1}{2} \\ 0 & 0 & 1 & 2 \end{bmatrix}
$$

This shows that $\mathbf{p} = -\frac{3}{2}\mathbf{v}_1 + \frac{1}{2}\mathbf{v}_2 + 2\mathbf{v}_3$.

8.2 Affine Independence

This section continues to explore the relation between linear concepts and affine concepts. Consider first a set of three vectors in $\mathbb{R}^3$, say $S = \{\mathbf{v}_1, \mathbf{v}_2, \mathbf{v}_3\}$. If S is linearly dependent, then one of the vectors is a linear combination of the other two vectors. What happens when one of the vectors is an *affine* combination of the others? For instance, suppose that

$$
\mathbf{v}_3 = (1 - t)\mathbf{v}_1 + t\mathbf{v}_2, \quad \text{for some } t \text{ in } \mathbb{R}.
$$

Then

$$
(1 - t)\mathbf{v}_1 + t\mathbf{v}_2 - \mathbf{v}_3 = \mathbf{0}.
$$

This is a linear dependence relation because not all the weights are zero. But more is true—the weights in the dependence relation sum to 0:

$$
(1 - t) + t + (-1) = 0.
$$

This is the additional property needed to define *affine dependence*.

DEFINITION

> An indexed set of points $\{\mathbf{v}_1, \ldots, \mathbf{v}_p\}$ in $\mathbb{R}^n$ is **affinely dependent** if there exist real numbers $c_1, \ldots, c_p$, not all zero, such that
>
> $$
> c_1 + \cdots + c_p = 0 \quad \text{and} \quad c_1\mathbf{v}_1 + \cdots + c_p\mathbf{v}_p = \mathbf{0} \tag{1}
> $$
>
> Otherwise, the set is **affinely independent**.

An affine combination is a special type of linear combination, and affine dependence is a restricted type of linear dependence. Thus, each affinely dependent set is automatically linearly dependent.

A set $\{\mathbf{v}_1\}$ of only one point (even the zero vector) must be affinely independent because the required properties of the coefficients c_i cannot be satisfied when there is only one coefficient. For $\{\mathbf{v}_1\}$, the first equation in (1) is just $c_1 = 0$, and yet at least one (the only one) coefficient must be nonzero.

Exercise 21 asks you to show that an indexed set $\{\mathbf{v}_1, \mathbf{v}_2\}$ is affinely dependent if and only if $\mathbf{v}_1 = \mathbf{v}_2$. The following theorem handles the general case and shows how the concept of affine dependence is analogous to that of linear dependence. Parts (c) and (d) give useful methods for determining whether a set is affinely dependent. Recall from Section 8.1 that if $\mathbf{v}$ is in $\mathbb{R}^n$, then the vector $\tilde{\mathbf{v}}$ in $\mathbb{R}^{n+1}$ denotes the homogeneous form of $\mathbf{v}$.

THEOREM 5

Given an indexed set $S = \{\mathbf{v}_1, \ldots, \mathbf{v}_p\}$ in $\mathbb{R}^n$, with $p \geq 2$, the following statements are logically equivalent. That is, either they are all true statements or they are all false.

a. S is affinely dependent.
b. One of the points in S is an affine combination of the other points in S.
c. The set $\{\mathbf{v}_2 - \mathbf{v}_1, \ldots, \mathbf{v}_p - \mathbf{v}_1\}$ in $\mathbb{R}^n$ is linearly dependent.
d. The set $\{\tilde{\mathbf{v}}_1, \ldots, \tilde{\mathbf{v}}_p\}$ of homogeneous forms in $\mathbb{R}^{n+1}$ is linearly dependent.

PROOF Suppose statement (a) is true, and let $c_1, \ldots, c_p$ satisfy (1). By renaming the points if necessary, one may assume that $c_1 \neq 0$ and divide both equations in (1) by c_1, so that $1 + (c_2/c_1) + \cdots + (c_p/c_1) = 0$ and

$$\mathbf{v}_1 = (-c_2/c_1)\mathbf{v}_2 + \cdots + (-c_p/c_1)\mathbf{v}_p \tag{2}$$

Note that the coefficients on the right side of (2) sum to 1. Thus (a) implies (b). Now, suppose that (b) is true. By renaming the points if necessary, one may assume that $\mathbf{v}_1 = c_2\mathbf{v}_2 + \cdots + c_p\mathbf{v}_p$, where $c_2 + \cdots + c_p = 1$. Then

$$(c_2 + \cdots + c_p)\mathbf{v}_1 = c_2\mathbf{v}_2 + \cdots + c_p\mathbf{v}_p \tag{3}$$

and

$$c_2(\mathbf{v}_2 - \mathbf{v}_1) + \cdots + c_p(\mathbf{v}_p - \mathbf{v}_1) = \mathbf{0} \tag{4}$$

Not all of $c_2, \ldots, c_p$ can be zero because they sum to 1. So (b) implies (c).

Next, if (c) is true, then there exist weights $c_2, \ldots, c_p$, not all zero, such that (4) holds. Rewrite (4) as (3) and set $c_1 = -(c_2 + \cdots + c_p)$. Then $c_1 + \cdots + c_p = 0$. Thus, (3) shows that (1) is true. So (c) implies (a), which proves that (a), (b), and (c) are logically equivalent. Finally, (d) is equivalent to (a) because the two equations in (1) are equivalent to the following equation involving the homogeneous forms of the points in S:

$$c_1 \begin{bmatrix} \mathbf{v}_1 \\ 1 \end{bmatrix} + \cdots + c_p \begin{bmatrix} \mathbf{v}_p \\ 1 \end{bmatrix} = \begin{bmatrix} \mathbf{0} \\ 0 \end{bmatrix} \qquad \blacksquare$$

In statement (c) of Theorem 5, $\mathbf{v}_1$ could be replaced by any of the other points in the list $\mathbf{v}_1, \ldots, \mathbf{v}_p$. Only the notation in the proof would change. So, to test whether a set is affinely dependent, subtract one point in the set from the other points, and check whether the translated set of $p - 1$ points is linearly dependent.

EXAMPLE 1 The affine hull of two distinct points $\mathbf{p}$ and $\mathbf{q}$ is a line. If a third point $\mathbf{r}$ is on the line, then $\{\mathbf{p}, \mathbf{q}, \mathbf{r}\}$ is an affinely dependent set. If a point $\mathbf{s}$ is not on the line through $\mathbf{p}$ and $\mathbf{q}$, then these three points are not collinear and $\{\mathbf{p}, \mathbf{q}, \mathbf{s}\}$ is an affinely independent set. See Figure 1. $\blacksquare$

FIGURE 1 $\{\mathbf{p}, \mathbf{q}, \mathbf{r}\}$ is affinely dependent.

EXAMPLE 2 Let $\mathbf{v}_1 = \begin{bmatrix} 1 \\ 3 \\ 7 \end{bmatrix}$, $\mathbf{v}_2 = \begin{bmatrix} 2 \\ 7 \\ 6.5 \end{bmatrix}$, $\mathbf{v}_3 = \begin{bmatrix} 0 \\ 4 \\ 7 \end{bmatrix}$, and $S = \{\mathbf{v}_1, \mathbf{v}_2, \mathbf{v}_3\}$.
Determine whether S is affinely independent.

SOLUTION Compute $\mathbf{v}_2 - \mathbf{v}_1 = \begin{bmatrix} 1 \\ 4 \\ -.5 \end{bmatrix}$ and $\mathbf{v}_3 - \mathbf{v}_1 = \begin{bmatrix} -1 \\ 1 \\ 0 \end{bmatrix}$. These two points
are not multiples and hence form a linearly independent set, S'. So all statements in
Theorem 5 are false, and S is affinely independent. Figure 2 shows S and the translated
set S'. Notice that Span S' is a plane through the origin and aff S is a parallel plane
through $\mathbf{v}_1$, $\mathbf{v}_2$, and $\mathbf{v}_3$. (Only a portion of each plane is shown here, of course.) ∎

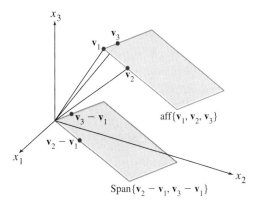

FIGURE 2 An affinely independent set
$\{\mathbf{v}_1, \mathbf{v}_2, \mathbf{v}_3\}$.

EXAMPLE 3 Let $\mathbf{v}_1 = \begin{bmatrix} 1 \\ 3 \\ 7 \end{bmatrix}$, $\mathbf{v}_2 = \begin{bmatrix} 2 \\ 7 \\ 6.5 \end{bmatrix}$, $\mathbf{v}_3 = \begin{bmatrix} 0 \\ 4 \\ 7 \end{bmatrix}$, and $\mathbf{v}_4 = \begin{bmatrix} 0 \\ 14 \\ 6 \end{bmatrix}$, and let
$S = \{\mathbf{v}_1, \ldots, \mathbf{v}_4\}$. Is S affinely dependent?

SOLUTION Compute $\mathbf{v}_2 - \mathbf{v}_1 = \begin{bmatrix} 1 \\ 4 \\ -.5 \end{bmatrix}$, $\mathbf{v}_3 - \mathbf{v}_1 = \begin{bmatrix} -1 \\ 1 \\ 0 \end{bmatrix}$, and $\mathbf{v}_4 - \mathbf{v}_1 = \begin{bmatrix} -1 \\ 11 \\ -1 \end{bmatrix}$,
and row reduce the matrix:
$$\begin{bmatrix} 1 & -1 & -1 \\ 4 & 1 & 11 \\ -.5 & 0 & -1 \end{bmatrix} \sim \begin{bmatrix} 1 & -1 & -1 \\ 0 & 5 & 15 \\ 0 & -.5 & -1.5 \end{bmatrix} \sim \begin{bmatrix} 1 & -1 & -1 \\ 0 & 5 & 15 \\ 0 & 0 & 0 \end{bmatrix}$$

Recall from Section 4.5 (or Section 2.8) that the columns are linearly dependent be-
cause not every column is a pivot column; so $\mathbf{v}_2 - \mathbf{v}_1$, $\mathbf{v}_3 - \mathbf{v}_1$, and $\mathbf{v}_4 - \mathbf{v}_1$ are linearly
dependent. By statement (c) in Theorem 5, $\{\mathbf{v}_1, \mathbf{v}_2, \mathbf{v}_3, \mathbf{v}_4\}$ is affinely dependent. This
dependence can also be established using (d) in Theorem 5 instead of (c). ∎

The calculations in Example 3 show that $\mathbf{v}_4 - \mathbf{v}_1$ is a linear combination of $\mathbf{v}_2 - \mathbf{v}_1$
and $\mathbf{v}_3 - \mathbf{v}_1$, which means that $\mathbf{v}_4 - \mathbf{v}_1$ is in Span $\{\mathbf{v}_2 - \mathbf{v}_1, \mathbf{v}_3 - \mathbf{v}_1\}$. By Theorem 1 in
Section 8.1, $\mathbf{v}_4$ is in aff $\{\mathbf{v}_1, \mathbf{v}_2, \mathbf{v}_3\}$. In fact, complete row reduction of the matrix in
Example 3 would show that

$$\mathbf{v}_4 - \mathbf{v}_1 = 2(\mathbf{v}_2 - \mathbf{v}_1) + 3(\mathbf{v}_3 - \mathbf{v}_1) \tag{5}$$
$$\mathbf{v}_4 = -4\mathbf{v}_1 + 2\mathbf{v}_2 + 3\mathbf{v}_3 \tag{6}$$

See Figure 3.

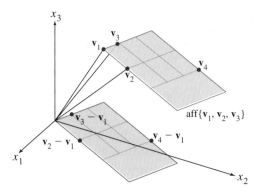

FIGURE 3 $\mathbf{v}_4$ is in the plane aff $\{\mathbf{v}_1, \mathbf{v}_2, \mathbf{v}_3\}$.

Figure 3 shows grids on both Span$\{\mathbf{v}_2 - \mathbf{v}_1, \mathbf{v}_3 - \mathbf{v}_1\}$ and aff $\{\mathbf{v}_1, \mathbf{v}_2, \mathbf{v}_3\}$. The grid on aff $\{\mathbf{v}_1, \mathbf{v}_2, \mathbf{v}_3\}$ is based on (5). Another "coordinate system" can be based on (6), in which the coefficients -4, 2, and 3 are called *affine* or *barycentric* coordinates of $\mathbf{v}_4$.

Barycentric Coordinates

The definition of barycentric coordinates depends on the following affine version of the Unique Representation Theorem in Section 4.4. See Exercise 25 in this section for the proof.

THEOREM 6

Let $S = \{\mathbf{v}_1, \ldots, \mathbf{v}_k\}$ be an affinely independent set in $\mathbb{R}^n$. Then each $\mathbf{p}$ in aff S has a unique representation as an affine combination of $\mathbf{v}_1, \ldots, \mathbf{v}_k$. That is, for each $\mathbf{p}$ there exists a unique set of scalars $c_1, \ldots, c_k$ such that

$$\mathbf{p} = c_1 \mathbf{v}_1 + \cdots + c_k \mathbf{v}_k \quad \text{and} \quad c_1 + \cdots + c_k = 1 \tag{7}$$

DEFINITION

Let $S = \{\mathbf{v}_1, \ldots, \mathbf{v}_k\}$ be an affinely independent set. Then for each point $\mathbf{p}$ in aff S, the coefficients $c_1, \ldots, c_k$ in the unique representation (7) of $\mathbf{p}$ are called the **barycentric** (or, sometimes, **affine**) **coordinates** of $\mathbf{p}$.

Observe that (7) is equivalent to the single equation

$$\begin{bmatrix} \mathbf{p} \\ 1 \end{bmatrix} = c_1 \begin{bmatrix} \mathbf{v}_1 \\ 1 \end{bmatrix} + \cdots + c_k \begin{bmatrix} \mathbf{v}_k \\ 1 \end{bmatrix} \tag{8}$$

involving the homogeneous forms of the points. Row reduction of the augmented matrix $\begin{bmatrix} \tilde{\mathbf{v}}_1 & \cdots & \tilde{\mathbf{v}}_k & \tilde{\mathbf{p}} \end{bmatrix}$ for (8) produces the barycentric coordinates of $\mathbf{p}$.

EXAMPLE 4 Let $\mathbf{a} = \begin{bmatrix} 1 \\ 7 \end{bmatrix}$, $\mathbf{b} = \begin{bmatrix} 3 \\ 0 \end{bmatrix}$, $\mathbf{c} = \begin{bmatrix} 9 \\ 3 \end{bmatrix}$, and $\mathbf{p} = \begin{bmatrix} 5 \\ 3 \end{bmatrix}$. Find the barycentric coordinates of $\mathbf{p}$ determined by the affinely independent set $\{\mathbf{a}, \mathbf{b}, \mathbf{c}\}$.

SOLUTION Row reduce the augmented matrix of points in homogeneous form, moving the last row of ones to the top to simplify the arithmetic:

$$\begin{bmatrix} \tilde{\mathbf{a}} & \tilde{\mathbf{b}} & \tilde{\mathbf{c}} & \tilde{\mathbf{p}} \end{bmatrix} = \begin{bmatrix} 1 & 3 & 9 & 5 \\ 7 & 0 & 3 & 3 \\ 1 & 1 & 1 & 1 \end{bmatrix} \sim \begin{bmatrix} 1 & 1 & 1 & 1 \\ 1 & 3 & 9 & 5 \\ 7 & 0 & 3 & 3 \end{bmatrix}$$

$$\sim \begin{bmatrix} 1 & 0 & 0 & \frac{1}{4} \\ 0 & 1 & 0 & \frac{1}{3} \\ 0 & 0 & 1 & \frac{5}{12} \end{bmatrix}$$

The coordinates are $\frac{1}{4}$, $\frac{1}{3}$, and $\frac{5}{12}$, so $\mathbf{p} = \frac{1}{4}\mathbf{a} + \frac{1}{3}\mathbf{b} + \frac{5}{12}\mathbf{c}$. ∎

Barycentric coordinates have both physical and geometric interpretations. They were originally defined by A. F. Moebius in 1827 for a point $\mathbf{p}$ inside a triangular region with vertices $\mathbf{a}$, $\mathbf{b}$, and $\mathbf{c}$. He wrote that the barycentric coordinates of $\mathbf{p}$ are three nonnegative numbers $m_{\mathbf{a}}, m_{\mathbf{b}}$, and $m_{\mathbf{c}}$ such that $\mathbf{p}$ is the center of mass of a system consisting of the triangle (with no mass) and masses $m_{\mathbf{a}}, m_{\mathbf{b}}$, and $m_{\mathbf{c}}$ at the corresponding vertices. The masses are uniquely determined by requiring that their sum be 1. This view is still useful in physics today.[1]

Figure 4 gives a geometric interpretation to the barycentric coordinates in Example 4, showing the triangle $\triangle\mathbf{abc}$ and three small triangles $\triangle\mathbf{pbc}$, $\triangle\mathbf{apc}$, and $\triangle\mathbf{abp}$. The areas of the small triangles are proportional to the barycentric coordinates of $\mathbf{p}$. In fact,

$$\text{area}(\triangle\mathbf{pbc}) = \frac{1}{4} \cdot \text{area}(\triangle\mathbf{abc})$$

$$\text{area}(\triangle\mathbf{apc}) = \frac{1}{3} \cdot \text{area}(\triangle\mathbf{abc}) \tag{9}$$

$$\text{area}(\triangle\mathbf{abp}) = \frac{5}{12} \cdot \text{area}(\triangle\mathbf{abc})$$

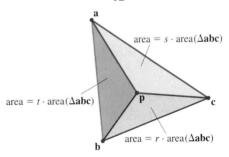

FIGURE 4 $\mathbf{p} = r\mathbf{a} + s\mathbf{b} + t\mathbf{c}$. Here, $r = \frac{1}{4}$, $s = \frac{1}{3}$, $t = \frac{5}{12}$.

The formulas in Figure 4 are verified in Exercises 29–31. Analogous equalities for volumes of tetrahedrons hold for the case when $\mathbf{p}$ is a point inside a tetrahedron in $\mathbb{R}^3$, with vertices $\mathbf{a}$, $\mathbf{b}$, $\mathbf{c}$, and $\mathbf{d}$.

When a point is not inside the triangle (or tetrahedron), some of the barycentric coordinates will be negative. The case of a triangle is illustrated in Figure 5, for vertices

[1] See Exercise 37 in Section 1.3. In astronomy, however, "barycentric coordinates" usually refer to ordinary $\mathbb{R}^3$ coordinates of points in what is now called the *International Celestial Reference System*, a Cartesian coordinate system for outer space, with the origin at the center of mass (the barycenter) of the solar system.

a, **b**, **c**, and coordinate values r, s, t, as above. The points on the line through **b** and **c**, for instance, have $r = 0$ because they are affine combinations of only **b** and **c**. The parallel line through **a** identifies points with $r = 1$.

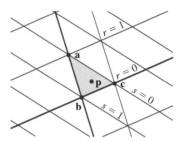

FIGURE 5 Barycentric coordinates for points in aff $\{\mathbf{a}, \mathbf{b}, \mathbf{c}\}$.

Barycentric Coordinates in Computer Graphics

When working with geometric objects in a computer graphics program, a designer may use a "wire-frame" approximation to an object at certain key points in the process of creating a realistic final image. For instance, if the surface of part of an object consists of small flat triangular surfaces, then a graphics program can easily add color, lighting, and shading to each small surface when that information is known only at the vertices. Barycentric coordinates provide the tool for smoothly interpolating the vertex information over the interior of a triangle. The interpolation at a point is simply the linear combination of the vertex values using the barycentric coordinates as weights.

Colors on a computer screen are often described by RGB coordinates. A triple (r, g, b) indicates the amount of each color—red, green, and blue—with the parameters varying from 0 to 1. For example, pure red is $(1, 0, 0)$, white is $(1, 1, 1)$, and black is $(0, 0, 0)$.

EXAMPLE 5 Let $\mathbf{v}_1 = \begin{bmatrix} 3 \\ 1 \\ 5 \end{bmatrix}$, $\mathbf{v}_2 = \begin{bmatrix} 4 \\ 3 \\ 4 \end{bmatrix}$, $\mathbf{v}_3 = \begin{bmatrix} 1 \\ 5 \\ 1 \end{bmatrix}$, and $\mathbf{p} = \begin{bmatrix} 3 \\ 3 \\ 3.5 \end{bmatrix}$. The colors at the vertices $\mathbf{v}_1$, $\mathbf{v}_2$, and $\mathbf{v}_3$ of a triangle are magenta $(1, 0, 1)$, light magenta $(1, .4, 1)$, and purple $(.6, 0, 1)$, respectively. Find the interpolated color at $\mathbf{p}$. See Figure 6.

FIGURE 6 Interpolated colors.

SOLUTION First, find the barycentric coordinates of $\mathbf{p}$. Here is the calculation using homogeneous forms of the points, with the first step moving row 4 to row 1:

$$\begin{bmatrix} \tilde{\mathbf{v}}_1 & \tilde{\mathbf{v}}_2 & \tilde{\mathbf{v}}_3 & \tilde{\mathbf{p}} \end{bmatrix} \sim \begin{bmatrix} 1 & 1 & 1 & 1 \\ 3 & 4 & 1 & 3 \\ 1 & 3 & 5 & 3 \\ 5 & 4 & 1 & 3.5 \end{bmatrix} \sim \begin{bmatrix} 1 & 0 & 0 & .25 \\ 0 & 1 & 0 & .50 \\ 0 & 0 & 1 & .25 \\ 0 & 0 & 0 & 0 \end{bmatrix}$$

So $\mathbf{p} = .25\mathbf{v}_1 + .5\mathbf{v}_2 + .25\mathbf{v}_3$. Use the barycentric coordinates of $\mathbf{p}$ to make a linear combination of the color data. The RGB values for $\mathbf{p}$ are

$$.25\begin{bmatrix} 1 \\ 0 \\ 1 \end{bmatrix} + .50\begin{bmatrix} 1 \\ .4 \\ 1 \end{bmatrix} + .25\begin{bmatrix} .6 \\ 0 \\ 1 \end{bmatrix} = \begin{bmatrix} .9 \\ .2 \\ 1 \end{bmatrix} \begin{matrix} \text{red} \\ \text{green} \\ \text{blue} \end{matrix}$$ ∎

One of the last steps in preparing a graphics scene for display on a computer screen is to remove "hidden surfaces" that should not be visible on the screen. Imagine the viewing screen as consisting of, say, a million pixels, and consider a ray or "line of sight" from the viewer's eye through a pixel and into the collection of objects that make up the 3D display. The color and other information displayed in the pixel on the screen should come from the object that the ray first intersects. See Figure 7. When the objects in the graphics scene are approximated by wire frames with triangular patches, the hidden surface problem can be solved using barycentric coordinates.

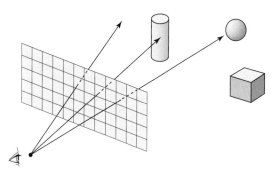

FIGURE 7 A ray from the eye through the screen to the nearest object.

The mathematics for finding the ray-triangle intersections can also be used to perform extremely realistic shading of objects. Currently, this *ray-tracing* method is too slow for real-time rendering, but recent advances in hardware implementation may change that in the future.[2]

EXAMPLE 6 Let

$$\mathbf{v}_1 = \begin{bmatrix} 1 \\ 1 \\ -6 \end{bmatrix}, \quad \mathbf{v}_2 = \begin{bmatrix} 8 \\ 1 \\ -4 \end{bmatrix}, \quad \mathbf{v}_3 = \begin{bmatrix} 5 \\ 11 \\ -2 \end{bmatrix}, \quad \mathbf{a} = \begin{bmatrix} 0 \\ 0 \\ 10 \end{bmatrix}, \quad \mathbf{b} = \begin{bmatrix} .7 \\ .4 \\ -3 \end{bmatrix},$$

and $\mathbf{x}(t) = \mathbf{a} + t\mathbf{b}$ for $t \geq 0$. Find the point where the ray $\mathbf{x}(t)$ intersects the plane that contains the triangle with vertices $\mathbf{v}_1$, $\mathbf{v}_2$, and $\mathbf{v}_3$. Is this point inside the triangle?

SOLUTION The plane is aff$\{\mathbf{v}_1, \mathbf{v}_2, \mathbf{v}_3\}$. A typical point in this plane may be written as $(1 - c_2 - c_3)\mathbf{v}_1 + c_2\mathbf{v}_2 + c_3\mathbf{v}_3$ for some c_2 and c_3. (The weights in this combination sum to 1.) The ray $\mathbf{x}(t)$ intersects the plane when c_2, c_3, and t satisfy

$$(1 - c_2 - c_3)\mathbf{v}_1 + c_2\mathbf{v}_2 + c_3\mathbf{v}_3 = \mathbf{a} + t\mathbf{b}$$

---

[2] See Joshua Fender and Jonathan Rose, "A High-Speed Ray Tracing Engine Built on a Field-Programmable System," in *Proc. Int. Conf on Field-Programmable Technology*, IEEE (2003). (A single processor can calculate 600 million ray-triangle intersections per second.)

Rearrange this as $c_2(\mathbf{v}_2 - \mathbf{v}_1) + c_3(\mathbf{v}_3 - \mathbf{v}_1) + t(-\mathbf{b}) = \mathbf{a} - \mathbf{v}_1$. In matrix form,

$$\begin{bmatrix} \mathbf{v}_2 - \mathbf{v}_1 & \mathbf{v}_3 - \mathbf{v}_1 & -\mathbf{b} \end{bmatrix} \begin{bmatrix} c_2 \\ c_3 \\ t \end{bmatrix} = \mathbf{a} - \mathbf{v}_1$$

For the specific points given here,

$$\mathbf{v}_2 - \mathbf{v}_1 = \begin{bmatrix} 7 \\ 0 \\ 2 \end{bmatrix}, \quad \mathbf{v}_3 - \mathbf{v}_1 = \begin{bmatrix} 4 \\ 10 \\ 4 \end{bmatrix}, \quad \mathbf{a} - \mathbf{v}_1 = \begin{bmatrix} -1 \\ -1 \\ 16 \end{bmatrix}$$

Row reduction of the augmented matrix above produces

$$\begin{bmatrix} 7 & 4 & -.7 & -1 \\ 0 & 10 & -.4 & -1 \\ 2 & 4 & 3 & 16 \end{bmatrix} \sim \begin{bmatrix} 1 & 0 & 0 & .3 \\ 0 & 1 & 0 & .1 \\ 0 & 0 & 1 & 5 \end{bmatrix}$$

Thus $c_2 = .3$, $c_3 = .1$, and $t = 5$. Therefore, the intersection point is

$$\mathbf{x}(5) = \mathbf{a} + 5\mathbf{b} = \begin{bmatrix} 0 \\ 0 \\ 10 \end{bmatrix} + 5 \begin{bmatrix} .7 \\ .4 \\ -3 \end{bmatrix} = \begin{bmatrix} 3.5 \\ 2.0 \\ -5.0 \end{bmatrix}$$

Also,

$$\mathbf{x}(5) = (1 - .3 - .1)\mathbf{v}_1 + .3\mathbf{v}_2 + .1\mathbf{v}_3$$

$$= .6 \begin{bmatrix} 1 \\ 1 \\ -6 \end{bmatrix} + .3 \begin{bmatrix} 8 \\ 1 \\ -4 \end{bmatrix} + .1 \begin{bmatrix} 5 \\ 11 \\ -2 \end{bmatrix} = \begin{bmatrix} 3.5 \\ 2.0 \\ -5.0 \end{bmatrix}$$

The intersection point is inside the triangle because the barycentric weights for $\mathbf{x}(5)$ are all positive. ∎

Practice Problems

1. Describe a fast way to determine when three points are collinear.

2. The points $\mathbf{v}_1 = \begin{bmatrix} 4 \\ 1 \end{bmatrix}$, $\mathbf{v}_2 = \begin{bmatrix} 1 \\ 0 \end{bmatrix}$, $\mathbf{v}_3 = \begin{bmatrix} 5 \\ 4 \end{bmatrix}$, and $\mathbf{v}_4 = \begin{bmatrix} 1 \\ 2 \end{bmatrix}$ form an affinely dependent set. Find weights $c_1, \ldots, c_4$ that produce an **affine dependence relation** $c_1\mathbf{v}_1 + \cdots + c_4\mathbf{v}_4 = \mathbf{0}$, where $c_1 + \cdots + c_4 = 0$ and not all c_i are zero. [*Hint:* See the end of the proof of Theorem 5.]

8.2 Exercises

In Exercises 1–6, determine if the set of points is affinely dependent. (See Practice Problem 2.) If so, construct an affine dependence relation for the points.

1. $\begin{bmatrix} 3 \\ -3 \end{bmatrix}, \begin{bmatrix} 0 \\ 6 \end{bmatrix}, \begin{bmatrix} 2 \\ 0 \end{bmatrix}$
   2. $\begin{bmatrix} 2 \\ 1 \end{bmatrix}, \begin{bmatrix} 5 \\ 4 \end{bmatrix}, \begin{bmatrix} -3 \\ -2 \end{bmatrix}$

3. $\begin{bmatrix} 1 \\ 2 \\ -1 \end{bmatrix}, \begin{bmatrix} -2 \\ -4 \\ 8 \end{bmatrix}, \begin{bmatrix} 2 \\ -1 \\ 11 \end{bmatrix}, \begin{bmatrix} 0 \\ 15 \\ -9 \end{bmatrix}$

4. $\begin{bmatrix} -2 \\ 5 \\ 3 \end{bmatrix}, \begin{bmatrix} 0 \\ -3 \\ 7 \end{bmatrix}, \begin{bmatrix} 1 \\ -2 \\ -6 \end{bmatrix}, \begin{bmatrix} -2 \\ 7 \\ -3 \end{bmatrix}$

5. $\begin{bmatrix} 1 \\ 0 \\ -2 \end{bmatrix}, \begin{bmatrix} 0 \\ 1 \\ 1 \end{bmatrix}, \begin{bmatrix} -1 \\ 5 \\ 1 \end{bmatrix}, \begin{bmatrix} 0 \\ 5 \\ -3 \end{bmatrix}$

6. $\begin{bmatrix} 1 \\ 3 \\ 1 \end{bmatrix}, \begin{bmatrix} 0 \\ -1 \\ -2 \end{bmatrix}, \begin{bmatrix} 2 \\ 5 \\ 2 \end{bmatrix}, \begin{bmatrix} 3 \\ 5 \\ 0 \end{bmatrix}$

In Exercises 7 and 8, find the barycentric coordinates of **p** with respect to the affinely independent set of points that precedes it.

7. $\begin{bmatrix} 1 \\ -1 \\ 2 \\ 1 \end{bmatrix}, \begin{bmatrix} 2 \\ 1 \\ 0 \\ 1 \end{bmatrix}, \begin{bmatrix} 1 \\ 2 \\ -2 \\ 0 \end{bmatrix}, \mathbf{p} = \begin{bmatrix} 5 \\ 4 \\ -2 \\ 2 \end{bmatrix}$

8. $\begin{bmatrix} 0 \\ 1 \\ -2 \\ 1 \end{bmatrix}, \begin{bmatrix} 1 \\ 1 \\ 0 \\ 2 \end{bmatrix}, \begin{bmatrix} 1 \\ 4 \\ -6 \\ 5 \end{bmatrix}, \mathbf{p} = \begin{bmatrix} -1 \\ 1 \\ -4 \\ 0 \end{bmatrix}$

In Exercises 9–18 mark each statement True or False (**T/F**). Justify each answer.

9. (**T/F**) If $\mathbf{v}_1, \ldots, \mathbf{v}_p$ are in $\mathbb{R}^n$ and if the set $\{\mathbf{v}_1 - \mathbf{v}_2, \mathbf{v}_3 - \mathbf{v}_2, \ldots, \mathbf{v}_p - \mathbf{v}_2\}$ is linearly dependent, then $\{\mathbf{v}_1, \ldots, \mathbf{v}_p\}$ is affinely dependent. (Read this carefully.)

10. (**T/F**) If $\{\mathbf{v}_1, \ldots, \mathbf{v}_p\}$ is an affinely dependent set in $\mathbb{R}^n$, then the set $\{\tilde{\mathbf{v}}_1, \ldots, \tilde{\mathbf{v}}_p\}$ in $\mathbb{R}^{n+1}$ of homogeneous forms may be linearly independent.

11. (**T/F**) If $\mathbf{v}_1, \ldots, \mathbf{v}_p$ are in $\mathbb{R}^n$ and if the set of homogeneous forms $\{\tilde{\mathbf{v}}_1, \ldots, \tilde{\mathbf{v}}_p\}$ in $\mathbb{R}^{n+1}$ is linearly independent, then $\{\mathbf{v}_1, \ldots, \mathbf{v}_p\}$ is affinely dependent.

12. (**T/F**) If $\mathbf{v}_1, \mathbf{v}_2, \mathbf{v}_3$, and $\mathbf{v}_4$ are in $\mathbb{R}^3$ and if the set $\{\mathbf{v}_2 - \mathbf{v}_1, \mathbf{v}_3 - \mathbf{v}_1, \mathbf{v}_4 - \mathbf{v}_1\}$ is linearly independent, then $\{\mathbf{v}_1, \ldots, \mathbf{v}_4\}$ is affinely independent.

13. (**T/F**) A finite set of points $\{\mathbf{v}_1, \ldots, \mathbf{v}_k\}$ is affinely dependent if there exist real numbers $c_1, \ldots, c_k$, not all zero, such that $c_1 + \cdots + c_k = 1$ and $c_1 \mathbf{v}_1 + \cdots + c_k \mathbf{v}_k = \mathbf{0}$.

14. (**T/F**) Given $S = \{\mathbf{b}_1, \ldots, \mathbf{b}_k\}$ in $\mathbb{R}^n$, each **p** in aff S has a unique representation as an affine combination of $\mathbf{b}_1, \ldots, \mathbf{b}_k$.

15. (**T/F**) If $S = \{\mathbf{v}_1, \ldots, \mathbf{v}_p\}$ is an affinely independent set in $\mathbb{R}^n$ and if **p** in $\mathbb{R}^n$ has a negative barycentric coordinate determined by S, then **p** is not in aff S.

16. (**T/F**) When color information is specified at each vertex $\mathbf{v}_1, \mathbf{v}_2, \mathbf{v}_3$ of a triangle in $\mathbb{R}^3$, then the color may be interpolated at a point **p** in aff $\{\mathbf{v}_1, \mathbf{v}_2, \mathbf{v}_3\}$ using the barycentric coordinates of **p**.

17. (**T/F**) If $\mathbf{v}_1, \mathbf{v}_2, \mathbf{v}_3, \mathbf{a}$, and **b** are in $\mathbb{R}^3$ and if a ray $\mathbf{a} + t\mathbf{b}$ for $t \geq 0$ intersects the triangle with vertices $\mathbf{v}_1, \mathbf{v}_2$, and $\mathbf{v}_3$, then the barycentric coordinates of the intersection point are all nonnegative.

18. (**T/F**) If T is a triangle in $\mathbb{R}^2$ and if a point **p** is on an edge of the triangle, then the barycentric coordinates of **p** (for this triangle) are not all positive.

19. Explain why any set of five or more points in $\mathbb{R}^3$ must be affinely dependent.

20. Show that a set $\{\mathbf{v}_1, \ldots, \mathbf{v}_p\}$ in $\mathbb{R}^n$ is affinely dependent when $p \geq n + 2$.

21. Use only the definition of affine dependence to show that an indexed set $\{\mathbf{v}_1, \mathbf{v}_2\}$ in $\mathbb{R}^n$ is affinely dependent if and only if $\mathbf{v}_1 = \mathbf{v}_2$.

22. The conditions for affine dependence are stronger than those for linear dependence, so an affinely dependent set is automatically linearly dependent. Also, a linearly independent set cannot be affinely dependent and therefore must be affinely independent. Construct two linearly dependent indexed sets S_1 and S_2 in $\mathbb{R}^2$ such that S_1 is affinely dependent and S_2 is affinely independent. In each case, the set should contain either one, two, or three nonzero points.

23. Let $\mathbf{v}_1 = \begin{bmatrix} -1 \\ 2 \end{bmatrix}$, $\mathbf{v}_2 = \begin{bmatrix} 0 \\ 4 \end{bmatrix}$, $\mathbf{v}_3 = \begin{bmatrix} 2 \\ 0 \end{bmatrix}$, and let $S = \{\mathbf{v}_1, \mathbf{v}_2, \mathbf{v}_3\}$.

   a. Show that the set S is affinely independent.

   b. Find the barycentric coordinates of $\mathbf{p}_1 = \begin{bmatrix} 2 \\ 3 \end{bmatrix}$, $\mathbf{p}_2 = \begin{bmatrix} 1 \\ 2 \end{bmatrix}$, $\mathbf{p}_3 = \begin{bmatrix} -2 \\ 1 \end{bmatrix}$, $\mathbf{p}_4 = \begin{bmatrix} 1 \\ -1 \end{bmatrix}$, and $\mathbf{p}_5 = \begin{bmatrix} 1 \\ 1 \end{bmatrix}$, with respect to S.

   c. Let T be the triangle with vertices $\mathbf{v}_1, \mathbf{v}_2$, and $\mathbf{v}_3$. When the sides of T are extended, the lines divide $\mathbb{R}^2$ into seven regions. See Figure 8. Note the signs of the barycentric coordinates of the points in each region. For example, $\mathbf{p}_5$ is inside the triangle T and all its barycentric coordinates are positive. Point $\mathbf{p}_1$ has coordinates $(-, +, +)$. Its third coordinate is positive because $\mathbf{p}_1$ is on the $\mathbf{v}_3$ side of the line through $\mathbf{v}_1$ and $\mathbf{v}_2$. Its first coordinate is negative because $\mathbf{p}_1$ is opposite the $\mathbf{v}_1$ side of the line through $\mathbf{v}_2$ and $\mathbf{v}_3$. Point $\mathbf{p}_2$ is on the $\mathbf{v}_2\mathbf{v}_3$ edge of T. Its coordinates are $(0, +, +)$. Without calculating the actual values, determine the signs of the barycentric coordinates of points $\mathbf{p}_6$, $\mathbf{p}_7$, and $\mathbf{p}_8$ as shown in Figure 8.

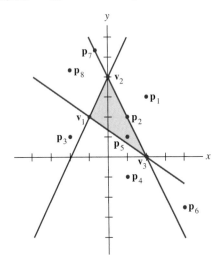

FIGURE 8

24. Let $\mathbf{v}_1 = \begin{bmatrix} 0 \\ 1 \end{bmatrix}$, $\mathbf{v}_2 = \begin{bmatrix} 1 \\ 5 \end{bmatrix}$, $\mathbf{v}_3 = \begin{bmatrix} 4 \\ 3 \end{bmatrix}$, $\mathbf{p}_1 = \begin{bmatrix} 3 \\ 5 \end{bmatrix}$,

$\mathbf{p}_2 = \begin{bmatrix} 5 \\ 1 \end{bmatrix}$, $\mathbf{p}_3 = \begin{bmatrix} 2 \\ 3 \end{bmatrix}$, $\mathbf{p}_4 = \begin{bmatrix} -1 \\ 0 \end{bmatrix}$, $\mathbf{p}_5 = \begin{bmatrix} 0 \\ 4 \end{bmatrix}$,

$\mathbf{p}_6 = \begin{bmatrix} 1 \\ 2 \end{bmatrix}$, $\mathbf{p}_7 = \begin{bmatrix} 6 \\ 4 \end{bmatrix}$, and $S = \{\mathbf{v}_1, \mathbf{v}_2, \mathbf{v}_3\}$.

a. Show that the set S is affinely independent.

b. Find the barycentric coordinates of $\mathbf{p}_1$, $\mathbf{p}_2$, and $\mathbf{p}_3$ with respect to S.

c. On graph paper, sketch the triangle T with vertices $\mathbf{v}_1$, $\mathbf{v}_2$, and $\mathbf{v}_3$, extend the sides as in Figure 8, and plot the points $\mathbf{p}_4$, $\mathbf{p}_5$, $\mathbf{p}_6$, and $\mathbf{p}_7$. Without calculating the actual values, determine the signs of the barycentric coordinates of points $\mathbf{p}_4$, $\mathbf{p}_5$, $\mathbf{p}_6$, and $\mathbf{p}_7$.

25. Prove Theorem 6 for an affinely independent set $S = \{\mathbf{v}_1, \ldots, \mathbf{v}_k\}$ in $\mathbb{R}^n$. [*Hint:* One method is to mimic the proof of Theorem 8 in Section 4.4.]

26. Let T be a tetrahedron in "standard" position, with three edges along the three positive coordinate axes in $\mathbb{R}^3$, and suppose the vertices are $a\mathbf{e}_1$, $b\mathbf{e}_2$, $c\mathbf{e}_3$, and $\mathbf{0}$, where $[\mathbf{e}_1 \ \mathbf{e}_2 \ \mathbf{e}_3] = I_3$. Find formulas for the barycentric coordinates of an arbitrary point $\mathbf{p}$ in $\mathbb{R}^3$.

27. Let $\{\mathbf{p}_1, \mathbf{p}_2, \mathbf{p}_3\}$ be an affinely dependent set of points in $\mathbb{R}^n$ and let $f: \mathbb{R}^n \to \mathbb{R}^m$ be a linear transformation. Show that $\{f(\mathbf{p}_1), f(\mathbf{p}_2), f(\mathbf{p}_3)\}$ is affinely dependent in $\mathbb{R}^m$.

28. Suppose that $\{\mathbf{p}_1, \mathbf{p}_2, \mathbf{p}_3\}$ is an affinely independent set in $\mathbb{R}^n$ and $\mathbf{q}$ is an arbitrary point in $\mathbb{R}^n$. Show that the translated set $\{\mathbf{p}_1 + \mathbf{q}, \mathbf{p}_2 + \mathbf{q}, \mathbf{p}_3 + \mathbf{q}\}$ is also affinely independent.

In Exercises 29–32, **a**, **b**, and **c** are noncollinear points in $\mathbb{R}^2$ and **p** is any other point in $\mathbb{R}^2$. Let $\triangle\mathbf{abc}$ denote the closed triangular region determined by **a**, **b**, and **c**, and let $\triangle\mathbf{pbc}$ be the region determined by **p**, **b**, and **c**. For convenience, assume that **a**, **b**, and **c** are arranged so that $\det[\tilde{\mathbf{a}} \ \tilde{\mathbf{b}} \ \tilde{\mathbf{c}}]$ is positive, where $\tilde{\mathbf{a}}$, $\tilde{\mathbf{b}}$, and $\tilde{\mathbf{c}}$ are the standard homogeneous forms for the points.

29. Show that the area of $\triangle\mathbf{abc}$ is $\det[\tilde{\mathbf{a}} \ \tilde{\mathbf{b}} \ \tilde{\mathbf{c}}]/2$. [*Hint:* Consult Sections 3.2 and 3.3, including the Exercises.]

30. Let **p** be a point on the line through **a** and **b**. Show that $\det[\tilde{\mathbf{a}} \ \tilde{\mathbf{b}} \ \tilde{\mathbf{p}}] = 0$.

31. Let **p** be any point in the interior of $\triangle\mathbf{abc}$, with barycentric coordinates (r, s, t), so that

$$\begin{bmatrix} \tilde{\mathbf{a}} & \tilde{\mathbf{b}} & \tilde{\mathbf{c}} \end{bmatrix} \begin{bmatrix} r \\ s \\ t \end{bmatrix} = \tilde{\mathbf{p}}$$

Use Exercise 29 and a fact about determinants (Chapter 3) to show that

$r = $ (area of $\triangle\mathbf{pbc}$)/(area of $\triangle\mathbf{abc}$)

$s = $ (area of $\triangle\mathbf{apc}$)/(area of $\triangle\mathbf{abc}$)

$t = $ (area of $\triangle\mathbf{abp}$)/(area of $\triangle\mathbf{abc}$)

32. Take **q** on the line segment from **b** to **c** and consider the line through **q** and **a**, which may be written as $\mathbf{p} = (1 - x)\mathbf{q} + x\mathbf{a}$ for all real x. Show that, for each x, $\det[\tilde{\mathbf{p}} \ \tilde{\mathbf{b}} \ \tilde{\mathbf{c}}] = x \cdot \det[\tilde{\mathbf{a}} \ \tilde{\mathbf{b}} \ \tilde{\mathbf{c}}]$. From this and earlier work, conclude that the parameter x is the first barycentric coordinate of **p**. However, by construction, the parameter x also determines the relative distance between **p** and **q** along the segment from **q** to **a**. (When $x = 1$, $\mathbf{p} = \mathbf{a}$.) When this fact is applied to Example 5, it shows that the colors at vertex **a** and the point **q** are smoothly interpolated as **p** moves along the line between **a** and **q**.

33. Let $\mathbf{v}_1 = \begin{bmatrix} 1 \\ 3 \\ -6 \end{bmatrix}$, $\mathbf{v}_2 = \begin{bmatrix} 7 \\ 3 \\ -5 \end{bmatrix}$, $\mathbf{v}_3 = \begin{bmatrix} 3 \\ 9 \\ -2 \end{bmatrix}$, $\mathbf{a} = \begin{bmatrix} 0 \\ 0 \\ 9 \end{bmatrix}$,

$\mathbf{b} = \begin{bmatrix} 1.4 \\ 1.5 \\ -3.1 \end{bmatrix}$, and $\mathbf{x}(t) = \mathbf{a} + t\mathbf{b}$ for $t \geq 0$. Find the point where the ray $\mathbf{x}(t)$ intersects the plane that contains the triangle with vertices $\mathbf{v}_1$, $\mathbf{v}_2$, and $\mathbf{v}_3$. Is this point inside the triangle?

34. Repeat Exercise 33 with $\mathbf{v}_1 = \begin{bmatrix} 1 \\ 2 \\ -4 \end{bmatrix}$, $\mathbf{v}_2 = \begin{bmatrix} 8 \\ 2 \\ -5 \end{bmatrix}$,

$\mathbf{v}_3 = \begin{bmatrix} 3 \\ 10 \\ -2 \end{bmatrix}$, $\mathbf{a} = \begin{bmatrix} 0 \\ 0 \\ 8 \end{bmatrix}$, and $\mathbf{b} = \begin{bmatrix} .9 \\ 2.0 \\ -3.7 \end{bmatrix}$.

Solutions to Practice Problems

1. From Example 1, the problem is to determine if the points are affinely dependent. Use the method of Example 2 and subtract one point from the other two. If one of these two new points is a multiple of the other, the original three points lie on a line.

2. The proof of Theorem 5 essentially points out that an affine dependence relation among points corresponds to a linear dependence relation among the homogeneous forms of the points, using the *same* weights. So, row reduce:

$$\begin{bmatrix} \tilde{\mathbf{v}}_1 & \tilde{\mathbf{v}}_2 & \tilde{\mathbf{v}}_3 & \tilde{\mathbf{v}}_4 \end{bmatrix} = \begin{bmatrix} 4 & 1 & 5 & 1 \\ 1 & 0 & 4 & 2 \\ 1 & 1 & 1 & 1 \end{bmatrix} \sim \begin{bmatrix} 1 & 1 & 1 & 1 \\ 4 & 1 & 5 & 1 \\ 1 & 0 & 4 & 2 \end{bmatrix}$$

$$\sim \begin{bmatrix} 1 & 0 & 0 & -1 \\ 0 & 1 & 0 & 1.25 \\ 0 & 0 & 1 & .75 \end{bmatrix}$$

View this matrix as the coefficient matrix for $A\mathbf{x} = \mathbf{0}$ with four variables. Then x_4 is free, $x_1 = x_4$, $x_2 = -1.25x_4$, and $x_3 = -.75x_4$. One solution is $x_1 = x_4 = 4$, $x_2 = -5$, and $x_3 = -3$. A linear dependence among the homogeneous forms is $4\tilde{\mathbf{v}}_1 - 5\tilde{\mathbf{v}}_2 - 3\tilde{\mathbf{v}}_3 + 4\tilde{\mathbf{v}}_4 = \mathbf{0}$. So $4\mathbf{v}_1 - 5\mathbf{v}_2 - 3\mathbf{v}_3 + 4\mathbf{v}_4 = \mathbf{0}$.

Another solution method is to translate the problem to the origin by subtracting $\mathbf{v}_1$ from the other points, find a linear dependence relation among the translated points, and then rearrange the terms. The amount of arithmetic involved is about the same as in the approach shown above.

8.3 Convex Combinations

Section 8.1 considered special linear combinations of the form

$$c_1\mathbf{v}_1 + c_2\mathbf{v}_2 + \cdots + c_k\mathbf{v}_k, \quad \text{where } c_1 + c_2 + \cdots + c_k = 1$$

This section further restricts the weights to be nonnegative.

DEFINITION

A **convex combination** of points $\mathbf{v}_1, \mathbf{v}_2, \ldots, \mathbf{v}_k$ in $\mathbb{R}^n$ is a linear combination of the form

$$c_1\mathbf{v}_1 + c_2\mathbf{v}_2 + \cdots + c_k\mathbf{v}_k$$

such that $c_1 + c_2 + \cdots + c_k = 1$ and $c_i \geq 0$ for all i. The set of all convex combinations of points in a set S is called the **convex hull** of S, denoted by conv S.

The convex hull of a single point $\mathbf{v}_1$ is just the set $\{\mathbf{v}_1\}$, the same as the affine hull. In other cases, the convex hull is properly contained in the affine hull. Recall that the affine hull of distinct points $\mathbf{v}_1$ and $\mathbf{v}_2$ is the line

$$\mathbf{y} = (1-t)\mathbf{v}_1 + t\mathbf{v}_2, \quad \text{with } t \text{ in } \mathbb{R}$$

Because the weights in a convex combination are nonnegative, the points in conv $\{\mathbf{v}_1, \mathbf{v}_2\}$ may be written as

$$\mathbf{y} = (1-t)\mathbf{v}_1 + t\mathbf{v}_2, \quad \text{with } 0 \leq t \leq 1$$

which is the **line segment** between $\mathbf{v}_1$ and $\mathbf{v}_2$, hereafter denoted by $\overline{\mathbf{v}_1\mathbf{v}_2}$.

If a set S is affinely independent and if $\mathbf{p} \in \text{aff } S$, then $\mathbf{p} \in \text{conv } S$ if and only if the barycentric coordinates of $\mathbf{p}$ are nonnegative. Example 1 shows a special situation in which S is much more than just affinely independent.

EXAMPLE 1 Let

$$\mathbf{v}_1 = \begin{bmatrix} 3 \\ 0 \\ 6 \\ -3 \end{bmatrix}, \quad \mathbf{v}_2 = \begin{bmatrix} -6 \\ 3 \\ 3 \\ 0 \end{bmatrix}, \quad \mathbf{v}_3 = \begin{bmatrix} 3 \\ 6 \\ 0 \\ 3 \end{bmatrix}, \quad \mathbf{p}_1 = \begin{bmatrix} 0 \\ 3 \\ 3 \\ 0 \end{bmatrix}, \quad \mathbf{p}_2 = \begin{bmatrix} -10 \\ 5 \\ 11 \\ -4 \end{bmatrix},$$

and $S = \{\mathbf{v}_1, \mathbf{v}_2, \mathbf{v}_3\}$. Note that S is an orthogonal set. Determine whether $\mathbf{p}_1$ is in Span S, aff S, and conv S. Then do the same for $\mathbf{p}_2$.

SOLUTION If $\mathbf{p}_1$ is at least a *linear* combination of the points in S, then the weights are easily found, because S is an orthogonal set. Let W be the subspace spanned by S. A calculation as in Section 6.3 shows that the orthogonal projection of $\mathbf{p}_1$ onto W is $\mathbf{p}_1$ itself:

$$\begin{aligned}
\text{proj}_W \mathbf{p}_1 &= \frac{\mathbf{p}_1 \cdot \mathbf{v}_1}{\mathbf{v}_1 \cdot \mathbf{v}_1}\mathbf{v}_1 + \frac{\mathbf{p}_1 \cdot \mathbf{v}_2}{\mathbf{v}_2 \cdot \mathbf{v}_2}\mathbf{v}_2 + \frac{\mathbf{p}_1 \cdot \mathbf{v}_3}{\mathbf{v}_3 \cdot \mathbf{v}_3}\mathbf{v}_3 \\
&= \frac{18}{54}\mathbf{v}_1 + \frac{18}{54}\mathbf{v}_2 + \frac{18}{54}\mathbf{v}_3 \\
&= \frac{1}{3}\begin{bmatrix} 3 \\ 0 \\ 6 \\ -3 \end{bmatrix} + \frac{1}{3}\begin{bmatrix} -6 \\ 3 \\ 3 \\ 0 \end{bmatrix} + \frac{1}{3}\begin{bmatrix} 3 \\ 6 \\ 0 \\ 3 \end{bmatrix} = \begin{bmatrix} 0 \\ 3 \\ 3 \\ 0 \end{bmatrix} = \mathbf{p}_1
\end{aligned}$$

This shows that $\mathbf{p}_1$ is *in* Span S. Also, since the coefficients sum to 1, $\mathbf{p}_1$ is in aff S. In fact, $\mathbf{p}_1$ is in conv S, because the coefficients are also nonnegative.

For $\mathbf{p}_2$, a similar calculation shows that $\text{proj}_W \mathbf{p}_2 \neq \mathbf{p}_2$. Since $\text{proj}_W \mathbf{p}_2$ is the closest point in Span S to $\mathbf{p}_2$, the point $\mathbf{p}_2$ is not in Span S. In particular, $\mathbf{p}_2$ cannot be in aff S or conv S. ∎

Recall that a set S is affine if it contains all lines determined by pairs of points in S. When attention is restricted to convex combinations, the appropriate condition involves line segments rather than lines.

DEFINITION   A set S is **convex** if for each $\mathbf{p}, \mathbf{q} \in S$, the line segment $\overline{\mathbf{pq}}$ is contained in S.

Intuitively, a set S is convex if every two points in the set can "see" each other without the line of sight leaving the set. Figure 1 illustrates this idea.

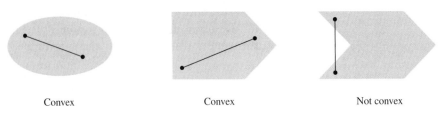

Convex   Convex   Not convex

FIGURE 1

The next result is analogous to Theorem 2 for affine sets.

THEOREM 7

A set S is convex if and only if every convex combination of points of S lies in S. That is, S is convex if and only if $S = \operatorname{conv} S$.

PROOF   The argument is similar to the proof of Theorem 2. The only difference is in the induction step. When taking a convex combination of $k + 1$ points, consider $\mathbf{y} = c_1\mathbf{v}_1 + \cdots + c_k\mathbf{v}_k + c_{k+1}\mathbf{v}_{k+1}$, where $c_1 + \cdots + c_{k+1} = 1$ and $0 \le c_i \le 1$ for all i. If $c_{k+1} = 1$, then $\mathbf{y} = \mathbf{v}_{k+1}$, which belongs to S, and there is nothing further to prove. If $c_{k+1} < 1$, let $t = c_1 + \cdots + c_k$. Then $t = 1 - c_{k+1} > 0$ and

$$\mathbf{y} = (1 - c_{k+1})\left(\frac{c_1}{t}\mathbf{v}_1 + \cdots + \frac{c_k}{t}\mathbf{v}_k\right) + c_{k+1}\mathbf{v}_{k+1} \tag{1}$$

By the induction hypothesis, the point $\mathbf{z} = (c_1/t)\mathbf{v}_1 + \cdots + (c_k/t)\mathbf{v}_k$ is in S, since the nonnegative coefficients sum to 1. Thus equation (1) displays $\mathbf{y}$ as a convex combination of two points in S. By the principle of induction, every convex combination of such points lies in S. ∎

Theorem 9 below provides a more geometric characterization of the convex hull of a set. It requires a preliminary result on intersections of sets. Recall from Section 4.1 (Exercise 40) that the intersection of two subspaces is itself a subspace. In fact, the intersection of any collection of subspaces is itself a subspace. A similar result holds for affine sets and convex sets.

THEOREM 8

Let $\{S_\alpha : \alpha \in \mathcal{A}\}$ be any collection of convex sets. Then $\bigcap_{\alpha \in \mathcal{A}} S_\alpha$ is convex. If $\{T_\beta : \beta \in \mathcal{B}\}$ is any collection of affine sets, then $\bigcap_{\beta \in \mathcal{B}} T_\beta$ is affine.

PROOF   If $\mathbf{p}$ and $\mathbf{q}$ are in $\cap S_\alpha$, then $\mathbf{p}$ and $\mathbf{q}$ are in each S_α. Since each S_α is convex, the line segment between $\mathbf{p}$ and $\mathbf{q}$ is in S_α for all α and hence that segment is contained in $\cap S_\alpha$. The proof of the affine case is similar. ∎

THEOREM 9

For any set S, the convex hull of S is the intersection of all the convex sets that contain S.

PROOF   Let T denote the intersection of all the convex sets containing S. Since $\operatorname{conv} S$ is a convex set containing S, it follows that $T \subseteq \operatorname{conv} S$. On the other hand, let C be any convex set containing S. Then C contains every convex combination of points of C (Theorem 7), and hence also contains every convex combination of points of the subset S. That is, $\operatorname{conv} S \subseteq C$. Since this is true for every convex set C containing S, it is also true for the intersection of them all. That is, $\operatorname{conv} S \subseteq T$. ∎

Theorem 9 shows that $\operatorname{conv} S$ is in a natural sense the "smallest" convex set containing S. For example, consider a set S that lies inside some large rectangle in $\mathbb{R}^2$, and imagine stretching a rubber band around the outside of S. As the rubber band contracts around S, it outlines the boundary of the convex hull of S. Or to use another analogy, the convex hull of S fills *in* all the holes in the inside of S and fills *out* all the dents in the boundary of S.

EXAMPLE 2

a. The convex hulls of sets S and T in $\mathbb{R}^2$ are shown below.

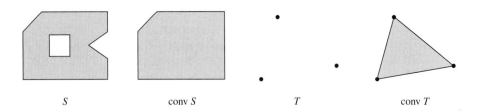

S conv S T conv T

b. Let S be the set consisting of the standard basis for $\mathbb{R}^3$, $S = \{\mathbf{e}_1, \mathbf{e}_2, \mathbf{e}_3\}$. Then conv S is a triangular surface in $\mathbb{R}^3$, with vertices $\mathbf{e}_1$, $\mathbf{e}_2$, and $\mathbf{e}_3$. See Figure 2. ∎

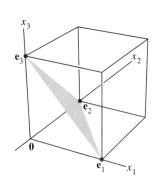

FIGURE 2

EXAMPLE 3 Let $S = \left\{ \begin{bmatrix} x \\ y \end{bmatrix} : x \geq 0 \text{ and } y = x^2 \right\}$. Show that the convex hull of S is the union of the origin and $\left\{ \begin{bmatrix} x \\ y \end{bmatrix} : x > 0 \text{ and } y \geq x^2 \right\}$. See Figure 3.

SOLUTION Every point in conv S must lie on a line segment that connects two points of S. The dashed line in Figure 3 indicates that, except for the origin, the positive y-axis is not in conv S, because the origin is the only point of S on the y-axis. It may seem reasonable that Figure 3 does show conv S, but how can you be sure that the point $(10^{-2}, 10^4)$, for example, is on a line segment from the origin to a point on the curve in S?

Consider any point $\mathbf{p}$ in the shaded region of Figure 3, say

$$\mathbf{p} = \begin{bmatrix} a \\ b \end{bmatrix}, \quad \text{with } a > 0 \text{ and } b \geq a^2$$

The line through $\mathbf{0}$ and $\mathbf{p}$ has the equation $y = (b/a)t$ for t real. That line intersects S where t satisfies $(b/a)t = t^2$, that is, when $t = b/a$. Thus, $\mathbf{p}$ is on the line segment from $\mathbf{0}$ to $\begin{bmatrix} b/a \\ b^2/a^2 \end{bmatrix}$, which shows that Figure 3 is correct. ∎

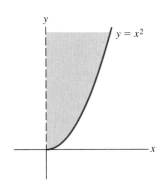

FIGURE 3

The following theorem is basic in the study of convex sets. It was first proved by Constantin Caratheodory in 1907. If $\mathbf{p}$ is in the convex hull of S, then, by definition, $\mathbf{p}$ must be a convex combination of points of S. But the definition makes no stipulation as to how many points of S are required to make the combination. Caratheodory's remarkable theorem says that in an n-dimensional space, the number of points of S in the convex combination never has to be more than $n + 1$.

THEOREM 10

(Caratheodory) If S is a nonempty subset of $\mathbb{R}^n$, then every point in conv S can be expressed as a convex combination of $n + 1$ or fewer points of S.

PROOF Given $\mathbf{p}$ in conv S, one may write $\mathbf{p} = c_1\mathbf{v}_1 + \cdots + c_k\mathbf{v}_k$, where $\mathbf{v}_i \in S$, $c_1 + \cdots + c_k = 1$, and $c_i \geq 0$, for some k and $i = 1, \ldots, k$. The goal is to show that such an expression exists for $\mathbf{p}$ with $k \leq n + 1$.

If $k > n + 1$, then $\{\mathbf{v}_1, \ldots, \mathbf{v}_k\}$ is affinely dependent, by Exercise 20 in Section 8.2. Thus there exist scalars $d_1, \ldots, d_k$, not all zero, such that

$$\sum_{i=1}^{k} d_i \mathbf{v}_i = \mathbf{0} \quad \text{and} \quad \sum_{i=1}^{k} d_i = 0$$

Consider the two equations

$$c_1 \mathbf{v}_1 + c_2 \mathbf{v}_2 + \cdots + c_k \mathbf{v}_k = \mathbf{p}$$

and

$$d_1 \mathbf{v}_1 + d_2 \mathbf{v}_2 + \cdots + d_k \mathbf{v}_k = \mathbf{0}$$

By subtracting an appropriate multiple of the second equation from the first, we now eliminate one of the $\mathbf{v}_i$ terms and obtain a convex combination of fewer than k elements of S that is equal to $\mathbf{p}$.

Since not all of the d_i coefficients are zero, we may assume (by reordering subscripts if necessary) that $d_k > 0$ and that $c_k/d_k \le c_i/d_i$ for all those i for which $d_i > 0$. For $i = 1, \ldots, k$, let $b_i = c_i - (c_k/d_k)d_i$. Then $b_k = 0$ and

$$\sum_{i=1}^{k} b_i = \sum_{i=1}^{k} c_i - \frac{c_k}{d_k} \sum_{i=1}^{k} d_i = 1 - 0 = 1$$

Furthermore, each $b_i \ge 0$. Indeed, if $d_i \le 0$, then $b_i \ge c_i \ge 0$. If $d_i > 0$, then $b_i = d_i(c_i/d_i - c_k/d_k) \ge 0$. By construction,

$$\sum_{i=1}^{k-1} b_i \mathbf{v}_i = \sum_{i=1}^{k} b_i \mathbf{v}_i = \sum_{i=1}^{k} \left(c_i - \frac{c_k}{d_k} d_i \right) \mathbf{v}_i$$

$$= \sum_{i=1}^{k} c_i \mathbf{v}_i - \frac{c_k}{d_k} \sum_{i=1}^{k} d_i \mathbf{v}_i = \sum_{i=1}^{k} c_i \mathbf{v}_i = \mathbf{p}$$

Thus $\mathbf{p}$ is now a convex combination of $k - 1$ of the points $\mathbf{v}_1, \ldots, \mathbf{v}_k$. This process may be repeated until $\mathbf{p}$ is expressed as a convex combination of at most $n + 1$ of the points of S. ∎

The following example illustrates the calculations in the proof above.

EXAMPLE 4 Let

$$\mathbf{v}_1 = \begin{bmatrix} 1 \\ 0 \end{bmatrix}, \quad \mathbf{v}_2 = \begin{bmatrix} 2 \\ 3 \end{bmatrix}, \quad \mathbf{v}_3 = \begin{bmatrix} 5 \\ 4 \end{bmatrix}, \quad \mathbf{v}_4 = \begin{bmatrix} 3 \\ 0 \end{bmatrix}, \quad \mathbf{p} = \begin{bmatrix} \frac{10}{3} \\ \frac{5}{2} \end{bmatrix},$$

and $S = \{\mathbf{v}_1, \mathbf{v}_2, \mathbf{v}_3, \mathbf{v}_4\}$. Then

$$\tfrac{1}{4}\mathbf{v}_1 + \tfrac{1}{6}\mathbf{v}_2 + \tfrac{1}{2}\mathbf{v}_3 + \tfrac{1}{12}\mathbf{v}_4 = \mathbf{p} \tag{2}$$

Use the procedure in the proof of Caratheodory's Theorem to express $\mathbf{p}$ as a convex combination of three points of S.

SOLUTION The set S is affinely dependent. Use the technique of Section 8.2 to obtain an affine dependence relation

$$-5\mathbf{v}_1 + 4\mathbf{v}_2 - 3\mathbf{v}_3 + 4\mathbf{v}_4 = \mathbf{0} \tag{3}$$

Next, choose the points $\mathbf{v}_2$ and $\mathbf{v}_4$ in (3), whose coefficients are positive. For each point, compute the ratio of the coefficients in equations (2) and (3). The ratio for $\mathbf{v}_2$ is $\frac{1}{6} \div 4 = \frac{1}{24}$, and that for $\mathbf{v}_4$ is $\frac{1}{12} \div 4 = \frac{1}{48}$. The ratio for $\mathbf{v}_4$ is smaller, so subtract $\frac{1}{48}$ times equation (3) from equation (2) to eliminate $\mathbf{v}_4$:

$$\left(\tfrac{1}{4} + \tfrac{5}{48}\right)\mathbf{v}_1 + \left(\tfrac{1}{6} - \tfrac{4}{48}\right)\mathbf{v}_2 + \left(\tfrac{1}{2} + \tfrac{3}{48}\right)\mathbf{v}_3 + \left(\tfrac{1}{12} - \tfrac{4}{48}\right)\mathbf{v}_4 = \mathbf{p}$$

$$\tfrac{17}{48}\mathbf{v}_1 + \tfrac{4}{48}\mathbf{v}_2 + \tfrac{27}{48}\mathbf{v}_3 = \mathbf{p} \qquad \blacksquare$$

This result cannot, in general, be improved by decreasing the required number of points. Indeed, given any three non-collinear points in $\mathbb{R}^2$, the centroid of the triangle formed by them is in the convex hull of all three, but is not in the convex hull of any two.

Practice Problems

1. Let $\mathbf{v}_1 = \begin{bmatrix} 6 \\ 2 \\ 2 \end{bmatrix}$, $\mathbf{v}_2 = \begin{bmatrix} 7 \\ 1 \\ 5 \end{bmatrix}$, $\mathbf{v}_3 = \begin{bmatrix} -2 \\ 4 \\ -1 \end{bmatrix}$, $\mathbf{p}_1 = \begin{bmatrix} 1 \\ 3 \\ 1 \end{bmatrix}$, and $\mathbf{p}_2 = \begin{bmatrix} 3 \\ 2 \\ 1 \end{bmatrix}$, and let $S = \{\mathbf{v}_1, \mathbf{v}_2, \mathbf{v}_3\}$. Determine whether $\mathbf{p}_1$ and $\mathbf{p}_2$ are in conv S.

2. Let S be the set of points on the curve $y = 1/x$ for $x > 0$. Explain geometrically why conv S consists of all points on and above the curve S.

8.3 Exercises

1. In $\mathbb{R}^2$, let $S = \left\{ \begin{bmatrix} 0 \\ y \end{bmatrix} : 0 \le y < 1 \right\} \cup \left\{ \begin{bmatrix} 2 \\ 0 \end{bmatrix} \right\}$. Describe (or sketch) the convex hull of S.

2. Describe the convex hull of the set S of points $\begin{bmatrix} x \\ y \end{bmatrix}$ in $\mathbb{R}^2$ that satisfy the given conditions. Justify your answers. (Show that an arbitrary point $\mathbf{p}$ in S belongs to conv S.)
   a. $y = 1/x$ and $x \ge 1/2$
   b. $y = \sin x$
   c. $y = x^{1/2}$ and $x \ge 0$

3. Consider the points in Exercise 5 in Section 8.1. Which of $\mathbf{p}_1$, $\mathbf{p}_2$, and $\mathbf{p}_3$ are in conv S?

4. Consider the points in Exercise 6 in Section 8.1. Which of $\mathbf{p}_1$, $\mathbf{p}_2$, and $\mathbf{p}_3$ are in conv S?

5. Let

$$\mathbf{v}_1 = \begin{bmatrix} -1 \\ -3 \\ 4 \end{bmatrix}, \mathbf{v}_2 = \begin{bmatrix} 0 \\ -3 \\ 1 \end{bmatrix}, \mathbf{v}_3 = \begin{bmatrix} 1 \\ -1 \\ 4 \end{bmatrix}, \mathbf{v}_4 = \begin{bmatrix} 1 \\ 1 \\ -2 \end{bmatrix},$$

$$\mathbf{p}_1 = \begin{bmatrix} 1 \\ -1 \\ 2 \end{bmatrix}, \mathbf{p}_2 = \begin{bmatrix} 0 \\ -2 \\ 2 \end{bmatrix},$$

and $S = \{\mathbf{v}_1, \mathbf{v}_2, \mathbf{v}_3, \mathbf{v}_4\}$. Determine whether $\mathbf{p}_1$ and $\mathbf{p}_2$ are in conv S.

6. Let $\mathbf{v}_1 = \begin{bmatrix} 2 \\ 0 \\ -1 \\ 2 \end{bmatrix}$, $\mathbf{v}_2 = \begin{bmatrix} 0 \\ -2 \\ 2 \\ 1 \end{bmatrix}$, $\mathbf{v}_3 = \begin{bmatrix} -2 \\ 1 \\ 0 \\ 2 \end{bmatrix}$, $\mathbf{p}_1 = \begin{bmatrix} -1 \\ 2 \\ -\frac{3}{2} \\ \frac{5}{2} \end{bmatrix}$,

$\mathbf{p}_2 = \begin{bmatrix} -\frac{1}{2} \\ 0 \\ \frac{1}{4} \\ \frac{7}{4} \end{bmatrix}$, $\mathbf{p}_3 = \begin{bmatrix} 6 \\ -4 \\ 1 \\ -1 \end{bmatrix}$, and $\mathbf{p}_4 = \begin{bmatrix} -1 \\ -2 \\ 0 \\ 4 \end{bmatrix}$, and let S be

the orthogonal set $\{\mathbf{v}_1, \mathbf{v}_2, \mathbf{v}_3\}$. Determine whether each $\mathbf{p}_i$ is in Span S, aff S, or conv S.
   a. $\mathbf{p}_1$ b. $\mathbf{p}_2$ c. $\mathbf{p}_3$ d. $\mathbf{p}_4$

Exercises 7–10 use the terminology from Section 8.2.

7. a. Let $T = \left\{ \begin{bmatrix} -1 \\ 0 \end{bmatrix}, \begin{bmatrix} 2 \\ 3 \end{bmatrix}, \begin{bmatrix} 4 \\ 1 \end{bmatrix} \right\}$, and let

   $\mathbf{p}_1 = \begin{bmatrix} 2 \\ 1 \end{bmatrix}$, $\mathbf{p}_2 = \begin{bmatrix} 3 \\ 2 \end{bmatrix}$, $\mathbf{p}_3 = \begin{bmatrix} 2 \\ 0 \end{bmatrix}$, and $\mathbf{p}_4 = \begin{bmatrix} 0 \\ 2 \end{bmatrix}$.

   Find the barycentric coordinates of $\mathbf{p}_1, \mathbf{p}_2, \mathbf{p}_3$, and $\mathbf{p}_4$ with respect to T.

   b. Use your answers in part (a) to determine whether each of $\mathbf{p}_1, \ldots, \mathbf{p}_4$ in part (a) is inside, outside, or on the edge of conv T, a triangular region.

8. Repeat Exercise 7 for $T = \left\{ \begin{bmatrix} 2 \\ 0 \end{bmatrix}, \begin{bmatrix} 0 \\ 5 \end{bmatrix}, \begin{bmatrix} -1 \\ 1 \end{bmatrix} \right\}$ and

$$\mathbf{p}_1 = \begin{bmatrix} 2 \\ 1 \end{bmatrix}, \ \mathbf{p}_2 = \begin{bmatrix} 1 \\ 1 \end{bmatrix}, \ \mathbf{p}_3 = \begin{bmatrix} 1 \\ \frac{1}{3} \end{bmatrix}, \ \text{and } \mathbf{p}_4 = \begin{bmatrix} 1 \\ 0 \end{bmatrix}.$$

9. Let $S = \{\mathbf{v}_1, \mathbf{v}_2, \mathbf{v}_3, \mathbf{v}_4\}$ be an affinely independent set. Consider the points $\mathbf{p}_1, \dots, \mathbf{p}_5$ whose barycentric coordinates with respect to S are given by $(2, 0, 0, -1)$, $\left(0, \frac{1}{2}, \frac{1}{4}, \frac{1}{4}\right)$, $\left(\frac{1}{2}, 0, \frac{3}{2}, -1\right)$, $\left(\frac{1}{3}, \frac{1}{4}, \frac{1}{4}, \frac{1}{6}\right)$, and $\left(\frac{1}{3}, 0, \frac{2}{3}, 0\right)$, respectively. Determine whether each of $\mathbf{p}_1, \dots, \mathbf{p}_5$ is inside, outside, or on the surface of conv S, a tetrahedron. Are any of these points on an edge of conv S?

10. Repeat Exercise 9 for the points $\mathbf{q}_1, \dots, \mathbf{q}_5$ whose barycentric coordinates with respect to S are given by $\left(\frac{1}{8}, \frac{1}{4}, \frac{1}{8}, \frac{1}{2}\right)$, $\left(\frac{3}{4}, -\frac{1}{4}, 0, \frac{1}{2}\right)$, $\left(0, \frac{3}{4}, \frac{1}{4}, 0\right)$, $(0, -2, 0, 3)$, and $\left(\frac{1}{3}, \frac{1}{3}, \frac{1}{3}, 0\right)$, respectively.

In Exercises 11–16, mark each statement True or False **(T/F)**. Justify each answer.

11. **(T/F)** If $\mathbf{y} = c_1 \mathbf{v}_1 + c_2 \mathbf{v}_2 + c_3 \mathbf{v}_3$ and $c_1 + c_2 + c_3 = 1$, then $\mathbf{y}$ is a convex combination of $\mathbf{v}_1, \mathbf{v}_2$, and $\mathbf{v}_3$.

12. **(T/F)** A set is convex if $\mathbf{x}, \mathbf{y} \in S$ implies that the line segment between $\mathbf{x}$ and $\mathbf{y}$ is contained in S.

13. **(T/F)** If S is a nonempty set, then conv S contains some points that are not in S.

14. **(T/F)** If S and T are convex sets, then $S \cap T$ is also convex.

15. **(T/F)** If S is a nonempty subset of $\mathbb{R}^5$ and $\mathbf{y} \in$ conv S, then there exist distinct points $\mathbf{v}_1, \dots, \mathbf{v}_6$ in S such that $\mathbf{y}$ is a convex combination of $\mathbf{v}_1, \dots, \mathbf{v}_6$.

16. **(T/F)** If S and T are convex sets, then $S \cup T$ is also convex.

17. Let S be a convex subset of $\mathbb{R}^n$ and suppose that $f : \mathbb{R}^n \to \mathbb{R}^m$ is a linear transformation. Prove that the set $f(S) = \{f(\mathbf{x}) : \mathbf{x} \in S\}$ is a convex subset of $\mathbb{R}^m$.

18. Let $f : \mathbb{R}^n \to \mathbb{R}^m$ be a linear transformation and let T be a convex subset of $\mathbb{R}^m$. Prove that the set $S = \{\mathbf{x} \in \mathbb{R}^n : f(\mathbf{x}) \in T\}$ is a convex subset of $\mathbb{R}^n$.

19. Let $\mathbf{v}_1 = \begin{bmatrix} 1 \\ 0 \end{bmatrix}$, $\mathbf{v}_2 = \begin{bmatrix} 1 \\ 2 \end{bmatrix}$, $\mathbf{v}_3 = \begin{bmatrix} 4 \\ 2 \end{bmatrix}$, $\mathbf{v}_4 = \begin{bmatrix} 4 \\ 0 \end{bmatrix}$, and $\mathbf{p} = \begin{bmatrix} 2 \\ 1 \end{bmatrix}$. Confirm that

$$\mathbf{p} = \tfrac{1}{3}\mathbf{v}_1 + \tfrac{1}{3}\mathbf{v}_2 + \tfrac{1}{6}\mathbf{v}_3 + \tfrac{1}{6}\mathbf{v}_4 \ \text{ and } \ \mathbf{v}_1 - \mathbf{v}_2 + \mathbf{v}_3 - \mathbf{v}_4 = \mathbf{0}.$$

Use the procedure in the proof of Caratheodory's Theorem to express $\mathbf{p}$ as a convex combination of three of the $\mathbf{v}_i$'s. Do this in *two* ways.

20. Repeat Exercise 19 for points $\mathbf{v}_1 = \begin{bmatrix} -1 \\ 0 \end{bmatrix}$, $\mathbf{v}_2 = \begin{bmatrix} 0 \\ 3 \end{bmatrix}$, $\mathbf{v}_3 = \begin{bmatrix} 3 \\ 1 \end{bmatrix}$, $\mathbf{v}_4 = \begin{bmatrix} 1 \\ -1 \end{bmatrix}$, and $\mathbf{p} = \begin{bmatrix} 1 \\ 2 \end{bmatrix}$, given that

$$\mathbf{p} = \tfrac{1}{121}\mathbf{v}_1 + \tfrac{72}{121}\mathbf{v}_2 + \tfrac{37}{121}\mathbf{v}_3 + \tfrac{1}{11}\mathbf{v}_4$$

and

$$10\mathbf{v}_1 - 6\mathbf{v}_2 + 7\mathbf{v}_3 - 11\mathbf{v}_4 = \mathbf{0}.$$

In Exercises 21–24, prove the given statement about subsets A and B of $\mathbb{R}^n$. A proof for an exercise may use results of earlier exercises.

21. If $A \subseteq B$ and B is convex, then conv $A \subseteq B$.

22. If $A \subseteq B$, then conv $A \subseteq$ conv B.

23. a. $[(\text{conv } A) \cup (\text{conv } B)] \subseteq \text{conv } (A \cup B)$
    b. Find an example in $\mathbb{R}^2$ to show that equality need not hold in part (a).

24. a. $\text{conv } (A \cap B) \subseteq [(\text{conv } A) \cap (\text{conv } B)]$
    b. Find an example in $\mathbb{R}^2$ to show that equality need not hold in part (a).

25. Let $\mathbf{p}_0$, $\mathbf{p}_1$, and $\mathbf{p}_2$ be points in $\mathbb{R}^n$, and define $\mathbf{f}_0(t) = (1-t)\mathbf{p}_0 + t\mathbf{p}_1$, $\mathbf{f}_1(t) = (1-t)\mathbf{p}_1 + t\mathbf{p}_2$, and $\mathbf{g}(t) = (1-t)\mathbf{f}_0(t) + t\mathbf{f}_1(t)$ for $0 \le t \le 1$. For the points as shown below, draw a picture that shows $\mathbf{f}_0\left(\frac{1}{2}\right), \mathbf{f}_1\left(\frac{1}{2}\right)$, and $\mathbf{g}\left(\frac{1}{2}\right)$.

26. Repeat Exercise 25 for $\mathbf{f}_0\left(\frac{3}{4}\right), \mathbf{f}_1\left(\frac{3}{4}\right)$, and $\mathbf{g}\left(\frac{3}{4}\right)$.

27. Let $\mathbf{g}(t)$ be defined as in Exercise 25. Its graph is called a *quadratic Bézier curve*, and it is used in some computer graphics designs. The points $\mathbf{p}_0$, $\mathbf{p}_1$, and $\mathbf{p}_2$ are called the *control points* for the curve. Compute a formula for $\mathbf{g}(t)$ that involves only $\mathbf{p}_0$, $\mathbf{p}_1$, and $\mathbf{p}_2$. Then show that $\mathbf{g}(t)$ is in conv $\{\mathbf{p}_0, \mathbf{p}_1, \mathbf{p}_2\}$ for $0 \le t \le 1$.

28. Given control points $\mathbf{p}_0$, $\mathbf{p}_1$, $\mathbf{p}_2$, and $\mathbf{p}_3$ in $\mathbb{R}^n$, let $\mathbf{g}_1(t)$ for $0 \le t \le 1$ be the quadratic Bézier curve from Exercise 27 determined by $\mathbf{p}_0$, $\mathbf{p}_1$, and $\mathbf{p}_2$, and let $\mathbf{g}_2(t)$ be defined similarly for $\mathbf{p}_1$, $\mathbf{p}_2$, and $\mathbf{p}_3$. For $0 \le t \le 1$, define $\mathbf{h}(t) = (1-t)\mathbf{g}_1(t) + t\mathbf{g}_2(t)$. Show that the graph of $\mathbf{h}(t)$ lies in the convex hull of the four control points. This curve is called a *cubic Bézier curve*, and its definition here is one step in an algorithm for constructing Bézier curves (discussed later in Section 8.6). A Bézier curve of degree k is determined by $k + 1$ control points, and its graph lies in the convex hull of these control points.

Solutions to Practice Problems

1. The points $\mathbf{v}_1$, $\mathbf{v}_2$, and $\mathbf{v}_3$ are not orthogonal, so compute

$$\mathbf{v}_2 - \mathbf{v}_1 = \begin{bmatrix} 1 \\ -1 \\ 3 \end{bmatrix}, \ \mathbf{v}_3 - \mathbf{v}_1 = \begin{bmatrix} -8 \\ 2 \\ -3 \end{bmatrix}, \ \mathbf{p}_1 - \mathbf{v}_1 = \begin{bmatrix} -5 \\ 1 \\ -1 \end{bmatrix}, \text{ and } \mathbf{p}_2 - \mathbf{v}_1 = \begin{bmatrix} -3 \\ 0 \\ -1 \end{bmatrix}$$

Augment the matrix $[\,\mathbf{v}_2 - \mathbf{v}_1 \quad \mathbf{v}_3 - \mathbf{v}_1\,]$ with both $\mathbf{p}_1 - \mathbf{v}_1$ and $\mathbf{p}_2 - \mathbf{v}_1$, and row reduce:

$$\begin{bmatrix} 1 & -8 & -5 & -3 \\ -1 & 2 & 1 & 0 \\ 3 & -3 & -1 & -1 \end{bmatrix} \sim \begin{bmatrix} 1 & 0 & \frac{1}{3} & 1 \\ 0 & 1 & \frac{2}{3} & \frac{1}{2} \\ 0 & 0 & 0 & -\frac{5}{2} \end{bmatrix}$$

The third column shows that $\mathbf{p}_1 - \mathbf{v}_1 = \frac{1}{3}(\mathbf{v}_2 - \mathbf{v}_1) + \frac{2}{3}(\mathbf{v}_3 - \mathbf{v}_1)$, which leads to $\mathbf{p}_1 = 0\mathbf{v}_1 + \frac{1}{3}\mathbf{v}_2 + \frac{2}{3}\mathbf{v}_3$. Thus $\mathbf{p}_1$ is in conv S. In fact, $\mathbf{p}_1$ is in conv $\{\mathbf{v}_2, \mathbf{v}_3\}$.

The last column of the matrix shows that $\mathbf{p}_2 - \mathbf{v}_1$ is not a linear combination of $\mathbf{v}_2 - \mathbf{v}_1$ and $\mathbf{v}_3 - \mathbf{v}_1$. Thus $\mathbf{p}_2$ is not an affine combination of $\mathbf{v}_1$, $\mathbf{v}_2$, and $\mathbf{v}_3$, so $\mathbf{p}_2$ cannot possibly be in conv S.

An alternative method of solution is to row reduce the augmented matrix of homogeneous forms:

$$\begin{bmatrix} \tilde{\mathbf{v}}_1 & \tilde{\mathbf{v}}_2 & \tilde{\mathbf{v}}_3 & \tilde{\mathbf{p}}_1 & \tilde{\mathbf{p}}_2 \end{bmatrix} \sim \begin{bmatrix} 1 & 0 & 0 & 0 & 0 \\ 0 & 1 & 0 & \frac{1}{3} & 0 \\ 0 & 0 & 1 & \frac{2}{3} & 0 \\ 0 & 0 & 0 & 0 & 1 \end{bmatrix}$$

2. If $\mathbf{p}$ is a point above S, then the line through $\mathbf{p}$ with slope -1 will intersect S at two points before it reaches the positive x- and y-axes.

8.4 Hyperplanes

Hyperplanes play a special role in the geometry of $\mathbb{R}^n$ because they divide the space into two disjoint pieces, just as a plane separates $\mathbb{R}^3$ into two parts and a line cuts through $\mathbb{R}^2$. The key to working with hyperplanes is to use simple *implicit* descriptions, rather than the *explicit* or parametric representations of lines and planes used in the earlier work with affine sets.[1]

An implicit equation of a line in $\mathbb{R}^2$ has the form $ax + by = d$. An implicit equation of a plane in $\mathbb{R}^3$ has the form $ax + by + cz = d$. Both equations describe the line or plane as the set of all points at which a linear expression (also called a *linear functional*) has a fixed value, d.

DEFINITION

A **linear functional** on $\mathbb{R}^n$ is a linear transformation f from $\mathbb{R}^n$ into $\mathbb{R}$. For each scalar d in $\mathbb{R}$, the symbol $[f:d]$ denotes the set of all $\mathbf{x}$ in $\mathbb{R}^n$ at which the value of f is d. That is,

---

[1] Parametric representations were introduced in Section 1.5.

$[f\!:\!d]$ is the set $\{\mathbf{x} \in \mathbb{R}^n\!: f(\mathbf{x}) = d\}$

The **zero functional** is the transformation such that $f(\mathbf{x}) = 0$ for all $\mathbf{x}$ in $\mathbb{R}^n$. All other linear functionals on $\mathbb{R}^n$ are said to be **nonzero**.

EXAMPLE 1 In $\mathbb{R}^2$, the line $x - 4y = 13$ is a hyperplane in $\mathbb{R}^2$, and it is the set of points at which the linear functional $f(x, y) = x - 4y$ has the value 13. That is, the line is the set $[f\!:\!13]$. ∎

EXAMPLE 2 In $\mathbb{R}^3$, the plane $5x - 2y + 3z = 21$ is a hyperplane, the set of points at which the linear functional $g(x, y, z) = 5x - 2y + 3z$ has the value 21. This hyperplane is the set $[g\!:\!21]$. ∎

If f is a linear functional on $\mathbb{R}^n$, then the standard matrix of this linear transformation f is a $1 \times n$ matrix A, say $A = [\,a_1 \;\; a_2 \;\; \cdots \;\; a_n\,]$. So

$$[f\!:\!0] \quad \text{is the same as} \quad \{\mathbf{x} \in \mathbb{R}^n\!: A\mathbf{x} = 0\} = \text{Nul } A \tag{1}$$

If f is a nonzero functional, then rank $A = 1$, and dim Nul $A = n - 1$, by the Rank Theorem.[2] Thus, the subspace $[f:0]$ has dimension $n - 1$ and so is a hyperplane. Also, if d is any number in $\mathbb{R}$, then

$$[f\!:\!d] \quad \text{is the same as} \quad \{\mathbf{x} \in \mathbb{R}^n\!: A\mathbf{x} = d\} \tag{2}$$

Recall from Theorem 6 in Section 1.5 that the set of solutions of $A\mathbf{x} = \mathbf{b}$ is obtained by translating the solution set of $A\mathbf{x} = \mathbf{0}$, using any particular solution $\mathbf{p}$ of $A\mathbf{x} = \mathbf{b}$. When A is the standard matrix of the transformation f, this theorem says that

$$[f\!:\!d] = [f:0] + \mathbf{p} \quad \text{for any } \mathbf{p} \text{ in } [f:d] \tag{3}$$

Thus the sets $[f:d]$ are hyperplanes parallel to $[f:0]$. See Figure 1.

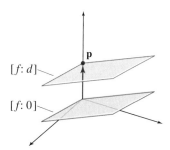

FIGURE 1 Parallel hyperplanes, with $f(\mathbf{p}) = d$.

When A is a $1 \times n$ matrix, the equation $A\mathbf{x} = d$ may be written with an inner product $\mathbf{n} \cdot \mathbf{x}$, using $\mathbf{n}$ in $\mathbb{R}^n$ with the same entries as A. Thus, from (2),

$$[f\!:\!d] \quad \text{is the same as} \quad \{\mathbf{x} \in \mathbb{R}^n\!: \mathbf{n} \cdot \mathbf{x} = d\} \tag{4}$$

---

[2] See Theorem 14 in Section 2.9 or Theorem 14 in Section 4.5.

Then $[f:0] = \{\mathbf{x} \in \mathbb{R}^n : \mathbf{n} \cdot \mathbf{x} = 0\}$, which shows that $[f:0]$ is the orthogonal complement of the subspace spanned by $\mathbf{n}$. In the terminology of calculus and geometry for $\mathbb{R}^3$, $\mathbf{n}$ is called a **normal** vector to $[f:0]$. (A "normal" vector in this sense need not have unit length.) Also, $\mathbf{n}$ is said to be **normal** to each parallel hyperplane $[f:d]$, even though $\mathbf{n} \cdot \mathbf{x}$ is not zero when $d \neq 0$.

Another name for $[f:d]$ is a *level set* of f, and $\mathbf{n}$ is sometimes called the *gradient* of f when $f(\mathbf{x}) = \mathbf{n} \cdot \mathbf{x}$ for each $\mathbf{x}$.

EXAMPLE 3 Let $\mathbf{n} = \begin{bmatrix} 3 \\ 4 \end{bmatrix}$ and $\mathbf{v} = \begin{bmatrix} 1 \\ -6 \end{bmatrix}$, and let $H = \{\mathbf{x} : \mathbf{n} \cdot \mathbf{x} = 12\}$, so $H = [f:12]$, where $f(x, y) = 3x + 4y$. Thus H is the line $3x + 4y = 12$. Find an implicit description of the parallel hyperplane (line) $H_1 = H + \mathbf{v}$.

SOLUTION First, find a point $\mathbf{p}$ in H_1. To do this, find a point in H and add $\mathbf{v}$ to it. For instance, $\begin{bmatrix} 0 \\ 3 \end{bmatrix}$ is in H, so $\mathbf{p} = \begin{bmatrix} 1 \\ -6 \end{bmatrix} + \begin{bmatrix} 0 \\ 3 \end{bmatrix} = \begin{bmatrix} 1 \\ -3 \end{bmatrix}$ is in H_1. Now, compute $\mathbf{n} \cdot \mathbf{p} = -9$. This shows that $H_1 = [f: -9]$. See Figure 2, which also shows the subspace $H_0 = \{\mathbf{x} : \mathbf{n} \cdot \mathbf{x} = 0\}$. ∎

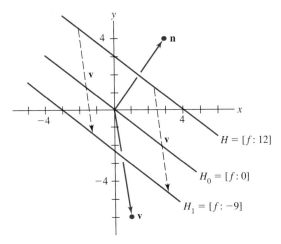

FIGURE 2

The next three examples show connections between implicit and explicit descriptions of hyperplanes. Example 4 begins with an implicit form.

EXAMPLE 4 In $\mathbb{R}^2$, give an explicit description of the line $x - 4y = 13$ in parametric vector form.

SOLUTION This amounts to solving a nonhomogeneous equation $A\mathbf{x} = \mathbf{b}$, where $A = [1 \quad -4]$ and $\mathbf{b}$ is the number 13 in $\mathbb{R}$. Write $x = 13 + 4y$, where y is a free variable. In parametric form, the solution is

$$\mathbf{x} = \begin{bmatrix} x \\ y \end{bmatrix} = \begin{bmatrix} 13 + 4y \\ y \end{bmatrix} = \begin{bmatrix} 13 \\ 0 \end{bmatrix} + y \begin{bmatrix} 4 \\ 1 \end{bmatrix} = \mathbf{p} + y\mathbf{q}, \quad y \in \mathbb{R} \qquad ∎$$

Converting an explicit description of a line into implicit form is more involved. The basic idea is to construct $[f:0]$ and then find d for $[f:d]$.

EXAMPLE 5 Let $\mathbf{v}_1 = \begin{bmatrix} 1 \\ 2 \end{bmatrix}$ and $\mathbf{v}_2 = \begin{bmatrix} 6 \\ 0 \end{bmatrix}$, and let L_1 be the line through $\mathbf{v}_1$ and $\mathbf{v}_2$. Find a linear functional f and a constant d such that $L_1 = [f:d]$.

SOLUTION   The line L_1 is parallel to the translated line L_0 through $\mathbf{v}_2 - \mathbf{v}_1$ and the origin. The defining equation for L_0 has the form

$$[a \quad b]\begin{bmatrix} x \\ y \end{bmatrix} = 0 \quad \text{or} \quad \mathbf{n} \cdot \mathbf{x} = 0, \quad \text{where} \quad \mathbf{n} = \begin{bmatrix} a \\ b \end{bmatrix} \tag{5}$$

Since $\mathbf{n}$ is orthogonal to the subspace L_0, which contains $\mathbf{v}_2 - \mathbf{v}_1$, compute

$$\mathbf{v}_2 - \mathbf{v}_1 = \begin{bmatrix} 6 \\ 0 \end{bmatrix} - \begin{bmatrix} 1 \\ 2 \end{bmatrix} = \begin{bmatrix} 5 \\ -2 \end{bmatrix}$$

and solve

$$[a \quad b]\begin{bmatrix} 5 \\ -2 \end{bmatrix} = 0$$

By inspection, a solution is $[a \quad b] = [2 \quad 5]$. Let $f(x, y) = 2x + 5y$. From (5), $L_0 = [f:0]$, and $L_1 = [f:d]$ for some d. Since $\mathbf{v}_1$ is on line L_1, $d = f(\mathbf{v}_1) = 2(1) + 5(2) = 12$. Thus, the equation for L_1 is $2x + 5y = 12$. As a check, note that $f(\mathbf{v}_2) = f(6, 0) = 2(6) + 5(0) = 12$, so $\mathbf{v}_2$ is on L_1, too.   ∎

EXAMPLE 6 Let $\mathbf{v}_1 = \begin{bmatrix} 1 \\ 1 \\ 1 \end{bmatrix}$, $\mathbf{v}_2 = \begin{bmatrix} 2 \\ -1 \\ 4 \end{bmatrix}$, and $\mathbf{v}_3 = \begin{bmatrix} 3 \\ 1 \\ 2 \end{bmatrix}$. Find an implicit description $[f:d]$ of the plane H_1 that passes through $\mathbf{v}_1$, $\mathbf{v}_2$, and $\mathbf{v}_3$.

SOLUTION H_1 is parallel to a plane H_0 through the origin that contains the translated points

$$\mathbf{v}_2 - \mathbf{v}_1 = \begin{bmatrix} 1 \\ -2 \\ 3 \end{bmatrix} \quad \text{and} \quad \mathbf{v}_3 - \mathbf{v}_1 = \begin{bmatrix} 2 \\ 0 \\ 1 \end{bmatrix}$$

Since these two points are linearly independent, $H_0 = \text{Span}\{\mathbf{v}_2 - \mathbf{v}_1, \mathbf{v}_3 - \mathbf{v}_1\}$. Let $\mathbf{n} = \begin{bmatrix} a \\ b \\ c \end{bmatrix}$ be the normal to H_0. Then $\mathbf{v}_2 - \mathbf{v}_1$ and $\mathbf{v}_3 - \mathbf{v}_1$ are each orthogonal to $\mathbf{n}$. That is, $(\mathbf{v}_2 - \mathbf{v}_1) \cdot \mathbf{n} = 0$ and $(\mathbf{v}_3 - \mathbf{v}_1) \cdot \mathbf{n} = 0$. These two equations form a system whose augmented matrix can be row reduced:

$$[1 \quad -2 \quad 3]\begin{bmatrix} a \\ b \\ c \end{bmatrix} = 0, \quad [2 \quad 0 \quad 1]\begin{bmatrix} a \\ b \\ c \end{bmatrix} = 0, \quad \begin{bmatrix} 1 & -2 & 3 & 0 \\ 2 & 0 & 1 & 0 \end{bmatrix}$$

Row operations yield $a = (-\frac{2}{4})c$, $b = (\frac{5}{4})c$, with c free. Set $c = 4$, for instance. Then $\mathbf{n} = \begin{bmatrix} -2 \\ 5 \\ 4 \end{bmatrix}$ and $H_0 = [f:0]$, where $f(\mathbf{x}) = -2x_1 + 5x_2 + 4x_3$.

The parallel hyperplane H_1 is $[f:d]$. To find d, use the fact that $\mathbf{v}_1$ is in H_1, and compute $d = f(\mathbf{v}_1) = f(1, 1, 1) = -2(1) + 5(1) + 4(1) = 7$. As a check, compute $f(\mathbf{v}_2) = f(2, -1, 4) = -2(2) + 5(-1) + 4(4) = 16 - 9 = 7$. Observe $f(\mathbf{v}_3) = 7$ also.   ∎

The procedure in Example 6 generalizes to higher dimensions. However, for the special case of $\mathbb{R}^3$, one can also use the **cross-product** formula to compute $\mathbf{n}$, using a symbolic determinant as a mnemonic device:

$$\mathbf{n} = (\mathbf{v}_2 - \mathbf{v}_1) \times (\mathbf{v}_3 - \mathbf{v}_1)$$

$$= \begin{vmatrix} 1 & 2 & \mathbf{i} \\ -2 & 0 & \mathbf{j} \\ 3 & 1 & \mathbf{k} \end{vmatrix} = \begin{vmatrix} -2 & 0 \\ 3 & 1 \end{vmatrix} \mathbf{i} - \begin{vmatrix} 1 & 2 \\ 3 & 1 \end{vmatrix} \mathbf{j} + \begin{vmatrix} 1 & 2 \\ -2 & 0 \end{vmatrix} \mathbf{k}$$

$$= -2\mathbf{i} + 5\mathbf{j} + 4\mathbf{k} = \begin{bmatrix} -2 \\ 5 \\ 4 \end{bmatrix}$$

If only the formula for f is needed, the cross-product calculation may be written as an ordinary determinant:

$$f(x_1, x_2, x_3) = \begin{vmatrix} 1 & 2 & x_1 \\ -2 & 0 & x_2 \\ 3 & 1 & x_3 \end{vmatrix} = \begin{vmatrix} -2 & 0 \\ 3 & 1 \end{vmatrix} x_1 - \begin{vmatrix} 1 & 2 \\ 3 & 1 \end{vmatrix} x_2 + \begin{vmatrix} 1 & 2 \\ -2 & 0 \end{vmatrix} x_3$$

$$= -2x_1 + 5x_2 + 4x_3$$

So far, every hyperplane examined has been described as $[f:d]$ for some linear functional f and some d in $\mathbb{R}$, or equivalently as $\{\mathbf{x} \in \mathbb{R}^n : \mathbf{n} \cdot \mathbf{x} = d\}$ for some $\mathbf{n}$ in $\mathbb{R}^n$. The following theorem shows that *every* hyperplane has these equivalent descriptions.

THEOREM 11

A subset H of $\mathbb{R}^n$ is a hyperplane if and only if $H = [f:d]$ for some nonzero linear functional f and some scalar d in $\mathbb{R}$. Thus, if H is a hyperplane, there exist a nonzero vector $\mathbf{n}$ and a real number d such that $H = \{\mathbf{x} : \mathbf{n} \cdot \mathbf{x} = d\}$.

PROOF Suppose that H is a hyperplane, take $\mathbf{p} \in H$, and let $H_0 = H - \mathbf{p}$. Then H_0 is an $(n-1)$-dimensional subspace. Next, take any point $\mathbf{y}$ that is not in H_0. By the Orthogonal Decomposition Theorem in Section 6.3,

$$\mathbf{y} = \mathbf{y}_1 + \mathbf{n}$$

where $\mathbf{y}_1$ is a vector in H_0 and $\mathbf{n}$ is orthogonal to every vector in H_0. The function f defined by

$$f(\mathbf{x}) = \mathbf{n} \cdot \mathbf{x} \quad \text{for } \mathbf{x} \in \mathbb{R}^n$$

is a linear functional, by properties of the inner product. Now, $[f:0]$ is a hyperplane that contains H_0, by construction of $\mathbf{n}$. It follows that

$$H_0 = [f:0]$$

[*Argument:* H_0 contains a basis S of $n-1$ vectors, and since S is in the $(n-1)$-dimensional subspace $[f:0]$, S must also be a basis for $[f:0]$, by the Basis Theorem.] Finally, let $d = f(\mathbf{p}) = \mathbf{n} \cdot \mathbf{p}$. Then, as in (3) shown earlier,

$$[f:d] = [f:0] + \mathbf{p} = H_0 + \mathbf{p} = H$$

The converse statement that $[f:d]$ is a hyperplane follows from (1) and (3) above. ∎

Many important applications of hyperplanes depend on the possibility of "separating" two sets by a hyperplane. Intuitively, this means that one of the sets is on one side of the hyperplane and the other set is on the other side. The following terminology and notation will help to make this idea more precise.

TOPOLOGY IN $\mathbb{R}^n$: TERMS AND FACTS

For any point $\mathbf{p}$ in $\mathbb{R}^n$ and any real $\delta > 0$, the **open ball** $B(\mathbf{p}, \delta)$ with center $\mathbf{p}$ and radius δ is given by
$$B(\mathbf{p}, \delta) = \{\mathbf{x} : \|\mathbf{x} - \mathbf{p}\| < \delta\}$$

Given a set S in $\mathbb{R}^n$, a point $\mathbf{p}$ is an **interior point** of S if there exists a $\delta > 0$ such that $B(\mathbf{p}, \delta) \subseteq S$. If every open ball centered at $\mathbf{p}$ intersects both S and the complement of S, then $\mathbf{p}$ is called a **boundary point** of S. A set is **open** if it contains none of its boundary points. (This is equivalent to saying that all of its points are interior points.) A set is **closed** if it contains all of its boundary points. (If S contains some but not all of its boundary points, then S is neither open nor closed.) A set S is **bounded** if there exists a $\delta > 0$ such that $S \subseteq B(\mathbf{0}, \delta)$. A set in $\mathbb{R}^n$ is **compact** if it is closed and bounded.

    Theorem: The convex hull of an open set is open, and the convex hull of a compact set is compact. (The convex hull of a closed set need not be closed. See Exercise 33.)

FIGURE 3

The set S is closed and bounded.

EXAMPLE 7 Let

$$S = \text{conv}\left\{\begin{bmatrix} -2 \\ 2 \end{bmatrix}, \begin{bmatrix} -2 \\ -2 \end{bmatrix}, \begin{bmatrix} 2 \\ -2 \end{bmatrix}, \begin{bmatrix} 2 \\ 2 \end{bmatrix}\right\}, \quad \mathbf{p}_1 = \begin{bmatrix} -1 \\ 0 \end{bmatrix}, \quad \text{and} \quad \mathbf{p}_2 = \begin{bmatrix} 2 \\ 1 \end{bmatrix},$$

as shown in Figure 3. Then $\mathbf{p}_1$ is an interior point since $B\left(\mathbf{p}_1, \frac{3}{4}\right) \subseteq S$. The point $\mathbf{p}_2$ is a boundary point since every open ball centered at $\mathbf{p}_2$ intersects both S and the complement of S. The set S is closed since it contains all its boundary points. The set S is bounded since $S \subseteq B(\mathbf{0}, 3)$. Thus S is also compact. ∎

Notation: If f is a linear functional, then $f(A) \leq d$ means $f(\mathbf{x}) \leq d$ for each $\mathbf{x} \in A$. Corresponding notations will be used when the inequalities are reversed or when they are strict.

DEFINITION

The hyperplane $H = [f : d]$ **separates** two sets A and B if one of the following holds:

(i) $f(A) \leq d$ and $f(B) \geq d$, or
(ii) $f(A) \geq d$ and $f(B) \leq d$.

If in the conditions above all the weak inequalities are replaced by strict inequalities, then H is said to **strictly separate** A and B.

Notice that strict separation requires that the two sets be disjoint, while mere separation does not. Indeed, if two circles in the plane are externally tangent, then their common tangent line separates them (but does not separate them strictly).

Although it is necessary that two sets be disjoint in order to strictly separate them, this condition is not sufficient, even for closed convex sets. For example, let

$$A = \left\{ \begin{bmatrix} x \\ y \end{bmatrix} : x \geq \frac{1}{2} \text{ and } \frac{1}{x} \leq y \leq 2 \right\} \quad \text{and} \quad B = \left\{ \begin{bmatrix} x \\ y \end{bmatrix} : x \geq 0 \text{ and } y = 0 \right\}$$

Then A and B are disjoint closed convex sets, but they cannot be strictly separated by a hyperplane (line in $\mathbb{R}^2$). See Figure 4. Thus the problem of separating (or strictly separating) two sets by a hyperplane is more complex than it might at first appear.

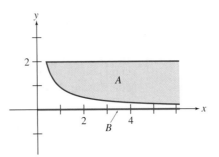

FIGURE 4 Disjoint closed convex sets.

There are many interesting conditions on the sets A and B that imply the existence of a separating hyperplane, but the following two theorems are sufficient for this section. The proof of the first theorem requires quite a bit of preliminary material,[3] but the second theorem follows easily from the first.

THEOREM 12

Suppose A and B are nonempty convex sets such that A is compact and B is closed. Then there exists a hyperplane H that strictly separates A and B if and only if $A \cap B = \varnothing$.

THEOREM 13

Suppose A and B are nonempty compact sets. Then there exists a hyperplane that strictly separates A and B if and only if $(\text{conv } A) \cap (\text{conv } B) = \varnothing$.

PROOF Suppose that $(\text{conv } A) \cap (\text{conv } B) = \varnothing$. Since the convex hull of a compact set is compact, Theorem 12 ensures that there is a hyperplane H that strictly separates conv A and conv B. Clearly, H also strictly separates the smaller sets A and B.

Conversely, suppose the hyperplane $H = [f : d]$ strictly separates A and B. Without loss of generality, assume that $f(A) < d$ and $f(B) > d$. Let $\mathbf{x} = c_1 \mathbf{x}_1 + \cdots + c_k \mathbf{x}_k$ be any convex combination of elements of A. Then

$$f(\mathbf{x}) = c_1 f(\mathbf{x}_1) + \cdots + c_k f(\mathbf{x}_k) < c_1 d + \cdots + c_k d = d$$

since $c_1 + \cdots + c_k = 1$. Thus $f(\text{conv } A) < d$. Likewise, $f(\text{conv } B) > d$, so $H = [f : d]$ strictly separates conv A and conv B. By Theorem 12, conv A and conv B must be disjoint. ∎

---

[3] A proof of Theorem 12 is given in Steven R. Lay, *Convex Sets and Their Applications* (New York: John Wiley & Sons, 1982; Mineola, NY: Dover Publications, 2007), pp. 34–39.

EXAMPLE 8   Let

$$\mathbf{a}_1 = \begin{bmatrix} 2 \\ 1 \\ 1 \end{bmatrix}, \quad \mathbf{a}_2 = \begin{bmatrix} -3 \\ 2 \\ 1 \end{bmatrix}, \quad \mathbf{a}_3 = \begin{bmatrix} 3 \\ 4 \\ 0 \end{bmatrix}, \quad \mathbf{b}_1 = \begin{bmatrix} 1 \\ 0 \\ 2 \end{bmatrix}, \quad \text{and} \quad \mathbf{b}_2 = \begin{bmatrix} 2 \\ -1 \\ 5 \end{bmatrix},$$

and let $A = \{\mathbf{a}_1, \mathbf{a}_2, \mathbf{a}_3\}$ and $B = \{\mathbf{b}_1, \mathbf{b}_2\}$. Show that the hyperplane $H = [f : 5]$, where $f(x_1, x_2, x_3) = 2x_1 - 3x_2 + x_3$, does not separate A and B. Is there a hyperplane parallel to H that does separate A and B? Do the convex hulls of A and B intersect?

SOLUTION   Evaluate the linear functional f at each of the points in A and B:

$$f(\mathbf{a}_1) = 2, \quad f(\mathbf{a}_2) = -11, \quad f(\mathbf{a}_3) = -6, \quad f(\mathbf{b}_1) = 4, \quad \text{and} \quad f(\mathbf{b}_2) = 12$$

Since $f(\mathbf{b}_1) = 4$ is less than 5 and $f(\mathbf{b}_2) = 12$ is greater than 5, points of B lie on both sides of $H = [f : 5]$ and so H does not separate A and B.

Since $f(A) < 3$ and $f(B) > 3$, the parallel hyperplane $[f : 3]$ strictly separates A and B. By Theorem 13, (conv A) $\cap$ (conv B) $= \varnothing$.

Caution: If there were no hyperplane parallel to H that strictly separated A and B, this would *not* necessarily imply that their convex hulls intersect. It might be that some other hyperplane not parallel to H would strictly separate them.   ∎

Practice Problem

Let $\mathbf{p}_1 = \begin{bmatrix} 1 \\ 0 \\ 2 \end{bmatrix}$, $\mathbf{p}_2 = \begin{bmatrix} -1 \\ 2 \\ 1 \end{bmatrix}$, $\mathbf{n}_1 = \begin{bmatrix} 1 \\ 1 \\ -2 \end{bmatrix}$, and $\mathbf{n}_2 = \begin{bmatrix} -2 \\ 1 \\ 3 \end{bmatrix}$; let H_1 be the hyperplane (plane) in $\mathbb{R}^3$ passing through the point $\mathbf{p}_1$ and having normal vector $\mathbf{n}_1$; and let H_2 be the hyperplane passing through the point $\mathbf{p}_2$ and having normal vector $\mathbf{n}_2$. Give an explicit description of $H_1 \cap H_2$ by a formula that shows how to generate all points in $H_1 \cap H_2$.

8.4 Exercises

1. Let L be the line in $\mathbb{R}^2$ through the points $\begin{bmatrix} -1 \\ 4 \end{bmatrix}$ and $\begin{bmatrix} 3 \\ 1 \end{bmatrix}$.
   Find a linear functional f and a real number d such that $L = [f : d]$.

2. Let L be the line in $\mathbb{R}^2$ through the points $\begin{bmatrix} 1 \\ 4 \end{bmatrix}$ and $\begin{bmatrix} -2 \\ -1 \end{bmatrix}$.
   Find a linear functional f and a real number d such that $L = [f : d]$.

In Exercises 3 and 4, determine whether each set is open or closed or neither open nor closed.

3. a. $\{(x, y) : y > 0\}$
   b. $\{(x, y) : x = 2 \text{ and } 1 \le y \le 3\}$
   c. $\{(x, y) : x = 2 \text{ and } 1 < y < 3\}$
   d. $\{(x, y) : xy = 1 \text{ and } x > 0\}$
   e. $\{(x, y) : xy \ge 1 \text{ and } x > 0\}$

4. a. $\{(x, y) : x^2 + y^2 = 1\}$
   b. $\{(x, y) : x^2 + y^2 > 1\}$
   c. $\{(x, y) : x^2 + y^2 \le 1 \text{ and } y > 0\}$
   d. $\{(x, y) : y \ge x^2\}$
   e. $\{(x, y) : y < x^2\}$

In Exercises 5 and 6, determine whether or not each set is compact and whether or not it is convex.

5. Use the sets from Exercise 3.

6. Use the sets from Exercise 4.

In Exercises 7–10, let H be the hyperplane through the listed points. (a) Find a vector **n** that is normal to the hyperplane. (b) Find a linear functional f and a real number d such that $H = [f : d]$.

7. $\begin{bmatrix} 1 \\ 1 \\ 3 \end{bmatrix}, \begin{bmatrix} 2 \\ 4 \\ 1 \end{bmatrix}, \begin{bmatrix} -1 \\ -2 \\ 5 \end{bmatrix}$ 8. $\begin{bmatrix} 1 \\ -2 \\ 1 \end{bmatrix}, \begin{bmatrix} 4 \\ -2 \\ 3 \end{bmatrix}, \begin{bmatrix} 7 \\ -4 \\ 4 \end{bmatrix}$

9. $\begin{bmatrix} 1 \\ 0 \\ 1 \\ 0 \end{bmatrix}, \begin{bmatrix} 2 \\ 3 \\ 1 \\ 0 \end{bmatrix}, \begin{bmatrix} 1 \\ 2 \\ 2 \\ 0 \end{bmatrix}, \begin{bmatrix} 1 \\ 1 \\ 1 \\ 1 \end{bmatrix}$

10. $\begin{bmatrix} 1 \\ 2 \\ 0 \\ 0 \end{bmatrix}, \begin{bmatrix} 2 \\ 2 \\ -1 \\ -3 \end{bmatrix}, \begin{bmatrix} 1 \\ 3 \\ 2 \\ 7 \end{bmatrix}, \begin{bmatrix} 3 \\ 2 \\ -1 \\ -1 \end{bmatrix}$

11. Let $\mathbf{p} = \begin{bmatrix} 1 \\ -3 \\ 1 \\ 2 \end{bmatrix}$, $\mathbf{n} = \begin{bmatrix} 2 \\ 1 \\ 5 \\ -1 \end{bmatrix}$, $\mathbf{v}_1 = \begin{bmatrix} 0 \\ 1 \\ 1 \\ 1 \end{bmatrix}$, $\mathbf{v}_2 = \begin{bmatrix} -2 \\ 0 \\ 1 \\ 3 \end{bmatrix}$,

and $\mathbf{v}_3 = \begin{bmatrix} 1 \\ 4 \\ 0 \\ 4 \end{bmatrix}$, and let H be the hyperplane in $\mathbb{R}^4$ with

normal $\mathbf{n}$ and passing through $\mathbf{p}$. Which of the points $\mathbf{v}_1$, $\mathbf{v}_2$, and $\mathbf{v}_3$ are on the same side of H as the origin, and which are not?

12. Let $\mathbf{a}_1 = \begin{bmatrix} 2 \\ -1 \\ 5 \end{bmatrix}$, $\mathbf{a}_2 = \begin{bmatrix} 3 \\ 1 \\ 3 \end{bmatrix}$, $\mathbf{a}_3 = \begin{bmatrix} -1 \\ 6 \\ 0 \end{bmatrix}$, $\mathbf{b}_1 = \begin{bmatrix} 0 \\ 5 \\ -1 \end{bmatrix}$,

$\mathbf{b}_2 = \begin{bmatrix} 1 \\ -3 \\ -2 \end{bmatrix}$, $\mathbf{b}_3 = \begin{bmatrix} 2 \\ 2 \\ 1 \end{bmatrix}$, and $\mathbf{n} = \begin{bmatrix} 3 \\ 1 \\ -2 \end{bmatrix}$, and let

$A = \{\mathbf{a}_1, \mathbf{a}_2, \mathbf{a}_3\}$ and $B = \{\mathbf{b}_1, \mathbf{b}_2, \mathbf{b}_3\}$. Find a hyperplane H with normal $\mathbf{n}$ that separates A and B. Is there a hyperplane parallel to H that strictly separates A and B?

13. Let $\mathbf{p}_1 = \begin{bmatrix} 2 \\ -3 \\ 1 \\ 2 \end{bmatrix}$, $\mathbf{p}_2 = \begin{bmatrix} 1 \\ 2 \\ -1 \\ 3 \end{bmatrix}$, $\mathbf{n}_1 = \begin{bmatrix} 1 \\ 2 \\ 4 \\ 2 \end{bmatrix}$, and

$\mathbf{n}_2 = \begin{bmatrix} 2 \\ 3 \\ 1 \\ 5 \end{bmatrix}$; let H_1 be the hyperplane in $\mathbb{R}^4$ through $\mathbf{p}_1$ with

normal $\mathbf{n}_1$; and let H_2 be the hyperplane through $\mathbf{p}_2$ with normal $\mathbf{n}_2$. Give an explicit description of $H_1 \cap H_2$. [*Hint:* Find a point $\mathbf{p}$ in $H_1 \cap H_2$ and two linearly independent vectors $\mathbf{v}_1$ and $\mathbf{v}_2$ that span a subspace parallel to the 2-dimensional flat $H_1 \cap H_2$.]

14. Let F_1 and F_2 be 4-dimensional flats in $\mathbb{R}^6$, and suppose that $F_1 \cap F_2 \neq \emptyset$. What are the possible dimensions of $F_1 \cap F_2$?

In Exercises 15–20, write a formula for a linear functional f and specify a number d, so that $[f : d]$ is the hyperplane H described in the exercise.

15. Let A be the 1×4 matrix $\begin{bmatrix} 1 & -3 & 4 & -2 \end{bmatrix}$ and let $b = 5$. Let $H = \{\mathbf{x} \text{ in } \mathbb{R}^4 : A\mathbf{x} = b\}$.

16. Let A be the 1×5 matrix $\begin{bmatrix} 2 & 5 & -3 & 0 & 6 \end{bmatrix}$. Note that Nul A is in $\mathbb{R}^5$. Let $H = $ Nul A.

17. Let H be the plane in $\mathbb{R}^3$ spanned by the rows of $B = \begin{bmatrix} 1 & 3 & 5 \\ 0 & 2 & 4 \end{bmatrix}$. That is, $H = $ Row B. [*Hint:* How is H related to Nul B? See Section 6.1.]

18. Let H be the plane in $\mathbb{R}^3$ spanned by the rows of $B = \begin{bmatrix} 1 & 4 & -5 \\ 0 & -2 & 8 \end{bmatrix}$. That is, $H = $ Row B.

19. Let H be the column space of the matrix $B = \begin{bmatrix} 1 & 0 \\ 4 & 2 \\ -7 & -6 \end{bmatrix}$.

That is, $H = $ Col B. [*Hint:* How is Col B related to Nul B^T? See Section 6.1.]

20. Let H be the column space of the matrix $B = \begin{bmatrix} 1 & 0 \\ 5 & 2 \\ -4 & -4 \end{bmatrix}$.

That is, $H = $ Col B.

In Exercises 21–28, mark each statement True or False (**T/F**). Justify each answer.

21. (**T/F**) A linear transformation from $\mathbb{R}$ to $\mathbb{R}^n$ is called a linear functional.

22. (**T/F**) If d is a real number and f is a nonzero linear functional defined on $\mathbb{R}^n$, then $[f : d]$ is a hyperplane in $\mathbb{R}^n$.

23. (**T/F**) If f is a linear functional defined on $\mathbb{R}^n$, then there exists a real number k such that $f(\mathbf{x}) = k\mathbf{x}$ for all $\mathbf{x}$ in $\mathbb{R}^n$.

24. (**T/F**) Given any vector $\mathbf{n}$ and any real number d, the set $\{\mathbf{x} : \mathbf{n} \cdot \mathbf{x} = d\}$ is a hyperplane.

25. (**T/F**) If a hyperplane strictly separates sets A and B, then $A \cap B = \emptyset$.

26. (**T/F**) If A and B are nonempty disjoint sets such that A is compact and B is closed, then there exists a hyperplane that strictly separates A and B.

27. (**T/F**) If A and B are closed convex sets and $A \cap B = \emptyset$, then there exists a hyperplane that strictly separates A and B.

28. (**T/F**) If there exists a hyperplane H such that H does not strictly separate two sets A and B, then (conv A) $\cap$ (conv B) $\neq \emptyset$.

29. Let $\mathbf{v}_1 = \begin{bmatrix} 1 \\ 1 \end{bmatrix}$, $\mathbf{v}_2 = \begin{bmatrix} 3 \\ 0 \end{bmatrix}$, $\mathbf{v}_3 = \begin{bmatrix} 5 \\ 3 \end{bmatrix}$, and $\mathbf{p} = \begin{bmatrix} 4 \\ 1 \end{bmatrix}$. Find

a hyperplane $[f : d]$ (in this case, a line) that strictly separates $\mathbf{p}$ from conv $\{\mathbf{v}_1, \mathbf{v}_2, \mathbf{v}_3\}$.

30. Repeat Exercise 29 for $\mathbf{v}_1 = \begin{bmatrix} 1 \\ 2 \end{bmatrix}$, $\mathbf{v}_2 = \begin{bmatrix} 5 \\ 1 \end{bmatrix}$, $\mathbf{v}_3 = \begin{bmatrix} 4 \\ 4 \end{bmatrix}$,

and $\mathbf{p} = \begin{bmatrix} 2 \\ 3 \end{bmatrix}$.

31. Let $\mathbf{p} = \begin{bmatrix} 4 \\ 1 \end{bmatrix}$, let $A = \{\mathbf{x} : \|\mathbf{x}\| \leq 3\}$, and let $B = \{\mathbf{x} : \|\mathbf{x} - \mathbf{p}\| \leq 1\}$. Find a hyperplane $[f : d]$ that strictly separates A and B. [*Hint:* Note that the point $\mathbf{v} = .75\mathbf{p}$ is neither in A nor in B].

32. Let $\mathbf{p} = \begin{bmatrix} 6 \\ 1 \end{bmatrix}$, $\mathbf{q} = \begin{bmatrix} 2 \\ 3 \end{bmatrix}$, $A = \{\mathbf{x} : \|\mathbf{x} - \mathbf{p}\| \le 1\}$, and $B = \{\mathbf{x} : \|\mathbf{x} - \mathbf{q}\| \le 3\}$. Find a hyperplane $[f:d]$ that strictly separates A and B.

33. Give an example of a closed subset S of $\mathbb{R}^2$ such that conv S is not closed.

34. Give an example of a compact set A and a closed set B in $\mathbb{R}^2$ such that $(\text{conv } A) \cap (\text{conv } B) = \varnothing$ but A and B cannot be strictly separated by a hyperplane.

35. Prove that the open ball $B(\mathbf{p}, \delta) = \{\mathbf{x} : \|\mathbf{x} - \mathbf{p}\| < \delta\}$ is a convex set. [*Hint:* Use the Triangle Inequality.]

36. Prove that the convex hull of a bounded set is bounded.

Solution to Practice Problem

First, compute $\mathbf{n}_1 \cdot \mathbf{p}_1 = -3$ and $\mathbf{n}_2 \cdot \mathbf{p}_2 = 7$. The hyperplane H_1 is the solution set of the equation $x_1 + x_2 - 2x_3 = -3$, and H_2 is the solution set of the equation $-2x_1 + x_2 + 3x_3 = 7$. Then

$$H_1 \cap H_2 = \{\mathbf{x} : x_1 + x_2 - 2x_3 = -3 \text{ and } -2x_1 + x_2 + 3x_3 = 7\}$$

This is an implicit description of $H_1 \cap H_2$. To find an explicit description, solve the system of equations by row reduction:

$$\begin{bmatrix} 1 & 1 & -2 & -3 \\ -2 & 1 & 3 & 7 \end{bmatrix} \sim \begin{bmatrix} 1 & 0 & -\frac{5}{3} & -\frac{10}{3} \\ 0 & 1 & -\frac{1}{3} & \frac{1}{3} \end{bmatrix}$$

Thus $x_1 = -\frac{10}{3} + \frac{5}{3}x_3, x_2 = \frac{1}{3} + \frac{1}{3}x_3, x_3 = x_3$. Let $\mathbf{p} = \begin{bmatrix} -\frac{10}{3} \\ \frac{1}{3} \\ 0 \end{bmatrix}$ and $\mathbf{v} = \begin{bmatrix} \frac{5}{3} \\ \frac{1}{3} \\ 1 \end{bmatrix}$. The general solution can be written as $\mathbf{x} = \mathbf{p} + x_3\mathbf{v}$. Thus $H_1 \cap H_2$ is the line through $\mathbf{p}$ in the direction of $\mathbf{v}$. Note that $\mathbf{v}$ is orthogonal to both $\mathbf{n}_1$ and $\mathbf{n}_2$.

8.5 Polytopes

This section studies geometric properties of an important class of compact convex sets called polytopes. These sets arise in all sorts of applications, including game theory, linear programming, and more general optimization problems, such as the design of feedback controls for engineering systems.

A **polytope** in $\mathbb{R}^n$ is the convex hull of a finite set of points. In $\mathbb{R}^2$, a polytope is simply a polygon. In $\mathbb{R}^3$, a polytope is called a polyhedron. Important features of a polyhedron are its faces, edges, and vertices. For example, the cube has 6 square faces, 12 edges, and 8 vertices. The following definitions provide terminology for higher dimensions as well as $\mathbb{R}^2$ and $\mathbb{R}^3$. Recall that the dimension of a set in $\mathbb{R}^n$ is the dimension of the smallest flat that contains it. Also, note that a polytope is a special type of compact convex set, because a finite set in $\mathbb{R}^n$ is compact and the convex hull of this set is compact, by the theorem in the topology terms and facts box in Section 8.4.

DEFINITION

Let S be a compact convex subset of $\mathbb{R}^n$. A nonempty subset F of S is called a (proper) **face** of S if $F \ne S$ and there exists a hyperplane $H = [f:d]$ such that $F = S \cap H$ and either $f(S) \le d$ or $f(S) \ge d$. The hyperplane H is called a

supporting hyperplane to S. If the dimension of F is k, then F is called a ***k*-face** of S.

If P is a polytope of dimension k, then P is called a ***k*-polytope**. A 0-face of P is called a **vertex** (plural: **vertices**), a 1-face is an **edge**, and a $(k-1)$-dimensional face is a **facet** of S.

EXAMPLE 1   Suppose S is a cube in $\mathbb{R}^3$. When a plane H is translated through $\mathbb{R}^3$ until it just touches (supports) the cube but does not cut through the interior of the cube, there are three possibilities for $H \cap S$, depending on the orientation of H. (See Figure 1.)

$H \cap S$ may be a 2-dimensional square face (facet) of the cube.

$H \cap S$ may be a 1-dimensional edge of the cube.

$H \cap S$ may be a 0-dimensional vertex of the cube.   ∎

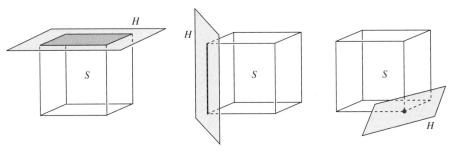

$H \cap S$ is 2-dimensional. $H \cap S$ is 1-dimensional. $H \cap S$ is 0-dimensional.

FIGURE 1

Most applications of polytopes involve the vertices in some way, because they have a special property that is identified in the following definition.

DEFINITION

Let S be a convex set. A point $\mathbf{p}$ in S is called an **extreme point** of S if $\mathbf{p}$ is not in the interior of any line segment that lies in S. More precisely, if $\mathbf{x}, \mathbf{y} \in S$ and $\mathbf{p} \in \overline{\mathbf{xy}}$, then $\mathbf{p} = \mathbf{x}$ or $\mathbf{p} = \mathbf{y}$. The set of all extreme points of S is called the **profile** of S.

A vertex of any compact convex set S is automatically an extreme point of S. This fact is proved during the proof of Theorem 14. In working with a polytope, say $P = \operatorname{conv}\{\mathbf{v}_1, \ldots, \mathbf{v}_k\}$ for $\mathbf{v}_1, \ldots, \mathbf{v}_k$ in $\mathbb{R}^n$, it is usually helpful to know that $\mathbf{v}_1, \ldots, \mathbf{v}_k$ are the extreme points of P. However, such a list might contain extraneous points. For example, some vector $\mathbf{v}_i$ could be the midpoint of an edge of the polytope. Of course, in this case $\mathbf{v}_i$ is not really needed to generate the convex hull. The following definition describes the property of the vertices that will make them all extreme points.

DEFINITION

The set $\{\mathbf{v}_1, \ldots, \mathbf{v}_k\}$ is a **minimal representation** of the polytope P if $P = \operatorname{conv}\{\mathbf{v}_1, \ldots, \mathbf{v}_k\}$ and for each $i = 1, \ldots, k$, $\mathbf{v}_i \notin \operatorname{conv}\{\mathbf{v}_j : j \neq i\}$.

Every polytope has a minimal representation. For if $P = \text{conv}\{\mathbf{v}_1, \ldots, \mathbf{v}_k\}$ and if some $\mathbf{v}_i$ is a convex combination of the other points, then $\mathbf{v}_i$ may be deleted from the set of points without changing the convex hull. This process may be repeated until the minimal representation is left. It can be shown that the minimal representation is unique.

THEOREM 14

Suppose $M = \{\mathbf{v}_1, \ldots, \mathbf{v}_k\}$ is the minimal representation of the polytope P. Then the following three statements are equivalent:

a. $\mathbf{p} \in M$.

b. $\mathbf{p}$ is a vertex of P.

c. $\mathbf{p}$ is an extreme point of P.

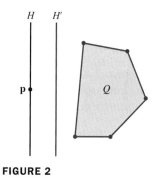

FIGURE 2

PROOF   (a) $\Rightarrow$ (b) Suppose $\mathbf{p} \in M$ and let $Q = \text{conv}\{\mathbf{v} : \mathbf{v} \in M \text{ and } \mathbf{v} \neq \mathbf{p}\}$. It follows from the definition of M that $\mathbf{p} \notin Q$, and since Q is compact, Theorem 13 implies the existence of a hyperplane H' that strictly separates $\{\mathbf{p}\}$ and Q. Let H be the hyperplane through $\mathbf{p}$ parallel to H'. See Figure 2.

Then Q lies in one of the closed half-spaces H^+ bounded by H and so $P \subseteq H^+$. Thus H supports P at $\mathbf{p}$. Furthermore, $\mathbf{p}$ is the only point of P that can lie on H, so $H \cap P = \{\mathbf{p}\}$ and $\mathbf{p}$ is a vertex of P.

(b) $\Rightarrow$ (c) Let $\mathbf{p}$ be a vertex of P. Then there exists a hyperplane $H = [f : d]$ such that $H \cap P = \{\mathbf{p}\}$ and $f(P) \geq d$. If $\mathbf{p}$ were not an extreme point, then there would exist distinct points $\mathbf{x}$ and $\mathbf{y}$ in P such that $\mathbf{p} = (1 - c)\mathbf{x} + c\mathbf{y}$ with $0 < c < 1$. That is,

$$(1 - c)\mathbf{x} = \mathbf{p} - c\mathbf{y} \text{ and } (1 - c)f(\mathbf{x}) = d - cf(\mathbf{y}) \text{ since } f(\mathbf{p}) = d$$

It follows that

$$f(\mathbf{x}) = \frac{d - cf(\mathbf{y})}{1 - c} \geq d \text{ since } f(\mathbf{x}) \geq d$$

But then $d - cf(\mathbf{y}) \geq d(1 - c)$ and $cf(\mathbf{y}) \leq d - d(1 - c) = cd$, so $f(\mathbf{y}) \leq d$. On the other hand, $\mathbf{y} \in P$, so $f(\mathbf{y}) \geq d$. It follows that $f(\mathbf{y}) = d$ and that $\mathbf{y} \in H \cap P$. This contradicts the fact that $\mathbf{p}$ is a vertex. So $\mathbf{p}$ must be an extreme point. (Note that this part of the proof does not depend on P being a polytope. It holds for any compact convex set.)

(c) $\Rightarrow$ (a) It is clear that any extreme point of P must be a member of M. ∎

EXAMPLE 2   Recall that the profile of a set S is the set of extreme points of S. Theorem 14 shows that the profile of a polygon in $\mathbb{R}^2$ is the set of vertices. (See Figure 3.) The profile of a closed ball is its boundary. An open set has no extreme points, so its profile is empty. A closed half-space has no extreme points, so its profile is empty. ∎

FIGURE 3

Exercise 24 asks you to show that a point **p** in a convex set S is an extreme point of S if and only if, when **p** is removed from S, the remaining points still form a convex set. It follows that if S^* is any subset of S such that conv S^* is equal to S, then S^* must contain the profile of S. The sets in Example 2 show that in general S^* may have to be larger than the profile of S. It is true, however, that when S is compact, we may actually take S^* to be the profile of S, as Theorem 15 will show. Thus every nonempty compact convex set S has an extreme point, and the set of all extreme points is the smallest subset of S whose convex hull is equal to S.

THEOREM 15

Let S be a nonempty compact convex set. Then S is the convex hull of its profile (the set of extreme points of S).

PROOF The proof is by induction on the dimension of the set S.[1] ∎

One important application of Theorem 15 is the following theorem. It is one of the key theoretical results in the development of linear programming. Linear functionals are continuous, and continuous functions always attain their maximum and minimum on a compact set. The significance of Theorem 16 is that for compact convex sets, the maximum (and minimum) is actually attained at an extreme point of S.

THEOREM 16

Let f be a linear functional defined on a nonempty compact convex set S. Then there exist extreme points $\hat{\mathbf{v}}$ and $\hat{\mathbf{w}}$ of S such that

$$f(\hat{\mathbf{v}}) = \max_{\mathbf{v} \in S} f(\mathbf{v}) \quad \text{and} \quad f(\hat{\mathbf{w}}) = \min_{\mathbf{v} \in S} f(\mathbf{v})$$

PROOF Assume that f attains its maximum m on S at some point $\mathbf{v}'$ in S. That is, $f(\mathbf{v}') = m$. We wish to show that there exists an extreme point in S with the same property. By Theorem 15, $\mathbf{v}'$ is a convex combination of the extreme points of S. That is, there exist extreme points $\mathbf{v}_1, \ldots, \mathbf{v}_k$ of S and nonnegative $c_1, \ldots, c_k$ such that

$$\mathbf{v}' = c_1 \mathbf{v}_1 + \cdots + c_k \mathbf{v}_k \quad \text{with } c_1 + \cdots + c_k = 1$$

If none of the extreme points of S satisfies $f(\mathbf{v}) = m$, then

$$f(\mathbf{v}_i) < m \quad \text{for } i = 1, \ldots, k$$

since m is the maximum of f on S. But then, because f is linear,

$$\begin{aligned} m = f(\mathbf{v}') &= f(c_1 \mathbf{v}_1 + \cdots + c_k \mathbf{v}_k) \\ &= c_1 f(\mathbf{v}_1) + \cdots + c_k f(\mathbf{v}_k) \\ &< c_1 m + \cdots + c_k m = m(c_1 + \cdots + c_k) = m \end{aligned}$$

This contradiction implies that some extreme point $\hat{\mathbf{v}}$ of S must satisfy $f(\hat{\mathbf{v}}) = m$. The proof for $\hat{\mathbf{w}}$ is similar. ∎

---

[1] The details may be found in Steven R. Lay, *Convex Sets and Their Applications* (New York: John Wiley & Sons, 1982; Mineola, NY: Dover Publications, 2007), p. 43.

EXAMPLE 3   Given points $\mathbf{p}_1 = \begin{bmatrix} -1 \\ 0 \end{bmatrix}$, $\mathbf{p}_2 = \begin{bmatrix} 3 \\ 1 \end{bmatrix}$, and $\mathbf{p}_3 = \begin{bmatrix} 1 \\ 2 \end{bmatrix}$ in $\mathbb{R}^2$, let $S = \text{conv}\{\mathbf{p}_1, \mathbf{p}_2, \mathbf{p}_3\}$. For each linear functional f, find the maximum value m of f on the set S, and find all points $\mathbf{x}$ in S at which $f(\mathbf{x}) = m$.

a. $f_1(x_1, x_2) = x_1 + x_2$ b. $f_2(x_1, x_2) = -3x_1 + x_2$ c. $f_3(x_1, x_2) = x_1 + 2x_2$

SOLUTION   By Theorem 16, the maximum value is attained at one of the extreme points of S. So to find m, evaluate f at each extreme point and select the largest value.

a. $f_1(\mathbf{p}_1) = -1$, $f_1(\mathbf{p}_2) = 4$, and $f_1(\mathbf{p}_3) = 3$, so $m_1 = 4$. Graph the line $f_1(x_1, x_2) = m_1$, that is, $x_1 + x_2 = 4$, and note that $\mathbf{x} = \mathbf{p}_2$ is the only point in S at which $f_1(\mathbf{x}) = 4$. See Figure 4(a).

b. $f_2(\mathbf{p}_1) = 3$, $f_2(\mathbf{p}_2) = -8$, and $f_2(\mathbf{p}_3) = -1$, so $m_2 = 3$. Graph the line $f_2(x_1, x_2) = m_2$, that is, $-3x_1 + x_2 = 3$, and note that $\mathbf{x} = \mathbf{p}_1$ is the only point in S at which $f_2(\mathbf{x}) = 3$. See Figure 4(b).

c. $f_3(\mathbf{p}_1) = -1$, $f_3(\mathbf{p}_2) = 5$, and $f_3(\mathbf{p}_3) = 5$, so $m_3 = 5$. Graph the line $f_3(x_1, x_2) = m_3$, that is, $x_1 + 2x_2 = 5$. Here, f_3 attains its maximum value at $\mathbf{p}_2$, at $\mathbf{p}_3$, and at every point in the convex hull of $\mathbf{p}_2$ and $\mathbf{p}_3$. See Figure 4(c).   ∎

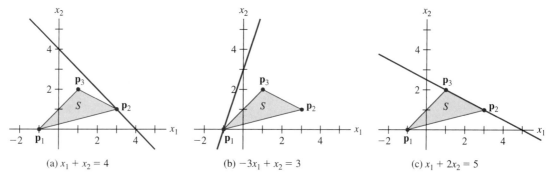

(a) $x_1 + x_2 = 4$ (b) $-3x_1 + x_2 = 3$ (c) $x_1 + 2x_2 = 5$

FIGURE 4

The situation illustrated in Example 3 for $\mathbb{R}^2$ also applies in higher dimensions. The maximum value of a linear functional f on a polytope P occurs at the intersection of a supporting hyperplane and P. This intersection is either a single extreme point of P, or the convex hull of 2 or more extreme points of P. In either case, the intersection is a polytope, and its extreme points form a subset of the extreme points of P.

By definition, a polytope is the convex hull of a finite set of points. This is an explicit representation of the polytope since it identifies points in the set. A polytope may also be represented implicitly as the intersection of a finite number of closed half-spaces. Example 4 illustrates this in $\mathbb{R}^2$.

EXAMPLE 4   Let

$$\mathbf{p}_1 = \begin{bmatrix} 0 \\ 1 \end{bmatrix}, \quad \mathbf{p}_2 = \begin{bmatrix} 1 \\ 0 \end{bmatrix}, \quad \text{and} \quad \mathbf{p}_3 = \begin{bmatrix} 3 \\ 2 \end{bmatrix}$$

in $\mathbb{R}^2$, and let $S = \text{conv}\{\mathbf{p}_1, \mathbf{p}_2, \mathbf{p}_3\}$. Simple algebra shows that the line through $\mathbf{p}_1$ and $\mathbf{p}_2$ is given by $x_1 + x_2 = 1$, and S is on the side of this line where

$$x_1 + x_2 \geq 1 \quad \text{or, equivalently,} \quad -x_1 - x_2 \leq -1.$$

Similarly, the line through $\mathbf{p}_2$ and $\mathbf{p}_3$ is $x_1 - x_2 = 1$, and S is on the side where

$$x_1 - x_2 \leq 1$$

Also, the line through $\mathbf{p}_3$ and $\mathbf{p}_1$ is $-x_1 + 3x_2 = 3$, and S is on the side where

$$-x_1 + 3x_2 \le 3$$

See Figure 5. It follows that S can be described as the solution set of the system of linear inequalities

$$-x_1 - x_2 \le -1$$
$$x_1 - x_2 \le 1$$
$$-x_1 + 3x_2 \le 3$$

This system may be written as $A\mathbf{x} \le \mathbf{b}$, where

$$A = \begin{bmatrix} -1 & -1 \\ 1 & -1 \\ -1 & 3 \end{bmatrix}, \quad \mathbf{x} = \begin{bmatrix} x_1 \\ x_2 \end{bmatrix}, \quad \text{and} \quad \mathbf{b} = \begin{bmatrix} -1 \\ 1 \\ 3 \end{bmatrix}.$$

Note that an inequality between two vectors, such as $A\mathbf{x}$ and $\mathbf{b}$, applies to each of the corresponding coordinates in those vectors. ∎

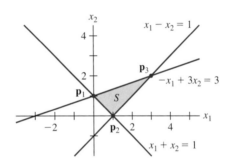

FIGURE 5

In Chapter 9, it will be necessary to replace an implicit description of a polytope by a minimal representation of the polytope, listing all the extreme points of the polytope. In simple cases, a graphical solution is feasible. The following example shows how to handle the situation when several points of interest are too close to identify easily on a graph.

EXAMPLE 5 Let P be the set of points in $\mathbb{R}^2$ that satisfy $A\mathbf{x} \le \mathbf{b}$, where

$$A = \begin{bmatrix} 1 & 3 \\ 1 & 1 \\ 3 & 2 \end{bmatrix} \quad \text{and} \quad \mathbf{b} = \begin{bmatrix} 18 \\ 8 \\ 21 \end{bmatrix}$$

and $\mathbf{x} \ge \mathbf{0}$. Find the minimal representation of P.

SOLUTION The condition $\mathbf{x} \ge \mathbf{0}$ places P in the first quadrant of $\mathbb{R}^2$, a typical condition in linear programming problems. The three inequalities in $A\mathbf{x} \le \mathbf{b}$ involve three boundary lines:

(1) $x_1 + 3x_2 = 18$ (2) $x_1 + x_2 = 8$ (3) $3x_1 + 2x_2 = 21$

All three lines have negative slopes, so a general idea of the shape of P is easy to visualize. Even a rough sketch of the graphs of these lines will reveal that $(0, 0)$, $(7, 0)$, and $(0, 6)$ are vertices of the polytope P.

What about the intersections of the lines (1), (2), and (3)? Sometimes it is clear from the graph which intersections to include. But if not, then the following algebraic procedure will work well:

> When an intersection point is found that corresponds to two inequalities, test it in the other inequalities to see whether the point is in the polytope.

The intersection of (1) and (2) is $\mathbf{p}_{12} = (3, 5)$. Both coordinates are nonnegative, so $\mathbf{p}_{12}$ satisfies all inequalities except possibly the third inequality. Test this:

$$3(3) + 2(5) = 19 < 21$$

This intersection point satisfies the inequality for (3), so $\mathbf{p}_{12}$ is in the polytope.

The intersection of (2) and (3) is $\mathbf{p}_{23} = (5, 3)$. This satisfies all inequalities except possibly the inequality for (1). Test this:

$$1(5) + 3(3) = 14 < 18$$

This shows that $\mathbf{p}_{23}$ is in the polytope.

Finally, the intersection of (1) and (3) is $\mathbf{p}_{13} = \left(\frac{27}{7}, \frac{33}{7}\right)$. Test this in the inequality for (2):

$$1\left(\tfrac{27}{7}\right) + 1\left(\tfrac{33}{7}\right) = \tfrac{60}{7} \approx 8.6 > 8$$

Thus $\mathbf{p}_{13}$ does **not** satisfy the second inequality, which shows that $\mathbf{p}_{13}$ is **not** in P. In conclusion, the minimal representation of the polytope P is

$$\left\{ \begin{bmatrix} 0 \\ 0 \end{bmatrix}, \begin{bmatrix} 7 \\ 0 \end{bmatrix}, \begin{bmatrix} 3 \\ 5 \end{bmatrix}, \begin{bmatrix} 5 \\ 3 \end{bmatrix}, \begin{bmatrix} 0 \\ 6 \end{bmatrix} \right\}.$$ ∎

The remainder of this section discusses the construction of two basic polytopes in $\mathbb{R}^3$ (and higher dimensions). The first appears in linear programming problems, the subject of Chapter 9. Both polytopes provide opportunities to visualize $\mathbb{R}^4$ in a remarkable way.

Simplex

A **simplex** is the convex hull of an affinely independent finite set of vectors. To construct a k-dimensional simplex (or k-simplex), proceed as follows:

0-simplex S^0:	a single point $\{\mathbf{v}_1\}$
1-simplex S^1:	$\mathrm{conv}(S^0 \cup \{\mathbf{v}_2\})$, with $\mathbf{v}_2$ not in aff S^0
2-simplex S^2:	$\mathrm{conv}(S^1 \cup \{\mathbf{v}_3\})$, with $\mathbf{v}_3$ not in aff S^1

$$\vdots$$

$$k\text{-simplex } S^k: \quad \mathrm{conv}(S^{k-1} \cup \{\mathbf{v}_{k+1}\}), \text{ with } \mathbf{v}_{k+1} \text{ not in aff } S^{k-1}$$

The simplex S^1 is a line segment. The triangle S^2 comes from choosing a point $\mathbf{v}_3$ that is not in the line containing S^1 and then forming the convex hull with S^1. (See Figure 6.) The tetrahedron S^3 is produced by choosing a point $\mathbf{v}_4$ not in the plane of S^2 and then forming the convex hull with S^2.

Before continuing, consider some of the patterns that are appearing. The triangle S^2 has three edges. Each of these edges is a line segment like S^1. Where do these three line segments come from? One of them is S^1. One of them comes by joining the endpoint $\mathbf{v}_2$ to the new point $\mathbf{v}_3$. The third comes from joining the other endpoint $\mathbf{v}_1$ to $\mathbf{v}_3$. You might say that each endpoint in S^1 is stretched out into a line segment in S^2.

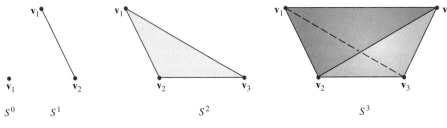

FIGURE 6

The tetrahedron S^3 in Figure 6 has four triangular faces. One of these is the original triangle S^2, and the other three come from stretching the edges of S^2 out to the new point $\mathbf{v}_4$. Notice too that the vertices of S^2 get stretched out into edges in S^3. The other edges in S^3 come from the edges in S^2. This suggests how to "visualize" the four-dimensional S^4.

The construction of S^4, called a pentatope, involves forming the convex hull of S^3 with a point $\mathbf{v}_5$ not in the 3-space of S^3. A complete picture is impossible, of course, but Figure 7 is suggestive: S^4 has five vertices, and any four of the vertices determine a facet in the shape of a tetrahedron. For example, the figure emphasizes the facet with vertices $\mathbf{v}_1$, $\mathbf{v}_2$, $\mathbf{v}_4$, and $\mathbf{v}_5$ and the facet with vertices $\mathbf{v}_2$, $\mathbf{v}_3$, $\mathbf{v}_4$, and $\mathbf{v}_5$. There are five such facets. Figure 7 identifies all ten edges of S^4, and these can be used to visualize the ten triangular faces.

Figure 8 shows another representation of the 4-dimensional simplex S^4. This time the fifth vertex appears "inside" the tetrahedron S^3. The highlighted tetrahedral facets also appear to be "inside" S^3.

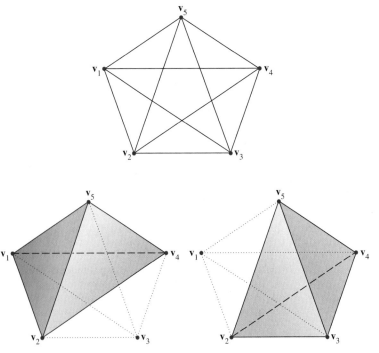

FIGURE 7 The 4-dimensional simplex S^4 projected onto $\mathbb{R}^2$, with two tetrahedral facets emphasized.

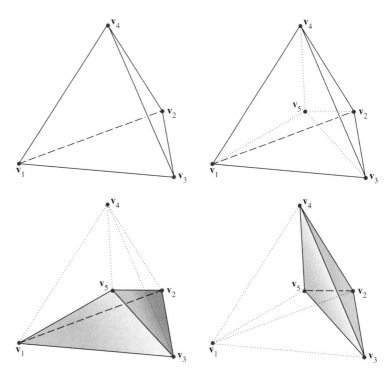

FIGURE 8 The fifth vertex of S^4 is "inside" S^3.

Hypercube

Let $I_i = \overline{\mathbf{0}\mathbf{e}_i}$ be the line segment from the origin $\mathbf{0}$ to the standard basis vector $\mathbf{e}_i$ in $\mathbb{R}^n$. Then for k such that $1 \leq k \leq n$, the vector sum[2]

$$C^k = I_1 + I_2 + \cdots + I_k$$

is called a k-dimensional **hypercube**.

To visualize the construction of C^k, start with the simple cases. The hypercube C^1 is the line segment I_1. If C^1 is translated by $\mathbf{e}_2$, the convex hull of its initial and final positions describes a square C^2. (See Figure 9.) Translating C^2 by $\mathbf{e}_3$ creates the cube C^3. A similar translation of C^3 by the vector $\mathbf{e}_4$ yields the 4-dimensional hypercube C^4.

C^1 C^2 C^3

FIGURE 9 Constructing the cube C^3.

Again, this is hard to visualize, but Figure 10 shows a 2-dimensional projection of C^4. Each of the edges of C^3 is stretched into a square facet of C^4. And each of the square

---

[2] The vector sum of two sets A and B is defined by $A + B = \{\mathbf{c} : \mathbf{c} = \mathbf{a} + \mathbf{b} \text{ for some } \mathbf{a} \in A \text{ and } \mathbf{b} \in B\}$.

FIGURE 10 C^4 projected onto $\mathbb{R}^2$.

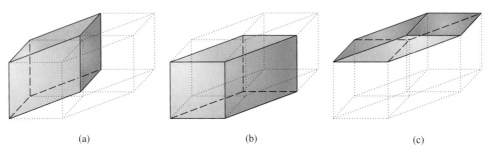

(a)   (b)   (c)

FIGURE 11 Three of the cubic facets of C^4.

faces of C^3 is stretched into a cubic facet of C^4. Figure 11 shows three facets of C^4. Part (a) highlights the cube that comes from the left square face of C^3. Part (b) shows the cube that comes from the front square face of C^3. And part (c) emphasizes the cube that comes from the top square face of C^3.

Figure 12 shows another representation of C^4 in which the translated cube is placed "inside" C^3. This makes it easier to visualize the cubic facets of C^4, since there is less distortion.

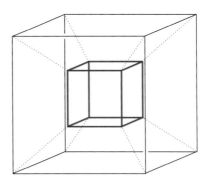

FIGURE 12 The translated image of C^3 is placed "inside" C^3 to obtain C^4.

Altogether, the 4-dimensional cube C^4 has eight cubic facets. Two come from the original and translated images of C^3, and six come from the square faces of C^3 that are stretched into cubes. The square 2-dimensional faces of C^4 come from the square faces of C^3 and its translate, and the edges of C^3 that are stretched into squares. Thus there are $2 \times 6 + 12 = 24$ square faces. To count the edges, take 2 times the number of edges in C^3 and add the number of vertices in C^3. This makes $2 \times 12 + 8 = 32$ edges in C^4. The vertices in C^4 all come from C^3 and its translate, so there are $2 \times 8 = 16$ vertices.

One of the truly remarkable results in the study of polytopes is the following formula, first proved by Leonard Euler (1707–1783). It establishes a simple relationship between the number of faces of different dimensions in a polytope. To simplify the statement of the formula, let $f_k(P)$ denote the number of k-dimensional faces of an n-dimensional polytope P.[3]

$$\text{Euler's formula:} \qquad \sum_{k=0}^{n-1}(-1)^k f_k(P) = 1 + (-1)^{n-1}$$

In particular, when $n = 3$, $v - e + f = 2$, where v, e, and f denote the number of vertices, edges, and facets (respectively) of P.

Practice Problem

Find the minimal representation of the polytope P defined by the inequalities $A\mathbf{x} \le \mathbf{b}$ and $\mathbf{x} \ge \mathbf{0}$, when $A = \begin{bmatrix} 1 & 3 \\ 1 & 2 \\ 2 & 1 \end{bmatrix}$ and $\mathbf{b} = \begin{bmatrix} 12 \\ 9 \\ 12 \end{bmatrix}$.

8.5 Exercises

1. Given points $\mathbf{p}_1 = \begin{bmatrix} 1 \\ 0 \end{bmatrix}$, $\mathbf{p}_2 = \begin{bmatrix} 2 \\ 3 \end{bmatrix}$, and $\mathbf{p}_3 = \begin{bmatrix} -1 \\ 2 \end{bmatrix}$ in $\mathbb{R}^2$, let $S = \text{conv}\{\mathbf{p}_1, \mathbf{p}_2, \mathbf{p}_3\}$. For each linear functional f, find the maximum value m of f on the set S, and find all points $\mathbf{x}$ in S at which $f(\mathbf{x}) = m$.
  a. $f(x_1, x_2) = x_1 - x_2$ b. $f(x_1, x_2) = x_1 + x_2$
  c. $f(x_1, x_2) = -3x_1 + x_2$

2. Given points $\mathbf{p}_1 = \begin{bmatrix} 0 \\ -1 \end{bmatrix}$, $\mathbf{p}_2 = \begin{bmatrix} 2 \\ 1 \end{bmatrix}$, and $\mathbf{p}_3 = \begin{bmatrix} 1 \\ 2 \end{bmatrix}$ in $\mathbb{R}^2$, let $S = \text{conv}\{\mathbf{p}_1, \mathbf{p}_2, \mathbf{p}_3\}$. For each linear functional f, find the maximum value m of f on the set S, and find all points $\mathbf{x}$ in S at which $f(\mathbf{x}) = m$.
  a. $f(x_1, x_2) = x_1 + x_2$ b. $f(x_1, x_2) = x_1 - x_2$
  c. $f(x_1, x_2) = -2x_1 + x_2$

3. Repeat Exercise 1 where m is the *minimum* value of f on S instead of the maximum value.

4. Repeat Exercise 2 where m is the *minimum* value of f on S instead of the maximum value.

In Exercises 5–8, find the minimal representation of the polytope defined by the inequalities $A\mathbf{x} \le \mathbf{b}$ and $\mathbf{x} \ge \mathbf{0}$.

5. $A = \begin{bmatrix} 1 & 2 \\ 3 & 1 \end{bmatrix}$, $\mathbf{b} = \begin{bmatrix} 10 \\ 15 \end{bmatrix}$

6. $A = \begin{bmatrix} 2 & 3 \\ 4 & 1 \end{bmatrix}$, $\mathbf{b} = \begin{bmatrix} 18 \\ 16 \end{bmatrix}$

7. $A = \begin{bmatrix} 1 & 3 \\ 1 & 1 \\ 4 & 1 \end{bmatrix}$, $\mathbf{b} = \begin{bmatrix} 18 \\ 10 \\ 28 \end{bmatrix}$

8. $A = \begin{bmatrix} 2 & 1 \\ 1 & 1 \\ 1 & 2 \end{bmatrix}$, $\mathbf{b} = \begin{bmatrix} 8 \\ 6 \\ 7 \end{bmatrix}$

9. Let $S = \{(x, y) : x^2 + (y - 1)^2 \le 1\} \cup \{(3, 0)\}$. Is the origin an extreme point of conv S? Is the origin a vertex of conv S?

10. Find an example of a closed convex set S in $\mathbb{R}^2$ such that its profile P is nonempty but conv $P \ne S$.

11. Find an example of a bounded convex set S in $\mathbb{R}^2$ such that its profile P is nonempty but conv $P \ne S$.

12. a. Determine the number of k-faces of the 5-dimensional simplex S^5 for $k = 0, 1, \ldots, 4$. Verify that your answer satisfies Euler's formula.

  b. Make a chart of the values of $f_k(S^n)$ for $n = 1, \ldots, 5$ and $k = 0, 1, \ldots, 4$. Can you see a pattern? Guess a general formula for $f_k(S^n)$.

13. a. Determine the number of k-faces of the 5-dimensional hypercube C^5 for $k = 0, 1, \ldots, 4$. Verify that your answer satisfies Euler's formula.

  b. Make a chart of the values of $f_k(C^n)$ for $n = 1, \ldots, 5$ and $k = 0, 1, \ldots, 4$. Can you see a pattern? Guess a general formula for $f_k(C^n)$.

14. Suppose $\mathbf{v}_1, \ldots, \mathbf{v}_k$ are linearly independent vectors in $\mathbb{R}^n$ ($1 \le k \le n$). Then the set $X^k = \text{conv}\{\pm\mathbf{v}_1, \ldots, \pm\mathbf{v}_k\}$ is called a ***k*-crosspolytope**.
  a. Sketch X^1 and X^2.

---

[3] A proof when $n = 3$ is presented in Steven R. Lay, *Convex Sets and Their Applications* (New York: John Wiley & Sons, 1982; Mineola, NY: Dover Publications, 2007), p. 131.

b. Determine the number of k-faces of the 3-dimensional crosspolytope X^3 for $k = 0, 1, 2$. What is another name for X^3?

c. Determine the number of k-faces of the 4-dimensional crosspolytope X^4 for $k = 0, 1, 2, 3$. Verify that your answer satisfies Euler's formula.

d. Find a formula for $f_k(X^n)$, the number of k-faces of X^n, for $0 \leq k \leq n - 1$.

15. A **k-pyramid** P^k is the convex hull of a $(k - 1)$-polytope Q and a point $\mathbf{x} \notin \text{aff } Q$. Find a formula for each of the following in terms of $f_j(Q), j = 0, \ldots, n - 1$.

a. The number of vertices of P^n: $f_0(P^n)$.

b. The number of k-faces of P^n: $f_k(P^n)$, for $1 \leq k \leq n - 2$.

c. The number of $(n - 1)$-dimensional facets of P^n: $f_{n-1}(P^n)$.

In Exercises 16–23, mark each statement True or False **(T/F)**. Justify each answer.

16. **(T/F)** A polytope is the convex hull of a finite set of points.

17. **(T/F)** A cube in $\mathbb{R}^3$ has exactly five facets.

18. **(T/F)** Let $\mathbf{p}$ be an extreme point of a convex set S. If $\mathbf{u}, \mathbf{v} \in S$, $\mathbf{p} \in \overline{\mathbf{uv}}$, and $\mathbf{p} \neq \mathbf{u}$, then $\mathbf{p} = \mathbf{v}$.

19. **(T/F)** A point $\mathbf{p}$ is an extreme point of a polytope P if and only if $\mathbf{p}$ is a vertex of P.

20. **(T/F)** If S is a nonempty convex subset of $\mathbb{R}^n$, then S is the convex hull of its profile.

21. **(T/F)** If S is a nonempty compact convex set and a linear functional attains its maximum at a point $\mathbf{p}$, then $\mathbf{p}$ is an extreme point of S.

22. **(T/F)** The 4-dimensional simplex S^4 has exactly five facets, each of which is a 3-dimensional tetrahedron.

23. **(T/F)** A 2-dimensional polytope always has the same number of vertices and edges.

24. Let $\mathbf{v}$ be an element of the convex set S. Prove that $\mathbf{v}$ is an extreme point of S if and only if the set $\{\mathbf{x} \in S : \mathbf{x} \neq \mathbf{v}\}$ is convex.

25. If $c \in \mathbb{R}$ and S is a set, define $cS = \{c\mathbf{x} : \mathbf{x} \in S\}$. Let S be a convex set and suppose $c > 0$ and $d > 0$. Prove that $cS + dS = (c + d)S$.

26. Find an example to show that the convexity of S is necessary in Exercise 25.

27. If A and B are convex sets, prove that $A + B$ is convex.

28. A polyhedron (3-polytope) is called **regular** if all its facets are congruent regular polygons and all the angles at the vertices are equal. Supply the details in the following proof that there are only five regular polyhedra.

a. Suppose that a regular polyhedron has r facets, each of which is a k-sided regular polygon, and that s edges meet at each vertex. Letting v and e denote the numbers of vertices and edges in the polyhedron, explain why $kr = 2e$ and $sv = 2e$.

b. Use Euler's formula to show that $\dfrac{1}{s} + \dfrac{1}{k} = \dfrac{1}{2} + \dfrac{1}{e}$.

c. Find all the integral solutions of the equation in part (b) that satisfy the geometric constraints of the problem. (How small can k and s be?)

For your information, the five regular polyhedra are the tetrahedron (4, 6, 4), the cube (8, 12, 6), the octahedron (6, 12, 8), the dodecahedron (20, 30, 12), and the icosahedron (12, 30, 20). (The numbers in parentheses indicate the numbers of vertices, edges, and faces, respectively.)

Solution to Practice Problem

The matrix inequality $A\mathbf{x} \leq \mathbf{b}$ yields the following system of inequalities:

$$\text{(a) } x_1 + 3x_2 \leq 12$$
$$\text{(b) } x_1 + 2x_2 \leq 9$$
$$\text{(c) } 2x_1 + x_2 \leq 12$$

The condition $\mathbf{x} \geq \mathbf{0}$, places the polytope in the first quadrant of the plane. One vertex is $(0, 0)$. The x_1-intercepts of the three lines (when $x_2 = 0$) are 12, 9, and 6, so $(6, 0)$ is a vertex. The x_2-intercepts of the three lines (when $x_1 = 0$) are 4, 4.5, and 12, so $(0, 4)$ is a vertex.

How do the three boundary lines intersect for positive values of x_1 and x_2? The intersection of (a) and (b) is at $\mathbf{p}_{ab} = (3, 3)$. Testing $\mathbf{p}_{ab}$ in (c) gives $2(3) + 1(3) = 9 < 12$, so $\mathbf{p}_{ab}$ is in P. The intersection of (b) and (c) is at $\mathbf{p}_{bc} = (5, 2)$. Testing $\mathbf{p}_{bc}$ in (a) gives $1(5) + 3(2) = 11 < 12$, so $\mathbf{p}_{bc}$ is in P. The intersection of (a) and (c) is at $\mathbf{p}_{ac} = (4.8, 2.4)$. Testing $\mathbf{p}_{ac}$ in (b) gives $1(4.8) + 2(2.4) = 9.6 > 9$. So $\mathbf{p}_{ac}$ is not in P.

Finally, the five vertices (extreme points) of the polytope are $(0, 0)$, $(6, 0)$, $(5, 2)$ $(3, 3)$, and $(0, 4)$. These points form the minimal representation of P. This is displayed graphically in Figure 13.

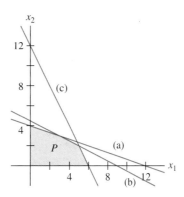

FIGURE 13

8.6 Curves and Surfaces

For thousands of years, builders used long thin strips of wood to create the hull of a boat. In more recent times, designers used long, flexible metal strips to lay out the surfaces of cars and airplanes. Weights and pegs shaped the strips into smooth curves called *natural cubic splines*. The curve between two successive control points (pegs or weights) has a parametric representation using cubic polynomials. Unfortunately, such curves have the property that moving one control point affects the shape of the entire curve, because of physical forces that the pegs and weights exert on the strip. Design engineers had long wanted local control of the curve—in which movement of one control point would affect only a small portion of the curve. In 1962, a French automotive engineer, Pierre Bézier, solved this problem by adding extra control points and using a class of curves now called by his name.

Bézier Curves

The curves described in this section play an important role in computer graphics as well as engineering. For example, they are used in Adobe Illustrator and Macromedia Freehand, and in application programming languages such as OpenGL. These curves permit a program to store exact information about curved segments and surfaces in a relatively small number of control points. All graphics commands for the segments and surfaces have only to be computed for the control points. The special structure of these curves also speeds up other calculations in the "graphics pipeline" that creates the final display on the viewing screen.

Exercises in Section 8.3 introduced quadratic Bézier curves and showed one method for constructing Bézier curves of higher degree. The discussion here focuses on quadratic and cubic Bézier curves, which are determined by three or four control points, denoted by $\mathbf{p}_0$, $\mathbf{p}_1$, $\mathbf{p}_2$, and $\mathbf{p}_3$. These points can be in $\mathbb{R}^2$ or $\mathbb{R}^3$, or they can be represented by homogeneous forms in $\mathbb{R}^3$ or $\mathbb{R}^4$. The standard parametric descriptions of these curves, for $0 \le t \le 1$, are

$$\mathbf{w}(t) = (1-t)^2\mathbf{p}_0 + 2t(1-t)\mathbf{p}_1 + t^2\mathbf{p}_2 \tag{1}$$

$$\mathbf{x}(t) = (1-t)^3\mathbf{p}_0 + 3t(1-t)^2\mathbf{p}_1 + 3t^2(1-t)\mathbf{p}_2 + t^3\mathbf{p}_3 \tag{2}$$

Figure 1 shows two typical curves. Usually, the curves pass through only the initial and terminal control points, but a Bézier curve is always in the convex hull of its control points. (See Exercises 25–28 in Section 8.3.)

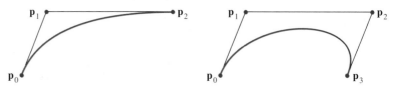

FIGURE 1 Quadratic and cubic Bézier curves.

Bézier curves are useful in computer graphics because their essential properties are preserved under the action of linear transformations and translations. For instance, if A is a matrix of appropriate size, then from the linearity of matrix multiplication, for $0 \le t \le 1$,

$$\begin{aligned} A\mathbf{x}(t) &= A[(1-t)^3\mathbf{p}_0 + 3t(1-t)^2\mathbf{p}_1 + 3t^2(1-t)\mathbf{p}_2 + t^3\mathbf{p}_3] \\ &= (1-t)^3 A\mathbf{p}_0 + 3t(1-t)^2 A\mathbf{p}_1 + 3t^2(1-t)A\mathbf{p}_2 + t^3 A\mathbf{p}_3 \end{aligned}$$

The new control points are $A\mathbf{p}_0, \ldots, A\mathbf{p}_3$. Translations of Bézier curves are considered in Exercise 1.

The curves in Figure 1 suggest that the control points determine the tangent lines to the curves at the initial and terminal control points. Recall from calculus that for any parametric curve, say $\mathbf{y}(t)$, the direction of the tangent line to the curve at a point $\mathbf{y}(t)$ is given by the derivative $\mathbf{y}'(t)$, called the **tangent vector** of the curve. (This derivative is computed entry by entry.)

EXAMPLE 1 Determine how the tangent vector of the quadratic Bézier curve $\mathbf{w}(t)$ is related to the control points of the curve, at $t = 0$ and $t = 1$.

SOLUTION Write the weights in equation (1) as simple polynomials

$$\mathbf{w}(t) = (1 - 2t + t^2)\mathbf{p}_0 + (2t - 2t^2)\mathbf{p}_1 + t^2\mathbf{p}_2$$

Then, because differentiation is a linear transformation on functions,

$$\mathbf{w}'(t) = (-2 + 2t)\mathbf{p}_0 + (2 - 4t)\mathbf{p}_1 + 2t\mathbf{p}_2$$

So

$$\mathbf{w}'(0) = -2\mathbf{p}_0 + 2\mathbf{p}_1 = 2(\mathbf{p}_1 - \mathbf{p}_0)$$
$$\mathbf{w}'(1) = -2\mathbf{p}_1 + 2\mathbf{p}_2 = 2(\mathbf{p}_2 - \mathbf{p}_1)$$

The tangent vector at $\mathbf{p}_0$, for instance, points from $\mathbf{p}_0$ to $\mathbf{p}_1$, but it is twice as long as the segment from $\mathbf{p}_0$ to $\mathbf{p}_1$. Notice that $\mathbf{w}'(0) = \mathbf{0}$ when $\mathbf{p}_1 = \mathbf{p}_0$. In this case, $\mathbf{w}(t) = (1 - t^2)\mathbf{p}_1 + t^2\mathbf{p}_2$, and the graph of $\mathbf{w}(t)$ is the line segment from $\mathbf{p}_1$ to $\mathbf{p}_2$. ∎

Connecting Two Bézier Curves

Two basic Bézier curves can be joined end to end, with the terminal point of the first curve $\mathbf{x}(t)$ being the initial point $\mathbf{p}_2$ of the second curve $\mathbf{y}(t)$. The combined curve is said to have G^0 *geometric continuity* (at $\mathbf{p}_2$) because the two segments join at $\mathbf{p}_2$. If the tangent line to curve 1 at $\mathbf{p}_2$ has a different direction than the tangent line to curve 2, then a "corner," or abrupt change of direction, may be apparent at $\mathbf{p}_2$. See Figure 2.

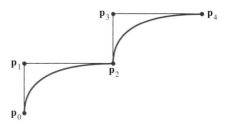

FIGURE 2 G^0 continuity at $\mathbf{p}_2$.

To avoid a sharp bend, it usually suffices to adjust the curves to have what is called G^1 *geometric continuity*, where both tangent vectors at $\mathbf{p}_2$ point in the same direction. That is, the derivatives $\mathbf{x}'(1)$ and $\mathbf{y}'(0)$ point in the same direction, even though their magnitudes may be different. When the tangent vectors are actually equal at $\mathbf{p}_2$, the tangent vector is continuous at $\mathbf{p}_2$, and the combined curve is said to have C^1 continuity, or C^1 *parametric continuity*. Figure 3 shows G^1 continuity in (a) and C^1 continuity in (b).

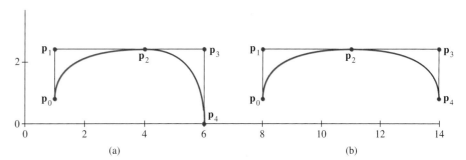

FIGURE 3 (a) G^1 continuity and (b) C^1 continuity.

EXAMPLE 2 Let $\mathbf{x}(t)$ and $\mathbf{y}(t)$ determine two quadratic Bézier curves, with control points $\{\mathbf{p}_0, \mathbf{p}_1, \mathbf{p}_2\}$ and $\{\mathbf{p}_2, \mathbf{p}_3, \mathbf{p}_4\}$, respectively. The curves are joined at $\mathbf{p}_2 = \mathbf{x}(1) = \mathbf{y}(0)$.

a. Suppose the combined curve has G^1 continuity (at $\mathbf{p}_2$). What algebraic restriction does this condition impose on the control points? Express this restriction in geometric language.

b. Repeat part (a) for C^1 continuity.

SOLUTION

a. From Example 1, $\mathbf{x}'(1) = 2(\mathbf{p}_2 - \mathbf{p}_1)$. Also, using the control points for $\mathbf{y}(t)$ in place of $\mathbf{w}(t)$, Example 1 shows that $\mathbf{y}'(0) = 2(\mathbf{p}_3 - \mathbf{p}_2)$. G^1 continuity means that $\mathbf{y}'(0) = k\mathbf{x}'(1)$ for some positive constant k. Equivalently,

$$\mathbf{p}_3 - \mathbf{p}_2 = k(\mathbf{p}_2 - \mathbf{p}_1), \quad \text{with } k > 0 \tag{3}$$

Geometrically, (3) implies that $\mathbf{p}_2$ lies on the line segment from $\mathbf{p}_1$ to $\mathbf{p}_3$. To prove this, let $t = (k + 1)^{-1}$, and note that $0 < t < 1$. Solve for k to obtain $k = (1 - t)/t$. When this expression is used for k in (3), a rearrangement shows that $\mathbf{p}_2 = (1 - t)\mathbf{p}_1 + t\mathbf{p}_3$, which verifies the assertion about $\mathbf{p}_2$.

b. C^1 continuity means that $\mathbf{y}'(0) = \mathbf{x}'(1)$. Thus $2(\mathbf{p}_3 - \mathbf{p}_2) = 2(\mathbf{p}_2 - \mathbf{p}_1)$, so $\mathbf{p}_3 - \mathbf{p}_2 = \mathbf{p}_2 - \mathbf{p}_1$, and $\mathbf{p}_2 = (\mathbf{p}_1 + \mathbf{p}_3)/2$. Geometrically, $\mathbf{p}_2$ is the midpoint of the line segment from $\mathbf{p}_1$ to $\mathbf{p}_3$. See Figure 3. ∎

Figure 4 shows C^1 continuity for two cubic Bézier curves. Notice how the point joining the two segments lies in the middle of the line segment between the adjacent control points.

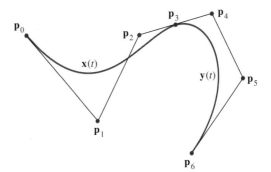

FIGURE 4 Two cubic Bézier curves.

Two curves have C^2 (parametric) continuity when they have C^1 continuity and the *second* derivatives $\mathbf{x}''(1)$ and $\mathbf{y}''(0)$ are equal. This is possible for cubic Bézier curves, but it severely limits the positions of the control points. Another class of cubic curves, called *B-splines*, always have C^2 continuity because each pair of curves share three control points rather than one. Graphics figures using B-splines have more control points and consequently require more computations. Some exercises for this section examine these curves.

Surprisingly, if $\mathbf{x}(t)$ and $\mathbf{y}(t)$ join at $\mathbf{p}_3$, the apparent smoothness of the curve at $\mathbf{p}_3$ is usually the same for both G^1 continuity and C^1 continuity. This is because the magnitude of $\mathbf{x}'(t)$ is not related to the physical shape of the curve. The magnitude reflects only the mathematical parameterization of the curve. For instance, if a new vector function $\mathbf{z}(t)$ equals $\mathbf{x}(2t)$, then the point $\mathbf{z}(t)$ traverses the curve from $\mathbf{p}_0$ to $\mathbf{p}_3$ twice as fast as the original version, because $2t$ reaches 1 when t is .5. But, by the chain rule of calculus, $\mathbf{z}'(t) = 2 \cdot \mathbf{x}'(2t)$, so the tangent vector to $\mathbf{z}(t)$ at $\mathbf{p}_3$ is twice the tangent vector to $\mathbf{x}(t)$ at $\mathbf{p}_3$.

In practice, many simple Bézier curves are joined to create graphics objects. Typesetting programs provide one important application, because many letters in a type font involve curved segments. Each letter in a PostScript® font, for example, is stored as a set of control points, along with information on how to construct the "outline" of the letter using line segments and Bézier curves. Enlarging such a letter basically requires multiplying the coordinates of each control point by one constant scale factor. Once the outline of the letter has been computed, the appropriate solid parts of the letter are filled in. Figure 5 illustrates this for a character in a PostScript font. Note the control points.

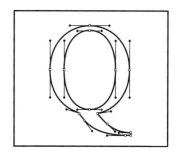

FIGURE 5 A PostScript character.

Matrix Equations for Bézier Curves

Since a Bézier curve is a linear combination of control points using polynomials as weights, the formula for $\mathbf{x}(t)$ may be written as

$$
\mathbf{x}(t) = \begin{bmatrix} \mathbf{p}_0 & \mathbf{p}_1 & \mathbf{p}_2 & \mathbf{p}_3 \end{bmatrix} \begin{bmatrix} (1-t)^3 \\ 3t(1-t)^2 \\ 3t^2(1-t) \\ t^3 \end{bmatrix}
$$

$$
= \begin{bmatrix} \mathbf{p}_0 & \mathbf{p}_1 & \mathbf{p}_2 & \mathbf{p}_3 \end{bmatrix} \begin{bmatrix} 1 - 3t + 3t^2 - t^3 \\ 3t - 6t^2 + 3t^3 \\ 3t^2 - 3t^3 \\ t^3 \end{bmatrix}
$$

$$
= \begin{bmatrix} \mathbf{p}_0 & \mathbf{p}_1 & \mathbf{p}_2 & \mathbf{p}_3 \end{bmatrix} \begin{bmatrix} 1 & -3 & 3 & -1 \\ 0 & 3 & -6 & 3 \\ 0 & 0 & 3 & -3 \\ 0 & 0 & 0 & 1 \end{bmatrix} \begin{bmatrix} 1 \\ t \\ t^2 \\ t^3 \end{bmatrix}
$$

The matrix whose columns are the four control points is called a **geometry matrix**, G. The 4×4 matrix of polynomial coefficients is the **Bézier basis matrix**, M_B. If $\mathbf{u}(t)$ is the column vector of powers of t, then the Bézier curve is given by

$$
\mathbf{x}(t) = G M_B \mathbf{u}(t) \tag{4}
$$

Other parametric cubic curves in computer graphics are written in this form, too. For instance, if the entries in the matrix M_B are changed appropriately, the resulting curves are B-splines. They are "smoother" than Bézier curves, but they do not pass through any of the control points. A **Hermite** cubic curve arises when the matrix M_B is replaced by a Hermite basis matrix. In this case, the columns of the geometry matrix consist of the starting and ending points of the curves and the tangent vectors to the curves at those points.[1]

The Bézier curve in equation (4) can also be "factored" in another way, to be used in the discussion of Bézier surfaces. For convenience later, the parameter t is replaced

---

[1] The term *basis matrix* comes from the rows of the matrix that list the coefficients of the *blending* polynomials used to define the curve. For a cubic Bézier curve, the four polynomials are $(1-t)^3$, $3t(1-t)^2$, $3t^2(1-t)$, and t^3. They form a basis for the space $\mathbb{P}_3$ of polynomials of degree 3 or less. Each entry in the vector $\mathbf{x}(t)$ is a linear combination of these polynomials. The weights come from the rows of the geometry matrix G in (4).

by a parameter s:

$$\mathbf{x}(s) = \mathbf{u}(s)^T M_B^T \begin{bmatrix} \mathbf{p}_0 \\ \mathbf{p}_1 \\ \mathbf{p}_2 \\ \mathbf{p}_3 \end{bmatrix} = \begin{bmatrix} 1 & s & s^2 & s^3 \end{bmatrix} \begin{bmatrix} 1 & 0 & 0 & 0 \\ -3 & 3 & 0 & 0 \\ 3 & -6 & 3 & 0 \\ -1 & 3 & -3 & 1 \end{bmatrix} \begin{bmatrix} \mathbf{p}_0 \\ \mathbf{p}_1 \\ \mathbf{p}_2 \\ \mathbf{p}_3 \end{bmatrix}$$

$$= \begin{bmatrix} (1-s)^3 & 3s(1-s)^2 & 3s^2(1-s) & s^3 \end{bmatrix} \begin{bmatrix} \mathbf{p}_0 \\ \mathbf{p}_1 \\ \mathbf{p}_2 \\ \mathbf{p}_3 \end{bmatrix} \tag{5}$$

This formula is not quite the same as the transpose of the product on the right of (4), because $\mathbf{x}(s)$ and the control points appear in (5) without transpose symbols. The matrix of control points in (5) is called a **geometry vector**. This should be viewed as a 4×1 block (partitioned) matrix whose entries are column vectors. The matrix to the left of the geometry vector, in the second part of (5), can be viewed as a block matrix, too, with a scalar in each block. The partitioned matrix multiplication makes sense, because each (vector) entry in the geometry vector can be left-multiplied by a scalar as well as by a matrix. Thus, the column vector $\mathbf{x}(s)$ is represented by (5).

Bézier Surfaces

A 3D bicubic surface patch can be constructed from a set of four Bézier curves. Consider the four geometry matrices

$$\begin{bmatrix} \mathbf{p}_{11} & \mathbf{p}_{12} & \mathbf{p}_{13} & \mathbf{p}_{14} \end{bmatrix}$$
$$\begin{bmatrix} \mathbf{p}_{21} & \mathbf{p}_{22} & \mathbf{p}_{23} & \mathbf{p}_{24} \end{bmatrix}$$
$$\begin{bmatrix} \mathbf{p}_{31} & \mathbf{p}_{32} & \mathbf{p}_{33} & \mathbf{p}_{34} \end{bmatrix}$$
$$\begin{bmatrix} \mathbf{p}_{41} & \mathbf{p}_{42} & \mathbf{p}_{43} & \mathbf{p}_{44} \end{bmatrix}$$

and recall from equation (4) that a Bézier curve is produced when any one of these matrices is multiplied on the right by the following vector of weights:

$$M_B \mathbf{u}(t) = \begin{bmatrix} (1-t)^3 \\ 3t(1-t)^2 \\ 3t^2(1-t) \\ t^3 \end{bmatrix}$$

Let G be the block (partitioned) 4×4 matrix whose entries are the control points $\mathbf{p}_{ij}$ displayed above. Then the following product is a block 4×1 matrix, and each entry is a Bézier curve:

$$GM_B \mathbf{u}(t) = \begin{bmatrix} \mathbf{p}_{11} & \mathbf{p}_{12} & \mathbf{p}_{13} & \mathbf{p}_{14} \\ \mathbf{p}_{21} & \mathbf{p}_{22} & \mathbf{p}_{23} & \mathbf{p}_{24} \\ \mathbf{p}_{31} & \mathbf{p}_{32} & \mathbf{p}_{33} & \mathbf{p}_{34} \\ \mathbf{p}_{41} & \mathbf{p}_{42} & \mathbf{p}_{43} & \mathbf{p}_{44} \end{bmatrix} \begin{bmatrix} (1-t)^3 \\ 3t(1-t)^2 \\ 3t^2(1-t) \\ t^3 \end{bmatrix}$$

In fact,

$$GM_B \mathbf{u}(t) = \begin{bmatrix} (1-t)^3\mathbf{p}_{11} + 3t(1-t)^2\mathbf{p}_{12} + 3t^2(1-t)\mathbf{p}_{13} + t^3\mathbf{p}_{14} \\ (1-t)^3\mathbf{p}_{21} + 3t(1-t)^2\mathbf{p}_{22} + 3t^2(1-t)\mathbf{p}_{23} + t^3\mathbf{p}_{24} \\ (1-t)^3\mathbf{p}_{31} + 3t(1-t)^2\mathbf{p}_{32} + 3t^2(1-t)\mathbf{p}_{33} + t^3\mathbf{p}_{34} \\ (1-t)^3\mathbf{p}_{41} + 3t(1-t)^2\mathbf{p}_{42} + 3t^2(1-t)\mathbf{p}_{43} + t^3\mathbf{p}_{44} \end{bmatrix}$$

Now fix t. Then $GM_B\mathbf{u}(t)$ is a column vector that can be used as a geometry vector in equation (5) for a Bézier curve in another variable s. This observation produces the **Bézier bicubic surface**:

$$\mathbf{x}(s, t) = \mathbf{u}(s)^T M_B^T GM_B\mathbf{u}(t), \quad \text{where } 0 \le s, t \le 1 \tag{6}$$

The formula for $\mathbf{x}(s, t)$ is a linear combination of the sixteen control points. If one imagines that these control points are arranged in a fairly uniform rectangular array, as in Figure 6, then the Bézier surface is controlled by a web of eight Bézier curves, four in the "s-direction" and four in the "t-direction." The surface actually passes through the four control points at its "corners." When it is in the middle of a larger surface, the sixteen-point surface shares its twelve boundary control points with its neighbors.

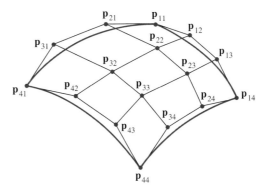

FIGURE 6 Sixteen control points for a Bézier bicubic surface patch.

Approximations to Curves and Surfaces

In CAD programs and in programs used to create realistic computer games, the designer often works at a graphics workstation to compose a "scene" involving various geometric structures. This process requires interaction between the designer and the geometric objects. Each slight repositioning of an object requires new mathematical computations by the graphics program. Bézier curves and surfaces can be useful in this process because they involve fewer control points than objects approximated by many polygons. This dramatically reduces the computation time and speeds up the designer's work.

After the scene composition, however, the final image preparation has different computational demands that are more easily met by objects consisting of flat surfaces and straight edges, such as polyhedra. The designer needs to *render* the scene, by introducing light sources, adding color and texture to surfaces, and simulating reflections from the surfaces.

Computing the direction of a reflected light at a point $\mathbf{p}$ on a surface, for instance, requires knowing the directions of both the incoming light and the *surface normal*—the vector perpendicular to the tangent plane at $\mathbf{p}$. Computing such normal vectors is much easier on a surface composed of, say, tiny flat polygons than on a curved surface whose normal vector changes continuously as $\mathbf{p}$ moves. If $\mathbf{p}_1$, $\mathbf{p}_2$, and $\mathbf{p}_3$ are adjacent vertices of a flat polygon, then the surface normal is just plus or minus the cross product $(\mathbf{p}_2 - \mathbf{p}_1) \times (\mathbf{p}_2 - \mathbf{p}_3)$. When the polygon is small, only one normal vector is needed for rendering the entire polygon. Also, two widely used shading routines, Gouraud shading and Phong shading, both require a surface to be defined by polygons.

As a result of these needs for flat surfaces, the Bézier curves and surfaces from the scene composition stage now are usually approximated by straight line segments and polyhedral surfaces. The basic idea for approximating a Bézier curve or surface is to divide the curve or surface into smaller pieces, with more and more control points.

Recursive Subdivision of Bézier Curves and Surfaces

Figure 7 shows the four control points $\mathbf{p}_0, \ldots, \mathbf{p}_3$ for a Bézier curve, along with control points for two new curves, each coinciding with half of the original curve. The "left" curve begins at $\mathbf{q}_0 = \mathbf{p}_0$ and ends at $\mathbf{q}_3$, at the midpoint of the original curve. The "right" curve begins at $\mathbf{r}_0 = \mathbf{q}_3$ and ends at $\mathbf{r}_3 = \mathbf{p}_3$.

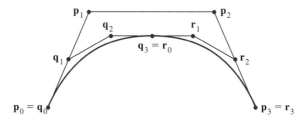

FIGURE 7 Subdivision of a Bézier curve.

Figure 8 shows how the new control points enclose regions that are "thinner" than the region enclosed by the original control points. As the distances between the control points decrease, the control points of each curve segment also move closer to a line segment. This *variation-diminishing property* of Bézier curves depends on the fact that a Bézier curve always lies in the convex hull of the control points.

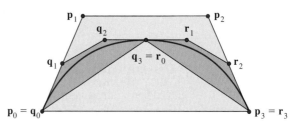

FIGURE 8 Convex hulls of the control points.

The new control points are related to the original control points by simple formulas. Of course, $\mathbf{q}_0 = \mathbf{p}_0$ and $\mathbf{r}_3 = \mathbf{p}_3$. The midpoint of the original curve $\mathbf{x}(t)$ occurs at $\mathbf{x}(.5)$ when $\mathbf{x}(t)$ has the standard parameterization,

$$\mathbf{x}(t) = (1 - 3t + 3t^2 - t^3)\mathbf{p}_0 + (3t - 6t^2 + 3t^3)\mathbf{p}_1 + (3t^2 - 3t^3)\mathbf{p}_2 + t^3\mathbf{p}_3 \qquad (7)$$

for $0 \le t \le 1$. Thus, the new control points $\mathbf{q}_3$ and $\mathbf{r}_0$ are given by

$$\mathbf{q}_3 = \mathbf{r}_0 = \mathbf{x}(.5) = \tfrac{1}{8}(\mathbf{p}_0 + 3\mathbf{p}_1 + 3\mathbf{p}_2 + \mathbf{p}_3) \qquad (8)$$

The formulas for the remaining "interior" control points are also simple, but the derivation of the formulas requires some work involving the tangent vectors of the curves. By definition, the tangent vector to a parameterized curve $\mathbf{x}(t)$ is the derivative $\mathbf{x}'(t)$. This vector shows the direction of the line tangent to the curve at $\mathbf{x}(t)$. For the Bézier curve in (7),

$$\mathbf{x}'(t) = (-3 + 6t - 3t^2)\mathbf{p}_0 + (3 - 12t + 9t^2)\mathbf{p}_1 + (6t - 9t^2)\mathbf{p}_2 + 3t^2\mathbf{p}_3$$

for $0 \le t \le 1$. In particular,

$$\mathbf{x}'(0) = 3(\mathbf{p}_1 - \mathbf{p}_0) \quad \text{and} \quad \mathbf{x}'(1) = 3(\mathbf{p}_3 - \mathbf{p}_2) \qquad (9)$$

Geometrically, $\mathbf{p}_1$ is on the line tangent to the curve at $\mathbf{p}_0$, and $\mathbf{p}_2$ is on the line tangent to the curve at $\mathbf{p}_3$. See Figure 8. Also, from $\mathbf{x}'(t)$, compute

$$\mathbf{x}'(.5) = \tfrac{3}{4}(-\mathbf{p}_0 - \mathbf{p}_1 + \mathbf{p}_2 + \mathbf{p}_3) \tag{10}$$

Let $\mathbf{y}(t)$ be the Bézier curve determined by $\mathbf{q}_0, \ldots, \mathbf{q}_3$, and let $\mathbf{z}(t)$ be the Bézier curve determined by $\mathbf{r}_0, \ldots, \mathbf{r}_3$. Since $\mathbf{y}(t)$ traverses the same path as $\mathbf{x}(t)$ but only gets to $\mathbf{x}(.5)$ as t goes from 0 to 1, $\mathbf{y}(t) = \mathbf{x}(.5t)$ for $0 \leq t \leq 1$. Similarly, since $\mathbf{z}(t)$ starts at $\mathbf{x}(.5)$ when $t = 0$, $\mathbf{z}(t) = \mathbf{x}(.5 + .5t)$ for $0 \leq t \leq 1$. By the chain rule for derivatives,

$$\mathbf{y}'(t) = .5\mathbf{x}'(.5t) \quad \text{and} \quad \mathbf{z}'(t) = .5\mathbf{x}'(.5 + .5t) \quad \text{for } 0 \leq t \leq 1 \tag{11}$$

From (9) with $\mathbf{y}'(0)$ in place of $\mathbf{x}'(0)$, from (11) with $t = 0$, and from (9), the control points for $\mathbf{y}(t)$ satisfy

$$3(\mathbf{q}_1 - \mathbf{q}_0) = \mathbf{y}'(0) = .5\mathbf{x}'(0) = \tfrac{3}{2}(\mathbf{p}_1 - \mathbf{p}_0) \tag{12}$$

From (9) with $\mathbf{y}'(1)$ in place of $\mathbf{x}'(1)$, from (11) with $t = 1$, and from (10),

$$3(\mathbf{q}_3 - \mathbf{q}_2) = \mathbf{y}'(1) = .5\mathbf{x}'(.5) = \tfrac{3}{8}(-\mathbf{p}_0 - \mathbf{p}_1 + \mathbf{p}_2 + \mathbf{p}_3) \tag{13}$$

Equations (8), (9), (10), (12), and (13) can be solved to produce the formulas for $\mathbf{q}_0, \ldots, \mathbf{q}_3$ shown in Exercise 17. Geometrically, the formulas are displayed in Figure 9. The interior control points $\mathbf{q}_1$ and $\mathbf{r}_2$ are the midpoints, respectively, of the segment from $\mathbf{p}_0$ to $\mathbf{p}_1$ and the segment from $\mathbf{p}_2$ to $\mathbf{p}_3$. When the midpoint of the segment from $\mathbf{p}_1$ to $\mathbf{p}_2$ is connected to $\mathbf{q}_1$, the resulting line segment has $\mathbf{q}_2$ in the middle!

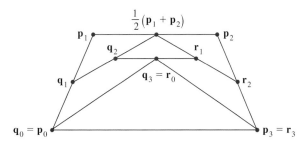

FIGURE 9 Geometric structure of new control points.

This completes one step of the subdivision process. The "recursion" begins, and both new curves are subdivided. The recursion continues to a depth at which all curves are sufficiently straight. Alternatively, at each step the recursion can be "adaptive" and not subdivide one of the two new curves if that curve is sufficiently straight. Once the subdivision completely stops, the endpoints of each curve are joined by line segments, and the scene is ready for the next step in the final image preparation.

A Bézier bicubic surface has the same variation-diminishing property as the Bézier curves that make up each cross-section of the surface, so the process described above can be applied in each cross-section. With the details omitted, here is the basic strategy. Consider the four "parallel" Bézier curves whose parameter is s, and apply the subdivision process to each of them. This produces four sets of eight control points; each set determines a curve as s varies from 0 to 1. As t varies, however, there are eight curves, each with four control points. Apply the subdivision process to each of these sets of four points, creating a total of 64 control points. Adaptive recursion is possible in this setting, too, but there are some subtleties involved.[2]

---

[2] See Foley, van Dam, Feiner, and Hughes, *Computer Graphics—Principles and Practice*, 2nd Ed. (Boston: Addison-Wesley, 1996), pp. 527–528.

Practice Problems

A *spline* usually refers to a curve that passes through specified points. A B-spline, however, usually does not pass through its control points. A single segment has the parametric form

$$\mathbf{x}(t) = \tfrac{1}{6}\big[(1-t)^3\mathbf{p}_0 + (3t^3 - 6t^2 + 4)\mathbf{p}_1 \\ + (-3t^3 + 3t^2 + 3t + 1)\mathbf{p}_2 + t^3\mathbf{p}_3\big] \tag{14}$$

for $0 \le t \le 1$, where $\mathbf{p}_0, \mathbf{p}_1, \mathbf{p}_2$, and $\mathbf{p}_3$ are the control points. When t varies from 0 to 1, $\mathbf{x}(t)$ creates a short curve that lies close to $\overline{\mathbf{p}_1\mathbf{p}_2}$. Basic algebra shows that the B-spline formula can also be written as

$$\mathbf{x}(t) = \tfrac{1}{6}\big[(1-t)^3\mathbf{p}_0 + (3t(1-t)^2 - 3t + 4)\mathbf{p}_1 \\ + (3t^2(1-t) + 3t + 1)\mathbf{p}_2 + t^3\mathbf{p}_3\big] \tag{15}$$

This shows the similarity with the Bézier curve. Except for the $1/6$ factor at the front, the $\mathbf{p}_0$ and $\mathbf{p}_3$ terms are the same. The $\mathbf{p}_1$ component has been increased by $-3t + 4$ and the $\mathbf{p}_2$ component has been increased by $3t + 1$. These components move the curve closer to $\overline{\mathbf{p}_1\mathbf{p}_2}$ than the Bézier curve. The $1/6$ factor is necessary to keep the sum of the coefficients equal to 1. Figure 10 compares a B-spline with a Bézier curve that has the same control points.

FIGURE 10 A B-spline segment and a Bézier curve.

1. Show that the B-spline does not begin at $\mathbf{p}_0$, but $\mathbf{x}(0)$ is in conv $\{\mathbf{p}_0, \mathbf{p}_1, \mathbf{p}_2\}$. Assuming that $\mathbf{p}_0, \mathbf{p}_1$, and $\mathbf{p}_2$ are affinely independent, find the affine coordinates of $\mathbf{x}(0)$ with respect to $\{\mathbf{p}_0, \mathbf{p}_1, \mathbf{p}_2\}$.

2. Show that the B-spline does not end at $\mathbf{p}_3$, but $\mathbf{x}(1)$ is in conv $\{\mathbf{p}_1, \mathbf{p}_2, \mathbf{p}_3\}$. Assuming that $\mathbf{p}_1, \mathbf{p}_2$, and $\mathbf{p}_3$ are affinely independent, find the affine coordinates of $\mathbf{x}(1)$ with respect to $\{\mathbf{p}_1, \mathbf{p}_2, \mathbf{p}_3\}$.

8.6 Exercises

1. Suppose a Bézier curve is translated to $\mathbf{x}(t) + \mathbf{b}$. That is, for $0 \le t \le 1$, the new curve is

$$\mathbf{x}(t) = (1-t)^3\mathbf{p}_0 + 3t(1-t)^2\mathbf{p}_1 \\ + 3t^2(1-t)\mathbf{p}_2 + t^3\mathbf{p}_3 + \mathbf{b}$$

Show that this new curve is again a Bézier curve. [*Hint:* Where are the new control points?]

2. The parametric vector form of a B-spline curve was defined in the Practice Problems as

$$\mathbf{x}(t) = \tfrac{1}{6}\big[(1-t)^3\mathbf{p}_0 + (3t(1-t)^2 - 3t + 4)\mathbf{p}_1 \\ + (3t^2(1-t) + 3t + 1)\mathbf{p}_2 + t^3\mathbf{p}_3\big] \quad \text{for } 0 \le t \le 1,$$

where $\mathbf{p}_0, \mathbf{p}_1, \mathbf{p}_2$, and $\mathbf{p}_3$ are the control points.

a. Show that for $0 \le t \le 1$, $\mathbf{x}(t)$ is in the convex hull of the control points.

b. Suppose that a B-spline curve $\mathbf{x}(t)$ is translated to $\mathbf{x}(t) + \mathbf{b}$ (as in Exercise 1). Show that this new curve is again a B-spline.

3. Let $\mathbf{x}(t)$ be a cubic Bézier curve determined by points $\mathbf{p}_0, \mathbf{p}_1, \mathbf{p}_2$, and $\mathbf{p}_3$.

a. Compute the *tangent* vector $\mathbf{x}'(t)$. Determine how $\mathbf{x}'(0)$ and $\mathbf{x}'(1)$ are related to the control points, and give geometric descriptions of the *directions* of these tangent vectors. Is it possible to have $\mathbf{x}'(1) = \mathbf{0}$?

b. Compute the second derivative $\mathbf{x}''(t)$ and determine how $\mathbf{x}''(0)$ and $\mathbf{x}''(1)$ are related to the control points. Draw a figure based on Figure 10, and construct a line segment that points in the direction of $\mathbf{x}''(0)$. [*Hint:* Use $\mathbf{p}_1$ as the origin of the coordinate system.]

4. Let $\mathbf{x}(t)$ be the B-spline in Exercise 2, with control points $\mathbf{p}_0$, $\mathbf{p}_1$, $\mathbf{p}_2$, and $\mathbf{p}_3$.

   a. Compute the tangent vector $\mathbf{x}'(t)$ and determine how the derivatives $\mathbf{x}'(0)$ and $\mathbf{x}'(1)$ are related to the control points. Give geometric descriptions of the *directions* of these tangent vectors. Explore what happens when both $\mathbf{x}'(0)$ and $\mathbf{x}'(1)$ equal $\mathbf{0}$. Justify your assertions.

   b. Compute the second derivative $\mathbf{x}''(t)$ and determine how $\mathbf{x}''(0)$ and $\mathbf{x}''(1)$ are related to the control points. Draw a figure based on Figure 10, and construct a line segment that points in the direction of $\mathbf{x}''(1)$. [*Hint:* Use $\mathbf{p}_2$ as the origin of the coordinate system.]

5. Let $\mathbf{x}(t)$ and $\mathbf{y}(t)$ be cubic Bézier curves with control points $\{\mathbf{p}_0, \mathbf{p}_1, \mathbf{p}_2, \mathbf{p}_3\}$ and $\{\mathbf{p}_3, \mathbf{p}_4, \mathbf{p}_5, \mathbf{p}_6\}$, respectively, so that $\mathbf{x}(t)$ and $\mathbf{y}(t)$ are joined at $\mathbf{p}_3$. The following questions refer to the curve consisting of $\mathbf{x}(t)$ followed by $\mathbf{y}(t)$. For simplicity, assume that the curve is in $\mathbb{R}^2$.

   a. What condition on the control points will guarantee that the curve has C^1 continuity at $\mathbf{p}_3$? Justify your answer.

   b. What happens when $\mathbf{x}'(1)$ and $\mathbf{y}'(0)$ are both the zero vector?

6. A B-spline is built out of B-spline segments, described in Exercise 2. Let $\mathbf{p}_0, \ldots, \mathbf{p}_4$ be control points. For $0 \le t \le 1$, let $\mathbf{x}(t)$ and $\mathbf{y}(t)$ be determined by the geometry matrices $[\,\mathbf{p}_0 \ \ \mathbf{p}_1 \ \ \mathbf{p}_2 \ \ \mathbf{p}_3\,]$ and $[\,\mathbf{p}_1 \ \ \mathbf{p}_2 \ \ \mathbf{p}_3 \ \ \mathbf{p}_4\,]$, respectively. Notice how the two segments share three control points. The two segments do not overlap, however—they join at a common endpoint, close to $\mathbf{p}_2$.

   a. Show that the combined curve has G^0 continuity—that is, $\mathbf{x}(1) = \mathbf{y}(0)$.

   b. Show that the curve has C^1 continuity at the join point, $\mathbf{x}(1)$. That is, show that $\mathbf{x}'(1) = \mathbf{y}'(0)$.

7. Let $\mathbf{x}(t)$ and $\mathbf{y}(t)$ be Bézier curves from Exercise 5, and suppose the combined curve has C^2 continuity (which includes C^1 continuity) at $\mathbf{p}_3$. Set $\mathbf{x}''(1) = \mathbf{y}''(0)$ and show that $\mathbf{p}_5$ is completely determined by $\mathbf{p}_1$, $\mathbf{p}_2$, and $\mathbf{p}_3$. Thus, the points $\mathbf{p}_0, \ldots, \mathbf{p}_3$ and the C^2 condition determine all but one of the control points for $\mathbf{y}(t)$.

8. Let $\mathbf{x}(t)$ and $\mathbf{y}(t)$ be segments of a B-spline as in Exercise 6. Show that the curve has C^2 continuity (as well as C^1 continuity) at $\mathbf{x}(1)$. That is, show that $\mathbf{x}''(1) = \mathbf{y}''(0)$. This higher-order continuity is desirable in CAD applications such as automotive body design, since the curves and surfaces appear much smoother. However, B-splines require three times the computation of Bézier curves, for curves of comparable length. For surfaces, B-splines require nine times the computation of Bézier surfaces. Programmers often

choose Bézier surfaces for applications (such as an airplane cockpit simulator) that require real-time rendering.

9. A quartic Bézier curve is determined by five control points, $\mathbf{p}_0$, $\mathbf{p}_1$, $\mathbf{p}_2$, $\mathbf{p}_3$, and $\mathbf{p}_4$:

$$\mathbf{x}(t) = (1-t)^4 \mathbf{p}_0 + 4t(1-t)^3 \mathbf{p}_1 + 6t^2(1-t)^2 \mathbf{p}_2$$
$$+ 4t^3(1-t)\mathbf{p}_3 + t^4 \mathbf{p}_4 \quad \text{for } 0 \le t \le 1$$

Construct the quartic basis matrix M_B for $\mathbf{x}(t)$.

10. The "B" in B-spline refers to the fact that a segment $\mathbf{x}(t)$ may be written in terms of a basis matrix, M_S, in a form similar to a Bézier curve. That is,

$$\mathbf{x}(t) = G M_S \mathbf{u}(t) \quad \text{for } 0 \le t \le 1$$

where G is the geometry matrix $[\,\mathbf{p}_0 \ \ \mathbf{p}_1 \ \ \mathbf{p}_2 \ \ \mathbf{p}_3\,]$ and $\mathbf{u}(t)$ is the column vector $(1, t, t^2, t^3)$. In a *uniform* B-spline, each segment uses the same basis matrix, but the geometry matrix changes. Construct the basis matrix M_S for $\mathbf{x}(t)$.

In Exercises 11–16, mark each statement True or False (**T/F**). Justify each answer.

11. (**T/F**) The cubic Bézier curve is based on four control points.

12. (**T/F**) The essential properties of Bézier curves are preserved under the action of linear transformations, but not translations.

13. (**T/F**) Given a quadratic Bézier curve $\mathbf{x}(t)$ with control points $\mathbf{p}_0$, $\mathbf{p}_1$, and $\mathbf{p}_2$, the directed line segment $\mathbf{p}_1 - \mathbf{p}_0$ (from $\mathbf{p}_0$ to $\mathbf{p}_1$) is the tangent vector to the curve at $\mathbf{p}_0$.

14. (**T/F**) When two Bézier curves $\mathbf{x}(t)$ and $\mathbf{y}(t)$ are joined at the point where $\mathbf{x}(1) = \mathbf{y}(0)$, the combined curve has G^0 continuity at that point.

15. (**T/F**) When two quadratic Bézier curves with control points $\{\mathbf{p}_0, \mathbf{p}_1, \mathbf{p}_2\}$ and $\{\mathbf{p}_2, \mathbf{p}_3, \mathbf{p}_4\}$ are joined at $\mathbf{p}_2$, the combined Bézier curve will have C^1 continuity at $\mathbf{p}_2$ if $\mathbf{p}_2$ is the midpoint of the line segment between $\mathbf{p}_1$ and $\mathbf{p}_3$.

16. (**T/F**) The Bézier basis matrix is a matrix whose columns are the control points of the curve.

Exercises 17–19 concern the subdivision of a Bézier curve shown in Figure 7. Let $\mathbf{x}(t)$ be the Bézier curve, with control points $\mathbf{p}_0, \ldots, \mathbf{p}_3$, and let $\mathbf{y}(t)$ and $\mathbf{z}(t)$ be the subdividing Bézier curves as in the text, with control points $\mathbf{q}_0, \ldots, \mathbf{q}_3$ and $\mathbf{r}_0, \ldots, \mathbf{r}_3$, respectively.

17. a. Use equation (12) to show that $\mathbf{q}_1$ is the midpoint of the segment from $\mathbf{p}_0$ to $\mathbf{p}_1$.

   b. Use equation (13) to show that

$$8\mathbf{q}_2 = 8\mathbf{q}_3 + \mathbf{p}_0 + \mathbf{p}_1 - \mathbf{p}_2 - \mathbf{p}_3.$$

   c. Use part (b), equation (8), and part (a) to show that $\mathbf{q}_2$ is the midpoint of the segment from $\mathbf{q}_1$ to the midpoint of the segment from $\mathbf{p}_1$ to $\mathbf{p}_2$. That is, $\mathbf{q}_2 = \frac{1}{2}[\mathbf{q}_1 + \frac{1}{2}(\mathbf{p}_1 + \mathbf{p}_2)]$.

18. a. Justify each equal sign:

$$3(\mathbf{r}_3 - \mathbf{r}_2) = \mathbf{z}'(1) = .5\mathbf{x}'(1) = \tfrac{3}{2}(\mathbf{p}_3 - \mathbf{p}_2).$$

b. Show that $\mathbf{r}_2$ is the midpoint of the segment from $\mathbf{p}_2$ to $\mathbf{p}_3$.

c. Justify each equal sign: $3(\mathbf{r}_1 - \mathbf{r}_0) = \mathbf{z}'(0) = .5\mathbf{x}'(.5)$.

d. Use part (c) to show that $8\mathbf{r}_1 = -\mathbf{p}_0 - \mathbf{p}_1 + \mathbf{p}_2 + \mathbf{p}_3 + 8\mathbf{r}_0$.

e. Use part (d), equation (8), and part (a) to show that $\mathbf{r}_1$ is the midpoint of the segment from $\mathbf{r}_2$ to the midpoint of the segment from $\mathbf{p}_1$ to $\mathbf{p}_2$. That is, $\mathbf{r}_1 = \tfrac{1}{2}[\mathbf{r}_2 + \tfrac{1}{2}(\mathbf{p}_1 + \mathbf{p}_2)]$.

19. Sometimes only one half of a Bézier curve needs further subdividing. For example, subdivision of the "left" side is accomplished with parts (a) and (c) of Exercise 17 and equation (8). When both halves of the curve $\mathbf{x}(t)$ are divided, it is possible to organize calculations efficiently to calculate both left and right control points concurrently, without using equation (8) directly.

a. Show that the tangent vectors $\mathbf{y}'(1)$ and $\mathbf{z}'(0)$ are equal.

b. Use part (a) to show that $\mathbf{q}_3$ (which equals $\mathbf{r}_0$) is the midpoint of the segment from $\mathbf{q}_2$ to $\mathbf{r}_1$.

c. Using part (b) and the results of Exercises 17 and 18, write an algorithm that computes the control points for both $\mathbf{y}(t)$ and $\mathbf{z}(t)$ in an efficient manner. The only operations needed are sums and division by 2.

20. Explain why a cubic Bézier curve is completely determined by $\mathbf{x}(0)$, $\mathbf{x}'(0)$, $\mathbf{x}(1)$, and $\mathbf{x}'(1)$.

21. TrueType® fonts, created by Apple Computer and Adobe Systems, use quadratic Bézier curves, while PostScript® fonts, created by Microsoft, use cubic Bézier curves. The cubic curves provide more flexibility for typeface design, but it is important to Microsoft that every typeface using quadratic curves can be transformed into one that uses cubic curves. Suppose that $\mathbf{w}(t)$ is a quadratic curve, with control points $\mathbf{p}_0$, $\mathbf{p}_1$, and $\mathbf{p}_2$.

a. Find control points $\mathbf{r}_0$, $\mathbf{r}_1$, $\mathbf{r}_2$, and $\mathbf{r}_3$ such that the cubic Bézier curve $\mathbf{x}(t)$ with these control points has the property that $\mathbf{x}(t)$ and $\mathbf{w}(t)$ have the same initial and terminal points and the same tangent vectors at $t = 0$ and $t = 1$. (See Exercise 20.)

b. Show that if $\mathbf{x}(t)$ is constructed as in part (a), then $\mathbf{x}(t) = \mathbf{w}(t)$ for $0 \le t \le 1$.

22. Use partitioned matrix multiplication to compute the following matrix product, which appears in the alternative formula (5) for a Bézier curve:

$$\begin{bmatrix} 1 & 0 & 0 & 0 \\ -3 & 3 & 0 & 0 \\ 3 & -6 & 3 & 0 \\ -1 & 3 & -3 & 1 \end{bmatrix} \begin{bmatrix} \mathbf{p}_0 \\ \mathbf{p}_1 \\ \mathbf{p}_2 \\ \mathbf{p}_3 \end{bmatrix}$$

Solutions to Practice Problems

1. From equation (14) with $t = 0$, $\mathbf{x}(0) \ne \mathbf{p}_0$ because

$$\mathbf{x}(0) = \tfrac{1}{6}[\mathbf{p}_0 + 4\mathbf{p}_1 + \mathbf{p}_2] = \tfrac{1}{6}\mathbf{p}_0 + \tfrac{2}{3}\mathbf{p}_1 + \tfrac{1}{6}\mathbf{p}_2.$$

The coefficients are nonnegative and sum to 1, so $\mathbf{x}(0)$ is in conv $\{\mathbf{p}_0, \mathbf{p}_1, \mathbf{p}_2\}$, and the affine coordinates with respect to $\{\mathbf{p}_0, \mathbf{p}_1, \mathbf{p}_2\}$ are $\left(\tfrac{1}{6}, \tfrac{2}{3}, \tfrac{1}{6}\right)$.

2. From equation (14) with $t = 1$, $\mathbf{x}(1) \ne \mathbf{p}_3$ because

$$\mathbf{x}(1) = \tfrac{1}{6}[\mathbf{p}_1 + 4\mathbf{p}_2 + \mathbf{p}_3] = \tfrac{1}{6}\mathbf{p}_1 + \tfrac{2}{3}\mathbf{p}_2 + \tfrac{1}{6}\mathbf{p}_3.$$

The coefficients are nonnegative and sum to 1, so $\mathbf{x}(1)$ is in conv $\{\mathbf{p}_1, \mathbf{p}_2, \mathbf{p}_3\}$, and the affine coordinates with respect to $\{\mathbf{p}_1, \mathbf{p}_2, \mathbf{p}_3\}$ are $\left(\tfrac{1}{6}, \tfrac{2}{3}, \tfrac{1}{6}\right)$.

CHAPTER 8 PROJECT

- -

The Chapter 8 project is available online at `bit.ly/30IM8gT`.

A. *Affine Combinations*: This project explores affine combinations of a given set of points.

CHAPTER 8 SUPPLEMENTARY EXERCISES

In Exercises 1–21, mark each statement True or False (**T/F**). Justify each answer.

1. (**T/F**) Given $\mathbf{v}_1, \mathbf{v}_2, \ldots, \mathbf{v}_p$ in $\mathbb{R}^n$ and scalars $c_1, \ldots, c_p$, an affine combination of $\mathbf{v}_1, \mathbf{v}_2, \ldots, \mathbf{v}_p$ is a linear combination $c_1 \mathbf{v}_1 + \cdots + c_p \mathbf{v}_p$ such that the weights satisfy $c_1 + \cdots + c_p = 1$.

2. (**T/F**) The affine hull of two points $\mathbf{v}_1$ and $\mathbf{v}_2$ is the set of all points $\mathbf{y} = t\mathbf{v}_1 + (1 - t)\mathbf{v}_2$, with t in $\mathbb{R}$.

3. (**T/F**) A hyperplane is a 4-dimensional flat.

4. (**T/F**) Two flats are parallel if their intersection is empty.

5. (**T/F**) Every subspace is a flat.

6. (**T/F**) Every subspace is an affine set.

7. (**T/F**) Every affinely dependent set is linearly dependent.

8. (**T/F**) Every affinely independent set is linearly independent.

9. (**T/F**) The barycentric coordinates of a point in $\mathbb{R}^2$ are always nonnegative.

10. (**T/F**) Let $S = \{\mathbf{v}_1, \ldots, \mathbf{v}_k\}$ be an affinely independent set in $\mathbb{R}^n$. Then each point $\mathbf{p}$ in $\mathbb{R}^n$ has a unique representation as an affine combination of $\mathbf{v}_1, \ldots, \mathbf{v}_k$.

11. (**T/F**) A convex combination of points $\{\mathbf{v}_1, \mathbf{v}_2, \ldots, \mathbf{v}_k\}$ in $\mathbb{R}^n$ is a linear combination of the form $c_1 \mathbf{v}_1 + c_2 \mathbf{v}_2 + \cdots + c_k \mathbf{v}_k$ such that $c_1 + c_2 + \cdots + c_k = 1$ and $c_i \geq 0$ for all i.

12. (**T/F**) If a set S is affinely independent and if $\mathbf{p} \in \text{aff } S$, then $\mathbf{p} \in \text{conv } S$ if and only if the barycentric coordinates of $\mathbf{p}$ with respect to S are nonnegative.

13. (**T/F**) The line segment between $\mathbf{x}$ and $\mathbf{y}$ is the set of all points of the form $(1 - t)\mathbf{x} + t\mathbf{y}$, where t is in $\mathbb{R}$.

14. (**T/F**) Every affine set is a convex set.

15. (**T/F**) For some $\mathbb{R}^n$, the dimension of a hyperplane can be the same as the dimension of a line.

16. (**T/F**) For some $\mathbb{R}^n$, the dimension of a hyperplane can be less than the dimension of a line.

17. (**T/F**) Suppose A and B are nonempty compact convex sets. Then there exists a hyperplane that strictly separates A and B if and only if $A \cap B = \varnothing$.

18. (**T/F**) A polytope can be the convex hull of infinitely many points.

19. (**T/F**) Every nonempty compact convex set S has an extreme point, and the set of all extreme points is the smallest subset of S whose convex hull is equal to S.

20. (**T/F**) If $\mathbf{w}(t)$ is a quadratic Bézier curve with control points $\mathbf{p}_0$, $\mathbf{p}_1$, and $\mathbf{p}_2$, then $\mathbf{w}'(0)$ has the same direction as the tangent to the curve at $\mathbf{p}_0$ and $\mathbf{w}'(1)$ has the same direction as the tangent to the curve at $\mathbf{p}_1$.

21. (**T/F**) When two Bézier curves are connected with G^1 geometric continuity, then the tangent vectors for the two curves at the common control point have the same direction.

22. If $S = \{\mathbf{v}_1, \ldots, \mathbf{v}_k\}$ is an affinely independent subset of $\mathbb{R}^n$, prove that $k \leq n + 1$.

23. Suppose that F and G are k-dimensional flats ($0 \leq k \leq n - 1$) in $\mathbb{R}^n$ with $F \subseteq G$. Prove that $F = G$.

24. Prove or give a counterexample: A set S is convex if and only if for each $\mathbf{p}$, $\mathbf{q}$ in S, the set of points of the form $(1 - t)\mathbf{p} + t\mathbf{q}$, where $0 < t < 1$, is contained in S.

25. Let V be a k-dimensional subspace ($0 \leq k \leq n - 1$) of $\mathbb{R}^n$ and let $F_1 = \mathbf{x}_1 + V$ and $F_2 = \mathbf{x}_2 + V$ for vectors $\mathbf{x}_1, \mathbf{x}_2$ in $\mathbb{R}^n$. Prove that either $F_1 = F_2$ or $F_1 \cap F_2 = \varnothing$. Thus two parallel flats either coincide or are disjoint.

26. Let f be a nonzero linear functional on $\mathbb{R}^n$ and suppose $H = [f : 7]$. If $\mathbf{p} \in \mathbb{R}^n$, $f(\mathbf{p}) = 2$, and $H_1 = H + 3\mathbf{p}$, then find d such that $H_1 = [f : d]$.

27. Let V be an $(n - 1)$-dimensional subspace of $\mathbb{R}^n$ and suppose $\mathbf{p} \in \mathbb{R}^n$ but $\mathbf{p} \notin V$. Prove that each vector $\mathbf{x}$ in $\mathbb{R}^n$ has a unique representation as $\mathbf{x} = \mathbf{v} + c\mathbf{p}$, where $\mathbf{v} \in V$ and $c \in \mathbb{R}$.

28. If m is the maximum value of the linear functional f on the convex set S, and $\mathbf{p}$, $\mathbf{q}$ are points in S such that $f(\mathbf{p}) = f(\mathbf{q}) = m$, show that $f(\mathbf{x}) = m$ for all $\mathbf{x}$ in $\overline{\mathbf{pq}}$.

29. If $B(\mathbf{p}, \delta)$ is the open ball with center $\mathbf{p}$ and radius δ in $\mathbb{R}^n$, prove that $\lambda B(\mathbf{p}, \delta) = B(\lambda \mathbf{p}, \lambda \delta)$, where $\delta > 0$ and $\lambda > 0$. This means that dilations and nonzero contractions map circles in $\mathbb{R}^2$ onto circles and balls in $\mathbb{R}^n$ onto balls.

30. In $\mathbb{R}^4$, let $\mathbf{v}_1 = (1, -1, 2, -1)$, $\mathbf{v}_2 = (2, -1, 2, 0)$, $\mathbf{v}_3 = (1, 0, 2, 0)$, and $\mathbf{v}_4 = (1, 0, 3, 1)$.
   a. Show that the set $\{\mathbf{v}_1, \mathbf{v}_2, \mathbf{v}_3, \mathbf{v}_4\}$ is affinely independent.
   b. Let $A = \text{aff}\{\mathbf{v}_1, \mathbf{v}_2, \mathbf{v}_3, \mathbf{v}_4\}$ and $B = [f : 3]$, where f is the linear functional defined by $f(x_1, x_2, x_3, x_4) = x_1 + x_2 + x_3 - x_4$. Prove that $A = B$. *Hint:* Use Exercise 23.

Exercises 31–35 deal with the following concepts: A point $\mathbf{p}$ is called a **positive combination** of the points $\mathbf{v}_1, \ldots, \mathbf{v}_k$ if $\mathbf{p} = c_1 \mathbf{v}_1 + \cdots + c_k \mathbf{v}_k$, with all $c_i \geq 0$. The set of all positive combinations of points of a set S is called the **positive hull** of S and is denoted by pos S.

31. Let $S = \{(-1, 1), (1, 1)\}$ in $\mathbb{R}^2$. Describe the set pos S geometrically.

32. Observe that in Exercise 31 we have pos $S \cap \text{aff } S = \text{conv } S$. Show that this is not true in general by verifying the following example: Let $T = \{\mathbf{v}_1, \mathbf{v}_2, \mathbf{v}_3\}$, where $\mathbf{v}_1 = (0, 1)$,

$\mathbf{v}_2 = (1, 1)$, and $\mathbf{v}_3 = (1, 0)$. And let $\mathbf{p} = (3, 2)$. Show that $\mathbf{p} \in \text{pos } T \cap \text{aff } T$, but $\mathbf{p} \notin \text{conv } T$.

33. What special property does set S in Exercise 31 have that makes pos $S \cap$ aff $S = \text{conv } S$?

34. Let S be a nonempty subset of $\mathbb{R}^n$. Show that pos $S = \text{pos (conv } S)$.

35. Let S be a nonempty convex subset of $\mathbb{R}^n$. Prove $\mathbf{x} \in \text{pos } S$ if and only if $\mathbf{x} = \lambda\mathbf{s}$ for some $\lambda \geq 0$ and some $\mathbf{x}$ in S.

9 Optimization

THE BERLIN AIRLIFT

After World War II, the city of Berlin was an "island" surrounded by the Soviet zone of occupied Germany. The city was divided into four sections, with the British, French, and Americans having jurisdiction over West Berlin and the Soviets over East Berlin. But the Russians were eager for the other three nations to abandon Berlin. After months of harassment, on June 24, 1948, they imposed a blockade on West Berlin, cutting off all access by land and rail. With a civilian population of about 2.5 million people, the isolated western sectors became dependent on reserve stocks and airlift replacements.

Four days later, the first American planes landed in Berlin with supplies of food, and "Operation Vittles" had begun. At first the airlift seemed doomed to failure because the needs of the city were overwhelming. The Russians had cut off all electricity and coal shipments, and the city

was literally under siege. But the Western Allies responded by flying in thousands of tons of food, coal, medicine, and other supplies on a daily basis. In May 1949, Stalin relented, and the blockade was lifted. The airlift, however, continued for another four months.

The Berlin Airlift was unbelievably successful in using relatively few aircraft to deliver an enormous amount of supplies. The design and conduct of this operation required intensive planning and calculations, which led to the theoretical development of linear programming, and the invention of the simplex method by George Dantzig. The potential of this new tool was quickly recognized by business and industry, where it is now used to allocate resources, plan production, schedule workers, organize investment portfolios, formulate marketing strategies, and perform many other tasks involving optimization.

There are many situations in business, politics, economics, military strategy, and other areas where one tries to optimize a certain benefit. This may involve maximizing a profit or the payoff in a contest or minimizing a cost or other loss. This chapter presents two mathematical models that deal with optimization problems. The fundamental results in both cases depend on properties of convex sets and hyperplanes. Section 9.1 introduces the theory of games and develops strategies based on probability. Sections 9.2–9.4 explore techniques of linear programming and use them to solve a variety of problems, including matrix games larger than those in Section 9.1.

9.1 Matrix Games

The theory of games analyzes competitive phenomena and seeks to provide a basis for rational decision-making. Its growing importance was highlighted in 1994 when the Nobel Prize in Economics was awarded to John Harsanyi, John Nash, and Reinhard Selten, for their pioneering work in the theory of noncooperative games.[1]

The games in this section are **matrix games** whose various outcomes are listed in a payoff matrix. Two players in a game compete according to a fixed set of rules. Player R (for *row*) has a choice of m possible moves (or choices of action), and player C (for *column*) has n moves. By convention, the **payoff matrix** $A = [a_{ij}]$ lists the amounts that the **row** player R wins **from** the column player C, depending on the choices R and C make. Entry a_{ij} shows the amount R wins when R chooses action i and C chooses action j. A negative value for a_{ij} indicates a loss for R, the amount R has to pay to C. The games are often called **two-person zero-sum games** because the algebraic sum of the amounts gained by R and C is zero.

EXAMPLE 1 Each player has a supply of pennies, nickels, and dimes. At a given signal, both players display (or "play") one coin. If the displayed coins are not the same, then the player showing the higher-valued coin gets to keep both. If they are both pennies or both nickels, then player C keeps both; but if they are both dimes, then player R keeps them. Construct a payoff matrix, using p for display of a penny, n for a nickel, and d for a dime.

SOLUTION   Each player has three choices, p, n, and d, so the payoff matrix is 3×3:

$$
\begin{array}{cc}
 & \begin{array}{ccc} p & n & d \end{array} \\
\text{Player } R \begin{array}{c} p \\ n \\ d \end{array} & \left[\begin{array}{ccc} & & \\ & & \\ & & \end{array} \right]
\end{array}
$$

Player C

Consider a row for R and fill in what R receives (or pays), depending on the choice C makes. First, suppose R plays a penny. If C also plays a penny, R loses 1 cent, because the coins match. The $(1, 1)$ entry is -1. If C plays either a nickel or a dime, R also loses 1 cent, because C displays the higher-valued coin. This information goes in row 1:

$$
\begin{array}{cc}
 & \begin{array}{ccc} p & n & d \end{array} \\
\text{Player } R \begin{array}{c} p \\ n \\ d \end{array} & \left[\begin{array}{ccc} -1 & -1 & -1 \\ & & \\ & & \end{array} \right]
\end{array}
$$

Player C

Next, suppose R plays a nickel. If C plays a penny, R wins the penny. Otherwise, R loses the nickel, because either C matches the nickel or shows the higher-value dime. Finally, when R plays a dime, R gains either a penny or a nickel, whichever is shown

---

[1] The popular 2002 movie *A Beautiful Mind* tells a poignant story of the life of John Nash.

by C, because R's dime is of higher value. Also, when both players display a dime, R wins the dime from C because of the special rule for that case.

$$
\begin{array}{cc}
 & \text{Player } C \\
 & \begin{array}{ccc} p & n & d \end{array} \\
\text{Player } R \begin{array}{c} p \\ n \\ d \end{array} &
\left[\begin{array}{ccc}
-1 & -1 & -1 \\
1 & -5 & -5 \\
1 & 5 & 10
\end{array}\right]
\end{array}
$$

■

By looking at the payoff matrix in Example 1, the players discover that some plays are better than others. Both players know that R is likely to choose a row that has positive entries, while C is likely to choose a column that has negative entries (a payment from R to C). Player R notes that every entry in row 3 is positive and chooses to play a dime. No matter what C may do, the worst that can happen to R is to win a penny. Player C notes that every column contains a positive entry and therefore C cannot be certain of winning anything. So player C chooses to play a penny, which will minimize the potential loss.

From a mathematical point of view, what has each player done? Player R has found the minimum of each row (the worst that could happen for that play) and has chosen the row for which this minimum is largest. (See Figure 1.) That is, R has computed

$$
\max_i \left[\min_j a_{ij} \right]
$$

$$
\begin{array}{ccc}
 & \text{Player } C & \text{Row minima} \\
\text{Player } R &
\left[\begin{array}{ccc}
-1 & -1 & -1 \\
1 & -5 & -5 \\
1 & 5 & 10
\end{array}\right] &
\begin{array}{c} -1 \\ -5 \\ 1 \leftarrow \text{Max of the minima} \end{array} \\
\text{Column maxima} & \begin{array}{ccc} 1 & 5 & 10 \end{array} \\
 & \uparrow\!\!\!\rule{0pt}{0pt}\text{Min of the maxima}
\end{array}
$$

FIGURE 1

Observe that for C, a large positive payment to R is worse than a small positive payment. Thus C has found the maximum of each column (the worst that can happen to C for that play) and has chosen the column for which this maximum is smallest. Player C has found

$$
\min_j \left[\max_i a_{ij} \right]
$$

For this payoff matrix $[a_{ij}]$,

$$
\max_i \min_j a_{ij} = \min_j \max_i a_{ij} = 1
$$

DEFINITION

> If the payoff matrix of a matrix game contains an entry a_{ij} that is both the minimum of row i and the maximum of column j, then a_{ij} is called a **saddle point**.

In Example 1, the entry a_{31} is a saddle point for the payoff matrix. As long as both players continue to seek their best advantage, player R will always display a dime (row 3) and player C will always display a penny (column 1). Some games may have more than one saddle point.

The situation is not quite so simple in the next example.

EXAMPLE 2 Again suppose that each player has a supply of pennies, nickels, and dimes to play, but this time the payoff matrix is given as follows:

$$
\begin{array}{cc}
& \text{Player } C \\
\end{array}
$$

		p	n	d	Row minima
	p	10	−5	5	−5
Player R	n	1	1	−1	−1 ←Max of the minima
	d	0	−10	−5	−10
Column maxima		10	1	5	

└—Min of the maxima

If player R reasons as in the first example and looks at the row minima, R will choose to play a nickel, thereby maximizing the minimum gain (in this case a loss of 1). Player C, looking at the column maxima (the greatest payment to R), will also select a nickel to minimize the loss to R.

Thus, as the game begins, R and C both continue to play a nickel. After a while, however, C begins to reason, "If R is going to play a nickel, then I'll play a dime so that I can win a penny." However, when C starts to play a dime repeatedly, R begins to reason, "If C is going to play a dime, then I'll play a penny so that I can win a nickel." Once R has done this, C switches to a nickel (to win a nickel) and then R starts playing a nickel ... and so on. It seems that neither player can develop a winning strategy. ■

Mathematically speaking, the payoff matrix for the game in Example 2 does not have a saddle point. Indeed,

$$\max_i \min_j a_{ij} = -1$$

while

$$\min_j \max_i a_{ij} = 1$$

This means that neither player can play the same coin repeatedly and be assured of optimizing the winnings. In fact, any predictable strategy can be countered by the opponent. But is it possible to formulate some combination of plays that over the long run will produce an optimal return? The answer is *yes* (as Theorem 3 later will show), when each move is made at random, but with a certain probability attached to each possible choice.

Here is a way to imagine how player R could develop a strategy for playing a matrix game. Suppose that R has a device consisting of a horizontal metal arrow whose center of gravity is supported on a vertical rod in the middle of a flat circular region. The region is cut into pie-shaped sectors, one for each of the rows in the payoff matrix. Player R gives the arrow an initial spin and waits for it to come to rest. The position of the arrowhead at rest determines one play for R in the matrix game.

If the area of the circle is taken as 1 unit, then the areas of the various sectors sum to 1, and these areas give the relative frequencies, or *probabilities*, of selecting the various plays in the matrix game, when the game is played many times. For instance, if there are five sectors of equal area and if the arrow is spun many times, player R will select each of the five plays about $1/5$ of the time. This strategy is specified by the vector in $\mathbb{R}^5$ whose entries all equal $1/5$. If the five sectors of the circle are unequal in size, then in the long run some game plays will be chosen more frequently than the others. The corresponding strategy for R is specified by a vector in $\mathbb{R}^5$ that lists the areas of the five sectors.

DEFINITION

A **probability vector** in $\mathbb{R}^m$ is a vector $\mathbf{x}$ in $\mathbb{R}^m$ whose entries are nonnegative and sum to 1. Such an $\mathbf{x}$ has the form

$$\mathbf{x} = \begin{bmatrix} x_1 \\ \vdots \\ x_m \end{bmatrix}, \quad x_i \geq 0 \text{ for } i = 1, \ldots, m \text{ and } \sum_{i=1}^{m} x_i = 1$$

Let A be an $m \times n$ payoff matrix for a game. The **strategy space** for player R is the set of all probability vectors in $\mathbb{R}^m$, and the **strategy space** for player C is the set of all probability vectors in $\mathbb{R}^n$. A point in a strategy space is called a **strategy**. If one entry in a strategy is 1 (and the other entries are zeros), the strategy is called a **pure strategy**.

The pure strategies in $\mathbb{R}^m$ are the standard basis vectors for $\mathbb{R}^m$, $\mathbf{e}_1, \ldots, \mathbf{e}_m$. In general, each strategy $\mathbf{x}$ is a linear combination, $x_1\mathbf{e}_1 + \cdots + x_m\mathbf{e}_m$, of these pure strategies with nonnegative weights that sum to 1.[2]

Suppose now that R and C are playing the $m \times n$ matrix game $A = [a_{ij}]$, where a_{ij} is the entry in the ith row and the jth column of A. There are mn possible outcomes of the game, depending on the row R chooses and the column C chooses. Suppose R uses strategy $\mathbf{x}$ and C uses strategy $\mathbf{y}$, where

$$\mathbf{x} = \begin{bmatrix} x_1 \\ \vdots \\ x_m \end{bmatrix} \quad \text{and} \quad \mathbf{y} = \begin{bmatrix} y_1 \\ \vdots \\ y_n \end{bmatrix}$$

Since R plays the first row with probability x_1 and C plays the first column with probability y_1 and since their choices are made independently, it can be shown that the probability is $x_1 y_1$ that R chooses the first row *and* C chooses the first column. Over the course of many games, the expected payoff to R for this outcome is $a_{11}x_1y_1$ for one game. A similar computation holds for each possible pair of choices that R and C can make. The sum of the expected payoffs to R over all possible pairs of choices is called the **expected payoff**, $E(\mathbf{x}, \mathbf{y})$, of the game to player R for strategies $\mathbf{x}$ and $\mathbf{y}$. That is,

$$E(\mathbf{x}, \mathbf{y}) = \sum_{i=1}^{m} \sum_{j=1}^{n} x_i a_{ij} y_j = \mathbf{x}^T A \mathbf{y}$$

Roughly speaking, the number $E(\mathbf{x}, \mathbf{y})$ is the average amount that C will pay to R per game, when R and C play a large number of games using the strategies $\mathbf{x}$ and $\mathbf{y}$, respectively.

---

[2] More precisely, each strategy is a convex combination of the set of pure strategies—that is, a point in the convex hull of the set of standard basis vectors. This fact connects the theory of convex sets to the study of matrix games. The strategy space for R is an $(m-1)$-dimensional simplex in $\mathbb{R}^m$, and the strategy space for C is an $(n-1)$-dimensional simplex in $\mathbb{R}^n$. See Sections 8.3 and 8.5 for definitions.

Let X denote the strategy space for R and Y the strategy space for C. If R were to choose a particular strategy, say $\tilde{\mathbf{x}}$, and if C were to discover this strategy, then C would certainly choose $\mathbf{y}$ to minimize

$$E(\tilde{\mathbf{x}}, \mathbf{y}) = \tilde{\mathbf{x}}^T A \mathbf{y}$$

The **value** of using strategy $\tilde{\mathbf{x}}$ is the number $v(\tilde{\mathbf{x}})$ defined by

$$v(\tilde{\mathbf{x}}) = \min_{\mathbf{y} \in Y} E(\tilde{\mathbf{x}}, \mathbf{y}) = \min_{\mathbf{y} \in Y} \tilde{\mathbf{x}}^T A \mathbf{y} \tag{1}$$

Since $\tilde{\mathbf{x}}^T A$ is a $1 \times n$ matrix, the mapping $\mathbf{y} \mapsto E(\tilde{\mathbf{x}}, \mathbf{y}) = \tilde{\mathbf{x}}^T A \mathbf{y}$ is a linear functional on the strategy space Y. From this, it can be shown that $E(\tilde{\mathbf{x}}, \mathbf{y})$ attains its minimum when $\mathbf{y}$ is one of the pure strategies, $\mathbf{e}_1, \ldots, \mathbf{e}_n$, for C.[3]

Recall that $A\mathbf{e}_j$ is the jth column of the matrix A, usually denoted by $\mathbf{a}_j$. Since the minimum in (1) is attained when $\mathbf{y} = \mathbf{e}_j$ for some j, (1) may be written, with $\mathbf{x}$ in place of $\tilde{\mathbf{x}}$, as

$$v(\mathbf{x}) = \min_j E(\mathbf{x}, \mathbf{e}_j) = \min_j \mathbf{x}^T A \mathbf{e}_j = \min_j \mathbf{x}^T \mathbf{a}_j = \min_j \mathbf{x} \cdot \mathbf{a}_j \tag{2}$$

That is, $v(\mathbf{x})$ is the minimum of the inner product of $\mathbf{x}$ with each of the columns of A. The goal of R is to choose $\mathbf{x}$ to maximize $v(\mathbf{x})$.

DEFINITION

The number v_R, defined by

$$v_R = \max_{\mathbf{x} \in X} v(\mathbf{x}) = \max_{\mathbf{x} \in X} \min_{\mathbf{y} \in Y} E(\mathbf{x}, \mathbf{y}) = \max_{\mathbf{x} \in X} \min_j \mathbf{x} \cdot \mathbf{a}_j$$

with the notation as described above, is called the **value of the game to row player** R. A strategy $\hat{\mathbf{x}}$ for R is called **optimal** if $v(\hat{\mathbf{x}}) = v_R$.

Of course, $E(\mathbf{x}, \mathbf{y})$ may exceed v_R for some $\mathbf{x}$ and $\mathbf{y}$ if C plays poorly. Thus, $\hat{\mathbf{x}}$ is optimal for R if $E(\hat{\mathbf{x}}, \mathbf{y}) \geq v_R$ for all $\mathbf{y} \in Y$. This value v_R can be thought of as the most that player R can be *sure* to receive from C, independent of what player C may do.

A similar analysis for player C, using the pure strategies for $\mathbf{x}$, shows that a particular strategy $\mathbf{y}$ will have a value $v(\mathbf{y})$ given by

$$v(\mathbf{y}) = \max_{\mathbf{x} \in X} E(\mathbf{x}, \mathbf{y}) = \max_i E(\mathbf{e}_i, \mathbf{y}) = \max_i \text{row}_i(A)\mathbf{y} \tag{3}$$

because $\mathbf{e}_i^T A = \text{row}_i(A)$. That is, the value of strategy $\mathbf{y}$ to C is the maximum of the inner product of $\mathbf{y}$ with each of the rows of A. The number v_C, defined by

$$v_C = \min_{\mathbf{y} \in Y} v(\mathbf{y}) = \min_{\mathbf{y} \in Y} \max_i \text{row}_i(A)\mathbf{y}$$

is called the **value of the game to column player** C. This is the least that C will have to lose regardless of what R may do. A strategy $\hat{\mathbf{y}}$ for C is called **optimal** if $v(\hat{\mathbf{y}}) = v_C$. Equivalently, $\hat{\mathbf{y}}$ is optimal if $E(\mathbf{x}, \hat{\mathbf{y}}) \leq v_C$ for all $\mathbf{x}$ in X.

---

[3] A linear functional on Y is a linear transformation from Y into $\mathbb{R}$. The pure strategies are the extreme points of the strategy space for a player. The stated result follows directly from Theorem 16 in Section 8.5.

THEOREM I

In any matrix game, $v_R \leq v_C$.

PROOF For any $\mathbf{x}$ in X, the definition $v(\mathbf{x}) = \min_{\mathbf{y} \in Y} E(\mathbf{x}, \mathbf{y})$ implies that $v(\mathbf{x}) \leq E(\mathbf{x}, \mathbf{y})$ for each $\mathbf{y}$ in Y. Also, since $v(\mathbf{y})$ is the maximum of $E(\mathbf{x}, \mathbf{y})$ over all $\mathbf{x}$, $v(\mathbf{y}) \geq E(\mathbf{x}, \mathbf{y})$ for each individual $\mathbf{x}$. These two inequalities show that

$$v(\mathbf{x}) \leq E(\mathbf{x}, \mathbf{y}) \leq v(\mathbf{y})$$

for all $\mathbf{x} \in X$ and for all $\mathbf{y} \in Y$. For any fixed $\mathbf{y}$, the left inequality above implies that $\max_{\mathbf{x} \in X} v(\mathbf{x}) \leq E(\mathbf{x}, \mathbf{y})$. Similarly, for each $\mathbf{x}$, $E(\mathbf{x}, \mathbf{y}) \leq \min_{\mathbf{y} \in Y} v(\mathbf{y})$. Thus,

$$\max_{\mathbf{x} \in X} v(\mathbf{x}) \leq \min_{\mathbf{y} \in Y} v(\mathbf{y})$$

which proves the theorem. ∎

EXAMPLE 3 Let $A = \begin{bmatrix} 10 & -5 & 5 \\ 1 & 1 & -1 \\ 0 & -10 & -5 \end{bmatrix}$, $\mathbf{x} = \begin{bmatrix} \frac{1}{4} \\ \frac{1}{2} \\ \frac{1}{4} \end{bmatrix}$, and $\mathbf{y} = \begin{bmatrix} \frac{1}{4} \\ \frac{1}{4} \\ \frac{1}{2} \end{bmatrix}$, where A

comes from Example 2. Compute $E(\mathbf{x}, \mathbf{y})$ and verify that this number lies between $v(\mathbf{x})$ and $v(\mathbf{y})$.

SOLUTION Compute

$$E(\mathbf{x}, \mathbf{y}) = \mathbf{x}^T A \mathbf{y} = \begin{bmatrix} \frac{1}{4} & \frac{1}{2} & \frac{1}{4} \end{bmatrix} \begin{bmatrix} 10 & -5 & 5 \\ 1 & 1 & -1 \\ 0 & -10 & -5 \end{bmatrix} \begin{bmatrix} \frac{1}{4} \\ \frac{1}{4} \\ \frac{1}{2} \end{bmatrix}$$

$$= \begin{bmatrix} \frac{1}{4} & \frac{1}{2} & \frac{1}{4} \end{bmatrix} \begin{bmatrix} \frac{15}{4} \\ 0 \\ -5 \end{bmatrix} = -\frac{5}{16}$$

Next, from (2), $v(\mathbf{x})$ is the minimum of $E(\mathbf{x}, \mathbf{e}_j)$ for $1 \leq j \leq 3$. So compute

$$E(\mathbf{x}, \mathbf{e}_1) = \frac{10}{4} + \frac{1}{2} + 0 = 3$$
$$E(\mathbf{x}, \mathbf{e}_2) = -\frac{5}{4} + \frac{1}{2} - \frac{10}{4} = -\frac{13}{4}$$
$$E(\mathbf{x}, \mathbf{e}_3) = \frac{5}{4} - \frac{1}{2} - \frac{5}{4} = -\frac{1}{2}$$

Then $v(\mathbf{x}) = \min \left\{ 3, -\frac{13}{4}, -\frac{1}{2} \right\} = -\frac{13}{4} < -\frac{5}{16} = E(\mathbf{x}, \mathbf{y})$. Similarly, $E(\mathbf{e}_1, \mathbf{y}) = \frac{15}{4}$, $E(\mathbf{e}_2, \mathbf{y}) = 0$, and $E(\mathbf{e}_3, \mathbf{y}) = -5$, and so $v(\mathbf{y}) = \max \left\{ \frac{15}{4}, 0, -5 \right\} = \frac{15}{4}$. Thus $E(\mathbf{x}, \mathbf{y}) \leq v(\mathbf{y})$, as expected. ∎

In Theorem 1, the proof that $v_R \leq v_C$ was simple. A fundamental result in game theory is that $v_R = v_C$, but this is not easy to prove. The first proof by John von Neumann in 1928 was technically difficult. Perhaps the best-known proof depends strongly on certain properties of convex sets and hyperplanes. It appeared in the classic 1944 book *Theory of Games and Economic Behavior*, by von Neumann and Oskar Morgenstern.[4]

---

[4] More precisely, the proof involves finding a hyperplane that strictly separates the origin $\mathbf{0}$ from the convex hull of $\{\mathbf{a}_1, \ldots, \mathbf{a}_n, \mathbf{e}_1, \ldots, \mathbf{e}_m\}$, where $\mathbf{a}_1, \ldots, \mathbf{a}_n$ are the columns of A and $\mathbf{e}_1, \ldots, \mathbf{e}_m$ are the standard basis vectors in $\mathbb{R}^m$. The details are in Steven R. Lay, *Convex Sets and Their Applications* (New York: John Wiley & Sons, 1982; Mineola, NY: Dover Publications, 2007), pp. 159–163.

THEOREM 2

Minimax Theorem

In any matrix game, $v_R = v_C$. That is,

$$\max_{\mathbf{x} \in X} \min_{\mathbf{y} \in Y} E(\mathbf{x}, \mathbf{y}) = \min_{\mathbf{y} \in Y} \max_{\mathbf{x} \in X} E(\mathbf{x}, \mathbf{y})$$

DEFINITION

The common value $v = v_R = v_C$ is called the **value of the game**. Any pair of optimal strategies $(\hat{\mathbf{x}}, \hat{\mathbf{y}})$ is called a **solution** to the game.

When $(\hat{\mathbf{x}}, \hat{\mathbf{y}})$ is a solution to the game, $v_R = v(\hat{\mathbf{x}}) \leq E(\hat{\mathbf{x}}, \hat{\mathbf{y}}) \leq v(\hat{\mathbf{y}}) = v_C$, which shows that $E(\hat{\mathbf{x}}, \hat{\mathbf{y}}) = v$.

The next theorem is the main theoretical result of this section. A proof can be based either on the Minimax Theorem or on the theory of linear programming (in Section 9.4).[5]

THEOREM 3

Fundamental Theorem for Matrix Games

In any matrix game, there are always optimal strategies. That is, every matrix game has a solution.

2 x n Matrix Games

When a game matrix A has 2 rows and n columns, an optimal row strategy and v_R are fairly easy to compute. Suppose

$$A = \begin{bmatrix} a_{11} & a_{12} & \cdots & a_{1n} \\ a_{21} & a_{22} & \cdots & a_{2n} \end{bmatrix}$$

The objective of player R is to choose $\mathbf{x}$ in $\mathbb{R}^2$ to maximize $v(\mathbf{x})$. Since $\mathbf{x}$ has only two entries, the strategy space X for R may be parameterized by a variable t, with a typical $\mathbf{x}$ in X having the form $\mathbf{x}(t) = \begin{bmatrix} 1 - t \\ t \end{bmatrix}$ for $0 \leq t \leq 1$. From formula (2), $v(\mathbf{x}(t))$ is the minimum of the inner product of $\mathbf{x}(t)$ with each of the columns of A. That is,

$$v(\mathbf{x}(t)) = \min \left\{ \mathbf{x}(t)^T \begin{bmatrix} a_{1j} \\ a_{2j} \end{bmatrix} : j = 1, \ldots, n \right\}$$

$$= \min \left\{ a_{1j}(1 - t) + a_{2j}t : j = 1, \ldots, n \right\} \tag{4}$$

---

[5] The proof based on the Minimax Theorem goes as follows: The function $v(\mathbf{x})$ is continuous on the compact set X, so there exists a point $\hat{\mathbf{x}}$ in X such that

$$v(\hat{\mathbf{x}}) = \max_{\mathbf{x} \in X} v(\mathbf{x}) = v_R$$

Similarly, there exists $\hat{\mathbf{y}}$ in Y such that

$$v(\hat{\mathbf{y}}) = \min_{\mathbf{y} \in Y} v(\mathbf{y}) = v_C$$

According to the Minimax Theorem, $v_R = v_C = v$.

Thus $v(\mathbf{x}(t))$ is the minimum value of n linear functions of t. When these functions are graphed on one coordinate system for $0 \le t \le 1$, the graph of $z = v(\mathbf{x}(t))$ as a function of t becomes evident, and the maximum value of $v(\mathbf{x}(t))$ is easy to find. The process is illustrated best by an example.

EXAMPLE 4   Consider the game whose payoff matrix is

$$A = \begin{bmatrix} 1 & 5 & 3 & 6 \\ 4 & 0 & 1 & 2 \end{bmatrix}$$

a. On a t-z coordinate system, sketch the four lines $z = a_{1j}(1-t) + a_{2j}t$ for $0 \le t \le 1$, and darken the line segments that correspond to the graph of $z = v(\mathbf{x}(t))$, from (4).

b. Identify the highest point $M = (t, z)$ on the graph of $v(\mathbf{x}(t))$. The z-coordinate of M is the value v_R of the game for R, and the t-coordinate determines an optimal strategy $\hat{\mathbf{x}}(t)$ for R.

SOLUTION

a. The four lines are

$$z = 1 \cdot (1-t) + 4 \cdot t = 3t + 1$$
$$z = 5 \cdot (1-t) + 0 \cdot t = -5t + 5$$
$$z = 3 \cdot (1-t) + 1 \cdot t = -2t + 3$$
$$z = 6 \cdot (1-t) + 2 \cdot t = -4t + 6$$

See Figure 2. Notice that the line $z = a_{1j} \cdot (1-t) + a_{2j} \cdot t$ goes through the points $(0, a_{1j})$ and $(1, a_{2j})$. For instance, the line $z = 6 \cdot (1-t) + 2 \cdot t$ for column 4 goes through the points $(0, 6)$ and $(1, 2)$. The heavy polygonal path in Figure 2 represents $v(\mathbf{x})$ as a function of t, because the z-coordinate of a point on this path is the minimum of the corresponding z-coordinates of points on the four lines in Figure 2.

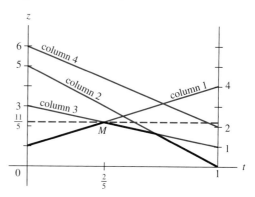

FIGURE 2

b. The highest point, M, on the graph of $v(\mathbf{x})$ is the intersection of the lines corresponding to the first and third columns of A. The coordinates of M are $\left(\frac{2}{5}, \frac{11}{5}\right)$.[6] The value

---

[6] Solve the equations for columns 1 and 3 simultaneously:

$$\left. \begin{array}{l} \text{(column 1) } z = 3t + 1 \\ \text{(column 3) } z = -2t + 3 \end{array} \right\} \Rightarrow t = \frac{2}{5}, z = \frac{11}{5}$$

of the game for R is $\frac{11}{5}$. This value is attained at $t = \frac{2}{5}$, so the optimal strategy for

$$R \text{ is } \hat{\mathbf{x}} = \begin{bmatrix} 1 - \frac{2}{5} \\ \frac{2}{5} \end{bmatrix} = \begin{bmatrix} \frac{3}{5} \\ \frac{2}{5} \end{bmatrix}. \qquad \blacksquare$$

For any $2 \times n$ matrix game, Example 4 illustrates the method for finding an optimal solution for player R. Theorem 3 guarantees that there also exists an optimal strategy for player C, and the value of the game is the same for C as for R. With this value available, an analysis of the graphical solution for R, as in Figure 2, will reveal how to produce an optimal strategy $\hat{\mathbf{y}}$ for C. The next theorem supplies the key information about $\hat{\mathbf{y}}$.

THEOREM 4

Let $\hat{\mathbf{x}}$ and $\hat{\mathbf{y}}$ be optimal strategies for an $m \times n$ matrix game whose value is v, and suppose that

$$\hat{\mathbf{x}} = \hat{x}_1 \mathbf{e}_1 + \cdots + \hat{x}_m \mathbf{e}_m \quad \text{in } \mathbb{R}^m \tag{5}$$

Then $\hat{\mathbf{y}}$ is a convex combination of the pure strategies $\mathbf{e}_j$ in $\mathbb{R}^n$ for which $E(\hat{\mathbf{x}}, \mathbf{e}_j) = v$. In addition, $\hat{\mathbf{y}}$ satisfies the equation

$$E(\mathbf{e}_i, \hat{\mathbf{y}}) = v \tag{6}$$

for each i such that $\hat{x}_i \neq 0$.

PROOF Write $\hat{\mathbf{y}} = \hat{y}_1 \mathbf{e}_1 + \cdots + \hat{y}_n \mathbf{e}_n$ in $\mathbb{R}^n$, and note that $v = E(\hat{\mathbf{x}}, \hat{\mathbf{y}}) = v(\hat{\mathbf{x}}) \leq E(\hat{\mathbf{x}}, \mathbf{e}_j)$ for $j = 1, \ldots, n$. So there exist nonnegative numbers ε_j such that

$$E(\hat{\mathbf{x}}, \mathbf{e}_j) = v + \varepsilon_j \quad (j = 1, \ldots, n)$$

Then

$$v = E(\hat{\mathbf{x}}, \hat{\mathbf{y}}) = E(\hat{\mathbf{x}}, \hat{y}_1 \mathbf{e}_1 + \cdots + \hat{y}_n \mathbf{e}_n)$$

$$= \sum_{j=1}^{n} \hat{y}_j E(\hat{\mathbf{x}}, \mathbf{e}_j) = \sum_{j=1}^{n} \hat{y}_j (v + \varepsilon_j)$$

$$= v + \sum_{j=1}^{n} \hat{y}_j \varepsilon_j$$

because the $\hat{y}_j$ sum to 1. This equality is possible only if $\hat{y}_j = 0$ whenever $\varepsilon_j > 0$. Thus $\hat{\mathbf{y}}$ is a linear combination of the $\mathbf{e}_j$ for which $\varepsilon_j = 0$. For such j, $E(\hat{\mathbf{x}}, \mathbf{e}_j) = v$.

Next, observe that $E(\mathbf{e}_i, \hat{\mathbf{y}}) \leq v(\hat{\mathbf{y}}) = E(\hat{\mathbf{x}}, \hat{\mathbf{y}})$ for $i = 1, \ldots, m$. So there exist nonnegative numbers δ_i such that

$$E(\mathbf{e}_i, \hat{\mathbf{y}}) + \delta_i = v \quad (i = 1, \ldots, m) \tag{7}$$

Then, using (5) gives

$$v = E(\hat{\mathbf{x}}, \hat{\mathbf{y}}) = \sum_{i=1}^{m} \hat{x}_i E(\mathbf{e}_i, \hat{\mathbf{y}})$$

$$= \sum_{i=1}^{m} \hat{x}_i (v - \delta_i) = v - \sum_{i=1}^{m} \hat{x}_i \delta_i$$

since the $\hat{x}_i$ sum to 1. This equality is possible only if $\delta_i = 0$ when $\hat{x}_i \neq 0$. By (7), $E(\mathbf{e}_i, \hat{\mathbf{y}}) = v$ for each i such that $\hat{x}_i \neq 0$. $\blacksquare$

EXAMPLE 5   The value of the game in Example 4 is $\frac{11}{5}$, attained when $\hat{\mathbf{x}} = \begin{bmatrix} \frac{3}{5} \\ \frac{2}{5} \end{bmatrix}$.
Use this fact to find an optimal strategy for the column player C.

SOLUTION   The z-coordinate of the maximum point M in Figure 2 is the value of
the game, and the t-coordinate identifies the optimal strategy $\mathbf{x}(\frac{2}{5}) = \hat{\mathbf{x}}$. Recall that the
z-coordinates of the lines in Figure 2 represent $E(\mathbf{x}(t), \mathbf{e}_j)$ for $j = 1, \ldots, 4$. Only the
lines for columns 1 and 3 pass through the point M, which means that

$$E(\hat{\mathbf{x}}, \mathbf{e}_1) = \tfrac{11}{5} \quad \text{and} \quad E(\hat{\mathbf{x}}, \mathbf{e}_3) = \tfrac{11}{5}$$

while $E(\hat{\mathbf{x}}, \mathbf{e}_2)$ and $E(\hat{\mathbf{x}}, \mathbf{e}_4)$ are greater than $\frac{11}{5}$. By Theorem 4, the optimal column
strategy $\hat{\mathbf{y}}$ for C is a linear combination of the pure strategies $\mathbf{e}_1$ and $\mathbf{e}_3$ in $\mathbb{R}^2$. Thus, $\hat{\mathbf{y}}$
has the form

$$\hat{\mathbf{y}} = c_1 \begin{bmatrix} 1 \\ 0 \\ 0 \\ 0 \end{bmatrix} + c_3 \begin{bmatrix} 0 \\ 0 \\ 1 \\ 0 \end{bmatrix} = \begin{bmatrix} c_1 \\ 0 \\ c_3 \\ 0 \end{bmatrix}$$

where $c_1 + c_3 = 1$. Since both coordinates of the optimal $\hat{\mathbf{x}}$ are nonzero, Theorem 4
shows that $E(\mathbf{e}_1, \hat{\mathbf{y}}) = \frac{11}{5}$ and $E(\mathbf{e}_2, \hat{\mathbf{y}}) = \frac{11}{5}$. Each condition, by itself, determines $\hat{\mathbf{y}}$.
For example,

$$E(\mathbf{e}_1, \hat{\mathbf{y}}) = \mathbf{e}_1^T A \hat{\mathbf{y}} = \begin{bmatrix} 1 & 0 \end{bmatrix} \begin{bmatrix} 1 & 5 & 3 & 6 \\ 4 & 0 & 1 & 2 \end{bmatrix} \begin{bmatrix} c_1 \\ 0 \\ c_3 \\ 0 \end{bmatrix} = c_1 + 3c_3 = \tfrac{11}{5}$$

Substitute $c_3 = 1 - c_1$, and obtain $c_1 + 3(1 - c_1) = \frac{11}{5}$, $c_1 = \frac{2}{5}$ and $c_3 = \frac{3}{5}$. The
optimal strategy for C is $\hat{\mathbf{y}} = \begin{bmatrix} \frac{2}{5} \\ 0 \\ \frac{3}{5} \\ 0 \end{bmatrix}$.   ∎

Reducing the Size of a Game

The general $m \times n$ matrix game can be solved using linear programming techniques, and
Section 9.4 describes one method for doing this. In some cases, however, a matrix game
can be reduced to a "smaller" game whose matrix has only two rows. If this happens,
the graphical method of Examples 4 and 5 is available.

DEFINITION

> Given $\mathbf{a}$ and $\mathbf{b}$ in $\mathbb{R}^n$, with entries a_i and b_i, respectively, vector $\mathbf{a}$ is said to
> **dominate** vector $\mathbf{b}$ if $a_i \geq b_i$ for all $i = 1, \ldots, n$ and $a_i > b_i$ for at least one i.
> If $\mathbf{a}$ dominates $\mathbf{b}$, then $\mathbf{b}$ is said to be **recessive** to $\mathbf{a}$.

Suppose that in the matrix game A, row r dominates row s. This means that for R
the pure strategy of choosing row r is at least as good as the pure strategy of choosing
row s, no matter what C may choose, and for some choice by C, r is better than s.
It follows that the recessive row s (the "smaller" one) can be ignored by R without
hurting R's expected payoff. A similar analysis applies to the columns of A, in which

case the dominating "larger" column is ignored. These observations are summarized in the following theorem.

THEOREM 5

Let A be an $m \times n$ matrix game. If row s in the matrix A is recessive to some other row, then let A_1 be the $(m-1) \times n$ matrix obtained by deleting row s from A. Similarly, if column t of matrix A dominates some other column, let A_2 be the $m \times (n-1)$ matrix obtained by deleting column t from A. In either case, any optimal strategy of the reduced matrix game A_1 or A_2 will determine an optimal strategy for A.

EXAMPLE 6 Use the process described in Theorem 5 to reduce the following matrix game to a smaller size. Then find the value of the game and optimal strategies for both players in the original game.

$$A = \begin{bmatrix} 7 & 1 & 6 & 7 \\ 8 & 3 & 1 & 0 \\ 4 & 5 & 3 & 3 \end{bmatrix}$$

SOLUTION Since the first column dominates the third, player C will never want to use the first pure strategy. So delete column 1 and obtain

$$\begin{bmatrix} * & 1 & 6 & 7 \\ * & 3 & 1 & 0 \\ * & 5 & 3 & 3 \end{bmatrix}$$

In this matrix, row 2 is recessive to row 3. Delete row 2 and obtain

$$\begin{bmatrix} * & 1 & 6 & 7 \\ * & * & * & * \\ * & 5 & 3 & 3 \end{bmatrix}$$

This reduced 2×3 matrix can be reduced further by dropping the last column, since it dominates column 2. Thus, the original matrix game A has been reduced to

$$B = \begin{bmatrix} 1 & 6 \\ 5 & 3 \end{bmatrix} \quad \text{when } A = \begin{bmatrix} 7 & 1 & 6 & 7 \\ 8 & 3 & 1 & 0 \\ 4 & 5 & 3 & 3 \end{bmatrix} \tag{8}$$

and any optimal strategy for B will produce an optimal strategy for A, with zeros as entries corresponding to deleted rows or columns.

A quick check of matrix B shows that the game has no saddle point (because 3 is the max of the row minima and 5 is the min of the column maxima). So the graphical solution method is needed. Figure 3 shows the lines corresponding to the two columns of B, whose equations are $z = 4t + 1$ and $z = -3t + 6$. They intersect where $t = \frac{5}{7}$; the value of the game is $\frac{27}{7}$, and the optimal row strategy for matrix B is

$$\hat{\mathbf{x}} = \mathbf{x}\left(\tfrac{5}{7}\right) = \begin{bmatrix} 1 - \frac{5}{7} \\ \frac{5}{7} \end{bmatrix} = \begin{bmatrix} \frac{2}{7} \\ \frac{5}{7} \end{bmatrix}$$

Since the game has no saddle point, the optimal column strategy must be a linear combination of the two pure strategies. Set $\hat{\mathbf{y}} = c_1\mathbf{e}_1 + c_2\mathbf{e}_2$, and use the second part of Theorem 4 to write

$$\tfrac{27}{7} = E(\mathbf{e}_1, \hat{\mathbf{y}}) = \begin{bmatrix} 1 & 0 \end{bmatrix} \begin{bmatrix} 1 & 6 \\ 5 & 3 \end{bmatrix} \begin{bmatrix} c_1 \\ c_2 \end{bmatrix} = c_1 + 6c_2 = (1 - c_2) + 6c_2$$

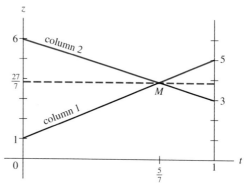

FIGURE 3

Solving gives $5c_2 = \frac{20}{7}$, $c_2 = \frac{4}{7}$, and $c_1 = 1 - c_2 = \frac{3}{7}$. Thus $\hat{\mathbf{y}} = \begin{bmatrix} \frac{3}{7} \\ \frac{4}{7} \end{bmatrix}$. As a check,

compute $E(\mathbf{e}_2, \hat{\mathbf{y}}) = 5(\frac{3}{7}) + 3(\frac{4}{7}) = \frac{27}{7} = v$.

The final step is to construct the solution for matrix A from the solution for matrix B (given by $\hat{\mathbf{x}}$ and $\hat{\mathbf{y}}$ above). Look at the matrices in (8) to see where the extra zeros go. The row and column strategies for A are, respectively,

$$\hat{\mathbf{x}} = \begin{bmatrix} \frac{2}{7} \\ 0 \\ \frac{5}{7} \end{bmatrix} \quad \text{and} \quad \hat{\mathbf{y}} = \begin{bmatrix} 0 \\ \frac{3}{7} \\ \frac{4}{7} \\ 0 \end{bmatrix}$$ ∎

Practice Problem

Find the optimal strategies and the value of the matrix game

$$\begin{bmatrix} -3 & 4 & 1 & 3 \\ 2 & 2 & -1 & 0 \\ 1 & 5 & 2 & 3 \end{bmatrix}$$

9.1 Exercises

In Exercises 1–4, write the payoff matrix for each game.

1. Player R has a supply of dimes and quarters. Player R chooses one of the coins, and player C must guess which coin R has chosen. If the guess is correct, C takes the coin. If the guess is incorrect, C gives R an amount equal to R's chosen coin.

2. Players R and C each show one, two, or three fingers. If the total number N of fingers shown is even, then C pays N dollars to R. If N is odd, R pays N dollars to C.

3. In the traditional Japanese children's game *janken* (or "rock, scissors, paper"), at a given signal, each of two players shows either no fingers (rock), two fingers (scissors), or all five

(paper). Rock beats scissors, scissors beats paper, and paper beats rock. In the case of a tie, there is no payoff. In the case of a win, suppose the winner collects 5 yen.

4. Player R has three cards: a red 3, a red 6, and a black 7. Player C has two cards: a red 4 and a black 9. They each show one of their cards. If the cards are the same color, R receives the larger of the two numbers. If the cards are of different colors, C receives the sum of the two numbers.

Find all saddle points for the matrix games in Exercises 5–8.

5. $\begin{bmatrix} 4 & 3 \\ 1 & -1 \end{bmatrix}$

6. $\begin{bmatrix} 2 & 1 & 3 \\ 4 & -2 & 1 \end{bmatrix}$

7. $\begin{bmatrix} 5 & 3 & 4 & 3 \\ -2 & 1 & -5 & 2 \\ 4 & 3 & 7 & 3 \end{bmatrix}$ 8. $\begin{bmatrix} -2 & 4 & 1 & -1 \\ 3 & 5 & 2 & 2 \\ 1 & -3 & 0 & 2 \end{bmatrix}$

9. Let M be the matrix game having payoff matrix
$\begin{bmatrix} 1 & 2 & -2 \\ 0 & 1 & 4 \\ 3 & -1 & 1 \end{bmatrix}$. Find $E(\mathbf{x}, \mathbf{y})$, $v(\mathbf{x})$, and $v(\mathbf{y})$ when $\mathbf{x}$ and $\mathbf{y}$ have the given values.

a. $\mathbf{x} = \begin{bmatrix} \frac{1}{3} \\ \frac{1}{2} \\ \frac{1}{6} \end{bmatrix}$ and $\mathbf{y} = \begin{bmatrix} \frac{1}{4} \\ \frac{1}{2} \\ \frac{1}{4} \end{bmatrix}$

b. $\mathbf{x} = \begin{bmatrix} \frac{1}{4} \\ \frac{1}{2} \\ \frac{1}{4} \end{bmatrix}$ and $\mathbf{y} = \begin{bmatrix} \frac{1}{2} \\ \frac{1}{4} \\ \frac{1}{4} \end{bmatrix}$

10. Let M be the matrix game having payoff matrix
$\begin{bmatrix} 2 & 0 & 1 & -1 \\ -1 & 1 & -2 & 0 \\ 1 & -2 & 2 & 1 \end{bmatrix}$. Find $E(\mathbf{x}, \mathbf{y})$, $v(\mathbf{x})$, and $v(\mathbf{y})$ when $\mathbf{x}$ and $\mathbf{y}$ have the given values.

a. $\mathbf{x} = \begin{bmatrix} \frac{1}{3} \\ 0 \\ \frac{2}{3} \end{bmatrix}$ and $\mathbf{y} = \begin{bmatrix} \frac{1}{4} \\ \frac{1}{2} \\ 0 \\ \frac{1}{4} \end{bmatrix}$

b. $\mathbf{x} = \begin{bmatrix} \frac{1}{2} \\ \frac{1}{4} \\ \frac{1}{4} \end{bmatrix}$ and $\mathbf{y} = \begin{bmatrix} 0 \\ \frac{1}{4} \\ \frac{1}{2} \\ \frac{1}{4} \end{bmatrix}$

In Exercises 11–18, find the optimal row and column strategies and the value of each matrix game.

11. $\begin{bmatrix} 3 & -2 \\ 0 & 1 \end{bmatrix}$ 12. $\begin{bmatrix} 2 & -2 \\ -3 & 6 \end{bmatrix}$

13. $\begin{bmatrix} 3 & 5 \\ 4 & 1 \end{bmatrix}$ 14. $\begin{bmatrix} 3 & 5 & 3 & 2 \\ -1 & 9 & 1 & 8 \end{bmatrix}$

15. $\begin{bmatrix} 4 & 6 & 2 & 0 \\ 1 & 3 & 2 & 5 \end{bmatrix}$ 16. $\begin{bmatrix} 5 & -1 & 1 \\ 4 & 2 & 3 \\ -2 & -3 & 1 \end{bmatrix}$

17. $\begin{bmatrix} 0 & 1 & -1 & 4 & 3 \\ 1 & -1 & 3 & -1 & -3 \\ 2 & -1 & 4 & 0 & -2 \\ -1 & 0 & -2 & 5 & 1 \end{bmatrix}$

18. $\begin{bmatrix} 6 & 4 & 5 & 5 \\ 0 & 4 & 2 & 7 \\ 6 & 3 & 5 & 2 \\ 2 & 5 & 3 & 7 \end{bmatrix}$

19. A certain army is engaged in guerrilla warfare. It has two ways of getting supplies to its troops: it can send a convoy up the river road or it can send a convoy overland through the jungle. On a given day, the guerrillas can watch only one of the two roads. If the convoy goes along the river and the guerrillas are there, the convoy will have to turn back and 4 army soldiers will be lost. If the convoy goes overland and encounters the guerrillas, half the supplies will get through, but 7 army soldiers will be lost. Each day a supply convoy travels one of the roads, and if the guerrillas are watching the other road, the convoy gets through with no losses. Set up and solve the following as matrix games, with R being the army.

a. What is the optimal strategy for the army if it wants to maximize the amount of supplies it gets to its troops? What is the optimal strategy for the guerrillas if they want to prevent the most supplies from getting through? If these strategies are followed, what portion of the supplies gets through?

b. What is the optimal strategy for the army if it wants to minimize its casualties? What is the optimal strategy for the guerrillas if they want to inflict maximum losses on the army? If these strategies are followed, what portion of the supplies gets through?

20. Suppose in Exercise 19 that whenever the convoy goes overland two soldiers are lost to land mines, whether they are attacked or not. Thus, if the army encounters the guerrillas, there will be 9 casualties. If it does not encounter the guerrillas, there will be 2 casualties.

a. Find the optimal strategies for the army and the guerrillas with respect to the number of army casualties.

b. In part (a), what is the "value" of the game? What does this represent in terms of the troops?

In Exercises 21–30, mark each statement True or False (**T/F**). Justify each answer.

21. (**T/F**) The payoff matrix for a matrix game indicates what R wins for each combination of moves.

22. (**T/F**) If a_{ij} is a saddle point, then a_{ij} is the smallest entry in row i and the largest entry in column j.

23. (**T/F**) With a pure strategy, a player makes the same choice each time the game is played.

24. (**T/F**) Each pure strategy is an optimal strategy.

25. (**T/F**) The value $v(\mathbf{x})$ of a particular strategy $\mathbf{x}$ to player R is equal to the maximum of the inner product of $\mathbf{x}$ with each of the columns of the payoff matrix.

26. (**T/F**) The value v_R of the game to player R is the maximum of the values of the various possible strategies for R.

27. (**T/F**) The Minimax Theorem says that every matrix game has a solution.

28. (**T/F**) The Fundamental Theorem for Matrix Games shows how to solve every matrix game.

29. (T/F) If row s is recessive to some other row in payoff matrix A, then row s will not be used (that is, have probability zero) in an optimal strategy for (row) player R.

30. (T/F) If column t dominates some other column in a payoff matrix A, then column t will not be used (that is, have probability zero) in an optimal strategy for (column) player C.

31. Find the optimal strategies and the value of the game in Example 2.

32. Bill and Wayne are playing a game in which each player has a choice of two colors: red or blue. The payoff matrix with Bill as the row player is given below.

$$\begin{array}{c} \\ \text{red} \\ \text{blue} \end{array} \begin{array}{cc} \text{red} & \text{blue} \\ \begin{bmatrix} -1 & 2 \\ 3 & -4 \end{bmatrix} \end{array}$$

For example, this means that if both people choose red, then Bill pays Wayne one unit.

a. Using the same payoffs for Bill and Wayne, write the matrix that shows the winnings with Wayne as the row player.

b. If A is the matrix with Bill as the row player, write your answer to (a) in terms of A.

33. Consider the matrix game $A = \begin{bmatrix} a & b \\ c & d \end{bmatrix}$, where A has no saddle point.

a. Find a formula for the optimal strategies $\hat{\mathbf{x}}$ for R and $\hat{\mathbf{y}}$ for C. What is the value of the game?

b. Let $J = \begin{bmatrix} 1 & 1 \\ 1 & 1 \end{bmatrix}$, and let α and β be real numbers with $\alpha \neq 0$. Use your answer in part (a) to show that the optimal strategies for the matrix game $B = \alpha A + \beta J$ are the same as for A. In particular, note that the optimal strategies for A and $A + \beta J$ are the same.

34. Let A be a matrix game having value v. Find an example to show that $E(\mathbf{x}, \mathbf{y}) = v$ does not necessarily imply that $\mathbf{x}$ and $\mathbf{y}$ are optimal strategies.

Solution to Practice Problem

The first row is recessive to the third row, so the first row may be eliminated. The second and fourth columns dominate the first and third columns, respectively. Deletion of the second and fourth columns leaves the matrix B as

$$B = \begin{bmatrix} 2 & -1 \\ 1 & 2 \end{bmatrix} \quad \text{when } A = \begin{bmatrix} -3 & 4 & 1 & 3 \\ 2 & 2 & -1 & 0 \\ 1 & 5 & 2 & 3 \end{bmatrix}$$

The game for B has no saddle point, but a graphical analysis will work. The two columns of B determine the two lines shown below, whose equations are $z = 2 \cdot (1 - t) + 1 \cdot t$ and $z = -1 \cdot (1 - t) + 2 \cdot t$.

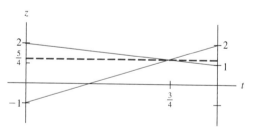

These lines intersect at the point $\left(\frac{3}{4}, \frac{5}{4}\right)$. The value of the game is $\frac{5}{4}$, and the optimal row strategy for the matrix game B is

$$\hat{\mathbf{x}} = \mathbf{x}\left(\tfrac{3}{4}\right) = \begin{bmatrix} 1 - \frac{3}{4} \\ \frac{3}{4} \end{bmatrix} = \begin{bmatrix} \frac{1}{4} \\ \frac{3}{4} \end{bmatrix}$$

By Theorem 4, the optimal column strategy, $\hat{\mathbf{y}} = \begin{bmatrix} c_1 \\ c_2 \end{bmatrix}$, satisfies two equations $E(\mathbf{e}_1, \hat{\mathbf{y}}) = \frac{5}{4}$ and $E(\mathbf{e}_2, \hat{\mathbf{y}}) = \frac{5}{4}$, because $\hat{\mathbf{x}}$ is a linear combination of both $\mathbf{e}_1$ and $\mathbf{e}_2$.

Solution to Practice Problem (Continued)

Each of these equations determines $\hat{\mathbf{y}}$. For example,

$$\frac{5}{4} = E(\mathbf{e}_1, \hat{\mathbf{y}}) = \begin{bmatrix} 1 & 0 \end{bmatrix}\begin{bmatrix} 2 & -1 \\ 1 & 2 \end{bmatrix}\begin{bmatrix} c_1 \\ c_2 \end{bmatrix} = 2c_1 - c_2 = 2c_1 - (1 - c_1) = 3c_1 - 1$$

Thus, $c_1 = \frac{3}{4}$, and so $c_2 = \frac{1}{4}$, and $\hat{\mathbf{y}} = \begin{bmatrix} \frac{3}{4} \\ \frac{1}{4} \end{bmatrix}$. As a check, compute

$$E(\mathbf{e}_2, \hat{\mathbf{y}}) = \begin{bmatrix} 0 & 1 \end{bmatrix}\begin{bmatrix} 2 & -1 \\ 1 & 2 \end{bmatrix}\begin{bmatrix} \frac{3}{4} \\ \frac{1}{4} \end{bmatrix} = \begin{bmatrix} 1 & 2 \end{bmatrix}\begin{bmatrix} \frac{3}{4} \\ \frac{1}{4} \end{bmatrix} = \frac{5}{4}$$

This solves the game for B. The optimal row strategy $\hat{\mathbf{x}}$ for A needs a 0 in the first entry (for the deleted first row); the optimal column strategy $\hat{\mathbf{y}}$ for A needs 0's in entries 2 and 4 (for the two deleted columns). Thus

$$\hat{\mathbf{x}} = \begin{bmatrix} 0 \\ \frac{1}{4} \\ \frac{3}{4} \end{bmatrix} \quad \text{and} \quad \hat{\mathbf{y}} = \begin{bmatrix} \frac{3}{4} \\ 0 \\ \frac{1}{4} \\ 0 \end{bmatrix}$$

9.2 Linear Programming—Geometric Method

Since the 1950s, the variety and size of industrial linear programming problems have grown along with the dramatic increase in computing power. Still, at their core, linear programming problems have a concise mathematical description, discussed in this section. The final example in the section presents a geometric view of linear programming that is important for visualizing the algebraic approach needed for larger problems.

Generally speaking, a linear programming problem involves a system of linear inequalities in variables $x_1, \ldots, x_n$ and a linear functional f from $\mathbb{R}^n$ into $\mathbb{R}$. The system typically has many free variables, and the problem is to find a solution $\mathbf{x}$ that maximizes or minimizes $f(\mathbf{x})$.

EXAMPLE 1 The Shady-Lane grass seed company blends two types of seed mixtures, EverGreen and QuickGreen. Each bag of EverGreen contains 3 pounds of fescue seed, 1 pound of rye seed, and 1 pound of bluegrass seed. Each bag of QuickGreen contains 2 pounds of fescue seed, 2 pounds of rye seed, and 1 pound of bluegrass seed. The company has 1,200 pounds of fescue seed, 800 pounds of rye seed, and 450 pounds of bluegrass seed available to put into its mixtures. The company makes a profit of \$2 on each bag of EverGreen and \$3 on each bag of QuickGreen that it produces. Set up the mathematical problem that determines the number of bags of each mixture that Shady-Lane should make in order to maximize its profit.

SOLUTION The phrase "maximize ... profit" identifies the goal or objective of the problem. The first step, then, is to create a formula for the profit. Begin by naming the quantities that can vary. Let x_1 be the number of bags of EverGreen and x_2 the number of bags of QuickGreen that are produced. Since the profit on each bag of EverGreen is $2 and the profit on each bag of QuickGreen is $3, the total profit (in dollars) is

$$2x_1 + 3x_2 \qquad \text{(profit function)}$$

The next step is to write inequalities or equalities that x_1 and x_2 must satisfy, one for each of the ingredients that are in limited supply. Notice that each bag of EverGreen requires 3 pounds of fescue seed and each bag of QuickGreen requires 2 pounds of fescue seed. So the total amount of fescue seed required is $3x_1 + 2x_2$ pounds. Since only 1,200 pounds are available, x_1 and x_2 must satisfy

$$3x_1 + 2x_2 \leq 1,200 \qquad \text{(fescue)}$$

Similarly, EverGreen needs 1 pound of rye seed per bag, QuickGreen needs 2 pounds per bag, and only 800 pounds of rye seed are available. Thus, the total amount of rye seed required is $x_1 + 2x_2$, and x_1 and x_2 must satisfy

$$x_1 + 2x_2 \leq 800 \qquad \text{(rye)}$$

As for the bluegrass seed, EverGreen requires 1 pound per bag and QuickGreen requires 1 pound per bag. Since 450 pounds are available,

$$x_1 + x_2 \leq 450 \qquad \text{(bluegrass)}$$

Of course, x_1 and x_2 cannot be negative, so x_1 and x_2 must also satisfy

$$x_1 \geq 0 \quad \text{and} \quad x_2 \geq 0$$

The problem is summarized mathematically as

$$
\begin{aligned}
\text{Maximize} \quad & 2x_1 + 3x_2 & \text{(profit function)} \\
\text{subject to} \quad & 3x_1 + 2x_2 \leq 1,200 & \text{(fescue)} \\
& x_1 + 2x_2 \leq 800 & \text{(rye)} \\
& x_1 + x_2 \leq 450 & \text{(bluegrass)} \\
\text{and } & x_1 \geq 0, x_2 \geq 0.
\end{aligned}
$$

■

EXAMPLE 2 An oil refining company has two refineries that produce three grades of unleaded gasoline. Each day refinery A produces 12,000 gallons of regular, 4,000 gallons of premium, and 1,000 gallons of super, at a cost of $3,500. Each day refinery B produces 4,000 gallons of regular, 4,000 gallons of premium, and 5,000 gallons of super, at a cost of $3,000. An order is received for 48,000 gallons of regular, 32,000 gallons of premium, and 20,000 gallons of super. Set up a mathematical problem that determines the number of days each refinery should operate in order to fill the order at the least cost.

SOLUTION Suppose that refinery A operates x_1 days and refinery B operates x_2 days. The cost of doing this is $3,500x_1 + 3,000x_2$ dollars. The problem is to find a production schedule (x_1, x_2) that minimizes this cost and also ensures that the required gasoline is produced.

Since refinery A produces 12,000 gallons of regular each day and refinery B produces 4,000 gallons of regular each day, the total produced is $12,000x_1 + 4,000x_2$. The total should be at least 48,000 gallons. That is,

$$12,000x_1 + 4,000x_2 \geq 48,000$$

Similarly, for the premium,

$$4{,}000x_1 + 4{,}000x_2 \geq 32{,}000$$

and, for the super,

$$1{,}000x_1 + 5{,}000x_2 \geq 20{,}000$$

As in Example 1, x_1 and x_2 cannot be negative, so $x_1 \geq 0$ and $x_2 \geq 0$.
The problem is summarized mathematically as

Minimize	$3{,}500x_1 + 3{,}000x_2$	(cost function)
subject to	$12{,}000x_1 + 4{,}000x_2 \geq 48{,}000$	(regular)
	$4{,}000x_1 + 4{,}000x_2 \geq 32{,}000$	(premium)
	$1{,}000x_1 + 5{,}000x_2 \geq 20{,}000$	(super)

and $x_1 \geq 0, x_2 \geq 0$. ∎

The examples show how a linear programming problem involves finding the maximum (or minimum) of a linear function, called the **objective function**, subject to certain linear constraints. In many situations, the constraints take the form of linear inequalities and the variables are restricted to nonnegative values. Here is a precise statement of the so-called canonical form of a linear programming problem.

DEFINITION

Given $\mathbf{b} = \begin{bmatrix} b_1 \\ \vdots \\ b_m \end{bmatrix}$ in $\mathbb{R}^m$, $\mathbf{c} = \begin{bmatrix} c_1 \\ \vdots \\ c_n \end{bmatrix}$ in $\mathbb{R}^n$, and an $m \times n$ matrix $A = \begin{bmatrix} a_{ij} \end{bmatrix}$, the **canonical linear programming problem** is the following:

Find an n-tuple $\mathbf{x} = \begin{bmatrix} x_1 \\ \vdots \\ x_n \end{bmatrix}$ in $\mathbb{R}^n$ to maximize

$$f(x_1, \ldots, x_n) = c_1 x_1 + c_2 x_2 + \cdots + c_n x_n$$

subject to the constraints

$$a_{11}x_1 + a_{12}x_2 + \cdots + a_{1n}x_n \leq b_1$$
$$a_{21}x_1 + a_{22}x_2 + \cdots + a_{2n}x_n \leq b_2$$
$$\vdots$$
$$a_{m1}x_1 + a_{m2}x_2 + \cdots + a_{mn}x_n \leq b_m$$

and

$$x_j \geq 0 \quad \text{for } j = 1, \ldots, n$$

This may be restated in vector-matrix notation[1] as follows:

Maximize $f(\mathbf{x}) = \mathbf{c}^T\mathbf{x}$	(1)
subject to the constraints $A\mathbf{x} \leq \mathbf{b}$	(2)
and $\mathbf{x} \geq \mathbf{0}$	(3)

where an inequality between two vectors applies to each of their coordinates.
Any vector $\mathbf{x}$ that satisfies (2) and (3) is called a **feasible solution**, and the set of all feasible solutions, denoted by $\mathcal{F}$, is called the **feasible set**. A vector $\bar{\mathbf{x}}$ in $\mathcal{F}$ is an **optimal solution** if $f(\bar{\mathbf{x}}) = \max_{\mathbf{x} \in \mathcal{F}} f(\mathbf{x})$.

---

[1] Some authors write $\mathbf{c}^T\mathbf{x}$ as $\mathbf{c} \cdot \mathbf{x}$, using the dot product.

The canonical statement of the problem is really not as restrictive as it might seem. To minimize a function $h(\mathbf{x})$, replace it with the problem of maximizing the function $-h(\mathbf{x})$. A constraint inequality of the sort

$$a_{i1}x_1 + \cdots + a_{in}x_n \geq b_i$$

can be replaced by

$$-a_{i1}x_1 - \cdots - a_{in}x_n \leq -b_i$$

An equality constraint

$$a_{i1}x_1 + \cdots + a_{in}x_n = b_i$$

can be replaced by two inequalities

$$a_{i1}x_1 + \cdots + a_{in}x_n \leq b_i$$
$$-a_{i1}x_1 - \cdots - a_{in}x_n \leq -b_i$$

With an arbitrary canonical linear programming problem, two things can go wrong. If the constraint inequalities are inconsistent, then $\mathcal{F}$ is the empty set. If the objective function takes on arbitrarily large values in $\mathcal{F}$, then the desired maximum does not exist. In the former case, the problem is said to be **infeasible**; in the latter case, the problem is called **unbounded**.

EXAMPLE 3 The problem

$$\text{Maximize} \quad 5x$$
$$\text{subject to} \quad x \leq 3$$
$$-x \leq -4$$
$$x \geq 0$$

is infeasible, since there is no x such that $x \leq 3$ and $x \geq 4$. ∎

EXAMPLE 4 The problem

$$\text{Maximize} \quad 5x$$
$$\text{subject to} \quad -x \leq 3$$
$$x \geq 0$$

is unbounded. The values of $5x$ may be arbitrarily large, as x is only required to satisfy $x \geq 0$ (and $x \geq -3$). ∎

Fortunately, these are the only two things that can go wrong.

THEOREM 6 If the feasible set $\mathcal{F}$ is nonempty and if the objective function is bounded above on $\mathcal{F}$, then the canonical linear programming problem has at least one optimal solution. Furthermore, at least one of the optimal solutions is an extreme point of $\mathcal{F}$.[2]

---

[2] The feasible set is the solution of a system of linear inequalities. Geometrically, this corresponds to the intersection of a finite number of (closed) half-spaces, sometimes called a polyhedral set. Intuitively, the extreme points correspond to the "corner points," or vertices, of this polyhedral set. The notion of an extreme point is discussed more fully in Section 8.5.

A proof of Theorem 6 is in Steven R. Lay, *Convex Sets and Their Applications* (New York: John Wiley & Sons, 1982; Mineola, NY: Dover Publications, 2007), p. 171.

Theorem 6 describes when an optimal solution exists, and it suggests a possible technique for finding one. That is, evaluate the objective function at each of the extreme points of $\mathcal{F}$ and select the point that gives the largest value. This works well in simple cases such as the next two examples. The geometric approach is limited to two or three dimensions, but it provides an important visualization of the nature of the solution set and how the objective function interacts with the feasible set to identify extreme points.

EXAMPLE 5 Maximize $f(x_1, x_2) = 2x_1 + 3x_2$

$$\text{subject to} \quad x_1 \qquad\qquad \leq 30$$
$$x_2 \leq 20$$
$$x_1 + 2x_2 \leq 54$$
$$\text{and } x_1 \geq 0, x_2 \geq 0.$$

SOLUTION Figure 1 shows the shaded pentagonal feasible set, obtained by graphing each of the constraint inequalities. (For simplicity, points in this section are displayed as ordered pairs or triples.) There are five extreme points, corresponding to the five vertices of the feasible set. They are found by solving the appropriate pairs of linear equations. For example, the extreme point $(14, 20)$ is found by solving the linear system $x_1 + 2x_2 = 54$ and $x_2 = 20$. The table below shows the value of the objective function at each extreme point. Evidently, the maximum is 96 at $x_1 = 30$ and $x_2 = 12$.

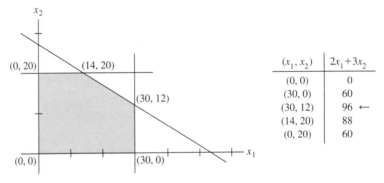

(x_1, x_2)	$2x_1 + 3x_2$
$(0, 0)$	0
$(30, 0)$	60
$(30, 12)$	96 ←
$(14, 20)$	88
$(0, 20)$	60

FIGURE 1

Another geometric technique that can be used when the problem involves two variables is to graph several *level lines* for the objective function. These are parallel lines, and the objective function has a constant value on each line. (See Figure 2.) The

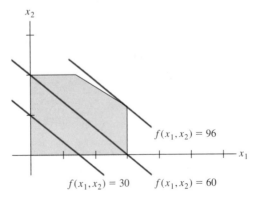

FIGURE 2

values of the objective function $f(x_1, x_2)$ increase as (x_1, x_2) moves from left to right. The level line farthest to the right that still intersects the feasible set is the line through the vertex $(30, 12)$. Thus, the point $(30, 12)$ yields the maximum value of $f(x_1, x_2)$ over the feasible set. ■

EXAMPLE 6   Maximize $f(x_1, x_2, x_3) = 2x_1 + 3x_2 + 4x_3$

$$\text{subject to} \qquad x_1 + \; x_2 + \; x_3 \le 50$$
$$x_1 + 2x_2 + 4x_3 \le 80$$
$$\text{and } x_1 \ge 0, x_2 \ge 0, x_3 \ge 0.$$

SOLUTION   Each of the five inequalities above determines a "half-space" in $\mathbb{R}^3$—a plane together with all points on one side of the plane. The feasible set of this linear programming problem is the intersection of these half-spaces, which is a convex set in the first octant of $\mathbb{R}^3$.

When the first inequality is changed to an equality, the graph is a plane that intercepts each coordinate axis 50 units from the origin and determines the equilateral triangular region shown in Figure 3. Since $(0, 0, 0)$ satisfies the inequality, so do all the other points "below" the plane. In a similar fashion, the second (in)equality determines a triangular region on a plane (shown in Figure 4) that passes somewhat closer to the origin. The two planes intersect in a line that contains segment EB.

The quadrilateral surface $BCDE$ forms a boundary of the feasible set, because it is below the equilateral triangular region. Beyond EB, however, the two planes change position relative to the origin, so the planar region ABE forms another bounding surface for the feasible set. The vertices of the feasible set are the points A, B, C, D, E, and 0 (the origin). See Figure 5, which has all sides of the feasible set shaded except the large "top" piece. To find the coordinates of B, solve the system

$$\begin{cases} x_1 + \; x_2 + \; x_3 = 50 \\ x_1 + 2x_2 + 4x_3 = 80 \\ \qquad\qquad\quad x_3 = \; 0 \end{cases} \Rightarrow \begin{cases} x_1 + \; x_2 = 50 \\ x_1 + 2x_2 = 80 \end{cases}$$

FIGURE 3

FIGURE 4

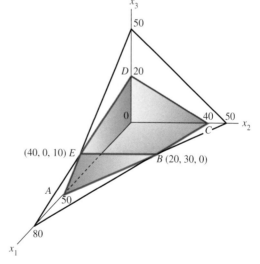

FIGURE 5

Obtain $x_2 = 30$, and find that B is $(20, 30, 0)$. For E, solve

$$\begin{cases} x_1 + x_2 + x_3 = 50 \\ x_1 + 2x_2 + 4x_3 = 80 \\ x_2 = 0 \end{cases} \Rightarrow \begin{cases} x_1 + x_3 = 50 \\ x_1 + 4x_3 = 80 \end{cases}$$

Obtain $x_3 = 10$, and find that $E = (40, 0, 10)$.

Now that the feasible set and its extreme points are clearly seen, the next step is to examine the objective function $f(x_1, x_2, x_3) = 2x_1 + 3x_2 + 4x_3$. The sets on which f is constant are planes, rather than lines, all having $(2, 3, 4)$ as a normal vector to the plane. This normal vector has a direction different from the normal vectors $(1, 1, 1)$ and $(1, 2, 4)$ to the two faces $BCDE$ and ABE. So the level sets of f are not parallel to any of the bounding surfaces of the feasible set. Figure 6 shows just the feasible set and a level set on which f has the value 120. This plane passes through C, E, and the point $(30, 20, 0)$ on the edge of the feasible set between A and B, which shows that the vertex B is "above" this level plane. In fact, $f(20, 30, 0) = 130$. Thus the unique solution of the linear programming problem is at $B = (20, 30, 0)$. ∎

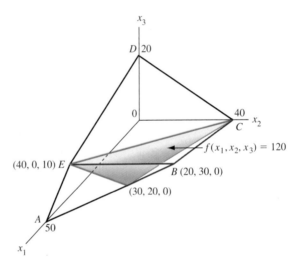

FIGURE 6

Practice Problems

1. Consider the following problem:

$$\begin{aligned} \text{Maximize} \quad & 2x_1 + x_2 \\ \text{subject to} \quad & x_1 - 2x_2 \geq -8 \\ & 3x_1 + 2x_2 \leq 24 \end{aligned}$$

   and $x_1 \geq 0, x_2 \geq 0$.

   Write this problem in the form of a canonical linear programming problem: Maximize $\mathbf{c}^T\mathbf{x}$ subject to $A\mathbf{x} \leq \mathbf{b}$ and $\mathbf{x} \geq 0$. Specify A, $\mathbf{b}$, and $\mathbf{c}$.

2. Graph the feasible set for Practice Problem 1.

3. Find the extreme points of the feasible set in Practice Problem 2.

4. Use the answer to Practice Problem 3 to find the solution to the linear programming problem in Practice Problem 1.

9.2 Exercises

1. Betty plans to invest a total of \$12,000 in mutual funds, certificates of deposit (CDs), and a high-yield savings account. Because of the risk involved in mutual funds, she wants to invest no more in mutual funds than the sum of her CDs and savings. She also wants the amount in savings to be at least half the amount in CDs. Her expected returns are 11% on the mutual funds, 8% on the CDs and 6% on savings. How much money should Betty invest in each area in order to have the largest return on her investments? Set this up as a linear programming problem in the following form: Maximize $\mathbf{c}^T\mathbf{x}$ subject to $A\mathbf{x} \le \mathbf{b}$ and $\mathbf{x} \ge \mathbf{0}$. Do *not* find the solution.

2. A dog breeder decides to feed his dogs a combination of two dog foods: Pixie Power and Misty Might. He wants the dogs to receive four nutritional factors each month. The amounts of these factors (a, b, c, and d) contained in 1 bag of each dog food are shown in the following chart, together with the total amounts needed.

	a	b	c	d
Pixie Power	3	2	1	2
Misty Might	2	4	3	1
Needed	28	30	20	25

The costs per bag are \$50 for Pixie Power and \$40 for Misty Might. How many bags of each dog food should be blended to meet the nutritional requirements at the lowest cost? Set this up as a linear programming problem in the following form: Minimize $\mathbf{c}^T\mathbf{x}$ subject to $A\mathbf{x} \ge \mathbf{b}$ and $\mathbf{x} \ge \mathbf{0}$. Do *not* find the solution.

In Exercises 3–6, find vectors $\mathbf{b}$ and $\mathbf{c}$ and matrix A so that each problem is set up as a canonical linear programming problem: Maximize $\mathbf{c}^T\mathbf{x}$ subject to $A\mathbf{x} \le \mathbf{b}$ and $\mathbf{x} \ge \mathbf{0}$. Do *not* find the solution.

3. Maximize $3x_1 + 4x_2 - 2x_3$
   subject to $x_1 + 2x_2 \qquad \le 20$
   $\qquad\qquad 3x_2 + 5x_3 \ge 10$
   and $x_1 \ge 0, x_2 \ge 0, x_3 \ge 0$.

4. Maximize $3x_1 + x_2 + 5x_3$
   subject to $5x_1 + 7x_2 + x_3 \le 25$
   $\qquad\quad 2x_1 + 3x_2 + 4x_3 = 40$
   and $x_1 \ge 0, x_2 \ge 0, x_3 \ge 0$.

5. Minimize $7x_1 - 3x_2 + x_3$
   subject to $x_1 - 4x_2 \qquad \ge 35$
   $\qquad\qquad\quad x_2 - 2x_3 = 20$
   and $x_1 \ge 0, x_2 \ge 0, x_3 \ge 0$.

6. Minimize $x_1 + 5x_2 - 2x_3$
   subject to $2x_1 + x_2 + 4x_3 \le 27$
   $\qquad\quad x_1 - 6x_2 + 3x_3 \ge 40$
   and $x_1 \ge 0, x_2 \ge 0, x_3 \ge 0$.

In Exercises 7–10, solve the linear programming problems.

7. Maximize $80x_1 + 65x_2$
   subject to $2x_1 + x_2 \le 32$
   $\qquad\quad x_1 + x_2 \le 18$
   $\qquad\quad x_1 + 3x_2 \le 24$
   and $x_1 \ge 0, x_2 \ge 0$.

8. Minimize $5x_1 + 3x_2$
   subject to $2x_1 + 5x_2 \ge 10$
   $\qquad\quad 3x_1 + x_2 \ge 6$
   $\qquad\quad x_1 + 7x_2 \ge 7$
   and $x_1 \ge 0, x_2 \ge 0$.

9. Maximize $2x_1 + 7x_2$
   subject to $-2x_1 + x_2 \le -4$
   $\qquad\quad x_1 - 2x_2 \le -4$
   and $x_1 \ge 0, x_2 \ge 0$.

10. Maximize $5x_1 + 12x_2$
    subject to $x_1 - x_2 \le 3$
    $\qquad\quad -x_1 + 2x_2 \le -4$
    and $x_1 \ge 0, x_2 \ge 0$.

In Exercises 11–14, mark each statement True or False (**T/F**). Justify each answer.

11. (**T/F**) In a canonical linear programming problem, a nonnegative vector $\mathbf{x}$ is a feasible solution if it satisfies $A\mathbf{x} \le \mathbf{b}$.

12. (**T/F**) If a canonical linear programming problem does not have an optimal solution, then either the objective function is not bounded on the feasible set $\mathcal{F}$ or $\mathcal{F}$ is the empty set.

13. (**T/F**) A vector $\overline{\mathbf{x}}$ is an optimal solution of a canonical linear programming problem if $f(\overline{\mathbf{x}})$ is equal to the maximum value of the linear functional f on the feasible set $\mathcal{F}$.

14. (**T/F**) If $\overline{\mathbf{x}}$ is an optimal solution of a canonical linear programming problem, then $\overline{\mathbf{x}}$ is an extreme point of the feasible set.

15. Solve the linear programming problem in Example 1.

16. Solve the linear programming problem in Example 2.

17. The Benri Company manufactures two kinds of kitchen gadgets: invertible widgets and collapsible whammies. The production process is divided into three departments: fabricating, packing, and shipping. The hours of labor required for each operation and the hours available in each department each day are shown below.

	Widgets	Whammies	Time available
Fabricating	5.0	2.0	200
Packing	0.2	0.4	16
Shipping	0.2	0.2	10

Suppose that the profit on each widget is $20 and the profit on each whammy is $26. How many widgets and how many whammies should be made each day to maximize the company's profit?

Exercises 18–21 use the notion of a convex set, studied in Section 8.3. A set S in $\mathbb{R}^n$ is convex if, for each $\mathbf{p}$ and $\mathbf{q}$ in S, the line segment between $\mathbf{p}$ and $\mathbf{q}$ lies in S. [This line segment is the set of points of the form $(1-t)\mathbf{p} + t\mathbf{q}$ for $0 \le t \le 1$.]

18. Let $\mathcal{F}$ be the feasible set of all solutions $\mathbf{x}$ of a linear programming problem $A\mathbf{x} \le \mathbf{b}$ with $\mathbf{x} \ge \mathbf{0}$. Assume that $\mathcal{F}$ is nonempty. Show that $\mathcal{F}$ is a convex set in $\mathbb{R}^n$. [*Hint:* Consider points $\mathbf{p}$ and $\mathbf{q}$ in $\mathcal{F}$ and t such that $0 \le t \le 1$. Show that $(1-t)\mathbf{p} + t\mathbf{q}$ is in $\mathcal{F}$.]

19. Let $\mathbf{v} = \begin{bmatrix} a \\ b \end{bmatrix}$ and $\mathbf{x} = \begin{bmatrix} x_1 \\ x_2 \end{bmatrix}$. The inequality $ax_1 + bx_2 \le c$ for some real number c may be written as $\mathbf{v}^T\mathbf{x} \le c$. The set S of all $\mathbf{x}$ that satisfy this inequality is called a **closed half-space** of $\mathbb{R}^2$. Show that S is convex. [See the Hint for Exercise 18.]

20. The feasible set in Example 5 is the intersection of five closed half-spaces. By Exercise 19, these half-spaces are convex sets. Show that the intersection of any five convex sets $S_1, \ldots, S_5$ in $\mathbb{R}^n$ is a convex set.

21. If c is in $\mathbb{R}^n$ and if f is defined on $\mathbb{R}^n$ by $f(\mathbf{x}) = \mathbf{c}^T\mathbf{x}$, then f is called a linear functional, and for any real number d, $\{\mathbf{x}: f(\mathbf{x}) = d\}$ is called a level set of f. (See level sets in Figure 2 of Example 5.) Show that any such level set is convex.

Solutions to Practice Problems

1. The first inequality has the wrong direction, so multiply by -1. This gives the following problem:

$$\begin{array}{ll} \text{Maximize} & 2x_1 + x_2 \\ \text{subject to} & -x_1 + 2x_2 \le 8 \\ & 3x_1 + 2x_2 \le 24 \end{array}$$

and $x_1 \ge 0, x_2 \ge 0$.

This corresponds to the canonical form

Maximize $\mathbf{c}^T\mathbf{x}$ subject to $A\mathbf{x} \le \mathbf{b}$ and $\mathbf{x} \ge \mathbf{0}$

when

$$\mathbf{b} = \begin{bmatrix} 8 \\ 24 \end{bmatrix}, \quad \mathbf{x} = \begin{bmatrix} x_1 \\ x_2 \end{bmatrix}, \quad \mathbf{c} = \begin{bmatrix} 2 \\ 1 \end{bmatrix}, \quad \text{and} \quad A = \begin{bmatrix} -1 & 2 \\ 3 & 2 \end{bmatrix}$$

2. To graph the inequality $-x_1 + 2x_2 \le 8$, first graph the corresponding equality $-x_1 + 2x_2 = 8$. The intercepts are easy to find: $(0, 4)$ and $(-8, 0)$. Figure 7 shows the straight line through these two points.

The graph of the inequality consists of this line together with all points on one side of the line. To determine which side, pick a point not on the line to see if its coordinates satisfy the inequality. For example, try the origin, $(0, 0)$. The inequality

$$-(0) + 2(0) \le 8$$

is a true statement. Thus the origin and all other points below the line satisfy the inequality. As another example, substituting the coordinates of the point $(0, 8)$ into the inequality produces a false statement:

$$-(0) + 2(8) \le 8$$

Thus $(0, 8)$ and all other points above the line do not satisfy the inequality. Figure 7 shows small arrows beneath the graph of $-x_1 + 2x_2 = 8$, to indicate which side is to be included.

For the inequality

$$3x_1 + 2x_2 \le 24$$

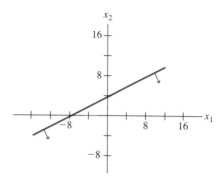

FIGURE 7 Graph of $-x_1 + 2x_2 \le 8$.

draw the graph of $3x_1 + 2x_2 = 24$, using the intercepts $(0, 12)$ and $(8, 0)$ or two other convenient points. Since $(0, 0)$ satisfies the inequality, the feasible set is on the side of the line containing the origin. The inequality $x_1 \ge 0$ gives the right half-plane, and the inequality $x_2 \ge 0$ gives the upper half-plane. All of these are graphed in Figure 8, and their common solution is the shaded feasible set.

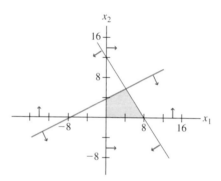

FIGURE 8 Graph of the feasible set.

3. There are four extreme points in the feasible set:

   1. The origin: $(0, 0)$

   2. The x_2-intercept of the first equality: $(0, 4)$

   3. The x_1-intercept of the second equality: $(8, 0)$

   4. The intersection of the two equalities.

For the fourth extreme point, solve the system of equations $-x_1 + 2x_2 = 8$ and $3x_1 + 2x_2 = 24$ to obtain $x_1 = 4$ and $x_2 = 6$.

4. To find the maximum value of the objective function $2x_1 + x_2$, evaluate it at each of the four extreme points of the feasible set.

	$2x_1 + x_2$
$(0, 0)$	$2(0) + 1(0) = 0$
$(0, 4)$	$2(0) + 1(4) = 4$
$(8, 0)$	$2(8) + 1(0) = 16$ ← max
$(4, 6)$	$2(4) + 1(6) = 14$

The maximum value is 16, attained when $x_1 = 8$ and $x_2 = 0$.

9.3 Linear Programming—Simplex Method

Transportation problems played an important role in the early days of linear programming, including the Berlin Airlift described in this chapter's Introductory Example. They are even more important today. The first example is simple, but it suggests how a problem of this type could involve hundreds, if not thousands, of variables and equations.

EXAMPLE 1 A retail sales company has two warehouses and four stores. A particular model of outdoor hot tub is sold at all four stores, and each store has placed an order with company headquarters for a certain number of these hot tubs. Headquarters determines that the warehouses have enough hot tubs and can ship them immediately. The distances from the warehouses to the stores vary, and the cost of transporting a hot tub from a warehouse to a store depends on the distance. The problem is to decide on a shipping schedule that minimizes the total cost of shipping. Let x_{ij} be the number of units (hot tubs) to ship from warehouse i to store j.

Let a_1 and a_2 be the numbers of units available at warehouses 1 and 2, and let $r_1, \ldots, r_4$ be the numbers of units requested by the various stores. Then the x_{ij} must satisfy the equations

$$
\begin{aligned}
x_{11} + x_{12} + x_{13} + x_{14} & & & \leq a_1 \\
& x_{21} + x_{22} + x_{23} + x_{24} & & \leq a_2 \\
x_{11} & \quad + x_{21} & & = r_1 \\
x_{12} & \quad + x_{22} & & = r_2 \\
x_{13} & \quad + x_{23} & & = r_3 \\
x_{14} & \quad + x_{24} & & = r_4
\end{aligned}
$$

and $x_{ij} \geq 0$ for $i = 1, 2$ and $j = 1, \ldots, 4$. If the cost of shipping one unit from warehouse i to store j is c_{ij}, then the problem is to minimize the function

$$
c_{11}x_{11} + c_{12}x_{12} + c_{13}x_{13} + c_{14}x_{14} + c_{21}x_{21} + c_{22}x_{22} + c_{23}x_{23} + c_{24}x_{24}
$$

subject to the four equalities and ten inequalities listed above. ∎

The *simplex method*, discussed below, can easily handle problems the size of Example 1. To introduce the method, however, this section focuses mainly on the canonical linear programming problem from Section 9.2, in which the objective function must be maximized. Here is an outline of the steps in the simplex method.

1. Select an extreme point $\mathbf{x}$ of the feasible set $\mathcal{F}$.

2. Consider all the edges of $\mathcal{F}$ that join at $\mathbf{x}$. If the objective function f cannot be increased by moving along any of these edges, then $\mathbf{x}$ is an optimal solution.

3. If f can be increased by moving along one or more of the edges, then follow the path that gives the largest increase and move to the extreme point of $\mathcal{F}$ at the opposite end.

4. Repeat the process, beginning at step 2.

Since the value of f increases at each step, the path will not go through the same extreme point twice. Since there are only a finite number of extreme points, this process will end at an optimal solution (if there is one) in a finite number of steps. If the problem is unbounded, then eventually the path will reach an unbounded edge at step 3 along which f increases without bound.

The next five examples concern canonical linear programming problems in which each of the entries in the m-tuple $\mathbf{b}$ is *positive*:

$$\text{Maximize } f(\mathbf{x}) = \mathbf{c}^T \mathbf{x}$$

$$\text{subject to the constraints } A\mathbf{x} \le \mathbf{b} \text{ and } \mathbf{x} \ge \mathbf{0}$$

Here $\mathbf{c}$ and $\mathbf{x}$ are in $\mathbb{R}^n$, A is an $m \times n$ matrix, and $\mathbf{b}$ is in $\mathbb{R}^m$.

The simplex method begins by changing each constraint inequality into an *equality*. This is done by adding one new variable to each inequality. These new variables are not part of the final solution; they appear only in the intermediate calculations.

DEFINITION

> A **slack variable** is a nonnegative variable that is added to the smaller side of an inequality to convert it to an equality.

EXAMPLE 2 Change the inequality

$$5x_1 + 7x_2 \le 80$$

into the equality

$$5x_1 + 7x_2 + x_3 = 80$$

by adding the slack variable x_3. Note that $x_3 = 80 - (5x_1 + 7x_2) \ge 0$. ∎

If A is $m \times n$, the addition of m slack variables in $A\mathbf{x} \le \mathbf{b}$ produces a linear system with m equations and $n + m$ variables. A solution to this system is called a **basic solution** if no more than m of the variables are nonzero. As in Section 9.2, a solution to the system is called **feasible** if each variable is nonnegative. Thus, in a **basic feasible solution**, each variable must be nonnegative and at most m of them can be positive. Geometrically, these basic feasible solutions correspond to the extreme points of the feasible set.

EXAMPLE 3 Find a basic feasible solution for the system

$$2x_1 + 3x_2 + 4x_3 \le 60$$
$$3x_1 + x_2 + 5x_3 \le 46$$
$$x_1 + 2x_2 + x_3 \le 50$$

SOLUTION   Add slack variables to obtain a system of three *equations*:

$$
\begin{aligned}
2x_1 + 3x_2 + 4x_3 + x_4 \qquad\qquad &= 60 \\
3x_1 + x_2 + 5x_3 \qquad\quad + x_5 \quad &= 46 \\
x_1 + 2x_2 + x_3 \qquad\qquad\; + x_6 &= 50
\end{aligned}
\tag{1}
$$

There were three inequalities in the original system, so a basic solution of (1) has at most three nonzero values for the variables. The following simple solution is called the basic feasible solution associated with (1):

$$
x_1 = x_2 = x_3 = 0, \quad x_4 = 60, \quad x_5 = 46, \quad \text{and} \quad x_6 = 50
$$

This solution corresponds to the extreme point **0** in the feasible set (in $\mathbb{R}^3$).   ∎

It is customary to refer to the nonzero variables x_4, x_5, and x_6 in system (1) as **basic variables** because each has a coefficient of 1 and occurs in only one equation.[1] The basic variables are said to be "in" the solution of (1). The variables x_1, x_2, and x_3 are said to be "out" of the solution. In a linear programming problem, this particular solution would probably not be optimal since only the slack variables are nonzero.

A standard procedure in the simplex method is to change the role a variable plays in a solution. For example, although x_2 is out of the solution in (1), it can be introduced "into" a solution by using elementary row operations. The goal is to **pivot** on the x_2 entry in the third equation of (1) to create a new system in which x_2 appears *only* in the third equation.[2]

First, divide the third equation in (1) by the coefficient of x_2 to obtain a new third equation:

$$
\tfrac{1}{2}x_1 + x_2 + \tfrac{1}{2}x_3 \qquad\qquad + \tfrac{1}{2}x_6 = \quad 25
$$

Second, to equations 1 and 2 of (1) add multiples of this new equation that will eliminate x_2 from those equations. This produces the system

$$
\begin{aligned}
\tfrac{1}{2}x_1 \qquad + \tfrac{5}{2}x_3 + x_4 \qquad - \tfrac{3}{2}x_6 &= -15 \\
\tfrac{5}{2}x_1 \qquad + \tfrac{9}{2}x_3 \qquad + x_5 - \tfrac{1}{2}x_6 &= \quad 21 \\
\tfrac{1}{2}x_1 + x_2 + \tfrac{1}{2}x_3 \qquad\qquad + \tfrac{1}{2}x_6 &= \quad 25
\end{aligned}
$$

The basic solution associated with this new system is

$$
x_1 = x_3 = x_6 = 0, \quad x_2 = 25, \quad x_4 = -15, \quad x_5 = 21
$$

The variable x_2 has come into the solution, and the variable x_6 has gone out. Unfortunately, this basic solution is not feasible since $x_4 < 0$. This lack of feasibility was caused by an improper choice of a pivot equation. The next paragraph shows how to avoid this problem.

---

[1] This terminology generalizes that used in Section 1.2, where basic variables also had to correspond to pivot positions in a matrix *echelon* form. Here, the goal is not to solve for basic variables in terms of free variables, but to obtain a particular solution of the system when the nonbasic (free) variables are zero.

[2] To "pivot" on a particular term here means to transform its coefficient into a 1 and then use it to eliminate corresponding terms in *all* the other equations, not just the equations below it, as was done in Section 1.2.

In general, consider the system

$$a_{11}x_1 + \cdots + a_{1k}x_k + \cdots + a_{1n}x_n = b_1$$

$$\vdots$$

$$a_{i1}x_1 + \cdots + a_{ik}x_k + \cdots + a_{in}x_n = b_i$$

$$\vdots$$

$$a_{m1}x_1 + \cdots + a_{mk}x_k + \cdots + a_{mn}x_n = b_n$$

and suppose the next step is to bring the variable x_k into the solution by using equation p to pivot on entry $a_{pk}x_k$. The basic solution corresponding to the resulting system will be feasible if the following two conditions are satisfied:

1. The coefficient a_{pk} of x_k must be positive. (When the pth equation is divided by a_{pk}, the new b_p term must be positive.)
2. The ratio b_p/a_{pk} must be the smallest among all the ratios b_i/a_{ik} for which $a_{ik} > 0$. (This will guarantee that when the pth equation is used to eliminate the x_k term from the ith equation, the resulting b_i term will be positive.)

EXAMPLE 4 Determine which row to use as a pivot in order to bring x_2 into the solution in Example 3.

SOLUTION Compute the ratios b_i/a_{i2}:

$$\frac{b_1}{a_{12}} = \frac{60}{3} = 20, \qquad \frac{b_2}{a_{22}} = 46, \qquad \text{and} \qquad \frac{b_3}{a_{32}} = \frac{50}{2} = 25$$

Since the first ratio is the smallest, pivot on the x_2 term in the first equation. This produces the system

$$\begin{aligned}
\tfrac{2}{3}x_1 + x_2 + \tfrac{4}{3}x_3 + \tfrac{1}{3}x_4 &= 20 \\
\tfrac{7}{3}x_1 \qquad + \tfrac{11}{3}x_3 - \tfrac{1}{3}x_4 + x_5 &= 26 \\
-\tfrac{1}{3}x_1 \qquad - \tfrac{5}{3}x_3 - \tfrac{2}{3}x_4 \qquad + x_6 &= 10
\end{aligned}$$

Now the basic feasible solution is

$$x_1 = x_3 = x_4 = 0, \quad x_2 = 20, \quad x_5 = 26, \quad x_6 = 10 \qquad \blacksquare$$

A matrix format greatly simplifies calculations of this type. For instance, system (1) in Example 3 is represented by the augmented matrix

$$\begin{array}{cccccc}
x_1 & x_2 & x_3 & x_4 & x_5 & x_6 \\
\end{array}$$
$$\left[\begin{array}{cccccc|c}
2 & ③ & 4 & 1 & 0 & 0 & 60 \\
3 & 1 & 5 & 0 & 1 & 0 & 46 \\
1 & 2 & 1 & 0 & 0 & 1 & 50
\end{array} \right]$$

The variables are used as column labels, with the slack variables in color. Recall that the basic feasible solution associated with this matrix is

$$x_1 = x_2 = x_3 = 0, \quad x_4 = 60, \quad x_5 = 46, \quad x_6 = 50$$

The circled 3 in the x_2 column indicates that this entry will be used as a pivot to bring x_2 into the solution. (The ratio calculations in Example 4 identified this entry as the appropriate pivot.) Complete row reduction in column 2 produces the new matrix that corresponds to the new system in Example 4:

$$
\begin{array}{cccccc}
x_1 & x_2 & x_3 & x_4 & x_5 & x_6 \\
\end{array}
$$

$$
\left[
\begin{array}{cccccc|c}
\frac{2}{3} & 1 & \frac{4}{3} & \frac{1}{3} & 0 & 0 & 20 \\
\frac{7}{3} & 0 & \frac{11}{3} & -\frac{1}{3} & 1 & 0 & 26 \\
-\frac{1}{3} & 0 & -\frac{5}{3} & -\frac{2}{3} & 0 & 1 & 10 \\
\end{array}
\right]
$$

As in Example 4, the new basic feasible solution is

$$
x_1 = x_3 = x_4 = 0, \quad x_2 = 20, \quad x_5 = 26, \quad x_6 = 10
$$

The preceding discussion has prepared the way for a full demonstration of the simplex method, based on the constraints in Example 3. At each step, the objective function in Example 5 will drive the choice of which variable to bring into the solution of the system.

EXAMPLE 5 Maximize $25x_1 + 33x_2 + 18x_3$

$$
\begin{aligned}
\text{subject to} \quad & 2x_1 + 3x_2 + 4x_3 \le 60 \\
& 3x_1 + x_2 + 5x_3 \le 46 \\
& x_1 + 2x_2 + x_3 \le 50
\end{aligned}
$$

and $x_j \ge 0$ for $j = 1, \ldots, 3$.

SOLUTION First, add slack variables, as before. Then change the objective *function* $25x_1 + 33x_2 + 18x_3$ into an *equation* by introducing a new variable M given by $M = 25x_1 + 33x_2 + 18x_3$. Now the goal is to maximize the variable M, where M satisfies the equation

$$
-25x_1 - 33x_2 - 18x_3 + M = 0
$$

The original problem is now restated as follows: Among all the solutions of the system of equations

$$
\begin{aligned}
2x_1 + 3x_2 + 4x_3 + x_4 &= 60 \\
3x_1 + x_2 + 5x_3 + x_5 &= 46 \\
x_1 + 2x_2 + x_3 + x_6 &= 50 \\
-25x_1 - 33x_2 - 18x_3 + M &= 0
\end{aligned}
$$

find a solution for which $x_j \ge 0$ $(j = 1, \ldots, 6)$ and for which M is as large as possible.

The augmented matrix for this new system is called the **initial simplex tableau**. It is written with two ruled lines in the matrix:

$$
\begin{array}{ccccccc}
x_1 & x_2 & x_3 & x_4 & x_5 & x_6 & M \\
\end{array}
$$

$$
\left[
\begin{array}{ccccccc|c}
2 & 3 & 4 & 1 & 0 & 0 & 0 & 60 \\
3 & 1 & 5 & 0 & 1 & 0 & 0 & 46 \\
1 & 2 & 1 & 0 & 0 & 1 & 0 & 50 \\
\hline
-25 & -33 & -18 & 0 & 0 & 0 & 1 & 0 \\
\end{array}
\right]
$$

The horizontal line above the bottom row isolates the equation corresponding to the objective function. This last row will play a special role in what follows. (The bottom row is used only to decide which variable to bring into the solution. Pivot positions are never chosen from the bottom row.) The column headings for the slack variables are in color, as a reminder that at the end of the calculations only the original variables are part of the final solution of the problem.

Look in rows 1 to 3 of the tableau above to find the basic feasible solution. The *columns of the* 3×3 *identity matrix* in these three rows *identify the basic variables*—namely, x_4, x_5, and x_6. The basic solution is

$$x_1 = x_2 = x_3 = 0, \quad x_4 = 60, \quad x_5 = 46, \quad x_6 = 50, \quad M = 0$$

This solution is not optimal, however, since only the slack variables are nonzero. However, the bottom row implies that

$$M = 25x_1 + 33x_2 + 18x_3$$

The value of M will rise when any of the variables $x_1, x_2,$ or x_3 rises. Since the coefficient of x_2 is the largest of the three coefficients, bringing x_2 into the solution will cause the greatest increase in M.

To bring x_2 into the solution, follow the pivoting procedure outlined earlier. In the tableau above, compare the ratios b_i / a_{i2} for each row except the last. They are $60/3$, $46/1$, and $50/2$. The smallest is $60/3$, so the pivot should be the entry 3 that is circled in the first row.

$$\begin{array}{ccccccc|c} x_1 & x_2 & x_3 & x_4 & x_5 & x_6 & M & \\ 2 & ③ & 4 & 1 & 0 & 0 & 0 & 60 \\ 3 & 1 & 5 & 0 & 1 & 0 & 0 & 46 \\ 1 & 2 & 1 & 0 & 0 & 1 & 0 & 50 \\ \hline -25 & -33 & -18 & 0 & 0 & 0 & 1 & 0 \end{array}$$

The result of the pivot operation is

$$\begin{array}{ccccccc|c} x_1 & x_2 & x_3 & x_4 & x_5 & x_6 & M & \\ \frac{2}{3} & 1 & \frac{4}{3} & \frac{1}{3} & 0 & 0 & 0 & 20 \\ \frac{7}{3} & 0 & \frac{11}{3} & -\frac{1}{3} & 1 & 0 & 0 & 26 \\ -\frac{1}{3} & 0 & -\frac{5}{3} & -\frac{2}{3} & 0 & 1 & 0 & 10 \\ \hline -3 & 0 & 26 & 11 & 0 & 0 & 1 & 660 \end{array} \tag{2}$$

Now the columns of the 3×3 identity matrix are in columns 2, 5, and 6 of the tableau. So the basic feasible solution is

$$x_1 = x_3 = x_4 = 0, \quad x_2 = 20, \quad x_5 = 26, \quad x_6 = 10, \quad M = 660$$

Thus M has increased from 0 to 660. To see if M can be increased further, look at the bottom row of the tableau and solve the equation for M:

$$M = 660 + 3x_1 - 26x_3 - 11x_4 \tag{3}$$

Since each of the variables x_j is nonnegative, the value of M will increase only if x_1 increases (from 0). (Since the coefficients of x_3 and x_4 are both negative at this point, increasing one of them would *decrease* M.) So x_1 needs to come into the solution. Compare the ratios (of the augmented column to column 1):

$$\frac{20}{\frac{2}{3}} = 30 \qquad \text{and} \qquad \frac{26}{\frac{7}{3}} = \frac{78}{7}$$

The second ratio is smaller, so the next pivot should be $\frac{7}{3}$ in row 2.

$$
\begin{array}{ccccccc}
x_1 & x_2 & x_3 & x_4 & x_5 & x_6 & M \\
\end{array}
$$

$$
\left[
\begin{array}{ccccccc|c}
\frac{2}{3} & 1 & \frac{4}{3} & \frac{1}{3} & 0 & 0 & 0 & 20 \\
\boxed{\frac{7}{3}} & 0 & \frac{11}{3} & -\frac{1}{3} & 1 & 0 & 0 & 26 \\
-\frac{1}{3} & 0 & -\frac{5}{3} & -\frac{2}{3} & 0 & 1 & 0 & 10 \\
\hline
-3 & 0 & 26 & 11 & 0 & 0 & 1 & 660
\end{array}
\right]
$$

After pivoting, the resulting tableau is

$$
\begin{array}{ccccccc}
x_1 & x_2 & x_3 & x_4 & x_5 & x_6 & M \\
\end{array}
$$

$$
\left[
\begin{array}{ccccccc|c}
0 & 1 & \frac{2}{7} & \frac{3}{7} & -\frac{2}{7} & 0 & 0 & \frac{88}{7} \\
1 & 0 & \frac{11}{7} & -\frac{1}{7} & \frac{3}{7} & 0 & 0 & \frac{78}{7} \\
0 & 0 & -\frac{8}{7} & -\frac{5}{7} & \frac{1}{7} & 1 & 0 & \frac{96}{7} \\
\hline
0 & 0 & \frac{215}{7} & \frac{74}{7} & \frac{9}{7} & 0 & 1 & \frac{4854}{7}
\end{array}
\right]
$$

The corresponding basic feasible solution is

$$x_3 = x_4 = x_5 = 0, \quad x_1 = \tfrac{78}{7}, \quad x_2 = \tfrac{88}{7}, \quad x_6 = \tfrac{96}{7}, \quad M = \tfrac{4854}{7}$$

The bottom row shows that

$$M = \tfrac{4854}{7} - \tfrac{215}{7}x_3 - \tfrac{74}{7}x_4 - \tfrac{9}{7}x_5$$

The negative coefficients of the variables here show that M can be no larger than $\frac{4854}{7}$ (because x_3, x_4, and x_5 are nonnegative), so the solution is optimal. The maximum value of $25x_1 + 33x_2 + 18x_3$ is $\frac{4854}{7}$, and this maximum occurs when $x_1 = \frac{78}{7}$, $x_2 = \frac{88}{7}$, and $x_3 = 0$. The variable x_3 is zero because in the optimal solution x_3 is a free variable, not a basic variable. Note that the value of x_6 is not part of the solution of the original problem, because x_6 is a slack variable. The fact that the slack variables x_4 and x_5 are zero means that the first two inequalities listed at the beginning of this example are both *equalities* at the optimal values of x_1, x_2, and x_3. ∎

Example 5 is worth reading carefully several times. In particular, notice that a *negative* entry in the bottom row of any x_j column will become a *positive* coefficient when that equation is solved for M, indicating that M has not reached its maximum. See tableau (2) and equation (3).

In summary, here is the simplex method for solving a canonical maximizing problem when each entry in the vector **b** is positive.

THE SIMPLEX ALGORITHM FOR A CANONICAL LINEAR PROGRAMMING PROBLEM

1. Change the inequality constraints into equalities by adding slack variables. Let M be a variable equal to the objective function, and below the constraint equations write an equation of the form

$$- \text{(objective function)} + M = 0$$

2. Set up the initial simplex tableau. The slack variables (and M) provide the initial basic feasible solution.

3. Check the bottom row of the tableau for optimality. If all the entries to the left of the vertical line are nonnegative, then the solution is optimal. If some are negative, then choose the variable x_k for which the entry in the bottom row is as negative as possible.[3]

4. Bring the variable x_k into the solution. Do this by pivoting on the positive entry a_{pk} for which the nonnegative ratio b_i / a_{ik} is the smallest. The new basic feasible solution includes an increased value for M.

5. Repeat the process, beginning at step 3, until all the entries in the bottom row are nonnegative.

Two things can go wrong in the simplex algorithm. At step 4, there might be a negative entry in the bottom row of the x_k column, but no positive entry a_{ik} above it. In this case, it will not be possible to find a pivot to bring x_k into the solution. This corresponds to the case where the objective function is unbounded and no optimal solution exists.

The second potential problem also occurs at step 4. The smallest ratio b_i / a_{ik} may occur in more than one row. When this happens, the next tableau will have at least one basic variable equal to zero, and in subsequent tableaus the value of M may remain constant. Theoretically it is possible for an infinite sequence of pivots to occur and fail to lead to an optimal solution. Such a phenomenon is called **cycling**. Fortunately, cycling occurs only rarely in practical applications. In most cases, one may arbitrarily choose either row with a minimum ratio as the pivot.

EXAMPLE 6 A health food store sells two different mixtures of nuts. A box of the first mixture contains 1 pound of cashews and 1 pound of peanuts. A box of the second mixture contains 1 pound of filberts and 2 pounds of peanuts. The store has available 30 pounds of cashews, 20 pounds of filberts, and 54 pounds of peanuts. Suppose the profit on each box of the first mixture is $2 and on each box of the second mixture is $3. If the store can sell all of the boxes it mixes, how many boxes of each mixture should be made in order to maximize the profit?

---

[3] The goal of step 3 is to produce the greatest increase possible in the value of M. This happens when only one variable x_k satisfies the conditions. Suppose, however, that the most negative entry in the bottom row appears in both columns j and k. Step 3 says that either x_j or x_k should be brought into the solution, and that is correct. Occasionally, a few computations can be avoided by first using step 4 to compute the "smallest ratio" for both columns j and k, and then choosing the column for which this "smallest ratio" is larger. This situation will arise in Section 9.4.

SOLUTION Let x_1 be the number of boxes of the first mixture, and let x_2 be the number of boxes of the second mixture. The problem can be expressed mathematically as

$$
\begin{aligned}
\text{Maximize} \quad & 2x_1 + 3x_2 \\
\text{subject to} \quad & x_1 \leq 30 \quad \text{(cashews)} \\
& x_2 \leq 20 \quad \text{(filberts)} \\
& x_1 + 2x_2 \leq 54 \quad \text{(peanuts)}
\end{aligned}
$$

and $x_1 \geq 0, x_2 \geq 0$.

This turns out to be the same problem solved graphically in Example 5 of Section 9.2. When it is solved by the simplex method, the basic feasible solution from each tableau corresponds to an extreme point of the feasible region. See Figure 1.

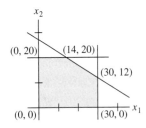

FIGURE 1

To construct the initial tableau, add slack variables and rewrite the objective function as an equation. The problem now is to find a nonnegative solution to the system

$$
\begin{aligned}
x_1 \qquad + x_3 \qquad\qquad\qquad &= 30 \\
x_2 + \quad x_4 \qquad\qquad &= 20 \\
x_1 + 2x_2 \qquad\quad + x_5 \qquad &= 54 \\
-2x_1 - 3x_2 \qquad\qquad\quad + M &= 0
\end{aligned}
$$

for which M is a maximum. The initial simplex tableau is

$$
\begin{array}{cccccc}
x_1 & x_2 & x_3 & x_4 & x_5 & M \\
\end{array}
$$
$$
\left[
\begin{array}{cccccc|c}
1 & 0 & 1 & 0 & 0 & 0 & 30 \\
0 & 1 & 0 & 1 & 0 & 0 & 20 \\
1 & 2 & 0 & 0 & 1 & 0 & 54 \\
\hline
-2 & -3 & 0 & 0 & 0 & 1 & 0
\end{array}
\right]
$$

The basic feasible solution, where x_1, x_2, and M are 0, corresponds to the extreme point $(x_1, x_2) = (0, 0)$ of the feasible region in Figure 1. In the bottom row of the tableau, the most negative entry is -3, so the first pivot should be in the x_2 column. The ratios $20/1$ and $54/2$ show that the pivot should be the 1 in the x_2 column:

$$
\begin{array}{cccccc}
x_1 & x_2 & x_3 & x_4 & x_5 & M \\
\end{array}
$$
$$
\left[
\begin{array}{cccccc|c}
1 & 0 & 1 & 0 & 0 & 0 & 30 \\
0 & ① & 0 & 1 & 0 & 0 & 20 \\
1 & 2 & 0 & 0 & 1 & 0 & 54 \\
\hline
-2 & -3 & 0 & 0 & 0 & 1 & 0
\end{array}
\right]
$$

After pivoting, the tableau becomes

$$\begin{array}{cccccc|c}
x_1 & x_2 & x_3 & x_4 & x_5 & M & \\
1 & 0 & 1 & 0 & 0 & 0 & 30 \\
0 & 1 & 0 & 1 & 0 & 0 & 20 \\
① & 0 & 0 & -2 & 1 & 0 & 14 \\
\hline
-2 & 0 & 0 & 3 & 0 & 1 & 60
\end{array}$$

The basic feasible solution is now

$$x_1 = x_4 = 0, \quad x_2 = 20, \quad x_3 = 30, \quad x_5 = 14, \quad M = 60$$

The new solution is at the extreme point $(x_1, x_2) = (0, 20)$ in Figure 1. The -2 in the bottom row of the tableau shows that the next pivot is in column 1, which produces

$$\begin{array}{cccccc|c}
x_1 & x_2 & x_3 & x_4 & x_5 & M & \\
0 & 0 & 1 & ② & -1 & 0 & 16 \\
0 & 1 & 0 & 1 & 0 & 0 & 20 \\
1 & 0 & 0 & -2 & 1 & 0 & 14 \\
\hline
0 & 0 & 0 & -1 & 2 & 1 & 88
\end{array}$$

This time $x_1 = 14$ and $x_2 = 20$, so the solution has moved across to the extreme point $(14, 20)$ in Figure 1, and the objective function has increased from 60 to 88. Finally, the -1 in the bottom row shows that the next pivot is in column 4. Pivoting on the 2 in the first row produces the final tableau:

$$\begin{array}{cccccc|c}
x_1 & x_2 & x_3 & x_4 & x_5 & M & \\
0 & 0 & \frac{1}{2} & 1 & -\frac{1}{2} & 0 & 8 \\
0 & 1 & -\frac{1}{2} & 0 & \frac{1}{2} & 0 & 12 \\
1 & 0 & 1 & 0 & 0 & 0 & 30 \\
\hline
0 & 0 & \frac{1}{2} & 0 & \frac{3}{2} & 1 & 96
\end{array}$$

Since all the entries in the bottom row are nonnegative, the solution now is optimal, with $x_1 = 30$ and $x_2 = 12$, corresponding to the extreme point $(30, 12)$. The maximum profit of \$96 is attained by making 30 boxes of the first mixture and 12 boxes of the second. Note that although x_4 is part of the basic feasible solution for this tableau, its value is not included in the solution of the original problem, because x_4 is a slack variable. ∎

Minimization Problems

So far, each canonical maximizing problem involved a vector **b** whose coordinates were positive. But what happens when some of the coordinates of **b** are zero or negative? And what about a minimizing problem?

If some of the coordinates of **b** are zero, then it is possible for cycling to occur and the simplex method to fail to terminate at an optimal solution. As mentioned earlier, however, cycling does not generally happen in practical applications, and so the presence of zero entries in the right-hand column seldom causes difficulty in the operation of the simplex method.

The case when one of the coordinates of **b** is negative can occur in practice and requires some special consideration. The difficulty is that all the b_i terms must be nonnegative in order for the slack variables to provide an initial basic feasible solution. One way to change a negative b_i term into a positive term would be to multiply the inequality by -1 (before introducing slack variables). But this would change the direction of the inequality. For example,

$$x_1 - 3x_2 + 2x_3 \leq -4$$

would become

$$-x_1 + 3x_2 - 2x_3 \geq 4$$

Thus a negative b_i term causes the same problem as a reversed inequality. Since reversed inequalities often occur in minimization problems, the following example discusses this case.

EXAMPLE 7 Minimize $x_1 + 2x_2$

$$\text{subject to} \qquad x_1 + x_2 \geq 14$$
$$x_1 - x_2 \leq 2$$
and $x_1 \geq 0, x_2 \geq 0$.

SOLUTION The minimum of $f(x_1, x_2)$ over a set occurs at the same point as the maximum of $-f(x_1, x_2)$ over the *same* set. However, in order to use the simplex algorithm, the canonical *description* of the feasible set must use $\leq$ signs. So the first inequality above must be rewritten. The second inequality is already in canonical form. Thus the original problem is equivalent to the following:

$$\text{Maximize} \qquad -x_1 - 2x_2$$
$$\text{subject to} \qquad -x_1 - x_2 \leq -14$$
$$x_1 - x_2 \leq 2$$
and $x_1 \geq 0, x_2 \geq 0$.

To solve this, let $M = -x_1 - 2x_2$ and add slack variables to the inequalities, as before. This creates the linear system

$$-x_1 - x_2 + x_3 \qquad\qquad = -14$$
$$x_1 - x_2 \qquad + x_4 \qquad = 2$$
$$x_1 + 2x_2 \qquad\qquad + M = 0$$

To find a nonnegative solution to this system for which M is a maximum, construct the initial simplex tableau:

$$
\begin{array}{ccccc}
x_1 & x_2 & x_3 & x_4 & M \\
\end{array}
$$
$$
\left[
\begin{array}{ccccc|c}
-1 & -1 & 1 & 0 & 0 & -14 \\
1 & -1 & 0 & 1 & 0 & 2 \\
\hline
1 & 2 & 0 & 0 & 1 & 0 \\
\end{array}
\right]
$$

The corresponding basic solution is

$$x_1 = x_2 = 0, \quad x_3 = -14, \quad x_4 = 2, \quad M = 0$$

However, since x_3 is negative, this basic solution is not feasible. *Before the standard simplex method can begin, each term in the augmented column above the horizontal line must be a nonnegative number.* This is accomplished by pivoting on a negative entry.

In order to replace a negative b_i entry by a positive number, find another negative entry in the same row. (If all the other entries in the row are nonnegative, then the problem has no feasible solution.) This negative entry is in the column corresponding to the variable that should now come into the solution. In this example, the first two columns both have negative entries, so either x_1 or x_2 should be brought into the solution.

For example, to bring x_2 into the solution, select as a pivot the entry a_{i2} in column 2 for which the ratio b_i/a_{i2} is the smallest nonnegative number. (The ratio is positive when both b_i and a_{i2} are negative.) In this case, only the ratio $(-14)/(-1)$ is nonnegative, so the -1 in the first row must be the pivot. After the pivot operations on column 2, the resulting tableau is

$$
\begin{array}{ccccc}
x_1 & x_2 & x_3 & x_4 & M \\
\end{array}
$$

$$
\left[\begin{array}{ccccc|c}
1 & 1 & -1 & 0 & 0 & 14 \\
2 & 0 & -1 & 1 & 0 & 16 \\
\hline
-1 & 0 & 2 & 0 & 1 & -28
\end{array}\right]
$$

Now each entry in the augmented column (except the bottom entry) is positive, and the simplex method can begin. (Sometimes it may be necessary to pivot more than once in order to make each of these terms nonnegative. See Exercise 19.) The next tableau turns out to be optimal:

$$
\begin{array}{ccccc}
x_1 & x_2 & x_3 & x_4 & M \\
\end{array}
$$

$$
\left[\begin{array}{ccccc|c}
0 & 1 & -\frac{1}{2} & -\frac{1}{2} & 0 & 6 \\
1 & 0 & -\frac{1}{2} & \frac{1}{2} & 0 & 8 \\
\hline
0 & 0 & \frac{3}{2} & \frac{1}{2} & 1 & -20
\end{array}\right]
$$

The maximum feasible value of $-x_1 - 2x_2$ is -20, when $x_1 = 8$ and $x_2 = 6$. So the *minimum* value of $x_1 + 2x_2$ is 20.   ∎

The final example uses the technique of Example 7, but the simplex tableau requires more preprocessing before the standard maximization operations can begin.

EXAMPLE 8   Minimize $5x_1 + 3x_2$

$$\text{subject to} \qquad 4x_1 + x_2 \geq 12$$
$$x_1 + 2x_2 \geq 10$$
$$x_1 + 4x_2 \geq 16$$
$$\text{and } x_1 \geq 0, x_2 \geq 0.$$

SOLUTION   Convert the problem into a maximization problem, setting $M = -5x_1 - 3x_2$ and reversing the three main constraint inequalities:

$$-4x_1 - x_2 \leq -12, \quad -x_1 - 2x_2 \leq -10, \quad -x_1 - 4x_2 \leq -16$$

Add nonnegative slack variables, and construct the initial simplex tableau:

$$
\begin{aligned}
-4x_1 - x_2 + x_3 && &= -12 \\
-x_1 - 2x_2 &&+ x_4 &= -10 \\
-x_1 - 4x_2 &&+ x_5 &= -16 \\
5x_1 + 3x_2 &&+ M &= 0
\end{aligned}
$$

$$
\begin{array}{cccccc}
x_1 & x_2 & x_3 & x_4 & x_5 & M \\
\end{array}
$$

$$
\left[\begin{array}{cccccc|c}
-4 & -1 & 1 & 0 & 0 & 0 & -12 \\
-1 & -2 & 0 & 1 & 0 & 0 & -10 \\
-1 & -4 & 0 & 0 & 1 & 0 & -16 \\
\hline
5 & 3 & 0 & 0 & 0 & 1 & 0
\end{array}\right]
$$

Before the simplex maximization process can begin, the top three entries in the augmented column must be nonnegative (to make the basic solution feasible). Pivoting on a negative entry to bring x_1 or x_2 into the solution will help. Trial and error will work. However, the fastest method is to compute the usual ratios b_i/a_{ij} for all negative entries in rows 1 to 3 of columns 1 and 2. Choose as the pivot the entry with the *largest* ratio. That will make *all* the augmented entries change sign (because the pivot operation will *add* multiples of the pivot row to the other rows). In this example, the pivot should be a_{31}, and the new tableau is

$$
\begin{array}{cccccc}
x_1 & x_2 & x_3 & x_4 & x_5 & M \\
\left[\begin{array}{cccccc|c}
0 & 15 & 1 & 0 & -4 & 0 & 52 \\
0 & 2 & 0 & 1 & -1 & 0 & 6 \\
1 & 4 & 0 & 0 & -1 & 0 & 16 \\
\hline
0 & -17 & 0 & 0 & 5 & 1 & -80
\end{array}\right]
\end{array}
$$

Now the simplex maximization algorithm is available. The -17 in the last row shows that x_2 must be brought into the solution. The smallest of the ratios $52/15$, $6/2$, and $16/4$ is $6/2$. A pivot on the 2 in column 2 produces

$$
\begin{array}{cccccc}
x_1 & x_2 & x_3 & x_4 & x_5 & M \\
\left[\begin{array}{cccccc|c}
0 & 0 & 1 & -\frac{15}{2} & \frac{7}{2} & 0 & 7 \\
0 & 1 & 0 & \frac{1}{2} & -\frac{1}{2} & 0 & 3 \\
1 & 0 & 0 & -2 & 1 & 0 & 4 \\
\hline
0 & 0 & 0 & \frac{17}{2} & -\frac{7}{2} & 1 & -29
\end{array}\right]
\end{array}
$$

The $-\frac{7}{2}$ in the last row shows that x_5 must be brought into the solution. The pivot is $\frac{7}{2}$ in column 5, and the new (and final) tableau is

$$
\begin{array}{cccccc}
x_1 & x_2 & x_3 & x_4 & x_5 & M \\
\left[\begin{array}{cccccc|c}
0 & 0 & \frac{2}{7} & -\frac{15}{7} & 1 & 0 & 2 \\
0 & 1 & \frac{1}{7} & -\frac{4}{7} & 0 & 0 & 4 \\
1 & 0 & -\frac{2}{7} & \frac{1}{7} & 0 & 0 & 2 \\
\hline
0 & 0 & 1 & 1 & 0 & 1 & -22
\end{array}\right]
\end{array}
$$

The solution occurs when $x_1 = 2$ (from row 3), $x_2 = 4$, and $M = -22$, so the minimum of the original objective function is 22. ∎

The "Simplex" in the Simplex Algorithm

The geometric approach in Section 9.2 focused on the *rows* of an $m \times 2$ matrix A, graphing each inequality as a half-space in $\mathbb{R}^2$, and viewing the solution set as the intersection of half-spaces. In higher-dimensional problems, the solution set is again an intersection of half-spaces, but this geometric view does not lead to an efficient algorithm for finding the optimal solution.

The simplex algorithm focuses on the *columns* of A instead of the rows. Suppose that A is $m \times n$ and denote the columns by $\mathbf{a}_1, \ldots, \mathbf{a}_n$. The addition of m slack variables creates an $m \times (n+m)$ system of equations of the form

$$x_1\mathbf{a}_1 + \cdots + x_n\mathbf{a}_n + x_{n+1}\mathbf{e}_1 + \cdots + x_{n+m}\mathbf{e}_m = \mathbf{b}$$

where $x_1, \ldots, x_{n+m}$ are nonnegative and $\{\mathbf{e}_1, \ldots, \mathbf{e}_m\}$ is the standard basis for $\mathbb{R}^m$. The initial basic feasible solution is obtained when $x_1, \ldots, x_n$ are zero and $b_1 \mathbf{e}_1 + \cdots + b_m \mathbf{e}_m = \mathbf{b}$. If $s = b_1 + \cdots + b_m$, then the equation

$$\mathbf{0} + \left(\frac{b_1}{s}\right) s \mathbf{e}_1 + \cdots + \left(\frac{b_m}{s}\right) s \mathbf{e}_m = \mathbf{b}$$

shows that $\mathbf{b}$ is in what is called the *simplex* generated by $\mathbf{0}, s\mathbf{e}_1, \ldots, s\mathbf{e}_m$. For simplicity, we say that "$\mathbf{b}$ is in an m-dimensional simplex determined by $\mathbf{e}_1, \ldots, \mathbf{e}_m$." This is the first simplex in the simplex algorithm.[4]

In general, if $\mathbf{v}_1, \ldots, \mathbf{v}_m$ is any basis of $\mathbb{R}^m$, selected from the columns of the matrix $P = [\, \mathbf{a}_1 \cdots \mathbf{a}_n \;\; \mathbf{e}_1 \cdots \mathbf{e}_m \,]$, and if $\mathbf{b}$ is a linear combination of these vectors with nonnegative weights, then $\mathbf{b}$ is in an m-dimensional simplex determined by $\mathbf{v}_1, \ldots, \mathbf{v}_m$. A *basic* feasible solution of the linear programming problem corresponds to a particular *basis* from the columns of P. The simplex algorithm changes this basis and hence the corresponding simplex that contains $\mathbf{b}$, one column at a time. The various ratios computed during the algorithm drive the choice of columns. Since row operations do not change the linear dependence relations among the columns, each basic feasible solution tells how to build $\mathbf{b}$ from the corresponding columns of P.

Practice Problem

Use the simplex method to solve the following linear programming problem:

$$\begin{array}{ll} \text{Maximize} & 2x_1 + x_2 \\ \text{subject to} & -x_1 + 2x_2 \le 8 \\ & 3x_1 + 2x_2 \le 24 \end{array}$$

and $x_1 \ge 0, x_2 \ge 0$.

9.3 Exercises

In Exercises 1 and 2, set up the initial simplex tableau for the given linear programming problem.

1. Maximize $\quad 21x_1 + 25x_2 + 15x_3$
   subject to $\quad 2x_1 + 7x_2 + 10x_3 \le 20$
   $\quad\quad\quad\quad\quad 3x_1 + 4x_2 + 18x_3 \le 25$
   and $x_1 \ge 0, x_2 \ge 0, x_3 \ge 0$.

2. Maximize $\quad 22x_1 + 14x_2$
   subject to $\quad 3x_1 + 5x_2 \le 30$
   $\quad\quad\quad\quad\quad 2x_1 + 7x_2 \le 24$
   $\quad\quad\quad\quad\quad 6x_1 + x_2 \le 42$
   and $x_1 \ge 0, x_2 \ge 0$.

For each simplex tableau in Exercises 3–6, do the following:

a. Determine which variable should be brought into the solution.

b. Compute the next tableau.

c. Identify the basic feasible solution corresponding to the tableau in part (b).

d. Determine if the answer in part (c) is optimal.

3.

x_1	x_2	x_3	x_4	M	
5	1	1	0	0	20
3	2	0	1	0	30
-4	-10	0	0	1	0

4.

x_1	x_2	x_3	x_4	M	
-1	1	2	0	0	4
1	0	5	1	0	6
-5	0	3	0	1	17

---

[4] If $\mathbf{v}_1, \ldots, \mathbf{v}_m$ are linearly independent vectors in $\mathbb{R}^m$, then the convex hull of the set $\{\mathbf{0}, \mathbf{v}_1, \ldots, \mathbf{v}_m\}$ is an m-dimensional simplex, S. (See Section 8.5.) A typical vector in S has the form $c_0 \mathbf{0} + c_1 \mathbf{v}_1 + \cdots + c_m \mathbf{v}_m$, where the weights are nonnegative and sum to one. (Equivalently, vectors in S have the form $c_1 \mathbf{v}_1 + \cdots + c_m \mathbf{v}_m$, where the weights are nonnegative and their sum is at most one.) Any set formed by translating such a set S is also called an m-dimensional simplex, but such sets do not appear in the simplex algorithm.

5.

$$\begin{array}{ccccc} x_1 & x_2 & x_3 & x_4 & M \end{array}$$
$$\begin{bmatrix} 2 & 3 & 1 & 0 & 0 & | & 20 \\ 2 & 1 & 0 & 1 & 0 & | & 16 \\ \hline -6 & -5 & 0 & 0 & 1 & | & 0 \end{bmatrix}$$

6.

$$\begin{array}{ccccc} x_1 & x_2 & x_3 & x_4 & M \end{array}$$
$$\begin{bmatrix} 5 & 8 & 1 & 0 & 0 & | & 80 \\ 12 & 6 & 0 & 1 & 0 & | & 30 \\ \hline 2 & -3 & 0 & 0 & 1 & | & 0 \end{bmatrix}$$

Exercises 7–12 relate to a canonical linear programming problem with an $m \times n$ coefficient matrix A in the constraint inequality $A\mathbf{x} \leq \mathbf{b}$. Mark each statement True or False **(T/F)**. Justify each answer.

7. **(T/F)** A slack variable is used to change an equality into an inequality.

8. **(T/F)** A solution is called a basic solution if m or fewer of the variables are nonzero.

9. **(T/F)** A solution is feasible if each variable is nonnegative.

10. **(T/F)** The basic feasible solutions correspond to the extreme points of the feasible set.

11. **(T/F)** If one of the coordinates in vector $\mathbf{b}$ is negative, then the problem is infeasible.

12. **(T/F)** The bottom entry in the right column of a simplex tableau gives the maximum value of the objective function.

Solve Exercises 13–18 by using the simplex method.

13. Maximize $\quad 10x_1 + 12x_2$
subject to $\quad 2x_1 + 3x_2 \leq 36$
$\quad\quad\quad\quad\quad 5x_1 + 4x_2 \leq 55$
and $x_1 \geq 0, x_2 \geq 0$.

14. Maximize $\quad 5x_1 + 4x_2$
subject to $\quad x_1 + 5x_2 \leq 70$
$\quad\quad\quad\quad\quad 3x_1 + 2x_2 \leq 54$
and $x_1 \geq 0, x_2 \geq 0$.

15. Maximize $\quad 4x_1 + 5x_2$
subject to $\quad x_1 + 2x_2 \leq 26$
$\quad\quad\quad\quad\quad 2x_1 + 3x_2 \leq 30$
$\quad\quad\quad\quad\quad x_1 + x_2 \leq 13$
and $x_1 \geq 0, x_2 \geq 0$.

16. Maximize $\quad 2x_1 + 5x_2 + 3x_3$
subject to $\quad x_1 + 2x_2 \quad\quad \leq 28$
$\quad\quad\quad\quad\quad 2x_1 \quad\quad + 4x_3 \leq 16$
$\quad\quad\quad\quad\quad\quad x_2 + x_3 \leq 12$
and $x_1 \geq 0, x_2 \geq 0, x_3 \geq 0$.

17. Minimize $\quad 12x_1 + 5x_2$
subject to $\quad 2x_1 + x_2 \geq 32$
$\quad\quad\quad\quad\quad -3x_1 + 5x_2 \leq 30$
and $x_1 \geq 0, x_2 \geq 0$.

18. Minimize $\quad 2x_1 + 3x_2 + 3x_3$
subject to $\quad x_1 - 2x_2 \quad\quad \geq -8$
$\quad\quad\quad\quad\quad 2x_2 + x_3 \geq 15$
$\quad\quad\quad\quad\quad 2x_1 - x_2 + x_3 \leq 25$
and $x_1 \geq 0, x_2 \geq 0, x_3 \geq 0$.

19. Solve Example 7 by bringing x_1 into the solution (instead of x_2) in the initial tableau.

20. Use the simplex method to solve the linear programming problem in Section 9.2, Exercise 1.

21. Use the simplex method to solve the linear programming problem in Section 9.2, Exercise 17.

22. Use the simplex method to solve the linear programming problem in Section 9.2, Example 1.

Solution to Practice Problem

Introduce slack variables x_3 and x_4 to rewrite the problem:

$$\begin{array}{ll} \text{Maximize} & 2x_1 + x_2 \\ \text{subject to} & -x_1 + 2x_2 + x_3 \quad\quad = 8 \\ & 3x_1 + 2x_2 \quad\quad + x_4 = 24 \end{array}$$

and $x_1 \geq 0, x_2 \geq 0$.

Then let $M = 2x_1 + x_2$, so that $-2x_1 - x_2 + M = 0$ provides the bottom row in the initial simplex tableau.

$$\begin{array}{ccccc} x_1 & x_2 & x_3 & x_4 & M \end{array}$$
$$\begin{bmatrix} -1 & 2 & 1 & 0 & 0 & | & 8 \\ ③ & 2 & 0 & 1 & 0 & | & 24 \\ \hline -2 & -1 & 0 & 0 & 1 & | & 0 \end{bmatrix}$$

Bring x_1 into the solution (because of the -2 entry in the bottom row), and pivot on the second row (because it is the only row with a positive entry in the first column). The second tableau turns out to be optimal, since all the entries in the bottom row are positive. Remember that the slack variables (in color) are never part of the solution.

$$
\begin{array}{ccccc}
x_1 & x_2 & x_3 & x_4 & M \\
\end{array}
$$
$$
\left[
\begin{array}{ccccc|c}
0 & \frac{8}{3} & 1 & \frac{1}{3} & 0 & 16 \\
1 & \frac{2}{3} & 0 & \frac{1}{3} & 0 & 8 \\
0 & \frac{1}{3} & 0 & \frac{2}{3} & 1 & 16 \\
\end{array}
\right]
$$

The maximum value is 16, when $x_1 = 8$ and $x_2 = 0$. Note that this problem was solved geometrically in the Practice Problem for Section 9.2.

9.4 Duality

Associated with each canonical (maximization) linear programming problem is a related minimization problem, called the *dual* problem. In this setting, the canonical problem is called the *primal* problem. This section describes the dual problem and how it is solved, along with an interesting economic interpretation of the dual variables. The section concludes by showing how any matrix game can be solved using the primal and dual versions of a suitable linear programming problem.

Given vectors $\mathbf{c}$ in $\mathbb{R}^n$ and $\mathbf{b}$ in $\mathbb{R}^m$, and given an $m \times n$ matrix A, the canonical (primal) problem is to find $\mathbf{x}$ in $\mathbb{R}^n$ so as to maximize $f(\mathbf{x}) = \mathbf{c}^T\mathbf{x}$ subject to the constraints $A\mathbf{x} \leq \mathbf{b}$ and $\mathbf{x} \geq \mathbf{0}$. The dual (minimization) problem is to find $\mathbf{y}$ in $\mathbb{R}^m$ so as to minimize $g(\mathbf{y}) = \mathbf{b}^T\mathbf{y}$ subject to $A^T\mathbf{y} \geq \mathbf{c}$ and $\mathbf{y} \geq \mathbf{0}$:

Primal Problem P		Dual Problem $P*$	
Maximize	$f(\mathbf{x}) = \mathbf{c}^T\mathbf{x}$	Minimize	$g(\mathbf{y}) = \mathbf{b}^T\mathbf{y}$
subject to	$A\mathbf{x} \leq \mathbf{b}$	subject to	$A^T\mathbf{y} \geq \mathbf{c}$
	$\mathbf{x} \geq \mathbf{0}$		$\mathbf{y} \geq \mathbf{0}$

Observe that in forming the dual problem, the c_i coefficients of x_i in the objective function of the primal problem become the constants on the right-hand side of the constraint inequalities in the dual. Likewise, the numbers in the right-hand side of the constraint inequalities in the primal problem become the coefficients b_j of y_j in the objective function in the dual. Also, note that the direction of the constraint inequalities is reversed from $A\mathbf{x} \leq \mathbf{b}$ to $A^T\mathbf{y} \geq \mathbf{c}$. In both cases, the variables $\mathbf{x}$ and $\mathbf{y}$ are nonnegative.

EXAMPLE 1 Find the dual of the following primal problem:

$$
\begin{array}{ll}
\text{Maximize} & 5x_1 + 7x_2 \\
\text{subject to} & 2x_1 + 3x_2 \leq 25 \\
& 7x_1 + 4x_2 \leq 16 \\
& x_1 + 9x_2 \leq 21 \\
\end{array}
$$
and $x_1 \geq 0, x_2 \geq 0$.

SOLUTION

$$\text{Minimize} \quad 25y_1 + 16y_2 + 21y_3$$
$$\text{subject to} \quad 2y_1 + 7y_2 + y_3 \geq 5$$
$$3y_1 + 4y_2 + 9y_3 \geq 7$$
$$\text{and } y_1 \geq 0, y_2 \geq 0, y_3 \geq 0. \qquad \blacksquare$$

Suppose that the dual problem above is rewritten as a canonical maximization problem:

$$\text{Maximize} \quad h(\mathbf{y}) = -\mathbf{b}^T\mathbf{y}$$
$$\text{subject to} \quad -A^T\mathbf{y} \leq -\mathbf{c} \quad \text{and} \quad \mathbf{y} \geq \mathbf{0}.$$

Then the dual of *this* problem is

$$\text{Minimize} \quad F(\mathbf{w}) = -\mathbf{c}^T\mathbf{w}$$
$$\text{subject to} \quad (-A^T)^T\mathbf{w} \geq -\mathbf{b} \quad \text{and} \quad \mathbf{w} \geq \mathbf{0}.$$

In canonical form, this minimization problem is equivalent to

$$\text{Maximize} \quad G(\mathbf{w}) = \mathbf{c}^T\mathbf{w}$$
$$\text{subject to} \quad A\mathbf{w} \leq \mathbf{b} \quad \text{and} \quad \mathbf{w} \geq \mathbf{0}.$$

If $\mathbf{w}$ is replaced by $\mathbf{x}$, this problem is precisely the primal problem. So the dual of the dual problem is the original primal problem.

Theorem 7 below is a fundamental result in linear programming. As with the Minimax Theorem in game theory, the proof depends on certain properties of convex sets and hyperplanes.[1]

THEOREM 7

The Duality Theorem

Let P be a (primal) linear programming problem with feasible set $\mathcal{F}$, and let P^* be the dual problem with feasible set $\mathcal{F}^*$.

a. If $\mathcal{F}$ and $\mathcal{F}^*$ are both nonempty, then P and P^* both have optimal solutions, say $\bar{\mathbf{x}}$ and $\bar{\mathbf{y}}$, respectively, and $f(\bar{\mathbf{x}}) = g(\bar{\mathbf{y}})$.

b. If one of the problems P or P^* has an optimal solution $\bar{\mathbf{x}}$ or $\bar{\mathbf{y}}$, respectively, then so does the other, and $f(\bar{\mathbf{x}}) = g(\bar{\mathbf{y}})$.

EXAMPLE 2 Set up and solve the dual to the problem in Example 5 of Section 9.2.

SOLUTION The original problem is to

$$\text{Maximize} \quad f(x_1, x_2) = 2x_1 + 3x_2$$
$$\text{subject to} \quad x_1 \qquad \leq 30$$
$$x_2 \leq 20$$
$$x_1 + 2x_2 \leq 54$$
$$\text{and } x_1 \geq 0, x_2 \geq 0.$$

---

[1] If the equation $A\mathbf{x} = \mathbf{b}$ has no nonnegative solution, then the sets $\{\mathbf{b}\}$ and $S = \{\mathbf{z} \in \mathbb{R}^m : \mathbf{z} = A\mathbf{x}, \mathbf{x} \geq \mathbf{0}\}$ are disjoint. It is not hard to show that S is a closed convex set, so Theorem 12 in Chapter 8 implies that there exists a hyperplane strictly separating $\{\mathbf{b}\}$ and S. This hyperplane plays a key role in the proof. For details, see Steven R. Lay, *Convex Sets and Their Applications* (New York: John Wiley & Sons, 1982; Mineola, NY: Dover Publications, 2007), pp. 174–178.

Calculations in Example 5 of Section 9.2 showed that the optimal solution of this problem is $\bar{\mathbf{x}} = \begin{bmatrix} 30 \\ 12 \end{bmatrix}$ with $f(\bar{\mathbf{x}}) = 96$. The dual problem is to

Minimize $g(y_1, y_2, y_3) = 30y_1 + 20y_2 + 54y_3$
subject to $y_1 \quad + \quad y_3 \geq 2$
$y_2 + 2y_3 \geq 3$
and $y_1 \geq 0, y_2 \geq 0, y_3 \geq 0$.

The simplex method could be used here, but the geometric method of Section 9.2 is not too difficult. Graphs of the constraint inequalities (Figure 1) reveal that $\mathscr{F}^*$ has three extreme points and that $\bar{\mathbf{y}} = \begin{bmatrix} \frac{1}{2} \\ 0 \\ \frac{3}{2} \end{bmatrix}$ is the optimal solution. Indeed, $g(\bar{\mathbf{y}}) = 30(\frac{1}{2}) + 20(0) + 54(\frac{3}{2}) = 96$, as expected.   ■

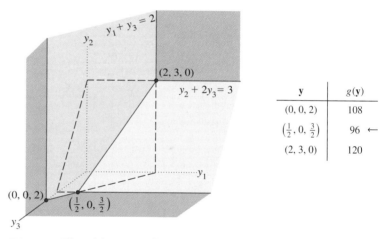

$\mathbf{y}$	$g(\mathbf{y})$
$(0, 0, 2)$	108
$(\frac{1}{2}, 0, \frac{3}{2})$	96 $\leftarrow$
$(2, 3, 0)$	120

FIGURE 1 The minimum of $g(y_1, y_2, y_3) = 30y_1 + 20y_2 + 54y_3$.

Example 2 illustrates another important property of duality and the simplex method. Recall that Example 6 of Section 9.3 solved this same maximizing problem using the simplex method. Here is the final tableau:

$$
\begin{array}{cccccc|c}
x_1 & x_2 & x_3 & x_4 & x_5 & M & \\
\hline
0 & 0 & \frac{1}{2} & 1 & -\frac{1}{2} & 0 & 8 \\
0 & 1 & -\frac{1}{2} & 0 & \frac{1}{2} & 0 & 12 \\
1 & 0 & 1 & 0 & 0 & 0 & 30 \\
\hline
0 & 0 & \frac{1}{2} & 0 & \frac{3}{2} & 1 & 96
\end{array}
$$

Notice that the optimal solution to the dual problem appears in the bottom row. The variables x_3, x_4, and x_5 are the slack variables for the first, second, and third equations, respectively. The bottom entry in each of these columns gives the optimal solution $\bar{\mathbf{y}} = \begin{bmatrix} \frac{1}{2} \\ 0 \\ \frac{3}{2} \end{bmatrix}$ to the dual problem. This is not a coincidence, as the following theorem shows.

THEOREM 7

The Duality Theorem (Continued)

Let P be a (primal) linear programming problem and let P^* be its dual problem. Suppose P (or P^*) has an optimal solution.

c. If either P or P^* is solved by the simplex method, then the solution of its dual is displayed in the bottom row of the final tableau in the columns associated with the slack variables.

EXAMPLE 3 Set up and solve the dual to the problem in Example 5 in Section 9.3.

SOLUTION The primal problem P is to

$$
\begin{aligned}
\text{Maximize} \quad & f(x_1, x_2, x_3) = 25x_1 + 33x_2 + 18x_3 \\
\text{subject to} \quad & 2x_1 + 3x_2 + 4x_3 \le 60 \\
& 3x_1 + x_2 + 5x_3 \le 46 \\
& x_1 + 2x_2 + x_3 \le 50
\end{aligned}
$$

and $x_1 \ge 0, x_2 \ge 0, x_3 \ge 0$.

The dual problem P^* is to

$$
\begin{aligned}
\text{Minimize} \quad & g(y_1, y_2, y_3) = 60y_1 + 46y_2 + 50y_3 \\
\text{subject to} \quad & 2y_1 + 3y_2 + y_3 \ge 25 \\
& 3y_1 + y_2 + 2y_3 \ge 33 \\
& 4y_1 + 5y_2 + y_3 \ge 18
\end{aligned}
$$

and $y_1 \ge 0, y_2 \ge 0, y_3 \ge 0$.

The final tableau for the solution of the primal problem was found to be

$$
\begin{array}{cccccccc}
x_1 & x_2 & x_3 & x_4 & x_5 & x_6 & M & \\
\left[\begin{array}{ccccccc|c}
0 & 1 & \frac{2}{7} & \frac{3}{7} & -\frac{2}{7} & 0 & 0 & \frac{88}{7} \\
1 & 0 & \frac{11}{7} & -\frac{1}{7} & \frac{3}{7} & 0 & 0 & \frac{78}{7} \\
0 & 0 & -\frac{8}{7} & -\frac{5}{7} & \frac{1}{7} & 1 & 0 & \frac{96}{7} \\
\hline
0 & 0 & \frac{215}{7} & \frac{74}{7} & \frac{9}{7} & 0 & 1 & \frac{4854}{7}
\end{array}\right]
\end{array}
$$

The slack variables are x_4, x_5, and x_6. They give the optimal solution to the dual problem P^*. Thus,

$$
y_1 = \tfrac{74}{7}, \quad y_2 = \tfrac{9}{7}, \quad \text{and} \quad y_3 = 0
$$

Note that the optimal value of the objective function in the dual problem is

$$
g\left(\tfrac{74}{7}, \tfrac{9}{7}, 0\right) = 60\left(\tfrac{74}{7}\right) + 46\left(\tfrac{9}{7}\right) + 50(0) = \tfrac{4854}{7}
$$

This agrees with the optimal value of the objective function in the primal problem. ∎

The variables in the dual problem have useful economic interpretations. For example, consider the problem of mixing nuts studied in Example 5 of Section 9.2 and Example 6 of Section 9.3:

$$
\begin{aligned}
\text{Maximize} \quad & f(x_1, x_2) = 2x_1 + 3x_2 \\
\text{subject to} \quad & x_1 \le 30 && \text{(cashews)} \\
& x_2 \le 20 && \text{(filberts)} \\
& x_1 + 2x_2 \le 54 && \text{(peanuts)}
\end{aligned}
$$

and $x_1 \ge 0, x_2 \ge 0$.

Recall that x_1 is the number of boxes of the first mixture and x_2 is the number of boxes of the second mixture. Example 2 displayed the following dual problem:

$$\begin{aligned}
\text{Minimize} \quad & g(y_1, y_2, y_3) = 30y_1 + 20y_2 + 54y_3 \\
\text{subject to} \quad & y_1 \quad + \quad y_3 \geq 2 \\
& \quad\quad y_2 + 2y_3 \geq 3
\end{aligned}$$

and $y_1 \geq 0, y_2 \geq 0, y_3 \geq 0$.

If $\bar{\mathbf{x}}$ and $\bar{\mathbf{y}}$ are optimal solutions of these problems, then by the Duality Theorem, the maximum profit $f(\bar{\mathbf{x}})$ satisfies the equation

$$f(\bar{\mathbf{x}}) = g(\bar{\mathbf{y}}) = 30\bar{y}_1 + 20\bar{y}_2 + 54\bar{y}_3$$

Suppose, for example, that the amount of cashews available was increased from 30 pounds to $30 + h$ pounds. Then the profit would increase by $h\bar{y}_1$. Likewise, if the amount of cashews was decreased by h pounds, then the profit would decrease by $h\bar{y}_1$. So $\bar{y}_1$ represents the value (per pound) of increasing or decreasing the amount of cashews available. This is usually referred to as the **marginal value** of the cashews. Similarly, $\bar{y}_2$ and $\bar{y}_3$ are the marginal values of the filberts and peanuts, respectively. These values indicate how much the company might be willing to pay for additional supplies of the various nuts.[2]

EXAMPLE 4 The final simplex tableau for the problem of mixing nuts was found (in Example 6 of Section 9.3) to be

$$\begin{array}{cccccc}
x_1 & x_2 & x_3 & x_4 & x_5 & M \\
\end{array}$$
$$\left[\begin{array}{cccccc|c}
0 & 0 & \frac{1}{2} & 1 & -\frac{1}{2} & 0 & 8 \\
0 & 1 & -\frac{1}{2} & 0 & \frac{1}{2} & 0 & 12 \\
1 & 0 & 1 & 0 & 0 & 0 & 30 \\
\hline
0 & 0 & \frac{1}{2} & 0 & \frac{3}{2} & 1 & 96
\end{array}\right]$$

so the optimal solution of the dual is $\bar{\mathbf{y}} = \begin{bmatrix} \frac{1}{2} \\ 0 \\ \frac{3}{2} \end{bmatrix}$. Thus the marginal value of the cashews is $\frac{1}{2}$, the marginal value of the filberts is 0, and the marginal value of the peanuts is $\frac{3}{2}$.

Note that the optimal production schedule $\bar{\mathbf{x}} = \begin{bmatrix} 30 \\ 12 \end{bmatrix}$ uses only 12 of the 20 pounds of filberts. (This corresponds to the slack variable x_4 for the filbert constraint inequality having value 8 in the final tableau.) This means that not all the available filberts are used, so there is no increase in profit from increasing the number of filberts available. That is, their marginal value is zero. ∎

Linear Programming and Matrix Games

Let A be an $m \times n$ payoff matrix for a matrix game, as in Section 9.1, and assume at first that each entry in A is positive. Let $\mathbf{u}$ in $\mathbb{R}^m$ and $\mathbf{v}$ in $\mathbb{R}^n$ be the vectors whose coordinates

---

[2] The other entries in the final tableau can also be given an economic interpretation. See Saul I. Gass, *Linear Programming Methods and Applications*, 5th Ed. (Danvers, MA: Boyd & Fraser Publishing, 1985), pp. 173–177. Also see Goldstein, Schneider, and Siegel, *Finite Mathematics and Its Applications*, 6th Ed. (Upper Saddle River, NJ: Prentice Hall, 1998), pp. 166–185.

are all equal to one, and consider the following linear programming problem P and its dual P^*. (Notice that the roles of $\mathbf{x}$ and $\mathbf{y}$ are reversed, with $\mathbf{x}$ in $\mathbb{R}^m$ and $\mathbf{y}$ in $\mathbb{R}^n$.)

$$P: \quad \text{Maximize} \quad \mathbf{v}^T\mathbf{y} \qquad P^*: \quad \text{Minimize} \quad \mathbf{u}^T\mathbf{x}$$
$$\text{subject to} \quad A\mathbf{y} \le \mathbf{u} \qquad\qquad \text{subject to} \quad A^T\mathbf{x} \ge \mathbf{v}$$
$$\mathbf{y} \ge \mathbf{0} \qquad\qquad\qquad\qquad \mathbf{x} \ge \mathbf{0}$$

The primal problem P is feasible since $\mathbf{y} = \mathbf{0}$ satisfies the constraints. The dual problem P^* is feasible since all the entries in A^T are positive and $\mathbf{v}$ is a vector of 1's. By the Duality Theorem, there exist optimal solutions $\bar{\mathbf{y}}$ and $\bar{\mathbf{x}}$ such that $\mathbf{v}^T\bar{\mathbf{y}} = \mathbf{u}^T\bar{\mathbf{x}}$. Set

$$\lambda = \mathbf{v}^T\bar{\mathbf{y}} = \mathbf{u}^T\bar{\mathbf{x}}$$

Since the entries in A and $\mathbf{u}$ are positive, the inequality $A\mathbf{y} \le \mathbf{u}$ has a nonzero solution $\mathbf{y}$ with $\mathbf{y} \ge \mathbf{0}$. As a result, the λ for the primal problem is positive. Let

$$\hat{\mathbf{y}} = \bar{\mathbf{y}}/\lambda \qquad \text{and} \qquad \hat{\mathbf{x}} = \bar{\mathbf{x}}/\lambda$$

It can be shown (Exercise 29) that $\hat{\mathbf{y}}$ is the optimal mixed strategy for the column player C and $\hat{\mathbf{x}}$ is the optimal mixed strategy for the row player R. Furthermore, the value of the game is equal to $1/\lambda$.

Finally, if the payoff matrix A has some entries that are not positive, add a fixed number, say k, to each entry to make the entries all positive. This will not change the optimal mixed strategies for the two players, and it will add an amount k to the value of the game. [See Exercise 33(b) in Section 9.1.]

EXAMPLE 5   Solve the game whose payoff matrix is $A = \begin{bmatrix} -2 & 1 & 2 \\ 3 & 2 & 0 \end{bmatrix}$.

SOLUTION   To produce a matrix B with positive entries, add 3 to each entry:

$$B = \begin{bmatrix} 1 & 4 & 5 \\ 6 & 5 & 3 \end{bmatrix}$$

The optimal strategy for the column player C is found by solving the linear programming problem

$$\begin{array}{ll} \text{Maximize} & y_1 + y_2 + y_3 \\ \text{subject to} & y_1 + 4y_2 + 5y_3 \le 1 \\ & 6y_1 + 5y_2 + 3y_3 \le 1 \end{array}$$

and $y_1 \ge 0$, $y_2 \ge 0$, $y_3 \ge 0$.

Introduce slack variables y_4 and y_5, let M be the objective function, and construct the initial simplex tableau:

$$\begin{array}{ccccccc} y_1 & y_2 & y_3 & y_4 & y_5 & M & \\ \left[\begin{array}{cccccc|c} 1 & 4 & 5 & 1 & 0 & 0 & 1 \\ 6 & 5 & 3 & 0 & 1 & 0 & 1 \\ \hline -1 & -1 & -1 & 0 & 0 & 1 & 0 \end{array}\right] \end{array}$$

The three -1 entries in the bottom row are equal, so any of columns 1 to 3 can be the first pivot column. Choose column 1 and check the ratios b_i/a_{i1}. To bring variable y_1 into the solution, pivot on the 6 in the second row.

$$\begin{array}{ccccccc} y_1 & y_2 & y_3 & y_4 & y_5 & M & \\ \left[\begin{array}{cccccc|c} 0 & \frac{19}{6} & \frac{9}{2} & 1 & -\frac{1}{6} & 0 & \frac{5}{6} \\ 1 & \frac{5}{6} & \frac{1}{2} & 0 & \frac{1}{6} & 0 & \frac{1}{6} \\ \hline 0 & -\frac{1}{6} & -\frac{1}{2} & 0 & \frac{1}{6} & 1 & \frac{1}{6} \end{array}\right] \end{array}$$

In the bottom row, the third entry is the most negative, so bring y_3 into the solution. The ratios b_i/a_{i3} are $\frac{5}{6}/\frac{9}{2} = \frac{5}{27}$ and $\frac{1}{6}/\frac{1}{2} = \frac{1}{3} = \frac{9}{27}$. The first ratio is smaller, so pivot on the $\frac{9}{2}$ in the first row.

$$
\begin{array}{cccccc}
y_1 & y_2 & y_3 & y_4 & y_5 & M \\
\end{array}
$$

$$
\left[
\begin{array}{ccccc|c}
0 & \frac{19}{27} & 1 & \frac{2}{9} & -\frac{1}{27} & 0 & \frac{5}{27} \\
1 & \frac{13}{27} & 0 & -\frac{1}{9} & \frac{5}{27} & 0 & \frac{2}{27} \\
\hline
0 & \frac{5}{27} & 0 & \frac{1}{9} & \frac{4}{27} & 1 & \frac{7}{27}
\end{array}
\right]
$$

The optimal solution of the primal problem is

$$
\bar{y}_1 = \tfrac{2}{27}, \quad \bar{y}_2 = 0, \quad \bar{y}_3 = \tfrac{5}{27}, \quad \text{with } \lambda = \bar{y}_1 + \bar{y}_2 + \bar{y}_3 = \tfrac{7}{27}
$$

The corresponding optimal mixed strategy for C is

$$
\hat{\mathbf{y}} = \bar{\mathbf{y}}/\lambda = \begin{bmatrix} \frac{2}{7} \\ 0 \\ \frac{5}{7} \end{bmatrix}
$$

The optimal solution of the dual problem comes from the bottom entries under the slack variables

$$
\bar{x}_1 = \tfrac{1}{9} = \tfrac{3}{27} \quad \text{and} \quad \bar{x}_2 = \tfrac{4}{27}, \quad \text{with } \lambda = \bar{x}_1 + \bar{x}_2 = \tfrac{7}{27}
$$

which shows that the optimal mixed strategy for R is

$$
\hat{\mathbf{x}} = \bar{\mathbf{x}}/\lambda = \begin{bmatrix} \frac{3}{7} \\ \frac{4}{7} \end{bmatrix}
$$

The value of the game with payoff matrix B is $v = \frac{1}{\lambda} = \frac{27}{7}$, so the value of the original matrix game A is $\frac{27}{7} - 3 = \frac{6}{7}$. ∎

Although matrix games are usually solved via linear programming, it is interesting that a linear programming problem can be reduced to a matrix game. If the programming problem has an optimal solution, then this solution is reflected in the solution of the matrix game. Suppose the problem is to maximize $\mathbf{c}^T\mathbf{x}$ subject to $A\mathbf{x} \leq \mathbf{b}$ and $\mathbf{x} \geq \mathbf{0}$, where A is $m \times n$ with $m \leq n$. Let

$$
M = \begin{bmatrix} 0 & A & -\mathbf{b} \\ -A^T & 0 & \mathbf{c} \\ \mathbf{b}^T & -\mathbf{c}^T & 0 \end{bmatrix} \quad \text{and} \quad \mathbf{s} = \begin{bmatrix} \bar{\mathbf{y}} \\ \bar{\mathbf{x}} \\ z \end{bmatrix}
$$

and suppose that M represents a matrix game and $\mathbf{s}$ is an optimal column strategy for M. The $(n + m + 1) \times (n + m + 1)$ matrix M is skew-symmetric; that is, $M^T = -M$. It can be shown that in this case the optimal row strategy equals the optimal column strategy, the value of the game is 0, and the maximum of the entries in the vector $M\mathbf{s}$ is 0. Observe that

$$
M\mathbf{s} = \begin{bmatrix} 0 & A & -\mathbf{b} \\ -A^T & 0 & \mathbf{c} \\ \mathbf{b}^T & -\mathbf{c}^T & 0 \end{bmatrix} \begin{bmatrix} \bar{\mathbf{y}} \\ \bar{\mathbf{x}} \\ z \end{bmatrix} = \begin{bmatrix} A\bar{\mathbf{x}} - z\mathbf{b} \\ -A^T\bar{\mathbf{y}} + z\mathbf{c} \\ \mathbf{b}^T\bar{\mathbf{y}} - \mathbf{c}^T\bar{\mathbf{x}} \end{bmatrix} \leq \begin{bmatrix} \mathbf{0} \\ \mathbf{0} \\ 0 \end{bmatrix}
$$

Thus $A\bar{\mathbf{x}} \le z\mathbf{b}$, $A^T\mathbf{y} \ge z\mathbf{c}$, and $\mathbf{b}^T\bar{\mathbf{y}} \le \mathbf{c}^T\bar{\mathbf{x}}$. Since the column strategy $\mathbf{s}$ is a probability vector, $z \ge 0$. It can be shown that if $z > 0$, then $\bar{\mathbf{x}}/z$ is an optimal solution for the primal (maximization) problem for $A\mathbf{x} \le \mathbf{b}$, and $\bar{\mathbf{y}}/z$ is an optimal solution for the dual problem for $A^T\mathbf{y} \ge \mathbf{c}$. Also, if $z = 0$, then the primal and dual problems have no optimal solutions.

In conclusion, the simplex method is a powerful tool in solving linear programming problems. Because a fixed procedure is followed, it lends itself well to using a computer for the tedious calculations involved. The algorithm presented here is not optimal for a computer, but many computer programs implement variants of the simplex method, and some programs even seek integer solutions. New methods developed in recent years take shortcuts through the interior of the feasible region instead of going from extreme point to extreme point. They are somewhat faster in certain situations (typically involving thousands of variables and constraints), but the simplex method is still the approach most widely used.

Practice Problems

The following questions relate to the Shady-Lane grass seed company from Example 1 in Section 9.2. The canonical linear programming problem can be stated as follows:

$$
\begin{array}{ll}
\text{Maximize} & 2x_1 + 3x_2 \\
\text{subject to} & 3x_1 + 2x_2 \le 1200 \quad \text{(fescue)} \\
& x_1 + 2x_2 \le 800 \quad \text{(rye)} \\
& x_1 + x_2 \le 450 \quad \text{(bluegrass)}
\end{array}
$$

and $x_1 \ge 0, x_2 \ge 0$.

1. State the dual problem.

2. Find the optimal solution to the dual problem, given that the final tableau in the simplex method for solving the primal problem is

x_1	x_2	x_3	x_4	x_5	M	
0	0	1	1	-4	0	200
0	1	0	1	-1	0	350
1	0	0	-1	1	0	100
0	0	0	1	1	1	1250

3. What are the marginal values of fescue, rye, and bluegrass at the optimal solution?

9.4 Exercises

In Exercises 1–4, state the dual of the given linear programming problem.

1. Exercise 13 in Section 9.3
2. Exercise 14 in Section 9.3
3. Exercise 15 in Section 9.3
4. Exercise 16 in Section 9.3

In Exercises 5–8, use the final tableau in the solution of the given exercise to solve its dual.

5. Exercise 13 in Section 9.3
6. Exercise 14 in Section 9.3
7. Exercise 15 in Section 9.3
8. Exercise 16 in Section 9.3

Exercises 9–16 relate to a primal linear programming problem of finding $\mathbf{x}$ in $\mathbb{R}^n$ so as to maximize $f(\mathbf{x}) = \mathbf{c}^T\mathbf{x}$ subject to $A\mathbf{x} \le \mathbf{b}$ and $\mathbf{x} \ge 0$. Mark each statement True or False (**T/F**). Justify each answer.

9. (**T/F**) The dual problem is to minimize $\mathbf{y}$ in $\mathbb{R}^m$ subject to $A\mathbf{y} \ge \mathbf{c}$ and $\mathbf{y} \ge \mathbf{0}$.

10. (**T/F**) The dual of the dual problem is the original primal problem.

11. (T/F) If both the primal and the dual problems are feasible, then they both have optimal solutions.

12. (T/F) If either the primal or the dual problem has an optimal solution, then they both do.

13. (T/F) If $\bar{x}$ is an optimal solution to the primal problem and $\hat{y}$ is a feasible solution to the dual problem such that $g(\hat{y}) = f(\bar{x})$, then $g(\hat{y})$ is an optimal solution to the dual problem.

14. (T/F) If the primal problem has an optimal solution, then the final tableau in the simplex method also gives the optimal solution to the dual problem.

15. (T/F) If a slack variable is in an optimal solution, then the marginal value of the item corresponding to its equation is positive.

16. (T/F) When a linear programming problem and its dual are used to solve a matrix game, the vectors **u** and **v** are unit.

Sometimes a minimization problem has inequalities only of the "≥" type. In this case, replace the problem by its dual. (Multiplying the original inequalities by -1 to reverse their direction will not work, because the basic solution of the initial simplex tableau in this case will be infeasible.) In Exercises 17–20, use the simplex method to solve the dual, and from this solve the original problem (the dual of the dual).

17. Minimize $\quad 16x_1 + 10x_2 + 20x_3$
subject to $\qquad x_1 + \quad x_2 + 3x_3 \geq 4$
$\qquad\qquad 2x_1 + \quad x_2 + 2x_3 \geq 5$
and $x_1 \geq 0, x_2 \geq 0, x_3 \geq 0$.

18. Minimize $\quad 10x_1 + 14x_2$
subject to $\qquad x_1 + 2x_2 \geq 3$
$\qquad\qquad 2x_1 + \quad x_2 \geq 4$
$\qquad\qquad 3x_1 + \quad x_2 \geq 2$
and $x_1 \geq 0, x_2 \geq 0$.

19. Solve Exercise 2 in Section 9.2.

20. Solve Example 2 in Section 9.2.

Exercises 21 and 22 refer to Exercise 17 in Section 9.2. This exercise was solved using the simplex method in Exercise 21 of Section 9.3. Use the final simplex tableau for that exercise to answer the following questions.

21. What is the marginal value of additional labor in the fabricating department? Give an economic interpretation to your answer.

22. If an extra hour of labor were available, to which department should it be allocated? Why?

Solve the matrix games in Exercises 23 and 24 by using linear programming.

23. $\begin{bmatrix} 2 & 0 \\ -4 & 5 \\ -1 & 3 \end{bmatrix}$ **24.** $\begin{bmatrix} 1 & -2 \\ 0 & 1 \\ -3 & 2 \end{bmatrix}$

25. Solve the matrix game in Exercise 9 in Section 9.1 using linear programming. This game and the one in Exercise 10 cannot be solved by the methods of Section 9.1.

26. Solve the matrix game in Exercise 10 in Section 9.1 using linear programming.

27. Bob wishes to invest \$35,000 in stocks, bonds, and gold coins. He knows that his rate of return will depend on the economic climate of the country, which is, of course, difficult to predict. After careful analysis, he determines the annual profit in dollars he would expect per hundred dollars on each type of investment, depending on whether the economy is strong, stable, or weak:

	Strong	Stable	Weak
Stocks	4	1	-2
Bonds	1	3	0
Gold	-1	0	4

How should Bob invest his money in order to maximize his profit regardless of what the economy does? That is, consider the problem as a matrix game in which Bob, the row player, is playing against the "economy." What is the expected value of his portfolio at the end of the year?

28. Let P be a (primal) linear programming problem with feasible set $\mathcal{F}$, and let P^* be the dual problem with feasible set $\mathcal{F}^*$. Prove the following:

a. If **x** is in $\mathcal{F}$ and **y** is in $\mathcal{F}^*$, then $f(\mathbf{x}) \leq g(\mathbf{y})$. [*Hint:* Write $f(\mathbf{x})$ as $\mathbf{x}^T \mathbf{c}$ and $g(\mathbf{y})$ as $\mathbf{y}^T \mathbf{b}$. Then begin with the inequality $\mathbf{c} \leq A^T \mathbf{y}$.]

b. If $f(\hat{\mathbf{x}}) = g(\hat{\mathbf{y}})$ for some $\hat{\mathbf{x}}$ in $\mathcal{F}$ and $\hat{\mathbf{y}}$ in $\mathcal{F}^*$, then $\hat{\mathbf{x}}$ is an optimal solution to P and $\hat{\mathbf{y}}$ is an optimal solution to P^*.

29. Let A be an $m \times n$ matrix game. Let $\bar{\mathbf{y}}$ and $\bar{\mathbf{x}}$ be the optimal solutions to the related primal and dual linear programming problems, respectively, as in the discussion prior to Example 5. Let $\lambda = \mathbf{u}^T \bar{\mathbf{x}} = \mathbf{v}^T \bar{\mathbf{y}}$, and define $\hat{\mathbf{x}} = \bar{\mathbf{x}}/\lambda$ and $\hat{\mathbf{y}} = \bar{\mathbf{y}}/\lambda$. Let R and C, respectively, denote the row and column players.

a. Show that $\hat{\mathbf{x}}$ and $\hat{\mathbf{y}}$ are mixed strategies for R and C, respectively.

b. If **y** is any mixed strategy for C, show that $E(\hat{\mathbf{x}}, \mathbf{y}) \geq 1/\lambda$.

c. If **x** is any mixed strategy for R, show that $E(\mathbf{x}, \hat{\mathbf{y}}) \leq 1/\lambda$.

d. Conclude that $\hat{\mathbf{x}}$ and $\hat{\mathbf{y}}$ are optimal mixed strategies for R and C, respectively, and that the value of the game is $1/\lambda$.

Solutions to Practice Problems

1. Minimize $\quad 1200y_1 + 800y_2 + 450y_3$
   subject to $\quad 3y_1 + \;\; y_2 + \;\; y_3 \geq 2$
   $\qquad\qquad\quad 2y_1 + \;\; 2y_2 + \;\; y_3 \geq 3$
   and $y_1 \geq 0$, $y_2 \geq 0$, $y_3 \geq 0$.

2. The slack variables are x_3, x_4, and x_5. The bottom row entries in these columns of the final simplex tableau give the optimal solution to the dual problem. Thus

$$\bar{\mathbf{y}} = \begin{bmatrix} 0 \\ 1 \\ 1 \end{bmatrix}.$$

3. Slack variable x_3 comes from the constraint inequality for fescue. This corresponds to variable y_1 in the dual problem, so the marginal value of fescue is 0. Similarly, x_4 and x_5 come from rye and bluegrass, respectively, so their marginal values are both equal to 1.

CHAPTER 9 PROJECT

The Chapter 9 project is available online at `bit.ly/30IM8gT`.

A. *Cycling*: This project investigates cycling in the Simplex Method.

CHAPTER 9 SUPPLEMENTARY EXERCISES

In Exercises 1–24, mark each statement True or False **(T/F)**. Justify each answer.

1. **(T/F)** A negative entry a_{ij} in a payoff matrix indicates the amount player R has to pay player C when R choses action i and C choses action j.

2. **(T/F)** Every payoff matrix has at least one saddle point.

3. **(T/F)** If $\mathbf{x}$ is a vector whose entries sum to 1, then $\mathbf{x}$ is a probability vector.

4. **(T/F)** If $\mathbf{x}$ is a pure strategy in a matrix game, then all the coordinates in $\mathbf{x}$ have the same value.

5. **(T/F)** Each strategy for player R in a matrix game is a convex combination of the set of pure strategies for R.

6. **(T/F)** If A is an $m \times n$ payoff matrix, then the strategy space for R is the set of all probability vectors in $\mathbb{R}^n$.

7. **(T/F)** A strategy $\hat{\mathbf{x}}$ for row player R is optimal if the value of $\hat{\mathbf{x}}$ is equal to the value of the game to R.

8. **(T/F)** If A is the payoff matrix for a matrix game, then the value of strategy $\mathbf{x}$ to player R, denoted $v(\mathbf{x})$, is the minimum of the inner product of $\mathbf{x}$ with each of the columns of A.

9. **(T/F)** If $\hat{\mathbf{x}}$ and $\hat{\mathbf{y}}$ are optimal strategies for an $m \times n$ matrix game whose value is v, then $\hat{\mathbf{y}}$ is a convex combination of the pure strategies $\mathbf{e}_j$ in $\mathbb{R}^n$ for which $E(\hat{\mathbf{x}}, \mathbf{e}_j) = v$.

10. **(T/F)** If A is the payoff matrix for a $2 \times n$ matrix game, then the value of strategy $\mathbf{x}(t)$ to player R, denoted $v(\mathbf{x}(t))$, is the maximum value of n linear functions of t.

11. **(T/F)** If a canonical linear programming problem has a feasible solution but no optimal solution, then the objective function must be unbounded on the feasible set.

12. **(T/F)** If $\mathbf{x}$ is an extreme point of the feasible set of a canonical linear programming problem, then $\mathbf{x}$ is an optimal solution.

13. **(T/F)** If the objective function in a canonical linear programming problem takes on arbitrarily large values in the feasible set, then the problem is infeasible.

14. **(T/F)** If a canonical linear programming problem is unbounded, then it must be feasible.

15. **(T/F)** If the feasible set of a canonical linear programming problem is unbounded, then the program has no optimal solution.

16. **(T/F)** The simplex method of solving a canonical linear programming problem begins by changing each constraint inequality into an equality.

17. **(T/F)** If A is $m \times n$, then it will require n slack variables to change $A\mathbf{x} \leq \mathbf{b}$ into a system of linear equations.

18. **(T/F)** In following the simplex method, when a variable goes "out" of a basic feasible solution, it stays out.

19. (T/F) In order to begin the standard simplex method, each term in the augmented column above the horizontal line must be a nonnegative number.

20. (T/F) When setting up the initial simplex tableau for a canonical linear programming problem, the coefficients of the objective function go in the bottom row.

21. (T/F) In setting up the dual linear programming problem, the matrix A in the primal problem is replaced by A^{-1} in the dual problem.

22. (T/F) If a primal linear programming problem has an optimal solution, then its dual program is bounded.

23. (T/F) Let P be a maximizing linear programming problem and let P^* be its dual. If P has an optimal solution, then the maximum value of the objective function of P in its feasible set is equal to the minimum value of the objective function of P^* in its feasible set.

24. (T/F) Let A be an $m \times n$ matrix game where all the entries in A are positive. Let $\bar{\mathbf{y}}$ and $\bar{\mathbf{x}}$ be the optimal solutions to the related primal and dual linear programming problems, respectively, as defined in Section 9.4. If λ is equal to the sum of the coordinates in $\bar{\mathbf{x}}$, then the value of the matrix game is equal to λ.

25. Consider the following problem:

$$\text{Maximize} \quad -x_1 + 2x_2$$
$$\text{subject to} \quad -x_1 + x_2 \le 1$$
$$x_2 \le 2$$
$$\text{and } x_1 \ge 0, x_2 \ge 0.$$

a. Graph the feasible set $\mathscr{F}$.

b. Find the extreme points of $\mathscr{F}$.

c. Draw some level lines of the objective function to show that the objective function is bounded on $\mathscr{F}$ even though $\mathscr{F}$ is not bounded.

d. Find an optimal solution to the problem.

26. Use the simplex method to find an optimal solution to the problem in Exercise 25.

27. Consider the following problem:

$$\text{Maximize} \quad x_1 + x_2$$
$$\text{subject to} \quad -x_1 + x_2 \le 1$$
$$x_2 \le 2$$

and $x_1 \ge 0, x_2 \ge 0$.

Note that this has the same constraint inequalities as Exercise 25 (and hence the same feasible set), but the objective function is different.

a. Draw the feasible set and some level lines for this new objective function to show that it is not bounded on $\mathscr{F}$.

b. Evaluate the objective function at each of the extreme points.

28. Try to use the simplex method to solve the problem in Exercise 27. Explain why it doesn't work.

29. Consider the following problem:

$$\text{Maximize} \quad 3x_1 + 4x_2$$
$$\text{subject to} \quad x_1 - x_2 \le 4$$
$$-2x_1 + 5x_2 \le -10$$
$$\text{and } x_1 \ge 0, x_2 \ge 0.$$

a. Set up the initial simplex tableau.

b. Try to apply the simplex method and explain why it does not work.

c. Graph the constraint inequalities and explain how they relate to your answer to part **b**.

30. The bottom row of the final tableau for a linear programming problem will have zeros as entries in the columns corresponding to the basic variables that are "in" the solution. There may also be zeros in some of the other columns in this bottom row. When this happens, the optimal solution will not be unique since these other variables could be brought into the solution without changing the value of the objective function.

a. Find *all* of the optimal solutions to the following problem:

$$\text{Maximize} \quad 4x_1 + 5x_2 - x_3$$
$$\text{Subject to} \quad x_1 + 2x_2 - x_3 \le 16$$
$$x_1 + x_2 \le 12$$
$$2x_1 + 2x_2 + x_3 \le 36$$
$$\text{and } x_1 \ge 0, x_2 \ge 0, x_3 \ge 0.$$

b. Describe the set of solutions geometrically.

31. Consider the matrix game having payoff matrix $A = \begin{bmatrix} -1 & 2 & 3 \\ 4 & 3 & -2 \end{bmatrix}$. Find the optimal mixed strategies and the value of the game by using the method of Example 4 in Section 9.1.

32. Find the optimal mixed strategies and the value of the game in Exercise 31 by using linear programming as in Example 5 in Section 9.4.

Appendix A
Uniqueness of the Reduced Echelon Form

THEOREM

Uniqueness of the Reduced Echelon Form

Each $m \times n$ matrix A is row equivalent to a unique reduced echelon matrix U.

PROOF   The proof uses the idea from Section 4.3 that the columns of row-equivalent matrices have exactly the same linear dependence relations.

The row reduction algorithm shows that there exists at least one such matrix U. Suppose that A is row equivalent to matrices U and V in reduced echelon form. The leftmost nonzero entry in a row of U is a "leading 1." Call the location of such a leading 1 a pivot position, and call the column that contains it a pivot column. (This definition uses only the echelon nature of U and V and does not assume the uniqueness of the reduced echelon form.)

The pivot columns of U and V are precisely the nonzero columns that are *not* linearly dependent on the columns to their left. (This condition is satisfied automatically by a *first* column if it is nonzero.) Since U and V are row equivalent (both being row equivalent to A), their columns have the same linear dependence relations. Hence, the pivot columns of U and V appear in the same locations. If there are r such columns, then since U and V are in reduced echelon form, their pivot columns are the first r columns of the $m \times m$ identity matrix. Thus, *corresponding pivot columns of U and V are equal.*

Finally, consider any nonpivot column of U, say column j. This column is either zero or a linear combination of the pivot columns to its left (because those pivot columns are a basis for the space spanned by the columns to the left of column j). Either case can be expressed by writing $U\mathbf{x} = \mathbf{0}$ for some $\mathbf{x}$ whose jth entry is 1. Then $V\mathbf{x} = \mathbf{0}$, too, which says that column j of V is either zero or the *same* linear combination of the pivot columns of V to *its* left. Since corresponding pivot columns of U and V are equal, columns j of U and V are also equal. This holds for all nonpivot columns, so $V = U$, which proves that U is unique.

Appendix B
Complex Numbers

A **complex number** is a number written in the form

$$z = a + bi$$

where a and b are real numbers and i is a formal symbol satisfying the relation $i^2 = -1$. The number a is the **real part** of z, denoted by $\mathrm{Re}\, z$, and b is the **imaginary part** of z, denoted by $\mathrm{Im}\, z$. Two complex numbers are considered equal if and only if their real and imaginary parts are equal. For example, if $z = 5 + (-2)i$, then $\mathrm{Re}\, z = 5$ and $\mathrm{Im}\, z = -2$. For simplicity, we write $z = 5 - 2i$.

A real number a is considered as a special type of complex number, by identifying a with $a + 0i$. Furthermore, arithmetic operations on real numbers can be extended to the set of complex numbers.

The **complex number system**, denoted by $\mathbb{C}$, is the set of all complex numbers, together with the following operations of addition and multiplication:

$$(a + bi) + (c + di) = (a + c) + (b + d)i \tag{1}$$

$$(a + bi)(c + di) = (ac - bd) + (ad + bc)i \tag{2}$$

These rules reduce to ordinary addition and multiplication of real numbers when b and d are zero in (1) and (2). It is readily checked that the usual laws of arithmetic for $\mathbb{R}$ also hold for $\mathbb{C}$. For this reason, multiplication is usually computed by algebraic expansion, as in the following example.

EXAMPLE 1
$$\begin{aligned}
(5 - 2i)(3 + 4i) &= 15 + 20i - 6i - 8i^2 \\
&= 15 + 14i - 8(-1) \\
&= 23 + 14i
\end{aligned}$$

That is, multiply each term of $5 - 2i$ by each term of $3 + 4i$, use $i^2 = -1$, and write the result in the form $a + bi$. ∎

Subtraction of complex numbers z_1 and z_2 is defined by

$$z_1 - z_2 = z_1 + (-1)z_2$$

In particular, we write $-z$ in place of $(-1)z$.

The **conjugate** of $z = a + bi$ is the complex number $\bar{z}$ (read as "z bar"), defined by

$$\bar{z} = a - bi$$

Obtain $\bar{z}$ from z by reversing the sign of the imaginary part.

EXAMPLE 2   The conjugate of $-3 + 4i$ is $-3 - 4i$; write $\overline{-3 + 4i} = -3 - 4i$.   ∎

Observe that if $z = a + bi$, then

$$z\bar{z} = (a + bi)(a - bi) = a^2 - abi + bai - b^2i^2 = a^2 + b^2 \qquad (3)$$

Since $z\bar{z}$ is real and nonnegative, it has a square root. The **absolute value** (or **modulus**) of z is the real number $|z|$ defined by

$$|z| = \sqrt{z\bar{z}} = \sqrt{a^2 + b^2}$$

If z is a real number, then $z = a + 0i$, and $|z| = \sqrt{a^2}$, which equals the ordinary absolute value of a.

Some useful properties of conjugates and absolute value are listed below; w and z denote complex numbers.

1. $\bar{z} = z$ if and only if z is a real number.
2. $\overline{w + z} = \bar{w} + \bar{z}$.
3. $\overline{wz} = \bar{w}\,\bar{z}$; in particular, $\overline{rz} = r\bar{z}$ if r is a real number.
4. $z\bar{z} = |z|^2 \geq 0$.
5. $|wz| = |w||z|$.
6. $|w + z| \leq |w| + |z|$.

If $z \neq 0$, then $|z| > 0$ and z has a multiplicative inverse, denoted by $1/z$ or z^{-1} and given by

$$\frac{1}{z} = z^{-1} = \frac{\bar{z}}{|z|^2}$$

Of course, a quotient w/z simply means $w \cdot (1/z)$.

EXAMPLE 3   Let $w = 3 + 4i$ and $z = 5 - 2i$. Compute $z\bar{z}$, $|z|$, and w/z.

SOLUTION   From equation (3),

$$z\bar{z} = 5^2 + (-2)^2 = 25 + 4 = 29$$

For the absolute value, $|z| = \sqrt{z\bar{z}} = \sqrt{29}$. To compute w/z, first multiply both the numerator and the denominator by $\bar{z}$, the conjugate of the denominator. Because of (3), this eliminates the i in the denominator:

$$\frac{w}{z} = \frac{3 + 4i}{5 - 2i}$$

$$= \frac{3 + 4i}{5 - 2i} \cdot \frac{5 + 2i}{5 + 2i}$$

$$= \frac{15 + 6i + 20i - 8}{5^2 + (-2)^2}$$

$$= \frac{7 + 26i}{29}$$

$$= \frac{7}{29} + \frac{26}{29}i \qquad\qquad ∎$$

Geometric Interpretation

Each complex number $z = a + bi$ corresponds to a point (a, b) in the plane $\mathbb{R}^2$, as in Figure 1. The horizontal axis is called the **real axis** because the points $(a, 0)$ on it correspond to the real numbers. The vertical axis is the **imaginary axis** because the points $(0, b)$ on it correspond to the **pure imaginary numbers** of the form $0 + bi$, or simply bi. The conjugate of z is the mirror image of z in the real axis. The absolute value of z is the distance from (a, b) to the origin.

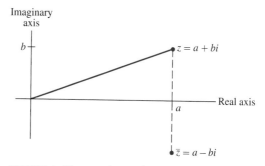

FIGURE 1 The complex conjugate is a mirror image.

Addition of complex numbers $z = a + bi$ and $w = c + di$ corresponds to vector addition of (a, b) and (c, d) in $\mathbb{R}^2$, as in Figure 2.

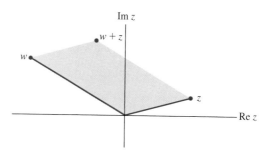

FIGURE 2 Addition of complex numbers.

To give a graphical representation of complex multiplication, we use **polar coordinates** in $\mathbb{R}^2$. Given a nonzero complex number $z = a + bi$, let φ be the angle between the positive real axis and the point (a, b), as in Figure 3 where $-\pi < \varphi \le \pi$. The angle φ is called the **argument** of z; we write $\varphi = \arg z$. From trigonometry,

$$a = |z| \cos \varphi, \qquad b = |z| \sin \varphi$$

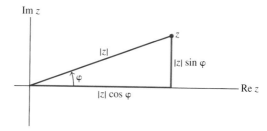

FIGURE 3 Polar coordinates of z.

and so
$$z = a + bi = |z|(\cos \varphi + i \sin \varphi)$$

If w is another nonzero complex number, say,
$$w = |w| (\cos \vartheta + i \sin \vartheta)$$

then, using standard trigonometric identities for the sine and cosine of the sum of two angles, one can verify that

$$wz = |w| |z| [\cos(\vartheta + \varphi) + i \sin(\vartheta + \varphi)] \tag{4}$$

See Figure 4. A similar formula may be written for quotients in polar form. The formulas for products and quotients can be stated in words as follows.

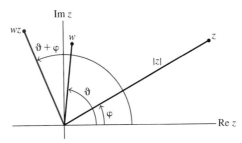

FIGURE 4 Multiplication with polar coordinates.

The product of two nonzero complex numbers is given in polar form by the product of their absolute values and the sum of their arguments. The quotient of two nonzero complex numbers is given by the quotient of their absolute values and the difference of their arguments.

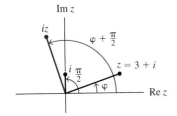

Multiplication by i.

EXAMPLE 4

a. If w has absolute value 1, then $w = \cos \vartheta + i \sin \vartheta$, where ϑ is the argument of w. Multiplication of any nonzero number z by w simply rotates z through the angle ϑ.

b. The argument of i itself is $\pi/2$ radians, so multiplication of z by i rotates z through an angle of $\pi/2$ radians. For example, $3 + i$ is rotated into $(3 + i)i = -1 + 3i$. ■

Powers of a Complex Number

Formula (4) applies when $z = w = r(\cos \varphi + i \sin \varphi)$. In this case
$$z^2 = r^2(\cos 2\varphi + i \sin 2\varphi)$$

and
$$
\begin{aligned}
z^3 &= z \cdot z^2 \\
&= r(\cos \varphi + i \sin \varphi) \cdot r^2(\cos 2\varphi + i \sin 2\varphi) \\
&= r^3(\cos 3\varphi + i \sin 3\varphi)
\end{aligned}
$$

In general, for any positive integer k,
$$z^k = r^k(\cos k\varphi + i \sin k\varphi)$$

This fact is known as *De Moivre's Theorem*.

Complex Numbers and $\mathbb{R}^2$

Although the elements of $\mathbb{R}^2$ and $\mathbb{C}$ are in one-to-one correspondence, and the operations of addition are essentially the same, there is a logical distinction between $\mathbb{R}^2$ and $\mathbb{C}$. In $\mathbb{R}^2$ we can only multiply a vector by a real scalar, whereas in $\mathbb{C}$ we can multiply any two complex numbers to obtain a third complex number. (The dot product in $\mathbb{R}^2$ doesn't count, because it produces a scalar, not an element of $\mathbb{R}^2$.) We use scalar notation for elements in $\mathbb{C}$ to emphasize this distinction.

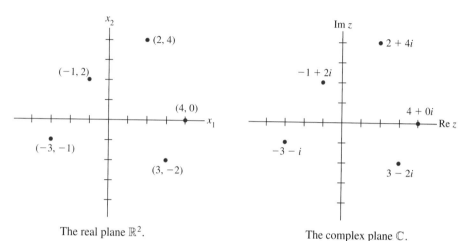

The real plane $\mathbb{R}^2$.   The complex plane $\mathbb{C}$.

Credits

Cover

DigiPub/Moment/Getty Images

Chapter 1

1, Hulton Archive/Keystone/Stringer/Getty Images; **54**, Olivier Le Queinec/Shutterstock; **54**, AButyrin 22/Shutterstock; **54**, SasinTipchai/Shutterstock; **58**, Shutterstock; Kurhan/Shutterstock; **86**, Ingredients in the diet as of 1984; nutrient data for ingredients adapted from USDA Agricultural Handbooks No. 8-1 and 8-6, 1976; **90**, Josef Hanus/Shutterstock; **90**, Gary Blakeley/Shutterstock

Chapter 2

97, LANBO/Shutterstock; **98**, Stocktrek Images, Inc/Alamy Stock Photo; **126**, Radu Bercan/Shutterstock; **131**, NASA; **145**, Fotokostic/Shutterstock; **145**, Rainer Plendl/Shutterstock; **145**, Designpics/123RF; **145**, Dotshock/Shutterstock; **146**, Wassily W. Leontief, "The Structure of the U.S. Economy," *Scientific American*, April 1965, pp. 30–32; **150**, Alexander Raths/Shutterstock

Chapter 3

171, Akkaradet Bangchun/Shutterstock; **171**, Zhengzaishuru/Shutterstock

Chapter 4

201, Interneteable/Shutterstock; **236**, Ifong/Shutterstock

Chapter 5

273, Digitalmedia.fws.gov; **336**, Archana bhartia/Shutterstock; **336**, Noah Strycker/Shutterstock

Chapter 6

349, Courtesy of Lester McDonald; **397**, Valerio Pardi/123RF; **397**, Jess Alford/Photodisc/Getty Images

Chapter 7

417, Landsat Data/U.S. Geological Survey; **418**, (Left): Landsat Data/U.S. Geological Survey; (Middle): Landsat Data/U.S. Geological Survey; (Right): Landsat Data/U.S. Geological Survey; (Left 2): Landsat Data/U.S. Geological Survey; (Middle 2): Landsat Data/U.S. Geological Survey; (Right 2): Landsat Data/U.S. Geological Survey; **436**, Onixxino/Getty Images; **436**, Dmitry Kalinovsky/Shutterstock; **436**, Viki2win/Shutterstock

Chapter 8

459, Canali/Index/Heritage Image Partnership Ltd/Alamy Stock Photo; **459**, City of Athens in 387 B.C., the Greek philosopher Plato

Chapter 9

521, CSU Archives/Everett Collection Inc/Alamy Stock Photo

Chapter 10 (online)

C-1, Dpa picture alliance/Alamy Stock Photo

Glossary

A

adjugate (or **classical adjoint**): The matrix adj A formed from a square matrix A by replacing the (i, j)-entry of A by the (i, j)-cofactor, for all i and j, and then transposing the resulting matrix.

affine combination: A linear combination of vectors (points in $\mathbb{R}^n$) in which the sum of the weights involved is 1.

affine dependence relation: An equation of the form $c_1\mathbf{v}_1 + \cdots + c_p\mathbf{v}_p = \mathbf{0}$, where the weights $c_1, \ldots, c_p$ are not all zero, and $c_1 + \cdots + c_p = 0$.

affine hull (or **affine span**) of a set S: The set of all affine combinations of points in S, denoted by aff S.

affinely dependent set: A set $\{\mathbf{v}_1, \ldots, \mathbf{v}_p\}$ in $\mathbb{R}^n$ such that there are real numbers $c_1, \ldots, c_p$, not all zero, such that $c_1 + \cdots + c_p = 0$ and $c_1\mathbf{v}_1 + \cdots + c_p\mathbf{v}_p = \mathbf{0}$.

affinely independent set: A set $\{\mathbf{v}_1, \ldots, \mathbf{v}_p\}$ in $\mathbb{R}^n$ that is not affinely dependent.

affine set (or **affine subset**): A set S of points such that if $\mathbf{p}$ and $\mathbf{q}$ are in S, then $(1 - t)\mathbf{p} + t\mathbf{q} \in S$ for each real number t.

affine transformation: A mapping $T: \mathbb{R}^n \to \mathbb{R}^m$ of the form $T(\mathbf{x}) = A\mathbf{x} + \mathbf{b}$, with A an $m \times n$ matrix and $\mathbf{b}$ in $\mathbb{R}^m$.

algebraic multiplicity: The multiplicity of an eigenvalue as a root of the characteristic equation.

angle (between nonzero vectors $\mathbf{u}$ and $\mathbf{v}$ in $\mathbb{R}^2$ or $\mathbb{R}^3$): The angle ϑ between the two directed line segments from the origin to the points $\mathbf{u}$ and $\mathbf{v}$. Related to the scalar product by

$$\mathbf{u} \cdot \mathbf{v} = \|\mathbf{u}\| \|\mathbf{v}\| \cos \vartheta$$

associative law of multiplication: $A(BC) = (AB)C$, for all A, B, C.

attractor (of a dynamical system in $\mathbb{R}^2$): The origin when all trajectories tend toward $\mathbf{0}$.

augmented matrix: A matrix made up of a coefficient matrix for a linear system and one or more columns to the right. Each extra column contains the constants from the right side of a system with the given coefficient matrix.

auxiliary equation: A polynomial equation in a variable r, created from the coefficients of a homogeneous difference equation.

B

back-substitution (with matrix notation): The backward phase of row reduction of an augmented matrix that transforms an echelon matrix into a reduced echelon matrix; used to find the solution(s) of a system of linear equations.

backward phase (of row reduction): The last part of the algorithm that reduces a matrix in echelon form to a reduced echelon form.

band matrix: A matrix whose nonzero entries lie within a band along the main diagonal.

barycentric coordinates (of a point $\mathbf{p}$ with respect to an affinely independent set $S = \{\mathbf{v}_1, \ldots, \mathbf{v}_k\}$): The (unique) set of weights $c_1, \ldots, c_k$ such that $\mathbf{p} = c_1\mathbf{v}_1 + \cdots + c_k\mathbf{v}_k$ and $c_1 + \cdots + c_k = 1$. (Sometimes also called the **affine coordinates** of $\mathbf{p}$ with respect to S.)

basic variable: A variable in a linear system that corresponds to a pivot column in the coefficient matrix.

basis (for a nontrivial subspace H of a vector space V): An indexed set $\mathcal{B} = \{\mathbf{v}_1, \ldots, \mathbf{v}_p\}$ in V such that: (i) $\mathcal{B}$ is a linearly independent set and (ii) the subspace spanned by $\mathcal{B}$ coincides with H, that is, $H = \text{Span}\{\mathbf{v}_1, \ldots, \mathbf{v}_p\}$.

$\mathcal{B}$-coordinates of x: *See* coordinates of $\mathbf{x}$ relative to the basis $\mathcal{B}$.

best approximation: The closest point in a given subspace to a given vector.

bidiagonal matrix: A matrix whose nonzero entries lie on the main diagonal and on one diagonal adjacent to the main diagonal.

block diagonal (matrix): A partitioned matrix $A = [A_{ij}]$ such that each block A_{ij} is a zero matrix for $i \neq j$.

block matrix: *See* partitioned matrix.

block matrix multiplication: The row–column multiplication of partitioned matrices as if the block entries were scalars.

block upper triangular (matrix): A partitioned matrix $A = [A_{ij}]$ such that each block A_{ij} is a zero matrix for $i > j$.

boundary point of a set S in $\mathbb{R}^n$: A point $\mathbf{p}$ such that every open ball in $\mathbb{R}^n$ centered at $\mathbf{p}$ intersects both S and the complement of S.

bounded set in $\mathbb{R}^n$: A set that is contained in an open ball $B(\mathbf{0}, \delta)$ for some $\delta > 0$.

$\mathcal{B}$-matrix (for T): A matrix $[T]_{\mathcal{B}}$ for a linear transformation $T : V \to V$ relative to a basis $\mathcal{B}$ for V, with the property that $[T(\mathbf{x})]_{\mathcal{B}} = [T]_{\mathcal{B}}[\mathbf{x}]_{\mathcal{B}}$ for all $\mathbf{x}$ in V.

C

Cauchy–Schwarz inequality: $|\langle \mathbf{u}, \mathbf{v} \rangle| \leq \|u\| \cdot \|v\|$ for all $\mathbf{u}, \mathbf{v}$.

change of basis: *See* change-of-coordinates matrix.

change-of-coordinates matrix (from a basis $\mathcal{B}$ to a basis $\mathcal{C}$): A matrix $\underset{\mathcal{C} \leftarrow \mathcal{B}}{P}$ that transforms $\mathcal{B}$-coordinate vectors into $\mathcal{C}$-coordinate vectors: $[\mathbf{x}]_{\mathcal{C}} = \underset{\mathcal{C} \leftarrow \mathcal{B}}{P}[\mathbf{x}]_{\mathcal{B}}$. If $\mathcal{C}$ is the standard basis for $\mathbb{R}^n$, then $\underset{\mathcal{C} \leftarrow \mathcal{B}}{P}$ is sometimes written as $P_{\mathcal{B}}$.

characteristic equation (of A): $\det(A - \lambda I) = 0$.

characteristic polynomial (of A): $\det(A - \lambda I)$ or, in some texts, $\det(\lambda I - A)$.

Cholesky factorization: A factorization $A = R^T R$, where R is an invertible upper triangular matrix whose diagonal entries are all positive.

closed ball (in $\mathbb{R}^n$): A set $\{\mathbf{x} : \|\mathbf{x} - \mathbf{p}\| < \delta\}$ in $\mathbb{R}^n$, where $\mathbf{p}$ is in $\mathbb{R}^n$ and $\delta > 0$.

closed set (in $\mathbb{R}^n$): A set that contains all of its boundary points.

codomain (of a transformation $T : \mathbb{R}^n \to \mathbb{R}^m$): The set $\mathbb{R}^m$ that contains the range of T. In general, if T maps a vector space V into a vector space W, then W is called the codomain of T.

coefficient matrix: A matrix whose entries are the coefficients of a system of linear equations.

cofactor: A number $C_{ij} = (-1)^{i+j} \det A_{ij}$, called the (i, j)-cofactor of A, where A_{ij} is the submatrix formed by deleting the ith row and the jth column of A.

cofactor expansion: A formula for $\det A$ using cofactors associated with one row or one column, such as for row 1:

$$\det A = a_{11}C_{11} + \cdots + a_{1n}C_{1n}$$

column–row expansion: The expression of a product AB as a sum of outer products: $\operatorname{col}_1(A)\operatorname{row}_1(B) + \cdots + \operatorname{col}_n(A)\operatorname{row}_n(B)$, where n is the number of columns of A.

column space (of an $m \times n$ matrix A): The set $\operatorname{Col} A$ of all linear combinations of the columns of A. If $A = [\mathbf{a}_1 \cdots \mathbf{a}_n]$, then $\operatorname{Col} A = \operatorname{Span}\{\mathbf{a}_1, \ldots, \mathbf{a}_n\}$. Equivalently,

$$\operatorname{Col} A = \{\mathbf{y} : \mathbf{y} = A\mathbf{x} \text{ for some } \mathbf{x} \text{ in } \mathbb{R}^n\}$$

column sum: The sum of the entries in a column of a matrix.

column vector: A matrix with only one column, or a single column of a matrix that has several columns.

commuting matrices: Two matrices A and B such that $AB = BA$.

compact set (in $\mathbb{R}^n$): A set in $\mathbb{R}^n$ that is both closed and bounded.

companion matrix: A special form of matrix whose characteristic polynomial is $(-1)^n p(\lambda)$ when $p(\lambda)$ is a specified polynomial whose leading term is λ^n.

complex eigenvalue: A nonreal root of the characteristic equation of an $n \times n$ matrix.

complex eigenvector: A nonzero vector $\mathbf{x}$ in $\mathbb{C}^n$ such that $A\mathbf{x} = \lambda\mathbf{x}$, where A is an $n \times n$ matrix and λ is a complex eigenvalue.

component of y orthogonal to u (for $\mathbf{u} \neq \mathbf{0}$): The vector $\mathbf{y} - \dfrac{\mathbf{y} \cdot \mathbf{u}}{\mathbf{u} \cdot \mathbf{u}}\mathbf{u}$.

composition of linear transformations: A mapping produced by applying two or more linear transformations in succession. If the transformations are matrix transformations, say left-multiplication by B followed by left-multiplication by A, then the composition is the mapping $\mathbf{x} \mapsto A(B\mathbf{x})$.

condition number (of A): The quotient σ_1/σ_n, where σ_1 is the largest singular value of A and σ_n is the smallest singular value. The condition number is $+\infty$ when σ_n is zero.

conformable for block multiplication: Two partitioned matrices A and B such that the block product AB is defined: The column partition of A must match the row partition of B.

consistent linear system: A linear system with at least one solution.

constrained optimization: The problem of maximizing a quantity such as $\mathbf{x}^T A\mathbf{x}$ or $\|A\mathbf{x}\|$ when $\mathbf{x}$ is subject to one or more constraints, such as $\mathbf{x}^T \mathbf{x} = 1$ or $\mathbf{x}^T \mathbf{v} = 0$.

consumption matrix: A matrix in the Leontief input–output model whose columns are the unit consumption vectors for the various sectors of an economy.

contraction: A mapping $\mathbf{x} \mapsto r\mathbf{x}$ for some scalar r, with $0 \leq r \leq 1$.

controllable (pair of matrices): A matrix pair (A, B) where A is $n \times n$, B has n rows, and

$$\operatorname{rank}\begin{bmatrix} B & AB & A^2B & \cdots & A^{n-1}B \end{bmatrix} = n$$

Related to a state-space model of a control system and the difference equation $\mathbf{x}_{k+1} = A\mathbf{x}_k + B\mathbf{u}_k$ ($k = 0, 1, \ldots$).

convergent (sequence of vectors): A sequence $\{\mathbf{x}_k\}$ such that the entries in $\mathbf{x}_k$ can be made as close as desired to the entries in some fixed vector for all k sufficiently large.

convex combination (of points $\mathbf{v}_1, \ldots, \mathbf{v}_k$ in $\mathbb{R}^n$): A linear combination of vectors (points) in which the weights in the combination are nonnegative and the sum of the weights is 1.

convex hull (of a set S): The set of all convex combinations of points in S, denoted by: conv S.

convex set: A set S with the property that for each $\mathbf{p}$ and $\mathbf{q}$ in S, the line segment $\overline{\mathbf{pq}}$ is contained in S.

coordinate mapping (determined by an ordered basis $\mathcal{B}$ in a vector space V): A mapping that associates to each $\mathbf{x}$ in V its coordinate vector $[\mathbf{x}]_\mathcal{B}$.

coordinates of x relative to the basis $\mathcal{B} = \{\mathbf{b}_1, \ldots, \mathbf{b}_n\}$: The weights $c_1, \ldots, c_n$ in the equation $\mathbf{x} = c_1\mathbf{b}_1 + \cdots + c_n\mathbf{b}_n$.

coordinate vector of x relative to $\mathcal{B}$: The vector $[\mathbf{x}]_\mathcal{B}$ whose entries are the coordinates of $\mathbf{x}$ relative to the basis $\mathcal{B}$.

covariance (of variables x_i and x_j, for $i \neq j$): The entry s_{ij} in the covariance matrix S for a matrix of observations, where x_i and x_j vary over the ith and jth coordinates, respectively, of the observation vectors.

covariance matrix (or **sample covariance matrix**): The $p \times p$ matrix S defined by $S = (N-1)^{-1}BB^T$, where B is a $p \times N$ matrix of observations in mean-deviation form.

Cramer's rule: A formula for each entry in the solution $\mathbf{x}$ of the equation $A\mathbf{x} = \mathbf{b}$ when A is an invertible matrix.

cross-product term: A term cx_ix_j in a quadratic form, with $i \neq j$.

cube: A three-dimensional solid object bounded by six square faces, with three faces meeting at each vertex.

D

decoupled system: A difference equation $\mathbf{y}_{k+1} = A\mathbf{y}_k$, or a differential equation $\mathbf{y}'(t) = A\mathbf{y}(t)$, in which A is a diagonal matrix. The discrete evolution of each entry in $\mathbf{y}_k$ (as a function of k), or the continuous evolution of each entry in the vector-valued function $\mathbf{y}(t)$, is unaffected by what happens to the other entries as $k \to \infty$ or $t \to \infty$.

design matrix: The matrix X in the linear model $\mathbf{y} = X\boldsymbol{\beta} + \boldsymbol{\epsilon}$, where the columns of X are determined in some way by the observed values of some independent variables.

determinant (of a square matrix A): The number $\det A$ defined inductively by a cofactor expansion along the first row of A. Also, $(-1)^r$ times the product of the diagonal entries in any echelon form U obtained from A by row replacements and r row interchanges (but no scaling operations).

diagonal entries (in a matrix): Entries having equal row and column indices.

diagonalizable (matrix): A matrix that can be written in factored form as PDP^{-1}, where D is a diagonal matrix and P is an invertible matrix.

diagonal matrix: A square matrix whose entries *not* on the main diagonal are all zero.

difference equation (or **linear recurrence relation**): An equation of the form $\mathbf{x}_{k+1} = A\mathbf{x}_k$ ($k = 0, 1, 2, \ldots$) whose solution is a sequence of vectors, $\mathbf{x}_0, \mathbf{x}_1, \ldots$.

dilation: A mapping $\mathbf{x} \mapsto r\mathbf{x}$ for some scalar r, with $1 < r$.

dimension:

of a flat S: The dimension of the corresponding parallel subspace.

of a set S: The dimension of the smallest flat containing S.

of a subspace S: The number of vectors in a basis for S, written as dim S.

of a vector space V: The number of vectors in a basis for V, written as dim V. The dimension of the zero space is 0.

discrete linear dynamical system: A difference equation of the form $\mathbf{x}_{k+1} = A\mathbf{x}_k$ that describes the changes in a system (usually a physical system) as time passes. The physical system is measured at discrete times, when $k = 0, 1, 2, \ldots$, and the **state** of the system at time k is a vector $\mathbf{x}_k$ whose entries provide certain facts of interest about the system.

distance between u and v: The length of the vector $\mathbf{u} - \mathbf{v}$, denoted by dist $(\mathbf{u}, \mathbf{v})$.

distance to a subspace: The distance from a given point (vector) $\mathbf{v}$ to the nearest point in the subspace.

distributive laws: (left) $A(B + C) = AB + AC$, and (right) $(B + C)A = BA + CA$, for all A, B, C.

domain (of a transformation T): The set of all vectors $\mathbf{x}$ for which $T(\mathbf{x})$ is defined.

dot product: *See* inner product.

dynamical system: *See* discrete linear dynamical system.

E

echelon form (or **row echelon form**, of a matrix): An echelon matrix that is row equivalent to the given matrix.

echelon matrix (or **row echelon matrix**): A rectangular matrix that has three properties: (1) All nonzero rows are above any row of all zeros. (2) Each leading entry of a row is in a column to the right of the leading entry of the row above it. (3) All entries in a column below a leading entry are zero.

eigenfunctions (of a differential equation $\mathbf{x}'(t) = A\mathbf{x}(t)$): A function $\mathbf{x}(t) = \mathbf{v}e^{\lambda t}$, where $\mathbf{v}$ is an eigenvector of A and λ is the corresponding eigenvalue.

eigenspace (of A corresponding to λ): The set of *all* solutions of $A\mathbf{x} = \lambda\mathbf{x}$, where λ is an eigenvalue of A. Consists of the zero vector and all eigenvectors corresponding to λ.

eigenvalue (of A): A scalar λ such that the equation $A\mathbf{x} = \lambda\mathbf{x}$ has a solution for some nonzero vector $\mathbf{x}$.

eigenvector (of A): A *nonzero* vector $\mathbf{x}$ such that $A\mathbf{x} = \lambda\mathbf{x}$ for some scalar λ.

eigenvector basis: A basis consisting entirely of eigenvectors of a given matrix.

eigenvector decomposition (of $\mathbf{x}$): An equation, $\mathbf{x} = c_1\mathbf{v}_1 + \cdots + c_n\mathbf{v}_n$, expressing $\mathbf{x}$ as a linear combination of eigenvectors of a matrix.

elementary matrix: An invertible matrix that results by performing one elementary row operation on an identity matrix.

elementary row operations: (1) (Replacement) Replace one row by the sum of itself and a multiple of another row. (2) Interchange two rows. (3) (Scaling) Multiply all entries in a row by a nonzero constant.

equal vectors: Vectors in $\mathbb{R}^n$ whose corresponding entries are the same.

equilibrium prices: A set of prices for the total output of the various sectors in an economy, such that the income of each sector exactly balances its expenses.

equilibrium vector: *See* steady-state vector.

equivalent (linear) systems: Linear systems with the same solution set.

exchange model: *See* Leontief exchange model.

existence question: Asks, "Does a solution to the system exist?" That is, "Is the system consistent?" Also, "Does a solution of $A\mathbf{x} = \mathbf{b}$ exist for *all* possible $\mathbf{b}$?"

expansion by cofactors: *See* cofactor expansion.

explicit description (of a subspace W of $\mathbb{R}^n$): A parametric representation of W as the set of all linear combinations of a set of specified vectors.

extreme point (of a convex set S): A point $\mathbf{p}$ in S such that $\mathbf{p}$ is not in the interior of any line segment that lies in S. (That is, if $\mathbf{x}, \mathbf{y}$ are in S and $\mathbf{p}$ is on the line segment $\overline{\mathbf{xy}}$, then $\mathbf{p} = \mathbf{x}$ or $\mathbf{p} = \mathbf{y}$.)

F

factorization (of A): An equation that expresses A as a product of two or more matrices.

final demand vector (or **bill of final demands**): The vector $\mathbf{d}$ in the Leontief input–output model that lists the dollar values of the goods and services demanded from the various sectors by the nonproductive part of the economy. The vector $\mathbf{d}$ can represent consumer demand, government consumption, surplus production, exports, or other external demand.

finite-dimensional (vector space): A vector space that is spanned by a finite set of vectors.

flat (in $\mathbb{R}^n$): A translate of a subspace of $\mathbb{R}^n$.

flexibility matrix: A matrix whose jth column gives the deflections of an elastic beam at specified points when a unit force is applied at the jth point on the beam.

floating point arithmetic: Arithmetic with numbers represented as decimals $\pm .d_1 \cdots d_p \times 10^r$, where r is an integer and the number p of digits to the right of the decimal point is usually between 8 and 16.

flop: One arithmetic operation $(+, -, *, /)$ on two real floating point numbers.

forward phase (of row reduction): The first part of the algorithm that reduces a matrix to echelon form.

Fourier approximation (of order n): The closest point in the subspace of nth-order trigonometric polynomials to a given function in $C[0, 2\pi]$.

Fourier coefficients: The weights used to make a trigonometric polynomial as a Fourier approximation to a function.

Fourier series: An infinite series that converges to a function in the inner product space $C[0, 2\pi]$, with the inner product given by a definite integral.

free variable: Any variable in a linear system that is not a basic variable.

full rank (matrix): An $m \times n$ matrix whose rank is the smaller of m and n.

fundamental set of solutions: A basis for the set of all solutions of a homogeneous linear difference or differential equation.

fundamental subspaces (determined by A): The null space and column space of A, and the null space and column space of A^T, with Col A^T commonly called the row space of A.

G

Gaussian elimination: *See* row reduction algorithm.

general least-squares problem: Given an $m \times n$ matrix A and a vector $\mathbf{b}$ in $\mathbb{R}^m$, find $\hat{\mathbf{x}}$ in $\mathbb{R}^n$ such that $\|\mathbf{b} - A\hat{\mathbf{x}}\| \leq \|\mathbf{b} - A\mathbf{x}\|$ for all $\mathbf{x}$ in $\mathbb{R}^n$.

general solution (of a linear system): A parametric description of a solution set that expresses the basic variables in terms of the free variables (the parameters), if any. After Section 1.5, the parametric description is written in vector form.

Givens rotation: A linear transformation from $\mathbb{R}^n$ to $\mathbb{R}^n$ used in computer programs to create zero entries in a vector (usually a column of a matrix).

Gram matrix (of A): The matrix $A^T A$.

Gram–Schmidt process: An algorithm for producing an orthogonal or orthonormal basis for a subspace that is spanned by a given set of vectors.

H

homogeneous coordinates: In $\mathbb{R}^3$, the representation of (x, y, z) as (X, Y, Z, H) for any $H \neq 0$, where $x = X/H$, $y = Y/H$, and $z = Z/H$. In $\mathbb{R}^2$, H is usually taken as 1, and the homogeneous coordinates of (x, y) are written as $(x, y, 1)$.

homogeneous equation: An equation of the form $A\mathbf{x} = \mathbf{0}$, possibly written as a vector equation or as a system of linear equations.

homogeneous form of (a vector) $\mathbf{v}$ in $\mathbb{R}^n$: The point $\tilde{\mathbf{v}} = \begin{bmatrix} \mathbf{v} \\ 1 \end{bmatrix}$ in $\mathbb{R}^{n+1}$.

Householder reflection: A transformation $\mathbf{x} \mapsto Q\mathbf{x}$, where $Q = I - 2\mathbf{u}\mathbf{u}^T$ and $\mathbf{u}$ is a unit vector ($\mathbf{u}^T\mathbf{u} = 1$).

hyperplane (in $\mathbb{R}^n$): A flat in $\mathbb{R}^n$ of dimension $n - 1$. Also: a translate of a subspace of dimension $n - 1$.

I

identity matrix (denoted by I or I_n): A square matrix with ones on the diagonal and zeros elsewhere.

ill-conditioned matrix: A square matrix with a large (or possibly infinite) condition number; a matrix that is singular or can become singular if some of its entries are changed ever so slightly.

image (of a vector $\mathbf{x}$ under a transformation T): The vector $T(\mathbf{x})$ assigned to $\mathbf{x}$ by T.

implicit description (of a subspace W of $\mathbb{R}^n$): A set of one or more homogeneous equations that characterize the points of W.

Im x: The vector in $\mathbb{R}^n$ formed from the imaginary parts of the entries of a vector $\mathbf{x}$ in $\mathbb{C}^n$.

inconsistent linear system: A linear system with no solution.

indefinite matrix: A symmetric matrix A such that $\mathbf{x}^T A \mathbf{x}$ assumes both positive and negative values.

indefinite quadratic form: A quadratic form Q such that $Q(\mathbf{x})$ assumes both positive and negative values.

infinite-dimensional (vector space): A nonzero vector space V that has no finite basis.

inner product: The scalar $\mathbf{u}^T \mathbf{v}$, usually written as $\mathbf{u} \cdot \mathbf{v}$, where $\mathbf{u}$ and $\mathbf{v}$ are vectors in $\mathbb{R}^n$ viewed as $n \times 1$ matrices. Also called the **dot product** of $\mathbf{u}$ and $\mathbf{v}$. In general, a function on a vector space that assigns to each pair of vectors $\mathbf{u}$ and $\mathbf{v}$ a number $\langle \mathbf{u}, \mathbf{v} \rangle$, subject to certain axioms. See Section 6.7.

inner product space: A vector space on which is defined an inner product.

input–output matrix: *See* consumption matrix.

input–output model: *See* Leontief input–output model.

interior point (of a set S in $\mathbb{R}^n$): A point $\mathbf{p}$ in S such that for some $\delta > 0$, the open ball $\mathbf{B}(\mathbf{p}, \delta)$ centered at $\mathbf{p}$ is contained in S.

intermediate demands: Demands for goods or services that will be consumed in the process of producing other goods and services for consumers. If $\mathbf{x}$ is the production level and C is the consumption matrix, then $C\mathbf{x}$ lists the intermediate demands.

interpolating polynomial: A polynomial whose graph passes through every point in a set of data points in $\mathbb{R}^2$.

invariant subspace (for A): A subspace H such that $A\mathbf{x}$ is in H whenever $\mathbf{x}$ is in H.

inverse (of an $n \times n$ matrix A): An $n \times n$ matrix A^{-1} such that $AA^{-1} = A^{-1}A = I_n$.

inverse power method: An algorithm for estimating an eigenvalue λ of a square matrix, when a good initial estimate of λ is available.

invertible linear transformation: A linear transformation $T : \mathbb{R}^n \to \mathbb{R}^n$ such that there exists a function $S : \mathbb{R}^n \to \mathbb{R}^n$ satisfying both $T(S(\mathbf{x})) = \mathbf{x}$ and $S(T(\mathbf{x})) = \mathbf{x}$ for all $\mathbf{x}$ in $\mathbb{R}^n$.

invertible matrix: A square matrix that possesses an inverse.

isomorphic vector spaces: Two vector spaces V and W for which there is a one-to-one linear transformation T that maps V onto W.

isomorphism: A one-to-one linear mapping from one vector space onto another.

K

kernel (of a linear transformation $T : V \to W$): The set of $\mathbf{x}$ in V such that $T(\mathbf{x}) = \mathbf{0}$.

Kirchhoff's laws: (1) (**voltage law**) The algebraic sum of the RI voltage drops in one direction around a loop equals the algebraic sum of the voltage sources in the same direction around the loop. (2) (**current law**) The current in a branch is the algebraic sum of the loop currents flowing through that branch.

L

ladder network: An electrical network assembled by connecting in series two or more electrical circuits.

leading entry: The leftmost nonzero entry in a row of a matrix.

least-squares error: The distance $\|\mathbf{b} - A\hat{\mathbf{x}}\|$ from $\mathbf{b}$ to $A\hat{\mathbf{x}}$, when $\hat{\mathbf{x}}$ is a least-squares solution of $A\mathbf{x} = \mathbf{b}$.

least-squares line: The line $y = \hat{\beta}_0 + \hat{\beta}_1 x$ that minimizes the least-squares error in the equation $\mathbf{y} = X\boldsymbol{\beta} + \boldsymbol{\epsilon}$.

least-squares solution (of $A\mathbf{x} = \mathbf{b}$): A vector $\hat{\mathbf{x}}$ such that $\|\mathbf{b} - A\hat{\mathbf{x}}\| \leq \|\mathbf{b} - A\mathbf{x}\|$ for all $\mathbf{x}$ in $\mathbb{R}^n$.

left inverse (of A): Any rectangular matrix C such that $CA = I$.

left-multiplication (by A): Multiplication of a vector or matrix on the left by A.

left singular vectors (of A): The columns of U in the singular value decomposition $A = U\Sigma V^T$.

length (or **norm**, of $\mathbf{v}$): The scalar $\|\mathbf{v}\| = \sqrt{\mathbf{v} \cdot \mathbf{v}} = \sqrt{\langle \mathbf{v}, \mathbf{v} \rangle}$.

Leontief exchange (or **closed**) **model**: A model of an economy where inputs and outputs are fixed, and where a set of prices for the outputs of the sectors is sought such that the income of each sector equals its expenditures. This "equilibrium" condition is expressed as a system of linear equations, with the prices as the unknowns.

Leontief input–output model (or **Leontief production equation**): The equation $\mathbf{x} = C\mathbf{x} + \mathbf{d}$, where $\mathbf{x}$ is production, $\mathbf{d}$ is final demand, and C is the consumption (or input–output) matrix. The jth column of C lists the inputs that sector j consumes per unit of output.

level set (or **gradient**) of a linear functional f on $\mathbb{R}^n$: A set $[f : d] = \{\mathbf{x} \in \mathbb{R}^n : f(\mathbf{x}) = d\}$

linear combination: A sum of scalar multiples of vectors. The scalars are called the *weights*.

linear dependence relation: A homogeneous vector equation where the weights are all specified and at least one weight is nonzero.

linear equation (in the variables $x_1, \ldots, x_n$): An equation that can be written in the form $a_1 x_1 + a_2 x_2 + \cdots + a_n x_n = b$, where b and the coefficients $a_1, \ldots, a_n$ are real or complex numbers.

linear filter: A linear difference equation used to transform discrete-time signals.

linear functional (on $\mathbb{R}^n$): A linear transformation f from $\mathbb{R}^n$ into $\mathbb{R}$.

linearly dependent (vectors): An indexed set $\{\mathbf{v}_1, \ldots, \mathbf{v}_p\}$ with the property that there exist weights $c_1, \ldots, c_p$, not all zero, such that $c_1\mathbf{v}_1 + \cdots + c_p\mathbf{v}_p = \mathbf{0}$. That is, the vector equation $c_1\mathbf{v}_1 + c_2\mathbf{v}_2 + \cdots + c_p\mathbf{v}_p = \mathbf{0}$ has a *nontrivial* solution.

linearly independent (vectors): An indexed set $\{\mathbf{v}_1, \ldots, \mathbf{v}_p\}$ with the property that the vector equation $c_1\mathbf{v}_1 + c_2\mathbf{v}_2 + \cdots + c_p\mathbf{v}_p = \mathbf{0}$ has *only* the trivial solution, $c_1 = \cdots = c_p = 0$.

linear model (in statistics): Any equation of the form $\mathbf{y} = X\boldsymbol{\beta} + \boldsymbol{\epsilon}$, where X and $\mathbf{y}$ are known and $\boldsymbol{\beta}$ is to be chosen to minimize the length of the **residual vector**, $\boldsymbol{\epsilon}$.

linear system: A collection of one or more linear equations involving the same variables, say, $x_1, \ldots, x_n$.

linear transformation T (from a vector space V into a vector space W): A rule T that assigns to each vector $\mathbf{x}$ in V a unique vector $T(\mathbf{x})$ in W, such that (i) $T(\mathbf{u} + \mathbf{v}) = T(\mathbf{u}) + T(\mathbf{v})$ for all $\mathbf{u}, \mathbf{v}$ in V, and (ii) $T(c\mathbf{u}) = cT(\mathbf{u})$ for all $\mathbf{u}$ in V and all scalars c. Notation: $T: V \to W$; also, $\mathbf{x} \mapsto A\mathbf{x}$ when $T: \mathbb{R}^n \to \mathbb{R}^m$ and A is the standard matrix for T.

line through p parallel to v: The set $\{\mathbf{p} + t\mathbf{v} : t \text{ in } \mathbb{R}\}$.

loop current: The amount of electric current flowing through a loop that makes the algebraic sum of the RI voltage drops around the loop equal to the algebraic sum of the voltage sources in the loop.

lower triangular matrix: A matrix with zeros above the main diagonal.

lower triangular part (of A): A lower triangular matrix whose entries on the main diagonal and below agree with those in A.

LU factorization: The representation of a matrix A in the form $A = LU$ where L is a square lower triangular matrix with ones on the diagonal (a unit lower triangular matrix) and U is an echelon form of A.

M

magnitude (of a vector): *See* norm.

main diagonal (of a matrix): The entries with equal row and column indices.

mapping: *See* transformation.

Markov chain: A sequence of probability vectors $\mathbf{x}_0$, $\mathbf{x}_1$, $\mathbf{x}_2, \ldots$, together with a stochastic matrix P such that $\mathbf{x}_{k+1} = P\mathbf{x}_k$ for $k = 0, 1, 2, \ldots$.

matrix: A rectangular array of numbers.

matrix equation: An equation that involves at least one matrix; for instance, $A\mathbf{x} = \mathbf{b}$.

matrix for T relative to bases $\mathcal{B}$ and $\mathcal{C}$: A matrix M for a linear transformation $T: V \to W$ with the property that $[T(\mathbf{x})]_\mathcal{C} = M[\mathbf{x}]_\mathcal{B}$ for all $\mathbf{x}$ in V, where $\mathcal{B}$ is a basis for V and $\mathcal{C}$ is a basis for W. When $W = V$ and $\mathcal{C} = \mathcal{B}$, the matrix M is called the $\mathcal{B}$-matrix for T and is denoted by $[T]_\mathcal{B}$.

matrix of observations: A $p \times N$ matrix whose columns are observation vectors, each column listing p measurements made on an individual or object in a specified population or set.

matrix transformation: A mapping $\mathbf{x} \mapsto A\mathbf{x}$, where A is an $m \times n$ matrix and $\mathbf{x}$ represents any vector in $\mathbb{R}^n$.

maximal linearly independent set (in V): A linearly independent set $\mathcal{B}$ in V such that if a vector $\mathbf{v}$ in V but not in $\mathcal{B}$ is added to $\mathcal{B}$, then the new set is linearly dependent.

mean-deviation form (of a matrix of observations): A matrix whose row vectors are in mean-deviation form. For each row, the entries sum to zero.

mean-deviation form (of a vector): A vector whose entries sum to zero.

mean square error: The error of an approximation in an inner product space, where the inner product is defined by a definite integral.

migration matrix: A matrix that gives the percentage movement between different locations, from one period to the next.

minimal spanning set (for a subspace H): A set $\mathcal{B}$ that spans H and has the property that if one of the elements of $\mathcal{B}$ is removed from $\mathcal{B}$, then the new set does not span H.

$m \times n$ matrix: A matrix with m rows and n columns.

Moore–Penrose inverse: *See* pseudoinverse.

multiple regression: A linear model involving several independent variables and one dependent variable.

N

nearly singular matrix: An ill-conditioned matrix.

negative definite matrix: A symmetric matrix A such that $\mathbf{x}^T A\mathbf{x} < 0$ for all $\mathbf{x} \neq \mathbf{0}$.

negative definite quadratic form: A quadratic form Q such that $Q(\mathbf{x}) < 0$ for all $\mathbf{x} \neq \mathbf{0}$.

negative semidefinite matrix: A symmetric matrix A such that $\mathbf{x}^T A\mathbf{x} \leq 0$ for all $\mathbf{x}$.

negative semidefinite quadratic form: A quadratic form Q such that $Q(\mathbf{x}) \leq 0$ for all $\mathbf{x}$.

nonhomogeneous equation: An equation of the form $A\mathbf{x} = \mathbf{b}$ with $\mathbf{b} \neq \mathbf{0}$, possibly written as a vector equation or as a system of linear equations.

nonsingular (matrix): An invertible matrix.

nontrivial solution: A nonzero solution of a homogeneous equation or system of homogeneous equations.

nonzero (matrix or vector): A matrix (with possibly only one row or column) that contains at least one nonzero entry.

norm (or **length**, of **v**): The scalar $\|\mathbf{v}\| = \sqrt{\mathbf{v} \cdot \mathbf{v}} = \sqrt{\langle \mathbf{v}, \mathbf{v} \rangle}$.

normal equations: The system of equations represented by $A^T A\mathbf{x} = A^T\mathbf{b}$, whose solution yields all least-squares solutions of $A\mathbf{x} = \mathbf{b}$. In statistics, a common notation is $X^T X\boldsymbol{\beta} = X^T\mathbf{y}$.

normalizing (a nonzero vector **v**): The process of creating a unit vector $\mathbf{u}$ that is a positive multiple of $\mathbf{v}$.

normal vector (to a subspace V of $\mathbb{R}^n$): A vector $\mathbf{n}$ in $\mathbb{R}^n$ such that $\mathbf{n} \cdot \mathbf{x} = 0$ for all $\mathbf{x}$ in V.

null space (of an $m \times n$ matrix A):   The set Nul A of all solutions to the homogeneous equation $A\mathbf{x} = \mathbf{0}$. Nul $A = \{\mathbf{x} : \mathbf{x}$ is in $\mathbb{R}^n$ and $A\mathbf{x} = \mathbf{0}\}$.

O

observation vector: The vector $\mathbf{y}$ in the linear model $\mathbf{y} = X\boldsymbol{\beta} + \boldsymbol{\epsilon}$, where the entries in $\mathbf{y}$ are the observed values of a dependent variable.

one-to-one (mapping):   A mapping $T : \mathbb{R}^n \to \mathbb{R}^m$ such that each $\mathbf{b}$ in R^m is the image of *at most* one $\mathbf{x}$ in $\mathbb{R}^n$.

onto (mapping):   A mapping $T : \mathbb{R}^n \to \mathbb{R}^m$ such that each $\mathbf{b}$ in R^m is the image of *at least* one $\mathbf{x}$ in $\mathbb{R}^n$.

open ball B(p, δ) in $\mathbb{R}^n$:   The set $\{\mathbf{x} : \|\mathbf{x} - \mathbf{p}\| < \delta\}$ in $\mathbb{R}^n$, where $\delta > 0$.

open set S in $\mathbb{R}^n$:   A set that contains none of its boundary points. (Equivalently, S is open if every point of S is an interior point.)

origin: The zero vector.

orthogonal basis: A basis that is also an orthogonal set.

orthogonal complement (of W):   The set $W^\perp$ of all vectors orthogonal to W.

orthogonal decomposition: The representation of a vector $\mathbf{y}$ as the sum of two vectors, one in a specified subspace W and the other in $W^\perp$. In general, a decomposition $\mathbf{y} = c_1\mathbf{u}_1 + \cdots + c_p\mathbf{u}_p$, where $\{\mathbf{u}_1, \ldots, \mathbf{u}_p\}$ is an orthogonal basis for a subspace that contains $\mathbf{y}$.

orthogonally diagonalizable (matrix):   A matrix A that admits a factorization, $A = PDP^{-1}$, with P an orthogonal matrix $(P^{-1} = P^T)$ and D diagonal.

orthogonal matrix: A square invertible matrix U such that $U^{-1} = U^T$.

orthogonal projection of y onto u (or onto the line through $\mathbf{u}$ and the origin, for $\mathbf{u} \neq \mathbf{0}$):   The vector $\hat{\mathbf{y}}$ defined by $\hat{\mathbf{y}} = \dfrac{\mathbf{y} \cdot \mathbf{u}}{\mathbf{u} \cdot \mathbf{u}} \mathbf{u}$.

orthogonal projection of y onto W: The unique vector $\hat{\mathbf{y}}$ in W such that $\mathbf{y} - \hat{\mathbf{y}}$ is orthogonal to W. Notation: $\hat{\mathbf{y}} = \text{proj}_W \mathbf{y}$.

orthogonal set: A set S of vectors such that $\mathbf{u} \cdot \mathbf{v} = 0$ for each distinct pair $\mathbf{u}, \mathbf{v}$ in S.

orthogonal to W: Orthogonal to every vector in W.

orthonormal basis: A basis that is an orthogonal set of unit vectors.

orthonormal set: An orthogonal set of unit vectors.

outer product: A matrix product $\mathbf{u}\mathbf{v}^T$ where $\mathbf{u}$ and $\mathbf{v}$ are vectors in $\mathbb{R}^n$ viewed as $n \times 1$ matrices. (The transpose symbol is on the "outside" of the symbols $\mathbf{u}$ and $\mathbf{v}$.)

overdetermined system: A system of equations with more equations than unknowns.

P

parallel flats: Two or more flats such that each flat is a translate of the other flats.

parallelogram rule for addition: A geometric interpretation of the sum of two vectors $\mathbf{u}, \mathbf{v}$ as the diagonal of the parallelogram determined by $\mathbf{u}, \mathbf{v}$, and $\mathbf{0}$.

parameter vector: The unknown vector $\boldsymbol{\beta}$ in the linear model $\mathbf{y} = X\boldsymbol{\beta} + \boldsymbol{\epsilon}$.

parametric equation of a line: An equation of the form $\mathbf{x} = \mathbf{p} + t\mathbf{v} \, (t \text{ in } \mathbb{R})$.

parametric equation of a plane: An equation of the form $\mathbf{x} = \mathbf{p} + s\mathbf{u} + t\mathbf{v} \, (s, t \text{ in } \mathbb{R})$, with $\mathbf{u}$ and $\mathbf{v}$ linearly independent.

partitioned matrix (or **block matrix**):   A matrix whose entries are themselves matrices of appropriate sizes.

permuted lower triangular matrix: A matrix such that a permutation of its rows will form a lower triangular matrix.

permuted LU factorization: The representation of a matrix A in the form $A = LU$ where L is a square matrix such that a permutation of its rows will form a unit lower triangular matrix, and U is an echelon form of A.

pivot: A nonzero number that either is used in a pivot position to create zeros through row operations or is changed into a leading 1, which in turn is used to create zeros.

pivot column: A column that contains a pivot position.

pivot position: A position in a matrix A that corresponds to a leading entry in an echelon form of A.

plane through u, v, and the origin: A set whose parametric equation is $\mathbf{x} = s\mathbf{u} + t\mathbf{v} \, (s, t \text{ in } \mathbb{R})$, with $\mathbf{u}$ and $\mathbf{v}$ linearly independent.

polar decomposition (of A):   A factorization $A = PQ$, where P is an $n \times n$ positive semidefinite matrix with the same rank as A, and Q is an $n \times n$ orthogonal matrix.

polygon: A polytope in $\mathbb{R}^2$.

polyhedron: A polytope in $\mathbb{R}^3$.

polytope: The convex hull of a finite set of points in $\mathbb{R}^n$ (a special type of compact convex set).

positive combination (of points $\mathbf{v}_1, \ldots, \mathbf{v}_m$ in $\mathbb{R}^n$):   A linear combination $c_1\mathbf{v}_1 + \cdots + c_m\mathbf{v}_m$, where all $c_i \geq 0$.

positive definite matrix: A symmetric matrix A such that $\mathbf{x}^T A \mathbf{x} > 0$ for all $\mathbf{x} \neq \mathbf{0}$.

positive definite quadratic form: A quadratic form Q such that $Q(\mathbf{x}) > 0$ for all $\mathbf{x} \neq \mathbf{0}$.

positive hull (of a set S):   The set of all positive combinations of points in S, denoted by pos S.

positive semidefinite matrix: A symmetric matrix A such that $\mathbf{x}^T A \mathbf{x} \geq 0$ for all $\mathbf{x}$.

positive semidefinite quadratic form: A quadratic form Q such that $Q(\mathbf{x}) \geq 0$ for all $\mathbf{x}$.

power method: An algorithm for estimating a strictly dominant eigenvalue of a square matrix.

principal axes (of a quadratic form $\mathbf{x}^T A\mathbf{x}$):   The orthonormal columns of an orthogonal matrix P such that $P^{-1}AP$ is diagonal. (These columns are unit eigenvectors of A.) Usually the columns of P are ordered in such a way that the corresponding eigenvalues of A are arranged in decreasing order of magnitude.

principal components (of the data in a matrix B of observations): The unit eigenvectors of a sample covariance matrix S for B, with the eigenvectors arranged so that the corresponding eigenvalues of S decrease in magnitude. If B is in mean-deviation form, then the principal components are the right singular vectors in a singular value decomposition of B^T.

probability vector: A vector in $\mathbb{R}^n$ whose entries are nonnegative and sum to one.

product Ax: The linear combination of the columns of A using the corresponding entries in $\mathbf{x}$ as weights.

production vector: The vector in the Leontief input–output model that lists the amounts that are to be produced by the various sectors of an economy.

profile (of a set S in $\mathbb{R}^n$): The set of extreme points of S.

projection matrix (or **orthogonal projection matrix**): A symmetric matrix B such that $B^2 = B$. A simple example is $B = \mathbf{v}\mathbf{v}^T$, where $\mathbf{v}$ is a unit vector.

proper subset of a set S: A subset of S that does not equal S itself.

proper subspace: Any subspace of a vector space V other than V itself.

pseudoinverse (of A): The matrix $VD^{-1}U^T$, when UDV^T is a reduced singular value decomposition of A.

Q

QR factorization: A factorization of an $m \times n$ matrix A with linearly independent columns, $A = QR$, where Q is an $m \times n$ matrix whose columns form an orthonormal basis for Col A, and R is an $n \times n$ upper triangular invertible matrix with positive entries on its diagonal.

quadratic Bézier curve: A curve whose description may be written in the form $\mathbf{g}(t) = (1 - t)\mathbf{f}_0(t) + t\mathbf{f}_1(t)$ for $0 \le t \le 1$, where $\mathbf{f}_0(t) = (1 - t)\mathbf{p}_0 + t\mathbf{p}_1$ and $\mathbf{f}_1(t) = (1 - t)\mathbf{p}_1 + t\mathbf{p}_2$. The points $\mathbf{p}_0, \mathbf{p}_1, \mathbf{p}_2$ are called the *control points* for the curve.

quadratic form: A function Q defined for $\mathbf{x}$ in $\mathbb{R}^n$ by $Q(\mathbf{x}) = \mathbf{x}^T A\mathbf{x}$, where A is an $n \times n$ symmetric matrix (called the **matrix of the quadratic form**).

R

range (of a linear transformation T): The set of all vectors of the form $T(\mathbf{x})$ for some $\mathbf{x}$ in the domain of T.

rank (of a matrix A): The dimension of the column space of A, denoted by rank A.

Rayleigh quotient: $R(\mathbf{x}) = (\mathbf{x}^T A\mathbf{x})/(\mathbf{x}^T\mathbf{x})$. An estimate of an eigenvalue of A (usually a symmetric matrix).

recurrence relation: *See* difference equation.

reduced echelon form (or **reduced row echelon form**): A reduced echelon matrix that is row equivalent to a given matrix.

reduced echelon matrix: A rectangular matrix in echelon form that has these additional properties: The leading entry in each nonzero row is 1, and each leading 1 is the only nonzero entry in its column.

reduced singular value decomposition: A factorization $A = UDV^T$, for an $m \times n$ matrix A of rank r, where U is $m \times r$ with orthonormal columns, D is an $r \times r$ diagonal matrix with the r nonzero singular values of A on its diagonal, and V is $n \times r$ with orthonormal columns.

regression coefficients: The coefficients β_0 and β_1 in the least-squares line $y = \beta_0 + \beta_1 x$.

regular solid: One of the five possible regular polyhedrons in $\mathbb{R}^3$: the tetrahedron (4 equal triangular faces), the cube (6 square faces), the octahedron (8 equal triangular faces), the dodecahedron (12 equal pentagonal faces), and the icosahedron (20 equal triangular faces).

regular stochastic matrix: A stochastic matrix P such that some matrix power P^k contains only strictly positive entries.

relative change or **relative error** (in $\mathbf{b}$): The quantity $\|\Delta\mathbf{b}\|/\|\mathbf{b}\|$ when $\mathbf{b}$ is changed to $\mathbf{b} + \Delta\mathbf{b}$.

repellor (of a dynamical system in $\mathbb{R}^2$): The origin when all trajectories except the constant zero sequence or function tend away from $\mathbf{0}$.

residual vector: The quantity $\boldsymbol{\epsilon}$ that appears in the general linear model: $\mathbf{y} = X\boldsymbol{\beta} + \boldsymbol{\epsilon}$; that is, $\boldsymbol{\epsilon} = \mathbf{y} - X\boldsymbol{\beta}$, the difference between the observed values and the predicted values (of y).

Re x: The vector in $\mathbb{R}^n$ formed from the real parts of the entries of a vector $\mathbf{x}$ in $\mathbb{C}^n$.

right inverse (of A): Any rectangular matrix C such that $AC = I$.

right-multiplication (by A): Multiplication of a matrix on the right by A.

right singular vectors (of A): The columns of V in the singular value decomposition $A = U\Sigma V^T$.

roundoff error: Error in floating point arithmetic caused when the result of a calculation is rounded (or truncated) to the number of floating point digits stored. Also, the error that results when the decimal representation of a number such as 1/3 is approximated by a floating point number with a finite number of digits.

row–column rule: The rule for computing a product AB in which the (i, j)-entry of AB is the sum of the products of corresponding entries from row i of A and column j of B.

row equivalent (matrices): Two matrices for which there exists a (finite) sequence of row operations that transforms one matrix into the other.

row reduction algorithm: A systematic method using elementary row operations that reduces a matrix to echelon form or reduced echelon form.

row replacement: An elementary row operation that replaces one row of a matrix by the sum of the row and a multiple of another row.

row space (of a matrix A): The set Row A of all linear combinations of the vectors formed from the rows of A; also denoted by Col A^T.

row sum: The sum of the entries in a row of a matrix.

row vector: A matrix with only one row, or a single row of a matrix that has several rows.

row–vector rule for computing $A\mathbf{x}$: The rule for computing a product $A\mathbf{x}$ in which the ith entry of $A\mathbf{x}$ is the sum of the products of corresponding entries from row i of A and from the vector $\mathbf{x}$.

S

saddle point (of a dynamical system in $\mathbb{R}^2$): The origin when some trajectories are attracted to $\mathbf{0}$ and other trajectories are repelled from $\mathbf{0}$.

same direction (as a vector $\mathbf{v}$): A vector that is a positive multiple of $\mathbf{v}$.

sample mean: The average M of a set of vectors, $\mathbf{X}_1, \ldots, \mathbf{X}_N$, given by $M = (1/N)(\mathbf{X}_1 + \cdots + \mathbf{X}_N)$.

scalar: A (real) number used to multiply either a vector or a matrix.

scalar multiple of $\mathbf{u}$ by c: The vector $c\mathbf{u}$ obtained by multiplying each entry in $\mathbf{u}$ by c.

scale (a vector): Multiply a vector (or a row or column of a matrix) by a nonzero scalar.

Schur complement: A certain matrix formed from the blocks of a 2×2 partitioned matrix $A = [A_{ij}]$. If A_{11} is invertible, its Schur complement is given by $A_{22} - A_{21}A_{11}^{-1}A_{12}$. If A_{22} is invertible, its Schur complement is given by $A_{11} - A_{12}A_{22}^{-1}A_{21}$.

Schur factorization (of A, for real scalars): A factorization $A = URU^T$ of an $n \times n$ matrix A having n real eigenvalues, where U is an $n \times n$ orthogonal matrix and R is an upper triangular matrix.

set spanned by $\{\mathbf{v}_1, \ldots, \mathbf{v}_p\}$: The set Span $\{\mathbf{v}_1, \ldots, \mathbf{v}_p\}$.

signal (or **discrete-time signal**): A doubly infinite sequence of numbers, $\{y_k\}$; a function defined on the integers; belongs to the vector space $\mathbb{S}$.

similar (matrices): Matrices A and B such that $P^{-1}AP = B$, or equivalently, $A = PBP^{-1}$, for some invertible matrix P.

similarity transformation: A transformation that changes A into $P^{-1}AP$.

simplex: The convex hull of an affinely independent finite set of vectors in $\mathbb{R}^n$.

singular (matrix): A square matrix that has no inverse.

singular value decomposition (of an $m \times n$ matrix A): $A = U\Sigma V^T$, where U is an $m \times m$ orthogonal matrix, V is an $n \times n$ orthogonal matrix, and Σ is an $m \times n$ matrix with non-negative entries on the main diagonal (arranged in decreasing

order of magnitude) and zeros elsewhere. If rank $A = r$, then Σ has exactly r positive entries (the nonzero singular values of A) on the diagonal.

singular values (of A): The (positive) square roots of the eigenvalues of A^TA, arranged in decreasing order of magnitude.

size (of a matrix): Two numbers, written in the form $m \times n$, that specify the number of rows (m) and columns (n) in the matrix.

solution (of a linear system involving variables $x_1, \ldots, x_n$): A list $(s_1, s_2, \ldots, s_n)$ of numbers that makes each equation in the system a true statement when the values $s_1, \ldots, s_n$ are substituted for $x_1, \ldots, x_n$, respectively.

solution set: The set of all possible solutions of a linear system. The solution set is empty when the linear system is inconsistent.

Span $\{\mathbf{v}_1, \ldots, \mathbf{v}_p\}$: The set of all linear combinations of $\mathbf{v}_1, \ldots, \mathbf{v}_p$. Also, the *subspace spanned* (or *generated*) by $\mathbf{v}_1, \ldots, \mathbf{v}_p$.

spanning set (for a subspace H): Any set $\{\mathbf{v}_1, \ldots, \mathbf{v}_p\}$ in H such that $H = \text{Span}\{\mathbf{v}_1, \ldots, \mathbf{v}_p\}$.

spectral decomposition (of A): A representation

$$A = \lambda_1 \mathbf{u}_1 \mathbf{u}_1^T + \cdots + \lambda_n \mathbf{u}_n \mathbf{u}_n^T$$

where $\{\mathbf{u}_1, \ldots, \mathbf{u}_n\}$ is an orthonormal basis of eigenvectors of A, and $\lambda_1, \ldots, \lambda_n$ are the corresponding eigenvalues of A.

spiral point (of a dynamical system in $\mathbb{R}^2$): The origin when the trajectories spiral about $\mathbf{0}$.

stage-matrix model: A difference equation $\mathbf{x}_{k+1} = A\mathbf{x}_k$ where $\mathbf{x}_k$ lists the number of females in a population at time k, with the females classified by various stages of development (such as juvenile, subadult, and adult).

standard basis: The basis $\mathcal{E} = \{\mathbf{e}_1, \ldots, \mathbf{e}_n\}$ for $\mathbb{R}^n$ consisting of the columns of the $n \times n$ identity matrix, or the basis $\{1, t, \ldots, t^n\}$ for $\mathbb{P}_n$.

standard matrix (for a linear transformation T): The matrix A such that $T(\mathbf{x}) = A\mathbf{x}$ for all $\mathbf{x}$ in the domain of T.

standard position: The position of the graph of an equation $\mathbf{x}^T A\mathbf{x} = c$, when A is a diagonal matrix.

state vector: A probability vector. In general, a vector that describes the "state" of a physical system, often in connection with a difference equation $\mathbf{x}_{k+1} = A\mathbf{x}_k$.

steady-state vector (for a stochastic matrix P): A probability vector $\mathbf{q}$ such that $P\mathbf{q} = \mathbf{q}$.

stiffness matrix: The inverse of a flexibility matrix. The jth column of a stiffness matrix gives the loads that must be applied at specified points on an elastic beam in order to produce a unit deflection at the jth point on the beam.

stochastic matrix: A square matrix whose columns are probability vectors.

strictly dominant eigenvalue: An eigenvalue λ_1 of a matrix A with the property that $|\lambda_1| > |\lambda_k|$ for all other eigenvalues λ_k of A.

submatrix (of A): Any matrix obtained by deleting some rows and/or columns of A; also, A itself.

subspace: A subset H of some vector space V such that H has these properties: (1) the zero vector of V is in H; (2) H is closed under vector addition; and (3) H is closed under multiplication by scalars.

supporting hyperplane (to a compact convex set S in $\mathbb{R}^n$): A hyperplane $H = [f:d]$ such that $H \cap S \neq \varnothing$ and either $f(x) \leq d$ for all x in S or $f(x) \geq d$ for all x in S.

symmetric matrix: A matrix A such that $A^T = A$.

system of linear equations (or a **linear system**): A collection of one or more linear equations involving the same set of variables, say, $x_1, \ldots, x_n$.

T

tetrahedron: A three-dimensional solid object bounded by four equal triangular faces, with three faces meeting at each vertex.

total variance: The trace of the covariance matrix S of a matrix of observations.

trace (of a square matrix A): The sum of the diagonal entries in A, denoted by tr A.

trajectory: The graph of a solution $\{\mathbf{x}_0, \mathbf{x}_1, \mathbf{x}_2, \ldots\}$ of a dynamical system $\mathbf{x}_{k+1} = A\mathbf{x}_k$, often connected by a thin curve to make the trajectory easier to see. Also, the graph of $\mathbf{x}(t)$ for $t \geq 0$, when $\mathbf{x}(t)$ is a solution of a differential equation $\mathbf{x}'(t) = A\mathbf{x}(t)$.

transfer matrix: A matrix A associated with an electrical circuit having input and output terminals, such that the output vector is A times the input vector.

transformation (or **function**, or **mapping**) T from $\mathbb{R}^n$ to $\mathbb{R}^m$: A rule that assigns to each vector $\mathbf{x}$ in $\mathbb{R}^n$ a unique vector $T(\mathbf{x})$ in $\mathbb{R}^m$. Notation: $T: \mathbb{R}^n \to \mathbb{R}^m$. Also, $T: V \to W$ denotes a rule that assigns to each $\mathbf{x}$ in V a unique vector $T(\mathbf{x})$ in W.

translation (by a vector $\mathbf{p}$): The operation of adding $\mathbf{p}$ to a vector or to each vector in a given set.

transpose (of A): An $n \times m$ matrix A^T whose columns are the corresponding rows of the $m \times n$ matrix A.

trend analysis: The use of orthogonal polynomials to fit data, with the inner product given by evaluation at a finite set of points.

triangle inequality: $\|\mathbf{u} + \mathbf{v}\| \leq \|\mathbf{u}\| + \|\mathbf{v}\|$ for all $\mathbf{u}, \mathbf{v}$.

triangular matrix: A matrix A with either zeros above or zeros below the diagonal entries.

trigonometric polynomial: A linear combination of the constant function 1 and sine and cosine functions such as $\cos nt$ and $\sin nt$.

trivial solution: The solution $\mathbf{x} = \mathbf{0}$ of a homogeneous equation $A\mathbf{x} = \mathbf{0}$.

U

uncorrelated variables: Any two variables x_i and x_j (with $i \neq j$) that range over the ith and jth coordinates of the observation vectors in an observation matrix, such that the covariance s_{ij} is zero.

underdetermined system: A system of equations with fewer equations than unknowns.

uniqueness question: Asks, "If a solution of a system exists, is it unique—that is, is it the only one?"

unit consumption vector: A column vector in the Leontief input–output model that lists the inputs a sector needs for each unit of its output; a column of the consumption matrix.

unit lower triangular matrix: A square lower triangular matrix with ones on the main diagonal.

unit vector: A vector $\mathbf{v}$ such that $\|\mathbf{v}\| = 1$.

upper triangular matrix: A matrix U (not necessarily square) with zeros below the diagonal entries $u_{11}, u_{22}, \ldots$.

V

Vandermonde matrix: An $n \times n$ matrix V or its transpose, when V has the form

$$
V = \begin{bmatrix} 1 & x_1 & x_1^2 & \cdots & x_1^{n-1} \\ 1 & x_2 & x_2^2 & \cdots & x_2^{n-1} \\ \vdots & \vdots & \vdots & & \vdots \\ 1 & x_n & x_n^2 & \cdots & x_n^{n-1} \end{bmatrix}
$$

variance (of a variable x_j): The diagonal entry s_{jj} in the covariance matrix S for a matrix of observations, where x_j varies over the jth coordinates of the observation vectors.

vector: A list of numbers; a matrix with only one column. In general, any element of a vector space.

vector addition: Adding vectors by adding corresponding entries.

vector equation: An equation involving a linear combination of vectors with undetermined weights.

vector space: A set of objects, called vectors, on which two operations are defined, called addition and multiplication by scalars. Ten axioms must be satisfied. See the first definition in Section 4.1.

vector subtraction: Computing $\mathbf{u} + (-1)\mathbf{v}$ and writing the result as $\mathbf{u} - \mathbf{v}$.

W

weighted least squares: Least-squares problems with a weighted inner product such as

$$
\langle \mathbf{x}, \mathbf{y} \rangle = w_1^2 x_1 y_1 + \cdots + w_n^2 x_n y_n.
$$

weights: The scalars used in a linear combination.

Z

zero subspace: The subspace $\{\mathbf{0}\}$ consisting of only the zero vector.

zero vector: The unique vector, denoted by $\mathbf{0}$, such that $\mathbf{u} + \mathbf{0} = \mathbf{u}$ for all $\mathbf{u}$. In $\mathbb{R}^n$, $\mathbf{0}$ is the vector whose entries are all zeros.

Answers to Odd-Numbered Exercises

Chapter 1

Section 1.1, page 10

1. The solution is $(x_1, x_2) = (-8, 3)$, or simply $(-8, 3)$.

3. $(4/7, 9/7)$

5. Replace row 2 by its sum with 3 times row 3, and then replace row 1 by its sum with -5 times row 3.

7. The solution set is empty.

9. No solutions   **11.** $(19, -8, 1)$

13. $(5, 3, -1)$

15.
$$
\begin{array}{rcrcrcr}
& -8 & + & 4(1) & = & -4 \\
19 & + & 3(-8) & + & 3(1) & = & -2 \\
3(19) & + & 7(-8) & + & 5(1) & = & 6
\end{array}
$$

17.
$$
\begin{array}{rcrcrcr}
(5) & & & - & 3(-1) & = & 8 \\
2(5) & + & 2(3) & + & 9(-1) & = & 7 \\
& & (3) & + & 5(-1) & = & -2
\end{array}
$$

19. Consistent

21. The three lines have one point in common.

23. $h \neq 2$ **25.** All h

27–33. Mark a statement True only if the statement is *always* true. Giving you the answers here would defeat the purpose of the true-false questions, which is to help you learn to read the text carefully. The *Study Guide* will tell you where to look for the answers, but you should not consult it until you have made an honest attempt to find the answers yourself.

35. $k + 2g + h = 0$

37. The row reduction of $\begin{bmatrix} 1 & 3 & f \\ c & d & g \end{bmatrix}$ to $\begin{bmatrix} 1 & 3 & f \\ 0 & d-3c & g-cf \end{bmatrix}$ shows that $d - 3c$ must be nonzero, since f and g are arbitrary. Otherwise, for some choices of f and g the second row could correspond to an equation of the form $0 = b$, where b is nonzero. Thus $d \neq 3c$.

39. Swap row 1 and row 2; swap row 1 and row 2.

41. Replace row 3 by row $3 + (-4)$ row 1; replace row 3 by row $3 + (4)$ row 1.

43.
$$
\begin{array}{rcrcrcrcr}
4T_1 & - & T_2 & & & - & T_4 & = & 30 \\
-T_1 & + & 4T_2 & - & T_3 & & & = & 60 \\
& & -T_2 & + & 4T_3 & - & T_4 & = & 70 \\
-T_1 & & & - & T_3 & + & 4T_4 & = & 40
\end{array}
$$

Section 1.2, page 23

1. Reduced echelon form: a and c. Echelon form: b and d.

3. $\begin{bmatrix} 1 & 0 & -1 & -2 \\ 0 & 1 & 2 & 3 \\ 0 & 0 & 0 & 0 \end{bmatrix}$. Pivot cols 1 and 2:
$\begin{bmatrix} 1 & 2 & 3 & 4 \\ 4 & 5 & 6 & 7 \\ 6 & 7 & 8 & 9 \end{bmatrix}$.

5. $\begin{bmatrix} \blacksquare & * \\ 0 & \blacksquare \end{bmatrix}, \begin{bmatrix} \blacksquare & * \\ 0 & 0 \end{bmatrix}, \begin{bmatrix} 0 & \blacksquare \\ 0 & 0 \end{bmatrix}$

7. $\begin{cases} x_1 = -5 - 3x_2 \\ x_2 \text{ is free} \\ x_3 = 3 \end{cases}$ **9.** $\begin{cases} x_1 = 6 + 5x_3 \\ x_2 = 5 + 6x_3 \\ x_3 \text{ is free} \end{cases}$

11. $\begin{cases} x_1 = \dfrac{4}{3}x_2 - \dfrac{2}{3}x_3 \\ x_2 \text{ is free} \\ x_3 \text{ is free} \end{cases}$

13. $\begin{cases} x_1 = -3 + 3x_5 \\ x_2 = 1 + 4x_5 \\ x_3 \text{ is free} \\ x_4 = -4 - 9x_5 \\ x_5 \text{ is free} \end{cases}$

Note: The *Study Guide* discusses the common mistake $x_3 = 0$.

15.
$$\begin{array}{rcrcrcr} & & x_2 & - & 6x_3 & = & 5 \\ x_1 & - & 2x_2 & + & 7x_3 & = & -4 \end{array} \text{, verify}$$

$$\begin{array}{rcrcrcr} & & 5 + 6x_3 & - & 6x_3 & = & 5 \\ (6 + 5x_3) & - & 2(5 + 6x_3) & + & 7x_3 & = & -4 \end{array}$$

17.
$$\begin{array}{rcrcrcr} 3x_1 & - & 4x_2 & + & 2x_3 & = & 0 \\ -9x_1 & + & 12x_2 & - & 6x_3 & = & 0 \\ -6x_1 & + & 8x_2 & - & 4x_3 & = & 0 \end{array} \text{, verify}$$

$$\begin{array}{rcrcrcr} 3\left(\frac{4}{3}x_2 - \frac{2}{3}x_3\right) & - & 4x_2 & + & 2x_3 & = & 0 \\ -9\left(\frac{4}{3}x_2 - \frac{2}{3}x_3\right) & + & 12x_2 & - & 6x_3 & = & 0 \\ -6\left(\frac{4}{3}x_2 - \frac{2}{3}x_3\right) & + & 8x_2 & - & 4x_3 & = & 0 \end{array}$$

19. a. Consistent, with a unique solution

b. Inconsistent

21. $h = 7/2$

23. a. Inconsistent when $h = 2$ and $k \neq 8$

b. A unique solution when $h \neq 2$

c. Many solutions when $h = 2$ and $k = 8$

25–33. Read the text carefully, and write your answers before you consult the *Study Guide*. Remember, a statement is true only if it is true in all cases.

35. Yes. The system is consistent because with three pivots, there must be a pivot in the third (bottom) row of the coefficient matrix. The reduced echelon form cannot contain a row of the form $[0 \ \ 0 \ \ 0 \ \ 0 \ \ 0 \ \ 1]$.

37. If the coefficient matrix has a pivot position in every row, then there is a pivot position in the bottom row, and there is no room for a pivot in the augmented column. So, the system is consistent, by Theorem 2.

39. If a linear system is consistent, then the solution is unique if and only if *every column in the coefficient matrix is a pivot column; otherwise, there are infinitely many solutions.*

41. An underdetermined system always has more variables than equations. There cannot be more basic variables than there are equations, so there must be at least one free variable. Such a variable may be assigned infinitely many different values. If the system is consistent, each different value of a free variable will produce a different solution.

43. Yes, a system of linear equations with more equations than unknowns can be consistent. The following system has a solution ($x_1 = x_2 = 1$):

$$\begin{array}{rcl} x_1 + x_2 & = & 2 \\ x_1 - x_2 & = & 0 \\ 3x_1 + 2x_2 & = & 5 \end{array}$$

45. $p(t) = 7 + 6t - t^2$

Section 1.3, page 34

1. $\begin{bmatrix} -4 \\ 5 \end{bmatrix}, \begin{bmatrix} 5 \\ -4 \end{bmatrix}$

3.

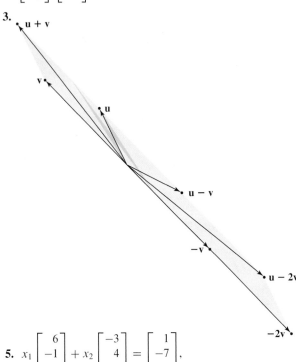

5. $x_1 \begin{bmatrix} 6 \\ -1 \\ 5 \end{bmatrix} + x_2 \begin{bmatrix} -3 \\ 4 \\ 0 \end{bmatrix} = \begin{bmatrix} 1 \\ -7 \\ -5 \end{bmatrix}$,

$$\begin{bmatrix} 6x_1 \\ -x_1 \\ 5x_1 \end{bmatrix} + \begin{bmatrix} -3x_2 \\ 4x_2 \\ 0 \end{bmatrix} = \begin{bmatrix} 1 \\ -7 \\ -5 \end{bmatrix}, \begin{bmatrix} 6x_1 - 3x_2 \\ -x_1 + 4x_2 \\ 5x_1 \end{bmatrix} = \begin{bmatrix} 1 \\ -7 \\ -5 \end{bmatrix}$$

$$\begin{array}{rcr} 6x_1 - 3x_2 & = & 1 \\ -x_1 + 4x_2 & = & -7 \\ 5x_1 & = & -5 \end{array}$$

Usually the intermediate steps are not displayed.

7. $\mathbf{a} = \mathbf{u} - 2\mathbf{v}, \mathbf{b} = 2\mathbf{u} - 2\mathbf{v}, \mathbf{c} = 2\mathbf{u} - 3.5\mathbf{v}, \mathbf{d} = 3\mathbf{u} - 4\mathbf{v}$

9. $x_1 \begin{bmatrix} 0 \\ 4 \\ -1 \end{bmatrix} + x_2 \begin{bmatrix} 1 \\ 6 \\ 3 \end{bmatrix} + x_3 \begin{bmatrix} 5 \\ -1 \\ -8 \end{bmatrix} = \begin{bmatrix} 0 \\ 0 \\ 0 \end{bmatrix}$

11. Yes, $\mathbf{b}$ is a linear combination of $\mathbf{a}_1$, $\mathbf{a}_2$, and $\mathbf{a}_3$.

13. No, $\mathbf{b}$ is *not* a linear combination of the columns of A.

15. Noninteger weights are acceptable, of course, but some simple choices are $0 \cdot \mathbf{v}_1 + 0 \cdot \mathbf{v}_2 = \mathbf{0}$, and

$$1 \cdot \mathbf{v}_1 + 0 \cdot \mathbf{v}_2 = \begin{bmatrix} 7 \\ 1 \\ -6 \end{bmatrix}, 0 \cdot \mathbf{v}_1 + 1 \cdot \mathbf{v}_2 = \begin{bmatrix} -5 \\ 3 \\ 0 \end{bmatrix}$$

$$1 \cdot \mathbf{v}_1 + 1 \cdot \mathbf{v}_2 = \begin{bmatrix} 2 \\ 4 \\ -6 \end{bmatrix}, 1 \cdot \mathbf{v}_1 - 1 \cdot \mathbf{v}_2 = \begin{bmatrix} 12 \\ -2 \\ -6 \end{bmatrix}$$

17. $h = -17$

19. Span $\{\mathbf{v}_1, \mathbf{v}_2\}$ is the set of points on the line through $\mathbf{v}_1$ and $\mathbf{0}$.

21. *Hint:* Show that $\begin{bmatrix} 2 & 2 & h \\ -1 & 1 & k \end{bmatrix}$ is consistent for all h and k. Explain what this calculation shows about Span $\{\mathbf{u}, \mathbf{v}\}$.

23–31. Before you consult your *Study Guide*, read the entire section carefully. Pay special attention to definitions and theorem statements, and note any remarks that precede or follow them.

33. a. No, three   **b.** Yes, infinitely many
   c. $\mathbf{a}_1 = 1 \cdot \mathbf{a}_1 + 0 \cdot \mathbf{a}_2 + 0 \cdot \mathbf{a}_3$

35. a. $5\mathbf{v}_1$ is the output of 5 day's operation of mine #1.
   b. The total output is $x_1\mathbf{v}_1 + x_2\mathbf{v}_2$, so x_1 and x_2 should satisfy $x_1\mathbf{v}_1 + x_2\mathbf{v}_2 = \begin{bmatrix} 150 \\ 2825 \end{bmatrix}$.
   c. 1.5 days for mine #1 and 4 days for mine #2

37. $(1.3, .9, 0)$

39. a. $\begin{bmatrix} 10/3 \\ 2 \end{bmatrix}$
   b. Add 3.5 g at $(0, 1)$, add .5 g at $(8, 1)$, and add 2 g at $(2, 4)$.

41. Review Practice Problem 1 and then *write* a solution. The *Study Guide* has a solution.

Section 1.4, page 42

1. The product is not defined because the number of columns (2) in the 3×2 matrix does not match the number of entries (3) in the vector.

3. $A\mathbf{x} = \begin{bmatrix} 6 & 5 \\ -4 & -3 \\ 7 & 6 \end{bmatrix}\begin{bmatrix} 1 \\ -3 \end{bmatrix} = 1\begin{bmatrix} 6 \\ -4 \\ 7 \end{bmatrix} - 3\begin{bmatrix} 5 \\ -3 \\ 6 \end{bmatrix}$

$= \begin{bmatrix} 6 \\ -4 \\ 7 \end{bmatrix} + \begin{bmatrix} -15 \\ 9 \\ -18 \end{bmatrix} = \begin{bmatrix} -9 \\ 5 \\ -11 \end{bmatrix}$, and

$A\mathbf{x} = \begin{bmatrix} 6 & 5 \\ -4 & -3 \\ 7 & 6 \end{bmatrix}\begin{bmatrix} 1 \\ -3 \end{bmatrix} = \begin{bmatrix} 6(1) + 5(-3) \\ (-4)(1) + (-3)(-3) \\ 7(1) + 6(-3) \end{bmatrix}$

$= \begin{bmatrix} -9 \\ 5 \\ -11 \end{bmatrix}$. Show your work here and for Exercises 4–6, but thereafter perform the calculations mentally.

5. $5\begin{bmatrix} 5 \\ -2 \end{bmatrix} - 1\begin{bmatrix} 1 \\ -7 \end{bmatrix} + 3\begin{bmatrix} -8 \\ 3 \end{bmatrix} - 2\begin{bmatrix} 4 \\ -5 \end{bmatrix} = \begin{bmatrix} -8 \\ 16 \end{bmatrix}$

7. $\begin{bmatrix} 4 & -5 & 7 \\ -1 & 3 & -8 \\ 7 & -5 & 0 \\ -4 & 1 & 2 \end{bmatrix}\begin{bmatrix} x_1 \\ x_2 \\ x_3 \end{bmatrix} = \begin{bmatrix} 6 \\ -8 \\ 0 \\ -7 \end{bmatrix}$

9. $x_1\begin{bmatrix} 3 \\ 0 \end{bmatrix} + x_2\begin{bmatrix} 1 \\ 1 \end{bmatrix} + x_3\begin{bmatrix} -5 \\ 4 \end{bmatrix} = \begin{bmatrix} 9 \\ 0 \end{bmatrix}$ and

$\begin{bmatrix} 3 & 1 & -5 \\ 0 & 1 & 4 \end{bmatrix}\begin{bmatrix} x_1 \\ x_2 \\ x_3 \end{bmatrix} = \begin{bmatrix} 9 \\ 0 \end{bmatrix}$

11. $\begin{bmatrix} 1 & 2 & 4 & -2 \\ 0 & 1 & 5 & 2 \\ -2 & -4 & -3 & 9 \end{bmatrix}, \mathbf{x} = \begin{bmatrix} x_1 \\ x_2 \\ x_3 \end{bmatrix} = \begin{bmatrix} 0 \\ -3 \\ 1 \end{bmatrix}$

13. Yes. (Justify your answer.)

$\mathbf{u}$ are here

15. The equation $A\mathbf{x} = \mathbf{b}$ is not consistent when $3b_1 + b_2$ is nonzero. (Show your work.) The set of $\mathbf{b}$ for which the equation *is* consistent is a line through the origin—the set of all points (b_1, b_2) satisfying $b_2 = -3b_1$.

17. Only three rows contain a pivot position. The equation $A\mathbf{x} = \mathbf{b}$ does *not* have a solution for each $\mathbf{b}$ in $\mathbb{R}^4$, by Theorem 4.

19. The work in Exercise 17 shows that statement (d) in Theorem 4 is false. So all four statements in Theorem 4 are false. Thus, not all vectors in $\mathbb{R}^4$ can be written as a linear combination of the columns of A. Also, the columns of A do *not* span $\mathbb{R}^4$.

21. The matrix $[\mathbf{v}_1 \quad \mathbf{v}_2 \quad \mathbf{v}_3]$ does not have a pivot in each row, so the columns of the matrix do not span $\mathbb{R}^4$, by Theorem 4. That is, $\{\mathbf{v}_1, \mathbf{v}_2, \mathbf{v}_3\}$ does not span $\mathbb{R}^4$.

23–33. Read the text carefully and try to mark each exercise statement True or False before you consult the *Study Guide*. Several parts of Exercises 23–24 are *implications* of the form

"If ⟨statement 1⟩, then ⟨statement 2⟩"

or equivalently,

"⟨statement 2⟩, if ⟨statement 1⟩"

Mark such an implication as True if ⟨statement 2⟩ is true in all cases when ⟨statement 1⟩ is true.

35. $c_1 = -3, c_2 = -1, c_3 = 2$

37. $Q\mathbf{x} = \mathbf{v}$, where $Q = [\mathbf{q}_1 \quad \mathbf{q}_2 \quad \mathbf{q}_3]$ and $\mathbf{x} = \begin{bmatrix} x_1 \\ x_2 \\ x_3 \end{bmatrix}$

Note: If your answer is the equation $A\mathbf{x} = \mathbf{b}$, you must specify what A and $\mathbf{b}$ are.

39. *Hint:* Start with any 3×3 matrix B in echelon form that has three pivot positions.

41. *Write* your solution before you check the *Study Guide*.

43. *Hint:* How many pivot columns does A have? Why?

45. Given $A\mathbf{x}_1 = \mathbf{y}_1$ and $A\mathbf{x}_2 = \mathbf{y}_2$, you are asked to show that the equation $A\mathbf{x} = \mathbf{w}$ has a solution, where $\mathbf{w} = \mathbf{y}_1 + \mathbf{y}_2$. Observe that $\mathbf{w} = A\mathbf{x}_1 + A\mathbf{x}_2$ and use Theorem 5(a) with $\mathbf{x}_1$ and $\mathbf{x}_2$ in place of $\mathbf{u}$ and $\mathbf{v}$, respectively. That is, $\mathbf{w} = A\mathbf{x}_1 + A\mathbf{x}_2 = A(\mathbf{x}_1 + \mathbf{x}_2)$. So the vector $\mathbf{x} = \mathbf{x}_1 + \mathbf{x}_2$ is a solution of $\mathbf{w} = A\mathbf{x}$.

47. The columns do not span $\mathbb{R}^4$.

49. The columns span $\mathbb{R}^4$.

51. Delete column 4 of the matrix in Exercise 49. It is also possible to delete column 3 instead of column 4.

Section 1.5, page 51

1. The system has a nontrivial solution because there is a free variable, x_3.

3. The system has a nontrivial solution because there is a free variable, x_3.

5. $\mathbf{x} = \begin{bmatrix} x_1 \\ x_2 \\ x_3 \end{bmatrix} = x_3 \begin{bmatrix} 5 \\ -2 \\ 1 \end{bmatrix}$

7. $\mathbf{x} = \begin{bmatrix} x_1 \\ x_2 \\ x_3 \\ x_4 \end{bmatrix} = x_3 \begin{bmatrix} -9 \\ 4 \\ 1 \\ 0 \end{bmatrix} + x_4 \begin{bmatrix} 8 \\ -5 \\ 0 \\ 1 \end{bmatrix}$

9. $\mathbf{x} = x_2 \begin{bmatrix} 3 \\ 1 \\ 0 \end{bmatrix} + x_3 \begin{bmatrix} -2 \\ 0 \\ 1 \end{bmatrix}$

11. *Hint:* The system derived from the *reduced* echelon form is

$$
\begin{aligned}
x_1 - 4x_2 \quad\quad + 5x_6 &= 0 \\
x_3 \quad - x_6 &= 0 \\
x_5 - 4x_6 &= 0 \\
0 &= 0
\end{aligned}
$$

The basic variables are x_1, x_3, and x_5. The remaining variables are free. The *Study Guide* discusses two mistakes that are often made on this type of problem.

13. $\begin{bmatrix} 3 & -9 & 6 \\ -1 & 3 & -2 \end{bmatrix} \left(x_2 \begin{bmatrix} 3 \\ 1 \\ 0 \end{bmatrix} + x_3 \begin{bmatrix} -2 \\ 0 \\ 1 \end{bmatrix} \right) =$

$x_2 \begin{bmatrix} 3 & -9 & 6 \\ -1 & 3 & -2 \end{bmatrix} \begin{bmatrix} 3 \\ 1 \\ 0 \end{bmatrix} +$

$x_3 \begin{bmatrix} 3 & -9 & 6 \\ -1 & 3 & -2 \end{bmatrix} \begin{bmatrix} -2 \\ 0 \\ 1 \end{bmatrix}$

$= x_2 \begin{bmatrix} 0 \\ 0 \end{bmatrix} + x_3 \begin{bmatrix} 0 \\ 0 \end{bmatrix} = \begin{bmatrix} 0 \\ 0 \end{bmatrix}$

15. $\begin{bmatrix} 1 & -4 & -2 & 0 & 3 & -5 \\ 0 & 0 & 1 & 0 & 0 & -1 \\ 0 & 0 & 0 & 0 & 1 & -4 \\ 0 & 0 & 0 & 0 & 0 & 0 \end{bmatrix} \left(x_2 \begin{bmatrix} 4 \\ 1 \\ 0 \\ 0 \\ 0 \\ 0 \end{bmatrix} + x_4 \begin{bmatrix} 0 \\ 0 \\ 0 \\ 1 \\ 0 \\ 0 \end{bmatrix} + x_6 \begin{bmatrix} -5 \\ 0 \\ 1 \\ 0 \\ 4 \\ 1 \end{bmatrix} \right)$

$= x_2 \begin{bmatrix} 1 & -4 & -2 & 0 & 3 & -5 \\ 0 & 0 & 1 & 0 & 0 & -1 \\ 0 & 0 & 0 & 0 & 1 & -4 \\ 0 & 0 & 0 & 0 & 0 & 0 \end{bmatrix} \begin{bmatrix} 4 \\ 1 \\ 0 \\ 0 \\ 0 \\ 0 \end{bmatrix} +$

$x_4 \begin{bmatrix} 1 & -4 & -2 & 0 & 3 & -5 \\ 0 & 0 & 1 & 0 & 0 & -1 \\ 0 & 0 & 0 & 0 & 1 & -4 \\ 0 & 0 & 0 & 0 & 0 & 0 \end{bmatrix} \begin{bmatrix} 0 \\ 0 \\ 0 \\ 1 \\ 0 \\ 0 \end{bmatrix} +$

$x_6 \begin{bmatrix} 1 & -4 & -2 & 0 & 3 & -5 \\ 0 & 0 & 1 & 0 & 0 & -1 \\ 0 & 0 & 0 & 0 & 1 & -4 \\ 0 & 0 & 0 & 0 & 0 & 0 \end{bmatrix} \begin{bmatrix} -5 \\ 0 \\ 1 \\ 0 \\ 4 \\ 1 \end{bmatrix}$

$= x_2 \begin{bmatrix} 0 \\ 0 \\ 0 \\ 0 \end{bmatrix} + x_4 \begin{bmatrix} 0 \\ 0 \\ 0 \\ 0 \end{bmatrix} + x_6 \begin{bmatrix} 0 \\ 0 \\ 0 \\ 0 \end{bmatrix} = \begin{bmatrix} 0 \\ 0 \\ 0 \\ 0 \end{bmatrix}$

17. $\mathbf{x} = \begin{bmatrix} 5 \\ -2 \\ 0 \end{bmatrix} + x_3 \begin{bmatrix} 4 \\ -7 \\ 1 \end{bmatrix} = \mathbf{p} + x_3\mathbf{q}$. Geometrically, the solution set is the line through $\begin{bmatrix} 5 \\ -2 \\ 0 \end{bmatrix}$ parallel to $\begin{bmatrix} 4 \\ -7 \\ 1 \end{bmatrix}$.

19. $\mathbf{x} = \begin{bmatrix} x_1 \\ x_2 \\ x_3 \end{bmatrix} = \begin{bmatrix} -2 \\ 1 \\ 0 \end{bmatrix} + x_3 \begin{bmatrix} 5 \\ -2 \\ 1 \end{bmatrix}$. The solution set is the line through $\begin{bmatrix} -2 \\ 1 \\ 0 \end{bmatrix}$, parallel to the line that is the solution set of the homogeneous system in Exercise 5.

21. Let $\mathbf{u} = \begin{bmatrix} -9 \\ 1 \\ 0 \end{bmatrix}$, $\mathbf{v} = \begin{bmatrix} 4 \\ 0 \\ 1 \end{bmatrix}$, $\mathbf{p} = \begin{bmatrix} -2 \\ 0 \\ 0 \end{bmatrix}$. The solution of the homogeneous equation is $\mathbf{x} = x_2\mathbf{u} + x_3\mathbf{v}$, the plane through the origin spanned by $\mathbf{u}$ and $\mathbf{v}$. The solution set of the nonhomogeneous system is $\mathbf{x} = \mathbf{p} + x_2\mathbf{u} + x_3\mathbf{v}$, the plane through $\mathbf{p}$ parallel to the solution set of the homogeneous equation.

23. $\mathbf{x} = \mathbf{a} + t\mathbf{b}$, where t represents a parameter, or $\mathbf{x} = \begin{bmatrix} x_1 \\ x_2 \end{bmatrix} = \begin{bmatrix} -2 \\ 0 \end{bmatrix} + t \begin{bmatrix} -5 \\ 3 \end{bmatrix}$, or $\begin{cases} x_1 = -2 - 5t \\ x_2 = 3t \end{cases}$

25. $\mathbf{x} = \mathbf{p} + t(\mathbf{q} - \mathbf{p}) = \begin{bmatrix} 2 \\ -5 \end{bmatrix} + t \begin{bmatrix} -5 \\ 6 \end{bmatrix}$

27–35. It is important to read the text carefully and write your answers. After that, check the *Study Guide*, if necessary.

37. $A\mathbf{v}_h = A(\mathbf{w} - \mathbf{p}) = A\mathbf{w} - A\mathbf{p} = \mathbf{b} - \mathbf{b} = \mathbf{0}$

39. When A is the 3×3 zero matrix, *every* $\mathbf{x}$ in $\mathbb{R}^3$ satisfies $A\mathbf{x} = \mathbf{0}$. So the solution set is all vectors in $\mathbb{R}^3$.

41. a. When A is a 3×3 matrix with three pivot positions, the equation $A\mathbf{x} = \mathbf{0}$ has no free variables and hence has no nontrivial solution.

    b. With three pivot positions, A has a pivot position in each of its three rows. By Theorem 4 in Section 1.4, the equation $A\mathbf{x} = \mathbf{b}$ has a solution for every possible $\mathbf{b}$. The word "possible" in the exercise means that the only vectors considered in this case are those in $\mathbb{R}^3$, because A has three rows.

43. a. When A is a 3×2 matrix with two pivot positions, each column is a pivot column. So the equation $A\mathbf{x} = \mathbf{0}$ has no free variables and hence no nontrivial solution.

    b. With two pivot positions and three rows, A cannot have a pivot in every row. So the equation $A\mathbf{x} = \mathbf{b}$ cannot have a solution for every possible $\mathbf{b}$ (in $\mathbb{R}^3$), by Theorem 4 in Section 1.4.

45. One answer: $\mathbf{x} = \begin{bmatrix} 3 \\ -1 \end{bmatrix}$

47. Your example should have the property that the sum of the entries in each row is zero. Why?

49. One answer is $A = \begin{bmatrix} 1 & -4 \\ 1 & -4 \end{bmatrix}$. The *Study Guide* shows how to analyze the problem in order to construct A. If $\mathbf{b}$ is any vector *not* a multiple of the first column of A, then the solution set of $A\mathbf{x} = \mathbf{b}$ is empty and thus cannot be formed by translating the solution set of $A\mathbf{x} = \mathbf{b}$. This does not contradict Theorem 6, because that theorem applies when the equation $A\mathbf{x} = \mathbf{b}$ has a nonempty solution set.

51. If c is a scalar, then $A(c\mathbf{u}) = cA\mathbf{u}$, by Theorem 5(b) in Section 1.4. If $\mathbf{u}$ satisfies $A\mathbf{x} = \mathbf{0}$, then $A\mathbf{u} = \mathbf{0}$, $cA\mathbf{u} = c \cdot \mathbf{0} = \mathbf{0}$, and so $A(c\mathbf{u}) = \mathbf{0}$.

Section 1.6, page 58

1. The general solution is $p_{Goods} = .875p_{Services}$, with $p_{Services}$ free. One equilibrium solution is $p_{Services} = 1000$ and $p_{Goods} = 875$. Using fractions, the general solution could be written $p_{Goods} = (7/8)p_{Services}$, and a natural choice of prices might be $p_{Services} = 80$ and $p_{Goods} = 70$. Only the *ratio* of the prices is important. The economic equilibrium is unaffected by a proportional change in prices.

3. a.

Distribution of Output From

Output	C&M ↓	F&P ↓	Mach. ↓	Input	Purchased By
	.2	.8	.4	→	C&M
	.3	.1	.4	→	F&P
	.5	.1	.2	→	Mach.

    b. $\begin{bmatrix} .8 & -.8 & -.4 & 0 \\ -.3 & .9 & -.4 & 0 \\ -.5 & -.1 & .8 & 0 \end{bmatrix}$

    c. $p_{Chemicals} = 141.7$, $p_{Fuels} = 91.7$, $p_{Machinery} = 100$. To two significant figures, $p_{Chemicals} = 140$, $p_{Fuels} = 92$, $p_{Machinery} = 100$.

5. $B_2S_3 + 6H_2O \rightarrow 2H_3BO_3 + 3H_2S$

7. $3NaHCO_3 + H_3C_6H_5O_7 \rightarrow Na_3C_6H_5O_7 + 3H_2O + 3CO_2$

9. $15PbN_6 + 44CrMn_2O_8 \rightarrow$
$5Pb_3O_4 + 22Cr_2O_3 + 88MnO_2 + 90NO$

11. $\begin{cases} x_1 = 20 - x_3 \\ x_2 = 60 + x_3 \\ x_3 \text{ is free} \\ x_4 = 60 \end{cases}$ The largest value of x_3 is 20.

13. a. $\begin{cases} x_1 = x_3 - 40 \\ x_2 = x_3 + 10 \\ x_3 \text{ is free} \\ x_4 = x_6 + 50 \\ x_5 = x_6 + 60 \\ x_6 \text{ is free} \end{cases}$ **b.** $\begin{cases} x_2 = 50 \\ x_3 = 40 \\ x_4 = 50 \\ x_5 = 60 \end{cases}$

Section 1.7, page 65

Justify your answers to Exercises 1–22.

1. Lin. dep.   **3.** Lin. indep.

5. Lin. indep.   **7.** Lin. depen.

9. a. $h = 4$ **b.** $h = 4$

11. $h = 6$ **13.** All h

15. Lin. depen.   **17.** Lin. depen.   **19.** Lin. indep.

21–27. Read through the definitions, examples, and theorems for this section before you consult the *Study Guide*.

29. $\begin{bmatrix} \blacksquare & * & * \\ 0 & \blacksquare & * \\ 0 & 0 & \blacksquare \end{bmatrix}$ **31.** $\begin{bmatrix} \blacksquare & * \\ 0 & \blacksquare \\ 0 & 0 \\ 0 & 0 \end{bmatrix}$ and $\begin{bmatrix} 0 & \blacksquare \\ 0 & 0 \\ 0 & 0 \\ 0 & 0 \end{bmatrix}$

33. All five columns of the 7×5 matrix A must be pivot columns. Otherwise, the equation $A\mathbf{x} = \mathbf{0}$ would have a free variable, in which case the columns of A would be linearly dependent.

35. A: Any 3×2 matrix with two nonzero columns such that neither column is a multiple of the other. In this case, the columns are linearly independent, and so the equation $A\mathbf{x} = \mathbf{0}$ has only the trivial solution.
B: Any 3×2 matrix with one column a multiple of the other.

37. $\mathbf{x} = \begin{bmatrix} 1 \\ 1 \\ -1 \end{bmatrix}$

39. True, by Theorem 7. (The *Study Guide* adds another justification.)

41. False. The vector $\mathbf{v}_1$ could be the zero vector.

43. True. A linear dependence relation among $\mathbf{v}_1, \mathbf{v}_2, \mathbf{v}_3$ may be extended to a linear dependence relation among $\mathbf{v}_1, \mathbf{v}_2, \mathbf{v}_3$, $\mathbf{v}_4$ by placing a zero weight on $\mathbf{v}_4$.

45. You should be able to work this important problem without help. *Write* your solution before you consult the *Study Guide*.

47. $B = \begin{bmatrix} 8 & -3 & 2 \\ -9 & 4 & -7 \\ 6 & -2 & 4 \\ 5 & -1 & 10 \end{bmatrix}$. Other choices are possible.

49. Each column of A that is not a column of B is in the set spanned by the columns of B.

Section 1.8, page 73

1. $\begin{bmatrix} 2 \\ -6 \end{bmatrix}, \begin{bmatrix} 2a \\ 2b \end{bmatrix}$ **3.** $\mathbf{x} = \begin{bmatrix} 3 \\ 1 \\ 2 \end{bmatrix}$, unique solution

5. $\mathbf{x} = \begin{bmatrix} 3 \\ 1 \\ 0 \end{bmatrix}$, not unique   **7.** $a = 5, b = 6$

9. $\mathbf{x} = x_3 \begin{bmatrix} 9 \\ 4 \\ 1 \\ 0 \end{bmatrix} + x_4 \begin{bmatrix} -7 \\ -3 \\ 0 \\ 1 \end{bmatrix}$

11. Yes, because the system represented by $[A \quad \mathbf{b}]$ is consistent.

13.

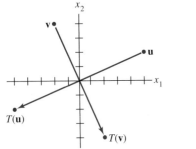

A reflection through the origin

15.

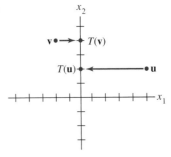

A projection onto the x_2-axis.

17. $\begin{bmatrix} 6 \\ 3 \end{bmatrix}, \begin{bmatrix} -2 \\ 6 \end{bmatrix}, \begin{bmatrix} 4 \\ 9 \end{bmatrix}$ **19.** $\begin{bmatrix} 13 \\ 7 \end{bmatrix}, \begin{bmatrix} 2x_1 - x_2 \\ 5x_1 + 6x_2 \end{bmatrix}$

21–29. If you consult your *Study Guide* before you make a good effort to answer the true-false questions, you will destroy most of their value.

31.

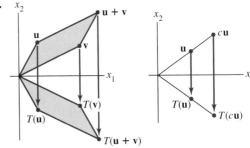

33. *Hint:* Show that the image of a line (that is, the set of images of all points on a line) can be represented by the parametric equation of a line.

35. a. The line through $\mathbf{p}$ and $\mathbf{q}$ is parallel to $\mathbf{q} - \mathbf{p}$. (See Exercises 25 and 26 in Section 1.5.) Since $\mathbf{p}$ is on the line, the equation of the line is $\mathbf{x} = \mathbf{p} + t(\mathbf{q} - \mathbf{p})$. Rewrite this as $\mathbf{x} = \mathbf{p} - t\mathbf{p} + t\mathbf{q}$ and $\mathbf{x} = (1 - t)\mathbf{p} + t\mathbf{q}$.

 b. Consider $\mathbf{x} = (1 - t)\mathbf{p} + t\mathbf{q}$ for t such that $0 \leq t \leq 1$. Then, by linearity of T, for $0 \leq t \leq 1$

$$T(\mathbf{x}) = T((1 - t)\mathbf{p} + t\mathbf{q}) = (1 - t)T(\mathbf{p}) + tT(\mathbf{q}) \quad (*)$$

 If $T(\mathbf{p})$ and $T(\mathbf{q})$ are distinct, then (*) is the equation for the line segment between $T(\mathbf{p})$ and $T(\mathbf{q})$, as shown in part (a). Otherwise, the set of images is just the single point $T(\mathbf{p})$, because

$$(1 - t)T(\mathbf{p}) + tT(\mathbf{q}) = (1 - t)T(\mathbf{p}) + tT(\mathbf{p}) = T(\mathbf{p})$$

37. a. When $b = 0$, $f(x) = mx$. In this case, for all x, y in $\mathbb{R}$ and all scalars c and d,

$$\begin{aligned} f(cx + dy) &= m(cx + dy) = mcx + mdy \\ &= c(mx) + d(my) = c \cdot f(x) + d \cdot f(y) \end{aligned}$$

 This shows that f is linear.

 b. When $f(x) = mx + b$, with b nonzero, $f(0) = m(0) + b = b \neq 0$.

c. In calculus, f is called a "linear function" because the graph of f is a line.

39. *Hint:* Since $\{\mathbf{v}_1, \mathbf{v}_2, \mathbf{v}_3\}$ is linearly dependent, you can write a certain equation and work with it.

41. One possibility is to show that T does not map the zero vector into the zero vector, something that every linear transformation *does* do: $T(0,0) = (0,4,0)$.

43. Take $\mathbf{u}$ and $\mathbf{v}$ in $\mathbb{R}^3$ and let c and d be scalars. Then

$$c\mathbf{u} + d\mathbf{v} = (cu_1 + dv_1, cu_2 + dv_2, cu_3 + dv_3)$$

The transformation T is linear because

$$
\begin{aligned}
T(c\mathbf{u} + d\mathbf{v}) &= (cu_1 + dv_1, cu_2 + dv_2, -(cu_3 + dv_3)) \\
&= (cu_1 + dv_1, cu_2 + dv_2, -cu_3 - dv_3) \\
&= (cu_1, cu_2, -cu_3) + (dv_1, dv_2, -dv_3) \\
&= c(u_1, u_2, -u_3) + d(v_1, v_2, -v_3) \\
&= cT(\mathbf{u}) + dT(\mathbf{v})
\end{aligned}
$$

45. All multiples of $(7, 9, 0, 2)$

47. Yes. One choice for $\mathbf{x}$ is $(4, 7, 1, 0)$.

Section 1.9, page 82

1. $\begin{bmatrix} 2 & -5 \\ 1 & 2 \\ 2 & 0 \\ 1 & 0 \end{bmatrix}$ **3.** $\begin{bmatrix} 0 & 1 \\ -1 & 0 \end{bmatrix}$ **5.** $\begin{bmatrix} 1 & 0 \\ -2 & 1 \end{bmatrix}$

7. $\begin{bmatrix} -1/\sqrt{2} & 1/\sqrt{2} \\ 1/\sqrt{2} & 1/\sqrt{2} \end{bmatrix}$ **9.** $\begin{bmatrix} 0 & -1 \\ -1 & 3 \end{bmatrix}$

11. The described transformation T maps $\mathbf{e}_1$ into $-\mathbf{e}_1$ and maps $\mathbf{e}_2$ into $-\mathbf{e}_2$. A rotation through π radians also maps $\mathbf{e}_1$ into $-\mathbf{e}_1$ and maps $\mathbf{e}_2$ into $-\mathbf{e}_2$. Since a linear transformation is completely determined by what it does to the columns of the identity matrix, the rotation transformation has the same effect as T on every vector in $\mathbb{R}^2$.

13.

15. $\begin{bmatrix} 2 & 0 & -3 \\ 4 & 0 & 0 \\ 1 & -1 & 1 \end{bmatrix}$ **17.** $\begin{bmatrix} 0 & 0 & 0 & 0 \\ 1 & 1 & 0 & 0 \\ 0 & 1 & 1 & 0 \\ 0 & 0 & 1 & 1 \end{bmatrix}$

19. $\begin{bmatrix} 1 & -5 & 4 \\ 0 & 1 & -6 \end{bmatrix}$ **21.** $\mathbf{x} = \begin{bmatrix} 7 \\ -4 \end{bmatrix}$

23–31. Read the text carefully, and write your answers before you consult the *Study Guide*. Remember, a statement is true only if it is true in all cases.

Justify your answers to Exercises 33–35.

33. Not one-to-one and does not map $\mathbb{R}^4$ onto $\mathbb{R}^4$

35. Not one-to-one but maps $\mathbb{R}^3$ onto $\mathbb{R}^2$

37. $\begin{bmatrix} \blacksquare & * & * \\ 0 & \blacksquare & * \\ 0 & 0 & \blacksquare \\ 0 & 0 & 0 \end{bmatrix}$

39. n. (Explain why, and then check the *Study Guide*).

41. *Hint:* If $\mathbf{e}_j$ is the jth column of I_n, then $B\mathbf{e}_j$ is the jth column of B.

43. *Hint:* Is it possible that $m > n$? What about $m < n$?

45. No. (Explain why.)

47. No. (Explain why.)

Section 1.10, page 91

1. a. $x_1 \begin{bmatrix} 110 \\ 4 \\ 20 \\ 2 \end{bmatrix} + x_2 \begin{bmatrix} 130 \\ 3 \\ 18 \\ 5 \end{bmatrix} = \begin{bmatrix} 295 \\ 9 \\ 48 \\ 8 \end{bmatrix}$, where x_1 is the number of servings of Cheerios and x_2 is the number of servings of 100% Natural Cereal.

b. $\begin{bmatrix} 110 & 130 \\ 4 & 3 \\ 20 & 18 \\ 2 & 5 \end{bmatrix} \begin{bmatrix} x_1 \\ x_2 \end{bmatrix} = \begin{bmatrix} 295 \\ 9 \\ 48 \\ 8 \end{bmatrix}$. Mix 1.5 servings of Cheerios together with 1 serving of 100% Natural Cereal.

3. a. She should mix .99 serving of Mac and Cheese, 1.54 servings of broccoli, and .79 serving of chicken to get her desired nutritional content.

b. She should mix 1.09 servings of shells and white cheddar, .88 serving of broccoli, and 1.03 servings of chicken to get her desired nutritional content. Notice that this mix contains significantly less broccoli, so she should like it better.

5. $R\mathbf{i} = \mathbf{v}$, $\begin{bmatrix} 11 & -5 & 0 & 0 \\ -5 & 10 & -1 & 0 \\ 0 & -1 & 9 & -2 \\ 0 & 0 & -2 & 10 \end{bmatrix} \begin{bmatrix} I_1 \\ I_2 \\ I_3 \\ I_4 \end{bmatrix} = \begin{bmatrix} 50 \\ -40 \\ 30 \\ -30 \end{bmatrix}$

$\mathbf{i} = \begin{bmatrix} I_1 \\ I_2 \\ I_3 \\ I_4 \end{bmatrix} = \begin{bmatrix} 3.68 \\ -1.90 \\ 2.57 \\ -2.49 \end{bmatrix}$

7. $R\mathbf{i} = \mathbf{v}$, $\begin{bmatrix} 12 & -7 & 0 & -4 \\ -7 & 15 & -6 & 0 \\ 0 & -6 & 14 & -5 \\ -4 & 0 & -5 & 13 \end{bmatrix} \begin{bmatrix} I_1 \\ I_2 \\ I_3 \\ I_4 \end{bmatrix} = \begin{bmatrix} 40 \\ 30 \\ 20 \\ -10 \end{bmatrix}$

$\mathbf{i} = \begin{bmatrix} I_1 \\ I_2 \\ I_3 \\ I_4 \end{bmatrix} = \begin{bmatrix} 11.43 \\ 10.55 \\ 8.04 \\ 5.84 \end{bmatrix}$

9. $\mathbf{x}_{k+1} = M\mathbf{x}_k$ for $k = 0, 1, 2, \ldots$, where

$$M = \begin{bmatrix} .93 & .05 \\ .07 & .95 \end{bmatrix} \quad \text{and} \quad \mathbf{x}_0 = \begin{bmatrix} 800{,}000 \\ 500{,}000 \end{bmatrix}.$$

The population in 2022 (for $k = 2$) is $\mathbf{x}_2 = \begin{bmatrix} 741{,}720 \\ 558{,}280 \end{bmatrix}$.

11. 32 in Pullman, 76 in Spokane, and 212 in Seattle.

13. a. The population of the city decreases. After 7 years, the populations are about equal, but the city population continues to decline. After 20 years, there are only 417,000 persons in the city (417,456 rounded off). However, the changes in population seem to grow smaller each year.

b. The city population is increasing slowly, and the suburban population is decreasing. After 20 years, the city population has grown from 350,000 to about 370,000.

Chapter 1 Supplementary Exercises, page 93

1. F   **2.** F   **3.** T   **4.** F   **5.** T

6. T   **7.** F   **8.** F   **9.** T   **10.** F

11. T   **12.** F   **13.** T   **14.** T   **15.** T

16. T   **17.** F   **18.** T   **19.** F   **20.** T

21. F   **22.** F   **23.** F   **24.** T   **25.** T

27. a. Any consistent linear system whose echelon form is

$$\begin{bmatrix} \blacksquare & * & * & * \\ 0 & \blacksquare & * & * \\ 0 & 0 & 0 & 0 \end{bmatrix} \text{ or } \begin{bmatrix} \blacksquare & * & * & * \\ 0 & 0 & \blacksquare & * \\ 0 & 0 & 0 & 0 \end{bmatrix}$$

$$\text{or } \begin{bmatrix} 0 & \blacksquare & * & * \\ 0 & 0 & \blacksquare & * \\ 0 & 0 & 0 & 0 \end{bmatrix}$$

b. Any consistent linear system whose reduced echelon form is I_3.

c. Any inconsistent linear system of three equations in three variables.

29. a. The solution set: (i) is empty if $h = 12$ and $k \neq 2$; (ii) contains a unique solution if $h \neq 12$; (iii) contains infinitely many solutions if $h = 12$ and $k = 2$.

b. The solution set is empty if $k + 3h = 0$; otherwise, the solution set contains a unique solution.

31. a. Set $\mathbf{v}_1 = \begin{bmatrix} 2 \\ -5 \\ 7 \end{bmatrix}$, $\mathbf{v}_2 = \begin{bmatrix} -4 \\ 1 \\ -5 \end{bmatrix}$, $\mathbf{v}_3 = \begin{bmatrix} -2 \\ 1 \\ -3 \end{bmatrix}$, and

$\mathbf{b} = \begin{bmatrix} b_1 \\ b_2 \\ b_3 \end{bmatrix}$. "Determine if $\mathbf{v}_1, \mathbf{v}_2, \mathbf{v}_3$ span $\mathbb{R}^3$."

Solution: No.

b. Set $A = \begin{bmatrix} 2 & -4 & -2 \\ -5 & 1 & 1 \\ 7 & -5 & -3 \end{bmatrix}$. "Determine if the columns of A span $\mathbb{R}^3$."

c. Define $T(\mathbf{x}) = A\mathbf{x}$. "Determine if T maps $\mathbb{R}^3$ onto $\mathbb{R}^3$."

33. $\begin{bmatrix} 5 \\ 6 \end{bmatrix} = \frac{4}{3}\begin{bmatrix} 2 \\ 1 \end{bmatrix} + \frac{7}{3}\begin{bmatrix} 1 \\ 2 \end{bmatrix}$ or $\begin{bmatrix} 5 \\ 6 \end{bmatrix} =$

$\begin{bmatrix} 8/3 \\ 4/3 \end{bmatrix} + \begin{bmatrix} 7/3 \\ 14/3 \end{bmatrix}$

34. *Hint:* Construct a "grid" on the $x_1 x_2$-plane determined by $\mathbf{a}_1$ and $\mathbf{a}_2$.

35. A solution set is a line when the system has one free variable. If the coefficient matrix is 2×3, then two of the columns should be pivot columns. For instance, take $\begin{bmatrix} 1 & 2 & * \\ 0 & 3 & * \end{bmatrix}$. Put anything in column 3. The resulting matrix will be in echelon form. Make one row replacement operation on the second row to create a matrix *not* in echelon form, such as $\begin{bmatrix} 1 & 2 & 1 \\ 0 & 3 & 1 \end{bmatrix} \sim \begin{bmatrix} 1 & 2 & 1 \\ 1 & 5 & 2 \end{bmatrix}$.

36. *Hint:* How many free variables are in the equation $A\mathbf{x} = \mathbf{0}$?

37. $E = \begin{bmatrix} 1 & 0 & -3 \\ 0 & 1 & 2 \\ 0 & 0 & 0 \end{bmatrix}$

39. a. If the three vectors are linearly independent, then a, c, and f must all be nonzero.

b. The numbers $a, \ldots, f$ can have any values.

40. *Hint:* List the columns from right to left as $\mathbf{v}_1, \ldots, \mathbf{v}_4$.

41. *Hint:* Use Theorem 7.

43. Let M be the line through the origin that is parallel to the line through $\mathbf{v}_1, \mathbf{v}_2$, and $\mathbf{v}_3$. Then $\mathbf{v}_2 - \mathbf{v}_1$ and $\mathbf{v}_3 - \mathbf{v}_1$ are both on M. So one of these two vectors is a multiple of the other, say $\mathbf{v}_2 - \mathbf{v}_1 = k(\mathbf{v}_3 - \mathbf{v}_1)$. This equation produces a linear dependence relation: $(k - 1)\mathbf{v}_1 + \mathbf{v}_2 - k\mathbf{v}_3 = \mathbf{0}$.

A second solution: A parametric equation of the line is $\mathbf{x} = \mathbf{v}_1 + t(\mathbf{v}_2 - \mathbf{v}_1)$. Since $\mathbf{v}_3$ is on the line, there is some t_0 such that $\mathbf{v}_3 = \mathbf{v}_1 + t_0(\mathbf{v}_2 - \mathbf{v}_1) = (1 - t_0)\mathbf{v}_1 + t_0\mathbf{v}_2$. So $\mathbf{v}_3$ is a linear combination of $\mathbf{v}_1$ and $\mathbf{v}_2$, and $\{\mathbf{v}_1, \mathbf{v}_2, \mathbf{v}_3\}$ is linearly dependent.

45. $\begin{bmatrix} 1 & 0 & 0 \\ 0 & -1 & 0 \\ 0 & 0 & 1 \end{bmatrix}$ **47.** $a = 4/5$ and $b = -3/5$

49. a. The vector lists the number of three-, two-, and one-bedroom apartments provided when x_1 floors of plan A are constructed.

b. $x_1\begin{bmatrix} 3 \\ 7 \\ 8 \end{bmatrix} + x_2\begin{bmatrix} 4 \\ 4 \\ 8 \end{bmatrix} + x_3\begin{bmatrix} 5 \\ 3 \\ 9 \end{bmatrix}$

c. Use 2 floors of plan A and 15 floors of plan B. Or, use 6 floors of plan A, 2 floors of plan B, and 8 floors of plan C. These are the only feasible solutions. There are other mathematical solutions, but they require a negative number of floors of one or two of the plans, which makes no physical sense.

Chapter 2

Section 2.1, page 108

1. $\begin{bmatrix} -4 & 0 & 2 \\ -8 & 6 & -4 \end{bmatrix}$, $\begin{bmatrix} 3 & -5 & 3 \\ -7 & 2 & -7 \end{bmatrix}$, not defined,

$\begin{bmatrix} 1 & 13 \\ -7 & -6 \end{bmatrix}$

3. $\begin{bmatrix} -1 & 1 \\ -5 & 5 \end{bmatrix}$, $\begin{bmatrix} 12 & -3 \\ 15 & -6 \end{bmatrix}$

5. a. $A\mathbf{b}_1 = \begin{bmatrix} -7 \\ 7 \\ 12 \end{bmatrix}$, $A\mathbf{b}_2 = \begin{bmatrix} 6 \\ -16 \\ -11 \end{bmatrix}$,

$AB = \begin{bmatrix} -7 & 6 \\ 7 & -16 \\ 12 & -11 \end{bmatrix}$

b. $AB = \begin{bmatrix} -1(3)+2(-2) & -1(-4)+2(1) \\ 5(3)+4(-2) & 5(-4)+4(1) \\ 2(3)-3(-2) & 2(-4)-3(1) \end{bmatrix}$

$= \begin{bmatrix} -7 & 6 \\ 7 & -16 \\ 12 & -11 \end{bmatrix}$

7. 3×7 **9.** $k = 5$

11. $AD = \begin{bmatrix} 2 & 3 & 5 \\ 2 & 6 & 15 \\ 2 & 12 & 25 \end{bmatrix}$, $DA = \begin{bmatrix} 2 & 2 & 2 \\ 3 & 6 & 9 \\ 5 & 20 & 25 \end{bmatrix}$

Right-multiplication (that is, multiplication on the right) by D multiplies each *column* of A by the corresponding diagonal entry of D. Left-multiplication by D multiplies each *row* of A by the corresponding diagonal entry of D. The *Study Guide* tells how to make $AB = BA$, but you should try this yourself before looking there.

13. *Hint:* One of the two matrices is Q.

15–23. Answer the questions before looking in the *Study Guide*.

25. $\mathbf{b}_1 = \begin{bmatrix} 7 \\ 4 \end{bmatrix}$, $\mathbf{b}_2 = \begin{bmatrix} -8 \\ -5 \end{bmatrix}$

27. The third column of AB is the sum of the first two columns of AB. Here's why. Write $B = [\ \mathbf{b}_1 \quad \mathbf{b}_2 \quad \mathbf{b}_3\ ]$. By definition, the third column of AB is $A\mathbf{b}_3$. If $\mathbf{b}_3 = \mathbf{b}_1 + \mathbf{b}_2$, then $A\mathbf{b}_3 = A(\mathbf{b}_1 + \mathbf{b}_2) = A\mathbf{b}_1 + A\mathbf{b}_2$, by a property of matrix-vector multiplication.

29. The columns of A are linearly dependent. Why?

31. *Hint:* Suppose $\mathbf{x}$ satisfies $A\mathbf{x} = \mathbf{0}$, and show that $\mathbf{x}$ must be $\mathbf{0}$.

33. *Hint:* Use the results of Exercises 31 and 32, and apply the associative law of multiplication to the product CAD.

35. $\mathbf{u}^T\mathbf{v} = \mathbf{v}^T\mathbf{u} = -2a + 3b - 4c$,

$\mathbf{u}\mathbf{v}^T = \begin{bmatrix} -2a & -2b & -2c \\ 3a & 3b & 3c \\ -4a & -4b & -4c \end{bmatrix}$,

$\mathbf{v}\mathbf{u}^T = \begin{bmatrix} -2a & 3a & -4a \\ -2b & 3b & -4b \\ -2c & 3c & -4c \end{bmatrix}$

37. *Hint:* For Theorem 2(b), show that the (i, j)-entry of $A(B + C)$ equals the (i, j)-entry of $AB + AC$.

39. *Hint:* Use the definition of the product $I_m A$ and the fact that $I_m\mathbf{x} = \mathbf{x}$ for $\mathbf{x}$ in $\mathbb{R}^m$.

41. *Hint:* First write the (i, j)-entry of $(AB)^T$, which is the (j, i)-entry of AB. Then, to compute the (i, j)-entry in $B^T A^T$, use the facts that the entries in row i of B^T are $b_{1i}, \ldots, b_{ni}$, because they come from column i of B, and the entries in column j of A^T are $a_{j1}, \ldots, a_{jn}$, because they come from row j of A.

43. The answer here depends on the choice of matrix program. For MATLAB, use the `help` command to read about `zeros`, `ones`, `eye`, and `diag`.

45. Display your results and report your conclusions.

47. The matrix S "shifts" the entries in a vector (a, b, c, d, e) to yield $(b, c, d, e, 0)$. S^5 is the 5×5 zero matrix. So is S^6.

49. $x = \begin{bmatrix} 1 \\ 0 \\ 1 \\ 0 \end{bmatrix}$

51. $\begin{bmatrix} 1 & 1 & 1 & 1 & 1 & 1 & 2 & 2 & 2 & 2 \\ 2 & 3 & 16 & 24 & 25 & 26 & 6 & 7 & 19 & 26 \end{bmatrix}$

Section 2.2, page 118

1. $\begin{bmatrix} 2 & -3 \\ -5 & 8 \end{bmatrix}$ **3.** $\dfrac{1}{3}\begin{bmatrix} 3 & 3 \\ -7 & -8 \end{bmatrix}$ or $\begin{bmatrix} 1 & 1 \\ -7/3 & -8/3 \end{bmatrix}$

5. $\begin{bmatrix} 8 & 3 \\ 5 & 2 \end{bmatrix}\begin{bmatrix} 2 & -3 \\ -5 & 8 \end{bmatrix} = \begin{bmatrix} 1 & 0 \\ 0 & 1 \end{bmatrix}$

7. $x_1 = 7$ and $x_2 = -18$

9. a and b: $\begin{bmatrix} -9 \\ 4 \end{bmatrix}$, $\begin{bmatrix} 11 \\ -5 \end{bmatrix}$, $\begin{bmatrix} 6 \\ -2 \end{bmatrix}$, and $\begin{bmatrix} 13 \\ -5 \end{bmatrix}$

11–19. Write out your answers before checking the *Study Guide*.

21. The proof can be modeled after the proof of Theorem 5.

23. $AB = AC \Rightarrow A^{-1}AB = A^{-1}AC \Rightarrow IB = IC \Rightarrow B = C$. No, in general, B and C can be different when A is not invertible. See Exercise 10 in Section 2.1.

25. $D = C^{-1}B^{-1}A^{-1}$. Show that D works.

27. $A = BCB^{-1}$

29. After you find $X = CB - A$, show that X is a solution.

31. *Hint:* Consider the equation $A\mathbf{x} = \mathbf{0}$.

33. *Hint:* If $A\mathbf{x} = \mathbf{0}$ has only the trivial solution, then there are no free variables in the equation $A\mathbf{x} = \mathbf{0}$, and each column of A is a pivot column.

35. *Hint:* Consider the case $a = b = 0$. Then consider the vector $\begin{bmatrix} -b \\ a \end{bmatrix}$, and use the fact that $ad - bc = 0$.

37. *Hint:* For part (a), interchange A and B in the box following Example 6 in Section 2.1, and then replace B by the identity matrix. For parts (b) and (c), begin by writing

$$A = \begin{bmatrix} \text{row}_1(A) \\ \text{row}_2(A) \\ \text{row}_3(A) \end{bmatrix}$$

39. $\begin{bmatrix} -7 & 2 \\ 4 & -1 \end{bmatrix}$ **41.** $\begin{bmatrix} 8 & 3 & 1 \\ 10 & 4 & 1 \\ 7/2 & 3/2 & 1/2 \end{bmatrix}$

43. $A^{-1} = B = \begin{bmatrix} 1 & 0 & 0 & \cdots & 0 \\ -1 & 1 & 0 & & 0 \\ 0 & -1 & 1 & & \\ \vdots & & & \ddots & \vdots \\ 0 & 0 & \cdots & -1 & 1 \end{bmatrix}$. *Hint:* For

$j = 1, \ldots, n$, let $\mathbf{a}_j$, $\mathbf{b}_j$, and $\mathbf{e}_j$ denote the jth columns of A, B, and I, respectively. Use the facts that $\mathbf{a}_j - \mathbf{a}_{j+1} = \mathbf{e}_j$ and $\mathbf{b}_j = \mathbf{e}_j - \mathbf{e}_{j+1}$ for $j = 1, \ldots, n - 1$, and $\mathbf{a}_n = \mathbf{b}_n = \mathbf{e}_n$.

45. $\begin{bmatrix} 3 \\ -6 \\ 4 \end{bmatrix}$. Find this by row reducing $[\ A \quad \mathbf{e}_3\ ]$.

47. $C = \begin{bmatrix} 1 & 1 & -1 \\ -1 & 1 & 0 \end{bmatrix}$

49. .27, .30, and .23 inch, respectively

51. 12, 1.5, 21.5, and 12 newtons, respectively

Section 2.3, page 124
The abbreviation IMT (here and in the *Study Guide*) denotes the Invertible Matrix Theorem (Theorem 8).

1. Invertible, by the IMT. Neither column of the matrix is a multiple of the other column, so they are linearly independent. Also, the matrix is invertible by Theorem 4 in Section 2.2 because the determinant is nonzero.

3. Invertible, by the IMT. The matrix row reduces to $\begin{bmatrix} 5 & 0 & 0 \\ 0 & -7 & 0 \\ 0 & 0 & -1 \end{bmatrix}$ and has 3 pivot positions.

5. Not invertible, by the IMT. The matrix row reduces to $\begin{bmatrix} 1 & 0 & 2 \\ 0 & 3 & -5 \\ 0 & 0 & 0 \end{bmatrix}$ and is not row equivalent to I_3.

7. Invertible, by the IMT. The matrix row reduces to $\begin{bmatrix} -1 & -3 & 0 & 1 \\ 0 & -4 & 8 & 0 \\ 0 & 0 & 3 & 0 \\ 0 & 0 & 0 & 1 \end{bmatrix}$ and has four pivot positions.

9. The 4×4 matrix has four pivot positions, so it is invertible by the IMT.

11–19. The *Study Guide* will help, but first try to answer the questions based on your careful reading of the text.

21. A square upper triangular matrix is invertible if and only if all the entries on the diagonal are nonzero. Why?

Note: The answers below for Exercises 15–29 mention the IMT. In many cases, part or all of an acceptable answer could also be based on results that were used to establish the IMT.

23. If A has two identical columns then its columns are linearly dependent. Part (e) of the IMT shows that A cannot be invertible.

25. If A is invertible, so is A^{-1}, by Theorem 6 in Section 2.2. By (e) of the IMT applied to A^{-1}, the columns of A^{-1} are linearly independent.

27. By (e) of the IMT, D is invertible. Thus the equation $D\mathbf{x} = \mathbf{b}$ has a solution for each $\mathbf{b}$ in $\mathbb{R}^7$, by (g) of the IMT. Can you say more?

29. The matrix G cannot be invertible, by Theorem 5 in Section 2.2 or by the paragraph following the IMT. So (g) of the IMT is false and so is (h). The columns of G do not span $\mathbb{R}^n$.

31. Statement (b) of the IMT is false for K, so statements (e) and (h) are also false. That is, the columns of K are linearly *dependent* and the columns do *not* span $\mathbb{R}^n$.

33. *Hint:* Use the IMT first.

35. Let W be the inverse of AB. Then $ABW = I$ and $A(BW) = I$. Unfortunately, this equation by itself does not prove that A is invertible. Why not? Finish the proof before you check the *Study Guide*.

37. Since the transformation $\mathbf{x} \mapsto A\mathbf{x}$ is not one-to-one, statement (f) of the IMT is false. Then (i) is also false and the transformation $\mathbf{x} \mapsto A\mathbf{x}$ does not map $\mathbb{R}^n$ onto $\mathbb{R}^n$. Also, A is not invertible, which implies that the transformation $\mathbf{x} \mapsto A\mathbf{x}$ is not invertible, by Theorem 9.

39. *Hint:* If the equation $A\mathbf{x} = \mathbf{b}$ has a solution for each $\mathbf{b}$, then A has a pivot in each row (Theorem 4 in Section 1.4). Could there be free variables in an equation $A\mathbf{x} = \mathbf{b}$?

41. *Hint:* First show that the standard matrix of T is invertible. Then use a theorem or theorems to show that $T^{-1}(\mathbf{x}) = B\mathbf{x}$, where $B = \begin{bmatrix} 7 & 9 \\ 4 & 5 \end{bmatrix}$.

43. *Hint:* To show that T is one-to-one, suppose that $T(\mathbf{u}) = T(\mathbf{v})$ for some vectors $\mathbf{u}$ and $\mathbf{v}$ in $\mathbb{R}^n$. Deduce that $\mathbf{u} = \mathbf{v}$. To show that T is onto, suppose $\mathbf{y}$ represents an arbitrary vector in $\mathbb{R}^n$ and use the inverse S to produce an $\mathbf{x}$ such that $T(\mathbf{x}) = \mathbf{y}$. A second proof can be given using Theorem 9 together with a theorem from Section 1.9.

45. *Hint:* Consider the standard matrices of T and U.

47. Given any $\mathbf{v}$ in $\mathbb{R}^n$, we may write $\mathbf{v} = T(\mathbf{x})$ for some $\mathbf{x}$, because T is an onto mapping. Then, the assumed properties of S and U show that $S(\mathbf{v}) = S(T(\mathbf{x})) = \mathbf{x}$ and $U(\mathbf{v}) = U(T(\mathbf{x})) = \mathbf{x}$. So $S(\mathbf{v})$ and $U(\mathbf{v})$ are equal for each $\mathbf{v}$. That is, S and U are the same function from $\mathbb{R}^n$ into $\mathbb{R}^n$.

49. a. The exact solution of (3) is $x_1 = 3.94$ and $x_2 = .49$. The exact solution of (4) is $x_1 = 2.90$ and $x_2 = 2.00$.

   b. When the solution of (4) is used as an approximation for the solution in (3), the error in using the value of 2.90 for x_1 is about 26%, and the error in using 2.0 for x_2 is about 308%.

   c. The condition number of the coefficient matrix is 3363. The percentage change in the solution from (3) to (4) is about 7700 times the percentage change in the right side of the equation. This is the same order of magnitude as the condition number. The condition number gives a rough measure of how sensitive the solution of $A\mathbf{x} = \mathbf{b}$ can be to changes in $\mathbf{b}$. Further information about the condition number is given at the end of Chapter 6 and in Chapter 7.

51. $\text{cond}(A) \approx 69{,}000$, which is between 10^4 and 10^5. So about 4 or 5 digits of accuracy may be lost. Several experiments with MATLAB should verify that $\mathbf{x}$ and $\mathbf{x}_1$ agree to 11 or 12 digits.

53. Some versions of MATLAB issue a warning when asked to invert a Hilbert matrix of order about 12 or larger using floating-point arithmetic. The product AA^{-1} should have several off-diagonal entries that are far from being zero. If not, try a larger matrix.

Section 2.4, page 130

1. $\begin{bmatrix} A & B \\ EA + C & EB + D \end{bmatrix}$ **3.** $\begin{bmatrix} Y & Z \\ W & X \end{bmatrix}$

5. $Y = B^{-1}$ (explain why), $X = -B^{-1}A$, $Z = C$

7. $X = A^{-1}$ (why?), $Y = -BA^{-1}$, $Z = 0$ (why?)

9. $X = -A_{21}A_{11}^{-1}$, $Y = -A_{31}A_{11}^{-1}$, $B_{22} = A_{22} - A_{21}A_{11}^{-1}A_{12}$

11–13. You can check your answers in the *Study Guide*.

15. *Hint:* Suppose A is invertible, and let $A^{-1} = \begin{bmatrix} D & E \\ F & G \end{bmatrix}$. Show that $BD = I$ and $CG = I$. This implies that B and C are invertible. (Explain why!) Conversely, suppose B and C are invertible. To prove that A is invertible, guess what A^{-1} must be and check that it works.

17. $\begin{bmatrix} A_{11} & A_{12} \\ A_{21} & A_{22} \end{bmatrix} = $

$\begin{bmatrix} I & 0 \\ A_{21}A_{11}^{-1} & I \end{bmatrix} \begin{bmatrix} A_{11} & 0 \\ 0 & S \end{bmatrix} \begin{bmatrix} I & A_{11}^{-1}A_{12} \\ 0 & I \end{bmatrix}$

with $S = A_{22} - A_{21}A_{11}^{-1}A_{12}$.

19. $G_{k+1} = \begin{bmatrix} X_k & \mathbf{x}_{k+1} \end{bmatrix} \begin{bmatrix} X_k^T \\ \mathbf{x}_{k+1}^T \end{bmatrix} = X_k X_k^T + \mathbf{x}_{k+1}\mathbf{x}_{k+1}^T$
$= G_k + \mathbf{x}_{k+1}\mathbf{x}_{k+1}^T$

Only the outer product matrix $\mathbf{x}_{k+1}\mathbf{x}_{k+1}^T$ needs to be computed (and then added to G_k).

21. $W(s) = I_m - C(A - sI_n)^{-1}B$. This is the Schur complement of $A - sI_n$ in the system matrix.

23. a. $A^2 = \begin{bmatrix} 1 & 0 \\ 3 & -1 \end{bmatrix}\begin{bmatrix} 1 & 0 \\ 3 & -1 \end{bmatrix}$

$= \begin{bmatrix} 1 + 0 & 0 + 0 \\ 3 - 3 & 0 + (-1)^2 \end{bmatrix} = \begin{bmatrix} 1 & 0 \\ 0 & 1 \end{bmatrix}$

   b. $M^2 = \begin{bmatrix} A & 0 \\ I & -A \end{bmatrix}\begin{bmatrix} A & 0 \\ I & -A \end{bmatrix}$

$= \begin{bmatrix} A^2 + 0 & 0 + 0 \\ A - A & 0 + (-A)^2 \end{bmatrix} = \begin{bmatrix} I & 0 \\ 0 & I \end{bmatrix}$

25. If A_1 and B_1 are $(k + 1) \times (k + 1)$ and lower triangular, then we can write $A_1 = \begin{bmatrix} a & \mathbf{0}^T \\ \mathbf{v} & A \end{bmatrix}$ and $B_1 = \begin{bmatrix} b & \mathbf{0}^T \\ \mathbf{w} & B \end{bmatrix}$, where A and B are $k \times k$ and lower triangular, $\mathbf{v}$ and $\mathbf{w}$ are in $\mathbb{R}^k$, and a and b are suitable scalars. Assume that the product of $k \times k$ lower triangular matrices is lower triangular, and compute the product $A_1 B_1$. What do you conclude?

27. Use Example 5 to find the inverse of a matrix of the form $B = \begin{bmatrix} B_{11} & 0 \\ 0 & B_{22} \end{bmatrix}$, where B_{11} is $p \times p$, B_{22} is $q \times q$ and B is invertible. Partition the matrix A, and apply your result twice to find that

$A^{-1} = \begin{bmatrix} -5 & 2 & 0 & 0 & 0 \\ 3 & -1 & 0 & 0 & 0 \\ 0 & 0 & 1/2 & 0 & 0 \\ 0 & 0 & 0 & 3 & -4 \\ 0 & 0 & 0 & -5/2 & 7/2 \end{bmatrix}$

29. a, b. The commands to be used in these exercises will depend on the matrix program.

   c. The algebra needed comes from the block matrix equation

$\begin{bmatrix} A_{11} & 0 \\ A_{21} & A_{22} \end{bmatrix}\begin{bmatrix} \mathbf{x}_1 \\ \mathbf{x}_2 \end{bmatrix} = \begin{bmatrix} \mathbf{b}_1 \\ \mathbf{b}_2 \end{bmatrix}$

where $\mathbf{x}_1$ and $\mathbf{b}_1$ are in $\mathbb{R}^{20}$ and $\mathbf{x}_2$ and $\mathbf{b}_2$ are in $\mathbb{R}^{30}$. Then $A_{11}\mathbf{x}_1 = \mathbf{b}_1$, which can be solved to produce $\mathbf{x}_1$. The equation $A_{21}\mathbf{x}_1 + A_{22}\mathbf{x}_2 = \mathbf{b}_2$ yields $A_{22}\mathbf{x}_2 = \mathbf{b}_2 - A_{21}\mathbf{x}_1$, which can be solved for $\mathbf{x}_2$ by row reducing the matrix $[A_{22} \ \ \mathbf{c}]$, where $\mathbf{c} = \mathbf{b}_2 - A_{21}\mathbf{x}_1$.

Section 2.5, page 138

1. $L\mathbf{y} = \mathbf{b} \Rightarrow \mathbf{y} = \begin{bmatrix} -7 \\ -2 \\ 6 \end{bmatrix}$, $U\mathbf{x} = \mathbf{y} \Rightarrow \mathbf{x} = \begin{bmatrix} 3 \\ 4 \\ -6 \end{bmatrix}$

3. $\mathbf{y} = \begin{bmatrix} 1 \\ 3 \\ 3 \end{bmatrix}$, $\mathbf{x} = \begin{bmatrix} -1 \\ 3 \\ 3 \end{bmatrix}$ **5.** $\mathbf{y} = \begin{bmatrix} 1 \\ 5 \\ 1 \\ -3 \end{bmatrix}$, $\mathbf{x} = \begin{bmatrix} -2 \\ -1 \\ 2 \\ -3 \end{bmatrix}$

7. $LU = \begin{bmatrix} 1 & 0 \\ -3/2 & 1 \end{bmatrix} \begin{bmatrix} 2 & 5 \\ 0 & 7/2 \end{bmatrix}$

9. $\begin{bmatrix} 1 & 0 & 0 \\ -1 & 1 & 0 \\ 3 & 2/3 & 1 \end{bmatrix} \begin{bmatrix} 3 & -1 & 2 \\ 0 & -3 & 12 \\ 0 & 0 & -8 \end{bmatrix}$

11. $\begin{bmatrix} 1 & 0 & 0 \\ 2 & 1 & 0 \\ -1/3 & 1 & 1 \end{bmatrix} \begin{bmatrix} 3 & -6 & 3 \\ 0 & 5 & -4 \\ 0 & 0 & 5 \end{bmatrix}$

13. $\begin{bmatrix} 1 & 0 & 0 & 0 \\ -1 & 1 & 0 & 0 \\ 4 & 5 & 1 & 0 \\ -2 & -1 & 0 & 1 \end{bmatrix} \begin{bmatrix} 1 & 3 & -5 & -3 \\ 0 & -2 & 3 & 1 \\ 0 & 0 & 0 & 0 \\ 0 & 0 & 0 & 0 \end{bmatrix}$

15. $\begin{bmatrix} 1 & 0 & 0 \\ 3 & 1 & 0 \\ -1/2 & -2 & 1 \end{bmatrix} \begin{bmatrix} 2 & -4 & 4 & -2 \\ 0 & 3 & -5 & 3 \\ 0 & 0 & 0 & 5 \end{bmatrix}$

17. $U^{-1} = \begin{bmatrix} 1/4 & 3/8 & 1/4 \\ 0 & -1/2 & 1/2 \\ 0 & 0 & 1/2 \end{bmatrix}$,

$L^{-1} = \begin{bmatrix} 1 & 0 & 0 \\ 1 & 1 & 0 \\ -2 & 0 & 1 \end{bmatrix}$,

$A^{-1} = \begin{bmatrix} 1/8 & 3/8 & 1/4 \\ -3/2 & -1/2 & 1/2 \\ -1 & 0 & 1/2 \end{bmatrix}$

19. *Hint:* Think about row reducing $\begin{bmatrix} A & I \end{bmatrix}$.

21. *Hint:* Represent the row operations by a sequence of elementary matrices.

23. a. Denote the rows of D as transposes of column vectors. Then partitioned matrix multiplication yields

$$A = CD = \begin{bmatrix} \mathbf{c}_1 & \cdots & \mathbf{c}_4 \end{bmatrix} \begin{bmatrix} \mathbf{d}_1^T \\ \vdots \\ \mathbf{d}_4^T \end{bmatrix}$$
$$= \mathbf{c}_1 \mathbf{d}_1^T + \cdots + \mathbf{c}_4 \mathbf{d}_4^T$$

b. A has 40,000 entries. Since C has 1600 entries and D has 400 entries, together they occupy only 5% of the memory needed to store A.

25. Explain why U, D, and V^T are invertible. Then use a theorem on the inverse of a product of invertible matrices.

27. a.

b.

29. a. $\begin{bmatrix} 1 + R_2/R_1 & -R_2 \\ -1/R_1 - R_2/(R_1 R_3) - 1/R_3 & 1 + R_2/R_3 \end{bmatrix}$

b. $A = \begin{bmatrix} 1 & 0 \\ -1/6 & 1 \end{bmatrix} \begin{bmatrix} 1 & -12 \\ 0 & 1 \end{bmatrix} \begin{bmatrix} 1 & 0 \\ -1/36 & 1 \end{bmatrix}$

31. a. $L = \begin{bmatrix} 1 & 0 & 0 & 0 & 0 & 0 & 0 & 0 \\ -.25 & 1 & 0 & 0 & 0 & 0 & 0 & 0 \\ -.25 & -.0667 & 1 & 0 & 0 & 0 & 0 & 0 \\ 0 & -.2667 & -.2857 & 1 & 0 & 0 & 0 & 0 \\ 0 & 0 & -.2679 & -.0833 & 1 & 0 & 0 & 0 \\ 0 & 0 & 0 & -.2917 & -.2921 & 1 & 0 & 0 \\ 0 & 0 & 0 & 0 & -.2697 & -.0861 & 1 & 0 \\ 0 & 0 & 0 & 0 & 0 & -.2948 & -.2931 & 1 \end{bmatrix}$

$U = \begin{bmatrix} 4 & -1 & -1 & 0 & 0 & 0 & 0 & 0 \\ 0 & 3.75 & -.25 & -1 & 0 & 0 & 0 & 0 \\ 0 & 0 & 3.7333 & -1.0667 & -1 & 0 & 0 & 0 \\ 0 & 0 & 0 & 3.4286 & -.2857 & -1 & 0 & 0 \\ 0 & 0 & 0 & 0 & 3.7083 & -1.0833 & -1 & 0 \\ 0 & 0 & 0 & 0 & 0 & 3.3919 & -.2921 & -1 \\ 0 & 0 & 0 & 0 & 0 & 0 & 3.7052 & -1.0861 \\ 0 & 0 & 0 & 0 & 0 & 0 & 0 & 3.3868 \end{bmatrix}$

b. $\mathbf{x} = (3.9569, 6.5885, 4.2392, 7.3971, 5.6029, 8.7608, 9.4115, 12.0431)$

c. $A^{-1} = \begin{bmatrix} .2953 & .0866 & .0945 & .0509 & .0318 & .0227 & .0100 & .0082 \\ .0866 & .2953 & .0509 & .0945 & .0227 & .0318 & .0082 & .0100 \\ .0945 & .0509 & .3271 & .1093 & .1045 & .0591 & .0318 & .0227 \\ .0509 & .0945 & .1093 & .3271 & .0591 & .1045 & .0227 & .0318 \\ .0318 & .0227 & .1045 & .0591 & .3271 & .1093 & .0945 & .0509 \\ .0227 & .0318 & .0591 & .1045 & .1093 & .3271 & .0509 & .0945 \\ .0100 & .0082 & .0318 & .0227 & .0945 & .0509 & .2953 & .0866 \\ .0082 & .0100 & .0227 & .0318 & .0509 & .0945 & .0866 & .2953 \end{bmatrix}$

Obtain A^{-1} directly and then compute $A^{-1} - U^{-1}L^{-1}$ to compare the two methods for inverting a matrix.

Section 2.6, page 145

1. $C = \begin{bmatrix} .10 & .60 & .60 \\ .30 & .20 & 0 \\ .30 & .10 & .10 \end{bmatrix}$, $\left\{ \begin{array}{c} \text{intermediate} \\ \text{demand} \end{array} \right\} = \begin{bmatrix} 60 \\ 20 \\ 10 \end{bmatrix}$

3. $\mathbf{x} = \begin{bmatrix} 40 \\ 15 \\ 15 \end{bmatrix}$ **5.** $\mathbf{x} = \begin{bmatrix} 110 \\ 120 \end{bmatrix}$

7. a. $\begin{bmatrix} 1.6 \\ 1.2 \end{bmatrix}$ **b.** $\begin{bmatrix} 111.6 \\ 121.2 \end{bmatrix}$

9. $\mathbf{x} = \begin{bmatrix} 82.8 \\ 131.0 \\ 110.3 \end{bmatrix}$

11. *Hint:* Use properties of transposes to obtain
$\mathbf{p}^T = \mathbf{p}^T C + \mathbf{v}^T$, so that $\mathbf{p}^T \mathbf{x} = (\mathbf{p}^T C + \mathbf{v}^T)\mathbf{x} = \mathbf{p}^T C \mathbf{x} + \mathbf{v}^T \mathbf{x}$. Now compute $\mathbf{p}^T \mathbf{x}$ from the production equation.

13. $\mathbf{x} = (99576, 97703, 51231, 131570, 49488, 329554, 13835)$. The entries in $\mathbf{x}$ suggest more precision in the answer than is warranted by the entries in $\mathbf{d}$, which appear to be accurate only to perhaps the nearest thousand. So a more realistic answer for $\mathbf{x}$ might be
$\mathbf{x} = 1000 \times (100, 98, 51, 132, 49, 330, 14)$.

15. $\mathbf{x}^{(12)}$ is the first vector whose entries are accurate to the nearest thousand. The calculation of $\mathbf{x}^{(12)}$ takes about 1260 flops, while row reduction of $\begin{bmatrix} (I - C) & \mathbf{d} \end{bmatrix}$ takes only about 550 flops. If C is larger than 20×20, then fewer flops are needed to compute $\mathbf{x}^{(12)}$ by iteration than to compute the equilibrium vector $\mathbf{x}$ by row reduction. As the size of C grows, the advantage of the iterative method increases. Also, because C becomes more sparse for larger models of the economy, fewer iterations are needed for reasonable accuracy.

Section 2.7, page 153

1. $\begin{bmatrix} 1 & .25 & 0 \\ 0 & 1 & 0 \\ 0 & 0 & 1 \end{bmatrix}$ **3.** $\begin{bmatrix} \sqrt{2}/2 & -\sqrt{2}/2 & \sqrt{2} \\ \sqrt{2}/2 & \sqrt{2}/2 & 2\sqrt{2} \\ 0 & 0 & 1 \end{bmatrix}$

5. $\begin{bmatrix} \sqrt{3}/2 & 1/2 & 0 \\ 1/2 & -\sqrt{3}/2 & 0 \\ 0 & 0 & 1 \end{bmatrix}$

7. $\begin{bmatrix} 1/2 & -\sqrt{3}/2 & 3 + 4\sqrt{3} \\ \sqrt{3}/2 & 1/2 & 4 - 3\sqrt{3} \\ 0 & 0 & 1 \end{bmatrix}$
See the Practice Problem.

9. $A(BD)$ requires 1600 multiplications. $(AB)D$ requires 808 multiplications. The first method uses about twice as many multiplications. If D had 20,000 columns, the counts would be 160,000 and 80,008, respectively.

11. Use the fact that
$$\sec \varphi - \tan \varphi \sin \varphi = \frac{1}{\cos \varphi} - \frac{\sin^2 \varphi}{\cos \varphi} = \cos \varphi$$

13. $\begin{bmatrix} A & \mathbf{p} \\ \mathbf{0}^T & 1 \end{bmatrix} = \begin{bmatrix} I & \mathbf{p} \\ \mathbf{0}^T & 1 \end{bmatrix} \begin{bmatrix} A & \mathbf{0} \\ \mathbf{0}^T & 1 \end{bmatrix}$. First apply the linear transformation A, and then translate by $\mathbf{p}$.

15. $(12, -6, 3)$ **17.** $\begin{bmatrix} 1 & 0 & 0 & 0 \\ 0 & 1/2 & -\sqrt{3}/2 & 0 \\ 0 & \sqrt{3}/2 & 1/2 & 0 \\ 0 & 0 & 0 & 1 \end{bmatrix}$

19. The triangle with vertices at $(7, 2, 0), (7.5, 5, 0), (5, 5, 0)$

21. $\begin{bmatrix} 2.2586 & -1.0395 & -.3473 \\ -1.3495 & 2.3441 & .0696 \\ .0910 & -.3046 & 1.2777 \end{bmatrix} \begin{bmatrix} X \\ Y \\ Z \end{bmatrix} = \begin{bmatrix} R \\ G \\ B \end{bmatrix}$

Section 2.8, page 160

1. The set is closed under sums but not under multiplication by a negative scalar. (Sketch an example.)

3. The set is not closed under sums or scalar multiples. The subset consisting of the points on the line $x_2 = x_1$ is a subspace, so any "counterexample" must use at least one point not on this line.

5. No. The system corresponding to $\begin{bmatrix} \mathbf{v}_1 & \mathbf{v}_2 & \mathbf{w} \end{bmatrix}$ is inconsistent.

7. **a.** The three vectors $\mathbf{v}_1, \mathbf{v}_2$, and $\mathbf{v}_3$

  b. Infinitely many vectors

  c. Yes, because $A\mathbf{x} = \mathbf{p}$ has a solution.

9. No, because $A\mathbf{p} \neq \mathbf{0}$.

11. $p = 4$ and $q = 3$. Nul A is a subspace of $\mathbb{R}^4$ because solutions of $A\mathbf{x} = \mathbf{0}$ must have four entries, to match the columns of A. Col A is a subspace of $\mathbb{R}^3$ because each column vector has three entries.

13. For Nul A, choose $(1, -2, 1, 0)$ or $(-1, 4, 0, 1)$, for example. For Col A, select any column of A.

15. Yes. Let A be the matrix whose columns are the vectors given. Then A is invertible because its determinant is nonzero, and so its columns form a basis for $\mathbb{R}^2$, by the IMT (or by Example 5). (Other reasons for the invertibility of A could be given.)

17. Yes. Let A be the matrix whose columns are the vectors given. Row reduction shows three pivots, so A is invertible. By the IMT, the columns of A form a basis for $\mathbb{R}^3$.

19. No. Let A be the 3×2 matrix whose columns are the vectors given. The columns of A cannot possibly span $\mathbb{R}^3$ because A cannot have a pivot in every row. So the columns are not a basis for $\mathbb{R}^3$. (They are a basis for a plane in $\mathbb{R}^3$.)

21–29. Read the section carefully, and write your answers before checking the *Study Guide*. This section has terms and key concepts that you must learn now before going on.

31. Basis for Col A: $\begin{bmatrix} 4 \\ 6 \\ 3 \end{bmatrix}, \begin{bmatrix} 5 \\ 5 \\ 4 \end{bmatrix}$

Basis for Nul A: $\begin{bmatrix} 4 \\ -5 \\ 1 \\ 0 \end{bmatrix}, \begin{bmatrix} -7 \\ 6 \\ 0 \\ 1 \end{bmatrix}$

33. Basis for Col A: $\begin{bmatrix} 1 \\ -1 \\ -2 \\ 3 \end{bmatrix}, \begin{bmatrix} 4 \\ 2 \\ 2 \\ 6 \end{bmatrix}, \begin{bmatrix} -3 \\ 3 \\ 5 \\ -5 \end{bmatrix}$

Basis for Nul A: $\begin{bmatrix} 2 \\ -2.5 \\ 1 \\ 0 \\ 0 \end{bmatrix}, \begin{bmatrix} -7 \\ .5 \\ 0 \\ -4 \\ 1 \end{bmatrix}$

35. Construct a nonzero 3×3 matrix A, and construct $\mathbf{b}$ to be almost any convenient linear combination of the columns of A.

37. *Hint:* You need a nonzero matrix whose columns are linearly dependent.

39. If Col $F \neq \mathbb{R}^5$, then the columns of F do not span $\mathbb{R}^5$. Since F is square, the IMT shows that F is not invertible and the equation $F\mathbf{x} = \mathbf{0}$ has a nontrivial solution. That is, Nul F contains a nonzero vector. Another way to describe this is to write Nul $F \neq \{\mathbf{0}\}$.

41. If Col $Q = \mathbb{R}^4$, then the columns of Q span $\mathbb{R}^4$. Since Q is square, the IMT shows that Q is invertible and the equation $Q\mathbf{x} = \mathbf{b}$ has a solution for each $\mathbf{b}$ in $\mathbb{R}^4$. Also, each solution is unique, by Theorem 5 in Section 2.2.

43. If the columns of B are linearly independent, then the equation $B\mathbf{x} = \mathbf{0}$ has only the trivial (zero) solution. That is, Nul $B = \{\mathbf{0}\}$.

45. Display the reduced echelon form of A, and select the pivot columns of A as a basis for Col A. For Nul A, write the solution of $A\mathbf{x} = \mathbf{0}$ in parametric vector form.

Basis for Col A : $\begin{bmatrix} 3 \\ -7 \\ -5 \\ 3 \end{bmatrix}, \begin{bmatrix} -5 \\ 9 \\ 7 \\ -7 \end{bmatrix}$

Basis for Nul A : $\begin{bmatrix} -2.5 \\ -1.5 \\ 1 \\ 0 \\ 0 \end{bmatrix}, \begin{bmatrix} 4.5 \\ 2.5 \\ 0 \\ 1 \\ 0 \end{bmatrix}, \begin{bmatrix} -3.5 \\ -1.5 \\ 0 \\ 0 \\ 1 \end{bmatrix}$

Section 2.9, page 166

1. $\mathbf{x} = 3\mathbf{b}_1 + 2\mathbf{b}_2 = 3\begin{bmatrix} 1 \\ 1 \end{bmatrix} + 2\begin{bmatrix} 2 \\ -1 \end{bmatrix} = \begin{bmatrix} 7 \\ 1 \end{bmatrix}$

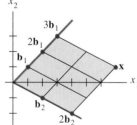

3. $\begin{bmatrix} 7 \\ 5 \end{bmatrix}$ **5.** $\begin{bmatrix} 1/4 \\ -5/4 \end{bmatrix}$

7. $[\mathbf{w}]_\mathcal{B} = \begin{bmatrix} 2 \\ -1 \end{bmatrix}, [\mathbf{x}]_\mathcal{B} = \begin{bmatrix} 1.5 \\ .5 \end{bmatrix}$

9. Basis for Col A: $\begin{bmatrix} 1 \\ -3 \\ 2 \\ -4 \end{bmatrix}, \begin{bmatrix} 2 \\ -1 \\ 4 \\ 2 \end{bmatrix}, \begin{bmatrix} -4 \\ 5 \\ -3 \\ 7 \end{bmatrix}$; dim Col $A = 3$

Basis for Nul A: $\begin{bmatrix} 3 \\ 1 \\ 0 \\ 0 \end{bmatrix}$; dim Nul $A = 1$

11. Basis for Col A: $\begin{bmatrix} 1 \\ 2 \\ -3 \\ 3 \end{bmatrix}, \begin{bmatrix} 2 \\ 5 \\ -9 \\ 10 \end{bmatrix}, \begin{bmatrix} 0 \\ 4 \\ -7 \\ 11 \end{bmatrix}$;

dim Col $A = 3$; Basis for Nul A: $\begin{bmatrix} 9 \\ -2 \\ 1 \\ 0 \\ 0 \end{bmatrix}, \begin{bmatrix} -5 \\ 3 \\ 0 \\ -2 \\ 1 \end{bmatrix}$;

dim Nul $A = 2$

13. Columns 1, 3, and 4 of the original matrix form a basis for H, so dim $H = 3$.

15. Col $A = \mathbb{R}^3$, because A has a pivot in each row, and so the columns of A span $\mathbb{R}^3$. Nul A *cannot* equal $\mathbb{R}^2$, because Nul A is a subspace of $\mathbb{R}^5$. It is true, however, that Nul A is two-dimensional. Reason: The equation $A\mathbf{x} = \mathbf{0}$ has two free variables, because A has five columns and only three of them are pivot columns.

17–25. See the *Study Guide* after you write your justifications.

27. The fact that the solution space of $A\mathbf{x} = \mathbf{0}$ has a basis of three vectors means that dim Nul $A = 3$. Since a 5×7 matrix A has seven columns, the Rank Theorem shows that rank $A = 7 - \text{dim Nul } A = 4$. See the *Study Guide* for a justification that does not explicitly mention the Rank Theorem.

29. A 7×6 matrix has six columns. By the Rank Theorem, dim Nul $A = 6 - \text{rank } A$. Since the rank is four, dim Nul $A = 2$. That is, the dimension of the solution space of $A\mathbf{x} = \mathbf{0}$ is two.

31. A 3×4 matrix A with a two-dimensional column space has two pivot columns. The remaining two columns will correspond to free variables in the equation $A\mathbf{x} = \mathbf{0}$. So the desired construction is possible. There are six possible locations for the two pivot columns, one of which is $\begin{bmatrix} \blacksquare & * & * & * \\ 0 & \blacksquare & * & * \\ 0 & 0 & 0 & 0 \end{bmatrix}$. A simple construction is to take two vectors in $\mathbb{R}^3$ that are obviously not linearly dependent and place them in a matrix along with a copy of each vector, in any order. The resulting matrix will obviously have a two-dimensional column space. There is no need to worry about whether Nul A has the correct dimension, since this is guaranteed by the Rank Theorem: dim Nul $A = 4 - \text{rank } A$.

33. The p columns of A span Col A by definition. If dim Col $A = p$, then the spanning set of p columns is automatically a basis for Col A, by the Basis Theorem. In particular, the columns are linearly independent.

35. a. *Hint:* The columns of B span W, and each vector $\mathbf{a}_j$ is in W. The vector $\mathbf{c}_j$ is in $\mathbb{R}^p$ because B has p columns.

   b. *Hint:* What is the size of C?

   c. *Hint:* How are B and C related to A?

37. Your calculations should show that the matrix $[\,\mathbf{v}_1 \quad \mathbf{v}_2 \quad \mathbf{x}\,]$ corresponds to a consistent system. The $\mathcal{B}$-coordinate vector of $\mathbf{x}$ is $(-5/3, 8/3)$.

Chapter 2 Supplementary Exercises, page 169

1. T   **2.** F   **3.** T   **4.** F

5. F   **6.** F   **7.** T   **8.** T

9. T   **10.** F   **11.** T   **12.** F

13. F   **14.** T   **15.** F

17. I

19. $A^2 = 2A - I$. Multiply by A: $A^3 = 2A^2 - A$.
Substitute $A^2 = 2A - I$: $A^3 = 2(2A - I) - A = 3A - 2I$.
Multiply by A again: $A^4 = A(3A - 2I) = 3A^2 - 2A$.
Substitute the identity $A^2 = 2A - I$ again:
$A^4 = 3(2A - I) - 2A = 4A - 3I$.

21. $\begin{bmatrix} 10 & -1 \\ 9 & 10 \\ -5 & -3 \end{bmatrix}$ **23.** $\begin{bmatrix} -3 & 13 \\ -8 & 27 \end{bmatrix}$

25. a. $p(x_i) = c_0 + c_1 x_i + \cdots + c_{n-1} x_i^{n-1}$

$= \text{row}_i(V) \begin{bmatrix} c_0 \\ \vdots \\ c_{n-1} \end{bmatrix} = \text{row}_i(V\mathbf{c}) = y_i$

   b. Suppose $x_1, \ldots, x_n$ are distinct, and suppose $V\mathbf{c} = \mathbf{0}$ for some vector $\mathbf{c}$. Then the entries in $\mathbf{c}$ are the coefficients of a polynomial whose value is zero at the distinct points $x_1, \ldots, x_n$. However, a nonzero polynomial of degree $n - 1$ cannot have n zeros, so the polynomial must be identically zero. That is, the entries in $\mathbf{c}$ must all be zero. This shows that the columns of V are linearly independent.

   c. *Hint:* When $x_1, \ldots, x_n$ are distinct, there is a vector $\mathbf{c}$ such that $V\mathbf{c} = \mathbf{y}$. Why?

27. a. $P^2 = (\mathbf{u}\mathbf{u}^T)(\mathbf{u}\mathbf{u}^T) = \mathbf{u}(\mathbf{u}^T\mathbf{u})\mathbf{u}^T = \mathbf{u}(1)\mathbf{u}^T = P$

   b. $P^T = (\mathbf{u}\mathbf{u}^T)^T = \mathbf{u}^{TT}\mathbf{u}^T = \mathbf{u}\mathbf{u}^T = P$

   c. $Q^2 = (I - 2P)(I - 2P)$
   $= I - I(2P) - 2PI + 2P(2P)$
   $= I - 4P + 4P^2 = I$, because of part (a).

29. Left-multiplication by an elementary matrix produces an elementary row operation:

$B \sim E_1 B \sim E_2 E_1 B \sim E_3 E_2 E_1 B = C$

So B is row equivalent to C. Since row operations are reversible, C is row equivalent to B. (Alternatively, show C

being changed into B by row operations using the inverses of the E_i.)

31. Since B is 4×6 (with more columns than rows), its six columns are linearly dependent and there is a nonzero $\mathbf{x}$ such that $B\mathbf{x} = \mathbf{0}$. Thus $AB\mathbf{x} = A\mathbf{0} = \mathbf{0}$, which shows that the matrix AB is not invertible, by the Invertible Matrix Theorem.

33. To four decimal places, as k increases,

$A^k \rightarrow \begin{bmatrix} .2857 & .2857 & .2857 \\ .4286 & .4286 & .4286 \\ .2857 & .2857 & .2857 \end{bmatrix}$ and

$B^k \rightarrow \begin{bmatrix} .2022 & .2022 & .2022 \\ .3708 & .3708 & .3708 \\ .4270 & .4270 & .4270 \end{bmatrix}$

or, in rational format,

$A^k \rightarrow \begin{bmatrix} 2/7 & 2/7 & 2/7 \\ 3/7 & 3/7 & 3/7 \\ 2/7 & 2/7 & 2/7 \end{bmatrix}$ and

$B^k \rightarrow \begin{bmatrix} 18/89 & 18/89 & 18/89 \\ 33/89 & 33/89 & 33/89 \\ 38/89 & 38/89 & 38/89 \end{bmatrix}$

Chapter 3

Section 3.1, page 177

1. 1   **3.** 0   **5.** -24 **7.** 4

9. 15. Start with row 3.

11. -18. Start with column 1 or row 4.

13. 6. Start with row 2 or column 2.

15. 24   **17.** -10

19. $ad - bc, cb - da$. Interchanging two rows changes the sign of the determinant.

21. 2; $3(4 + 2k) - 2(5 + 3k) = 12 + 6k - 10 - 6k$. Row replacement does not change a determinant.

23. $7a - 14b + 7c. -7a + 14b - 7c$. Interchanging two rows changes the sign of the determinant.

25. 1   **27.** 1   **29.** k

31. 1. The matrix is upper or lower triangular, with only 1's on the diagonal. The determinant is 1, the product of the diagonal entries.

33. $\det EA = \det \begin{bmatrix} a + kc & b + kd \\ c & d \end{bmatrix}$
$= (a + kc)d - (b + kd)c$
$= ad + kcd - bc - kdc = (+1)(ad - bc)$
$= (\det E)(\det A)$

35. $\det EA = \det \begin{bmatrix} c & d \\ a & b \end{bmatrix} = cb - ad = (-1)(ad - bc)$
$= (\det E)(\det A)$

37. $5A = \begin{bmatrix} 15 & 5 \\ 20 & 10 \end{bmatrix}$; no

39–41. Hints are in the *Study Guide*.

43. The area of the parallelogram and the determinant of $\begin{bmatrix} \mathbf{u} & \mathbf{v} \end{bmatrix}$ both equal 6. If $\mathbf{v} = \begin{bmatrix} x \\ 2 \end{bmatrix}$ for any x, the area is still 6. In each case the base of the parallelogram is unchanged, and the altitude remains 2 because the second coordinate of $\mathbf{v}$ is always 2.

45. a. yes **b.** no **c.** yes **d.** no

47. In general, det $A^{-1} = 1/\det A$ as long as det A is nonzero.

49. You can check your conjectures when you get to Section 3.2.

51. b. *llll; rllr; lrlr; llrr;*

Section 3.2, page 185

1. Interchanging two rows reverses the sign of the determinant.

3. A row replacement operation does not change the determinant.

5. -3 **7.** 0 **9.** -28 **11.** -48

13. 6 **15.** 21 **17.** 7 **19.** 14

21. Not invertible **23.** Invertible

25. Linearly independent

27–33. See the *Study Guide*.

35. 16

37. *Hint:* Show that $(\det A)(\det A^{-1}) = 1$.

39. *Hint:* Use Theorem 6.

41. *Hint:* Use Theorem 6 and another theorem.

43. $\det AB = \det \begin{bmatrix} 6 & 0 \\ 17 & 4 \end{bmatrix} = 24$; $(\det A)(\det B) = 3 \cdot 8 = 24$

45. a. -6 **b.** -250 **c.** 3 **d.** $-1/2$ **e.** -8

47. $\det A = (a + e)d - (b + f)c = ad + ed - bc - fc$
$= (ad - bc) + (ed - fc) = \det B + \det C$

49. *Hint:* Compute det A by a cofactor expansion down column 3.

51. No. det A det A^{-1} should equal 1.

53. See the *Study Guide* after you have made a conjecture about $A^T A$ and $A A^T$.

Section 3.3, page 195

1. $\begin{bmatrix} 5/6 \\ -1/6 \end{bmatrix}$ **3.** $\begin{bmatrix} 4/5 \\ -3/10 \end{bmatrix}$ **5.** $\begin{bmatrix} 1/4 \\ 11/4 \\ 3/8 \end{bmatrix}$

7. $s \neq \pm\sqrt{3}$; $x_1 = \dfrac{5s + 4}{6(s^2 - 3)}, x_2 = \dfrac{-4s - 15}{4(s^2 - 3)}$

9. $s \neq 0, 1$; $x_1 = \dfrac{-7}{3(s - 1)}, x_2 = \dfrac{4s + 3}{6s(s - 1)}$

11. adj $A = \begin{bmatrix} 0 & 1 & 0 \\ -5 & -1 & -5 \\ 5 & 2 & 10 \end{bmatrix}$, $A^{-1} = \dfrac{1}{5}\begin{bmatrix} 0 & 1 & 0 \\ -5 & -1 & -5 \\ 5 & 2 & 10 \end{bmatrix}$

13. adj $A = \begin{bmatrix} -1 & -1 & 5 \\ 1 & -5 & 1 \\ 1 & 7 & -5 \end{bmatrix}$, $A^{-1} = \dfrac{1}{6}\begin{bmatrix} -1 & -1 & 5 \\ 1 & -5 & 1 \\ 1 & 7 & -5 \end{bmatrix}$

15. adj $A = \begin{bmatrix} -1 & 0 & 0 \\ -1 & -5 & 0 \\ -1 & -15 & 5 \end{bmatrix}$, $A^{-1} = \dfrac{-1}{5}\begin{bmatrix} -1 & 0 & 0 \\ -1 & -5 & 0 \\ -1 & -15 & 5 \end{bmatrix}$

17. If $A = \begin{bmatrix} a & b \\ c & d \end{bmatrix}$, then $C_{11} = d$, $C_{12} = -c$, $C_{21} = -b$, $C_{22} = a$. The adjugate matrix is the transpose of cofactors:

$$\text{adj } A = \begin{bmatrix} d & -b \\ -c & a \end{bmatrix}$$

Following Theorem 8, we divide by det A; this produces the formula from Section 2.2.

19. 8 **21.** 3 **23.** 23

25. A 3×3 matrix A is not invertible if and only if its columns are linearly dependent (by the Invertible Matrix Theorem). This happens if and only if one of the columns is in the plane spanned by the other two columns, which is equivalent to the condition that the parallelepiped determined by these columns has zero volume, which in turn is equivalent to the condition that det $A = 0$.

27. 12 **29.** $\frac{1}{2}|\det \begin{bmatrix} \mathbf{v}_1 & \mathbf{v}_2 \end{bmatrix}|$

31. a. See Example 5. **b.** $4\pi abc/3$

33. I.

35–37. By now you know to try these before you look in the *Study Guide*.

39. In MATLAB, the entries in $B - \text{inv}(A)$ are approximately 10^{-15} or smaller. See the *Study Guide* for suggestions that may save you keystrokes as you work.

41. MATLAB Student Version 4.0 uses 57,771 flops for inv(A), and 14,269,045 flops for the inverse formula. The `inv (A)` command requires only about 0.4% of the operations for the inverse formula. The *Study Guide* shows how to use the `flops` command.

Chapter 3 Supplementary Exercises, page 197

1. T **2.** T **3.** F **4.** F

5. F **6.** F **7.** T **8.** T

9. F **10.** F **11.** T **12.** F

13. F **14.** T **15.** F

The solution for Exercise 17 is based on the fact that if a matrix contains two rows (or two columns) that are multiples of each other, then the determinant of the matrix is zero, by Theorem.

17. Make two row replacement operations, and then factor out a common multiple in row 2 and a common multiple in row 3.

$$\begin{vmatrix} 1 & a & b+c \\ 1 & b & a+c \\ 1 & c & a+b \end{vmatrix} = \begin{vmatrix} 1 & a & b+c \\ 0 & b-a & a-b \\ 0 & c-a & a-c \end{vmatrix}$$

$$= (b-a)(c-a)\begin{vmatrix} 1 & a & b+c \\ 0 & 1 & -1 \\ 0 & 1 & -1 \end{vmatrix}$$

$$= 0$$

19. -12

21. When the determinant is expanded by cofactors of the first row, the equation has the form $ax + by + c = 0$, where at least one of a and b is not zero. This is the equation of a line. It is clear that (x_1, y_1) and (x_2, y_2) are on the line, because when the coordinates of one of the points are substituted for x and y, two rows of the matrix are equal and so the determinant is zero.

23. $T \sim \begin{bmatrix} 1 & a & a^2 \\ 0 & b-a & b^2-a^2 \\ 0 & c-a & c^2-a^2 \end{bmatrix}$. Thus, by Theorem 3,

$$\det T = (b-a)(c-a)\det\begin{bmatrix} 1 & a & a^2 \\ 0 & 1 & b+a \\ 0 & 1 & c+a \end{bmatrix}$$

$$= (b-a)(c-a)\det\begin{bmatrix} 1 & a & a^2 \\ 0 & 1 & b+a \\ 0 & 0 & c-b \end{bmatrix}$$

$$= (b-a)(c-a)(c-b)$$

25. Area $= 12$. If one vertex is subtracted from all four vertices, and if the new vertices are $\mathbf{0}$, $\mathbf{v}_1$, $\mathbf{v}_2$, and $\mathbf{v}_3$, then the translated figure (and hence the original figure) will be a parallelogram if and only if one of $\mathbf{v}_1$, $\mathbf{v}_2$, and $\mathbf{v}_3$ is the sum of the other two vectors.

27. By the Inverse Formula, $(\operatorname{adj} A) \cdot \dfrac{1}{\det A} A = A^{-1}A = I$. By the Invertible Matrix Theorem, $\operatorname{adj} A$ is invertible

$(\operatorname{adj} A)^{-1} = \dfrac{1}{\det A} A$.

29. a. $X = CA^{-1}$, $Y = D - CA^{-1}B$. Now use Exercise 28(c).

b. From part (a), and the property of determinants,

$$\det\begin{bmatrix} A & B \\ C & D \end{bmatrix} = \det[A(D - CA^{-1}B)]$$

$$= \det[AD - ACA^{-1}B]$$

$$= \det[AD - CAA^{-1}B]$$

$$= \det[AD - CB]$$

where the equality $AC = CA$ was used in the third step.

31. First consider the case $n = 2$, and prove that the result holds by directly computing the determinants of B and C. Now assume that the formula holds for all $(k-1) \times (k-1)$ matrices, and let A, B, and C be $k \times k$ matrices. Use a cofactor expansion along the first column and the inductive

hypothesis to find $\det B$. Use row replacement operations on C to create zeros below the first pivot and produce a triangular matrix. Find the determinant of this matrix and add to $\det B$ to get the result.

33. Compute:

$$\begin{vmatrix} 1 & 1 & 1 \\ 1 & 2 & 2 \\ 1 & 2 & 3 \end{vmatrix} = 1, \quad \begin{vmatrix} 1 & 1 & 1 & 1 \\ 1 & 2 & 2 & 2 \\ 1 & 2 & 3 & 3 \\ 1 & 2 & 3 & 4 \end{vmatrix} = 1,$$

$$\begin{vmatrix} 1 & 1 & 1 & 1 & 1 \\ 1 & 2 & 2 & 2 & 2 \\ 1 & 2 & 3 & 3 & 3 \\ 1 & 2 & 3 & 4 & 4 \\ 1 & 2 & 3 & 4 & 5 \end{vmatrix} = 1$$

Conjecture:

$$\begin{vmatrix} 1 & 1 & 1 & \cdots & 1 \\ 1 & 2 & 2 & & 2 \\ 1 & 2 & 3 & & 3 \\ \vdots & & & \ddots & \vdots \\ 1 & 2 & 3 & \cdots & n \end{vmatrix} = 1$$

To confirm the conjecture, use row replacement operations to create zeros below the first pivot, then the second pivot, and so on. The resulting matrix is

$$\begin{vmatrix} 1 & 1 & 1 & \cdots & 1 \\ 0 & 1 & 1 & & 1 \\ 0 & 0 & 1 & & 1 \\ \vdots & & & \ddots & \vdots \\ 0 & 0 & 0 & \cdots & 1 \end{vmatrix}$$

which is an upper triangular matrix with determinant 1.

Chapter 4

Section 4.1, page 208

1. a. $\mathbf{u} + \mathbf{v}$ is in V because its entries will both be nonnegative.

b. *Example:* If $\mathbf{u} = \begin{bmatrix} 2 \\ 2 \end{bmatrix}$ and $c = -1$, then $\mathbf{u}$ is in V, but $c\mathbf{u}$ is not in V.

3. *Example:* If $\mathbf{u} = \begin{bmatrix} .5 \\ .5 \end{bmatrix}$ and $c = 4$, then $\mathbf{u}$ is in H, but $c\mathbf{u}$ is not in H.

5. Yes, by Theorem 1, because the set is Span $\{t^2\}$.

7. No, the set is not closed under multiplication by scalars that are not integers.

9. $H = $ Span $\{\mathbf{v}\}$, where $\mathbf{v} = \begin{bmatrix} 1 \\ 3 \\ 2 \end{bmatrix}$. By Theorem 1, H is a subspace of $\mathbb{R}^3$.

11. $W = \text{Span}\{\mathbf{u}, \mathbf{v}\}$, where $\mathbf{u} = \begin{bmatrix} 5 \\ 1 \\ 0 \end{bmatrix}$, $\mathbf{v} = \begin{bmatrix} 2 \\ 0 \\ 1 \end{bmatrix}$. By

Theorem 1, W is a subspace of $\mathbb{R}^3$.

13. a. There are only three vectors in $\{\mathbf{v}_1, \mathbf{v}_2, \mathbf{v}_3\}$, and $\mathbf{w}$ is not one of them.

b. There are infinitely many vectors in $\text{Span}\{\mathbf{v}_1, \mathbf{v}_2, \mathbf{v}_3\}$.

c. $\mathbf{w}$ is in $\text{Span}\{\mathbf{v}_1, \mathbf{v}_2, \mathbf{v}_3\}$.

15. Not a vector space because the zero vector is not in W

17. $S = \left\{ \begin{bmatrix} 1 \\ 0 \\ -1 \\ 0 \end{bmatrix}, \begin{bmatrix} -1 \\ 1 \\ 0 \\ 1 \end{bmatrix}, \begin{bmatrix} 0 \\ -1 \\ 1 \\ 0 \end{bmatrix} \right\}$

19. *Hint:* Use Theorem 1.

Warning: Although the *Study Guide* has complete solutions for *every* odd-numbered exercise whose answer here is only a "Hint," you *must* really try to work the solution yourself. Otherwise, you will not benefit from the exercise.

21. Yes. The conditions for a subspace are obviously satisfied: The zero matrix is in H, the sum of two upper triangular matrices is upper triangular, and any scalar multiple of an upper triangular matrix is again upper triangular.

23–31. See the *Study Guide* after you have written your answers.

33. 4   **35. a.** 8   **b.** 3   **c.** 5   **d.** 4

37. $\mathbf{u} + (-1)\mathbf{u} = 1\mathbf{u} + (-1)\mathbf{u}$ Axiom 10
$= [1 + (-1)]\mathbf{u}$ Axiom 8
$= 0\mathbf{u} = \mathbf{0}$ Exercise 35
From Exercise 34, it follows that $(-1)\mathbf{u} = -\mathbf{u}$.

39. Any subspace H that contains $\mathbf{u}$ and $\mathbf{v}$ must also contain all scalar multiples of $\mathbf{u}$ and $\mathbf{v}$ and hence must contain all sums of scalar multiples of $\mathbf{u}$ and $\mathbf{v}$. Thus H must contain $\text{Span}\{\mathbf{u}, \mathbf{v}\}$.

41. *Hint:* For part of the solution, consider $\mathbf{w}_1$ and $\mathbf{w}_2$ in $H + K$, and write $\mathbf{w}_1$ and $\mathbf{w}_2$ in the form $\mathbf{w}_1 = \mathbf{u}_1 + \mathbf{v}_1$ and $\mathbf{w}_2 = \mathbf{u}_2 + \mathbf{v}_2$, where $\mathbf{u}_1$ and $\mathbf{u}_2$ are in H, and $\mathbf{v}_1$ and $\mathbf{v}_2$ are in K.

43. The reduced echelon form of $[\,\mathbf{v}_1 \quad \mathbf{v}_2 \quad \mathbf{v}_3 \quad \mathbf{w}\,]$ shows that $\mathbf{w} = \mathbf{v}_1 - 2\mathbf{v}_2 + \mathbf{v}_3$.

45. The functions are $\cos 4t$ and $\cos 6t$. See Exercise 54 in Section 4.5.

Section 4.2, page 219

1. $\begin{bmatrix} 3 & -5 & -3 \\ 6 & -2 & 0 \\ -8 & 4 & 1 \end{bmatrix} \begin{bmatrix} 1 \\ 3 \\ -4 \end{bmatrix} = \begin{bmatrix} 0 \\ 0 \\ 0 \end{bmatrix}$, so $\mathbf{w}$ is in Nul A.

3. $\begin{bmatrix} 7 \\ -4 \\ 1 \\ 0 \end{bmatrix}, \begin{bmatrix} -6 \\ 2 \\ 0 \\ 1 \end{bmatrix}$ **5.** $\begin{bmatrix} 2 \\ 1 \\ 0 \\ 0 \\ 0 \end{bmatrix}, \begin{bmatrix} -4 \\ 0 \\ 9 \\ 1 \\ 0 \end{bmatrix}$

7. W is not a subspace of $\mathbb{R}^3$ because the zero vector $(0, 0, 0)$ is not in W.

9. W is a subspace of $\mathbb{R}^4$, by Theorem 2, because W is the set of solutions of the homogeneous system

$$a - 2b - 4c \qquad\quad = 0$$
$$2a \qquad\quad - c - 3d = 0$$

11. W is not a subspace because $\mathbf{0}$ is not in W. *Justification:* If a typical element $(b - 2d, 5 + d, b + 3d, d)$ were zero, then $5 + d = 0$ and $d = 0$, which is impossible.

13. $W = \text{Col } A$ for $A = \begin{bmatrix} 1 & -6 \\ 0 & 1 \\ 1 & 0 \end{bmatrix}$, so W is a vector space by Theorem 3.

15. $\begin{bmatrix} 0 & 2 & 3 \\ 1 & 1 & -2 \\ 4 & 1 & 0 \\ 3 & -1 & -1 \end{bmatrix}$

17. a. 2   **b.** 4   **19. a.** 5   **b.** 2

21. $\begin{bmatrix} 3 \\ 1 \end{bmatrix}$ in Nul A, $\begin{bmatrix} 2 \\ -1 \\ -4 \\ 3 \end{bmatrix}$ in Col A, and $[2 \ -6]$ is in Row A.

Other answers possible.

23. $\mathbf{w}$ is in both Nul A and Col A.

25–37. See the *Study Guide*. By now you should know how to use it properly.

39. Let $\mathbf{x} = \begin{bmatrix} 3 \\ 2 \\ -1 \end{bmatrix}$ and $A = \begin{bmatrix} 1 & -3 & -3 \\ -2 & 4 & 2 \\ -1 & 5 & 7 \end{bmatrix}$. Then $\mathbf{x}$ is in Nul A. Since Nul A is a subspace of $\mathbb{R}^3$, $10\mathbf{x}$ is in Nul A.

41. a. $A\mathbf{0} = \mathbf{0}$, so the zero vector is in Col A.

b. By a property of matrix multiplication, $A\mathbf{x} + A\mathbf{w} = A(\mathbf{x} + \mathbf{w})$, which shows that $A\mathbf{x} + A\mathbf{w}$ is a linear combination of the columns of A and hence is in Col A.

c. $c(A\mathbf{x}) = A(c\mathbf{x})$, which shows that $c(A\mathbf{x})$ is in Col A for all scalars c.

43. a. For arbitrary polynomials $\mathbf{p}, \mathbf{q}$ in $\mathbb{P}_2$ and any scalar c,

$$T(\mathbf{p} + \mathbf{q}) = \begin{bmatrix} (\mathbf{p} + \mathbf{q})(0) \\ (\mathbf{p} + \mathbf{q})(1) \end{bmatrix} = \begin{bmatrix} \mathbf{p}(0) + \mathbf{q}(0) \\ \mathbf{p}(1) + \mathbf{q}(1) \end{bmatrix}$$

$$= \begin{bmatrix} \mathbf{p}(0) \\ \mathbf{p}(1) \end{bmatrix} + \begin{bmatrix} \mathbf{q}(0) \\ \mathbf{q}(1) \end{bmatrix} = T(\mathbf{p}) + T(\mathbf{q})$$

$$T(c\mathbf{p}) = \begin{bmatrix} c\mathbf{p}(0) \\ c\mathbf{p}(1) \end{bmatrix} = c\begin{bmatrix} \mathbf{p}(0) \\ \mathbf{p}(1) \end{bmatrix} = cT(\mathbf{p})$$

So T is a linear transformation from $\mathbb{P}_2$ into $\mathbb{P}_2$.

b. Any quadratic polynomial that vanishes at 0 and 1 must be a multiple of $\mathbf{p}(t) = t(t - 1)$. The range of T is $\mathbb{R}^2$.

45. a. For A, B in $M_{2\times 2}$ and any scalar c,

$$T(A + B) = (A + B) + (A + B)^T$$
$$= A + B + A^T + B^T \quad \text{Transpose property}$$
$$= (A + A^T) + (B + B^T) = T(A) + T(B)$$
$$T(cA) = (cA) + (cA)^T = cA + cA^T$$
$$= c(A + A^T) = cT(A)$$

So T is a linear transformation from $M_{2\times 2}$ into $M_{2\times 2}$.

b. If B is any element in $M_{2\times 2}$ with the property that $B^T = B$, and if $A = \frac{1}{2}B$, then

$$T(A) = \frac{1}{2}B + \left(\frac{1}{2}B\right)^T = \frac{1}{2}B + \frac{1}{2}B = B$$

c. Part (b) showed that the range of T contains all B such that $B^T = B$. So it suffices to show that any B in the range of T has this property. If $B = T(A)$, then by properties of transposes,

$$B^T = (A + A^T)^T = A^T + A^{TT} = A^T + A = B$$

d. The kernel of T is $\left\{ \begin{bmatrix} 0 & b \\ -b & 0 \end{bmatrix} : b \text{ real} \right\}$.

47. *Hint:* Check the three conditions for a subspace. Typical elements of $T(U)$ have the form $T(\mathbf{u}_1)$ and $T(\mathbf{u}_2)$, where $\mathbf{u}_1$ and $\mathbf{u}_2$ are in U.

49. $\mathbf{w}$ is in Col A but not in Nul A. (Explain why.)

51. The reduced echelon form of A is

$$\begin{bmatrix} 1 & 0 & 1/3 & 0 & 10/3 \\ 0 & 1 & 1/3 & 0 & -26/3 \\ 0 & 0 & 0 & 1 & -4 \\ 0 & 0 & 0 & 0 & 0 \end{bmatrix}$$

Section 4.3, page 228

1. Yes, the 3×3 matrix $A = \begin{bmatrix} 1 & 1 & 1 \\ 0 & 1 & 1 \\ 0 & 0 & 1 \end{bmatrix}$ has 3 pivot positions. By the Invertible Matrix Theorem, A is invertible and its columns form a basis for $\mathbb{R}^3$. (See Example 3.)

3. No, the vectors are linearly dependent and do not span $\mathbb{R}^3$.

5. No, the set is linearly dependent because the zero vector is in the set. However,

$$\begin{bmatrix} 1 & -2 & 0 & 0 \\ -3 & 9 & 0 & -3 \\ 0 & 0 & 0 & 5 \end{bmatrix} \sim \begin{bmatrix} 1 & -2 & 0 & 0 \\ 0 & 3 & 0 & -3 \\ 0 & 0 & 0 & 5 \end{bmatrix}$$

The matrix has pivots in each row and hence its columns span $\mathbb{R}^3$.

7. No, the vectors are linearly independent because they are not multiples. (More precisely, neither vector is a multiple of the other.) However, the vectors do not span $\mathbb{R}^3$. The

matrix $\begin{bmatrix} -2 & 6 \\ 3 & -1 \\ 0 & 5 \end{bmatrix}$ can have at most two pivots since it has only two columns. So there will not be a pivot in each row.

9. $\begin{bmatrix} 3 \\ 5 \\ 1 \\ 0 \end{bmatrix}, \begin{bmatrix} -2 \\ -4 \\ 0 \\ 1 \end{bmatrix}$ **11.** $\begin{bmatrix} -2 \\ 1 \\ 0 \end{bmatrix}, \begin{bmatrix} -1 \\ 0 \\ 1 \end{bmatrix}$

13. Basis for Nul A: $\begin{bmatrix} -6 \\ -5/2 \\ 1 \\ 0 \end{bmatrix}, \begin{bmatrix} -5 \\ -3/2 \\ 0 \\ 1 \end{bmatrix}$

Basis for Col A: $\begin{bmatrix} -2 \\ 2 \\ -3 \end{bmatrix}, \begin{bmatrix} 4 \\ -6 \\ 8 \end{bmatrix}$

Basis for Row A: $\begin{bmatrix} 1 & 0 & 6 & 5 \end{bmatrix}, \begin{bmatrix} 0 & 2 & 5 & 3 \end{bmatrix}$

15. $\{\mathbf{v}_1, \mathbf{v}_2, \mathbf{v}_4\}$ **17.** $\{\mathbf{v}_1, \mathbf{v}_2, \mathbf{v}_3\}$

19. The three simplest answers are $\{\mathbf{v}_1, \mathbf{v}_2\}$ or $\{\mathbf{v}_1, \mathbf{v}_3\}$ or $\{\mathbf{v}_2, \mathbf{v}_3\}$. Other answers are possible.

21–31. See the *Study Guide* for hints.

33. *Hint:* Use the Invertible Matrix Theorem.

35. No. (Why is the set not a basis for H?)

37. $\{\cos \omega t, \sin \omega t\}$

39. Let A be the $n \times k$ matrix $\begin{bmatrix} \mathbf{v}_1 & \cdots & \mathbf{v}_k \end{bmatrix}$. Since A has fewer columns than rows, there cannot be a pivot position in each row of A. By Theorem 4 in Section 1.4, the columns of A do not span $\mathbb{R}^n$ and hence are not a basis for $\mathbb{R}^n$.

41. *Hint:* If $\{\mathbf{v}_1, \ldots, \mathbf{v}_p\}$ is linearly dependent, then there exist $c_1, \ldots, c_p$, not all zero, such that $c_1\mathbf{v}_1 + \cdots + c_p\mathbf{v}_p = \mathbf{0}$. Use this equation.

43. Neither polynomial is a multiple of the other polynomial, so $\{\mathbf{p}_1, \mathbf{p}_2\}$ is a linearly independent set in $\mathbb{P}_3$.

45. Let $\{\mathbf{v}_1, \mathbf{v}_3\}$ be any linearly independent set in the vector space V, and let $\mathbf{v}_2$ and $\mathbf{v}_4$ be linear combinations of $\mathbf{v}_1$ and $\mathbf{v}_3$. Then $\{\mathbf{v}_1, \mathbf{v}_3\}$ is a basis for Span$\{\mathbf{v}_1, \mathbf{v}_2, \mathbf{v}_3, \mathbf{v}_4\}$.

47. You could be clever and find special values of t that produce several zeros in (5), and thereby create a system of equations that can be solved easily by hand. Or, you could use values of t such as $t = 0, .1, .2, \ldots$ to create a system of equations that you can solve with a matrix program.

Section 4.4, page 238

1. $\begin{bmatrix} 3 \\ -7 \end{bmatrix}$ **3.** $\begin{bmatrix} -1 \\ -5 \\ 9 \end{bmatrix}$ **5.** $\begin{bmatrix} 8 \\ -5 \end{bmatrix}$ **7.** $\begin{bmatrix} -1 \\ -1 \\ 3 \end{bmatrix}$

9. $\begin{bmatrix} 2 & 1 \\ -9 & 8 \end{bmatrix}$ **11.** $\begin{bmatrix} 6 \\ 4 \end{bmatrix}$ **13.** $\begin{bmatrix} 2 \\ 6 \\ -1 \end{bmatrix}$

15–19. The *Study Guide* has hints.

21. $\begin{bmatrix} 1 \\ 1 \end{bmatrix} = 5\mathbf{v}_1 - 2\mathbf{v}_2 = 10\mathbf{v}_1 - 3\mathbf{v}_2 + \mathbf{v}_3$ (infinitely many answers)

23. *Hint:* By hypothesis, the zero vector has a unique representation as a linear combination of elements of S.

25. $\begin{bmatrix} 9 & 2 \\ 4 & 1 \end{bmatrix}$

27. *Hint:* Suppose that $[\mathbf{u}]_{\mathcal{B}} = [\mathbf{w}]_{\mathcal{B}}$ for some $\mathbf{u}$ and $\mathbf{w}$ in V, and denote the entries in $[\mathbf{u}]_{\mathcal{B}}$ by $c_1, \ldots, c_n$. Use the definition of $[\mathbf{u}]_{\mathcal{B}}$.

29. One possible approach: First, show that if $\mathbf{u}_1, \ldots, \mathbf{u}_p$ are linearly *dependent*, then $[\mathbf{u}_1]_{\mathcal{B}}, \ldots, [\mathbf{u}_p]_{\mathcal{B}}$ are linearly dependent. Second, show that if $[\mathbf{u}_1]_{\mathcal{B}}, \ldots, [\mathbf{u}_p]_{\mathcal{B}}$ are linearly dependent, then $\mathbf{u}_1, \ldots, \mathbf{u}_p$ are linearly *dependent*. Use the two equations displayed in the exercise. A slightly different proof is given in the *Study Guide*.

31. Linearly independent. (Justify answers to Exercises 31–38.)

33. Linearly dependent

35. a. The coordinate vectors $\begin{bmatrix} 1 \\ -3 \\ 5 \end{bmatrix}, \begin{bmatrix} -3 \\ 5 \\ -7 \end{bmatrix}, \begin{bmatrix} -4 \\ 5 \\ -6 \end{bmatrix}, \begin{bmatrix} 1 \\ 0 \\ -1 \end{bmatrix}$ do not span $\mathbb{R}^3$. Because of the isomorphism between $\mathbb{R}^3$ and $\mathbb{P}_2$, the corresponding polynomials do not span $\mathbb{P}_2$.

b. The coordinate vectors $\begin{bmatrix} 0 \\ 5 \\ 1 \end{bmatrix}, \begin{bmatrix} 1 \\ -8 \\ -2 \end{bmatrix}, \begin{bmatrix} -3 \\ 4 \\ 2 \end{bmatrix}, \begin{bmatrix} 2 \\ -3 \\ 0 \end{bmatrix}$ span $\mathbb{R}^3$. Because of the isomorphism between $\mathbb{R}^3$ and $\mathbb{P}_2$, the corresponding polynomials span $\mathbb{P}_2$.

37. The coordinate vectors $\begin{bmatrix} 3 \\ 7 \\ 0 \\ 0 \end{bmatrix}, \begin{bmatrix} 5 \\ 1 \\ 0 \\ -2 \end{bmatrix}, \begin{bmatrix} 0 \\ 1 \\ -2 \\ 0 \end{bmatrix}, \begin{bmatrix} 1 \\ 16 \\ -6 \\ 2 \end{bmatrix}$ are a linearly dependent subset of $\mathbb{R}^4$. Because of the isomorphism between $\mathbb{R}^4$ and $\mathbb{P}_3$, the corresponding polynomials form a linearly dependent subset of $\mathbb{P}_3$, and thus cannot be a basis for $\mathbb{P}_3$.

39. $[\mathbf{x}]_{\mathcal{B}} = \begin{bmatrix} -5/3 \\ 8/3 \end{bmatrix}$ **41.** $\begin{bmatrix} 1.3 \\ 0 \\ 0.8 \end{bmatrix}$

Section 4.5, page 247

1. $\begin{bmatrix} 1 \\ 1 \\ 0 \end{bmatrix}, \begin{bmatrix} -2 \\ 1 \\ 3 \end{bmatrix}$; dim is 2

3. $\begin{bmatrix} 0 \\ 1 \\ 0 \\ 1 \end{bmatrix}, \begin{bmatrix} 0 \\ -1 \\ 1 \\ 2 \end{bmatrix}, \begin{bmatrix} 2 \\ 0 \\ -3 \\ 0 \end{bmatrix}$; dim is 3

5. $\begin{bmatrix} 1 \\ 2 \\ -1 \\ -3 \end{bmatrix}, \begin{bmatrix} -4 \\ 5 \\ 0 \\ 7 \end{bmatrix}$; dim is 2

7. No basis; dim is 0   **9.** 2   **11.** 2, 3, 3

13. 2, 2, 2   **15.** 0, 3, 3

17–25. See the *Study Guide*.

27. *Hint:* You need only show that the first four Hermite polynomials are linearly independent. Why?

29. $[\mathbf{p}]_{\mathcal{B}} = (3, 3, -2, \frac{3}{2})$

31. *Hint:* Suppose S does span V, and use the Spanning Set Theorem. This leads to a contradiction, which shows that the spanning hypothesis is false.

33. 5, 3, 3

35. Yes; no. Since Col A is a four-dimensional subspace of $\mathbb{R}^4$, it coincides with $\mathbb{R}^4$. The null space cannot be $\mathbb{R}^3$, because the vectors in Nul A have 7 entries. Nul A is a three-dimensional subspace of $\mathbb{R}^7$, by the Rank Theorem.

37. 2

39. 5, 5. In both cases, the number of pivots cannot exceed the number of columns or the number of rows.

41. The functions $\{1, x, x^2, \ldots\}$ are a linearly independent set with infinitely many vectors.

43–47. Consult the *Study Guide*.

49. dim Row $A =$ dim Col $A =$ rank$A$, so the result follows from the Rank Theorem.

51. *Hint:* Since H is a nonzero subspace of a finite-dimensional space, H is finite-dimensional and has a basis, say, $\mathbf{v}_1, \ldots, \mathbf{v}_p$. First show that $\{T(\mathbf{v}_1), \ldots, T(\mathbf{v}_p)\}$ spans $T(H)$.

53. a. One basis is $\{\mathbf{v}_1, \mathbf{v}_2, \mathbf{v}_3, \mathbf{e}_2, \mathbf{e}_3\}$. In fact, any two of the vectors $\mathbf{e}_2, \ldots, \mathbf{e}_5$ will extend $\{\mathbf{v}_1, \mathbf{v}_2, \mathbf{v}_3\}$ to a basis of $\mathbb{R}^5$.

Section 4.6, page 253

1. a. $\begin{bmatrix} 6 & 9 \\ -2 & -4 \end{bmatrix}$ **b.** $\begin{bmatrix} 0 \\ -2 \end{bmatrix}$ **3.** (ii)

5. a. $\begin{bmatrix} 4 & -1 & 0 \\ -1 & 1 & 1 \\ 0 & 1 & -2 \end{bmatrix}$ **b.** $\begin{bmatrix} 8 \\ 2 \\ 2 \end{bmatrix}$

7. $\underset{C \leftarrow B}{P} = \begin{bmatrix} -3 & 1 \\ -5 & 2 \end{bmatrix}$, $\underset{B \leftarrow C}{P} = \begin{bmatrix} -2 & 1 \\ -5 & 3 \end{bmatrix}$

9. $\underset{C \leftarrow B}{P} = \begin{bmatrix} 9 & -2 \\ -4 & 1 \end{bmatrix}$, $\underset{B \leftarrow C}{P} = \begin{bmatrix} 1 & 2 \\ 4 & 9 \end{bmatrix}$

11–13. See the *Study Guide*.

15. $\underset{C \leftarrow B}{P} = \begin{bmatrix} 1 & 3 & 0 \\ -2 & -5 & 2 \\ 1 & 4 & 3 \end{bmatrix}$, $[-1 + 2t]_{\mathcal{B}} = \begin{bmatrix} 5 \\ -2 \\ 1 \end{bmatrix}$

17. a. $\mathcal{B}$ is a basis for V.

b. The coordinate mapping is a linear transformation.

c. The product of a matrix and a vector

d. The coordinate vector of $\mathbf{v}$ relative to $\mathcal{B}$

19. a. $P^{-1} = \dfrac{1}{32}\begin{bmatrix} 32 & 0 & 16 & 0 & 12 & 0 & 10 \\ 0 & 32 & 0 & 24 & 0 & 20 & 0 \\ 0 & 0 & 16 & 0 & 16 & 0 & 15 \\ 0 & 0 & 0 & 8 & 0 & 10 & 0 \\ 0 & 0 & 0 & 0 & 4 & 0 & 6 \\ 0 & 0 & 0 & 0 & 0 & 2 & 0 \\ 0 & 0 & 0 & 0 & 0 & 0 & 1 \end{bmatrix}$

b. P is the change-of-coordinates matrix from $\mathcal{C}$ to $\mathcal{B}$. So P^{-1} is the change-of-coordinates matrix from $\mathcal{B}$ to $\mathcal{C}$, by equation (5), and the columns of this matrix are the $\mathcal{C}$-coordinate vectors of the basis vectors in $\mathcal{B}$, by Theorem 15.

21. *Hint:* Let $\mathcal{C}$ be the basis $\{\mathbf{v}_1, \mathbf{v}_2, \mathbf{v}_3\}$. Then the columns of P are $[\mathbf{u}_1]_{\mathcal{C}}$, $[\mathbf{u}_2]_{\mathcal{C}}$, and $[\mathbf{u}_3]_{\mathcal{C}}$. Use the definition of $\mathcal{C}$-coordinate vectors and matrix algebra to compute $\mathbf{u}_1, \mathbf{u}_2$, and $\mathbf{u}_3$. The solution method is discussed in the *Study Guide*. Here are the numerical answers:

a. $\mathbf{u}_1 = \begin{bmatrix} -6 \\ -5 \\ 21 \end{bmatrix}$, $\mathbf{u}_2 = \begin{bmatrix} -6 \\ -9 \\ 32 \end{bmatrix}$, $\mathbf{u}_3 = \begin{bmatrix} -5 \\ 0 \\ 3 \end{bmatrix}$

b. $\mathbf{w}_1 = \begin{bmatrix} 28 \\ -9 \\ -3 \end{bmatrix}$, $\mathbf{w}_2 = \begin{bmatrix} 38 \\ -13 \\ 2 \end{bmatrix}$, $\mathbf{w}_3 = \begin{bmatrix} 21 \\ -7 \\ 3 \end{bmatrix}$

Section 4.7, Page 260

1. $(\ldots, 0, 2, 0, 2, 0, 2, 0, \ldots)$

3. $(\ldots, -2, 2, -2, 3, -1, 3, -1, \ldots)$

5. α

7. ϵ_c

9. Verify that the three properties in the definition of a LTI transformation are satisfied.

11. χ

13. Apply T to any signal to get a signal in the range of T.

15–21. See the *Study Guide*.

23. $I - \frac{3}{4}S$.

25. Show that W satisfies the three properties of a subspace.

27. $\{\chi - \alpha\}, 1$

29. Show that W satisfies the three properties of a subspace.

31. $\{S^{2m-1}(\delta)|$ where m is any integer$\}$. Yes W is an infinite dimensional subspace. Justify your answer.

Section 4.8, page 268

1. If $y_k = 2^k$, then $y_{k+1} = 2^{k+1}$ and $y_{k+2} = 2^{k+2}$. Substituting these formulas into the left side of the equation gives

$$\begin{aligned} y_{k+2} + 2y_{k+1} - 8y_k &= 2^{k+2} + 2 \cdot 2^{k+1} - 8 \cdot 2^k \\ &= 2^k(2^2 + 2 \cdot 2 - 8) \\ &= 2^k(0) = 0 \quad \text{for all } k \end{aligned}$$

Since the difference equation holds for all k, 2^k is a solution. A similar calculation works for $y_k = (-4)^k$.

3. The signals 2^k and $(-4)^k$ are linearly independent because neither is a multiple of the other. For instance, there is no scalar c such that $2^k = c(-4)^k$ *for all k*. By Theorem 17, the solution set H of the difference equation in Exercise 1 is two-dimensional. By the Basis Theorem in Section 4.5, the two linearly independent signals 2^k and $(-4)^k$ form a basis for H.

5. If $y_k = (-3)^k$, then

$$\begin{aligned} y_{k+2} + 6y_{k+1} + 9y_k &= (-3)^{k+2} + 6(-3)^{k+1} + 9(-3)^k \\ &= (-3)^k[(-3)^2 + 6(-3) + 9] \\ &= (-3)^k(0) = 0 \quad \text{for all } k \end{aligned}$$

Similarly, if $y_k = k(-3)^k$, then

$$\begin{aligned} &y_{k+2} + 6y_{k+1} + 9y_k \\ &= (k+2)(-3)^{k+2} + 6(k+1)(-3)^{k+1} + 9k(-3)^k \\ &= (-3)^k[(k+2)(-3)^2 + 6(k+1)(-3) + 9k] \\ &= (-3)^k[9k + 18 - 18k - 18 + 9k] \\ &= (-3)^k(0) \quad \text{for all } k \end{aligned}$$

Thus both $(-3)^k$ and $k(-3)^k$ are in the solution space H of the difference equation. Also, there is no scalar c such that $k(-3)^k = c(-3)^k$ *for all k*, because c must be chosen independently of k. Likewise, there is no scalar c such that $(-3)^k = ck(-3)^k$ for all k. So the two signals are linearly independent. Since dim $H = 2$, the signals form a basis for H, by the Basis Theorem.

7. Yes   **9.** Yes

11. No, two signals cannot span the three-dimensional solution space.

13. $\left(\frac{1}{3}\right)^k, \left(\frac{2}{3}\right)^k$ **15.** $5^k, (-5)^k$

17. $y_k = \frac{1}{\sqrt{5}}\left(\frac{1+\sqrt{5}}{2}\right)^k - \frac{1}{\sqrt{5}}\left(\frac{1-\sqrt{5}}{2}\right)^k$

19. $Y_k = c_1(.8)^k + c_2(.5)^k + 10 \to 10 \quad$ as $k \to \infty$

21. $y_k = c_1(-2+\sqrt{3})^k + c_2(-2-\sqrt{3})^k$

23. 7, 5, 4, 3, 4, 5, 6, 6, 7, 8, 9, 8, 7; see figure below.

25. a. $y_{k+1} - 1.01y_k = -450$, $y_0 = 10,000$

b. MATLAB code:

```
pay = 450, y = 10000, m = 0
table = [0 ; y]
while y > 450
    y = 1.01*y - pay
    m = m + 1
    table = [table [m ; y] ]
            %append new column
end
m, y
```

c. At month 26, the last payment is $114.88. The total paid by the borrower is $11,364.88.

27. $k^2 + c_1 \cdot (-4)^k + c_2$ **29.** $2 - 2k + c_1 \cdot 4^k + c_2 \cdot 2^{-k}$

31. $\mathbf{x}_{k+1} = A\mathbf{x}_k$, where

$$A = \begin{bmatrix} 0 & 1 & 0 & 0 \\ 0 & 0 & 1 & 0 \\ 0 & 0 & 0 & 1 \\ 9 & -6 & -8 & 6 \end{bmatrix}, \mathbf{x} = \begin{bmatrix} y_k \\ y_{k+1} \\ y_{k+2} \\ y_{k+3} \end{bmatrix}$$

33. The equation holds for all k, so it holds with k replaced by $k - 1$, which transforms the equation into

$$y_{k+2} + 5y_{k+1} + 6y_k = 0 \quad \text{for all } k$$

The equation is of order 2.

35. For all k, the Casorati matrix $C(k)$ is not invertible. In this case, the Casorati matrix gives no information about the linear independence/dependence of the set of signals. In fact, neither signal is a multiple of the other, so they are linearly independent.

Chapter 4 Supplementary Exercises, page 271

1. T   **2.** T   **3.** F   **4.** F   **5.** T   **6.** T

7. F   **8.** F   **9.** T   **10.** F   **11.** F   **12.** F

13. T   **14.** F   **15.** T   **16.** T   **17.** F   **18.** T

19. T

21. The set of all (b_1, b_2, b_3) satisfying $b_1 + 2b_2 + b_3 = 0$.

23. The vector $\mathbf{p}_1$ is not zero and $\mathbf{p}_2$ is not a multiple of $\mathbf{p}_1$, so keep both of these vectors. Since $\mathbf{p}_3 = 2\mathbf{p}_1 + 2\mathbf{p}_2$, discard $\mathbf{p}_3$. Since $\mathbf{p}_4$ has a t^2 term, it cannot be a linear combination of $\mathbf{p}_1$ and $\mathbf{p}_2$, so keep $\mathbf{p}_4$. Finally, $\mathbf{p}_5 = \mathbf{p}_1 + \mathbf{p}_4$, so discard $\mathbf{p}_5$. The resulting basis is $\{\mathbf{p}_1, \mathbf{p}_2, \mathbf{p}_4\}$.

25. You would have to know that the solution set of the homogeneous system is spanned by two solutions. In this case, the null space of the 18×20 coefficient matrix A is at most two-dimensional. By the Rank Theorem, $\dim \text{Col } A \geq 20 - 2 = 18$, which means that $\text{Col } A = \mathbb{R}^{18}$, because A has 18 rows, and every equation $A\mathbf{x} = \mathbf{b}$ is consistent.

27. Let A be the standard $m \times n$ matrix of the transformation T.
  a. If T is one-to-one, then the columns of A are linearly independent (Theorem 12 in Section 1.9), so $\dim \text{Nul } A = 0$. By the Rank Theorem, $\dim \text{Col } A = \text{rank } A = n$. Since the range of T is Col A, the dimension of the range of T is n.
  b. If T is onto, then the columns of A span $\mathbb{R}^m$ (Theorem 12 in Section 1.9), so $\dim \text{Col } A = m$. By the Rank Theorem, $\dim \text{Nul } A = n - \dim \text{Col } A = n - m$. Since the kernel of T is Nul A, the dimension of the kernel of T is $n - m$.

29. If S is a finite spanning set for V, then a subset of S—say S'—is a basis for V. Since S' must span V, S' cannot be a

proper subset of S because of the minimality of S. Thus $S' = S$, which proves that S is a basis for V.

30. **a.** *Hint:* Any $\mathbf{y}$ in Col AB has the form $\mathbf{y} = AB\mathbf{x}$ for some $\mathbf{x}$.

31. By Exercise 12, rank $PA \leq$ rank A, and rank $A = $ rank $P^{-1}PA \leq$ rank PA. Thus rank $PA = $ rank A.

33. The equation $AB = 0$ shows that each column of B is in Nul A. Since Nul A is a subspace, all linear combinations of the columns of B are in Nul A, so Col B is a subspace of Nul A. By Theorem 12 in Section 4.5, $\dim \text{Col } B \leq \dim \text{Nul } A$. Applying the Rank Theorem, we find that

$$n = \text{rank } A + \dim \text{Nul } A \geq \text{rank } A + \text{rank } B$$

35. **a.** Let A_1 consist of the r pivot columns in A. The columns of A_1 are linearly independent. So A_1 is an $m \times r$ with rank r.
  b. By the Rank Theorem applied to A_1, the dimension of Row A is r, so A_1 has r linearly independent rows. Use them to form A_2. Then A_2 is $r \times r$ with linearly independent rows. By the Invertible Matrix Theorem, A_2 is invertible.

37. $[\,B \quad AB \quad A^2B\,] = \begin{bmatrix} 0 & 1 & 0 \\ 1 & -.9 & .81 \\ 1 & .5 & .25 \end{bmatrix}$

$$\sim \begin{bmatrix} 1 & -.9 & .81 \\ 0 & 1 & 0 \\ 0 & 0 & -.56 \end{bmatrix}$$

This matrix has rank 3, so the pair (A, B) is controllable.

39. rank $[\,B \quad AB \quad A^2B \quad A^3B\,] = 3$. The pair (A, B) is not controllable.

Chapter 5

Section 5.1, page 280

1. Yes   **3.** No   **5.** Yes, $\lambda = 0$ **7.** Yes, $\begin{bmatrix} 1 \\ 1 \\ -1 \end{bmatrix}$

9. $\lambda = 1: \begin{bmatrix} 0 \\ 1 \end{bmatrix}$; $\lambda = 5: \begin{bmatrix} 2 \\ 1 \end{bmatrix}$ **11.** $\begin{bmatrix} -1 \\ 3 \end{bmatrix}$

13. $\lambda = 1: \begin{bmatrix} 0 \\ 1 \\ 0 \end{bmatrix}$; $\lambda = 2: \begin{bmatrix} -1 \\ 2 \\ 2 \end{bmatrix}$; $\lambda = 3: \begin{bmatrix} -1 \\ 1 \\ 1 \end{bmatrix}$

15. $\begin{bmatrix} -2 \\ 1 \\ 0 \end{bmatrix}, \begin{bmatrix} -3 \\ 0 \\ 1 \end{bmatrix}$ **17.** $0, 2, -1$

19. 0. Justify your answer.

21–29. See the *Study Guide*, after you have written your answers.

31. *Hint:* Use Theorem 2.

33. *Hint:* Use the equation $A\mathbf{x} = \lambda\mathbf{x}$ to find an equation involving A^{-1}.

35. *Hint:* For any λ, $(A - \lambda I)^T = A^T - \lambda I$. By a theorem (which one?), $A^T - \lambda I$ is invertible if and only if $A - \lambda I$ is invertible.

37. Let $\mathbf{v}$ be the vector in $\mathbb{R}^n$ whose entries are all 1's. Then $A\mathbf{v} = s\mathbf{v}$.

39. *Hint:* If A is the standard matrix of T, look for a nonzero vector $\mathbf{v}$ (a point in the plane) such that $A\mathbf{v} = \mathbf{v}$.

41. **a.** $\mathbf{x}_{k+1} = c_1\lambda^{k+1}\mathbf{u} + c_2\mu^{k+1}\mathbf{v}$

b. $A\mathbf{x}_k = A(c_1\lambda^k\mathbf{u} + c_2\mu^k\mathbf{v})$
$\qquad = c_1\lambda^k A\mathbf{u} + c_2\mu^k A\mathbf{v}$ Linearity
$\qquad = c_1\lambda^k\lambda\mathbf{u} + c_2\mu^k\mu\mathbf{v}$ $\mathbf{u}$ and $\mathbf{v}$ are eigenvectors.
$\qquad = \mathbf{x}_{k+1}$

43.

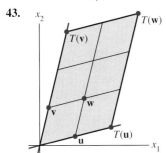

45. $\lambda = 3$: $\begin{bmatrix} 5 \\ -2 \\ 9 \end{bmatrix}$; $\lambda = 13$: $\begin{bmatrix} -2 \\ 1 \\ 0 \end{bmatrix}, \begin{bmatrix} -1 \\ 0 \\ 1 \end{bmatrix}$. You can speed up your calculations with the program `nulbasis` discussed in the *Study Guide*.

47. $\lambda = -2$: $\begin{bmatrix} -2 \\ 7 \\ -5 \\ 5 \\ 0 \end{bmatrix}, \begin{bmatrix} 3 \\ 7 \\ -5 \\ 0 \\ 5 \end{bmatrix}$;

$\lambda = 5$: $\begin{bmatrix} 2 \\ -1 \\ 1 \\ 0 \\ 0 \end{bmatrix}, \begin{bmatrix} -1 \\ 1 \\ 0 \\ 1 \\ 0 \end{bmatrix}, \begin{bmatrix} 2 \\ 0 \\ 0 \\ 0 \\ 1 \end{bmatrix}$

Section 5.2, page 288

1. $\lambda^2 - 4\lambda - 45$; $9, -5$ **3.** $\lambda^2 - 2\lambda - 1$; $1 \pm \sqrt{2}$

5. $\lambda^2 - 6\lambda + 9$; 3 **7.** $\lambda^2 - 9\lambda + 32$; no real eigenvalues

9. $-\lambda^3 + 4\lambda^2 - 9\lambda - 6$ **11.** $-\lambda^3 + 9\lambda^2 - 26\lambda + 24$

13. $-\lambda^3 + 18\lambda^2 - 95\lambda + 150$ **15.** $4, 3, 3, 1$

17. $3, 3, 1, 1, 0$

19. *Hint:* The equation given holds for all λ.

21–29. The *Study Guide* has hints.

31. *Hint:* Find an invertible matrix P so that $RQ = P^{-1}AP$.

33. In general, the eigenvectors of A are not the same as the eigenvectors of A^T, unless, of course, $A^T = A$.

35. $a = 32$: $\lambda = 1, 1, 2$
$a = 31.9$: $\lambda = .2958, 1, 2.7042$
$a = 31.8$: $\lambda = -.1279, 1, 3.1279$
$a = 32.1$: $\lambda = 1, 1.5 \pm .9747i$
$a = 32.2$: $\lambda = 1; 1.5 \pm 1.4663i$

Section 5.3, page 295

1. $\begin{bmatrix} 226 & -525 \\ 90 & -209 \end{bmatrix}$ **3.** $\begin{bmatrix} a^k & 0 \\ 3(a^k - b^k) & b^k \end{bmatrix}$

5. $\lambda = 5$: $\begin{bmatrix} 1 \\ 1 \\ 1 \end{bmatrix}$; $\lambda = 1$: $\begin{bmatrix} 1 \\ 0 \\ -1 \end{bmatrix}, \begin{bmatrix} 2 \\ -1 \\ 0 \end{bmatrix}$

When an answer involves a diagonalization, $A = PDP^{-1}$, the factors P and D are not unique, so your answer may differ from that given here.

7. $P = \begin{bmatrix} 1 & 0 \\ 3 & 1 \end{bmatrix}$, $D = \begin{bmatrix} 1 & 0 \\ 0 & -1 \end{bmatrix}$ **9.** Not diagonalizable

11. $P = \begin{bmatrix} 1 & 2 & 1 \\ 3 & 3 & 1 \\ 4 & 3 & 1 \end{bmatrix}$, $D = \begin{bmatrix} 3 & 0 & 0 \\ 0 & 2 & 0 \\ 0 & 0 & 1 \end{bmatrix}$

13. $P = \begin{bmatrix} -1 & 2 & 1 \\ -1 & -1 & 0 \\ 1 & 0 & 1 \end{bmatrix}$, $D = \begin{bmatrix} 5 & 0 & 0 \\ 0 & 1 & 0 \\ 0 & 0 & 1 \end{bmatrix}$

15. $P = \begin{bmatrix} -1 & -4 & -2 \\ 1 & 0 & -1 \\ 0 & 1 & 1 \end{bmatrix}$, $D = \begin{bmatrix} 3 & 0 & 0 \\ 0 & 3 & 0 \\ 0 & 0 & 1 \end{bmatrix}$

17. Not diagonalizable

19. $P = \begin{bmatrix} 1 & 3 & -1 & -1 \\ 0 & 2 & -1 & 2 \\ 0 & 0 & 1 & 0 \\ 0 & 0 & 0 & 1 \end{bmatrix}$, $D = \begin{bmatrix} 5 & 0 & 0 & 0 \\ 0 & 3 & 0 & 0 \\ 0 & 0 & 2 & 0 \\ 0 & 0 & 0 & 2 \end{bmatrix}$

21–27. See the *Study Guide*.

29. Yes. (Explain why.)

31. No, A must be diagonalizable. (Explain why.)

33. *Hint:* Write $A = PDP^{-1}$. Since A is invertible, 0 is not an eigenvalue of A, so D has nonzero entries on its diagonal.

35. One answer is $P_1 = \begin{bmatrix} 1 & 1 \\ -2 & -1 \end{bmatrix}$, whose columns are eigenvectors corresponding to the eigenvalues in D_1.

37. *Hint:* Construct a suitable 2×2 triangular matrix.

39. $P = \begin{bmatrix} 2 & 2 & 1 & 6 \\ 1 & -1 & 1 & -3 \\ -1 & -7 & 1 & 0 \\ 2 & 2 & 0 & 4 \end{bmatrix}$,

$D = \begin{bmatrix} 5 & 0 & 0 & 0 \\ 0 & 1 & 0 & 0 \\ 0 & 0 & -2 & 0 \\ 0 & 0 & 0 & -2 \end{bmatrix}$

41. $P = \begin{bmatrix} 6 & 3 & 2 & 4 & 3 \\ -1 & -1 & -1 & -3 & -1 \\ -3 & -3 & -4 & -2 & -4 \\ 3 & 0 & -1 & 5 & 0 \\ 0 & 3 & 4 & 0 & 5 \end{bmatrix}$,

$D = \begin{bmatrix} 5 & 0 & 0 & 0 & 0 \\ 0 & 5 & 0 & 0 & 0 \\ 0 & 0 & 3 & 0 & 0 \\ 0 & 0 & 0 & 1 & 0 \\ 0 & 0 & 0 & 0 & 1 \end{bmatrix}$

Section 5.4, page 303

1. $\begin{bmatrix} 3 & -1 & 0 \\ -5 & 6 & 4 \\ 0 & 0 & 0 \end{bmatrix}$

3. $\begin{bmatrix} 3 & 0 & 0 \\ 5 & -2 & 0 \\ 0 & 4 & 1 \end{bmatrix}$

5. $24\mathbf{b}_1 - 20\mathbf{b}_2 + 11\mathbf{b}_3$

7. $\begin{bmatrix} 1 & 5 \\ 0 & 1 \end{bmatrix}$ **9.** $\mathbf{b}_1 = \begin{bmatrix} 1 \\ 1 \end{bmatrix}, \mathbf{b}_2 = \begin{bmatrix} 1 \\ 3 \end{bmatrix}$

11. $\mathbf{b}_1 = \begin{bmatrix} -2 \\ 1 \end{bmatrix}, \mathbf{b}_2 = \begin{bmatrix} 1 \\ 1 \end{bmatrix}$

13. a. $A\mathbf{b}_1 = 2\mathbf{b}_1$, so $\mathbf{b}_1$ is an eigenvector of A. However, A has only one eigenvalue, $\lambda = 2$, and the eigenspace is only one-dimensional, so A is not diagonalizable.

b. $\begin{bmatrix} 2 & -1 \\ 0 & 2 \end{bmatrix}$

15. a. $T(\mathbf{p}) = 3 + 3t + 3t^2 = 3\mathbf{p}$ so $\mathbf{p}$ is an eigenvector of T with eigenvalue 3.

b. $T(\mathbf{p}) = -1 - t - t^2$ so $\mathbf{p}$ is not an eigenvector.

17–19. See the *Study Guide*.

21. By definition, if A is similar to B, there exists an invertible matrix P such that $P^{-1}AP = B$. (See Section 5.2.) Then B is invertible because it is the product of invertible matrices. To show that A^{-1} is similar to B^{-1}, use the equation $P^{-1}AP = B$. See the *Study Guide*.

23. *Hint:* Review Practice Problem 2.

25. *Hint:* Compute $B(P^{-1}\mathbf{x})$.

27. *Hint:* Write $A = PBP^{-1} = (PB)P^{-1}$, and use the trace property.

29. $S(\chi) = \chi$ so χ is an eigenvector of S with eigenvalue 1.

31. $M_2(\alpha) = 0$ so α is an eigenvector of M_2 with eigenvalue 0.

33. $P^{-1}AP = \begin{bmatrix} 8 & 3 & -6 \\ 0 & 1 & 3 \\ 0 & 0 & -3 \end{bmatrix}$

35. $\lambda = 2$: $\mathbf{b}_1 = \begin{bmatrix} 0 \\ -3 \\ 3 \\ 2 \end{bmatrix}$; $\lambda = 4$: $\mathbf{b}_2 = \begin{bmatrix} -30 \\ -7 \\ 3 \\ 0 \end{bmatrix}$,

$\mathbf{b}_3 = \begin{bmatrix} 39 \\ 5 \\ 0 \\ 3 \end{bmatrix}$; $\lambda = 5$: $\mathbf{b}_4 = \begin{bmatrix} 11 \\ -3 \\ 4 \\ 4 \end{bmatrix}$;

basis: $\mathcal{B} = \{\mathbf{b}_1, \mathbf{b}_2, \mathbf{b}_3, \mathbf{b}_4\}$

Section 5.5, page 310

1. $\lambda = 2 + i, \begin{bmatrix} -1 + i \\ 1 \end{bmatrix}$; $\lambda = 2 - i, \begin{bmatrix} -1 - i \\ 1 \end{bmatrix}$

3. $\lambda = 2 + 3i, \begin{bmatrix} 1 - 3i \\ 2 \end{bmatrix}$; $\lambda = 2 - 3i, \begin{bmatrix} 1 + 3i \\ 2 \end{bmatrix}$

5. $\lambda = 2 + 2i, \begin{bmatrix} 1 \\ 2 + 2i \end{bmatrix}$; $\lambda = 2 - 2i, \begin{bmatrix} 1 \\ 2 - 2i \end{bmatrix}$

7. $\lambda = \sqrt{3} \pm i, \varphi = \pi/6$ radian, $r = 2$

9. $\lambda = -\sqrt{3}/2 \pm (1/2)i, \varphi = -5\pi/6$ radians, $r = 1$

11. $\lambda = .1 \pm .1i, \varphi = -\pi/4$ radian, $r = \sqrt{2}/10$

In Exercises 13–20, other answers are possible. Any P that makes $P^{-1}AP$ equal to the given C or to C^T is a satisfactory answer. First find P; then compute $P^{-1}AP$.

13. $P = \begin{bmatrix} -1 & -1 \\ 1 & 0 \end{bmatrix}, C = \begin{bmatrix} 2 & -1 \\ 1 & 2 \end{bmatrix}$

15. $P = \begin{bmatrix} 1 & 3 \\ 2 & 0 \end{bmatrix}, C = \begin{bmatrix} 2 & -3 \\ 3 & 2 \end{bmatrix}$

17. $P = \begin{bmatrix} 2 & -1 \\ 5 & 0 \end{bmatrix}, C = \begin{bmatrix} -.6 & -.8 \\ .8 & -.6 \end{bmatrix}$

19. $P = \begin{bmatrix} 2 & -1 \\ 2 & 0 \end{bmatrix}, C = \begin{bmatrix} .96 & -.28 \\ .28 & .96 \end{bmatrix}$

21. $\mathbf{y} = \begin{bmatrix} 2 \\ -1 + 2i \end{bmatrix} = \dfrac{-1 + 2i}{5} \begin{bmatrix} -2 - 4i \\ 5 \end{bmatrix}$

23–25. See the *Study Guide*.

27. (a) Properties of conjugates and the fact that $\overline{\mathbf{x}}^T = \overline{\mathbf{x}^T}$; **(b)** $\overline{A\mathbf{x}} = A\overline{\mathbf{x}}$ and A is real; **(c)** because $\mathbf{x}^T A\overline{\mathbf{x}}$ is a scalar and hence may be viewed as a 1×1 matrix; **(d)** properties of transposes; **(e)** $A^T = A$, definition of q

29. *Hint:* First write $\mathbf{x} = \text{Re }\mathbf{x} + i(\text{Im }\mathbf{x})$.

31. $P = \begin{bmatrix} 1 & -1 & -2 & 0 \\ -4 & 0 & 0 & 2 \\ 0 & 0 & -3 & -1 \\ 2 & 0 & 4 & 0 \end{bmatrix}$, $C = \begin{bmatrix} .2 & -.5 & 0 & 0 \\ .5 & .2 & 0 & 0 \\ 0 & 0 & .3 & -.1 \\ 0 & 0 & .1 & .3 \end{bmatrix}$.

Other choices are possible, but C must equal $P^{-1}AP$.

Section 5.6, page 320

1. a. *Hint:* Find c_1, c_2 such that $\mathbf{x}_0 = c_1\mathbf{v}_1 + c_2\mathbf{v}_2$. Use this representation and the fact that $\mathbf{v}_1$ and $\mathbf{v}_2$ are eigenvectors of A to compute $\mathbf{x}_1 = \begin{bmatrix} 49/3 \\ 41/3 \end{bmatrix}$.

   b. In general, $\mathbf{x}_k = 5(3)^k\mathbf{v}_1 - 4(\frac{1}{3})^k\mathbf{v}_2$ for $k \geq 0$.

3. When $p = .2$, the eigenvalues of A are .9 and .7, and

$$\mathbf{x}_k = c_1(.9)^k\begin{bmatrix} 1 \\ 1 \end{bmatrix} + c_2(.7)^k\begin{bmatrix} 2 \\ 1 \end{bmatrix} \to \mathbf{0} \quad \text{as } k \to \infty$$

The higher predation rate cuts down the owls' food supply, and eventually both predator and prey populations perish.

5. If $p = .325$, the eigenvalues are 1.05 and .55. Since $1.05 > 1$, both populations will grow at 5% per year. An eigenvector for 1.05 is $(6, 13)$, so eventually there will be approximately 6 spotted owls to every 13 (thousand) flying squirrels.

7. a. The origin is a saddle point because A has one eigenvalue larger than 1 and one smaller than 1 (in absolute value).

   b. The direction of greatest attraction is given by the eigenvector corresponding to the eigenvalue $1/3$, namely, $\mathbf{v}_2$. All vectors that are multiples of $\mathbf{v}_2$ are attracted to the origin. The direction of greatest repulsion is given by the eigenvector $\mathbf{v}_1$. All multiples of $\mathbf{v}_1$ are repelled.

   c. See the *Study Guide.*

9. Saddle point; eigenvalues: 2, .5; direction of greatest repulsion: the line through $(0, 0)$ and $(-1, 1)$; direction of greatest attraction: the line through $(0, 0)$ and $(1, 4)$

11. Attractor; eigenvalues: .9, .8; greatest attraction: line through $(0, 0)$ and $(5, 4)$

13. Repellor; eigenvalues: 1.2, 1.1; greatest repulsion: line through $(0, 0)$ and $(3, 4)$

15. $\mathbf{x}_k = \mathbf{v}_1 + .1(.5)^k\begin{bmatrix} 2 \\ -3 \\ 1 \end{bmatrix} + .3(.2)^k\begin{bmatrix} -1 \\ 0 \\ 1 \end{bmatrix} \to \mathbf{v}_1$ as $k \to \infty$

17. a. $A = \begin{bmatrix} 0 & 1.6 \\ .3 & .8 \end{bmatrix}$

   b. The population is growing because the largest eigenvalue of A is 1.2, which is larger than 1 in magnitude. The eventual growth rate is 1.2, which is 20% per year. The eigenvector $(4, 3)$ for $\lambda_1 = 1.2$ shows that there will be 4 juveniles for every 3 adults.

   c. The juvenile–adult ratio seems to stabilize after about 5 or 6 years. The *Study Guide* describes how to construct a matrix program to generate a data matrix whose columns list the numbers of juveniles and adults each year. Graphing the data is also discussed.

Section 5.7, page 328

1. $\mathbf{x}(t) = \frac{5}{2}\begin{bmatrix} -3 \\ 1 \end{bmatrix}e^{4t} - \frac{3}{2}\begin{bmatrix} -1 \\ 1 \end{bmatrix}e^{2t}$

3. $-\frac{5}{2}\begin{bmatrix} -3 \\ 1 \end{bmatrix}e^t + \frac{9}{2}\begin{bmatrix} -1 \\ 1 \end{bmatrix}e^{-t}$. The origin is a saddle point. The direction of greatest attraction is the line through $(-1, 1)$ and the origin. The direction of greatest repulsion is the line through $(-3, 1)$ and the origin.

5. $-\frac{1}{2}\begin{bmatrix} 1 \\ 3 \end{bmatrix}e^{4t} + \frac{7}{2}\begin{bmatrix} 1 \\ 1 \end{bmatrix}e^{6t}$. The origin is a repellor. The direction of greatest repulsion is the line through $(1, 1)$ and the origin.

7. Set $P = \begin{bmatrix} 1 & 1 \\ 3 & 1 \end{bmatrix}$ and $D = \begin{bmatrix} 4 & 0 \\ 0 & 6 \end{bmatrix}$. Then $A = PDP^{-1}$. Substituting $\mathbf{x} = P\mathbf{y}$ into $\mathbf{x}' = A\mathbf{x}$, we have

$$\frac{d}{dt}(P\mathbf{y}) = A(P\mathbf{y})$$
$$P\mathbf{y}' = PDP^{-1}(P\mathbf{y}) = PD\mathbf{y}$$

Left-multiplying by P^{-1} gives

$$\mathbf{y}' = D\mathbf{y}, \quad \text{or} \quad \begin{bmatrix} y_1'(t) \\ y_2'(t) \end{bmatrix} = \begin{bmatrix} 4 & 0 \\ 0 & 6 \end{bmatrix}\begin{bmatrix} y_1(t) \\ y_2(t) \end{bmatrix}$$

9. (complex solution):

$$c_1\begin{bmatrix} 1 - i \\ 1 \end{bmatrix}e^{(-2+i)t} + c_2\begin{bmatrix} 1 + i \\ 1 \end{bmatrix}e^{(-2-i)t}$$

(real solution):

$$c_1\begin{bmatrix} \cos t + \sin t \\ \cos t \end{bmatrix}e^{-2t} + c_2\begin{bmatrix} \sin t - \cos t \\ \sin t \end{bmatrix}e^{-2t}$$

The trajectories spiral in toward the origin.

11. (complex): $c_1\begin{bmatrix} -3 + 3i \\ 2 \end{bmatrix}e^{3it} + c_2\begin{bmatrix} -3 - 3i \\ 2 \end{bmatrix}e^{-3it}$

(real):

$$c_1\begin{bmatrix} -3\cos 3t - 3\sin 3t \\ 2\cos 3t \end{bmatrix} + c_2\begin{bmatrix} -3\sin 3t + 3\cos 3t \\ 2\sin 3t \end{bmatrix}$$

The trajectories are ellipses about the origin.

13. (complex): $c_1\begin{bmatrix} 1 + i \\ 2 \end{bmatrix}e^{(1+3i)t} + c_2\begin{bmatrix} 1 - i \\ 2 \end{bmatrix}e^{(1-3i)t}$

(real): $c_1\begin{bmatrix} \cos 3t - \sin 3t \\ 2\cos 3t \end{bmatrix}e^t + c_2\begin{bmatrix} \sin 3t + \cos 3t \\ 2\sin 3t \end{bmatrix}e^t$

The trajectories spiral out, away from the origin.

15. $\mathbf{x}(t) = c_1 \begin{bmatrix} -1 \\ 0 \\ 1 \end{bmatrix} e^{-2t} + c_2 \begin{bmatrix} -6 \\ 1 \\ 5 \end{bmatrix} e^{-t} + c_3 \begin{bmatrix} -4 \\ 1 \\ 4 \end{bmatrix} e^{t}$

The origin is a saddle point. A solution with $c_3 = 0$ is attracted to the origin. A solution with $c_1 = c_2 = 0$ is repelled.

17. (complex):

$c_1 \begin{bmatrix} -3 \\ 1 \\ 1 \end{bmatrix} e^{t} + c_2 \begin{bmatrix} 23 - 34i \\ -9 + 14i \\ 3 \end{bmatrix} e^{(5+2i)t} +$

$c_3 \begin{bmatrix} 23 + 34i \\ -9 - 14i \\ 3 \end{bmatrix} e^{(5-2i)t}$

(real): $c_1 \begin{bmatrix} -3 \\ 1 \\ 1 \end{bmatrix} e^{t} + c_2 \begin{bmatrix} 23\cos 2t + 34\sin 2t \\ -9\cos 2t - 14\sin 2t \\ 3\cos 2t \end{bmatrix} e^{5t} +$

$c_3 \begin{bmatrix} 23\sin 2t - 34\cos 2t \\ -9\sin 2t + 14\cos 2t \\ 3\sin 2t \end{bmatrix} e^{5t}$

The origin is a repellor. The trajectories spiral outward, away from the origin.

19. $A = \begin{bmatrix} -2 & 3/4 \\ 1 & -1 \end{bmatrix}$,

$\begin{bmatrix} v_1(t) \\ v_2(t) \end{bmatrix} = \dfrac{5}{2} \begin{bmatrix} 1 \\ 2 \end{bmatrix} e^{-.5t} - \dfrac{1}{2} \begin{bmatrix} -3 \\ 2 \end{bmatrix} e^{-2.5t}$

21. $A = \begin{bmatrix} -1 & -8 \\ 5 & -5 \end{bmatrix}$,

$\begin{bmatrix} i_L(t) \\ v_C(t) \end{bmatrix} = \begin{bmatrix} -20\sin 6t \\ 15\cos 6t - 5\sin 6t \end{bmatrix} e^{-3t}$

Section 5.8, page 335

1. Eigenvector: $\mathbf{x}_4 = \begin{bmatrix} 1 \\ .3326 \end{bmatrix}$, or $A\mathbf{x}_4 = \begin{bmatrix} 4.9978 \\ 1.6652 \end{bmatrix}$; $\lambda \approx 4.9978$

3. Eigenvector: $\mathbf{x}_4 = \begin{bmatrix} .5188 \\ 1 \end{bmatrix}$, or $A\mathbf{x}_4 = \begin{bmatrix} .4594 \\ .9075 \end{bmatrix}$; $\lambda \approx .9075$

5. $\mathbf{x} = \begin{bmatrix} -.7999 \\ 1 \end{bmatrix}$, $A\mathbf{x} = \begin{bmatrix} 4.0015 \\ -5.0020 \end{bmatrix}$; estimated $\lambda = -5.0020$

7. $\mathbf{x}_k$: $\begin{bmatrix} .75 \\ 1 \end{bmatrix}$, $\begin{bmatrix} 1 \\ .9565 \end{bmatrix}$, $\begin{bmatrix} .9932 \\ 1 \end{bmatrix}$, $\begin{bmatrix} 1 \\ .9990 \end{bmatrix}$, $\begin{bmatrix} .9998 \\ 1 \end{bmatrix}$
μ_k: 11.5, 12.78, 12.96, 12.9948, 12.9990

9. $\mu_5 = 8.4233$, $\mu_6 = 8.4246$; actual value: 8.42443 (accurate to 5 places)

11. μ_k: 5.8000, 5.9655, 5.9942, 5.9990 ($k = 1, 2, 3, 4$); $R(\mathbf{x}_k)$: 5.9655, 5.9990, 5.99997, 5.9999993

13. Yes, but the sequences may converge very slowly.

15. *Hint:* Write $A\mathbf{x} - \alpha\mathbf{x} = (A - \alpha I)\mathbf{x}$, and use the fact that $(A - \alpha I)$ is invertible when α is *not* an eigenvalue of A.

17. $v_0 = 3.3384$, $v_1 = 3.32119$ (accurate to 4 places with rounding), $v_2 = 3.3212209$. Actual value: 3.3212201 (accurate to 7 places)

19. a. $\mu_6 = 30.2887 = \mu_7$ to four decimal places. To six places, the largest eigenvalue is 30.288685, with eigenvector $(.957629, .688937, 1, .943782)$.

b. The inverse power method (with $\alpha = 0$) produces $\mu_1^{-1} = .010141$, $\mu_2^{-1} = .010150$. To seven places, the smallest eigenvalue is .0101500, with eigenvector $(-.603972, 1, -.251135, .148953)$. The reason for the rapid convergence is that the next-to-smallest eigenvalue is near .85.

21. a. If the eigenvalues of A are all less than 1 in magnitude, and if $\mathbf{x} \neq \mathbf{0}$, then $A^k\mathbf{x}$ is approximately an eigenvector for large k.

b. If the strictly dominant eigenvalue is 1, and if $\mathbf{x}$ has a component in the direction of the corresponding eigenvector, then $\{A^k\mathbf{x}\}$ will converge to a multiple of that eigenvector.

c. If the eigenvalues of A are all greater than 1 in magnitude, and if $\mathbf{x}$ is not an eigenvector, then the distance from $A^k\mathbf{x}$ to the nearest eigenvector will *increase* as $k \to \infty$.

Section 5.9, page 344

1. a. From
	N	M	To
	.7	.6	News
	.3	.4	Music

b. $\begin{bmatrix} 1 \\ 0 \end{bmatrix}$ **c.** 33%

3. a. From
	H	I	To
	.95	.45	Healthy
	.05	.55	Ill

b. 15%, 12.5%

c. .925; use $\mathbf{x}_0 = \begin{bmatrix} 1 \\ 0 \end{bmatrix}$.

5. $\begin{bmatrix} .4 \\ .6 \end{bmatrix}$ **7.** $\begin{bmatrix} 1/4 \\ 1/2 \\ 1/4 \end{bmatrix}$

9. Yes, because P^2 has all positive entries.

11. a. $\begin{bmatrix} 2/3 \\ 1/3 \end{bmatrix}$ **b.** 2/3

13. a. $\begin{bmatrix} .9 \\ .1 \end{bmatrix}$ **b.** .10, no

15–19. See the *Study Guide*.

21. No. $\mathbf{q}$ is not a probability vector since its entries do not add to 1.

23. No. $A\mathbf{q}$ does not equal $\mathbf{q}$.

25. 67%

27. a. The entries in a column of P sum to 1. A column in the matrix $P - I$ has the same entries as in P except that one of the entries is decreased by 1. Hence each column sum is 0.

   b. By (a), the bottom row of $P - I$ is the negative of the sum of the other rows.

   c. By (b) and the Spanning Set Theorem, the bottom row of $P - I$ can be removed and the remaining $(n - 1)$ rows will still span the row space. Alternatively, use (a) and the fact that row operations do not change the row space. Let A be the matrix obtained from $P - I$ by adding to the bottom row all the other rows. By (a), the row space is spanned by the first $(n - 1)$ rows of A.

   d. By the Rank Theorem and (c), the dimension of the column space of $P - I$ is less than n, and hence the null space is nontrivial. Instead of the Rank Theorem, you may use the Invertible Matrix Theorem, since $P - I$ is a square matrix.

29. a. The product $S\mathbf{x}$ equals the sum of the entries in $\mathbf{x}$. For a probability vector, this sum must be 1.

   b. $P = [\,\mathbf{p}_1 \;\; \mathbf{p}_2 \;\; \cdots \;\; \mathbf{p}_n\,]$, where the $\mathbf{p}_i$ are probability vectors. By matrix multiplication and part (a),

$$SP = [\, S\mathbf{p}_1 \;\; S\mathbf{p}_2 \;\; \cdots \;\; S\mathbf{p}_n\,] = [\,1 \;\; 1 \;\; \cdots \;\; 1\,] = S$$

   c. By part (b), $S(P\mathbf{x}) = (SP)\mathbf{x} = S\mathbf{x} = 1$. Also, the entries in $P\mathbf{x}$ are nonnegative (because P and $\mathbf{x}$ have nonnegative entries). Hence, by (a), $P\mathbf{x}$ is a probability vector.

31. a. To four decimal places,

$$P^4 = P^5 = \begin{bmatrix} .2816 & .2816 & .2816 & .2816 \\ .3355 & .3355 & .3355 & .3355 \\ .1819 & .1819 & .1819 & .1819 \\ .2009 & .2009 & .2009 & .2009 \end{bmatrix},$$

$$\mathbf{q} = \begin{bmatrix} .2816 \\ .3355 \\ .1819 \\ .2009 \end{bmatrix}$$

   Note that, due to round-off, the column sums are not 1.

   b. To four decimal places,

$$Q^{80} = \begin{bmatrix} .7354 & .7348 & .7351 \\ .0881 & .0887 & .0884 \\ .1764 & .1766 & .1765 \end{bmatrix},$$

$$Q^{116} = Q^{117} = \begin{bmatrix} .7353 & .7353 & .7353 \\ .0882 & .0882 & .0882 \\ .1765 & .1765 & .1765 \end{bmatrix},$$

$$\mathbf{q} = \begin{bmatrix} .7353 \\ .0882 \\ .1765 \end{bmatrix}$$

   c. Let P be an $n \times n$ regular stochastic matrix, $\mathbf{q}$ the steady-state vector of P, and $\mathbf{e}_1$ the first column of the

identity matrix. Then $P^k\mathbf{e}_1$ is the first column of P^k. By Theorem 11, $P^k\mathbf{e}_1 \to \mathbf{q}$ as $k \to \infty$. Replacing $\mathbf{e}_1$ by the other columns of the identity matrix, we conclude that each column of P^k converges to $\mathbf{q}$ as $k \to \infty$. Thus $P^k \to [\,\mathbf{q} \;\; \mathbf{q} \;\; \cdots \;\; \mathbf{q}\,]$.

Chapter 5 Supplementary Exercises, page 346

1. T **2.** F **3.** T **4.** F **5.** T

6. T **7.** F **8.** T **9.** F **10.** T

11. F **12.** F **13.** F **14.** T **15.** F

16. T **17.** F **18.** T **19.** F **20.** T

21. T **22.** T **23.** F

25. a. Suppose $A\mathbf{x} = \lambda\mathbf{x}$, with $\mathbf{x} \neq \mathbf{0}$. Then $(5I - A)\mathbf{x} = 5\mathbf{x} - A\mathbf{x} = 5\mathbf{x} - \lambda\mathbf{x} = (5 - \lambda)\mathbf{x}$. The eigenvalue is $5 - \lambda$.

   b. $(5I - 3A + A^2)\mathbf{x} = 5\mathbf{x} - 3A\mathbf{x} + A(A\mathbf{x}) = 5\mathbf{x} - 3\lambda\mathbf{x} + \lambda^2\mathbf{x} = (5 - 3\lambda + \lambda^2)\mathbf{x}$. The eigenvalue is $5 - 3\lambda + \lambda^2$.

27. Suppose $A\mathbf{x} = \lambda\mathbf{x}$, with $\mathbf{x} \neq \mathbf{0}$. Then

$$\begin{aligned} p(A)\mathbf{x} &= (c_0 I + c_1 A + c_2 A^2 + \cdots + c_n A^n)\mathbf{x} \\ &= c_0\mathbf{x} + c_1 A\mathbf{x} + c_2 A^2\mathbf{x} + \cdots + c_n A^n\mathbf{x} \\ &= c_0\mathbf{x} + c_1\lambda\mathbf{x} + c_2\lambda^2\mathbf{x} + \cdots + c_n\lambda^n\mathbf{x} = p(\lambda)\mathbf{x} \end{aligned}$$

So $p(\lambda)$ is an eigenvalue of the matrix $p(A)$.

29. If $A = PDP^{-1}$, then $p(A) = Pp(D)P^{-1}$, as shown in Exercise 28. If the (j, j) entry in D is λ, then the (j, j) entry in D^k is λ^k, and so the (j, j) entry in $p(D)$ is $p(\lambda)$. If p is the characteristic polynomial of A, then $p(\lambda) = 0$ for each diagonal entry of D, because these entries in D are the eigenvalues of A. Thus $p(D)$ is the zero matrix. Thus $p(A) = P0P^{-1} = 0$.

31. If $I - A$ were not invertible, then the equation $(I - A)\mathbf{x} = \mathbf{0}$ would have a nontrivial solution $\mathbf{x}$. Then $\mathbf{x} - A\mathbf{x} = \mathbf{0}$ and $A\mathbf{x} = 1 \cdot \mathbf{x}$, which shows that A would have 1 as an eigenvalue. This cannot happen if all the eigenvalues are less than 1 in magnitude. So $I - A$ must be invertible.

33. a. Take $\mathbf{x}$ in H. Then $\mathbf{x} = c\mathbf{u}$ for some scalar c. So $A\mathbf{x} = A(c\mathbf{u}) = c(A\mathbf{u}) = c(\lambda\mathbf{u}) = (c\lambda)\mathbf{u}$, which shows that $A\mathbf{x}$ is in H.

   b. Let $\mathbf{x}$ be a nonzero vector in K. Since K is one-dimensional, K must be the set of all scalar multiples of $\mathbf{x}$. If K is invariant under A, then $A\mathbf{x}$ is in K and hence $A\mathbf{x}$ is a multiple of $\mathbf{x}$. Thus $\mathbf{x}$ is an eigenvector of A.

35. 1, 3, 7

37. Replace a by $a - \lambda$ in the determinant formula from Exercise 30 in Chapter 3 Supplementary Exercises:

$$\det(A - \lambda I) = (a - b - \lambda)^{n-1}[a - \lambda + (n - 1)b]$$

This determinant is zero only if $a - b - \lambda = 0$ or $a - \lambda + (n - 1)b = 0$. Thus λ is an eigenvalue of A if and only if $\lambda = a - b$ or $\lambda = a + (n - 1)b$. From the formula for $\det(A - \lambda I)$ above, the algebraic multiplicity is $n - 1$ for $a - b$ and 1 for $a + (n - 1)b$.

39. $\det(A - \lambda I) = (a_{11} - \lambda)(a_{22} - \lambda) - a_{12}a_{21} =$
$\lambda^2 - (a_{11} + a_{22})\lambda + (a_{11}a_{22} - a_{12}a_{21}) =$
$\lambda^2 - (\text{tr } A)\lambda + \det A$. Use the quadratic formula to solve the characteristic equation:

$$\lambda = \frac{\text{tr } A \pm \sqrt{(\text{tr } A)^2 - 4 \det A}}{2}$$

The eigenvalues are both real if and only if the discriminant is nonnegative, that is, $(\text{tr } A)^2 - 4 \det A \geq 0$. This inequality simplifies to $(\text{tr } A)^2 \geq 4 \det A$ and $\left(\dfrac{\text{tr } A}{2}\right)^2 \geq \det A$.

41. $C_p = \begin{bmatrix} 0 & 1 \\ -6 & 5 \end{bmatrix}$; $\det(C_p - \lambda I) = 6 - 5\lambda + \lambda^2 = p(\lambda)$

43. If p is a polynomial of order 2, then a calculation such as in Exercise 41 shows that the characteristic polynomial of C_p is $p(\lambda) = (-1)^2 p(\lambda)$, so the result is true for $n = 2$. Suppose the result is true for $n = k$ for some $k \geq 2$, and consider a polynomial p of degree $k + 1$. Then expanding $\det(C_p - \lambda I)$ by cofactors down the first column, the determinant of $C_p - \lambda I$ equals

$$(-\lambda) \det \begin{bmatrix} -\lambda & 1 & \cdots & 0 \\ \vdots & & & \vdots \\ 0 & & & 1 \\ -a_1 & -a_2 & \cdots & -a_k - \lambda \end{bmatrix} + (-1)^{k+1} a_0$$

The $k \times k$ matrix shown is $C_q - \lambda I$, where $q(t) = a_1 + a_2 t + \cdots + a_k t^{k-1} + t^k$. By the induction assumption, the determinant of $C_q - \lambda I$ is $(-1)^k q(\lambda)$. Thus

$\det(C_p - \lambda I) = (-1)^{k+1} a_0 + (-\lambda)(-1)^k q(\lambda)$
$\qquad = (-1)^{k+1}[a_0 + \lambda(a_1 + \cdots + a_k \lambda^{k-1} + \lambda^k)]$
$\qquad = (-1)^{k+1} p(\lambda)$

So the formula holds for $n = k + 1$ when it holds for $n = k$. By the principle of induction, the formula for $\det(C_p - \lambda I)$ is true for all $n \geq 2$.

45. From Exercise 44, the columns of the Vandermonde matrix V are eigenvectors of C_p, corresponding to the eigenvalues $\lambda_1, \lambda_2, \lambda_3$ (the roots of the polynomial p). Since these eigenvalues are distinct, the eigenvectors form a linearly independent set, by Theorem 2 in Section 5.1. Thus V has linearly independent columns and hence is invertible, by the Invertible Matrix Theorem. Finally, since the columns of V are eigenvectors of C_p, the Diagonalization Theorem (Theorem 5 in Section 5.3) shows that $V^{-1} C_p V$ is diagonal.

47. If your matrix program computes eigenvalues and eigenvectors by iterative methods rather than symbolic calculations, you may have some difficulties. You should find that $AP - PD$ has extremely small entries and PDP^{-1} is close to A. (This was true just a few years ago, but the situation could change as matrix programs continue

to improve.) If you constructed P from the program's eigenvectors, check the condition number of P. This may indicate that you do not really have three linearly independent eigenvectors.

Chapter 6

Section 6.1, page 356

1. $5, 4, \dfrac{4}{5}$ **3.** $\begin{bmatrix} 3/35 \\ -1/35 \\ -1/7 \end{bmatrix}$ **5.** $\begin{bmatrix} 8/13 \\ 12/13 \end{bmatrix}$

7. $\sqrt{35}$ **9.** $\begin{bmatrix} -.6 \\ .8 \end{bmatrix}$ **11.** $\begin{bmatrix} 7/\sqrt{69} \\ 2/\sqrt{69} \\ 4/\sqrt{69} \end{bmatrix}$

13. $5\sqrt{5}$ **15.** Not orthogonal **17.** Orthogonal

19–27. Refer to the *Study Guide* after you have written your answers.

29. *Hint:* Use Theorems 3 and 2 from Section 2.1.

31. $\mathbf{u} \cdot \mathbf{v} = 0$, $\|\mathbf{u}\|^2 = 30$, $\|\mathbf{v}\|^2 = 101$,
$\quad \|\mathbf{u} + \mathbf{v}\|^2 = (-5)^2 + (-9)^2 + 5^2 = 131 = 30 + 101$

33. The set of all multiples of $\begin{bmatrix} -b \\ a \end{bmatrix}$ (when $\mathbf{v} \neq \mathbf{0}$)

35. *Hint:* Use the definition of orthogonality.

37. *Hint:* Consider a typical vector $\mathbf{w} = c_1\mathbf{v}_1 + \cdots + c_p\mathbf{v}_p$ in W.

39. *Hint:* If $\mathbf{x}$ is in $W^\perp$, then $\mathbf{x}$ is orthogonal to every vector in W.

41. State your conjecture and verify it algebraically.

Section 6.2, page 364

1. Not orthogonal **3.** Not orthogonal **5.** Orthogonal

7. Show $\mathbf{u}_1 \cdot \mathbf{u}_2 = 0$, mention Theorem 4, and observe that two linearly independent vectors in $\mathbb{R}^2$ form a basis. Then obtain

$$\mathbf{x} = \frac{39}{13}\begin{bmatrix} 2 \\ -3 \end{bmatrix} + \frac{26}{52}\begin{bmatrix} 6 \\ 4 \end{bmatrix} = 3\begin{bmatrix} 2 \\ -3 \end{bmatrix} + \frac{1}{2}\begin{bmatrix} 6 \\ 4 \end{bmatrix}$$

9. Show $\mathbf{u}_1 \cdot \mathbf{u}_2 = 0$, $\mathbf{u}_1 \cdot \mathbf{u}_3 = 0$, and $\mathbf{u}_2 \cdot \mathbf{u}_3 = 0$. Mention Theorem 4, and observe that three linearly independent vectors in $\mathbb{R}^3$ form a basis. Then obtain

$$\mathbf{x} = \frac{5}{2}\mathbf{u}_1 - \frac{27}{18}\mathbf{u}_2 + \frac{18}{9}\mathbf{u}_3 = \frac{5}{2}\mathbf{u}_1 - \frac{3}{2}\mathbf{u}_2 + 2\mathbf{u}_3$$

11. $\begin{bmatrix} -2 \\ 1 \end{bmatrix}$ **13.** $\mathbf{y} = \begin{bmatrix} -4/5 \\ 7/5 \end{bmatrix} + \begin{bmatrix} 14/5 \\ 8/5 \end{bmatrix}$

15. $\mathbf{y} - \hat{\mathbf{y}} = \begin{bmatrix} .6 \\ -.8 \end{bmatrix}$, distance is 1

17. $\begin{bmatrix} 1/\sqrt{3} \\ 1/\sqrt{3} \\ 1/\sqrt{3} \end{bmatrix}, \begin{bmatrix} -1/\sqrt{2} \\ 0 \\ 1/\sqrt{2} \end{bmatrix}$

19. Orthonormal **21.** Orthonormal

23–31. See the *Study Guide*.

33. *Hint:* $\|U\mathbf{x}\|^2 = (U\mathbf{x})^T(U\mathbf{x})$. Also, parts (a) and (c) follow from (b).

35. *Hint:* You need two theorems, one of which applies only to *square* matrices.

37. *Hint:* If you have a candidate for an inverse, you can check to see whether the candidate works.

39. Suppose $\hat{\mathbf{y}} = \dfrac{\mathbf{y} \cdot \mathbf{u}}{\mathbf{u} \cdot \mathbf{u}}\mathbf{u}$. Replace $\mathbf{u}$ by $c\mathbf{u}$ with $c \neq 0$; then

$$\frac{\mathbf{y} \cdot (c\mathbf{u})}{(c\mathbf{u}) \cdot (c\mathbf{u})}(c\mathbf{u}) = \frac{c(\mathbf{y} \cdot \mathbf{u})}{c^2 \mathbf{u} \cdot \mathbf{u}}(c)\mathbf{u} = \hat{\mathbf{y}}$$

41. Let $L = \text{Span}\{\mathbf{u}\}$, where $\mathbf{u}$ is nonzero, and let $T(\mathbf{x}) = \text{proj}_L \mathbf{x}$. By definition,

$$T(\mathbf{x}) = \frac{\mathbf{x} \cdot \mathbf{u}}{\mathbf{u} \cdot \mathbf{u}}\mathbf{u} = (\mathbf{x} \cdot \mathbf{u})(\mathbf{u} \cdot \mathbf{u})^{-1}\mathbf{u}$$

For $\mathbf{x}$ and $\mathbf{y}$ in $\mathbb{R}^n$ and any scalars c and d, properties of the inner product (Theorem 1) show that

$$\begin{aligned}
T(c\mathbf{x} + d\mathbf{y}) &= [(c\mathbf{x} + d\mathbf{y}) \cdot \mathbf{u}](\mathbf{u} \cdot \mathbf{u})^{-1}\mathbf{u} \\
&= [c(\mathbf{x} \cdot \mathbf{u}) + d(\mathbf{y} \cdot \mathbf{u})](\mathbf{u} \cdot \mathbf{u})^{-1}\mathbf{u} \\
&= c(\mathbf{x} \cdot \mathbf{u})(\mathbf{u} \cdot \mathbf{u})^{-1}\mathbf{u} + d(\mathbf{y} \cdot \mathbf{u})(\mathbf{u} \cdot \mathbf{u})^{-1}\mathbf{u} \\
&= cT(\mathbf{x}) + dT(\mathbf{y})
\end{aligned}$$

Thus T is linear.

43. The proof of Theorem 6 shows that the inner products to be checked are actually entries in the matrix product $A^T A$. A calculation shows that $A^T A = 100 I_4$. Since the off-diagonal entries in $A^T A$ are zero, the columns of A are orthogonal.

Section 6.3, page 374

1. $\mathbf{x} = -\dfrac{8}{9}\mathbf{u}_1 - \dfrac{2}{9}\mathbf{u}_2 + \dfrac{2}{3}\mathbf{u}_3 + 2\mathbf{u}_4;$ $\mathbf{x} = \begin{bmatrix} 0 \\ -2 \\ 4 \\ -2 \end{bmatrix} + \begin{bmatrix} 10 \\ -6 \\ -2 \\ 2 \end{bmatrix}$

3. $\begin{bmatrix} -1 \\ 4 \\ 0 \end{bmatrix}$ **5.** $\begin{bmatrix} -1 \\ 2 \\ 6 \end{bmatrix} = \mathbf{y}$

7. $\mathbf{y} = \begin{bmatrix} 10/3 \\ 2/3 \\ 8/3 \end{bmatrix} + \begin{bmatrix} -7/3 \\ 7/3 \\ 7/3 \end{bmatrix}$ **9.** $\mathbf{y} = \begin{bmatrix} 2 \\ 4 \\ 0 \\ 0 \end{bmatrix} + \begin{bmatrix} 2 \\ -1 \\ 3 \\ -1 \end{bmatrix}$

11. $\begin{bmatrix} 3 \\ -1 \\ 1 \\ -1 \end{bmatrix}$ **13.** $\begin{bmatrix} -1 \\ -3 \\ -2 \\ 3 \end{bmatrix}$ **15.** $\sqrt{40}$

17. a. $U^T U = \begin{bmatrix} 1 & 0 \\ 0 & 1 \end{bmatrix}$,

$$U U^T = \begin{bmatrix} 8/9 & -2/9 & 2/9 \\ -2/9 & 5/9 & 4/9 \\ 2/9 & 4/9 & 5/9 \end{bmatrix}$$

b. $\text{proj}_W \mathbf{y} = 6\mathbf{u}_1 + 3\mathbf{u}_2 = \begin{bmatrix} 2 \\ 4 \\ 5 \end{bmatrix}$, and $(U U^T)\mathbf{y} = \begin{bmatrix} 2 \\ 4 \\ 5 \end{bmatrix}$

19. Any multiple of $\begin{bmatrix} 0 \\ 2/5 \\ 1/5 \end{bmatrix}$, such as $\begin{bmatrix} 0 \\ 2 \\ 1 \end{bmatrix}$

21–29. Write your answers before checking the *Study Guide*.

31. *Hint:* Use Theorem 3 and the Orthogonal Decomposition Theorem. For the uniqueness, suppose $A\mathbf{p} = \mathbf{b}$ and $A\mathbf{p}_1 = \mathbf{b}$, and consider the equations $\mathbf{p} = \mathbf{p}_1 + (\mathbf{p} - \mathbf{p}_1)$ and $\mathbf{p} = \mathbf{p} + \mathbf{0}$.

33. $\mathbf{w} = \begin{bmatrix} 1 \\ 0 \\ 0 \\ 1 \end{bmatrix}$ $M = \begin{bmatrix} 1 & 0 & 0 & -1 \\ 0 & 1 & 0 & 0 \\ 0 & 0 & 1 & 0 \\ -1 & 0 & 0 & 1 \end{bmatrix}$

35. $\mathbf{w} = \begin{bmatrix} 1 \\ 1 \\ 1 \\ 1 \\ 0 \\ 0 \\ 1 \\ 1 \\ 1 \end{bmatrix}$;

$$M = \begin{bmatrix} 6 & -1 & -1 & -1 & 0 & 0 & -1 & -1 & -1 \\ -1 & 1 & 0 & 0 & 0 & 0 & 0 & 0 & 0 \\ -1 & 0 & 1 & 0 & 0 & 0 & 0 & 0 & 0 \\ -1 & 0 & 0 & 1 & 0 & 0 & 0 & 0 & 0 \\ 0 & 0 & 0 & 0 & 1 & 0 & 0 & 0 & 0 \\ 0 & 0 & 0 & 0 & 0 & 1 & 0 & 0 & 0 \\ -1 & 0 & 0 & 0 & 0 & 0 & 1 & 0 & 0 \\ -1 & 0 & 0 & 0 & 0 & 0 & 0 & 1 & 0 \\ -1 & 0 & 0 & 0 & 0 & 0 & 0 & 0 & 1 \end{bmatrix}$$

37. U has orthonormal columns, by Theorem 6 in Section 6.2, because $U^T U = I_4$. The closest point to $\mathbf{y}$ in $\text{Col}\,U$ is the orthogonal projection $\hat{\mathbf{y}}$ of $\mathbf{y}$ onto $\text{Col}\,U$. From Theorem 10,

$$\hat{\mathbf{y}} = U U^T \mathbf{y} = (1.2, .4, 1.2, 1.2, .4, 1.2, .4, .4)$$

Section 6.4, page 380

1. $\begin{bmatrix} 3 \\ 0 \\ -1 \end{bmatrix}$, $\begin{bmatrix} -1 \\ 5 \\ -3 \end{bmatrix}$ **3.** $\begin{bmatrix} 2 \\ -5 \\ 1 \end{bmatrix}$, $\begin{bmatrix} 3 \\ 3/2 \\ 3/2 \end{bmatrix}$

5. $\begin{bmatrix} 1 \\ -4 \\ 0 \\ 1 \end{bmatrix}$, $\begin{bmatrix} 5 \\ 1 \\ -4 \\ -1 \end{bmatrix}$ **7.** $\begin{bmatrix} 2/\sqrt{30} \\ -5/\sqrt{30} \\ 1/\sqrt{30} \end{bmatrix}$, $\begin{bmatrix} 2/\sqrt{6} \\ 1/\sqrt{6} \\ 1/\sqrt{6} \end{bmatrix}$

9. $\begin{bmatrix} 3 \\ 1 \\ -1 \\ 3 \end{bmatrix}$, $\begin{bmatrix} 1 \\ 3 \\ 3 \\ -1 \end{bmatrix}$, $\begin{bmatrix} -3 \\ 1 \\ 1 \\ 3 \end{bmatrix}$ **11.** $\begin{bmatrix} 1 \\ -1 \\ -1 \\ 1 \\ 1 \end{bmatrix}$, $\begin{bmatrix} 3 \\ 0 \\ 3 \\ -3 \\ 3 \end{bmatrix}$, $\begin{bmatrix} 2 \\ 0 \\ 2 \\ 2 \\ -2 \end{bmatrix}$

13. $R = \begin{bmatrix} 6 & 12 \\ 0 & 6 \end{bmatrix}$

15. $Q = \begin{bmatrix} 1/\sqrt{5} & 1/2 & 1/2 \\ -1/\sqrt{5} & 0 & 0 \\ -1/\sqrt{5} & 1/2 & 1/2 \\ 1/\sqrt{5} & -1/2 & 1/2 \\ 1/\sqrt{5} & 1/2 & -1/2 \end{bmatrix}$,

$R = \begin{bmatrix} \sqrt{5} & -\sqrt{5} & 4\sqrt{5} \\ 0 & 6 & -2 \\ 0 & 0 & 4 \end{bmatrix}$

17–21. See the *Study Guide*.

23. Suppose **x** satisfies $R\mathbf{x} = \mathbf{0}$; then $QR\mathbf{x} = Q\mathbf{0} = \mathbf{0}$, and $A\mathbf{x} = \mathbf{0}$. Since the columns of A are linearly independent, **x** must be zero. This fact, in turn, shows that the columns of R are linearly independent. Since R is square, it is invertible, by the Invertible Matrix Theorem.

25. Denote the columns of Q by $\mathbf{q}_1, \ldots, \mathbf{q}_n$. Note that $n \leq m$, because A is $m \times n$ and has linearly independent columns. Use the fact that the columns of Q can be extended to an orthonormal basis for $\mathbb{R}^m$, say, $\{\mathbf{q}_1, \ldots, \mathbf{q}_m\}$. (The *Study Guide* describes one method.) Let $Q_0 = [\mathbf{q}_{n+1} \cdots \mathbf{q}_m]$ and $Q_1 = [Q \quad Q_0]$. Then, using partitioned matrix multiplication, $Q_1 \begin{bmatrix} R \\ 0 \end{bmatrix} = QR = A$.

27. *Hint:* Partition R as a 2×2 block matrix.

29. The diagonal entries of R are 20, 6, 10.3923, and 7.0711, to four decimal places.

Section 6.5, page 388

1. a. $\begin{bmatrix} 6 & -11 \\ -11 & 22 \end{bmatrix}\begin{bmatrix} x_1 \\ x_2 \end{bmatrix} = \begin{bmatrix} -4 \\ 11 \end{bmatrix}$ **b.** $\hat{\mathbf{x}} = \begin{bmatrix} 3 \\ 2 \end{bmatrix}$

3. a. $\begin{bmatrix} 6 & 6 \\ 6 & 42 \end{bmatrix}\begin{bmatrix} x_1 \\ x_2 \end{bmatrix} = \begin{bmatrix} 6 \\ -6 \end{bmatrix}$ **b.** $\hat{\mathbf{x}} = \begin{bmatrix} 4/3 \\ -1/3 \end{bmatrix}$

5. $\hat{\mathbf{x}} = \begin{bmatrix} 5 \\ -3 \\ 0 \end{bmatrix} + x_3\begin{bmatrix} -1 \\ 1 \\ 1 \end{bmatrix}$ **7.** $2\sqrt{5}$

9. a. $\hat{\mathbf{b}} = \begin{bmatrix} 1 \\ 1 \\ 0 \end{bmatrix}$ **b.** $\hat{\mathbf{x}} = \begin{bmatrix} 2/7 \\ 1/7 \end{bmatrix}$

11. a. $\hat{\mathbf{b}} = \begin{bmatrix} 3 \\ 1 \\ 4 \\ -1 \end{bmatrix}$ **b.** $\hat{\mathbf{x}} = \begin{bmatrix} 2/3 \\ 0 \\ 1/3 \end{bmatrix}$

13. $A\mathbf{u} = \begin{bmatrix} 11 \\ -11 \\ 11 \end{bmatrix}$, $A\mathbf{v} = \begin{bmatrix} 7 \\ -12 \\ 7 \end{bmatrix}$,

$\mathbf{b} - A\mathbf{u} = \begin{bmatrix} 0 \\ 2 \\ -6 \end{bmatrix}$, $\mathbf{b} - A\mathbf{v} = \begin{bmatrix} 4 \\ 3 \\ -2 \end{bmatrix}$. No, **u** could not possibly be a least-squares solution of $A\mathbf{x} = \mathbf{b}$. Why?

15. $\hat{\mathbf{x}} = \begin{bmatrix} 4 \\ -1 \end{bmatrix}$

17–25. See the *Study Guide*.

27. a. If $A\mathbf{x} = \mathbf{0}$, then $A^TA\mathbf{x} = A^T\mathbf{0} = \mathbf{0}$. This shows that Nul A is contained in Nul A^TA.

b. If $A^TA\mathbf{x} = \mathbf{0}$, then $\mathbf{x}^TA^TA\mathbf{x} = \mathbf{x}^T\mathbf{0} = 0$. So $(A\mathbf{x})^T(A\mathbf{x}) = 0$ (which means that $\|A\mathbf{x}\|^2 = 0$), and hence $A\mathbf{x} = \mathbf{0}$. This shows that Nul A^TA is contained in Nul A.

29. *Hint:* For part (a), use an important theorem from Chapter 2.

31. By Theorem 14, $\hat{\mathbf{b}} = A\hat{\mathbf{x}} = A(A^TA)^{-1}A^T\mathbf{b}$. The matrix $A(A^TA)^{-1}A^T$ occurs frequently in statistics, where it is sometimes called the *hat-matrix*.

33. The normal equations are $\begin{bmatrix} 2 & 2 \\ 2 & 2 \end{bmatrix}\begin{bmatrix} x \\ y \end{bmatrix} = \begin{bmatrix} 6 \\ 6 \end{bmatrix}$, whose solution is the set of (x, y) such that $x + y = 3$. The solutions correspond to points on the line midway between the lines $x + y = 2$ and $x + y = 4$.

Section 6.6, page 397

1. $y = .9 + .4x$ **3.** $y = 1.1 + 1.3x$

5. 2.5

7. 2.1, a difference of .1 is reasonable.

9. No. A y-value of 20 is quite far from the other y-values.

11. If two data points have different x-coordinates, then the two columns of the design matrix X cannot be multiples of each other and hence are linearly independent. By Theorem 14 in Section 6.5, the normal equations have a unique solution.

13. a. $\mathbf{y} = X\boldsymbol{\beta} + \boldsymbol{\epsilon}$, where $\mathbf{y} = \begin{bmatrix} 1.8 \\ 2.7 \\ 3.4 \\ 3.8 \\ 3.9 \end{bmatrix}$, $X = \begin{bmatrix} 1 & 1 \\ 2 & 4 \\ 3 & 9 \\ 4 & 16 \\ 5 & 25 \end{bmatrix}$,

$\boldsymbol{\beta} = \begin{bmatrix} \beta_1 \\ \beta_2 \end{bmatrix}$, $\boldsymbol{\epsilon} = \begin{bmatrix} \epsilon_1 \\ \epsilon_2 \\ \epsilon_3 \\ \epsilon_4 \\ \epsilon_5 \end{bmatrix}$

b. $y = 1.76x - .20x^2$

c. $y = 3.36$

15. $\mathbf{y} = X\boldsymbol{\beta} + \boldsymbol{\epsilon}$, where $\mathbf{y} = \begin{bmatrix} 7.9 \\ 5.4 \\ -.9 \end{bmatrix}$, $X = \begin{bmatrix} \cos 1 & \sin 1 \\ \cos 2 & \sin 2 \\ \cos 3 & \sin 3 \end{bmatrix}$,

$\boldsymbol{\beta} = \begin{bmatrix} A \\ B \end{bmatrix}$, $\boldsymbol{\epsilon} = \begin{bmatrix} \epsilon_1 \\ \epsilon_2 \\ \epsilon_3 \end{bmatrix}$

17. $\beta = 1.45$ and $e = .811$; the orbit is an ellipse. The equation $r = \beta/(1 - e \cdot \cos \vartheta)$ produces $r = 1.33$ when $\vartheta = 4.6$.

19. a. $y = -.8558 + 4.7025t + 5.5554t^2 - .0274t^3$

b. The velocity function is $v(t) = 4.7025 + 11.1108t - .0822t^2$, and $v(4.5) = 53.0$ ft/sec.

21. *Hint:* Write X and **y** as in equation (1), and compute X^TX and $X^T\mathbf{y}$.

23. a. The mean of the x-data is $\bar{x} = 5.5$. The data in mean-deviation form are $(-3.5, 1)$, $(-.5, 2)$, $(1.5, 3)$, and $(2.5, 3)$. The columns of X are orthogonal because the entries in the second column sum to 0.

b. $\begin{bmatrix} 4 & 0 \\ 0 & 21 \end{bmatrix} \begin{bmatrix} \beta_0 \\ \beta_1 \end{bmatrix} = \begin{bmatrix} 9 \\ 7.5 \end{bmatrix}$,

$y = \frac{9}{4} + \frac{5}{14} x^* = \frac{9}{4} + \frac{5}{14}(x - 5.5)$

25. *Hint:* The equation has a nice geometric interpretation.

Section 6.7, page 406

1. a. $3, \sqrt{105}, 225$ **b.** All multiples of $\begin{bmatrix} 1 \\ 4 \end{bmatrix}$

3. 28   **5.** $5\sqrt{2}, 3\sqrt{3}$ **7.** $\frac{56}{25} + \frac{14}{25} t$

9. a. Constant polynomial, $p(t) = 5$

b. $t^2 - 5$ is orthogonal to p_0 and p_1; values: $(4, -4, -4, 4)$; answer: $q(t) = \frac{1}{4}(t^2 - 5)$

11. $\frac{17}{5} t$

13. Verify each of the four axioms. For instance:

$\begin{aligned} 1. \quad \langle \mathbf{u}, \mathbf{v} \rangle &= (A\mathbf{u}) \cdot (A\mathbf{v}) & \text{Definition} \\ &= (A\mathbf{v}) \cdot (A\mathbf{u}) & \text{Property of the dot product} \\ &= \langle \mathbf{v}, \mathbf{u} \rangle & \text{Definition} \end{aligned}$

15. $\begin{aligned} \langle \mathbf{u}, c\mathbf{v} \rangle &= \langle c\mathbf{v}, \mathbf{u} \rangle & \text{Axiom 1} \\ &= c \langle \mathbf{v}, \mathbf{u} \rangle & \text{Axiom 3} \\ &= c \langle \mathbf{u}, \mathbf{v} \rangle & \text{Axiom 1} \end{aligned}$

17. *Hint:* Compute 4 times the right-hand side.

19–23. See the Study Guide.

25. $\langle \mathbf{u}, \mathbf{v} \rangle = \sqrt{a}\sqrt{b} + \sqrt{b}\sqrt{a} = 2\sqrt{ab}$, $\|\mathbf{u}\|^2 = (\sqrt{a})^2 + (\sqrt{b})^2 = a + b$. Since a and b are nonnegative, $\|\mathbf{u}\| = \sqrt{a + b}$. Similarly, $\|\mathbf{v}\| = \sqrt{b + a}$. By Cauchy–Schwarz, $2\sqrt{ab} \le \sqrt{a + b}\sqrt{b + a} = a + b$. Hence, $\sqrt{ab} \le \dfrac{a + b}{2}$.

27. 0   **29.** $2/\sqrt{5}$ **31.** $1, t, 3t^2 - 1$

33. The new orthogonal polynomials are multiples of $-17t + 5t^3$ and $72 - 155t^2 + 35t^4$. Scale these polynomials so their values at $-2, -1, 0, 1$, and 2 are small integers.

Section 6.8, page 412

1. $y = 2 + \frac{3}{2} t$

3. $\begin{aligned} p(t) &= 4p_0 - .1p_1 - .5p_2 + .2p_3 \\ &= 4 - .1t - .5(t^2 - 2) + .2\left(\frac{5}{6}t^3 - \frac{17}{6}t\right) \end{aligned}$
(This polynomial happens to fit the data exactly.)

5. Use the identity $\sin mt \sin nt = \frac{1}{2}[\cos(mt - nt) - \cos(mt + nt)]$

7. Use the identity $\cos^2 kt = \dfrac{1 + \cos 2kt}{2}$.

9. $\pi + 2\sin t + \sin 2t + \frac{2}{3}\sin 3t$ [*Hint:* Save time by using the results from Example 4.]

11. $\frac{1}{2} - \frac{1}{2}\cos 2t$ (Why?)

13. *Hint:* Take functions f and g in $C[0, 2\pi]$, and fix an integer $m \ge 0$. Write the Fourier coefficient of $f + g$ that involves $\cos mt$, and write the Fourier coefficient that involves $\sin mt (m > 0)$.

15. The cubic curve is the graph of $g(t) = -.2685 + 3.6095t + 5.8576t^2 - .0477t^3$. The velocity at $t = 4.5$ seconds is $g'(4.5) = 53.4$ ft/sec. This is about .7% faster than the estimate obtained in Exercise 19 in Section 6.6.

Chapter 6 Supplementary Exercises, page 414

1. F   **2.** T   **3.** T   **4.** F   **5.** F   **6.** T

7. T   **8.** T   **9.** F   **10.** T   **11.** T   **12.** F

13. T   **14.** F   **15.** F   **16.** T   **17.** T   **18.** F

19. F

20. *Hint:* If $\{\mathbf{v}_1, \mathbf{v}_2\}$ is an orthonormal set and $\mathbf{x} = c_1\mathbf{v}_1 + c_2\mathbf{v}_2$, then the vectors $c_1\mathbf{v}_1$ and $c_2\mathbf{v}_2$ are orthogonal, and

$\begin{aligned} \|\mathbf{x}\|^2 &= \|c_1\mathbf{v}_1 + c_2\mathbf{v}_2\|^2 = \|c_1\mathbf{v}_1\|^2 + \|c_2\mathbf{v}_2\|^2 \\ &= (|c_1|\|\mathbf{v}_1\|)^2 + (|c_2|\|\mathbf{v}_2\|)^2 = |c_1|^2 + |c_2|^2 \end{aligned}$

(Explain why.) So the stated equality holds for $p = 2$. Suppose that the equality holds for $p = k$, with $k \ge 2$, let $\{\mathbf{v}_1, \ldots, \mathbf{v}_{k+1}\}$ be an orthonormal set, and consider $\mathbf{x} = c_1\mathbf{v}_1 + \cdots + c_k\mathbf{v}_k + c_{k+1}\mathbf{v}_{k+1} = \mathbf{u}_k + c_{k+1}\mathbf{v}_{k+1}$, where $\mathbf{u}_k = c_1\mathbf{v}_1 + \cdots + c_k\mathbf{v}_k$.

21. Given $\mathbf{x}$ and an orthonormal set $\{\mathbf{v}_1, \ldots, \mathbf{v}_p\}$ in $\mathbb{R}^n$, let $\hat{\mathbf{x}}$ be the orthogonal projection of $\mathbf{x}$ onto the subspace spanned by $\mathbf{v}_1, \ldots, \mathbf{v}_p$. By Theorem 10 in Section 6.3,

$\hat{\mathbf{x}} = (\mathbf{x} \cdot \mathbf{v}_1)\mathbf{v}_1 + \cdots + (\mathbf{x} \cdot \mathbf{v}_p)\mathbf{v}_p$

By Exercise 20, $\|\hat{\mathbf{x}}\|^2 = |\mathbf{x} \cdot \mathbf{v}_1|^2 + \cdots + |\mathbf{x} \cdot \mathbf{v}_p|^2$. Bessel's inequality follows from the fact that $\|\hat{\mathbf{x}}\|^2 \le \|\mathbf{x}\|^2$, noted before the statement of the Cauchy–Schwarz inequality, in Section 6.7.

23. Suppose $(U\mathbf{x}) \cdot (U\mathbf{y}) = \mathbf{x} \cdot \mathbf{y}$ for all $\mathbf{x}, \mathbf{y}$ in $\mathbb{R}^n$, and let $\mathbf{e}_1, \ldots, \mathbf{e}_n$ be the standard basis for $\mathbb{R}^n$. For $j = 1, \ldots, n$, $U\mathbf{e}_j$ is the jth column of U. Since $\|U\mathbf{e}_j\|^2 = (U\mathbf{e}_j) \cdot (U\mathbf{e}_j) = \mathbf{e}_j \cdot \mathbf{e}_j = 1$, the columns of U are unit vectors; since $(U\mathbf{e}_j) \cdot (U\mathbf{e}_k) = \mathbf{e}_j \cdot \mathbf{e}_k = 0$ for $j \ne k$, the columns are pairwise orthogonal.

25. *Hint:* Compute $Q^T Q$, using the fact that $(\mathbf{u}\mathbf{u}^T)^T = \mathbf{u}^{TT}\mathbf{u}^T = \mathbf{u}\mathbf{u}^T$.

27. Let $W = \text{Span}\{\mathbf{u}, \mathbf{v}\}$. Given $\mathbf{z}$ in $\mathbb{R}^n$, let $\hat{\mathbf{z}} = \text{proj}_W \mathbf{z}$. Then $\hat{\mathbf{z}}$ is in Col A, where $A = [\mathbf{u} \quad \mathbf{v}]$, say, $\hat{\mathbf{z}} = A\hat{\mathbf{x}}$ for some $\hat{\mathbf{x}}$ in $\mathbb{R}^2$. So $\hat{\mathbf{x}}$ is a least-squares solution of $A\mathbf{x} = \mathbf{z}$. The normal equations can be solved to produce $\hat{\mathbf{x}}$, and then $\hat{\mathbf{z}}$ is found by computing $A\hat{\mathbf{x}}$.

29. *Hint:* Let $\mathbf{x} = \begin{bmatrix} x \\ y \\ z \end{bmatrix}$, $\mathbf{b} = \begin{bmatrix} a \\ b \\ c \end{bmatrix}$, $\mathbf{v} = \begin{bmatrix} 1 \\ -2 \\ 5 \end{bmatrix}$, and

$A = \begin{bmatrix} \mathbf{v}^T \\ \mathbf{v}^T \\ \mathbf{v}^T \end{bmatrix} = \begin{bmatrix} 1 & -2 & 5 \\ 1 & -2 & 5 \\ 1 & -2 & 5 \end{bmatrix}$. The given set of

equations is $A\mathbf{x} = \mathbf{b}$, and the set of all least-squares solutions coincides with the set of solutions of $A^T A\mathbf{x} = A^T \mathbf{b}$ (Theorem 13 in Section 6.5). Study this equation, and use the fact that $(\mathbf{vv}^T)\mathbf{x} = \mathbf{v}(\mathbf{v}^T\mathbf{x}) = (\mathbf{v}^T\mathbf{x})\mathbf{v}$, because $\mathbf{v}^T\mathbf{x}$ is a scalar.

31. a. The row–column calculation of $A\mathbf{u}$ shows that each row of A is orthogonal to every $\mathbf{u}$ in Nul A. So each row of A is in $(\text{Nul } A)^\perp$. Since $(\text{Nul } A)^\perp$ is a subspace, it must contain all linear combinations of the rows of A; hence $(\text{Nul } A)^\perp$ contains Row A.

b. If rank $A = r$, then dim Nul $A = n - r$, by the Rank Theorem. By Exercise 32(c) in Section 6.3,

$$\dim \text{Nul } A + \dim(\text{Nul } A)^\perp = n$$

So $\dim(\text{Nul } A)^\perp$ must be r. But Row A is an r-dimensional subspace of $(\text{Nul } A)^\perp$, by the Rank Theorem and part (a). Therefore, Row A must coincide with $(\text{Nul } A)^\perp$.

c. Replace A by A^T in part (b) and conclude that Row A^T coincides with $(\text{Nul } A^T)^\perp$. Since Row $A^T = \text{Col } A$, this proves (c).

33. If $A = URU^T$ with U orthogonal, then A is similar to R (because U is invertible and $U^T = U^{-1}$) and so A has the same eigenvalues as R (by Theorem 4 in Section 5.2), namely, the n real numbers on the diagonal of R.

35. $\dfrac{\|\Delta\mathbf{x}\|}{\|\mathbf{x}\|} = .4618$,

$\text{cond}(A) \times \dfrac{\|\Delta\mathbf{b}\|}{\|\mathbf{b}\|} = 3363 \times (1.548 \times 10^{-4}) = .5206$.

Observe that $\|\Delta\mathbf{x}\|/\|\mathbf{x}\|$ almost equals $\text{cond}(A)$ times $\|\Delta\mathbf{b}\|/\|\mathbf{b}\|$.

37. $\dfrac{\|\Delta\mathbf{x}\|}{\|\mathbf{x}\|} = 7.178 \times 10^{-8}$, $\dfrac{\|\Delta\mathbf{b}\|}{\|\mathbf{b}\|} = 2.832 \times 10^{-4}$. Observe that the relative change in $\mathbf{x}$ is *much* smaller than the relative change in $\mathbf{b}$. In fact, since

$$\text{cond}(A) \times \dfrac{\|\Delta\mathbf{b}\|}{\|\mathbf{b}\|} = 23{,}683 \times (2.832 \times 10^{-4}) = 6.707$$

the theoretical bound on the relative change in $\mathbf{x}$ is 6.707 (to four significant figures). This exercise shows that even when a condition number is large, the relative error in a solution need not be as large as you might expect.

Chapter 7

Section 7.1, page 423

1. Symmetric   **3.** Not symmetric   **5.** Symmetric

7. Orthogonal, $\begin{bmatrix} .6 & .8 \\ .8 & -.6 \end{bmatrix}$

9. Orthogonal, $\begin{bmatrix} -4/5 & 3/5 \\ 3/5 & 4/5 \end{bmatrix}$

11. Not orthogonal

13. $P = \begin{bmatrix} 1/\sqrt{2} & -1/\sqrt{2} \\ 1/\sqrt{2} & 1/\sqrt{2} \end{bmatrix}$, $D = \begin{bmatrix} 4 & 0 \\ 0 & 2 \end{bmatrix}$

15. $P = \begin{bmatrix} -2/\sqrt{5} & 1/\sqrt{5} \\ 1/\sqrt{5} & 2/\sqrt{5} \end{bmatrix}$, $D = \begin{bmatrix} 1 & 0 \\ 0 & 11 \end{bmatrix}$

17. $P = \begin{bmatrix} -1/\sqrt{2} & 1/\sqrt{6} & 1/\sqrt{3} \\ 0 & -2/\sqrt{6} & 1/\sqrt{3} \\ 1/\sqrt{2} & 1/\sqrt{6} & 1/\sqrt{3} \end{bmatrix}$,

$D = \begin{bmatrix} -4 & 0 & 0 \\ 0 & 4 & 0 \\ 0 & 0 & 7 \end{bmatrix}$

19. $P = \begin{bmatrix} -1/\sqrt{5} & 4/\sqrt{45} & -2/3 \\ 2/\sqrt{5} & 2/\sqrt{45} & -1/3 \\ 0 & 5/\sqrt{45} & 2/3 \end{bmatrix}$,

$D = \begin{bmatrix} 7 & 0 & 0 \\ 0 & 7 & 0 \\ 0 & 0 & -2 \end{bmatrix}$

21. $P = \begin{bmatrix} 0 & 1/\sqrt{2} & 1/2 & 1/2 \\ 0 & -1/\sqrt{2} & 1/2 & 1/2 \\ 1/\sqrt{2} & 0 & -1/2 & 1/2 \\ -1/\sqrt{2} & 0 & -1/2 & 1/2 \end{bmatrix}$,

$D = \begin{bmatrix} 1 & 0 & 0 & 0 \\ 0 & 1 & 0 & 0 \\ 0 & 0 & 5 & 0 \\ 0 & 0 & 0 & 9 \end{bmatrix}$

23. $P = \begin{bmatrix} 1/\sqrt{3} & 1/\sqrt{2} & -1/\sqrt{6} \\ 1/\sqrt{3} & -1/\sqrt{2} & -1/\sqrt{6} \\ 1/\sqrt{3} & 0 & 2/\sqrt{6} \end{bmatrix}$,

$D = \begin{bmatrix} 2 & 0 & 0 \\ 0 & 5 & 0 \\ 0 & 0 & 5 \end{bmatrix}$

25–31. See the *Study Guide*.

33. $(A\mathbf{x})\cdot\mathbf{y} = (A\mathbf{x})^T\mathbf{y} = \mathbf{x}^T A^T \mathbf{y} = \mathbf{x}^T A\mathbf{y} = \mathbf{x}\cdot(A\mathbf{y})$, because $A^T = A$.

35. *Hint:* Use an orthogonal diagonalization of A, or appeal to Theorem 2.

37. The Diagonalization Theorem in Section 5.3 says that the columns of P are (linearly independent) eigenvectors corresponding to the eigenvalues of A listed on the diagonal of D. So P has exactly k columns of eigenvectors corresponding to λ. These k columns form a basis for the eigenspace.

39. $A = 8\mathbf{u}_1\mathbf{u}_1^T + 6\mathbf{u}_2\mathbf{u}_2^T + 3\mathbf{u}_3\mathbf{u}_3^T$

$$= 8\begin{bmatrix} 1/2 & -1/2 & 0 \\ -1/2 & 1/2 & 0 \\ 0 & 0 & 0 \end{bmatrix}$$

$$+ 6\begin{bmatrix} 1/6 & 1/6 & -2/6 \\ 1/6 & 1/6 & -2/6 \\ -2/6 & -2/6 & 4/6 \end{bmatrix}$$

$$+ 3\begin{bmatrix} 1/3 & 1/3 & 1/3 \\ 1/3 & 1/3 & 1/3 \\ 1/3 & 1/3 & 1/3 \end{bmatrix}$$

41. *Hint:* $(\mathbf{u}\mathbf{u}^T)\mathbf{x} = \mathbf{u}(\mathbf{u}^T\mathbf{x}) = (\mathbf{u}^T\mathbf{x})\mathbf{u}$, because $\mathbf{u}^T\mathbf{x}$ is a scalar.

43. $P = \dfrac{1}{2}\begin{bmatrix} -1 & 1 & 1 & 1 \\ 1 & 1 & 1 & -1 \\ -1 & 1 & -1 & -1 \\ 1 & 1 & -1 & 1 \end{bmatrix}$,

$$D = \begin{bmatrix} 19 & 0 & 0 & 0 \\ 0 & 11 & 0 & 0 \\ 0 & 0 & 5 & 0 \\ 0 & 0 & 0 & -11 \end{bmatrix}$$

45. $P = \begin{bmatrix} 1/\sqrt{2} & 3/\sqrt{50} & -2/5 & -2/5 \\ 0 & 4/\sqrt{50} & -1/5 & 4/5 \\ 0 & 4/\sqrt{50} & 4/5 & -1/5 \\ 1/\sqrt{2} & -3/\sqrt{50} & 2/5 & 2/5 \end{bmatrix}$

$$D = \begin{bmatrix} .75 & 0 & 0 & 0 \\ 0 & .75 & 0 & 0 \\ 0 & 0 & 0 & 0 \\ 0 & 0 & 0 & -1.25 \end{bmatrix}$$

Section 7.2, page 430

1. a. $5x_1^2 + \frac{2}{3}x_1x_2 + x_2^2$ **b.** 185   **c.** 16

3. a. $\begin{bmatrix} 3 & -2 \\ -2 & 5 \end{bmatrix}$ **b.** $\begin{bmatrix} 3 & 1 \\ 1 & 0 \end{bmatrix}$

5. a. $\begin{bmatrix} 3 & -3 & 4 \\ -3 & 2 & -2 \\ 4 & -2 & -5 \end{bmatrix}$ **b.** $\begin{bmatrix} 0 & 3 & 2 \\ 3 & 0 & -5 \\ 2 & -5 & 0 \end{bmatrix}$

7. $\mathbf{x} = P\mathbf{y}$, where $P = \dfrac{1}{\sqrt{2}}\begin{bmatrix} 1 & -1 \\ 1 & 1 \end{bmatrix}$, $\mathbf{y}^T D\mathbf{y} = 6y_1^2 - 4y_2^2$

In Exercises 9–14, other answers (change of variables and new quadratic form) are possible.

9. Positive definite; eigenvalues are 6 and 2

Change of variable: $\mathbf{x} = P\mathbf{y}$, with $P = \dfrac{1}{\sqrt{2}}\begin{bmatrix} -1 & 1 \\ 1 & 1 \end{bmatrix}$

New quadratic form: $6y_1^2 + 2y_2^2$

11. Indefinite; eigenvalues are 3 and -2

Change of variable: $\mathbf{x} = P\mathbf{y}$, with $P = \dfrac{1}{\sqrt{5}}\begin{bmatrix} -2 & 1 \\ 1 & 2 \end{bmatrix}$

New quadratic form: $3y_1^2 - 2y_2^2$

13. Positive semidefinite; eigenvalues are 10 and 0

Change of variable: $\mathbf{x} = P\mathbf{y}$, with $P = \dfrac{1}{\sqrt{10}}\begin{bmatrix} 1 & 3 \\ -3 & 1 \end{bmatrix}$

New quadratic form: $10y_1^2$

15. Negative definite; eigenvalues are $-13, -9, -7, -1$

Change of variable: $\mathbf{x} = P\mathbf{y}$;

$$P = \begin{bmatrix} 0 & -1/2 & 0 & 3/\sqrt{12} \\ 0 & 1/2 & -2/\sqrt{6} & 1/\sqrt{12} \\ -1/\sqrt{2} & 1/2 & 1/\sqrt{6} & 1/\sqrt{12} \\ 1/\sqrt{2} & 1/2 & 1/\sqrt{6} & 1/\sqrt{12} \end{bmatrix}$$

New quadratic form: $-13y_1^2 - 9y_2^2 - 7y_3^2 - y_4^2$

17. Positive definite; eigenvalues are 1 and 21.

Change of variable: $\mathbf{x} = P\mathbf{y}$;

$$P = \dfrac{1}{\sqrt{50}}\begin{bmatrix} 4 & 3 & 4 & -3 \\ -5 & 0 & 5 & 0 \\ 3 & -4 & 3 & 4 \\ 0 & 5 & 0 & 5 \end{bmatrix}$$

New quadratic form: $y_1^2 + y_2^2 + 21y_3^2 + 21y_4^2$

19. 8

21–29. See the *Study Guide.*

31. Write the characteristic polynomial in two ways:

$$\det(A - \lambda I) = \det\begin{bmatrix} a - \lambda & b \\ b & d - \lambda \end{bmatrix}$$
$$= \lambda^2 - (a + d)\lambda + ad - b^2$$

and

$$(\lambda - \lambda_1)(\lambda - \lambda_2) = \lambda^2 - (\lambda_1 + \lambda_2)\lambda + \lambda_1\lambda_2$$

Equate coefficients to obtain $\lambda_1 + \lambda_2 = a + d$ and $\lambda_1\lambda_2 = ad - b^2 = \det A$.

33. Exercise 34 in Section 7.1 showed that B^TB is symmetric. Also, $\mathbf{x}^TB^TB\mathbf{x} = (B\mathbf{x})^TB\mathbf{x} = \|B\mathbf{x}\|^2 \geq 0$, so the quadratic form is positive semidefinite, and we say that the matrix B^TB is positive semidefinite. *Hint:* To show that B^TB is positive definite when B is square and invertible, suppose that $\mathbf{x}^TB^TB\mathbf{x} = 0$ and deduce that $\mathbf{x} = \mathbf{0}$.

35. *Hint:* Show that $A + B$ is symmetric and the quadratic form $\mathbf{x}^T(A + B)\mathbf{x}$ is positive definite.

Section 7.3, page 438

1. $\mathbf{x} = P\mathbf{y}$, where $P = \begin{bmatrix} 1/3 & 2/3 & -2/3 \\ 2/3 & 1/3 & 2/3 \\ -2/3 & 2/3 & 1/3 \end{bmatrix}$

3. a. 9   **b.** $\pm\begin{bmatrix} 1/3 \\ 2/3 \\ -2/3 \end{bmatrix}$ **c.** 6

5. a. 6   **b.** $\pm\begin{bmatrix} 1/\sqrt{2} \\ -1/\sqrt{2} \end{bmatrix}$ **c.** -4

7. $\pm\begin{bmatrix} 1/3 \\ 2/3 \\ 2/3 \end{bmatrix}$ **9.** $5 + \sqrt{5}$ **11.** 3

13. *Hint:* If $m = M$, take $\alpha = 0$ in the formula for $\mathbf{x}$. That is, let $\mathbf{x} = \mathbf{u}_n$, and verify that $\mathbf{x}^T A \mathbf{x} = m$. If $m < M$ and if t is a number between m and M, then $0 \le t - m \le M - m$ and $0 \le (t - m)/(M - m) \le 1$. So let $\alpha = (t - m)/(M - m)$. Solve the expression for α to see that $t = (1 - \alpha)m + \alpha M$. As α goes from 0 to 1, t goes from m to M. Construct $\mathbf{x}$ as in the statement of the exercise, and verify its properties.

15. a. 9   **b.** $\begin{bmatrix} -2/\sqrt{6} \\ 0 \\ 1/\sqrt{6} \\ 1/\sqrt{6} \end{bmatrix}$ **c.** 3

17. a. 17   **b.** $\begin{bmatrix} 1/2 \\ 1/2 \\ 1/2 \\ 1/2 \end{bmatrix}$ **c.** 13

Section 7.4, page 448

1. 3, 1   **3.** 4, 1

The answers in Exercises 5–13 are not the only possibilities.

5. $\begin{bmatrix} -1 & 0 \\ 0 & 1 \end{bmatrix}\begin{bmatrix} 2 & 0 \\ 0 & 0 \end{bmatrix}\begin{bmatrix} 1 & 0 \\ 0 & 1 \end{bmatrix}$

7. $\begin{bmatrix} 1/\sqrt{5} & -2/\sqrt{5} \\ 2/\sqrt{5} & 1/\sqrt{5} \end{bmatrix}\begin{bmatrix} 3 & 0 \\ 0 & 2 \end{bmatrix}$
$\times \begin{bmatrix} 2/\sqrt{5} & 1/\sqrt{5} \\ -1/\sqrt{5} & 2/\sqrt{5} \end{bmatrix}$

9. $\begin{bmatrix} -1 & 0 & 0 \\ 0 & 0 & 1 \\ 0 & 1 & 0 \end{bmatrix}\begin{bmatrix} 3\sqrt{2} & 0 \\ 0 & \sqrt{2} \\ 0 & 0 \end{bmatrix}$
$\times \begin{bmatrix} -1/\sqrt{2} & 1/\sqrt{2} \\ 1/\sqrt{2} & 1/\sqrt{2} \end{bmatrix}$

11. $\begin{bmatrix} -1/3 & 2/3 & 2/3 \\ 2/3 & -1/3 & 2/3 \\ 2/3 & 2/3 & -1/3 \end{bmatrix}\begin{bmatrix} \sqrt{90} & 0 \\ 0 & 0 \\ 0 & 0 \end{bmatrix}$
$\times \begin{bmatrix} 3/\sqrt{10} & -1/\sqrt{10} \\ 1/\sqrt{10} & 3/\sqrt{10} \end{bmatrix}$

13. $\begin{bmatrix} 1/\sqrt{2} & -1/\sqrt{2} \\ 1/\sqrt{2} & 1/\sqrt{2} \end{bmatrix}\begin{bmatrix} 5 & 0 & 0 \\ 0 & 3 & 0 \end{bmatrix}$
$\times \begin{bmatrix} 1/\sqrt{2} & 1/\sqrt{2} & 0 \\ -1/\sqrt{18} & 1/\sqrt{18} & -4/\sqrt{18} \\ -2/3 & 2/3 & 1/3 \end{bmatrix}$

15. a. rank $A = 2$

b. Basis for Col A: $\begin{bmatrix} .40 \\ .37 \\ -.84 \end{bmatrix}$, $\begin{bmatrix} -.78 \\ -.33 \\ -.52 \end{bmatrix}$

Basis for Nul A: $\begin{bmatrix} .58 \\ -.58 \\ .58 \end{bmatrix}$

(Remember that V^T appears in the SVD.)

17. If U is an orthogonal matrix then det $U = \pm 1$. If $A = U\Sigma V^T$ and A is square, then so are U, Σ, and V. Hence det $A = $ det U det Σ det V^T $= \pm 1$ det $\Sigma = \pm\sigma_1 \cdots \sigma_n$

19. *Hint:* Since U and V are orthogonal,

$$A^T A = (U\Sigma V^T)^T U\Sigma V^T = V\Sigma^T U^T U\Sigma V^T$$
$$= V(\Sigma^T \Sigma)V^{-1}$$

Thus V diagonalizes $A^T A$. What does this tell you about V?

21. The right singular vector $\mathbf{v}_1$ is an eigenvector for the largest eigenvalue λ_1 of $A^T A$. By Theorem 7 in Section 7.3, the largest eigenvalue, λ_2, is the maximum of $\mathbf{x}^T (A^T A)\mathbf{x}$ over all unit vectors orthogonal to $\mathbf{v}_1$. Since $\mathbf{x}^T (A^T A)\mathbf{x} = \|A\mathbf{x}\|^2$, the square root of λ_2, which is the second largest eigenvalue, is the maximum of $\|A\mathbf{x}\|$ over all unit vectors orthogonal to $\mathbf{v}_1$.

23. *Hint:* Use a column–row expansion of $(U\Sigma)V^T$.

25. *Hint:* Consider the SVD for the standard matrix of T—say, $A = U\Sigma V^T = U\Sigma V^{-1}$. Let $\mathcal{B} = \{\mathbf{v}_1, \ldots, \mathbf{v}_n\}$ and $\mathcal{C} = \{\mathbf{u}_1, \ldots, \mathbf{u}_m\}$ be bases constructed from the columns of V and U, respectively. Compute the matrix for T relative to $\mathcal{B}$ and $\mathcal{C}$, as in Section 5.4. To do this, you must show that $V^{-1}\mathbf{v}_j = \mathbf{e}_j$, the jth column of I_n.

27. $\begin{bmatrix} -.57 & -.65 & -.42 & .27 \\ .63 & -.24 & -.68 & -.29 \\ .07 & -.63 & .53 & -.56 \\ -.51 & .34 & -.29 & -.73 \end{bmatrix}$
$\times \begin{bmatrix} 16.46 & 0 & 0 & 0 & 0 \\ 0 & 12.16 & 0 & 0 & 0 \\ 0 & 0 & 4.87 & 0 & 0 \\ 0 & 0 & 0 & 4.31 & 0 \end{bmatrix}$
$\times \begin{bmatrix} -.10 & .61 & -.21 & -.52 & .55 \\ -.39 & .29 & .84 & -.14 & -.19 \\ -.74 & -.27 & -.07 & .38 & .49 \\ .41 & -.50 & .45 & -.23 & .58 \\ -.36 & -.48 & -.19 & -.72 & -.29 \end{bmatrix}$

29. 25.9343, 16.7554, 11.2917, 1.0785, .00037793; $\sigma_1/\sigma_5 = 68{,}622$

Section 7.5, page 455

1. $M = \begin{bmatrix} 12 \\ 10 \end{bmatrix}$; $B = \begin{bmatrix} 7 & 10 & -6 & -9 & -10 & 8 \\ 2 & -4 & -1 & 5 & 3 & -5 \end{bmatrix}$;
$S = \begin{bmatrix} 86 & -27 \\ -27 & 16 \end{bmatrix}$

3. $\begin{bmatrix} .95 \\ -.32 \end{bmatrix}$ for $\lambda = 95.2$, $\begin{bmatrix} .32 \\ .95 \end{bmatrix}$ for $\lambda = 6.8$

5. (.130, .874, .468), 75.9% of the variance

7. $y_1 = .95x_1 - .32x_2$; y_1 explains 93.3% of the variance.

9. $c_1 = 1/3$, $c_2 = 2/3$, $c_3 = 2/3$; the variance of y is 9.

11. **a.** If $\mathbf{w}$ is the vector in $\mathbb{R}^N$ with a 1 in each position, then

$$\begin{bmatrix} \mathbf{X}_1 & \cdots & \mathbf{X}_N \end{bmatrix} \mathbf{w} = \mathbf{X}_1 + \cdots + \mathbf{X}_N = \mathbf{0}$$

because the $\mathbf{X}_k$ are in mean-deviation form. Then

$$\begin{bmatrix} \mathbf{Y}_1 & \cdots & \mathbf{Y}_N \end{bmatrix} \mathbf{w}$$
$$= \begin{bmatrix} P^T\mathbf{X}_1 & \cdots & P^T\mathbf{X}_N \end{bmatrix} \mathbf{w} \qquad \text{By definition}$$
$$= P^T\begin{bmatrix} \mathbf{X}_1 & \cdots & \mathbf{X}_N \end{bmatrix} \mathbf{w} = P^T\mathbf{0} = \mathbf{0}$$

That is, $\mathbf{Y}_1 + \cdots + \mathbf{Y}_N = \mathbf{0}$, so the $\mathbf{Y}_k$ are in mean-deviation form.

b. *Hint:* Because the $\mathbf{X}_j$ are in mean-deviation form, the covariance matrix of the $\mathbf{X}_j$ is

$$1/(N-1)\begin{bmatrix} \mathbf{X}_1 & \cdots & \mathbf{X}_N \end{bmatrix}\begin{bmatrix} \mathbf{X}_1 & \cdots & \mathbf{X}_N \end{bmatrix}^T$$

Compute the covariance matrix of the $\mathbf{Y}_j$, using part (a).

13. If $B = \begin{bmatrix} \hat{\mathbf{X}}_1 & \cdots & \hat{\mathbf{X}}_N \end{bmatrix}$, then

$$S = \frac{1}{N-1}BB^T = \frac{1}{N-1}\begin{bmatrix} \hat{\mathbf{X}}_1 & \cdots & \hat{\mathbf{X}}_n \end{bmatrix}\begin{bmatrix} \hat{\mathbf{X}}_1^T \\ \vdots \\ \hat{\mathbf{X}}_N^T \end{bmatrix}$$

$$= \frac{1}{N-1}\sum_1^N \hat{\mathbf{X}}_k\hat{\mathbf{X}}_k^T = \frac{1}{N-1}\sum_1^N (\mathbf{X}_k - \mathbf{M})(\mathbf{X}_k - \mathbf{M})^T$$

Chapter 7 Supplementary Exercises, page 457

1. T   2. F   3. T   4. F   5. F   6. F

7. F   8. T   9. F   10. F   11. F   12. F

13. T   14. F   15. T   16. T   17. F

19. If rank $A = r$, then $\dim \text{Nul } A = n - r$, by the Rank Theorem. So 0 is an eigenvalue of multiplicity $n - r$. Hence, of the n terms in the spectral decomposition of A, exactly $n - r$ are zero. The remaining r terms (corresponding to the nonzero eigenvalues) are all rank 1 matrices, as mentioned in the discussion of the spectral decomposition.

21. If $A\mathbf{v} = \lambda\mathbf{v}$ for some nonzero λ, then $\mathbf{v} = \lambda^{-1}A\mathbf{v} = A(\lambda^{-1}\mathbf{v})$, which shows that $\mathbf{v}$ is a linear combination of the columns of A.

23. *Hint:* If $A = R^T R$, where R is invertible, then A is positive definite, by Exercise 33 in Section 7.2. Conversely, suppose that A is positive definite. Then by Exercise 34 in Section 7.2, $A = B^T B$ for some positive definite matrix B. Explain why B admits a QR factorization, and use it to create the Cholesky factorization of A.

25. If A is $m \times n$ and $\mathbf{x}$ is in $\mathbb{R}^n$, then $\mathbf{x}^T A^T A\mathbf{x} = (A\mathbf{x})^T(A\mathbf{x}) = \|A\mathbf{x}\|^2 \geq 0$. Thus $A^T A$ is positive semidefinite. By Exercise 30 in Section 6.5, rank $A^T A = $ rank A.

27. *Hint:* Write an SVD of A in the form $A = U\Sigma V^T = PQ$, where $P = U\Sigma U^T$ and $Q = UV^T$. Show that P is symmetric and has the same eigenvalues as Σ. Explain why Q is an orthogonal matrix.

29. **a.** If $\mathbf{b} = A\mathbf{x}$, then $\mathbf{x}^+ = A^+\mathbf{b} = A^+A\mathbf{x}$. By Exercise 28(b), $\mathbf{x}^+$ is the orthogonal projection of $\mathbf{x}$ onto Row A.

b. From (a) and then Exercise 28(c), $A\mathbf{x}^+ = A(A^+A\mathbf{x}) = (AA^+A)\mathbf{x} = A\mathbf{x} = \mathbf{b}$.

c. Since $\mathbf{x}^+$ is the orthogonal projection onto Row A, the Pythagorean Theorem shows that $\|\mathbf{u}\|^2 = \|\mathbf{x}^+\|^2 + \|\mathbf{u} - \mathbf{x}^+\|^2$. Part (c) follows immediately.

31. $A^+ = \dfrac{1}{40} \cdot \begin{bmatrix} -2 & -14 & 13 & 13 \\ -2 & -14 & 13 & 13 \\ -2 & 6 & -7 & -7 \\ 2 & -6 & 7 & 7 \\ 4 & -12 & -6 & -6 \end{bmatrix}$, $\hat{\mathbf{x}} = \begin{bmatrix} .7 \\ .7 \\ -.8 \\ .8 \\ .6 \end{bmatrix}$

The reduced echelon form of $\begin{bmatrix} A \\ \mathbf{x}^T \end{bmatrix}$ is the same as the reduced echelon form of A, except for an extra row of zeros. So adding scalar multiples of the rows of A to $\mathbf{x}^T$ can produce the zero vector, which shows that $\mathbf{x}^T$ is in Row A.

Basis for Nul A: $\begin{bmatrix} -1 \\ 1 \\ 0 \\ 0 \\ 0 \end{bmatrix}, \begin{bmatrix} 0 \\ 0 \\ 1 \\ 1 \\ 0 \end{bmatrix}$

Chapter 8

Section 8.1, page 467

1. Some possible answers: $\mathbf{y} = 2\mathbf{v}_1 - 1.5\mathbf{v}_2 + .5\mathbf{v}_3$, $\mathbf{y} = 2\mathbf{v}_1 - 2\mathbf{v}_3 + \mathbf{v}_4$, $\mathbf{y} = 2\mathbf{v}_1 + 3\mathbf{v}_2 - 7\mathbf{v}_3 + 3\mathbf{v}_4$

3. $\mathbf{y} = -3\mathbf{v}_1 + 2\mathbf{v}_2 + 2\mathbf{v}_3$. The weights sum to 1, so this is an affine sum.

5. **a.** $\mathbf{p}_1 = 3\mathbf{b}_1 - \mathbf{b}_2 - \mathbf{b}_3 \in \text{aff } S$ since the coefficients sum to 1.

b. $\mathbf{p}_2 = 2\mathbf{b}_1 + 0\mathbf{b}_2 + \mathbf{b}_3 \notin \text{aff } S$ since the coefficients do not sum to 1.

c. $\mathbf{p}_3 = -\mathbf{b}_1 + 2\mathbf{b}_2 + 0\mathbf{b}_3 \in \text{aff } S$ since the coefficients sum to 1.

7. **a.** $\mathbf{p}_1 \in \text{Span } S$, but $\mathbf{p}_1 \notin \text{aff } S$

b. $\mathbf{p}_2 \in \text{Span } S$, and $\mathbf{p}_2 \in \text{aff } S$

c. $\mathbf{p}_3 \notin \text{Span } S$, so $\mathbf{p}_3 \notin \text{aff } S$

9. $\mathbf{v}_1 = \begin{bmatrix} -3 \\ 0 \end{bmatrix}$ and $\mathbf{v}_2 = \begin{bmatrix} 1 \\ -2 \end{bmatrix}$. Other answers are possible.

11–19. See the *Study Guide*.

21. Span $\{\mathbf{v}_2 - \mathbf{v}_1, \mathbf{v}_3 - \mathbf{v}_1\}$ is a plane if and only if $\{\mathbf{v}_2 - \mathbf{v}_1, \mathbf{v}_3 - \mathbf{v}_1\}$ is linearly independent. Suppose c_2 and c_3 satisfy $c_2(\mathbf{v}_2 - \mathbf{v}_1) + c_3(\mathbf{v}_3 - \mathbf{v}_1) = \mathbf{0}$. Show that this implies $c_2 = c_3 = 0$.

23. Let $S = \{\mathbf{x} : A\mathbf{x} = \mathbf{b}\}$. To show that S is affine, it suffices to show that S is a flat, by Theorem 3. Let $W = \{\mathbf{x} : A\mathbf{x} = \mathbf{0}\}$. Then W is a subspace of $\mathbb{R}^n$, by Theorem 2 in Section 4.2 (or Theorem 12 in Section 2.8). Since $S = W + \mathbf{p}$, where $\mathbf{p}$ satisfies $A\mathbf{p} = \mathbf{b}$, by Theorem 6 in Section 1.5, S is a translate of W, and hence S is a flat.

25. A suitable set consists of any three vectors that are not collinear and have 5 as their third entry. If 5 is their third entry, they lie in the plane $z = 5$. If the vectors are not collinear, their affine hull cannot be a line, so it must be the plane.

27. If $\mathbf{p}, \mathbf{q} \in f(S)$, then there exist $\mathbf{r}, \mathbf{s} \in S$ such that $f(\mathbf{r}) = \mathbf{p}$ and $f(\mathbf{s}) = \mathbf{q}$. Given any $t \in \mathbb{R}$, we must show that $\mathbf{z} = (1 - t)\mathbf{p} + t\mathbf{q}$ is in $f(S)$. Now use definitions of $\mathbf{p}$ and $\mathbf{q}$, and the fact that f is linear. The complete proof is presented in the *Study Guide*.

29. Since B is affine, Theorem 2 implies that B contains all affine combinations of points of B. Hence B contains all affine combinations of points of A. That is, aff $A \subseteq B$.

31. Since $A \subseteq (A \cup B)$, it follows from Exercise 30 that aff $A \subseteq$ aff $(A \cup B)$. Similarly, aff $B \subseteq$ aff $(A \cup B)$, so $[\text{aff } A \cup \text{aff } B] \subseteq \text{aff } (A \cup B)$.

33. To show that $D \subseteq E \cap F$, show that $D \subseteq E$ and $D \subseteq F$. The complete proof is presented in the *Study Guide*.

Section 8.2, page 477

1. Affinely dependent and $2\mathbf{v}_1 + \mathbf{v}_2 - 3\mathbf{v}_3 = \mathbf{0}$

3. The set is affinely independent. If the points are called $\mathbf{v}_1$, $\mathbf{v}_2$, $\mathbf{v}_3$, and $\mathbf{v}_4$, then $\{\mathbf{v}_1, \mathbf{v}_2, \mathbf{v}_3\}$ is a basis for $\mathbb{R}^3$ and $\mathbf{v}_4 = 16\mathbf{v}_1 + 5\mathbf{v}_2 - 3\mathbf{v}_3$, but the weights in the linear combination do not sum to 1.

5. $-4\mathbf{v}_1 + 5\mathbf{v}_2 - 4\mathbf{v}_3 + 3\mathbf{v}_4 = \mathbf{0}$

7. The barycentric coordinates are $(-2, 4, -1)$.

9–17. See the *Study Guide*.

19. When a set of five points is translated by subtracting, say, the first point, the new set of four points must be linearly dependent, by Theorem 8 in Section 1.7, because the four points are in $\mathbb{R}^3$. By Theorem 5, the original set of five points is affinely dependent.

21. If $\{\mathbf{v}_1, \mathbf{v}_2\}$ is affinely dependent, then there exist c_1 and c_2, not both zero, such that $c_1 + c_2 = 0$ and $c_1\mathbf{v}_1 + c_2\mathbf{v}_2 = \mathbf{0}$. Show that this implies $\mathbf{v}_1 = \mathbf{v}_2$. For the converse, suppose $\mathbf{v}_1 = \mathbf{v}_2$ and select specific c_1 and c_2 that show their affine dependence. The details are in the *Study Guide*.

23. a. The vectors $\mathbf{v}_2 - \mathbf{v}_1 = \begin{bmatrix} 1 \\ 2 \end{bmatrix}$ and $\mathbf{v}_3 - \mathbf{v}_1 = \begin{bmatrix} 3 \\ -2 \end{bmatrix}$ are not multiples and hence are linearly independent. By Theorem 5, S is affinely independent.

b. $\mathbf{p}_1 \leftrightarrow \left(-\frac{6}{8}, \frac{9}{8}, \frac{5}{8}\right)$, $\mathbf{p}_2 \leftrightarrow \left(0, \frac{1}{2}, \frac{1}{2}\right)$, $\mathbf{p}_3 \leftrightarrow \left(\frac{14}{8}, -\frac{5}{8}, -\frac{1}{8}\right)$, $\mathbf{p}_4 \leftrightarrow \left(\frac{6}{8}, -\frac{5}{8}, \frac{7}{8}\right)$, $\mathbf{p}_5 \leftrightarrow \left(\frac{1}{4}, \frac{1}{8}, \frac{5}{8}\right)$

c. $\mathbf{p}_6$ is $(-, -, +)$, $\mathbf{p}_7$ is $(0, +, -)$, and $\mathbf{p}_8$ is $(+, +, -)$.

25. Suppose $S = \{\mathbf{b}_1, \dots, \mathbf{b}_k\}$ is an affinely independent set. Then equation (7) has a solution, because $\mathbf{p}$ is in aff S. Hence equation (8) has a solution. By Theorem 5, the homogeneous forms of the points in S are linearly independent. Thus (8) has a unique solution. Then (7) also has a unique solution, because (8) encodes both equations that appear in (7).

The following argument mimics the proof of Theorem 8 in Section 4.4. If $S = \{\mathbf{b}_1, \dots, \mathbf{b}_k\}$ is an affinely independent set, then scalars $c_1, \dots, c_k$ exist that satisfy (7), by definition of aff S. Suppose $\mathbf{x}$ also has the representation

$$\mathbf{x} = d_1\mathbf{b}_1 + \cdots + d_k\mathbf{b}_k \quad \text{and} \quad d_1 + \cdots + d_k = 1 \quad (7a)$$

for scalars $d_1, \dots, d_k$. Then subtraction produces the equation

$$\mathbf{0} = \mathbf{x} - \mathbf{x} = (c_1 - d_1)\mathbf{b}_1 + \cdots + (c_k - d_k)\mathbf{b}_k \quad (7b)$$

The weights in (7b) sum to 0 because the c's and the d's separately sum to 1. This is impossible, unless each weight in (8) is 0, because S is an affinely independent set. This proves that $c_i = d_i$ for $i = 1, \dots, k$.

27. If $\{\mathbf{p}_1, \mathbf{p}_2, \mathbf{p}_3\}$ is an affinely dependent set, then there exist scalars c_1, c_2, and c_3, not all zero, such that $c_1\mathbf{p}_1 + c_2\mathbf{p}_2 + c_3\mathbf{p}_3 = \mathbf{0}$ and $c_1 + c_2 + c_3 = 0$. Now use the linearity of f.

29. Let $\mathbf{a} = \begin{bmatrix} a_1 \\ a_2 \end{bmatrix}$, $\mathbf{b} = \begin{bmatrix} b_1 \\ b_2 \end{bmatrix}$, and $\mathbf{c} = \begin{bmatrix} c_1 \\ c_2 \end{bmatrix}$. Then

$$\det[\tilde{\mathbf{a}} \quad \tilde{\mathbf{b}} \quad \tilde{\mathbf{c}}] = \det \begin{bmatrix} a_1 & b_1 & c_1 \\ a_2 & b_2 & c_2 \\ 1 & 1 & 1 \end{bmatrix} =$$

$\det \begin{bmatrix} a_1 & a_2 & 1 \\ b_1 & b_2 & 1 \\ c_1 & c_2 & 1 \end{bmatrix}$, by the transpose property of the determinant (Theorem 5 in Section 3.2). By Exercise 30 in Section 3.3, this determinant equals 2 times the area of the triangle with vertices at $\mathbf{a}$, $\mathbf{b}$, and $\mathbf{c}$.

31. If $[\tilde{\mathbf{a}} \quad \tilde{\mathbf{b}} \quad \tilde{\mathbf{c}}] \begin{bmatrix} r \\ s \\ t \end{bmatrix} = \tilde{\mathbf{p}}$, then Cramer's rule gives $r = \det[\tilde{\mathbf{p}} \quad \tilde{\mathbf{b}} \quad \tilde{\mathbf{c}}] / \det[\tilde{\mathbf{a}} \quad \tilde{\mathbf{b}} \quad \tilde{\mathbf{c}}]$. By Exercise 29, the numerator of this quotient is twice the area of $\triangle\mathbf{pbc}$, and the denominator is twice the area of $\triangle\mathbf{abc}$. This proves the formula for r. The other formulas are proved using Cramer's rule for s and t.

33. The intersection point is $\mathbf{x}(4) =$
$$-.1 \begin{bmatrix} 1 \\ 3 \\ -6 \end{bmatrix} + .6 \begin{bmatrix} 7 \\ 3 \\ -5 \end{bmatrix} + .5 \begin{bmatrix} 3 \\ 9 \\ -2 \end{bmatrix} = \begin{bmatrix} 5.6 \\ 6.0 \\ -3.4 \end{bmatrix}. \text{ It is}$$
not inside the triangle.

Section 8.3, page 484

1. See the *Study Guide*.

3. None are in conv S.

5. $\mathbf{p}_1 = -\frac{1}{6}\mathbf{v}_1 + \frac{1}{3}\mathbf{v}_2 + \frac{2}{3}\mathbf{v}_3 + \frac{1}{6}\mathbf{v}_4$, so $\mathbf{p}_1 \notin \text{conv } S$.
$\mathbf{p}_2 = \frac{1}{3}\mathbf{v}_1 + \frac{1}{3}\mathbf{v}_2 + \frac{1}{6}\mathbf{v}_3 + \frac{1}{6}\mathbf{v}_4$, so $\mathbf{p}_2 \in \text{conv } S$.

7. a. The barycentric coordinates of $\mathbf{p}_1$, $\mathbf{p}_2$, $\mathbf{p}_3$, and $\mathbf{p}_4$ are, respectively, $\left(\frac{1}{3}, \frac{1}{6}, \frac{1}{2}\right)$, $\left(0, \frac{1}{2}, \frac{1}{2}\right)$, $\left(\frac{1}{2}, -\frac{1}{4}, \frac{3}{4}\right)$, and $\left(\frac{1}{2}, \frac{3}{4}, -\frac{1}{4}\right)$.

   b. $\mathbf{p}_3$ and $\mathbf{p}_4$ are outside conv T. $\mathbf{p}_1$ is inside conv T. $\mathbf{p}_2$ is on the edge $\overline{\mathbf{v}_2\mathbf{v}_3}$ of conv T.

9. $\mathbf{p}_1$ and $\mathbf{p}_3$ are outside the tetrahedron conv S. $\mathbf{p}_2$ is on the face containing the vertices $\mathbf{v}_2$, $\mathbf{v}_3$, and $\mathbf{v}_4$. $\mathbf{p}_4$ is inside conv S. $\mathbf{p}_5$ is on the edge between $\mathbf{v}_1$ and $\mathbf{v}_3$.

11–15. See the *Study Guide*.

17. If $\mathbf{p}, \mathbf{q} \in f(S)$, then there exist $\mathbf{r}, \mathbf{s} \in S$ such that $f(\mathbf{r}) = \mathbf{p}$ and $f(\mathbf{s}) = \mathbf{q}$. The goal is to show that the line segment $\mathbf{y} = (1-t)\mathbf{p} + t\mathbf{q}$, for $0 \le t \le 1$, is in $f(S)$. Use the linearity of f and the convexity of S to show that $\mathbf{y} = f(\mathbf{w})$ for some $\mathbf{w}$ in S. This will show that $\mathbf{y}$ is in $f(S)$ and that $f(S)$ is convex.

19. $\mathbf{p} = \frac{1}{6}\mathbf{v}_1 + \frac{1}{2}\mathbf{v}_2 + \frac{1}{3}\mathbf{v}_4$ and $\mathbf{p} = \frac{1}{2}\mathbf{v}_1 + \frac{1}{6}\mathbf{v}_2 + \frac{1}{3}\mathbf{v}_3$.

21. Suppose $A \subseteq B$, where B is convex. Then, since B is convex, Theorem 7 implies that B contains all convex combinations of points of B. Hence B contains all convex combinations of points of A. That is, conv $A \subseteq B$.

23. a. Use Exercise 22 to show that conv A and conv B are both subsets of conv $(A \cup B)$. This will imply that their union is also a subset of conv $(A \cup B)$.

   b. One possibility is to let A be two adjacent corners of a square and let B be the other two corners. Then what is (conv A) $\cup$ (conv B), and what is conv $(A \cup B)$?

25.

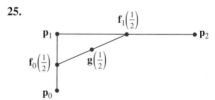

27. $\mathbf{g}(t) = (1-t)\mathbf{f}_0(t) + t\mathbf{f}_1(t)$
$= (1-t)[(1-t)\mathbf{p}_0 + t\mathbf{p}_1] + t[(1-t)\mathbf{p}_1 + t\mathbf{p}_2]$
$= (1-t)^2\mathbf{p}_0 + 2t(1-t)\mathbf{p}_1 + t^2\mathbf{p}_2$.
The sum of the weights in the linear combination for $\mathbf{g}$ is $(1-t)^2 + 2t(1-t) + t^2$, which equals $(1 - 2t + t^2) + (2t - 2t^2) + t^2 = 1$. The weights are each between 0 and 1 when $0 \le t \le 1$, so $\mathbf{g}(t)$ is in conv $\{\mathbf{p}_0, \mathbf{p}_1, \mathbf{p}_2\}$.

Section 8.4, page 493

1. $f(x_1, x_2) = 3x_1 + 4x_2$ and $d = 13$

3. a. Open   **b.** Closed   **c.** Neither
   d. Closed   **e.** Closed

5. a. Not compact, convex
   b. Compact, convex

c. Not compact, convex
d. Not compact, not convex
e. Not compact, convex

7. a. $\mathbf{n} = \begin{bmatrix} 0 \\ 2 \\ 3 \end{bmatrix}$ or a multiple

   b. $f(\mathbf{x}) = 2x_2 + 3x_3, d = 11$

9. a. $\mathbf{n} = \begin{bmatrix} 3 \\ -1 \\ 2 \\ 1 \end{bmatrix}$ or a multiple

   b. $f(\mathbf{x}) = 3x_1 - x_2 + 2x_3 + x_4, d = 5$

11. $\mathbf{v}_2$ is on the same side as $\mathbf{0}$, $\mathbf{v}_1$ is on the other side, and $\mathbf{v}_3$ is in H.

13. One possibility is $\mathbf{p} = \begin{bmatrix} 32 \\ -14 \\ 0 \\ 0 \end{bmatrix}$, $\mathbf{v}_1 = \begin{bmatrix} 10 \\ -7 \\ 1 \\ 0 \end{bmatrix}$,

$\mathbf{v}_2 = \begin{bmatrix} -4 \\ 1 \\ 0 \\ 1 \end{bmatrix}$.

15. $f(x_1, x_2, x_3, x_4) = x_1 - 3x_2 + 4x_3 - 2x_4$, and $d = 5$

17. $f(x_1, x_2, x_3) = x_1 - 2x_2 + x_3$, and $d = 0$

19. $f(x_1, x_2, x_3) = -5x_1 + 3x_2 + x_3$, and $d = 0$

21–27. See the *Study Guide*.

29. $f(x_1, x_2) = 3x_1 - 2x_2$ with d satisfying $9 < d < 10$ is one possibility.

31. $f(x, y) = 4x + y$. A natural choice for d is 12.75, which equals $f(3, .75)$. The point $(3, .75)$ is three-fourths of the distance between the center of A and the center of B.

33. Exercise 2(a) in Section 8.3 gives one possibility. Or let $S = \{(x, y) : x^2y^2 = 1 \text{ and } y > 0\}$. Then conv S is the upper (open) half-plane.

35. Let $\mathbf{x}, \mathbf{y} \in B(\mathbf{p}, \delta)$ and suppose $\mathbf{z} = (1 - t)\mathbf{x} + t\mathbf{y}$, where $0 \le t \le 1$. Then show that
$\|\mathbf{z} - \mathbf{p}\| = \|[(1 - t)\mathbf{x} + t\mathbf{y}] - \mathbf{p}\|$
$= \|(1 - t)(\mathbf{x} - \mathbf{p}) + t(\mathbf{y} - \mathbf{p})\| < \delta$.

Section 8.5, page 505

1. a. $m = 1$ at the point $\mathbf{p}_1$ **b.** $m = 5$ at the point $\mathbf{p}_2$
   c. $m = 5$ at the point $\mathbf{p}_3$

3. a. $m = -3$ at the point $\mathbf{p}_3$
   b. $m = 1$ on the set conv $\{\mathbf{p}_1, \mathbf{p}_3\}$
   c. $m = -3$ on the set conv $\{\mathbf{p}_1, \mathbf{p}_2\}$

5. $\left\{ \begin{bmatrix} 0 \\ 0 \end{bmatrix}, \begin{bmatrix} 5 \\ 0 \end{bmatrix}, \begin{bmatrix} 4 \\ 3 \end{bmatrix}, \begin{bmatrix} 0 \\ 5 \end{bmatrix} \right\}$

7. $\left\{ \begin{bmatrix} 0 \\ 0 \end{bmatrix}, \begin{bmatrix} 7 \\ 0 \end{bmatrix}, \begin{bmatrix} 6 \\ 4 \end{bmatrix}, \begin{bmatrix} 0 \\ 6 \end{bmatrix} \right\}$

9. The origin is an extreme point, but it is not a vertex. Explain why.

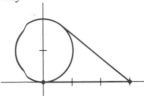

11. One possibility is to let S be a square that includes part of the boundary but not all of it. For example, include just two adjacent edges. The convex hull of the profile P is a triangular region.

13. **a.** $f_0(C^5) = 32$, $f_1(C^5) = 80$, $f_2(C^5) = 80$, $f_3(C^5) = 40$, $f_4(C^5) = 10$, and $32 - 80 + 80 - 40 + 10 = 2$.

b.

	f_0	f_1	f_2	f_3	f_4
C^1	2				
C^2	4	4			
C^3	8	12	6		
C^4	16	32	24	8	
C^5	32	80	80	40	10

For a general formula, see the *Study Guide*.

15. **a.** $f_0(P^n) = f_0(Q) + 1$

b. $f_k(P^n) = f_k(Q) + f_{k-1}(Q)$

c. $f_{n-1}(P^n) = f_{n-2}(Q) + 1$

17–23. See the *Study Guide*.

25. Let S be convex and let $\mathbf{x} \in cS + dS$, where $c > 0$ and $d > 0$. Then there exist $\mathbf{s}_1$ and $\mathbf{s}_2$ in S such that $\mathbf{x} = c\mathbf{s}_1 + d\mathbf{s}_2$. But then

$$\mathbf{x} = c\mathbf{s}_1 + d\mathbf{s}_2 = (c + d)\left(\frac{c}{c + d}\mathbf{s}_1 + \frac{d}{c + d}\mathbf{s}_2\right).$$

Now show that the expression on the right side is a member of $(c + d)S$.

For the converse, pick a typical point in $(c + d)S$ and show it is in $cS + dS$.

27. *Hint:* Suppose A and B are convex. Let $\mathbf{x}, \mathbf{y} \in A + B$. Then there exist $\mathbf{a}, \mathbf{c} \in A$ and $\mathbf{b}, \mathbf{d} \in B$ such that $\mathbf{x} = \mathbf{a} + \mathbf{b}$ and $\mathbf{y} = \mathbf{c} + \mathbf{d}$. For any t such that $0 \leq t \leq 1$, show that

$$\mathbf{w} = (1 - t)\mathbf{x} + t\mathbf{y} = (1 - t)(\mathbf{a} + \mathbf{b}) + t(\mathbf{c} + \mathbf{d})$$

represents a point in $A + B$.

Section 8.6, page 516

1. The control points for $\mathbf{x}(t) + \mathbf{b}$ should be $\mathbf{p}_0 + \mathbf{b}$, $\mathbf{p}_1 + \mathbf{b}$, and $\mathbf{p}_3 + \mathbf{b}$. Write the Bézier curve through these points, and show algebraically that this curve is $\mathbf{x}(t) + \mathbf{b}$. See the *Study Guide*.

3. **a.** $\mathbf{x}'(t) = (-3 + 6t - 3t^2)\mathbf{p}_0 + (3 - 12t + 9t^2)\mathbf{p}_1 + (6t - 9t^2)\mathbf{p}_2 + 3t^2\mathbf{p}_3$, so
$\mathbf{x}'(0) = -3\mathbf{p}_0 + 3\mathbf{p}_1 = 3(\mathbf{p}_1 - \mathbf{p}_0)$, and
$\mathbf{x}'(1) = -3\mathbf{p}_2 + 3\mathbf{p}_3 = 3(\mathbf{p}_3 - \mathbf{p}_2)$. This shows that the tangent vector $\mathbf{x}'(0)$ points in the direction from $\mathbf{p}_0$ to $\mathbf{p}_1$ and is three times the length of $\mathbf{p}_1 - \mathbf{p}_0$. Likewise, $\mathbf{x}'(1)$ points in the direction from $\mathbf{p}_2$ to $\mathbf{p}_3$ and is three times the length of $\mathbf{p}_3 - \mathbf{p}_2$. In particular, $\mathbf{x}'(1) = \mathbf{0}$ if and only if $\mathbf{p}_3 = \mathbf{p}_2$.

b. $\mathbf{x}''(t) = (6 - 6t)\mathbf{p}_0 + (-12 + 18t)\mathbf{p}_1$
$+ (6 - 18t)\mathbf{p}_2 + 6t\mathbf{p}_3$, so that
$\mathbf{x}''(0) = 6\mathbf{p}_0 - 12\mathbf{p}_1 + 6\mathbf{p}_2 = 6(\mathbf{p}_0 - \mathbf{p}_1) + 6(\mathbf{p}_2 - \mathbf{p}_1)$
and
$\mathbf{x}''(1) = 6\mathbf{p}_1 - 12\mathbf{p}_2 + 6\mathbf{p}_3 = 6(\mathbf{p}_1 - \mathbf{p}_2) + 6(\mathbf{p}_3 - \mathbf{p}_2)$
For a picture of $\mathbf{x}''(0)$, construct a coordinate system with the origin at $\mathbf{p}_1$, temporarily, label $\mathbf{p}_0$ as $\mathbf{p}_0 - \mathbf{p}_1$, and label $\mathbf{p}_2$ as $\mathbf{p}_2 - \mathbf{p}_1$. Finally, construct a line from this new origin through the sum of $\mathbf{p}_0 - \mathbf{p}_1$ and $\mathbf{p}_2 - \mathbf{p}_1$, extended out a bit. That line points in the direction of $\mathbf{x}''(0)$.

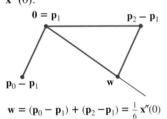

$$\mathbf{w} = (\mathbf{p}_0 - \mathbf{p}_1) + (\mathbf{p}_2 - \mathbf{p}_1) = \tfrac{1}{6}\mathbf{x}''(0)$$

5. **a.** From Exercise 3(a) or equation (9) in the text,

$$\mathbf{x}'(1) = 3(\mathbf{p}_3 - \mathbf{p}_2)$$

Use the formula for $\mathbf{x}'(0)$, with the control points from $\mathbf{y}(t)$, and obtain

$$\mathbf{y}'(0) = -3\mathbf{p}_3 + 3\mathbf{p}_4 = 3(\mathbf{p}_4 - \mathbf{p}_3)$$

For C^1 continuity, $3(\mathbf{p}_3 - \mathbf{p}_2) = 3(\mathbf{p}_4 - \mathbf{p}_3)$, so $\mathbf{p}_3 = (\mathbf{p}_4 + \mathbf{p}_2)/2$, and $\mathbf{p}_3$ is the midpoint of the line segment from $\mathbf{p}_2$ to $\mathbf{p}_4$.

b. If $\mathbf{x}'(1) = \mathbf{y}'(0) = \mathbf{0}$, then $\mathbf{p}_2 = \mathbf{p}_3$ and $\mathbf{p}_3 = \mathbf{p}_4$. Thus, the "line segment" from $\mathbf{p}_2$ to $\mathbf{p}_4$ is just the point $\mathbf{p}_3$. [*Note:* In this case, the combined curve is still C^1 continuous, by definition. However, some choices of the other "control" points, $\mathbf{p}_0$, $\mathbf{p}_1$, $\mathbf{p}_5$, and $\mathbf{p}_6$, can produce a curve with a visible corner at $\mathbf{p}_3$, in which case the curve is not G^1 continuous at $\mathbf{p}_3$.]

7. *Hint:* Use $\mathbf{x}''(t)$ from Exercise 3 and adapt this for the second curve to see that

$$\mathbf{y}''(t) = 6(1 - t)\mathbf{p}_3 + 6(-2 + 3t)\mathbf{p}_4 + 6(1 - 3t)\mathbf{p}_5 + 6t\mathbf{p}_6$$

Then set $\mathbf{x}''(1) = \mathbf{y}''(0)$. Since the curve is C^1 continuous at $\mathbf{p}_3$, Exercise 5(a) says that the point $\mathbf{p}_3$ is the midpoint of the segment from $\mathbf{p}_2$ to $\mathbf{p}_4$. This implies that $\mathbf{p}_4 - \mathbf{p}_3 = \mathbf{p}_3 - \mathbf{p}_2$. Use this substitution to show that $\mathbf{p}_4$ and $\mathbf{p}_5$ are uniquely determined by $\mathbf{p}_1, \mathbf{p}_2$, and $\mathbf{p}_3$. Only $\mathbf{p}_6$ can be chosen arbitrarily.

9. Write a vector of the polynomial weights for $\mathbf{x}(t)$, expand the polynomial weights, and factor the vector as $M_B \mathbf{u}(t)$:

$$\begin{bmatrix} 1 - 4t + 6t^2 - 4t^3 + t^4 \\ 4t - 12t^2 + 12t^3 - 4t^4 \\ 6t^2 - 12t^3 + 6t^4 \\ 4t^3 - 4t^4 \\ t^4 \end{bmatrix}$$

$$= \begin{bmatrix} 1 & -4 & 6 & -4 & 1 \\ 0 & 4 & -12 & 12 & -4 \\ 0 & 0 & 6 & -12 & 6 \\ 0 & 0 & 0 & 4 & -4 \\ 0 & 0 & 0 & 0 & 1 \end{bmatrix} \begin{bmatrix} 1 \\ t \\ t^2 \\ t^3 \\ t^4 \end{bmatrix},$$

$$M_B = \begin{bmatrix} 1 & -4 & 6 & -4 & 1 \\ 0 & 4 & -12 & 12 & -4 \\ 0 & 0 & 6 & -12 & 6 \\ 0 & 0 & 0 & 4 & -4 \\ 0 & 0 & 0 & 0 & 1 \end{bmatrix}$$

11–15. See the *Study Guide*.

17. a. *Hint:* Use the fact that $\mathbf{q}_0 = \mathbf{p}_0$.
   b. Multiply the first and last parts of equation (13) by $\frac{8}{3}$ and solve for $8\mathbf{q}_2$.
   c. Use equation (8) to substitute for $8\mathbf{q}_3$ and then apply part (a).

19. a. From equation (11), $\mathbf{y}'(1) = .5\mathbf{x}'(.5) = \mathbf{z}'(0)$.
   b. Observe that $\mathbf{y}'(1) = 3(\mathbf{q}_3 - \mathbf{q}_2)$. This follows from equation (9), with $\mathbf{y}(t)$ and its control points in place of $\mathbf{x}(t)$ and its control points. Similarly, for $\mathbf{z}(t)$ and its control points, $\mathbf{z}'(0) = 3(\mathbf{r}_1 - \mathbf{r}_0)$. By part (a), $3(\mathbf{q}_3 - \mathbf{q}_2) = 3(\mathbf{r}_1 - \mathbf{r}_0)$. Replace $\mathbf{r}_0$ by $\mathbf{q}_3$, and obtain $\mathbf{q}_3 - \mathbf{q}_2 = \mathbf{r}_1 - \mathbf{q}_3$, and hence $\mathbf{q}_3 = (\mathbf{q}_2 + \mathbf{r}_1)/2$.
   c. Set $\mathbf{q}_0 = \mathbf{p}_0$ and $\mathbf{r}_3 = \mathbf{p}_3$. Compute $\mathbf{q}_1 = (\mathbf{p}_0 + \mathbf{p}_1)/2$ and $\mathbf{r}_2 = (\mathbf{p}_2 + \mathbf{p}_3)/2$. Compute $\mathbf{m} = (\mathbf{p}_1 + \mathbf{p}_2)/2$. Compute $\mathbf{q}_2 = (\mathbf{q}_1 + \mathbf{m})/2$ and $\mathbf{r}_1 = (\mathbf{m} + \mathbf{r}_2)/2$. Compute $\mathbf{q}_3 = (\mathbf{q}_2 + \mathbf{r}_1)/2$ and set $\mathbf{r}_0 = \mathbf{q}_3$.

21. a. $\mathbf{r}_0 = \mathbf{p}_0, \mathbf{r}_1 = \dfrac{\mathbf{p}_0 + 2\mathbf{p}_1}{3}, \mathbf{r}_2 = \dfrac{2\mathbf{p}_1 + \mathbf{p}_2}{3}, \mathbf{r}_3 = \mathbf{p}_2$
   b. *Hint:* Write the standard formula (7) in this section, with $\mathbf{r}_i$ in place of $\mathbf{p}_i$ for $i = 0, \ldots, 3$, and then replace $\mathbf{r}_0$ and $\mathbf{r}_3$ by $\mathbf{p}_0$ and $\mathbf{p}_2$, respectively:

   $$\mathbf{x}(t) = (1 - 3t + 3t^2 - t^3)\mathbf{p}_0 \\ + (3t - 6t^2 + 3t^3)\mathbf{r}_1 \\ + (3t^2 - 3t^3)\mathbf{r}_2 + t^3\mathbf{p}_2$$

   Use the formulas for $\mathbf{r}_1$ and $\mathbf{r}_2$ from part (a) to examine the second and third terms in this expression for $\mathbf{x}(t)$.

Chapter 8 Supplementary Exercises, pages 519

1. T 2. T 3. F 4. F 5. T 6. T
7. T 8. F 9. F 10. F 11. T 12. T
13. F 14. T 15. T 16. T 17. T 18. T
19. T 20. F 21. T

23. Let $\mathbf{y} \in F$. Then $U = F - \mathbf{y}$ and $V = G - \mathbf{y}$ are k-dimensional subspaces with $U \subseteq V$. Let $B = \{\mathbf{x}_1, \ldots, \mathbf{x}_k\}$ be a basis for U. Since $\dim V = k$, B is also a basis for V. Hence $U = V$, and $F = U + \mathbf{y} = V + \mathbf{y} = G$.

25. *Hint:* Suppose $F_1 \cap F_2 \neq \varnothing$. Then there exist $\mathbf{v}_1$ and $\mathbf{v}_2$ in V such that $\mathbf{x}_1 + \mathbf{v}_1 = \mathbf{x}_2 + \mathbf{v}_2$. Use this and the properties of a subspace to show that for all $\mathbf{v}$ in V, $\mathbf{x}_1 + \mathbf{v} \in \mathbf{x}_2 + V$ and $\mathbf{x}_2 + \mathbf{v} \in \mathbf{x}_1 + V$.

27. *Hint:* Start with a basis for V and expand it by joining $\mathbf{p}$ to get a basis for $\mathbb{R}^n$.

29. *Hint:* Suppose $\mathbf{x} \in \lambda B(\mathbf{p}, \delta)$. This means that there exists $\mathbf{y} \in B(\mathbf{p}, \delta)$ such that $\mathbf{x} = \lambda\mathbf{y}$. Use the definition of $B(\mathbf{p}, \delta)$ to show that this implies $\mathbf{x} \in B(\lambda\mathbf{p}, \lambda\delta)$. The converse is similar.

31. The positive hull of S is a cone with vertex $(0, 0)$ containing the positive y axis and with sides on the lines $y = \pm x$.

33. *Hint:* It is significant that the set in Exercise 31 consists of exactly two non-collinear points. Explain why this is important.

35. *Hint:* Suppose $\mathbf{x} \in \text{pos } S$. Then $\mathbf{x} = c_1\mathbf{v}_1 + \cdots c_k\mathbf{v}_k$, where $\mathbf{v}_i \in S$ and all $c_i \geq 0$. Let $d = \sum_{i=1}^{k} c_i$. Consider two cases: $d = 0$ and $d \neq 0$.

Chapter 9

Section 9.1, page 533

1. $\begin{array}{c c} & \begin{array}{cc} d & q \end{array} \\ \begin{array}{c} d \\ q \end{array} & \begin{bmatrix} -10 & 10 \\ 25 & -25 \end{bmatrix} \end{array}$

3. $\begin{array}{c c} & \begin{array}{ccc} r & s & p \end{array} \\ \begin{array}{c} \text{rock} \\ \text{scissors} \\ \text{paper} \end{array} & \begin{bmatrix} 0 & 5 & -5 \\ -5 & 0 & 5 \\ 5 & -5 & 0 \end{bmatrix} \end{array}$

5. $\begin{bmatrix} 4 & ③ \\ 1 & -1 \end{bmatrix}$

7. $\begin{bmatrix} 5 & ③ & 4 & ③ \\ -2 & 1 & -5 & 2 \\ 4 & ③ & 7 & ③ \end{bmatrix}$

9. a. $E(\mathbf{x}, \mathbf{y}) = \frac{13}{12}, v(\mathbf{x}) = \min\{\frac{5}{6}, 1, \frac{9}{6}\} = \frac{5}{6}$, $v(\mathbf{y}) = \max\{\frac{3}{4}, \frac{3}{2}, \frac{1}{2}\} = \frac{3}{2}$
   b. $E(\mathbf{x}, \mathbf{y}) = \frac{9}{8}, v(\mathbf{x}) = \min\{1, \frac{3}{4}, \frac{7}{4}\} = \frac{3}{4}$, $v(\mathbf{y}) = \max\{\frac{1}{2}, \frac{5}{4}, \frac{3}{2}\} = \frac{3}{2}$

11. $\hat{\mathbf{x}} = \begin{bmatrix} \frac{1}{6} \\ \frac{5}{6} \end{bmatrix}$, $\hat{\mathbf{y}} = \begin{bmatrix} \frac{1}{2} \\ \frac{1}{2} \end{bmatrix}$, $v = \frac{1}{2}$

13. $\hat{\mathbf{x}} = \begin{bmatrix} \frac{3}{5} \\ \frac{2}{5} \end{bmatrix}$, $\hat{\mathbf{y}} = \begin{bmatrix} \frac{4}{5} \\ \frac{1}{5} \end{bmatrix}$, $v = \frac{17}{5}$

15. $\hat{\mathbf{x}} = \begin{bmatrix} \frac{1}{3} \\ \frac{2}{3} \end{bmatrix}$ or $\begin{bmatrix} \frac{3}{5} \\ \frac{2}{5} \end{bmatrix}$ or any convex combination of these row

strategies, $\hat{\mathbf{y}} = \begin{bmatrix} 0 \\ 0 \\ 1 \\ 0 \end{bmatrix}$, $v = 2$

17. $\hat{\mathbf{x}} = \begin{bmatrix} \frac{5}{7} \\ 0 \\ \frac{2}{7} \\ 0 \end{bmatrix}$, $\hat{\mathbf{y}} = \begin{bmatrix} 0 \\ \frac{5}{7} \\ \frac{2}{7} \\ 0 \\ 0 \end{bmatrix}$, $v = \frac{3}{7}$

19. a. Army: 1/3 river, 2/3 land; guerrillas: 1/3 river, 2/3 land; 2/3 of the supplies get through.

b. Army: 7/11 river, 4/11 land; guerrillas: 7/11 river, 4/11 land; 64/121 of the supplies get through.

21–29. See the *Study Guide*.

31. $\hat{\mathbf{x}} = \begin{bmatrix} \frac{1}{6} \\ \frac{5}{6} \\ 0 \end{bmatrix}$, $\hat{\mathbf{y}} = \begin{bmatrix} 0 \\ \frac{1}{2} \\ \frac{1}{2} \end{bmatrix}$, $v = 0$

33. $\hat{\mathbf{x}} = \left(\dfrac{d-c}{a-b+d-c}, \dfrac{a-b}{a-b+d-c} \right)$,

$\hat{\mathbf{y}} = \left(\dfrac{d-b}{a-b+d-c}, \dfrac{a-c}{a-b+d-c} \right)$,

$v = \dfrac{ad-bc}{a-b+d-c}$

Section 9.2, page 543

1. Let x_1 be the amount invested in mutual funds, x_2 the amount in CDs, and x_3 the amount in savings. Then

$\mathbf{b} = \begin{bmatrix} 12{,}000 \\ 0 \\ 0 \end{bmatrix}$, $\mathbf{x} = \begin{bmatrix} x_1 \\ x_2 \\ x_3 \end{bmatrix}$, $\mathbf{c} = \begin{bmatrix} .11 \\ .08 \\ .06 \end{bmatrix}$, and

$A = \begin{bmatrix} 1 & 1 & 1 \\ 1 & -1 & -1 \\ 0 & 1 & -2 \end{bmatrix}$.

3. $\mathbf{b} = \begin{bmatrix} 20 \\ -10 \end{bmatrix}$, $\mathbf{c} = \begin{bmatrix} 3 \\ 4 \\ -2 \end{bmatrix}$, $A = \begin{bmatrix} 1 & 2 & 0 \\ 0 & -3 & -5 \end{bmatrix}$

5. $\mathbf{b} = \begin{bmatrix} -35 \\ 20 \\ -20 \end{bmatrix}$, $\mathbf{c} = \begin{bmatrix} -7 \\ 3 \\ -1 \end{bmatrix}$, $A = \begin{bmatrix} -1 & 4 & 0 \\ 0 & 1 & -2 \\ 0 & -1 & 2 \end{bmatrix}$

7. max $= 1360$, when $x_1 = \frac{72}{5}$ and $x_2 = \frac{16}{5}$

9. unbounded

11–13. See the *Study Guide*.

15. max profit $= \$1250$, when $x_1 = 100$ bags of EverGreen and $x_2 = 350$ bags of QuickGreen

17. max profit $= \$1180$, for 20 widgets and 30 whammies

19. Take any $\mathbf{p}$ and $\mathbf{q}$ in S, with $\mathbf{p} = \begin{bmatrix} x_1 \\ x_2 \end{bmatrix}$ and $\mathbf{q} = \begin{bmatrix} y_1 \\ y_2 \end{bmatrix}$.

Then $\mathbf{v}^T \mathbf{p} \le c$ and $\mathbf{v}^T \mathbf{q} \le c$. Take any scalar t such that $0 \le t \le 1$. Then, by the linearity of matrix multiplication (or the dot product if $\mathbf{v}^T \mathbf{p}$ is written as $\mathbf{v} \cdot \mathbf{p}$, and so on),

$$\mathbf{v}^T[(1-t)\mathbf{p} + t\mathbf{q}] = (1-t)\mathbf{v}^T\mathbf{p} + t\mathbf{v}^T\mathbf{q}$$
$$\le (1-t)c + tc = c$$

because $(1-t)$ and t are both positive and $\mathbf{p}$ and $\mathbf{q}$ are in S. So the line segment between $\mathbf{p}$ and $\mathbf{q}$ is in S. Since $\mathbf{p}$ and $\mathbf{q}$ were any points in S, the set S is convex.

21. Let $S = \{\mathbf{x} : f(\mathbf{x}) = d\}$, and take $\mathbf{p}$ and $\mathbf{q}$ in S. Also, take t with $0 \le t \le 1$, and let $\mathbf{x} = (1-t)\mathbf{p} + t\mathbf{q}$. Then

$$f(\mathbf{x}) = \mathbf{c}^T\mathbf{x} = \mathbf{c}^T[(1-t)\mathbf{p} + t\mathbf{q}]$$
$$= (1-t)\mathbf{c}^T\mathbf{p} + t\mathbf{c}^T\mathbf{q} = (1-t)d + td = d$$

Thus, $\mathbf{x}$ is in S. This shows that S is convex.

Section 9.3, page 559

1.

x_1	x_2	x_3	x_4	x_5	M	
2	7	10	1	0	0	20
3	4	18	0	1	0	25
-21	-25	-15	0	0	1	0

3. a. x_2

b.

x_1	x_2	x_3	x_4	M	
$\frac{7}{2}$	0	1	$-\frac{1}{2}$	0	5
$\frac{3}{2}$	1	0	$\frac{1}{2}$	0	15
11	0	0	5	1	150

c. $x_1 = 0$, $x_2 = 15$, $x_3 = 5$, $x_4 = 0$, $M = 150$

d. optimal

5. a. x_1

b.

x_1	x_2	x_3	x_4	M	
0	2	1	-1	0	4
1	$\frac{1}{2}$	0	$\frac{1}{2}$	0	8
0	-2	0	3	1	48

c. $x_1 = 8$, $x_2 = 0$, $x_3 = 4$, $x_4 = 0$, $M = 48$

d. not optimal

7–11. See the *Study Guide*.

13. The maximum is 150, when $x_1 = 3$ and $x_2 = 10$.

15. The maximum is 56, when $x_1 = 9$ and $x_2 = 4$.

17. The minimum is 180, when $x_1 = 10$ and $x_2 = 12$.

19. The answer matches that in Example 7. The minimum is 20, when $x_1 = 8$ and $x_2 = 6$.

21. The maximum profit is \$1180, achieved by making 20 widgets and 30 whammies each day.

Section 9.4, page 568

1. Minimize $\quad 36y_1 + 55y_2$
subject to $\quad 2y_1 + 5y_2 \geq 10$
$\qquad\qquad\quad 3y_1 + 4y_2 \geq 12$
and $y_1 \geq 0, y_2 \geq 0$.

3. Minimize $\quad 26y_1 + 30y_2 + 13y_3$
subject to $\quad y_1 + 2y_2 + y_3 \geq 4$
$\qquad\qquad\quad 2y_1 + 3y_2 + y_3 \geq 5$
and $y_1 \geq 0, y_2 \geq 0, y_3 \geq 0$.

5. The minimum is $M = 150$, attained when $y_1 = \frac{20}{7}$ and $y_2 = \frac{6}{7}$.

7. The minimum is $M = 56$, attained when $y_1 = 0$, $y_2 = 1$, and $y_3 = 2$.

9–15. See the *Study Guide*.

17. The minimum is 43, when $x_1 = \frac{7}{4}$, $x_2 = 0$, and $x_3 = \frac{3}{4}$.

19. The minimum cost is \$670, using 11 bags of Pixie Power and 3 bags of Misty Might.

21. The marginal value is zero. This corresponds to labor in the fabricating department being underutilized. That is, at the optimal production schedule with $x_1 = 20$ and $x_2 = 30$, only 160 of the 200 available hours in fabricating are needed. The extra labor is wasted, and so it has value zero.

23. $\hat{\mathbf{x}} = \begin{bmatrix} \frac{2}{3} \\ 0 \\ \frac{1}{3} \end{bmatrix}$, $\hat{\mathbf{y}} = \begin{bmatrix} \frac{1}{2} \\ \frac{1}{2} \end{bmatrix}$, $v = 1$

25. $\hat{\mathbf{x}} = \begin{bmatrix} \frac{2}{5} \\ \frac{2}{5} \\ \frac{1}{5} \end{bmatrix}$, $\hat{\mathbf{y}} = \begin{bmatrix} \frac{3}{7} \\ \frac{3}{7} \\ \frac{1}{7} \end{bmatrix}$, $v = 1$

27. Change this "game" into a linear programming problem and use the simplex method to analyze the game. The expected value of the game is $\frac{38}{35}$, based on a payoff matrix for an investment of \$100. With \$35,000 to invest, Bob "plays"

this game 350 times. Thus, he expects to gain \$380, and the expected value of his portfolio at the end of the year is \$35,380. Using the optimal game strategy, Bob should invest \$11,000 in stocks, \$9000 in bonds, and \$15,000 in gold.

29. a. The coordinates of $\bar{\mathbf{x}}$ are all nonnegative. From the definition of $\mathbf{u}$, λ is equal to the sum of these coordinates. It follows that the coordinates of $\hat{\mathbf{x}}$ are nonnegative and sum to 1. Thus, $\hat{\mathbf{x}}$ is a mixed strategy for the row player R. A similar argument holds for $\hat{\mathbf{y}}$ and the column player C.

b. If $\mathbf{y}$ is any mixed strategy for C, then

$$E(\hat{\mathbf{x}}, \mathbf{y}) = \hat{\mathbf{x}}^T A \mathbf{y} = \frac{1}{\lambda} \left(\bar{\mathbf{x}}^T A \mathbf{y} \right) = \frac{1}{\lambda} \left[\left(A^T \bar{\mathbf{x}} \right) \cdot \mathbf{y} \right]$$
$$\geq \frac{1}{\lambda} (\mathbf{v} \cdot \mathbf{y}) = \frac{1}{\lambda}$$

c. If $\mathbf{x}$ is any mixed strategy for R, then

$$E(\mathbf{x}, \hat{\mathbf{y}}) = \mathbf{x}^T A \hat{\mathbf{y}} = \frac{1}{\lambda} \left(\mathbf{x}^T A \bar{\mathbf{y}} \right) = \frac{1}{\lambda} \left[\mathbf{x} \cdot A \bar{\mathbf{y}} \right]$$
$$\leq \frac{1}{\lambda} (\mathbf{x} \cdot \mathbf{u}) = \frac{1}{\lambda}$$

d. Part (b) implies $v(\hat{\mathbf{x}}) \geq 1/\lambda$, so $v_R \geq 1/\lambda$. Part (c) implies $v(\hat{\mathbf{y}}) \leq 1/\lambda$, so $v_C \leq 1/\lambda$. It follows from the Minimax Theorem in Section 9.1 that $\hat{\mathbf{x}}$ and $\hat{\mathbf{y}}$ are optimal mixed strategies for R and C, respectively, and that the value of the game is $1/\lambda$.

Chapter 9 Supplementary Exercises, page 570

1. T **2.** F **3.** F **4.** F **5.** T **6.** F

7. T **8.** T **9.** T **10.** F **11.** T **12.** F

13. F **14.** T **15.** F **16.** T **17.** F **18.** F

19. T **20.** F **21.** F **22.** T **23.** T **24.** F

25. b. The extreme points are $(0, 0)$, $(0, 1)$, and $(1, 2)$.

27. $f(x_1, x_2) = x_1 + x_2$, $f(0, 0) = 0$, $f(0, 1) = 1$, and $f(1, 2) = 3$.

29. *Hint:* there are no feasible solutions.

31. $\hat{\mathbf{x}} = \begin{bmatrix} \frac{3}{5} \\ \frac{2}{5} \end{bmatrix}$, $\hat{\mathbf{y}} = \begin{bmatrix} \frac{1}{2} \\ 0 \\ \frac{1}{2} \end{bmatrix}$, and $v = 1$.

Index

Page numbers denoted with "C" are found within the online chapter 10 as available at `bit.ly/2nj1HhO`.

I-1

Advice on Reading Linear Algebra

When you are reading linear algebra, you will encounter several different types of mathematical objects. Here are some conventions that will help you to keep track of which objects are which.

- Pay attention to case and font. Mathematical objects are usually **bold** or *italic*, and "**A**," "*A*," "*a*," and "**a**" may be used to represent four different mathematical objects at the same time, while "A" and "a" are one-letter words.

- Pay attention to sub$_{\text{scripts}}$ and super^{scripts}, which also give you clues about the objects you are reading about.

- Repeated letters in the same font and case are significant. An $n \times n$ matrix has the same number of rows as columns, whereas an $m \times n$ matrix may or may not have the same number of rows as columns.

- As you read, it may help to imagine shapes and pictures. I think of a matrix as a large, wide rectangle full of numbers and a vector as a tall, thin rectangle of numbers.

Finally, read the surrounding text carefully as the type of object you are reading (or writing) about should be clearly stated, and the conventions in this list are *not* always followed. When you are writing by hand, it is particularly important to indicate what each letter represents.

c, k, i, j	Italicized lowercase letters usually represent numbers. Pay attention to the context to determine whether they are integers, real numbers, or complex numbers.
A, B	Italicized capital letters near the beginning of the alphabet are frequently used to represent matrices.
a_{ij}, b_{ij}	When a matrix is listed as an italicized capital letter, the entries in the matrix are frequently represented using the same letter, but in lowercase italic with two subscripts. The first subscript tells you the row number. The second subscript tells you the column number. So a_{23} is the entry in the second row and third column of matrix A, whereas b_{31} is the entry in the third row and first column of the matrix B.
A_{ij}, B_{ij}	When you see an italicized capital letter with two subscripts, you need to check the surrounding text to see how this notation is being used. For example, in the context of Section 2.4, a matrix A has been partitioned into submatrices and A_{34} donates the submatrix in the third row and fourth column of this block partitioning. In contrast, in Section 3.1, A_{34} represents the matrix that was created by deleting the third row and fourth column of the matrix A.
E_1, E_2	When we have a list of matrices, we may use the same italicized capital letter with a single subscript. A different subscript indicates a different matrix. The letter E is used for special matrices such as the elementary matrices introduced in Chapter 2. Other special matrices may also be represented by the first letter of their description—so a diagonal matrix might be called D.

(continued)

T	An italicized capital T usually represents a linear transformation. If more than one transformation is needed, then S or U is a frequent second choice.
$\mathbf{v}, \mathbf{x}$	Vectors are frequently listed in this text with bold, but not italic, letters. They are often chosen to be near the end of the alphabet; however, $\mathbf{b}$ is used for vectors that represent the right-hand side of equations. In higher-level linear algebra or math textbooks, they may be italicized, but not bold. In calculus and physics books, vectors may be written with an arrow over top of them.
v_i, x_i	The entries in a vector may be represented using the same letter as the vector, but these letters will be italicized (not bold), with one subscript. The subscript describes the position of the entry. For example, v_4 is the fourth entry in the vector $\mathbf{v}$.
$\mathbf{v}_1, \mathbf{v}_2$	This one gets a little tricky too. To list several vectors, bold letters with a single subscript are often used. Thus, $\mathbf{v}_1$, $\mathbf{v}_2$, and $\mathbf{v}_3$ are three different vectors, whereas v_1, v_2, and v_3 are the first three entries in a vector named $\mathbf{v}$. Clearly, we need to move to a different convention if we want to describe the entries of $\mathbf{v}_2$.
$\mathbf{e}_j$	A bold $\mathbf{e}$ is usually associated with the standard basis vectors, with the subscript indicating which row contains a 1. Thus, $\mathbf{e}_2$ is a vector with a 1 in the second row and zeros in all the other rows. You will have to read the surrounding text to tell how many zero entries are in $\mathbf{e}_2$.
I	An italicized I always represents an identity matrix. Sometimes we indicate its size using a subscript. For example, I_3 is the identity matrix with three rows and columns, while I_n represents an $n \times n$ identity matrix. But if the letter has not been italicized, I just represents me!
A^{-1}	An $n \times n$ matrix with superscript -1 denotes the inverse of that matrix. When matrix A has an inverse, it has the special property that $A^{-1}A = I = AA^{-1}$. What may be different from what you have seen before is that we do *not* write $1/A$, when A is a matrix.
A^k	An $n \times n$ matrix with a superscript positive integer works much like the exponents you have encountered in other math classes. For example, $A^1 = A, A^2 = A(A), A^3 = A(A)(A)$, and so on. We also define $A^0 = I$, the identity matrix. For negative superscripts, use the inverse. For example $A^{-2} = (A^{-1})^2 = A^{-1}(A^{-1}), A^{-3} = (A^{-1})^3 = A^{-1}(A^{-1})(A^{-1})$, and so on.
A^T	Be careful. When the superscript is an italicized capital T, it represents the transpose of the matrix named with the same letter. Thus, A^T is the matrix formed from the matrix A by switching the rows with the columns.
$\mathbb{R}$	Letters where parts of the font are double barred represent specific number or mathematical systems. In particular, $\mathbb{R}$ always represents the real numbers.

(*continued*)